Y. Tanida · H. Miyashiro (Eds.)

Flow Visualization VI

Proceedings of the
Sixth International Symposium on Flow Visualization,
October 5 - 9, 1992, Yokohama, Japan

Springer-Verlag Berlin Heidelberg GmbH

Prof. Yoshimichi Tanida
Dept. of Mechanical Engineering
Tokai University
Kitakaname, Hiratsuka
Kanagawa 259-12
Japan

Prof. Hiroshi Miyashiro
Dept. of Mechanical Systems Engineering
Toyama Prefectural University
Kosugi-machi
Toyama 939-03
Japan

ISBN 978-3-642-84826-1 ISBN 978-3-642-84824-7 (eBook)
DOI 10.1007/ 978-3-642-84824-7

Originally published by Springer-Verlag Berlin Heidelberg New York 1992
Softcover reprint of the hardcover 1st edition 1992

Typesetting: Camera ready by author

61/3020-5 4 3 2 1 0 - Printed on acid -free paper

Preface

Since 1977, when the International Symposium on Flow Visualization was born at Tokyo under the leadership of Prof.T. Asanuma, it has been held every three years at Bochum, Ann Arbor, Paris and Prague. It is now held in Japan again, and we are very happy to have many of the old and new friends here from abroad.

Over the past two decades, the flow visualization has made marvelous advance and got definite appraisal in many special fields. Particularly, invention of Laser has led to a revolutional progress in flow visualization, creating new mthods such as holography and speckle method. Lasers are now widely used in flow visualization as light sources and as signal carriers in laser optical systems. The flow visualization is, however, originally to give information of the whole flow field observed, that makes it easy to grasp the physical characteristics of phenomena qualitatively but leads to the difficulty in quantitative analysis of flow.

These ten years advances in computer technologies have brought great changes to the field of flow visualization, specifically in the forms of image processing and computer graphics, that is conputer-aided flow visualization(CAFV). These technologies are now being used to great effect for quantitative analysis in flow visualization, a capability which this field had previously lacked. They have also made it possible to visualize nonvisual informaion in living bodies, sea etc., that is called computer tomography(CT). Numerical calculation, too, has come to be recognized as a partner of flow visualization with regard to computational fluid dynamics(CFD). And now we are even seeing flow visualization expand to cover a range of information generally thought of as nonvisual, such as heat and mass transfer, acoustic energy flow, electromagnetic flow etc. Hence, the flow visualization is not only be concerned with the fields of engineering and physics but also with other fields such as medical science, agriculture, marine products, oceanography, meterology and sports.

On such a background, the Sixth International Symposium on Flow Visualization is organized to present about two hundred papers as well as exhibition of advanced technologies of flow visualization. I am convinced that the ISFV-6 brings about fruitful results for the development of flow visualization and the international exchange of friendship.

Finally, on behalf of the Organizaing Committee, I would give my hearty acknowledgement to the sponsoring societies and to the authors who prestent their brilliant papers and photographs, and hope that much more interdisciplinary and worldwide cooperation will be promoted through this and following symposia.

Y.Tanida
Chairman
Organizing Committee
ISFV6

The Sixth International Symposium on Flow Visualization is Sponsored by

The Visualiztion Society of Japan and

Cosponsored by

The International Flow Visualization Society,
The Institution of Mechanical Engineers,
The International Association for Hydraulic Research,
The Japan Society for Aeronautical and Space Siences, and
The Japan Society of Mechanical Engineers.

History of the International Symposium on Flow Visualization

First Symposium
1977, Tokyo, Japan
Chairman: Asanuma, T.

Second Symposium
1980, Bochum, Germany
Chairman: Merzkirch, W.

Third Symposium
1983, Ann Arbor, U.S.A.
Chairman: Yang, W.-J.

Fourth Symposium
1986, Paris, France
Chairman: Veret, C.

Fifth Symposium
1989, Prague, Czechoslovakia
Chairman: Reznicek, R.

International Organizing Committee

Asanuma, T., Honorary Chairman, Japan
Tanida, Y., Chairman, Japan
Kobayashi, T., Secretary General, Japan
Boucher, R.F., U.K.
Carlomagno, G.M., Italy
Cheng, K.C., Canada
Crowder, J., U.S.A.
Dewey, J.M., Canada
Doi, J., Japan
Dracos, T., Switzerland
Fomin, N.A., C.I.S.
Gostelow, J.P., Australia
Lee, T.S., Korea
Merzkirch, W., Germany
Miyashiro, H., Japan
Reznicek, R., Czechoslovakia
Stasiek, J., Poland
Takagi, M., Japan
Tanaka, H., Japan
Veret, C., France
Wei, Q.D., China
Weinberg, F., U.K.
Yang, W.-J., U.S.A.

Organizing Committee

Tanida, Y., Chairman
Kobayashi, T., Secretary
Akiyama, M.
Ayukawa, K.
Doi, J.
Hiroyasu, H.
Imaichi, K.
Inoue, T.
Kamemoto, T.
Kameoka, T.
Kasagi, N.
Kato, H.
Kiya, M.
Matsuo, K.
Miyashiro, H.
Mochizuki, S.
Murakami, S.
Nakajima, Y.
Nakamura, I.
Okajima, A.
Okamoto, Y.
Takagi, M.
Takayama, K.
Tanaka, H.
Tanaka, M.
Yano, M.
Yoshida, T.

Executive Committee

Tanida, Y., Chairman
Kobayashi, T., Secretary

General Affairs
Aoki, K.
Fujii, K.
Hibi, K.
Ishizuka, M.
Iwanaga, M.
Kamemoto, T.
Kasagi, N.
Kobayashi, T.
Mochizuki, S.
Yoshida, T.

Financial Affairs
Inoue, T.
Kameoka, T.
Nakayama, Y.
Obokata, T.
Tanaka, H.
Yano, M.

Scientific Program
Akiyama, M.
Kawahashi, M.
Matsumoto, Y.
Miyashiro, H.
Tanahashi, T.

Social Program
Arakawa, T.
Doi, J.
Fukase, T.
Hosoi, Y.
Nagumo, S.
Nakajima, Y.
Ohashi, M.
Okamoto, Y.
Sawada, T.

Exhibition
Akino, N.
Saga, T.
Takagi, M.
Takakura, S.

Contents

2.2 Vortices

2.3 Turbulent Flows

3. Flows Around Bodies

3.1 Flows Around Moving Bodies

3.2 Flows Around Columns

Author Index

Special Lectures

Quantitative Flow Visualization of Three-Dimensional Flows

CHING-JEN CHEN[1], LUKE J. CHEN[2], AND YOU-GON KIM[3]

Department of Mechanical Engineering
Iowa Institute of Hydraulic Research
The University of Iowa, Iowa City, Iowa 52242, U.S.A.

Summary

Quantitative flow visualization has the advantage of providing instantaneous full field velocity measurement of complex flows. In this article, methods of quantitative velocity measurement are briefly reviewed. Then two benchmark experiments of three-dimensional flow conducted by the authors are presented. The purpose of the experiments is to study the complex three-dimensional motion and to measure quantitative velocity profiles and stream patterns for verification of the accuracy of the three-dimensional numerical simulation. From the investigation of flow past a cubic chamber with offset inlet and outlet flows and the lid driven rotating flow, complex three-dimensional flows are elucidated and quantitative data are documented.

Methods of Quantitative Velocity Measurement

Flow visualization has long been used by researchers to investigate fluid motion because of its ability to reveal fluid dynamics at multiple points and to provide direct visualization of particles or streak motion. Recently, many methods of quantitative flow visualization have been proposed which not only achieve the visualization of fluid motion, but also provide the full field quantitative velocity measurement. Good descriptions of quantitative flow visualization can be found in review articles by Adrian (1986, 1991) and Hesselink (1988), and the Flow Visualization Handbook by Yang (1989).

The fundamental principle for quantitative flow visualization at multipoints of fluid flow is simply the definitions of velocity $\vec{V} = \frac{\Delta \vec{s}}{\Delta t}$. Here Δt is the time interval in which the vector displacement $\Delta \vec{s}$ is measured. $\vec{V}$ is the velocity vector of the particle, or the fluid element if the particle moves with the fluid, at the point of measurement. The shorter the measurement time, Δt, the more precisely the measurement may be designated at a position in the measuring field. However, the shorter the measuring time, the greater the error that may be introduced into the velocity measurement due to the uncertainty in the displacement length measurement and the finite size of the particle marker. On the other hand, the longer the displacement measurement is allowed, the lesser the precision with which one may designate the position at which the velocity is measured.

---

1 Currently, Professor and Dean, College of Engineering, Florida A&M University/Florida State University, Tallahassee, Florida 32316-2175, U.S.A.

2 Currently, Research Engineer, Industrial Technology Research Institute, Bldg. 64, Hsinchu, Taiwan 310, Republic of China

3 Currently, Senior Researcher, Fast Breeder Reactor Coolant Department, Korea Atomic Energy Research Institute, P.O. Box 76, Daeduk-danji, Daejon, Korea 305-353

Whatever the methods used, the quantitative velocity measurement requires measuring the distance between two end points of the particle displacement and the time interval in measuring the displacement distance. According to how the displacement distance, $\Delta \vec{s}$, is measured, various methods are proposed and investigated. In general, fluid particles or seeded particles are made visible by an optical source. For quantitative visualization of a three-dimensional flow in a plane, a laser light is often used since it can be easily managed by an optical system to generate the necessary illumination in a narrow sheet. To visualize the three-dimensional flow field, a reconstruction can be made from the series of planer quantitative measurements. Another approach is to measure the three-dimensional velocity vector directly by determining the three-dimensional vector at each position of the measuring field. This approach, in general, requires a holographic recording or a stereoscopic imaging or tracking three-dimensional motion of the seeded particles.

In the following, several methods of quantitative velocity measurement are briefly described.

1. Particle Streak Velocimetry (PSV)

In the particle streak velocimetry (PSV) method, the displacement of seeded particles or the length of the particle streaks, Δs, are measured from the recording media. Knowing the time interval of the measurements, Δt, quantitative velocities are determined from $\Delta s/\Delta t$. To apply the PSV method, a light sheet is formed in the test section to illuminate the seeded particles. The particles scatter the light from the light sheet and form the flow image. The recording device is then exposed for a time interval such that the illuminated particles travel more than their own diameter during the exposure time. The resulting streaks of the particles in the image plane are measured for length. Given the exposure time of the imaging system, one may determine the particle velocity. To determine the direction of flow velocity in the image recorded, one may use a streak coding scheme to identify the heads and tails of the streaks. Further details on the PSV method and the experimental system can be found, for example, in the articles by Khalighi (1989) and Walter and Chen (1992). A disadvantage of the particle streak velocimetry method is that streaks may overlap if the seeding of particles is dense and there may be difficulty in measuring the streak lengths accurately. Also, the PSV method has difficulty in recording rapid changes in the velocity vector, such as fluctuations in turbulence.

2. Particle Tracking Velocimetry (PTV)

The particle tracking velocimetry (PTV) method essentially uses the same principle as the particle streak velocimetry method by finding the streak length. However, the streak lengths are obtained by the superposition of many instantaneous particle images recorded in sequence. It is important that the sequential positions of the same particle are properly identified and tracked. The velocity is thus measure from the streak created by superimposing multiframes on the image processor with each image having a time interval of prescribed duration. The particle tracking velocimetry (PTV) method is sometimes referred to as the low image density mode of the particle image velocimetry (PIV) method (Adrian, 1991). Further details of the particle tracking velocimetry method can be found in the work of Kobayashi et al. (1987) and Ohmi et al. (1991), Boucher and Kamala (1991).

3. Laser-Speckle Velocimetry (LSV)

This method has also recently been called particle image displacement velocimetry(PIDV) or the high image density mode of particle image velocimetry (PIV). This method involves the illumination of a flow densely seeded with small tracer particles within a thin laser light sheet. The images of the particles are recorded with a multiple exposure photographic technique with short time intervals. The distance between corresponding particle images can be determined by focusing the laser beam to interrogate a small area of the multiply exposed photographic film. The diffraction pattern produced by the coherent illumination of the multiple images in the photograph transparency generates Young's fringe. This fringe has an orientation that is perpendicular to the direction of the local displacement and a spacing inversely proportional to the displacement. The velocity vector is thus determined knowing the displacement, the direction, and the time interval. The interrogation of local images can be repeated in an automated process for the entire flow field.

One of the limitations of the laser-speckle velocimetry method is that it is not possible to determine the direction of the velocity vector. Thus, there is ambiguity in the direction of the flow. However, this can be resolved using a "velocity bias technique" in which a uniform reference motion is added to the flow, thus superimposing a velocity shift to the real flow field. A properly chosen shift can be introduced to insure that all image displacement occurs in the same direction thereby eliminating the ambiguity. The true flow field can then be recovered later by removing this artificial shift from the measured velocity data. The details the LSV or PIDV technique are described by Shih et al. (1992), Lourenco and Krothapalli (1987), Lourenco et al. (1988), and Adrian (1991).

4. Digital Particle Imaging Velocimetry (DPIV)

In the digital particle imaging velocimetry (DPIV) method, the velocity measurement is achieved by processing two sequential digital flow images, each with a single exposure of flow particles. This differs from the laser-speckle velocimetry method in that the determination of velocity measurement does not rely on an optical diffraction pattern of Young's fringe. Instead, in the DPIV method, the relation between two sequential digital images is modeled as input and output, such that in the interrogating area a correlation is found between the first and second image for the most probable displacement. This leads to the determination of the velocity. This DPIV method was originally proposed by Willert and Gharib (1991). This method was further developed by Chen and Chen (1991) and was called the digital correlation velocimetry (DCV) method. The above two methods use the idea of a correlation between the first and second image to determine the most probable displacement. Recently, Chen, Kim and Walter (1992) proposed to treat the two sequential images as a digital vector displacement. Instead of conducting correlations between the sequential images, a direct count of all possible velocity vectors formed between the two successive images are digitally constructed on the digital vectorgram. The digital vector velocimetry (DVV) method invokes the idea of a direct count of the most populated state of the digital vectors. In other words, the most repeated or intense digital vector in the interrogating area is the velocity vector of the fluid motion. Therefore, the DVV method can be implemented without optical diffraction pattern and computation of correlations. Okamoto, et al. (1991) modified the DVV method in their flow visualization. Further details of the digital imaging velocimetry method (DIV) or digital correlation velocimetry method (DCV) and digital vector velocimetry method (DVV) can be found in the above cited references.

5. Three-Dimensional Particle Tracking Velocimetry (3D PTV)

The term three-dimensional particle tracking velocimetry (3D PTV) method is used here to describe the most recent development of quantitative flow visualization of three-dimensional velocity vectors. Different names are used by various authors. For example, Kimura and Kohno (1991) use the term spatio-temporal imaging correlation to describe their direct measurement of three-dimensional velocity vectors in a flow field. On the other hand, Yamamoto et al (1991) describes a new particle identification method for three-dimensional particle tracking velocimetry (3D PTV). In the early attempts of 3D PTV by Chang, Watson, and Tatterson (1985) and Nishino, Kasagi, and Hirata (1989), each tracer particle is tracked along the flow field by calculating its three-dimensional position using a stereoscopic measurement method. This method requires rather elaborate imaging processing for the spatio and temporal identification of hundreds of tracer particles in the flow field. For the temporal identification, some known flow characteristics, such as mean velocity, are necessary. And also, the tracer particle density in the flow field cannot be dense. Kimura and Kohno (1991) proposed the spatio-temporal correlation method as shown in Fig. 1. In a small n x n pixel interrogating area, a close correlation is performed to find the δx, δy, and time interval, such that maximum correlation is achieved. The δx, δy, and the time interval found are the displacement of the tracer particle group in the interrogating area. Consequently, the three-dimensional vector from the initial image to the subsequent image is identified. Further details of the three-dimensional particle tracking velocimetry (3D PTV) method can be found in the above cited references or in work by Papantoniou and Mass (1990).

The three-dimensional particle tracking velocimetry (3D PTV) method is just beginning to develop. It may eventually lead to a useful quantitative flow visualization for three-dimensional flow. However, at the present time, most of the three-dimensional quantitative visualizations are done with the method where two-dimensional projections are made for the three-dimensional flow. Therefore, a series of two-dimensional quantitative visualizations can be used to reconstruct the three-dimensional flow field. In the following two examples conducted by the authors, using the digital correlation velocimetry (DCV) method and the digital velocity vector (DVV) method are illustrated.

Three Dimensional Flow in Cubic Chamber With Offset Inlet and Outlet

The flow system in this study is shown in Fig. 2. Water is made to circulate through the cubic chamber (10 x 10 x 10 cm^3) with offset inlet and outlet. The inlet and outlet duct has a cross section of 2 x 2 cm^2. The inlet flow velocity of 0.015 m/s is set in the present study. This gives the Reynolds number of 298, based on the inlet height and inlet velocity. Details of the experimental setup and the explanations of the physics of fluid in the cubic chamber are given by Chen and Chen (1991). Figures 3 and 4 illustrate the qualitative and quantitative flow visualization of a flow passing through the cubic chamber in three planes, each respectively viewed from axial and side views. The measurements were made by the digital correlation velocimetry (DCV) method. The particle streak velocimetry (PSV) method is used to obtain the qualitative view. The study was made for the purpose of providing researchers with a comprehensive understanding of complex flow phenomena in a cubic chamber, and the quantitative measurement as benchmark data for verification of numerical simulation of three-dimensional flow problems.

Quantitative Visualization of Three-Dimensional Flow in Lid-Driven Rotating Flow

This study was made by Kim and Chen (1992) where the lid-driven cylindrical cavity flow in the laminar region was visualized by the digital vector velocimetry (DVV) method. Figure 5 shows the schematics of test sections. The lid, with a 2.5 inch radius and a depth of 5 inches, is driven at the angular speed of 10 rpm. The Reynolds number, based on the angular speed and radius of the cylinder, is 3200. Water was used as a fluid medium. Figure 6 shows the typical particle streak view and the quantitative measurements obtained by the DVV method. Interesting secondary flow is observed from the side view of the cylinder, a flow moving along the rotating disk toward the cylinder wall and turning downward to the bottom and returning to the center of the axis. However, the secondary flow quickly turns in a radical direction to move up from the bottom of the cylinder to the top of the lid at approximately one-half of the cylinder radius instead of moving along the axis. This pattern was also numerically verified by Kim (1991). Further details of three-dimensional flow in lid-driven rotating flow can be found in Kim's (1991) work.

Conclusion

Recently, quantitative flow visualization has become a very popular and powerful tool for researchers and engineers to analyze complex fluid motions. Because quantitative flow visualization provide the full field velocity measurement instead of point-wise measurement, such as laser-Doppler anemometry and hot-wire or hot film anemometry, it greatly simplifies the data acquisition of complex flow fields. Therefore, the quantitative flow visualization can be used to investigate complex three-dimensional flow patterns. In this paper, two complex three-dimensional flows are illustrated using different methods of quantitative flow visualization, namely the particle streak velocimetry method, the digital correlation velocimetry method, and the digital velocity vector method. It is anticipated that the three-dimensional particle tracking velocimetry method will be further developed and become a practical tool to investigate complex three-dimensional flows.

References

Adrian, R.J., "Particle-Imaging Techniques for Experimental Fluid Mechanics," *Annual Review of Fluid Mechanics,* Vol. 23, pp. 261-304, 1991.

Boucher, R.F. and Kamala, M.A., "Near Real-time particle Image Velocimetry (PIV) by Streak Analysis," FLUCOME 91, 3rd Triennial International Symposium on Fluid Control, Measurement, and Visualization, pp. 443-454, 1991.

Chang, T.P.K., Watson, A.T., and Tatterson, G.B., "Image Processing of Tracer Particle Motions as Applied to Mixing and Turbulent Flows," *Chemical Engineering Science,* Vol. 40, No. 2, pp. 269-285, 1985.

Chen, C.J., Kim, Y.G., and Walter, J.A.. "Progress in Quantitative Flow Visualization and Imaging Process," *Proc. 4th Int. Symp. on Refined Flow Modelling and Turbulence Measurement,* Wuhan, pp. KA-15, Sept. 20-23, 1990. Also *Progress in Visualization,* Pergamom Press, Vol. 1, 1992.

Chen, L.J. and Chen, C.J., "Visualization of Three-Dimensional Flow in a Cubic Chamber with Digital Correlation Method," FLUCOME '91, 3rd Triennial International Symposium on Fluid Control, MNeasurement, and Visualization, pp. 407-414, 1991.

Hesselink, L., "Digital Image Processing in Flow Visualization," *Annual Review of Fluid Mechanics,* Vol. 20, pp. 421-485, 1988.

Khalighi, B., "Quantitative Fluid Velocity Measurements by Automatic Analysis of Flow Visualization Images," *Experiments in Fluids,* Vol. 7, No. 2, pp. 142-144, 1989.

Kim, Y.G., "Development of Digital Vector Velocimetry Method and Its Application to Rotating Flows," Ph.D. Dissertation, Department of Mechanical Engineering, The University of Iowa, Iowa City, Iowa 52242, 1991.

Kimura, I., and Kohno, Y., "Measurement of Three-Dimensional Velocity Vectors in a Flow Field Based on Spatio-Temporal Image Correlation," FLUCOME '91, 3rd Triennial International Symposium on Fluid Control, Measurement, and Visualization, pp. 609-615, 1991.

Kobayashi, T., Saga, T., and Segawa, S., "Some Considerations on Automated Image Processing of Pathline Photographs," *Flow Visualization IV* (C. Véret, ed.), Hemisphere Publishing Corporation, Washington, D.C. pp. 241-246, 1987.

Lourenco, L., an Krothapalli, A., "The Role of Photographic Parameters in Laser Speckle or Particle Image Displacement Velocimetry," *Journal of Experiments in Fluids,* Vol. 5, No. 1, pp. 29-32, 1987.

Lourenco, L.M., Krothapalli, A., Buchlin, J.M., and Riethmuller, M.L., "A Non-Invasive Technique for the Measurement of Unsteady Velocity and Vorticity Fields," in von Karman Institute for Fluid Dynamics Lecture Series 1988-06, *Particle Image Displacement Velocimetry,* Rhode St. Genèse, Belgium, 1988.

Nishino, K., Kasagi, N., and Hirata, M., "Three-Dimensional Particle Tracking Velocimetry Based on Automated Digital Image Processing," *ASME Journal of Fluids Engineering,* Vol. 111, pp. 384-391, 1989.

Ohmi, K., Vemura, T., Yamamoto, R., and Coutarrceau, M., "Comparative Measurement between Particle Streak and Particle Tracking Velocimetry," FLUCOME '91, 3rd Triennial International Symposium on Fluid Control, Measurement, and Visualization, pp. 435-442, 1991.

Okamoto, K., Madarame, H., and Shioya, R., "Measurement of Velocity Distribution in a Tank by Velocity Vector Histogram," FLUCOME '91, 3rd Triennial International Symposium on Fluid Control, Measurement, and Visualization, pp. 601-607, 1991.

Papantoniou, D., and Mass, H.G., "Recent Advances in 3-D Particle Tracking Velocimetry," *Proc. 5th Intl. Symp. On Application of Laser Techniques to Fluid Mechanics and Workshop on the Use of Computers in Flow Measurements,* Lisbon, pp. (18.4) 1-5, July 1990.

Shih, C., Lourenco, L., Van Dommelen, L., and Krothapalli, A., "Unsteady Flow Past an Airfoil Pitching at a Constant Rate," *AIAA Journal,* Vol. 30, No. 5, May 1992.

Walter, J.A. and Chen, C.J., "Visualization and Analysis of Flow in an Offset Channel," *Journal of Heat Transfer* (to appear) Vol. 114, 1992.

Willert, C., and Gharib, M., "Digital Particle Image Velocimetry," *Experiments in Fluid,* No. 10, pp. 181-193, 1991.

Yamamoto, F., et al., "A New Particle Identification Method for 3D PTV," FLUCOME '91, 3rd Triennial International Symposium on Fluid Control, Measurement, and Visualization, pp. 617-622, 1991.

Yang, W.J., *Handbook of Flow Visualization,* Hemisphere Publishing Co., New York, 1989.

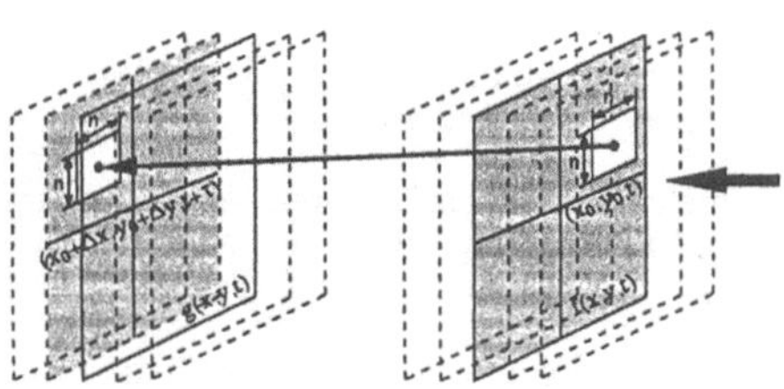

Figure 1.

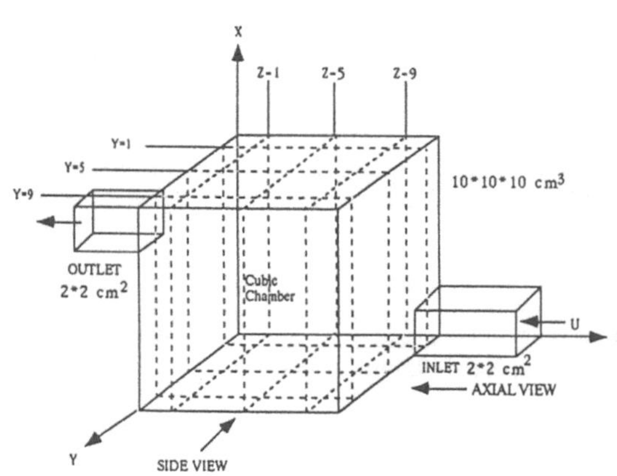

Figure 2

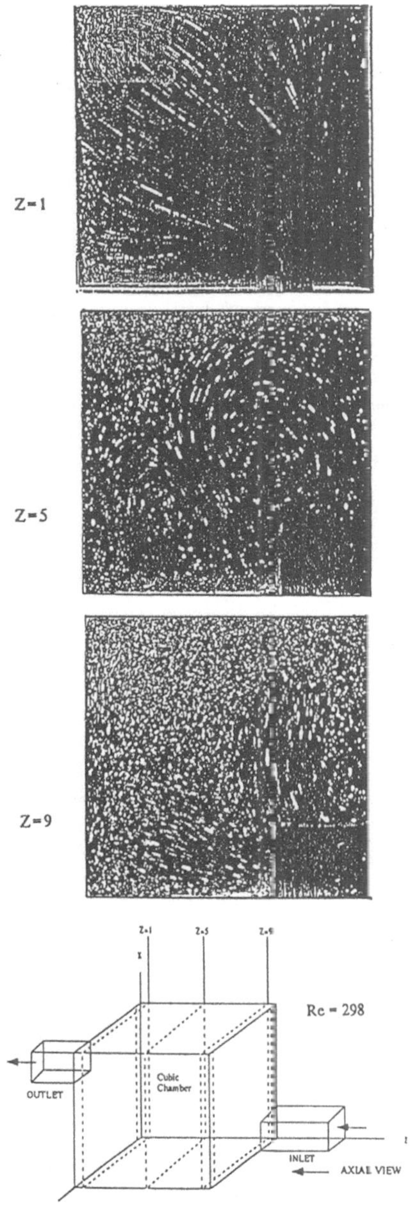

Qualitative Flow Visualization at the Plane

Z = 1, 5, 9 of the Axial View

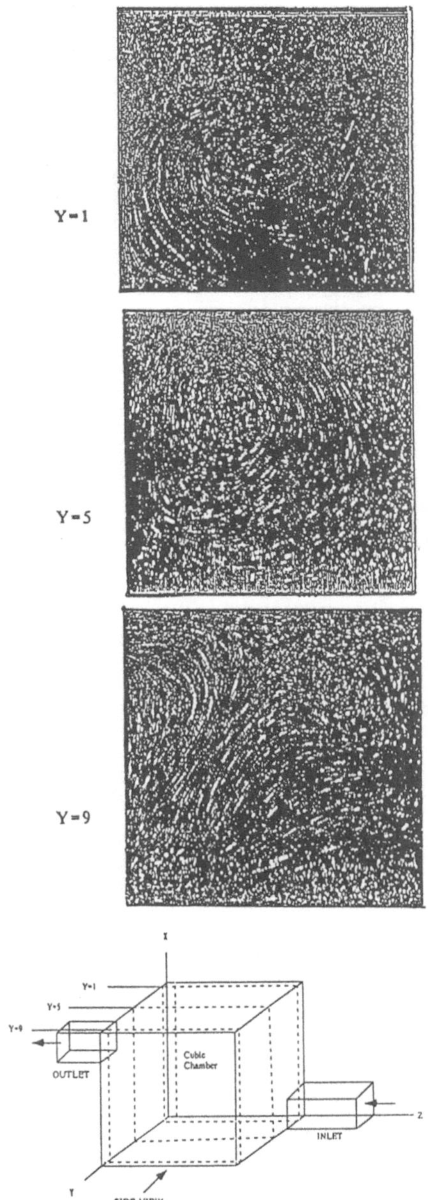

Qualitative Flow Visualization at the Plane

Y = 1, 5, 9 of the Side View

Figure 3

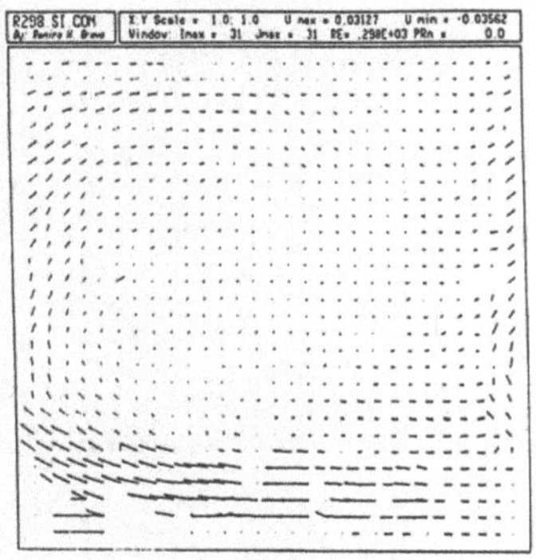

Velocity Profiles of X-Component on Y=1 Plane

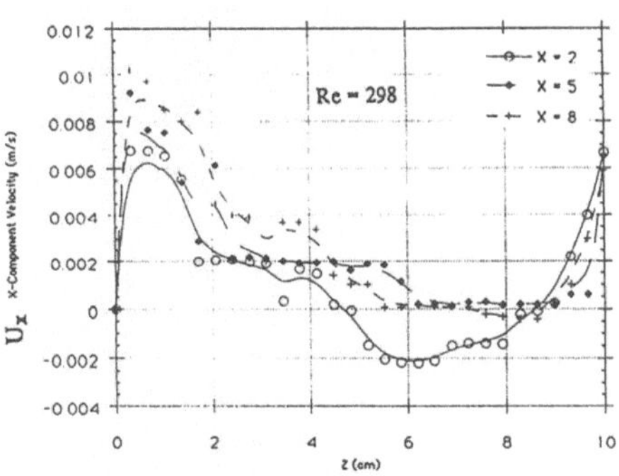

Velocity Profiles of Z -Component on Y=1 Plane

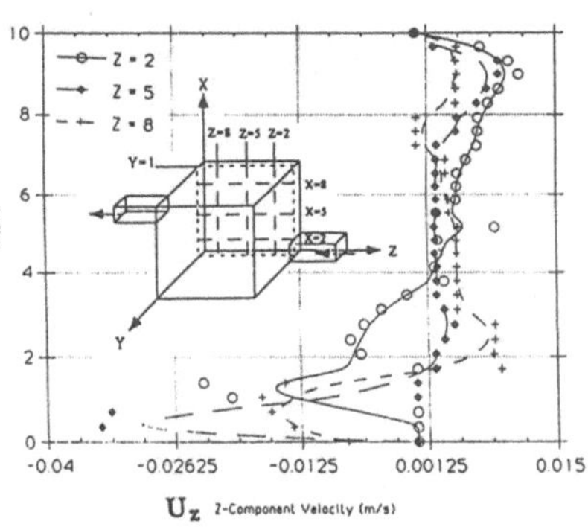

Full-Field Velocity Measurement at the plane Y = 1 of the Side View

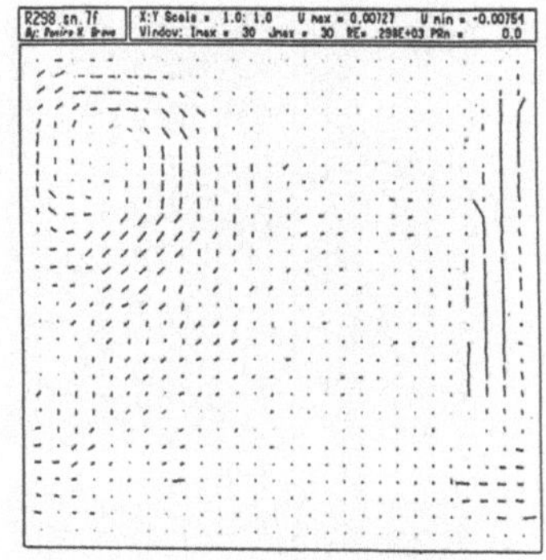

Velocity Profiles of X-Component on Y=5 Plane

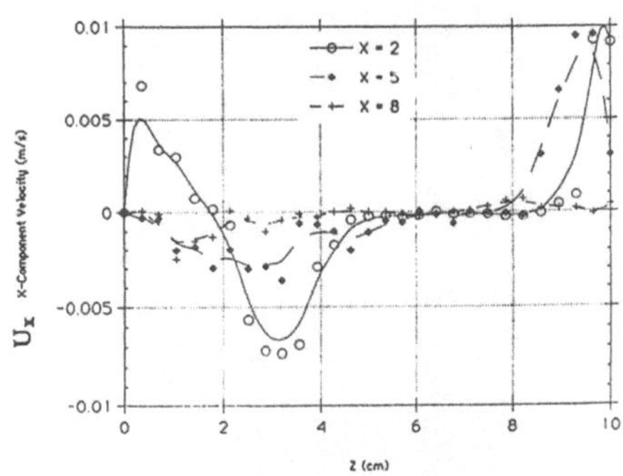

Velocity Profiles of Z-Component on Y=5 Plane

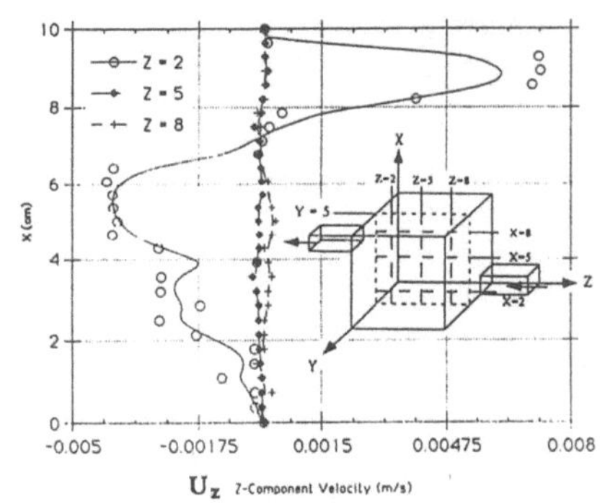

Full-Field Velocity Measurement at the plane Y = 5 of the Side View

Figure 4

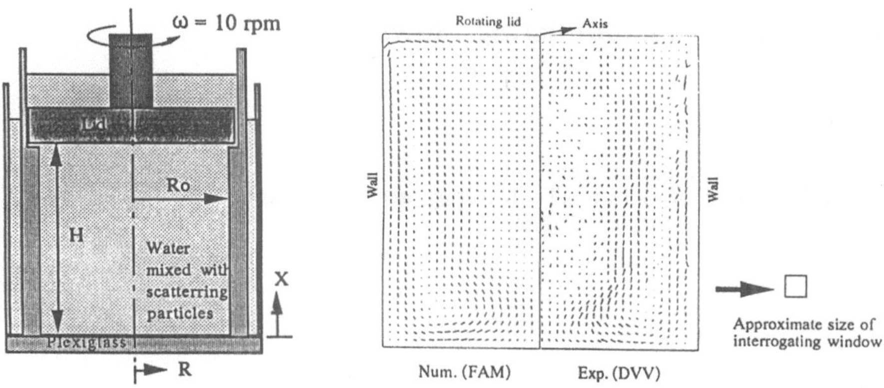

Figure 5 Schematic test cell and comparison of experimental (DVV) data and computational (FAM) result

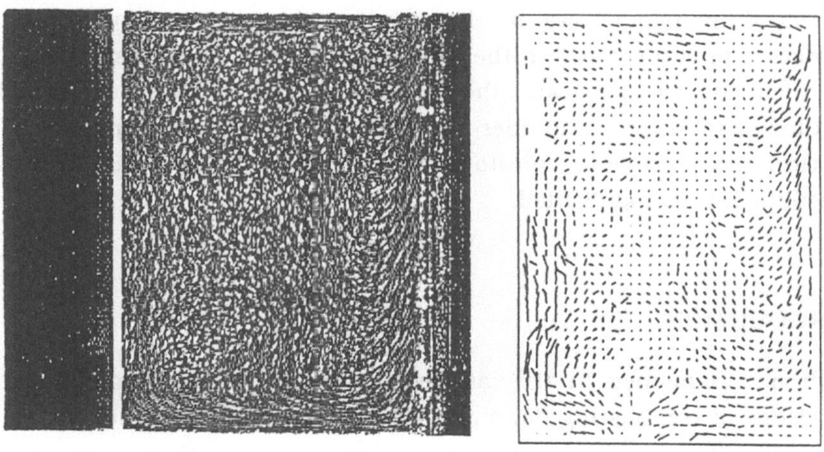

Figure 6 Particle streak view and result of DVV measurement

Visualization of Combusting Flows

J.H. Whitelaw

Head, Thermofluids Section,
Imperial College of Science,Technology and Medicine,
Exhibition Road, London, England.

SUMMARY

The flow characteristics of internal-combustion and gas-turbine engines can be determined in terms of local velocities, temperatures and concentrations and these measurements can be greatly facilitated by visualisation so that the general features of the flow are known prior to more detailed examination. This lecture considers some small flames, aspects of impinging flows of relevance to fuel sprays striking a surface, the nature of vortices within the cylinders of internal combustion engines and their consequences and the distribution of heat release within the annular combustion chamber of a small gas turbine engine. The relationship between photographic visualisation and detailed measurements is highlighted.

INTRODUCTION

Much has happened in the last decade to allow improvement of our knowledge of combusting flows and the various measurement techniques now available can provide qualitative and quantitative information, usually with greater costs associated with the latter. As in all flows, therefore, cheap and rapid qualitative information is desirable prior to the use of the more complex instrumentation required for the quantitative measurements. In the case of combusting flows, the variations in colour provide a convenient way in which to identify some important features and this paper is concerned mainly with the use of visualisation and photography to guide more detailed investigations of configurations ranging from small and open flows to those confined within gas turbines and internal-combustion engines. This simple, and often preliminary, method of investigation should be considered in the context of more complete investigations and some aspects of those are described here. Also, the following paragraphs provide an indication of the available measurement methods.

Weinberg (1963) provides a clear impression of the extent to which visualisation, and shadowgraph and Schlieren photographs, can assist the interpretation of small flames. This simple approach has been complemented, and in some cases improved, by the use of diodes to provide signals proportional to amplitude and, therefore quantitative images with time response determined by the particular instrumentation. This approach has been used in a number of ways, including quantification of the intensity of the coherent structures in jet flow by Chao, Han and Jeng (1990), of laser-induced fluorescence to give information of scalar transport in turbulent jets by van Cruyingen, Lozano and Hanson (1990), of heat release in flames through fluorescence of the OH radical by Gutmark, Parr, Hanson-Parr and Schadow (1991) and to observe the location of droplets or particles as by Adrian and Landreth (1990) in forms of particle-image velocimetry to provide information of velocity, size and concentration of the discrete phase. These methods have the common feature of observation by an array of diodes, digital processing of amplitude signals and laser sheets to ensure near two- dimensional images. The extent to which they measure two-dimensional images in real time depends on the thickness of the sheet, the means of focussing the images and the time-response of the interface, computer and related software.

Measurements at single locations, often with better spatial resolution and time response than the methods described above, can be obtained with a number of optical techniques including laser-Doppler and phase-Doppler velocimetry for the measurement of velocity and droplet or particle size, concentration and flux as described for example by Durst, Melling and Whitelaw (1981) and Hardalupas, Taylor and Whitelaw (1989,1990a,b); Rayleigh and Raman scattering for the measurement of density, temperature and concentration as for example by Dibble and Hollenbach (1981), Pitts and Kashiwagi (1984), Kadota, Zhao and Miyoshi (1990) and Enotiadis (1991) and by Drake (1988) respectively; and where Mie-scattered signals corrupt the Rayleigh and Raman signals, by CARS as successfully used in a few laboratories, for example by Eckbreth (1992).

Two-colour pyrometry has also been used to measure temperature but with uncertainties which remain to be established. Further details of these techniques can be found, for example in Taylor (1992).

The flow-visualisation results of the following sections are complemented by velocity and droplet-size measurements, using Doppler-shifted light, and by temperatures measured with small-diameter, digitally-compensated thermocouples, as described by Heitor, Taylor and Whitelaw (1985, 1987). They are presented in a sequence of increased complexity of flow, so that we begin with small gas flames, move to the gas, liquid and solid-fired flames of the type used in furnaces and end with consideration of the recirculating combusting flows within gas-turbines and internal-combustion engines. In all cases, photographs of the flames are used to provide an impression of the physical processes which are subsequently confirmed and elaborated by local values of flow properties. The examples are taken exclusively from work performed at Imperial College over a number of years.

Small Gas Flames

Two examples are considered here the low Reynolds number flames of Durst (1972) and the opposed-jet flames of Mastorakos, Taylor and Whitelaw (1991a,b). Both flames burned methane gas in air with low heat release and with important characteristics which were readily observed by eye.

The small diffusion flames of Durst were pulsed at low frequencies and used, in the early days of laser-Doppler anemometry, to show that this instrumentation was able to measure in combusting flows with fluctuations. Thus, signals from a laser-Doppler anemometer were obtained from a sequence of flames corresponding to increasing velocities of methane gas in the exit plane of a burner pipe with an imposed frequency of 15Hz and examined to ensure that they corresponded to the visual observation. At the same time, the probability density distributions of velocity were shown to be bimodal, as expected with signals which corresponded nearly but not quite to sine waves. This comparison between the results of, at that time, a new technique for combustion studies and visualisation provided a convincing justification for a substantial number of subsequent investigations as shown, in small part, in the following descriptions.

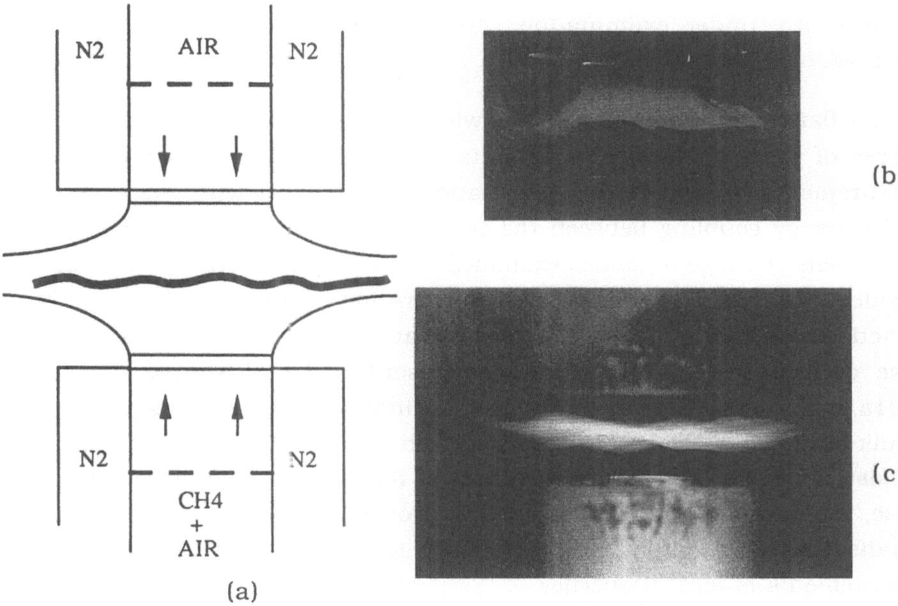

FIGURE 1 a) Opposed-jets burner; b) Isothermal flow with the upper jet seeded by AlO_2 particles; c) Diffusion stagnation flame.

The opposed jet flames of Mastorakos *et al.* are considered in detail elsewhere in this Symposium and are used here to demonstrate the link between visualisation and local measurements of velocity and temperature. They involved a jet of methane and air opposing a jet of air and Figure 1 shows the geometrical arrangement, isothermal opposed jets with one stream seeded and a turbulent diffusion flame where the fuel jet is methane alone, with jets separated by 20 mm or two pipe diameters. In this case, and mindful of the dangers of relaminarisation due to the temperature rise, turbulence-promoting screens were located in the pipes carrying the opposed flows which had average velocities of 1.5 m/s . The photograph shows that the isothermal jets, with the upper jet seeded with aluminum oxide particles, impinge and turn through 90 degrees with a tendency for waviness of the interface with a length scale larger than that of the expected turbulent integral scale. Corresponding velocity measurements, by laser-Doppler anemometry, confirmed an increase in the axial fluctuations above those of the turbulent jet without impingement and the implications for correlations between the two jets are presently under examination with the visualisation providing the incentive.

The flat flame, and others like it but with different momentum ratios and degrees of premixedness, also exhibits a waviness which has prompted measurements of temperature correlation and spectra with evidence of low-frequency coupling between the two sides of the opposed flame. The more extensive range of flames examined visually by Mastorakos *et al.* has provided the basis for choosing configurations with which to study extinction and the influence of strain of various scales. First results of these quantitative investigations are described by Mastorakos *et al.* (1991a,b) and show, for example, that extinction requires consideration of bulk strain as well as that associated with small scales. The flat flames will also allow investigation of the effect of the concentration of a second phase, such as a liquid or solid fuel, and once again it is expected that visualisation will guide the application of instrumentation for the measurements of local properties.

Burners for Furnaces

The flames discussed here formed after the mixing of fuel and air directed vertically upwards with various arrangements and all of which involved flow from an annulus of inner and outer diameters 18 and 50.8 mm, with a corresponding Reynolds number of 53700. The fuel, which comprised natural gas, kerosene oil and pulverised coal, was added to the annulus or injected along the axis, depending upon the configuration, and the heat-release rates ranged to more than 100 kW and considerably greater than the flows of the previous paragraphs.

Figure 2a shown one of the burner arrangements, that of Hardalupas, Liu and Whitelaw (1991), fuelled centrally with a kerosene spray and air in the annulus. The sequence of three photographs of Figure 2b correspond to the spray without and with the annular air and no combustion and the spray with annular air and combustion. The quarl diffuser is absent, a 43.8 mm stabiliser disk was used and sheet lighting was required to observe the non-combusting flows. It is evident that the presence of the annulus air tended to disperse the kerosene droplets from the centreline and thereby to shorten the spray and increase its diameter. The separated flow caused by the disk ensured that some of the droplets assumed negative velocities and moved towards the disk so that, in the

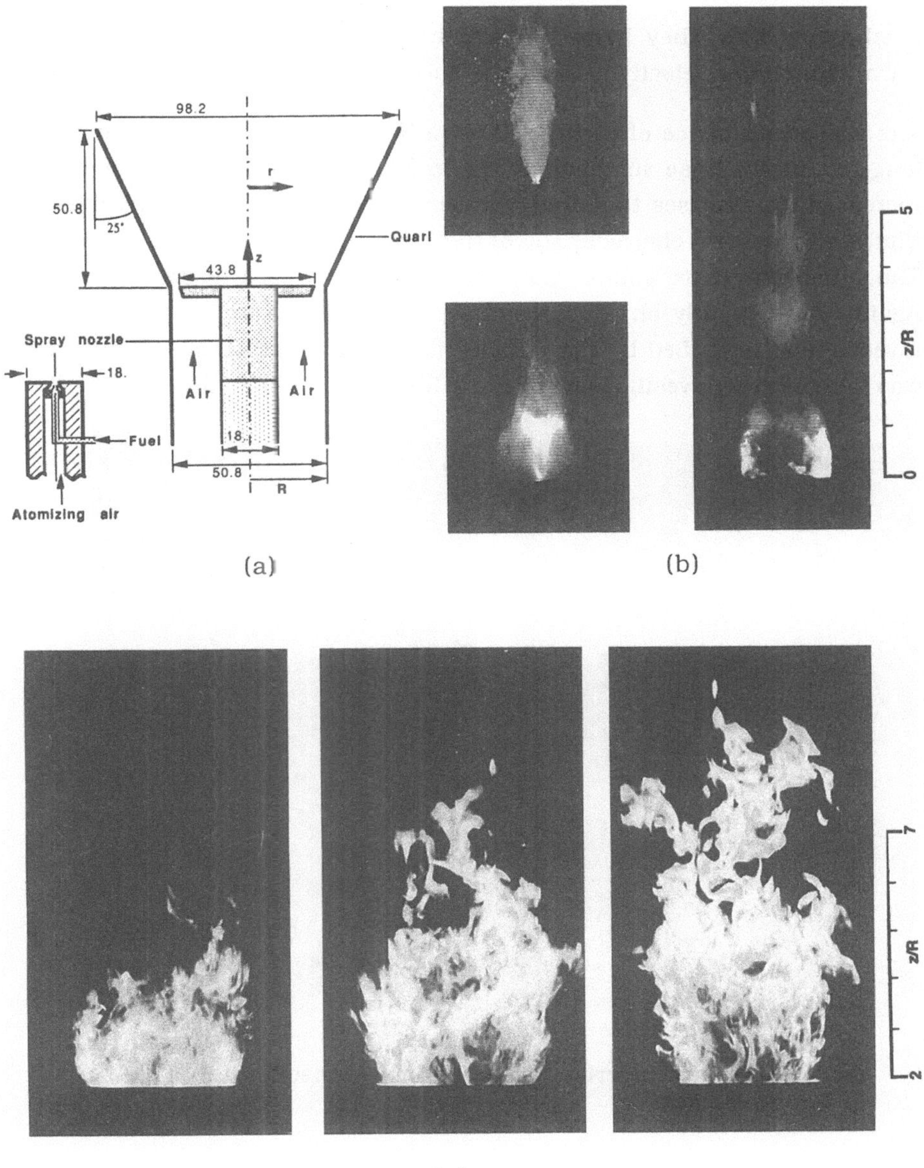

FIGURE 2 a) Kerosene-fuelled burner arrangement; b) Isothermal spray without (top left) and with (bottom left) annulus air and spray flame (right); c) Spray flames stabilised with disks of 43.8, 38.2 and 31.6 mm diameters (left to right).

combusting flow, they helped to stabilise the flame by evaporating and burning with low velocity close to the surface.

The second sequence of pictures, Figure 2c, shows the decrease in flame length and increase in combustion intensity as the disk diameter is increased; in all cases the quarl diffuser is present in the photographs so that the flames are characteristic of the burners found in furnace practice. The flame length is almost three times longer with the smallest disk. Again, these rapidly obtained results provided guidance for the detailed investigation described by Hardalupas *et al.* (1991) and a similar approach was taken in the investigations of Hardalupas *et al.* (1990a,b).

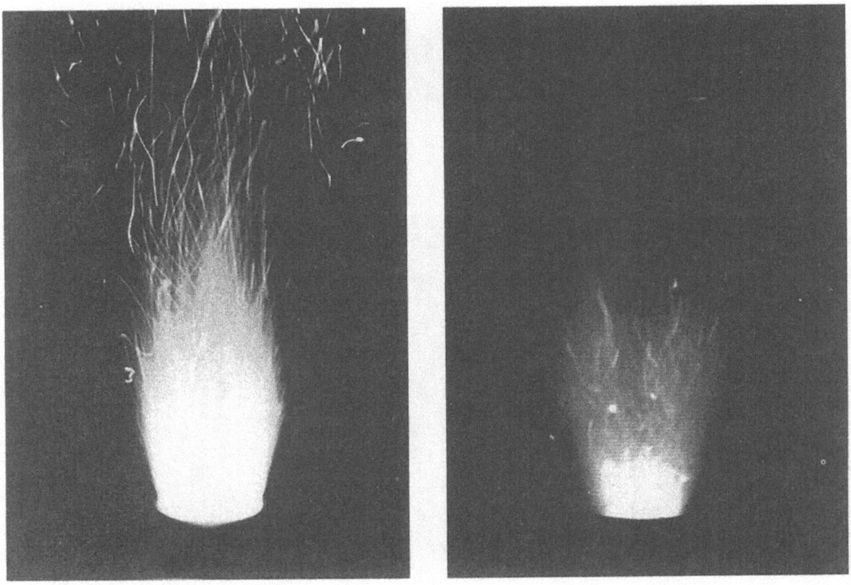

(a) (b)

FIGURE 3 a) Pulverised coal- and b) kerosene-fuelled flames in a quarl burner.

The sequence of Figure 3 represents the difference between atomised kerosene and pulverised coal added to the annular air flow and without a kerosene spray, as described by Milosavljevic, Taylor and Whitelaw (1990). The pathlines of burning kerosene droplets and coal particles can be clearly seen within the swirl-stabilised flames. As expected, the kerosene droplets evaporate rapidly and the pathlines are limited to regions near the quarl. In contrast, the coal pathlines over the entire length of the photograph. Both flows show potential for the measurement of temperature by two-colour pyrometry, with the measurement corresponding to the soot mantle burning around the droplet or to the particle or its surrounding volatiles.

As a final comment on quarl burners, it is possible to arrange a flame which is lifted away from the burner and stabilised, though not firmly, in a region of low velocity gas as shown in Figure 4a. The addition of a low-energy pressure signal at an appropriate frequency to the air in the upstream settling chamber leads to the stable and attached flame of Figure 4b, as demonstrated by Milosavljevic (1992). The reasons for this occurrence are unknown though it can be conjectured that the shear layer is energised to provide eddies of size and frequency appropriate for combustion and that this process is strongest close to the exit of the burner.

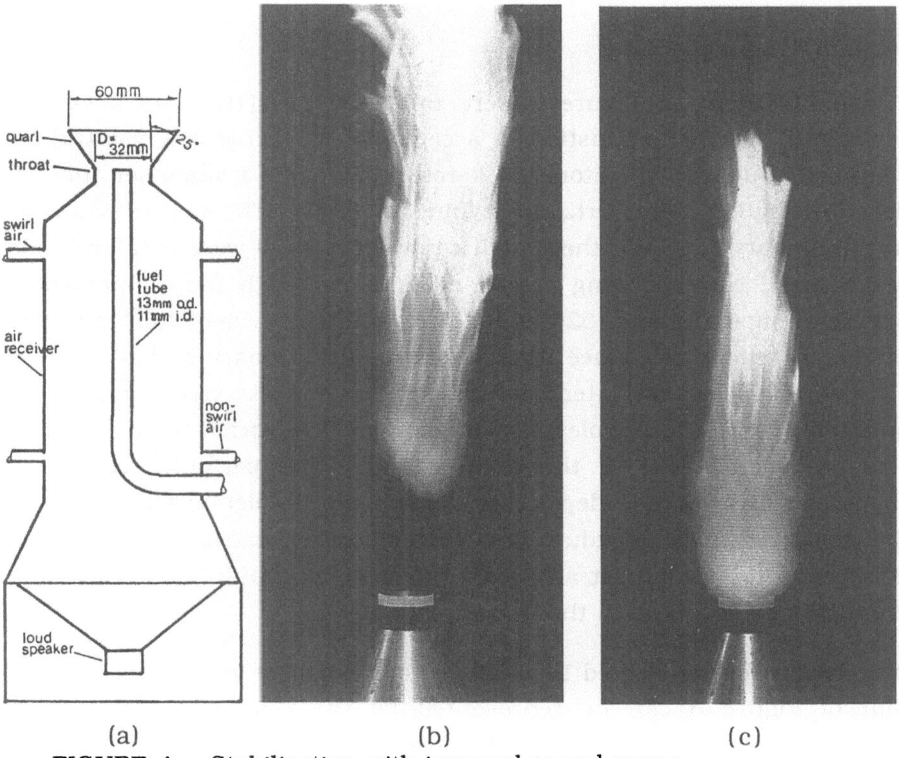

FIGURE 4 Stabilisation with imposed sound waves.
a) Burner arrangement; b) Flame with swirl and no sound input;
c) Flame with sound input, 30W and 190 Hz, and no swirl.

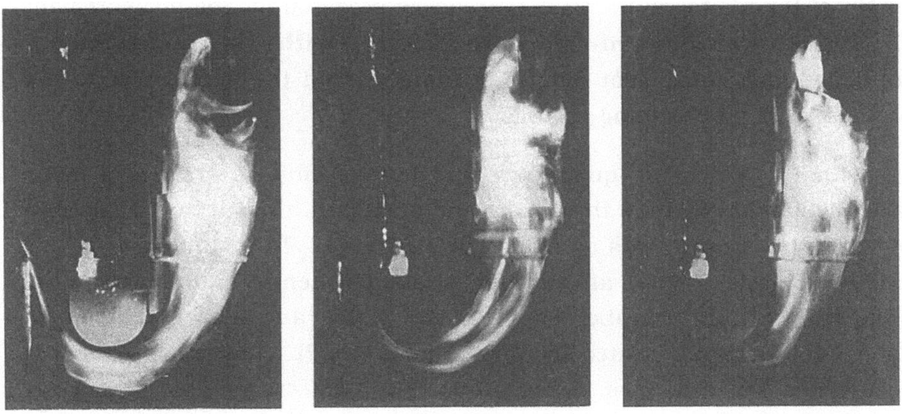

(a) (b) (c)

FIGURE 5 Flames in a reverse-flow gas-turbine combustor at air-to-fuel ratio of 42. Air preheat temperatures of a) 523, b) 473 and c) 423 °K.

Gas-Turbine Combustor

The photographs of Figure 5 were taken by Liu (1992) to show the regions of intense combustion in a sector of a reverse-flow gas-turbine combustor fuelled with atomised kerosene through a vaporiser tube as described by Liu, Perez-Ortiz and Whitelaw (1991). The equivalence ratio in the primary zone and the overall equivalence ratio were 0.72 and 0.35 respectively, corresponding to take-off condition with the air preheated between temperatures of 523 and 423°K, and it is evident that the more intense reaction took place downstream of the primary region where a cloud of droplets was later confirmed by sheet-light visualisation and quantified by phase-Doppler measurements. The vortex in the primary zone is also evident in the photographs. The performance of the combustor was found to depend upon the liquid droplets discharged from the vapouriser, and the reduction of air temperature caused an increase of liquid from the vapouriser and consequent increase of luminous flame in the primary region and in the downstream.

It is sometimes considered that the flows in complex geometries such as that of Figure 5 can be represented by the numerical solution of differential equations in finite-difference form and Chow and McGuirk

(1991), for example, have described attempts to calculate gaseous flames similar to the kerosene flames of Figure 5. As one example of the need to interact with experiment, they found that the trajectory of the flow from the upper dilution holes influenced the flow to an extent that it was necessary to determine the angle experimentally. The flow-visualisation results of Tse (1988), based on the method described by Sivasegaram and Whitelaw (1986), provided this information.

Internal-Combustion Engines

Many sequences of photographs have been obtained in internal-combustion engines, though all required skill and care, and involved some expense. In general, they were obtained in order to improve knowledge of the physical and chemical processes within the engine and, on some occasions, to guide subsequent measurements by alternative optical measuring methods. Margary *et al.* (1990) have described one such investigation in a spark-ignition engine where velocities in the region of the flame front were measured subsequently but here the example of a Diesel engine is chosen and the results of Arcoumanis *et al.* (1991). In these and other cases, optical access is essential to the use of photography and both of the above single-cylinder engines were designed and constructed with optical measurements in mind.

The Diesel engine of Arcoumanis *et al.* was a single-cylinder Hydra engine manufactured by Ricardo and mounted on an instrumented test bed. The combustion chamber was similar to that of the Ford York 2.5L HSDI engine and optical access was provided through the piston and, where required, through windows in an annular steel ring in an elongated cylinder. The piston arrangement was complex to avoid oil leakage and involved a 30 mm thick insert of quartz. High-speed photography was employed with light emitted by the combustion process. A Photec high speed camera was used at 6000 frames/s corresponding to an effective filming speed of 12000 frames/s as a result of the half-frame 16-sided rotating prism. The aperture of the 45 mm objective lens remained open at f/2.8 and the film was 16 mm Ektachrome at ASA 400. When loaded with 30 m of film, the camera had some 250 ms of run time which was sufficient to capture two consecutive combustion events at an engine speed of 1000 rpm. The engine also required careful operation to allow measurements without oil films and the procedures are described by Arcoumanis *et al.*

Figure 6 presents a sequence of photographs for which the end of fuel injection overlapped by some 4 degrees of crank angle the beginning of combustion so that some of the fuel burned before it reached the wall, which was not the case at lower loads. The four sprays are visible in the photographs up to, and perhaps beyond, 10 degrees ATDC and the presence of combustion prior to the sprays striking the wall is evident in earlier pictures. The swirl motion was responsible for the macroscopic mixing which tended to make the combustion homogeneous and almost all fuel had burned by 50 degrees ATDC.

The photographic study revealed many features which influenced local measurements and deductions about performance and emission tests. They include the variation in location of ignition sites, combustion duration and spray characteristics as a function of load, speed and timing.

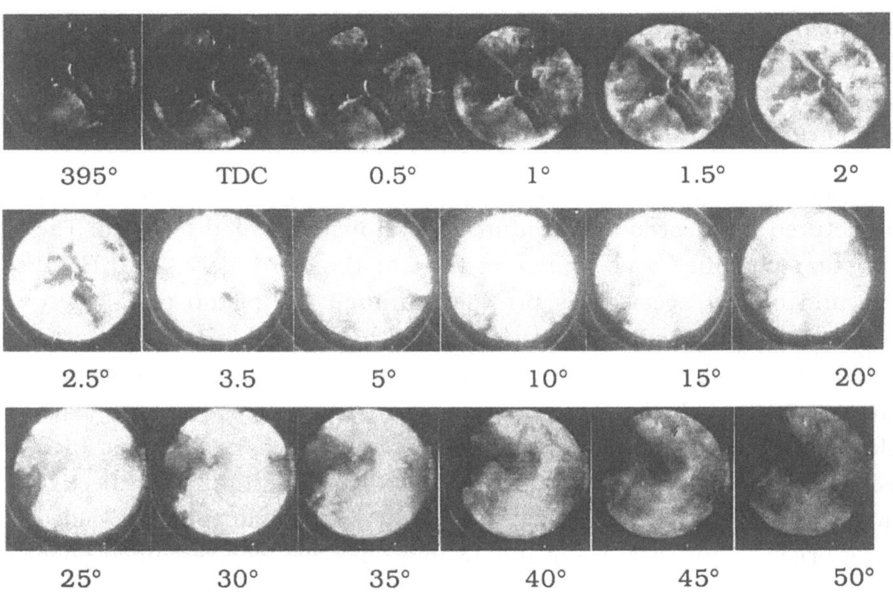

FIGURE 6 High-speed photographs of flame in the combustion chamber of a diesel engine running at 1000 rpm. Numbers are crank angles beginning from top dead centre.

ACKNOWLEDGEMENTS

The author is grateful to many colleagues and particularly to those whose results are used in this paper The results quoted were obtained with financial support from the Science and Engineering Research Council, the European Commission and Rolls-Royce plc. Mr. C.H. Liu was of considerable help in assembling the paper in its final form, as well as by his contributions to the results.

REFERENCES

Adrian, R.J. and Landreth, C.C. (1990) Impingement of low Reynolds number turbulent circular jet onto a flat plate at normal incidence. *Exp. in Fluids*, **9**, 74.

Arcoumanis, C., Hadjiapostolou, A. and Whitelaw, J.H. (1991) Flow and combustion in a Hydra direct-injection Diesel engine. SAE-paper-910177.

Chao, Y.C., Han, J.M. and Jeng, M.S. (1990) A quantitative laser sheet image processing method for the study of the coherent structure of a circular jet flow. *Exp. in Fluids*, **9**, 323.

Chow, S.K. and McGuirk, J.J. (1991) Numerical prediction of flow and combustion characteristics of a model annular combustor. ASME paper 91-GT-364.

van Cruyingen, I., Lozano, A. and Hanson, R.K. (1990) Quantitative imaging of concentration by planar laser-induced fluorescence. *Exp. in Fluids*, **10**, 41.

Dibble, R.W. and Hollenbach, R.E. (1981) Laser Rayleigh thermometry in turbulent flames. Proc. 19th (Int'l) Symp. on Combustion, 1489.

Drake, M.C. (1988) Stretched laminar flamelet analysis of turbulent hydrogen and CO/H2/N2 diffusion flames. in Proc. 21st Symp. (Int'l) on Combustion, Combustion Institute, 1579.

Durst, F. (1972) Development and application of optical anemometry. Ph.D. thesis, University of London.

Durst, F., Melling, A. and Whitelaw, J.H. (1981) Principles and Practice of Laser-Doppler Anemometry. 2nd edition, Academic Press.

Eckbreth, A.C. (1992) Coherent laser diagnostics for temperature/species measurements in advanced engines. in Combusting Flow Diagnostics., edited by D.F.L. Durao, M.V. Heitor, J.H. Whitelaw and P.D. Witze, Kluwer Academic Publishers, 399.

Enotiadis, A. (1992) Velocity and scalar variations in spark-ignition engines. Ph.D. thesis, University of London. (in preparation)

Gutmark, E., Parr, T.P., Hanson-Parr, D.M. and Schadow, K.C. (1990) Coherent and random structure in reacting jets. *Exp. in Fluids*, **10**, 147.

Hardalupas, Y., Liu, C.H. and Whitelaw, J.H. (1991) Experiments with disk stabilised kerosene fuelled flames. Imperial College, Thermofluids Section Report TF/91/17. Submitted to *Comb. Sci. and Tech.*

Hardalupas, Y., Milosavljevic, V.D., Taylor, A.M.K.P. and Whitelaw, J.H. (1991) In-quarl measurements of liquid fuel velocity, mass flux and concentration distributions. Imperial College, Thermofluids Section Report TF/91/03.

Hardalupas, Y., Taylor, A.M.K.P. and Whitelaw, J.H. (1989) Velocity and particle flux characteristics of a turbulent particle-laden jet. *Proc. Roy. Soc. Lond. A*, **426**, 31.

Hardalupas, Y., Taylor, A.M.K.P. and Whitelaw, J.H. (1990a) Velocity and size characteristics of liquid-fuelled flames stabilised by a swirl burner. *Proc. Roy. Soc.*, Lond. A, **428**, 129.

Hardalupas, Y., Taylor, A.M.K.P. and Whitelaw, J.H. (1990b) Liquid fuel mass flux, and concentration distributions in a swirl-stabilised flame. Imperial College, Fluids Section Report FS/90/09.

Heitor, M., Taylor, A.M.K.P. and Whitelaw, J.H. (1985) Simultaneous velocity and temperature measurements in a premixed flame. *Exp. in Fluids*, **3**, 323.

Heitor, M., Taylor, A.M.K.P. and Whitelaw, J.H. (1987) The interaction of turbulence and pressure gradients in a baffle-stabilised premixed flame. *J. Fluid Mech.*, **181**, 387.

Liu, C.H. (1992) Applications of laser- and phase-Doppler velocimetry, Ph.D. thesis, University of London. (in preparation)

Liu, C.H., Perez-Ortiz, R. and Whitelaw, J.H. (1991) Vaporiser performance. Imperial College, Thermofluids Section Report TF/91/01. Submitted to *Proc. IMechE.*

Kadota, T., Zhao, F.Q. and Miyoshi, K. (1990) Rayleigh scattering measurements of transient fuel vapour concentration in a motored spark-ignition engine. SAE-paper-900481.

Margary, R., Nino, E. and Whitelaw, J.H. (1990) LDA measurements of the flow field and flame visualisation inside an internal combustion engine. in Proc. FISITA.

Mastorakos, E. and Taylor, A.M.K.P. (1992) Visualisation of counterflow flames. in Proc. 6th Symp. on Flow Visualisation.

Mastorakos, E., Taylor, A.M.K.P. and Whitelaw, J.H. (1991a) Extinction and temperature characteristics of turbulent counterflow diffusion flames with partial premixing. Imperial College, Thermofluids Section Report TF/91/24. Submitted to *Comb. and Flame.*

Mastorakos, E., Taylor, A.M.K.P. and Whitelaw, J.H. (1991b) Scalar dissipation rate and extinction of turbulent counterflow non-premixed flames. Imperial College, Thermofluids Section Report TF/91/34. Submitted to *Comb. and Flame.*

Milosavljevic, V. (1992) Ph.D. thesis, University of London. (in preparation)

Milosavljevic, V.D., Taylor, A.M.K.P. and Whitelaw, J.H. (1990) Stability of pulverised coal burners. in Proc. 23rd Symp. (Int'l) on Combustion, 957.

Pitts, W.M. and Kashiwagi, T. (1984) The application of laser-induced Rayleigh scattering to the study of turbulent mixing. *J. Fluid Mech.*, **141**, 394.

Sivasegaram, S. and Whitelaw, J.H. (1986) Flow characteristics of opposed rows of jets in a confined space. *Proc. IMechE*, **200**, 71.

Taylor, A.M.K.P. (editor) (1992) Experimental Methods for Flows with Combustion. Academic Press.

Tse, D.G.N. (1988) Flow and combustion characteristics of annular and can-type combustors. Ph.D. Thesis, University of London.

Weinberg, F.J. (1963) Optics of Flames. Butterworths.

Advances in Computer-Aided Flow Visualization

T. KOBAYASHI

Institute of Industrial Science,
University of Tokyo, Tokyo 106 Japan

Summary

Because of the value of computer aided flow visualization for quantitative measurement, it is already being used for a variety of research activities of both basic and applied engineering. Flow visualization image analysis supported by computers are divided into three branches; flow image analysis using digital image processing techniques, flow image production from the collected data of a probe measurements, and an animated graphics using computational fluid dynamics.

Some recent developments in flow field analysis are reviewed, with particular attention to particle-imaging velocimetry, computed tomography and computational fluid dynamics.

Introduction

The ability to visualize dynamic phenomena in flows has two major benefits. Firstly, through visualization techniques, we can create images of these phenomena, making them easier to comprehend. Secondly, visualization makes available to us volumes of quantitative information which can be harnessed provided that processing is fast and accurate.

The availability of supercomputer, computer graphics, and improved software has spurred the growth of computer-aided engineering in the broad fields. The analytical technique in fluid engineering changes its style, and the capability of "computer aided flow visualization" is becoming crucial[1]-[3].

Flow visualization images supported by computers can be generated in different ways and divided into three branches;animated graphics using computational fluid dynamics[4][5], flow image production from the collected data of probe measurements[6]-[8], and flow image analysis using digital image processing techniques[6]-[22].

In the following, some recent developments in flow field analysis are reviewed with particular attention to computational fluid dynamics, computed tomography and particle-imaging velocimetry.

Computational Fluid Dynamics

The analysis and control of a fluid flow have a big impact upon various fields of science and technology. Especially, the numerical analysis will make such a large contribution to their development that it can break through the limits of experimental research, and a lot of information offered by it will stimulate the expansion of the theory of fluid flow, which is opening the new field of computational fluid dynamics(CFD).

CFD enables us to estimate quantitatively the unsteady and/or local structures of turbulent flow which were difficult to observe in previous experiments, because, in CFD, the conditions of the flow field can be easily controlled: for example, initial and boundary conditions, time and spatial scale of measurements, or physical properties of fluid. Also, flows which are difficult to reproduce in the laboratory can also be investigated, such as flows in nuclear fusion or high-vacuum vessels, high-temperature or high-velocity flows, or flows under exactly symmetrical or periodical conditions. In other words, CFD expands widely and deeply the range of investigation possible, by way of "experiments" on computer to explore unknown fields in fluid dynamics.

There are many forms of flow phenomena, classified by parameters such as Reynolds number, Mach number, or other physical quantities, and for each form, a different numerical simulation analysis technique has been proposed or is being researched. Perhaps the most actively researched of these at present are the turbulent flow analysis and the analysis of high-velocity flow including shock waves. In the following, we will look at a few examples of recent studies on turbulent flow analysis[18].

Direct simulation which one might expect to be the simplest means of turbulent flow analysis is, in practice, the most difficult of numerical simulation processes. To obtain an accurate solution of the fundamental equations (the equation of continuity and the Navier-Stokes equation), grid points are needed numbering as many as about $10^{9/4}$ times the Reynolds number, that is, in the case of engineering flow more than 10^9 grid points. In this regards, attention is now focused on a technique known as large eddy simulation (LES), which mitigates this problem and is widely used for the study of turbulent flow structure. In view of the rapid progress being seen in computer performance (Fig.1), it seems that this technique will come to play a dominant role in turbulent flow analysis in the future. Figure 2 shows an example of its application to a comparatively complex field of flow called a back-facing step flow[19].

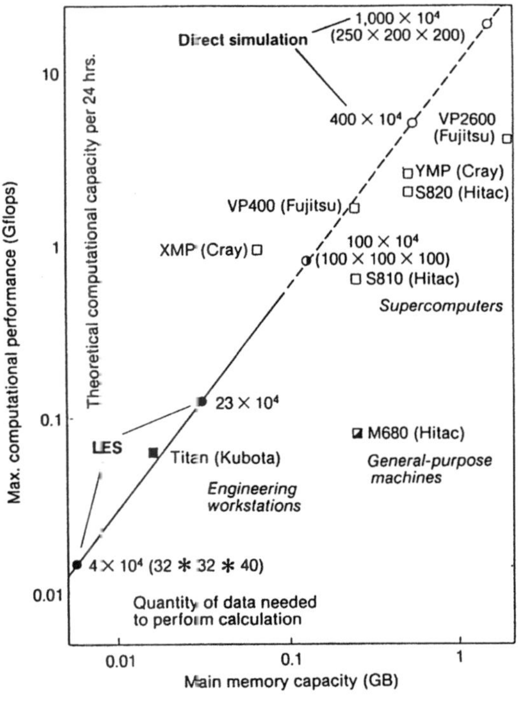

Fig.1 Computer power and scale of flow simulation

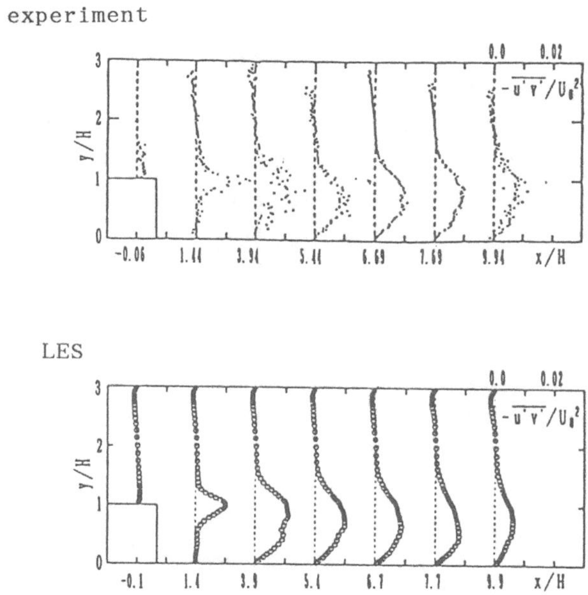

Fig.2 Turbulent shear stress profiles

In the engineering field, an extensive research effort has been directed at developing into practical tools for averaged flow analysis processes based on various models of fundamental equations. Typical one of such models is the k-ε model and the Reynolds stress model. These models are based on the Reynolds equation, which uses velocity and pressure averaged over a specified span of time, and the effects of complex changes of turbulent flow are represented by parameters such as turbulence kinetic energy (k) and dissipation of turbulence kinetic energy (ε). Figure 3 shows an example of k-ε model applied to air conditioning flow in a vehicle passenger compartment[20].

On the other hand, several calculation results by quasi-direct simulation with third order upwind scheme are proposed. Figure 4 shows air flows around a passenger car. However, the solutions obtained by this method depend strongly on the computational grid and the reliability of turbulence related quantities is not argued.

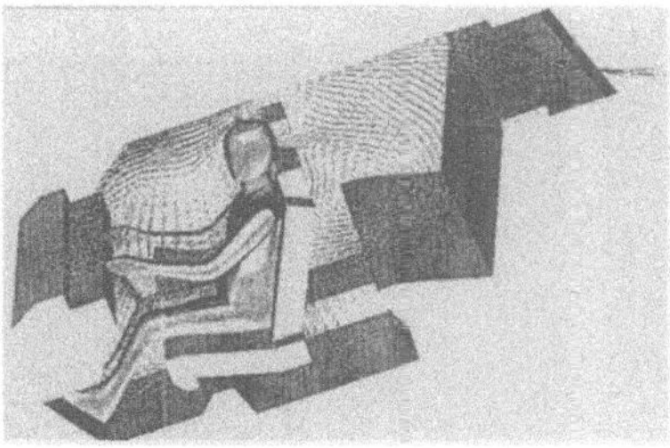

Fig.3 Air flow in a passenger compartment

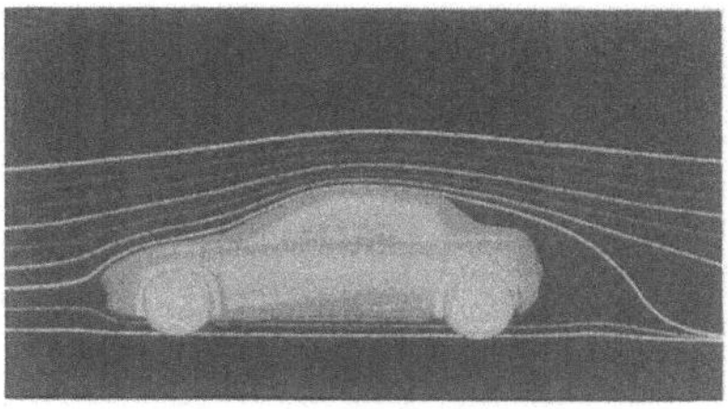

Fig.4 Streamlines around a passenger car

These analytical processes use bolder assumptions in modeling than direct simulation or LES. Consequently, verification of models developed in this way is of more important than the technical problems of numerical simulation. Reliable simulation codes, if available at low cost, will contribute enormously to the promotion of turbulent flow analysis. In the same way that certain instruments are invaluable to wind tunnel tests, numerical simulation will prove a powerful tool for CFD studies.

Figure 5 shows the schematic illustration of analysis model of Czochralski melt growth[22]. The analysis is performed for the melt in which crystal radius Re is 7.14 cm, Crucible radius Rc is 20.0 cm, and melt depth is 12.0 cm. In the analysis, the shapes of free surface and crystal/melt interface are assumed to be flat. For the velocity boundary conditions, the crystal is assumed to be rotated at the rotation rate Ws, 15.0 rpm in the clockwise direction, while the crucible is assumed to rotated at rotation rate Wc, 5 rpm in the counterclockwise. On the free surface, the free slip condition is given. For the temperature boundary conditions, the temperature profile on the crucible bottom is given by a linear function which varies the temperature in terms of the crucible radius, r between Tcm, 1700 Kelvin the temperature at the middle of the crucible, and Twb, 1745 Kelvin the temperature at the bottom edge of the crucible. Similarly, the temperature profile is given by a linear function of melt height z, varying the temperature between Twb, 1745 Kelvin and Twt, 1694 Kelvin the temperature at the upper edge of the crucible. The temperature on the crystal/melt interface is assumed to be at the melting point temperature Ts, 1685.0 Kelvin. And also the free surface is assumed to be adiabatic.

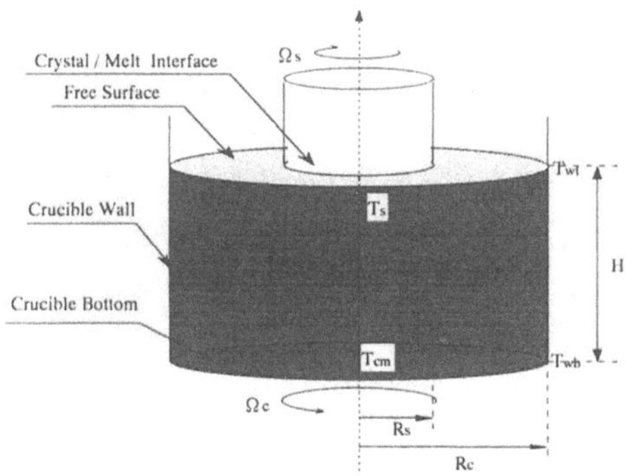

Fig.5 Analysis model of Czochralski melt growth

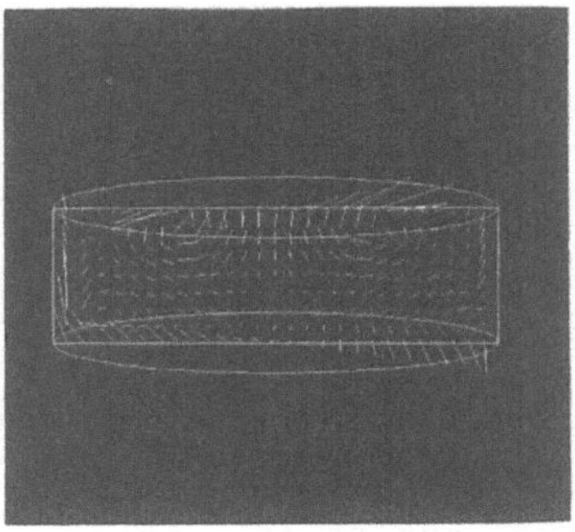

Fig.6 Velocity profile in crucible

Figure 6 shows the velocity profile resulted from the substructure method with eight substructures using finite element mesh with 25,000,000 number of degrees of freedom.

While the above figures appear to provide a satisfactory visualization of flow phenomena, it may be argued from the standpoint of CFD that quantitative comparison of these visualized data with experimental data (such as time-averaged values) carries far more importance. In the sense that numerical data can provide concrete answers to equations for flow fields, which themselves are totally unvisual, the numerical data should be viewed as the most fundamentally important products of visualization by numerical simulation. Also important, therefore, are displays produced by computational graphics (CG) techniques, as they help us greatly to understand flow.

Computed Tomography

Computed tomography (CT)is an advance computer-aided visualization technique, which enables us to view an object in cross section without having to cut the object. The essential invention, for medical use, was made in UK, and CT systems are used widely as a powerful tool for medical diagnoses. This same technology can also be used for non-destructive inspection and flow visualization.

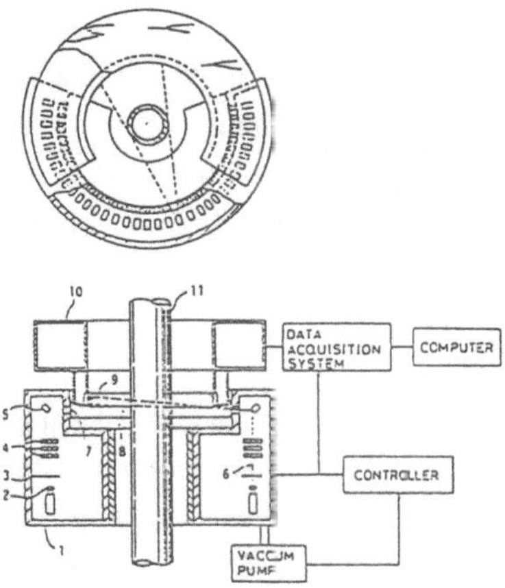

Fig.7 Layout of the high speed X-ray CT system

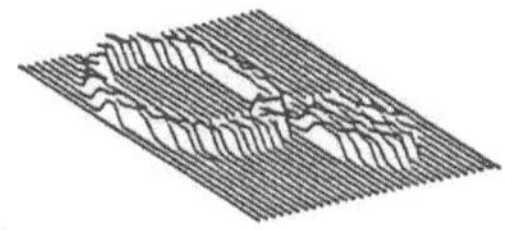

(a)Tomographic image

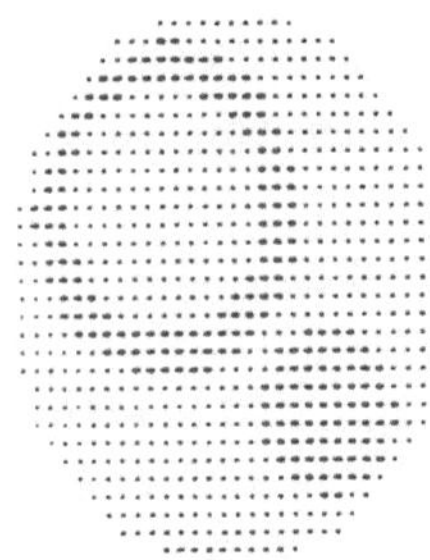

(b)Binary image

Fig.8 Tomographic image of acrylic phantoms

High speed X-ray scanner system is developed for measuring the instantaneous cross sectional structures of unsteady multi phase flows in a pipe [23]. The high speed x-ray CT scanner is shown in Fig.7. The system uses seventeen x-ray generating units arranged on an semicircle surrounding an object, and each unit has a cathode, a grid, a three stage electron beam acceleration/ converging electrode, and a tungsten target saving as an anode. A high speed accelerating voltage is applied between the cathode and anode electrodes to obtain x-rays having the required intensity. In the presence of high voltage, a bias voltage, which is applied to the individual grid to suppress an electron beam directed towards the anode from the cathode, is sequentially temporarily removed to obtain scanning x-rays. Since the electron beam on-off control is carried out electronically, the scanning of x-ray can be conducted continuously at an high speed of 3,5 milliseconds per one slice, thus permitting to obtain a time-dependent sectional view so that the dynamic state object is determined accurately. Figure 8 shows an example of tomographic image and its binary image of the phantom consisting of a pipe and a rod. The void fractions of the object phantoms, determined by dimensions of pipes and rod, are well reproduced by the CT system.

Simultaneous Measurement of Temperature and Velocity Fields

In the case of analyzing the phenomenon of unsteady thermal fluid flows, it is very efficient to measure the velocity and temperature distributions of flow field simultaneously and to understand their time dependent variation. The image processing technique has been developing to draw out the composite information in the thermal flow fields.

Figure 9 shows the flow pattern of jet flow with temperature different from that of environment visualized by the use of fine cupsled thermo-sensing liquid crystals and nylon tracer particles. The simultaneous measurement of velocity and temperature is carried out using a color image processing on color TV image. The color images are consecutively taken by a color CCD camera. The temperature distribution of liquid crystal is automatically analyzed by the identification of intensifies of three components (R, G and B) on liquid crystal particles. On the other hand, the velocity of flow field is measured by the displacement of nylon particles in a short time interval.

The simultaneous measurement of temperature and velocity combined by correlation method and laser induced fluorescence method is also available approach to measure the thermal flows[24]. The system is composed of a set of two monochrome TV cameras, and argon laser illumination and personal computer. The laser light sheet is controlled to be synchronously illuminated with the set of two TV cameras. In this case, the tracer of the sawdust for velocity measurement and fluorescence dye for the temperature measurement are mixed into fluid flow in the same time. The velocity distribution can be obtained by the correlation method using the instantaneous density pattern images taken from camera 1. In the same time, fluoresces dye is illuminated by Ar laser and the pattern of the intensity of fluorescence is recorded in the camera; 2. The temperature distribution is reconstructed automatically from the relation between the intensity of fluorescence and temperature. Figure 10 shows the velocity and temperature distribution of the turbulent mixing state of thermal stratified fluids.

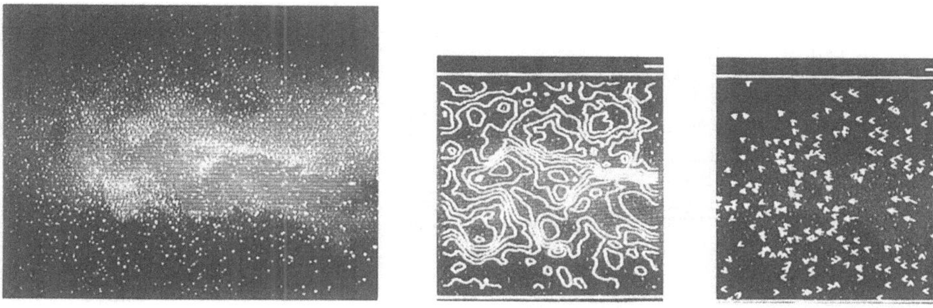

(a)Input image (b)Temperature distribution (c)velocity vectors

Fig.9 Simultaneous measurement of temperature and velocity fields by color image processing

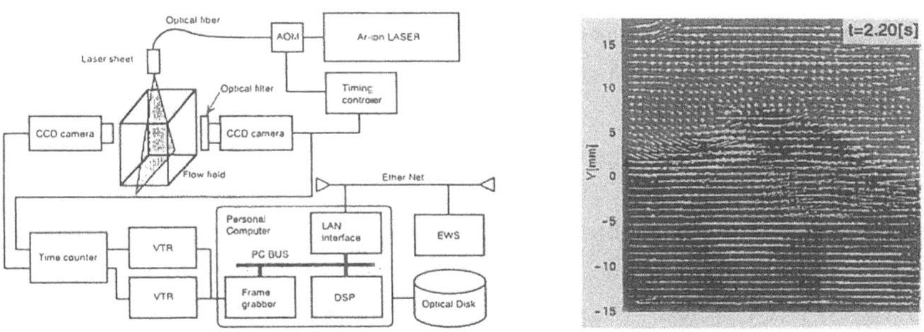

Fig.10 Temperature and velocity measurement by LIF and correlation method

<u>Particle-Imaging Velocimetry for High Reynolds Fluid Flow</u>

Particle-imaging velocimetry (PIV) is a method by which movements of fine particles injected into a flow are traced in real time to enable dynamic measurement of flow velocity.

Certain outstanding characteristics have generated great interest in PIV: 1)two dimensional and three-dimensional information can be logged in real-time for multiple points over a broad field of flow; 2)from this information we can interpret flow structure; 3)the need to calibrate velocities is eliminated; 4)the method is applicable to complex flows; 5)measurement is non-contact; and 6)data can be automatically processed with case.

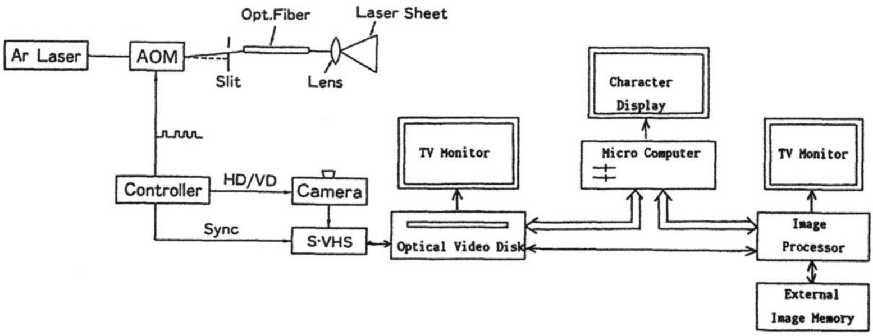

Fig.11 Optical and image processing system

The PIV using the video camera is often limited to low velocity flow fields. When the flow speed is high, it is very difficult to track tracer particles on the consecutive NTSC video images. The major limitation is the sampling rate of image which is 30 frames per second (minimum frame rate of NTSC video format). In such a case, the tracking of high-speed particles is usually required to use a high-speed video camera. This way is followed not only by the complicated process of frequent video frame controls but also by the reduction of the image quality and resolution. To overcome this shortcomings, we have developed a video capture technique and a fully automated image processing algorithm for high Reynolds fluid flow. This technique allows rapid extraction of two-dimensional velocity vectors with a certain accuracy.

Figure 11 shows schematic diagram of the optical and the image processing system [11]. The system consists of a 4W Argon laser, an acousto-optic modulator(AOM) and CCD camera (SONY XC 007). In order to visualize the high-speed flow, particles are illuminated with a laser light sheet(LLS). The working fluids is water which is seeded with small amount of neutrally buoyant nylon particles with sizes of about 50 μ m in diameter. The laser beam is passed through a series of optical components to produce a modulated sheet of laser light (2mm thickness). The temporal light modulation is obtained by using scattered light of the first-order which is produced by the AOM. Figure 12 shows the typical pulsing pattern used in the present system. The illuminations is synchronized with the vertical and horizontal blanking signals of the CCD camera. In this case, the LLS is illuminated one time at the first field and three times at the second field. When the flow speed is high, the illumination time T_1 and T_2, T_3, T_4 are settled close to the vertical blanking of second field. The illumination time T_i and exposure time dt_i can be adjusted by 1H unit (63.56 μ s) which could be adapted to various flow conditions. The illumination times of LLS also can be changed. Scattered light from particles at times T_1, T_2, T_3 and T_4 are sensed by the photosensors of the CCD camera which generated the electric charge proportionally to the light intensity. The electric charge of photosensors by scattered light are transferred to the output amplifier of the CCD camera as an interline-transfer organization. Frame image is made up of 2:1 interlaced video format and is the composite video signal (VBS:video burst signal). This composite video image has 480 vertical effective lines (240 vertical lines for the first field and the second field, respectively). In this system, the frame integration charge mode of the CCD camera is used. In this mode, the electric charge of each interline CCD photosensor is integrated as shown in Fig.12. This is, the electric charge of the first field photosensors is integrated between the beginning of the first field and the end of the second field, and the charge for the second field photosensors also integrate between the beginning of the second field and the end of the first field of next frame. As a result of the multi-pulse illumination and the frame integration, the particle positions at T_1 to T_5 are obtained on a frame image.

The ability and the usefulness of the system are confirmed by the measurement for complex flow fields.

The system was applied to the turbulent water flow in a continuous casting model. In Fig. 13, the results by LDV measurement are compared with that of present method. The mean velocity profiles by PIV have excellent agreement with LDV results. It is expected that the technique has good measurement accuracy, at least concerning the mean flow velocity.

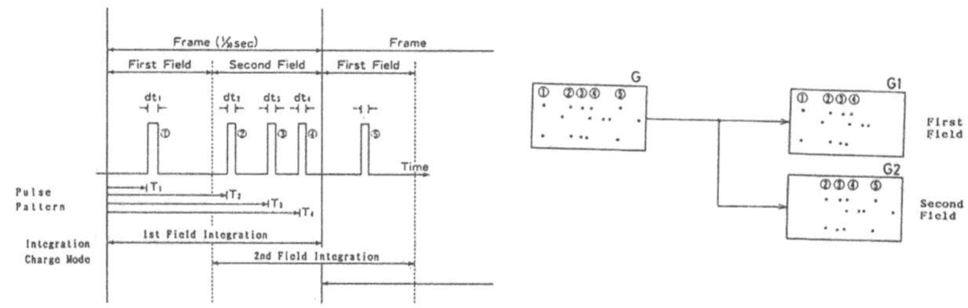

Fig.12 Typical illumination coding pattern of LLS and frame images

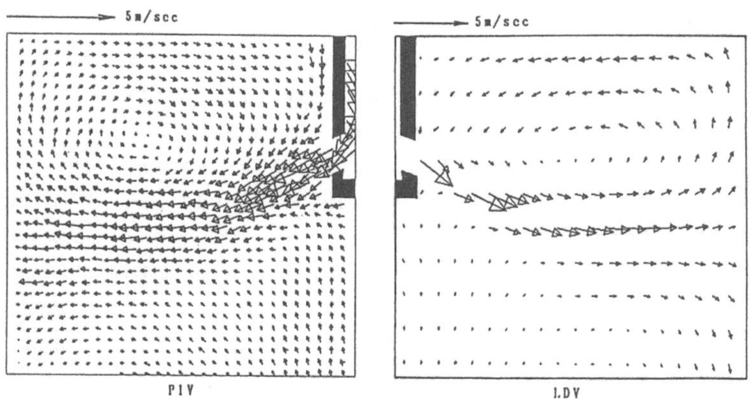

Fig.13 Comparison PIV and LDV results

Conclusion

Computer aided flow visualization for quantitative measurement is already being used for a variety of research activities of both basic and applied nature. In most cases of flow image analysis using digital image processing techniques, accuracy of the analysis essentially depends on correctness of the flow visualization. Efforts will be necessary to consider both the flow visualization techniques and the image processing techniques, to enhance the functions and to make systems more popular.

CFD is beginning to play an important role in the engineering industry. However, there are a lot of problems to be solved in order that CFD can be applied as a really and truly useful tool to future engineering. Simply speaking, they are concerned with reliability and efficiency.

In addition, it should be emphasized that present delays in constructing networks will become a very serious obstacle in the future for the continued development of CG. As CG is a tool for analyzing enormous amounts of data, it requires high-speed and general-purpose networks, and judging from the performance of present computers, a communication capability of at least MB/sec would seem appropriate.

References

1. Kobayashi, T.:Flow visualization image analysis review. Proc. FLUCOME '88 (1988) pp.545-556.
2. Hesselink, I.:Digital image processing in flow visualization. Ann. Rev. Fluid Mech. (1988) pp.421-485.
3. Adrian R.J.:Particle-imaging technique for experimental fluid mechanics. Ann. Rev. Fluid Mech. (1991) pp261-304.
4. Pitcher, E.J.(ed.):Science and engineering on supercomputers. Springer-Verlag (1990).
5. Fujii, K.:Supercomputers and workstations in fluid dynamic research. Visualization in Supercomputing (Mendez, R.H. ed.). Springer-Verlag (1990) pp.3-13.
6. Akino, N.; Kunugi.T.; Shiina, Y.; Ichimiya, K.; Kurosawa, A.: Fundamental study on visualization of temperature fields using thermosensitive liquid-crystals. Flow Visualization 5. Hemisphere (1990) pp.87-92.
7. Takeda, Y.:Velocity profile measurement by ultrasound doppler shift method. Int. J. Heat & Fluid Flow 7-4 (1986) pp.313-318.
8. Hibi, K.; Fujii, K.; Ueda, H.; Shimada, K.: Visualization of fluctuating surface pressure distribution on bluff body using electronically scanning pressure sensors. Flow Visualization 5. Hemisphere (1990) pp.215-220.
9. Landreth, C.C.; Adrian, R.J.: Impingement of a low Reynolds number turbulent circular jet onto flat plate at normal incidence. Exp. in Fluids 9 (1990) pp.74-84.
10. Kawahashi, M.; Hosoi, K.: Dual-beam-sweep laser speckle velocimetry. Exp. in Fluids 11 (1991) pp.278-289.
11. Kobayashi, T.; Saga, T.; Haeno, T.; Tsuda, N.: Development of a real-time velocity measurement system for high Reynolds fluid flow using a digital image processing technique. ASME FED-128 (1991) pp.9-14.
12. Kobayashi,T.; Saga,T.; Sekimoto, K.: Velocity measurement of three-dimensional flow around rotating parallel disks by digital image processing. ASME FED-85 (1989) pp.29-36.

13. Nishino,K.; Kasagi,N.: Hirata,M.: Three-dimensional particle tracking velocimetry based on digital image processing. Trans.ASME-111 (1989) PP.384-391.
14. Mass,H.G.: A high-resolution photogrammetric 3-D particle tracking velocimeter. ASME FED-128 (1991) PP.79-84.
15. Ni-imi,T.; Fujimoto,T.: Development of a CAFV system for gas dynamics (Synthesis of visualized picture at arbitrary cross sections). Flow Visualization 5. Hemisphere 1990 pp.240-247.
16 Hijikata,K.; Mimatsu,J.; Inoue,J.: A study of wall pressure structure in backward step flow by a holographic/velocity-pressure cross-correlation visualization. ASME FED-128 (1991) pp.61-68.
17. Watanabe,M.; Eguchi,M.: Kakimoto,K.; Hibiya,T.: Three-dimensional visualization of molten silicon convection. ASME FED-128 (1991) pp.255-260.
18. Kobayashi,T.: Supercomputing of incompressible fluid flow. Supercomputing (Kondo,J. ed.). Springer-Verlag (1991) pp.125-140.
19. Morinishi,Y.; Kobayashi,T.: Large eddy simulation of backward facing step flow. Elsevier Sci. Pub. (1991) pp.279-286.
20. Komoriya,T,; Kobayashi,T.; Taniguchi,N.: Numerical simulation of the air flow around a driver. JSAE Spring Convention Proc. 1 (1992) pp.25-28 (in Japanese).
21. Yamada,A.; Ito,S.; Okamoto,M.: Computational analysis of the flow around a car body. JSAE Spring Convention Proc. 1 (1992) pp.13-16 (in Japanese).
22. Oshima,M.: A substructure method for large scale magneto-thermal-hydrodynamic flows. Doctor Thesis, Univ. of Tokyo (1991).
23. Akai, M.; Takahashi,S.: High speed x-ray tomography for multiphase flow visualization . Proc. 9th Task Leader Meting. IEA Working Party on Conservation in Combustion (1987).
24. Kobayashi,K.; Sakakibara,J.; Hishida,K.; Maeda,M.:Time-series measurements of turbulent flow field using image processing system. ASME FED-128 (1991) pp.155-162.

Internal Flow

Flows in Chamber

Use of a Colored Water Surface Flow Visualization Technique in a Supersonic Wind Tunnel to Investigate Cavity Flow Fields

Floyd J. Wilcox, Jr.
NASA Langley Research Center
Hampton, VA 23665-5225
USA

Summary

A surface flow visualization technique was developed that uses colored water as a medium to visualize surface flows over wind tunnel models at supersonic speeds. Wind tunnel tests were conducted with this technique used to visualize surface flows in a rectangular box cavity that was mounted in a flat plate at Mach numbers of 1.50 and 2.16. The results showed that this technique was effective in visualizing the surface flows in the cavity and has produced a unique set of data that has increased the understanding of complex unsteady three-dimensional cavity flow fields.

Introduction

For many years, NASA Langley Research Center has been studying cavity flow fields at supersonic speeds. This research has included various flow visualization wind tunnel tests to document both the cavity surface and off-body flow fields. During one of these tests, water was injected through pressure measurement orifices located on the model surface in an attempt to visualize vortices that typically form above certain cavity configurations. Because of the low static pressure in the tunnel, it was anticipated that the water would vaporize as it exited the orifice and the resulting vapor would be entrained into the vortices and render them visible. This technique did not work, and the water instead flowed along the model surface. Although the water was difficult to discern against the model surface, it was obvious that the water was being entrained by the model surface flows and had a fast response to the cavity flow unsteadiness. This impetus led to the development of a technique that uses colored water as a medium to visualize surface flows over wind tunnel models.

The colored water surface flow visualization technique basically involves injecting colored water through the pressure measurement orifice tubes installed in a wind tunnel model . The colored water exits the orifices and flows along the model surface to form streaklines. The resulting flow patterns are then photographed for later analysis. This method has several advantages over the conventional oil flow and tuft techniques that can make it a useful procedure for routine testing. For example, it is sometimes difficult to interpret oil flow or tuft data to determine the origin of vortices that have formed from a merging of two or more flows. With the colored water technique this determination is simplified because water that is tinted with different colors can be injected from various locations on the model so that as the streaklines merge to form vortices the mixing of the different water colors creates a new color. An example of this aspect of the colored water method will be discussed later in this paper. Another example is that oil flows typically cannot quickly respond to either general flow unsteadiness or gross flow field changes. This slow response is a result of the oil viscosity being large enough to prevent the oil from quickly blowing off the model and thus requiring the tunnel to be shut down to recoat the model. The colored water, on the other hand, has a low viscosity and can respond to fast dynamically changing flow fields. It should be emphasized that the colored water technique complements the oil flow and tuft techniques and gives researchers another test technique to analyze complex flow fields. The remainder of this paper will discuss wind tunnel tests that were conducted to visualize surface flows in a rectangular box cavity at Mach numbers of 1.50 and 2.16.

Experimental methods

The apparatus used in the colored water surface flow visualization technique consists of three components: 1) a pressure instrumented wind tunnel model, 2) a set of control valves, and 3) containers of colored water (fig. 1). The model pressure tubes are routed outside of the tunnel and attached to a series of metering, shutoff, and switching valves and then routed to the containers of colored water. The water is tinted with ordinary food coloring and mixed at various concentrations depending on the surface flow field phenomenon to be visualized, e.g., tight vortex formations require that the water color be relatively light so that the vortex is not a dark mass of water. Typical concentrations are 25 to 75 ml of food coloring to 1000 ml of water. The natural pressure difference between the model surface pressure and atmospheric pressure draws the colored water from the containers and through the pressure tubes. The colored water then exits the pressure orifices on the model and flows along the surface to produce streaklines of the model surface flow. The series of control valves allow 1) the rate of water flow to be varied, 2) the water to be turned off and on, and 3) the water color to be varied or air to flow through the tubes (to clean the pressure tubes).

The wind tunnel model (fig. 2) consisted of a rectangular box cavity mounted in a flat plate. The rear face of the cavity was attached to a remotely controlled electrical drive mechanism that allowed the rear face to move and thus vary the length of the cavity while the tunnel was in operation. The cavity height (h) and width were 0.5 in. and 2.5 in., respectively, and were held constant throughout the entire test while the cavity length (L) was varied from 12 in. to approximately 0.5 in. The cavity leading edge was 10.4 in. aft of the flat plate leading edge. The boundary layer on the plate was turbulent, having been tripped at the plate leading edge with sand grit. Colored water was injected through pressures orifices located along the floor of the cavity and on the flat plate surface ahead of the cavity. All of the tests were conducted with the cavity at angle of attack and sideslip of 0°. The model was vertically mounted in the wind tunnel to allow video and still photographic data to be obtained through the test section windows. The model was painted white for good contrast with the colored water and was illuminated with three flood lamps.

The tests were conducted in the low Mach number test section of the Langley Unitary Plan Wind Tunnel (ref. 1), which is a continuous flow, variable pressure supersonic wind tunnel with a Mach number range from 1.5 to 4.6. The test section is approximately 4 ft square and 7 ft long. The tests were conducted at the following conditions:

Mach number	Reynolds number (1/ft)	Total pressure (psfa)	Total temperature (°F)
1.50	2.0×10^6	1051	125
2.16	2.0×10^6	1349	125

The basic procedure used during these tests was to bring the tunnel to the desired test conditions and then start the flow of colored water. Typically, the length of the cavity was varied over a specific range while the photographic data were obtained. The colored water had a tendency to freeze in certain locations on the model surface such as in the slot on the flat plate surface aft of the cavity where the rear face slides. To minimize this problem, the colored water flow would only be turned on for two or three minutes at a time, thus allowing any small ice buildup to dissipate. Although it was not attempted during this test, the addition of alcohol to the colored water mixture would probably lower the freezing point of the colored water enough to prevent any ice buildup for the short periods of time that the colored water is flowing.

Results

Before presenting some sample results, it would be helpful to briefly describe the two primary types of cavity flow fields that exist at supersonic speeds (fig. 3). These two flow fields are generally referred to as closed and open cavity flow (ref. 2 and 3). The type of flow field that exists for a given cavity depends primarily on the cavity length-to-height (L/h) ratio. Closed flow (L/h $\gtrsim$ 13) is characterized by the flow expanding over the cavity leading edge, impinging on the cavity floor, passing along the floor, and then exiting at the rear of the cavity. This flow produces two small recirculation regions at the forward and rear sections of the cavity. In contrast, open flow (L/h $\lesssim$ 10) is characterized by the flow passing over the cavity with little expansion into the cavity and a large recirculation region in the cavity such that the flow along the floor is in the opposite direction to the free-stream flow. The present flow visualization technique was used to investigate these two flow types.

Shown in figure 4 are a series of photographs that were acquired as the cavity L/h was decreased at a Mach number of 1.50. The free-stream flow is from left to right and the scale located on the flat plate above the cavity indicates the cavity length in inches. Unfortunately, a significant amount information is lost due to the use of black-and-white halftone reproductions in this paper rather than the original color photographs. Colored water was injected from two orifices located on the flat plate ahead of the cavity and from three orifices located on the cavity floor approximately 3 in. aft of the forward face. For the cavity with L/h = 24 (closed flow), the streaklines on the flat plate surface curve into the forward section of the cavity and indicate that the flow is expanding into the low pressure region. At the rear of the cavity, these streaklines curve away from the cavity as the flow exits from the high pressure region. On the cavity floor, two vortices have formed ahead of the rear face from a merging of the three cavity floor streaklines. The water color for these streaklines (yellow — middle orifice, blue — outside orifices) was chosen so that as the streaklines merged to form vortices, a new color (green) was formed. The formation of new colors from the merging of flows greatly enhances the data analysis. The upper vortex has a counterclockwise rotation and the lower vortex has a clockwise rotation. As the cavity length is reduced from L/h = 24 to L/h $\approx$ 13, these vortices keep the same rotation direction. At this point the vortices start to dissipate and begin to form in the opposite direction. As the cavity length is reduced still further, the flow field rapidly switches from closed to open flow (L/h $\approx$ 12) as indicated by the streaklines on the cavity floor flowing upstream and the streaklines on the flat plate surface not curving into the cavity.

Figure 5 shows the cavity surface flows at a Mach number of 2.16. The main difference between the results at Mach 1.50 and 2.16 is the formation of the vortices at the rear of the cavity. For L/h = 24 (closed flow), the two vortices ahead of the rear face have the same rotation direction as the Mach 1.50 case. As the cavity L/h is decreased to approximately 16, the upper and lower vortices have dissipated and two new vortices, which together span the entire cavity width, have formed in the opposite direction to the original vortices. These vortices continue to exist as the cavity L/h is decreased to approximately 13, where the flow rapidly switches from closed to open flow. The location, size, and rotation direction of the cavity floor vortices as shown in figures 4 and 5 were previously unknown. The visualization of these vortices using the colored water surface flow visualization technique has significantly improved the understanding of three-dimensional cavity flow fields.

Conclusions

A surface flow visualization technique was developed that uses colored water as a medium to visualize surface flows over wind tunnel models at supersonic speeds. Wind tunnel tests were conducted on a cavity model and the results showed that the colored water was effective in visualizing the cavity surface flows. Two main advantages of this method are that the colored water greatly enhances the data analysis of merging flows and that the low viscosity of the water allows the visualization of rapidly changing flow fields. The results from this test have led to a further understanding of the complex nature of three-dimensional cavity flow fields.

References
1. Jackson, Charlie M., Jr.; Corlett, William A.; and Monta, William J.: *Description and Calibration of the Langley Unitary Plan Wind Tunnel.* NASA TP-1905, 1981.
2. Charwat, A. F.; Roos, J. N.; Dewey, F. C., Jr.; and Hitz, J. A.: An Investigation of Separated Flows — Part I: The Pressure Field. *J. Aeosp. Sci.*, vol. 28, no. 6, June 1961, pp. 457-470.
3. Stallings, Robert L., Jr.; and Wilcox, Floyd J., Jr.: *Experimental Cavity Pressure Distributions at Supersonic Speeds.* NASA TP-2683, 1987.

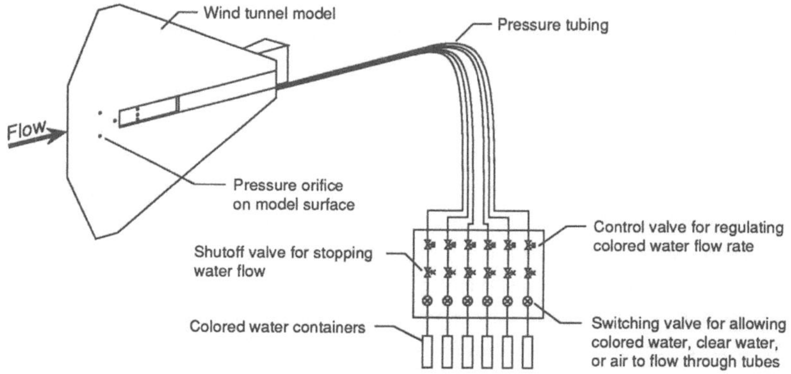

Figure 1. Schematic of experimental appratus.

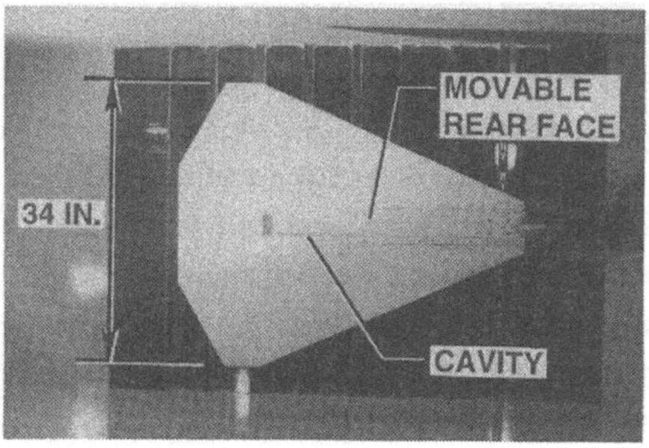

Figure 2. Photograph of flate plate cavity model vertically mounted in the wind tunnel.

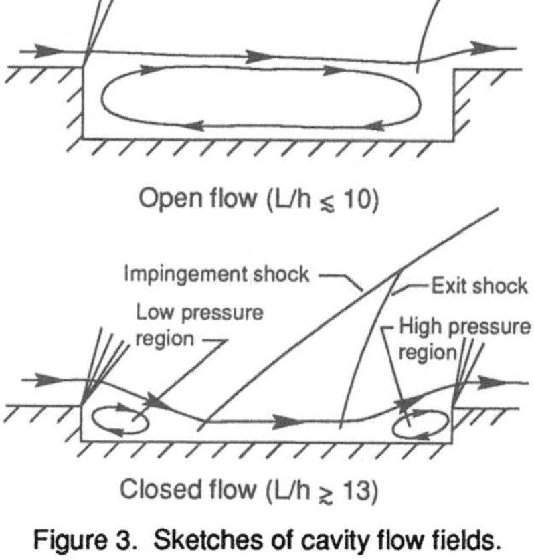

Figure 3. Sketches of cavity flow fields.

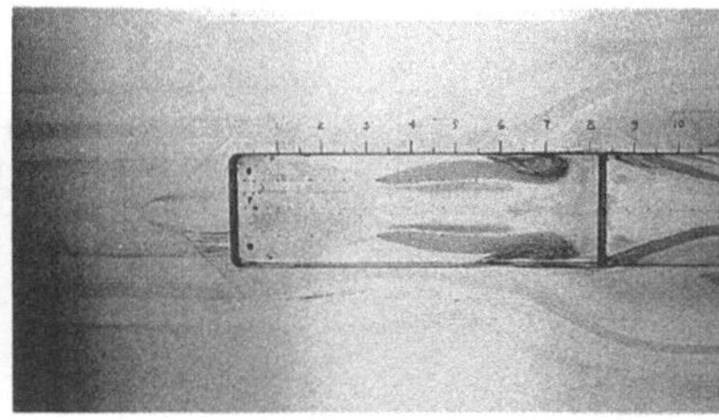

L/h = 24 L/h ≈ 17

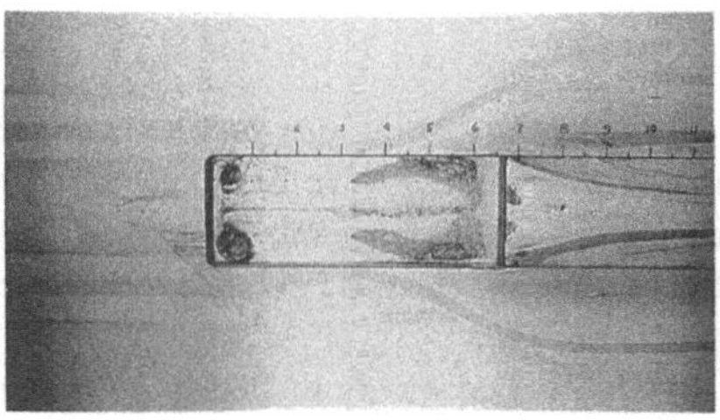
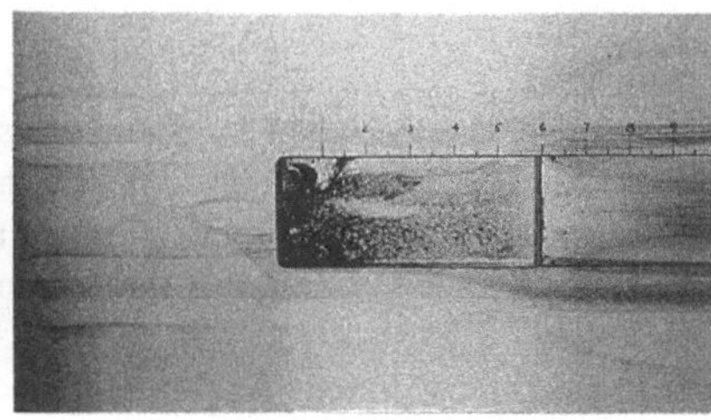

L/h ≈ 13 L/h ≈ 12

Figure 4. Cavity flow fields at various length-to-height ratios (L/h). M = 1.50.

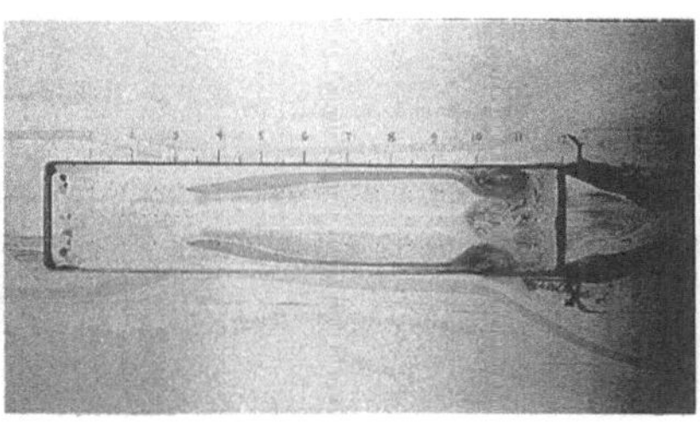
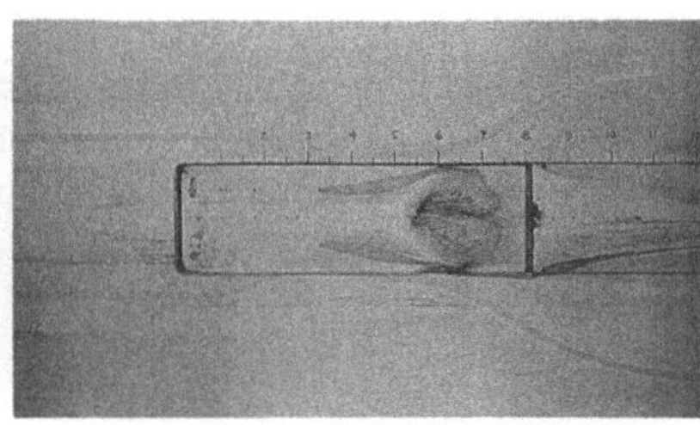

L/h = 24 L/h ≈ 16

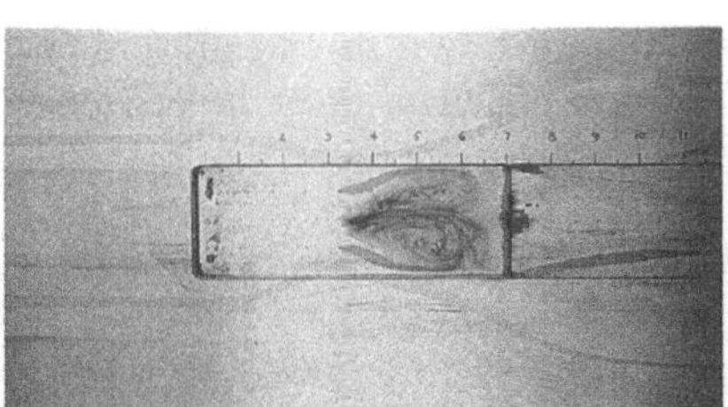

L/h = 14 L/h ≈ 13

Figure 5. Cavity flow fields at various length-to-height ratios (L/h). M = 2.16.

Visualization of Gravity Current Flow Using Laser-Induced Dye Fluorescence

W.D. Peters, C.R. Dutcher and J.E.S. Venart
Fire Science Centre, University of New Brunswick
Fredericton, New Brunswick, Canada, E3B 5A3

Abstract
Salt water modelling and laser-induced dye fluorescence are being used to study gravity current flows. Under specific geometric constraints, the dilution mechanisms of these flows are being analyzed to assess the mixing dynamics.

Introduction

Gravity currents arise when buoyant forces, due to density variations in the fluid, produce motions mainly in the direction normal to the gravitational field. The basic characteristics of the flow are a head, with a slightly elevated nose, at the leading edge of the flowing layer, and a stable mixed region of fluid that settles into the area behind the head and above the layer.

To evaluate gravity current dilution levels accurately, it is necessary to clearly understand the interchange processes involved. Several mixing mechanisms have been defined; 1) shearing at the current/ambient fluid interface, 2) mixing at the current/boundary interface due to the surface roughness, and 3) dilution within the head structure due to the ingestion of ambient fluid over-run by the advancing system. In the past, these last two mechanism have not been considered significant contributors to the mixing.

Simpson [1] and Simpson and Britter [2] studied gravity currents using salt water modelling techniques. They stated that the intense shear between the gravity current and the ambient fluid was the main mixing process at the front. Using lock exchange experiments, Winant and Bratkovich [3] also studied such flows. They stressed that an examination of the mixing sites and mechanisms within the gravity current head structure is required to fully understand the behaviour which governs the dynamics and mixing.

Peters et al. [4] examined gravity current flows using salt water modelling techniques. Ongoing work attempts to examine, from fundamental principles, the mixing mechanisms that occur within the head structure. Preliminary work indicates that significant mixing and dilution, initiated by the ingestion of ambient over-run fluid, occur in this region. This process has received little attention to date.

The flexibility of the small-scale experiments described here allows extension of the results to dense gas dispersion and smoke movement in corridors. In the spread of dense gases, the effect of terrain roughness and slope can, also, be considered. Steady-state and transient smoke transport down a corridor ceiling, as well as the transient modelling of the fire source plume dynamics can be modelled.

Empirical relations, used to predict the dilution and advance of a dense vapour cloud, have become the basis for the many dense gas dispersion models currently available [5]. They rely heavily on data gathered from limited large scale field tests and wind tunnel simulations. Unfortunately, this approach has not addressed the fundamental phenomena and basic principles of the mixing process. Thus far, the models, therefore, require significant empirical input. It is, therefore, desirable to develop more basic descriptions of the dilution processes.

Typically, in building fires, the largest losses can be attributed to the transport of large quantities of combustion products throughout the space. The interaction of these flows with ceilings, walls, openings, and other structural features induces mixing and entrainment [6]. The introduction of oxygen to otherwise oxygen-deficient and fuel-rich areas in a stratified flow can lead to further ignition, or flashover. An understanding of the flow dynamics, in such circumstances, can lead to a better evaluation of the hazards.

Scaling Laws

Fully developed gravity current flow, generally, requires Reynolds numbers above 1000. Gravity current Reynolds numbers have been defined using (1) the dense current depth and velocity [6] and (2) the dense current depth with the head frontal velocity [2]. The dense current layer velocity, however, is necessarily greater than that of the head to satisfy mass continuity. Since the source condition sets this velocity, a Reynolds number based on the current depth and velocity should be used as in the work of Zukoski and Kubota [6]. They suggest that the dynamics of the gravity current display a Reynolds number independence for values greater than 1000.

Simpson and Britter [2] characterized their flows with three dimensionless parameters. These were: a Froude number; a mixing parameter describing the flux of dense fluid mixed out of the current; and the ratio of mixed layer depth to current depth. All were given as functions of the fractional depth of the gravity current to the overall fluid height. These parameters, along with the Reynolds number, were used for scaling.

Reynolds numbers in the range 250-5000 and density variations, $\Delta\rho/\rho$, of 0.004-0.15 have all been used in the past [2,6]. The Froude numbers and ratios of mixed layer to current layer depth, thus, varied from 0.75-1.30 and 0.50-3.0, respectively. These values were obtained over a range of fractional depth of the gravity current to the overall fluid height of 0.05-0.30 [2]. The values of these parameters for our work, also, fall within these ranges.

For the tests analysed here, the ambient/current fluid specific gravity difference was 1.0%. The ambient fluid depth was 15.0 cm. A flow, per channel unit width, of 15.2 cm^2/s was used. These conditions resulted in a turbulent gravity current flow with a Reynolds number of 1500.

Experimental Facility

Experiments are carried out in an open plexiglass channel, 240 cm long and 20 cm wide with a maximum depth of 30 cm. Figure 1 schematically illustrates the facility. Saline fluid enters through an upstream entry section on the channel bottom and, after spilling over an inlet edge, flows down the channel length, forming a gravity current.

The visualization section is illuminated by passing a vertical laser sheet, approximately 20 cm wide, through the longitudinal mid-plane of the channel. It is produced using an argon ion laser with sheet generating optics and a collimating lens. Fluorescein sodium is contained in the saline so that it fluoresces in the presence of the laser sheet, illuminating the gravity current as it flows down the channel. Mixing can be seen, since the diluted regions fluoresce in proportion with the mixing.

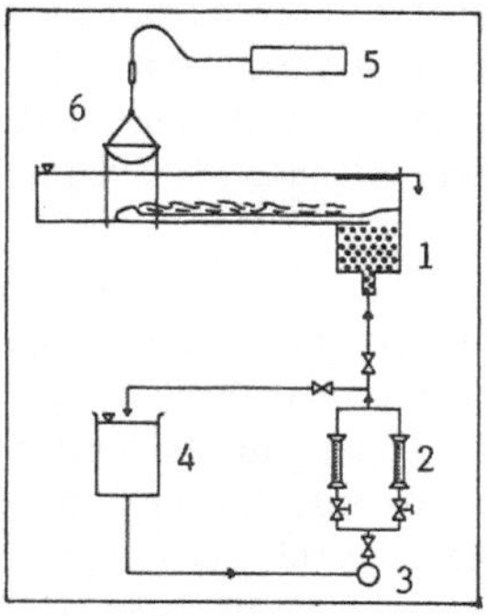

Fig.1. Schematic diagram of apparatus showing (1) entry box, (2) flowmeters, (3) pump, (4) saline reservoir, (5) argon laser and (6) sheet generating optics.

The flow illumination system and a video camera are mounted on a stepper motor driven traversing carriage. The speed, acceleration/deceleration and direction of the carriage can be maintained and/or altered while the gravity current head is tracked. Carriage position, speed and elapsed time are recorded, providing reference data for the current advance.

Tests are recorded using a SVHS/VHS video cassette recorder with an external timer and a B/W CCD video camera. Images are processed using a PC-based image processing system equipped with a frame grabber board (512 x 512 resolution). Use of this system, along with custom and commercial software packages for image processing, allows analysis of the flow.

Treatment of Video Images

Images are grabbed from the video using specific frame grabbing brightness and contrast values to optimize image offset and gain. 3 x 3 neighborhood averaging and low level thresholding are used to smooth the images and eliminate low level noise. Vertical bands of dark and light regions were apparent in the laser sheet, due primarily to non-uniformities in the optics. It was necessary, therefore, to apply a correction to achieve a uniform horizontal intensity profile. Further enhancement of the mixing regions within the structure is achieved using linear scaling. Pseudo-Colour can next be added to the images to improve interpretation. Contour maps of concentration can, also, be extracted.

Discussion

Figure 2 shows the general flow patterns observed within the gravity current structure. As the current progresses to the left, lighter ambient fluid is ingested under the head (a) until it combines with the advancing fluid of the feeding dense layer (b). Dilution occurs, with the mixed fluid being carried up and back through the current head (c).

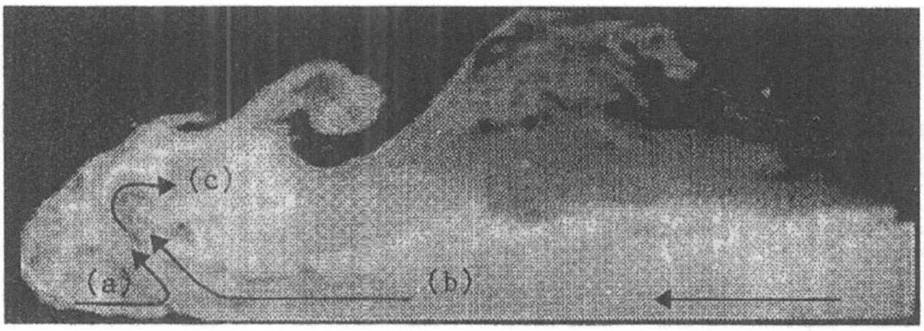

Fig.2. Internal flows of (a) lighter ingested fluid, (b) dense current fluid and (c) mixed fluid.

Concentration gradients induced by mixing within the gravity current head are apparent in Figure 3. A mixed layer exists in the region behind the head and above the current layer, fed, in part, by the shear mixing of the flow up and over the head. More importantly, however, is the substantial mixing that is apparent within the head structure that appears to be initiated by the ingestion of over-run fluid as was previously described.

The gravity current structure in Figure 3 has a head height of 6.0 cm. The observed frontal velocity of the nose is 4.0 cm/s. The feeding current layer has a thickness of 3.0 cm and is moving at a velocity of 5.0 cm/s. Above it, the mixed layer has a thickness of 2.0 cm.

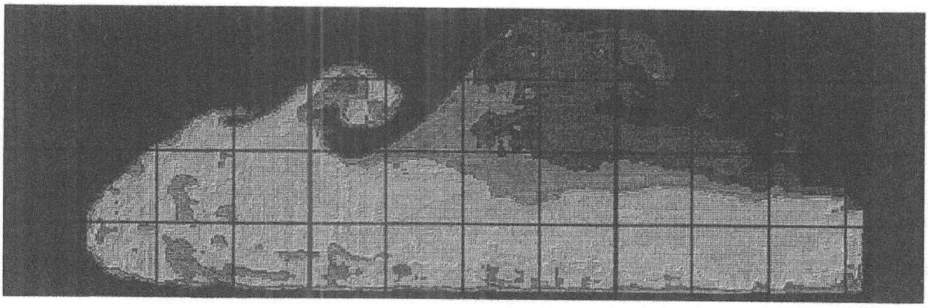

Fig.3. Mixing observed within the gravity current head structure (2 cm x 2 cm grid).

In Figure 4, concentration contours within the gravity current structure (Figure 3) are extracted. Contour maps of this type can highlight areas of greatest mixing. Significant mixing is evident in Figure 4 in both the trailing mixed layer and in the central regions of the head.

Simpson [1] states that the mixing process at the front is primarily due to the formation of billows and shearing over the head. He reports that the mixing caused by the over-run fluid is only a small fraction of the total. Analysis of the video sequences and images obtained here (Figure 3) indicates that there is, however, much greater mixing occurring intermittently inside the head. This, perhaps, is indicative of a larger role for the over-run fluid in destabilizing the current head resulting in its intermittent buckling, folding and collapse.

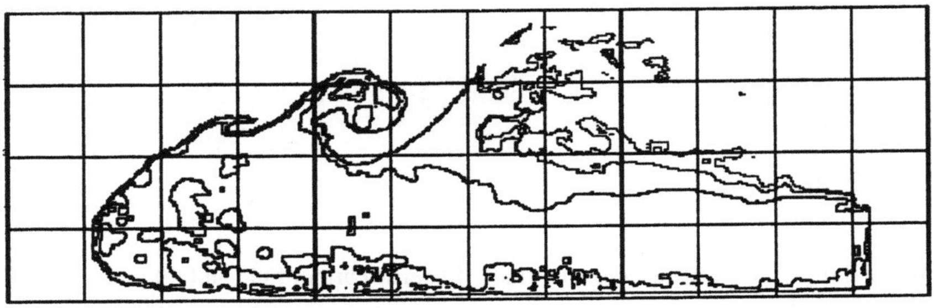

Fig.4. Contours of internal mixing extracted from Fig.3 (2 cm x 2 cm grid).

Conclusion

The present visualization technique, applied to gravity current flows, allows substantial qualitative analysis. It seems evident that the degree of mixing occurring within the head structure is more important than was previously noted. Continuing work will extend this technique to provide more detailed quantitative evaluation of the mixing processes.

Acknowledgements

The present work is sponsored in part by the Natural Sciences and Engineering Research Council (NSERC) and the Society of Fire Protection Engineers (SFPE).

References

1. Simpson, J.E., 1986, "Mixing at the Front of a Gravity Current", Acta Mechanica, Vol. 63, pp. 245-253.
2. Simpson, J.E., and Britter, R.E., 1979, "The Dynamics of the Head of a Gravity Current Advancing Over a Horizontal Surface", Journal of Fluid Mechanics, Vol. 94, Part 3, pp. 477-495.
3. Winant, C.D., and Bratkovich, A., 1977, "Structure and Mixing Within the Frontal Region of a Density Current", 6th Australasian Hydraulics and Fluid Mechanics Conference, Adelaide, Australia, Dec. 5-9, 1977.
4. Peters, W.D., Dutcher, C.R., and Venart, J.E.S., 1991, "Transient Flow Visualization of Gravity Currents Using Full-Field Laser-Induced Dye Fluorescence", Symposium on Experimental and Numerical Flow Visualization, FED-Vol. 128, pp. 209-215, ASME Winter Annual Meeting, Atlanta, GA., Dec. 1-6, 1991.
5. Blackmore, D.R., Herman, M.N., and Woodward, J.L., 1982, "Heavy Gas Dispersion Models", Journal of Hazardous Materials, Vol. 6, pp. 107-128.
6. Zukoski, E.E. and Kubota, T., 1988, "Experimental Study of Environment and Heat Transfer in a Room Fire," Report NIST-GCR-88-554, Center for Fire Research, National Institute of Standards and Technology, Gaithersburg, MD.

Flow Analysis Inside a Passenger Car Silencer by Laser Light-Sheet and Coded Particle Trace Techniques

W. Hentschel, Volkswagen Research, Wolfsburg, Germany

S. Hehn, Inst. of Applied Physics, Univ. of Frankfurt, Germany

Summary:

The well-known technique of flow field measurements by particle tracking (e.g.[1]) in a laser light-sheet has been improved by using a special coding function for the illumination, and by the development of algorithms for the automatic analysis of pictures of the entire flow field. The geometric boundary conditions of the object are taken into account. This technique was first tested under well known flow conditions in a rectangular tube with a ramp and afterwards used to measure the flow field inside a passenger car silencer under steady state conditions.

Introduction

The flow inside a passenger car silencer is expected to be turbulent within wide areas. In order to obtain more detailed information about the flow field, a plexiglass model of the original shape of a series silencer was made which allows optical access to the different chambers and pipes inside the silencer, Fig. 1. Visualization of the air flow in this model is performed by illuminating a plane of the flow field by means of a high power

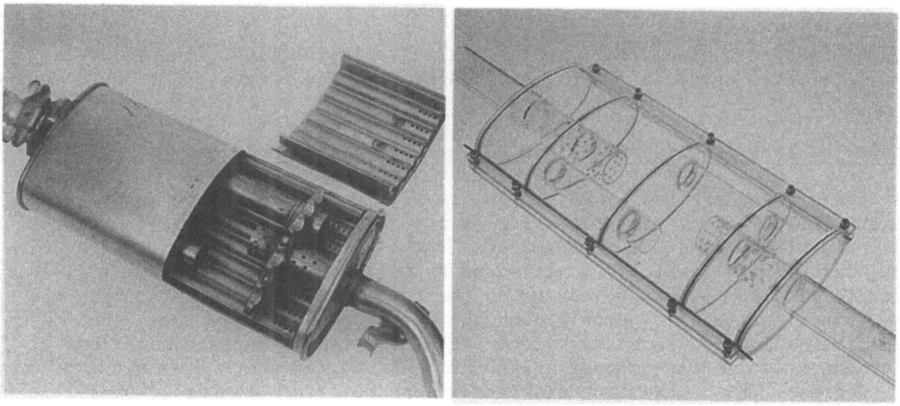

Fig. 1: Original passenger car silencer (left, opened for insight) and plexiglass model (right) for flow field visualization.

argon-ion laser, by seeding the air flow with lycopod spores as tracer particles (mean diameter 36 µm), and by registration of the pathlines onto black-and white film.

Due to the turbulent character of the flow simple stroboscopic illumination does not yield into quantitative information of the flow velocity including its direction. Therefore, we describe a method and its implementation for the automated analysis of 2-D flow fields using a coded sequence of light pulses for the light-sheet illumination [2].

Coded illumination

The modulation of the illumination is achieved by a rotating disk (chopper wheel) in the raypath (Fig. 2). The choosen code "short - long - short - short - short" allows the unambiguous identification of the flow direction even with the first or last pulse missing in the picture. In addition, the orientation of the long piece of a particle trace can be used as an approximate direction for the particle movement. This simplifies the assignment of the correct short pieces to a trajectory.

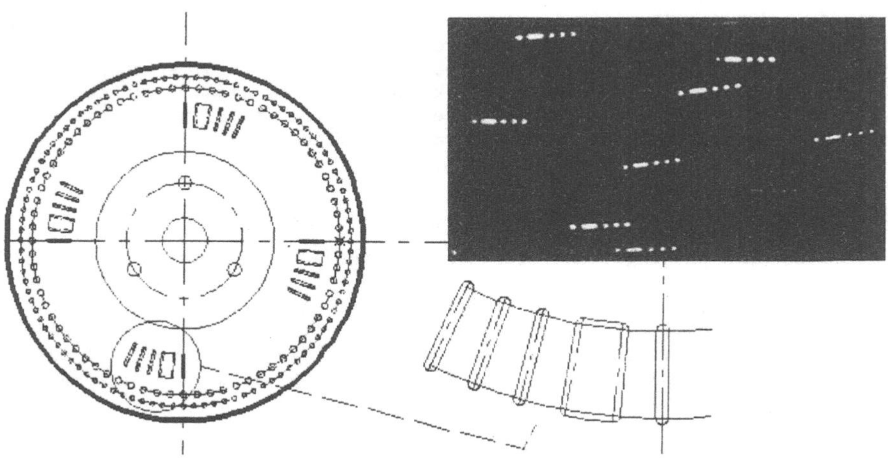

Fig. 2: Coded illumination of tracer particles in the flow. Chopper wheel (left), particle traces, and coding sequence (right).

Data extraction and reconstruction of the particle velocities

The photographs are fed into a high resolution digital image processing system (HDTV). After pre-processing of the pictures (contrast enhancement and noise reduction) the coordinates, the length, and the orientation of each part of a trajectory is extracted by new developed algorithms [3]. In order to find the pieces belonging to one common trajectory, these data are used to calculate all possible combinations of pieces. Only those combinations are allowed, where the directions of the longer exposed piece and of the connection to its nearest neighbours make angles which does not exceed ten degrees. This condition drastically reduces the number of possible combinations. The most probable combination is used to calculate the particle velocity.

Interpolation of the entire flow field

As the velocity information is distributed stochastically in the plane under investigation, the measured velocity values v_i are interpolated onto an equally spaced grid. The used interpolation function is an improved form of Shepards interpolation formula (1) [4] adapted to the special requirements of fluid dynamics:

$$v(x,y) = \begin{cases} \sum_{i=1}^{N} \frac{v_i}{r_i^{\mu}} \Big/ \sum_{i=1}^{N} \frac{1}{r_i^{\mu}} &, \text{ if } r_i \neq 0 \\ v_i &, \text{ if } r_i = 0 \end{cases} \qquad (1)$$

with r_i being the distance of the ith measured value to the point (x,y) and μ a parameter which was found to be optimal between 2.8 and 3.5 [2].

In a first step, in order to assure a tangential flow around obstacles and along walls in the flow, velocity vectors close to the walls are interpolated and only their tangential component with respect to the wall is kept.

The accuracy of formula (1) can be remarkably increased when v_i is replaced by

$$v_i^* = v_i + \delta v_i/\delta x\ (x-x_i) + \delta v_i/\delta y\ (y-y_i) \qquad , \qquad (2)$$

where $\delta v_i/\delta x$ and $\delta v_i/\delta y$ are the partial derivations of the velocity with respect to the x- and y- direction. Since the partial derivations cannot be measured, they have to be estimated in a suitable way. This is done by calculating the flow field around each measured velocity vector using formula (1) and computing the derivations $\delta v_i/\delta x$ and $\delta v_i/\delta y$ from these fields. The complete velocity map is than interpolated using formula (1) with the values of v_i^* received by formula (2).

Verification of the expected accuracy

Investigations with simulated flow fields show a mean error of less than 3 % per interpolation point, if a resonable number of velocity vectors is given. The method was tested by measuring the flow field in a rectangular tube with 30 mm x 50 mm cross section in which a ramp of 100 mm length and 30 mm hight was installed. The velocity above the ramp can be calculated analytically. Fig. 3 shows a photograph of the flow, the extracted velocity vectors, the interpolated flow field, and in comparison the theoretical and measured velocities above the ramp. The slight differences of less than 2 % between the measured and the calculated velocities show the good suitability of the presented method for measuring 2-D flow fields.

Results

As a first result from the passenger car silencer the air flow field in the center plane was investigated under steady flow conditions which corresponds to a low load engine operation. Fig. 4 shows the measured velocity vectors (top) and the interpolated flow field (middle). In some areas no velocities could be measured due to the fact that the flow cuts the plane of illumination nearly perpendicular. In those regions no velocity vectors were interpolated. The observations during the experiments in combination with the interpolated velocity map were used to sketch the

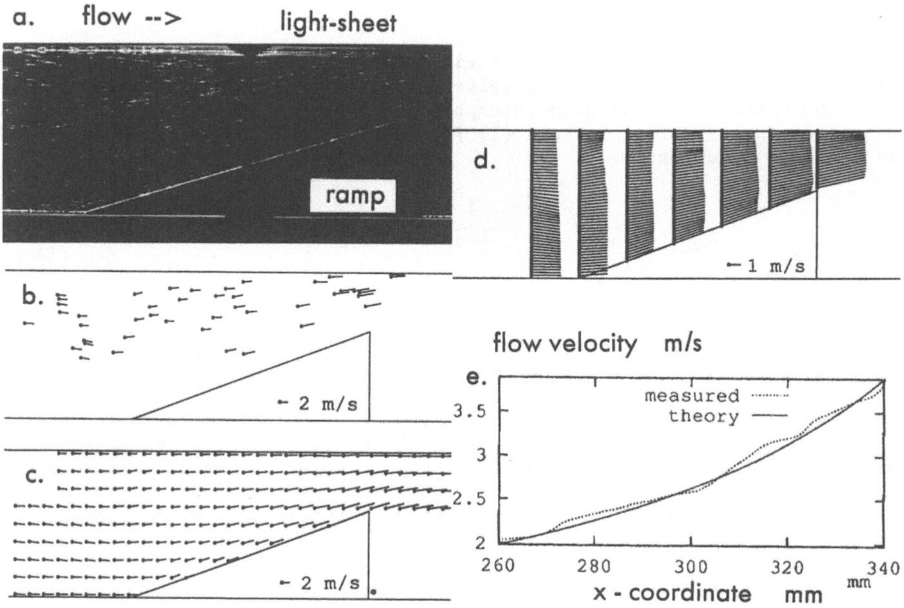

Fig. 3: Flow in a rectangular tube with ramp.
a. Photograhy of the particle movement in the light-sheet;
b. extracted velocity vectors from particle trajectory analysis;
c. interpolated flow field;
d. interpolated flow profiles;
e. comparison of theoretical and measured velocities.

3-D flow field within the car silencer (bottom). The flow is devided into two branches. In the first, the air leaves the inlet tube through the small holes to chamber I, passes chamber II through the inner tube, and enters the exit tube in chamber III. In the second branch, the air leaves the inlet tube in chamber II, passes chamber III through the other inner tube, and enters the exit tube in chamber IV through the small holes. Due to the symmetry of the design, the volume flows are equal in the two branches and this fact was used for a check of plausibility of the evaluated results.

Additionally, flow field studies in the fully transparent silencer operated behind a 1.8 l spark-ignition engine were carried out at an engine speed of 1200 rpm and part load. The motion of the tracer particles visualized by a high power laser light-sheet (20 W argon laser, without modulation) is registrated by a high-speed cine-camera (16 mm, 500 frames per second). One can clearly observe the instationary character of the silencer flow with an intermediate reverse flow in the different strokes of the engine operation. The condensation of steam (product of the combustion) and over-heating of the plexiglass model limits the time of operation.

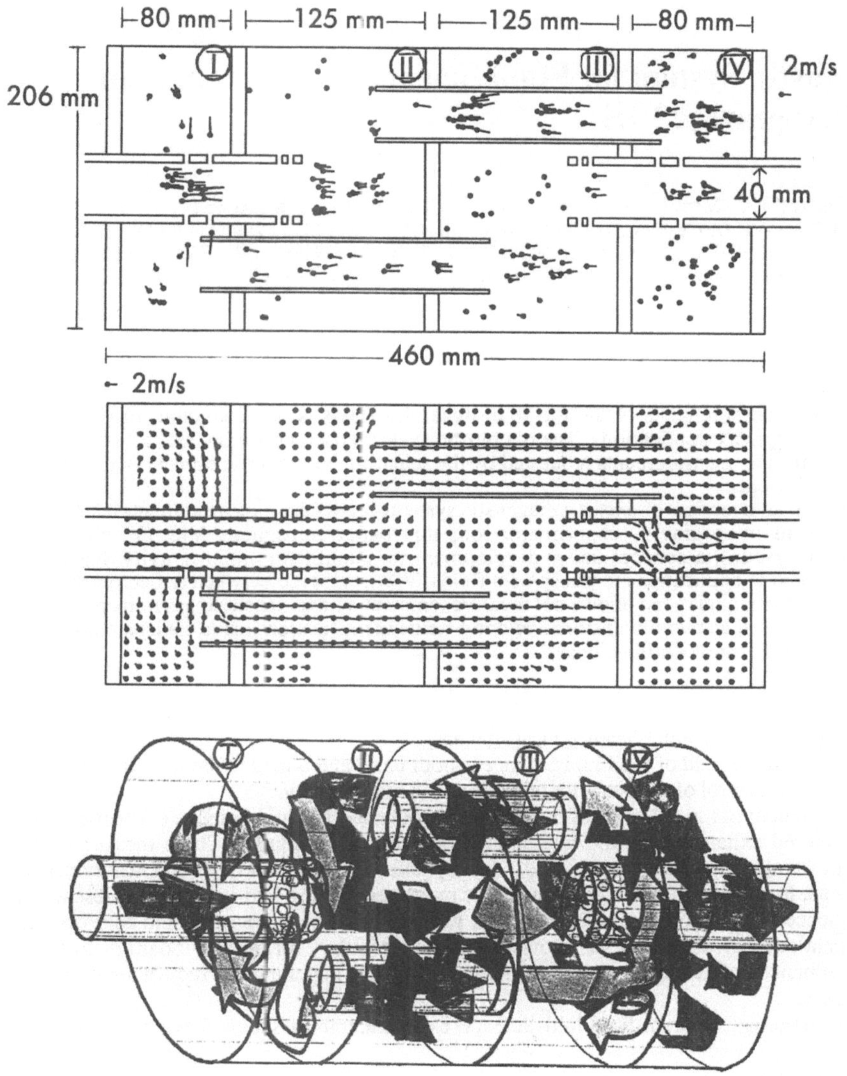

Fig. 4: Analysis of the flow field in the centre plane of a plexiglass model of a passenger car silencer by means of a coded particle trace technique under steady-state conditions.
top: measured flow velocity vectors by automated analysis of the light-sheet photographs;
middle: interpolated flow field on an equally spaced grid;
bottom: sketch of the 3-D flow in the silencer model.

References
[1] Merzkirch, W.: Flow Visualization. Academic Press, London 1987
[2] Hehn, S.: PhD thesis. Frankfurt 1992 (in German)
[3] Puffer, F.: Masters thesis. Frankfurt 1992 (in German)
[4] Lorentz, G.G. et al. (eds.): Approximation Theory. Acad. Press, 1973

Diagnostic System for Fluid Science Experimentation in Microgravity: VIS

V. De Chiara(°), C. Mirra(°), P.G. Falciani(^), M. Tacconi (^), S. Musazzi (*)
(°) MARS Center-Via Diocleziano, 328 - 80125 Napoli (Italy) Ph. +39817252772; FAX +39817252750
(^) Officine GALILEO-Via A. Einstein, 35 - 50013 Firenze (Italy) Ph. +39558950672; FAX +39558950613
(*) CISE-V.le Reggio Emilia, 39 - 20100 Milano (Italy) Ph. +39221671; FAX +39221672620

Summary

Not invasive measurements of thermo-fluid dynamics fields is of paramount importance in several microgravity applications. The diagnostic systems used until now are not completely exhaustive for the measurements requested by the scientific community, especially in the field of the Fluid Science.
Starting from an overview of existing diagnostic systems for accurate measurements of physical parameters in microgravity experimentation, this paper introduces VIS (Visualisation Integration System), a concept under development to manufacture microgravity diagnostic system tailored to applications to transparent fluids. VIS is a merge between deflectometry and photochromic techniques and allows not invasive temperature and density measurements (beam deflection) as well as two dimensional velocity fields identification inside transparent liquids (tracing methods - photochromic dyes).

Diagnostic systems for Fluid Science measurements.

The research in the field of the Fluid Science is both basic and aimed at the identification, the knowledge and control of the behavior of the fluids in microgravity.
As it was pointed out in [1], a lot of measurements obtained in Fluid Science experiments are not considered exhaustive by the international scientific community. These measurements regard the velocity and temperature fields, density and concentration, mass and heat transport phenomena, in general diffusive and convective processes. Taking into account the presence in Fluid Science of the various problems and the wide range of measurements and controls, the above mentioned investigation has showed how, in particulary for the on - board diagnostic, the performances of actual diagnostic systems are not full matching the requirements of the investigator.
An ideal diagnostic system for Fluid Science experimentation should have the following features:
1) not invasive;
2) to be related to all flow effects in both coordinate systems (Eulerian and Lagrangian system);
3) to provide quantitative measurements with an high degree of precision and accuracy;
4) to correlate continuosly the data in space and time.
From a design point of view it has to satisfy the following requirements:
1) to be convenient, simple and easy to use;
2) the final results have to be in the most simple, opportune and usable form in order to be able to understand, for instance, unexpected effects.
The existing techniques do not satisfy all the requirements of an ideal diagnostic system, but an integration of techniques, for instance photochromic dyes and beam deflection, could provide one of the most complete diagnostic systems for Fluid Science experimentation.

Photochromic technique

The photochromic dyes method is a tracing technique which employs an UV laser as excitation source and a RGB (or CCD) camera, with a suitable illumination system, as visualization method (Fig.1).

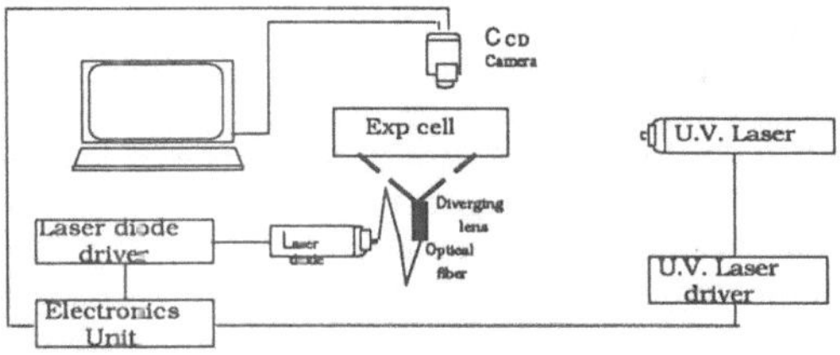

Fig.1. The photochromic dyes experimental set-up

A photochromic dye solution, in absence of UV illumination, is transparent but it changes its optical density when lightened with a suitable UV light. The most important feature of this technique is that the trace within the fluid is a portion of the fluid itself. The physical properties of the fluid within the trace and outside, are similar and so no intrusive bodies or substances are embedded in the test fluid.

In spite of the extremely low invasivity of this technique, it is necessary to point out two drawbacks: the liquid used must be an organic one (only mixtures of organic fluids with water can be used) and an intrusive thermal action of the UV laser exists, even if this is almost negligible.

In the photochromic dyes technique, a line of fluid, usually normal to the flow, is marked at a given instant because of its optical density change due to the photochromic effect when lightened with a UV radiation (Fig.2). Any later location of this line is called timeline.

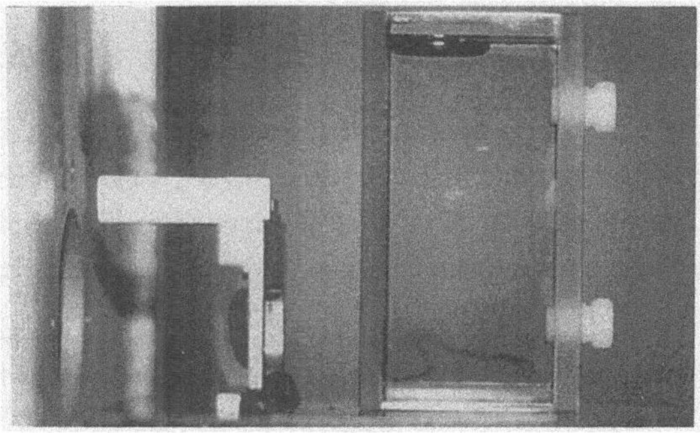

Fig.2. A line of fluid (a photochromic solution) marked by means of an U.V. ECCIMER laser

The visualization method of flow fields through photochromic dyes was introduced by Popovich and Hummel in 1967 [2]. In presence of UV laser illumination, a plot of dark lines is generated in the fluid matrix at once. A continous illumination without disturbance allows to visualize diffusive and turbulent phenomena. On the other hand, pulsed UV laser light is used to generate several centimeters long linear traces in quite instantaneous mode with respect to high velocity photography (10^6 images per second), giving a possibility to study high frequence phenomena of the order of MHz.

Features and constrains

The usual application of this technique regards the monitoring of the velocity field in single-phase fluid or fluid in liquid-liquid double-phase. This diagnostic method gives excellent results in interface phenomena experiments.

The constraints of the system are the following:

1) it is applicable only to liquids;

2) apart from the points where two traces intersect, it is not possible to measure directly the velocity, that has to be calculated starting from a digital image processing;

3) the system results invasive if too much high density energy is used to lightened the dyes;

4) adaptation of the laser system to the microgravity environment in terms of power, voltage, mass, dimensions, coupling, beam quality, robustness;

5) trace shape definition: the ideal trace for the visualization of 2-dimensional velocity field should be as thin as possible, regular and symmetrical, sharp and straight at time T_0 and sharp and regular at times $T_n > T_0$.

Visualization method

The target of the photochromic dyes method is a quantitative measurement of the velocity field, but neither the timeline patterns nor streakline (defined as locus of points connecting all fluid particles that have passed through a given fixed point in the flow) alone can give a complete information on the velocity flow field.

In order to obtain a good estimation of the velocity field the position of an identifiable fluid particles at the end of short enough time intervals is needed. This requires a combined time-streak photochromic dyes tracer method.

This target can be achieved with two distinct U.V. laser beams (or one splitted in two by means of a beam splitter and two suitable mirrors). The point of the photochromic solution in which both the beams pass through, will be darker than the other points of one or other excited trace. This fact permits the recognition of one point of the trace in all the subsequent timelines.

In reducing data from the dyes traces on digitized image to obtain velocity, the x-component of velocity, for example, is given by:

$$< v_x > = (\Delta x / M) * \nu \qquad (1)$$

where $< v_x >$ is the estimated x component of the local Eulerian velocity v(x,y,z,t) at some point (a time and space average taken during a short time Δt), Δx is the x component of the displacement during a known interval $\Delta t/\nu$, M is the scale factor converting displacements on the digital image Δx in the actual measurement units and ν is the RGB framing speed (frames number per second).

It is useful to note that the minimum distance between two recordable timelines depends on the flow velocity and the RGB framing speed. Since the framing speed of a RGB has an upper limit of 50 frames/sec., one can monitor the flow velocity field with a good spatial resolution only if the flow is slow enough. Otherwise special CCD cameras exist with a very high framing speed (≈5000 frames/sec.) but with a lower spatial resolution.

From the eq.(1) it is possible to estimate the uncertainty, that is the possible error in the velocity measurements with the photochromic dyes technique.

If one defines $\varepsilon_{<v_x>}$ the uncertainty interval in the $< v_x >(\Delta x, M, \nu)$ measurement, the result expressed in non dimensional form is:

$$\frac{\varepsilon_{<V_x>}}{<V_x>} = \left[\left(\frac{\varepsilon_{\Delta x}}{\Delta x}\right)^2 + \left(\frac{\varepsilon_M}{M}\right)^2 + \left(\frac{\varepsilon_\nu}{\nu}\right)^2 \right]^{\frac{1}{2}} \qquad (2)$$

The uncertainties in Δx obviously overshadow those in M and v. Infact with an RGB or a CCD camera the framing speed can be driven with a precision better than 0.1% and the scale factor could be well defined by the optical set-up.

In the photochromic dyes technique the uncertainty for the displacement Δx arises from:

1) Measurement of Δx on the digital image;
2) Averaging effects in changing velocity field;
3) Possible displacements of the trace out of the x-z plane of the Laser beam;
4) Resolution problems due to the finite trace size and finite averaging interval;
5) Blurring of the trace due to thermal effects within the trace.

Image processing

Regarding the uncertainties on the Δx measurement, the method can be improved by the use of digital image processing either in resolution limits or in optical distortions. As just explained above, the CCD or RGB cameras are not very fast but their sensitivity permits to stop the image on the target in 1.0 μsec by means of a suitable illumination system. This fact allows to obtain very high quality images in a wide range of flow velocities. The possible blurring in the image is due, in this case, to the movement of the trace during the recording time. To have an uncertainty on the position of the traces their edge must be recorded, on the direction of the velocity field, by two or more pixels during the exposure. If the exposure time is short enough, the possibility of a limited image resolution due to this effect is very low. If the velocity is low, the probability that the above described effect occurs decreases since to be very low for extremely slow flows like the thermoconvective flows. The suitable illumination for this kind of recording techniques is a low power diode laser, driven with a square wave of 50 Hz and 1-3 μsec of pulse duration. In the photochromic technique, the illumination must be sincronized with the laser and the RGB camera. In order to avoid errors due to an excessive diffusion of the trace shape, a good method consists in recording the trace just during the first milliseconds. The use of the laser diode permits the recording of both the first (straight line) and the second (deformed by the flow) traces. Moreover the rising time of the diode laser pulse, that is extremely shorter than those of usual flashes (≈0.1 μsec against 0.1-0.2 msec) allows a better trace definition and a lower timing error.

This kind of illumination can be used only with a CCD or RGB camera because of their very high sensitivity.

In this marker method the target is the prediction of the Eulerian velocity of the flow field in a fixed point in time and space by calculating the Lagrangian time average velocity of a darker fluid parcel of fixed identity. The difference between this method and those which use tracing particles (powder, bubbles, microspheres etc.) is that in these methods the particles have to be assumed istantaneously attaining the local field velocity around them. This is not always true because of the particle different density, with respect to the surrounding fluid, which leads to a rising or descending velocity component and a delay in responding to changes in fluid velocity.

The beam deflection technique

The beam deflection technique is a diagnostic tool utilized in ground based experiments to study temperature/concentration induced density gradients in fluids [3]. It is a powerful non invasive technique being actually more sensitive and less critical in terms of mechanical stability and overall design complexity than interferometric systems (although in the limit of 1-D analysis).

The principle of operation is quite simple and relies on the measurement of the angular deflection of a laser beam propagating through a region where a varying optical gradient $\delta n/\delta z$ (orthogonal to the beam) is present. The deflection angle is given by:

$$\vartheta = (\delta n/\delta z)*1 \quad (3)$$

where l is the length of the region where the gradient exists (Fig. 3).

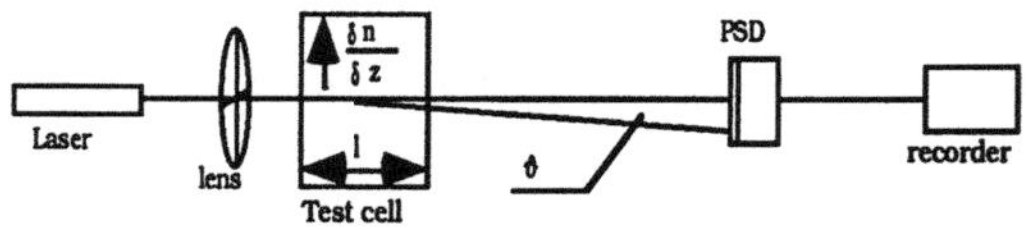

Fig.3. Mapping 1-D set-up

Notice the ϑ includes the refraction at the boundaries so that the difference between the average index inside the test fluid and the fluid outside the test region has not to be taken into account. The measurement of the deflection is usually performed by means of a Position Sensing Device (PSD) placed at a proper distance L from the test cell. The PSD is a sensor capable of providing an electrical output proportional to the displacement of the beam centroid from the unperturbed zero position, of course, since the larger L the larger the beam centroid displacement will be, one gains in sensitivity by selecting L adequately large. In the case the beam has to be focused in the test region so to probe the gradient across a very thin layer, the beam spot on the sensor will increase linearly with the distance L. Luckily, position sensors exist which give an output proportional to the centroid displacement irrespective of the actual beam spot, provided the spot diameter is only a fraction of the sensing element diameter. Typically, centroid displacements of a few microns can be detected, the range of measurement being a few millimeters. When 2-D mapping of the test region is requested a more complex scheme has to be considered [4]. It is schematically shown in Fig. 4.

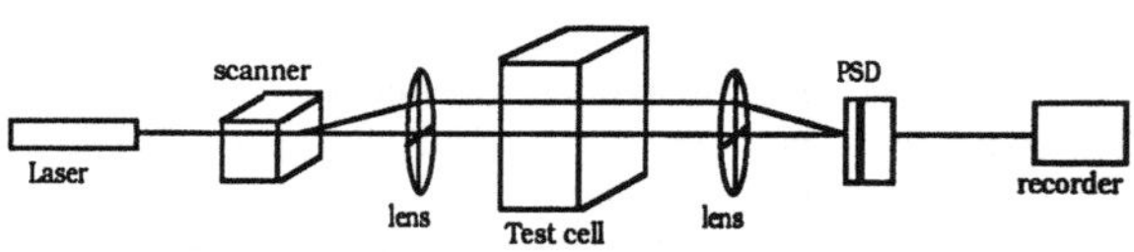

Fig.4. Mapping 2-D set-up

The beam coming from the laser source is made scan a two-dimensional angular region by means of a scanning device (utilizing either two mirrors mounted on a couple of galvanometers oscillating around two orthogonal axes or a single mirror mounted on a magnetically driven gimbal mount). The beam angular displacements are transformed into linear sweeps by means of a lens placed at the focal distance from the scanning device. In this way the test region (whose largest dimension depends on the lens diameter) is scanned by a probe beam propagating along directions parallels to the optical axes. A second lens, positioned after the test cell, is used to focus the probe beam onto the PSD.

VIS

As mentioned before, an integration of different tchniques can lead to the almost ideal diagnostic system. Our proposed system is the Visualization Integrated System that is a diagnostic system for Fluid Science experiments, utilizing the two already described techniques: the photochromic dyes technique and the laser beam deflection technique.
Tracing methods, in particular, allow the measurament of the velocity field by visual inspection or photographic recording of the flow field. All these methods have in common that portion of the fluid is marked or seeded with a foreign material, and that it is possible to draw conclusion on the velocity of the fluid from the motion of this observable foreign material, whose motion allows the identification of the flow pattern.

Among these techniques, the photochromic dyes method has been pointed out, being the best candidate for VIS, because of its very low invasivity and its high spatial resolution. In this technique the dyes may be selectively activated by an U.V. light source to produce bright colours.

On the other hand, the potentialities of the integration of this method with a beam deflection system is highlighted as a powerful opportunity to offer a complete measurement of the physical parameters of the whole flow field.

Infact, in the absence of index of refraction gradients in the test region the probe beam is focused in correspondance of the PSD center irrespective of the explored region in the test cell.

When a perturbation is created in the fluid, the probe beam is deflected by an angle ϑ and consequentely the position of the focussed spot on the PSD will change by a quantity:

$$d = f\tan\vartheta - f\vartheta \qquad (4)$$

being f the focal length of the focusing lens.

By recording the PSD outputs, the components of d along two orthogonal axes can be derived as a function of time and consequently the beam deflection can be calculated for each point explored during the 2-D scanning of the test region.

In respect of this, in our case, the photochromic traces will be the above mentioned perturbation, surely less invasive than others. Moreover, the temperature field visualization inside the test cell, utilizing the laser beam deflection technique, will allow to evaluate the invasivity degree of the photochromic system. It will be necessary to measure the temperature field before, during and after the excitation of the traces, obtaing information on the occurred modification and on the right proper way to utilize the system. On the other hand, the nucleation and coalescence phenomena in the presence of high intensity laser beam could be used like a stimulus or a generating bubbles system.

The VIS configuration is showed in Fig. 5

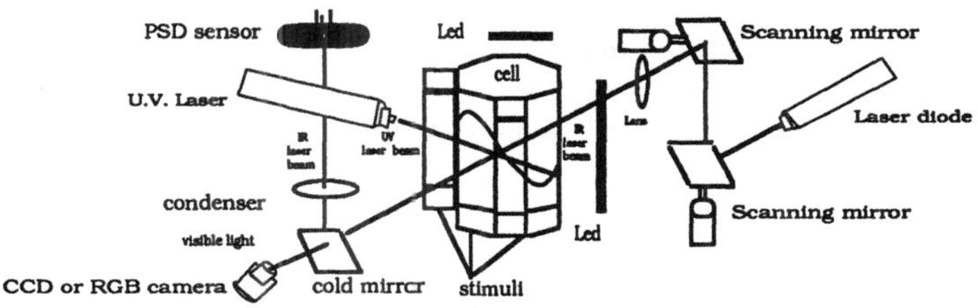

Fig.5. V I S configuration

The advantages of a diagnostic system like this are the following:

- to utilize a technique to measure the largest number of parameters as possible in Fluid Science in the less invasive way;
- to obtain in real time information about the evolution of the system;
- a real possibility to utilize it for microgravity Fluid Science experiments.

The first step will be the manufacturing of a bread-board for on-ground experimentation and later to focus on the integration and use of the same in microgravity environment.

References

1. GALILEO/MARS/CISE : Diagnostic Advanced System final report, ASI Contract 91/AF/1990

2. Popovich, Hummel : A new method for non-disturbing turbolent flow measurement very close to a wall. *Chem. Eng. Sci.*, 22 pp. 21-25

3. M. Giglio, A. Vendramini, *Phys. Rev. Lett.*, 38, 26 (1977)

4. S. Ciliberto, F. Francini, F. Simonelli, *Opt. Commun*, 54 (5), 251 (1985)

Visualization of Unsteady Flow in Rotating Drums

Wen-Jei Yang
Department of Mechanical Engineering and Applied Mechanics,
The University of Michigan,
Ann Arbor, Michigan 48109, U.S.A.

Genshi Kawashima
Department of Mechanical Engineering,
Musashi Institute of Technology,
Tokyo, Japan

Hiroshi Ohue
Musashi Institute of Technology

Summery

Flow visualization studies were conducted on the unsteady flow of heat and fluid in rotating drums resulting from a sudden acceleration/deceleration, both with and without a sudden application of inner surface heating. The drum consisted of an inner cylinder, an outer cylinder, and two circular plates for the side walls. The outer cylinder and side walls were transparent to allow observation. A flexible electrical heater was attached uniformly to the inner cylinder. Water and aqueous solution of glycerin at various concentrations were used as the fluid in the drum. Different methods were employed to observe unsteady recirculating flow: (1) a laser light sheet with double-exposure photography, (2) temperature-sensitive liquid crystal combined with image processing, (3) laser Doppler velocimetry, (4) temperature measurements with thermocouples. The flow patterns and temperature distributions thus obtained were compared with each other.

Introduction

Unsteady flow and heat transfer are induced by a suddenly accelerated or decelerated drum, possibly in combination with sudden heating on its inner surface, is usually a rather complicated phenomenon. Because of technical difficulty in both measurements and theoretical analysis, minimal effort has been directed toward studying this type of flow.
Recently, a series of studies were performed on unsteady drum flow theoretically[1]; by means of L.D.V.[2]; by visualization with double-exposure photography[3]; by the laser light sheet method [4,5]; by temperature measurements[6,7]; and using temperature-sensitive liquid crystal[8]. This study compares results obtained by the above methods to determine unsteady flow mechanisms in rotating the drum.

Experimental Apparatus and Prrocedure

A schematic diagram of the experimental apparatus is illustrated in Figure 1. The drum consisted of an inner cylinder, an outer cylinder and two circular plates. The two cylinders were placed concentrically with the two plates as the side walls to enclose the test fluid. Both

outer cylinder and side walls were made of transparent acrylic to allow visual observation.
First, double-exposure photography was employed with a He-Ne laser light sheet for illumination. The test liquid was seeded with plastic powder of 150 μm in diameter which had the same specific weight as the liquid. Flow direction was determined from the traces of long and short intermittent light pulses. The particle traces recorded on photographs may be analyzed to determine the radial and angular velocity components[3]. Next, a temperature-sensitive liquid crystal which was microencapsulated in gelatin of about 50 μm was used. Simultaneously, the relationship between the hue angle and the temperature was calibrated. In both methods, flow patterns were recorded from two directions, one being perpendicular to, and the other at a slight angle to the illuminated plane using a camera with ASA 3000 film (in double-exposure photography) and ASA 1600 color reversal films (in the case of liquid crystal). The liquid temperature in the drum was measured by means of three thermocouples, one each on a side wall and the inner and outer cylinders at the midplane.

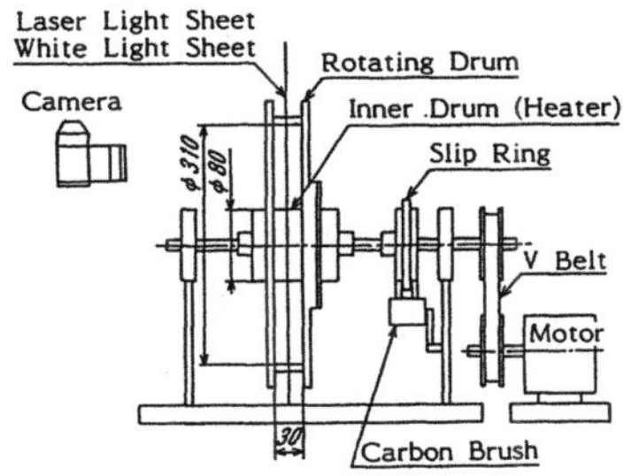

Fig.1. A schematic of experimental apparatus.

Experimental Results and Discussion

In the interest of brevity, only the main features of observed flow patterns are presented here. Figures 2 through 4 are sketches of flow patterns in double-exposure photographs. Figure 2 is for a stationary drum and Figures 3 and 4 are for drums rotating counter-clockwise at the same speed at 136 sec. per revolution. The fluid in the drum was water for Figures 2 and 3 and a 50 % water-glycerin mixture for Figure 4. Figures 5 and 6 are sketches based on photographs obtained using temperature-sensitive liquid crystal. Figure 5 is for a stationary drum and Figure 6 for a drum rotating counter-clockwise at the same speed as above. The liquid was water for both. The power supplied to the heater was 50 W in all cases.
Figure 7 shows temperature distributions in the flow field corresponding to Figure 5, obtained by image processing. One is a two-dimensional equivalent temperature map for (a) and the other is a three dimensional contour map with temperature being the vertical, or Z, axis. The abscissa and ordinate define the domain of observation, shown as rectangles in Figure 5. One very interesting and typical feature of flow in a stationary drum is the stratified layer with a horizontal temperature distribution near the top of the inner cylinder, as shown in Figure 5, within which the velocity is downward, as shown in Figure 2(a). In other words, even though the velocity is non-zero in the vertical direction, the temperature distribution is stable in the horizontal direction.
In a rotating drum, Figures 6(a)(b), a well-defined hot plume ascends upward toward the outer cylinder and slants slightly in the direction of rotation during the early stage at 60 seconds after a sudden initiation of both rotation and heating. Next, the hot plume slants in the counter-rotational direction and the stable, thermally stratified layer remains near the outer cylinder even though the drum is rotating. Sketches of these flow patterns based on photographs taken under the same conditions are shown in Figure 3. The slanting hot plume is shown in Figure 6. In this case also, the velocity was non-zero in the stable, thermally stratified layer with the horizontal temperature distribution.
The features of flow pattern for a different liquid, 50 % water-glycerin, are shown in Figure

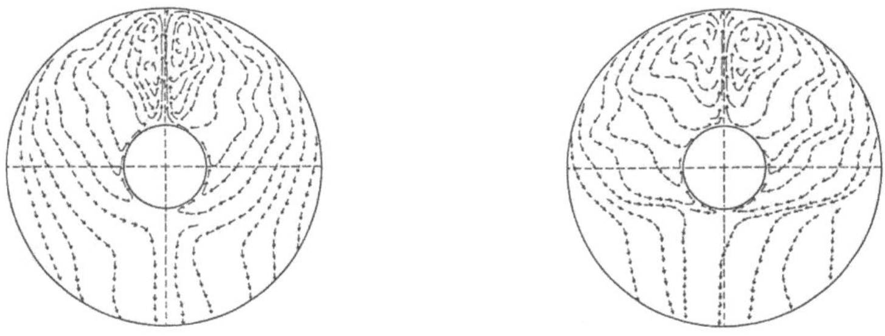

(a) at 240 sec. after initiation of the transient (b) at 1800 sec. after initiation of the transient

Fig.2. Unsteady flow patterns in water inside a stationary drum induced by sudden inner surface heating at 50 W obtained by double-exposure photography.

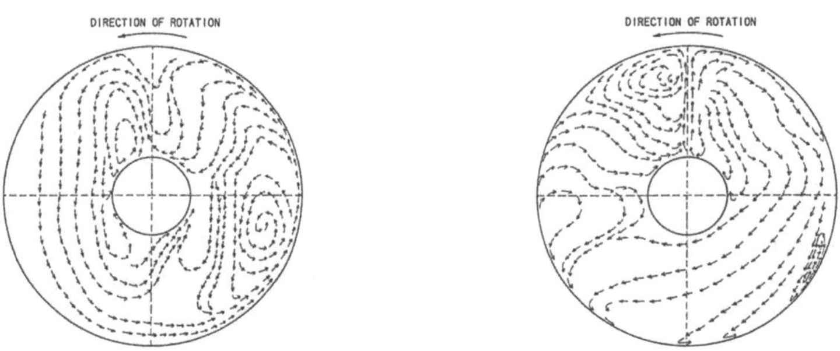

(a) at 30 sec. after initiation of the transient (b) at 240 sec. after initiation of the transient

Fig.3. Unsteady flow patterns in water inside a drum rotating at 136 sec/rev induced by sudden inner surface heating of 50 W and rotation obtained by double-exposure photography.

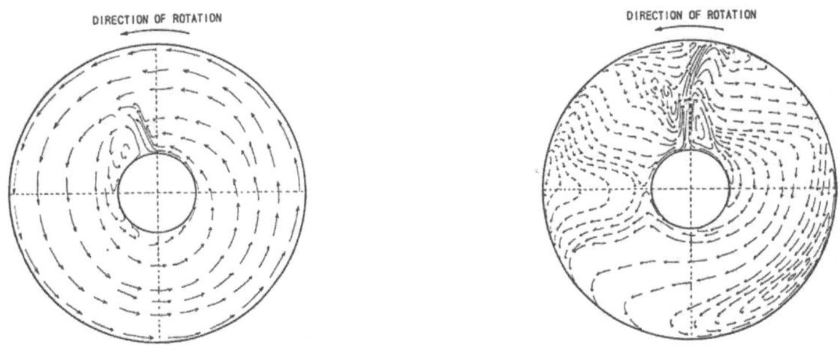

(a) at 30 sec. after initiation of the transient (b) at 240 sec. after initiation of the transient

Fig.4. Unsteady flow patterns in glycerin solution inside a drum rotating at 136 sec/rev induced by sudden inner surface heating of 50 W and rotation obtained by double-exposure photography.

4. A comparison of Figure 3 and Figure 4 reveals that the flow pattern at the beginning of the rotation and heating are different because of the difference in viscosity. As time passed, the difference diminished and the flow reached a steady state. These results are corroborated by the temperature measurements.

Figure 7 shows velocity distributions as measured by L.D.V. and double-exposure photogra-

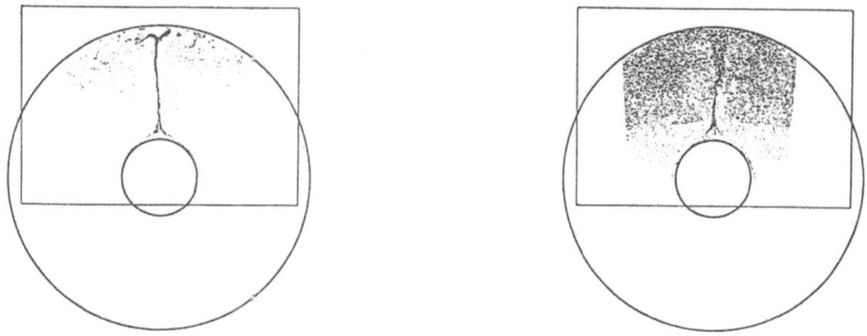

(a) at 60 sec. after initiation of the transient (b) at 230 sec. after initiation of the transient

Fig.5. A sketch of the photograph taken by the temperature sensitive liquid crystal method - stationary drum.

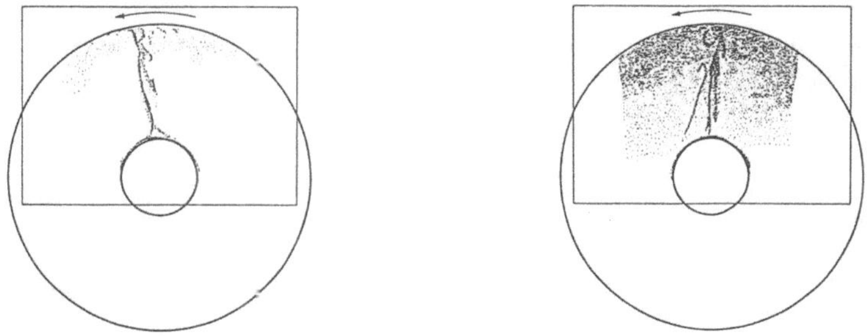

(a) at 60 sec. after initiation of the transient (b) at 230 sec. after initiation of the transient

Fig.6. A sketch of the photograph taken by the temperature sensitive liquid crystal method - rotating drum.

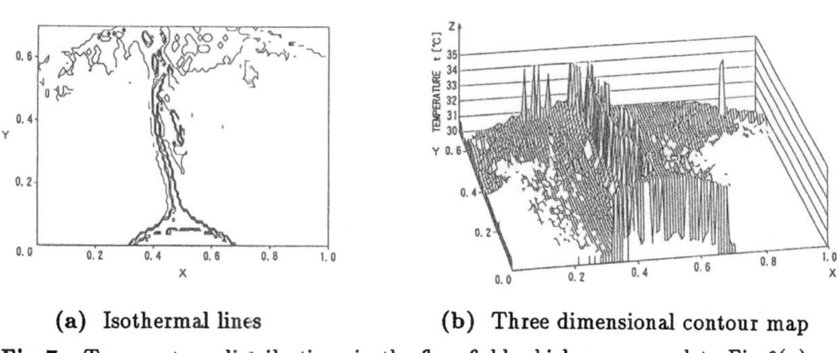

(a) Isothermal lines (b) Three dimensional contour map

Fig.7. Temperature distributions in the flow field which correspond to Fig.6(a).

phy. It is clear that double-exposure photography is an effective method of measuring very small and negative values.

The contour maps of the temperature distribution obtained by image processing clearly show the actual temperature situation, as shown in Figure 8.

The results of all four methods – temperature-sensitive liquid crystal, double-exposure photography, velocity measurements by L.D.V., and temperature measurements with thermocouples, – agree very well and provide a clear picture of the flow pattern in both stationary and rotating drums.

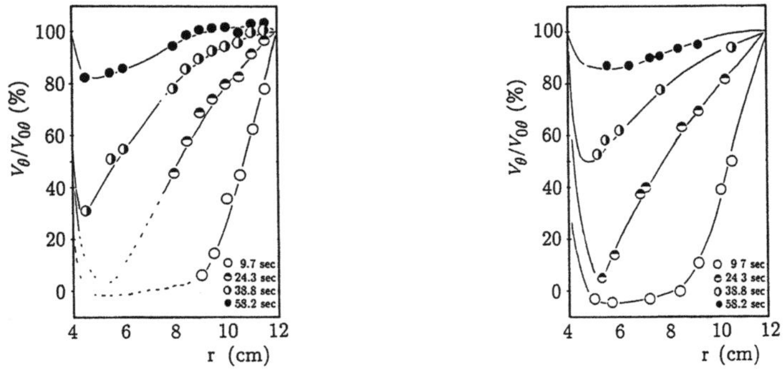

(a) L.D.V (b) Visualization

Fig.8. Transient angular velocity distribution in the radial direction in water inside Drum for a sudden acceleration to 10 sec/rev at the midplane.

References

1. Yang, W.J.; Kawashima,G.: Hydrodynamics in Unsteady Drums. Preprint of 2nd International Symposium on Transport Phenomena and Dynamics of Rotating Machinery Vol.1, Honolulu. (1988) 33-39.
2. Kawashima,G.; Yang, W.J.: Unsteady Flow in rotating Drums using Laser Doppler Velocimetry. Experiments in Fluids 9 (1988) 165-171.
3. Kawashima,G.; Yang, W.J.: Determination of Unsteady Flow Inside Rotating Drums by Flow Visualization with Double-exposure Photograph. Measuring and Matering of Unsteady Flows, ASME FED Vol.106, (1991) 41-46.
4. Kawashima,G.; Yang, W.J.: Light Sheet Visualization of Unsteady Recirculating Flow inside Rotating Drums. Flow Visualization V, Hemisphere Publishing Corporation, Washington D.C., (1990) 633-639.
5. Yang, W.J.; Ohue,H.; Kawashima,G.:Experimental Study by L.D.V. and Flow Visualization on the Unsteady Flow Pattern in Rotating Drums. Preprint of 4th International Symposium on Transport Phenomena and Dynamics of Rotating Machinery, Honolulu. (1992).
6. Kawashima,G.; Yang, W.J.: Experimental Study on Unsteady Recirculating Flow in Rotating and Stationary Drum Inner Surface Heating. International Symposium on Nonsteady Fluid Dynamics, ASME FED Vol.92, (1990) 219-224.
7. Yang, W.J.; Kawashima,G.: Experimental Study on Unsteady Convection in Rotating Drums with Inner Surface Heating. Preprint of 3ed International Symposium on Transport Phenomena and Dynamics of Rotating Machinery, Honolulu. (1990) 156-166.
8. Kawashima,G.; Yang, W.J.; Ohue, H.: Image Processing of Flow Patterns in Rotating Drum with Inner Surface Heating using Temperature Sensitive Liquid Crystal. Preprint of internal Seminar on Imaging in transport Processes, Greece. (1992).

Flow Visualization of Liquid Crystalline Polymer Solutions in Rectangular Channels

S. Guido, N.Grizzuti, and G.Marrucci

Dipartimento di Ingegneria Chimica, Università di Napoli Federico II
P.le V.Tecchio 80, 80125 Napoli- ITALY

Summary

Velocity profiles of aqueous solutions of Hydroxypropylcellulose flowing in rectangular channels were visualized by means of an electrochemical technique. Isotropic and anisotropic solutions were used. The velocity profiles of the liquid crystalline solution were found to be qualitatively different from those of the isotropic one. In particular, anomalous maxima in the velocity near the side walls were observed.

Introduction

One of the most attractive features of Liquid Crystalline Polymers (LCPs) is the high degree of flow-induced molecular orientation in view of their possible application as self-reinforced materials. However, the flow-induced orientation process is still poorly understood, due to the extremely complex rheological behaviour of such materials.

While optical techniques, such as polarized light microscopy and light scattering [1-5], have been extensively applied to study the microscopic structure exhibited by LCPs during and after the cessation of flow, so far little use has been made of direct flow visualization techniques, which are able in principle to give macroscopic information on the flow behaviour.

In this work, isotropic and anisotropic aqueous solutions of Hydroxypropylcellulose (HPC) were injected in a transparent mold. The relatively low opacity of the liquid crystalline solutions (and the full transparency of the isotropic ones) allowed us to use a coloured tracer technique to visualize the velocity profiles in a shallow channel geometry.

In the following, after a description of the experimental setup, the results of the flow visualization will be shown, and the differences in flow behaviour between the isotropic and anisotropic phases will be discussed.

Materials and Methods

A schematic view of the apparatus is shown in Fig.1a. A syringe pump, driven by a servo controlled motor or by an infusion pump (model 11, Harvard Apparatus), injects the polymer into a transparent mold. For observations and recording of images a CCD videocamera, which can be connected to a microscope tube, or a photographic camera equipped with a macro (short focus) objective can alternatively be used. Mold, lamp and camera are mounted on an optical bench, so that good alignment and fine focussing are possible. The mold (see Fig.1b) consists of two parallel plates of plexiglass (length 10 cm, width 5 cm), separated by a metallic spacer, suitably shaped, which defines the geometry of the flow field. The feeding gate is a circular hole through one of the confining plates. A more detailed description of the apparatus can be found elsewhere [6].

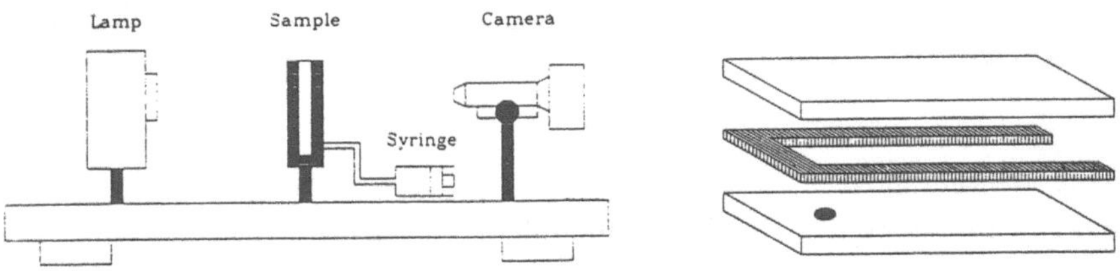

(a) (b)

Fig.1. (a) Schematic view of the experimental setup; (b) Design of the transparent mold.

Aqueous solutions of HPC (Aldrich, $M_W = 100{,}000$) were prepared by dissolution and long time mixing. The critical weight concentration for mesophase formation is reported to be ca. 41%. Above such a concentration the material is cholesteric at rest. The concentration of the anisotropic (A) and isotropic (I) solutions used in the experiments were 50%wt and 30% wt, respectively.

Velocity profiles were visualized by means of an electrochemical technique. A thin filament (platinum or tungsten) with diameter ranging from about 10 to 50 μm, was placed in the fluid, acting as a cathode. The anode was another filament, placed a few centimeters away from the cathode. The two electrodes were connected to a DC generator, which was able to produce constant tension pulses. If a tension of about 20 volts or more is applied, a line of tiny hydrogen bubbles is generated at the cathodic filament surface by water electrolysis. The bubble line displacement is then monitored as a function of time in order to determine the velocity profiles [7]. Alternatively, if a pH indicator is dissolved in the material, and a lower tension is applied, a coloured line is produced by the local change in pH due to electrolysis [8]. The high viscosity of the sample, in determining a slow diffusion rate of the coloured fluid, keeps the line sharp and well visible for long times. A thymol blue indicator, which turned from orange to blue upon electrolysis, was used in the experiments. In both cases, small amounts of the electrolyte NaCl were added to the solution, in order to increase the conductivity of the sample.

A particle tracer technique was also used, details of which are described elsewhere [9]. The motion of isolated glass microspheres inside the mold was followed by a microscope. Though reproducing qualitatively the same results obtained by means of the direct techniques, quantitative reproducibility was not excellent, due to difficulties in focussing the glass particles and tracing their motion over microscopic length scales.

Experimental results

The origin of the coordinate system used to describe the velocity profile is in the center of the channel section, where the filament resides. The x-axis corresponds to the channel length, and the z-axis runs along the filament, i.e. along the channel width, called 2W. The thickness is called 2H. The filament is placed at about 5 cm from the feeding gate.

The experiments were performed under different flow conditions. In particular, values of the volumetric flow rate Q ranging from about 2 to 30 mm^3/s, and mold thicknesses from 1 to 2 mm were used. For any choice of flow rate and thickness, the measurements were repeated several times in different runs. Reproducibility was always better than 5%.

Fig.2 shows a typical velocity profile for both the I and A solutions. Here the absolute velocity v(z) at $y=0$ is reported as a function of the dimensionless coordinate $1-z/W$. The mold geometry used in this case was a rectangular channel with a sudden enlargement of width (ratio 4:1), where the coloured generating filament resided some distance away from the enlargement. Due to geometrical symmetry, only one half of the channel is considered in Fig.2.

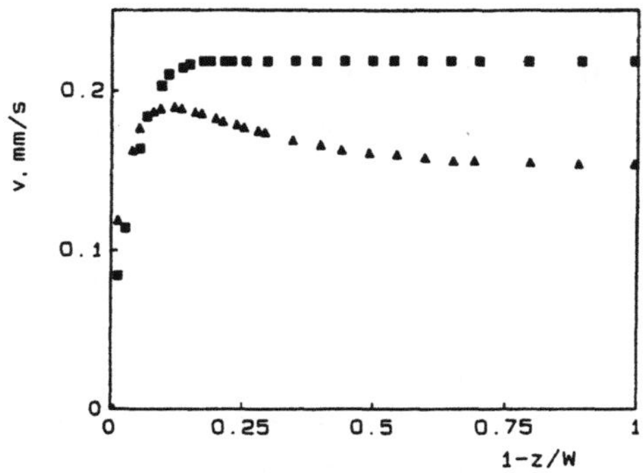

Fig.2. Midplane velocity profiles along the z direction (channel width). Q=2.9 mm^3/s, H=0.5 mm. (▲) Anisotropic sample; (■) Isotropic sample.

The I solution displays the expected profile, which is essentially flat except for a small region near the side wall, where the velocity sharply falls to zero. Conversely, the A sample is characterized by an anomalous behaviour: The velocity profile shows a shallow minimum at the center of the channel, i.e., it grows steadily from the center to the side, except for the extreme layers where it is bound to drop to zero at the side wall.

Conclusions

Two solutions of the same polymer, one isotropic and the other in the liquid crystalline phase, exhibited a qualitatively different flow behaviour during mold filling. The midplane velocity of the LCP sample, starting from a plateau value at the center of the mold, grows steadily towards the side walls, reaching a maximum before falling to zero at the boundary (due to the non-slip condition at the wall).

This behaviour is reminiscent of the flow of two immiscible liquids of different viscosities, the less viscous one flowing near the side walls. In such a case, a velocity profile similar to that of Fig.2 would be observed, provided that the two viscosities are sufficiently different. Thus, a profile such as that in Fig.2 would be explained by assuming that the viscosity of the anisotropic liquid steadily decreases from the center to the side walls. Which mechanism is responsible for such a viscosity change is unknown, however. In any event, optical observation of the anisotropic sample under flow rules out the occurrence of a flow induced phase separation. Another explanation of the anomalous maxima exhibited by the anisotropic solution might be a not fully developed velocity profile. More experimental work is in progress to shed some light on this effect. The aim is to determine velocity profiles at different distances from the feeding gate and at various depth along the thickness of the channel.

Acknowledgments.

Work supported by the Ministry of the University and Research (MURST).

References

1. Viney, C.; Donald, A.M.; Windle, A.H.: Molecular correlations in sheared thermotropic copolyesters showing banded textures. Polymer 26 (1985) 870-878.

2. Maffettone, P.L.; Grizzuti, N.; Marrucci, G.: Band formation in HPC solutions by consecutive shears along orthogonal directions. Liq. Cryst. 4 (1989) 385-391.

3. Onogi, Y.; White, J.L.; Fellers, J.F.: Rheo-optics of shear and elongational flow of liquid crystalline polymer solutions: hydroxypropyl cellulose/water and poly-p-phenylene terephthlamide/sulfuric acid. J. Non-Newt. Fluid Mech. 7 (1980) 121-151.

4. Meeten, G.H.; Navard, P.: Cholesteric hydroxypropylcellulose solutions: microscopy and small-angle light scattering. J. Polym. Sci.: Polym. Phys. 26 (1988) 413-419.

5. Takebe, T.; Takeji, T.; Ernst, B.; Navard, P.; Stein, R.S.: Small-angle light scattering of polymer liquid crystals under shear flow. J. Chem. Phys. 92 (1990) 1386-1392.

6. Grizzuti N., Guido S., Nastri, V. and Marrucci, G.: Velocity profiles in rectangular channel flow of liquid crystalline polymer solutions. Rheol. Acta 30 (1991) 71-76.

7. Schraub, F.A.; Kline, S.J.; Henry, J.; Runstadler, J.W., Jr.; Littell, A.: Use of hydrogen bubbles for quantitative determination of time-dependent velocity fields in low speed water flows. J. Basic Eng., Trans. ASME, Series D, 87 (1965) 429-444.

8. Baker, D.J.: A technique for the precise measurement of small fluid velocities. J. Fluid Mech. 26 (1966) 573-575.

9. Guido, S.; Marrucci, G.; Grizzuti, N.: S-shaped deformation profiles in sheared liquid-crystalline polymers. Liq. Cryst. 7 (1990) 279-281.

The Viscoelastic Rayleigh-Bénard Convection Investigated by Different Optical Methods

B.Gampert, J.Domjahn

Universität Essen, Angewandte Mechanik, 4300 Essen 1, Schützenbahn 70, Germany

1. Summary

The instabilities in horizontal layers of viscoelastic fluids heated from below and cooled from above have been investigated by the present authors in earlier papers [1] applying holographic interferometry. It was shown that the critical Rayleigh number Ra_c characterizing the onset of laminar steady two-dimensional Rayleigh-Bénard convection can be increased in viscoelastic polymer solutions as compared to Newtonian fluids. Subject of this paper are the transitions to three dimensional or time dependent stability regions.

2. Introduction. The Rayleigh-Bénard Convection

For fluids with Prandtl numbers smaller than 100, the Pr number is an additional parameter which influences the transition to higher instabilities (Fig.1). While the steady, two dimensional convection starts instantaneously from the state of pure heat conduction, no discrete transition to higher instabilities could be found by holographic interferometry, which is a two dimensional method and yields an integral signal over the (optical) path transmitted. On the contrary the change from the parallel rolls to the three dimensional, later time dependent or oscillating rolls seemed to develop quasi continuously. Therefore the three dimensional structure of the rolls has been investigated by a plane light sheet method (Fig.3). For steady or only slightly oscillating motions of the rolls the creation of three dimensional states could be successfully visualized applying the light sheet method. The polymer solutions investigated have Pr numbers of about 7, which is approximately the number of water, up to Pr $\approx$ 300, which corresponds to moderatly concentrated polymer solutions. The Xanthan - H_2O solution used in the experiments has a concentration of 100 ppm and a Pr number of 17.5. For a given geometry with a definite wave number the two dimensional convection first changes into three dimensional and later into time dependent three dimensional structures [3]. The viscosity of the polymer solutions investigated is stress dependent. Therefore it is necessary to determine the field of velocity gradients. The results of the Laser-Doppler measurements allow for strictly steady, laminar stability regimes the determination of the shear and the extention rates in the flow field.

In the example shown in Fig.2 a Xanthan solution of 100 ppm has been used and the Rayleigh number was 12 000. The largest velocity was v=0.33 mm/s, the highest shear rate normal to the velocity direction $\dot{\gamma}=0.145\ ^1/_s$ and the strongest extension rate in direction of the local velocity $\dot{\epsilon}=0.094\ ^1/_s$. In the buoyancy driven convection region between the rolls a relation between shear and extension rate of $(\dot{\epsilon}/\dot{\gamma})_{max} = 18.4$ appears.

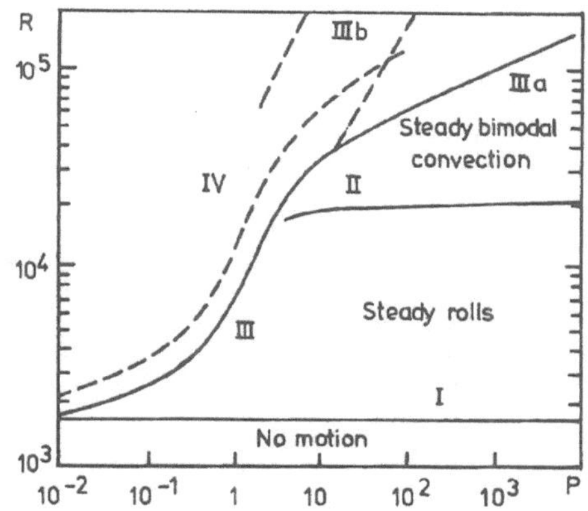

Fig. 1 The regions of stability of Rayleigh-Bénard convection [2]

3. Experimental Method

The holographic interferometry and the electronic parts of the experimental arrangement applied are described in detail in [1]. The light sheet has been build up for a transmission length of 70 mm and a transmission height of 6 mm up to 15 mm with a He-Ne Laser. Because of the heating from below and the cooling from above the chamber has to be transmitted laterally. For the LDV a two channel system from Dantec has been applied. For the measurement of the extremely small velocities of -2 mm/s up to +2 mm/s the signal had to be amplyfied by analog technique. Before each measurement a calibration concerning the determination of the zero point and the amplification has been performed.

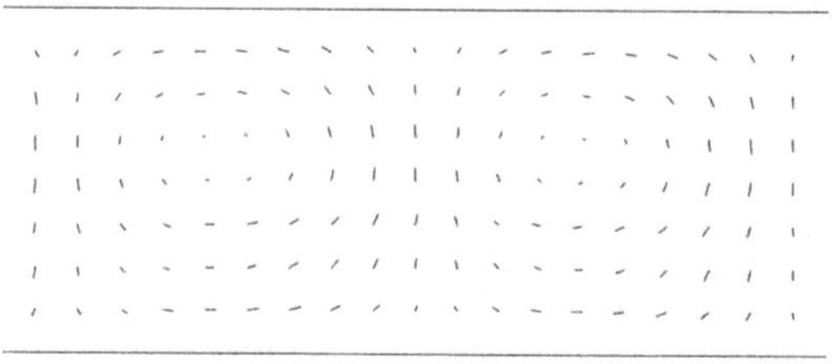

Fig. 2a Two convection rolls measured by LDV

Fig. 2b and visualized by light sheet technique

4. Results

In Fig.3 the results of the three optical methods applied have been compared for a Ra number of Ra ≈ 12 000. The data of the hight/length/depth of the test chamber are respectively 8.5/70/20 mm. Fig.3a shows the vertical velocity in the horizontal centerline of the test chamber measured by LDV. Fig.3b is a light sheet photography in the vertical center plane and Fig.3c shows the holographic interferogramm which is an integral signal taken over the depth of the test chamber. The flow is strictly laminar, two dimensional and steady. In Fig.4 and the following, the light sheets were taken

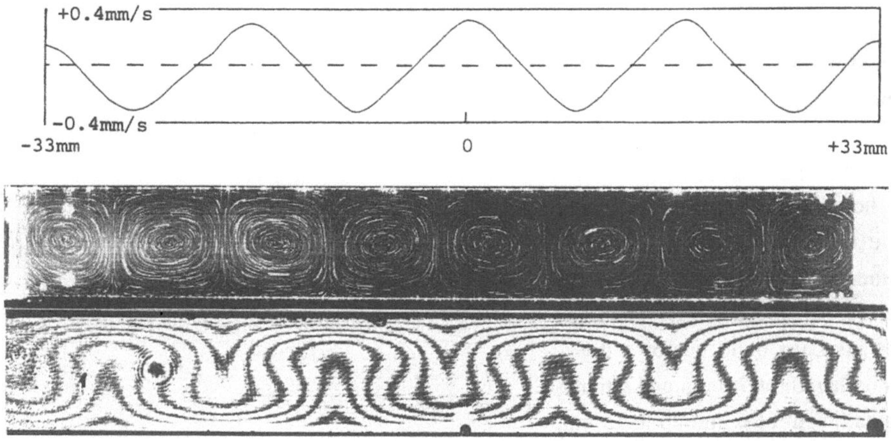

Fig. 3 Comparison of different optical techniques

in the symmetry plane M, the plane F which is 5 mm in front of M and in the plane B, 5 mm behind M.

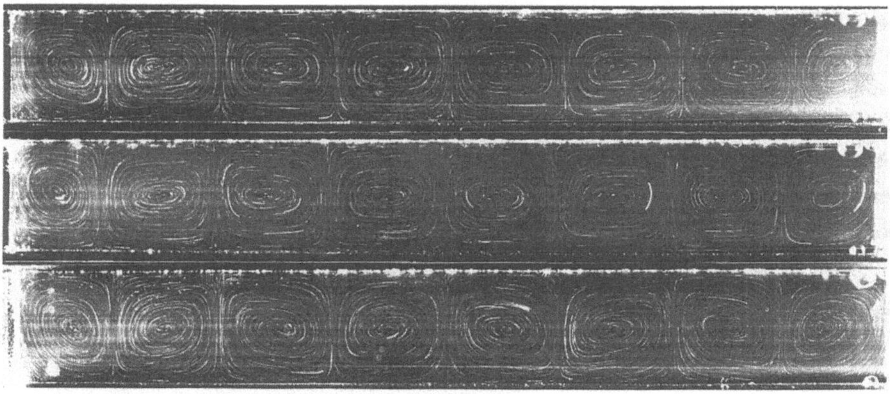

Fig. 4 Light sheets at Ra = 40 880 in plane F on the top, plane M in the middle and B at the button

For a Ra number of 40 880 only small three dimensional deformations of the rolls appeared and are visualized in Fig. 4. The rolls in the center of Fig. 4 M are placed symmetrically to the horizontal centerline. In the planes B and F the streamlines in the region of the rolls are transformed from circles into ellipses with axes turned off from the horizontal/vertical position. Following the vertical flow direction between the rolls the axes are moved upward or downward. After a continuing increase of the temperature difference in the test chamber pronounced three dimensional deformations can be seen at the right and left edge of the test chamber which finally result in a disappearence of the rolls present there, see Fig.5. Applying the light sheet method changes of this kind can be watched with greater accuracy as compared with holographic interferometry.

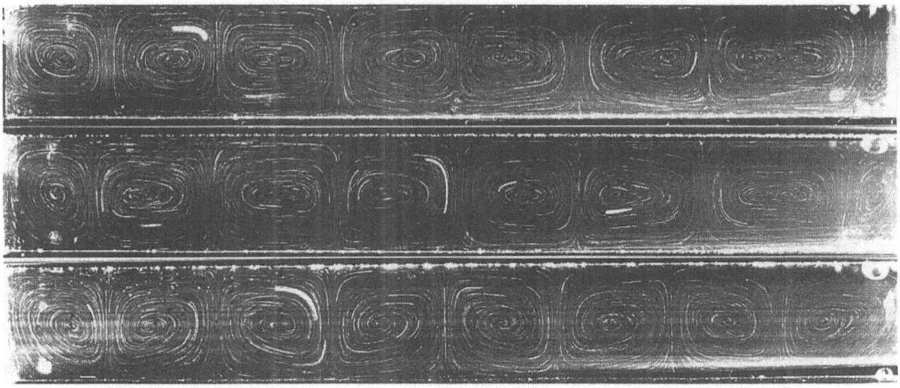

Fig. 5 Light sheets at Ra= 67 160

This convection form belongs to the region of steady bimodal convection of Fig.1. The largest velocity is $v_{max.}$ = 1.16 mm/s for a Ra number of Ra = 53 334, measured with the LDV system. In Fig. 6 a top view sketch of the test chamber is shown. The lines correspond to the upward and downward motions shown in the light sheet photography, see Fig.5.

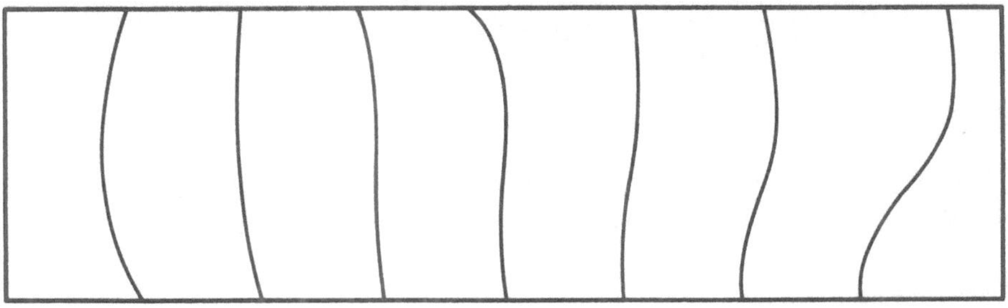

Fig. 6 Top view of the test chamber

After a further increase of the temperature difference a slightly oscillating motion of the rolls starts. At first the general structure of the rolls does not change (Fig.7). But the orientation of the rolls becomes increasingly irregular. In Fig.7 M a circular roll similar to those which can be observed in convection flows with free surfaces can be seen. In Fig.8 again rolls with axes orientated parallel to the shorter wall of the test chamber appear. The number of rolls

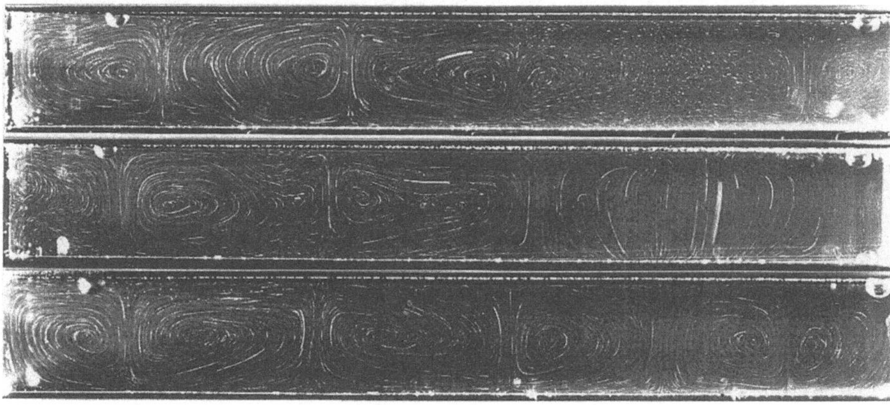

Fig. 7 Light sheets at Ra = 61 600

Fig. 8 Light sheets at Ra = 81 360

varies from four in the plane M up to eight in the plane F. In the plane F even two rolls placed above one other can be observed in the left hand part of the figure.

Acknowledgement

This research has been supported by the Forschungspool of the University of Essen

5. References

1. Gampert,B.; J.Domjahn: Visualization of Ranges of Stability in Viscoelastic Rayleigh-Bénard Convection, Flow Visualization V, (ed. R.Reznicek, Hemisphere, Washington)
2. Krishnamurti,R.: On the transition to turbulent convection. Part 2, The transition to time-dependent flow. J.Fluid.Mech.42, pp 309-320 (1970b)
3. Busse, F.H.: Transition to Turbulence in Rayleigh-Bénard convection in Hydrodynamic Instabilities and the Transition to Turbulence (eds. H.L.Swinney; J.P.Gollup, Springer Verlag Heidelberg Berlin New York)

Flows in Pipe and Channel

Secondary Flow Phenomena in Curved Pipes and Rotating Channels

K.C. CHENG

Department of Mechanical Engineering
University of Alberta
Edmonton, Alberta, Canada T6G 2G8

Summary

This paper demonstrates the applicability of the simple smoke injection method for visualization of secondary flow phenomena caused by such body forces as centrifugal and Coriolis forces in curved pipes and rotating channels. Examples are shown for the following secondary flow phenomena: (1) laminar entry flow in a curved pipe, (2) flow in a 180-degree sharp turn of square duct, (3) instability phenomena in a curved rectangular channel with aspect ratio 10, (4) relaminarization in a helical pipe, and (5) pitch effect on secondary flow in helical pipes. The unsteady nature of the secondary flow phenomena caused by such body forces as buoyancy, centrifugal and Coriolis forces at a higher physical parameter is revealed by flow visualization. The applicability of the smoke injection method in understanding many practical internal flow phenomena up to Reynolds number Re ≈ 5,000 is emphasized.

1. Introduction

Flow visualization contributes to the physical understanding of complex flow phenomena and structures and may be used for comparison with the results of numerical investigations and checking the physical model used. Many flow and heat transfer phenomena were investigated using steady state models in the past whereas the real flow phenomena in many internal or external flow problems may become time-dependent (unsteady) at higher physical parameters such as Reynolds and other parameters.

The purpose of this paper is to present some recent flow visualization results on secondary flow phenomena in curved pipes and rotating channels obtained by smoke injection method. Recent literature relating to flow and heat transfer in curved pipes and rotating channels is very extensive and is well reviewed in [1-7].

The smoke injection method is probably the most simple and inexpensive flow visualization technique and is applicable to many internal flow problems with low to intermediate Reynolds numbers. At present it is well known that secondary flow phenomena caused by such body forces as buoyancy, centrifugal, and Coriolis forces in heated, curved and rotating channels are known to be qualitatively similar. The secondary flow phenomena in heated, curved and rotating channels have received much attention during the past 25 years because of practical applications.

The experimental procedure relating to the smoke injection method is well discussed in recent works [8-17] and the details will be omitted entirely here in order to provide more space for presenting photographic results.

2. Developing Secondary Flow Patterns in the Hydrodynamic Entrance Region of a Curved Pipe

Photographic results are presented in Fig. 1 for developing secondary flow patterns in the hydrodynamic entrance region of a curved pipe (pipe inside diameter d = 25.4 mm, radius of curvature R_c = 127 mm, and curvature ratio $(d/2)/R_c = 0.1$) with fully developed laminar entry flow from a straight entrance pipe for the entrance bend angles ϕ = 0, 15, 45, 90 and 180°. The Dean number range is K = 91 to 653 ($K = Re\,(d/2\,R_c)^{1/2}$, Re = Reynolds number, $u_m d/\nu$, u_m = average velocity, ν = kinematic viscosity) and the corresponding Reynolds number range is Re = 287 to 2064.

In each photograph, the right-hand side corresponds to the outer concave wall and the left-hand side represents the inner convex wall of a curved pipe. At the inlet to the curved pipe ϕ = 0, one can see the concentric smoke traces which correspond to the parabolic entry flow for Re = 294 to 928. For larger Reynolds numbers (Re $\geq$ 1193) some disturbances can be observed. At entrance angle ϕ = 45°, a pair of vortices near the inner wall can be seen clearly. The changing secondary flow patterns with the increase of Dean number is of interest.

For K $\geq$ 470 at ϕ = 45°, an additional pair of very small vortices appear near the inner wall. At ϕ = 90°, rather complex secondary flow patterns appear near the inner wall with the increase of Dean (or Reynolds) number. At ϕ = 180°, rather complex secondary flow patterns appear near the outer wall and no symmetry exists for K $\geq$ 374.

It is of considerable interest to compare the developing secondary flow patterns with the available theoretical results in the literature. The photographs presented provide considerable physical insight into the complex developing secondary flow patterns for the present basic flow problem in curved pipe.

3. Secondary Flow Patterns in a 180-Degree Sharp Turn of a Square Duct [13]

The test section is shown in Fig. 2 and the entry flow is fully developed laminar flow. The secondary flow patterns for the case L = 1.5D and Re = 500 ($Re = u_m D/\nu$, u_m = average velocity) are shown in Fig. 3. It is seen that the secondary flow patterns are not symmetric and the flow is already unsteady at Re = 500. The unsteady nature of the secondary flow patterns at downstream positions z/D = -1, -2 and -3 for the case L = D and Re = 500 is shown in Fig. 4 at three different instants. For the case L = 1.5D and Re = 800, turbulent-like flow seems to appear at z/D = -3 as shown in Fig. 5. At larger downstream distance, a relaminarization phenomenon should appear after a sharp 180-degree turn. Further details can be found in [13].

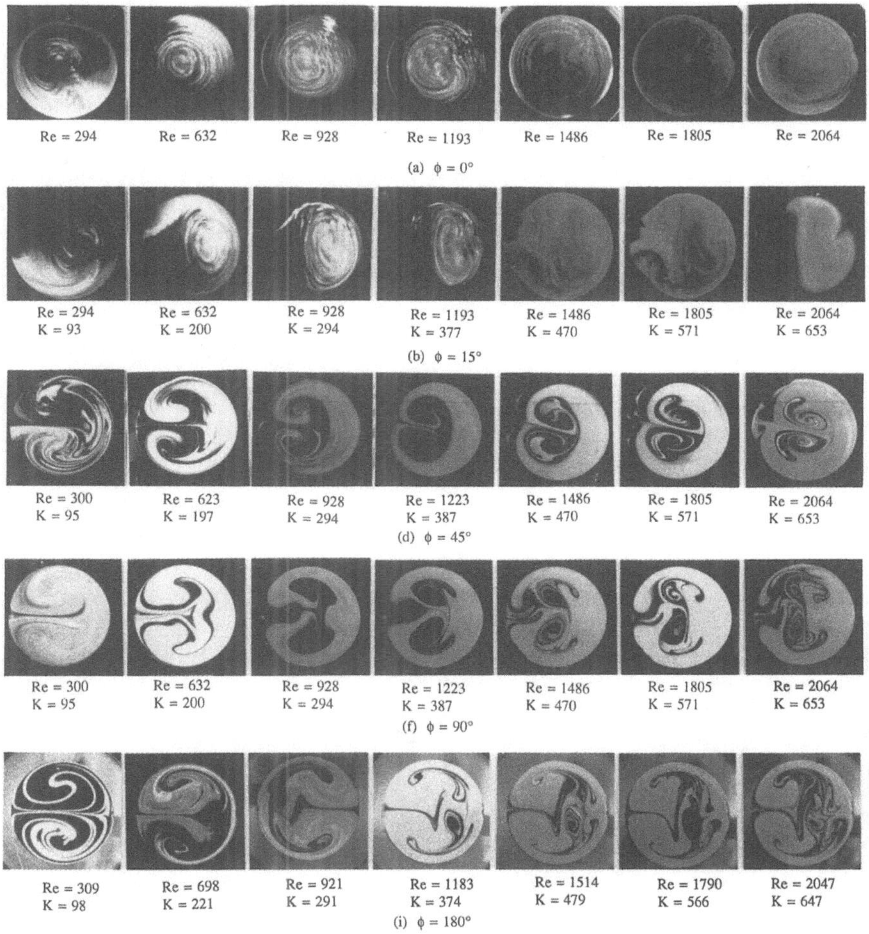

Fig. 1 Developing secondary flow patterns in a curved pipe at bend angles $\phi = 0, 15, 45, 90$ and $180°$

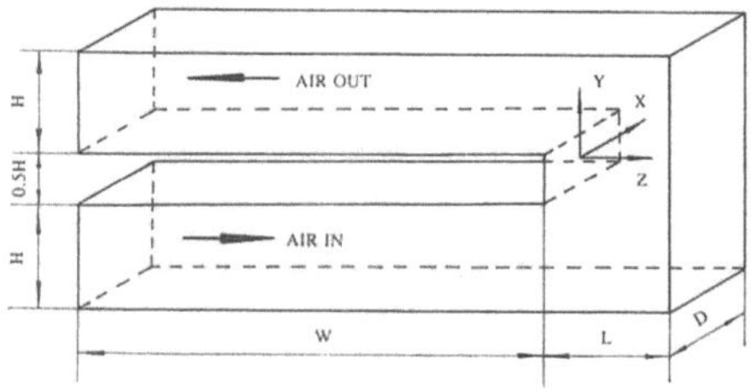

Fig. 2 The test section and coordinate system

Fig. 4 Unsteady secondary flow patterns at z/D = -1, -2 and -3 for the case L = D and Re = 500

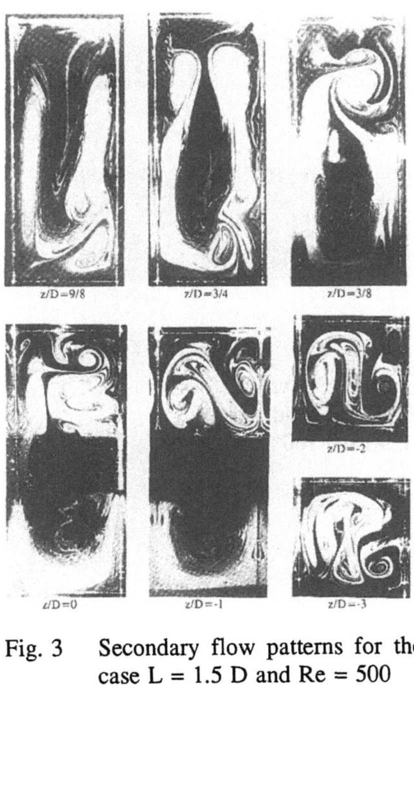

Fig. 3 Secondary flow patterns for the case L = 1.5 D and Re = 500

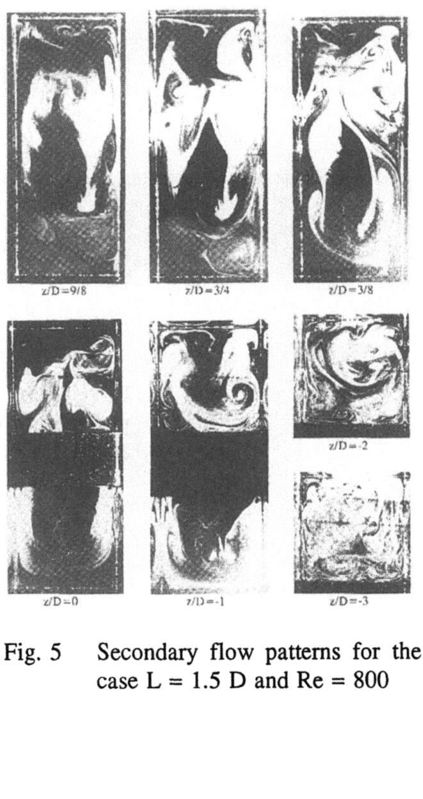

Fig. 5 Secondary flow patterns for the case L = 1.5 D and Re = 800

4. Instability Phenomena in a Rotating Curved Rectangular Channel with Aspect Ratio 10 [14]

A schematic diagram of the experimental apparatus is shown in Fig. 6 and the rotating test section is shown in Fig. 7.

The rotating test section installed on the turning table consists of an entrance spiral rectangular channel with axial length 1.25 m and a curved rectangular channel (270° bend) with constant radius of curvature R_c = 25.4 cm (see Fig. 7). The curved rectangular channel has cross-section with width a = 2.54 cm and height b = 25.4 cm. The air flows through a rotating straight tube (inside diameter = 4 cm) along the axis of rotation (see Fig. 7) and then enters the spiral rectangular channel before entering the test section. The top view and the exit cross-section of the test section are shown in Fig. 7. The test section was made from acrylic sheets.

The rotating table is driven by an electric motor with adjustable speed drive and the range of rotating speed is n = 0 ~ 500 rpm for this study. The rotational speed is measured by using an optical slot switch running on a disc with 60 holes on its perimeter. The signal from the switch is fed to a Hewlett Packard HP5314A universal counter. With 60 holes in the disc, the frequency in Hz equals the rotational speed in rpm.

A second disc with a single hole provides the signal for firing the General Radio 1540 Strobolume by way of a delay generator/single flash flip flop which allows visual observation using a slit light source with one flash per revolution and also permits a single, properly timed flash for photographing the whole flow field. The duration of the flash is 15 μs.

The air flow rate was measured by the Meriam laminar flow element with a calibrated differential pressure transducer. The smoke generated by burning Chinese incense sticks was injected through a dispersing tube before the test section as shown in Fig. 6. The smoke particles following the flow were photographed instantly at the exit of the test section revealing the secondary flow pattern. A Nikon FM2 single lens reflex camera with 55 mm micro-lens and the Kodak T Max black and white film P3200 were used.

The experimental parameters are: curvature ratio (d_e/R_c), aspect ratio (b/a), Reynolds number Re = $u_m d_e/\nu$ (d_e = 2ab/(a+b) and u_m = average velocity, ν = kinematic viscosity), Dean number (K_1,K_2) ($K_1 = Re(d_e/R_c)^{1/2}$, $K_2 = Re(a/R_c)^{1/2}$), rotational Reynolds number Re_ω = $\omega d_e^2/\nu$ (ω = angular velocity), and rotating speed n. The parametric values are listed in Table 1.

TABLE 1 Values of Experimental Parameters

Curvature ratio, d_e/R_c	0.1818
Aspect ratio, b/a	10
Reynolds number, Re	199 and 821
Rotating Reynolds number, Re_ω	0 ~ 5725
Dean number, K_1	350
Rotating speed n, rpm	0 ~ 413

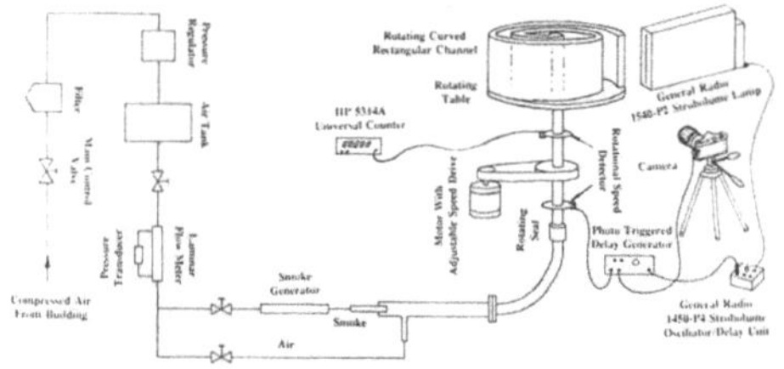

Fig. 6 Schematic diagram of experimental apparatus

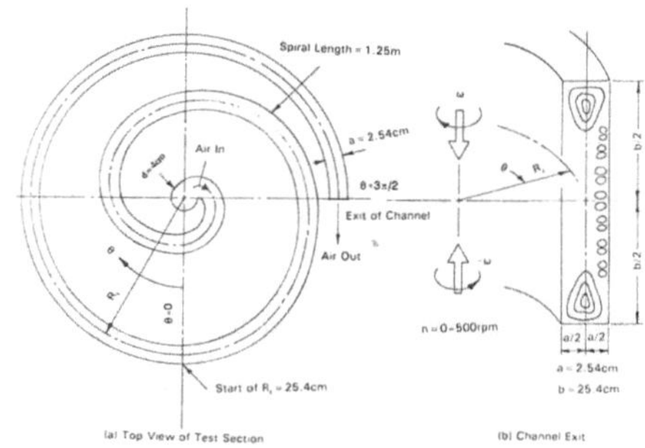

Fig. 7 The top view of rotating test section and configuration at the channel exit

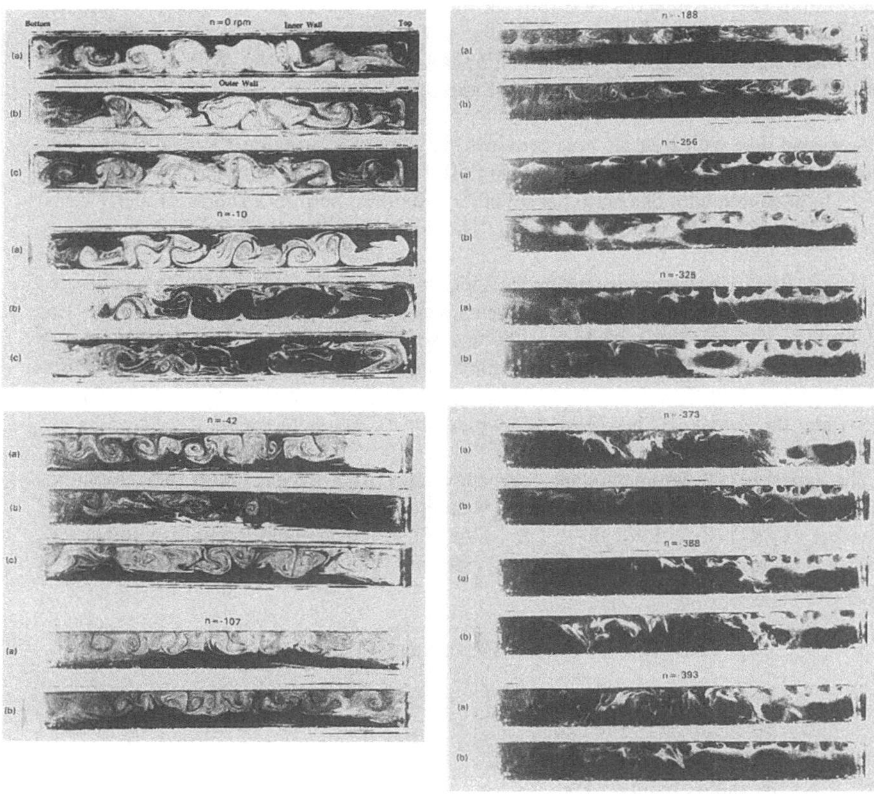

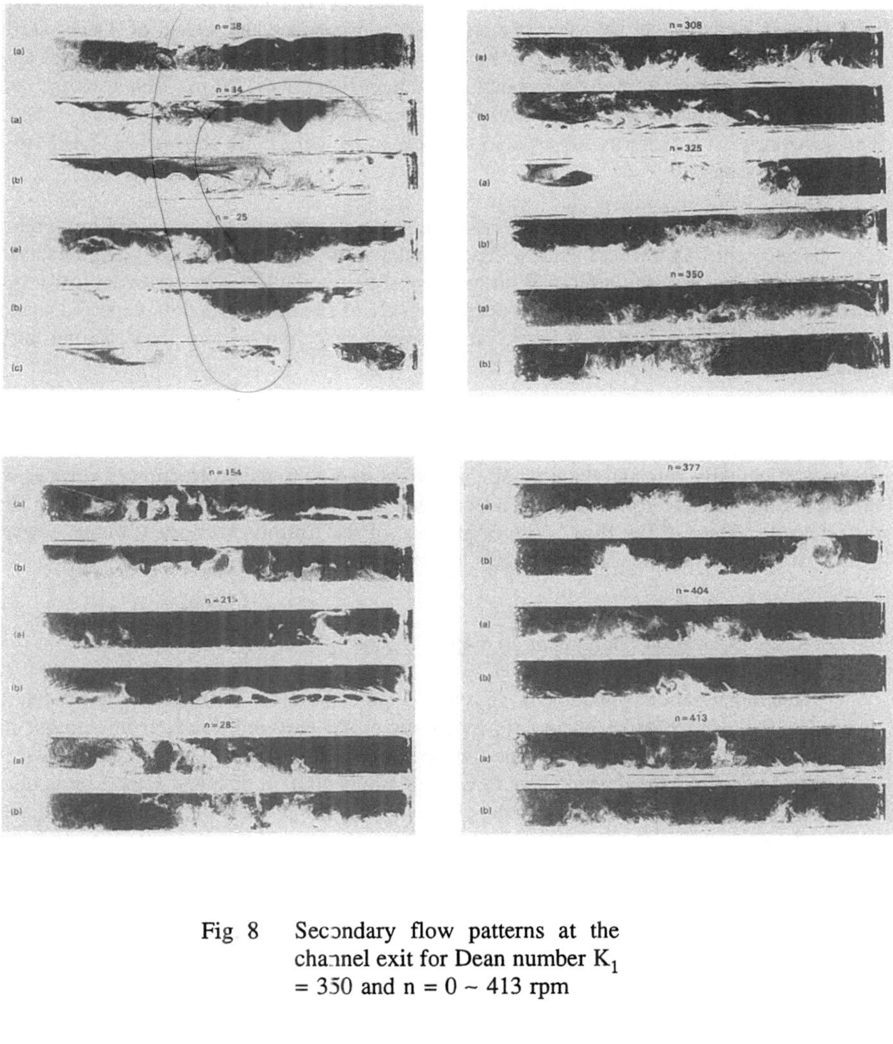

Fig 8 Secondary flow patterns at the channel exit for Dean number K_1 = 350 and n = 0 ~ 413 rpm

For fully developed laminar flow in a stationary curved rectangular channel (n = 0 rpm) with aspect ratio b/a = 10, the critical Dean number for the onset of Dean vortices (longitudinal vortex rolls) is known to be $K_1 = 100$. With an entrance spiral channel length 1.25 m and a subsequent curved rectangular channel (length = 1.20 m) with constant radius of curvature, the flow at the exit of the channel becomes fully developed flow ($\ell/d_e = 53.03$). The secondary flow patterns for aspect ratios b/a = 10 and 12 are shown in [17,18] for the case without rotation.

When the direction of rotation is in the same direction as the main airflow, both the Coriolis and centrifugal forces act towards the outer wall (high pressure side) and the Coriolis forces have a destabilizing effect. With negative rotation, the Coriolis forces act towards the inner wall (low pressure side) and a stabilizing effect results. It is thus expected that the rotational motion can lead to a very complex secondary flow pattern in a rotating curved rectangular channel.

The photographic results are shown in Fig. 8 for the case $K_1 = 350$ and n = 0 ~ 413 rpm. It is of interest to note the decreasing sizes of vortices near the inner wall with the increase of negative rpm and the destabilizing effect leading to the destruction of vortices with the increasing positive rpm. Different secondary flow patterns for the same rpm also show the unsteady nature of the flow. The occurrence of the Coriolis vortices at higher negative rpm is of special interest. Further details can be found in [14].

5. Relaminarization Phenomena in a Helical Pipe

Relaminarization phenomena in curved pipes are of considerable practical and theoretical interest. The secondary flow patterns at the exit and downstream straight pipe positions of a closely coiled helical pipe (inside diameter d = 37.5 mm, radius of curvature R_c = 381 mm) with fully developed entry flow from a long straight pipe and after 2 turns of coil are shown in Fig. 9 for Reynolds number ($Re = u_m\, d/\nu$) range Re = 2200 ~ 5300. The Dean number is $K = Re(d/2\, R_c)^{1/2}$. In each photo, the right-hand side represents the outer wall. When the flow becomes turbulent at downstream position in straight pipe, z/d, for higher Reynolds number ($Re \geq 4300$), the secondary flow pattern disappears or cannot be seen clearly due to smoke diffusion.

6. The Pitch Effect on Secondary Flow Patterns in Helical Pipes

In recent years the theoretical analysis for the pitch effect on developing laminar flow and heat transfer in loosely coiled helical pipes has received some attention in the literature. The secondary flow patterns observed at the exit of a single coiling (one turn) of vinyl tube (inside diameter d = 17.8 mm, outside diameter = 25.4 mm) loosely coiled around a long cylinder (outside diameter = 168 mm) are shown in Fig. 10 with pitches 80 and 120 cm. The entry flow to the helical pipe is fully developed laminar flow from a long straight tube. The Dean number is defined as $K = Re\, (d/2\, R_c)^{1/2}$ ($Re = u_m d/\nu$, u_m = average velocity). In each photo, the upper part represents the inner wall. The secondary flow patterns are seen to be complex depending on the Dean number and the pitch of coiling. For the Dean number range K = 100 to 600, no symmetry of secondary flow exists.

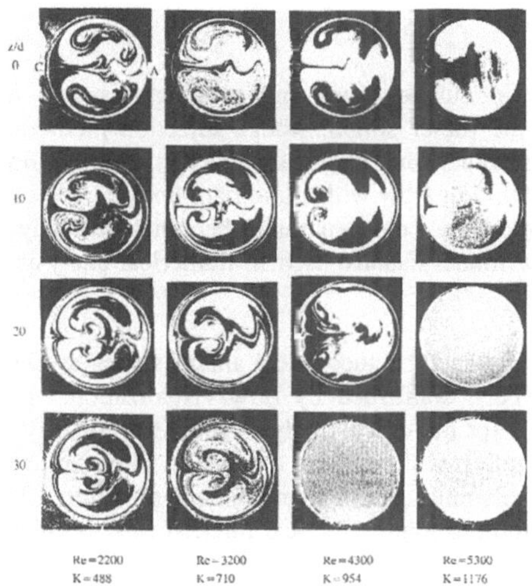

Fig. 9 Secondary flow patterns at the exit of a helical coil with 2 turns and downstream locations (relaminarization phenomena)

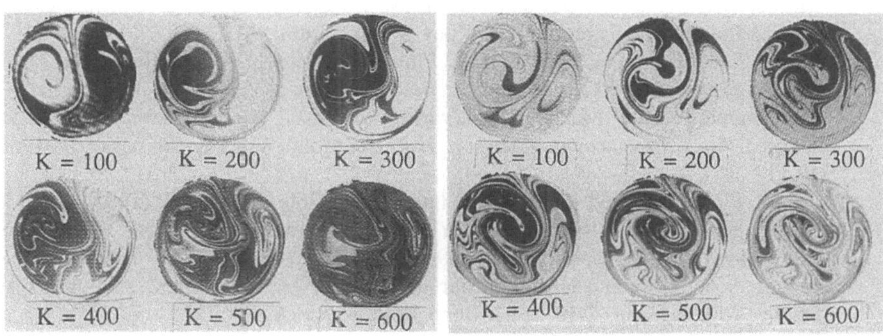

(a) Pitch = 80 cm

(b) Pitch = 120 cm

Fig. 10 The pitch effect on secondary flow patterns at the exit of a single coiling (one turn) of a helical pipe for (a) pitch = 80 cm and (b) pitch = 120 cm

7. Concluding Remarks

Examples of secondary flow patterns in curved pipes and rotating channels are presented to demonstrate the applicability of the rather simple smoke injection method in visualizing the secondary flow involving centrifugal and Coriolis forces in internal flow problems with Reynolds number up to Re = 5300. For flow with significant body forces such as buoyancy, centrifugal and Coriolis forces, the flow field eventually becomes unsteady at higher physical parameters and the steady flow model usually used in numerical analysis is not applicable.

As well, many secondary flow patterns obtained by theoretical analysis such as Dean vortices in curved circular pipe [6] have not been confirmed by flow visualization. It is believed that laminar and turbulent flow models for numerical analysis should eventually be confirmed by flow visualization. Flow visualization will continue to play an important role in understanding the flow phenomena including relaminarization phenomena in the future. The applicability of the simple smoke injection method to many practical internal flow problems is also obvious.

References

1. Morris, W.D., Heat Transfer and Fluid Flow in Rotating Coolant Channels, Research Studies Press, A Division of John Wiley & Sons, Ltd., 1981.
2. Berger, S.A., Talbot, L. and Yao, L.S., Flow in Curved Pipes, Ann. Rev. Fluid Mech., Vol. 15 (1983), 461-512.
3. Nandakumar, K. and Masliyah, J., Swirling Flow and Heat Transfer in Coiled and Twisted Pipes, Advances in Transport Processes, John Wiley & Sons, Vol. 4 (1986), 49-112.
4. Ito, H., Flow in Curved Pipes, JSME International Journal, Vol. 30 (1987), 543-552.
5. Cheng, K.C., Internal Flows, W.J. Yang (ed.), Handbook of Flow Visualization, Hemisphere Pub. Corp., 415-457, 1989.
6. Berger, S.A., Flow and Heat Transfer in Curved Pipes and Tubes, 29th Aerospace Science Meeting, Jan. 7-10, 1991, Reno Nevada, AIAA 91-0030.
7. Hwang, G.J. and Soong, C.Y., Convective Heat Transfer in Rotating Channels, 4th Int. Symp. on Transport Phenomena and Dynamics of Rotating Machinery, Honolulu, HI, April 5-8, 1992.
8. Cheng, K.C., Takuma, M. and Akiyama, M., Flow Visualization Studies on Developing Secondary Flow Patterns in the Simultaneous Hydrodynamic and Thermal Entrance Region of an Isothermally Heated Horizontal Curved Pipes, Flow Visualization V, ed., R. Reznicek, Hemisphere Pub. Corp., 1990, 456-463.
9. Cheng, K.C. and Takuma, M., Mixed Convection Effects on Laminar Air Flow in an Isothermally Heated Horizontal 90-Degree Bend with Offset Angles 0 and 180°, Procs. ASME/JSME Thermal Eng. Joint Conf., ASME, Vol. 1, 1991, 35-42.
10. Cheng, K.C. and Shi, L., Visualization of Developing Secondary Flow and Measurements of Velocity Profiles in a Curved Square Duct With and Without an Offset Bend, Procs. 3rd Int. Symp. on Fluid Control, Measurement and Visualization, ASME, Aug. 29-31, 1991, 415-422.

11. Cheng, K.C. and Takuma, M., Visualizations of Developing Secondary Flow Patterns and Temperature Measurements for Mixed Convection Effects in an Isothermally Heated Horizontal Curved Pipe, Mixed Convection Heat Transfer, HTD-Vol. 163, ASME, 1991, 53-60.
12. Cheng, K.C. and Takuma, M., Visualization of Centrifugal Phenomena in Parallel, Converging and Diverging Curved Channels, Procs. 3rd Int. Symp. on Transport Phenomena and Dynamics of Rotating Machinery, Honolulu, HI, Hemisphere Pub. Corp., Vol. 1, 1990, 391-403.
13. Cheng, K.C., Shi, L., Kurokawa, M. and Chyu, M.K., Visualization of Flow Patterns in a 180-degree Sharp Turn of a Square Duct, Preprints, 4th Int. Symp. on Transport Phenomena and Dynamics of Rotating Machinery, April 5-8, 1992, Honolulu, HI, Vol. A, 541-548.
14. Cheng, K.C., Wang, L. and Faulkner, B.G., Visualization of Instability Phenomena in a Rotating Curved Rectangular Channel with Aspect Ratio 10, Preprints, 4th Int. Symp. on Transport Phenomena and Dynamics of Rotating Machinery, Honolulu, HI, April 5-8, 1992, Vol. A, 549-560.
15. Cheng, K.C. and Yuen, F.P., Flow Visualization Studies on Secondary Flow Patterns in Straight Tubes Downstream of a 180 deg Bend and in Isothermally Heated Horizontal Tubes, ASME J. of Heat Transfer, Vol. 109, 1987, 49-54.
16. Cheng, K.C. and Yuen, F.P., Flow Visualization Experiments on Secondary Flow Patterns in an Isothermally Heated Curved Pipe, ASME J. of Heat Transfer, Vol. 109, 1987, 55-61.
17. Cheng, K.C., Nakayama, J. and Akiyama, M., Effect on Finite and Infinite Aspect Ratios on Flow Patterns in Curved Rectangular Channels, Flow Visualization (Procs. Int. Symp. on Flow Visualization, Tokyo), Hemisphere Pub. Corp., 1979, 181-186.
18. Akiyama, M., Cheng, K.C., Urai, I., Suzuki, M. and Nishiwaki, I., Effect of Curvature on the Hydrodynamic Instability in Curved Rectangular Channels, J. of Flow Visualization Soc. of Japan, Vol. 2, 1982, 553-558.

An Application of Hydrogen Bubble Method to Flow Networks

Nengli Zhang, Wen-Jei Yang, Youren Xu
Department of Mechanical Engineering & Applied Mechanics
The University of Michigan, Ann Arbor, MI 48109, U. S. A.
C. P. Lee
General Electric Company, Cincinnati, Ohio 45215, U. S. A.

Abstract

The flow at each intersection in a flow network is characterized by a collision of two flow streams, which results in complex flow patterns through the downstream holes. The hydrogen bubble method is employed to observe flow patterns and mixing behavior of water in the flow network. It is disclosed that (1) the entrance lengths of both the laminar and turbulent flows are much shorter than those predicted by conventional empirical formulas, (2) intermixing of the ramming flows in the intersection zone occurs in turbulent flow but not in laminar flow, (3) intermixing propagates upstream with an increase in the Reynolds number, (4) the streak lines downstream of the intersection are shifted toward the inner walls, forming long recirculation zones that stretch downstream until the exit with less flow through the other half of the channel, and (5) velocity profiles in stagnation-recirculation zones can be obtained through pulse hydrogen bubble technology.

Introduction

A flow network is a system of mutually intersecting holes inside a plate or an assembly of plates. Figure 1 depicts the cross-sectional view of a flow network with (a) a single intersection for each stream of throughflow, and (b) multiple intersections. A flow network of multiple intersections may be instituted inside a single plate or in an assembly of plates each of which has a single intersection, as shown in Fig. 1 (b). The flow at each intersection is characterized by a collision of two flow streams, resulting in complex flow patterns through the downstream holes. In the case of multiple intersections, the flow is periodically disrupted at each succeeding intersection, thus preventing the formation of a fully developed flow through the holes. Hence, it is expected that a flow network provides a better flow mixing, and consequently, a higher heat transfer performance.

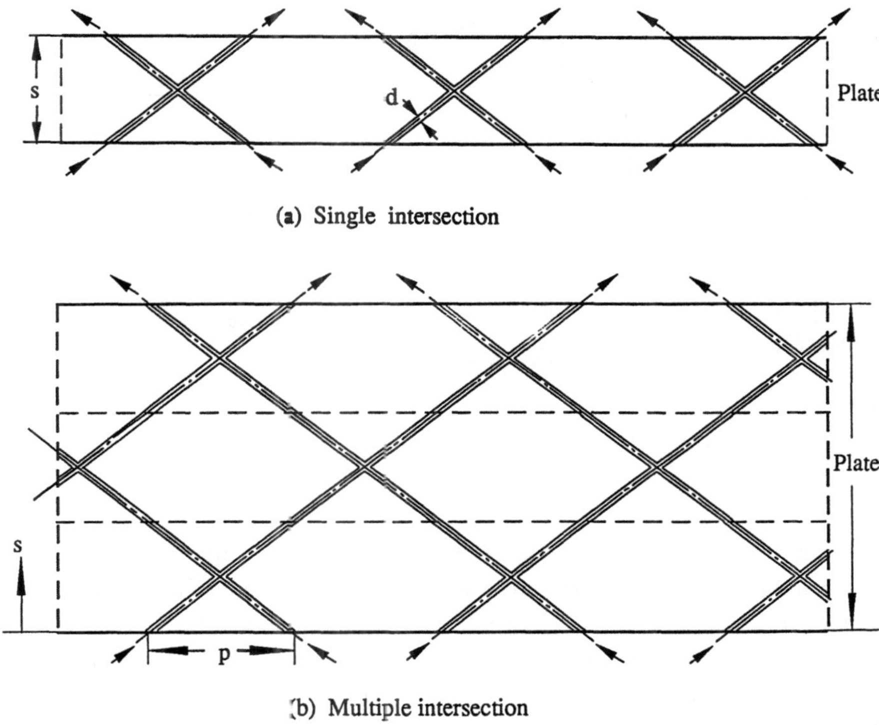

Fig. 1 Cross-sectional view of a flow network with (a) a single intersection and (b) multiple intersection

Flow characteristics, such as pressure drop and flow discharge coefficient, are important in sizing the network and predicting the flow rate through the network. The empirical flow and pressure drop data were obtained in various flow networks by systematically varying geometrical parameters and flow conditions. A physical model was developed to predict the intersection pressure loss in flow networks [1]. In this study, flow visualization using the hydrogen bubble method is conducted to clarify the mixing phenomena in the intersection zone and the flow pattern downstream from the intersection area .

Experimental Setup and Procedure

A water model was employed to observe flow patterns in a flow network. The test setup consisted of a water filter, a rotameter, a test section, two stabilizing plenums, an illumination system, a hydrogen-bubble generator and a camera, as shown in Fig. 2.

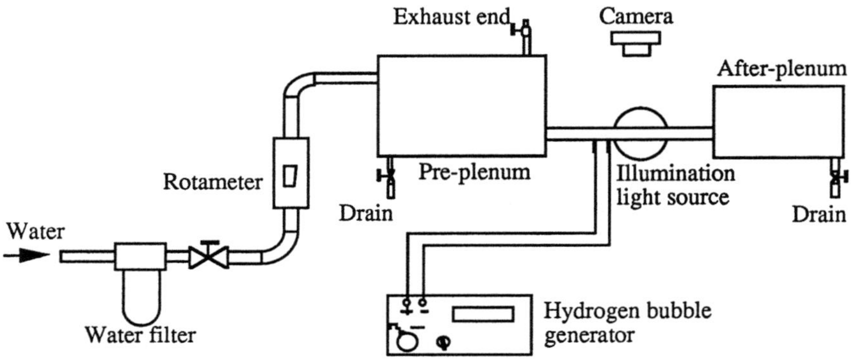

Fig. 2 Flow visualization experimental setup

A pre-plenum was installed to maintain constant pressure while a after-plenum was open to the ambient via an overflow rim. The test section was made of acrylic for observation. It consisted of two 15.9 mm X 15.9 mm square channels intersecting at an angle of 60 ° and with the length-to-diameter ratio, L/d, of 19.2. The hydrogen bubble generation system consisted of two pairs of copper plates as the anodes, two pairs of fine platinum wires as the cathodes, and a hydrogen bubble generator. One pair of the cathodes were placed at a distance of 127 mm from each entrance of the flow network, while the other pair were placed at a distance of 38 mm downstream from the intersection center of the network, as shown in Fig. 3. As the illumination light source, a slide projector (800 W) was placed downside from the test section and the angle between the illuminating beam and the viewing line were set at 110 ° ~ 120 ° to obtain high-quality photographs.

Results and Discussion

Single pulse generation of hydrogen bubbles provided velocity profiles in the channels. Figures 4 (a) and (b) show typical velocity profiles for laminar and turbulent flows at the distance of 127 mm from the entrance for Re = 2000 and 3200, respectively. It is observed that the flows are already full-developed and the entrance length-to-diameter ratios (L/d) of both laminar and turbulence flows are less than 8, much less than the conventional values of L/d = 0.06Re for laminar flow and L/d = 25 ~ 40 for turbulent flow through a pipe [2]. The results are in qualitative agreement with those obtained from the pressure distribution study [1], probably because the flow enters the channels at an angle of 30 ° rather than 90 °.

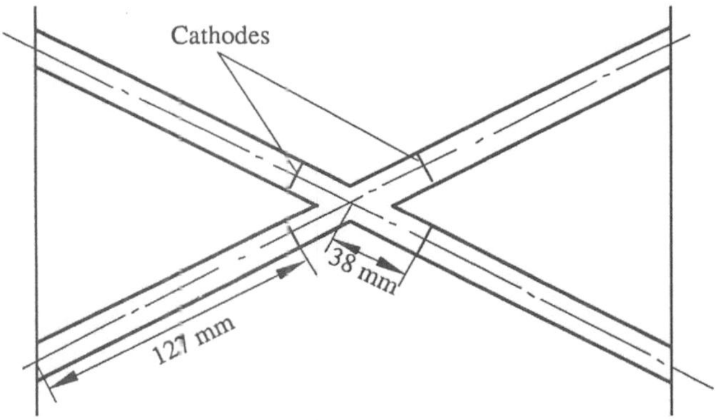

Fig. 3 Locations of cathodes

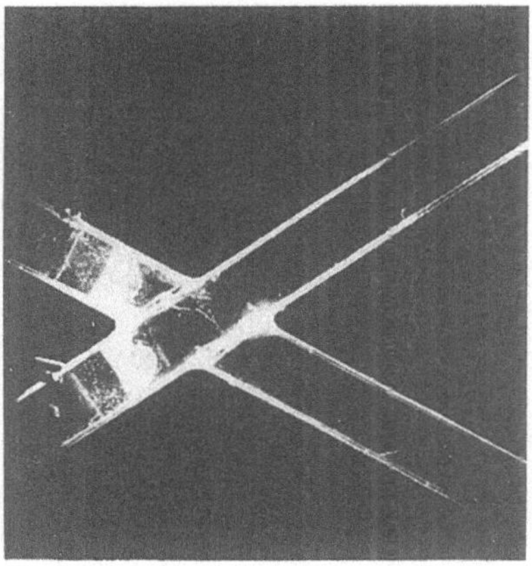

(a) Re = 2,000

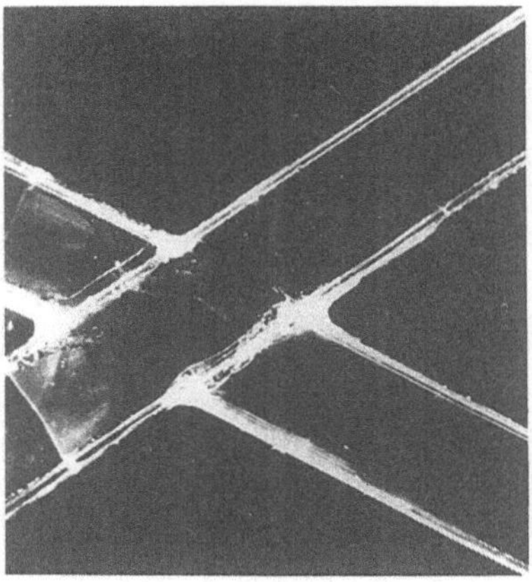

(b) Re = 3,200

Fig. 4 Flow visualization of velocity profiles in (a) laminar flow at Re=2,000 and (b)turbulent flow at Re=3,200

Figure 5 depicts a phenomenon of flow ramming in the intersecting zone. In the laminar-flow case, no intermixing occurs in the intersection area, as seen in Fig. 5 (a) and (b) which were taken at different times after hydrogen bubbles were generated. In contrast, turbulent flow at Re = 3200 is characterized by an intermixing in the downstream end of the intersecting zone, as shown in Fig. 5 (c). As the Reynolds

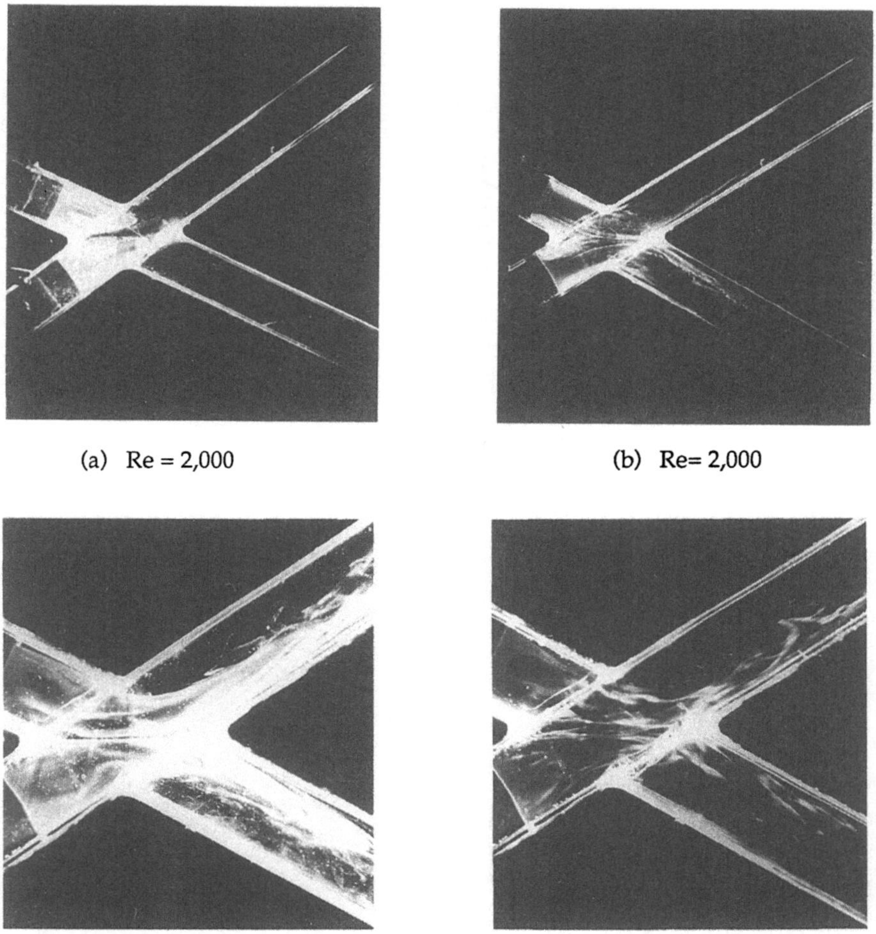

(a) Re = 2,000 (b) Re= 2,000

(c) Re = 3,200 (d) Re = 5,000

Fig. 5 Visualization of flow ramming in intersecting zone, (a) and (b) for laminar-flow and (c) and (d) for turbulent-flow

number is increased, the occurrence of intermixing migrates upstream. Figure 5 (d) shows an example for the Reynolds number of 5000 in which the intermixing starts at the mid-portion of the intersecting zone. Flow patterns in the post-intersecting region are shown in Fig. 6 with hydrogen bubbles being released from one of the two intersecting channels. It is observed at the Reynolds number of 3200 and 5000 that the streak lines downstream of the intersection areas shift toward the inner walls, forming long stagnation-recirculation zones which extend downstream to the exit (L/d = 9.6) with less flow through the other half of the channel. The velocity profiles in the stagnation-recirculation zone of the right channel is shown in Fig. 7 with a release of pulses of hydrogen bubbles (at a pulsating frequency of 20 $sec.^{-1}$ and a pulse duration of 10^{-4} sec.).

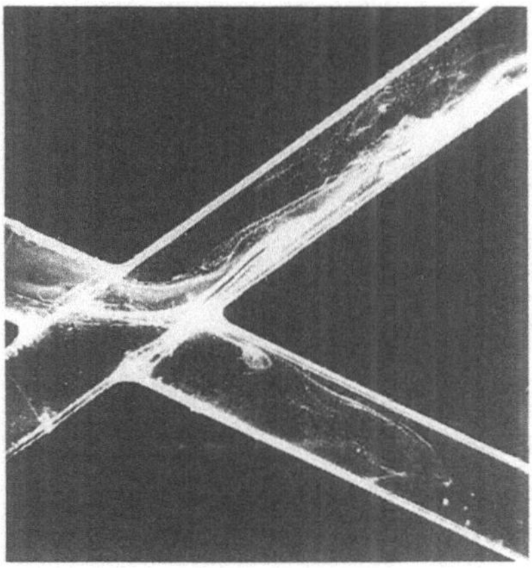

(a) Re = 3,200 with bubble release in left channel

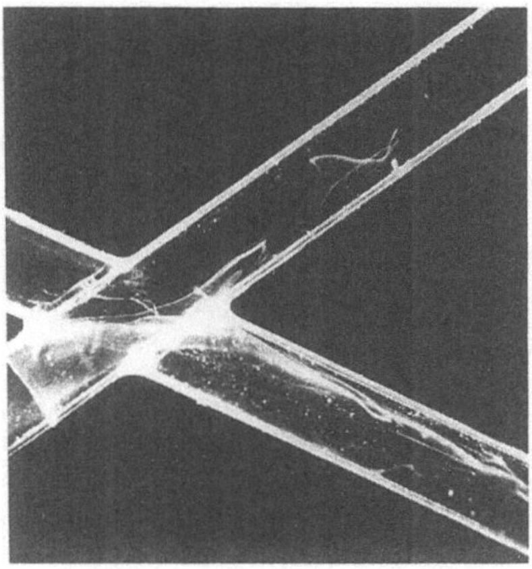

(b) Re = 3,200 with bubble release in right channel

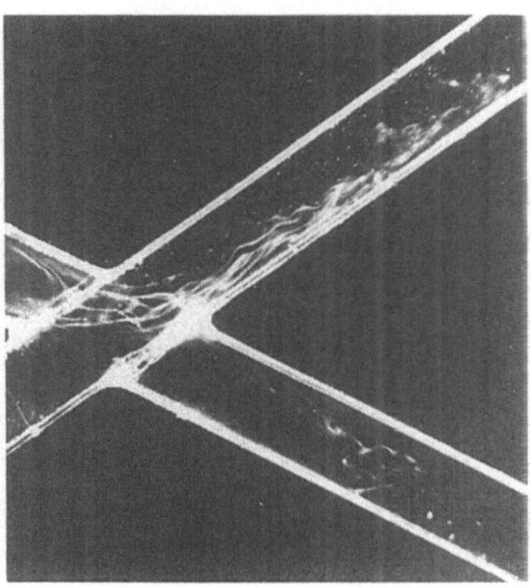

(c) Re = 5,000 with bubble release in left channel

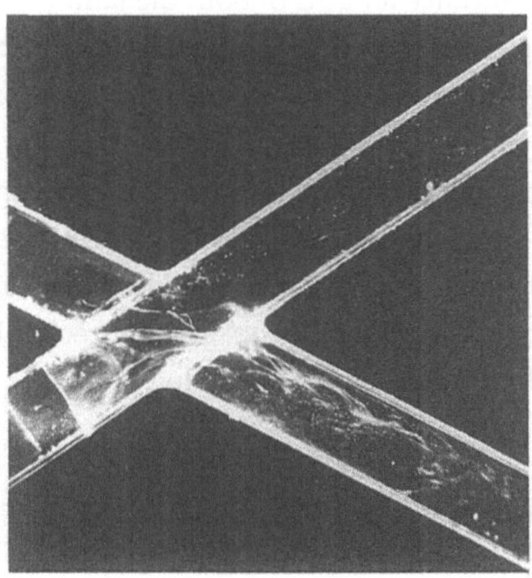

(d) Re = 5,000 with bubble release in right channel

Fig. 6 Flow patterns in post-intersecting region

Fig. 7 Velocity profiles in stagnation-recirculation zone

Conclusions

Hydrogen bubble method has been applied to visualize the flow patterns in a flow network. It is disclosed that

1. the entrance lengths of both the laminar and turbulent flows are less than 8d, much shorter than those in conventional tube flow,
2. flow ramming occurs in the intersection zone, with intermixing seen only in turbulent flow,
3. intermixing propagates upstream with an increase in the Reynolds number,
4. the streak lines downstream of the intersection area shift toward the inner walls, forming long stagnation-recirculation zones,
5. velocity profiles in the stagnation-recirculation zones can be obtained through pulse hydrogen bubble technology.

References

1. Zhang, N., Yang, W. J., Xu, Y., and Lee, C. P.: Flow characteristics in flow networks. Experiments in Fluids (in print) (1992).

2. Robert, W. F., and McDonald, A. T.: Introduction of Fluid Mechanics, Third Edition, Wiley, New York (1985).

Numerical Simulation of Viscoelastic Fluid in Two Dimensional Channel Junctions

S. GOTO and H. KATO

Department of Mechanical Engineering
Faculty of Technology
Tokyo Metropolitan University, Tokyo

Summary

The flow pattern for the viscoelastic fluid, the power law fluid and Newtonian fluid are simulated using the finite difference scheme. The Denn model is used as a constitutive equation. The influence of non-Newtonian viscosity, Reynolds number, Weissenberg number and the dividing flow rate on the sizes of the recirculation zones is studied. Shear thinning decreases the size of the recirculation zone in the lateral conduit and increases it in the main conduit, whereas shear thickening shows the reverse tendency. The size of the recirculation zone in the lateral conduit decreases with increasing Weissenberg number, and in the main conduit increases with increasing Weissenbrg number. The former agrees with the earlier experimental results, the latter does not agree with it.

Basic Equations

A Denn model is used as the constitutive equation:

$$\tau_{ij}=\mu\left[\frac{1}{2}\Pi\right]^{(n-1)/2}B_{(1)}^{ij}-\lambda\left[\frac{1}{2}\Pi\right]^{(s-2)/2}B_{(2)}^{ij} \qquad (1)$$

Here, $B_{(1)}^{ij}$ and $B_{(2)}^{ij}$ are the first and second acceleration tensors, respectively, and Π is the second invariant of $B_{(1)}^{ij}$.
The basic equations are expressed nondimensionally by the nondimensional qualities defined as follows:

$$u^*=u/V_1,\ v^*=v/V_1,\ x^*=x/H,\ y^*=y/H,$$

$$\zeta^*=H\zeta/V_1,\ \psi^*=\psi/HV_1,\ \tau_{ij}^*=\tau_{ij}/(1/2)\rho V_1^2$$

where u is the x-directional velocity component, v is the y-directional velocity component, V_1 is the mean velocity in the upstream main conduit, H is the width of the conduit, ζ is the vorticity, ψ is the stream function, and the symbol * is the nondimensional qualitities. These symbols are omitted hereafter. The velocities and the vorticity are:

$$u=\frac{\partial \psi}{\partial y} \quad , \quad v=-\frac{\partial \psi}{\partial x} \qquad (2) \qquad \zeta=\frac{\partial v}{\partial x}-\frac{\partial u}{\partial y} \qquad (3)$$

The relationship between the stream functiona and the vorticity is:

$$\frac{\partial^2 \psi}{\partial x^2}+\frac{\partial^2 \psi}{\partial y^2}=-\zeta \qquad (4)$$

The vorticity transport equation for the Denn model is:

$$\frac{\partial}{\partial x}(u\zeta)+\frac{\partial}{\partial y}(v\zeta)=\frac{1}{Re}\left[A\nabla^2\zeta+B\right]-C \qquad (5)$$

Where

$$\nabla^2\zeta=\frac{\partial^2\zeta}{\partial x^2}+\frac{\partial^2\zeta}{\partial y^2} \quad , \quad A=\left(\frac{1}{2}\Pi\right)^{(n-1)/2}$$

$$B=2\frac{\partial A}{\partial x}\frac{\partial \zeta}{\partial x}+2\frac{\partial A}{\partial x}\frac{\partial \zeta}{\partial x}+\frac{\partial^2 A}{\partial x^2}\left(\frac{\partial v}{\partial x}+\frac{\partial u}{\partial y}\right)-\frac{\partial^2 A}{\partial y^2}\left(\frac{\partial v}{\partial x}+\frac{\partial u}{\partial y}\right)-4\frac{\partial^2 A}{\partial x\partial y}\frac{\partial u}{\partial x}$$

$$C=\frac{\partial^2\tau_{xy}}{\partial x^2}-\frac{\partial^2\tau_{xy}}{\partial y^2}+\frac{\partial^2}{\partial x\partial y}(\tau_{yy}-\tau_{xx})$$

$$\tau_{xy}=2\frac{We}{Re}\left(\frac{1}{2}\Pi\right)^{(s-2)/2}\left\{u\frac{\partial}{\partial x}\left(\frac{\partial v}{\partial x}+\frac{\partial u}{\partial y}\right)+v\frac{\partial}{\partial y}\left(\frac{\partial v}{\partial x}+\frac{\partial u}{\partial y}\right)-2\frac{\partial u}{\partial y}\frac{\partial v}{\partial y}-2\frac{\partial v}{\partial x}\frac{\partial u}{\partial x}\right.$$

$$\tau_{xy}-\tau_{xy}=2\frac{We}{Re}\left(\frac{1}{2}\Pi\right)^{(s-2)/2}\left\{-2u\frac{\partial^2 u}{\partial x^2}+2v\frac{\partial^2 v}{\partial y^2}-\left(\frac{\partial v}{\partial x}-\frac{\partial u}{\partial y}\right)\left(\frac{\partial v}{\partial x}+\frac{\partial u}{\partial y}\right)\right.$$

Here, Re and We are the Reynolds number and Weissenberg number, respectively, defined as:

$$Re=\frac{\rho H^n V_1^{(2-n)}}{\mu} \quad , \quad We=\frac{\lambda}{\mu}\left(\frac{V_1}{H}\right)^{(s-n)} \qquad (6)$$

where, $\mu, \lambda,$ n and s are the rheological constants of the Denn model and ρ is the density of the fluid.

C=0 at We=0 represents the power low fluid, A=1, B=0 and C=0 at We=0 and n=1 represents the Newtonian fluid. In Eq.(1), the factor of $B^{ij}_{(1)}$ becomes μ for n=1. This represents the Newtonian viscosity. Also,the factor of $B^{ij}_{(2)}$ becomes λ for s=2. It represents a constant of the first normal stress factor.

Numerical analysis

Numerical soluitones are obtained by the central difference fomula which are substituted for the basic equations. The ratio of the widths of the main conduit to that of the lateral conduit is 1.0. The vorticity equation is caluculated by using the under relaxation method and the stream function by the over relaxation method. The fully developed flow is used at the entrance of the main conduit and at the exit of the main and lateral conduit, respectively. The vorticity at the corner of the junction is calculated by using the scheme used at the wall for the value of the stream function at the point of 45°. The condition of a convergent solution is $(A_{k+1}-A_k)/A_k < 10^{-4}$ at each grid point, where A is ζ, ψ, u and v.

Numerical Results

The numerical results are described for 90° junction only. Those for 45° junction are the same qualitatively. The two recirculation zones occur by the separation of the flow for Re > about 10. One is formed on the upstream wall in the lateral conduit and another on the the upper wall of the main conduit. We called the former vortex A and the latter vortex B, respectively. A schematic diagram of the 90° junction is shown in Fig.1.
Fig.2 shows examples of the simulation for Newtonian fluid. The sizes of both vortex grow with increasing Reynolds number(Re) for each fluid. Also the vortex A decreases with increasing dividing flow rate(Q_2/Q_1), whereas the vortex B increases with

Fig.1 Schematic diagram of 90°junction

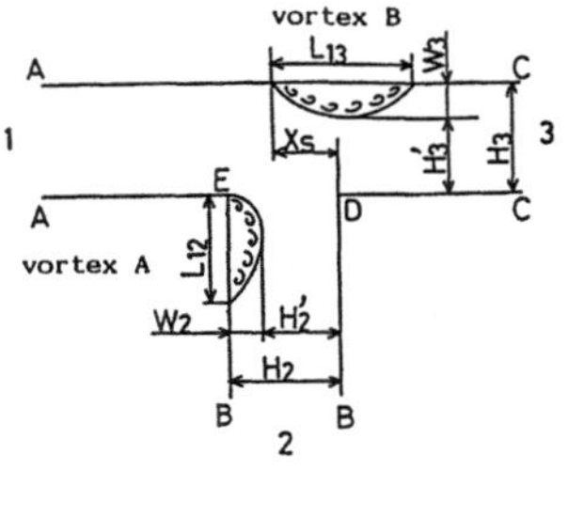

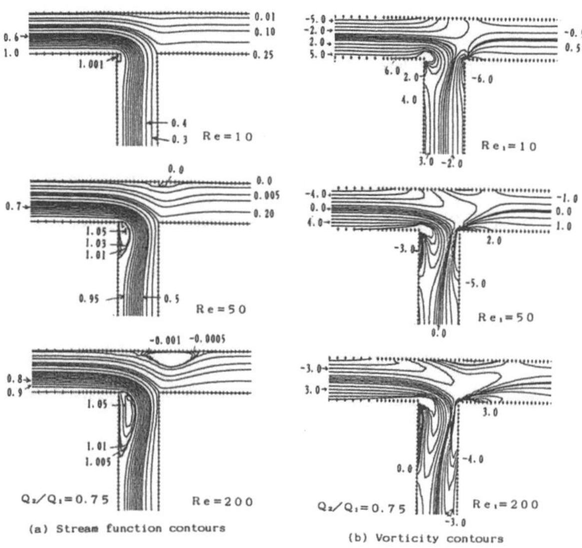

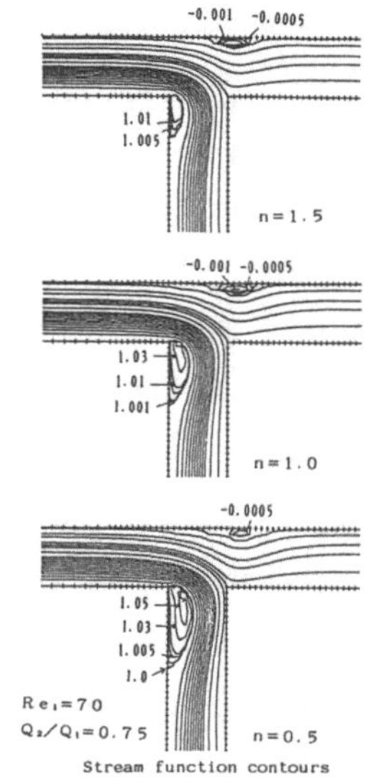

Fig. 2 Influence of Re on the sizes of the recirculation zones

Fig. 3 Influence of non-Newtonian viscosity on the sizes of the recirculation zones

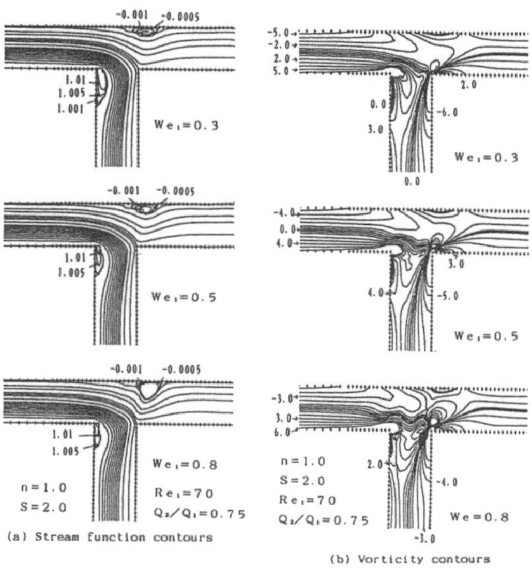

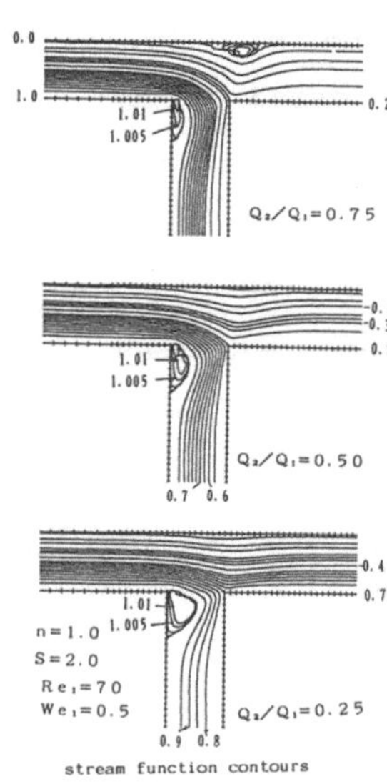

Fig. 4 Influence of We on the sizes of the recirculation zones

Fig. 5 Influence of the dividing flow rate on the sizes of the recirculation zones

it. These numerical results agree well with the earlier experimental results[1].

Fig.3 shows the influence of non-Newtonian viscosity on the size of the vortex. The shear thinning(n<1) increases the size of the vortex A and decreases the size of the vortex B, whereas the shear thickening(n>1) shows the reverse tendency. This numerical result cannot be confirmed because there is no exprimental result. Fig.4 shows the influence of the viscoelastic effect on the size of the vortex(n=1,s=2). The increase of the Weissenberg number(We) decreases the size of the vortex A and increases the size of the vortex B. This result for vortex A agree qualitatively well with our exprimental results[1] for the polymer soluiton (separan AP-30). But, the vortex B does not agrees with those. This may be because the converging solution cannot be obtained for the same rheological constants (n,s) and Wessenberg number,as those in the experiment. Fig.5 shows the influence of the dividing flow rate (Q_2/Q_1) on the sizes of the recirculation zones.

The decrease of the rheological constant s, which represents the non-linearity for the viscoelastic property, increases the size of the vortex A and decreases the size of the vortex B.

Conclusion

The influence of shear thinning and thickening on the sizes of the recirculation zones show the reverse tendency of each other. The increase of the Weissenberg number decreases the size of the recirculation zone in the lateral conduit and increases it in the main conduit.

Reference

1. Goto, S.; Kato, H.:Study on the Flow Viscoelastic fluid in Channel junctions (flow Ptterns). Preprint of JSME No.900-54 (1990) 140-142.

Vortex Formation in a T-Junction

BURGHARD BARTMANN, RUDOLF NEIKES, THOMAS LEWEKE, WOLFRAM LIMBERG, EGON KRAUSE

Aerodynamisches Institut, RWTH Aachen,
Wüllnerstr. 5-7, D-5100 Aachen, F.R.Germany

Summary

Flow visualization was used to study the jet in a crossflow problem in the bifurcating pipe flow of a T-junction. The visualizations revealed a great variety of vortex phenomena and -structures generated by the complex interaction between jet and crossflow. This paper reports on some of these and offers a qualitative insight into the very complicated three-dimensional nature of flow fields in a T-junction.

Bifurcating pipe flow, jet in a crossflow, flow visualization, coherent vortex structures, recirculation bubbles

Introduction

Internal flow in bifurcating pipes is of fundamental importance to mechanical engineering and is of great technical and scientific interest. In nearly all technical devices, in which fluids or gases are transported, bifurcating or branched pipes are used in many ways. In industrial application among others the following technical areas are concerned: motors, turbines, pumps, exhaust systems and air supply piping, hydraulics and pneumatics, aeration and deaeration, inlet- and outlet pipe connections, fittings, lubricant- and coolant circular courses, transport of gas and fluid in pipelines.

In order to increase physical understanding of incompressible internal flows in branched pipes and to obtain test data for numerical simulations of such highly three-dimensional and often time-dependent flows, an experimental program was performed at the Aerodynamisches Institut. The investigations included visualization studies in water flows and measurements of wall-pressure distributions, wall-shear-stresses and velocities in incompressible air flows of numerous bifurcations [1,2,3]. The results reported here are restricted to flow visualization of some vortex phenomena and -structures generated by the interaction of jet and crossflow in a T-junction.

Experimental Details

Flow visualization was carried out at a testing stand for water flows. The experimental setup is illustrated in Fig.1. The piping of the testing stand, made of plexiglass, has an inner diameter of D=80mm. The diameter of the branch is 2.3 times smaller than the diameter of the stem. To make an undistored flow observation possible, the T-junction was embedded in a water basin. Water was supplied from a water tank 5m high with a volume of $10m^3$. In

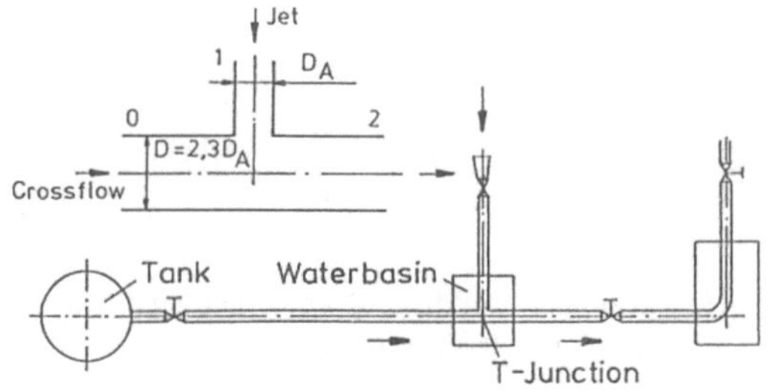

Fig.1 Experimental setup

order to keep the outflow constant, the tank could be filled up during the experiments. The mass fluxes were determined with standard orifices. The Reynolds number is defined with the average velocity calculated from mass flux at the outflow, the diameter of the stem and the viscosity of water. To avoid pollution by the additives for flow visualization, the water flows through the piping only once. Visualizations were carried out by addition of different fluorescent dyes and reflecting plastic particles for various velocity ratios v_1/v_0 and Reynolds numbers. V_0 is the mean velocity of the entering crossflow in the stem, v_1 is the mean velocity of the jet. The Reynolds numbers laid in the range between 10^2 and 10^5. The mean flow velocity for $Re=10^5$ was about 1m/s. The inflow profiles were fully developed.

Results

The visualisation offered a great variety of flow phenomena, e.g. pulsating deadwater, secondary motions, shedding of ring- or horseshoe vortices periodic in time, steady and timedependend stagnation point flows, revers-flow regions, formation and decay of shear-layer structures andlarge systems of vortices, periodical vortex formation with abrupt frequency changes, recirculation bubbles and vortex pairing. Some of them are illustrated in the following pictures. (The outflow is on the right. With exception of Fig.6b, Fig.6c and Fig.8 the flows are visualized by light sheets in the horizontal respectively vertical middle plane).

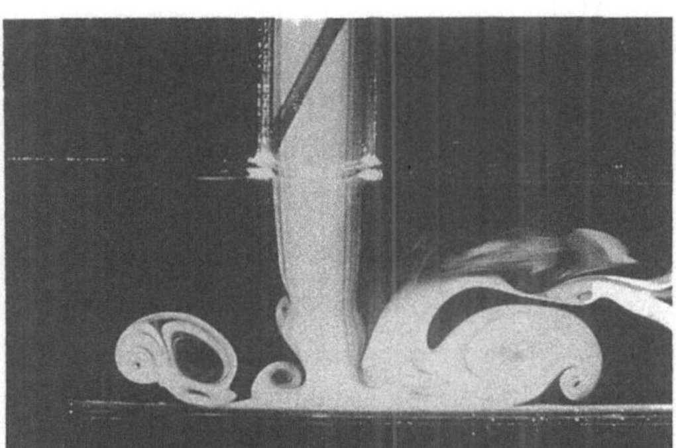

Fig.2 Bursting of ring-shaped jet structures at the stem wall. $v_1 >> v_0$.

Fig.2 shows a cut through ring-sha-

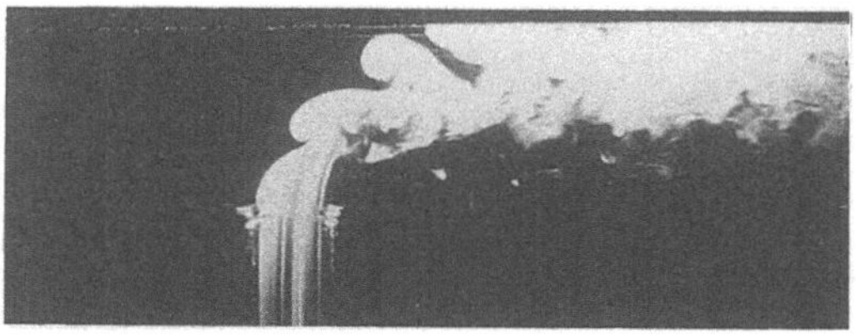

Fig.3 Periodical shedding of deformed ring vortices. Re = 3630; v_1/v_0 = 2.7.

ped jet structures bursting at the stem wall. The flow structure created by rapid increase of jet velocity is similar to a jet discharging into a quiescent ambient. Fig.3 reveals periodical formation of deformed ring vortices due to the interaction of crossflow and jet. The vortices carry vorticity of the same sign as the jet flow in the branch. Very close to the entrance of the jet the initial vortical rings are asymmetrically stretched by the different acceleration of the windward and lee side of the jet column. Dependent upon the velocity ratio vortical rings break down to turbulence before the bending over of the jet, burst at stem wall or move further downstream interacting with the crossflow. Fig.4 shows the largest one of a set of unsteady horseshoe vortices existing in the separation region on the left side of the stagnation point. Fig.5 presents periodical shedding of secondary horseshoe vortices in the unsteady stagnation region. In both cases vortex lines are bend around the jet column producing streamwise vorticity in the outflow.Fig.6a shows a region of two recirculation bubbles located symmetrically to the horizontal middle plane in the lee of the jet. Further downstream coherent vortex structures are formed carrying vorticity of opposite sign as the jet in the branch. In the sequence in Fig.6b the whole pipe is illuminated. The visualization shows the spiral motion of a vortex filament around a recirculation bubble and the temporal evolution of the reverse flow. The flow structure resembles the vortex break down phenomenon. The reverse flow is directed towards the pressure minimum at the trailing edge of the jet. Fig.6c shows a transversal cut through the bubbles and the jet. The middle pair of symmetrical counter-

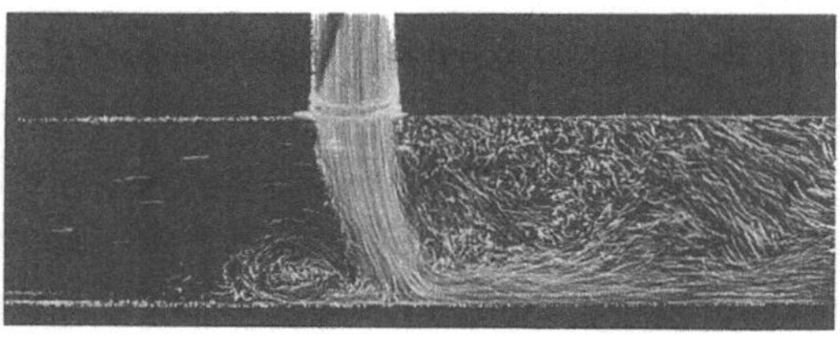

Fig.4 Unsteady horseshoe vortex in the stagnation point. Re = 15000; v_1/v_0 = 4.6.

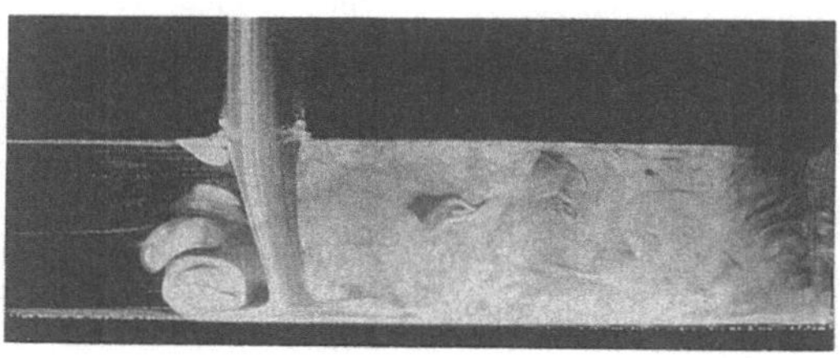

Fig.5 Periodical seperation of secondary horseshoe vortices in unsteady stagnation region. Re = 560; v_1/v_0 = 3.5.

rotating vortices reveals the structure of secondary motion in the bubbles, which are found in the concave region of the kidney shaped jet column. The visualized cross section reveals two further symmetrical vortex pairs: a pair of relative small vortices, located on the right, and the "bound vortex", typically of jets in cross-flows[4,5,6].

Fig.7 and Fig.8 offer a further insight into the very complicated three-dimensional nature of flows in a T-junction. In both cases the interaction of jet and crossflow creates complex coherent vortex structures further downstream. Fig.7 shows the wake of a jet resembling a three-dimensional version of Kármán vortex street. Fig.8 illustrates the formation of interlocking vortex loops. Coherent vortex structures of the type mentioned above were also found by Perry & Lim and Perry & Tan in coflowing jets and wakes produced by a laterally oscillating axissymmetric jet in a coflowing uniform stream [7,8].

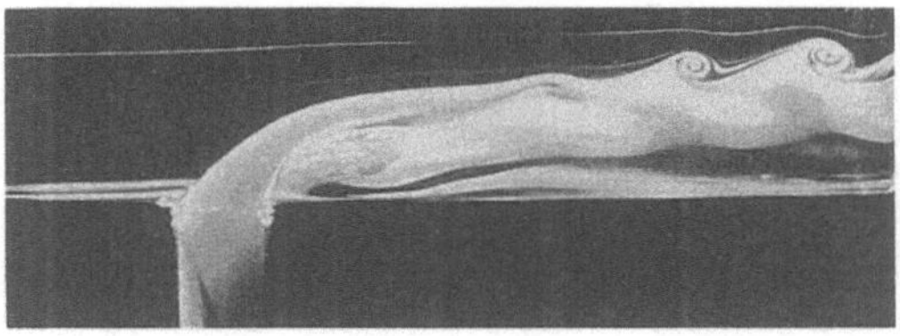

Fig.6a Region of recirculation bubbles in the lee of a jet and coherent vortex structures in the wake. Re = 1120; v_1/v_0 = 1.3.

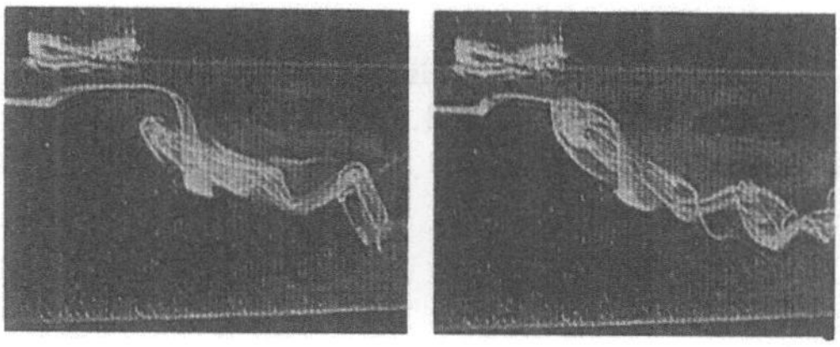

Fig.6b Spiral motion of vortex filament around recirculation bubbles resembling vortex breakdown. Re = 1120; v_1/v_0 = 2.7.

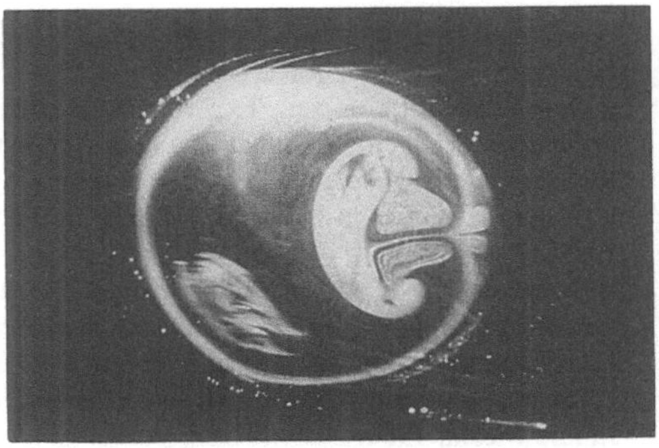

Fig.6c Vertical cut through jet and recirculation bubbles. Re = 1320; v_1/v_0 = 2.7.

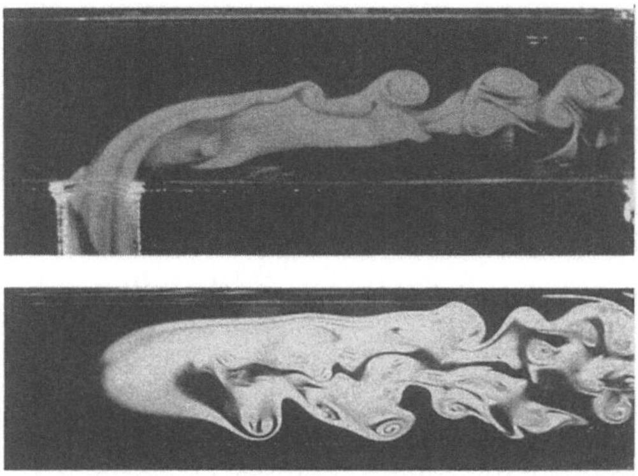

Fig.7 Wake of a jet resembling Kármán vortex street.
Above: Re = 1520; v_1/v_0 = 1.8.
Below: Re = 1240; v_1/v_0 = 1.8.

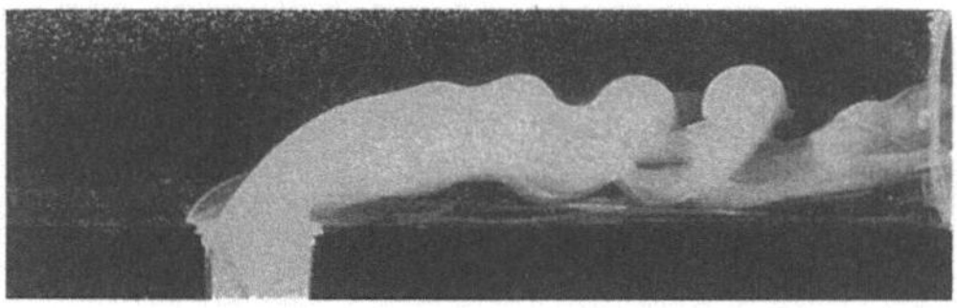

Fig.8 Formation of interlocking vortex loops in the wake of a jet.
Re = 1230; v_1/v_0 = 1.3.

References

1. R.Neikes und B.Bartmann, "Experimentelle Untersuchung turbulenter Strömungen in Rohrverzweigungen", Abhandl. aus dem Aerodyn. Institut, Heft 30, 1990.
2. R.Neikes and B.Bartmann, "Experimental Investigation of Three-dimensional, Turbulent, Internal Flows in Bifurcating Pipes", to be published in Experiments in Fluids, 1992.
3. B.Bartmann und R.Neikes, "Experimentelle Untersuchung inkompressibler verzweigender Innenströmungen", Abhandl. aus dem Aerodyn. Institut, Heft 31, 1992.
4. J.Andreopoulos, "On the Structure of Jets in a Crossflow", J.Fluid Mech., 1985
5. J.Andreopoulos and W.Rodi, "An Experimental Investigation of Jets in a Crossflow", J.Fluid Mech., 1984.
6. J.F.Foss, "Interaction Region Phenomena for the Jet in a Crossflow Problem", Rep. SFB 80/E/161, University of Karlsruhe.
7. A.E.Perry and D.K.M.Tan, "Simple Three-dimensional Vortex Motion in Coflowing Jets and Wakes", J.Fluid Mech., 1984.
8. A.E.Perry and T.T.Lim, "Coherent structures in Coflowing Jets and Wakes", J.Fluid Mech., 1978.

Steady and Oscillatory Flows in a Furrowed Channel with Arc-Shaped Walls

T. Nishimura
Department of Mechanical Engineering, Yamaguchi University, Ube, 755 Japan

Y. Kawamura
Department of Chemical Engineering, Hiroshima University, Higashi-Hiroshima 724 Japan

Flow patterns in a symmetric two-dimensional wavy-walled channel were investigated by flow visualizations at low Reynolds numbers and intermediate Strouhal numbers : Re=10-300 and St=0.008-0.05. For steady flow, we discovered a new flow structure consisting of a regular three-dimensional flow as the Reynolds number exceeds 27.5. For oscillatory flow, the critical value of the Reynolds number at the onset of three-dimensional flow increases with the Strouhal number.

Introduction

Previous our studies have been shown that it is possible to generate efficient vortex mixing with a symmetric sinusoidal wavy-walled channel by using an appropriate level of fluid oscillation at low Reynolds numbers and that the mass transfer from the channel wall to the bulk of the flow or vice versa is enhanced [1, 2]. The present work is concerned with the flow patterns in a furrowed channel with a sharp edge to examine the effect of wall geometories. The sharp edge is expected to lead to more efficient fluid mixing since the occurrence of flow separation is easy. The flow patterns are observed experimentally for steady and oscillatory flows.

Experimental Apparatus and Procedure

The experimental apparatus is the same as that used previously [1]. Flow oscillations were produced by a Scotch-yoke-driven piston/cylinder arrangement. The volumetric flow rate varies with time in a sinusoidal form. The channel consisted of a pair of undulated plates, which are principal walls. Figure 1 shows the dimensions of the channel. There are 14 furrows in the wall of the channel; each furrow is 14 mm long (λ), 3.5 mm deep (2a) and 80 mm wide (W). The minimum gap between the upper and lower walls H_{min} is 3 mm. The dimensions are the same as those of sinusoidal wavy walls used previously [1].

There are two flow parameters that characterize oscillatory flow, the Strouhal number

$$St=fH_{min}/(2U_p)$$

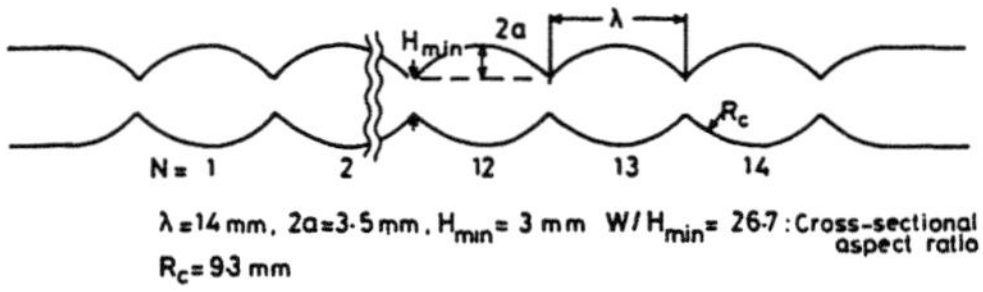

Fig.1. Details of test section

and the Reynolds number

$$Re=U_pH_{min}/(2\nu)$$

Where f is the frequency of flow oscillation, U_p is the peak velocity through the minimum gap and ν is the kinematic viscosity of the fluid.

Experiments were carried out for Reynolds numbers up to 300 and intermediate Strouhal numbers between 0.008 and 0.05. The ranges belong to the inertially dominated flow regime as indicated from the results for a sinusoidal wavy-walled channel [1].

Flow visualizations are by means of the aluminum dust method. Perfusion with a suspension of aluminum particles, about 40 μm in diameter, enabled us to observe path lines. Illumination was provided through a slit by a 500 W projector light source. To examine three-dimensionality of the flow we viewed the path lines of aluminum particles from the side and top of the channel. Photographs were taken using a motor-driven camera which was electrically triggered by the pulse of a photo sensor. In oscillatory flow, particle paths do not in general coincide with streamlines, but if the exposure time is short compared with the timescale for significant changes in the streamline pattern, there will be approximate coincidence over the exposure time. From this point of view it was desirable to use as short exposure as possible. However, a limit of the exposure time is imposed by the need for a clear photograph of the flow pattern and the frequency of flow oscillation. In this experiment, an exposure time of 1/30 to 1/8 s was selected for a fluid particle to trace a path sufficiently long. Many of the experiments were also recorded by a video camera.

Results and Discussion

Figure 2 shows photographs of the flow for steady flow. An exposure time of 0.5 s was used. In the photographs the mainstream is moving from left to right and the location is from the sixth to the eighth wave sections in the channel. At Re=10 (Fig. 2(a)) there is a stagnant region in the upstream part of each furrow. As the Reynolds number increases the separated vortex is clearly observed. At Re=25 (Fig. 2(b)), the separation point is located near the crest of the wall. The mainstream moves straight through the channel and thus the flow is symmetric at that stage. An unexpected flow pattern is observed at Re=50. The photograph of Fig. 2(g) represents the behavior of the aluminum particles accumulated on the lower wall. A periodic pattern with a constant wavelength λ_v is formed in the spanwise direction and also undergoes a 180° phase shift between adjacent furrows, which indicates the existence of three-dimensional flow with a regular structure. Figs. 2(c) and (d) show the flows at two different longitudinal cross sections (A-A' and B-B') separated by a distance of a half wave length $1/2\lambda_v$. From the comparison of both figures, the mainstream alternately deflects per one wave section through the channel, forming an asymmetric flow and also the deflection is reversed between the cross sections A-A' and B-B', but the flow behavior is identical. The details of the structure are described later. As the Reynolds number exceeds 50 the flow becomes unsteady. Figs. 2(e) and (f) show the results as the cross section B-B' for Re=100 and 150, respectively. Unsteady asymmetric flow develops and also turbulence exists. For arc-shaped walls, the onset of turbulence is found to occur at a much lower Reynolds number than for sinusoidal wavy walls reported previously.

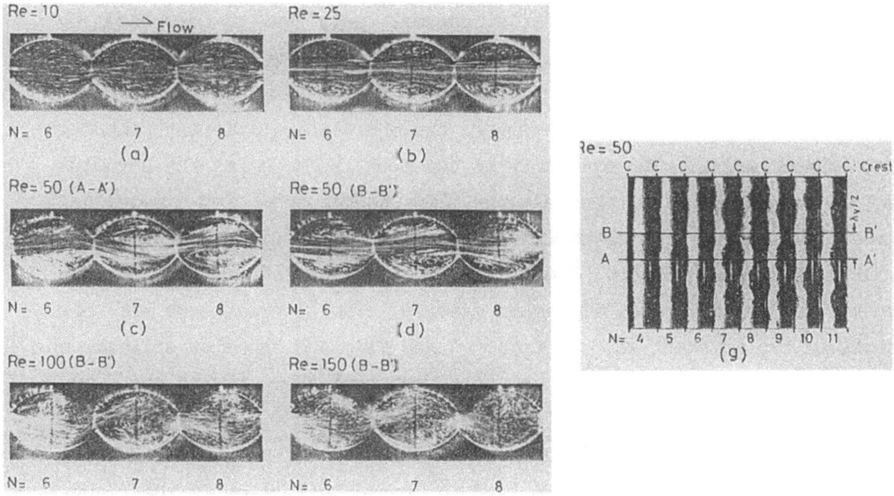

Fig.2. Flow patterns for steady flow

We furthermore examine the structure of a new flow pattern by the dye injection method. The Reynolds number at the onset of three-dimensional flow was 27.5. Figure 3 shows the streakline of dye in furrows. We call the region of one wavelength of λ_V a unit cell. The photographs reveal, firstly that the fluid in each furrow enters at the cell centerline and leaves at the cell boundary, i.e. inflow-outflow process. Secondly, the fluid within the unit cell exhibits a pair of tornado-like motions the axis of which is normal to the wall. Thus it is found that the tornado-like motion causes a periodic pattern of aluminum particles as shown in Fig. 2(g). A similar observation has been reported for a single cavity in turbulent boundary layer flows by Maul and East [3] and Kistler and Tan [4], but it is not evident whether the reason for the three-dimensional flow is identical.

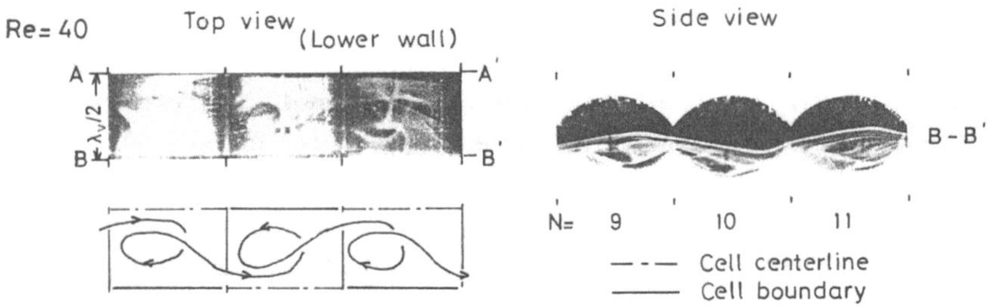

Fig.3. Flow patterns within furrows

The basic flow patterns which accompany unsteady separation in the present channel are similar to those for sinusoidal wavy walls described by Nishimura et al. [1]. The process entails formation of a separated vortex, followed by ejection of the vortex beginning prior to flow reversal, i.e. vortex formation/ejection cycle. However, there are some different features for arc-shaped walls. The symmetory of the flow is broken as the Reynolds number increases and simultaneously the flow changes from two-dimensional to three-dimensional. The asymmetric three-dimensional flow is kept at any time during flow oscillation. Figure 4 shows the flow patterns at various times for Re=78.7 and St=0.0284 as an example. The flow patterns are not identical in two adjacent wave sections at any time and the flow deflection alternates between the wave sections. Also the time asymmetry in each wave section can be seen immediately from the photographs for times t/T=0.0 and 0.5, of which the former shows a downward deflection of the separated shear layer in the left-hand wave section and an upward deflection in the right-hand wave section, while the latter shows the reverse situation.

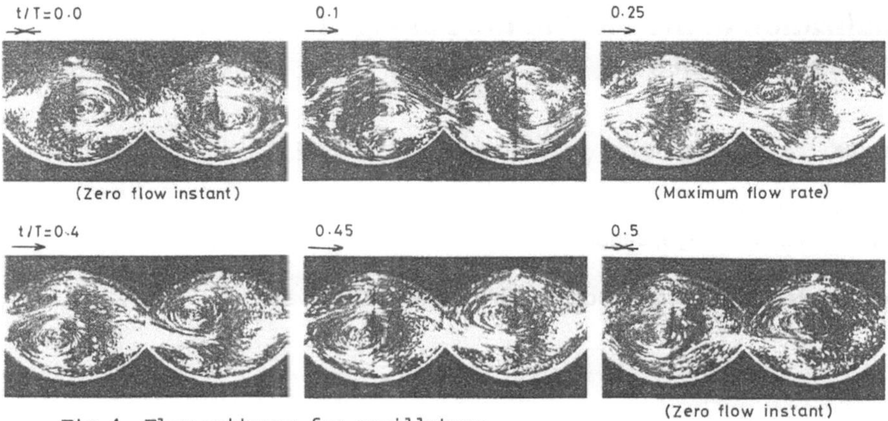

Fig.4. Flow patterns for oscillatory flow

St=0.0284, Re=78.7

Furthermore, the transition to three-dimensional flow for various flow parameters was examined as shown in Fig. 5. The critical Reynolds number at the onset of three-dimensional flow tends to increase as the Strouhal number increases. This behavior is similar to the suppression of turbulence in a straight-walled channel for oscillatory flow [5].

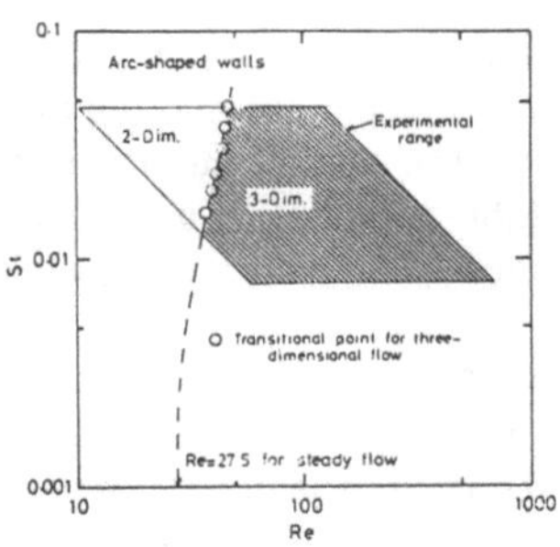

Fig.5. Formation region of three-dimensional flow

References

1. Nishimura, T.; Miyashita, H.; Murakami, S.; Kawamura, Y.: Oscillatory flow in a symmetric sinusoidal wavy-walled channel at intermediate Strouhal numbers. Chem. Engng Sci. 46 (1991) 757-771.

2. Nishimura, T.; Murakami, S.; Kawamura, Y.: Mass transfer in a symmetric sinusoidal wavy-walled channel for oscillatory flow. Submitted to Chem. Engng Sci. (1992).

3. Maul, D. J.; East, L. F.: Three-dimensional flow in cavities. J. Fluid Mech. 16 (1963) 620-632.

4. Kistler, A. L.; Tan, F. C.: Some properties of turbulent separated flows. Physics Fluids. (1967) S167-S173.

5. Ohmi, S.; Iguchi, M.; Akao, F.: Transition to turbulent flow and velocity profiles in a rectangular channel for oscillatory flow. Trans JSME Ser. B. 47 (1983) 2343-2353.

Visualization of the Flow in the Passage Between Two Coaxial Cones

K. Balatka, S. Mochizuki, A. Murata, ○ Y. Yagi

Tokyo University of Agriculture and Technology,
Koganei — shi, 183 Tokyo, Japan

○ Tokyo Polytechnic College, Kodaira — shi, Tokyo 187, Japan

Summary

The present paper deals with a flow behavior in an annular-conical passage. During the investigation, two steady states of the flow pattern were observed, that is, the streak flow pattern and the annular-vortex flow pattern.

The flow visualization was performed by using paraffin mist. A 1.5 W Argon laser light-sheet and a 500 W photographic flood-light were used as an illumination. The pictures were taken using a video camera and a still camera.

1. Introduction

The annular-conical passage is one of the fundamental flow passages where the flow behavior is not yet well known. Mochizuki et al. [1] investigated spray characteristics of a gasoline injection nozzle (poppet nozzle), that is one of the technical applications of the passage of concern. They reported that the flow in the outlet jet from the nozzle, forms multiple and regularly-spaced streaks instead of forming a circumferentially uniform liquid film. However, the origin and the physical mechanism of the streak flow pattern had not been known yet.

In order to clarify the flow behavior in the annular-conical passage of the poppet nozzle, the scale up model was built up and the flow visualization study was performed.

Fig.1. Outlet jet from the poppet nozzle.

2. Experimental Apparatus

Fig. 2 shows a test section. It is located inside an airtight chamber. To obtain as uniform and steady flow in the inlet section as possible, a suction arrangement is used to drive air through the passage. Air is sucked through a bell-mouth and an inlet pipe into the test section and exhausted from the airtight chamber by means of a blower.

A flow rate is set by controlling the blower R.P.M. and measured by a hot wire anemometer.

Two axially symmetrical cones, shown in Fig. 2, form the annular-conical passage. An apex angle is 90 ° and both the outer and the inner cones are made of 1 mm thick transparent PVC. The upper part of the test section (the outer cone, the mist chamber, and the inlet pipe) is supported by the airtight chamber and also hooked on steel wires with turnbuckles that are used to establish the axial symmetry. The cone spacing *s* can be varied by moving the inner cone vertically.

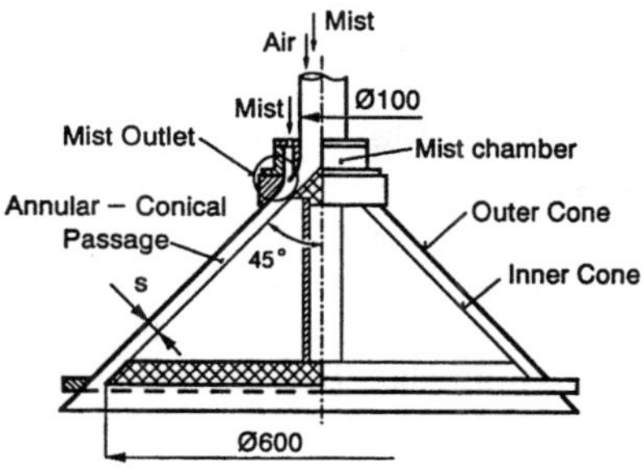

Fig. 2. Schematic of the test section.

Fig. 3 shows a definition of geometrical parameters. The Reynolds number is defined as $u_i \cdot 2s/\nu$, where u_i denotes the average inlet-flow velocity based on the diameter d_i of the inner cone. The value $2s$ represents the hydraulic diameter and ν is the kinematic viscosity.

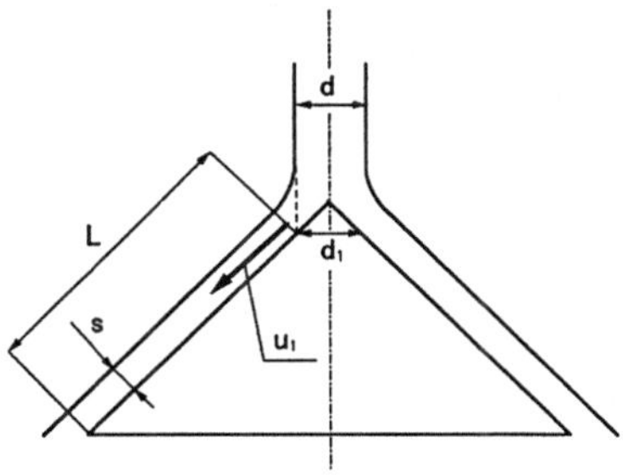

Fig. 3. Definition of the geometrical parameters.

3. Flow Visualization Method

The flow visualization is carried out by using mist produced by vaporization of liquid paraffin with the aid of an electrically-heated spiral wire. Figs. 4 and 5 illustrate the schematics of visualizing the flow and for taking photographs.

To visualize the streaks, the mist is injected first into a mist chamber, (see Fig. 2), that equally distributes the tracer around the perimeter and that injects it to the passage through a slot. The slot is built in around a rounded corner of the passage entrance. In order to visualize the annular-vortex flow pattern, an upper mist injector that injects the mist directly into the air inlet is used in addition to the mist injection described above. By using these two mist inlets it is possible to have a clear vision of the flow along the outer and the inner wall of the passage. To avoid a light reflection, a black mat paper covers the inner cone. A 500 W flood-light was used while taking black and white photographs of the flow patterns, Figs. 6 and 8. To

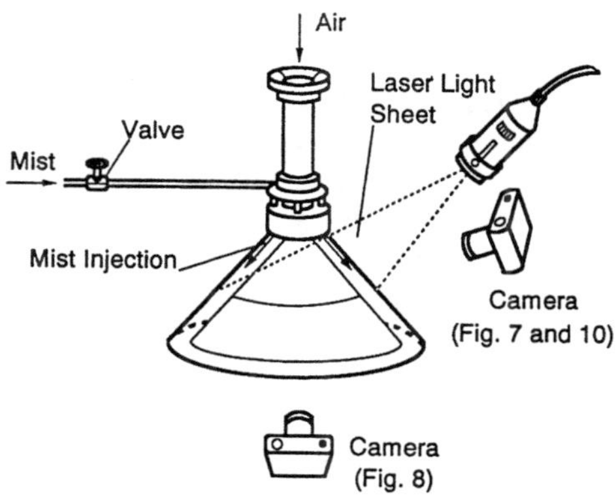

Fig. 4. Setup for visualizing the streak flow pattern.

photograph cross sections, Figs. 7 and 9, a 1.5 W Argon laser light-sheet was sufficient as an illumination.

4. Experimental results and discussion

Figs. 6 and 7 show representative results of the streak flow pattern visualization for the Reynolds number $Re=300$ and the gap $s=10$ mm.

Comparing Fig. 1 with 6, we believe that these two flow patterns are similar. In addition, we found that the streaks are formed by longitudinal vortices and Fig. 6 shows that these vortices originate in the entrance region of the passage. Fig. 7 shows the cross section "a" of the flow pattern in Fig. 6 and a sketch in Fig. 10 provides a cross sectional view on the history of longitudinal vortex development.

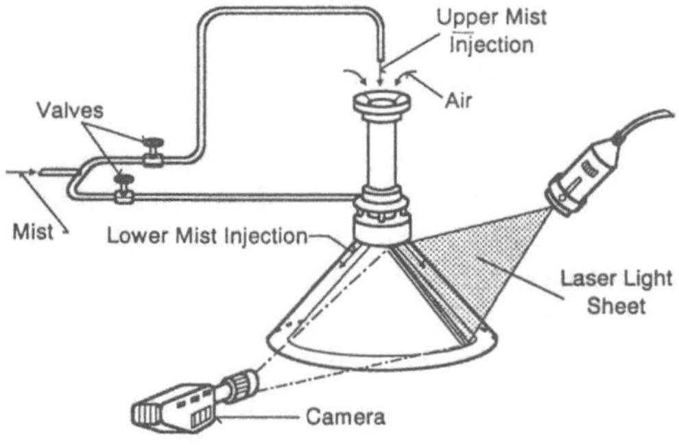

Fig.5. Setup for visualizing the annular – vortices.

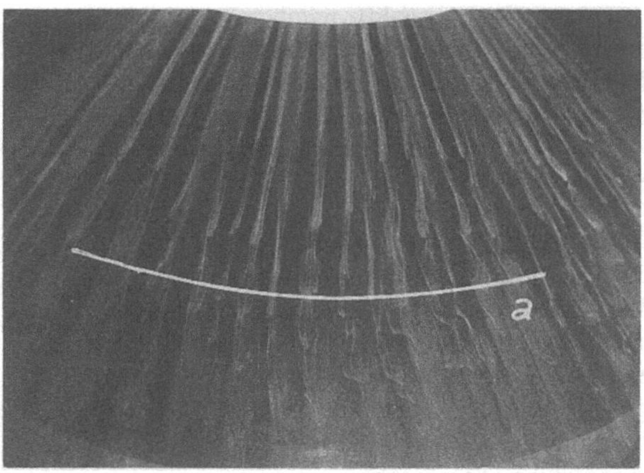

Fig.6. Streak flow pattern.

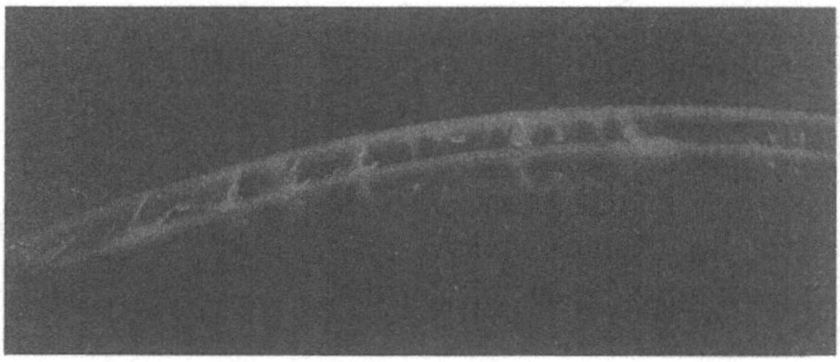

Fig.7. Cross section "a" (marked in Fig.6).

If the Reynolds number in the passage of concern is increased over a certain critical value, the self - sustained oscillating flow takes place. Figs. 8 and 9 show that these oscillations are caused by a vortex street consisting of annular vortices that are separated from the wall on both sides of the passage. Similarly as in the case of the von Karman vortex street, the flow separation occurs periodically and alternately.

In addition to the flow visualization, a frequency of the periodical flow separation was measured by using the hot wire anemometer and a FFT frequency analyzer. The representative result is shown in

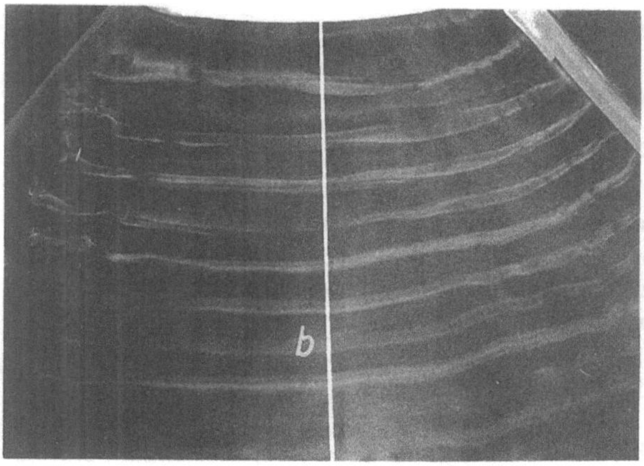

Fig.8. Annular – vortex flow pattern.

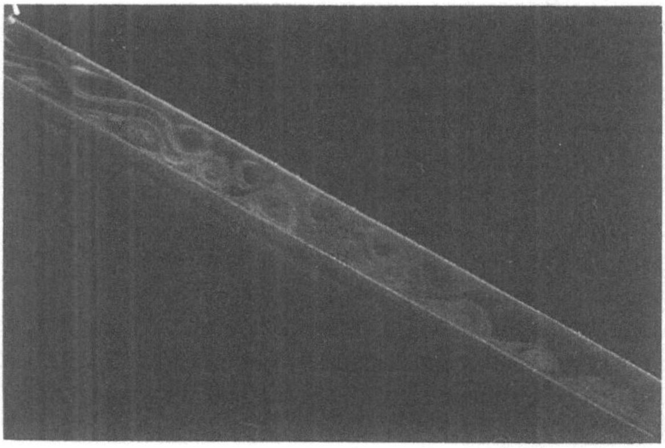

Fig.9. Cross section "b" (marked in Fig.8)

Fig. 11. The main peak designates the most significant frequency, that is, the frequency of the periodical flow separation.

From our observations we conclude that the smooth laminar flow, (see Fig. 10a), becomes unstable above a certain critical Reynolds number. We belive that disturbances grow due to a centrifugal force that is introduced to the flow in the entrance region of the passage. This changes the flow pattern and in the range *Re=200~400* the flow reaches a first level of equilibrium, resulting in a new steady state. We call this first level of equilibrium (steady state) the streak flow pattern, (Fig. 6). The physical mechanism is considered to be similar to the one of the Goertler vortices [2]. In other words, we think that this centrifugal instability results in a transverse secondary flow, that moves the tracer particles in the manner shown in Fig. 10b,c,d. Taking the shape of the annular- conical passage into consideration, it is clear that the flow conditions change along the passage and the force that drives the secondary flow is believed to be vanishing. This causes a change of the streak flow pattern from the origin, (Fig. 10b), through the developed longitudinal vortex, (Fig. 10c,d), and results in a vortex breakdown and turbulence that takes place near the passage outlet as illustrated in Fig. 10e.

The flow, that now is in temporary equilibrium, becomes unstable to other type of disturbances, while increasing the Reynolds number. The flow pattern thus starts to change again and reaches the second level of equilibrium in the range of the Reynolds number *Re=600~800*. That is, the annular vortices become to be the characteristic flow pattern. As in the previous case, the flow pattern changes downstream due to the change in flow conditions. Fig. 11 shows that the frequency of the periodical flow separation is approximately *f=3 Hz.*

When the Reynolds number increases further, the flow does not reach any other state of equilibrium. However, large and random disturbances that starts to prevail cause the flow to be turbulent. The flow becomes completely turbulent over the Reynolds number *Re=2000.*

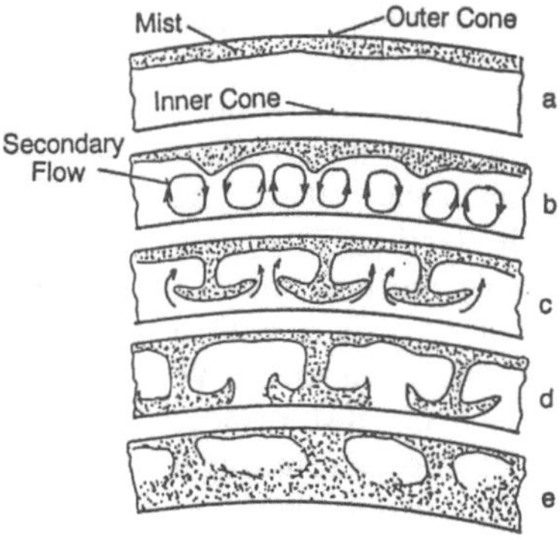

Fig.10. Streaks developement.

5. Conclusions

1. The flow between two coaxial cones exhibits two characteristic flow patterns; the streak flow pattern and the annular- vortex flow pattern.

2. The streak flow pattern originates in the entrance region of the test section and the streaks are formed by the transverse secondary flow. The secondary flow is believed to be caused by the centrifugal instability.

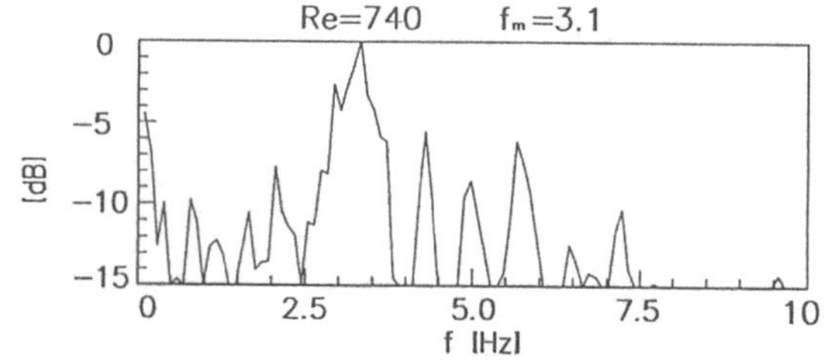

Fig.11. Example of a result of frequency measurement.

3. The first level of equilibrium, the streak flow pattern, takes place for the Reynolds number approx. *Re=200~400.* The second one, the annular- vortex flow pattern, occurs when the Reynolds number is approximately in the range *Re=600~800*.

4. The frequencies of the periodical flow separation *(Re=600~ 800)* slightly vary around the value 3 Hz.

References

1. Mochizuki, S.; Morita, D.; Sato, T.; Yang,W.J.: Flow Visualization in Conical Jets by Laser-Induced Fluorescence Method. Flow Visualization 5th ed., R. Reznicek Hemisphere (1989) 824-831.

2. Schlichting, H. (7th ed.) Boundary-Layer Theory. New York, London, Tokyo: McGraw-Hill 1979.

3. Mochizuki, S.; Yang, W.J.: Self Sustained Radial Oscillating Flows Between Parallel Disks. J. Fluid Mech. 154 (1985) 337-397.

4. Kundu, P. K. (1st ed.) Fluid Mechanics. San Diego, London: Academic Press 1990.

Flows in Bio-System

Visualization of Velocity Profile in Blood-Related Flow by Laser Technique

K. OHBA

Department of Mechanical and Systems Engineering,
Kansai University, Suita, Osaka 564, Japan

Summary

Laser techniques to visualize velocity profile in model blood flows are described. The availability of a laser–induced fluorescence method and a photochromic dye tracer method to get the quantitative information of the detailed velocity field in pulsatile flow through bends, bifurcations and stenoses is shown.

Introduction

In the biomedical field, knowledge of the detailed velocity field in the pulsating blood flow at stenoses, bends and bifurcations has been required, because such flow fields has been considered to play an important role in cardiovascular diseases like atherosclerosis. Several experimental studies have been made primarily to understand the changes to the blood flow pattern due to bends and bifurcations by Ku and Giddens [1], constrictions or stenoses by Ahmed and Giddens [2] and aortic valve prostheses by Yoganathan, Corcoran and Harrison [3] and Hanle *et al.* [4], and also to assess the potential clinical use of high frequency ventilator systems for use during surgery and in intensive care by Gaver and Grotberg [5]. It has only been recently, however, that the distribution of lesions in given anatomical regions of human subjects has been correlated with quantitative fluid dynamic measurements, thus allowing a sharper focus upon the potentially relevant hemodynamic factors such as high and/or low wall shear stress, high and/or locally low pressure, turbulence and oscillating wall shear *etc*. Thus, for the experimental study in this field the quantitative measurements of the local and temporal velocity field over the whole flow field are required. Therefore, qualitative methods of flow visualization such as a dye injection technique is not appropriate. Laser Doppler velocimetry is not necessarily complete for this purpose because it is primarily a point measurement method and to get the information of the whole flow field by using LDV is very tedious and elaborate. So, a method of time line visualization is considered to be very useful, especially non–invasive method using laser is most appropriate. In this paper several methods of time line (velocity profile) visualization will be described briefly.

Time Line Visualization by Laser

Conventional visualization methods for a liquid flow through a duct such as a dye injection method and another tracer injection methods can visualize the stream line, but not the time line. Hence they give only a qualitative information of flow field. There are several methods of time line visualization such as a hydrogen bubble method of Geller [6] and a photo–chromic dye tracing method of Hummel [7]. The former can make quantitative measurements to some extent, but it has some disadvantages such that flow is disturbed by a tracer generator in a duct, that the traceability of the hydrogen bubble is not sufficient and that there is some time delay, about 100 μs, in generating bubbles when stimulated. The photochromic dye tracing technique is based on the practically instantaneous (5–10 μs) reaction of a photochromic compound with ultraviolet light. It is a non–invasive method with fast response, but it needs a background light for visualizing the time line. A laser–induced fluorescence or phosphorescence (LIF or LIP) method of Nakatani [8] and Ohba [9,10] can be used to advantage to make quantitative measurements of a complex flow field non–invasively, because it can visualize the velocity distribution directly without any background light. A comparison among there these methods for time line visualization is made in Table 1.

Table 1. Visualization of Time Line (Velocity Profile)

Hydrogen bubble method
- Instantaneous (10-100 μs) bubble generation on a platinum wire by a pulse (width ~100 μs) electric current
- Life Time: $\gtrsim$ 5s
- Invasive method
- Background light necessary

Photochromic Dye Tracing
- Instantaneous (5-10 μs) reaction of photochromic compound with pulse (width ~5 ns) ultraviolet light
- Life Time: ~1 s
- Non-invasive method
- Background light needed

Laser Induced Fluorescence (Phosphorescence)
- Instantaneous (~1 fs) excitation of fluorescence by a pulse (width 5 ns) ultraviolet light
- Life Time: ~1.1 s
- Non-invasive method
- No Background light needed. Image Intensifier needed

Photochromic Dye Tracer Method

The photochromic tracer method makes use of the properties of a normally colorless indicator in the test fluid. Upon irradiation with UV light the indicator undergoes a reversible photochemical reaction resulting in a color change. The trace produced by a pulsed UV laser is photographed at a selected time after its formation, thereby enabling the average velocity within the time interval to be obtained. The computer–controlled measurement system used by Ojha *et al.* [11] is illustrated in Fig. 1.

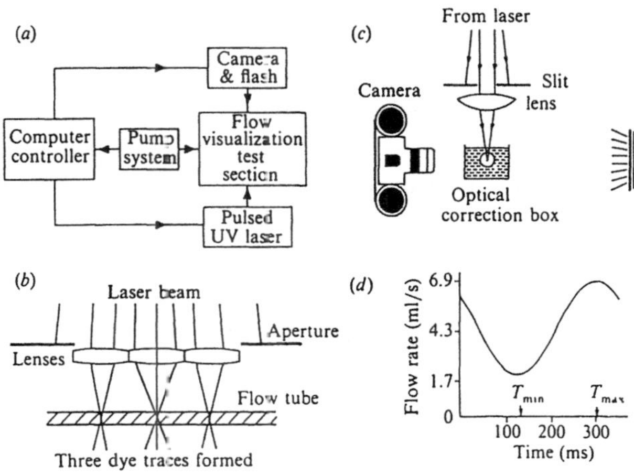

Fig. 1 Schematic representations of (a) the major components of the measurement system; (b) longitudinal view showing the lens arrangement for the formation of multiple traces (note, the focal length of the lens are 10 cm and the aperture width is 2.4 cm, so that the beam divergence is much less than that illustrated); (c) cross–sectional view of the test and optical systems; (d) volumetric flow waveform whose period is 345 ms. [11]

The photochromic solution consisted of 0.01% by weight of the indicator 1',3',3'–trimethylindoline–6–nitrobenzospiropyran (TNSB) dissolved in Shell–Sol 715. Such a low concentration does not affect the physical properties of the solvent, but the concentration is sufficient to ensure that a fairly uniform trace with satisfactory contrast is produced across the Pyrex tube. A dark blue trace develops in less than 10^{-6}s following the laser discharge (duration < 10^{-8}s). The trace is photographed at a selected time after its formation on a film through the use of a camera equipped with a micro lens and a flash. This selected time will be referred to as the flash delay and it actually represents the time interval between the triggering of the nitrogen laser and the discharge of the flash. Best contrast in photographing the blue dye trace is achieved using green light. This was achieved by using a green dichroic filter on the flash and a yellow filter on the camera.

Laser–Induced Fluorescence Method

Figure 2 shows an apparatus for the visualization and measurement of the flow velocity distribution using a laser–induced fluorescence method. A nitrogen pulse laser is used for exciting the fluorescence (phosphorescence) particle, thus making it fluoresce. The laser emits ultraviolet light of 5 nanosecond in pulse width, 337.1 nm in wave length and 1 MW in peak power. For the fluorescence particles, a zinc sulfide particle of about 8 μm in mean diameter and 4.1 g/cc in density is used, which is available in a commercial market. It is intermixed into water in a concentration of 0.1 to 0.2 % by weight. The ZnS particles, irradiated by a well collimated pulse ultraviolet laser light, emit a green fluorescence light of about 1 second in life time. From Franck–Condon principle, a fluorescence particle emits a fluorescence light without any time delay on being irradiated by pulse laser. As it is well known as Stokes' law, a frequency of a fluorescence light is lower than that of the stimulator light, so fluorescence is not absorbed by other fluorescence particles. As a row of fluorescing particles moves with a flow, a time line representing a velocity profile along the laser beam is formed. The motion of the time line so generated is photographed and tracked by a high speed video camera having a time resolving power of 200 frames per second in the direction perpendicular to both the laser beam and the flow. The camera is equipped with an image intensifier for amplifying the weak fluorescence light.

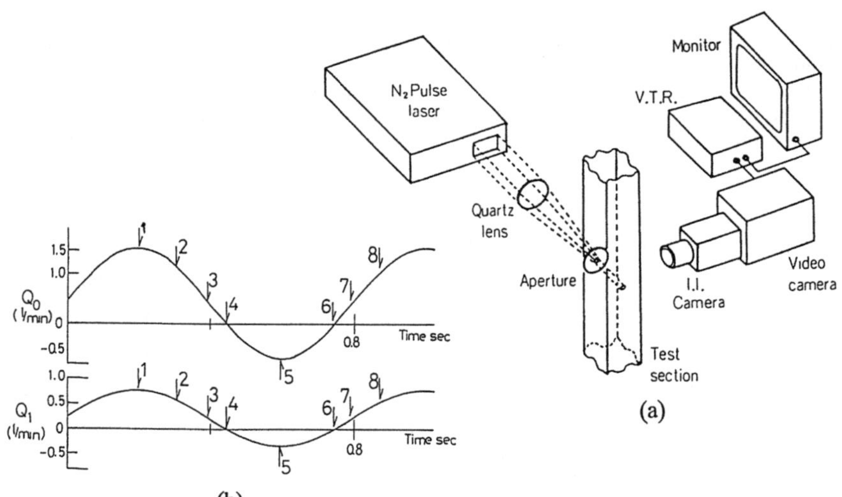

Fig. 2 Schematic diagram of (a) the apparatus for visualization and measurement of flow velocity distribution by the LIF method and (b) volumetric flow wave forms whose frequencies are 1.25 Hz. [9, 10]

Volumetric flow waveforms which are to be measured by the LIF are also shown in this figure. Where Q_0 is the flow rate in the parent tube and Q_1 is that in one of the daughter tubes. A target to be measured is a pulsating laminar water flow through a 45–degree branch tube of 11.5mm x 11.5mm square cross section. Some results of detailed velocity profiles measurement in a branched region are shown in Fig. 3. The mean Reynolds number is 290. At the peak of Q_0 a flow separation takes place in the outer wall region of the branch 2. During the deceleration of the flow (Phase 2 to 5), another flow separation also takes place at the outer wall of the branch 1. Eddies from these separated flow region gradually spread to the upperstream, and velocity profile at the upper stream from the branch are strongly disturbed. In the acceleration phase, such eddies are washed away to the downstream, and eddies and flow separation almost vanish.

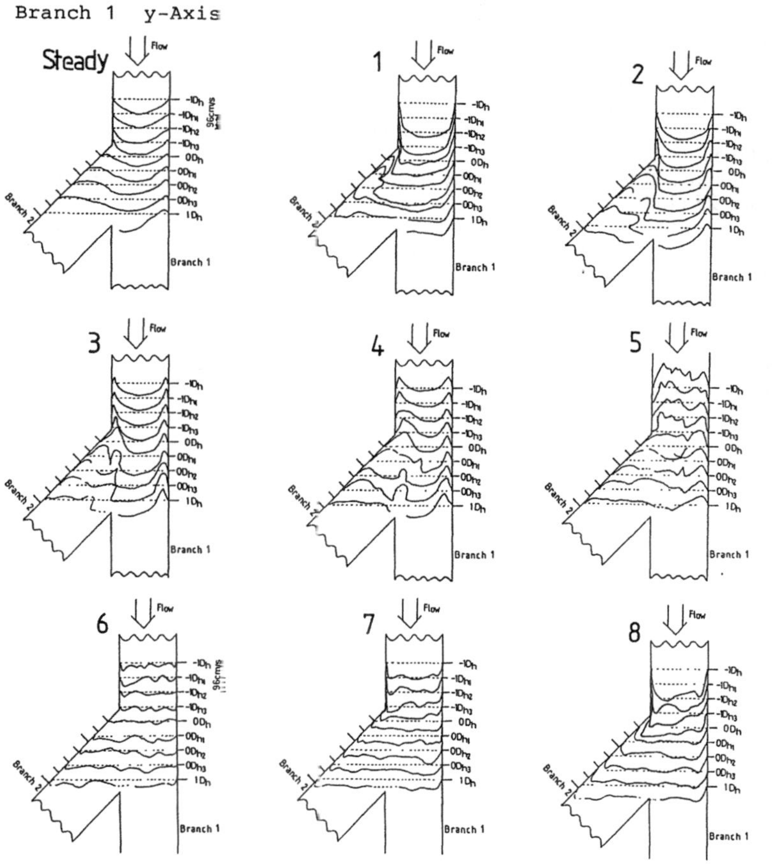

Fig. 3 Detailed velocity profiles in a pulsatile flow through a branched square tube. [10]

Important Remarks

Although a laser technique to visualize velocity profile is a powerful tool to get the quantitative information of the detailed velocity field over the whole flow field, it is not applicable to blood flow itself *in vivo*. So, the significance of any model blood flow experiments should be assessed.

References

1. Ku, D.N.; Giddens, D.P.: Laser Doppler anemometry measurements of pulsatile flow in a model carotid bifurcation. J. Biomech. **20** (1987) 407–431.
2. Ahmed, S.; Giddens, D.P.: Pulsatile poststenotic flow studies with laser Doppler anemometry. J. Biomech. **17** (1984) 695–705.
3. Yoganathan, A.P; Corcoran, W.H.; Harrison, E.C.: In vivo velocimetry measurements in the vicinity of aortic prostheses. J. Biomech. **12** (1979) 12 135–152.
4. Hanle, D.D; Harrison, E.C.; Yoganathan, A.P.; Corcoran, W.H.: Turbulence downstream from the Ionescu–Shiley bioprosthesis in steady and pulsatile flow. Med. Biol. Engng & Comput. **25** (1987) 645–649.
5. Gaver, D.P.; Grotberg, J.B.: An experimental investigation of oscillatory flow in a tapered channel. J. Fluid Mech. **172** (1986) 47–61.
6. Geller, E.W.: An Electrochemical method of visualizing the boundary layer. J. Aeronaut. Sci. **22** (1955) 869.
7. Ho, F.; Hummel, R.L: Average velocity distributions within falling liquid films. Chem. Engng Sci. **25** (1970) 1225.
8. Nakatani, N. *et al.*: Measurement of flow velocity distributions by pulse luminescence method. J. Phys. E: Sci. Instrnum. **8** (1975) 1042.
9. Ohba, K.; Sato M.; Sakaguchi, S.: Role of blood flow in atherogenesis. (Ed. Yoshida *et al.*) Spriger–Verlag 1988.
10. Ohba, K.; Sato M.; Sakaguchi, S.; Sakurai, A.: Applications of laser techniques to fluid mechanics (Ed. Durao *et al.*) Springer–Verlag 1991.
11. Ojha, M.: Pulsatile flow through constricted tubes: an experimental investigation using photochromic tracer methods. **203** (1987) 173–197

In-Vitro Flow Visualization of Mitral Valve Regurgitation: Laser Flow Visualization, Color Doppler Flow Mapping and Coated Albunex Particles

Edmond Rambod, Steven Khan, David Allen*, Shmuel Einav, Gerald Maurer, Jack Matloff
Cedars-Sinai Medical Center, Los Angeles, CA 90048, USA
UCLA-School of Engineering, Los Angeles, CA 90024, USA*

INTRODUCTION

The development of color Doppler echocardiography flow mapping (CDFM) represented a major technological breakthrough in the noninvasive investigation of flow fields in the heart. CDFM displays the spatial distribution of abnormal blood velocities in the heart and has been primarily used for evaluation of valvular regurgitation [1]. However, quantitation of regurgitation by CDFM remains difficult [2].

The failure of current analysis techniques may be due to the presence of vortices shed from the body of the jet and the three dimensional nature of the flow field in the heart. Theory suggests that these vortices can be responsible for a significant amount of energy dissipation in such jets. Clinically, these vortices may be of particular importance in understanding mitral regurgitation which represents a jet in a constrained chamber: the left atrium. However, these vortices create low velocity flow fields and may therefore not be visualized by standard CDFM methods.

In order to determine the role of these vortices in mitral regurgitation, we studied an in vitro model using laser light-sheet flow visualization and 2-*D* echo contrast using coated albunex particles. To evaluate the ability of color Doppler flow mapping we performed in vitro experiments with different Doppler settings to visualize velocity field and echo contrast to visualize vortices.

EXPERIMENTAL APPARATUS AND METHODOLOGY

IN VITRO MODEL

The flow imaging and CDFM studies were performed in the pulse duplicator system shown in Figure 1. The experiments were conducted in symmetric ventricular and atrial chambers constructed of lucite. The inner diameter of both chambers was 5.08 cm. Circular plates with central orifices constructed of lucite were mounted between the ventricular and atrial chambers. The lucite walls were transparent to allow photographic accessibility. Flow imaging studies can be performed throughout the atrium, including the orifice plane.

Flow was generated by a pulsatile blood pump with adjustable stroke volume, heart rate and systolic interval using a blood analog solution (glycerol-cornstarch adjusted to a bulk viscosity of 3.5 *cp*). The circulation was arranged as follows: the ventricular chamber had no aortic side and the pump drew fluid directly from the reservoir, via ventricular section and atrial chamber, and recirculate it through outgoing tubing (Figure 1). Atrial pressure was maintained at 10 mmHg throughout the experiments and a constant pressure gradient of 100 mmHg was maintained across the orifices.

Although pulse duplicators frequently produce water hammer effects that result in unphysiologic pressure spikes, the careful choice of tubing materials and judicious placement of a variable compliance chamber minimized these pressure spikes in our system. Regurgitant volumes across the orifices were created by adjusting the pump rates from 5 to 40 RPM and the stroke volumes from 5 to 15 *cc*. The ejection time was maintained at 30% of each cycle.

LASER FLOW VISUALIZATION

For laser imaging, an atrial chamber constructed of clear crystal glass with a hemispherical back wall was used. Flow visualization studies were performed by creating a planar light sheet in the model atrium and videotaping the trajectories of particles in the light sheet. The light sheet was generated by passing the output of an 8mW He-*Ne* laser through several lenses. The laser beam was diverged and directed through a cylindrical lens, creating laser light sheet of approximately 1.5 mm thick which could illuminate an entire cross section of the model atrium, (Figure 2). The width of the plane of light can be controlled by two variables: the lens diameter and the distance between the lens and the flow section. As the diameter of the lens decreases, the angle of divergence increases, permitting a closer distance between the laser and the flow section for a desired width of the plane of light. The intensity of the light is inversely proportional to its width, (Figure 2). In addition, this setup could be adjusted so that the sheet of light could pass through various cross sections of the flow channel, thereby facilitating flow visualization at different cross sections. Such cross sectional variations are useful in studying any asymmetric recirculating regions in the flow field. A mixture of alumina and glass beads of 3-50 micron was added to the fluid to act as tracer particles to observe the flow patterns downstream of the orifice. The reflective particles were illuminated by the laser light, and the flow patterns were photographed in real time with a Sony Hi8-video camera connected to a VCR, filming at 30 frames per second against a black background. Figure 3 demonstrate typical still photographs of the flow fields. Large scale flow structures, particularly vortices are readily apparent.

2-*D* COLOR DOPPLER FLOW MAPPING

CDFM was performed with an Aloka-870 color Doppler scanner using a 3.5-MHz phased array transducer placed on the back wall of the model left atrium opposite the inflow orifice. The echo sector was aimed at the inflow orifice to image the central plane of the flow jet. Images were obtained at a color gain setting just below that which produced noise in the color Doppler image, and this setting was kept constant throughout the study. A fixed pulse repetition frequency of 9 KHz was used, yielding a Nyquist limit of 1.2 m/s. Two regurgitant orifices were studied, 3.4 and 7.9 mm^2. Two different velocity filter settings were used for jet area measurements. The CD-clinical settings represented typical clinical filter settings which remove low velocity components of the jet. The CD-low settings allowed visualization of all velocity components including the low velocities. Figures 4,a,b illustrate typical flow jet images for CD-clinical and CD-low settings, respectively, for the 3.4 mm^2 orifice. Color Doppler images consist adjacent colors which represent different velocity distributions in the flow field including the core of the jet and the areas of recirculation or vortices at the edge of the jet. Images were obtained in one specific frame for each cycle and recorded on a 1/2-in videotape for further analysis.

2-*D* ECHO CONTRAST FLOW MAPPING

The flow jets were also mapped with 2-*D* echo cardiography using albumen coated air microspheres (Albunex particles) as an echo contrast agent. These microspheres are echogenic and are made in a sonication process in a sterile non-pyrogenic 5 percent solution of albumen. In the process, the sonication horn oscillates at 20,000 Hertz causing cavitation within the solution and the formation of air-filled microspheres, with a mean diameter of 3-5 microns. Figure 5 illustrates a typical jet mapping with 2-*D* echo contrast.

DATA MEASUREMENTS

Pulsatile flow rates ranging from 0.2 to 0.6 liters per minute were studied using two orifice sizes, 3.4 and 7.9 mm^2. The calculated orifice exit velocities at peak systole ranged from 0.98 to 2.94 m/s and from 0.42 to 1.27 m/s, respectively. Regurgitant jet velocities were calculated using the calibrated pump output (in cubic centimeters per minute) and the known cross sectional area of the inflow orifice (in square centimeters). The calculated velocities were then cross-checked against regurgitant flow velocity measurements by continous-wave Doppler. Reynolds numbers, based on orifice velocities obtained by pulsed wave Doppler immediately downstream of the orifices, ranged from 600 to 1800 for the 3.4 mm^2 orifice and from 440 to 1330 for the 7.9 mm^2 orifice. Resultant jet flow areas for each velocity filter setting (Figure 4) were measured using the built-in Aloka software. Jet area values were expressed as $Value+SD$ (in square centimeters) and regurgitant flow rates were expressed in liters per minute.

For each orifice in each condition, jet areas were measured and averaged in two frames. These two frames were chosen according to the largest color Doppler jet area as imaged in the velocity-variance mode. Intraobserver variabilities for jet area measurements were less than 6% for both CD-clinical settings and 2-*D* echo contrast and within 10% for CD-low settings due to the aliasing effect of low velocities that influenced the jet area measurements.

RESULTS

Figures 6 compares the color Doppler and echo contrast measured jet areas for the 3.4 mm^2 orifice. As revealed, CD-low jet areas are the largest followed by the CD-clinical jet areas. The echo contrast areas are consistently smaller than the color Doppler jet areas. Similar behavior of jet areas was observed for the 7.9 mm^2 orifice.

The jet areas, as illustrated, initially increase with increasing flow rates and then decrease at higher flow rates. These results demonstrate that the jet area is not linearly related to the regurgitant flow rate. In a confined test chamber, the laser flow visualization images are critical in understanding the biphasic response of CDFM areas. Laser flow visualization revealed the formation of large scale vortices. These vortices persisted from beat to beat under appropriate conditions indicating energy storage. As flow rates increased, the vortices could be seen to, progressively, narrow the central jet. Thus, the laser flow visualization studies demonstrate that the biphasic response of CDFM jet area is due to competition of recirculating vortices with central jet in a confined chamber.

CONCLUSIONS

CD-low jet areas are larger than the CD-clinical jet areas due to the inclusion of lower velocity components. The echo contrast areas are consistently smaller than the color Doppler areas due to the fact that echo contrast images represent only the flow entering the left atrium from the left ventricle. Color Doppler jet areas are larger than echo contrast areas because thay also visualize left atrial fluid entrained by the central jet. The biphasic characteristics of jet areas with increasing flow rates can be explained by competition between the vortices and the central core of the jet for the limited space in the left atrium, as demonstrated by laser flow visualization. As flow rates are increased, larger amounts of energy is stored in these vortices and progressively narrow the central jet.

REFERENCES

1. Sweitzer, D.F.; Yoganathan, A.P.; Nanda, N.C.; Woo, Y.R.; Ridgway. A.J.: Calibration of color Doppler flow mapping during extreme hemodynamic conditions in vitro: *A* foundation for a reliable quantitative grading system for aortic incompetence. Circulation 75 (1987) 837-846.
2. Baumgartner, H.; Khan, S.; DeRobertis, M.; Czer, L.; Maurer, G.: Color Doppler regurgitant characteristics of normal mechanical mitral valve prosthesis in vitro. Circulation 85 (1992), 323-332.

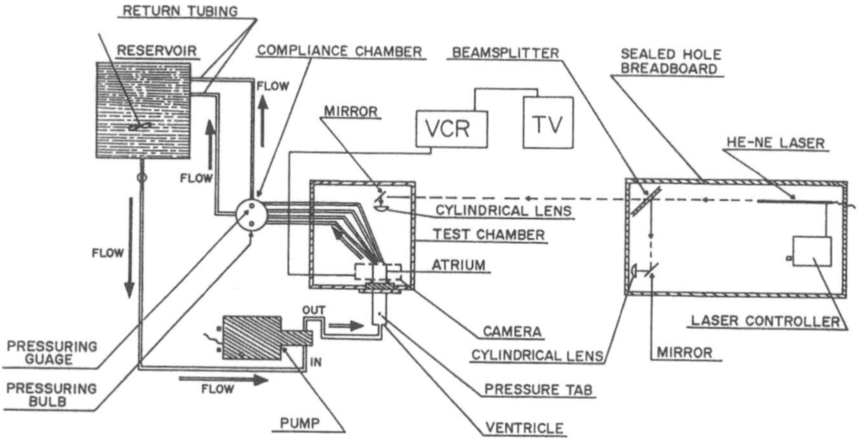

CARDIOVASCULAR FLOW IMAGING SYSTEM

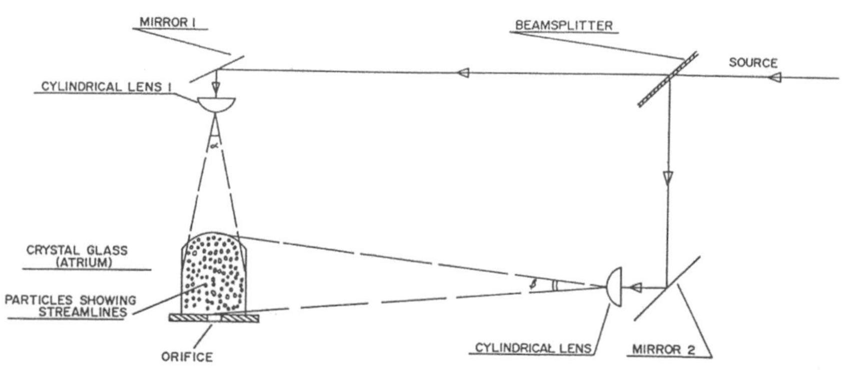

L. ATRIUM AND THE STRUCTURE OF LASER LIGHT SHEETS

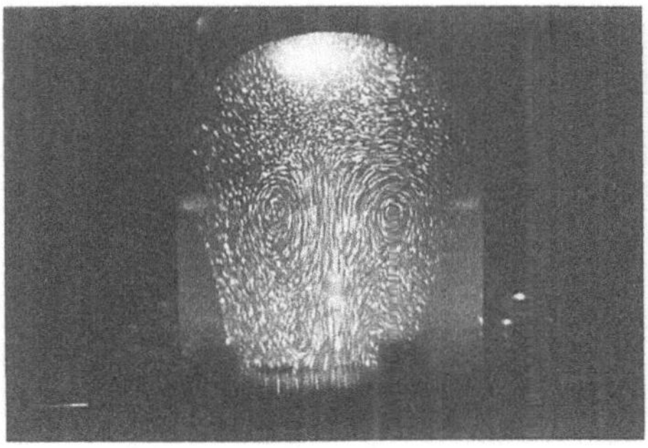

Figure 3: Typical Laser Flow Image Including Vortex Shedding

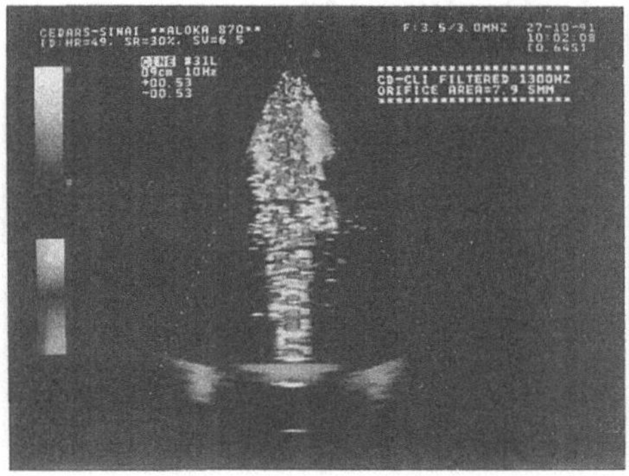

Figure 4a: Typical CDFM for CD-clinical Settings

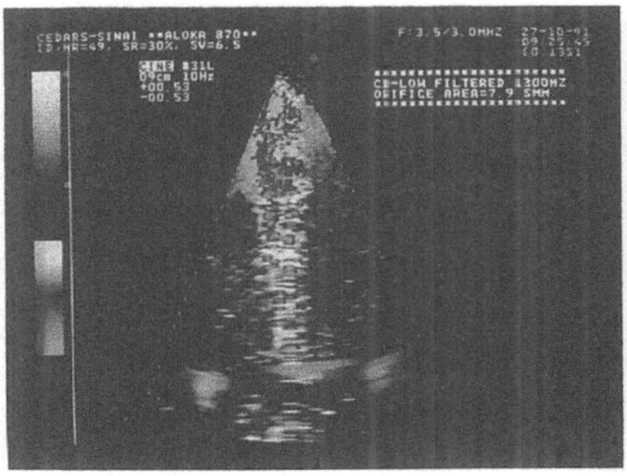

Figure 4b: Typical CDFM for CD-low Settings

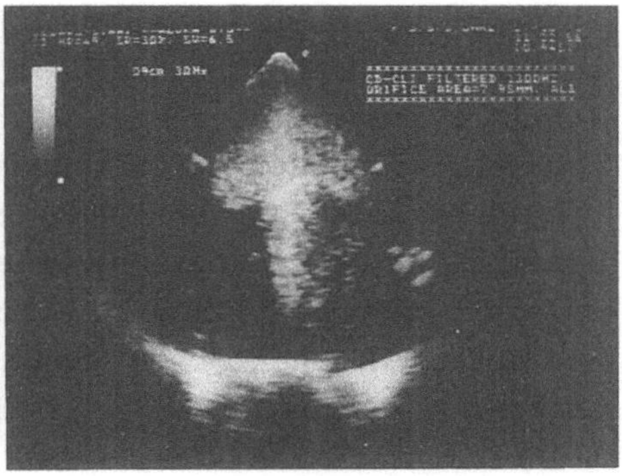

Figure 5: Typical Echo Contrast Flow Mapping

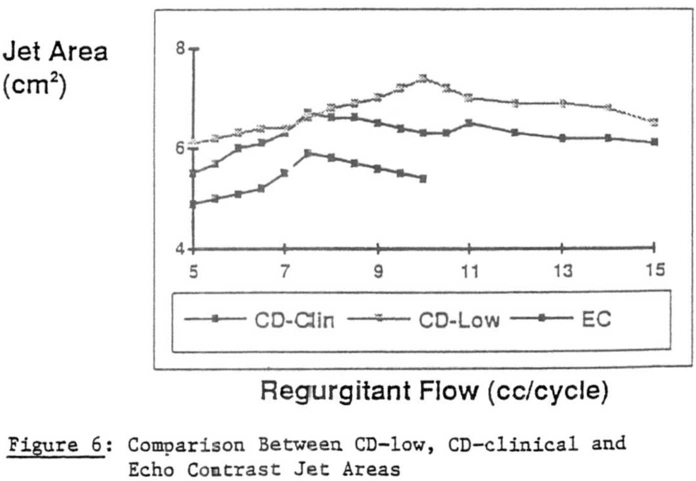

Figure 6: Comparison Between CD-low, CD-clinical and Echo Contrast Jet Areas

Holographic Diagnostics of Transfer Processes in Capillary Flows

O.G. LYSENKO and N.A. KARBALEVICH

Heat and Mass Transfer Institute,
Belarus Academy of Sciences, Minsk, Belarus

Summary

The results of use of the holographic interferometry for measuring of surface tension, diffusion coefficients, thermal diffusivity of liquid media in capillary are discussed.

Study of various medical and biological microobjects and processes (blood erythracites, protein diffusion) under the conditions ultimately close to natural ones is a highly urgent problem. The behaviour of these processes is much affected both by the material properties of the walls of the vessels that contain the media under study and by the surface characteristics of these media. Solving the problem stated requires development and use of contactless methods, in particular, of optical methods allowing total measurements of the main parameters of transfer processes accompanied by simultaneous control of structural, geometrical, and surface characteristics of capillaries.

A modified holographic microinterferometry-based method [1] was used to study transfer processes in liquid media. In infinite and finite fringes, records were made of a hologram of a liquid meniscus in a capillary placed into a specially arranged thin-wall cell filled with an immersion liquid. The specific features of an optical scheme enabled one to measure wetting angles and surface tension coefficients at the boundaries: solid body (capillary) - liquid-gas and solid body-liquid-liquid. A choice of liquids to be studied was determined by their wide use in organic synthesis and as blood substitutes.

The interface was visualized on the reconstructed hologram (Fig. 1). A wetting angle was calculated by the relation $\cos\Theta = 2m/(m^2+1)$ where $m = r/d$, r is the capillary radius and d is the meniscus depth. The obtained experimental findings have shown that the equilibrium wetting angles of n-alkanes grow with increasing the ordinal number of hydrocarbon.

A capillary rise method is most exact for measurements of surface tension of liquid media.

The holographic methods were adopted to measure interphase characteristics of liquid media: wetting angle and interphase tension of saturated hydrocarbons at the air (Fig. 2) and water (Fig. 3) interface. A value of an angle θ_{12} depends on a length, l, of a hydrocarbon column, time contact between the liquid and capillary, and on liquid viscosity. For fixed values, the angle at the interphase boundary grows with increasing the number of carbon atoms in a molecule, this growth being in a homological series. A dynamic wetting angle vs hydrocarbon surface tension is a monotonic function and is approximated by the relation $\cos\theta = 1 - a(\sigma - \sigma_c)$ where a constant a = 0.03. By extrapolating to the zero wetting angle the values of σ_c are obtained which characterize the solid surface properties. This enables one to analyze the wetting data and to determine wetting angles for different components of the interphase contact. An absolute measuring error of these angles does not exceed 0.5%. more pronounced than for the n-alkane-air boundary. The measured values of θ_{12} enabled one to calculate the surface tension at the interphase boundary by using the relation $\sigma_{12} = (rg(\rho_1 l_1 + \rho_2 l_2) - 2\sigma_1 \cos\theta_1)/2\cos\theta_{12}$. A difference between the experimental and tabular data did not exceed 0.3%. The advantage of the holographic methods is that during one experiment, it is possible, with a high accuracy, to measure all parameters required to calculate σ and σ_{12} and to control a capillary inner diameter and wall thickness along the entire length [1].

To correctly describe the capillary rise it is necessary to allow for the wetting angle relaxation effect on an impregnation rate. This factor is also of importance for studying the kinetics of displacing one liquid by another. Obviously, developing the methods to measure a dynamic wetting angle at the liquid--gas and liquid-liquid boundaries is an urgent problem.

A dynamic wetting angle vs meniscus rate at capillary impregnation is found. Over the impregnation rate range 10^{-5} to 10^{-2} cm/s, the dynamic angle weakly depends on its value and over the range 10^{-3} to 10^{-1} cm/s this dependence becomes exponential for n-alkanes.

Study of the displacement (saturated hydrocarbons and water) has shown that a displacement rate depends on a column length of n-alkane, its viscosity, and a capillary diameter.

The used experimental method enabled one also to study the wetting angle relaxation at a moving boundary of two liquids over the range 10^{-1} - 10^{-2} cm/s.

In infinite and finite fringes, the two-exposure method was u used to measure diffusion coefficients of saturated hydrocarbons in benzene (Fig. 4).

Experimental values well agree with those predicted by Vilke- - Chang 's equation [2]. The thus found diffusion coefficients of saturated hydrocarbons in benzene well lie on a straight line in the coordinates $\ln D - \ln \mu$. Determination of the diffusion coefficients of different terms of the homologic series and analysis of these coefficients vs molecular mass characterize the mobility of dissolved macromolecules and make it possible to calculate their hydrodynamic and geometric characteristics.

Thermal diffusivities of saturated hydrocarbons in capillaries measured by the proposed methods at pulse and continuous thermal effects well agree with those cited in [3]. Making such measurements is important in experiments outside the laboratory conditions, e.g., at space stations [4].

Thus, the modified method of holographic microinterferometry enables one, with a high accuracy, to measure surface tension and diffusion coefficients, thermal diffusivity of liquid media. It may be used for investigating biological solutions and different protein components under the conditions close to the natural ones (in capillaries and in glass tubes), for solving problems on designing capillaries and blood vessel grafts with assigned properties. Results for the processes to proceed in single capillaries may be also adopted while designing capillary evaporators, pumps, and heat pipes, whose operation principle is based on liquid motion in capillary-porous structures.

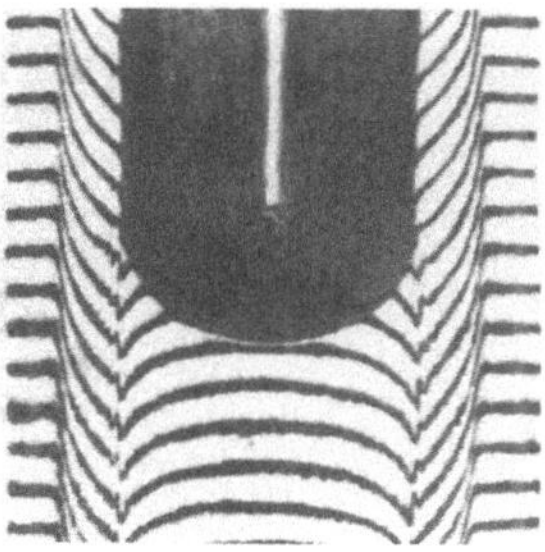

Fig. 1

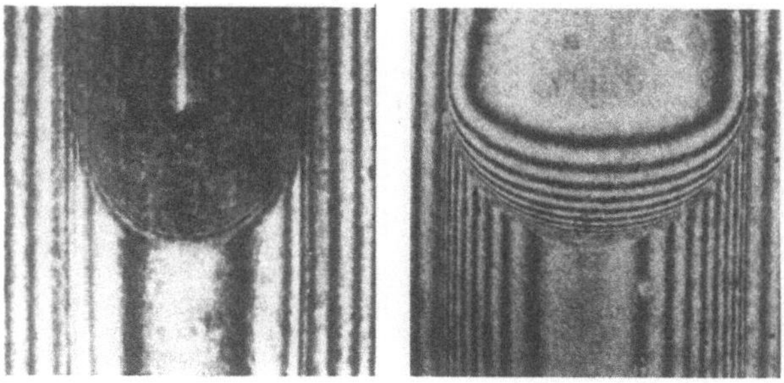

Fig. 2. Liquid-gas interface

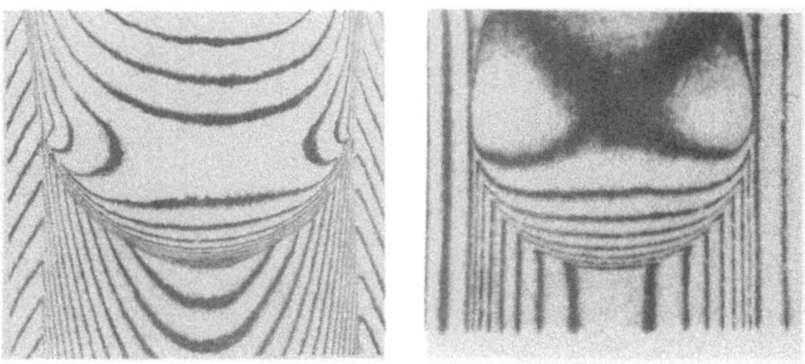

Fig. 3. Interface liquid-liquid

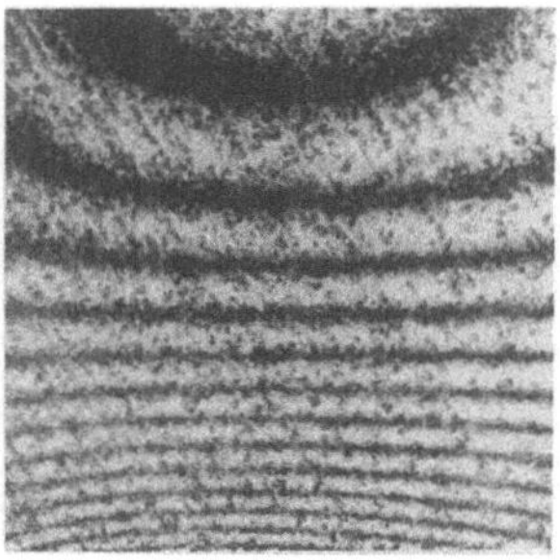

Fig. 4

REFERENCES

1. Lysenko O.G. Holographic diagnostics of capillary fluid flow//Proc. 5th Int. Symp. on Flow Visualization, 1989, Prague, pp. 167-169.

2. Reid R., Prausnicz J., Sherwood T. "The Properties of Gases and Liquids", Leningrad, Cham.Press, 1972.

3. Vargaftik N.B. Handboock on thermophysical properties of gases and liquids. Moscow, Nauka Press, 1972.

4. Gangerly N.M., Gurevich S.B. Study of kinetics of processes by means of compact holographic equipment / Holography: theoretical and applied problems, Leningrad, 1988, pp. 108--124.

Jets and Turbulence

Jets

Subsonic Jet Visualization

T.M.MAKARENKO

Central Aerohydrodynamic Institute (TsAGI), Acoustic Division
Radio street 17, 107005 Moscow, Russia

T.U.VOLNOVA, T.N.BEZMENOVA, V.I.RIBAKOV

Central Aerohydrodynamic Institute (TsAGI)

Summary

The visualization of the jet flowing out of a round nozzle with the diameter d = 20 mm at the velocity Uo = 20 m/s has been conducted with the aid of a shear interferometer connected with photo- and cinema-cameras. Transversal acoustic excitation influence on a coherent structure type (at Sh_S=0.125-2.5) was considered. Their qualitative and quantitative characteristics were obtained. The laser knife application permitted to obtain vortex generation structures in different jet sections.

Introduction

Recently there have been published many papers devoted to coherent structure investigations. It can be accounted for by their influence on the dynamics of turbulent flow development and the noise generation process. Knowing the contribution of these structures to noise generation one can evaluate noise reduction possibilities by effecting these structures.

Detection of basic coherent structure parameters is conducted by either a direct method or an indirect method of experimental investigations. The direct periodicity detection method is to single out the most probable oscillation either as the result of visual examination or with the aid of spectral or correlational measurements of velocity and pressure pulsation in the jet. The indirect method comes to periodic jet excitation and detection of jet responses to excitations at different frequencies. With the natural tendency for degree of order the periodic excitation may amplify regular structure developing against a background of small - scale turbulence. Measurements [1] of space - time correlation coefficients of

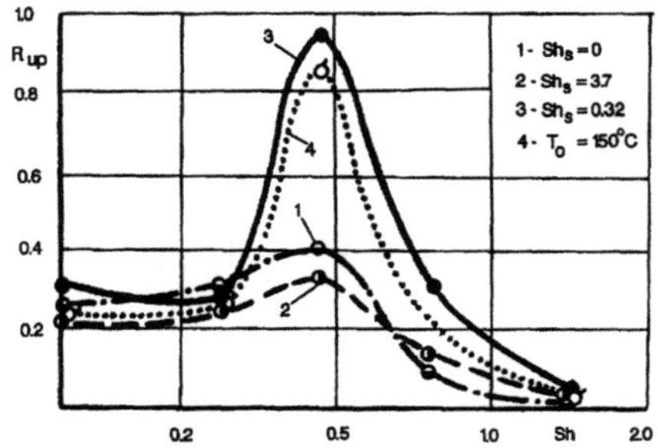

Fig.1 Correlation cofficient of velocity and pressure pulsations in the near field.

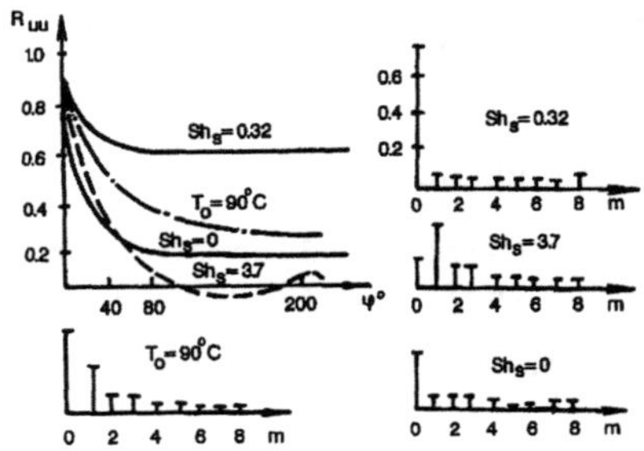

Fig.2 Azimutal correlation coefficients of velocity pulsations at different Sh_S

velocity pulsations in the jet core and of pressure pulsations in the near field showed a close relation of pressure pulsations to coherent structures, low - frequency acoustic excitation strengthening this relation. Heating provides the similar effect (Fig.1). We can make a conclusion about the shape of coherent structures analysing the measured cofficients of velocity pulsation azimuthal correlation (Fig.2). In Fourier cosine expansion the zero mode (m=0) corresponds to large toroidal vortex structures. The appearance of the higher order modes is related to the violation of axial structure symmetry. Low-frequency excitation (Sh_S=0.32) results in the considerable increase of the zero mode contribution and suppression of all other modes. High-frequency excitation (Sh_S= 3.7) results in decreasing of the zero mode share considerably increasing first mode contribution, and increasing a bit the share of the higher order modes corresponding to small - scale turbulence. Heating results in same increasing of contribution of the zero and first modes, that is similarly to the case of low-frequency jet excitation, it results in large-scale structure stabilization.

Experimental Arrangement

The investigation was carried out on the aerodynamic installation which had a round nozzle with the diameter d=20 mm with compression degree 400 and generator length 135 mm at the jet outflow velocity Uo = 20 m/s (M=0.06, Re=$2*10^4$), and the jet temperature at the nozzle exit t=45 C. Heating was used to

improve the conditions of visualization.
It has been already noticed that an acoustic excitation amplifies coherent structures so the acoustic excitation influence on coherent structure types at $Sh_S(f*d/Uo)$ = 0.125-2.5 was considered. The excitation was accomplished with the aid of two loudspeakers symmetrically distributed about the jet axis, at 1.5d distance therefrom. The loudspeakers' axes were normal to the jet axis , those centers being at three diameter distance from the nozzle exit plane. Three regimes of the loudspeaker functioning were examined : an electric signal was supplied to one of the dynamics, to both dynamics in phase and out of phase. In this case noise pressure levels near the nozzle section were L = 103 - 108 db. .

In the current study an experimental shear interferometer connected to photo- and cinema-cameras was used for visualization. This device has some advantages in comparision with other devices for flow visualization. The main advantage is the sensitivity increase due to two-fold passage of beam through non-uniformity. The device can function with mirrors of different diameters (250-1000 mm),i.e. it can change the visualization field size. High-speed pictures were done by a camera "PENTAZET-35" with 2000 pictures per second using an impulse light source (flash energy 4.6 joule,flash duration 3-5 mcsec).

Results and Discussion

Some photographs obtained during the work distinctly show that toroidal structures exist in a jet and they can be clearly observed behind the nozzle exit section across about 4 diameter's lengh starting at the distance of l = 0.5d (Fig.3). l - decreases under acoustic excitation (Fig. 4 - 6). It is worth noticing that when we use another device as well as in the works of other researchers [2, 3, 4] and in the case of shadow photography this distance was 1.5d. It is probably connected with outflow conditions.Low-frequency excitation,at the characteristic frequency of coherent structures in particular,results in their stabilization (Fig.4,5). At high frequency excitation the coherent structures are destroyed more rapidly (Fig.6) which agrees with the research data obtained by other ways.

In the study [1] on the basis of an average velosity measurement in a fixed point on the jet axis (x/d = 8) it was established that in the case of low - frequency (Sh_s= var) transversal jet excitation at the same frequency by two loudspeakers arranged symmetricly with regard to jet axis, exitation effect strongly depends on the phase shift between both acoustic signals . With acoustic signals being 180° out of phase the mixing intensification effect was the highest one (Sh_s=0.1 - 0.25). The visual investigation confirms supports this result (Fig.7).

The photographs give an idea about the jet structure in different separately considered moments of time. High - speed pictures permit to follow the successive development of vortex structures. One can also obtain quantative characteristics of the structures: geometrical vortex sizes and the convective velocity thereof as well. The cinema-frame analysis showed that

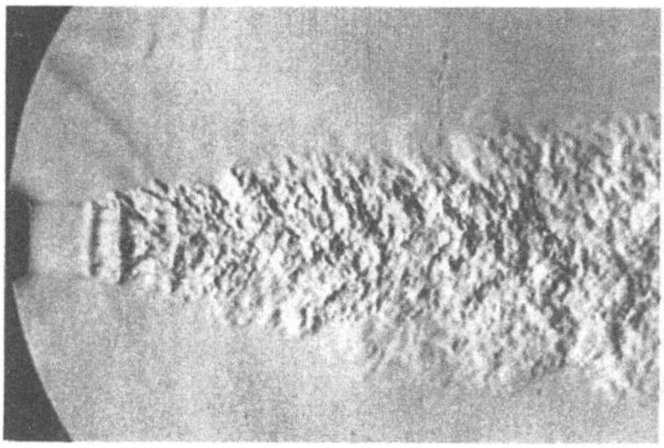

Fig.3 Photo of unexcited jet

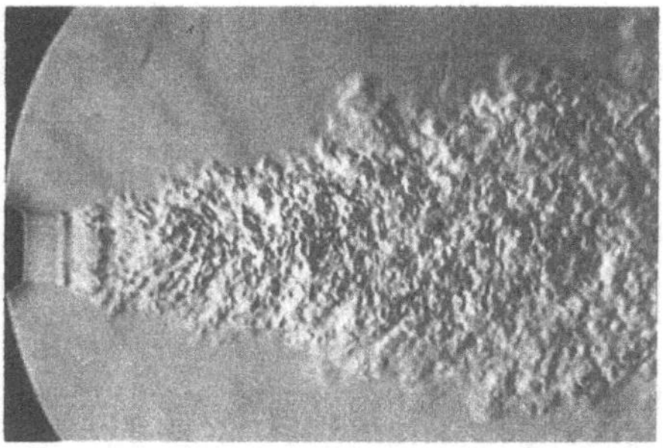

Fig.4 Sh_s=0.125 (1 loudsp.)

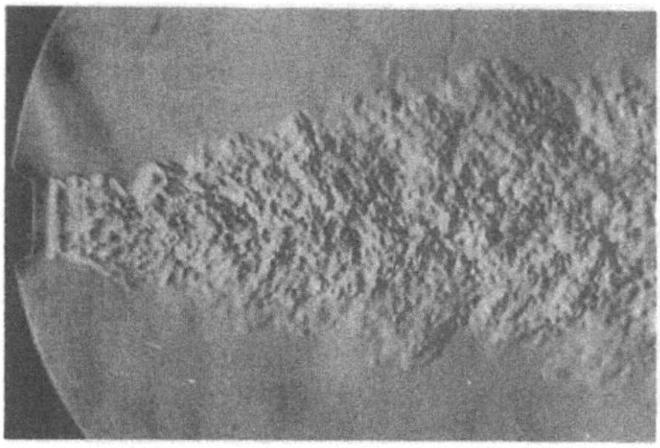

Fig.5 Sh_s=0.425 (1 loudsp.)

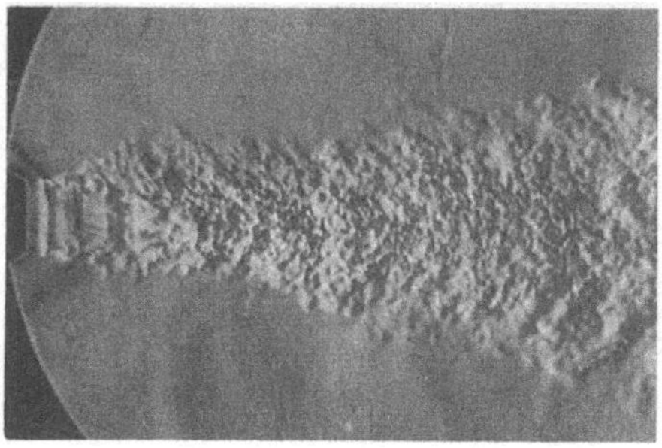

Fig.6 Sh_s=1.25 (1 loudsp.)

acoustic excitation influenced all the characteristics of large-scale structures. At low-frequence exitation the vortex thickness increases, and at high - frequence excitation it decreases. The convection velocity was analysed from the vortex center shear during the frame change. Maximum convection velocity equal to 0.58Uo, corresponds to characteristic frequency of coherent structures.

The laser knife application permitted to obtain vortex generation structure in different jet sections. In this case light knife thickness was 2 mm. Nitrogen vapours were inserted into the air jet.

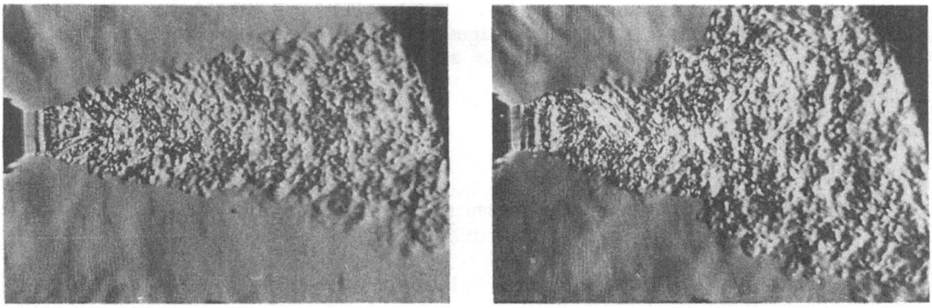

Fig.7 Photograph of excited jet (2 loudspeakers)
a) $Sh_S=0.125$ ($\Delta\varphi_{ac}=0$) b) $Sh_S=0.125$ ($\Delta\varphi_{ac}=180^{\circ}$)

Conclusions

Thus it is found that visualization helps to realise better the physics of process, confirms the results gained by other methods and substantially expands possibilities of achieving qualitative and quantitive characteristics of coherent sruc - tures.

References

1. Vlasov E.V., Ginevskey A.S., Karavosov R.K.: Coherent struc. in immersed jets and wakes. In: Papers of the 7 conference on aero- acoustics. Moscow, TsAGI, 1981, pp.61-62 .
2. Browand F.K., Laufer J. :The role of large scale structures in the initial development of circular jets.In:Turbulence in liquids. Proc.4 Biennial Symp. Turbulence Liquids. Sep., 1975 pp.333-334, Princeton, New Jersey, 1977.
3. Hussain A.K.M.F., Clarc A.R. :On the coherent structure of the axisymmetric mixing layer: a flow visualization study. -J.Fluid Mech., 1981, 104, pp.263-294.
4. Yule A.J. : Large-scale structure in the mixing layer of a round jet. -J.Fluid mech.,1978, 89, pp.413-432.

Visualizing and Charactering Free Jets Flow with Laser Velocimetries

A. Cenedese *, G. Doglia *, G.P.Romano *, G. Tanzini **

* Department of Mechanics and Aeronautics
University of Roma "*La Sapienza*"
Via Eudossiana 18 - 00184 Roma - ITALY
Phone: +39/6/44285218 Fax: +39/6/4881759

** CRTN - ENEL
Via A. Pisano 120 - Pisa - ITALY

1. Introduction

During last years the use of whole field velocimetry techniques in fluid flows overlays the limits of single point measurements (HWA, HFA, LDA etc.). The enhancements in image recording and processing gave advantages to optical over classical methods. If the interest is focalized on velocity measurements, several new methods can be employed:

- Holographic Interferometry [1];
- Doppler Global Velocimetry (DGV) [2];
- Particle Image Velocimetry (PIV) [3,4].

For temperature, density and species measurements also Liquid Crystals and methods based on molecular inelastic scattering (Raman Spettroscopy, Laser Induced Fluorescence (LIF) etc.) are used.

Among these optical methods, the PIV technique allows a simpler acquisition of images and a wide range of approaches for processing and recognizing the particles trajectories in the investigated region; it still suffers, however, of a higher complexity in measuring all three the velocity components. Whenever small seeding density are present, as frequently in air flows, the direct investigation of images [5] takes advantages over the usual Young fringes investigation [3]: it follows each trajectory of the seedi ng particles either using time markers [6] or validation criteria for the individuation of successive positions [5]. Moreover,using a videocamera, it allows a simple on-line acquisition of images [7].

In complex turbulent flows the PIV must overcome several problems:

- spatial extension of the phenomenon;
- temporal evolution of such phenomenon;
- velocity gradients and turbulence intensities.

Wide spatial and temporal ranges must be investigated for a detailed analysis of the flow configuration to be performed; this can be accomplished by using high energy light source, microscopes and high resolution cameras (up to 0.05 mm [7] as with LDA); furthermore the use of videocameras connected to videorecorders allows a fine temporal resolution (up to 10^{-3} s [8] comparable with single point techniques). High velocity gradients and turbulent intensities are severe tests for all image analysis methods and particulary for the validation criteria used for recognizing the real trajectories.

Beyond different turbulent flows, free jets must be investigated for both theirs basic properties and relevance as test cases: the former contribuiting to the understanding of complex fluid-dynamics phenomena for optimal designing of efficient burners, the latter for evaluating the limits of the measurement technique.

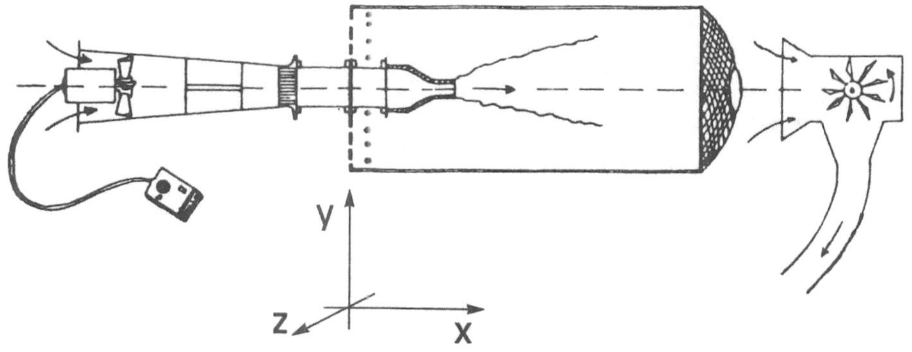

Figure 1: Experimental set-up and reference system.

2. Set up

A subsonic open loop wind tunnel, equipped with a motor fan with variable velocity (from 1 m/s to 40 m/s), is used for generating the turbulent jets. The air is captured onto the fan, stabilized in a square duct and then converged towards the outlet circular section (max. diameter $\simeq$ 3.6 cm) (figure 1). Two geometrical configurations were investigated: a circular and an anular jet, the latter beeing obtained by inserting, in the centre of the converger out section, a cone (max. diameter $\simeq$ 2.8 cm) supported at the wall with thin wires. These configurations are used for simulating the inner section of a burner.

The test section (from the outlet up to an axial distance of 21 cm) is illuminated with a light sheet, obtained from the beam of a pulsed *Nd-Yag* laser (max. power 300 mJ/ 50 ns, wavelenght 514.5 nm) crossing spherical and cilindrical lenses; one single pulse (mean duration $\simeq$ 1 ns) or three consecutive pulses (mean duration $\simeq$ 100 ns) separated by 1 ms can be used for taking images of the seeding particles on the plane. These images are acquired with a videocamera, to perform visualizations, and with a photocamera on slides for quantitative measurements.

The image analysis is performed by recognizing each particle and its trajectory in magnifyied parts of the total image; a microscope (max. enlargment 16 times), equipped with a CCD device connected to a frame grabber (512 * 512 pixels with 256 grey levels), is used for this purpose. Step motors (minimum displacement $\simeq 10\mu m$) are used for examining different regions of the slide. The acquired image appears as a set of points ("spots") representing the illuminated seeding particles. A suitable software applyies validation criteria on the detected trajectories to avoid erroneous connections between spots; it consists of comparisons on areas, distances and mean direction of consecutive spots [5,7]. The investigated region is about 20 cm * 15 cm, resulting a spatial resolution less than 1 mm. About 40 images of the field are superimposed to obtain a higher statistical accuracy.

3. Experimental results

3.1. Visualizations

As previously mentioned, in both geometrical configurations, visualizations were performed as a first approach to the problems. The single pulse laser configuration is used for "*freezing*" the field at very small time intervals; the pulse frequency is about 10 Hz. Smoke is used as seeding. The outlet velocity ranges from 2 m/s to 50 m/s corresponding to Reynolds number from about 5000 to 100000 (taking the outlet diameter). In figure 2 the flow configuration at $Re \simeq 6000$ for the circular jet is shown: vortical simmetrical

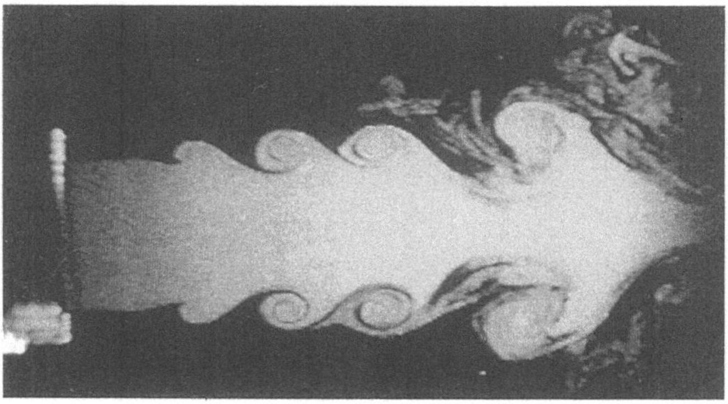

Figure 2: Flow visualization on circular jet ($Re \simeq 6000$); the real dimensions are about 1.5 times larger.

structures, whose dimensions increase with the distance from the outlet can be observed. Very close to the outlet, a region without vortices can be also noticed; this is the so called "self-preservation core" where the velocity profiles are scaled with axial and transverse coordinates [9]. It extends over two outlet diameter and is as small as higher the Reynolds number. Moreover an efficient mixing is obtained at higher Reynolds number.

The visualizations on anular jets show a higher asimmetry dependent on erroneous centering of the inner cone base; the mixing is more efficient than before.

3.2. Circular jet

The laser is used in the three pulses configuration. Flour particles are used for seed the flow. The reference system (figure 1) is taken with the X axis in the direction of the mean flow and the Y, Z axes on the transverse section. In figure 3 the original image with the relative spots (a) is shown at a velocity equal to 4 m/s ($Re \simeq 10000$). The vortical structures, already noticed in visualizations, are recognized: in the recirculating regions, however, there is a lack of measurements connected to the difficulty of detecting very small displacements. A rotating mirror overcomes this problem as well as the one of reverse velocities.

From such pictures velocities can be calculated using the time interval between successive laser pulses. To obtain measurements on a regular grid , interpolation algorithms must be employed. In figure 3 (b) the axial velocity, V_x along the maximum transverse diameter, Y direction, is shown at different axial distances from the outlet; an almost gaussian profile is obtained at all sections, scaling its vertical and horizontal extensions with the distance from the outlet. This can be better evidenciated by looking at the iso-velocity contours on the same plane (figure 3 (c)): the lines separates gradually, but the maximum velocity doesn't decrease considerably in the axial direction. Moreover the "self-preservation" region is shown to start after 2 times the outlet diameter as from visualizations.

3.3. Anular jet

In the same laser configuration, a jet in the anular geometrical configuration was investigated. In figure 4 (a) the detected spots from the original slide are shown (velocity equal to 3.5 m/s, $Re \simeq 10000$). The 3D velocity plot is represented in figure 4 (b): the velocity profile consists of two peaks merging a few centimeters after the outlet. The effect of mixing can be observed also from the iso-velocity contours (figure 4 (c)), where a lower simmetry than for the circular configuration can be noticed. Moreover the undisturbed velocity region is shown to be smaller; also in the axial direction the velocity decreases

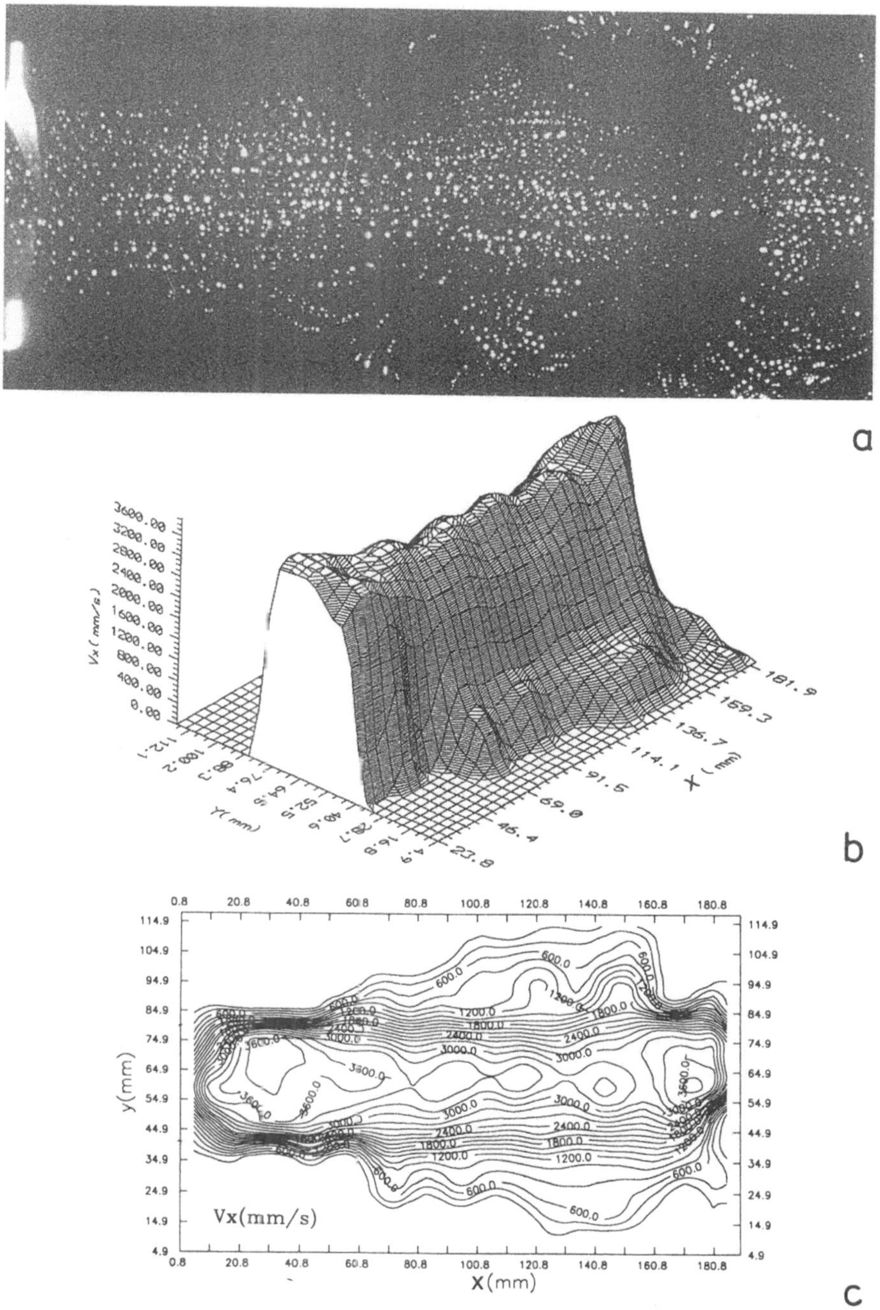

Figure 3: Acquired image with spots (a), axial velocities on plane (x, y) (b) and velocity contours (c): circular jet configuration with $Re \simeq 10000$.

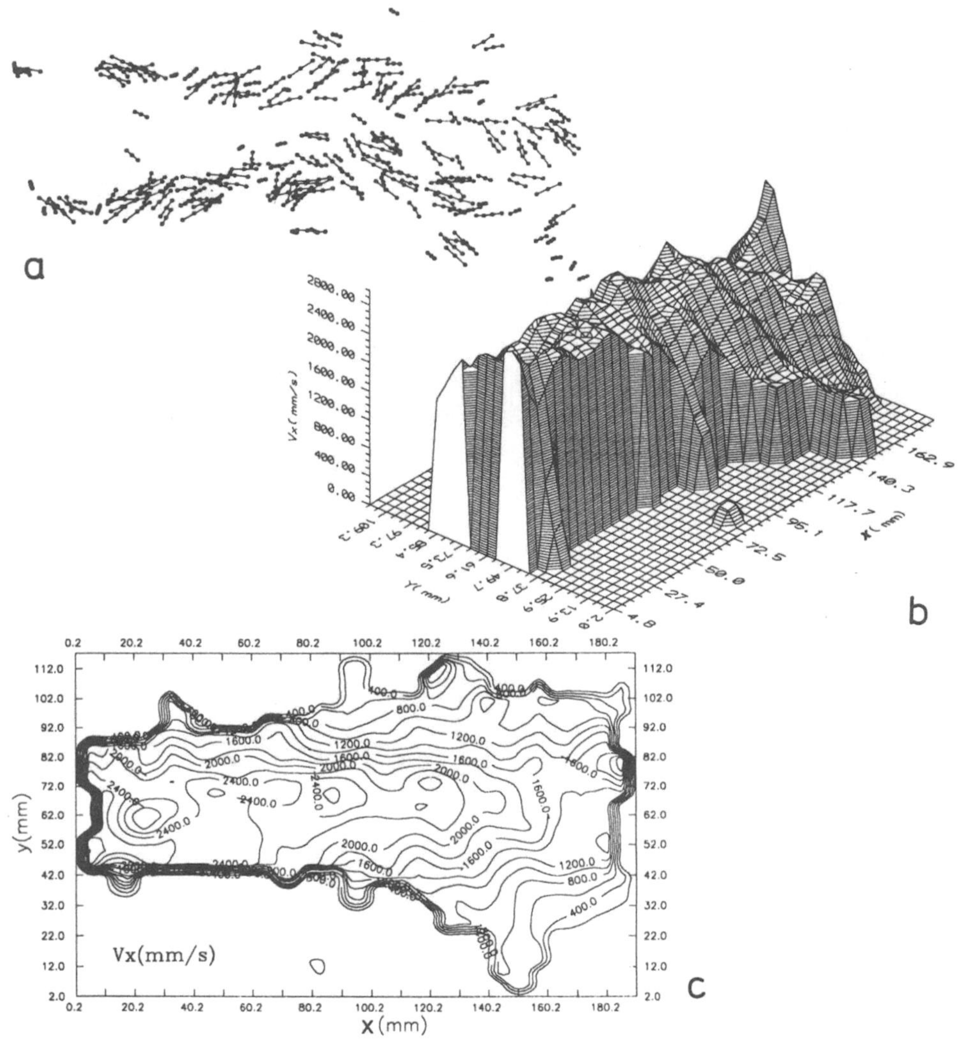

Figure 4: Detected spots (a), axial velocities on plane (x, y) (b) and velocity contours (c): anular jet configuration with $Re \simeq 10000$.

much faster than before, showing a more efficient mixing with the sorrounding flow at rest.

3.4. LDA-PIV comparison

In figure 5 the results from LDA and PIV measurements on the circular jet, at about 6 cm from the outlet, with an outlet velocity equal to 6 m/s ($Re \simeq 15000$) are shown. The velocity are normalized with the maximum velocity in that transverse section. The LDA data are obtained with a standard fiber optic system, in backscatter congiguration, connected to an *Ar-Ion* laser (volume dimensions 0.13 mm * 0.13 mm * 2.0 mm, fringe spacing 5.9μm, number of samples per point 5000). There is a good agreement between the two techniques except for the results near the border; as already mentioned this is a consequence of non-detecting negative velocities with PIV.

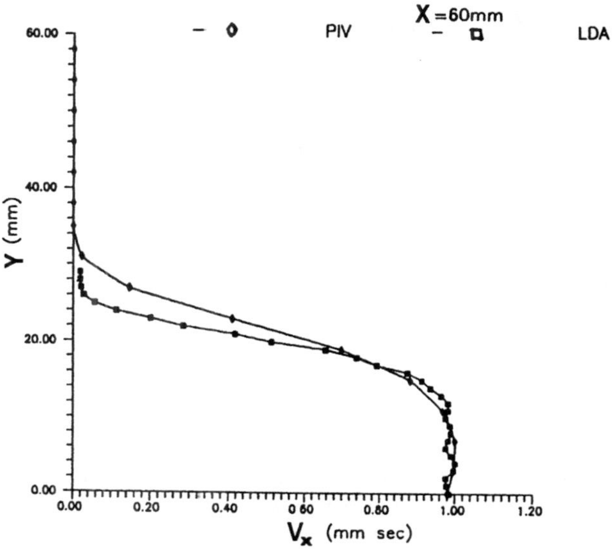

Figure 5: Comparison between LDA (□) and PIV (◇) results on the circular jet configuration at 6 cm from the outlet ($Re \simeq 15000$).

References

[1] PARKER R. *"New directions in Holographic Flow Visualizations"* Seminar on Optical Methods and Data Processing in Heat and Fluid Flow - Institution of Mechanical Engineers - London 1992

[2] MEYERS J., KOMINE H. *"Doppler Global Velocimetry: a new way to look at velocity"* 4th Conference on Laser Anemometry Advances and Applications - American Society of Mechanical Engineers - Cleveland 1991

[3] ADRIAN R., YAO C.S. *"Development of pulsed laser velocimetry for measurement of turbulent flow"* 8th Biennal Symposium on Turbulence - University of Missouri - Rolla 1983

[4] MERZKIRCH W. *"Flow visualizations and digital image processing"* Von Karman Lecture Series "Laser Speckle Methods in Fluid Mechanics"- Springer Verlag - 1986

[5] CENEDESE A., PAGLIALUNGA A. *"Direct digital analysis of a multiexposed photograph in PIV"* Experiments in Fluids 8:273-280 1990

[6] CENEDESE A., QUERZOLI G., ROMANO G.P. *"Comparison between PIV and LDV Velocity Measurements in a Convective Boundary Layer"* International Symposium on Imaging in Transport Processes - Athens 1992

[7] CENEDESE A., PALMIERI G., ROMANO G.P. *"Turbulent Intensity Evaluations with PIV"* 5th International Symposium on Applications of Laser Anemometry to Fluid Mechanics - Lisbon 1990

[8] KOBAYASHI K., SAKAKIBARA J., HISHIDA K., MAEDA M. *"Time series measurements of Turbulent Flow Field using Image Processing System"* Conference on Experimental and Numerical Flow Visualization - American Society of Mechanical Engineers - Atlanta 1991

[9] SCHLICHTING H. *"Boundary Layer Theory"* Mc Graw-Hill - 7th Edition 1979

Visualization of Sub-Nozzle Jet of Air Jet Looms

Ryuji SHINTANI
Researcher, Industrial Research Institute of Ishikawa,
Ro-1, Tomizu-machi, Kanazawa, Ishikawa, 920-02, Japan
Atsushi OKAJIMA
Professor, Faculty of Technology, Kanazawa University,
2-40-20, Kodatsuno, Kanazawa, Ishikawa, 920, Japan

Summary

This study has attempted to visualize an air flow in the filling weft yarn process in an air jet loom which consists of a special reed and sub-nozzles. A jet flow from one sub-nozzle and a flow through a reed channel have been experimented by using a hot-wire anemometer and a schlieren method with the carbonic acid gas. It is certified by a hot-wire measuring that an angle of jet issued from a nozzle exit with different shapes is different, and that a flow configuration of the *vena contracta* at the nozzle exit has influence on the angle of jet. Visualization by a schlieren method suggests that a leakage flow from the channel of a reed greatly depends on the thickness and the density of a special reed, also.

Introduction

An air jet loom (AJL) has a mechanism to carry a weft yarn to the end of the loom between warp yarns by an air flow. The AJL is generally divided into three different types of instruments with regard to a filling weft yarn: that is, the air guide type, the special reed and sub-nozzle type, and the type of the air guide with sub-nozzles [1]. In all types of the AJL, the characteristics of air flow are considered to be most important for the filling weft yarn.

The sub-nozzles of the AJL require the stable conditions of angle and velocity of the jet, independent of an air velocity. However, it is well known from experience that an angle of jet changes with increasing the supplied pressure or the velocity at the nozzle exit.

The air flows in the AJL have been studied and reported by many researchers [2]-[6]. These studies were on a main-nozzle, an air-guide and a reed, but there was a few studies which focused on the jet of a sub-nozzle and the relationships between a flow through a special reed and jets of sub-nozzles.

So, we have investigated on the air flows in the AJL which consists of a special reed and sub-nozzles, because this type of AJL is more popular than other types. Although this type of AJL practically used has ten and more sub-nozzles and the air flow through a reed channel in the filling weft yarn process is intermittent and unsteady, we have visualized the steady free jet from a sub-nozzle and the flows issued from three sub-nozzles jet into the reed channel.

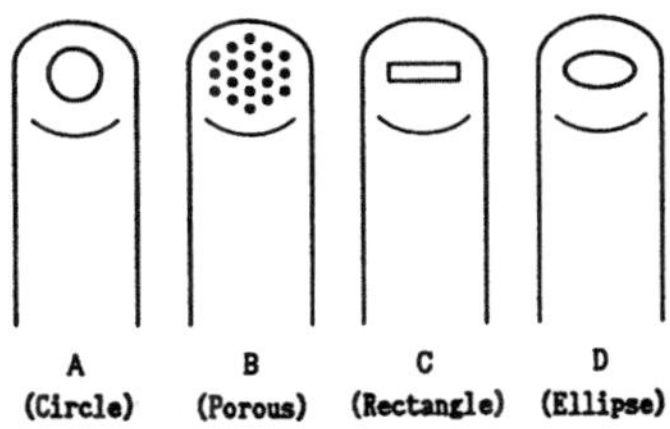

Fig.1 Commercial sub-nozzles.

Experimental apparatus and procedures

Figure 1 shows four commercial sub-nozzles that have different shapes of the nozzle exit, i.e. circle, porous, rectangle and ellipse. All of these nozzles are in practical and wide use.

A hot-wire anemometer system (SYSTEM 7201 Kanomax Japan Inc.) was employed to measure the velocity distributions of a sub-nozzles jet. A sensor length of a I-type hot-wire probe is 1mm, and a diameter is 5 μ m. The probe was mounted on a computer-controlled precise traverse mechanism enabling movements in all three axes.

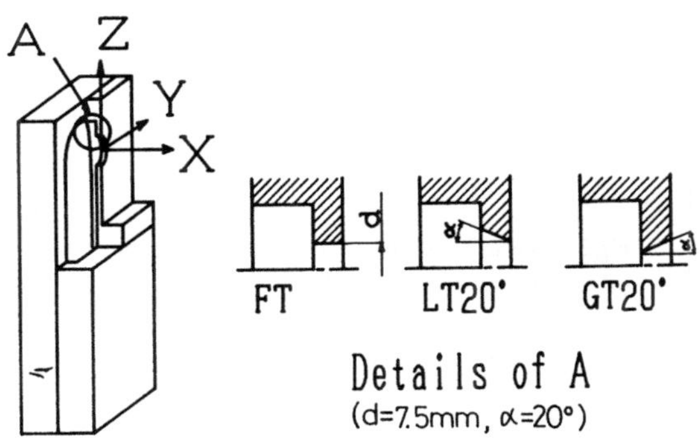

Fig.2 Models of the sub-nozzle.

Three models of sub-nozzle made of plastics were employed to examine effects of the shape of a cross section of a nozzle exit on the angle of jet. Their cross sections are shown in Fig. 2. The diameter of the nozzles is d=7.5mm, and the nozzles whose exit-shape are designed to be straight, convergent and divergent, are denoted by FT, LT20° and GT20°, respectively, and they have a common conduit.

To visualize a jet flow from the sub-nozzle, a schlieren compact system (SCL-200S Mizojiri) was employed with carbonic acid gas. It has a lens and a mirror whose diameters are 200mm. The analog signals of flow-patterns were stored on the video recorder system at a rate of 30pictures/s.

Results and discussion

Commercial sub-nozzles

The mean velocity contours of the jets from four commercial sub-nozzles are shown in Fig. 3. The contour value is 10m/s and several lines are measured at different X positions from a nozzle exit. The supplied pressure is 98kPa. This figure shows that all of the jets

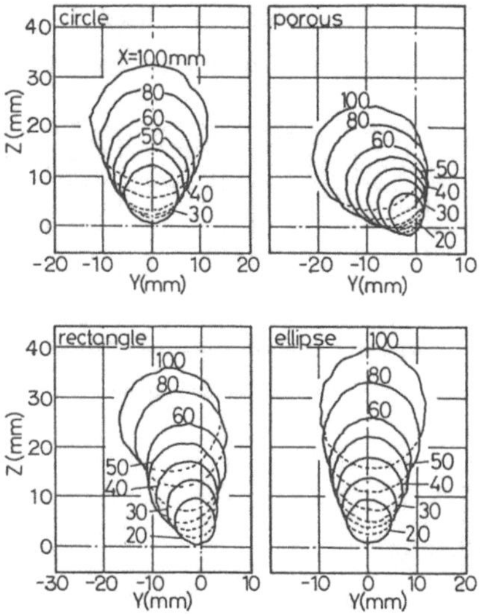

Fig.3 Mean velocity contours of four sub-nozzles. The contour value is 10m/s.

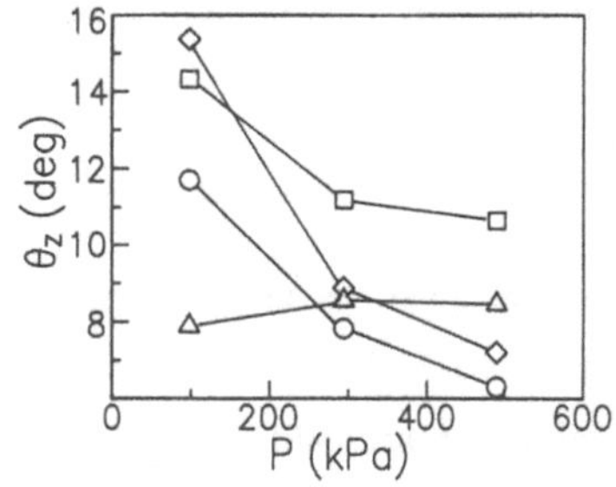

Fig.4 Jet angles against a pressure in a reservoir. ○: Circle, △: Porous, □: Rectangle and ◇: Ellipse.

issue upward and that the spread of jets is different for indivisual sub-nozzles. For the rectangle nozzle, note that the aspect ratio of the spread is not only strongly dependent on the X position, but also a trajectory of the jet center is not straight.

The jet angles against the supplied air pressure are shown in Fig. 4. The jet angles of circle-, ellipse- and rectangle-shaped nozzles are reduced with increasing the pressure supplied from a reservoir. This tendency of the jet angle agrees well with experience. The jet angle of a porous nozzle is almost constant, independent of the supplied pressure. It is concluded that the jet angles are influenced by the shapes of the nozzle exits and the conduit, and that the variation of the jet angles against the supplied pressure is seen to be caused by the differece of velocity distributions at the nozzle exit.

Models of sub-nozzle

Figure 5 shows a velocity distribution of the FT nozzle whose cross section is straight. From this figure, the region of a constant velocity contracts to be smaller than the exit cross area. Namely, note that the phenomenon of the *vena contracta* occurs at the exit of the nozzle. The jet angles of three sub-nozzle models against the mean velocity Uo at the nozzle exit are shown in Fig. 6. The jet angles are reduced with increasing the mean velocity, like the results for the

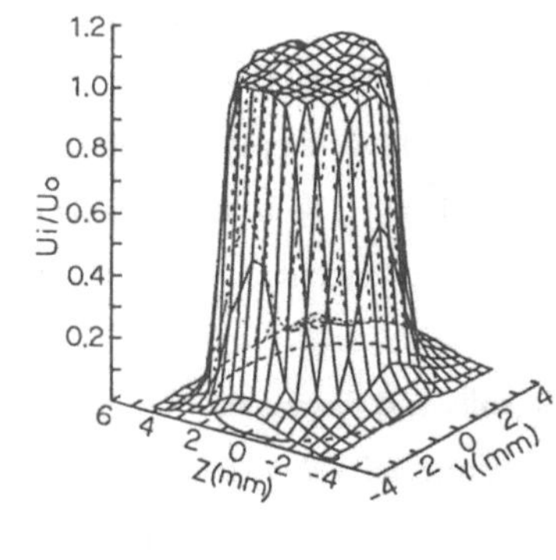

Fig.5 Velocity distribution at the nozzle exit. (FT-nozzle)

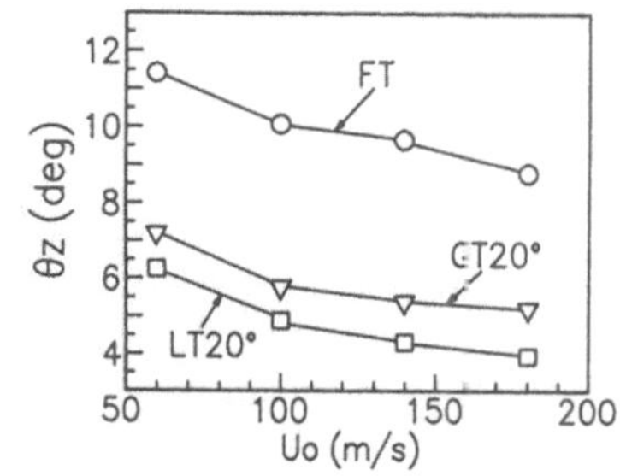

Fig.6 Jet angles of three models FT, LT20° and GT20° against the mean velocity.

commercial sub-nozzles. The flow configuration of the *vena contracta* which occurs at the nozzle exit, seems to change with the mean velocity.

Figures 7 (a), (b) and (c) show flow-patterns of the jets from three models of sub-nozzles, FT, LT20° and GT20°, respectively which are visualized by a schlieren method. From these figures, it is clear that the jet angle of the FT nozzle is larger than the others.

Flow through a special reed

An experiment was performed setting a reed parallel to the light axis of the schllieren method in order to examine how the thickness and density of reed blades affect the flow through a special reed. In the case in which a light axis is set parallel to the reed, the photograph by the schllieren method obviously indicates that the flow issues into the back and the upper sides of the reed. Figures 8 (a), (b) and (c) show the flows for the different reeds with variable density,153.8, 134.8 and 115.4 blades/m, respectively. The leakage flows at the back and the upper sides of the reed are seen to be different due to the density of the reed. The back leakage flow spreads widely, and the upper leakage flows go out along the surfaces of the reed with decreasing the density. We find an amount of flow leakage issued from the reed into the

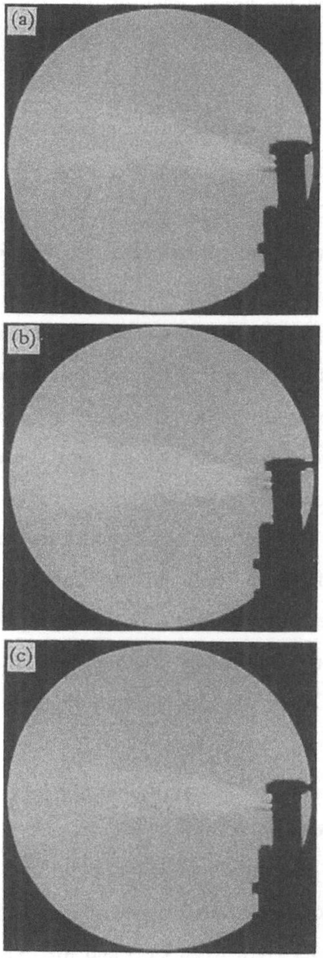

Fig.7 Jets of three models. (a) FT nozzle, (b) LT20° and (c) GT20° .

back and the upper sides of the reed by the schlieren method.

Conclusion

We try to estimate the performance of the sub-nozzles and special reed by the measurement of a hot-wire anemometer, and the flows in the filling weft yarn process are visualized by the schlieren method.

It is found that the angle of jet strongly depends on the exit shapes of the nozzle, and that the angle of jet is closely related with the *vena contracta* which occurs at the nozzle exit.

We find also an amount of flow leakage issued from the reed channel into the back and the upper sides of the reed by the schlieren visualization method, and that the leakage flow depends on the thickness and the density of the reed.

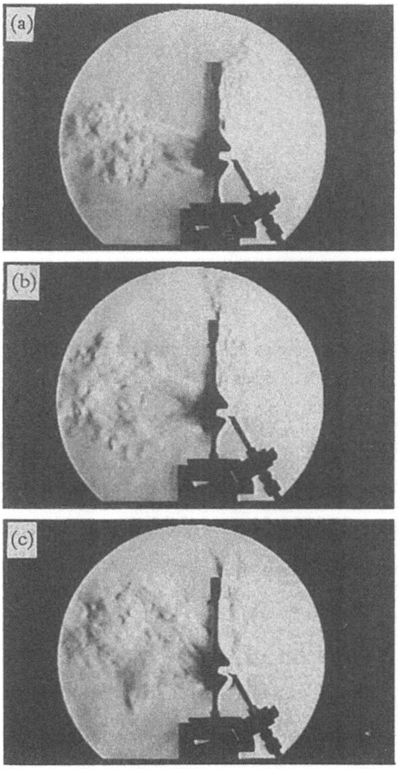

Fig.8 Flows around the different reeds. (a) Density of 153.8 blades/m, (b) 134.8 and (c) 115.4.

Acknowledgment

We gratefully acknowledge the interest and unsparingly support by Chief scientist Mr. K. Chikaoka, Senior staff scientist Mr. I. Donjou (IRII), Mr. S. Yano (Yano Labo. Co.), Mr. K. Tomita (Kiji Reed Co.) and Mr. T. Nakagawa (Tsudakoma Co.).

References

1. Text. Mach. Soc. Japan(ed.): The Jet Loom Handbook, part 1, 1982.
2. M.Uno: A Study on Air-Jet Loom with Substream Added. Part 3: Synthesis of Substreams, J. Text. Mach. Soc. Japan, Vol.25 (1972) T130-T139.
3. M.Muramatsu: A Study on the Air Flow and Weft Projection of Air Jet Looms, Nissan Tech. Review, No.24 (1984) 57-68.
4. K.Yoshida; F.Suzuki; S.Kawabata; J.Hasegawa: Jet Injected into Air Guides of an Air Jet Loom. Part 2: Special-Reed System, J. Text. Mach. Soc. Japan, Vol.40 (1987) T125-T136.
5. M.Ishida; A.Okajima; Y.Shimada; T.Kurata; H.Hoshiai: Experimental Study on the Air Jet Loom with an Air-Guide. Part 1: Characteristics of the Flow Injected into an Air-Guide, J. Text. Mach. Soc. Japan, Vol.36, No.4 (1990) 127-132.
6. M.Ishida; A.Okajima; Y.Shimada; T.Kurata; H.Hoshiai: Experimental Study on the Air Jet Loom with an Air-Guide. Part 2: A Comparison of the Air Flow Characteristics between in an Air-Guide and in Various Air Tubes, J. Text. Mach. Soc. Japan, Vol.37, No.1 (1991) 8-13.

Two-Dimensional Time-Resolving Concentration Measurement of Turbulent Mixing Layers

T. HASEGAWA, S. YAMAGUCHI, T. AMANO, and H. GOTOH

Department of Mechanical Engineering
Nagoya Institute of Technology, Nagoya, Japan

Abstracts
A visualization system that enabled two-dimensional time-resolving concentration measurement was constructed, and was applied to study the mixing process of a free circular jet and a confined planar shear layer. The visualized concentration images were digitally processed and analyzed to get quantitative properties such as mean value, fluctuation intensity and frequency spectra of the mixing layers. The system had been proved to be an capable tool for both visualization of vortical structure and quantitative measurement of turbulent mixing.

Visualization system and image analysis
An Ar-ion laser of 1.5 W at 514.5 nm was used to form a laser sheet of 0.5 mm thickness. Water vapor of about 10 μm diameter was seeded in the flow, and the intensity of the scattered

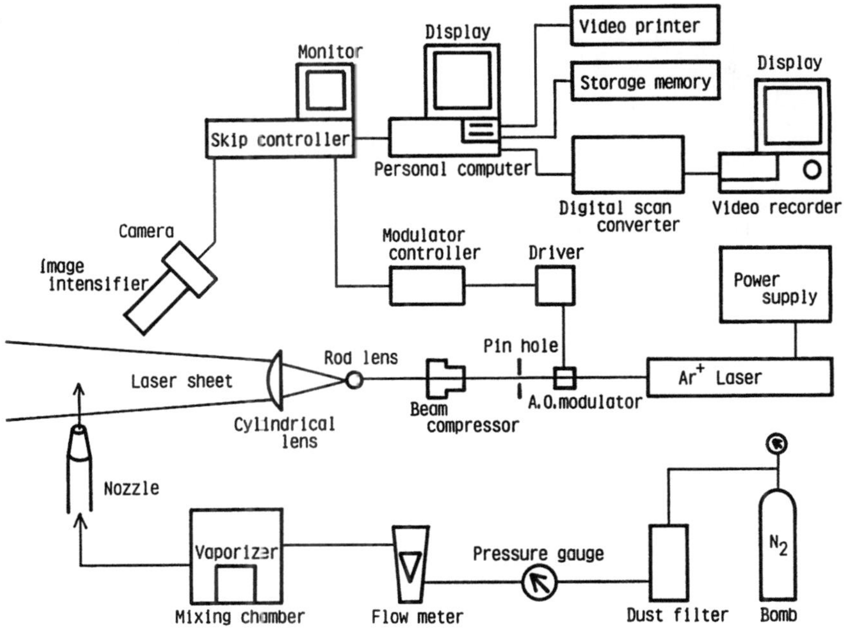

Fig. 1 Visualization system.

light was detected by a high-speed image sensor through an image intensifier. The framing rate was varied from 330 to 500 frames/s, and the exposure time was synchronously controlled by an acousto-optic modulator. More than one hundred sequential frames were taken into the storage memory and used for image analyses. A schematic diagram of the visualization system is shown in Fig. 1.

The scattered light images were transformed into the concentration images after correcting the non-uniformity of the laser sheet. The sequential concentration images were used to understand the dynamic process of mixing, and analyzed to obtain statistical quantities, such as mean value, fluctuation intensity, probability density function, intermittency, and spatial correlation. The frequency spectra was also analyzed by the maximum entropy method.

Visualization of a circular jet

As a verification of the visualization system, a circular jet at a velocity of 4.9 m/s (Re = 1,900) and with a nozzle diameter of 6 mm was visualized at a rate of 500 frames/s, at an exposure

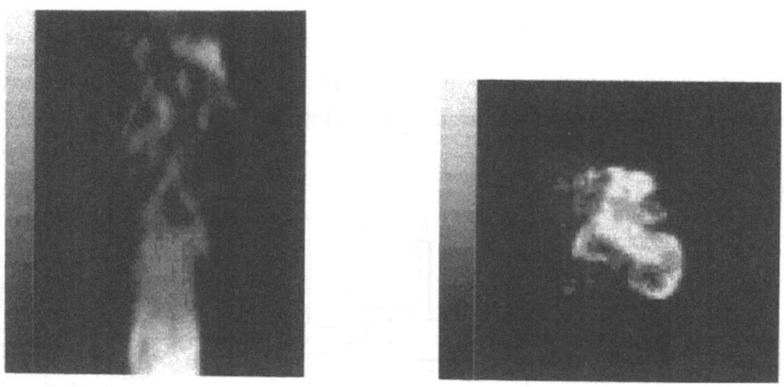

Fig. 2 Instantaneous concentration fields of a circular jet (Re=1,900). Longitudinal cross section between z/d = 5-10 and lateral cross section at z/d = 8.

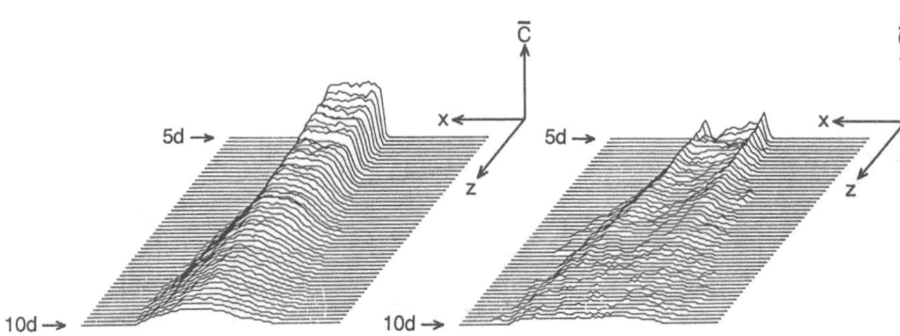

(a) mean concentration (b) fluctuation intensity

Fig. 3 Mean and fluctuation intensity of the concentration of a circular jet (Re=1,900).

time of 200 μs, and with a resolution of 0.3 mm × 0.3 mm for each pixel. Fig. 2 shows instantaneous concentration images in longitudinal and lateral cross sections. It is clearly found that longitudinal and lateral vortices exist and entrainment of the ambient gas happens in three-dimension [1, 2].

The longitudinal distribution of the mean concentration in Fig. 3a shows a trapezoid profile near the nozzle and a Gaussian profile in the developed region. The longitudinal fluctuation intensity distribution in Fig. 3b shows peaks near the mixing region. Concentration intermittency and probability density functions could also be obtained from the visualized data.

Visualization of a confined planar mixing layer

A planar mixing layer between parallel walls with a high-speed side velocity of 1.4 m/s and a low-speed side velocity of 1.0 m/s was visualized by seeding water vapor in the high-speed side. The velocity ratio was 0.71 and the Reynolds number defined by the distance between the walls of 80 mm and the mean velocity of the high-speed and the low-speed sides was 6,400. The turbulent intensity at the entrance of the test section were 0.4 % at high-speed side and 0.6 % at low-speed side.

Fig. 4 shows instantaneous concentration images between x = 70 mm - 120 mm, 120 mm - 170 mm, 170 mm - 220 mm, and 210 mm - 260 mm downstream from the splitter plate, mea-

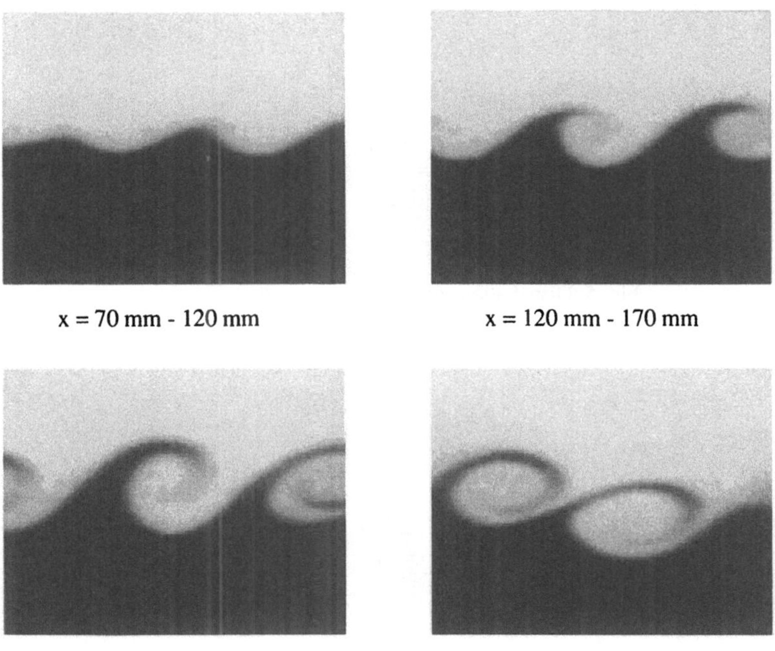

Fig. 4 Instantaneous concentration fields of a confined planar mixing layer.

suring at a rate of 330 frames/s, at an exposure time of 3.3 ms and with a resolution of 0.5 mm × 0.5 mm for each pixel. The figures indicate an instability wave of planar shear layer, rollup of two-dimensional vortices, mixing of the fluids in both sides, and vortex paring.

Figs. 5 and 6 show mean value and fluctuation intensity of concentration at x = 70 mm - 120 mm, 120 mm - 170 mm, 170 mm - 220 mm. Mean concentration shows a step profile near the splitter plate and a two-step profile at downstream regions as a result of development of rollup vortices [3]. The fluctuation intensity has a single peak near the splitter plate and two peaks at downstream regions corresponding to the mean concentration, because the fluctuation intensity is related to the transversal gradient of mean concentration as $\overline{v'C'}\,\partial\overline{C}/\partial y$. The transition of mean concentration and fluctuation intensity profiles are also confirmed by the numerical simulation of spatially developing mixing layer and its scalar transport process [4, 5].

Spatial correlation of concentration between x = 120 mm - 170 mm with a reference point of x = 130 mm is shown in Fig. 7. The correlation has clear peaks as a result of two-dimensional rollup vortices. The distance of the peaks can be regarded as a mean distance between the vortices, and measured to be 21.5 mm. Fig. 8 shows frequency spectra of concentration at x = 80

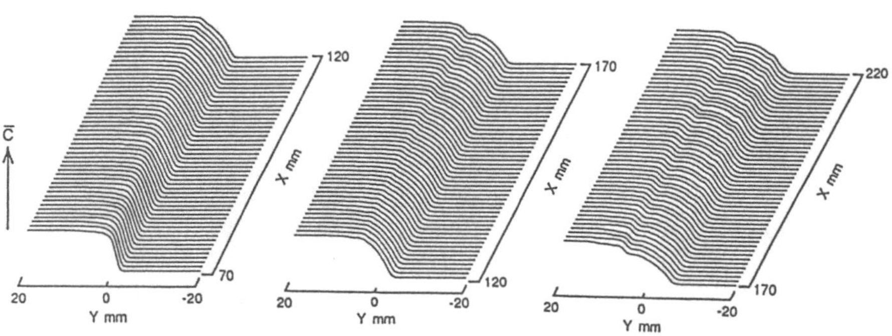

Fig. 5 Distributions of mean concentration.

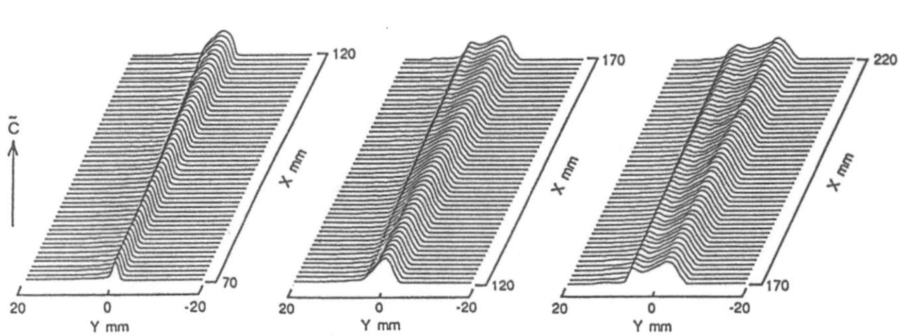

Fig. 6 Distributions of fluctuation intensity.

mm analyzed by the maximum entropy method. The range of frequency spectra is limited to 167 Hz due to the Nyquist's theorem applied to the rate of 333 frames/s. The frequency spectra show a peak of 53 Hz which is suspected to be an eigenvalue of the planar shear layer.

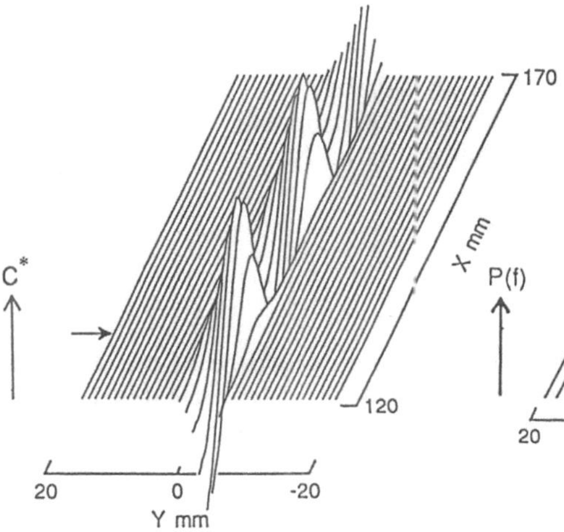

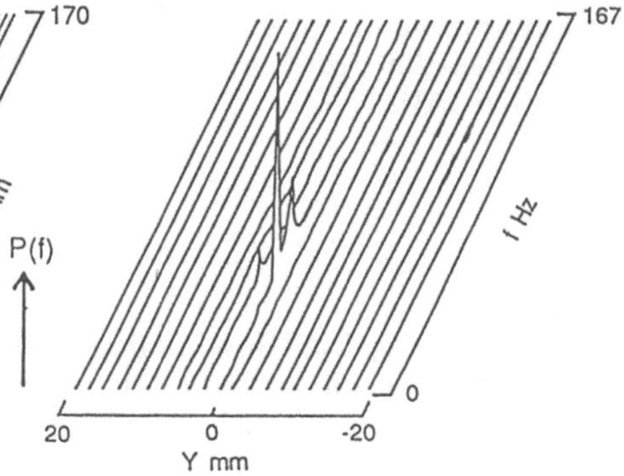

Fig. 7 Spatial correlation between x = 120 mm - 170 mm with a reference point of x = 130 mm.

Fig. 8 Frequency spectra at x = 80 mm.

Summary

The two-dimensional time-resolving visualization system was applied to study concentration fields of a circular jet and a confined mixing layer. It was demonstrated that the system gave us insights into the dynamics of vortices and quantitative properties of turbulent mixing.

Acknowledgement

This work was partially supported by the grant from the NAITO Foundation in 1991.

References

1. Long, M.B.; Chu, B.T.: Mixing mechanism and structure of an axisymmetric turbulent mixing layer. AIAA J. 19 (1981) 1158-1163.

2. Long, M.B; Tzeng, H.-M; Fourguette, D.: Time-resolved two-dimensional concentration measurements in an acoustically driven flow. AIAA J. 21 (1983) 1085-1088.

3. Koochesfahani, M.M.; Dimotakis, P.E.: Laser-induced fluorescence measurements of mixed fluid concentration in a liquid plane shear layer. AIAA J. 23 (1985) 1700-1707.

4. Hasegawa, T.; Yamaguchi, S.; Ohiwa, N.: A numerical analysis of mass transfer in a plane shear layer. JSME Int. J. 30 (1987) 785-791.

5. Hasegawa, T.; Yamaguchi, S.: Numerical study on the reaction process in a plane shear layer. Progress in Astronautics and Aeronautics vol. 113 ed. Kuhl, A.; Bowen, J.R.; Leyer J.-L.; Borisov, A. 289. Washington D.C.: AIAA 1988.

Flow Visualization of Jet Flow in the High Speed Switching Spool Valve

Tetsuhiro TSUKIJI
Dept. of Mechanical Engineering
Ashikaga Institute of Technology
268-1, Oomae-cho, Ashikaga-shi,
Tochigi, 326 JAPAN

Masahiko SOSHINO
Hitachi, Ltd
2520, Takaba,
Oowaza, Katsuta-shi,
Ibaragi, JAPAN

Summary

The two-dimensional jet flow issuing from the metering orifice of the spool valve commonly used for hydraulic power application is investigated by a experimental and a numerical flow visualization when the valve is instantly opened and closed repeatedly. Flow pattern of the jet flow, pressure distributions and flow forces are discussed.

Introduction

The high speed switching valve is sometimes used to obtain a high performance of the electro-hydraulic control system. It is necessary to investigate the flow near the orifice of the valve so as to improve the valve. In the present study, a two-dimensional model of the spool valve is designed to visualize the jet flow in a valve chamber by a hydrogen bubble method. The spool is directly moved by an oil hydraulic actuator to open and close the valve repeatedly. The illuminated flow fields are directly recorded on a video imaging system by a CCD monochrome camera. A digital image processing technique is employed. A numerical simulation[1] by a source distribution method and a discrete-vortex method is carried out to analyze the two-dimensional jet flow issuing from the spool valve orifice. A prediction of the flow pattern of the jet flow, the pressure distributions and the flow forces is undertaken. The behavior of the vortex shedding near the orifice outlet is discussed by comparing the numerical flow visualization results with the experimental flow visualization ones.

Spool valve model

Figure 1 shows the test valve that was used during these experiments. The flow through the metering orifice forms a two-dimensional jet. This model is made of transparent acrylic resins. The spool was made of the Teflon resin to move the spool easily with low friction force. And, the over-lap(2mm) was set up to decrease the leakage as shown in Fig.1 (part C). Furthermore, the spool was pressed by the springs and the bolts to decrease the clearance between the spool and the valve body. A hydrogen bubble method was applied to visualize the jet formation. The platinum wire 50 μm in diameter was mounted horizontally near the orifice. For the direct current(d.c.) technique, application of a fixed value of d.c. voltage(300Volts) produced a uniform boundary between the edge of the jet and the ambient water. The spool was directly moved by an oil hydraulic actuator.

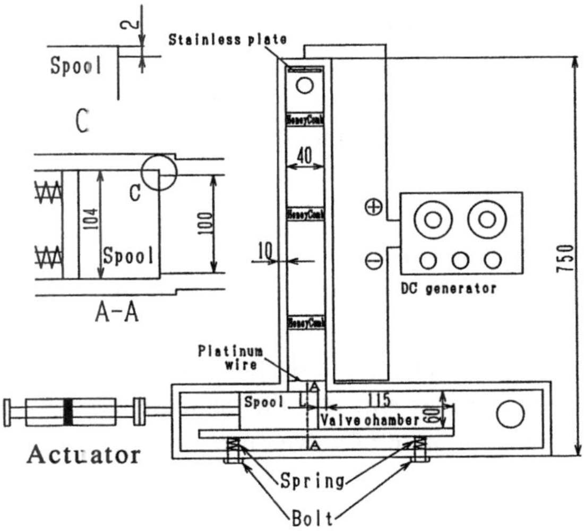

Fig.1 Spool valve model

<u>Experimental and image processing procedure</u>

The video images were recorded using a CCD camera and processed to yield quantitative flow field information. The first step in processing is to remove the image background(orifice image, etc.) except the jet image in an original image. This was achieved by subtracting the gray level of the original image before the jet starts from the gray level of the original image with the jet flow. Some jet image are overlapped to increase the difference of gray levels between the jet image and the image background. The contrast between the jet image and the image background can be increased. A suitable threshold level was chosen to classify each pixel as belonging to a jet or to the background. The binary images were obtained for the suitable threshold level and the orifice image is added. Then the digital images of jet edges are obtained from the binary image. The Reynolds nember at the orifice exit, Re , is defined as

$$Re = 4mU / \nu \qquad (1)$$

where ν, m and U are the kinematic viscosity, the hydraulic mean depth and the average velocity at the maximum valve opening.

Jet formation

The calculated results by a discrete vortex method and the experimental ones are shown in Fig.2. The maximum valve opening is 15mm. Figure2(a) shows the digital image of jet edge. Figure 2(b) shows the distribution of elemental vortices corresponding to the experimental results and ● shows the clockwise vortices and ○ the counterclockwise vortices. Figure 2(c) shows the time variation of valve opening L (see Fig.1). In these figures, the time T is non-dimensionalized. The signs A ~ AA below Figs.(a) and (b) show those in Fig.(c). The error of the dimensional time t is 0.06sec correspondind to the error of the dimensionless time 0.4. The computational results of the process of the jet formation are found to agree with the experimental results.

Fluctuations of the flow force

Calculated fluctuations of the forces obtained by integrating the pressure distributions on the surface wall $\overline{FG}$ (see the left side of Fig.4) are shown in Fig.3. The flow force decreases during the spool opening and increases during it closing. At the beginning and the end of the spool movement, the force fluctuates considerably.

Pressure fluctuation

The calculated pressure distributions on the wall $\overline{FG}$ are shown in Fig.4 . The pressure distributions on the wall $\overline{FG}$ almost increase gradually with the time. The force fluctuations in Fig.3 can be understood from the fluctuations of the pressure distributions. Especially the profile of the pressure distributions are different between the steady and the unsteady states.

Conclusions

A deeper understanding of the behavior of the vortex shedding near the orifice outlet for repeated spool-valve opening and closing has been acquired by the experimental and the numerical flow visualization. The results

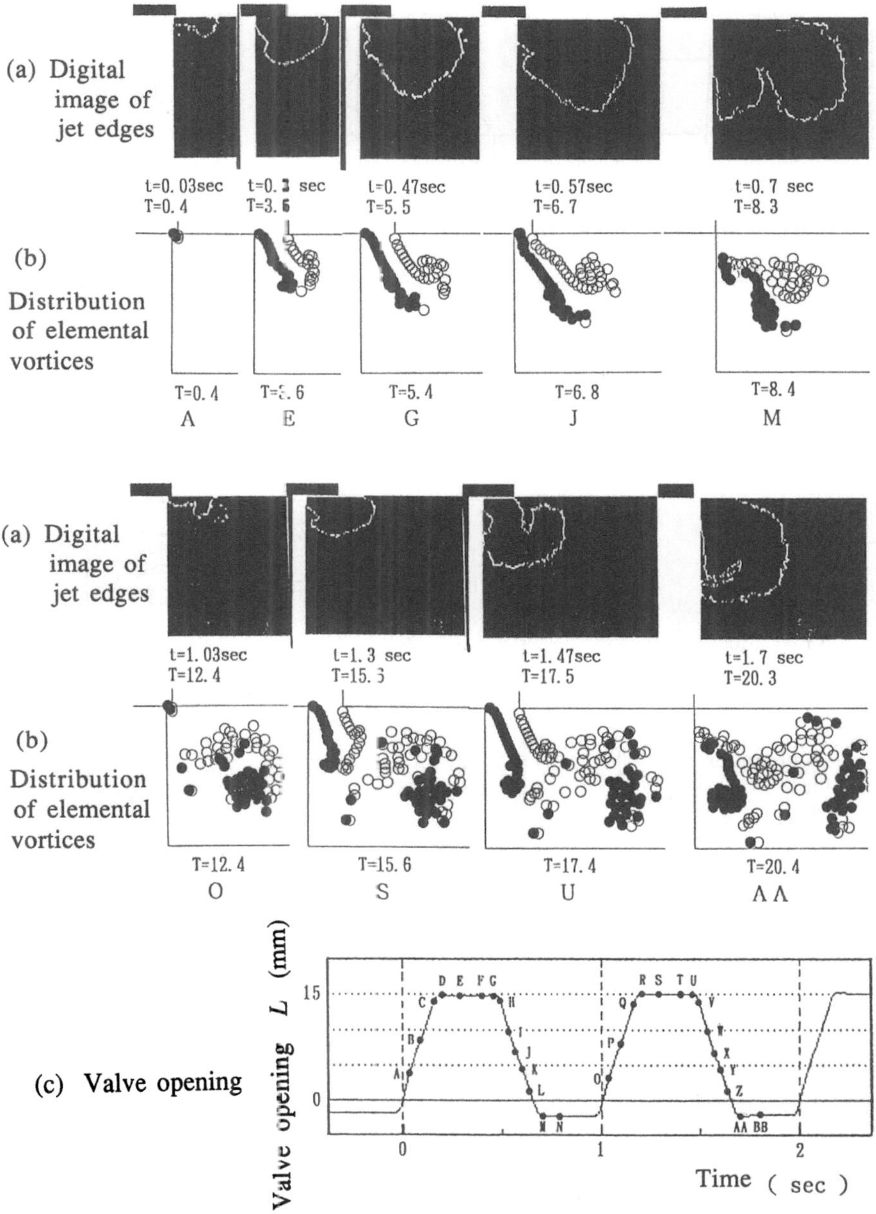

Fig.2 Jet formation

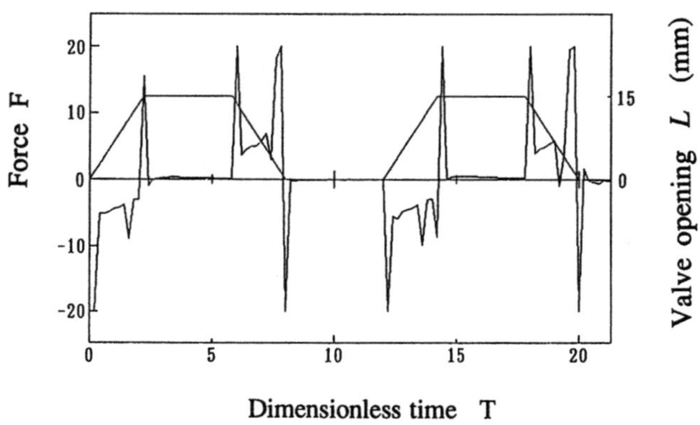

Fig.3 Force

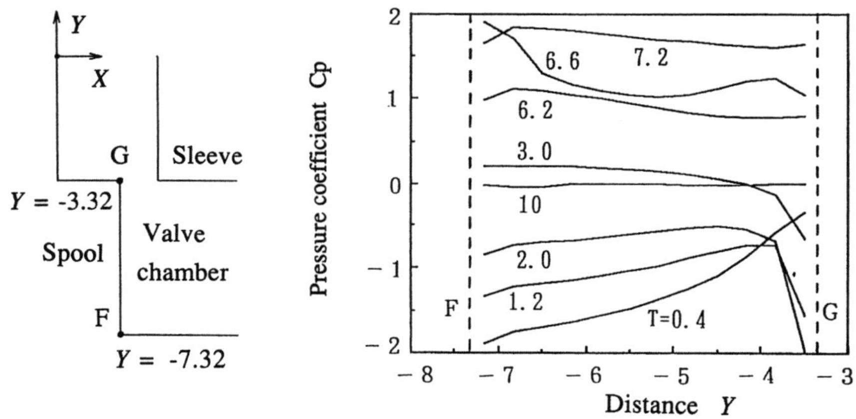

Fig.4 Pressure distributions

obtained by the present investigation are very important to develop the high speed switching spool valve.

Reference

1. Tsukiji, T.: Experimental and numerical flow visualization of impulsively started jet from spool valve orifices, Proceedings of the second international symposium on experimental and numerical flow visualization (ASME-FED-Vol.128),(1991)181-187.

Interaction Between Impulse Jet and Flat Plate

O.V.BULAT, V.V.GOLUB, V.N.LYAKHOV, A.M.SHULMEISTER

Institute for High Temperatures Academy
of Sciences, Moscow, 127412, CIS

Summary

Formation of the interaction between impulse jet and flat plate was investigated. Flow field between the nozzle and the plate was analized using shadowphotography, interferometry, numerical calculation, piezoceramic pressure pickups and thin-film temperature transmitters.

Introduction

In paper [1] the interaction of the stationary jet with a flat plate was experimentally studied and in [2] the flow pattern of the free impulse jet was revealed. The present paper is devoted to the investigation of the unsteady interaction between an impulse jet and flat plate.

Experimental set-up

At the end of the shock tube a sonic nozzle was installed with the diameter of the critical cross-section d =4.0 mm, the outflowing gas passes through the nozzle into a vacuum chamber. A flat plate was mounted in front of the nozzle at the dimensionless distances L=L/d =6-40. The shock tube and vacuum chamber were filled by nitrogen in the range from 4kPa up to 50 kPa. The Mach number of incident shock waves was varied in the range of M =2-3, that provided the pressure-ration n=10-28 ($n=p_a/p_\infty$, where p_a is the pressure at the nozzle exit, p_∞ is the pressure of the ambient gas). The flow was visualized by means of the shadow column instrument and interferometer with a rubi laser.

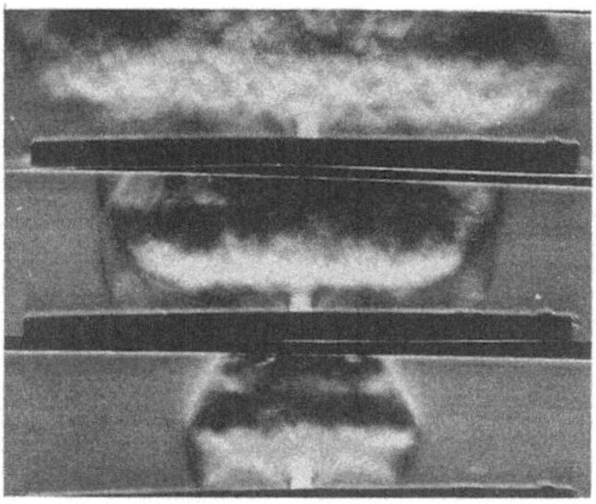

Fig.1.

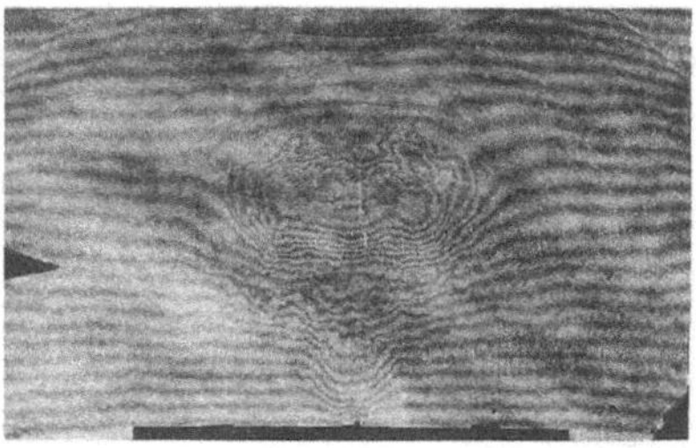

Fig.2.

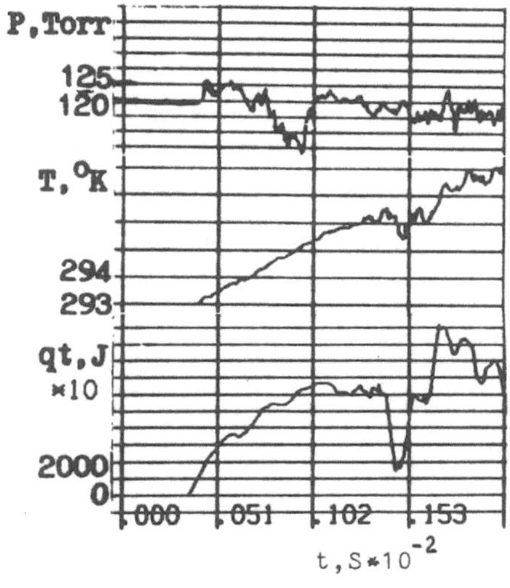

Fig.3.

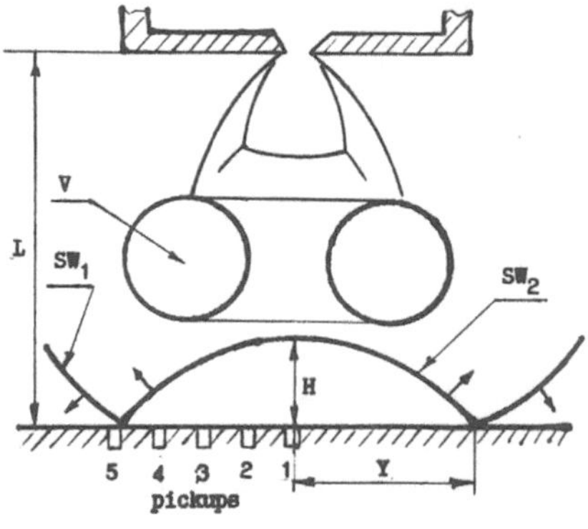

Fig.5.

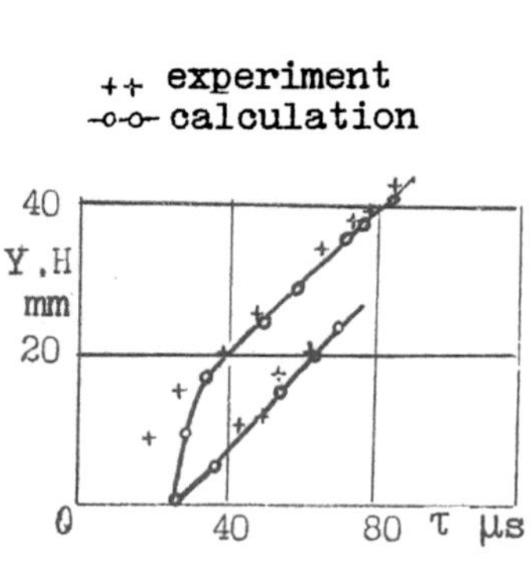

Fig.6.

Results and discussions

Fig.1 presents the shadowgrath pictures of the impulse jet impingement on a flat plate at the instans of t=47mcs(a),62mcs(b), 254mcs(c). Fig.2 presents the interferogram of interaction between the jet and plate at the instant of t=138mcs after starting of the outflow. The density field was obtaned by means of the image analizer IBAS followed by solving Abel equation. Mounted in the plate were 5 piezoceramic pressure pickups and 5 thin-film temperature transmitters, the distances between pickups are 8.5 mm. Fig.3 gives the oscillograms of pressure, temperature and heat flow that were obtaned in the interaction between the jet and plate installed at distance l/d =38 from the nozzle, the pickups being mounted at distance 8.5d from the axis of the jet. Concurrent with the experimental investigations the Euler equations were solved by McCormack method. Fig.4 presents the computational velociti vectors, density and pressure distributions in the impinging impulse jet on the flat plate. Given in Fig.5 is the diagram of the initial stage of interaction between the jet and plate. The starting shock wave SW_1 going in front of the hot gas flowing out of the nozzle, reaches the plate and is reflected from it, forming shock wave SW_2 mowing towards the nozzle. SW_2 has the shape of a semisphere, the radius of the latter increasing in time. Plotted in Fig.6 is the diagram x-t that characterizes the movement of the point of reflection of SW_1 from the plate and travel of the axial point of SW_2 to the nozzle. A good coordination of the physical and numerical experiment is noted. Fig.7 presents the experimental and computed distributions of the pressure in the plate at L/d=6 (numbers "1" and "2" correspond to pickups in Fig.5). As seen from interferograms of the process, vortex ring V (Fig.2) is formed in the vicinity of the wavefront of the outflowing gas. The arrival of the vortex to the plate is clearly traced by the considerable drop of the pressure recorded by the pickups (Fig.3).

Conclusions

Thus, the substantional difference of the initial stage of interaction between the supersonic jet and plate from the steady interaction consists in the following:

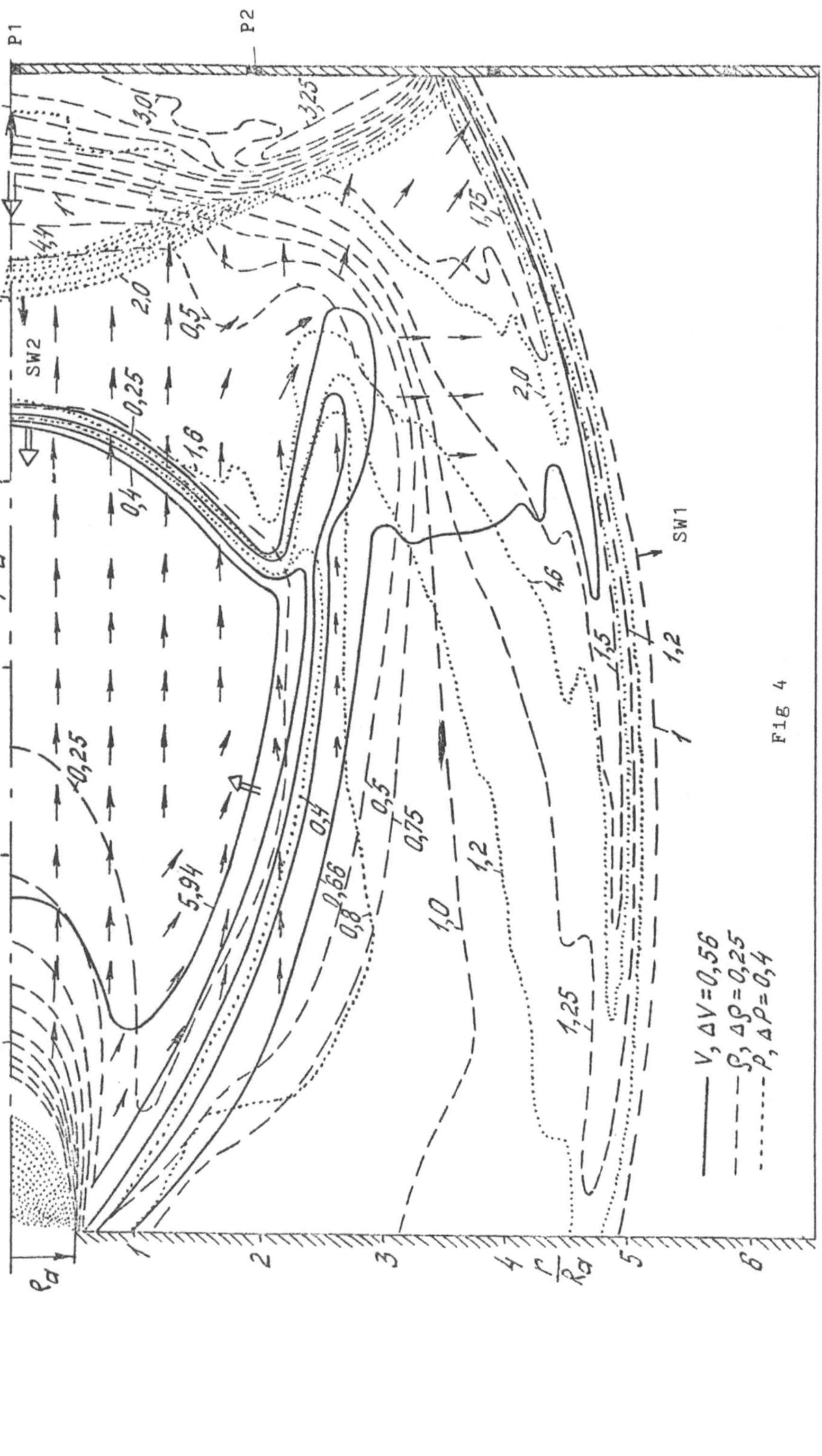

Fig 4

1. At the small distances between the nozzle and plate, the reflected wave plays an important part, bringing about the rise of the temperature and pressure (thus, at L/d =6, the pressure in the plate increases 2-3 times).

2. At the great distances between the nozzle and plate, the vortex ring plays the decisive part, bringing about the pressure drop (thus, at L/d =38, the pressure in the plate drops twofold).

3. The heat flow in the wall at the initial instant greatly exceeds the steady flow in the entire range of the investigated initial parameters.

References

1. Iwamoto,J.:The impingement of a choked jet on a flat plate. Proc. of the 5th Intern. Symp, on Flow Visualization. (1989) 422-427.

2.Golub, V.V.; Kharitonov, A.I.; Sharow, Y.L.; Shulmeister, F.M.: Two-direction visualization of vortex rings emerging in the cource of formation of the supersonic jet. Proc. of the 5th Intern. Symp. on Flow visualization. (1989) 556-561.

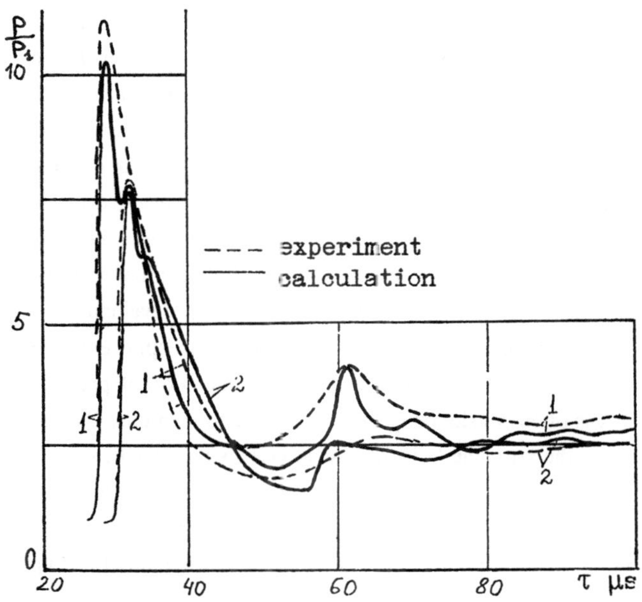

Fig 7

A Flow Visualization Study of Chaotic Vortical Motion in the Near Field of a Planar Jet

F. O. Thomas and J. W. Cutbirth*
Hessert Center for Aerospace Research
Department of Aerospace & Mechanical Engineering
University of Notre Dame
Notre Dame, IN 46556

Abstract
A flow visualization-based investigation of the transition of planar submerged water jet shear layers at moderate Reynolds number is presented.The development of the free shear layers under low level artificial excitation was captured on video tape by dye injection flow visualization. Despite "clean" initial conditions and the imposition of periodic excitation, examination of vortex trajectories revealed that the flow structure was not strictly periodic but exhibited "jitter" in the vortex roll-up location and subsequent vortex interaction events. This study focuses on this non-periodic character of the flow. Since even very simple nonlinear systems can exhibit chaotic behavior, the flow visualization images were examined for evidence of whether the observed non-periodic behavior could be associated with chaos. The flow visualization frames were utilized to derive return maps enabling examination of the flow visualization images in pseudo phase space. Results indicate that many aspects of the non-periodic vortical motion show evidence of chaos similar to that exhibited by lower dimension nonlinear dynamical systems.

Introduction

The important role of large-scale vorticity upon the dynamics of the evolving free shear layer has stimulated numerous investigations; too many to mention here but they are reviewed by Ho and Huerre[1], and Thomas[2]. From these studies an *idealized* picture of the evolution of vorticity in the free shear layer has emerged. The instability undergoes spatial growth leading to a predominantly two-dimensional "roll-up" into a periodic array of discrete spanwise vortices. The transition of the shear layer is in large part dominated by a sequence of discrete subharmonic modes and the selective growth of the local subharmonic gives rise to pairing of neighboring vortices. Vortex pairing doubles the local thickness of the layer, while the local characteristic passage frequency is halved. The streamwise distance between consecutive pairing events also approximately doubles. However, this picture of shear layer development is only an idealization that is seldom exactly realized in the laboratory. Despite extreme precautions taken to establish "clean" initial conditions and/or the imposition of periodic excitation, the actual development is often considerably more complicated than this. Careful examination of vortex trajectories in the x-t plane (x is the streamwise spatial coordinate, t is time) is shown in Figure 1 (which are derived from planar water jet shear layer flow visualization) and reveals that the flow structure is not truly periodic. This study focuses on this non-periodic character of the flow. From Figure 1 it may be noted that the initial roll-up

* Oklahoma Christian University of Science & Arts, Oklahoma City, OK

location varies spatially and the initial vortex passage frequency is not periodic in time. The subsequent vortex interactions can be quite complicated; the pairing locations and convective velocity of the vortices vary considerably as does the number of vortices involved in the interaction. In this investigation we attempt to characterize this complexity by asking whether recent developments in dynamical systems theory can be used to characterize the apparent randomness associated with the development of the large-scale vorticity field (see the review by Sen[3]).

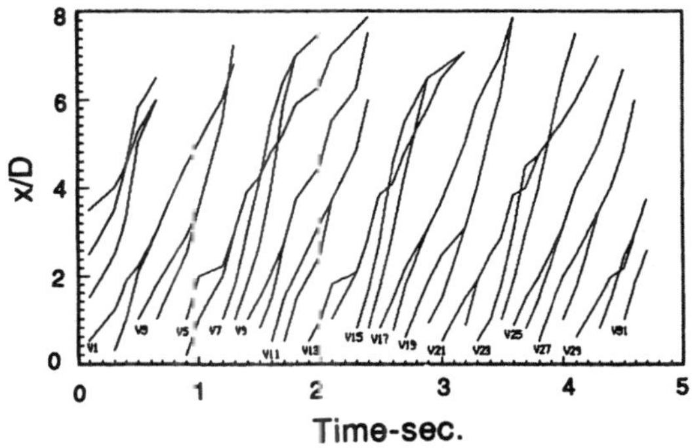

Figure 1. Sample vortex trajectories in the artificially excited jet shear layer

Experimental Facility and Instrumentation

The investigation was performed in a planar submerged water jet facility which is described in detail by Cutbirth[4]. Hence, only essential details are presented here. The two-dimensional jet nozzle has a contraction ratio of 16:1 and terminates in a rectangular slot with aspect ratio 10:1. The jet Reynolds number was 2200 (based on the jet slot width D=0.5 inch) corresponding to an exit velocity of U_o = 0.58 ft/s. The water jet has a flat exit profile and discharges into the flow field which is formed by a large, fully enclosed plexiglass tank. A constant head plenum supplies water to the nozzle by way of a duct containing flow straighteners and multiple turbulence management screens. Slots on the nozzle walls allow colored dye to be injected into the nozzle wall boundary layers for subsequent visualization of the jet free shear layers. In order to control the initial shear layer instability, a sinusoidal streamwise velocity perturbation was introduced into the base flow upstream of the nozzle exit by means of a variable speed transmission and cam operating in contact with a rubber disc assembly. Sinusoidal excitation was applied near the initially most unstable shear layer frequency of f_e=7Hz at an initial amplitude of $\sqrt{u^2(f)} / U_0 = 0.3\%$. The development of the free shear layers was captured on video tape by a video camera aligned either in the direction of mean flow homogeneity or in the lateral cross-stream direction. Digital computer analysis of the video images was accomplished by using a Data Translation frame grabber board and custom software in conjunction with a Macintosh II microcomputer system.

Results

On average roll-up occurred at x/D=1.0, the first pairing event involved two vortical structures merging near x/D=2.5 while a second paring of vortical structures (each of which resulted from an upstream pairing) occurred at x/D=5.5. However, this most common mode of structural interaction only occurred about 60% of the time. Numerous other modes of interaction were observed to occur randomly and for shorter periods of time, as shown in Figure 1. In order to examine this aspect of the shear layer development, the time interval T_n between two adjacent vortices at a selected x/D location was measured and plotted against T_{n+1} to obtain a one-dimensional map of form $T_{n+1}= f\{T_n\}$. Both ordinate and abscissa are normalized by the forcing period. The time resolution was 1/30 second. The resulting maps show a wide range of behavior as the streamwise distance x/D is varied. A few selected locations are shown in Figure 2. At x/D=0.75 the map indicates the flow structure is almost periodic but even here scatter occurs well outside the error band associated with the time interval measurement. More importantly, the scatter is symmetric with respect to the line $T_{n+1}=T_n$ indicating an invertible map. This would not be expected if the structural phase "jitter" was truly random. Increased broadening occurs near x/D=1.125 (not shown) and the single peak bifurcates near x/D=1.25 (not shown). At x/D=1.5 the map is dominated by two peaks (suggesting the intersection of a torus with a plane) and is still invertible as shown in Figure 2. For locations between 1.5 and 2.0 the maps indicate the flow to be non-periodic although the flow is dominated by two competing frequencies. The maps are observed to become noninvertible by x/D=2 as evidenced by the development of a non-symmetric off diagonal point. This leads to a second bifurcation at x/D=2.25 and the map at x/D=2.5 shown in Figure 2 is clearly noninvertible and chaotic. These results are summarized in Figure 3 where the number of events are plotted in the T_n , x/D plane resulting in a bifurcation diagram (where the streamwise distance x/D is taken as the bifurcation parameter, λ).This figure clearly shows two pitchfork type bifurcations that precede chaos. By knowing two consecutive values of the bifurcation parameter λ_{n+1} , λ_n associated with a quadratic map an estimate of the location of the onset of chaos λ_∞ can be made. This relationship is $\lambda_\infty = 1/(\delta-1) [\delta \lambda_{n+1} - \lambda_n]$ (Berge et al [5]) where δ is the Feigenbaum number. From the approximate bifurcation locations visible in Figure 3 an estimate for the onset of chaos is found to be $\lambda_\infty = x/D)_{chaos} \approx 2.5$.

A non integer fractal dimension is the hallmark of a strange attractor and implies chaotic motion. There are several measures of fractal dimension; the correlation dimension described by Berge et al[5]) appears best suited to the time series derived from this flow visualization study. This fractal dimension is defined by $d = \lim_{r\to 0} [\log C(r)/\log r]$ where r is the radius of a circle centered on the origin and $C(r) = \lim_{N\to\infty} 1/N^2$ [number of pairs (i,j) with distance $s_{ij} < r$]. Data sets were generated at selected x/D by calculating the time interval s_{ij} (i.e. $|T_i - T_j|$)

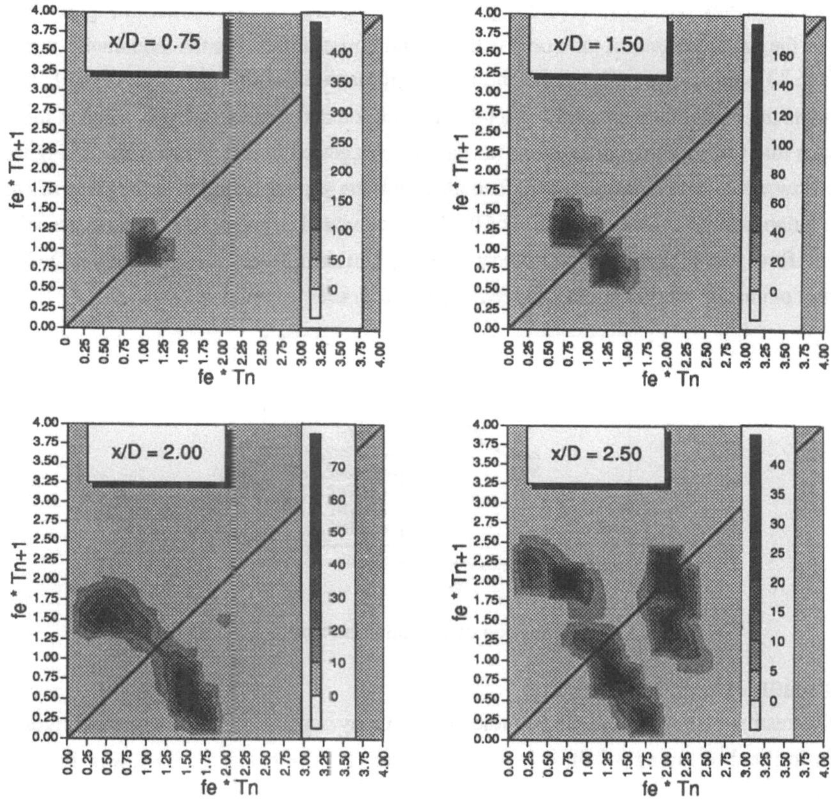

Figure 2. Sample one dimensional mappings

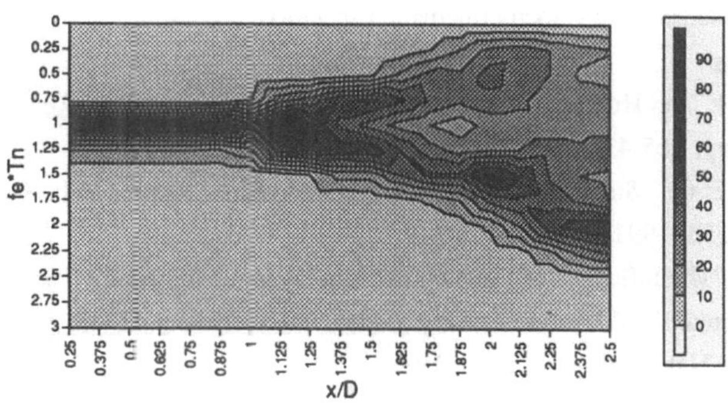

Figure 3. Bifurcation diagram

between all vortex passages. Resulting plots of C(r) vs. log r are presented in Figure 4. The slope of the linear portion of the curve is the fractal dimension d. As the streamwise distance x/D (i.e. λ) approaches x/D = 2.25 the slope approaches a limiting value of 0.58 which represents the fractal dimension. The fact that this limit exists and that the slope is non integer, indicates that the attractors are strange as was also suggested by the return maps of Figure 2. These flow visualization data compare favorably with the logistic equation at the onset of chaos where a non-integer fractal dimension of 0.538 is calculated. Lyapunov exponents associated with the flow visualization were also computed by Cutbirth[4] and these suggest x/D=2.4 as the onset of chaotic motion in close agreement with the other techniques.

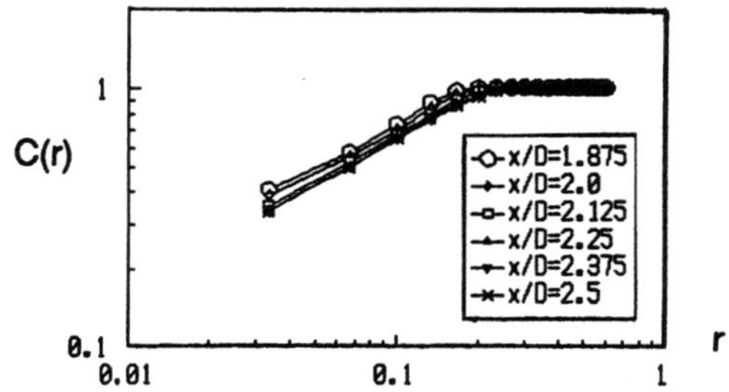

Figure 4. Estimation of Fractal Dimension

Conclusion

The results of this investigation indicate that many aspects of the non-periodic vortical motion in the developing jet shear layer are similar to those exhibited by lower dimension nonlinear dynamical systems. Hence, despite clean initial conditions and the use of periodic excitation, the results suggest that the non-periodic behavior exhibited is due to the inherently nonlinear nature of the flow. Further, onset of chaotic motion appears upstream of the location where the shear layer becomes turbulent according to traditional diagnostics (e.g. the existence of an inertial subrange in the velocity fluctuation spectra).

References

1. Ho, C. M. and Huerre, P.,"Perturbed Free Shear Layers," Ann. Rev. Fluid Mech., vol. 16, pp. 365-424, 1984.
2. Thomas, F. O. ,"Structure of Mixing Layers and Jets," Appl. Mech. Rev., vol 44, 3, pp. 119-153, 1991.
3. Sen, M., "The Influence of Developments in Dynamical Systems Theory on Experimental Fluid Mechanics," in Frontiers in Experimental Fluid Mechanics, ed. M.Gad-el-Hak, Springer-Verlag, 1989.
4. Cutbirth, J. W.,"Chaotic Vortical Motion in the Near Region of a Plane Jet," Ph.D. Thesis, Oklahoma State University, 1991.
5. Berge, P., Pomeau, Y., and Vidal, C.,"Order Within Chaos," John-Wiley, 1984.

A Visual Study of a Round Jet in Cross-Flow

R.M. Kelso, T.T. Lim & A.E. Perry
Department of Mechanical & Manufacturing Engineering
University of Melbourne, Parkville, VICTORIA 3052. AUSTRALIA.

Abstract
Dye visualization techniques have been used to study the flow structures of a round jet in a cross-flow at a velocity ratio ($R=V_{jet}/U_o$) of four. The study has revealed some interesting flow features including a small separation region which occurs just below the upstream lip of the pipe exit and that the jet shear layer rolls up to produce distorted ring vortices which tilt and fold with streamwise distance. These and other flow features are analyzed with the aid of critical point theory.

Introduction

When a jet discharges normal to a cross-flow there is a complex interaction, resulting in the deflection of the jet in the direction of the cross-flow. The jet flow forms a pair of counter-rotating vortices which seem to originate near the pipe exit and which dominate the flow far downstream. The structure near the pipe exit is dominated by ring-like vortices which become distorted with streamwise distance. Even though this phenomenon has also been noted by many workers, including [1,2], very little is understood about the topological features of the flow. The flow pattern adjacent to the the pipe exit [3,4] is also not fully understood. The aim of this paper is to address some of these problems. It should be pointed out that this work forms part of an ongoing research program into the structure of jets in cross-flow.

Apparatus

The experiments were conducted in the closed-return water channel at the University of Melbourne. Figure 1 shows a schematic diagram of the apparatus used. The pipe was very short with the fluid entering it from a 36:1 contraction, thus giving a top-hat velocity profile in the case of no cross-flow. No external forcing was deliberately applied to this flow.

Results

Figure 2 shows the effect when (red) dye is injected from a small port below the pipe exit into the upstream side of the pipe (see figure 1). Figure 3 shows a similar flow case but with the blue dye injected from a circumferential slit in the pipe. Here the Reynolds number (U_oD/ν) is about 1000 and the velocity ratio (Vj/Uo) is about 4. These conditions were chosen because they show the features of the flow more clearly than at higher Reynolds numbers.

When (red) dye is released from the injection port, it seems to fill a small separation region which occurs just below the upstream lip of the pipe exit. Consequently, all the red dye seen in the photographs originates from this small separation region. The red dye marks the jet shear layer which rolls up naturally to produce distorted ring vortices. The sides of these vortices appear to tilt with the bending of the jet. These structures appear to pair regularly downstream. Figure 3 shows that at about 2D above the exit, the upstream and downstream sides of the vortex rings converge and the blue-marked fluid from the lee side appears to be convected around the red-marked fluid on the upstream side. This behaviour may give a clue as to the particle transport and mixing mechanisms between the jet and free-stream fluids.

It can also be seen in figure 3 that the blue dye is swept away from the front, thus indicating a very strong secondary flow developing near the exit of the pipe. In this figure, a second shear layer roll-up can also be seen. This roll-up appears to originate from the upstream separation and contributes to the formation of the counter-rotating vortex pair in the flow downstream.

To further aid our understanding of the flow at the upstream edge of the pipe, filaments of dye were injected from far upstream. Figure 4 shows that the dye filament impinging on the corner splits into two with one part flowing upstream along the channel floor and the other part flowing down the front side of the pipe against the main jet flow. The dye flowing onto the channel floor was swept upstream into the horse-shoe vortex system, whereas the dye flowing into the pipe was swept into the separation region within the pipe. Careful observation of the filament splitting at the corner revealed that a second separation region may exist immediately inside the pipe, below the corner. The size of this separation was very small indeed and may be related to the radius of the corner of the pipe and the channel floor.

Discussion and Conclusions

Figure 5 shows the authors' interpretation of the evolution of the jet shear layer vortex rings as they convect downstream. A filament of dye indicates the behaviour of the flow pattern in the centreplane. This interpretation suggests that the rings fold at the sides of the jet. The upstream sides tilt with the mean curvature of the jet, and the downstream sides become aligned with the direction of the jet, thus contributing to the circulation of the counter-rotating vortex pair. Similar mechanisms have been suggested by other workers [1, 5] for jets in cross-flow and Perry & Lim [6] for the structure of buoyant co-flowing jets. The present process is, however, affected by the presence of other vortex systems (not described in this paper). It should be noted that, unlike the flow of Perry & Lim, the jet shear layer rolls up on the lee side of the jet.

Figure 6 shows the authors' interpretation of the time-averaged flow pattern in the centreplane as seen by a stationary observer. Features such as the 'separation vortex' and the horse-shoe vortex system have been included, but detailed discussion of them is beyond the scope of this paper. The pattern is represented by critical points such as stable foci and saddles. This interpretation is topologically similar to the pattern described in [7] for a rectangular jet, although no upstream separation in the pipe was reported.

To complete the picture, the surface streamline pattern inside the pipe can be interpreted and related to the rest of the flow pattern. Figure 7 shows this pattern, presented as though the pipe was split at the lee side ($\theta=180^o$) and opened out flat. The upstream separation within the pipe occurs at a negative streamsurface bifurcation line which extends from a saddle point S at the centreline, then around the sides of the pipe to the edge of the pipe. When dye was injected from the port into the upstream side of the pipe, the dye was convected downwards (upstream) towards this bifurcation line, then across the front of the pipe before lifting away from the pipe edge at approximately $\theta = \pm 60^o$. The extent of the separation region decreased with increasing R. At the upstream edge of the pipe, a node is indicated. This point corresponds to point A where the filament of dye impinges on the pipe corner in figure 4. The existence of a second separation at the front edge of the pipe, as discussed above, would only affect the topology in the immediate vicinity of the node. In the large, the pattern would be as shown here. Other flow features have been included in this diagram, but discussion of them is beyond the scope of this paper.

The authors acknowledge the support of the Australian Research Council.

References

1. Andreopoulos, J. 1985 On the structure of jets in a cross flow. J. Fluid Mech. **157,** 163-197

2. Fric, T. F. & Roshko, A. 1989 Structure in the near field of the transverse jet. Seventh Symposium on Turbulent Shear Flows. 6.4.1-6.4.6.

3. Foss, J. F. 1980 Interaction region phenomena for the jet in a cross-flow problem. Rep. SFB 80/E/161, University of Karlsruhe

4. Andreopoulos, J. 1982 Measurements in a jet-pipe flow issuing perpendicularly into a cross stream. J. Fluids Engr. **104,** 493-499

5. Moussa, Z. M., Trischka, J. W. & Eskinazi, S. 1977 The near field in the mixing of a round jet with a cross-stream. J. Fluid Mech. **80,** 49-80

6. Perry, A. E. & Lim, T. T. 1978 Coherent structures in co-flowing jets and wakes. J. Fluid Mech. **88 part 3,** 451-463

7. Krothapali, A., Lourenco, L. & Buchlin, J. M. 1990 Separated flow upstream of a jet in a crossflow. AIAA J. **28,** 414-420

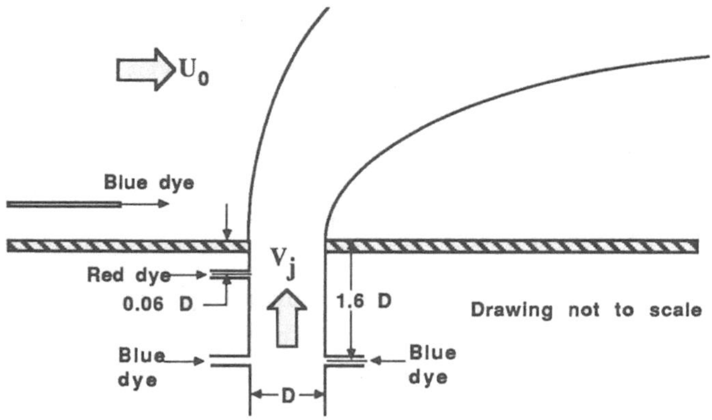

Figure 1. Schematic daigram of jet apparatus

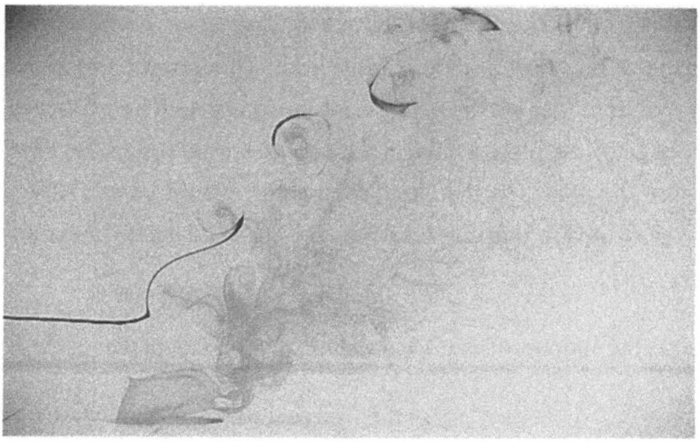

Figure 2.

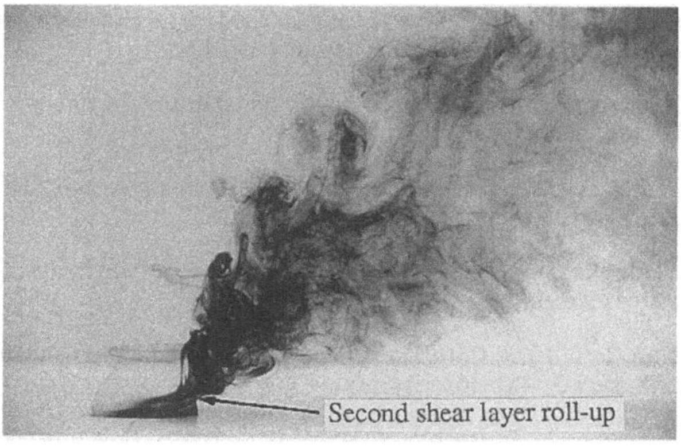

Figure 3.

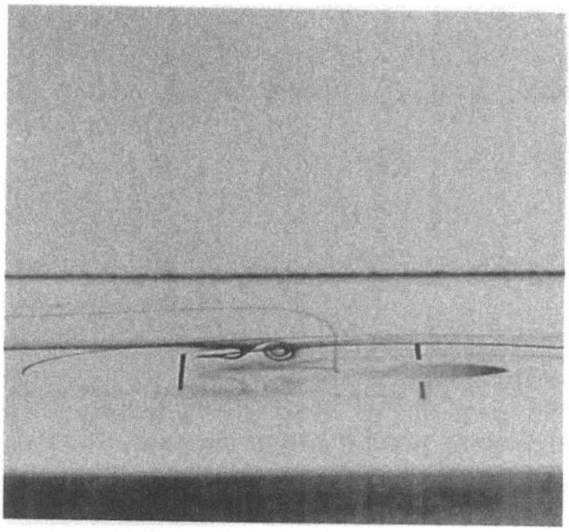

Figure 4.

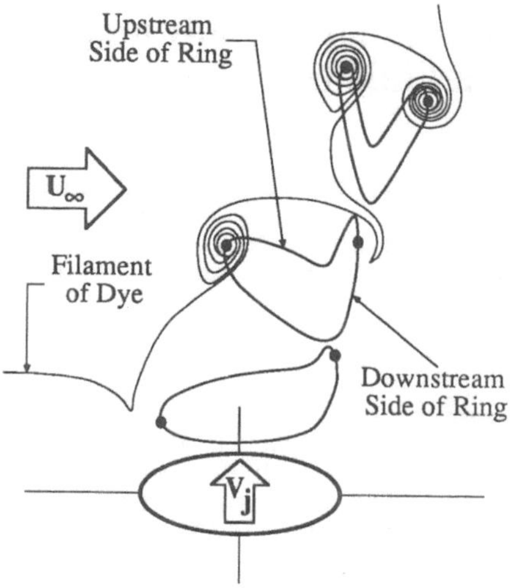

Figure 5. Perspective view of authors' interpretation of the evolution of the jet shear layer vortex rings as they convect downstream.

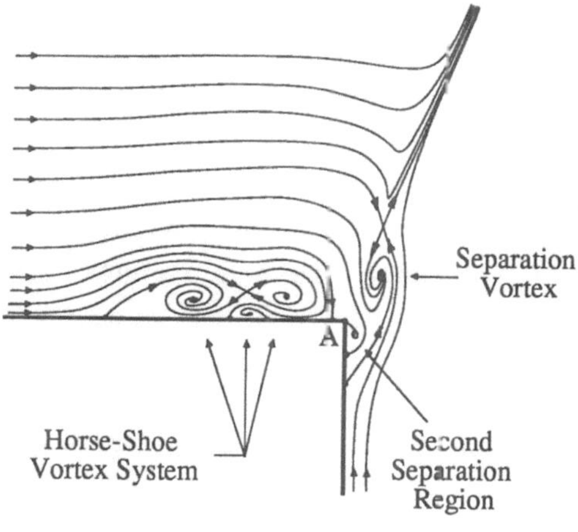

Figure 6. Authors' interpretation of the time-averaged flow pattern in the centreplane as seen by a stationary observer.

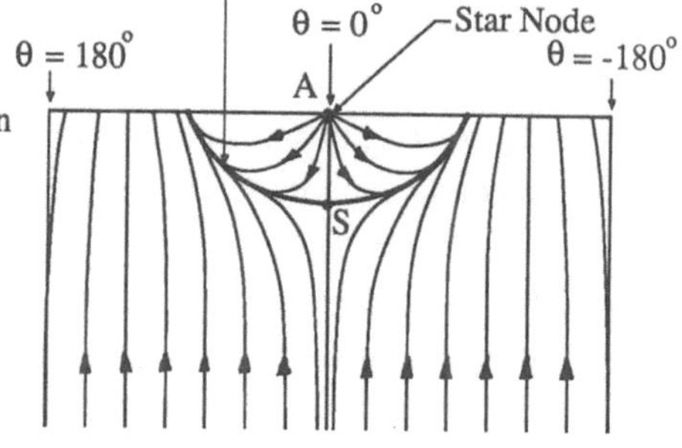

Figure 7. Surface streamline pattern within the pipe. The pipe surface is represented as a developed surface i.e the pipe has been split at the rear, then opened out flat.

Aerodynamic Study on Flow Characteristics of a Fluidics

Y.Oshima, Y.Watanabe, Y.Ishii, H.Kondoh and T.Takamiya

Research and Development Center, Ricoh Co.Ltd.
Kohoku-ku, Yokohama 223, Japan

Summary

Fluid flow in fluidics which composed of a jet nozzle, an oscillation inducer and an endblock arranged in one framework, is visualized by smoke in air and by hydrogen bubble in water. It is found that the frequency of the self-sustaining oscillation of the flow is linearly proportional to the fluid velocity at the jet nozzle and that the non-dimensional parameter, the Strouhal number, is almost constant regardless the Reynolds number.

1. Introduction

The self-sustaining fluid oscillation in the fluidics have been attracting much interests among the fluid engineers for many years. Recent studies are focused on the linear dependence of its frequency on the flow rate with the application as a flow-meter in mind.[1)-3)] Numerical simulation of such phenomena have been carried out in order to clarify the mechanism of the oscillation.[4)]

In this paper, the flow field in a fluidics, which is composed of a jet nozzle, an oscillation inducer and an endblock arranged in one framework without a particular return circuit, is investigated. Visualization by smoke in air and by hydrogen bubble in water is carried out, and the onset of the fluidic oscillation is detected.

A non-dimensional parameters of the flow rate, Reynolds number, Re is defined as U*L/Nu, where U is the mean flow velocity at the jet nozzle, L is the width of the nozzle exit, and Nu is the kinematic viscosity of the fluid. Also the non-dimensional parameter of the oscillation frequency, Strouhal number St, is defined as n*D/U, where n is the frequency of the oscillation, and D is the width of the oscillation inducer.

2. Experimental Apparatus and Method

A fluidics used in the experiment is made of thick acrylic

plate by whittling for visualization and the name of each element is shown in Fig.1. Air and water are used for the experiment. The schematic diagram of the experimental apparatus for air is shown in Fig.2. Two kinds of flow generators, an air compressor (the flow rate<600L/H) or a blower (the flow rate>400L/H) is used according to the flow rate.

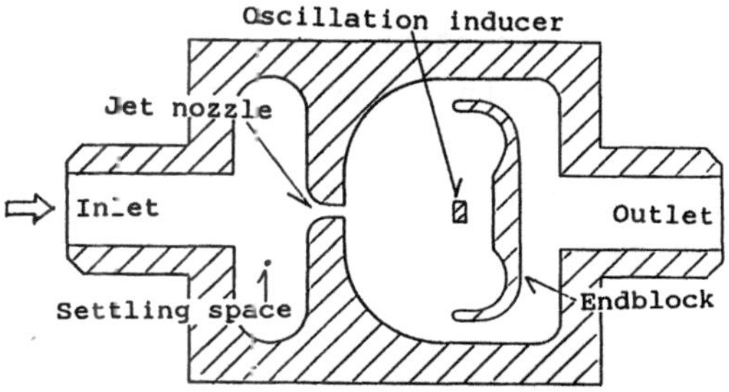

Fig.1 Fluidics and its elements.

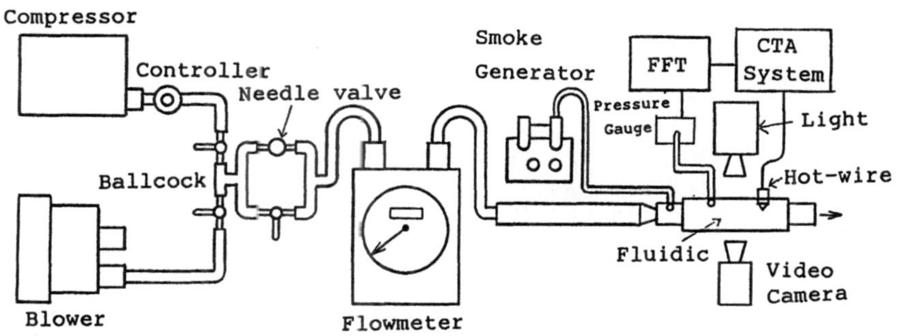

Fig.2 Schematic diagram of experimental apparatus.

Flow field is visualized by smoke injected into the inlet part of the fluidics. The smoke generator ejects mist of isoparafin continuously at a controlled the smoke density. Hydrogen bubble method is used in water. Video recorder is used to record these flow fields, and high speed video system is also adopted at the case of higher flow rates.

Both hot-wire anemometer and pressure gauge are used to detect the flow oscillation, and their output are analyzed by frequency analyzer using FFT.

3. Results and Discussions

In the low flow rate range, the flow ejected from the jet nozzle is evenly divided into two directions against the oscillation inducer, and they join together behind the endblock after flowing around it, as shown in Fig.3. Two vortices are formed between the inner wall and jet flow. As the flow rate increases, the flow leans toward one direction and looses balance due to the vortex formation behind the inducer. At this flow rate, the jet flow changes its direction due to the disturbance caused by returning flow along the inner wall of the framework, and it changes to the opposite direction. Thus, the sustaining oscillation begins. Figure 4 is the flow field visualized by hydrogen bubble in water at a flow rate of 15(L/H), in which a vortex behind the inducer is clearly seen. Figure 5 shows visualized flow field by smoke at a rate of 150(L/H).

Fig.3 Flow pattern of the lower flow rate.

Fig.4 Oscillation pattern of water visualized by hydrogen bubble method. Vortex is formed behind the inducer. Flow rate=15L/H.

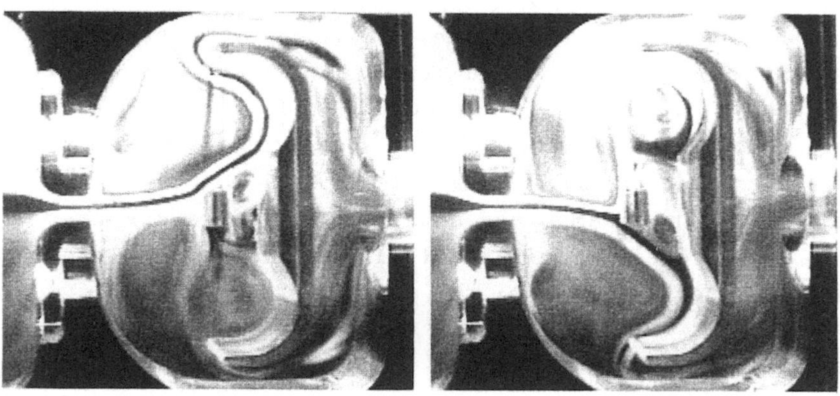

Fig.5 Examples of oscillation pattern visualized by smoke. Flow rate=150L/H.

The measurements of the frequency of the oscillation are carried out using a hot-wire and a pressure gauge in air and water. Figure 6 shows an example of hot-wire output of the air at flow rate of 150(L/H). As the flow rate increases, the frequency of the fluidic oscillation increases linearly. The outputs of them contain ripples caused by the small vortices. The output of the pressure gauge shows no ripples at the same flow condition, while the low level of output due to the large sensitive area.

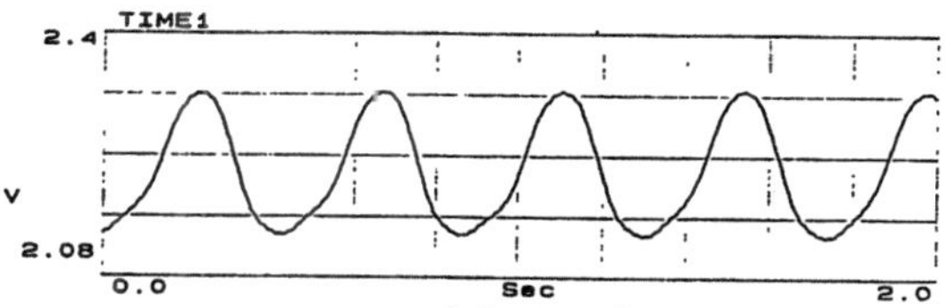

Fig.6 Typical wave pattern of hot-wire output. Flow rate=150L/H.

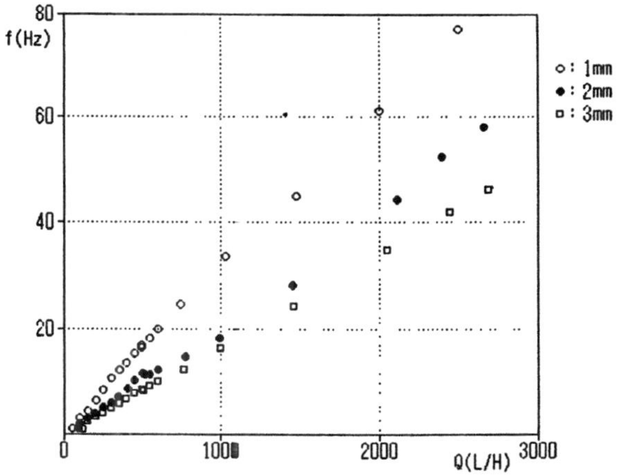

Fig.7 Relation between oscillation frequency and flow rate of air flow for various width of jet nozzle.

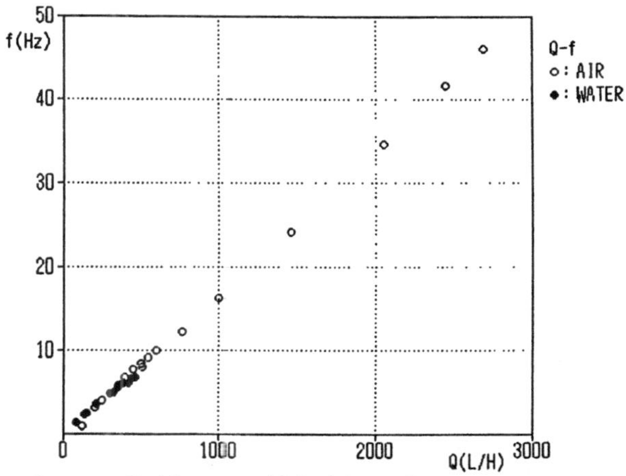

Fig.8 Comparison of the oscillation frequency vs. flow rate of air and water.

Figure 7 shows the relation between the oscillation frequency and the flow rate for various widths of jet-nozzle, and the relation for the flow of air and water with 3mm width of jet nozzle are shown in Fig.8. The frequency is linearly proportional to the flow rate. The relation in non-dimensional parameters, St and Re, is shown in Fig.9. The Strouhal number holds a constant value in the wide range of the Reynolds number.

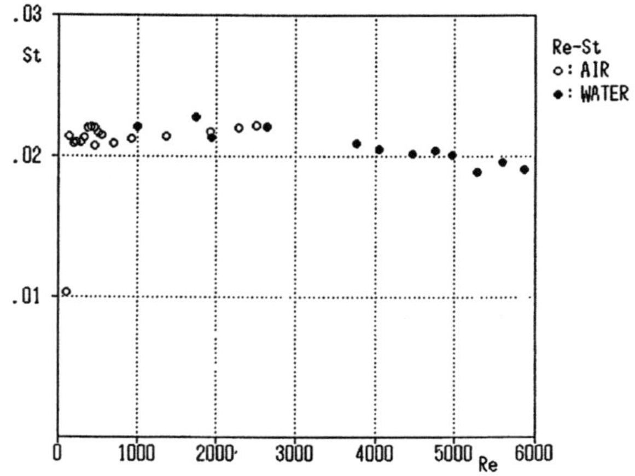

Fig.9 Relation between Reynolds number Re and Strouhal number St.

5. Conclusion

1. The frequency of the self-sustaining oscillation is linearly proportional to the flow rate.

2. The Strouhal number holds a constant value of 0.021 not depending upon the geometrical sizes and the species of the fluid.

3. The lower limit of the self-sustained oscillation is determined by the Reynolds number.

References

[1] H.Yasuda, K.Sakai and M.Okabayashi: Development of small-sized fluidic gas flowmeter, Int.Gas Res.Conf.(1989),261.

[2] M.Okabayashi and H.Yamasaki: Feasibility study of new fluidic gas meters, FLUCOME (1991),313.

[3] H.S.Kalsi, E.Markland, N.P.Samuel and G.M.Tofield: Factors affecting choice of fluidic flowmeters, 2nd Int.Symp.on Fluid (1988),1.

[4] A.J.Parry, S.G.Chiwanga, H.S.Kalsi and P.Jepson: Numerical and experimental visualization of flow through a target fluidic oscillator, ASME Conf.(1991),1.

Direct Shadow Method Application to Aero-Acoustic Interaction Investigation

V. G. PIMSCHTEIN
Central Aerohydrodynamic Institute (TsAGI),
Moscow Branch, Moscow, Russia.

Summary

Some investigation results for interaction process of saw-like sound waves of high intensity (160-170 dB) with axisymmetrical subsonic and supersonic jets are presented. Flow and sound waves visualization was performed by the direct shadowgraph method using an impulse source with exposure duration $2*10^{-7}$s. Peculiarities of disturbance emergence and development in a subsonic air jet and in a helium jet under acoustic excitation are considered. It is shown that large-scale disturbances arise on the acoustically excited side of a supersonic jet and the jet radiates directional sound at the frequency of the external excitation (Mach waves).

Results and Discussion

Experiments were carried out with air and helium jets flowing out of conic subsonic nozzles with diameters d=10-60 mm at pressure ratios in the nozzle $\overline{p}$ =0.2-7.8. Hartman's generators (HG) with frequencies f=1-11kHz and respectively radiated acoustic power 30-0.4kWt, were used as sound sources.External transversal sound excitation of the jet, including excitation with applications of parabolic and elliptic reflectors, was applied as well as internal longitudinal excitation. HG was located in the receiver of the investigated nozzle during the longitudinal internal excitation. Photographs of the aero-acoustic interaction process were taken by the direct shadow method using a spark impulse light with the size of an illuminatory body 0.8 mm, exposure duration 0.2 mcsec and discharge energy 2 Jl. Registration was done on a film with the sensitivity 1200 ASA, the size of one frame 200x500 mm, distance from the illumination source to the object of photography was 1200-1500mm. Application of the direct shadow method to investigation of the processes which take place in sound-jet interactions is based on the possibility of visualization not only ofjets but also of saw-like sound waves, intensive enough.

1. Subsonic jets

A toroidal vortex appears at the edge of the nozzle during a longitudinal internal acoustic excitation of the jet, when the maximum compression phase of a sound wave passes the nozzle outlet. Its dimensions increase exponentially at the initial length of the flow. This process of vortex generation is shown on Fig.1, in its development, together with the exciting sound wave.

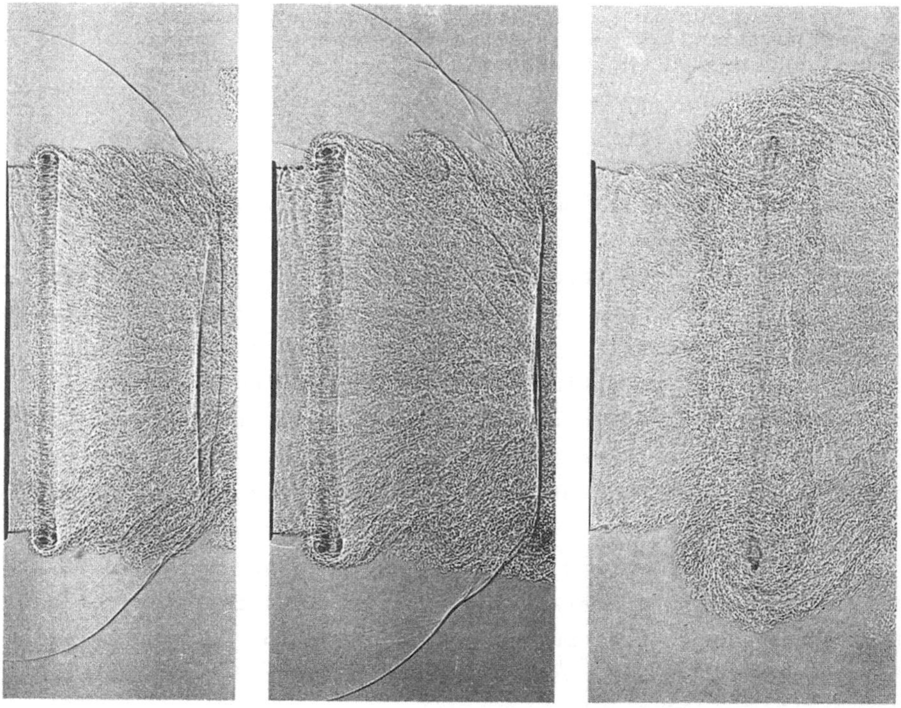

Fig.1.Vortex generation and development in a subsonic jet ($\bar{p}$=1.2; f=1.6 kHz)

The direction of propagation of any sound wave visible in the photograph here and below corresponds to the direction from light lines to dark ones in the picture of the sound wave front. All the waves visible in the photographs can be identified this way, including those reflected by the nozzle, radiated by the jet itself, radiated during vortex interaction and so on. Vortices increase in dimensions moving downstream only to a certain limit and then they change insignificantly: a chain of four vortices emerged during the consequent passage of four wave fronts through the nozzle outlet is shown in Fig.2.

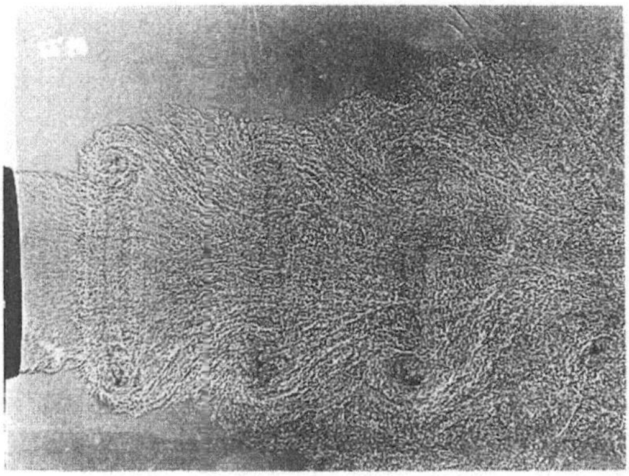

Fig.2. Vortex chain origin and sound wave interaction with earlier generated vortices ($\bar{p}$=1.4; f=2.5 kHz)

The same figure shows that the sound wave of considerably high intensity (170 dB) does not actually interact with the vortices already generated: the sound wave visible in the photograph generates a vortex nearest to the nozzle outlet and passes through the second and the third ones without producing any changes in them. The fourth vortex is in the phase of destruction.

It is known that the jet under a sufficiently intensive sound excitation expands much more rapidly than an unexcited jet. An especially considerable effect is achieved under external excitation of a helium jet with the high intensity sound (170dB). The half-opening angle of the jet attains $\alpha = 50^\circ$ (Fig.3).

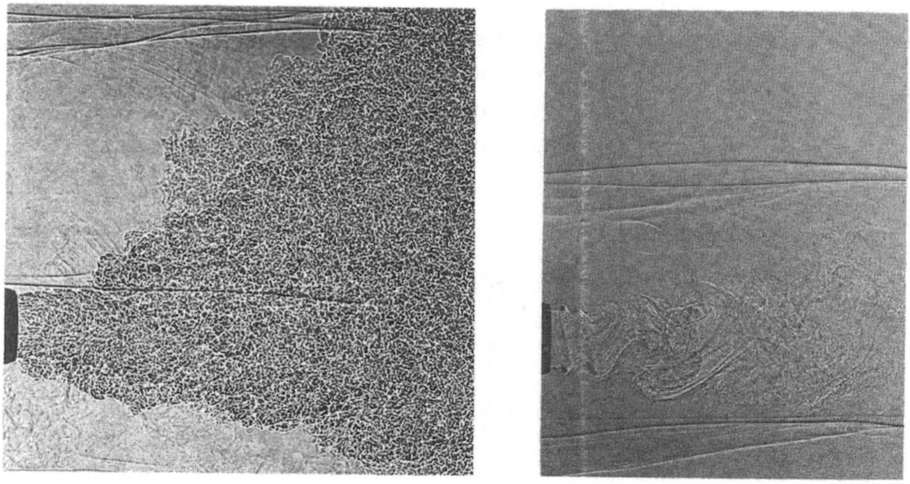

Fig.3. Helium and air jets under sound excitation ($\bar{p}$=0.8)

A photograph of an air jet excited under the same sound pressure level and pressure ratio on the nozzle are shown in Fig.3.

This result points out an absolutely new aspect in the question of sound intensity necessary to achieve a considerable jet expansion.

2.Supersonic jets

Disturbances arise on the irradiated side of a jet under sound excitation from an external sound source the same way as it takes place in the case of a sùbsonic jet, when the maximum compression phase of a sound wave passes through the nozzle outlet. These disturbances (if their convection velocity is supersonic) radiate Mach waves (Fig.4).

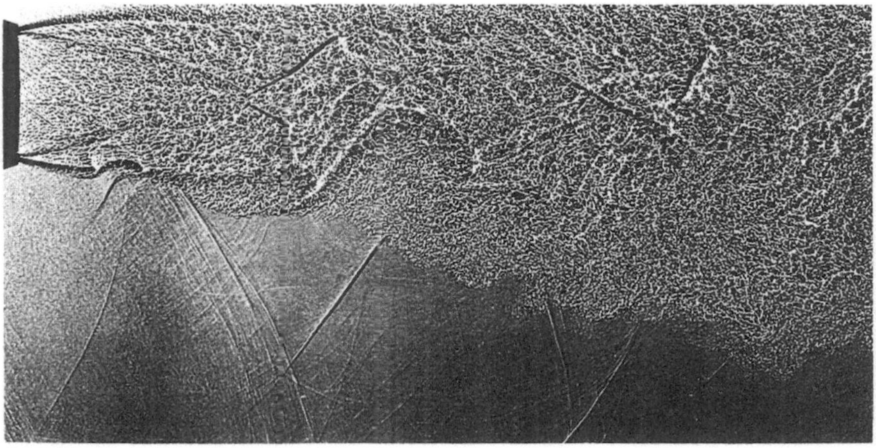

Fig.4. Disturbance generation in supersonic jet under sound excitation (M=2.0; r=0.8; f=8.5 kHz; α=30°; L=162 dB)

Disturbances posessing their own space scales and radiating Mach waves at corresponding frequencies (Fig.5) appear at the jet boundaries during the acoustic excitation of a supersonic jet by two independent sources at different frequencies on the opposite sides of the jet.

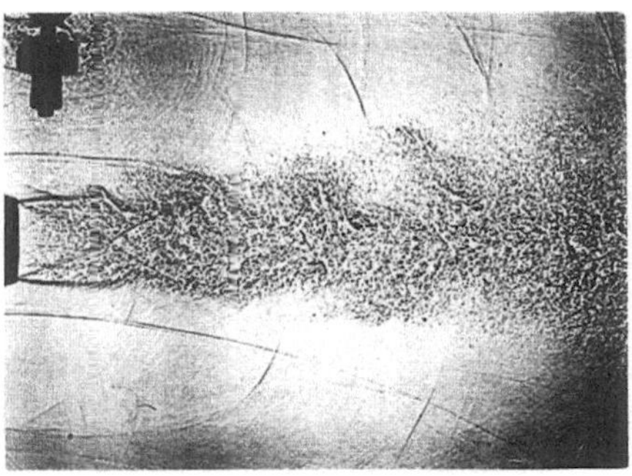

Fig.5.Jet excitation from two sides at different frequencies (M=2.0; n=0,8; f=8.5 kHz; f=11.8 kHz; L=164 dB)

It is possible to investigate a number of phenomena connected with simultaneous longitudinal and transversal sound excitation of jets, sound influence on vortices, sound emission during vortex interaction, etc., if one considers aero-acoustic processes in subsonic and supersonic jets by the method applied here.

Vortices

Visualization of the Turbulent Vortex Ring Oscillations

V. F. KOPIEV
Central Aerohydrodynamic Institute (TsAGI),
Acoustic Division, Radio str. 17, 107005, Moscow, Russia.

M. Yu. ZAITSEV , V. A. YAKOVLEV , L. P. GURIASHKIN
Central Aerohydrodynamic Institute (TsAGI)

Summary

The experimental investigation directed to visualization method development for fast proceeding processes in the turbulent vortex ring core is carried out. The data obtained support the existence and the possibility of registration of the theoretically predicted separate oscillation modes.

Introduction

The subject of the present study is a gas turbulent flow well known in the field of hydrodynamics as a vortex ring [1]. A vortex ring structure allows to distinguish between three characteristic flow fields: a toroidal vortex core, an ellipsoidal field surrounding the core and moving together with it (the so called ring "atmosphere ") and an outer flow. At high Reynolds number (Re>1000) the flow appears turbulent throughout the whole ring "atmosphere" , for the exception of the slowly altering vortex core in which the turbulence is suppresed [2]. The fact of the turbulence suppresion in the core is principal in the vortex ring dynamics and allows to apply the laminar theory which predicts the possibility of regular vortex oscillations [3-5] (Fig.1).

It was experimentally found [6], that vortex ring movements are followed by the narrow-band sound radiation which can be associated with its core regular oscillations (Fig.2). For vortex ring production the piston vortex generator with a removable nozzle set was used. The characteristic Reynolds number, predicted with reference to the initial vortex ring

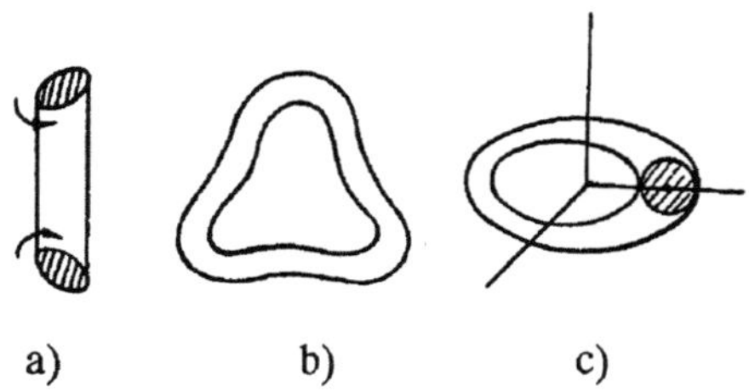

Fig.1. Vortex ring core oscillations: a- Kelvin's mode, b- flexural mode, c-barrel-like mode.

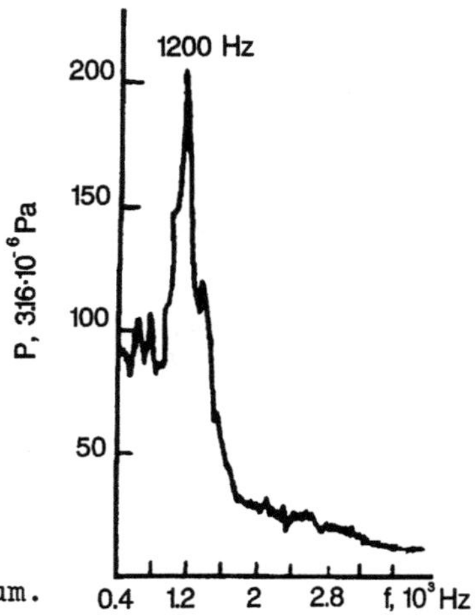

Fig.2. Averaged sound pressure spectrum.

velocity U=25m/s and the generator nozzle diameter d=5cm was Re=83000 which corresponds to turbulent flow conditions.

For direct investigation of the unsteady processes in a vortex ring core and for the possibility of comparison between real and ideal vortex dynamics the visualization was necessary. The only reference for the selection of characteristic frequency of high-speed photography was the acoustic experiment mentioned above. Thus, the exposure time and the minimum acceptable image frequency is strictly limited by the frequency of about 1200Hz which corresponds to the peak in acoustic spectrum from Fig.2.

Experimental Arrangements

The present work studied possibilities of visualization of fast non-stationary processes in the vortex ring core and methods of their high-speed and flash photography were worked out. An equipment set used in optics and physics investigations for high-speed cinematography and flash photography was applied here for realization of the questions raised. These are the following: pulsed ruby laser (impulse duration 30 ns); specially developed flash white-light source of extended duration up to 3 ms for the work with high-speed cameras;

industrial high-speed drum camera with the frame rate up to 20000 frames/s; a mirror shearing interferometer; a system for digital image processing based on PC 2001 "Pericolor".

<u>Flow Visualization</u>

Flow field structure features assosiated with the existing of the distinct boundary between the laminar core and the turbulent "atmosphere" and with fluid particles transition from the "atmosphere" into the wake require the development of adequate visualization methods. Two visualization types were used:
a) smoke visualization in reflected light (Fig.3a,e),
b) visualization with helium or ethyl steam admixture by means of the shearing interferometer (Fig.3b,d).

In the first case heavy smoke particles are pushed out of the core and the contrast is determined by the difference of smoke concentrations in the uncoloured core and the vortex ring "atmosphere". This method allows to visualize the vortex core at a distance not more than 40 diameters from the nozzle exit plane of the vortex generator because at a greater distance the smoke concentration in the core and in the "atmosphere" is equalized due to smoke transition from the "atmosphere" into the wake. Visualization with the help of the interferometer when helium and ethyl steam are used is on the contrary effective at a distance more than 20-30 diameters from the nozzle exit plane because the core is masked by the helium coloured small scale turbulence of the "atmosphere" in the initial part of the vortex trajectory. In this case visualizating substance transition from the "atmosphere" into the wake is a positive factor as it reveals the helium-coloured core. It is possible to visualize pure rings without any admixtures at a distance up to 30 diameters, if one uses the naturally originating density reduction in the core (Fig.3c). Only the central line of the vortex core appears visible, where the density is minimal. The flow intensity weakens as moving away from the nozzle exit plane and the density ratio appears insufficient to receive an image even with the help of an interferometer.

Single photoes were taken according to an impulse light-screen technique (exposure time 30 ns). A smoke ring entering the registration field was cut by a laser "knife" in longitudinal or lateral directions which permitted to obtain characteristic ring sections. In the case of interferometer the vortex ring image is a projection of the whole ring volume onto the plane.

Cinema-grammes of the vortex core were obtained with the help of a flash source of the extended duration t=3ms similarly to single photoes. In this case the process registration time was determined by the flash duration. With the maximum achieved frame rate 10000 frames/s there are 32 pictures on the film. Digital processing pictures were orginized as films demonstrating dynamical processes in the vortex core on the computer display.

Conclusions

The photoes taken during the work give an idea of momentary states of the vortex ring core and the developed registration technique with the maximum achieved frame rate allows to investigate fast dynamic processes in the core which do not exceed the frequency 1,5-2 kHz. The visualization results confirmed the presence of the vortex core oscillation modes . The most characteristic films demonstrating different modes were copied on a video.

References

1. Maxworthy, T.: Some experimental studies of vortex rings. J. Fluid Mech. 81 (1977), pp.465-495.

2. Vladimirov, V.A.; Lugovtsov, B.A.; Tarasov,V.F.: Turbulence suppression in the cores of concentrated vortices. J. Appl. Mech. Tech. Phys. 5 (1980), pp.69-76.

3. Kopiev, V.F.; Leontiev, E.A.: Radiation and scattering of sound from a vortex ring. Izv. AN SSSR, Mech. Zidk. Gasa 22(3) (1987),pp.83-95.

4. Widnall, S.E.: The structure and dynamics of vortex filaments. Ann. Rev. Fluid Mech. 7 (1975), p.141-166.

5. Kopiev, V.F.; Chernyshov, S.A.: Small disturbances of the stationary vortices. Izv. AN SSSR, Mech. Zidk. Gaza. 26(5) (1991), pp. 99-109.

6. Zaitsev, M.Yu.; Kopiev, V.F.; Munin, A.G.; Potokin, A.A.: Sound emission by the turbulent vortex ring. Dokl. Akad. Nauk SSSR 312(5) (1990), pp.1080-1083.

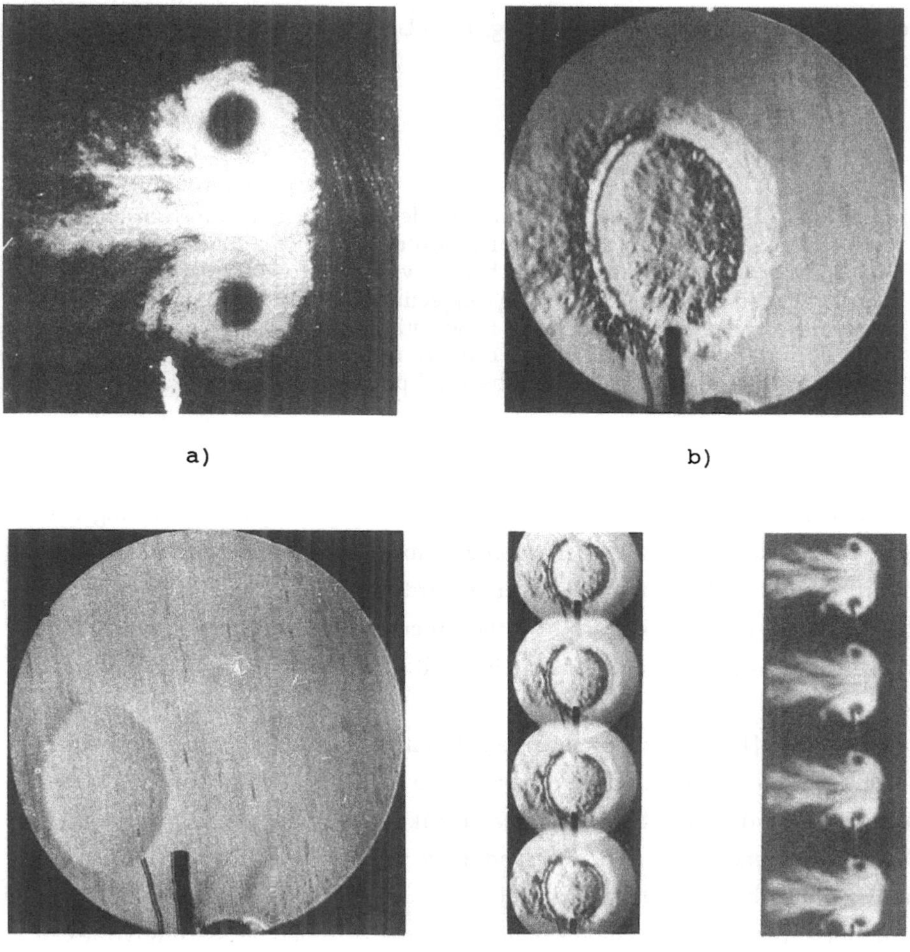

a) b)

c) d) e)

Fig.3. Vortex ring photographies: a- smoke ring, longitudinal section; b- vortex ring with helium admixture, a 3/4 front view; c- pure vortex ring. Cinema-grammes of vortex ring movements, 10000 frames/s: d- with helium admixtures, a front view; e- smoke ring, a side view. (There is a trigger microphone at the bottom of the picture)

Structure of a Starting Vortex

T. Leweke[1] and T. Sawada[2]

[1]Aerodynamisches Institut, RWTH Aachen, Germany
[2]Department of Mechanical Engineering, Keio University, Yokohama, Japan

The time-dependent flow structure inside a slender starting vortex, which was bounded in the axial direction either by two solid walls or by a wall and a free surface, has been investigated experimentally in water. Particle traces and the dye-injection method were used for velocity measurements and flow visualizations. Numerical simulations were also carried out in order to identify several mechanisms responsible for the observed phenomena.

Introduction

The main feature of the flow consisting of a concentrated vortex interacting with the boundary layer of a wall perpendicular to its axis, is an inward-spiralling motion of fluid in the boundary layer towards the vortex axis, driven by the radial pressure gradient imposed by the outer swirling flow, and which then turns into an axial flow away from the wall along the vortex core ("tornado effect").

Numerical (e.g. [1]) and experimental (e.g. [2]) investigations of this flow problem also revealed the possibility of a reversal of the axial flow on and near the vortex axis, as well as different types of vortex breakdown in the vicinity of the wall. These phenomena were also observed in a somewhat different situation: a vortex bounded by *two* parallel walls [3].

In the present study the time-dependent structure of a slender starting vortex of finite length is examined. Radial profiles of the swirl velocity at different times, measured in order to find suitable initial conditions for a numerical simulation, are presented. Attention is then focussed on the evolution of the axial motion inside the vortex, which is visualized for two cases: a vortex bounded by two parallel walls and one bounded on one side by a free surface. Finally, the flow structure in the region where the vortex axis meets a wall is shown.

Experimental methods

The experimental set-up is shown in Fig. 1. A flat metal vane of length $L = 500$ mm, width $B = 180$ mm, and thickness $d = 5$ mm was fixed at the rear wall of a rectangular plexiglass water container. The vane, which extended from one side wall of the container to the other, was free to rotate about the fixed edge. When the vane was impulsively (at time $t = 0$) brought to a rotation of constant angular velocity ω, a slender starting vortex was created along the sharpened free edge, the vortex being bounded by the container walls. The relevant parameter is the Reynolds number $Re = B^2\omega/\nu$, where ν is the kinematic viscosity of water.
Profiles of the azimuthal velocity in the vortex were obtained by Particle-Image-Velocimetry (see Fig. 2) using white Vestamid particles and a 10 mm light sheet from a 500 W xenon vapour lamp.

The axial flow inside the vortex was visualized by dyes injected just before the creation of the vortex at different positions along the free edge by means of thin metal cannulae taped onto the vane. For the investigation of a vortex bounded on one side by a free surface, the container was turned 90°, so that the vortex extended vertically, and it was only partially filled up.
The flow near a wall was visualized with the container turned 90°, by very carefully injecting dye or particles, which were slightly heavier than water, before the start of the vane rotation. The dye or particles spread along the wall and were then dragged along by the fluid motion.

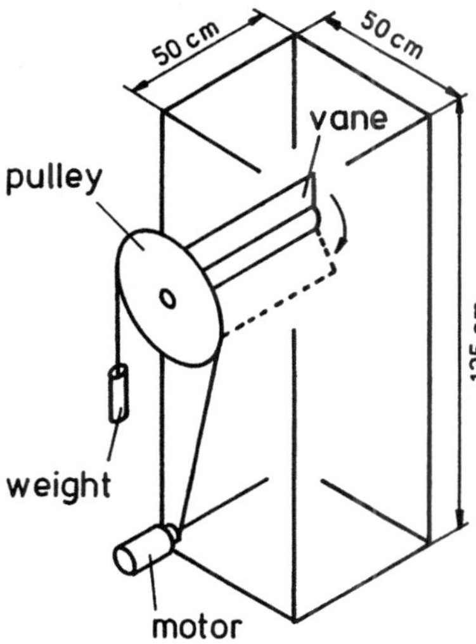

Fig. 1: Experimental set-up.

Results

Figure 2 shows particle traces in the symmetry plane for *Re* = 1900 and t = 5.7 s. The velocities obtained from these photographs were fitted to the fomula of Oseen in order to obtain the circulation Γ and the radius r_k of the vortex core. The result for Fig. 2 is shown in Fig. 3. During the motion of the vane ($0<t<24$ s) the circulation varies approx. with $t^{1/3}$, which corresponds to the theoretical result for the self-similar roll-up of an inviscid starting vortex at a semi-infinite plate [4]. After the stopping of the vane the circulation seems to decrease slowly. The evolution of the axial flow in the main part of the vortex is shown in Fig. 4. In the beginning, fluid is sucked into the vortex core from the wall boundary layer. Later on, this axial motion slows down and even reverses its direction in the immediate vicinity of the axis. The resulting 'inversion' of the conical dye patterns can well be seen in Fig. 5. This flow reversal is due to the modification of the pressure distribution on the axis caused by local changes of the azimuthal velocity profile by radial flow [5]. The same qualitative behaviour is observed in a vortex bounded on one side by a free surface (Fig. 6). This surface has the same properties as the symmetry plane in the first case, provided that the deformation caused by the the flow is weak, which is true for the range of *Re* considered here. Figure 7 shows the flow near the wall at *Re* = 1940, t = 15 s, which resembles the bubble-type vortex breakdown, i.e. a sudden expansion of the thin vortex core and the formation of a recirculation bubble (see e.g. [6]). The bubble in Fig. 7 is subject to spiral perturbations at later times. At higher *Re* the spiral-type vortex breakdown can be observed as well (Fig. 8). As shown by numerical simulations of the same flow [5], the breakdown is mainly caused by a sharp pressure minimum on the vortex axis near the wall, due to the advection of circulation towards the axis in the wall boundary layer.

This work was supported by EC-Contract No. SC1-0212-C (TT).

References

[1] R. Rotunno, J. Atmos. Sci. **36** (1979) 140-155.
[2] W.R.C. Phillips, Proc. R. Soc. Lond. A **400** (1985) 253-261.
[3] D. Auerbach, Exp. in Fluids **5** (1987) 385-392.
[4] E. Wedemeyer, Ing.-Arch. **30** (1961) 187-200.
[5] T. Leweke, to be published in Exp. in Fluids (1992).
[6] J.H. Faler & S. Leibovich, Phys. Fluids **14** (1977) 1385-1400.

Fig. 2: Particle traces in mid-plane at $Re = 1900$, $t = 5.7$ s.

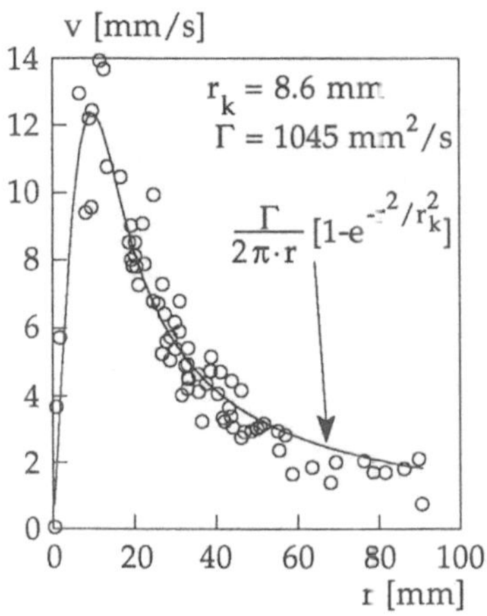

Fig. 3: Velocity profile in mid-plane at $Re = 1900$, $t = 5.7$ s.

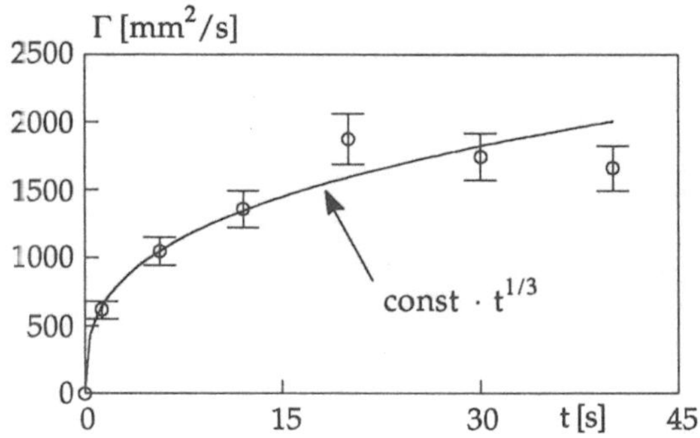

Fig. 4: Circulation in mid-plane at $Re = 1900$.

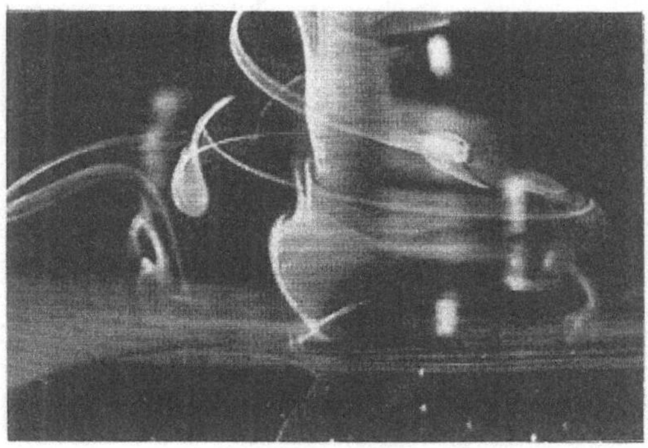

Fig. 7: Flow near a wall at $Re = 1940$, $t = 15$ s. bubble-type vortex breakdown.

Fig. 8: Flow near a wall at $Re = 4050$, spiral-type vortex breakdown.

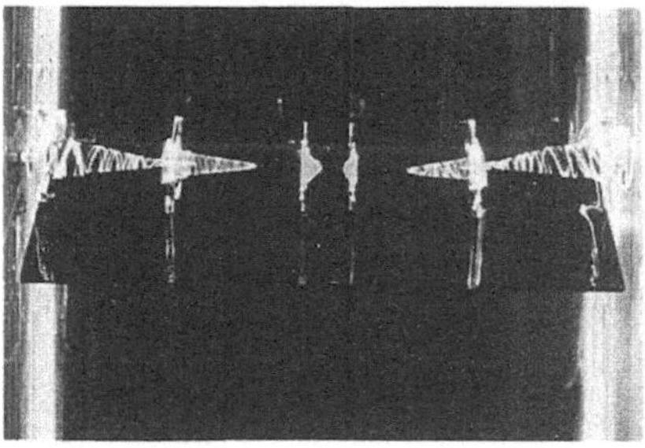

a) $t = 13$ s

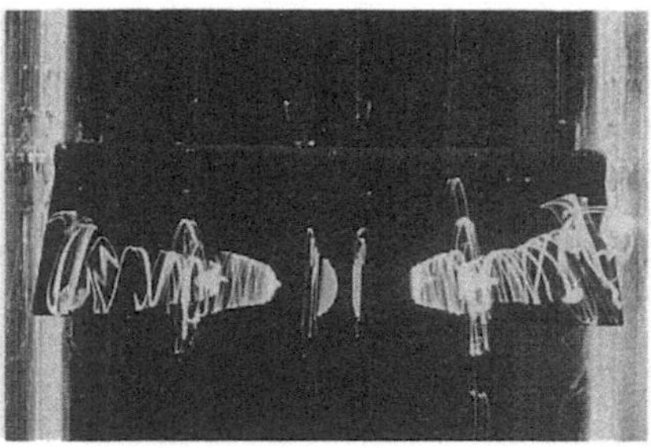

b) $t = 42$ s

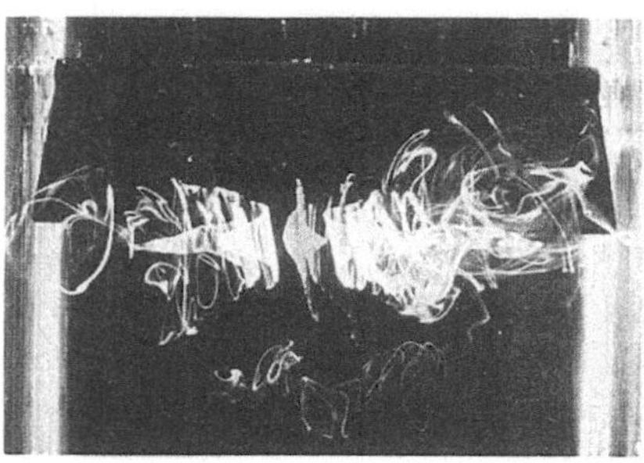

c) $t = 65$ s

Fig. 5: Axial flow in vortex bounded by two walls at $Re = 1900$.

a) $t = 7$ s

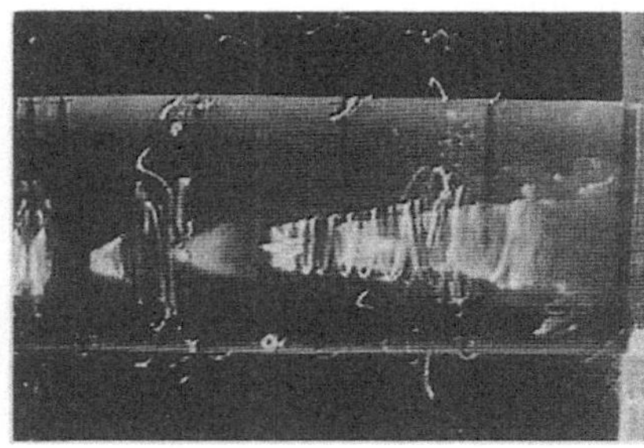

b) $t = 29$ s

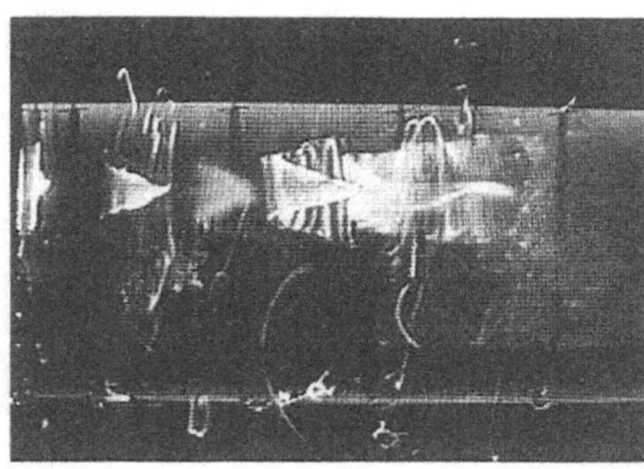

c) $t = 45$ s

Fig. 6: Axial flow in vortex bounded by a free surface (left) and a wall (right) at $Re = 4100$.

A Visualization Study on Necklace and Shedding Vortices Around a Three-Dimensional Protuberance

H. YAMADA and H. OSAKA

Mechanical Engineering,
Yamaguchi University, Ube, Japan

Summary

The behavior of a necklace vortex and a shedding vortex around a square plate protuberance standing vertically on a flat ground wall where the laminar boundary layer develops were studied using visualization techniques. Three different types of steady necklace vortex system and four different types of shedding vortex would appear against the various values of the Reynolds number and the boundary layer thickness. Furthermore, the interference pattern between the main necklace vortex and the shedding vortex in the wake region was examined.

Introduction

The aim of this investigation is to clarify the behavior of necklace and shedding vortices generated by a square plate protuberance. The structure of the shedding vortex formed behind the three-dimensional body of the various geometrical shapes such as a sphere[1, 2], a hemisphere[3] and a short circular cylinder[4] has been investigated in prior literature. On the other hand, the structure of the necklace vortex has hardly been investigated. Baker[5] shows that the pattern of the laminar necklace vortex systems formed in the frontal region of a effectively long circular cylinder standing on a flat ground wall would be determined by the value of the Reynolds number $U_0 \cdot d/\nu$ and the ratio of cylinder diameter to boundary layer displacement thickness. Because a thin square plate regarded as a three-dimensional protuberance has a finite height, the Reynolds number based on the plate height and the tip velocity must be considered to exhibit the behavior of the necklace vortex. It is also expected that the behavior of the necklace vortex will affect the structure of the separation vortex shed from the square plate.

In this paper the effect of both the rate of square plate height to boundary layer thickness h/δ and the Reynolds number $R_{eh} = U_h \cdot h/\nu$ on the behavior of the necklace vortex and the shedding vortex around a square plate standing vertically on the flat ground wall, where the laminar boundary layer develops, is studied using a visualization technique. Furthermore, the interference pattern between the main necklace vortex and the shedding vortex in the wake region was examined.

Experimental Apparatus and Techniques

The water open channel used in this experiment has an acrylic test section of 300mm width, 300mm height and 1100mm length. Owing to make the laminar boundary layer on the ground wall, a flat plate for the ground wall, having an elliptical nose, is submerged in the water flow at a position of 50mm height from the channel floor. Therefore, the velocity profile of the laminar boundary layer would correspond with Blasius's solution in the uniform velocity range of $U_0 \leqq 70$ mm/s. Fig. 1 shows the coordinate system and the nomenclature. The square plate for a three-dimensional protuberance has a thickness of 1.5 mm and a height of the range of h=5~56 mm, and it is set on the ground wall at the position of 150 mm downstream from the leading edge of the ground wall. The behavior of necklace vortex was examined by a dye injected from the tracer tube set in various positions from the ground wall. The behavior of vortex shed from the square plate was observed by issuing a dye continuously into the shear layer from its edge. Their behaviors were recorded by a still camera and a video camera. The binary image of the vortices structure is produced by processing the video signal sent to the personal computer.

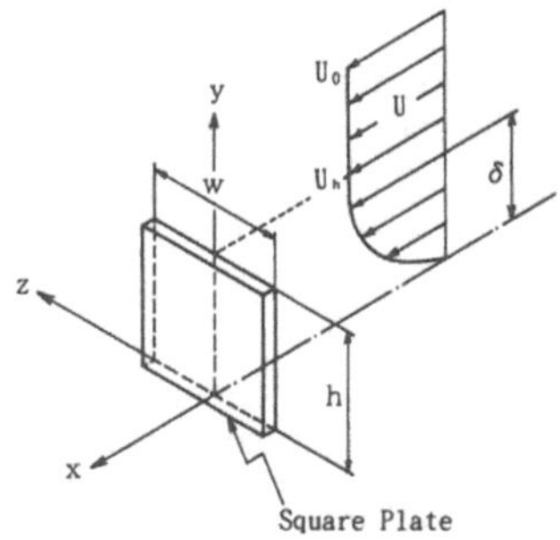

Fig. 1 Coordinate System

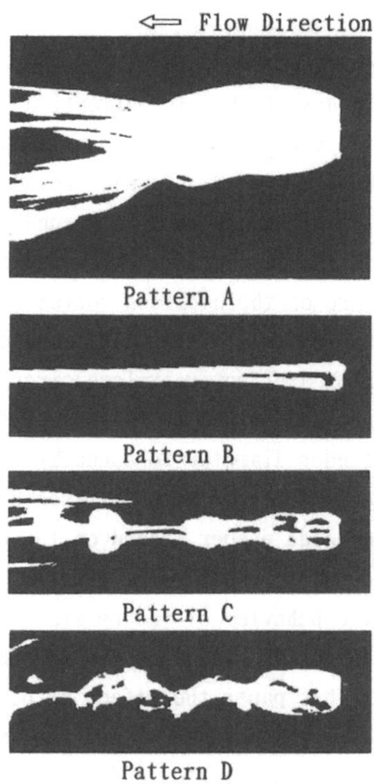

Fig. 2 Binary Image of Shedding Vortex

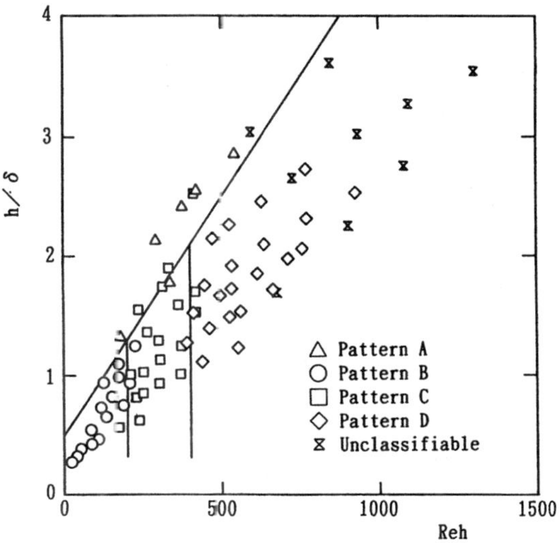

Fig. 3 Formation Condition of Shedding Vortex Patterns

Shedding Vortex Patterns

The behavior of the separation vortex shed from a square plate may be classified into four patterns, as shown in Fig. 2, which shows a horseshoe-like trailing vortex (Pattern A), an ordinary trailing vortex (Pattern B), a regu-ar hairpin vortex (Pattern C) and an irregular hairpin vortex (Pattern D). These patterns alter depending on both R_{eh} and h/δ as shown in Fig. 3. That is to say, in the range of about $h/\delta \geqq 4\times10^{-3}R_{eh}+0.5$ pattern A is observed. In the range of $h/\delta \leqq 4\times10^{-3}R_{eh}+0.5$, pattern B is observed in Reynolds numbers R_{eh} below 200, pattern C is observed in Reynolds numbers between 200 and 400, and pattern D is observed in Reynolds numbers beyond 400. In Reynolds numbers beyond about 10^3, however, the shedding vortex pattern becomes unclassifiable in this visualization study.

Necklace Vortex Patterns

In the range of the present R_{eh} and h/δ, three different types of the necklace vortex system, called patterns I ~ III, were formed in the frontal region of the square plate as shown in Fig. 4. Each pattern of necklace vortex systems is similar to the experimental results of Baker[5], therefore the patterns I, II and III can be regarded as a two vortex system, a four vortex sys-

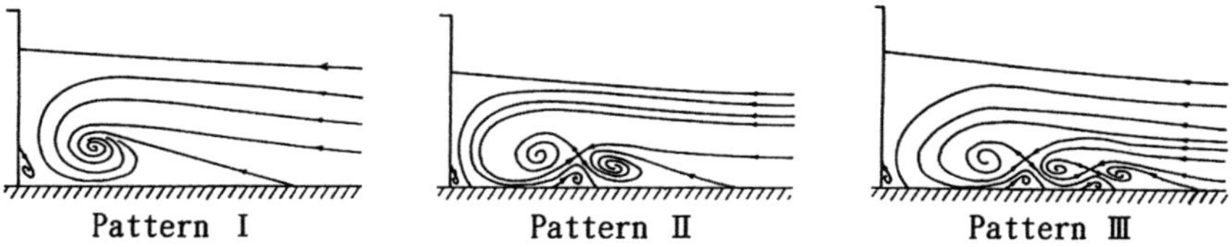

Fig. 4 Pattern of Necklace Vortex System

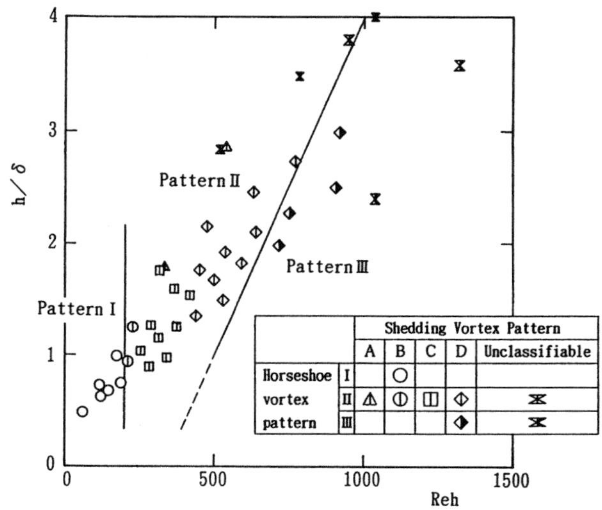

Fig. 5 Formation Condition of Necklace Vortex Patterns

tem and a six vortex system, respectively. Since a different vortex pattern appears in three condition zones of R_{eh} and h/δ as shown in Fig. 5, it is known that the pattern of necklace vortex system is determined by the variation of R_{eh} and h/δ. The pattern I is observed in the same Reynolds number range, of about $R_{eh} \leqq 200$, as the shedding vortex pattern B appears. The pattern II is observed in the range of about $R_{eh} \geqq 200$ and $h/\delta \geqq 6\times10^{-3}R_{eh}-2.0$ in which the shedding vortex indicates the patterns A, C or D, and the pattern III is observed in the range of about $h/\delta \leqq 6\times10^{-3}R_{eh}-2.0$ in which the shedding vortex is the patterns D. It is known that the necklace vortex pattern I and the shedding vortex pattern B are formed simultaneously, the necklace vortex pattern II appears whenever the shedding vortex is pattern A or C, and the necklace vortex pattern III appears whenever the shedding vortex is pattern D.

Interference between Necklace and Shedding Vortices

There is interference between the necklace vortex and the shedding vortex in the wake region downstream the square plate. Fig. 6(a)~(c) shows a sketch of simultaneous visualization of the main necklace vortex and the shedding vortex obtained using two kinds of color dye. The main necklace vortex is lifted up by the horseshoe-like trailing vortex in the wake when the necklace vortex pattern II and the shedding vortex pattern A appear at the same time as shown in Fig. 6(a). The main necklace vortex is located in a position below the ordinary trailing vortex when the vortex patterns I and B appear as shown in Fig. 6(b).
The main necklace vortex is entrained by the regular or the irregular hairpin vortices when they appear as shown in Fig. 6(c).

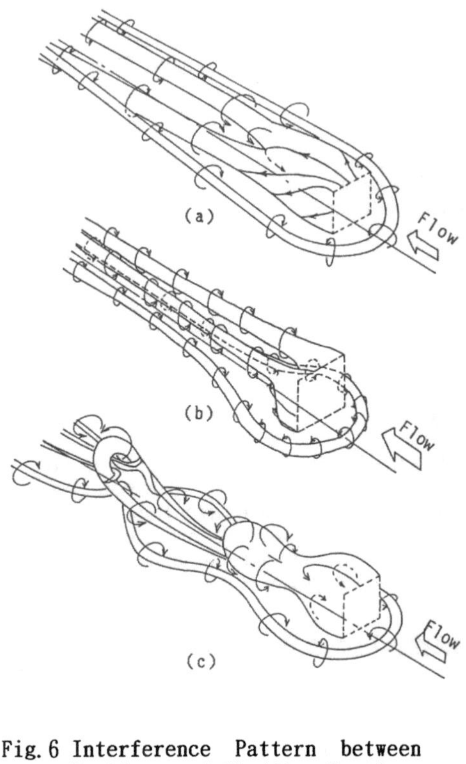

Fig. 6 Interference Pattern between Necklace and Shedding Vortices

References

1. Mochizuki, M.: Smoke observation on boundary layer transition caused by aspherical roughness element, J. Phys. Soc. Japan, 16-5 (1961) 995-1008.
2. Furuya, Y. and Miyata, M.: Visual studies on the wake of a roughness element proximate to a wall, Mem. Fac. Eng., Nagoya Univ., 24-2 (1972) 278-293.
3. Acarlar, M. S. and Smith, C. R.: A study of hairpin vortices in a laminar boundary layer (Part 1. Hairpin vortices generated by a hemisphere protuberance), J. Fluid Mech., 175 (1987) 1.
4. Gregory, N. and Walker, W. S.: The effect on transition of isolated surface excrescence in the boundary layer, ARC R and M, No. 2779, Part I (1951) 1-23.
5. Baker, C. J.: The laminar horseshoe vortex, J. Fluid Mech., 95-2 (1979) 347-367.

In-Depth Analysis of Couette-Taylor Flows by Means of Rotating Laser Beams

G. SEMON, J.P. PRENEL, R. PORCAR

Institut de Génie Energétique
2, Avenue Jean Moulin 90000 Belfort France
Tel. : (33) 84 57 82 00   Fax. : (33) 84 57 00 32

G. COGNET

Institut National Polytechnique de Grenoble
Avenue Felix Viallet
38000 Grenoble France

KEYWORDS : Light Sheet, Couette-Taylor flow, Laser Beam Sweep.

Summary

To visualize the Couette-Taylor flows it is essential to know about various observable singularities from which we shall set out an original device carried at permitting the observation of revolution symmetry flow. The system based on the rotation of a laser beam allows multiple adjustments thus offering a wide range of the exploration of flows. Our first results are set out and discussed on the hypothesis of an even more comprehensive development of the system in the future.

The Couette-Taylor flows

For about one century, the fluid mechanics have been fascinated and questionned themselves about the issue of the revolution symmetry flow. This growing interest on the one hand, and the diversity of the techniques which are used, offer a considerably innovative sphere of investigation. In such a vast field as that of the rotating flows, we shall single out a particular case, a great number of works refering to it, that is to say ; the Couette-Taylor flows.

Our team given to specifity, finds in such a wide issue, the chance to apply the techniques mainly directed towards the methods of analysis of hydrodynamic phenomenon which are principally directed towards the developpment of new concepts of visualization.

It's advisable, before getting into considerations connected with the optical device of visualization, to give precisions about the stratified nature of the Couette-taylor flows, indeed ; the static and dynamic structural complexity of these flows permits an infinity of various observations whose main evolutive steps can be summed up as follows.

We observe for a low rotation speed of the inner cylinder, a rotation movement of the fluid similar to that of a solid, the stability of the flow is checked if the square of the circulation increases with the radius, which offers then a first criterion of stability. Then instabilities appear having a certain axisymmetrical stratification as well as a toric shape. By taking into account viscosity in the criterion of stability, we can determine the size of the cells. The first visualizations carried out in larger fields bring to the fore other kinds of instabilities, indeed, after a new increase from a critical number of Reynolds, we find a second threshold from which the lines of separation are no longer seen in plans orthogonal to the axis, but they are enriched by real ondulations on each vortex whose distinctive characteristic is to move in the rotation direction of the inner cylinder.
The movement which is then more complex, has a double periodicity identified by a number of axial waves and a number of azimuthal waves. We observe that, depending on whether we are in accelerating or decelerating phase, the number of waves can vary and, the phenomenon can be accompanied by hysteresis. Once of our important observations, shows the existence of disturbances called 'turbulent', despite an axial stratification which can still be seen in the flow. Now, numerous researches are carried out so as to have a better understanding of the evolution towards the turbulence, then towards the chaos of the flow. The techniques applied are all of a theoretical nature which lead us to great mathematical investigation, taking for exemple the theory of bifurcation and the fractal study of chaos, in the same manner and with the utilisation of computers, numerous dynamic simulations of the fluid can be effected. Because of these techniques, our device is not marginal but it allows to have a concrete approach to the theory. In fact the visualization of the phenomenon expected by calculations induces an adequacy with no equivalent.

Experimental Device

Let have a device with vertical symmetry including two coaxial cylinders of the same length and of diameter Φ1 and Φ2, so that, between them a free space called annular space, can exist. From a dynamic point of view, both cylinders can revolve independently or not to a driving mecanism around a vertical axis. A transparent liquid into which are plunged submicroscopic glass particules or nylon ones is put between the two cylinders becoming thus the stand of the hydrodynamic singularities one wishes to observe.

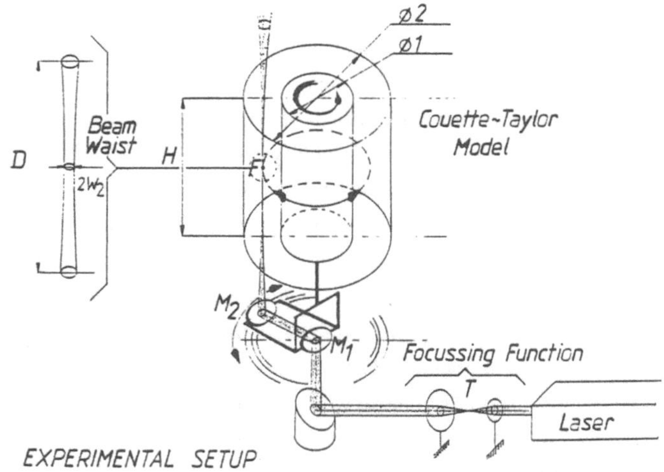

EXPERIMENTAL SETUP

The optical device connected with the previous system is made of two parts : the first one **T** is fixed including two spherical lenses making a telescope, the second one **M1** + **M2** is moving including two interdependent mirrors placed face to face. The laser beam after going through the optical device creates a generatrix of light, parallel to the axis of the two cylinders whose rotation induces a laser light pipe, the characteristics of which it will be advisable to be specified. An adjustment of the radial position of the generatrix permits a diametral adaptation of the cylindrical sheet of light to the space observation of the singularities in the whole annular space. It is obvious that the fineness and in particular the position of the beam waist depend on the range of the potential observations. We can say that with a shorter waist and an important depth of the field **D**, the precision of the tomographic cutting will be more in considering clearly the intern structure of the flow by the scattering of the laser light on the particles carried by the fluid.

Spotting The Waist With The Optical System

Initial waist (upstream) W_0, the more frequently of type " laser cavity "
Intermediate waist W_1, original point
Final waist (downstream) W_2, work zone

Parameter control

$$\Delta = d_1 + d_2$$

$$\Delta = (f_1 + f_2) + (\delta_1 + \delta_2)$$

The choice of δ_2 gives the work distance d_3

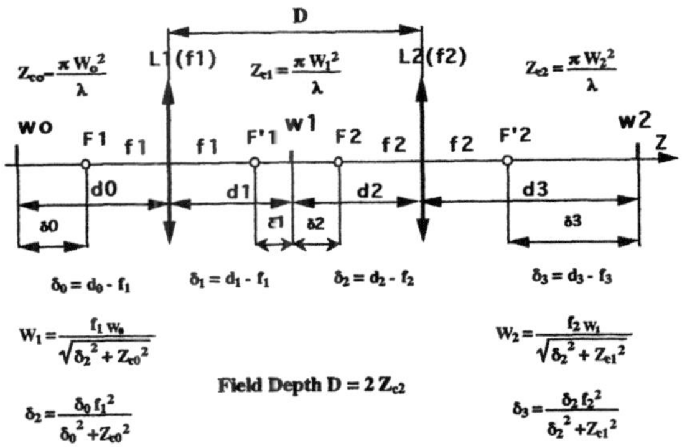

Results

We lay out the results of our first photographic investigations when the fluid is water or oil, the scattering particles are being small glass balls of 10 μm of diameter. Our device allowed cinematic exploration for the inner cylinder from 0 to 3200 rpm and 2500 rpm for the reflecting system **M1** + **M2**, let us remark that it is possible to integrate a metrologic criterion to the device by using the rotation of the beam which naturally induces a stroboscopic effect which we modify with a precision of a few Hz, the aim in this case is to fix the circumferential perturbations so as to calculate their rotation velocity. By taking a picture under the same conditions it is possible, by using the same units of measures to calculate the length of the azimuthal waves. Four types of information are then available, rotation speed of the singularities, photograph then dimensional study of the instabilities, and this study of the evolution of the instabilities on a wide scale of speed.

Conclusion

The first photographic results show a very good adaptation of the device to the observation of singularities which exist in the Couette-Taylor flows. After the improvement and the optimization of the various components of the system, we can point out within framework of the advanced development of the device, the possibility to bring a radial laser sheet contained in the annular space making then the possibility of the inner observation of the Taylor cells. The means of visualization can be completed by a device of laser velocimetry as well as a camera linked to the instabilities to be observed, the whole system already built is being tested at our laboratory.

Figure 2
Stratification of Couette-Taylor Flow

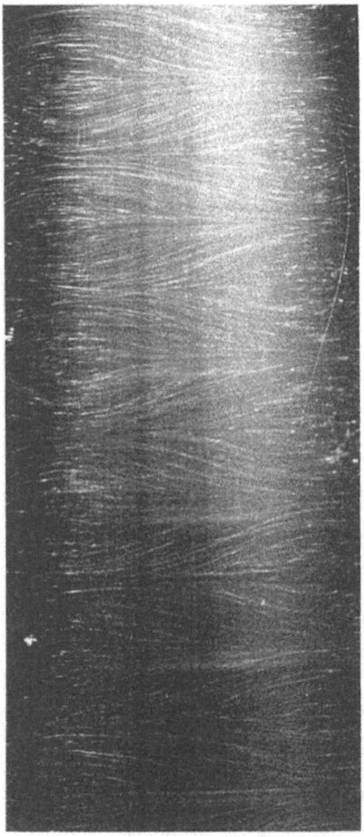

Figure 1
Internal Structure of Taylor Cells

(1) COGNET, G.
" The Way To Turbulence In Couette Flow Between Coaxial Cylinders "
Journal de Mécanique Théorique et Appliquée, spec, supp. , p. 7-44. 1984

(2) PORCAR. R, J.P. PRENEL, P. HAMELIN
Visualizations by means of coherent light sheets
Flow visualization III. Ann Arbor 1983 (USA)
Proceedings HPC Mac Graw Hill (Edit. W. J.Yang) . pp 123 - 27

(3) PORCAR. R, J.P. PRENEL
Un générateur de nappes cylindriques pour visualisations tridimensionnelles.
Colloque National Visualisation et Traitement d'Images.
Lille 1990 (France)
Proceedings ONERA - IMFL 1991 - (Edit. M. STANISLAS) pp 37 - 42

Visualization of Curtain Coating Flows

YUTAKA KASHIWABARA, SHOTARO OGAWA, KAZUO OZAKI

Development Group
Production Engineering & Development Division
Fujinomiya Factory Fuji Photo Film Co., LTD.

1. Summary

In photograph, paper, and steel industries, some products are made by coating a liquid on a web-shaped base material.

In the coating process, the condition of the coating bead affects the coating face. Keeping the coating bead stable is an important task. If we can visualize the shape and flow of the coating bead, we can begin to determine more stable coating conditions. However, the coating bead is very small, spread out along the web and the web moves quickly. Therefore it is difficult to visualize the coating bead.

Currently in Prof. Scriven's group at the University of Minnesota, visualization of the coating bead is being studied. Their method uses a glycerine-water solution, hydrogen bubbles and color dye. The coating types which they have tried are slide coating and slot coating.

In our experiment, we tried to visualize curtain coating flow. During curtain coating, the coating bead was less stable than that during slide coating or slot coating, because the coating die is farther apart from the coating bead. Therefore, its visualization needs a more highly clarifying technique. Moreover, we considered the influence of the web under several conditions of the coating bead.

In our results, we found some characteristic phenomena associated with the coating bead such as recirculating vortices and changing dynamic wetting lines.

2. Curtain coating

Curtain coating is the means by which a sheet of liquid falls on the surface of a solid sheet on which it is then carried away to be solidified into a permanent coating as shown in Fig.1. Fig.2(a) shows an ideal coating bead for steady flow. However, under some conditions coating flow can be unsteady as shown in Fig.2(b) and (c).

Therefore, visualization of the coating bead is very important in order to understand the actual coating flows and to jugde whether or not the conditions are ideal.

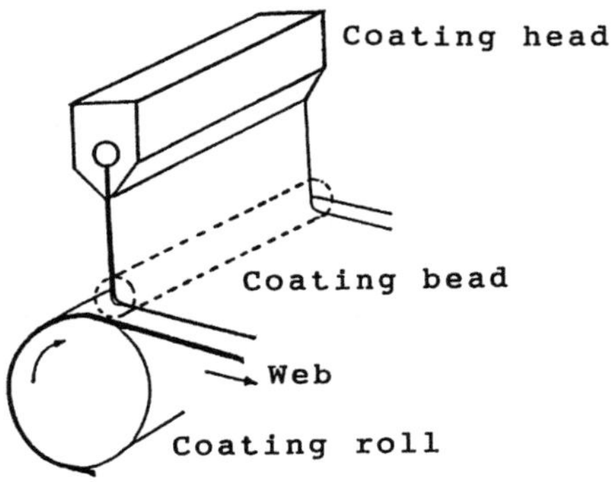

Fig.1 Curtain coating

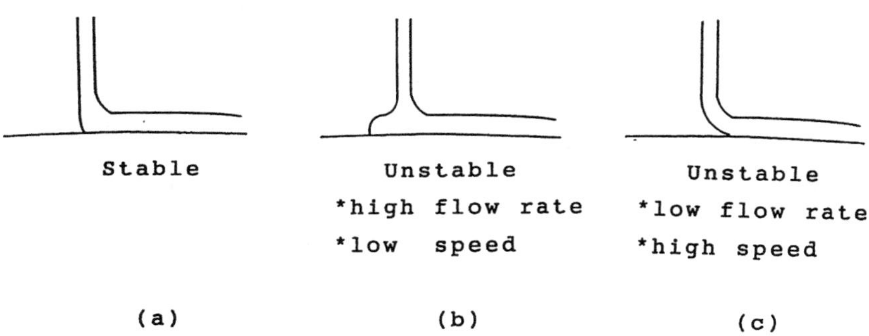

Fig.2 Coating bead

3. Visualization method

A transparent plate, tracers and optical instruments were used in the coating bead visualization as shown in Fig.3.

The transparent plate was set between the coating die and the coating roll. Hydrogen bubbles were used for the tracer in order to visualize the stream line in the coating bead. Color dye was used for the tracer in order to visualize the surface of the coating bead.

The visualization image is dark because the two-dimensional sections of the coating bead are visualized through the side plate and the liquid layer. Therefore, optical instruments which are able to visualize dark images were used. The optical section can be distorted or obscured by edge effects of the side plate.

A number of designs for the side plate were tried and a flat plate was chosen. It has good wettablity for the coating solution, to which was added a small amount of surfactant.

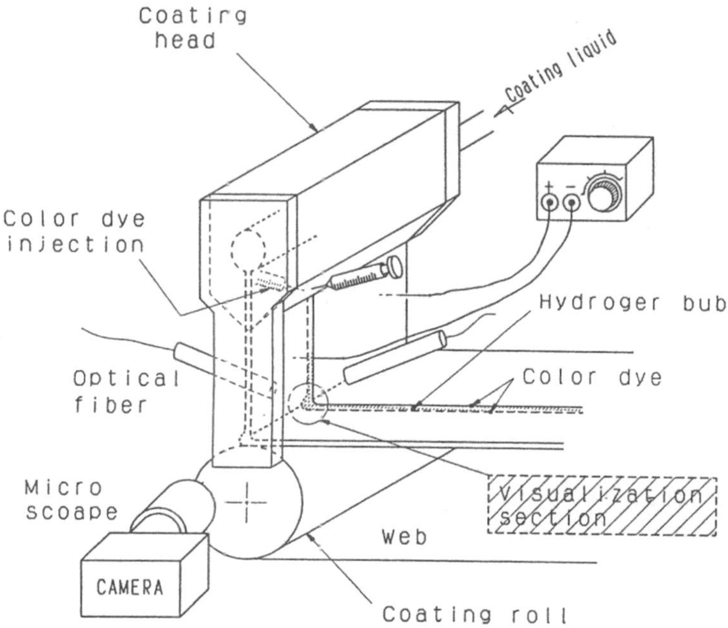

Fig.3 Visualization method

4.Result

One of the visualization examples of curtain coating flows is shown in Fig.4. The results show some characteristic phenom enaassociated with the coating bead such as recirculating vort i- ces and changing dynamic wetting lines.

These visualization results under certain coating conditions, illuminate characteristic behavior of curtain coating flow. They can help not only understand its flow mechanism but also identify optimal operating conditions and design for steady flow and uniform coating.

Fig.4 Visualization result

References

1.P.M.Schweizwer, J.Fluid Mech. 193(1988)285-302

2.L.E.Scriven, W.J.Suszynski, Chemical Eng.Progress, September (1990)24-29

3.S.Ogawa, L.E.Scriven, AIChE Spring National Mtg. (1992)41B

Turbulent Flows

Visual Representation of Single Point Measurements in Turbulent Spots

J.P. GOSTELOW

School of Mechanical Engineering
University of Technology, Sydney
Sydney, Australia

Summary

Measurements of triggered turbulent spots have been made in a flat plate boundary layer under zero and adverse streamwise pressure gradients. The paper gives samples of the velocity profiles throughout the spot and perspective views of integral parameters. Variations in disturbance level are displayed using graphics packages which facilitate representation of flow features.

Introduction

Use of single point measurements results in large quantities of data. For the data to be effectively analysed and presented visual representation is necessary. This paper describes some techniques in use for visual representation of the single point data.

Recent experiments have been oriented to provide turbomachinery designers with reliable correlations for boundary layer transition, thus addressing inadequacies in the prediction of transition length. Under adverse pressure gradient conditions, and for low free-stream turbulence levels, transition occurs rapidly; the time required for the velocity profile to adjust from laminar to turbulent may be greater than the transition length, resulting in a violation of similarity (1).

Linear computations of the transition region are predicated on a spatial division of the boundary layer into laminar and turbulent regions. It is assumed that transition occurs in discrete turbulent spots extending beyond the thickness of the laminar layer. Through the present project it is intended to provide information on the transition process by studying triggered spots.

Nomenclature

H	form factor
N	total number of phase-locked samples
q	disturbance level, %
t	time

U	free-stream velocity
u	local velocity
$\bar{u}^2(j)$	disturbance level at j'th sample point
$\bar{u}(j)$	ensemble-averaged velocity at j'th sample point
x	streamwise distance from leading edge
y	normal distance from plate surface
z	spanwise distance
δ^*	displacement thickness

Velocities

The experiments were performed in the 608 mm x 608 mm octagonal open circuit tunnel at UTS. The free stream velocity in these experiments was approximately 10 m/s with a turbulence level of 0.3%. Measurements were made on the upper surface of a 1500 x 608 x 25 mm flat aluminium plate having a high quality surface finish. The leading edge of the upper surface is of elliptical arc form and the plate has a 0.25° negative incidence to avoid leading edge separation.

The turbulent spot was triggered by a jet of air injected at a station 205 mm downstream of the leading edge. For adverse pressure gradient testing the jet was produced by a speaker driven by sharp-edged pulses of 0.4 ms duration having a 1 Hz repetition rate. The eleven 0.5 mm diameter injection ports were located at 8 mm spanwise intervals on the centre-line and to one side of it.

Velocities were measured using a Dantec 55M10 anemometer with a single wire platinum probe. The probe was traversed in streamwise and normal directions on the tunnel centre-line; its position was monitored by a dial gauge having a least count of 0.01 mm. A light source was focused on the probe tip, the reflection of which in the surface was used for accurate probe positioning.

The sharp edge of the square wave pulse driving the triggering jet was adopted as a phase reference. Each signal trace was digitised at 5 kHz using a Yokogawa DL3120 digital recording oscilloscope; 1000 readings were taken for each trace and stored on an IBM compatible computer. In order to obtain reliable velocity data a phase-lock averaging procedure was adopted for velocity traces; the set of traces corresponding to the spot's centre line are presented in Figure 1. Averaging over 128 cycles has resulted in traces which reveal the structural features consistently whilst obviously smoothing out most of the variability from one realisation to the next. Velocity variations have been measured at eight spanwise locations and for four different stages of spot development.

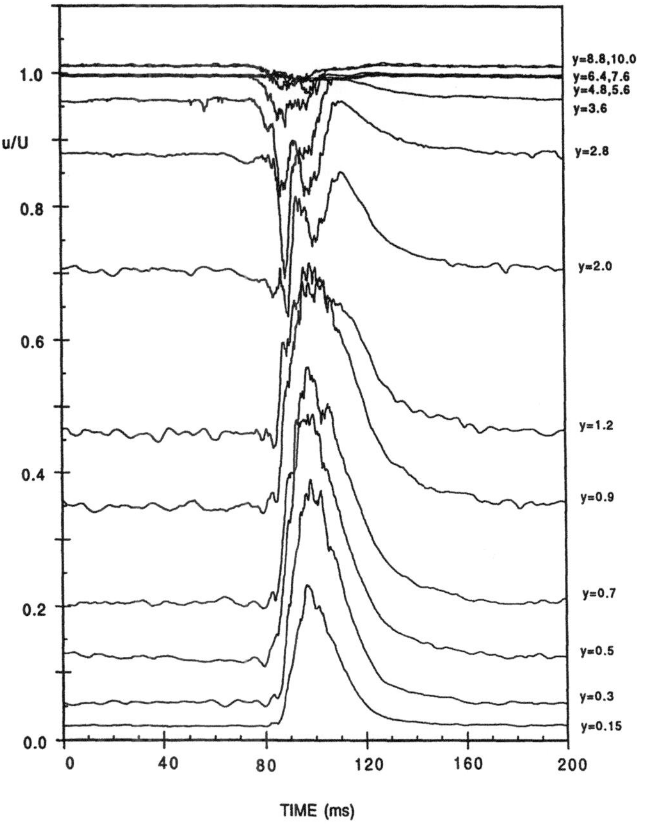

Fig. 1. Phase-averaged velocity traces for different heights through the boundary layer.

By interrogating the resulting y-t arrays of averaged velocity data, for each recorded time interval, velocity profiles throughout the spot could be assembled. These were integrated to give integral parameters of the boundary layer throughout the spot region; results for displacement thickness are presented in Figure 2 using the Surfer package (2). Boundary layers are considerably thinner immediately after the spot than at its leading edge.

Disturbance Levels

The array of velocity data was analyzed for variations in disturbance level in order to map the geometry of the spot and of any attendant disorderly activity. The variation of disturbance level throughout the spot was intended to assist physical understanding. The disturbance level of the streamwise velocity component is defined as

$$\bar{u}^{\,2} = \lim_{T \to \infty} (1/T) \int_0^T (u-\bar{u})^2 dt. \qquad (1)$$

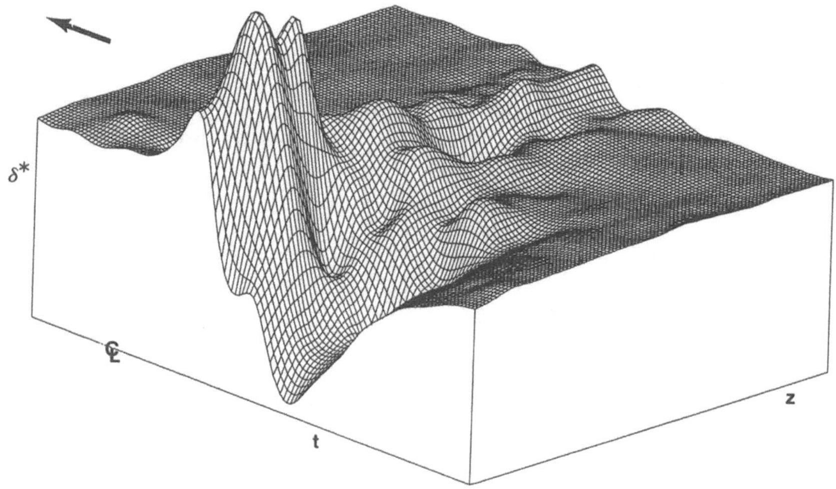

Fig. 2. Three-dimensional representation of displacement thickness.

Fig. 3. Two-dimensional slice of velocity disturbance in a spot.

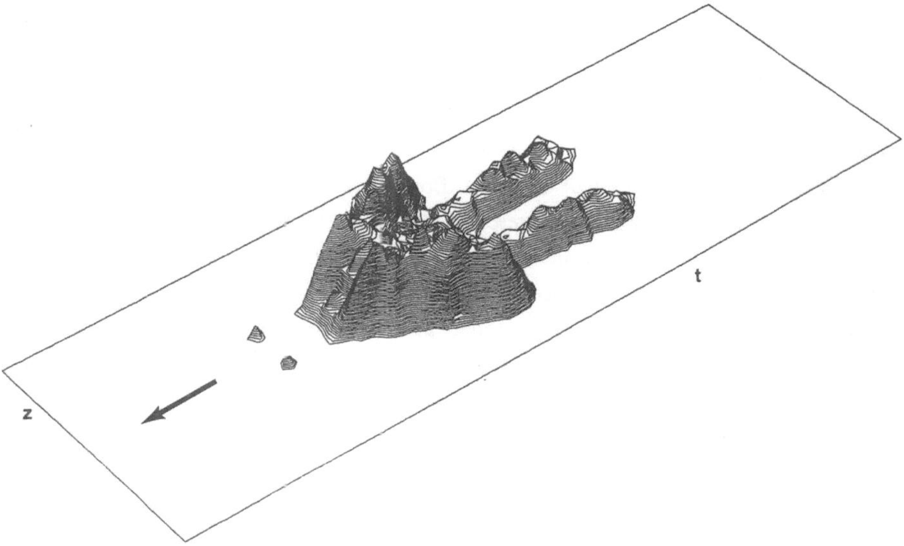

Fig. 4. Three-dimensional representation of disturbance level at critical height.

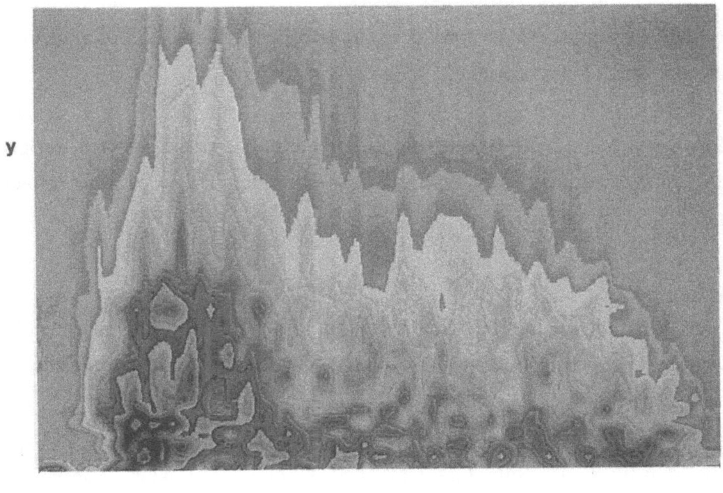

Fig. 5. Disturbance level at spanwise location 20% out from centre-line.

On the basis of the above data disturbance levels in the spot could be displayed in various ways. A representation of disturbance level on the centre-line using the Spyglass package (3) is given in Figure 3.

Spots tested under a zero pressure gradient and excited by a relatively energetic pulse exhibited extensive trailing lobes of disturbed flow. Contours of disturbance level at a height of 2.7 mm above the plate are presented in Figure 4. For presentation purposes the spot has been assumed to be symmetrical about the centre-line. A representation of disturbance level at a spanwise distance of 16 mm from the centre-line is given in Figure 5. On the basis of this example it was possible to identify saddle points associated with the vortex structure which moves into the spot from upstream. Further details of this case are given in Ref. (4).

Conclusions

These measurements of a turbulent spot under an adverse pressure gradient are a sample of a more extensive data set. The aim of this project is to obtain sufficient information to give a more accurate modeling of the spot in predictions of transitional boundary layers for turbomachinery design.

Phase-lock averaging was used; velocity profiles throughout the spot and its surrounding region were plotted and integral properties were derived throughout the region. The array of velocity data was analysed for variations in disturbance level in order to map the geometry of the spot and of any attendant disorderly activity.

The thin boundary layer behind the spot is complex in structure having, in part, properties usually associated with a laminar layer and yet also exhibiting zones of high turbulence close to the wall.

Acknowledgements

The author wishes to acknowledge the contributions of Dr. G. Hong, N. Melwani, J.K. Lee, W.T. Yu and D. Wan and the financial support of the Australian Research Council.

References

1. Gostelow, J.P. and Walker, G.J. Similarity Behavior in Transitional Boundary Layers Over a Range of Adverse Pressure Gradients and Turbulence Levels. *ASME, Journal of Turbomachinery,* Vol. 113, 617-625, 1991.
2. Surfer, Golden Software, 807 14th St., Golden, Colorado, 80401.
3. Spyglass Transform 2.1, Spyglass Inc., 701 Devonshire Drive, Champaign, Illinois 61820.
4. Gostelow, J.P., Hong, G. and Sheppeard, M.A. Investigations of a By-pass Spot with Reference to Boundary Layers on Turbomachinery Blading. *Proceedings 4th International Symposium on Rotating Machinery,* Honolulu, 1992.

Measurement of the Spatial, Stationary and Time Variable Velocity Distribution of Airflow Using Tracer Particles and Still Video Techniques

Dr.L.Besse, G. Gottschalk, Dr. A. Moser, Dr. P. Suter
Swiss Federal Institute of Technology (ETH)
Laboratory for Energy Systems
CH-8092 Zurich / Switzerland

Summary

This paper describes on a method to experimentally measure the three dimensional velocity distribution of the airflow in rooms. The velocity is measured at many random points by imaging small 'weightless' tracers (i.e soap bubbles with diameters of ca. 3mm) carried by the air stream. The area to be analyzed is illuminated along the direction parallel to the main velocity components (x-y plane) in three colored stripes. Color changes of a particular tracer trajectory delivers information in the third dimension. Visual information of the scene is captured with a still video camera in a timed multi-frame sequence (25 frames) and stored on a video floppy disk. This information in the form of images is subsequently processed in a computer to explore the spatial structure of the air velocity field in the observed air volume and time frame.

The length and the direction of the particle trajectory is needed to calculate the stationary velocity field. Each single frame produces a snapshot of the paths of the soap bubbles during 1/8 or 1/16sec. In the sequence shooting mode the camera can capture a total of 25 frames at the maximum speed of 8 images/second. Typically, this short time interval is required to analyze small airflow changes, i.e., deviation from the stationary observation, and experimentally predict the turbulence.

Keywords: Physically-based modeling of airflow, turbulence, image processing, pattern recognition, feature extraction, connectivity and frame matching, tracer particles in airflow, image capture with still video technique.

Introduction

Attempts to experimentally observe the dynamic behavior of airflow in rooms are often limited either by the tracer techniques (i.e. smoke particles) or the recording techniques (i.e single shot, long exposure). The development of modern techniques with timely and precise capture of image sequences, the use of computers for subsequent image processing, and finally the data analysis have made quantitative measurements in this field feasible. Displacement ventilation (i.e. air-conditioning) is still the most applied method to remove dissipated heat in office and in manufacturing space. The need for a good and efficient ventilation within the range of human comfort has always been considered important.

Model based computer simulation of the airflow in existing spaces often deviates from the realistic situation. The computational principle is often to simple; the room is divided into a number of volume cells (30 x 30 x 30), say, 27,000 cells, and the balance equations of mass, momentum and heat are solved iteratively for those cells. The typical computational cost to calculate one parameter-set is more than 2 hours on a CRAY type super-computer. Computational fluid dynamics methods are especially promising when used in large, minimally furnished or empty rooms. This includes gymnasiums, halls, and shopping centers or wherever experiments are difficult and the comfort requirements are low. Other areas such as office space with heat sources, furnitures, and humans with requirements for acceptable thermal climate need a precise simulation. These small details, such as a computer and a printer on the desk, cannot be handled in most practical simulations due to the limited computer capacity and the incomplete turbulence model. Therefore, it is important to have a tool to compare the effects of the simplifying assumptions with the real air flow situation.

An alternative approach is *visual observation* of the air flow using tracer elements, where the resolution of the observed area is much smaller, and hence more accurate than the computational grid.

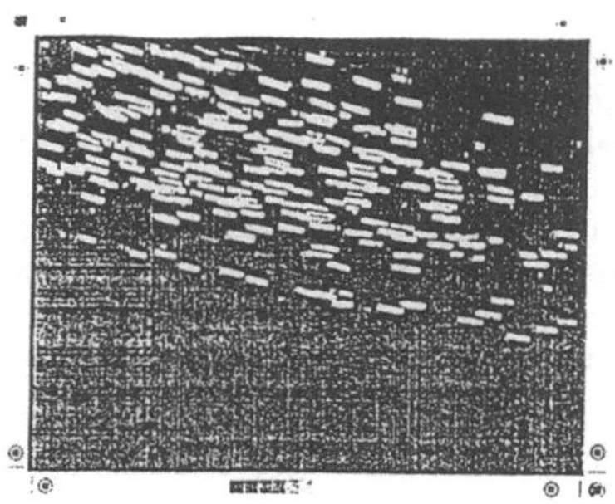

Figure 1 + 2: Soap bubbles travelling in the room and an example printout of the electronically recorded image sequence.

In this paper we describe the improved technique to calculate velocity variation in a low speed air field from a set of timely captured instantaneous vector information. We know from a practical point of view that most of the natural flows are turbulent (i.e. chaotic). Observation of the tracer particle motion over time has also indicated a practical instability in the flow structure. Here, we present a method to demonstrate quantitatively the turbulent characteristics in a room segment registered with 25 delayed snap-shot measurements.

Tracer Techniques

The basic principle of tracer particle image-velocimetry are well known and published by several authors [1][2]. These technique can be described as follows. The illuminated particle create an image on a photo-sensitive media (film or CCD array) in a specific time. This results in a particle path that indicates the speed and the direction of the flow field.

The soap bubble tracer method used for the air-flow has been experimentally developed in our ETH laboratory. The bubbles are produced by a generator in still air and injected into the experimental space. It has been observed that the neutral buoyancy results at a diameter of 3 - 4 mm. The average bubble life time exceeds several minutes [3].

Imaging Techniques

The velocity of soap bubble is calculated based on the particle displacement and the time interval between successive images (frames). In our velocity study, we use 25 consecutive time frames from the same observation area (1x0.7x0.2meters). The CANON RC-760 still video camera is used to record this area on a video floppy disk by using its high-resolution CCD-chip (600,000-pixel and color). In addition to every image on the floppy disk, the most important features such as shutter speed, aperture, time, etc. are also recorded. The observed field in the x-y plane spans over a 640x480 pixels area. Each pixel is stored with full color information (24bits) and the scene is illuminated with color light bands in the z-direction, which makes it easy to detect pathlines moving in direction perpendicular to the x-y plane.

Low-level Image Processing

Extraction of particle traces is based on a good visual track to noise relation. Even the best quality image frame will have a certain level of background noise and other pixel distortions within the context of the airflow information. Therefore, standard image enhancement strategies are applied first. The first step removes "salt and pepper noise" using a non-linear intensity transformation, and subsequently the "facet filter" removes single small distortions leaving only the tracer paths untouched.

Feature Extractions, Tracking

The velocity of the soap bubble is calculated based on the particle displacement and the time interval between successive image frames [4]. The feature extraction is performed by searching for connected pixels (pattern) in a single frame, starting at the upper left corner and ending in the lower right corner. If the path is found to be consistent, the position (x, y) and velocity components (vx, vy) are determined and stored into the vector list. Next, the time ordered frames are searched for traces corresponding to the same particle in order to model the airflow.

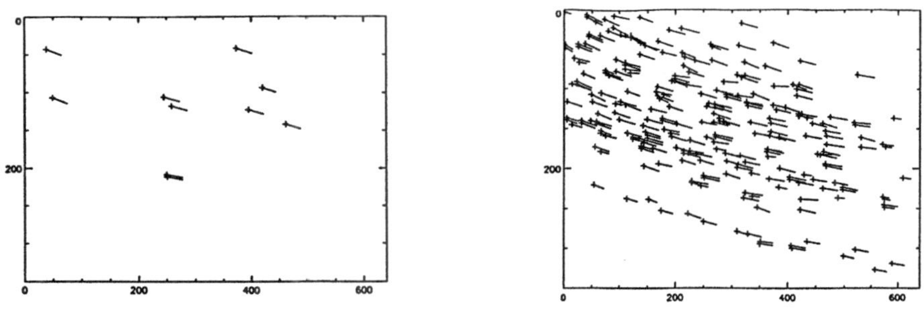

Figures 3 + 4: Single frame events (left) and events of all frames.

Vector Field Estimation

Velocity vectors obtained in this manner are scattered irregularly in the flow field, and must be rearranged by a surface fitting method. Consequently, the flow field is divided into a regular grid and the velocity vector split in x- and y-components. The velocity is then estimated on every grid point by the bivariate smooth surface fitting method earlier proposed by Akima [5].

Instantaneous Velocity Field

To calculate the instantaneous vector field there are two possible strategies to select. In the first strategy a few, e.g. 3 or 4 consecutive frames, are used to estimate the instantaneous vector field. In the second strategy all 25 frames are used to generate an average vector field. We use the first technique since we observe a continous, time dependent variation in our data set. Fig. 4 shows all bubble paths captured during 25 frames. This timedependent variation can be observed by close inspection from Fig. 4.

```
200(low_lim)      255(high-lim)
 12(min_length of tracks accepted)
----            250 tracks identified in 25 frames ----
```

x	y	vx	vy	iframe
373	42	28	7	1
38	43	27	8	1
420	94	24	6	1
245	106	28	6	1
49	107	26	8	1
259	118	27	6	1
395	123	27	6	1
462	143	26	6	1
250	209	28	4	1
251	212	27	4	1
167	97	28	6	2
310	108	27	3	2
379	126	26	4	2
521	145	27	3	2
532	173	27	3	2
169	194	25	4	2
423	239	31	1	2
264	64	27	7	3

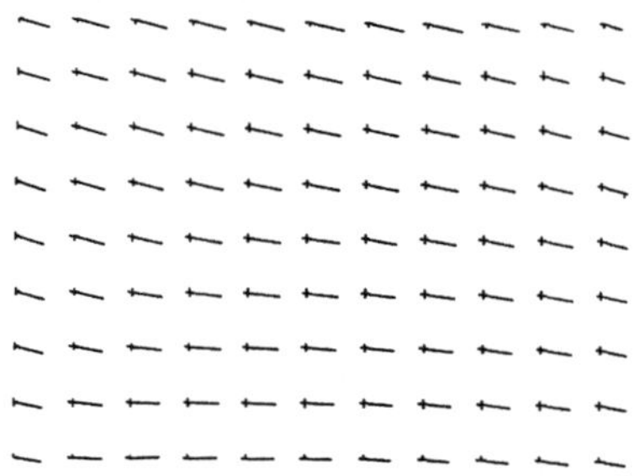

Figures 5 + 6: Extracted vector list (left) and the estimated vector field

Time Variant Field

The velocity information extracted at different time frames at the same location often deviates in magnitude and direction from the stationary values. This suggests that the soap bubble arriving later at the same geometrical position can provide crucial information about the structure of slow turbulences. This is especially true for rooms with displacement ventilation, where the air is supplied into the measured zone at low velocity from a wall mounted diffuser. It is evident that there are some low-turbulent chaotic effects in this situation, which can be theoretically predicted by the effects of the 'free jet' of the inlet (Karman) and by the vertical temperature gradient [6].

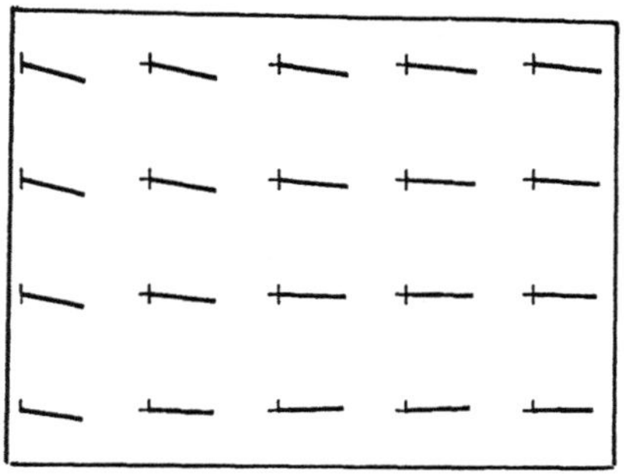

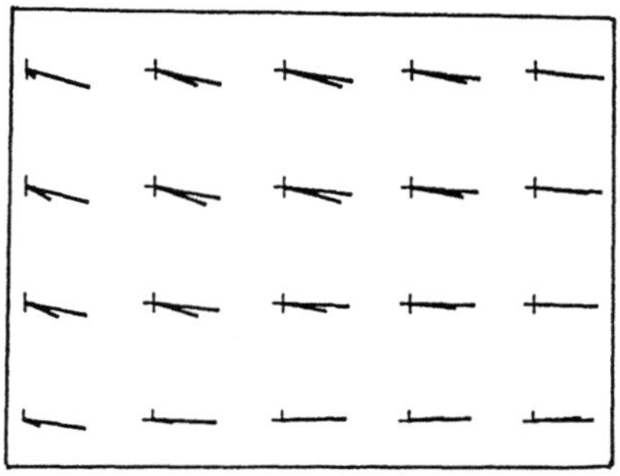

Figure 7 + 8: Vector field data based on the frame No.16 (left) and the vector variation 4 frames later.

During the examination procedure, every trajectory established by a new frame is compared against the mean velocity vector and against the instantaneous velocity components of the previous and succeeding frames to estimate the turbulence.

Concluding Remarks

The imaging technique applied to the slow speed air flow observation is based on automatized particle tracking. The method automatically extracts velocity vectors from a flow field captured by a still video camera. This information is stored in an exact defined time sequence of 25 time frames. Slow changes in the air flow can be observed and qualified if the sampling rate adapted to the airflow variation properly (i.e. Shannon's sampling theorem). As a next step, we plan to provide mutiple sequences with varying sampling rate from the same observed scene and automatically evaluating the frequency of the local air speed variation.

References

[1] A.Kaga, Y. Inoue, A. Yoshikawa: 'Velocity distribution measurement trough digital image processing of visualized flow images' Proceedings of the 2nd International Conference on Engineering Aero- and Thermodynamics of Ventilated Rooms. Section B1/2; Oslo Norway [1990]

[2] T. Kobayashi, T.Saga, S. Segawa: 'Some Consideration on Automated Image Processing of Pathline Photographs' Univ. of Tokyo, Inst. of Industrial Science Tokyo, 106 Japan

[3] P.Suter, Z. Revesz, P.Tanner, G. Gottschalk: 'Reconstruction of Velocity Vectors in a 3D Flow Field from Photographed Tracer Trajectories' Proceedings of the 5th International Symposium on Flow Visualization' Hemisphere Publishing Corp. [1989/8]

[4] L. Besse: 'Measurement and Computer Modeling of the Airflow Velocity Distribution' Proceedings of Compugraphics 91, Sesimbra Portugal.

[5] Hiroshi Akima: 'A method of bivariate interpolation and smooth surface fitting for irregularly distributed data points' ACM Transactions on Mathematical Software Vol. 4 [1978]

[6] F.T.M. Nieuwstadt, 'Direct and Large Eddy Simulation of Free Convection', 9th Int. Heat Transfer Conference Jerusalem [1990]

Study of a Turbulent Boundary Layer, Using High Speed Tomography

M.C. Hoyez†, J.C. Monnier, M. Stanislas‡.

IMFL-ONERA, 5 Bvd Paul Painlevé 59000 Lille FRANCE.

‡also LML, URA CNRS 1441, Cité Scientifique 59650 Villeneuve d'Ascq FRANCE

†ARIST, 77 rue Nationale, 59000 Lille FRANCE

Summary

Improved images of the coherent structures inside a turbulent boundary layer are obtained. Digital analysis of these images allows to obtain pertinent quantitative informations on the typical eddies for two values of the Reynolds number.

Introduction

In a paper presented at the previous conference by Stanislas & Hoyez [1], an experiment of visualization of a turbulent boundary layer developing on a flat plate was reported. Since then, the optical set-up has been improved to allow a better image quality. Moreover, quantitative results have been obtained from the visualizations using a digital analysis system.

Experimental set-up

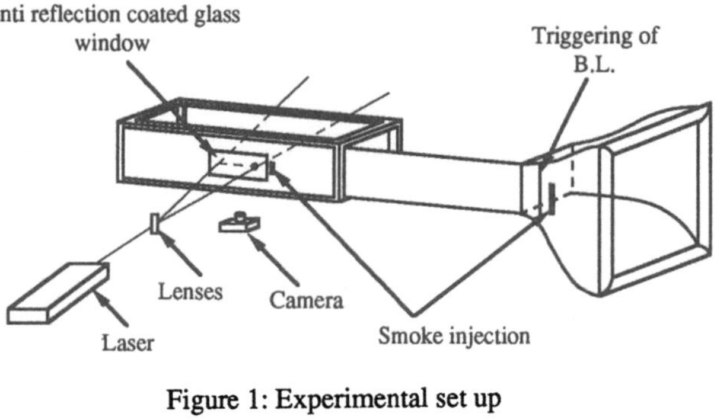

Figure 1: Experimental set up

The experiment which is described in detail by Hoyez [2] was conducted in a boundary layer wind tunnel specially designed for this study (figure 1). The test section is 2.5 m long and has a cross section of 0.3 x 0.3 m^2. It is divided into four parts, the last one of which is only represented in figure 1. The velocity in the test section can be varied from 0 to 40 m/s. The turbulence level at the inlet is 0.3 %.

The boundary layer under study is that on a flat plate which is as long as the test section. The plate can be mounted on each of the lateral walls of the wind tunnel. The boundary layer is triggered in the contraction, 0.20 m upstream of the inlet section. The origin of the turbulent boundary layer is considered to be at this trigger point, x being the longitudinal coordinate along the wall, y that normal to the wall and z the spanwise coordinate. It has been verified, using hot wire anemometry, that the boundary layer has usual turbulent characteristics [1,2].

Visualization method

The set-up used to visualize the boundary layer is also presented in figure 1. The light source consists of four Ruby lasers on the same optical bench. Each of them can emit a pulse of red light which has an energy of 220 mJ and a duration of 20 ηs. The four laser cavities are synchronised. The time intervals between the pulses can be varied independently between 5 μs and 10 s.

The light beams coming out of the lasers are transformed into plane light sheets using sphero-cylindrical lenses systems. The typical thickness of the light sheets varies between 2/10 and 4/10 mm. Their width is about 15 cm. They are superimposed and introduced in the test section horizontally, normal to the wall and parallel to the mean velocity. They enter through a glass window of good quality, which is anti-reflection coated on both sides and which is carefully adjusted in the boundary layer plate. To minimize light diffusion inside the test section, the light sheets go out through a plexiglas slit on the opposit wall. The slit dimensions are adjusted to that of the sheet of light.

The boundary layer is made visible by seeding it with smoke. Oil smoke with droplets diameter less than 1 micron is used. It can be injected in the contraction, 60 mm upstream of the triggering device, by a slit which is 5 mm by 10 cm. The angle of injection is 20°. The smoke can also be injected by a smaller slit (2mm x 40 mm) located underneath the turbulent boundary layer just upstream of the glass window mentioned above. It has been verified, using hot wire anemometry, that the injection at both stations does not perturb neither the mean nor the fluctuating velocities in the visualization section.

The images are recorded through the underneath tunnel wall which is also equiped with an anti-reflection coated glass window. Pictures are taken through a mirror at 45°, using a STROMBODRUM high speed drum camera, on 35 mm KODAK Recording 2475 film, which nominal sensitivity is 1000 ASA. The lens is a ZEISS 125 mm f/2.5. A synchronization electronic allows the recording of seven sequences of images. The triggering of the laser system through this electronic can be conditioned on an external voltage signal going over an adjustable threshold.

Visualizations

All the visualizations mentioned here have been obtained in the vicinity of x = 2550 mm. The light sheets were superimposed, giving a field of view of approximatly 100 mm in the streamwise direction. Images were recorded using one smoke injection through the most upstream slit of figure 1, for free stream velocities of 12.6 m/s and 22 m/s (corresponding to $R_\theta \simeq 3500$ and 6300), with time intervals Δt of respectively 1700 μs and 900 μs (corresponding to Δt^+ of 31 and 43). A second serie of images were recorded using a double smoke injection to visualize both internal and external structures. The time intervals for the two previous velocities are respectively 1000 μs (corresponding to

$\Delta t^+ = 18$) and 500 μs ($\Delta t^+ = 24$). Plate 1 gives an example of image obtained with a single smoke injection. In plate 2, a sequence of 4 images obtained with a double smoke injection is presented with a close-up view to the near wall region.

Analysis

The detailed visual analysis of the obtained images has been first performed using a NAC film analyser. It showed an overall behaviour which is in close agreement with the model proposed by Falco [3], but the information obtained was more or less qualitative. A more quantitative study has thus been undertaken.

The images were digitized using a THOMSON high definition vidicon camera (850 pts/line) connected to a Matrox MVP 512x512 frame grabber placed in a PC micro-computer. A software was developed to measure various characteristics of typical eddy structures as indicated in figure 2.

First, the Cx and Cy as defined by Falco [4] (figure 2a) were measured. Figure 3 compares the mean values obtained for the two characteristic length to the laws proposed by Falco[4]. The difference is within the measurement accuracy which is of the order of ± 5%.

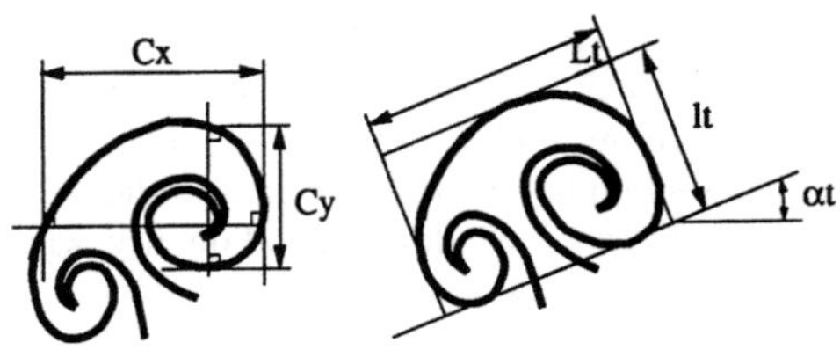

Figure 2a

Figure 2b

Figure 2: Dimensions of typical Eddies measured

On a second step, the parameters presented on figure 2b were measured. Figure 4 presents, as an example, the length L_t scaled with external variables as a function of the wall distance.

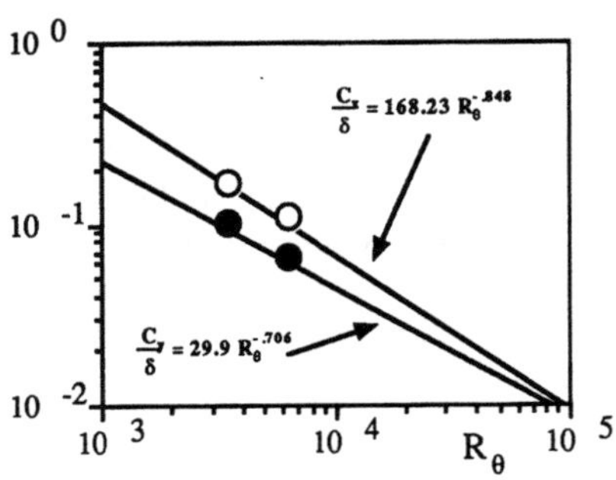

Figure 3: TE size, comparison with [4]

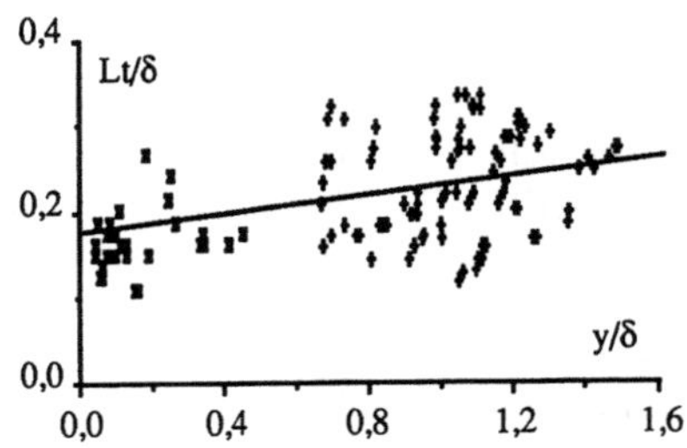

Figure 4: Typical Eddies Size

The quantitative analysis of the images allows also to obtain the velocities of the coherent structures. Knowing the position of a structure on each of the four images and the time interval between them, one can dedùce the convective longitudinal U and normal V velocity components of this structure. Knowing the local mean velocity $\overline{U}$ which was obtained by hot wire anemometry, it is possible to deduce the local transport velocity components.

Taking into account all the elements of the analysis system, the precision obtained on U and V is $\pm$ 4% of U_∞.

Figure 5 presents the transport velocity components $u_t = U - \overline{U}$ measured for the typical eddies, as a function of the wall distance and for R_θ = 3500 compared to the mean turbulence intensity. The results are scaled using external variables.

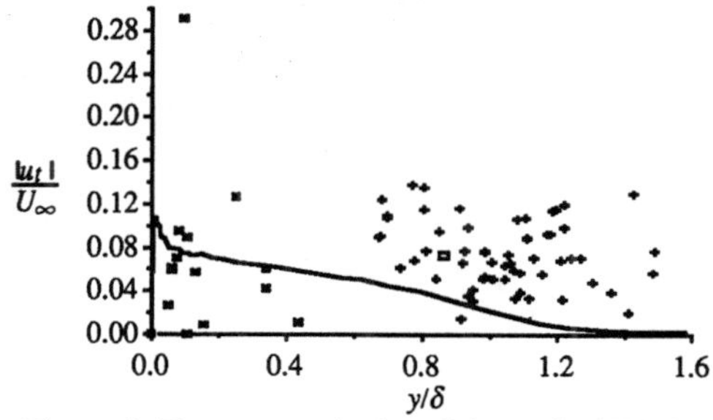

Figure 5: Transport velocity of the typical Eddies

Conclusion

The improvement of the optical set-up used in the previous study has allowed to obtain better images of the coherent structures in a turbulent boundary layer. A digital analysis of the recorded images has led to valuable quantitative informations which compare closely to the results of Falco [4]. These results have now to be completed and analysed carefully to try to get a better understanding of the turbulence production near a wall.

We are thankfull to the DRET G6 for the financial support to this study.

References

[1] Stanislas, M.; Hoyez, M.C.: Study of the structure of an equilibrium turbulent boundary layer using a rapid light sheet visualization technique. *Flow Visualization V, Prague, August 21-25, 1989.*

[2] Hoyez, M.C.: Etude des caractéristiques instationnaires d'une couche limite turbulente de plaque plane sans gradient de pression. *Thèse de Doctorat n° 610, Lille, 1990.*

[3] Falco, R.E.: New results, a review and synthesis of the mechanism of turbulence production in boundary layers and its modifications. *AIAA paper n° 83-0377, 1983.*

[4] Falco, R.E.: A coherent structure model of the turbulent boundary layer and its ability to predict Reynolds number dependence. *Phil. Soc. London, Series A, vol. 336, pp. 103-129, 1991.*

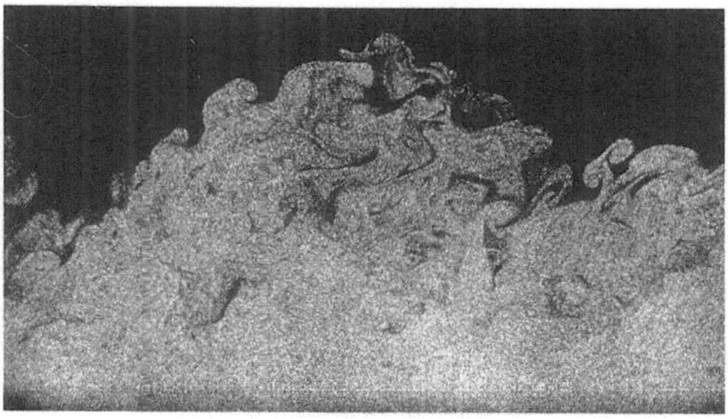

Plate 1

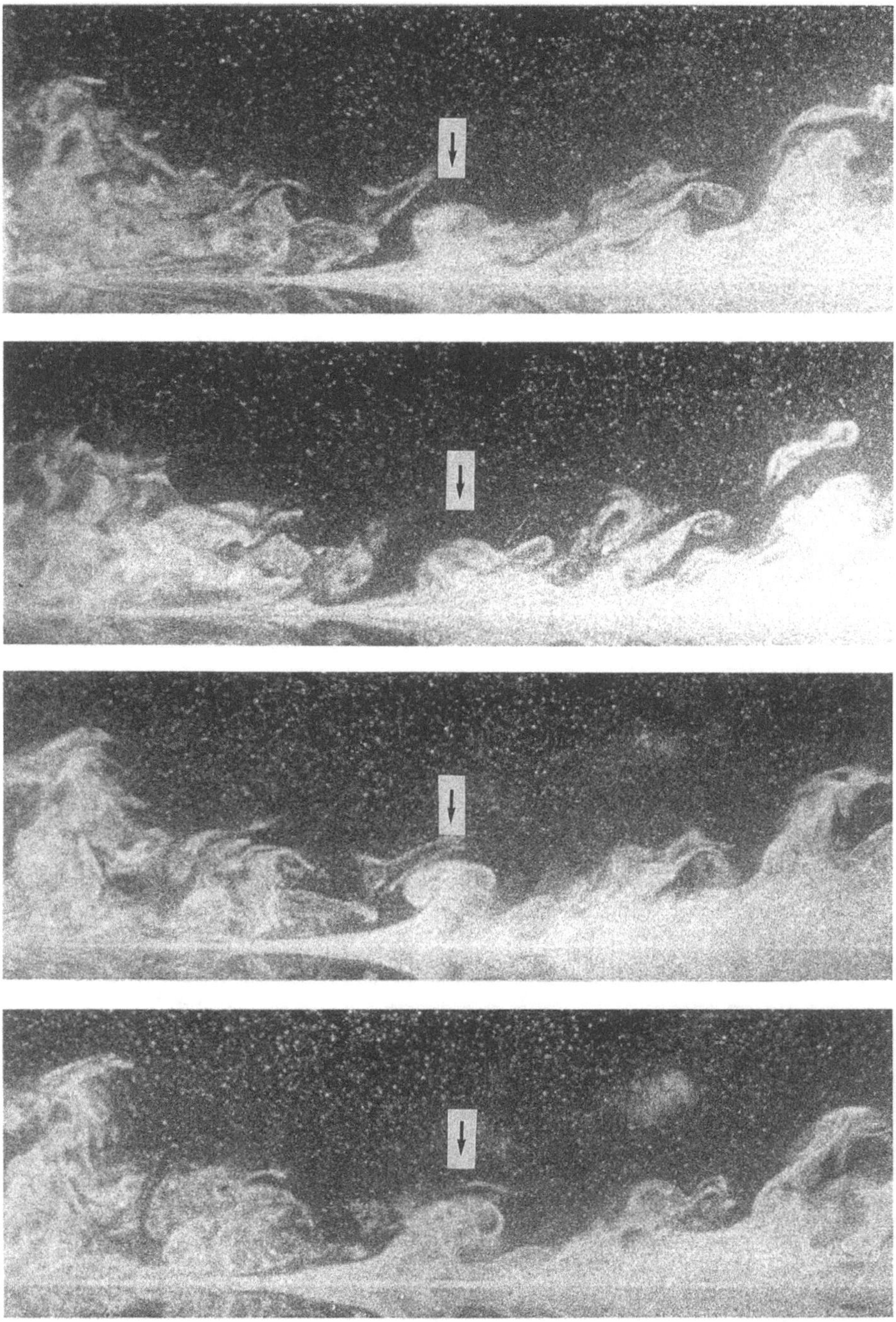

Plate 2

A Novel Soap-Film Tunnel for Studying Two-Dimensional Flows

M. GHARIB and M. BEIZAIE

Department of Applied Mechanics and Engineering Science
University of California, San Diego

Summary
The concept and operation of a novel device to study two-dimensional flows is presented. A suspended liquid film (soap film) is set to motion in a long frame by using the shearing action of an impinging planar water jet. By reshaping the frame's structure, various shear flows for qualitative (flow visualization) or quantitative (laser Doppler velocimetry) studies can be produced. Several examples of shear flows are presented. It is shown that this device can be a valuable tool in establishing a quantitative experimental basis for two-dimensional flows including wakes, jets and grid-generated flows.

Introduction

While two-dimensional flows have played a central role in the theoretical foundation of fluid mechanics, their experimental realization has been difficult if not impossible. However, large scale natural flows (geostrophic flows), due to their large span to depth ratio, can be considered as two-dimensional flows. The experimental simulation of these flows and verification of numerous theories for two-dimensional flows in the laboratory is difficult and seldom decisive, owing in part to the difficulty of isolating the primary two-dimensional flow from evolving three-dimensional instabilities.

Mysels [1], in his ingenious work on liquid films, proposed soap films as potential candidates to produce and study two-dimensional hydrodynamics. The work of Couder, Chomaz and Rabaud [2] on the dynamics of the vortex wakes behind a towed cylinder in a soap film frame, proved the feasibility of this idea. Unfortunately, the towing technique has severe limitations associated with the short period of observation and non-uniform film thickness.

Recently, we have developed a continuously running soap-film tunnel (Gharib and Derango [3], Beizaie [4]) which has eliminated the short-observation-period and non-uniform-film-thickness problems. This tunnel is essentially a two-dimensional counterpart of conventional wind and water tunnels.

Soap Film Tunnel

This device consists of a frame (4 inches wide and 12 inches long), one end of which is positioned in a diluted surfactant mixture while the other end is subjected to a film-pulling mechanism. The frame is constructed of either copper or steel (Fig. 1). The main portion of

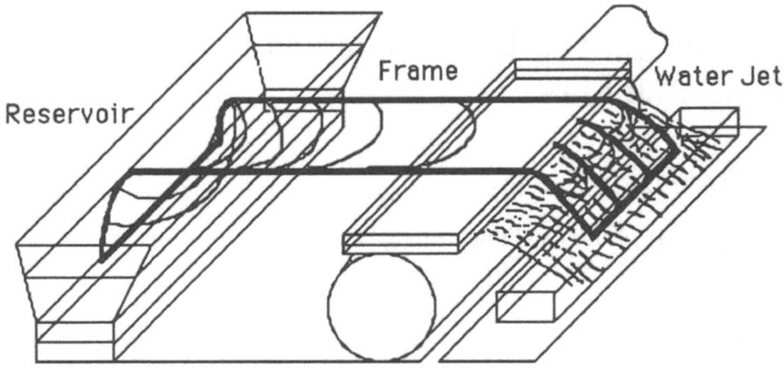

Fig. 1. The Liquid Film Tunnel.

the frame is a flat section consisting of two parallel rods. This flat section is supported by two legs, one at each end. The liquid films are pulled by shearing action of a two-dimensional water jet which is produced by flow of water through a rectangular opening, mounted on a horizontal pipe. By rotating this pipe, the shearing angle can be changed and by varying the water flow rate one can vary the pulling force. Notice that the leg at the film/water jet interface has fork-like structures. These structures help to stabilize the film/water jet interface.

The size of the frame is limited by the film's tendency to bow at flat sections of the tunnel. The maximum width of the frame to avoid this effect is 4". The streamwise length of the test section can be extended up to 1.5 ft without greatly reducing the life of the film. Several other factors, including dry air, severe ventilation, vibration and unsteadiness or turbulence in the water jet can shorten the film life. Fine cloth ribbons are glued to the top of the frame to help prevent rupture of the liquid films due to water evaporation. So, the films are actually supported by these thin wet ribbons rather than by the rods themselves. Therefore, the menisci along the ribbons have a constant thickness, independent of the surfactant concentration. The films are also quite uniform in thickness when running steadily, but this thickness can be changed by changing the surfactant concentration. With a steady water jet and in absence of severe ventilation and vibration, the films can run for hours, provided that fresh surfactant mixture is added continuously to the reservoir.

The liquid film tunnel is started by pulling a film from reservoir's surface and drawing it on the wet frame by a wet cloth and finally contacting it to the running water sheet. Once a film starts to run, it produces two-dimensional patterns depending on the frame's design. Examples include frames that produce flows such as jets, cavity flow, grid turbulence and vortex shedding.

Flow Visualization

Soap films can be visualized through light interference effect produced by the small variations in the film thickness. Such variations are produced during pulling of the film from the reservoir due to the gravitational effects. However, in the horizontal section these variations are simply transported and act like a scalar. This is true since the time scale of film restoration forces is much longer than residency of the film in the frame. The maximum variation between two consecutive fringe patterns is in the order of 100 nm for a monochromatic green light of 500 nm wavelength.

Usually, the distance between fringes in the horizontal section is about 5 to 10 cm. Also color interference method can be used to utilize the full effect of the visible color band for the visualization effect.

Representative Two-Dimensional Flow Patterns

In this section, two examples of two-dimensional flows are presented to demonstrate the capabilities of the soap film tunnel. The first flow visualization (Fig. 2) represents a laminar jet produced by a thin film of one micron thickness.

Fig. 2. Two-Dimensional Laminar Jet.

In Figure 3 we show a grid flow generated in the soap film tunnel.

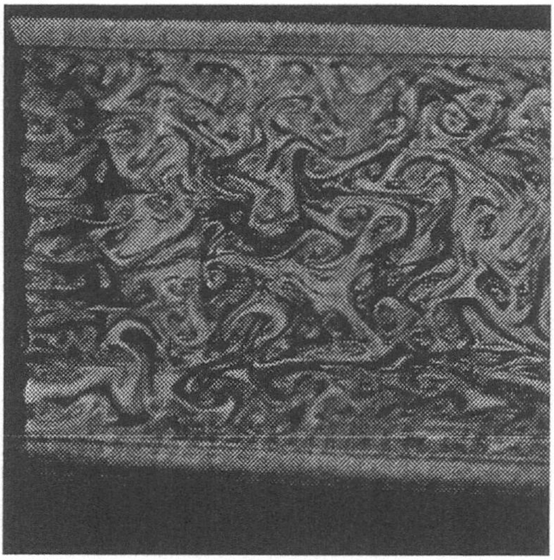

Fig. 3. Two-Dimensional Grid Turbulence.

Aside from fascinating flow patterns that can be generated in soap films, we employed laser Doppler velocimetry technique to quantify various flow situations. Fig. 4 gives an example of such a measurement where mean velocity profile is shown at a downstream position.

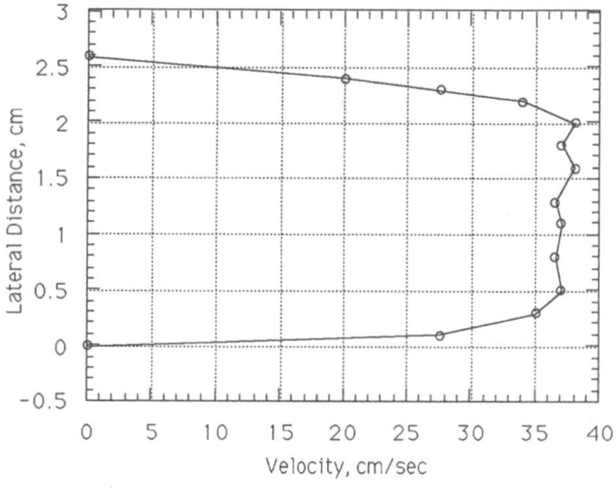

Fig. 4. Sample Velocity Profile of the Soap Film.

Conclusion

We have shown that the soap film tunnel can be considered as a serious tool in simulating and understanding two-dimensional flows. A combination of qualitative and quantitative methods can be used to study various shear flows.

Acknowledgement

We are grateful to Professor Karol Mysels for his support and many advices. This work was supported by National Science Foundation.

References

1. Mysels, K.: Soap films-studies of their thinning and a bibliography. New York, London, Paris, Los Angeles: Pergamon Press 1959.

2. Couder, Y.; Chomaz, J.M.; Rabaud, M.: On the hydrodynamics of soap films. Physica D 37 (1989) 384-405.

3. Gharib, M.; Derango, P.: A liquid film (soap film) tunnel to study two-dimensional laminar and turbulent shear flows. Physica D 37 (1989) 406-416.

4. Beizaie, M.: Physical fundamentals of the surfactant film tunnel. Bull. Am. Phys. Soc., 36 (1991), 2718.

Detection of Coherent Structures in LDV Data Using Combined Visualization in Square Duct Flow

G. KAWAHARA, K. AYUKAWA and J. OCHI

Department of Mechanical Engineering, Ehime University, Matsuyama, Japan

T. WATANABE

Nippon Telegraph and Telephone, Chiyoda-ku, Japan

Abstract

Near-wall streaky structures have been studied in a fully-developed turbulent square-duct flow by using combined dye flow visualization - LDV velocity measurement techniques. The velocity fluctuations are conditioned on the topological properties, such as coalescing and branching, and the transverse locations of the visually-identified low-speed streaks. It is shown that a symmetric pair of counter-rotating quasi-streamwise vortices is associated with one low-speed streak in a statistical sense. However, upstream of coalescing point and downstream of branching point of the low-speed streaks, a *single* quasi-streamwise vortex is associated with one low-speed streak.

Introduction

Coherent and large-scale structures exist in the near-wall region of wall-bounded turbulent flows, and they are believed to make a great contribution to momentum and mass transfer near the wall. A number of detection schemes have been applied to hot-wire or laser-Doppler velocimeter (LDV) signals in order to describe these structures quantitatively. However, flow structures corresponding to coherent events have been unclear because of large discrepancies [1] among the results obtained by the different schemes.

In this study, dye flow visualization has been used to detect coherent structures in LDV data for a fully-developed turbulent square-duct flow. Near-wall streaky structures are visualized, and then the visually-identified low-speed streaks are classified into three categories, i.e., coalescing, branching and the other ones. The axial, wall-normal and spanwise velocity fluctuations are conditioned on the topological properties and the transverse locations of the low-speed streaks. The most probable values of the velocity fluctuations are estimated from the resulting conditional probability density distributions. The flow structures associated with the low-speed streaks are discussed for each topological category.

Experimental procedure

Experiments were conducted in the fully-developed turbulent flow of a water duct. The duct had an internal cross-sectional dimension of 100×100mm, and was about 9m in length. A fluorescent dye was injected through ten dye taps in the bottom wall of the duct to mark the sublayer flow structures. Its specific gravity was 1.04 so that the dye spread over the bottom wall. The dye taps were located $60H$ (H : duct width) downstream of the duct inlet. The flow at $y < 25\nu/u_\tau$ (ν : kinematic viscosity, u_τ : friction velocity) was filmed from above at 18frames/s with 16mm movie camera. The film frames were synchronized with the LDV data.

A Dantec model 55x LDV system was used to obtain the axial, wall-normal and spanwise velocity fluctuations, u',v' and w' on the bottom-wall bisector, $(x, z) = (0, 0)$ (see Fig. 1). The velocity-measurement point was located at $y = 10\nu/u_\tau$ (for w'), $30\nu/u_\tau$ (for u' and v') and $8H$ downstream of the dye taps. Velocity component signals were digitized and registered on a magnetic disk.

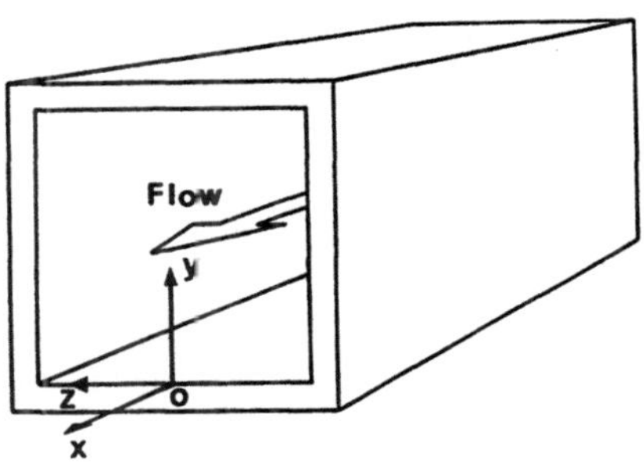

Fig. 1. Coordinate system.

Results and discussions

The basic flow conditions are given in Table 1. Here U_m is the bulk-mean velocity.

Table 1. Basic flow conditions

ν m^2/s	u_τ cm/s	$R_e = U_m H/\nu$
1.09×10^{-6}	0.547	6870

An example of the flow visualization is shown in Fig. 2. The visual data shows that the dye-filled streaky structures [2] appear near the wall, and that they exhibit the characteristic aspects, such as coalescing and branching [3]. We identify such dye-filled streaks as the low-speed streaks, and define T, which indicates the topological categories, i.e., coalescing, branching and the other ones, as follows. If the low-speed streak adjacent to the velocity-measurement point has the coalescing point less than $150\nu/u_\tau$ downstream of the velocity-measurement point, $T = C$, if it has the branching point less than $200\nu/u_\tau$ upstream of that, $T = B$, and $T = O$ for the other cases, as shown in Fig. 3. In addition, the abscissa λ of the adjacent streaks was fixed with its origin set at the velocity-measurement point, as shown in Fig. 4. The positive direction of w' and the negative direction of λ are also specified for each category in Fig. 3. The conditional probability density of the velocity fluctuation V is defined in the form

$$P_V(V \mid \lambda\ ;\ T) = P_j(V, \lambda \mid T)/P(\lambda \mid T),$$

where P_j is the joint probability density of V and λ, and P is the probability density of λ on condition, T.

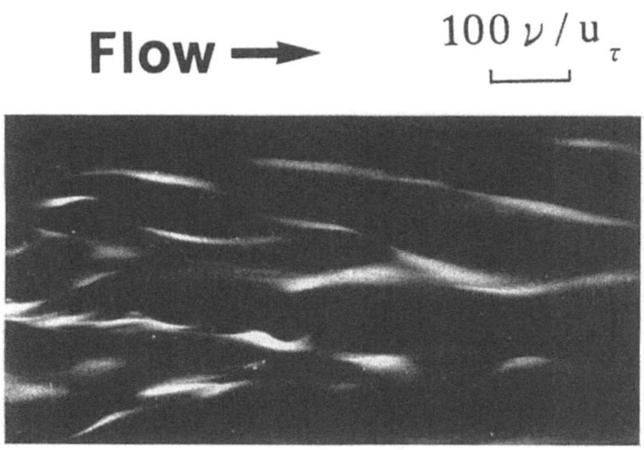

Fig. 2. Dye flow visualization in turbulent square-duct flow.

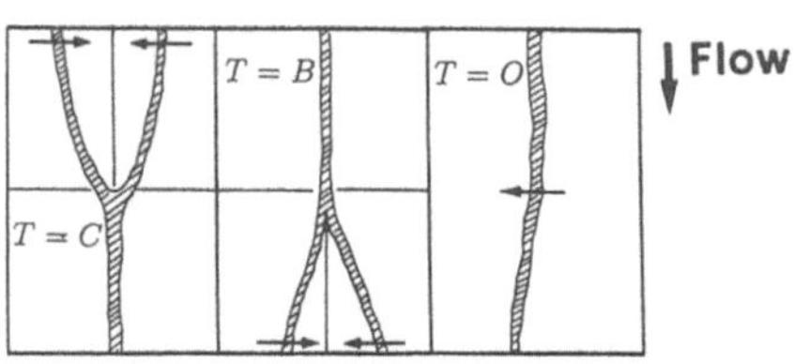

Fig. 3. Classification of low-speed streaks.

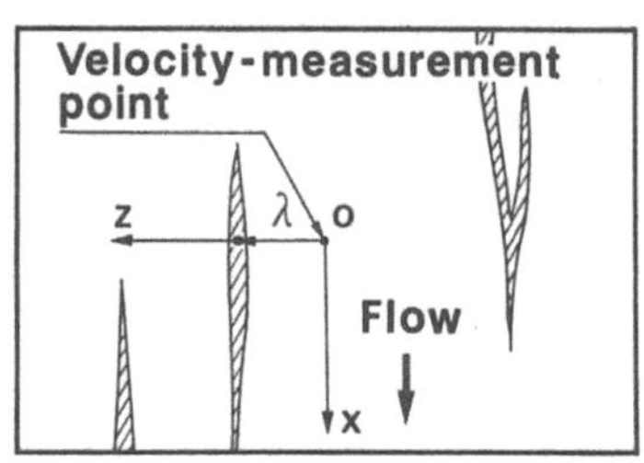

Fig. 4. Fix of λ.

We can estimate the most probable values of the axial, wall-normal and spanwise velocity fluctuations, u'_{mp}, v'_{mp} and w'_{mp} from the resulting conditional probability density distributions. The variations of u'_{mp}, v'_{mp} and w'_{mp} with λ are shown for $T = C$ and B in Fig. 5 (a)~(f) respectively. These figures include the results for $T = O$, corresponding to the same velocity components.

For $T = O$, the profile of u'_{mp} is symmetric about $\lambda = 0$, while for $T = C$, u'_{mp} is negative at $\lambda > 0$ and positive at $\lambda < 0$, as shown in Fig. 5 (a). For $T = B$, u'_{mp} is positive at $\lambda > 0$ and negative at $\lambda < 0$, as shown in Fig. 5 (b). For $T = O$, the profile of v'_{mp} is also symmetric about $\lambda = 0$, while for $T = C$, v'_{mp} is positive at $\lambda > 0$ and negative at $\lambda < 0$, as shown in Fig. 5 (c). For $T = B$, v'_{mp} is negative at $\lambda \geq 20\nu/u_\tau$ and positive at $\lambda \leq -20\nu/u_\tau$, as shown in Fig. 5 (d). For $T = O$, w'_{mp} is positive at $\lambda \leq -20\nu/u_\tau$ and negative at $\lambda = 20\nu/u_\tau$, while for $T = C$, w'_{mp} is positive at any λ-location, as shown in Fig. 5 (e). For $T = B$, w'_{mp} is almost negative, as shown in Fig. 5 (f).

These results show that a symmetric pair of counter-rotating quasi-streamwise vortices is associated with one low-speed streak without coalescing and branching in a statistical sense. However, upstream of the coalescing point and downstream of the branching point of the low-speed streaks, a *single* quasi-streamwise vortex [4] is associated with one low-speed streak.

Concluding remarks

The flow structures, associated with the low-speed streaks without coalescing and branching have symmetry in a statistical sense. On the other hand, upstream of the coalescing point and downstream of the branching point of the low-speed streaks, they have asymmetry. This asymmetry follows the fact that a *single* quasi-streamwise vortex is associated with one low-speed streak.

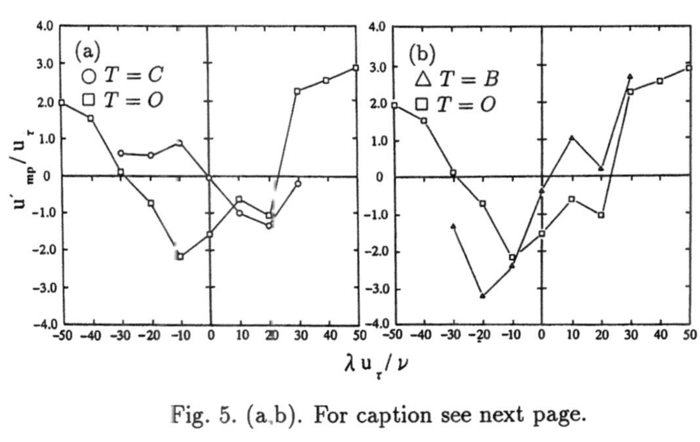

Fig. 5. (a,b). For caption see next page.

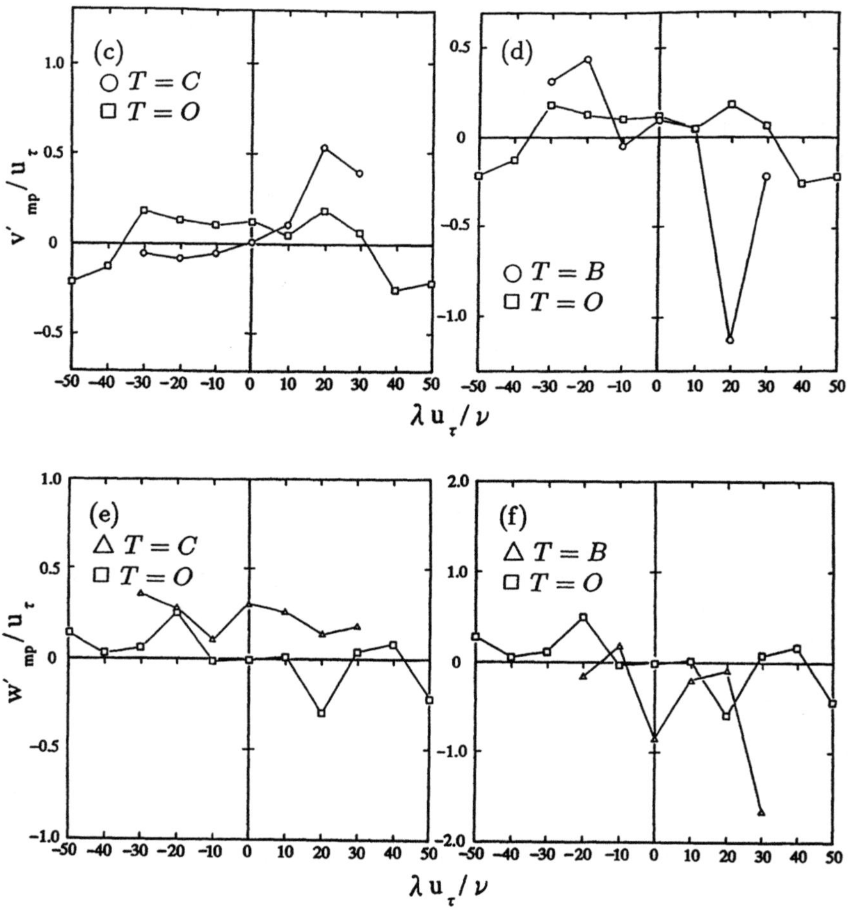

Fig. 5. The most probable value of velocity fluctuation versus λ: (a) u'-component for $T = C$; (b) u'-component for $T = B$; (c) v'-component for $T = C$; (d) v'-component for $T = B$; (e) w'-component for $T = C$; (f) w'-component for $T = B$. Note that the figures include the results for $T = O$ corresponding to the same component.

References

1. Bogard, D. G. and Tiederman, W. G., *J. Fluid Mech.*, **162** (1986), 389.
2. Zaric, Z., Falco, R. E. and Blackwelder, R. F., in *Turbulence and Chaotic Phenomena in Fluids* (1987), 439, North-Holland.
3. Smith, C. R. and Metzler, S. P., *J. Fluid Mech.*, **129** (1983), 27.
4. Guezennec, Y. G., Piomelli, U. and Kim, J., *Phys. Fluids*, A **1** (1989), 764.

Quantitative, Four-Dimensional, Laser Induced Fluorescence Imaging Studies of Scalar Mixing in Turbulent Flows

WERNER J.A. DAHM AND KENNETH B. SOUTHERLAND

Department of Aerospace Engineering
The University of Michigan
Ann Arbor, MI 48109-2140, USA

Summary

The small scale structure of the scalar energy dissipation rate field $\nabla\zeta \cdot \nabla\zeta(\mathbf{x},t)$ in turbulent shear flows is studied using quantitative, four-dimensional, spatio-temporal, laser induced fluorescence imaging of a large Schmidt number conserved scalar field $\zeta(\mathbf{x},t)$. Previous work [1–3] has examined the mixing process at the diffusive scale λ_D, and has shown that essentially all of the molecular mixing at $Sc \gg 1$ occurs in thin, strained, laminar diffusion layers of the Burgers and Townsend type, in which the internal structure is self-similar. This paper addresses the scalar mixing process at scales between the scalar diffusion limit λ_D and the vorticity diffusion limit λ_ν. The emphasis is on an approach for quantitatively characterizing certain similarity properties of the canonical patterns into which the scalar energy dissipation layers are arranged by the continual stretching and folding action of the vorticity and strain rate fields in turbulent flows.

Introduction

The structure associated with mixing of dynamically passive conserved scalar quantities in turbulent flows has been a problem that has received considerable attention over the past several decades. The smallest scalar gradient lengthscales are set by a balance between the thinning action due to the continual straining of scalar interfaces by velocity gradients and the thickening by molecular diffusion of the scalar. The resulting strain-diffusion equilibrium gives this minimum scalar gradient length as $\lambda_D \sim \delta \cdot Re_\delta^{-3/4} \cdot Sc^{-1/2}$, where $Re_\delta \equiv u\delta/\nu$ is the local outer-scale Reynolds number, with δ and u the local outer length and velocity scales characterizing the mean velocity gradient in the flow, and with $Sc \equiv \nu/D$ giving the ratio of the vorticity and scalar diffusivities. In the vorticity field, this balance between strain and diffusion leads to the strain-limited vorticity lengthscale $\lambda_\nu \sim \lambda_D \cdot Sc^{1/2}$. When the Schmidt number is large, the vorticity and strain rate are therefore essentially uniform over lengthscales significantly larger than λ_D. Burgers (1948, 1950) and Townsend (1951) used this to deduce the presence of thin dissipation sheets with a simple self-similar internal structure, in which essentially all the scalar energy dissipation occurs. Recent experimental measurements [1–3] have confirmed this structure. Over lengthscales somewhat larger than λ_D but sufficiently smaller than λ_ν, Batchelor (1959) argued on dimensional grounds that the essentially uniform straining action together with the action of molecular diffusion of the scalar will lead to a –1 power law form of the scalar energy spectrum. At lengthscales larger than λ_ν and but still sufficiently smaller than the local outer scale δ, the essentially nondiffusive distortion of the scalar dissipation sheets by the vorticity field leads to another relatively simple scaling regime in the scalar energy spectrum, having an at least approximately –5/3 power law form.

These power law scalings in the spectral domain indicate the dominance of potentially simple multiplicative processes associated with the deformation of the scalar dissipation sheets in the physical domain. An understanding of these processes may suggest relatively simple physically-based subgrid models for incorporating them in practical numerical simulations of turbulent flows. With this goal in mind, we have over the past few years undertaken a series of studies of the small scale structure of conserved scalar mixing in turbulent flows. Previous work has principally addressed the structure over lengthscales λ of the order of λ_D; here we investigate the range $\lambda_D < \lambda < \lambda_\nu$.

Experimental Technique

Traditionally, the detailed physical structure of large *Sc* scalar mixing in turbulent flows has proven extremely difficult to address experimentally, owing to the very small length and time scales involved. In relatively recent times, high resolution three- and four-dimensional spatio-temporal imaging techniques have been developed [1–3] which are capable of reaching the extremely small scales associated with large Schmidt number scalar mixing. These are based on high-speed quantitative laser induced fluorescence imaging techniques specifically designed for fully-resolved measurements of the fine scale structure of the space and time varying conserved scalar field $\zeta(\mathbf{x},t)$ and the associated scalar energy dissipation rate field $\nabla\zeta \cdot \nabla\zeta(\mathbf{x},t)$ in turbulent flows.

The present measurements are based on rapid acquisition of successive laser induced fluorescence data planes at sustained throughput rates up to 9.3 MB/sec, into gigabyte sized data sets using a 3.1 GB high-speed parallel transfer disk rank, to produce a four-dimensional spatio-temporal data space – see Figs. 2 and 3 of [1]. Each data space consists of a time sequence of three-dimensional spatial data volumes composed of a spatial array of parallel two-dimensional data planes, each comprising a 256 × 256 array of data points. The separation between adjacent points within each data plane, and between adjacent data planes within each data volume, is smaller than the local strain-limited diffusion lengthscale λ_D of the turbulent scalar field. Similarly, the temporal separation between adjacent data planes within each data volume, and between the same data point in successive data volumes, is shorter than the local diffusion scale advection time λ_D/u. This resolution, together with the high signal quality attained, allows accurate differentiation of the measured conserved scalar data in all three space dimensions and in time to determine the instantaneous scalar gradient vector field $\nabla\zeta(\mathbf{x},t)$ at every point in the data space. From this, the true instantaneous scalar energy dissipation rate field $\nabla\zeta\cdot\nabla\zeta(\mathbf{x},t)$ is obtained. Measurements are obtained in the self-similar far field of an axisymmetric turbulent jet, at outer-scale Reynolds numbers Re_δ as high as 6,000.

Results and Discussion

Results are typically presented in the form of fine structure maps of the scalar fields $\zeta(\mathbf{x},t)$ and the associated dissipation rate fields $log_e\nabla\zeta\cdot\nabla\zeta(\mathbf{x},t)$ – see Fig. 1. To begin to understand the structural topology of the dissipation sheets evident in these data over lengthscales in the range between λ_D and λ_v, it is useful to view the deformation in terms of the relative scales of stretching (strain) and folding (vorticity). For $Sc \gg 1$, the smallest scale on which folding of the sheets can occur is of order $Sc^{1/2}$ larger than the sheet thickness λ_D. As a result, within a region extending a few λ_D in any direction, the structure would locally undergo stretching but virtually no folding, because the scale λ_v of the local vorticity gradient requires such a region to be in essentially solid body rotation. This produces the locally sheet-like nature of the dissipation fields. On lengthscales somewhat larger than this, but still smaller than λ_v, differential rotation effects cause the dissipation sheets to deform and eventually become interleaved.

Figure 1 gives an example of the topologies associated with the arrangement of the dissipation layers relative to one another; other examples can be found in [1–3]. Data such as these show highly convoluted patterns into which the dissipation layers are kneaded by the continual stretching action of the strain field and the folding action due to differential rotations introduced by gradients in the vorticity field as well as by continuous reorientations of the local principal strain rate axes. The basic components of this continually repeated volume-preserving stretching and folding process suggest a self-similar form of a bakers transformation, somewhat similar to the familiar the Hénon map. Indeed, certain features can be identified in the scalar dissipation layer patterns that at least qualitatively resemble features of the Poincaré map of this classic example of stretching and folding transformations. This suggests a self-similarity in, for example, the distribution of dissipation layer separations. To assess this possibility, in Fig. 2 shows the locii of local sheet-normal scalar dissipation peaks. From these we compute the one-sided distance to the next neighboring dissipation sheet in these two-dimensional slices. The resulting distribution of layer separations is shown in Fig. 3. If the conjectures above are correct, then we can expect that distributions such as this will demonstrate certain self-similar scaling properties. Based on flow visualization measurements of this type, we are presently conducting a relatively detailed examination of the self-similarity properties of distributions like these and their relation to power-law scalings in the spectral domain.

While more precise statements must await results from these investigations, it is worth noting that both the underlying sheet-like topology of the scalar dissipation field in Fig. 1, as well as the canonical patterns by which these sheets are interwoven, are qualitatively reminiscent of results from experiments on mixing by chaotic advection in very low Reynolds number flows [4, 5]. In these chaotic flows, analytical techniques allow the scalings properties of distributions such as these to be deduced. The overall similarity with our turbulent flow results suggests a degree of universality in the structure of scalar mixing that extends across the boundaries of the Reynolds number regimes traditionally associated with low Reynolds number chaotic advection and the higher Reynolds number turbulent flows examined here.

Conclusions

Quantitative flow visualization measurements such as these are providing previously inaccessible details of the scalar mixing process in turbulent flows. Results such as those presented here suggest a simple self-similar foundation in the deformation of the scalar dissipation sheets, which ultimately produces the –1 power-law between the vorticity and scalar gradient cutoff scales in the spectral domain. Further studies are planned to examine these processes in quantitative detail. Moreover, the continual stretching and folding action of strain and differential rotation in the flow at these scales appears to lead to just a few canonical patterns into which the dissipation layers can form. Critical point analyses of these patterns using flow visualization data of this type are also underway.

References

1. Dahm, W.J.A., Southerland, K.B. and Buch, K.A. (1991) Direct, high resolution, four-dimensional measurements of the fine scale structure of Sc • 1 molecular mixing in turbulent flows, Phys. Fluids A **3**, 1115-1127.

2. Dahm, W.J.A., Southerland, K.B. and Buch, K.A. (1991) Four-dimensional laser induced fluorescence measurements of conserved scalar mixing in turbulent flows, Applications of Laser Techniques to Fluid Mechanics, R. Adrian, Ed., Springer Verlag, Berlin, in press.

3. Dahm, W.J.A. and Buch, K.A. (1991) Fine structure characteristics of large Schmidt number molecular mixing in turbulent flows, Chemical Reactions in Turbulent Liquids, J. Hunt, Ed., Cambridge University Press, in press.

4. Ottino, J.M. (1989) The kinematics of mixing: stretching, chaos, and transport. Cambridge Univ. Press.

5. Ottino, J.M. (1982) Description of mixing with diffusion and reaction in terms of the concept of material surfaces. J. Fluid Mech. **114**, 83-103.

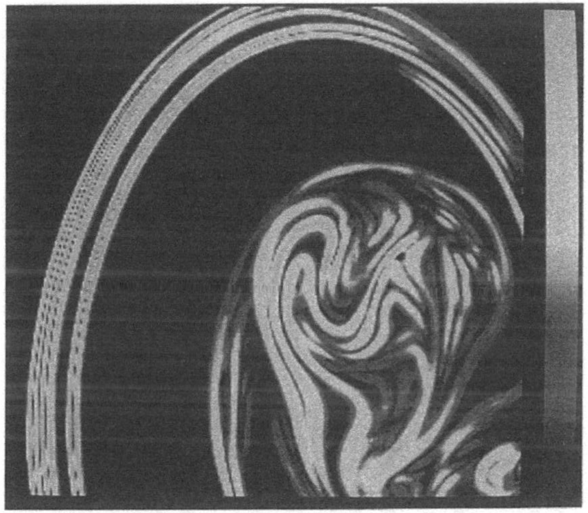

Fig. 1. Typical instantaneous scalar energy dissipation rate field $log_e \nabla\zeta \cdot \nabla\zeta(\mathbf{x},t)$..

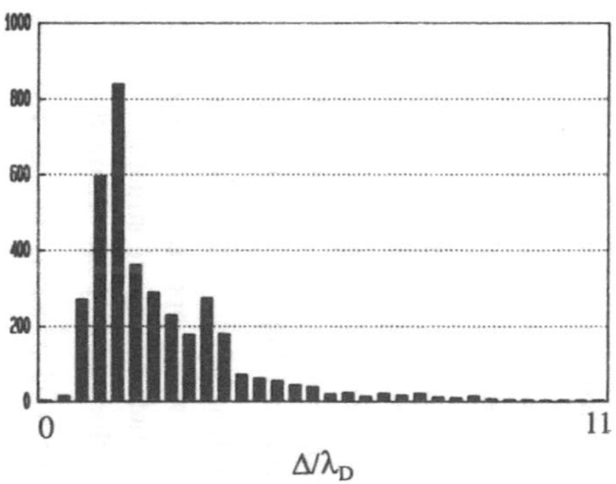

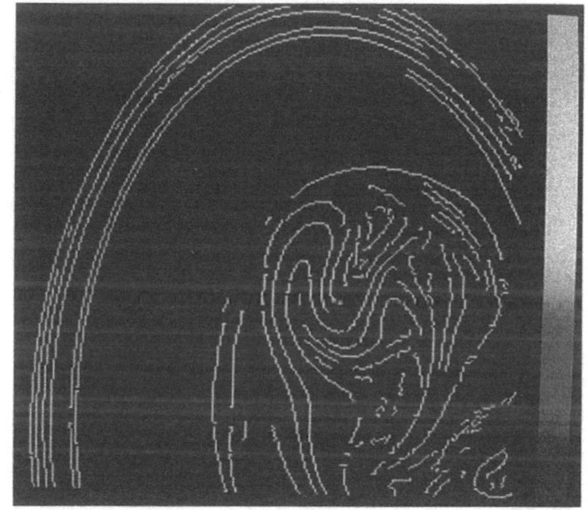

Fig. 2. The locii of sheet-normal scalar dissipation peaks from Fig. 1.

Fig. 3. The distribution of one-sided dissipation layer separations obtained from the data in Fig. 2. If the layer deformation results from simple multiplicative processes, then certain self-similar scaling properties of distributions such as these are implied, which ultimately lead to the simple −1 power-law Batchelor scaling in the spectral domain.

Visualization of Mixing and Residence Time Characteristics of the Flow in Static Mixers and Structured Packings

Gerd Gaiser

Institut für Chemische Verfahrenstechnik,
University of Stuttgart
D7000 Stuttgart 1, Böblinger Str. 72, FRG
Phone: 049-711-6412229, Fax: 049-711-6412242

SUMMARY

The flow and mixing behavior in structures used in static mixers and structured packings are visualized. Results show, that the geometrical parameters of these structures significantly affect the mixing behavior of the flow within these structures. By proper choice of the parameters compact mixers with a homogeneous mixing behavior can be achieved.

1. INTRODUCTION

Static mixers are well established in process engineering. Their particular advantages are in-line mixing, no moving parts, low power consumption and simultaneous homogenisation of residence time behavior.

Structured packings as well wind a wide field of application in chemical and process engineering. Examples of their use are distillation columns, extraction and absorption.

Several species of static mixers are commonly known /1,2/. One example of their design is based on single bars crosswise arranged in layers with opposing orientation. Fig. 1 shows a such a structure similar to a species of static mixers commonly used. Geometrical parameters of these structures are the inclination angle φ, that means the orientation of the bars to the main flow direcion, the spacing λ of the bars and of course their shape and size.

Another species of static mixers is based on corrugated passages formed by layers of single plates with opposing orientation of the corrugation lines as illustrated in fig. 2. Essential parameters of these structures are the inclination angle φ, the wavelength λ, the amplitude a and the shape of the corrugation.

Structured packings of this type are also commonly used as column-inserts for distillation, extraction or absorption applications. Other applications of these corrugated structures are plate heat exchangers, regenerators and catalysts.

Though several species of static mixers are known, they are commonly designed in a narrow range of the geometrical parameters according to few published experimental studies. The optimization of these structures however requires information on the effects of the structure parameters on local flow and transport phenomena. Results on flow and transport phenomena in corrugated passages have been previously reported /3,4/, some results on cross-type structures according to fig.1 will be discussed below.

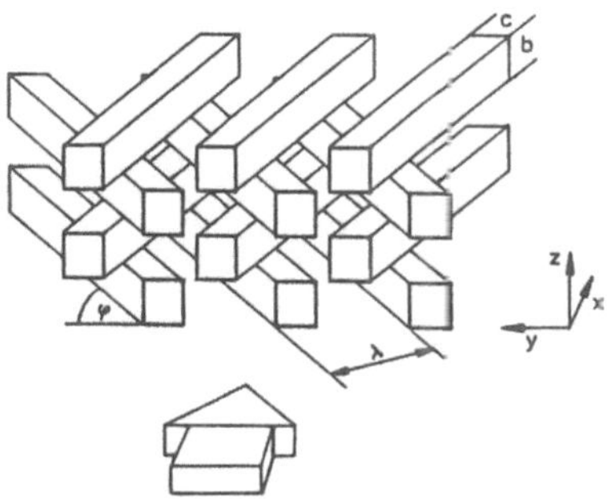

Fig. 1: Geometrical parameters of cross-type structure

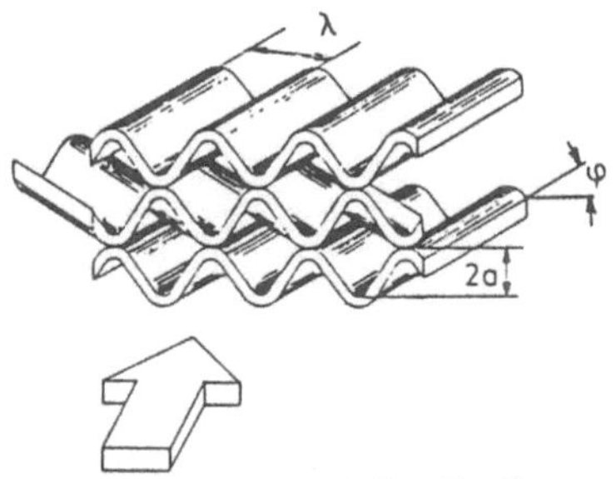

Fig. 2: Geometrical parameters of corrugated structure

2. FLOW BEHAVIOR IN CROSS-TYPE STRUCTURES

The experiments on flow visualization are performed in a water tunnel using transparent structures and dye technique. Pressure transducers as well as conductivity sensors for quantitative measurements of mixing and residence time characteristics can be placed in front of and behind the structures.

Experiments on flow phenomena and visualization of mixing behavior in cross-type structures have been performed using an entire stack of the structure. The effects of geometrical parameters have been investigated in single passages as well.

2.1 Entire Stack

Fig. 3 shows the visualization of the flow and mixing behavior in a structure with square bars at an inclination angle of $\varphi = 52°$. The structure is shown from top-view. The experiment shows a rather good mixing of the locally added tracer across the entire structure within a short distance. Obviously geometrical parameters like these lead to a good mixing in the x-y plane. Fig. 4 shows the same experiment at the same structure from the side-view. The experiment proves, that the mixing across the layers even at higher Re-numbers is significantly worse than the mixing in the layers. This is the reason why static mixers of this kind usually consist of several consecutive elements, each one turned at 90° to the previous one. Thereby mixing in two directions is achieved by two consecutive elements.

Fig. 3: Flow and mixing behavior in cross-type structure (top view; $\lambda/b = 2.75$, $\varphi = 52°$)

Fig. 4: Flow and mixing behavior in cross-type structure (side view) ($\lambda/b = 2.75$, $\varphi = 52°$)

2.2 Single Passage

Since the mixing along the layers (x-y plane) is the essential phenomenon, while the mixing across several layers (in z-direction) is of minor significance, the structure parameters may be designed primarily for a homogeneous mixing in the x-y plane.

Therefore the investigations on the mixing behavior in the x-y plane may be performed in a single passage consisting of two opposing bar layers. Of course the mixing behavior in a single passage does not exactly correspond to the mixing behavior in a x-y plane in entire stacks, but it gives a rather good approximation to it.

Fig. 5 shows the effects of the inclination angle on the mixing behavior in single passages. The geometrical parameters of the bars correspond to those of the structure in fig. 3. The experiments correspond to a Re-number of Re=2000. These experiments have been perfomed in a wind tunnel using a method of wall tracing described in /3/. With this method flow phenomena are visualized at the wall coated with a filter paper. Fig. 5 shows one side of the structure, the examined passage consists of two sides with opposing direction of the bars. The white areas correspond to the contact areas of the two adjacent layers.

At higher inclination angles of $\varphi = 62°$ the main part of the flow follows the path between the bars. Therefore structures with higher inclination angles need a long distance to generate a homogeneous concentration profile and therefore lead to rather long mixers. Decreasing inclination angles at this ration of λ/b lead to an improved mixing behavior. At small inclination angles homogeneous mixing can be achieved within a short distance resulting in very compact mixers.

3. RELATIONS TO CORRUGATED STRUCTURES

Corresponding investigations on flow phenomena, mixing behavior, pressure drop and local mass transfer in corrugated structures have been performed in previous investigations /3,4/. The results show some relations in flow behavior between corrugated structures and cross-type-structures in single passage arrangement. Corrugated structures show similar effects of the inclination angle on mixing behavior /3,4/. At small inclination angles φ (and certain wave-lengths λ) corrugated structures as well lead to a homogeneous mixing within a short distance resulting in compact mixers. This is however related to a relative high pressure drop /4/.

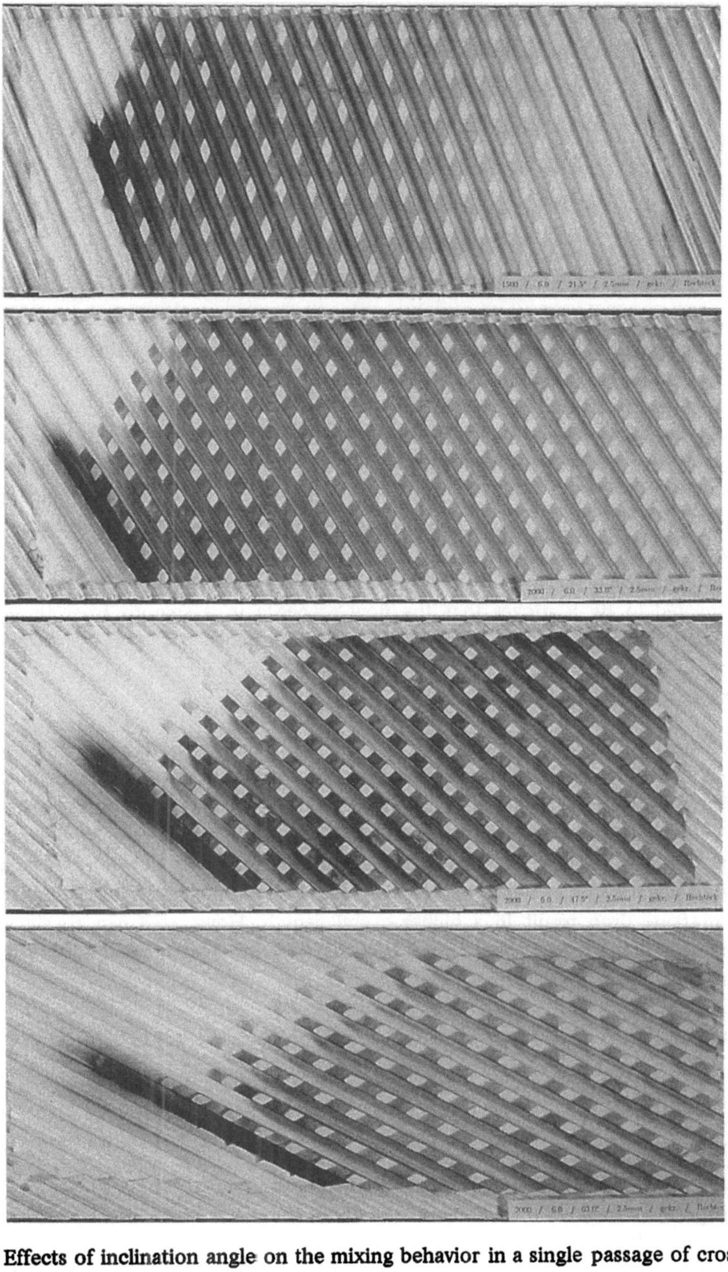

Fig. 5: **Effects of inclination angle on the mixing behavior in a single passage of cross-type structures**
(Re = 2000, λ/b = 2.75, φ = 63°, 47°, 33°, 22° from top to bottom)

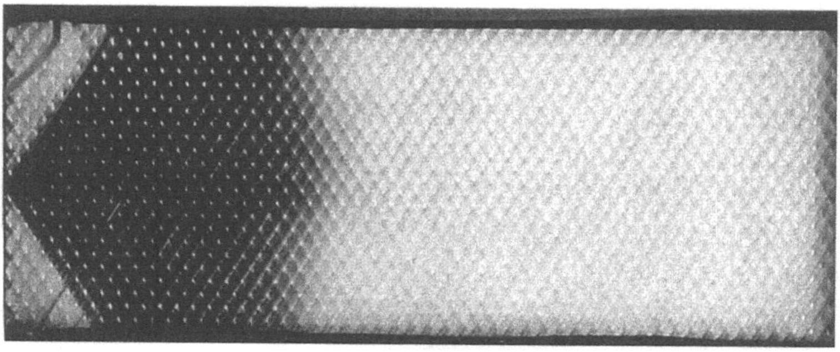

Fig. 6: Homogeneous mixing and residence time behavior in corrugated structure ($\lambda/a = 3.56$, $\varphi = 32°$)

Another point of interest with static mixers and structured packings is the residence time charateristic of the flow. Experiments with corrugated structures in the water tunnel have shown, that the residene time charateristic strongly depends on the geometrical parameters of the structure as well as on the Re-number. Some geometrical parameter combinations lead to a homogeneous mixing as well as to a homogeneous residence time behavior as illustrated in fig. 6 while other parameters lead to a rather inhomogeneous residence time behavior. The experiment related to fig 6 proves that the flow proceeds in a smooth front through the structure, leading to a real homogeneous residence time behavior.

4. CONCLUDING REMARKS

The flow and mixing behavior in structures, used in static mixers, have been visualized. The results prove, that the geometrical parameters of the structure, i.e. the inclination angle, strongly affect the mixing behavior in these structures. For technical applications motionless mixers can be designed with regard to a compact structure or with regard to a minimum in pressure drop. Additional investigations have shown, that compact structures leading to a homogeneous mixing within a short length usually cause a high pressure drop, while structures with a low pressure drop result in a higher overall length. However it can be shown, that parameter optimization for the same mixing task may lead to structures with a low pressure drop and an acceptable overall length.

5. REFERENCES

/1/ Pahl, M.H., Muschelknautz, E.: Einsatz und Auslegung Statischer Mischer. Chem.-Ing.-Tech. 51 (1979) 5, 347-364.

/2/ Bürgi, R., Tauscher, W., Streiff, F.: Vermischen von Gasen in statischen Mischern. Chem.-Ing.-Tech. 53 (1981) 1, 39-42.

/3/ Gaiser, G., Kottke,V.: Visualization of flow phenomena and local heat and mass transfer in corrugated passages. Proceedings of the 5th International Symposium on Flow Visualization, Prag, 1989, 835-842.

/4/ Gaiser, G.: Strömungs- und Transportvorgänge in gewellten Strukturen. Dissertation, Universität Stuttgart 1990.

Flows Around Bodies

Flows Around Moving Bodies

Flow Visualization of the Leading Edge Vortices on Slender Wings Undergoing Large Amplitude Unsteady Motions

DR. ROBERT C. NELSON
DR. SCOTT A. THOMPSON

Department of Aerospace and Mechanical Engineering
Hessert Center for Aerospace Research
University of Notre Dame, Notre Dame, Indiana, 46556
United States of America

Abstract

An experimental wind tunnel investigation was performed using a 70° sweep delta wing. Flow visualization and surface pressure tests were conducted in order to relate the aerodynamic loads to the vortex flow characteristics for both steady and unsteady cases. For ranges of motion precluding the occurrence of vortex breakdown, quasi-steady behavior was seen. For ranges including very high angles of attack, large unsteady effects were measured. The unsteady breakdown location was then correlated with the unsteady surface pressure data.

Introduction

The emphasis on maneuverability and agility of modern fighter aircraft has required controlled flight at angles of attack where the leeward flow field is mostly separated. This separated flow field is dominated by highly organized vortical flows that are generated from the separation of the flow around the slender forebody and the lifting surfaces such as the strake (leading edge extension) wing or control surfaces.

Only recently have experimental studies been conducted to examine the aerodynamics of slender wings undergoing large amplitude pitch oscillations.[1-8] These experiments have shown that for oscillations at low-angles-of-attack the aerodynamic forces act in an essentially quasi-steady manner. However, for oscillations at large-angles-of-attack significant aerodynamic hysteresis has been observed. The overshoot in lift force has been seen to exceed 50% of the steady state loads.[1,6]

In this paper the relationship between the vortical flow characteristics of an oscillating, highly-swept wing and its surface pressure field is examined. Both the steady and unsteady cases will be examined. Extensive testing was conducted that consisted primarily of the measurement of surface pressures and vortex breakdown locations as functions of angle of attack. The flow visualization data were then correlated to the surface pressure data. Additional data and discussion can be found in Reference 8.

Experimental Equipment and Procedure

The experimental equipment included an unsteady pitching mechanism, the flow visualization material, and the image processing equipment. The unsteady pitching mechanism consisted of a five-bar armature attached on one end to the wing and on the other end to a motor. The mechanism converted the rotational motion of the motor to a translational motion of the armature. This armature then drove the model about a pivot point, creating angle of attack excursions. The model was a 70° sweep, $\frac{3}{4}$" (1.9 cm) thick, $16\frac{7}{16}$" (0.42 m) root chord delta wing with the leading edge beveled on the pressure surface only. The unsteady motion used was a sinusoidal oscillation from approximately 2-60°.

Titanium tetrachloride ($TiCl_4$) was used as the flow visualization method. Inert Nitrogen gas was used to deliver $TiCl_4$ vapor to the smoke emission ports located on the wing surface near the apex. Upon being emitted from the wing, the vapor reacts with water vapor in the air to form a dense white smoke.[9] The emission port was placed so that the smoke was entrained into the leading edge vortex, effectively visualizing the core and the breakdown. Each experiment was then recorded with a charge-coupled device (CCD) high resolution super-VHS video camera.

The videotapes were then analyzed to obtain information on the vortex core and vortex breakdown location as a functions of both time and instantaneous angle of attack. The videotapes were analyzed using a professional video editing and analysis system. This system allowed for image enhancement, coordinate digitization, and extraction of time-dependent data. Unsteady data was obtained to examine unsteady effects such as lags in the flowfield development and vortex field recovery. Steady data was obtained as a reference for the unsteady data, and for use in comparing with data available in the literature.

Results

Figure 1 shows a photograph of an actual flow visualization experiment. Note that this photograph shows a wing with both an upper and lower surface bevel; the data shown will be for a wing with a lower surface bevel only. Figure 1 is shown to illustrate the flow visualization method. The $TiCl_4$ vapor is being emitted near the apex, where it is entrained into the vortex core. Vortex breakdown can be seen in Fig. 1 as a rapid expansion of the core diameter. The actual point of breakdown was taken as the point at which the core rapidly expands from a tightly rolled vortex to the wider breakdown region. This subjectivity, in addition to the steady state oscillation of breakdown, can account for a maximum uncertainty in the breakdown location of 3%.

Figure 2 shows unsteady breakdown location data obtained from a flow visualization experiment. The breakdown location is shown as a function of instantaneous angle of attack for four nondimensional pitching frequencies, where $k = 2\pi fc/U$ (where c is the root chord length,

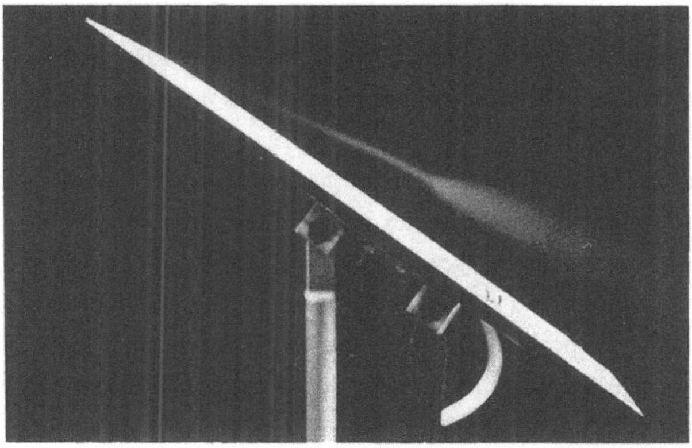

Figure 1 Vortex Breakdown over a Delta Wing. Flow Visualized by use of $TiCl_4$.

f the pitch frequency, and U the freestream velocity). The arrows in Fig. 2 indicate the direction of motion (angle of attack increasing or decreasing). The steady breakdown location data has not been shown in order to maintain clarity. However, it was seen that the unsteady data bracketed the steady data.

The hysteresis which is characteristic of unsteady high angle of attack delta wing aerodynamics can be seen in Fig. 2. At a given instantaneous angle of attack, the difference in breakdown location is as much as 0.45c, depending on the direction of motion. The size of the hysteresis loop increases with increasing reduced frequency. At very high angles of attack,

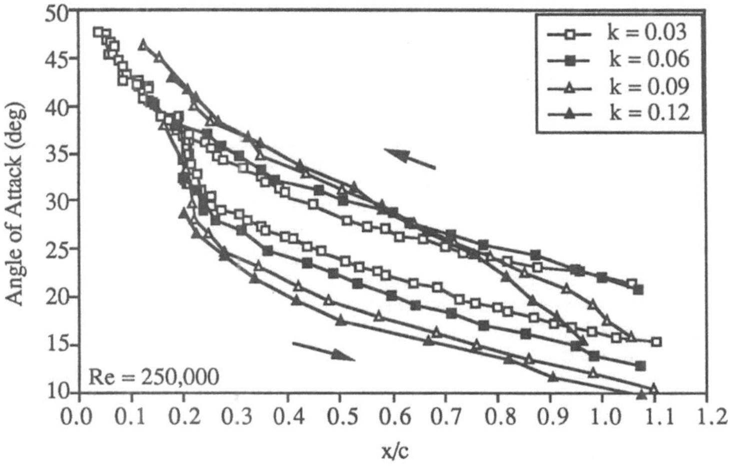

Figure 2 Breakdown Location for 2-60° Oscillation. Four Pitch Frequencies Shown.

over 50°, the flow over the wing becomes fully separated, similar to a bluff-body flow. Thus breakdown does not exist. As the wing pitches downs, the leading edge vortex system reforms with breakdown near the apex. Breakdown then moves aft as the angle of attack continues to decrease.

Figure 2 illustrates the lags that develop for the unsteady case. For the upward motion, there is a lag in the development of the vortex core and the upstream progression of vortex breakdown. For the downward motion there is a lag in reformation of the vortex system (from fully separated flow conditions) and thus in the downstream progression of vortex breakdown. These lags give rise to the drastic unsteady effects seen in the aerodynamic loads on the wing.

Figure 3 shows an example of the unsteady pressure data as a function of instantaneous angle of attack, along with the steady data. This data is for a single surface pressure tap location. The location of vortex breakdown has been noted at specific points along the curve.

As the angle of attack increases the pressures uniformly decrease. With the occurrence of breakdown on the wing the pressure gradient begins to decrease. Further increase in angle of attack moves the breakdown to the apex. This precedes the total separation of the leeward side,

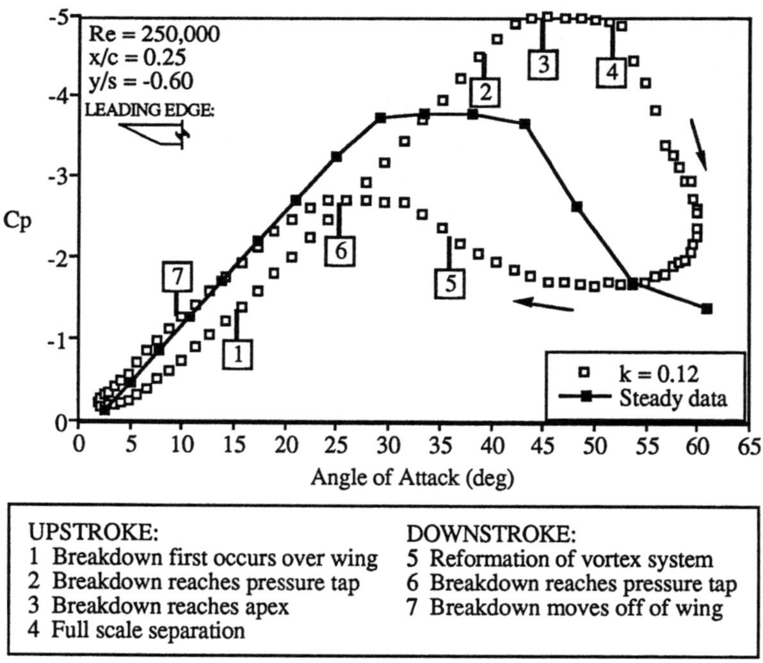

Figure 3 Unsteady Surface Pressure Data for 2-60° Oscillation. Chordwise Progress of Vortex Breakdown (from Flow Visualization Tests) also Noted.

and as a result the pressures begin increasing due to the lack of an organized flow structure over the wing. However, before the pressure field can become relatively uniform (as for the steady high angles of attack case), the wing pitches down again, until the point is reached at which the leading edge vortex structure reforms (with breakdown near the apex). Thus a pressure recovery begins and the pressures begin decreasing. Breakdown then moves down the length of the wing and into the wake. Concurrent with this behavior of the vortex is a partial recovery of the suction pressures. Further decrease in angle of attack results in a collapse of the vortex system and the vortex induced pressure field.

Conclusions

This research was one of the first experimental studies to obtain unsteady surface pressures for a pitching delta wing. The surface pressure data was then correlated to the unsteady location of the vortex breakdown during an oscillatory pitching motion encompassing both low and high angles of attack. For ranges of motion where breakdown moved onto or off of the wing, the unsteady effects were large. This was also the case for motions including full scale separation . During the unsteady motion, a lag was measured in the reformation of the vortex system from a fully separated flow condition (at very high angles). This lag is related to the decreased recovery of the suction pressure during the downstroke portion of the motion. The decreased pressure recovery contributes to the hysteresis seen in the unsteady loads.

Acknowledgments

This research was sponsored by the University of Notre Dame and the Air Force Office of Scientific Research under grant AFOSR-90-0321.

References

1. Brandon, Jay M., and Shah, G. H., "Effect of Large Amplitude Pitching Motions on the Unsteady Aerodynamic Characteristics of Flat-Plate Wings", AIAA-88-4331, Aug, 1988.
2. Reynolds, G.A., and Abtahi, A.A., "Instabilities in Leading-Edge Vortex Development", AIAA-87-2424, Aug, 1987.
3. LeMay, S.P., Batill, S.M., and Nelson, R.C, "Leading Edge Vortex Dynamics on a Pitching Delta Wing", AIAA-88-2559-CP, June, 1988.
4. Bragg, M.B., and Soltani, M.R., "An Experimental Study of the Effect of Asymmetrical Vortex Bursting on a Pitching Delta Wing", AIAA-88-4334, Aug, 1988.
5. Magness, C., Robinson, O., and Rockwell, D., "Control of Leading-Edge Vortices on a Delta Wing", AIAA-89-0999, Mar, 1989.
6. Jarrah, Mohammad-Ameen M., "Unsteady Aerodynamics of Delta Wings Performing Maneuvers to High Angle of Attack," Ph.D. Dissertation, Stanford University, Dec, 1988.
7. Thompson, S.A., Batill, S.M., and Nelson, R.C., "The Separated Flow Field on a Slender Wing Undergoing Transient Pitching Motions", AIAA-89-0194, Jan, 1989.
8. Thompson, S.A., and Nelson, R.C., "The Unsteady Aerodynamics of a Delta Wing Undergoing Large Amplitude Pitching Motiòns", Ph.D. Dissertation, University of Notre Dame, Apr, 1992.
9. Visser, K. D., Nelson, R. C., and Ng., T. T., "Method of Cold Smoke Generation for Vortex Core Tagging," AIAA Journal of Aircraft, Vol. 25, No. 11, Nov, 1988.

Flow Visualization Study of Vortex Breakdown on a 65° Delta Wing in Dynamic Motion Regime

G. Guglieri - Centro Studi Dinamica Fluidi - CNR, Torino , Italy
M. Onorato, F. Quagliotti - Aerospace Engineering Dept., Turin Polytechnic Institute - Technical University, Torino, Italy

Abstract

An application of helium-bubble technique has been attempted in order to investigate the behaviour of the vortical structures on a delta wing in oscillatory conditions.
The dynamic behaviour of the vortex breakdown phenomenon has been analyzed.

Introduction

The flow structure on the upper side of a delta wing is extremely complex. At moderate angle of attack the leeward flow field is dominated by a highly organized vortical flow structures emanating from the leading edge.
The pressure distribution created on the wing surface by these leading edge vortices causes an increment of lift that may be a significant percentage of the total wing lift.
Delta wing performance is limited, however, by a phenomenon known as vortex breakdown or vortex "bursting", that appears at higher angle of attack. This leads to a drastic change in the flow field.
Many experiments have been carried out in an effort to understand the factors which can influence the phenomenon of breakdown by changing all possible parameters, including wing geometry, angle of attack and Reynolds number.
As a consequence, the position of the breakdown is a function of the pressure gradient along the vortex, the initial axial core velocity and the angle of sideslip. Rounded leading edges are also influenced by the Reynolds number.
Although much has been learned about this phenomenon, no theory exists which can yield the flow details in the breakdown location which consistently compare with experimental results. For this reason, it is essential to continue examining the flow field experimentally in an effort to uncover information which may shed light on some unresolved aspects of this problem.

The physical understanding of the behaviour of the vortical flows is not as advanced for wings undergoing dynamic motions.
The unsteady motion of the wing can result in delays of flow separation, changes in vortex location and onset of breakdown at higher angle of attack.
During longitudinal oscillatory or transient motions, a hysteresis develops in the positions of the vortex core and the vortex breakdown, relative to the static locations. Due to this hysteresis in the flow field, there is a corresponding modification of the aerodynamic loads on the wing. In post-stall conditions, a substantial overshoot in the aerodynamic force is typically seen for oscillatory or transient pitching maneuvers, as a consequence of breakdown dynamics.
In references 1-8 recent studies on the subject are reported.
In an attempt to contribute to the understanding of the phenomenon of the vortex brekdown, a delta wing has been extensively tested, both in static and dynamic conditions, under forced pitching oscillation motion. Results from flow visualization are shown here.

The Experimental Apparatus

The wind tunnel used for this experimental program has a closed circuit with a 3 meter diameter circular test section. The maximum speed is about 80 m/s and the turbulence intensity is 0.3%. This tunnel is located in the "Modesto Panetti" Aeronautical Laboratory of the Turin Polytecnic Institute, Aerospace Department.
The model used is a sharp leading edge flat plate 65° delta wing with 0.793 m span and 0.850 m root chord. A fuselage under the wing can house the five component balance, the scanivalve system and the pressure trasducer.
The moving-support rig permits the setting of the model angle of attack in the range from -7° to 60° and may produce the oscillatory motion in pitch in the range +/- 3.5°, with a maximum frequency of 5 Hz. Details of the moving rig and support system, plus details of the motion control system may be found in ref.9.
The helium-bubble technique has been used for both static and dynamic flow visualization. The bubbles were injected in the flow near to the model apex, with an injecting pipe. The corrected proportion of helium-soap mixture generates bubbles which have the same specific weight as the air. Illumination of the flow was achieved using a light sheet, located downstream of the model support, in a low interference position.
Both in static and dynamic conditions, the flow visualizations have been filmed by a super VHS video-camera. In order to correlate the positions of the oscillating

model, a LED was suitably positioned on the model. The periodic lighting of the LED gave the model position as a function of time.

The flow visualization tests were performed at the flow velocity of 5 m/s. The vortex core and the vortex breakdown position can be clearly visualized. From the analysis of the video frames it was possible to obtain quantitative measurements of the vortex breakdown position and propagation speed, during the model oscillation, and its phase-lag with respect to the primary motion.

Experimental Results

The model has been forced to oscillate in pitch about its centre of gravity, following a harmonic motion, with an amplitude of +/- 3.5° for different mean angles of attack, α. Results are shown here for $\alpha = 15°$. In figures 1a and 1b the vortex breakdown position (non dimensional distance from the leading edge) during the oscillation is shown. For comparison the static results are also shown. The test Reynolds number and the reduced frequency, K, are shown on the figures. The reduced frequency is defined as $K = \omega C/Uo$, where ω is the oscillation frequency, C is the wing root chord and Uo the wind velocity.

Comparing the dynamic result with the static one, the flow hysteresis is evident. During the upstroke phase, the breakdown assumes a backshifted position with respect to the static case, at the same angle of attack. The opposite is observed during the dowstroke phase.

At the lowest value of the reduced frequency, K = 0.101, the hysteresis loop assumes the shape shown in Fig. 1a. This distribution is almost symmetrical about the static data. As the reduced frequency increases, the hysteresis effect becomes greater at a given angle of attack and the region of the model over which the breakdown occurs decreases.

In Fig. 2 the breakdown propagation speed is reported as a function of the angle of attack for different values of the reduced frequency, K. A positive velocity corresponds to the breakdown location moving downstream. The magnitude of the velocity increases with reduced frequency. The changes in velocity are more sudden at the lower and higher angles of attack, where the breakdown position is, respectively, closer to the trailing edge and to the leading edge.

Moreover it is interesting to remark that, at low values of K, the diagrams have a symmetrical shape, while at K = 1.068 the pattern is more twisted.

It is evident that sign inversion of the propagation speed is consistently delayed with respect to the maximum and minimum angle of attack position.

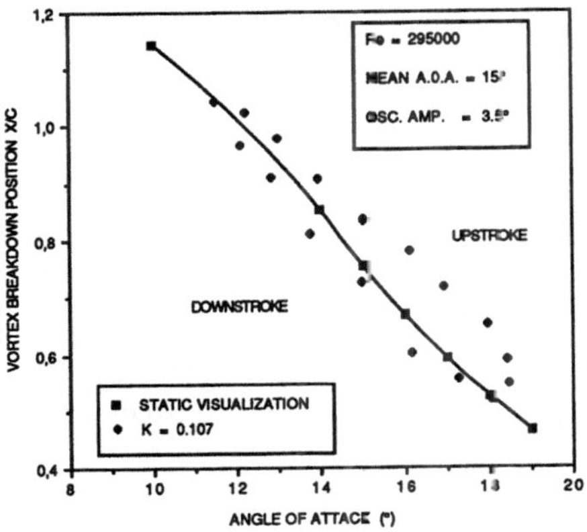

Fig. 1 a - Vortex Breakdown Position

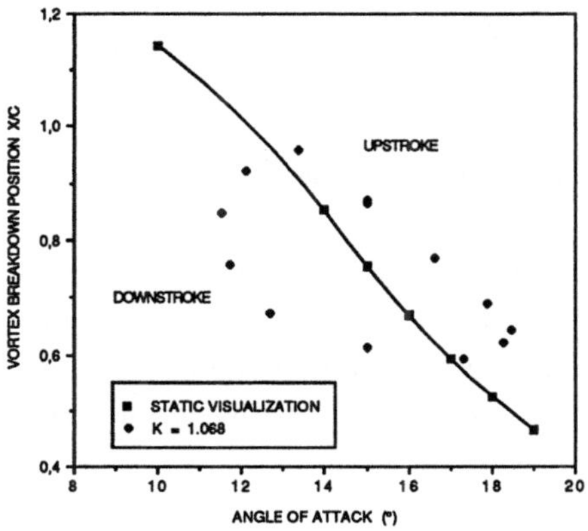

Fig. 1 b - Vortex Breakdown Position

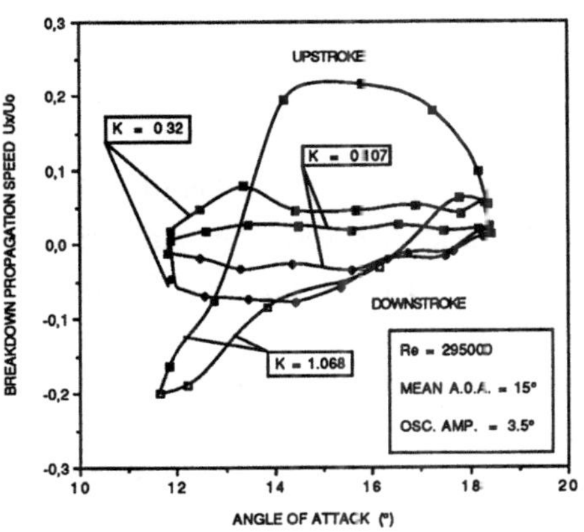

Fig. 2 - Breakdown Propagation Speed

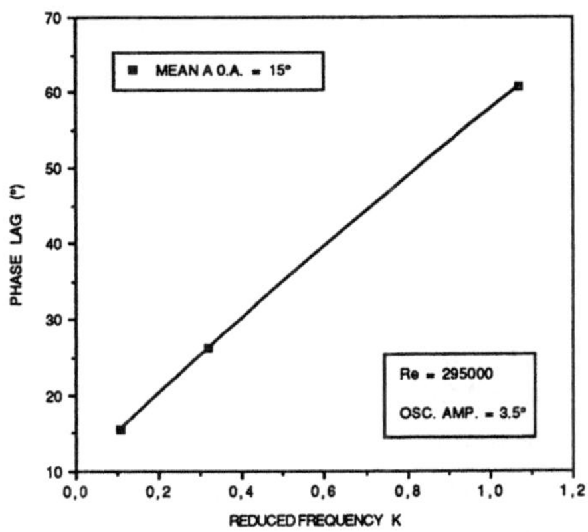

Fig. 3 - Phase Lag Effect

Finally, in Fig. 3, the phase lag between the dynamic vortex breakdown and the model primary motion is reported as a function of the reduced frequency. From the diagram the linear dependence is evident.

Similar results have been obtained for other values of the mean angle of attack.

Concluding Remarks

The helium bubble flow visualization technique has shown to be a reliable method to study vortex flow stuctures, even in dynamic conditions.
As the delta wing oscillate a hysteresis is observed in the chordwise position of vortex breakdown. This hysteresis effect becomes larger increasing the oscillation reduced frequency.
The flow visualization method used has been so effective as to permit quantitative measurements of the vortex breakdown propagation speed and of the relative phase lag.

Acknowledgments

The research has been supported by CNR (Italian National Research Council) and MURST (Italian Ministry of University and Scientific Research)

References

1. Jarrah M.A.M., "Low Speed Wind Tunnel Investigation of Flow About Delta Wing, Oscillating in Pitch to Very High Angle of Attack" AIAA 89-0295, Jan. 1989
2. Cunningham Jr. A.M., "Unsteady Low Speed Wind Tunnel Test of a Straked Delta Wing, Oscillating in Pitch" AFWAL TR-87-3098, April 1988
3. Bragg M.B., "Experimental Study of the Effect of Asymmetrical Vortex Bursting on a Pitching Delta Wing" AIAA 88-4334, August 1988
4. Brandon J.M., "Effect of Large Amplitude Pitching Motions on the Unsteady Aerodynamic Characteristics of Flat-plate Wings" AIAA 88-4331, August 1988
5. Lee M., "Vortex Dynamics of Delta Wings" L.N.E., Vol. 46, 1989 Springer Verlag, Berlin
6. Ashley H., "Unsteady Aerodynamic Loading of Delta Wings for Low and High Angles of Attack" Proc. Int. Symp. on Non-steady Fluid Dynamics, June1990
7. Hanff E.S., "Oscillatory Test Techniques" AGARD-R-776, April 1991
8. "Hanff E.S., "Large Amplitude Oscillations" AGARD-R-776, April 1991
9. Guglieri G., Onorato M., Quagliotti F. "Untersuchung des Wirbelaufplatzens an Deltaflugeln", Zeitschrift fur Flugwiss., n.16-4, 1992

In-Flight Flow Visualization Results from High Angle of Attack Tests on a Forward Swept Wing Aircraft with Comparisons to Water Tunnel Results

John H. Del Frate

NASA Dryden Flight Research Facility
Edwards, California, U.S.A. 93523-0273

Abstract

Asymmetries in the vortical flow generated by the long forebody of the X-29 aircraft were suspected of causing uncommanded yawing moments at high-angles of attack. A smoke flow visualization technique was used to visualize the forebody vortices and to correlate any asymmetries observed with flight yawing moment data. This paper presents a method of analyzing smoke flow visualization data for comparison with asymmetric aircraft moments or forces. Good agreement was found between the forebody vortex asymmetries and uncommanded yawing moments.

Nomenclature

C_{n_0}	yawing moment coefficient at 0° sideslip (positive is nose right)
α	angle of attack, deg
β	angle of sideslip, deg (positive is nose left)
θ_v	angular location of the mid-plane between the right and left forebody vortices (0° is top centerline, positive is to the right as viewed from back of aircraft, negative is to the left), deg

Introduction

Since 1984, the X-29 aircraft (Fig. 1) has been flight tested at the National Aeronautics and Space Administration's (NASA) Dryden Flight Research Facility located at Edwards, California, USA. In one of the test programs, the envelope of the X-29 was expanded to include the high-angle-of-attack regime where significant vortical flows exist. Uncommanded yawing moments at high-angles of attack at $\beta = 0°$ were discovered in the flight tests [1]. Asymmetries in the vortical flow generated by the long forebody of the X-29 were thought to be the cause of the yawing moments. A smoke flow visualization technique used previously on the F-18 High Alpha Research Vehicle (HARV) [2–4] was used to study the forebody vortices.

This paper describes a method of analyzing smoke flow visualization data for comparison with asymmetric aircraft moments or forces. A more extensive and in-depth report on this flight experiment can be found in reference 5.

Aircraft Description and Test Hardware

Figure 1 is a three-view sketch of the X-29 aircraft. The aircraft is a technology demonstrator and features forward-swept wings, close coupled canards, and relaxed static stability. The X-29 is powered by a single turbofan engine. The forebody on the X-29 is from an F-5A aircraft that was shortened 11 in. (27.9 cm). A nose strake and noseboom were added at the apex [6].

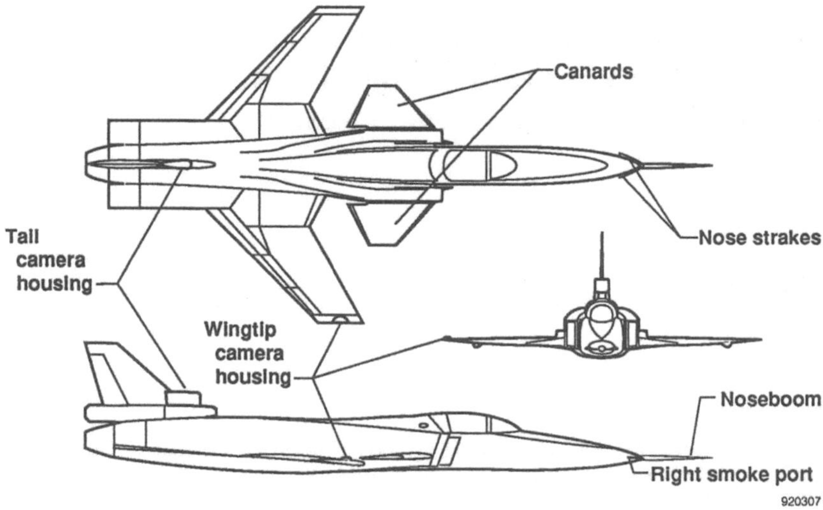

Fig. 1. Three view of X-29 configuration.

The smoke used to visualize the forebody vortices was introduced through a 1-in. diameter port near the nose apex just under the nose strakes as shown in Fig. 1. The port location was determined with a sub-scale model in a water tunnel flow visualization test. Smoke was generated in a forebody compartment using the same method used for the vortex flow visualization in the F-18 HARV program [2–4]. To produce smoke, the pilot remotely ignites a pyrotechnic device which provides approximately 30 sec. of dense white smoke.

The visualization results were recorded by an onboard camera system which consisted of a video camera and 35-mm still camera housed in the wingtip and an 8-mm camcorder and 35-mm still camera housed at the base of the vertical tail (Fig. 1). The wingtip video was telemetered to the ground, viewed, and recorded at the ground station in real time. The 35-mm cameras were controlled from the ground station. A local timing standard was used to coordinate all video and still camera data with the flight conditions.

Test Conditions

The flight-test results presented were documented in 1-g flight while stabilized at a constant angle of attack, in wings-level sideslip maneuvers where the sideslip was commanded by the pilot ($\alpha > 37°$) or in wingrock where the sideslip was uncommanded ($\alpha \geq 37°$). Data is presented for angles of attack from 25.5 to 50.5°. All test points were flown between 17,000- and 30,000-ft altitude.

Results and Discussion

The smoke entrained in the forebody vortex system was documented by the tail and wingtip cameras. Figure 2(a) is a photo of the forebody vortex pair at $\alpha = 49.7°$ and $\beta = -6.1°$ as seen from the wingtip 35-mm still camera. As angle of attack increased, the vortices seen from the wingtip camera became better defined and their distance from the top of the forebody increased. The forebody vortex pair visible in Fig. 2(b) was photographed by the tail camera at $\alpha = 33.2°$ and $\beta = 1.0°$. The vortices appear as a pair of white lobes with a dark region mid-plane between them. Similar tail video images were analyzed and the pair of lobes were observed shifting to the left and right on the upper surface of the aircraft. As the lobes shifted, the dark region remained between the two lobes and shifted with them as a unit. This dark region or mid-plane was considered representative of the vortex system angular position.

Based on the wingtip camera video, the image recorded by the tail cameras is believed to correspond to the longitudinal plane about mid-canopy.

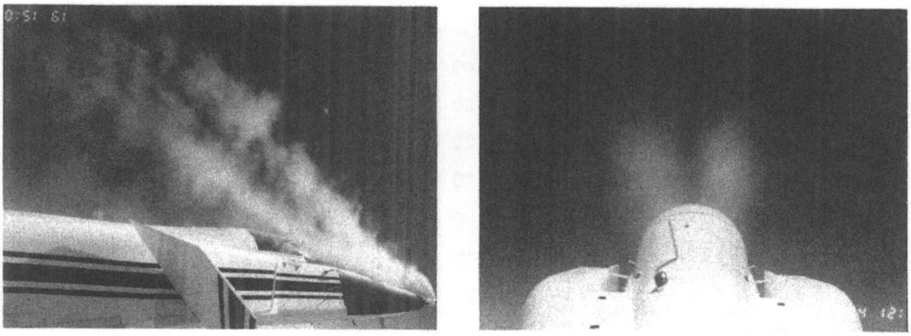

(a) Wingtip camera view of forebody vortices, $\alpha = 49.7°, \beta = -6.1°$.

(b) Tail camera view of forebody vortices, $\alpha = 33.2\,°, \beta = 1.0\,°$.

Fig. 2. Examples of camera views.

By measuring the angular position of the forebody vortices over a range of angle-of-attack and angle-of-sideslip conditions, the existence of asymmetries could be determined. Figure 3 is a sketch of the tail camera view with the measurement reference system in use. A line was drawn through the mid-plane (in this case, the vortex system position is shown shifted to the left). This was done for several angle-of-attack and angle-of-sideslip combinations. The lines drawn through the mid-plane tended to intersect at a point just below the canopy surface as shown in the sketch by a cross. This center provided a simple reference system for tracking the angular position of the vortex system. The angle between the mid-plane and the vertical plane was called θ_v. As noted in the figure, positive θ_v is to the right, as seen by the tail camera, and negative θ_v is to the left.

The forebody vortex system positions were plotted as a function of sideslip for several angles of attack and a linear approximation was drawn through each of the respective data sets (Fig. 4). In general, the line slopes are similar, however for some angles of attack, the bias of the line at $\beta = 0°$ can be slightly positive or negative. This indicates that at $\beta = 0°$, the forebody vortex positions varied. A positive θ_v indicated that the forebody vortices lay slightly to the right as seen from the tail camera and to the left for a negative θ_v. The lines for $\alpha = 25.5°$ and 41.3° cross through the origin indicating that at $\beta = 0°$ the forebody vortices are oriented symmetrically. The lines for $\alpha = 35.0°$ and 50.5° are above the origin indicating that the forebody vortices are shifted slightly to the right. At $\alpha = 46.0°$ the line lies below the origin indicating that the forebody vortices at this condition are shifted slightly to the left.

Wind-tunnel research has shown that forebody vortex system asymmetry can result in side forces. An orientation toward the left, as noted in the flight data for $\alpha = 46.0°$, would promote premature boundary layer separation on the left side of the forebody and delay separation on the right side. In this case, the right side would have a higher suction level than the left, yielding a net side force to the right [7–9].

Figure 5 shows the yawing moment of the X-29 at $\beta = 0° (C_{n_0})$ and the forebody vortex system positions, θ_v, at $\beta = 0°$ as a function of angle of attack. The scale on the left corresponds to the yawing moment data with nose right being positive. The yawing moment data set had been previously determined from parameter estimation and pilot comments in flight [1] and is currently used in the flight simulator. The scale on the right side of the plot corresponds to the θ_v data and reflects the proper yawing moment direction (for example, negative θ_v would yield a positive yawing moment). The X-29 was found to have a $C_{n_0} = 0$ up to $\alpha = 40°$ where a right yawing moment began and continued out to about 48°. At angles of attack of 48° or greater, the yawing moment reversed to the left. With the exception of the data point at $\alpha = 35°$, all of the vortex system data points agreed well with the yawing moment data.

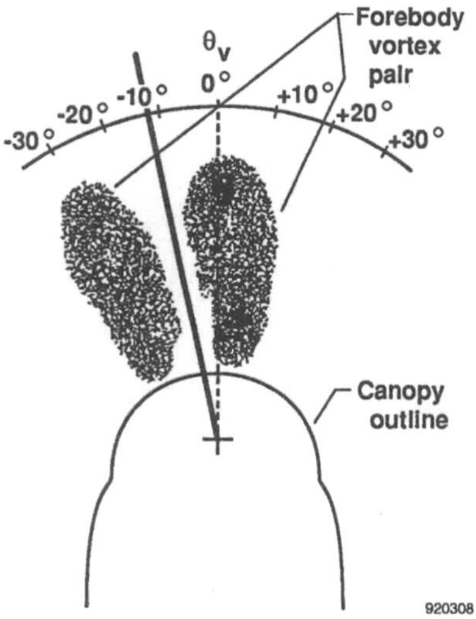

Fig. 3. Forebody vortex measurement technique.

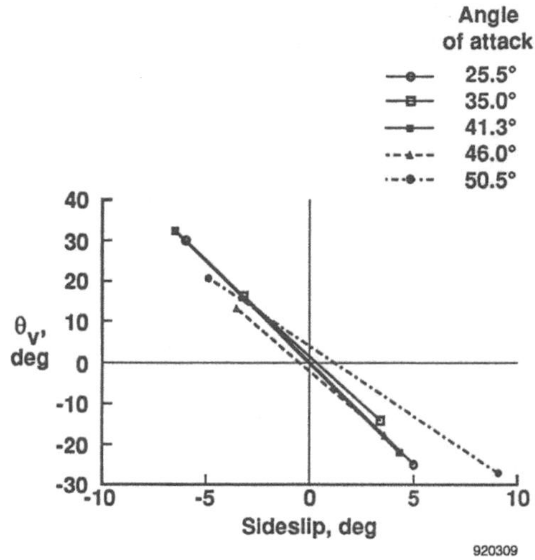

Fig. 4. Approximate forebody vortex system position (θ_v) vs. sideslip for various angles of attack.

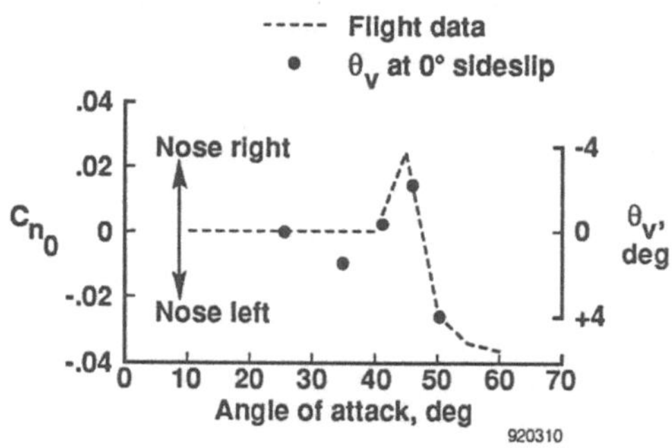

Fig. 5. Comparison of yawing moment at zero sideslip and forebody vortex system position, θ_v.

The generally good agreement between the uncommanded yawing moments at $\beta = 0°(C_{n_0})$ and the angular position of the forebody vortex system confirms the hypothesis that the yawing moments at $\beta = 0°(C_{n_0})$ at high-angles of attack are related to asymmetries in the vortex system [1, 10]. The cause of the asymmetries is not known at this time, however, because they are repeatable they are attributed to asymmetries or irregularities on the aircraft surface.

Concluding Remarks

Visualization of the forebody vortices on the X-29 was accomplished by using a smoke flow visualization technique during flight. Results were presented and comparisons between the flight yawing moment and forebody vortex angular position data were made. There was generally good agreement between the asymmetries in the angular position forebody vortex system and airplane yawing moments at high-angles of attack at $\beta = 0°$.

References

1. Webster, Frederick R. and Dana Purifoy, "X-29 High Angle-of-Attack Flying Qualities," AFFTC-TR-91-15, July 1991.
2. Fisher, D.F., R.E. Curry, J H. Del Frate, and D.M. Richwine, "In-Flight Flow Visualization Techniques on a Vortex-Lift Fighter Aircraft," Flow Visualization V; Proceedings of the 5th International Symposium, Prague, Czechoslovakia, Aug. 21–25, 1989. Hemisphere Publishing Corp., New York, NY, 1990.
3. Richwine, David M., Robert E. Curry, and Gene V. Tracy, "A Smoke Generator System for Aerodynamic Flight Research," NASA TM-4137, 1989.
4. Fisher, David F., John H. Del Frate, and David M. Richwine, "In-Flight Flow Visualization Characteristics of the NASA F-18 High Alpha Research Vehicle at High Angles of Attack," NASA TM-4193, 1990.
5. Del Frate, John H. and J.A. Saltzman, "In-Flight Flow Visualization Results from the X-29 Aircraft at High Angles of Attack," AIAA 92-4102, Aug. 1992.
6. Moore, M. and D. Frei, "X-29 Forward Swept Wing Aerodynamic Overview," AIAA 83-1834, July 1983.
7. Keener, Earl R., "Flow Separation Patterns on Symmetric Forebodies," NASA TM-86016, 1986.
8. Grafton, Sue B., Joseph R. Chambers, and Paul L. Coe, Jr., "Wind - Tunnel Free-Flight Investigation of a Model of a Spin-Resistant Fighter Configuration," NASA TN D-7716, 1974.
9. Skow, A. M. and G. E. Erickson, "Modern Fighter Aircraft Design for High Angle-of-Attack Maneuvering," AGARD-LS-121, Dec. 1982, pp. 4–1 to 4–59.
10. Klein, Vladislav, Brent R. Cobleigh, and Keith D. Noderer, "Lateral Aerodynamic Parameters of the X-29 Aircraft Estimated From Flight Data at Moderate to High Angles of Attack," NASA TM-104155, 1991.

On the Boundary-Layer Control Through Momentum Injection: Numerical, Flow Visualization, and Experimental Studies

V.J. MODI

Department of Mechanical Engineering
University of British Columbia
Vancouver, B.C., Canada V6T 1Z4

and

T. YOKOMIZO

Department of Mechanical Engineering
Kanto Gakuin University
Mutsuura, Kanazawa, Yokohama, Japan 236

Abstract

The concept of Moving Surface Boundary-layer Control (MSBC) as applied to two-dimensional airfoils is investigated through a planned numerical simulation program, complemented by wind tunnel tests and extensive flow visualization study. The moving surface was provided by rotating cylinders located at the leading edge and upper surface of the airfoil. The results suggest that the concept is quite promising and can provide a substantial increase in the lift and a delay of the stall. The maximum coefficient of lift realized was around 2.4, approximately three times that of the base airfoil. The stall was delayed from around 10° to 48°! Application of the MSBC procedure to bluff bodies showed a remarkable decrease in the pressure drag.

Introduction

Ever since the introduction of the boundary-layer concept by Prandtl, there has been a constant challenge faced by scientists and engineers to minimize its adverse effects and control it to advantage. Methods such as suction, blowing, vortex generators, turbulence promoters, etc. have been investigated at length and employed in practice with a varying degree of success. A vast body of literature accumulated over years has been reviewed rather effectively by Modi et al. [1,2]. However the use of moving wall for boundary-layer control has received relatively less attention.

Irrespective of the method used, the main objective of a control procedure is to prevent, or at least delay, the separation of boundary-layer from the wall. A moving surface attempts to accomplish this in two ways :

(i) it retards growth of the boundary–layer by minimizing relative motion between the surface and the freestream;

(ii) it injects momentum into the existing boundary layer.

Using rotating cylinders as momentum injecting elements (Fig. 1), the paper studies effectiveness of the MSBC as applied to the next generation of highly maneuverable airplanes negotiating steep gradients requiring increased lift, and operating at high angles of attack. Three different approaches are used and their validity assessed through comparison.

Numerical Simulation

Here the stream function–vorticity form of the Navier–Stokes equations were used, in conjunction with the variable grid–size ($\approx$ 3000 nodes, Fig. 2) finite element analysis. Such numerical simulation of the two and three element airfoils with momentum injection has not been reported in literature. Note, accuracy of the results is governed by the number of nodes which can be increased depending on the available computational facility. The parametric analysis, involving a systematic variation of the speed ratio (U_c, cylinder surface velocity; U, free stream velocity), angle of attack (α) and the Reynolds number (R_n, $10^4 - 10^6$), gave detailed information about the pressure loading, separation condition and the time dependent wake (Fig. 3). It also showed, rather spectacularly, effectiveness of the MSBC.

Flow Visualization

Various airfoil and bluff body configurations were subjected to an extensive flow visualization study to get better appreciation of the flow, particularly with reference to stagnation, separation, reattachment and wake conditions. The tests were carried out in a water channel using slit lighting with polyvinylchloride particles serving as tracers. The flow visualization pictures were obtained over a wide range of model geometry and orientation, cylinder location, surface condition and speed, cylinder projection into the boundary layer, etc. A video was also taken. The test Reynolds number, based on the freestream speed and the airfoil chord was around 4×10^4, which can be quite different from the real life situation. Hence, the results should be considered only qualitative in character, however, they do show effectiveness of the boundary–layer control procedure.

Fig. 5 shows a 15% thickness, symmetrical Joukowsky airfoil, with its nose replaced by a rotating element, at $\alpha = 25°$. As the reference unmodified airfoil stalls at around $\alpha = 10°$, a highly separated flow with a wide turbulent wake is expected as indicated in Fig. 4(a). However, with the momentum injection corresponding to $U_c/U = 4$, the flow remains essentially attached right up to the vicinity of the trailing edge, thus exhibiting near potential flow character (Fig. 4b). Note the similarity between the numerical predictions and the flow visualization data.

Wind Tunnel Tests

The numerical and flow visualization studies were complemented by an extensive wind tunnel test–program involving a family of airfoils with one or more momentum injecting elements (rotating cylinders) forming moving surfaces [2]. Fig. 5 shows a typical C_L vs α plot for the case where the cylinder is located slightly away from the symmetrical nose position (Upper L.E. Cylinder). In absence of the rotation ($U_c/U = 0$), there is a drop in C_L due to the modified nose geometry. However, with the cylinder rotation, the lift curve is extended, essentially at the same slope, beyond the stall reaching the $C_{L,max}$ of around 2.4, a 193 % increase! Note, the stall is now delayed to around 48°! The pressure distribution data also provided an estimate of the separation location which correlated quite well with the flow visualization results.

Concluding Remarks

Based on the numerical, flow visualization, and wind tunnel tests with a two-dimensional, 15% thickness ratio, Joukowsky airfoil, it can be concluded that the MSBC represents an effective procedure for increasing lift and delaying stall. The semi–passive character of the procedure makes it quite attractive for real–life prototype applications. When applied to two–dimensional prisms is resulted in a significant reduction in drag [3].

Acknowledgment

The investigation reported here was supported by the Natural Sciences and Engineering Research Council of Canada, Grant No. A–2181.

References

[1] Modi, V.J.; Sun, J.L.C.; Akutsu, T.; Lake, P.; Swinton, P.G.; and Mullins, D. : Moving surface boundary–layer control for aircraft operation at high incidence. Journal of Aircraft. 18 (1981) 963–968.

[2] Mokhtarian, F. : Fluid dynamics of airfoils with moving surface boundary–layer control. Ph.D. Thesis. University of British Columbia, Vancouver, B.C., Canada 1988.

[3] Modi, V.J.; Ying, B.; and Yokomizo, T. : Boundary–layer control of bluff bodies through momentum injection. 1990 SAE International Bus/Truck Meeting and Exposition, Detroit, U.S.A. 1990; also SAE Transactions, Journal of Commercial Vehicles, Section 2. 99 (1991) 778–794.

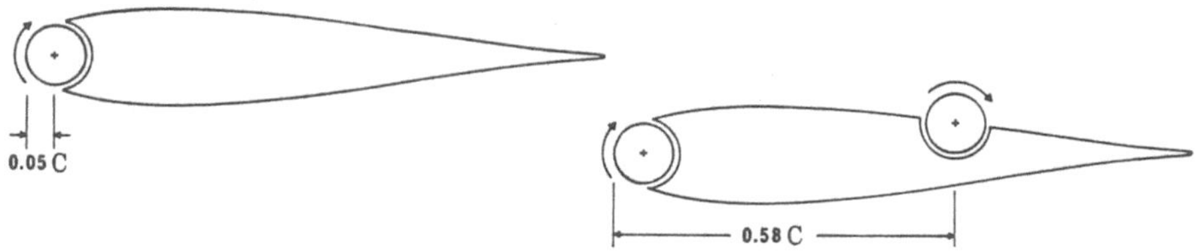

Fig. 1 Airfoils with rotating cylinders, serving as momentum injection units, used in the study.

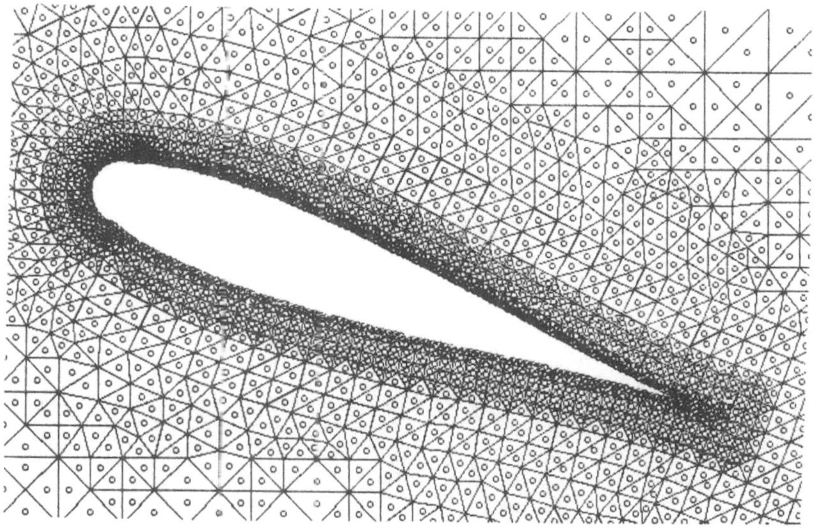

Fig. 2 Triangular elements with spatially varying grid–size used in the numerical simulation.

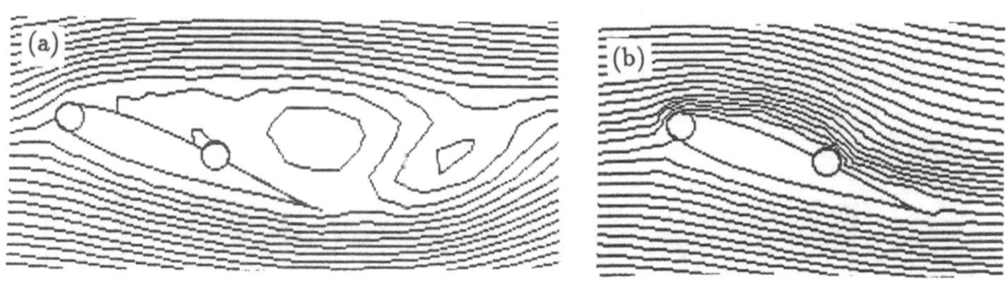

Fig. 3. Typical numerically obtained flow patterns, for an airfoil at α = 30°, showing effectiveness of the MSBC : (a) $U_c/U = 0$ for both the cylinders; (b) $U_c/U = 3$ for both the cylinders.

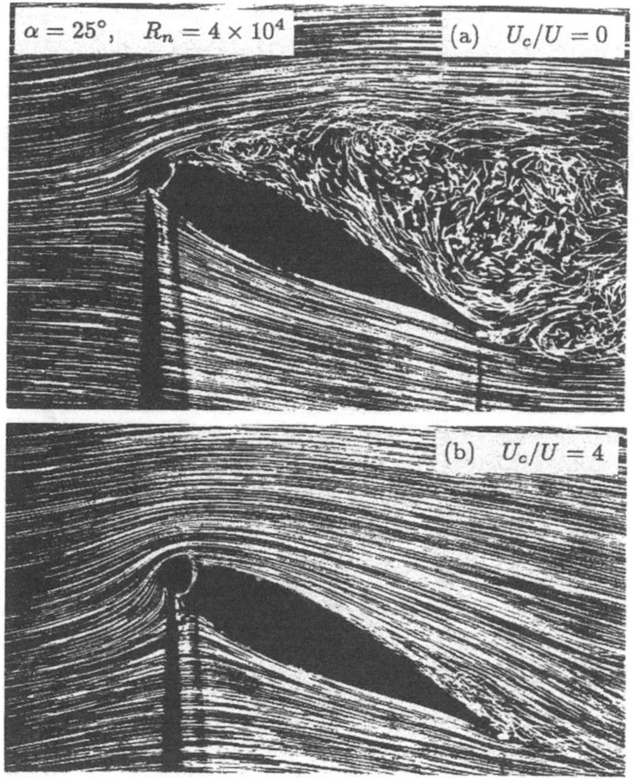

Fig. 4 Representative flow visualization pictures, showing rather dramatically, successful control of the boundary-layer separation through momentum injection.

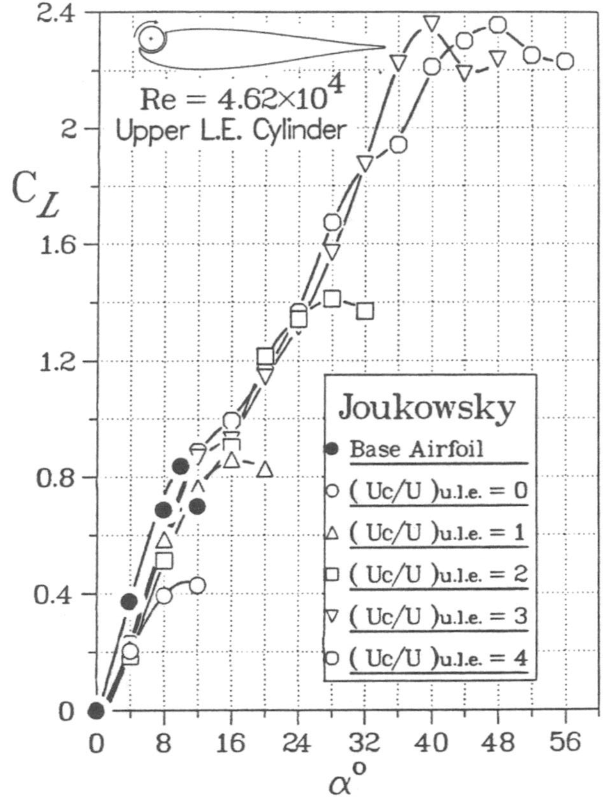

Fig. 5 Lift and stall characteristics of the Joukowsky airfoil as affected by the upper leading edge cylinder rotation.

The Effect of Car Configurations on the Flow Around Elevator Models (Oil Flow Pattern and Distribution of Pressure Fluctuation on the Model Wall Surface)

Hisashi Matsuda*, Yoshitaka Fukuyama*, Yasuyuki Yokono*
Kazunori Miyasako**
* Mechanical Engineering Lab., R & D Center TOSHIBA Corp.
4-1 Ukishima-cho, Kawasaki-ku, Kawasaki
** Fuchu Works, TOSHIBA Corp.

Summary
In order to have a clear understanding of the flow phenomena that are expected to cause the generation of aerodynamic noise from the elevator car and to examine the effect of car configurations on such flow behavior, flow around two elevator car models, one with a standard type apron and the other with an aerodynamic cover, was studied experimentally. Oil flow pattern technique was used for flow visualization. Distributions of pressure fluctuation on the wall surfaces of the models were also measured using a miniature pressure sensor and the RMS results were compared with the oil flow patterns.

1.Introduction

The recent trend in major Japanese cities is towards construction of increasing numbers of multistoried buildings, and this has generated demand for high speed (above 400m/min or so) elevators. High speed, however, has resulted in a new problem, namely aerodynamic noise[1,2,3].

Author et.al. investigated the flow structure around the conventional type of elevator (Fig.1) and demonstrated clearly that during downward motion flow is accelerated in front of the car door due to the generation of streamwise vortices along the apron edges (Fig.2), and as a result noise inside the car during downward motion increases[2,3].

To reduce the noise inside the car during high speed running, it is necessary to prevent the generation of the accelerated flow in front of the car door, and furthermore to control the flow separation and reattachment phenomena around the car. This is because this type of noise has a strong correlation with the pressure fluctuation on the wall and the above mentioned phenomena play an important role in causing the pressure fluctuation[4].

In this paper the effect of car configuration on the flow around the car was investigated experimentally using oil flow pattern technique and measuring distribution of wall pressure fluctuation.

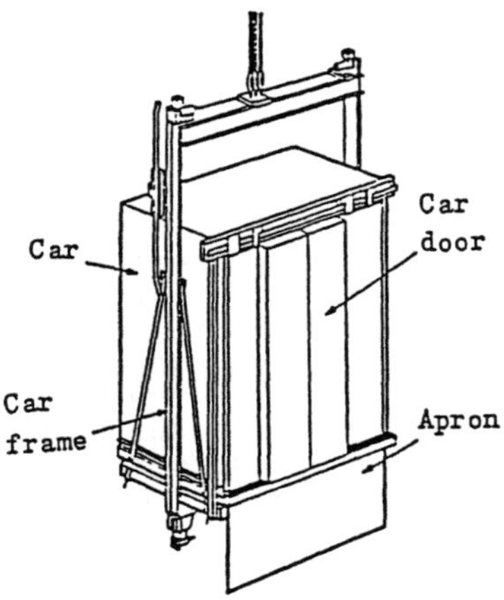

Fig.1 Schematic diagram of elevator

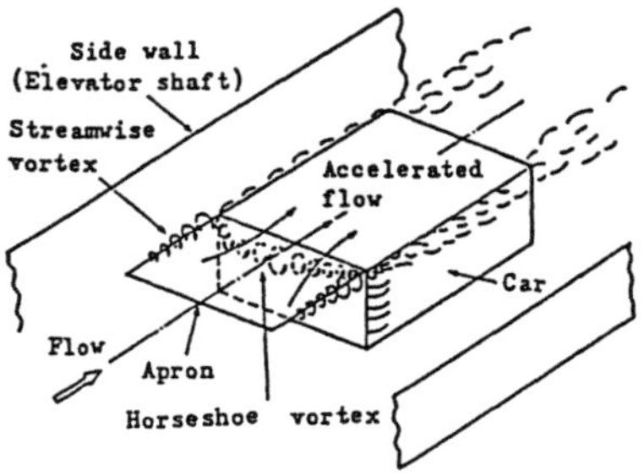

Fig.2 Flow around apron during downward motion of car

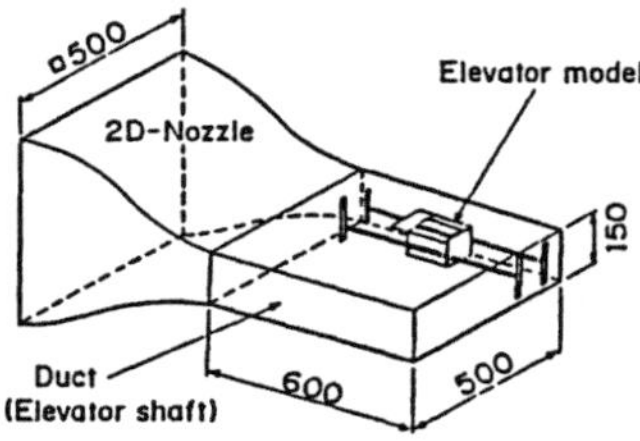

Fig.3 Wind tunnel experimental setup

2.Experiments

Experiments were carried out in a low-speed wind tunnel facility under the conditions of the Re number based on the car width and mean flow velocity equal to 1.3×10^5. A duct made of plexiglas which simulates the elevator shaft was placed at the end of the nozzle. A schematic diagram of the experimental setup is shown in Fig.3. Two scale models (approximately 1/20) identical in size which simulate the basic shape of an actual elevator were used. One model was for flow visualization and the other was to measure wall pressure fluctuation. An apron or the aerodynamic cover could be attached to both models.

For flow visualization, liquid paraffin was used as base oil and titanium oxide was used as pigment. To measure pressure fluctuation (from 50Hz to 15KHz) on the wall surfaces a miniature pressure sensor (EPI-080-2G, Entran Corp. make) was used.

3.Results

Oil flow pattern and its interpretation at the front wall of the model with standard type apron during downward motion at the center of the duct are shown in Figs.4(a,b), respectively. In Fig.4(b) "N", "S" and "A.L." mean the node point, the saddle point and the attachment line, respectively[5]. In Figs.4(a,b) it can be seen that the oil converges into the center portion of the car from the front half of the apron. This phenomenon corresponds to the above mentioned fact that the accelerated flow, generated by the streamwise vortices, occurs in front of the car door. Also clear reattachment lines are observed at the door step and both door pockets. These factors indicate that during downward motion a back-step flow was formed in this area.

By obstructing the generation of the streamwise vortices the accelerated flow can be suppressed. The oil flow pattern and its interpretation in the case of the aerodynamic cover are shown in Figs.5(a,b), respectively. Clear reattachment phenomena are observed at the leading edge of the cover. However, in front of the door region the oil flows parallel to the main flow without converging at the center. It is clear that with the cover the accelerated flow could be suppressed.

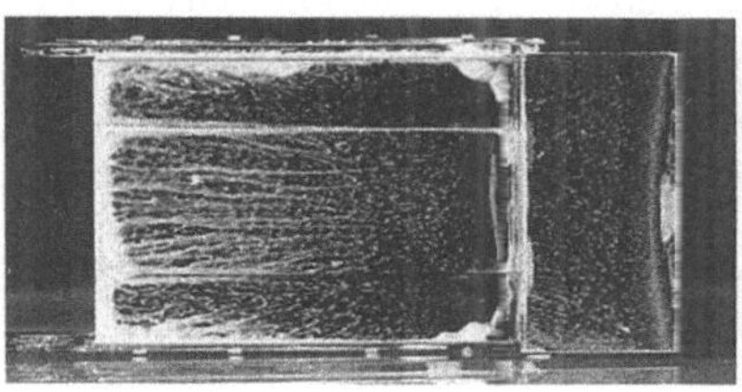

(a) Oil flow pattern

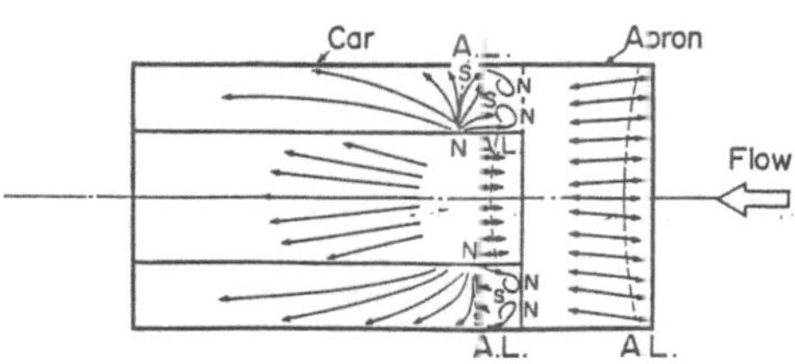

(b) Interpretation of the flow

Fig.4 Flow around the model with a standard type apron (Front view)

(a) Oil flow pattern

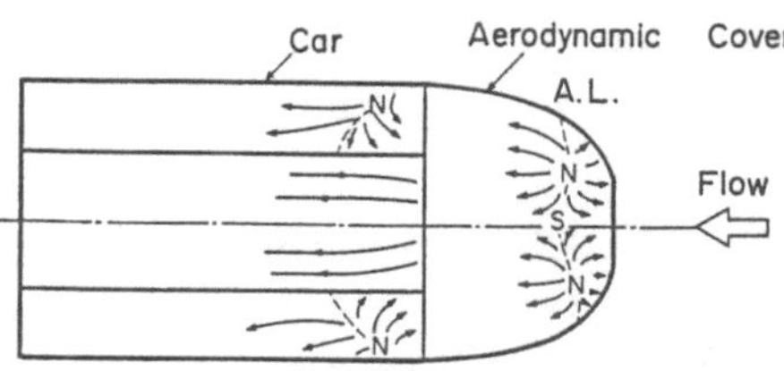

(b) Interpretation of the flow

Fig.5 Flow around the model with an aerodynamic cover (Front view)

Next, we consider the flow along another surface of the car. Figs.6(a,b.c) present the car side wall oil flow pattern(a) and its interpretation(b) as well as distribution of the wall pressure fluctuation(c) of the model with the standard type apron. In fig.(c) the relative size of the fluctuations are shown taking RMS value as a radius. At the side wall extremely complex flow is generated by the interference of separated flow at the lower side of the car with horseshoe shaped vortices generated at the mounting point of the apron and carframe surrounding. A comparison of Fig.6(b) and (c) shows that the locations of big pressure fluctuation correspond to the positions in the vicinity of the attachment node point ("N") where the oil spreads radially. A big pressure fluctuation is measured at the upper central part in the door side region (upper half of the figure) and at the middle part close to the guide rail and the left central part in the rear side region (lower half of the figure). It is not shown here but correspondence of the positions of the flow field interpretation and distribution of the pressure fluctuation is good on other measured surfaces also.

Results for the case of the model with an aerodynamic cover are presented in Figs.7(a,b,c). It seems that the flows once separated at the cover center and then reattached while interfering with the horseshoe shaped vortex generated along the carframe (Fig.7a). At the top end of the carframe a clear node point is recognized and one pair of node points exists along the guide rail (Fig.7b). However, as compared to Figs.6(a,b), it is evident that a very smooth flow field was realized by mounting the cover. Therefore, the wall pressure fluctuation is small along the whole side surface (Fig.7c). The relatively big pressure fluctuation at the right of the figure in the door region is presumed to be caused by irregularity in mounting the cover.

4.Concluding Remarks

The effect of the car configuration on the flow around the elevator models was investigated experimentally. The results are as follows. 1) By obstructing the generation of the streamwise vortices along the apron edges, the accelerated flow in front of the car door during downward motion can be suppressed. 2) From the oil flow pattern information, a region with a big wall pressure fluctuation can be assumed. A big wall pressure fluctuation occurs in the vicinity of the attachment node point. 3) Mounting of an aerodynamic cover can reduce wall pressure fluctuation around the car.

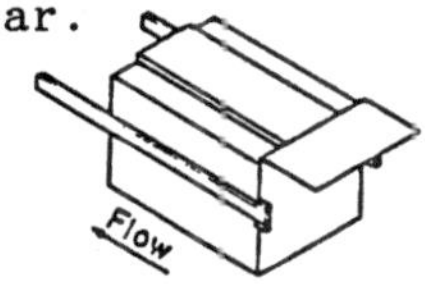

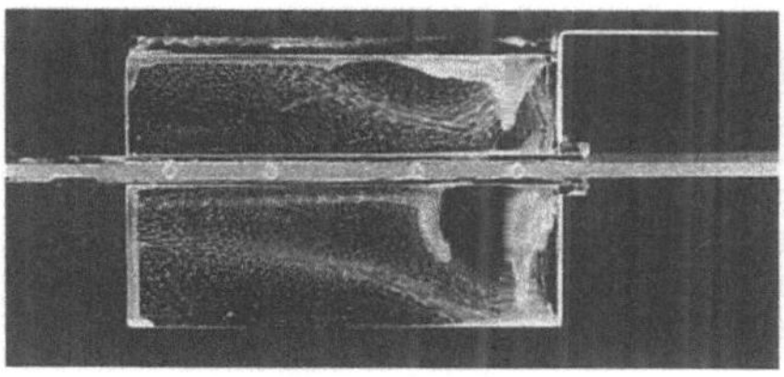

(a) Oil flow pattern

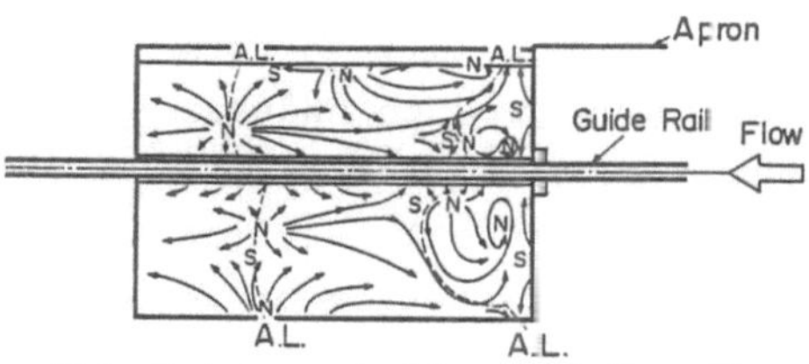

(b) Interpretation of the flow

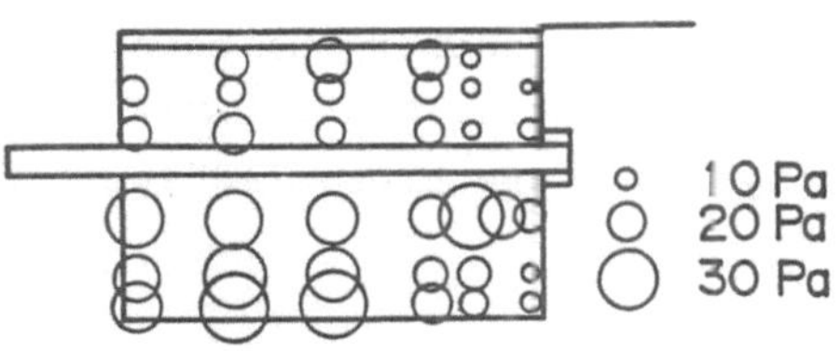

(c) Wall pressure fluctuation

Fig.6 Flow around the model with a standard type apron (Side view)

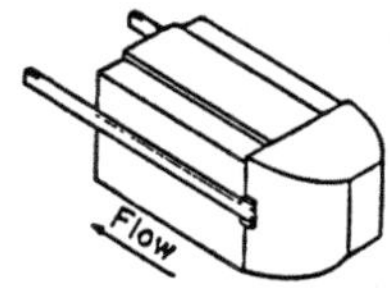

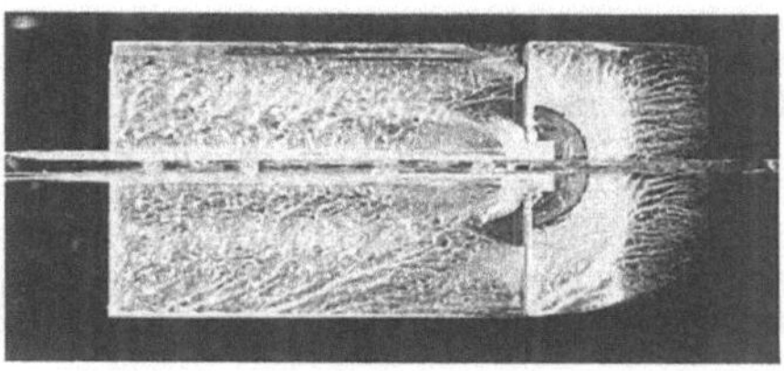

(a) Oil flow pattern

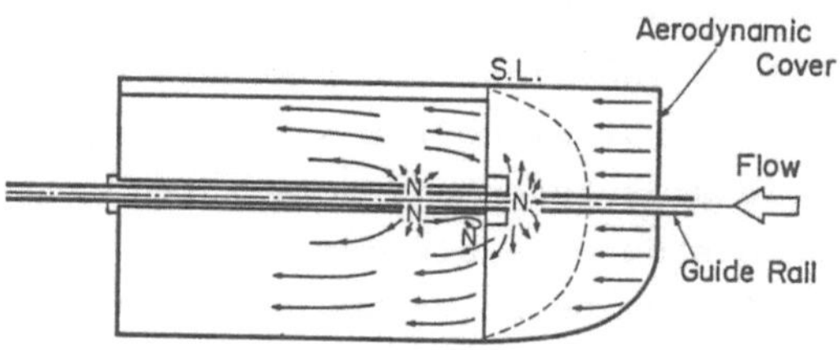

(b) Interpretation of the flow

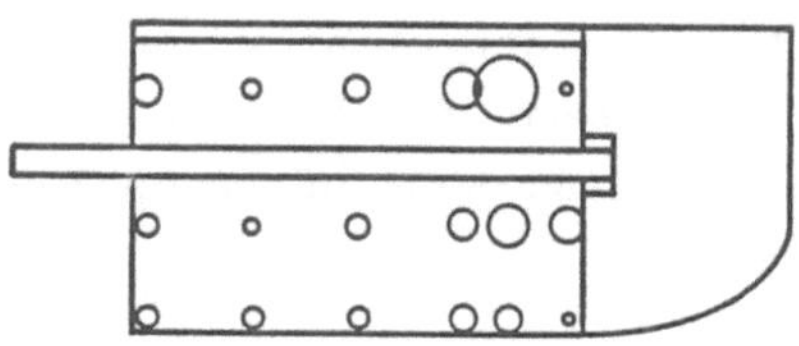

(c) Wall pressure fluctuation

Fig.7 Flow around the model with an aerodynamic cover (Side view)

References

1. Nakajima,S.and Kanamori,O.,Proceeding of JSME,No.910-3, (1991),pp.55-58.(in japanese)
2. Matsuda,H.et.al.,Proceeding of JSME,No910-62,(1991), pp.540-542.(in japanese)
3. Teshima,N.et.al.,Proceeding of ELEVCON'92,(1992),in press.
4. Haruna,S.et.al.,SAE Paper No.900316.(1990)
5. Peake,D.J. and Tobak,M.(1980),AGARDogrqaph,No.252.

Some New Applications of the Fluorescent Mini Tuft Technique

ZIXING WANG

Department of Aerodynamics, Nanjing Aeronautical Institute
Nanjing, China

Abstract

This paper briefly presents some engineering applications of the Fluorescent Mini Tuft Technique (FMTT) by author in recent years. The FMTT can simply, quickly, intuitionally and quantitatively measures and displays the traces, cross section shapes and circulation of a spatial vortex in an ordinary wind tunnel, so it is useful for author to investigate the physical mechanism of vortex flow and many engineering problems.

Introduction

It is inavoidable for modern vehicle of high mobility to form many spatial vortices. The vortices sometimes produce undesirable side force and moment (eg. the asymmetrical vortex on the lee side of a slender body at high angles of attack). But the lift can be augmented due to the favourable interaction of vortex. Therefore, it is important to understand the characteristics of a spatial vortex. The FMTT can be used to investigate any vortex flow field efficiently and quickly, so the author has been investigated many aerodynamic problems sucessfully by the FMTT in recent years. Some of these investigations are presented in this paper.

Description of Experimental Equipment

Tests were conducted in a low speed wind tunnel. The test velo-ocity is 30 meters per second. The Reynolds number was about 2-3 X 10^5. The fluorescent mini tuft is specially manufactured by author, its diameter is only about 0.01 mm., so the interference of mini tuft to the potential flow can be neglected. When the fluorescent mini tuft is stimulated by some ultra violate rays, the diameter of mini tuft seems about 1 mm. The fluorescent mini tuft rotates simultaneously with the vortex flow when it was immersed in a spatial vortex; so we can easily determine the circulation of a vortex by use of a Nitrogen Molecular Laser. This technique was invented by author and has been awarded a National Invention Prize.

Some New Applications of FMTT

1. The vortex system on the lee side of a slender body at high angles of attack.

Fig. 1&2 are the top view and side view of a pair of asymmetrical primary vortex on the lee side of a slender body at high angles of attack, which were visualized by the FMTT in a low speed wind tunnel.

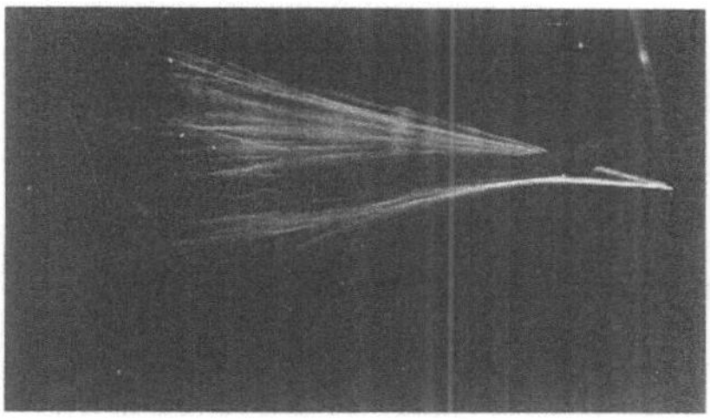

Fig. 1 Side view

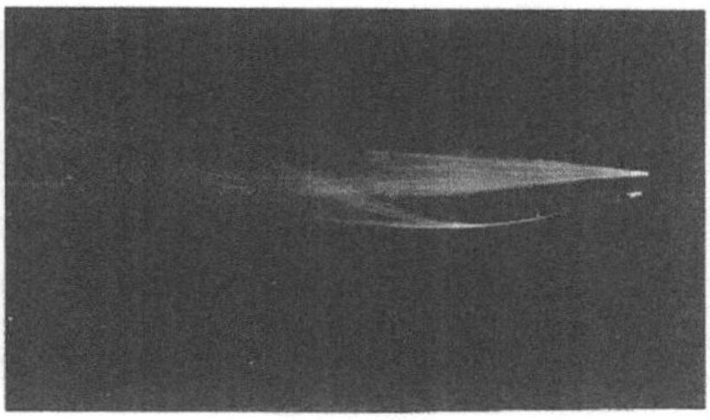

Fig. 2 Top view

Fig. 3&4 are two cross section of the above mentioned vortices.

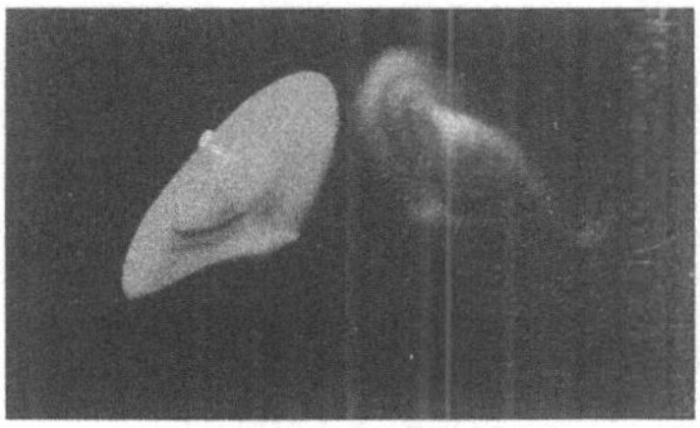

Fig. 3

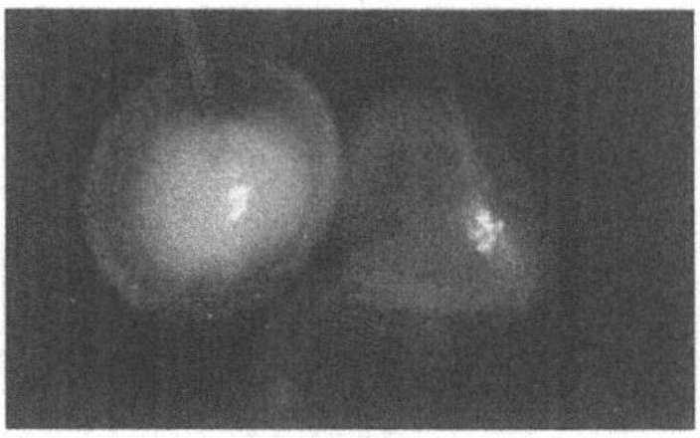

Fig. 4

2. The formation mechanism of two asymmetric spatial vortex on the lee side of a slender body at high angles of attack.

There are different viewpoints about the formation mechanism of asymmetric vortex in recent years, because all lack of sufficient experimental quantitave verification. When you use the FMTT the formation mechanism can be clearly shown, because one can know the circulation of two asymmetric vortex and the boundary condition (solid boundary and potential flow boundary). For details, see the ACTA AERODYNAMICA SINICA Vol.6 No.1. Fig. 5 shows the circulation of the vortex along the vortex track of a pair of asymmetric vortex.

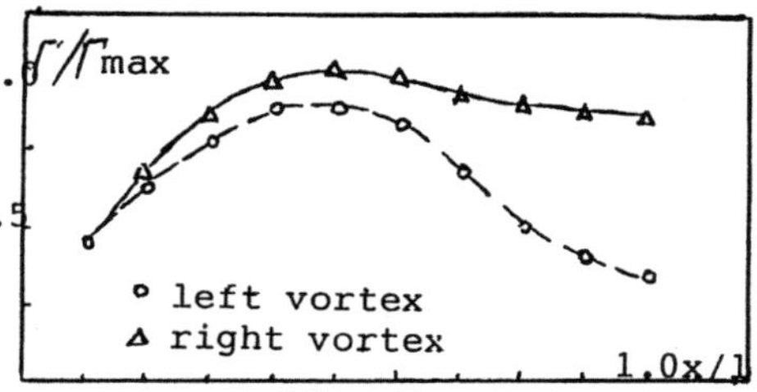

Fig. 5

3. The effect of winglets on the leeside spatial vortex of a slender body at high angles of attack.

The FMTT clearly shows the effect of a nose winglet to eliminate or delay the development of spatial vortex on the lee side of a slender body was due to the winglets can reduce the strength of spatial vortex and change the separation effectively. Fig. 6 shows the vortex track of a slender body at high angles of attack with and without a nose winglet. For details, please see AD-A 174599.

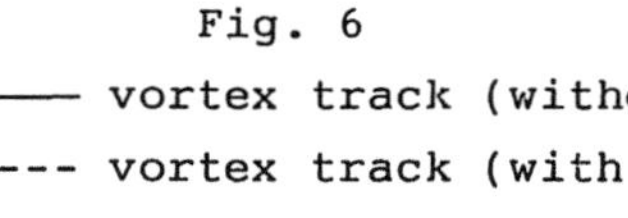

Fig. 6

——— vortex track (without winglet)

------ vortex track (with a nose winglet)

4. The lift distribution of a missile with "X" or "+" wing configulation.

When a missile with a "X" or "+" wing configulation, owing to the small aspect ratio wings and large missile body, there are serious interference between wings and body, strong downwash and sidewash were a further complication. It is very important in the project design stage to know accurately the lift and control surface efficiency. The FMTT can quickly determines the lift and induced drag distribution of wings or control surfaces at any angle of attack or angle of bank in some minutes. An example of the lift distribution of fcur wings at a certain angle of attack and angle of bank were shown in Fig. 7&8 respectively. The fundamental principle is: Lift is proportional to the strength of wing tip vortex, and the induced drag is proportional to the square of wing tip vortex: the strength of wing tip vortex can be easily determined by the FMTT.

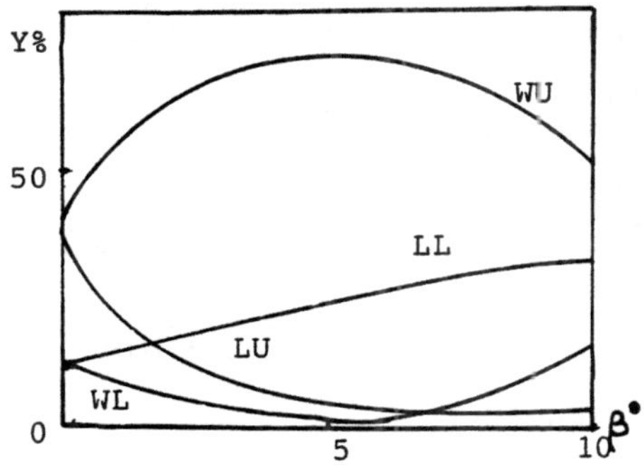

Fig.7 $\alpha=5°$

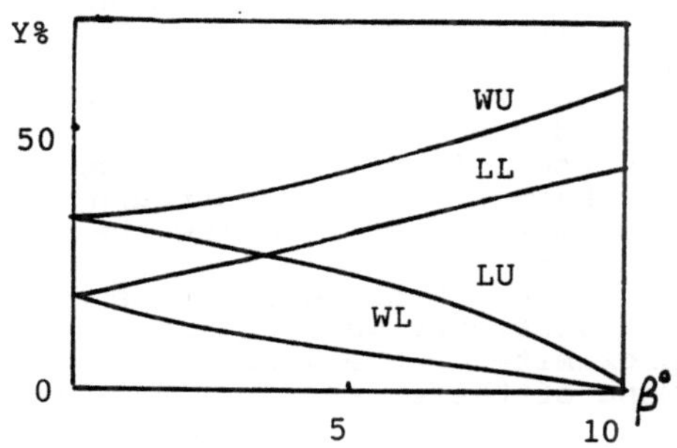

Fig. 8 $\alpha=10°$

where: WU-windward upper wing
WL-windward lower wing
LU-lee ward upper wing
LL-lee ward lower wing

Key words: Vortex flow, Light sheet display, Fluorescent tuft, Lift, Missile aerodynamics

Flow Field Around a Rotating Wheel

Antonello COGOTTI

Industrie Pininfarina s.p.a.
Wind Tunnel Department
Grugliasco (Torino) ITALY

Abstract

It is known that full-scale wind tunnel tests on passenger cars are usually carried out with fixed wheels and fixed ground plane and this represents the major difference between the wind tunnel simulation and the road running conditions.
In this exercise, the wheel rotation was simulated and at the same time, the floor boundary layer was reduced by using the existing boundary layer suction system.
The paper reports results of the investigation carried out around an isolated wheel, both rotating and stationary, with a fixed ground plane.
The wheel flow field was mapped by means of a 14-hole probe.
Furthermore, new software, developed for the new traversing system,was used to allow the probe to accurately follow the wheel contour.
With this technique, time averaged values of pressure, velocity and vorticity were measured and used to map the flow field in proximity of the wheel.
Some of these maps are shown in this paper.

WHEELS REPRESENT a relatively small part of a passenger car, however they are extremely important.
Their primary task is obviously to reduce to a minimum the rolling resistance of the car in the direction of the motion and, at the same time, to generate maximum friction during the various manoeuvres made by the car on the road (acceleration,braking,turns, etc.). Therefore, car wheels are usually optimized accordingly.
However, their shape and location in the car, contributes significantly towards increasing the drag coefficient of the car and, more in general, towards determining its overall aerodynamic characteristics.
Therefore, it is not surprising that, in the last 25 years,wheels have been investigated many times in order to achieve a better understanding of their aerodynamic characteristics. An excellent summary of these investigations was recently published by Mercker and Berneburg (1).
However, further research is needed to fully understand the aerodynamic behaviour of a rotating wheel and, to this aim, this paper

reports some further experimental results regarding the flow field of an isolated wheel, both stationary and rotating.

TEST CONDITIONS

- An isolated wheel with the drive unit and the relevant shaft at the right side and a dummy drive unit to produce a symmetric flow around the wheel, on the opposite side.

The wheel has diameter D= .480 m,width W= .140 m,area S= .064 m2.

The wheel rim is faired by a flat disk on each side.

The wheel tread is turned so that the tyre is slick (ref. 2).

- The ground is fixed. A brush is inserted under the wheel in order to avoid any through-flow between the tyre and the ground.
- The ground boundary layer is reduced to δ_{99} = 30 mm and δ^*= 5 mm by using the suction system at the beginning of the test section.

Test velocity is V = 38 m/s, which corresponds to 1500 rpm of the wheel.

FLOW FIELD VISUALIZATION BY THE 14-HOLE PROBE

The 14-hole probe technique is already described in ref. 3,4. However, the wind tunnel traversing system was recently replaced by a new one, its computer system was upgraded and a new software was installed. Therefore, the probe can now accurately follow a body contour in order to obtain information about pressure and velocities close to the surface.

Some of the flow surveys made in proximity of the wheel, both stationary and rotating, are shown in Fig. 1 - flow field behind the wheel (cross sections) and Fig. 2 - flow at the wheel side and in plan view.

COMMENTS ON THE RESULTS

- The flow differences between rotating and stationary wheels can be clearly seen from Figs. 1 and 2.

In the case of the rotating wheel,velocities in the lower wake are higher and pressure losses are lower (Figs. 2D,2E in plan view and 2B,2C at the side of the wheel). Furthermore (Figs.2A and 1B,1E) the flow over the rotating wheel separates earlier.

Therefore the vortex pattern and the wake shape are different, as well as the drag & lift coeffs. (Cd = .53 and Cl = .46 for the rotating wheel, .60 and .53 for the station. wheel,at Beta = 0 deg).
- The flow behind the wheel,at Beta = 0 deg,is symmetric,both for the stationary (Fig.1E) and rotating wheel (1B). However, a small yaw angle (5.0 deg) immediately produces an asymmetric flow (see Cp tot and vorticity maps) and therefore a larger wake.
This effect is more important for the stationary wheel than for the rotating wheel, because the wheel rotation delays the flow separation on the outer side of the wheel, at higher yaw angles.
This is confirmed by the force measurements at Beta = 0 and 5 deg: Cd grows with the yaw angle from 0.53 to 0.57 (rotat. wheel) and from 0.60 to 0.64 (stat. wheel),while Cs (side force coeff.) grows up to 0.12 for the rot. wheel and only to 0.04 for the stat. wheel. This means that the rotating wheel develops a stronger side force, produces a smaller wake and directs (for the same Beta) its wake further behind the wheel itself and less outwards.
Therefore, in passenger cars where front wheels are subjected to an asymmetric flow, the flow separation behind the wheels obviously increases noticeably the overall drag of the car. However, this increase is slightly lower for the car with rotating wheels,due to the lower Cd of the rotating wheel itself and to the different interference between the wake of the rotating wheel and the underbody of the car.

CONCLUSIONS

Although these results refer to an isolated wheel and not to a wheel within a wheelhousing on a vehicle, it seems clear that the wheel rotation may significantly modify the overall flow along the car underbody. This may result in a different local design,especially in the case of a well aerodynamically optimized underbody. Therefore, the simulation of the wheel rotation, during the wind tunnel development of new low-drag car models, is likely to become more and more important in the future.

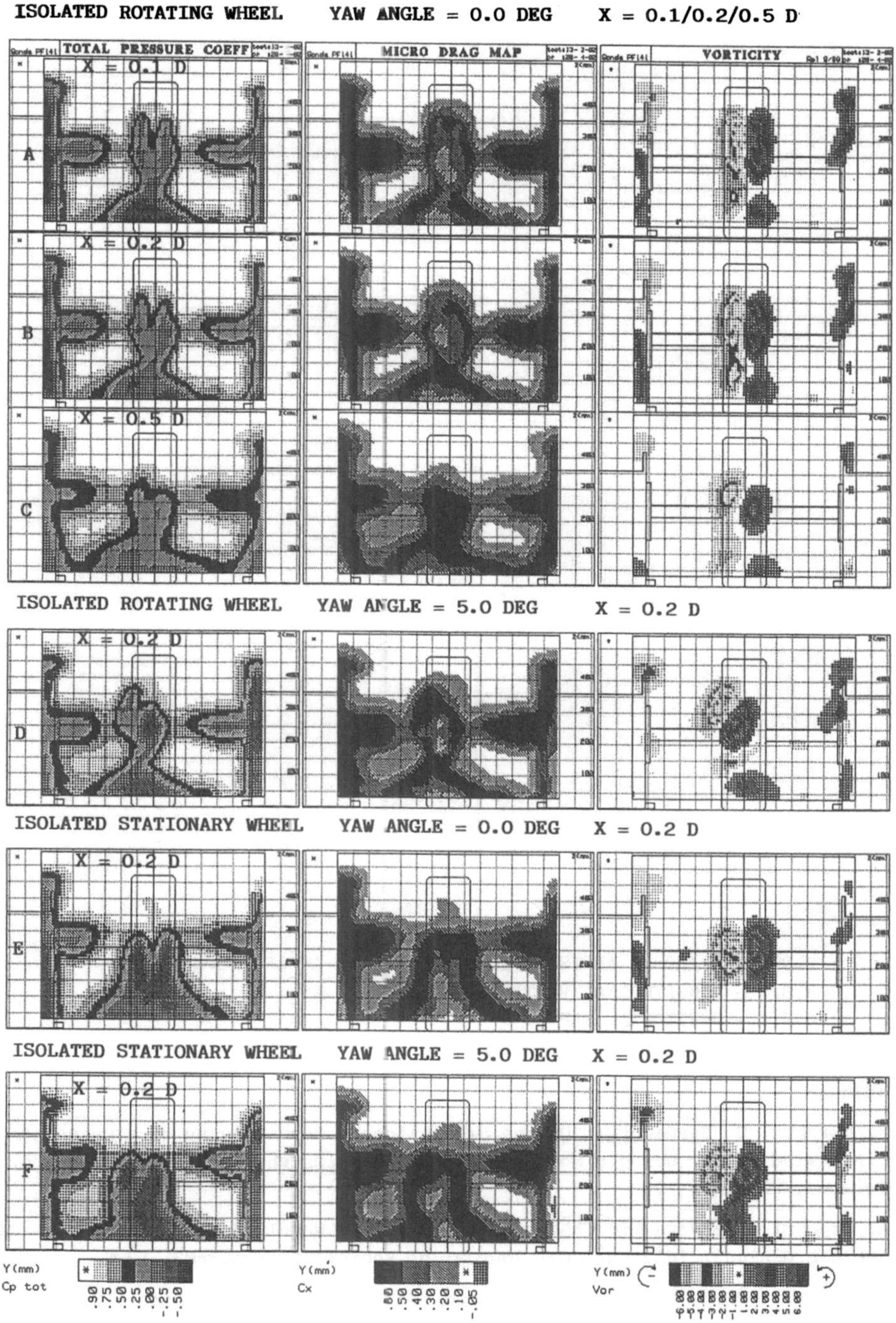

FIG. 1 : FLOW MAPS BEHIND AN ISOLATED WHEEL

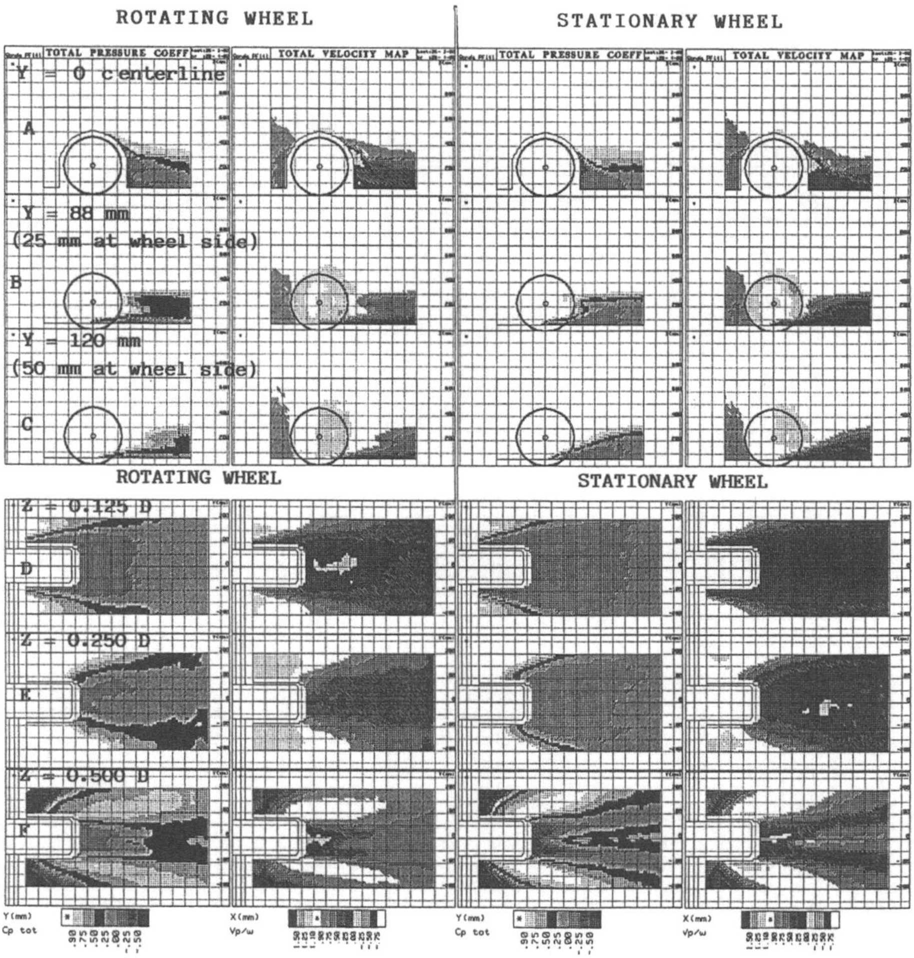

FIG. 2 : FLOW MAPS AROUND AN ISOLATED WHEEL AT YAW ANGLE = 0.0 DEG
A,B,C - LONGITUDINAL VERTICAL MAPS
D,E,F - HORIZONTAL MAPS

REFERENCES

1. E. Mercker, H. Berneburg, "On the Simulation of Road Driving of a Passenger Car in a Wind Tunnel Using a Moving Belt and Rotating Wheels", ATA 3rd Intern. Conference, Florence 1992

2. A. Cogotti, "Aerodynamic Characteristics of Car Wheel-Impact of Aerodynamics on Vehicle Design", International Journal of Vehicle Design, SP3, 1983

3. A. Cogotti, "Flow Field Surveys behind Three Squareback Car Models Using a New "Fourteen-Hole Probe", SAE paper 870243, 1987

4. A. Cogotti, "Prospects for Aerodynamic Research in the Pininfarina Wind Tunnel", FISITA paper 905149, 1990

Visualization of Flow Phenomena in and Around a Savonius Rotor

Nobuyuki FUJISAWA
Department of Mechanical Engineering
Gunma University, Kiryu, Japan

Summary

Flow phenomena occurring in and around a Savonius rotor have been studied by extensive flow visualizations and by the measurement of pressure distributions on the blade surfaces. Experiments are performed both for the stationary and rotating rotor with and without overlap. Based on the results, the variations of the torque mechanisms due to the effect of rotation and overlap are discussed by examining the flow patterns in and around the rotor and the points of flow separation and stagnation on the blades.

Introduction

The Savonius rotor has been applied to wind power generation systems and to pumping water systems. The global characteristics of the rotor such as the torque and the power performance are well established in a literature[1,2]. However, the mechanisms of torque and the power generation of the rotor are not well clarified because of the complexity of the flow field in and around the rotor, especially for the rotating rotor[3].

In the present paper, extensive flow visualization studies are carried out to gain an insight into the flow phenomena occurring in and around the stationary and rotating Savonius rotor with and without overlap, and the results of which are compared with the measured pressure distributions on the blade surfaces in order to examine the torque mechanisms quantitatively.

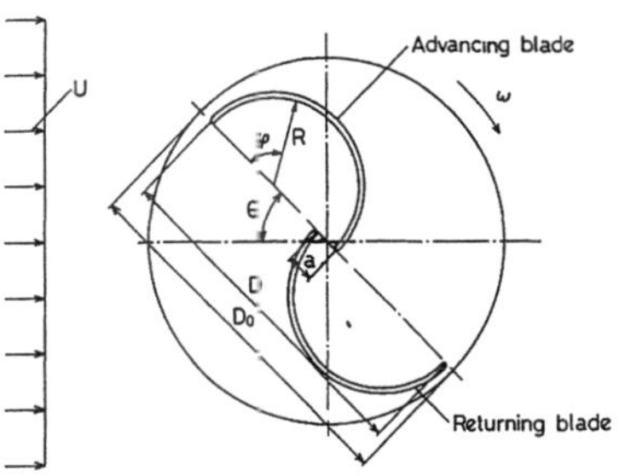

Fig.1 Rotor configuration and geometric parameters

Experimental Apparatus

A schematic diagram of the experimental apparatus is shown in Fig.1. Uniform flow is produced in an open circuit type wind tunnel by a squared nozzle with an outlet width of 900mm. The test Savonius rotor is placed at the tunnel center and 700mm downstream of the nozzle exit. The measured velocity distributions at this position with the rotor removed is uniform to within ±0.5% in the central area of 750mm×750mm.

The test Savonius rotor has two semi-circular blades with an overlap between them. The design parameters are selected such as dimensionless overlap distance G(=a/(2R))=0.15, end plate diameter D_0=1.1D and aspect ratio A_s(=H/D)=1, where D is the rotor diameter and H the rotor height. They are taken from optimization experiments[1,2]. While for comparative purposes, the rotor without overlap with the other parameters being fixed to the optimum values is also adopted. The rotor shafts between the blades are removed for both cases. The physical size of the rotor is D=H=300mm for the rotor with overlap and D=H=320mm for the rotor without overlap.

Flow Visualization Techniques

Figure 2 shows the experimental arrangement for visualizing the flow outside the rotor. The main flow is visualized by the smoke-wire technique, while the wake flow is observed simultaneously by injecting smoke through the nozzle located at the upstream end of the rotor. The smoke is produced by a smoke candle in the smoke tank. The observations are made at the mid-plane of the rotor using a light sheet from a 3W argon-ion laser and a CCD camera with a shutter speed of 0.002s. Experiments are carried out at a flow velocity U=1.5m/s.

Figure 3 shows the experimental arrangement for visualizing the internal flow field. The smoke-wire is set in the mid-plane of the Savonius rotor. Heating of the wire is provided by a direct current of 40-60V which is supplied through the rotating contacts at the end plate. The flow is observed by a CCD camera, and the illumination is provided by

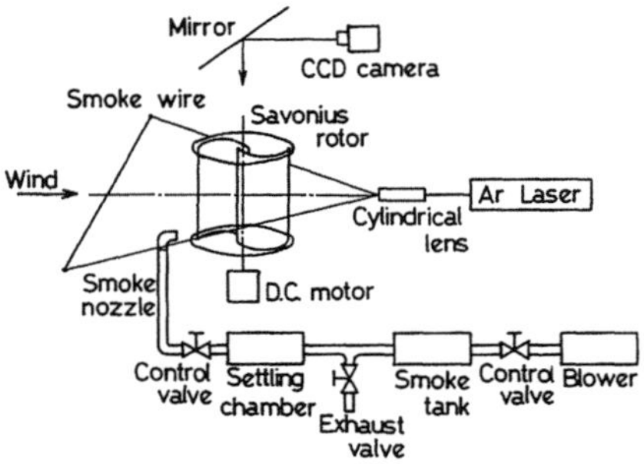

Fig.2 Experimental arrangement for visualizing external flow

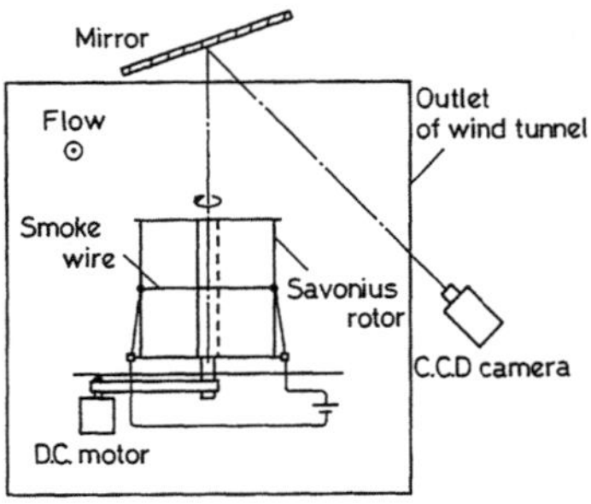

Fig.3 Experimental arrangement for visualizing internal flow

three lamps of 300W. The experiment is performed at a flow velocity U=0.7m/s.

The surface flow on the blade is visualized by a surface tuft technique. Observations at the same rotor angles are made for every cycle of rotor rotation, which is realized by a photosensor and a stroboscope. The tuft is made of silk thread for embroidery of 0.1mm in diameter with length 5mm, which is unraveled to increase the frequency response of the tuft motion. The separation point is judged from the tuft observation, where the centrifugal force is dominant compared with the surface flow without any fluctuations with time. The tufts are located every 10mm in both the streamwise and spanwise directions. The separation point is also evaluated by Thwaites's method for laminar boundary-layers with the measured pressure distributions on the stationary and rotating rotor[4]. Both the results are in close agreement with each other, suggesting the validity of the present tuft method for evaluating a separation point.

Results and Discussion

Typical results of the visualized flow fields in and around the Savonius rotor are shown in Figs. 4 and 5. Figure 4 shows the result of stationary rotor and Fig. 5 is that of the rotating rotor operating at the tip-speed

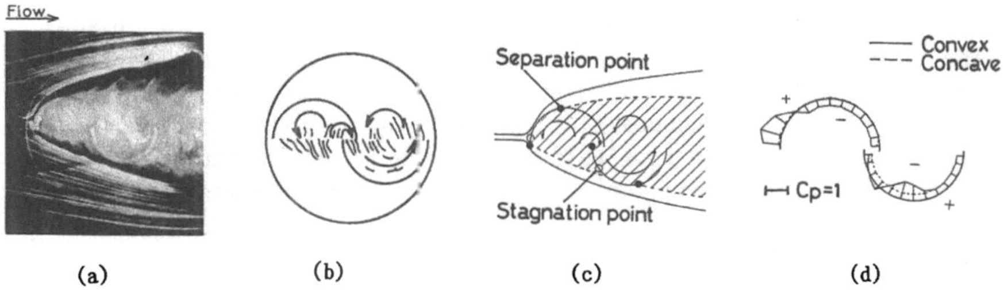

Fig.4 Flow in and around a stationary Savonius rotor

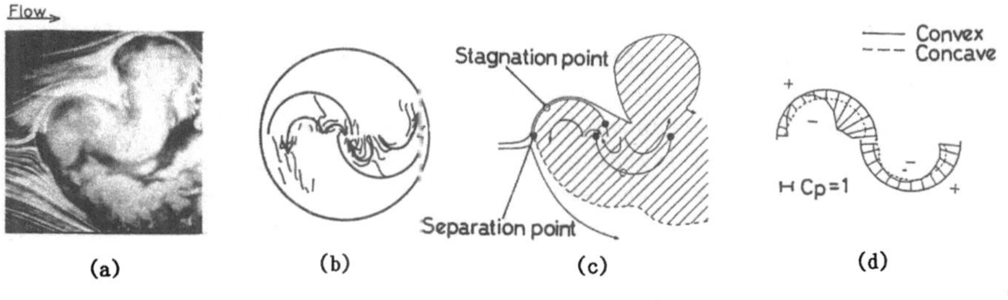

Fig.5 Flow in and around a rotating Savonius rotor

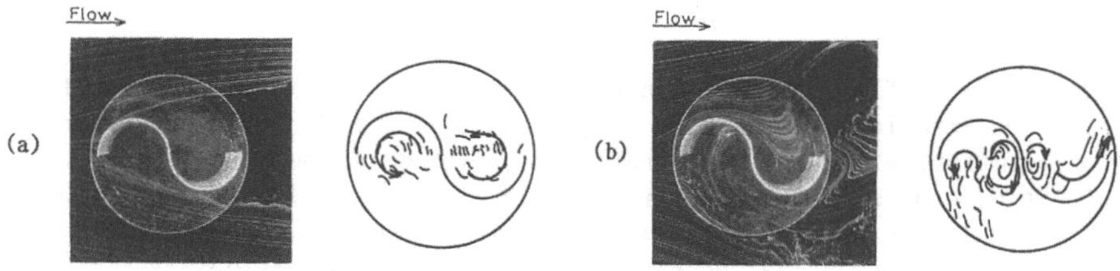

Fig.6 Flow in and around a Savonius rotor without overlap

ratio $X(=D\omega/(2U))=0.9$, where the maximum power performance is obtained. Each of the figures shows the visualized flow field outside the rotor (a), the flow inside the rotor (b), the schematic illustration of the flow field with the indications of separation and stagnation points (c) and the measured pressure distributions on the blade surfaces (d).

The flow over the stationary rotor separates at both sides of the rotor and hence roughly symmetric wake is produced in the downstream, as seen in Fig.4(a). However, the pressure distributions on the blade surfaces are not symmetric because of the wake effect of the upstream rotor. Therefore, a large magnitude of stagnation pressure appears on the upstream side of the advancing blade, but it is rather weak on the returning blade. The flow inside the rotor moves from the advancing side to the returning side of the rotor through the overlap and produces a pressure recovery effect on the concave side of the returning blade.

The flow field of the rotating rotor shown in Fig.5 is much different from the stationary case. It is clearly seen from Fig.5(a) that the flow separating region on the convex side of the rotor is fairly reduced by the formation of a Coanda-like flow pattern, which is due to the moving blade effect by the clockwise rotation of the rotor. This phenomenon is closely related to the appearance of the favorable pressure gradient on the convex side of the rotor, as seen in the pressure distributions. The appearance of the low pressure region on the convex side of the advancing blade contributes largely to the production of clockwise rotating torque. The Coanda-like flow together with the clockwise rotation of the advancing blade generates the vortex structure in the downstream of the advancing blade, which circulates in clockwise and grows in size in the downstream. The observation near the overlap shows the flow through the overlap, as was seen in the stationary rotor. However, the reverse flow is formed in the returning blade by the effect of rotation, which resists the flow through the overlap. Hence, the pressure recovery effect on the returning blade is expected to be reduced in the rotating rotor. This result is supported by the measured pressure distributions near the overlap.

Figure 6(a),(b) shows the smoke-wire visualization results of the Savonius rotor without overlap for both the stationary and rotating ($X=0.9$) case, respectively. It is noted that the smoke injection technique is not adopted here for visualizing the external flow. This flow field is very similar to that observed in Figs.4 and 5 for the rotor with overlap

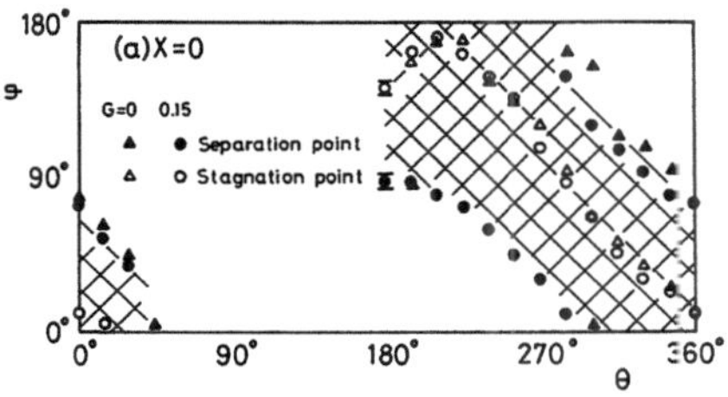

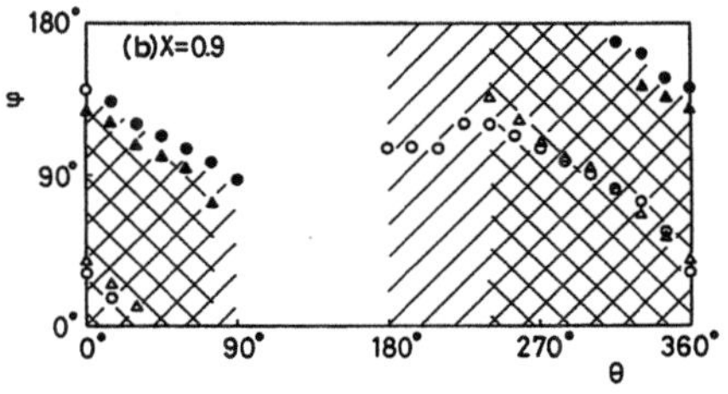

Fig.7 Variations of attached flow region with rotor angle θ

except for the flow near the overlap and the separating region of flow over convex side of advancing blade with rotation. The former is due to the direct effect of the overlap, while the latter can be explained by the appearance of secondary vortices near the center of the rotor. They are produced by the effect of rotor rotation and acts to separate the flow over the convex side of the advancing blade of the rotor without overlap.

Figure 7 gives the variations of the points of separation and stagnation for the stationary and rotating rotor, which are plotted against the rotor angle θ. The points of separation and stagnation are shown by the blade angle ϕ measured from the outer tip of the blade. It is seen that the attached flow region, which is shown by the shaded area in this figure, is increased when the rotor rotates. This is mainly due to the moving blade effect of the rotor, as was mentioned previously. It is also proved that the attached flow region is increased for the rotor with overlap and in rotation, as has been suggested in Fig.6. It can be concluded that the aerodynamic characteristics of the Savonius rotor is improved when the rotor with overlap is in rotation.

Conclusions

The aerodynamic characteristics of the rotating Savonius rotor is much improved by the moving blade effect, which generates the Coanda-like flow over the convex side of the advancing blade and results in a rotating torque. While it is also enhanced by the presence of overlap between the blades, which causes a reduced effect of secondary vortices produced near the rotor center.

References

1. Sheldahl, R.E; Blackwell, B.F.; Feltz, L.V.: Wind tunnel performance data for two- and three-bucket Savonius rotors. J. Energy 2(1978) 160-164.
2. Sivasegaram, S.:Secondary parameters affecting the performance of resistance-type vertical-axis wind rotors. Wind Eng. 2(1978) 49-58.
3. Fujisawa, N.; Shirai, H.: Experimental investigation on the unsteady flow field around a Savonius rotor at the maximum power performance. Wind Eng. 11(1987) 195-206.
4. Fujisawa, N.; Gotoh, F.: Pressure measurements and flow visualization study of a Savonius rotor. Proc. EWEA Conf., Amsterdam (1991) 199-203.

Flows Around Columns

Flow Around a Cylinder with Helical Surface Perturbations

Jose V. Nebres
Hani H. Nigim*
Stephen M. Batill

Department of Aerospace and Mechanical Engineering
Hessert Center for Aerospace Research
University of Notre Dame, Notre Dame, Indiana, 46556, USA
* Visiting Fulbright Fellow from Birzeit University, Birzeit, West Bank

Summary
Characteristics of the near wake of the flow around a circular cylinder with and without helical surface perturbations were studied. The tests were performed in a low speed wind tunnel and were conducted at a Reynolds number of 10,000 based upon the nominal cylinder diameter. The helix angle of the perturbations was varied in a range from 11° to 21°. The vortex wake formation length was estimated using smoke wire flow visualization and correlated with hot wire anemometry measurements. The smoke flow visualization was also used to qualitatively investigate the vortex formation process in the near wake.

Near Wake FlowField for a Cylinder

There are numerous situations in which the flow about a circular cylinder is modified by the presence of regular surface perturbations. One particular application deals with the use of helical perturbations for suppressing flow induced vibrations. Zdravkovich[1] reviewed several vortex suppression methods which included surface mounted helical wires. Optimal configurations for reducing vibrations have been proposed in terms of perturbation size, number of perturbations and helix angle from force transducer measurements by Weaver[2]. He recommended that for Reynolds numbers of 10,000 to 400,000 (based upon cylinder diameter), the most effective configuration would be 4 helical perturbations positioned at a helix angle of 14.7° with a perturbation size of 0.083D where D is the cylinder diameter. A schematic of this is shown in Figure 1. Relative to the boundary layer, the perturbation height is significantly greater than the boundary layer thickness near separation, approximately an order of magnitude greater. The helix geometry can be defined by the pitch, P, which is related to the helix angle, ϕ, and diameter, D, by $P = \pi D/\tan \phi$.

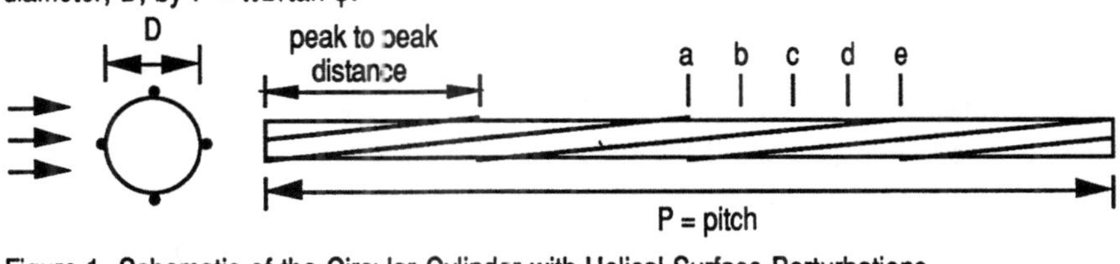

Figure 1 Schematic of the Circular Cylinder with Helical Surface Perturbations

Investigations by Nebres and Batill[3] revealed that vortex induced vibration suppression accomplished using surface perturbations based on Weaver's optimal configuration appeared to be due in part to an elongated vortex formation length. Hot wire measurements and flow visualization were performed to measure the formation length. At the Reynolds number of 10,000, they observed that the wake formation length of the cylinder with helical perturbations and P=12D was 2.5 times the formation length of the plain circular cylinder. The increased vortex formation length was correlated to periodic spanwise variation in boundary layer separation, shear layer transition, entrainment, diffusion and thickness. The measurements by Nebres and Batill[3] were only performed for the single model based on Weaver's[2] optimal configuration. The sensitivity of the formation length to the helix angle of the helical perturbations was not examined. The current study was intended to provide additional insight by varying the helix angle. With these results an improved understanding of the mechanics of the flow in the formation region of three dimensional bluff bodies can be obtained.

Wind Tunnel Experiments

All tests were conducted at the University of Notre Dame Hessert Center for Aerospace Research. An in-draft, subsonic wind tunnel was used. The wind tunnel has a contraction ratio of 20.6:1 based on area and 12 anti-turbulence screens. This wind tunnel has a 610 mm by 610 mm test section with a near uniform free-stream velocity profile and a turbulence intensity of about 0.05%. Four models were considered; a plain circular cylinder and three cylinders with 4 helical perturbations with pitches of 8D, 12D and 16D. The cylinder diameter was D=44.5 mm and the perturbations were cylindrical with a diameter, d=4.0 mm, and the span/D for the model was 13.7. No endplates were used for this experiment and the model spanned the width of the test section. The smoke wire technique of flow visualization was used; details are available in Reference 4. Wake flow visualization was achieved by positioning the smoke wire at different downstream stations in the wake at mid-span. This gave an indication of the formation length by observing reversed flow. Formation length measurements similar to Bloor[5] were conducted using a hot wire anemometery system. The single wire probe was traversed streamwise along the wake centerline from the base to approximately 6 diameters behind the model. The hot-wire data was filtered and the RMS of the AC coupled hot-wire signal at twice the shedding frequency was computed. The definition of formation length, as used in this paper, is that location is where this value of the RMS is at its maximum.

Results

An evaluation of the concept of formation length has been conducted using flow visualization and results have been compared to those obtained using hot wire anemometry. By placing the smoke-wire normal to a plane formed by the free-stream velocity vector and the model longitudinal axis, and viewing along the model's longitudinal axis, detailed information concerning the wake formation can be obtained. When placed behind the model, smoke in the wake region is entrained into the wake and marks the region enclosed by the free shear layers. The smoke entrained into the wake can also

provide an indication of the separation points on the bluff body and the spacing between the shear layers. As the smoke-wire is moved downstream, a range of smoke-wire locations can be identified where the tracer is no longer entrained into the near-base region and all of the tracer is convected downstream. This provides an indirect measurement of the length required for wake formation.
For the case of the circular cylinder at Re= 10,000, this region occurs between 1D and 2D (this distance is measured from the rear of the cylinder) as illustrated in Figure 2a and 2b. In Figure 2a the formation region is clearly defined. Figure 2b illustrates the wake after the formation region and provides detail on the flow as influenced by the shear layers and those regions still "potential" in nature. Figures 3, 4 and 5 illustrate the influence of the helical surface perturbations on the near wake flow. They represent the cases with pitches of 8D, 12D and 16D respectively. For the sake of space, only the 3D and 4D smoke wire locations are illustrated but these locations provide information necessary to define the wake formation length. Most obvious is the marked difference between the results for the helical perturbations and the smooth cylinder at the same Reynolds number. The wake formation has been significantly extended though qualitatively the results are comparable to the smooth cylinder case. Figures 6 a, b and c present the results from the hot-wire measurements for the same cases. The individual data sets which are labeled a, b, c, d, e correspond to various spanwise locations along the helix as indicated in Figure 1. The smooth curves, which are the result of data averaging, indicate the wake formation length based upon the peak RMS criteria. The solid vertical line indicates an estimation of the mean formation length. The pitch of 12D appears to have a slightly longer formation length than the other two cases based upon the hot-wire data. This is in agreement with the results of Weaver[2] and the results are consistent with the smoke-wire data. The P=16D was the only case where spanwise variations in the formation length were indicated and this could imply a more complex, three-dimensional wake development.
Only increments of 1D were used with the smoke-wire for the reasons indicated by Figure 7. These results illustrate the fact that the wake formation does not occur at a point. The wake closure is a periodic process and therefore the *formation length* is also time varying. For this example the P=8D and 16D case are shown and the smoke-wire was placed at 3D. Comparison with Figures 3a and 5a where flow was entrained for the same wire position illustrates the uncertainty in the formation length measurement and periodic nature of the wake formation process. It also implies that the formation length for these two cases is somewhat shorter since this did not occur for the P=12D case.

Conclusions

The combination of smoke-wire visualization and hot-wire anemometry has provided insight into the wake formation process and the influence of helical surface perturbations on a circular cylinder. The perturbations do extend the formation region which may influence the unsteady surface pressures and flow induced vibration. That point in the wake which separates fluid which is entrained into the formation region from that which is convected downstream varies periodically during the vortex wake formation process.

References
1. Zdravkovich, M., 1981: "Review and classification of various aerodynamic means for suppressing vortex shedding", Journal of Wind Engineering and Industrial Aerodynamics, vol. 7, pp. 145-189, Amsterdam: Elsevier Scientific Publishing Company.
2. Weaver, W., 1961: "Wind-induced vibrations in antenna members", Proceedings of the American Society of Civil Engineers, Journal of the Eng'g. Mechanics Division, Vol. 87, pp. 141–165.
3. Nebres, J. V. and Batill, S. M., 1992: "Flow about cylinders with helical surface protrusions", 30th Aerospace Sciences Meeting & Exhibit, Reno, Nevada, AIAA Paper No. 92-0540.
4. Batill, S. M. and Mueller, T. J.,1981: "Visualization of transition in the flow over an airfoil using the smoke-wire technique", AIAA Journal, Vol. 19, No. 3, March, pp. 340-345.
5. Bloor, M. S., 1963: "The transition to turbulence in the wake of a circular cylinder", Journal of Fluid Mechanics, Vol. 19, part 2, pp. 290-304.

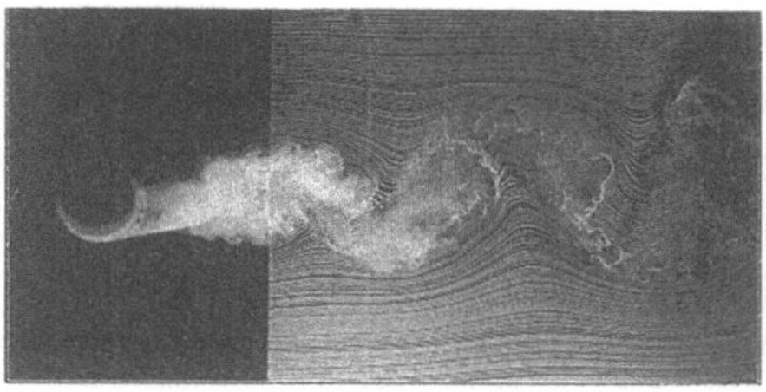

Figure 2a. Cylinder, Smoke-Wire at 1D

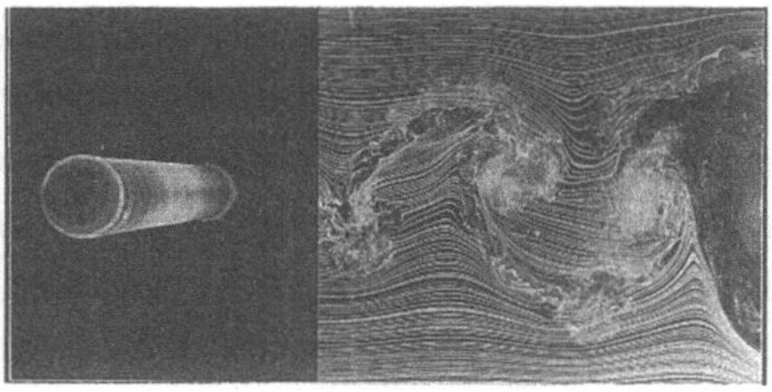

Figure 2b. Cylinder, Smoke-Wire at 2D

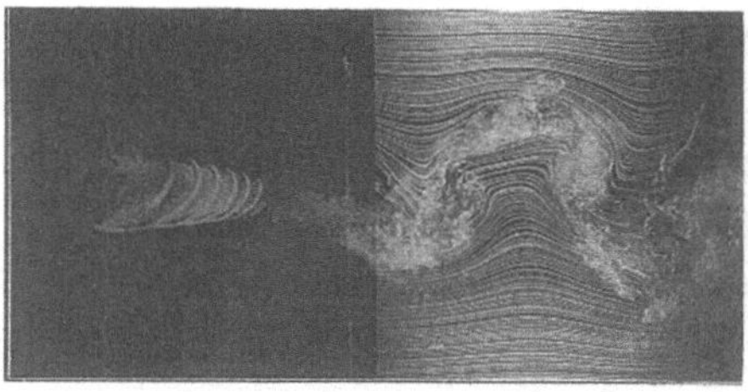

Figure 3a. Cylinder With Perturbations, P=8D, Smoke-Wire at 3D

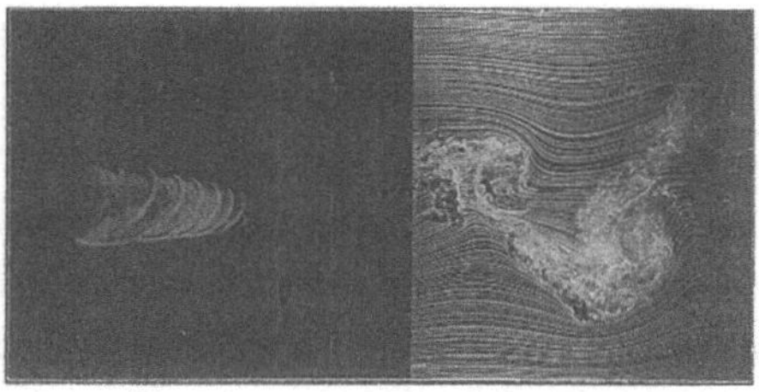

Figure 3b. Cylinder With Perturbations, P=8D, Smoke-Wire at 4D

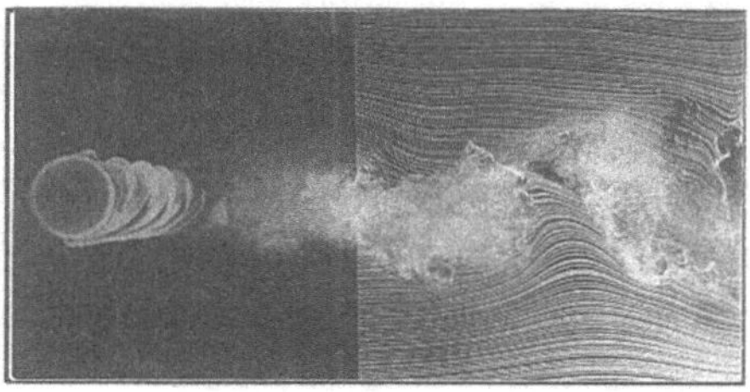

Figure 4a. Cylinder With Perturbations, P=12D, Smoke-Wire at 3D

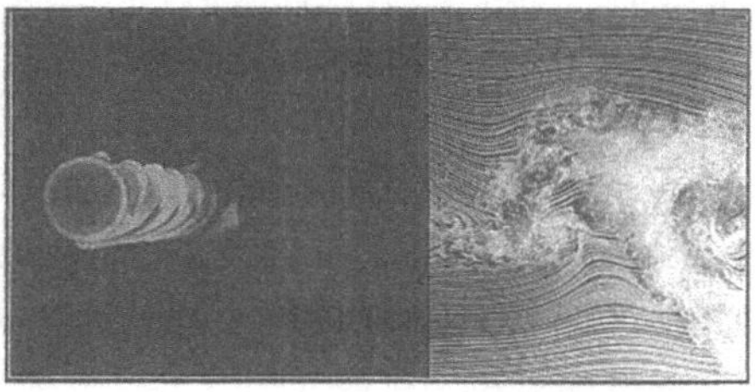

Figure 4b. Cylinder With Perturbations, P=12D, Smoke-Wire at 4D

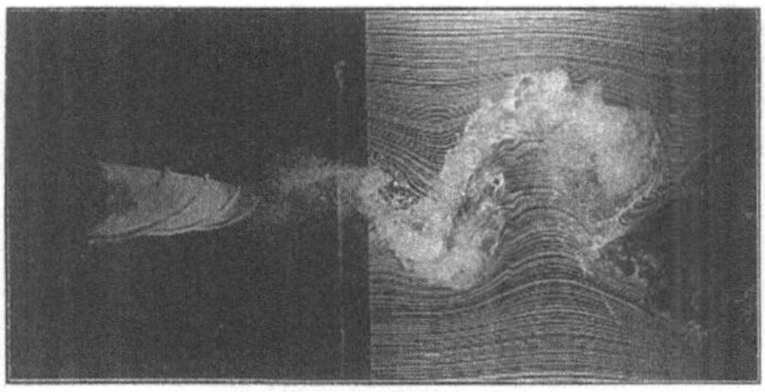

Figure 5a. Cylinder With Perturbation, P=16D, Smoke-Wire at 3D

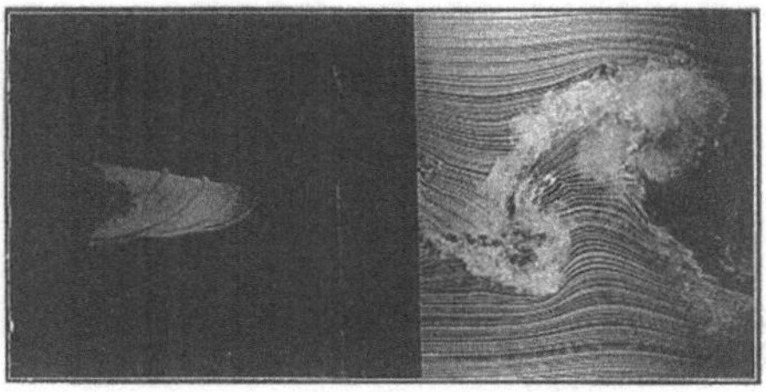

Figure 5b. Cylinder With Perturbation, P=16D, Smoke-Wire at 4D

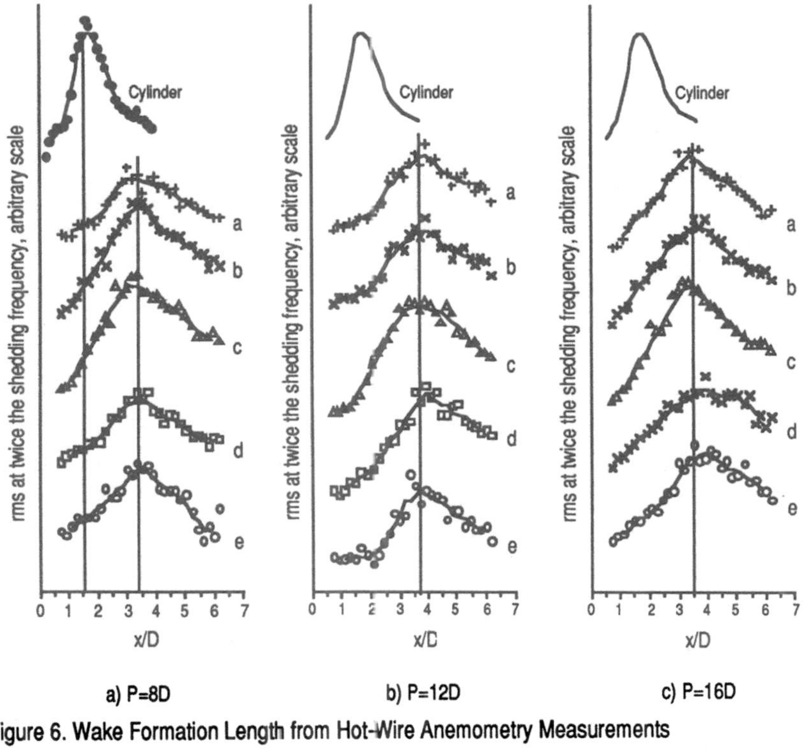

Figure 6. Wake Formation Length from Hot-Wire Anemometry Measurements

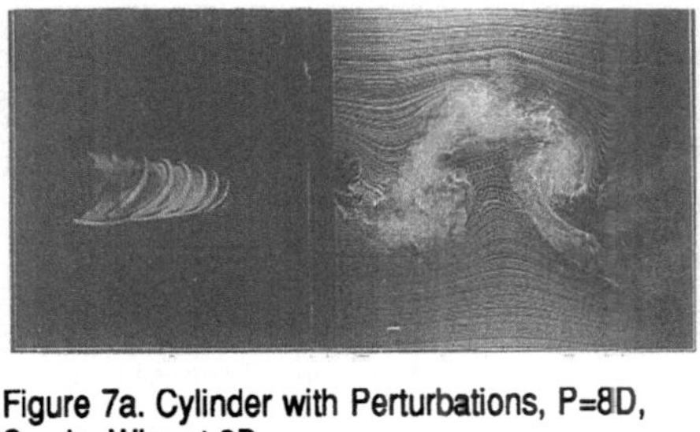

Figure 7a. Cylinder with Perturbations, P=8D, Smoke Wire at 3D

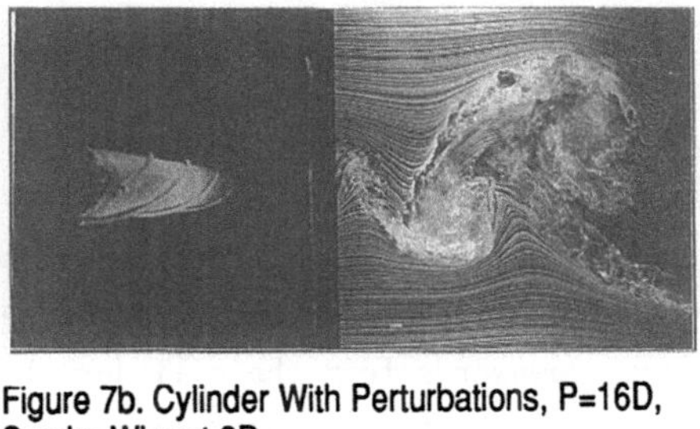

Figure 7b. Cylinder With Perturbations, P=16D, Smoke-Wire at 3D

Generation of the Karman Vortex Street at Low Reynolds Number by Cooling a Circular Cylinder

K.NOTO and T.NAKAJIMA

Department of Mechanical Engineering
Kobe University
Nada, Kobe 657
JAPAN

Abstract

This paper describes the generation of the Karman vortex street and the change of its vortex shedding due to cooling a circular cylinder submerged in an upward free-stream of air at the low Reynolds number 40-50, by showing streaklines obtained by computations of the time-dependent Navier-Stokes equations with the buoyancy term and flow visualizations using the smoke-wire methods. The shedding frequency of the vortex generated at the low Reynolds number is different from that in the isothermal wake.

Introduction

The Karman vortex street has been studied by many researchers[1]. However, their studies have been limited only in either the isothermal wake or the wake with negligibly buoyant force. The Karman vortex street does not occur at the Reynolds number less than 40[2], 45[3], 50-60[4], or 55-70[5] in these wakes behind a circular cylinder.

Noto[6] discovered the Karman vortex street is certainly generated at the low Reynolds number range, when the thermal operation is added to a circular cylinder. Noto[7,8] made clear the generation of the Karman vortex street due to cooling a cylinder. Nobody except Noto has reported this phenomenon.

This paper shows the Karman vortex street and its vortex shedding generated by cooling a horizontally circular cylinder in the upward free-stream.

Flow System

A circular cylinder is horizontally placed in an infinite space. A free-stream has the X-direction, where X is the coordinate with the anti-gravitational direction, Y is the horizontal coordinate, and the coordinate origin is at the cylinder center. When the cylinder is cooled, the buoyant force occurs downwards from the cylinder. This buoyant force and the free-stream are combined to form a complicated wake.

This wake phenomenon is assumed to be time-dependent, two-dimensional, and laminar. The Boussinesq's approximation can be applied to any thermal flow with small temperature difference and is employed in the present study.

Methods of Numerical Computation and Flow Visualization

The governing equations are the continuity equation, the time-dependent Navier-Stokes equations with the buoyancy term, and the time-dependent energy equation. The space is discretized by the central finite difference. The time step was determined by the Courant condition. The time integral was done by using the ADI method. The governing equations with the initial and boundary conditions were numerically solved.

The present computation was given attention as follows. One is the initial disturbance for occurrence of an asymmetric wake. A time-dependent, forced disturbance with a finite amplitude is given at the initial stage of the present computation. The second is a recognition of a vortex. The streakline is obtained in order to detect a vortex in the present computation, because a vortex cannot be detected with a velocity measurement and visualizations of the pathline, streamline, and velocity vector.

The streakline is visualized using the smoke-wire methods in the present experiment. The opposing flow is realized by heating a circular cylinder submerged in the downward free-stream in the wind tunnel.

Isothermal Wake

Figure 1(1) shows computed streaklines in the isothermal wake (Ri=0) at Re=44. The wake is not a symmetric flow but has not a vortex. The isothermal wake is a wavy flow.

Generation of the Karman Vortex Street Due to Cooling a Circular Cylinder

When the cylinder is cooled at the same velocity of the free-stream as Fig.1(1), WHAT DOES HAPPEN IN THE WAKE? Computational results for the cooled wake are shown in Figs. 1(2)-1(7).

The wake at Ri=-0.025 in Fig.1(2) is more remarkably wavy flow than that at Ri=0 in Fig.1(1), and generates vortex filaments. With a decrease of the cylinder temperature, a vortex filament becomes large and the wake has vortex spirals in Figs.1(3)-1(4). The wake at Ri=-0.1 becomes like the Karman vortex street in Fig.1(5). The Karman vortex street becomes large by cooling a cylinder in Fig.1(6).

Although not shown in figure, the time-averaged velocity in the X-direc-

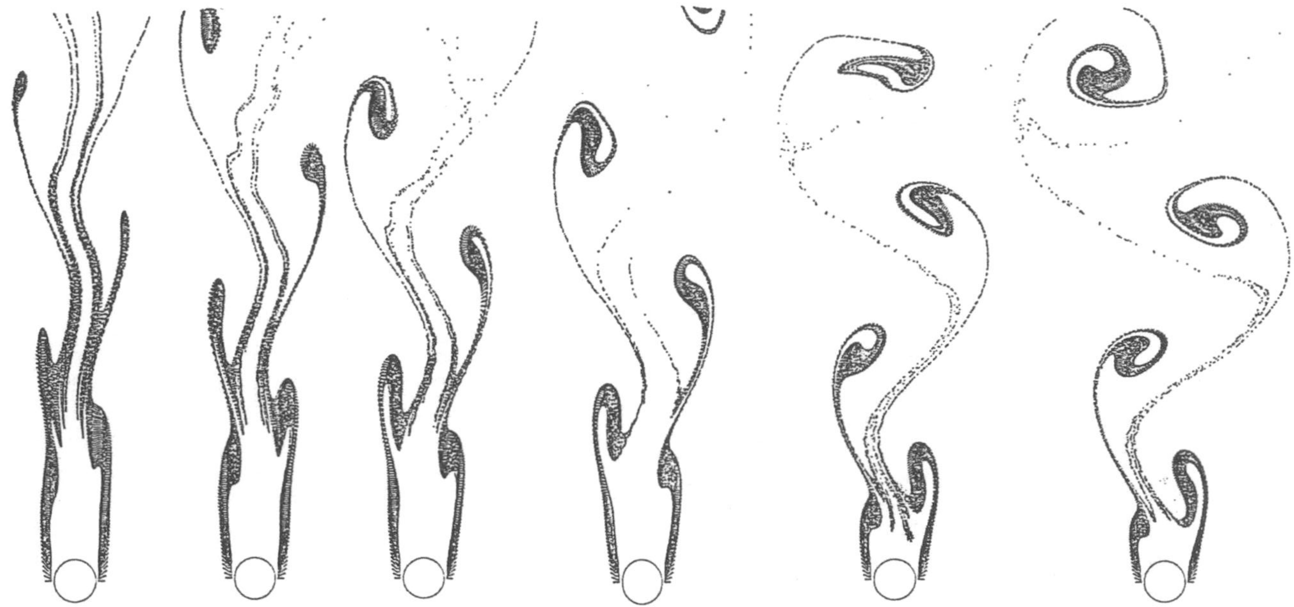

Fig.1. Generation of the Karman Vortex Street Due to Cooling a Circular Cylinder at Re=44 for Air.

(The gravitational direction is downward. The free-stream direction is upward. The free-stream velocity is constant and only the cylinder temperature changes in Fig.(1)-(6). The nomenclature Re is the Reynolds number defined as $U_\infty d/\nu$, U_∞ the free-stream velocity, d the cylinder diameter, ν the kinematic viscosity of fluid, Ri the Richardson number defined as Gr/Re^2, Gr the Grashof number defined as $d^3 g\beta(T_w - T_\infty)/\nu^2$, g the gravitational acceleration, β the expansion coefficient of fluid, T_w the temperature of cylinder surface, and T_∞ the free-stream temperature.)

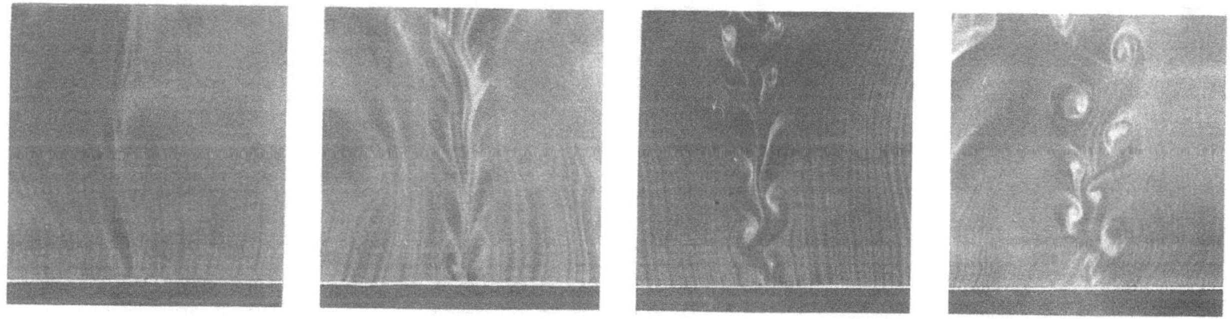

	(1)	(2)	(3)	(4)
Gr	= 0	-153	-348	-478
Re	=50	50	51	50
Ri	= 0	-0.061	-0.135	-0.188
Tw(^{o}C)	=38.7	92.4	156.2	200.7

Fig.2 Flow Visualization of Generation of the Karman Vortex Street at Re=50 for Air.

(The flow visualization experiment was performed in the wake behind a heated circular cylinder submerged in a downward free-stream. This opposing flow is the same as that behind a cooled circular cylinder submerged in an upward free-stream. The flow visualization results are shown in the same manner as Fig.1 with the upward free-stream. The circular cylinder with the diameter 3mm is at the center near the bottom in Fig.2. The horizontal line with white color is the nichrome wire generating smoke.

Figure 2 shows streaklines obtained by visualization experiment. The isothermal wake is a wavy flow in Fig.2(1). With a decrease of the Richardson number, the wake becomes asymmetric and to have vortices, as shown in Figs. 2(2) and 2(3). The Karman vortex street is generated in Fig.2(4).)

tion decreases with an increase of X. This fact means that the vortex street generated by cooling does not return to the isothermal wake in the fully downstream.

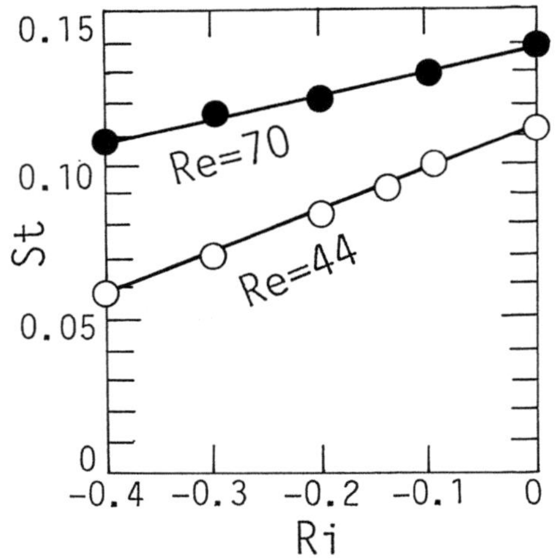

Fig.3. Strouhal Number

Flow Visualization

See Figure 2.

Vortex Shedding & Strouhal Number

Figure 3 shows a change of the Strouhal number St defined as fd/U_∞, where f is the wake frequency determined by the time-dependent velocity in the X-direction. The Strouhal number at Re=44 decreases by cooling a cylinder. This means that the shedding frequency of the Karman vortex generated by cooling is different from that in the isothermal vortex. That is, the generated vortex streets in Figs. 1(6) and 2(4) do not occur in any isothermal wake.

Figure 3 shows also the Strouhal number at Re=70. The wake at Ri=0 was a vortex street. This vortex street became to have a large spiral of vortex by cooling a cylinder. Although the vortical scale becomes dominant in the wake pattern, the shedding frequency of vortex decreases by cooling.

Conclusions

The present computations and flow visualizations show that the Karman vortex street is generated by cooling a cylinder at the low Reynolds number 40-50. This vortex street generated by cooling has the wake frequency less than that in the isothermal wake. The phenomenon discovered in this study occurs at any Reynolds number less than 50 and is applicable to the opposing flow with the downward free-stream.

References

1. e.g. Williamson,C.H.K.: Phys. Fluids. 30-10(1988), 2742-2744.
2. Roshko,A.: NACA Annual Report. No.1191(1954).
3. Taneda,S.: J. Phys. Soc. of Japan. 11-3(1956), 302-307.
4. Zdravkovich,M.M.: J. Fluid Mechanics. 37-3(1969), 491-496.
5. Gerrard,J.H.: Phil. Trans. of Royal Soc. of London. A-288(1978), 351-382.
6. Noto,K.: FLOW VISUALIZATION V. Springer-Verlag(1987), 649-654.
7. Noto,K.: J. Flow Visualization Japan, 8-Suppl(1988), 23-28.
8. Noto,K.: Numerical Methods in Laminar & Turbulent Flow, 7(1991), 671-684.

Flow Visualization of Separation Bubble on a Circular Cylinder

Y TOMONARI

Department of Aeronautical Engineering
Nippon Bunri University
Ichigi, Oita, 870-03 Japan

M M ZDRAVKOVICH

Department of Aeronautical and Mechanical Engineering
University of Salford
Salford M5 4WT, UK

Summary

A blockage tolerant test section based on a modified Parkinson's design allows tests up to $Re = 4.3 \times 10^6$. A schlieren method is used to visualise boundary and free shear layers emanating from a slightly heated circular cylinder. Colour photographs are presented to illustrate the separation of boundary layers and the breakaway of free shear layers in the range $0.4 \times 10^6 < Re < 4.3 \times 10^6$. The shift of the separation point for rising and falling *Re* exhibited a considerable hysteresis.

Introduction

Large size cylinders and huge wind or water tunnels are required to bring transition in free shear layers near to the separation line. The high cost of such tunnels forced most researchers to use medium size test sections and outsized models leading to unacceptably high blockage. The present tests are carried out in a medium blockage tolerant test section modified according to the Parkinson's design(1).

The interaction between the transition and separation results in a unique flow phenomenon known as a separation bubble (2, 3). The transitional boundary layer separates at *S1* and further transition in the free shear layer induces a reattachment at *RA* so that a separation bubble is formed. Further transition in the reattached boundary layer is presumably completed when the turbulent separation at *S2* takes place. Typical values of the angles *S1*, *RA* and *S2* are 97, 107 and 135 degrees, respectively, at $Re = 3 \times 10^6$ (4). The first separation *S1* is displaced more than 10 degrees in comparison with the laminar separation at 82-85 degrees. The extent of the separation bubble is about 10 degrees while the subsequent transition in the reattached boundary layer spreads over 30 degrees. The narrowing of the wake caused by *S2* considerably reduces the drag coefficient (4, 5, 6).

Object of Present Work

The separation bubble is a transitional phenomenon and as such sensitive to the effect of free stream turbulence (7), surface irregularities (3,4), surface roughness (8,9), etc. Hence it is important to avoid all flow visualisation techniques which might either affect any two-dimensionality or destabilise separation. The intrusive visualisation techniques, such as surface oil, smoke wire, hydrogen bubbles, dye release, etc, might affect the shape and stability of the separation bubble. The non-intrusive schlieren method is successfully used in this work to visualise the separation and free shear layers.

Experimental Arrangement

Figure 1 shows the outline of the blockage tolerant test section which was used in the present experiment. The upper and lower walls of the test section, originally 46 x 46 cm, were replaced by two arrays of aerofoil slats. Each slat had a chord of 5 cm and a span of 46 cm. Two outer plenum chambers were attached, each of 98 x 11 x 46 cm in size. A cylinder was placed horizontally at the centre of the test section. The cylinder diameter was 22 cm and the nominal blockage ratio was 48%. The air flow was allowed to pass through the gaps

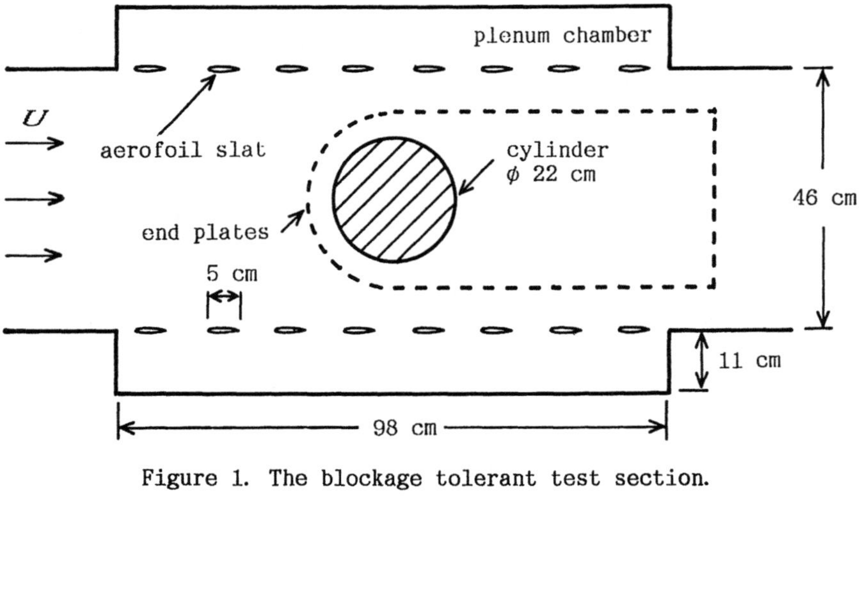

Figure 1. The blockage tolerant test section.

between the aerofoil slats and to simulate the blockage-free flow. The effectiveness of the blockage tolerant test section had been checked and described in (10).

The cylinder had a span to diameter ratio of 1.8. A circumferential groove was cut around the cylinder perimeter at the mid span. The groove had a depth of 5 mm, a width of 10 mm and it was covered by a thin brass plate 8 cm wide. Hot water was forced to pass through the groove from the rear to the front side of the cylinder. The average temperature of the hot water was 41℃. The cylinder was fitted with 3 mm -thick glass end-plates located 25 mm from the glass side walls so that the effect of wall boundary layers was excluded.

A standard schlieren optical equipment for high speed flow was used (20 cm diameter optical mirrors). The optical axis was positioned about 1/4 cylinder diameter above the cylinder axis and covered only the upper cylinder surface. The knife edge at the focus was horizontal and the adjustment was downwards. The flow visualisation was photographed by using a 35 mm SLR camera and colour films (ISO 100). The exposure time was 1 second and the aperture was f4.

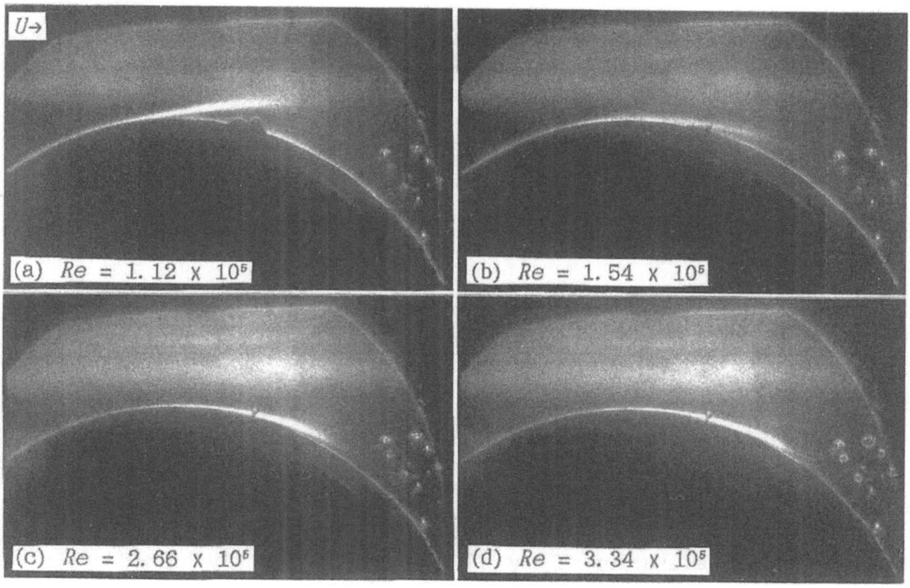

Figure 2. Photographs of boundary and free shear layers.

Results

Figure 2 shows some examples of the photographs obtained by the present method. Figure 2(a) shows a subcritical pattern when the boundary layer separates at around 90 degrees and the free shear layer moves deeply into the free stream (wide wake). Figure 2(b) shows a quite different pattern at a slightly higher *Re* = 1.54 x 10^5 when the separation point is displaced downstream. The free shear layer is not stationary and its image is somewhat obscured in Figure 2(b). At *Re* = 2.66 x 10^5, see figure 2(c), the free shear layer is stable again and it is located close to the cylinder surface. The separation point is displaced further downstream. Finally, figure 2(d) shows the pattern at *Re* = 3.34 x 10^5. The presence of a separation bubble can be deduced from the attached former free shear layer. The thickness of the separation bubble is so small that it cannot be estimated in figure 2(d).

The first separation angle *S1* is measured for various values of *Re* by using the film negatives. Figure 3 shows a large hysteresis for rising and falling *Re*. This result is in agreement with the previous measurements (5,6). The hysteretic effect is produced by a delay of the free shear layers to adjust to the changes in *Re*. The wide wake at low *Re* has to re-adjust to the narrow wake at high *Re* and vice versa.

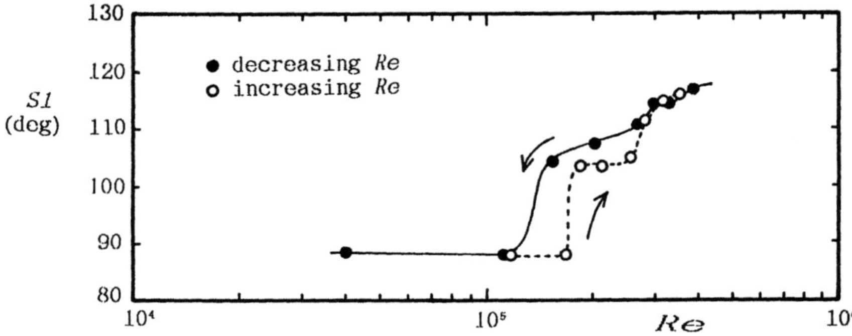

Figure 3. Measurements of first separation point.

Conclusion

The schlieren visualisation of the flow around a slightly heated cylinder shows the following:

i) the downstream shift of the separation point with rising *Re*.

ii) the gradual approach of the free shear layer toward the cylinder surface.

iii) the formation of a separation bubble.

iv) the hysteretic change from wide to narrow wake and vice versa.

Further work will examine the effect of the surface roughness and free stream turbulence on the hysteretic appearance and disappearance of the separation bubble.

Acknowledgement

The work reported in this paper has been carried out at the University of Salford while the first author was on sabbatical leave 1990/91.

References

1. Parkinson, G V and Hameury M: Performance of the tolerant tunnel for bluff body testing. Bluff Body Aerodynamics and Its Applications, Eds Ito, M et al, Amsterdam, Elsevier, (1990) 35-42.

2. Tani, I: Low speed flow involving bubble separations. Prog Aero Sci 5 (1964) 70-90.

3. Bearman, P W: On vortex shedding from a circular cylinder in the critical Reynolds number regime. J Fluid Mech 37 (1969) 577-585.

4. Achenbach, E: Distribution of local pressure and skin friction around a circular cylinder in cross flow up to $Re = 5 \times 10^6$. J Fluid Mech 34 (1968) 625-639.

5. Kamiya, N, Suzuki, S and Nishi, T: On the aerodynamic force acting on a circular cylinder in the critical range of the Reynolds number. AIAA report 79-1475 (1979).

6. Schewe, G: On the force fluctuations acting on a circular cylinder in cross flow from subcritical up to transcritical Reynolds numbers. J Fluid Mech 133 (1983) 265-285.

7. Kiya, M, Suzuki, Y, Arie, M and Hagino, M: A Contribution in the free stream turbulence effect on the flow past a circular cylinder. J Fluid Mech 115 (1982) 151-164.

8. Achenbach, E: Influence of surface roughness on the cross flow around a circular cylinder. J Fluid Mech 46 (1971) 321-335.

9. Okajima, A: Flow around two tandem circular cylinders at very high Reynolds numbers. Bull JSME 22 (1979) 504-511.

10. Zdravkovich, M M: On Parkinson's tolerant wind tunnel for circular cylinders. J Wind Eng and Ind Aerodyn (1992) (proofs sent).

Visualization of Flow Control Around a Circular Cylinder by a New Method

T. IGARASHI

Department of Mechanical Engineering,
The National Defense Academy
Yokosuka 239, Japan

Summary

Experimental studies on the control of the flow around a circular cylinder were conducted by a new method, setting up a small cylinder in the shear layer near the main cylinder. It was found that a forced reattachment of the shear layer separated from the cylinder was realized. The shear layer reattached onto the rear surface of the cylinder and the flow adhered beyond the rear stagnation point. The results obtained were as follows: The drag of the main cylinder decreased about 20 to 30% and the lift was generated with a value of C_L=1.0.

Introduction

According to the previous papers [1,2] of flow around a rocket composed of a main cylinder and two sub-cylinders, the shear layer separated from the main cylinder reattaches and adheres to the rear surface of the cylinder at an attack angle $\alpha=60°$ as shown in Fig. 1. This coanda effect suggests a new method of flow control around a circular cylinder by a sub-cylinder. This paper summarizes the interest results of previous papers [3∿5].

Experimental Apparatus and Procedure

The configuration of a model is illustrated in Fig. 2. The diameters of two cylinders were 40 and 2 mm, respectively, and

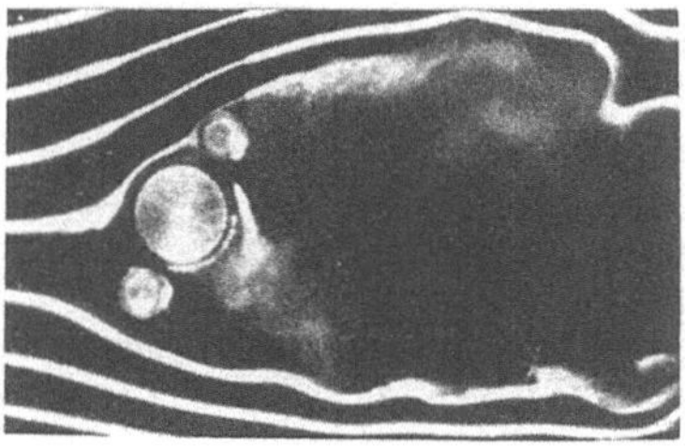

Fig. 1 Visualization of the flow around three circular cylinders

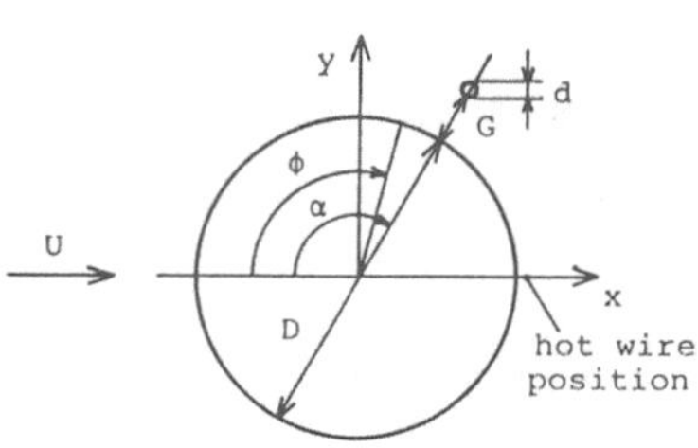

Fig. 2 Flow geometry

the Reynolds number, $Re=UD/\nu$, ranged from 5.1×10^3 to 5.1×10^4. The small cylinder was set up at various angular position α and with gap $G=2\sim7$ mm. The flow around the cylinders was visualized in a smoke tunnel and by incense stick smoke. The surface flow on the main cylinder was visualized by an oil-film method. The velocity in the near wake was measured, as well as the vortex shedding frequency was obtained from the frequency analysis of the velocity fluctuations. The pressure distributions around the cylinders were measured, and the drag and lift coefficients were obtained by integration of the pressure distributions.

Flow Visualization

Figures 3 (a) and (b) are smoke wind photographs obtained under long and instantaneous exposures, respectively. For $\alpha=120°$, the shear layer separated from the cylinder is divided at the small cylinder, and the lower shear layer reattaches onto the rear surface of the cylinder and adhered beyond the rear stagnation point. For $\alpha=124°$, the shear layer elongates, so that the vortex formation region moves downstream. The two are referred to as patterns A and B, respectively. The surface oil-flow

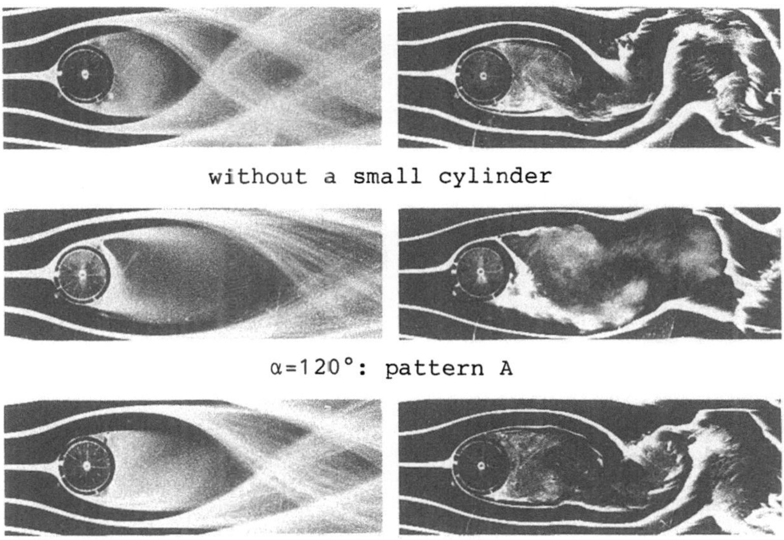

without a small cylinder

$\alpha=120°$: pattern A

$\alpha=124°$: pattern B

(a) long-exposure(1s) (b) instantaneous
[$G=6$ mm, $U=4$ m/s, $Re=10^4$]

Fig. 3 Visualization of the forced reattachment of the shear layer by a small cylinder patterns are shown in Fig. 4. The pattern A indicates laminar separation at $\phi=80°$, reattachment at $\phi=130°$ and turbulent separation at $\phi=160°$. The pattern B is laminar separation and looks like the case of without a small cylinder.

Strouhal Number and Near Wake Flow

Figure 5 shows the variations of the Strouhal number, S=fD/U, and the time-averaged velocity, u_b, near the rear stagnation point with angle of α. The symbols A and B are correspond to the patterns A and B, respectively. The two are represented the same profiles and the flow patterns change drastically. The values of u_b and 1/S have the maximum at the pattern A and the minimum at the pattern B. In the pattern A, the upper shear layer separated from the main cylinder was divided by the small cylinder, and the outer shear layer rolls up and forms the vortex street. The wake width defined by the width between the upper and lower shear layers increases, therefore the vortex shedding frequency decreaes. For the pattern B, the shear layer elongates and the wake width is reduced, so that the vortex shedding frequency increases. The velocity in the vicinity of the rear face exceed the undisturbed free stream velocity.

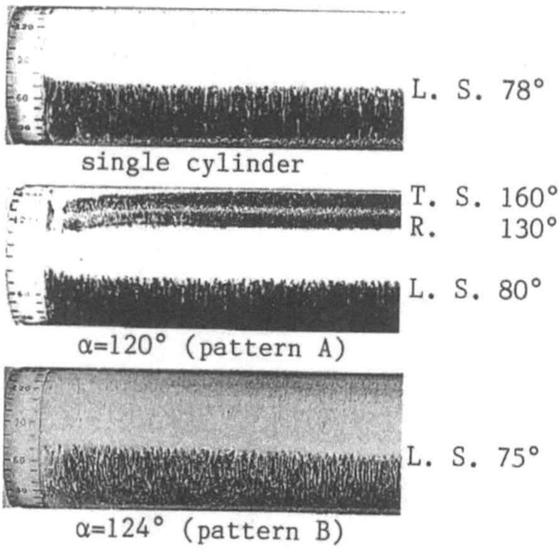

L. S. : laminar separation
T. S. : turbulent separation
R. : reattachment
[G=6mm, U=16m/s, Re=4.1×10^4]

Fig. 4 Oil-flow pattern

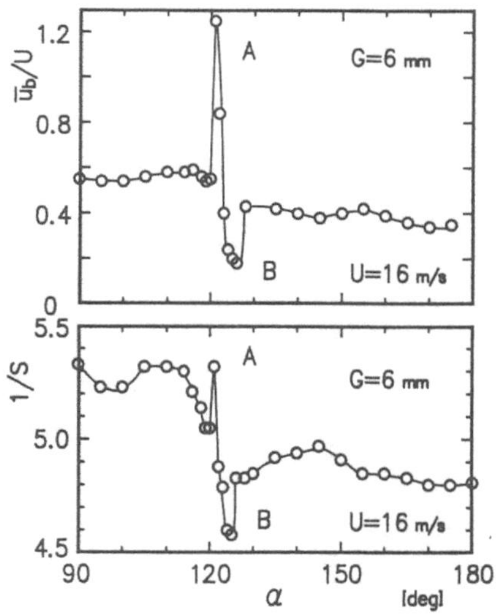

Fig. 5 Mean velocity and Strouhal number

Reduction of Drag and Induced Lift

Figures 6 and 7 show the drag and lift coefficients. The drag of the main cylinder decreases about 20 to 30% compared with that of the single cylinder. The lift of C_L=1.0 was generated by the forced reattachment. The decrease of the total drag force of the main cylinder including the small one was about 5% [4]. Therefore, the total drag force decreases about 25%. Further, it was found that the direction of the fluid force acting on the main cylinder is related to the deflection of the front stagnation point on the main cylinder [4].

Properties of the Reattachment Jet

Figures 8 and 9 show the contours of the mean velocity in the near wake of the main cylinder and the velocity distribution of

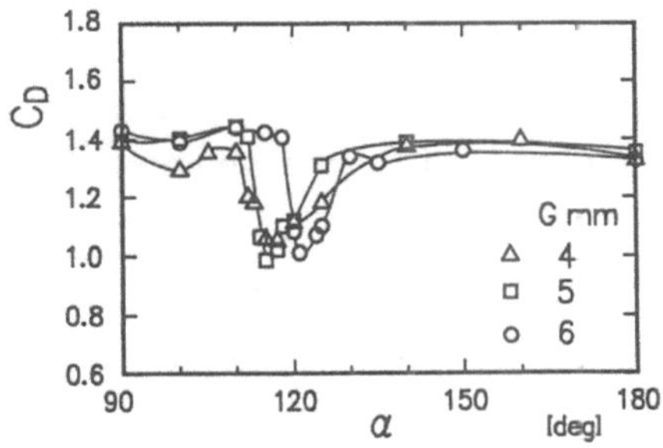

Fig. 6 Drag coefficient

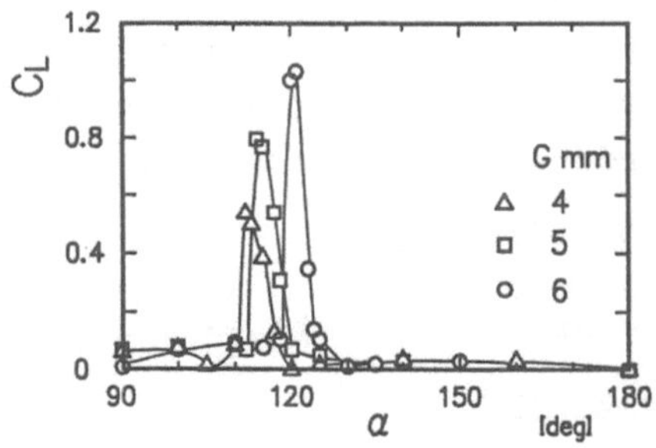

Fig. 7 Lift coefficient

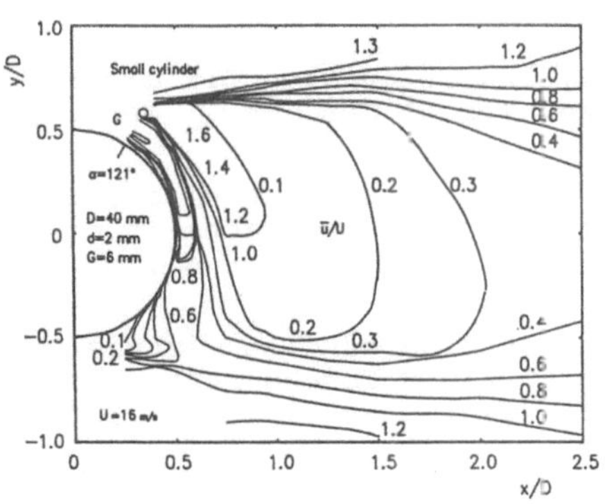

Fig. 8. Contours of mean velocity near wake

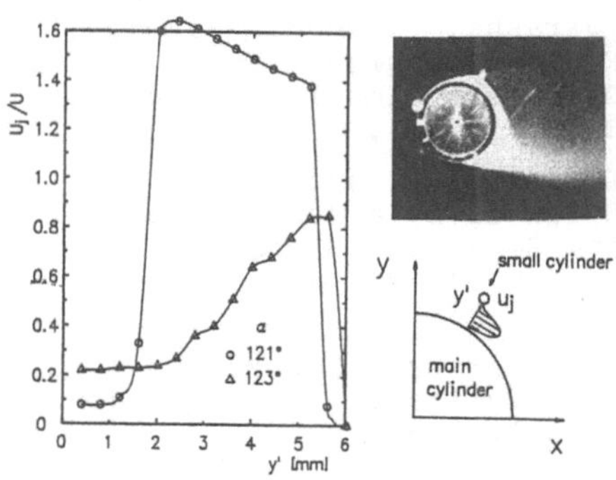

Fig. 9. Velocity distribution of the reattachment jet

the reattachment jet, respectively. The maximum velocity is 1.6 times as large as undisturbed free stream velocity. The near wake of $x/D \leqq 1.5$ is a stagnant region, the maximum value is $u/U = 0.2$. The forced reattachment is based on the same phenomena as a coanda effect. The critical condition, that is, $G/D \leqq 0.17$ and $C_j \geqq 0.045$, coincides with that of a turbulent wall jet along a circular cylinder and a jet on a circular cylinder immersed in uniform flow [5], where C_j is the momentum coefficient of the jet. The properties of the reattachment jet are similar to those of turbulent wall jet along a circular cylinder, namely, these distributions of mean velocity are nearly equal [5].

Conclusions

The flow control around a circular cylinder was realized by a new method, setting up a small cylinder in the separated shear layer near the main cylinder. That is to say, the shear layer reattached onto the rear surface of the cylinder and the flow adhered beyond the rear stagnation point. The drag of the main cylinder decreased about 20 to 30%, and the lift was generated.

Acknowledgement

The author wishes to express appreciate to Mr. T. Tsutsui for his assistance in the experiments.

References

1. Igarashi, T., K. Nishida and S. Mochizuki, Aerodynamic Forces Acting on Three Circular Cylinders Having Different Diameters Closely Arranged in Line, Trans. Jpn. Soc. Mech. Eng., 54-505, B (1988), 2349-2356.

2. Igarashi, T., K. Nishida, S. Mochizuki and H. Yamasaki, Fluid Flow around Three Circular Cylinders Having Different Diameters Closely Arranged in Line, Trans. Jpn. Soc. Mech. Eng., 55-511, B (1989), 693-700.

3. Igarashi, T. and T. Tsutsui, Flow Control Around a Circular Cylinder by a New Method (1st Report, Forced Reattachment of the Separated Shear Layer), Trans. Jpn. Soc. Mech. Eng., 55-511, B (1989), 701-707.

4. Igarashi, T. and T. Tsutsui, Flow Control Around a Circular Cylinder by a New Method (2nd Report, Fluid Forces Acting on the Cylinder), Trans. Jpn. Soc. Mech. Eng., 55-511, B (1989), 708-714.

5. Igarashi, T. and T. Tsutsui, Flow Control Around a Circular Cylinder by a New Method (3rd Report, Properties of the Reattachment Jet), Trans. Jpn. Soc. Mech. Eng., 57-533, B (1991), 8-13.

Visualization of Flow Past Inclined Bluff Cylinders

Luo, S. C., Lee, T. S., Chew, Y. T., Yazdani, M. G.

Department of Mechanical and Production Engineering
National University of Singapore, 10 Kent Ridge Crescent, Singapore 0511

Summary

Most of the engineering structures like buildings normally have either square, rectangular, trapezoidal or triangular cross-sectional shape. It is thus essential to know the aerodynamic loading, both in magnitude and frequency, on prismatic bodies. Some preliminary measurement had been carried out by the authors. The objective of the present paper is to conduct flow visualization to gain better understanding of the flow past prismatic bodies with different cross-sectional shape, which in turn will help the analysis of the experimental data.

Experimental Set-Up

To understand the effect of after body shapes and angle of attack on the flow structure and vortex shedding process, three cylinders of three different shapes , namely square, trapezium , and triangle were selected for flow visualization. All three models have the same stream wise (depth) and cross-stream (front face width) dimensions of d (=46 mm). The rear surface of the trapezoidal cylinder is d/2. For the triangular cylinder the apex always faces the downstream direction. The aspect ratio of each cylinder is 12. When placed vertically in the water tunnel, the blockage ratio is 5%. End plates were installed to ensure two-dimensional flow. They are 10 d (streamwise) X 8 d in dimensions. The models were installed such that the distance between the leading edge of the end plate and the front face of model was 2.5 d. The flow visualization experiment were carried out in the 1.8 m X 0.6 m X 0.9 m working section of the Singapore Aerospace Limited water tunnel. The experimental speed of the flow (U_∞) in the tunnel was 4.57×10^{-3} m/s. Outside the floor boundary layer the uniformity of the velocity profile of the tunnel is within ±2% and the mean flow direction is within $\pm 1^0$. The flow is visualized by dye injection from two needles placed upstream to the cylinders. Both video and still colour photographs were taken. Colour photographs were taken because two different colour dye were used for better visualization effect. Based on U_∞ and d, the Reynoulds number of the flow was 2.1×10^3.

Results and Discussion

From analyzing the video recording of the visualized flow (sometimes frame by frame) the following conclusions can be made :-

1. The Vortex shedding process associated with all the three shapes tested follows Gerrard's [1] proposed vortex shedding mechanism.

2. When the angle of attack (α) is altered , the interaction between the separated shear layers and the side of the body is also altered. When α increases from 0^0, it was observed that the vortex formation length (l_f)increases for all the three shapes up to a certain critical angle of attack (α_{crit}). There after l_f decreases as α further increases to the maximum value investigated which are 25^0 , 30^0 and 40^0 for the square, trapezoidal and triangular cylinder respectively (see plate 1 & 2). α_{crit} for the square, trapezoidal & triangular cylinder was found to be 12^0, 28^0 and 32^0 respectively. At α_{crit} , the value of l_f/d is estimated to be 2.4, 1.78 and 1.76 for the square,trapezoidal and triangular cylinder respectively. Since we know the video camera operates at 25 frames per second, Strouhal number (S) was also estimated from the video recording. The magnitude of S estimated agrees quite well with independent wind tunnel measurement. Variation of S with α is also dependent on afterbody shape. At $\alpha = 0^0$, S was estimated to be about 0.13 for the square and trapezoidal cylinder and about 0.15 for the triangular cylinder. S initially increases with α up to a maximum value of about 0.15, 0.18 and 0.23 for the square, trapezoidal and triangular cylinder respectively at their respective α_{crit} . Flow visualization suggests that α_{crit} exits because at $\alpha \sim \alpha_{crit}$, one separated shear layer begins to reattach onto the trailing corner of the cylinder. At $\alpha > \alpha_{crit}$, S begins to reduce in magnitude as α further increases. It is well known that the presence of an afterbody (the part of the cylinder that is downstream to the flow separation positions) delays or even suppress in certain cases the interaction of the two separated shear layers, resulting in the variation in the magnitudes of l_f . In the present investigation, despite keeping both the front face width and the depth of the three shapes identical, the variation of l_f/d and S with α still exists. This observation thus confirms the expectation that in addition to the depth, the afterbody shape, the proximity between the sides of the cylinder and the two separated shear layers also has a

strong influence on the interaction of the two separated shear layers and hence the formation of vortices.

3. When α varies, l_f also varies. However l_f varies in such a way that the magnitude of l_f associated with both the top and bottom shear layers remains the same as each other at any particular value of α despite the asymmetry of the cylinder with respect to the flow.

4. Bearman & Trueman [2] found that the distance between a vortex when it is shed and the base of the cylinder has a strong influence on the magnitude of the base pressure coefficient(C_{pb})(hence the drag coefficient C_D). Their findings were found to apply to the present investigation. It was found that the variation in C_{pb} measured in an independent experiment involving the same three cross-sectional shapes can be explained by the variation in l_f(i.e, $|C_{pb}|$ varies approximately inversely with l_f, with C_{pb} being a negative quantity all the time) observed in the present flow visualization experiment.

5. The variation of S estimated from flow visualization with α for square cylinder agrees quite well with the results of Vickery [3], and Lee [4].

6. The body axis lift coefficient (C_N) versus α had been measured in aseparate wind tunnel experiment. From the wind tunnel measurement,C_N for the square cylinder increase as α increases from 0^0 Whereas C_N for the other two shapes remains near zero up to α~ 10^0 & 11^0 for the trapezoidal & triangular cylinder respectively. By visualizing the flow it is confirmed, as expected, that the magnitude of C_N is related to the proximity between each separated shear layer and the corresponding side face of the cylinder. For example, for the square cylinder at a small α (e.g 5^0)shown in plate 1, the top shear layer is closer to the top face than the bottom shear layer is to the bottom face. Since shear layer causes entraintment of fluid, it is thus suggested that the top face is subjected to greater suction than the bottom face, hence C_N will be positive (i.e upwards). However, for the other two shapes, because of the tapering of the two side faces, at small α like 5^0 both the side faces are still relatively far away from the separated shear layers (plate 3 & 4) and hence C_N is still near zero. At even larger α, the effect of the separated shear layers began to influence the pressure on the side face and C_N began to become non-zero (plate 5 & 6).

References

1. Gerrard, J.H.: The mechanics of the formation region of vortices behind bluff bodies, J. Fld. Mech. 25 (1966) 401-413.

2. Bearman, P.W. & Trueman, D .M.: An investigation of flow around rectangular cylinders, Aero. Quart. 23 (1972) 229-236.

3. Vickery, B.J.: Fluctuating lift and drag on a long cylinder of square cross-section in a smooth and turbulent stream, J. Fld. Mech. 25 (1966) 481-494.

4. Lee, B.E.: The surface pressure field experienced by a two-dimensional square prism, Central Electricity Generating Board, RD/L/N17/74 (1974).

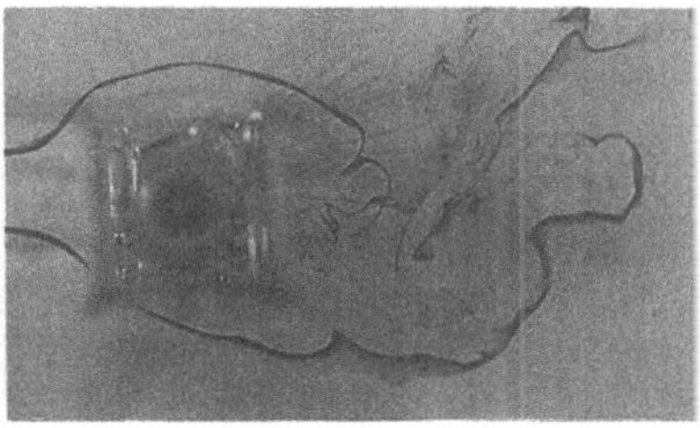

Plate 1: Square cylinder at $\alpha=5^{o}$.

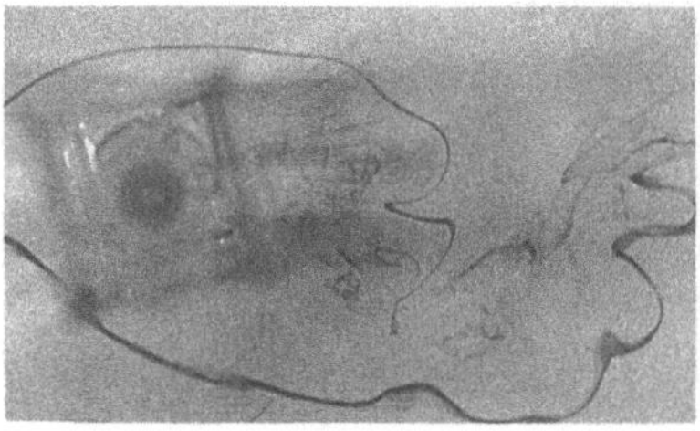

Plate 2: Square cylinder at $\alpha=10^{o}$.

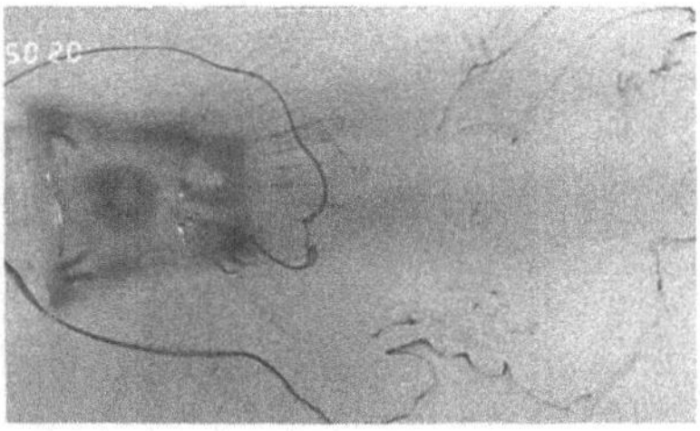

Plate 3: Trapezioidal cylinder at $\alpha=5^{o}$.

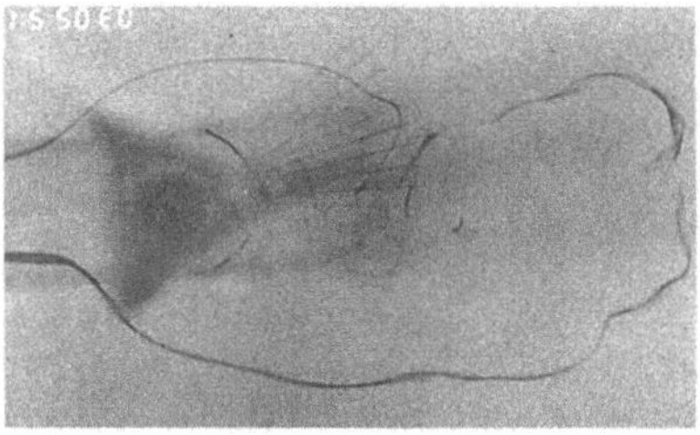

Plate 4: Triangular cylinder at $\alpha=5^{o}$.

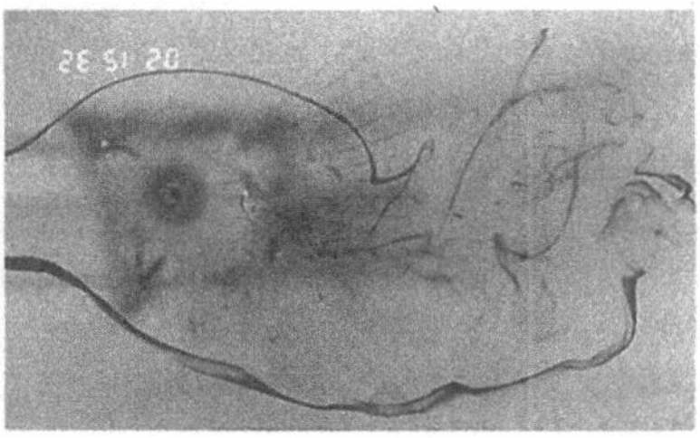

Plate 5: Trapezoidal cylinder at $\alpha=15^{o}$.

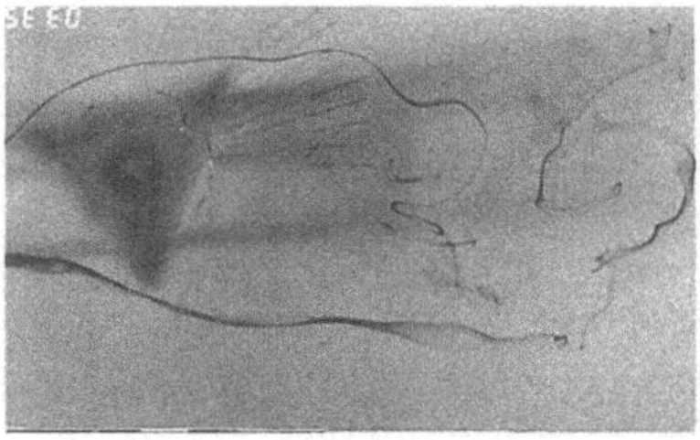

Plate 6: Triangular cylinder at $\alpha=35^{o}$.

Vortex Wake Visualizations of Two Circular Cylinders in Parallel

Kazuo OHMI

Osaka Sangyo University
Faculty of Engineering
Daito, Osaka 574, JAPAN

Kensaku IMAICHI

Osaka Hoki Co. Inc.
Fukushima-ku, Osaka 553, JAPAN

Summary

Vortex wakes of two parallel circular cylinders in tandem, side-by-side and staggered arrangements are examined by visualization experiments using a towing water tank. The cylinder spacing parameters are expressed in terms of longitudinal (or streamwise) spacing ratio L/d and transverse spacing ratio T/d, where L and T denote two Cartesian components of the distance between two cylinder centers and d the diameter. The flow analysis is generally made from direct observation of the visualization images as they are but, in some cases, complemented by image processing results relating to the estimated values of stream function, vorticity and wake frequency.

Introduction

Parallel cylinders in cross flow are often encountered in some practical situations of aerodynamical, architectural and civil engineering and have long been an object of intensive study in their respective fields. If the flow model is confined to two parallel circular cylinders, there still exists a good deal of literature, mostly derived from experimental work, which is timely followed by two review papers [1]-[2]. As point out both of these papers, the general flow character is varied depending on the spacing of the two cylinders and the most important feature is the existence of critical spacings, at which drastic change is observed in the aerodynamic force curves and those of wake frequency.

It has been reported that the wake of two tandem cylinders meets a drastic change at a critical spacing of about L/d = 3.5-3.9, where a marked jump is recognized on the drag and wake frequency charts. In the case of two side-by-side cylinders, the drag coefficients and wake frequencies present a bistable nature at a critical spacing of about T/d = 2.0 and flow asymmetry is observed between the two cylinders. If the two cylinders are in staggered arrangement, the drag, lift and frequency values are varied more complicatedly depending upon both L/d and T/d, inclusive of change-up of signs for the drag and lift coefficients.

The characteristics of the two circular cylinders flow have been thus revealed on a fluid measurement basis but the basic aspects of the flow phenomena corresponding to the above force and frequency variations have not been closely examined so far. On the other hand, at least as far as tandem and side-by-side arrangements are concerned, there is a growing interest in the wake behaviors at post-critical spacings [2]. Therefore, the present paper attempts to investigate the vortex wake behavior via visualization experiments for various arrangements and spacings of two cylinders and review the earlier understandings relative to the force and frequency measurements mentioned above.

Flow Visualization and Image Processing

The visualization experiment is conducted in a towing water tank of 250 x 30 x 30 cm^3 capacity. The test cylinders, 1.0 or 2.0 cm in diameter, are towed at a constant speed of 1.2 to 9.0 cm/s, the resultant Reynolds numbers being 40 to 500 (40 to 120 in most cases). The flow visualization is carried out under the water (not on the free surface) by using very fine aluminum particles in suspension and white tin hydrate dissolved out of the cylinder surface. These produce respectively streamlines and streak line profiles of the relevant flow.

The cylinder spacing parameter L/d is varied from 2.0 to 20.0 in the case of tandem arrangement, while another parameter T/d is from 1.0 to 5.0 in side-by-side arrangement. These spacing parameters are limited to a number of combinations in staggered arrangement, where $0 \leq L/d \leq 10.0$ and $0 \leq T/d \leq 4.0$. The generated vortex wakes are analyzed generally by direct observation of the visualization photographs and classified into some typical patterns depending on the spacing parameter. Then image processing techniques are introduced in order to extract quantitative results from the visualizations showing particle paths. The wake frequency is determined from the wave length and the displacement rate of the vortex street. These variables are estimated from the three sequential vortex centers located at about 10 d downstream of the leeward cylinder.

Another quantitative result of the image processing is the 2-D velocity profiles as defined as length and gradient of the particle paths. And no less important are the distributions of stream function and vorticity computed numerically from these velocity profiles. Every detail of this numerical method is described elsewhere by the present authors [3].

Results and Discussions

(A) Tandem arrangement The visualizations given in Fig.1 show three typical wake patterns observed at L/d = 4.0, 6.0 and 9.0 respectively. The difference between these spacings is relatively clear and can be described as follows: (1) Regime I, where the shear layers separating from the upstream cylinder reattach the downstream one without growing roll-up of vortices, and the two cylinders, approximately as a single body, shed alternate vortices towards downstream; (2) Regime II, in which roll-up of vortices originating from the upstream cylinder is steadily observed between the two cylinders, and the wake of the downstream cylinder forms regular vortex streets, where the wave length, amplitude and the phase of the upstream streets are almost maintained; (3) Regime III, where the downstream cylinder starts to shed its own vortices within the wake of the upstream cylinder and, as a result, the downstream wake experiences marked vortex interaction, such as beat, contraction or cancellation, the latter being often followed by reorganization of larger scale vortex streets.

The wake frequency, determined from the image processing, is normalized as Strouhal number S (= nd/U). The results of the present experiment is shown in Fig.2, where S is dependent on L/d and Re. The horizontal lines in this figure denote the Strouhal numbers of a single circular cylinder for comparison. The frequency curves in Fig.2, if compared with those of other earlier experiments (many of which are cited in Reference [2]) in the Reynolds number range of 10^3 to 10^5, follows similar variation of S with respect to L/d, but the values of

S, themselves, are decreased by 20 - 40 % in general. This is mainly because of the Reynolds number range of 40 to 120 in the present experiment, where S of a circular cylinder is not yet constant but variable linearly depending on Re.

The computed distributions of stream function and vorticity, determined also from the image processing, are shown in Fig.3 at three typical spacings of L/d = 3.0, 5.0, 8.0. The Reynolds number is fixed at 100 in these experiments. This figure demonstrates clearly that at L/d = 3.0, vorticity generated from the upstream cylinder does not roll up before the downstream one while at L/d = 5.0, it does roll up between the two cylinders. In the latter case, more precisely, some vorticity is supplied to upstream vortices during their passage over the downstream cylinder. Finally, at L/d = 8.0, the vortex interaction behind the downstream cylinder is clearly recognized by vorticity maps, where the periodic vortices coming from upstream are prone to deformation, division and/or unification during their passage over the downstream cylinder, because of complicity caused by the disturbed phase of vortex shedding and vortex wave length.

(B) Side-by-side arrangement The visualizations in Fig.4 show four typical wake patterns observed at T/d = 1.3, 1.9, 3.5 and 4.5 respectively. Since the last two patterns are possible variants of the same type, there are three flow regimes which are described as follows: (1) Regime I, where the two cylinders behave approximately as a single body and shed a single Karman vortex street; (2) Regime II, in which the flow passing through the two cylinders gap affects the downstream wake by shedding weak periodic vortices produced from the inner edges. These vortices do not last long but are merged into either of the outer vortex rows within about 10 d downstream of the two cylinders and, as a result, give rise to well known biased flow with respect to the wake center line; (3) Regime III, where the inner vortices originating from the gap flow are stable and last much longer (until at least 15 d downstream) and the whole wake consists of two sets of Karman vortex street. These two vortex streets are aligned often in phase or in counterphase (symmetric or counter-symmetric with respect to the center line) as shown by the last two patterns in Fig.4 but they are no less often irregularly out of phase and, in some cases, this phase lag is gradually changed during their downstream movement. It seems there are no special laws which may even explain the change up of these complicated patterns, though some tendencies can be pointed out as to the effects of T/d and Re number.

The variation of wake Strouhal number in this arrangement is shown in Fig.5, where the frequency is plotted by a slightly different symbols between the two cylinders. According to this figure, the frequencies in Regime I are about half the single cylinder value in the same Re number range. This is mainly because the effective length scale for estimation of Re number is doubled in this case. The frequency plots in Regime II are more or less scattered which is probably indicative of the wake asymmetry caused by biased gap flows. But the frequency curve does not make a bifurcation as in the case of hot wire measurements cited in reference [2]. The vortex frequency of the weak gap flow cannot be captured by the present method of measurement. Yet, different vortex frequencies can be visually recognized in some of the visualization frames (e.g. the second photograph in Fig.4). Obviously, the frequency curves in Regime III are more stable and likely to approach constant levels with increasing T/d. And each of these levels is nearly consistent to single cylinder frequency in the same Re range.

The distributions of stream function and vorticity are given in Fig.6 at two selected spacings of T/d = 1.5 and 3.0 (corresponding to Regimes II & III). The Re number is 100 in any case. The vorticity maps of this figure indicate that the biased gap flow in Regime II is caused by linking and merging of one of the corotative inner and outer vortex pairs. This takes place periodically in response to the vortex shedding from the outer rows, and the bias does not change up in the course of experiment unless the flow is once stopped and restarted.

<u>(C) Staggered arrangement</u> Four visualization frames are picked up in Fig.4 from many combinations of the spacing parameters. The Reynolds number is fixed at 80 in these experiments. If T/d = 0.5 or 1.0, the downstream wake bears essentially the same characteristics as in the tandem case (first two frames in Fig.7), whereas if T/d $\geq$ 2.0, it is evolved into the side-by-side case regardless of the values of L/d (last two frames in Fig.7). The criteria of this distinction are the onset of vortex shedding from the cylinders gap region and the absence of vortex interaction on the same side of the two cylinders. More precisely, the first two frames in Fig.7 correspond to Regimes II and III of the tandem case (vortex interaction in Regime III is not pronounced) and the last two frames to Regimes II and III of the side-by-side case.

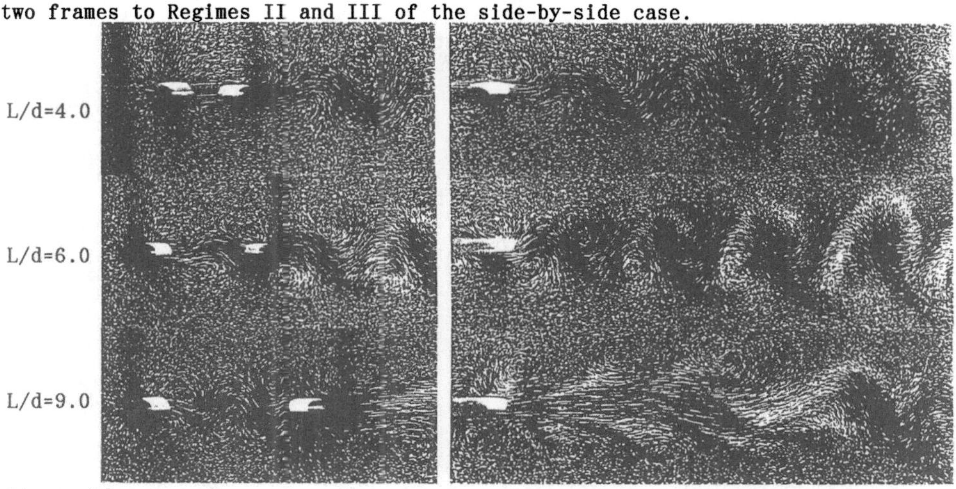

Fig.1 Visualizations of tandem cylinders (left:upstream ; right:downstream).

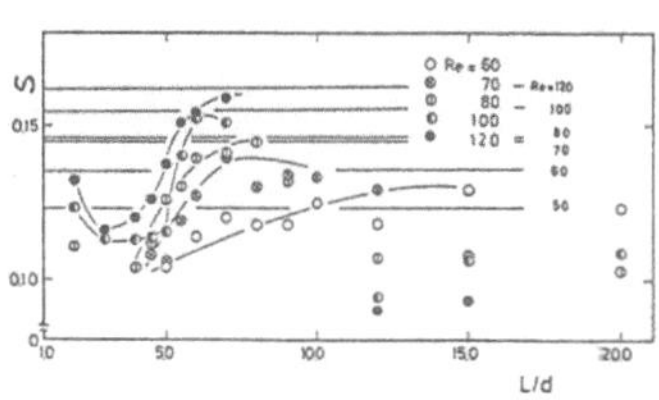

Fig.2 Strouhal number.

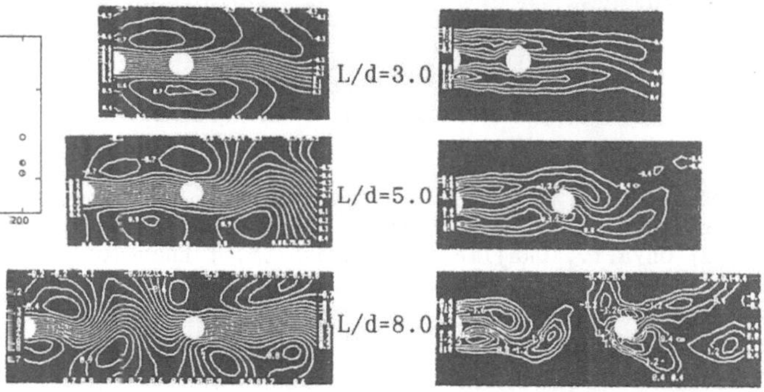

Fig.3 Stream function (left) and vorticity (right).

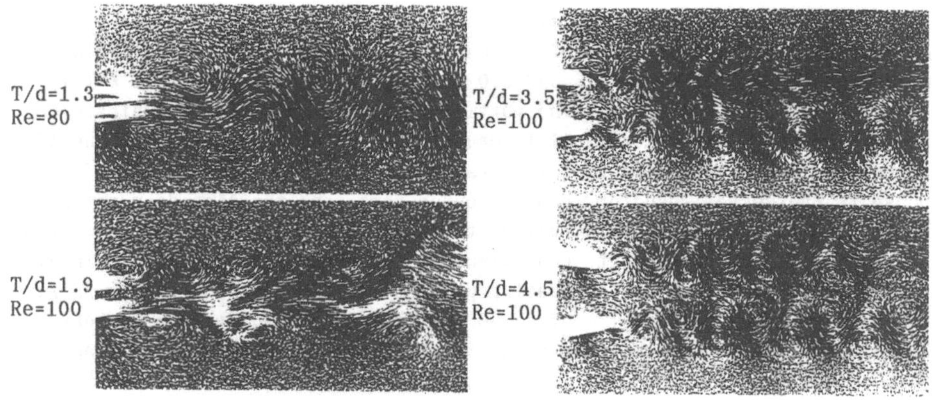

Fig.4 Visualizations of side-by-side cylinders.

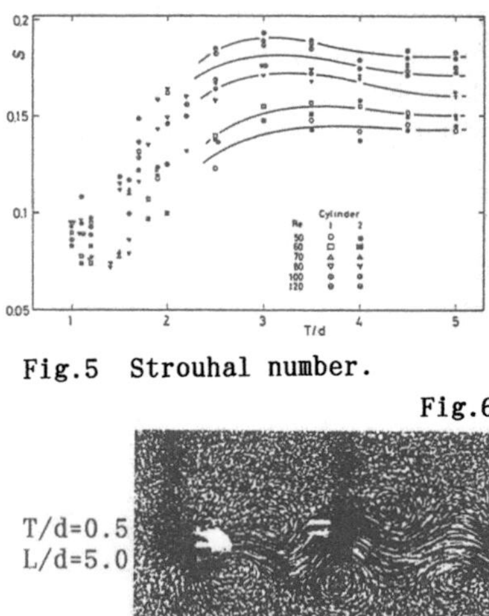

Fig.5 Strouhal number.

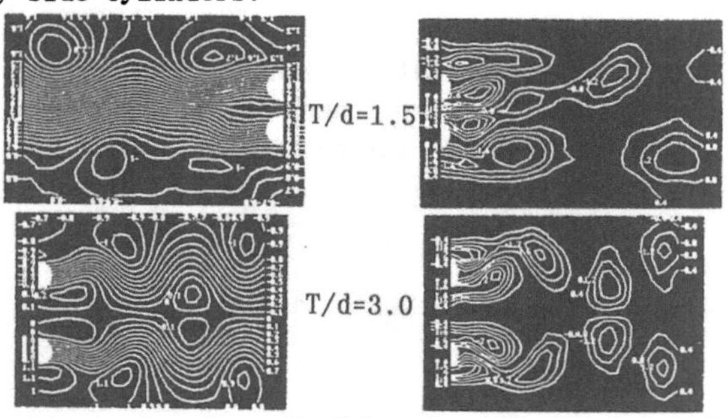

Fig.6 Stream function (left) and vorticity (right).

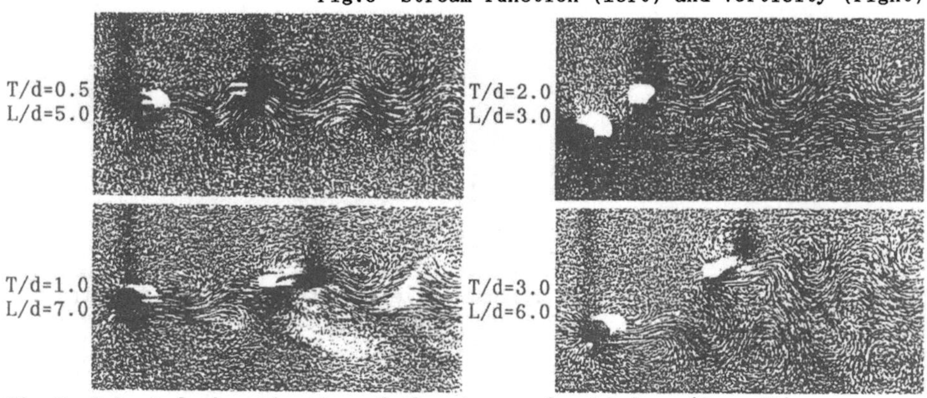

Fig.7 Selected visualization of the staggered cylinders (Re = 80).

References

[1] Zdravkovich,M.M. : Trans. ASME - J. Fluids Engineering **99**, 618 (1977).
[2] Ohya,Y., Okajima,A. & Hayashi,M. : Encyclop. Fluid Mech. **8**, 323 (1989).
[3] Imaichi,K. & Ohmi,K. : J. Fluid Mech. **129**, 283 (1983).

Flow Around a Circular Cylinder with and without Grooves

Makoto OKI
Faculty of High–Technology for Human Welfare,Tokai Univ., Japan

Masumitsu SUEHIRO
Mazda Co., Japan

Hiroo OKANAGA and Katsumi AOKI
Faculty of Engineering, Tokai Univ., Japan

Yasuki NAKAYAMA
Future Technology Research Institute, Japan

Takaharu OKUMOTO
Yokohama Rubber Co., Ltd., Japan

Summary

It is well-known the dimples on the surface of golf ball increase its flying distance. But the detailed mechanism of flow by dimples has not been made clear. The purpose of this research is to make clear the mechanism of this flow in the case of changing the number and shape of dimples by computation and experiment. As the first stage of this research, this paper deals with the effect of the grooves as modeled dimples that were put on the surface of a circular cylinder. The computation was performed by solving the Navier–Stokes equation using the QUICK method. The grooves were formed in the same shape as a golf ball and the number of dimples is 32. The results by computation of the flow pattern around the circular cylinder with and without grooves showed large difference.

Nomenclature

- u, v = flow velocities normalized by u_0
- u_0 = uniform velocity
- p = pressure normalized by ρu_0^2
- ν = kinematic viscosity
- d = diameter of cylinder
- t = time normalized by d/u_0
- Re = Reynolds number ($Re = u_0 d/\nu$)
- x, y = physical coordinates normalized by d
- ξ, η = computational coordinates normalized by d
- C_P = pressure coefficient
- C_D = drag coefficient
- θ = circumferential angle from stagnation point

Introduction

It is well-known the dimples on the surface of golf ball increase its flying distance. The boundary layer on the surface of golf ball is disturbed by dimples and changes from laminar to turbulent flow, so the separation point on the surface moves downstream and its drag decreases. But flow around high speed and spinning golf ball is complicated and its flow mechanism is not clear in full detail. Therefore, up to now, the number and shape of dimples have been determined by experimental data. The study on the surface roughness around a circular cylinder is investigated experimentally by Achenbach [1],[2], Güven and others [3], Adachi and others [4]. But, the surface roughness on a cylinder is similar to the shape of a pyramid or produced by wrapping a sandpaper, so the shape of roughness is different from a dimple. On the other hand, there are some studies on a circular cylinder having partial projections [5],[6]. But they have not been reported about a circular cylinder having dimples all over the surface.
The purpose of this research is to make clear the mechanism of this flow in the case of changing the number and shape of grooves as modeled dimples that were put on the surface of a circular cylinder by computation and experiment. As the first stage of this research this paper deals with the effect of grooves by computation.
And, this result was compared with the experimental result of reference. The computation was performed for Re=50000.

Numerical Analysis

Fundamental equations

The computation was performed under the following assumptions. Fluid is incompressible and viscous, two dimensional flow and isothermal. No external force is acting on the fluid. The fundamental equations are obtained from the continuity equation derived from conservation of mass and Navier–Stokes equation derived from equation of motion. These fundamental equations are normalized by diameter of cylinder d and uniform velocity u_0, and transformed from physical coordinates to computational coordinates as shown in next equations.

$$\hat{q}_t + \hat{E}_\xi + \hat{F}_\eta = \frac{1}{Re}(\hat{R}_\xi + \hat{S}_\eta) \tag{1}$$

$$\hat{q} = \frac{1}{J}\begin{bmatrix} 0 \\ u \\ v \end{bmatrix} \qquad \hat{E} = \frac{1}{J}\begin{bmatrix} U \\ uU + \xi_x p \\ vV + \xi_y p \end{bmatrix} \qquad \hat{F} = \frac{1}{J}\begin{bmatrix} V \\ uV + \eta_x p \\ vV + \eta_y p \end{bmatrix}$$

$$\hat{R} = \frac{1}{J}\begin{bmatrix} 0 \\ (\xi_x^2 + \xi_y^2)u_\xi + (\xi_x\eta_x + \xi_y\eta_y)u_\eta \\ (\xi_x^2 + \xi_y^2)v_\xi + (\xi_x\eta_x + \xi_y\eta_y)v_\eta \end{bmatrix} \qquad \hat{S} = \frac{1}{J}\begin{bmatrix} 0 \\ (\xi_x\eta_x + \xi_y\eta_y)u_\xi + (\eta_x^2 + \eta_y^2)u_\eta \\ (\xi_x\eta_x + \xi_y\eta_y)v_\xi + (\eta_x^2 + \eta_y^2)v_\eta \end{bmatrix}$$

where J is the Jacobian of coordinates transformation, U and V are the velocity components along the ξ and η computational coordinates.

Numerical analysis

In this study, the solution procedure used spatially the third-order accurate upwind finite difference technique based on the QUICK algorithm that has been extended to generalized coordinates[7]. The fundamental equations are integrated over the control-volume computing cell on a uniform regular grid, are numerically solved by means of a time integration using fractional step method.
The pressure field is simultaneously calculated from the Poisson equation. Finally, the velocity and pressure are iterated so as to satisfy continuity equation.

Computational region and meshes

The mesh of circular cylinder is shown in Fig.1. The same number and interval of meshes are designed for both circular cylinder without grooves and with grooves. The number of meshes is 224 in circular direction and 100 in radial direction. And, computational region takes the circular region to be 52 times the radius. The grooves were made as the same shape as a golf ball and the number of grooves is 32. The meshes were concentrated in near surface, and included 4 $\sim$ 5 lines into boundary layer. The two meshes of near surface are shown in Fig.2.

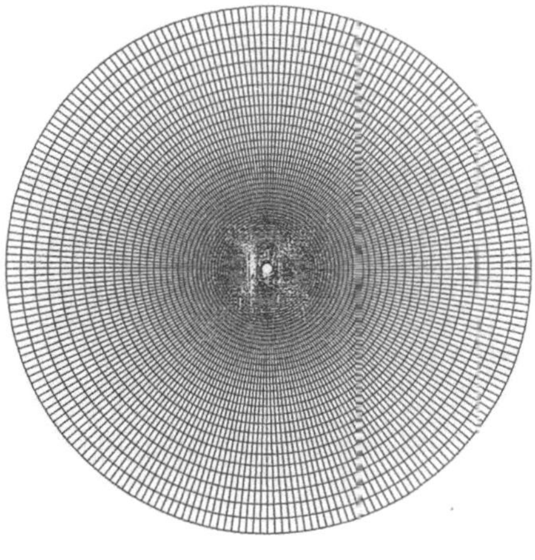

Fig.1 Mesh

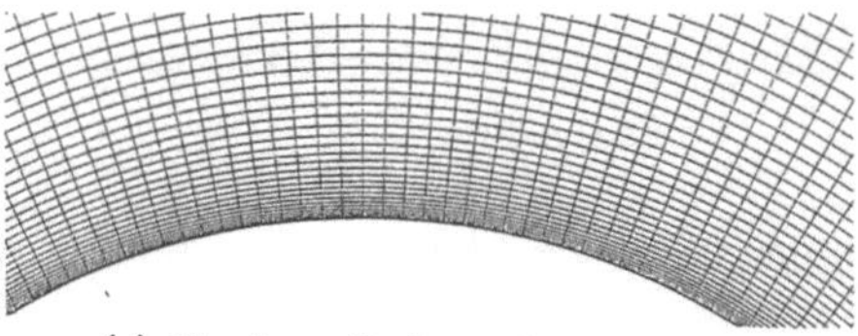

(a) Circular cylinder without grooves

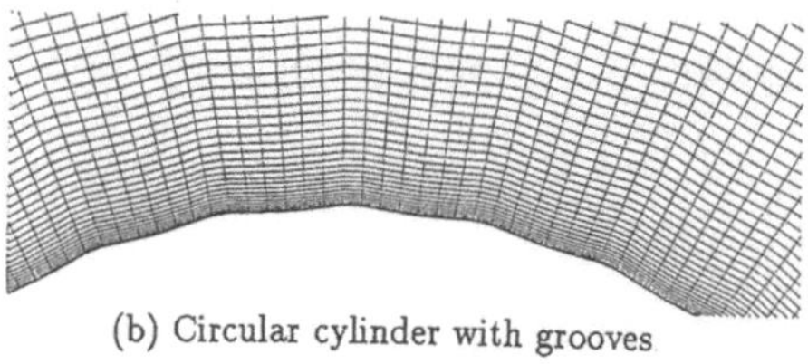

(b) Circular cylinder with grooves

Fig.2 Meshes of near surface

Boundary conditions

When the flow is viscous, it is assumed that the flow does not slip on the cylinder surface.

$$u=0, \qquad v=0, \qquad \partial p/\partial \eta=0$$

The velocities and pressure on the outside boundary are given as follows

$$u=1, \qquad v=0, \qquad p=0$$

Results and Discussions

Computation was carried out for flow around the circular cylinder at Reynolds number of 50000. Computation was performed from 0 to 100 in normalized time. The computational results at any times are shown in Fig.3. The intensity of shade shows the pressure distribution. The stronger the intensity of the shade lower the pressure.

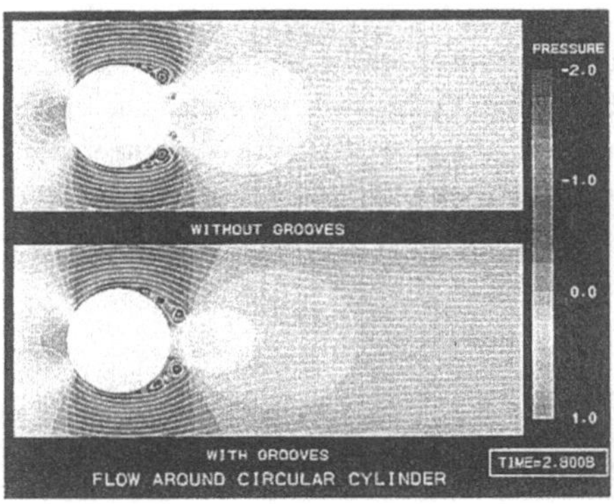

(a) t=2.8

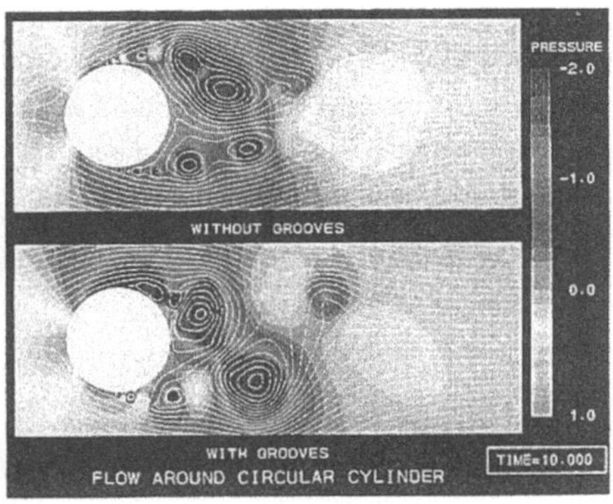

(c) t=10.0

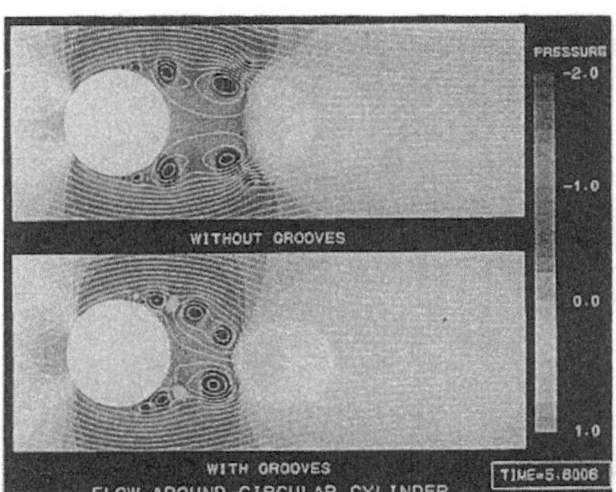

(b) t=5.6

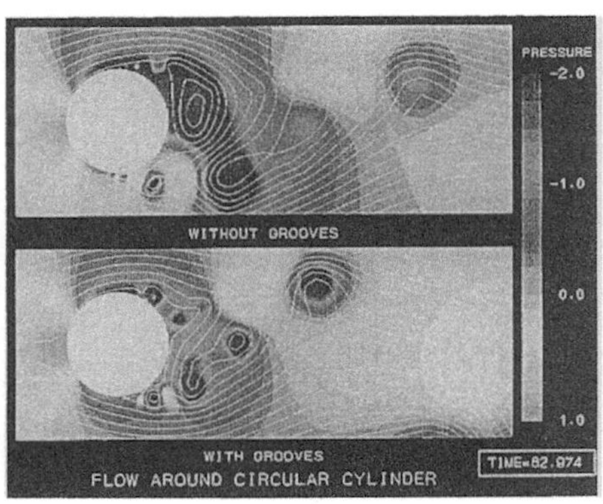

(d) t=63.0

Fig.3 Pressure distribution and streamlines of unsteady flow

The time average values of pressure and velocity were obtained in the normalized time of 50 to 90 for the circular cylinder without grooves and 50 to 100 for the circular cylinder with grooves. These results are shown in Fig.4. The streamlines behind the circular cylinder show the separation region. Both shade and separation areas behind the cylinder with grooves are smaller than the one without grooves. These depict the drag force of the former is smaller than the latter. Fig.5 is the zooming figure of near surface of Fig.4. From the streamline of near surface of Fig.5, because separation point of the circular cylinder with grooves was found to occur 10° (θ= 80° in circular cylinder without grooves, θ= 90° with grooves) downstream compared with the one without grooves, the effect of grooves was verified.

The pressure coefficient C_P on the surface of circular cylinder was calculated with and without grooves and computed results were compared as shown in Fig.6. The computed result without grooves and experimental result at *Re* of 1.1×10^5 [8] were compared. They were agreed each other as shown in Fig.6.

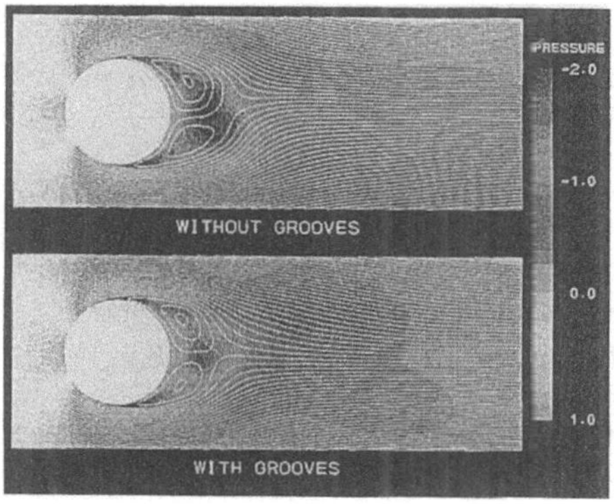

Fig.4 Pressure distribution and streamlines of time average flow

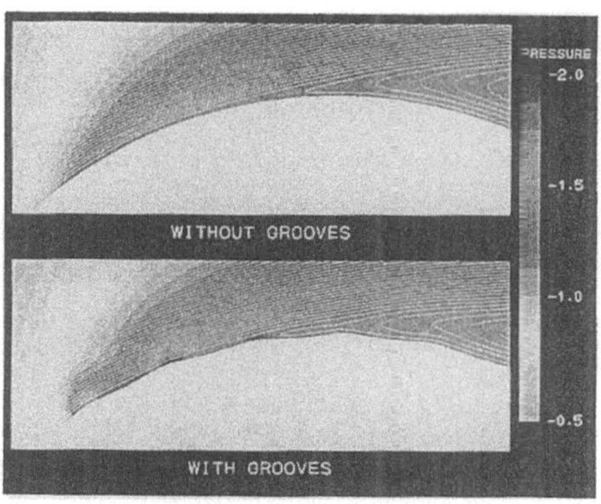

Fig.5 Pressure distribution and streamlines of near surface

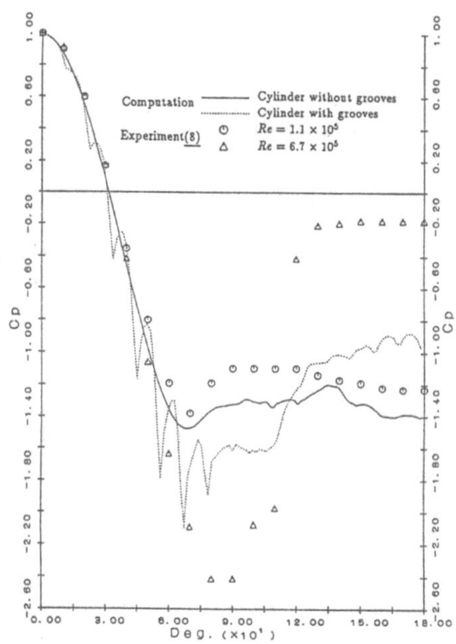

Fig.6 Pressure distributions on surface

Next, the pressure distribution on the surface of circular cylinder with grooves in Fig.5 and Fig.6 differs from the one of circular cylinder without grooves due to the pressure difference in the top and bottom part of grooves. The pressure at the top of grooves is lower than it's circumference.
The pressure distribution is similar to the one of supercritical range($Re = 6.7 \times 10^5$) by experiment [8] as shown in Fig.6. But, the flat part of pressure distribution from laminar separation point by computed result is longer than the one of experimental result.
Drag coefficient C_D is given by integrating pressure coefficient C_P. As the result, C_D is 1.243 for the circular cylinder without grooves, 1.004 for the circular cylinder with grooves. C_D of the circular cylinder with grooves is lower than one of the circular cylinder without grooves. This show that grooves reduce C_D value due to flow moved into critical range.

Conclusions

The results of flow simulation around a circular cylinder with and without grooves lead to the following conclusions:

(1) The large difference was shown in the flow pattern around the circular cylinder with and without grooves.

(2) The separation point moved 10° downstream when grooves were introduced.

(3) The pressure distribution of circular cylinder with grooves found from computation that pressure at the top of grooves was lower than it's circumference. This form was similar to the form of near critical range of circular cylinder Reynolds number.

(4) It was found from computation that the drag coefficient decreased by 0.24 with grooves.

References

1. Achenbach, E. and Heinecke, E.: On vortex shedding from smooth and rough cylinders in the range of reynolds numbers 6×10^3 to 5×10^6, Journal of fluid mechanics, vol.109, 1981, 239.

2. Achenbach, E.: Influence of surface roughness on the cross–flow around a circular cylinder, Journal of fluid mechanics, vol.46, part.2, 1971, 321.

3. Güven, O., Farell, C. and Patel, V.C.: Surface–roughness effects on the mean flow past circular cylinders, Journal of fluid mechanics, vol.98, part.4, 1980, 673.

4. Adachi, T., Ono, H., Matsuuchi, K., Kawai, T. and Cho, T.: Flow around a circular cylinder in the high reynolds number range, Transactions of the japan society of mechanical engineers (in Japanese), vol.55, no.511, B, 1989, 685.

5. Kawamura, T. and Kuwahara, K.: Computation of high reynolds number flow around a circular cylinder with surface roughness, AIAA paper, 84–0340, 1984.

6. Kimura, T., Tsutahara, M., Kamata, K. and Ikushima, Y.: Flow past a circular cylinder with circular–arc grooves on its surface, Journal of the visualization society of Japan, vol.10, suppl.no1, 1990, 73.

7. Nakamura, Y. and Takemoto, Y.: Solutions of incompressible flowes using a generalized QUICK method, Proceedings of international symposium on computational fluid dynamics–Tokyo, sept, 1985, 386.

8. Roshko, A.: Experiments on the flow past a circular cylinder at very high reynolds number, Journal of fluid mechanics, vol.10, 1961, 345.

Syncronous Visualization of Flow Patterns and Correlative Physical Parameters

WEI Qing-ding WANG Wen-bao LI Chang-Lin & DU Xiang-dong
Department of Mechanics, Peking University, Beijing, China 100871

Summary

A visualization method — syncronous visualization of flow patterns and correlative physical parameters is introduced. By the principle of image composition the image of flow patterns taken by video cameras and the image representing correlative physical parameters can be combined in composite image which can not only shows the flow patterns at some moment but also can shows syncronously the correlative physical parameters, such as vortisity, shear stress, pressure, force, moment, temperature and concentration etc. It is very useful for researching unsteady flow. As an example, the patterns of a flow past a two dimensional H-shape cross-section and the pitching moment acting on the H-shape cylinder are visualized syncronously. The results show a strong correlation between vortex shedding and fluctuating pitching moment.

Introduction

Flow can be visualized by means of many approaches. The visualized flow is represented by one or a few kinds of characteristic lines, such as stream lines, streak lines, path lines, time lines and so on, or by color, fringe or brilliance. In general a picture or a frame of flow visualization provides either the information of flow patterns or the information of correlative physical paramcters with the flow. It is hoped that flow patterns and correlative physical parameters, for example fluid forces acting on a body which is placed in the flow, can be visualized on a same picture or a same frame syncronously. It is not much difficult to steady flow, periodic flow or quasi steady flow. One method has been described in reference[1] by Haolian H. To unsteady flow an ordinary method is coordinating image pictures of flow patterns and figures of correlative physical parameters as shown in fig. 1 (by Sun R. [2]). This paper introduces a new method by which the flow patterns and the figures representing correlative physical parameters can be shown on a same picture or a same frame syncronously.

Visualization Facilities

1. *Image composition system*

To composite a frame of TV image and a frame of computer figures a image composition system was used. The block diagram of this system is shown in fig. 2. The information of flow patterns is taken by some video cameras and is sent to a video recorder, then past a adapter is sent to a syncro adjustment and image composition board. At the same time the information of correlative physical parameters taken by some detectors is sent to an A/D converter and a computer for processing and drawing, and then is also sent to the board. Every new frame of image is composited of two syncro frames of image at the image composition board.

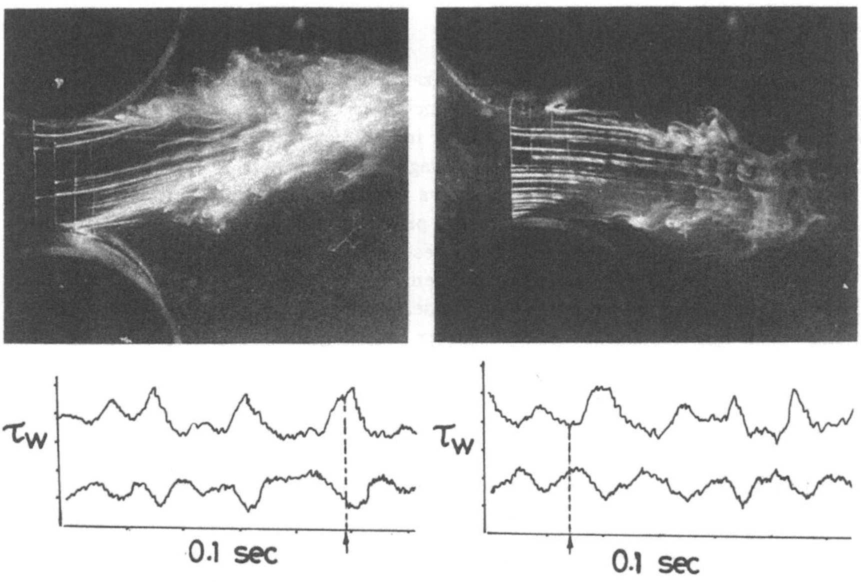

Fig. 1. The simultaneous visualization of the flow patters past two circular cylinders arranged side by side and the signals of surface shear stress at $\pm 90°$ (to lower and upper cylinder respectively), The arrows indicate the photo-time

2. *Water tunnel*

A water tunnel with test section 0. 4m high, 0. 4m wide and 6m long was used for testing. The Reynolds number is about 6×10^3 by characteristic dimension 15cm.

3. *Model*

A two dimensional cylinder with H-shape cross section was tested as a model. The scale of the model and the coordinate system are shown in fig. 3.

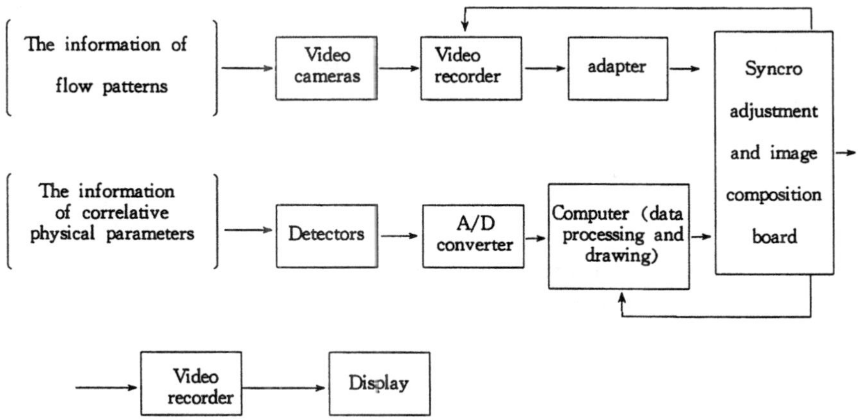

Fig. 2 The block diagram of image composition system

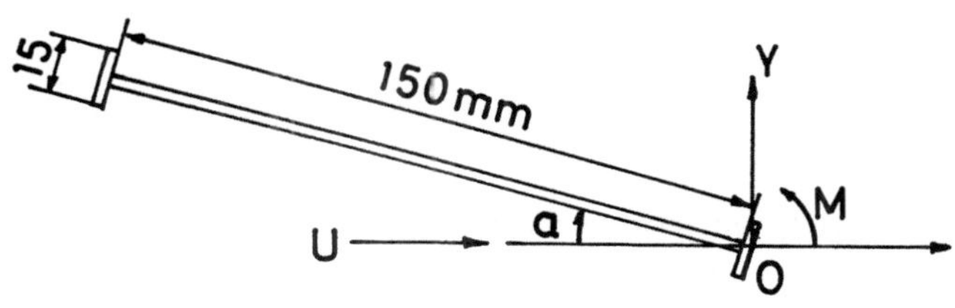

Fig. 3 The scale of the model and the coordinate system

Experimental Results

Using image composition technique the flow patterns past a two dimensional H-shape cross-section and the pitching moment acting on the cylider are visualized syncronously. fig. 4 shows the flow at the small attack angle $\alpha = -3°$. It can be seen that when the shedding vortex at leading edge is strong the pitching moment is positive (left), conversely is negative (right).

The correlation between the vortex shedding and the fluctuating pithing moment can be seen from fig. 5. The strong fluctuation of the pitching moment occures to follow the vortex shedding.

At the biger attack angle, $\alpha = 30°$, the separation region changes obviously. Along with it the pitching moment changes(see fig. 6)

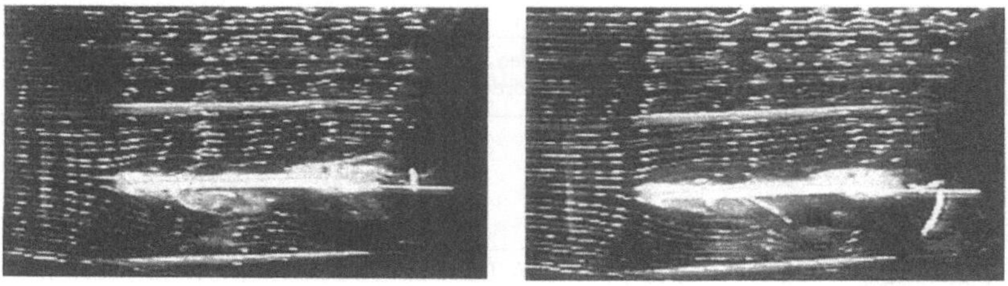

Fig. 4 The flow patterns past a two dimensional H-shape cross-section and the pitching moment at the attack angle $\alpha=-3°$

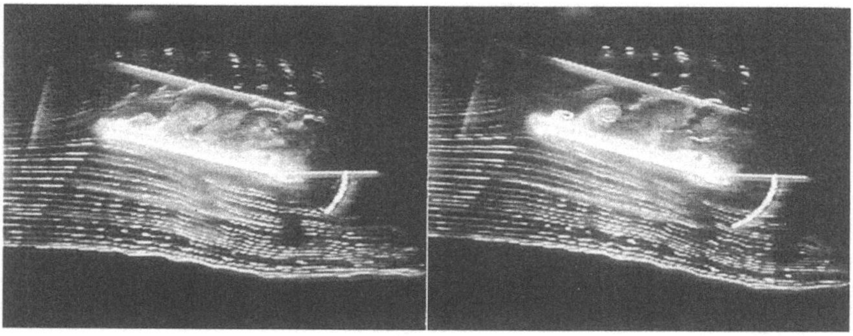

Fig. 5 The correlation between the vortex shedding and the fluctuation of pitching moment ($\alpha=14°$)

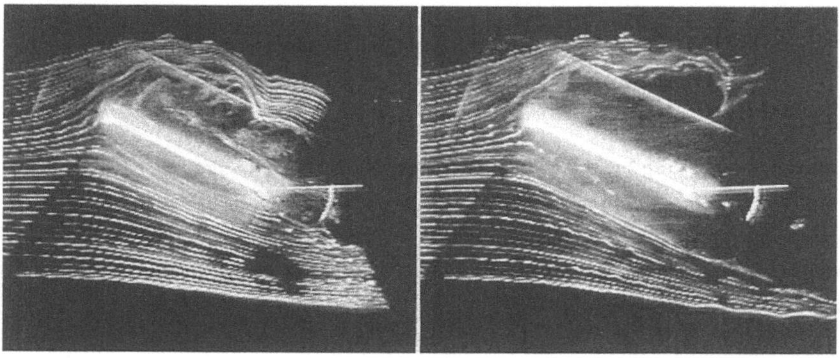

Fig. 6 The relationship between the change of separation region and the change of pitching moment.

Conclusions

1. The method of syncronous visualization of flow patterns and correlative physical parameters offers a new approach to recoganize the dynamical relationship between the flow patterns and the correlative physical parameters.

2. The syncronous visualization of the flow patterns past a two dimensional H-shape cross-section and the pitching moment acting on it shows that the vortex shedding and the change of separation region effect on the fluctuation of pitching moment.

References

1. Helian H. Z. : Simultaneous Visualization of Mutiple Parameters of Fluid Mechanics. Proceedings of the Second China-Japan Symposium of Visualization, Hangzhou, China (1992).

2. Sun R. & Wei Q. D. : Applications of Smoke Wire Method for Visualization. Proceedings of the Second National Symposium of Flow Visualization, Huangshan, China (1986).

Application of Flow Visualization to the Study of Very Long Cylinders in Near-Axial Flow

M.K.BULL and W.A.DEKKERS

Department of Mechanical Engineering
University of Adelaide
Adelaide, S.A. 5000.

1. Introduction

As part of a study directed towards the axisymmetric boundary layer on a very long cylinder, the flow over cylinders yawed to the main flow direction has been studied by several methods of flow visualisation.

For axial flow, the conditions of interest are those which lead to both the outer boundary-layer scale (the boundary layer thickness δ) and the inner scale (the wall scale ν/U_τ) being large in relation to the scale of transverse curvature of the boundary layer (the cylinder radius, a). Then δ/a is large (as high as 40) and $a_+ = aU_\tau/\nu$ is small (10-40), and the whole boundary layer is influenced by the curvature [1].

In the case of a cylinder inclined at an angle β to the flow direction, the boundary layer becomes highly asymmetric even at very small values of β (one or two degrees), and over a finite range of Reynolds number vortex shedding can take place [2].

These observations raise the possibility that vortex-shedding from a cylinder in axial flow, as a result of cross-flows within its own turbulent boundary layer, may play an important part in turbulence generation in very thick axisymmetric turbulent boundary layers [3]. One of the aims of the flow-visualisation studies is to explore this possibility.

2. Experimental Techniques.

The flow visualisation techniques which have been employed are as follows:

(i) Planar arrays of smoke streaks in an air-flow. These were obtained by electrically heating a fine stainless steel wire spanning the wind tunnel and smeared with oil (the thickness of the oil film varying periodically along the wire under the influence of surface tension).

(ii) Smoke generation, in an air flow, by a short streamwise segment of the cylinder itself. For this purpose a ceramic insert in the cylinder was wound with fine resistance wire; the resistance wire was smeared with oil which was then burned off by electrical heating.

(iii) Hydrogen bubble visualisation in a water flow.

The air flow experiments were conducted in a wind tunnel with a maximum value of flow velocity U of 25 m/s. The tunnel test-section (300 x 650 mm) is 3m long, with side walls which may be adjusted to vary the streamwise pressure gradient. The cylinder in the axial flow case was a tensioned 3mm-diameter teflon line; for the yawed flow case a 6mm-diameter steel rod was used. The cylinder Reynolds number $Re_a = Ua/\nu$ ranged from 250 to 500.

A special recirculating water tunnel is being constructed with a 3.6 metre long transparent working section to accommodate boundary layer development on cylinders with length-to-diameter ratios of several thousand, and with a depth of 375 mm to allow yaw angles of up to 5° to be obtained. The data presented here have been obtained either in a similar existing tunnel with a much shorter test section, or in a large open circuit water channel. In these water flows, Re_a was in the range 150 to 200.

The water tunnel has a 0.8-metre-long working section (100 x 200 mm cross section), and a maximum flow velocity of 0.15 m/s. Flow smoothing is achieved by a honeycomb section and two fine mesh screens followed by a 4 : 1 contraction. The water channel has a 3m-long test-section (200 x 400 mm cross section), and a maximum flow velocity of 0.18 m/s. Limited flow smoothing is obtained by a honeycomb section and two mesh screens at the upstream end of the test section.

Two video recording systems were used, a conventional video camera with a high definition recorder capable of variable speed replay, and a Kodak Ekta Pro 1000 high speed video system capable of recording up to 1000 frames per second. Still photographs were produced from the video recordings using a Sony colour video printer.

3. Comparison of Visualisation Techniques.

The results shown in figure 1 demonstrate that comparable still images can be obtained from both air and water flows. However, for the same cylinder Reynolds number and cylinder diameter, the continuity of the video recording, which requires a video framing rate higher than the vortex shedding frequency, is far better for the water flow. This results from the fact that with cylinder diameter and Re_a constant the vortex shedding frequency is proportional to the kinematic viscosity of the fluid. The kinematic viscosity of water is lower than that of air by a factor of about 15 leading to a lowering of the vortex shedding frequency by the same factor and hence to significant improvement in video continuity. The use of high speed video recording was severely limited by the amount of light which could be directed onto the area of flow under investigation, and by the density of visualisation particles within this area.

(a)

(b)

(c)

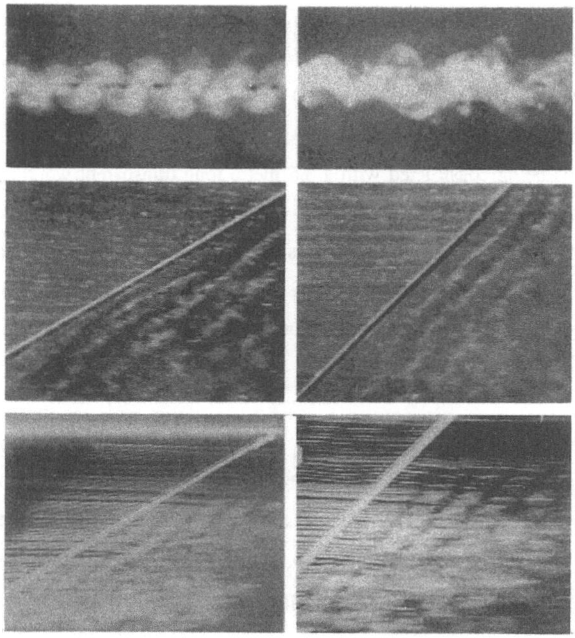

$\beta = 30°$ $\beta = 45°$

Figure 1. Vortex wake of a cylinder yawed at angle β in the vertical plane. (a) and (b) views of horizontal and vertical planes, water flow, $Re_a = 200$; (c) view of vertical plane, air flow, $Re_a = 260$. Flow from left to right.

The short duration of the smoke streaks requires repeated recording to determine the effects of varying the lighting and rate of smoke generation, whereas the continuous stream of hydrogen bubbles in the water flow allows experimental variations to be viewed directly. These differences suggest that, for the types of flow being investigated here, the advantages of flow visualisation in water rather than in air are sufficient to override the problems of producing a water flow of comparable quality to that readily available in a wind tunnel.

4. Discussion of Flow Details and Conclusions.

The inclination of the vortex axes to the cylinder axis is expected to increase as the yaw angle is decreased. As the yaw angle is reduced, the magnitude of the component of main-flow velocity parallel to the cylinder axis increases relative to the normal component; the resultant vorticity vector in the cylinder boundary layer, and consequently the axes of the vortices, therefore becomes increasingly inclined to the cylinder axis as β is reduced. The visible concentrated lines of hydrogen bubbles in figure 2 demonstrate this effect as the cylinder yaw is decreased from 45°, with vortex lines almost parallel to the cylinder, to 8° yaw where the vortex axes are inclined at 30° to the cylinder axis.

The views of a horizontal section of the wake of a cylinder yawed in the vertical plane (figure 3) indicate that the rolling-up of the separated shear layer in the wake of the cylinder takes place more quickly at the larger angles of yaw, i.e. that the vortex formation length increases as β is decreased. At β = 45° vortex-formation appears to be complete, whereas at β = 8° the wake shows a sinuous character and rolling-up is only just beginning to occur within the viewing area. The latter effect is accentuated at β = 3.5°; and at this angle the formation length appears to be in excess of 100 cylinder radii. These results suggest that at the small yaw angles induced by turbulent cross-flows in a thick

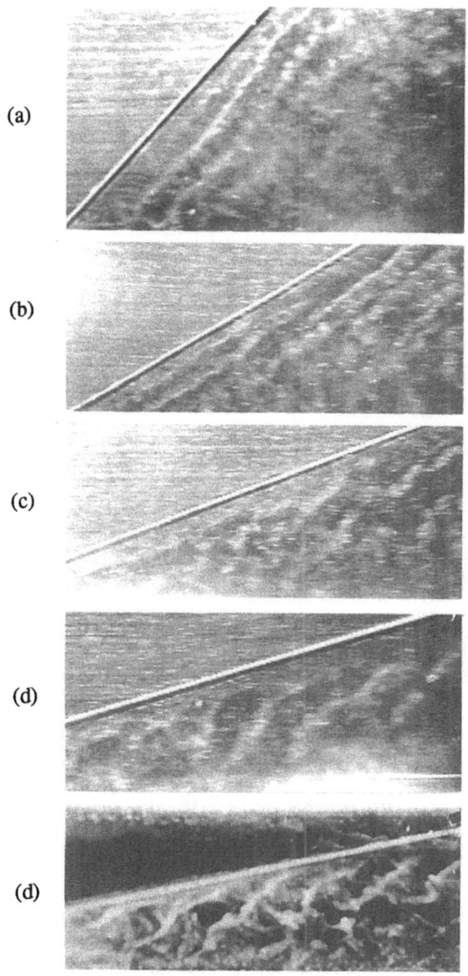

Figure 2. Hydrogen-bubble visualisation of the vortex wake of a cylinder yawed in the vertical plane at Re_a = 200. Views of vertical plane for (a) β = 45°; (b) 30°; (c) 20°; (d) 15°; (e) 8°. Flow from left to right.

axisymmetric turbulent boundary layer, with δ/a typically 30 or 40, the formation length may be of the order of several boundary layer thicknesses in which case it would seem that complete vortex formation is unlikely to occur. However, flow-visualisation from a cylinder segment in an axisymmetric turbulent boundary layer (not shown here) also shows sinuous flow near the cylinder surface, and the process in the axisymmetric layer clearly merits further investigation.

(a)

(b)

(c)

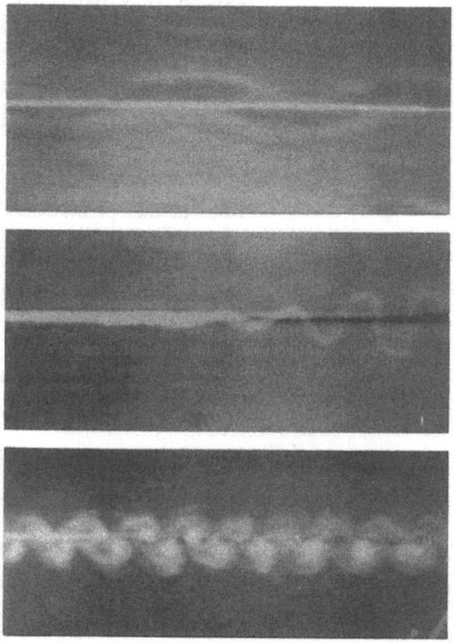

Figure 3. Hydrogen-bubble visualisation of the wake of a cylinder yawed in the vertical plane. Views of horizontal plane for (a) $\beta = 3.5°$, $Re_a = 150°$; (b) $\beta = 8°$, $Re_a = 200$; (c) $\beta = 30°$, $Re_a = 200$. Flow from left to right.

References.

1. Luxton, R.E.; Bull, M.K.; Rajagopalan, S.: The thick turbulent boundary layer on a long fine cylinder in axial flow. Aeron. Jl. 88 (1984) 186-199.

2. Bull, M.K.; Dekkers, W.A.: Vortex shedding from cylinders in near axial flow. Proc. 10th Australasian Fluid Mech. Conf. Melbourne. (1989) 6.41-6.44.

3. Bull,M.K.; Luxton,R.E.: Vortex-shedding and the maintenance of thick turbulent boundary layers on long cylinders in axial flow. Advances in turbulence 3. Springer Verlag 1991, 13-21.

Experimental and Numerical Visualization of the 3D Flow Around a Short Circular Cylinder Fitted with Endplates

G. PINEAU *, A. TEXIER *, M. COUTANCEAU * and TA PHUOC LOC **

(*) Laboratoire de Mécanique des Fluides, (URA CNRS 191), 40 av. du Recteur Pineau Université de Poitiers, France
(**) Laboratoire d'Informatique pour la Mécanique et les Sciences de l'Ingénieur, Orsay, France

Summary :

The 3D effects induced by two symmetric endplates upon the flow around an impulsively started circular cylinder are evidenced and analysed by means of two complementary experimental methods of visualization : dispersed and continuous tracers. The corresponding data are used as references to perfect a numerical process. The two sorts of results are presented as comparative cross and spanwise streamline or streakline patterns visualizing the details of the flow structure; velocity distributions are also deduced. Two values of the Reynolds numbers (1000 and 3000) and of the aspect ratio of the cylinder (5 and 10) have been studied.

Introduction :

Most of flow fields in nature and technology are unsteady, three-dimensional and influenced by viscosity. However, calculation of such flows is really a complex problem and mobilize the efforts of many reseachers around the world. As well, experimentation is also difficult, mainly if precise quantitative data are required.

As an example, we consider, in this work, the flow generated by the impulsive start from rest of a short circular cylinder (of diameter D and length L) fitted with two large symmetric endplates, translating with the constant velocity V_0 in a liquid of kinematic viscosity ν.

This relatively simple and well defined geometry can serve as a 3D model; indeed, several recent investigations [1-2-3-4-5] have shown the high sensibility of flow to the cylinder extremity effects especially for low Reynolds numbers.

The object of the present study is to evidence the way the the 3D effects induced by the endplates modify the flow field and to determine how they set up with time. Two parallel approaches have been developed. One is by means of an experimental flow visualization using dispersed or continuous tracers, the other by means of a numerical visualization resulting from the integration of the 3D unsteady Navier-Stokes equations. The first approach is used to help in the elaboration of the numerical process, after what the two sorts of results complement each other.

For this purpose, the time-evolution of the flow structure has been captured in different cross sections regularly distributed along the $\vec{z}$ spanwise direction and in the middle spanwise section. The results are presented as quasi-instantaneous 2D velocity field and vorticity propagation patterns which complement each other and facilitate the interpretation of the 3D time-development of the flow. Moreover, a quantitative comparison between calculation and experiment, especially by means of velocity distributions along selected directions, show a good agreement for the initial phase of the flow establishment which ensures the efficiency of the proposed numerical technique whose originality lies mainly in the treatment of the pressure equation.

The study has been made for two values of the aspect ratio Al of the cylinder (Al=D/L=5 and 10) and two values of the Reynolds number Re ($Re=V_0D/\nu=1000$ and 3000) ; the case presented in this paper, i.e. Al=5 and Re=1000, has served as a reference. The time-evolution has been followed up to t* (tV_0/D) of 12.0 with $\Delta t^*=0.5$.

<u>The experimental technique</u> :

The cylinder, attached to a heavy towing carrier, translates vertically, under gravity, in a tank filled with a quiescent liquid. Its impulsive start is assured by cutting off the current supplying an electro-magnet which maintains the cylinder in its upper fixed position. The selected cross or spanwise sections are illuminated by a plane light sheet coming from a 5W laser. This laser is put on a horizontal carrier movable on two rectified guides , so that the light sheet may be positioned precisely.

The photographs are taken sequentially by an automatic driving camera which accompanies the cylinder in its displacement.

For the experimental visualization, rilsan particles and dye emitted/electrochemical precipitation of hydrated tin washed off/from the wall are used as dispersed and continuous tracers respectively.

<u>The numerical method</u> :

The numerical integration of the 3D unsteady Navier-Stokes equations, in a velocity-vorticity formulation, is carried out using second-order space and time finite-difference schemes [6]; for the space discretization, a generalized coordinate system suitable for complex geometries is used. In a post treatment, the pressure is obtained by minimizing a discrete norm of the velocity divergence.

Two grid systems of 61*81*61 and 61*81*101 nodes with a time step of 0.02 have been used.

The results :

The influence of the endplates and its progressive spanwise development towards the middle cross section has been evidenced, in a first stage, for Re=1000 and Al=5, by means of complementary visualizations such as those presented in figs 1,2,3. The pictures of fig.1 show the quasi-instantaneous velocity field in three different cross sections ranging from Z/D = 0.5 (near the endplate) to Z/D = 2.5 (middle cross section) and this for normalized time t* of 3.0 (standing eddies) and 6 (outset of the shedding process). Taking into account that each time-evolution is captured from different downruns of the cylinder, it is either the lower or the upper vortex which loses its downstream extremity at first. Its appears that the influence of the endplates remains relatively weak up to t*=3.0, but develops clearly with time and modifies completely the shedding zone in the plate proximity (Z/D=0.5) for t*=6. This is well confirmed by spanwise visualizations like those of fig.2 (solid tracers) and fig.3 (dye filaments emitted from the plate and four other cross sections). The dye filaments explain how the plates interact on the flow by means of the spanwise propagation of the vorticity originating from the plate itself.

Finally quantitative evaluation of these 3D endplate effects in the velocity distribution along the downstream symmetry axis is given in fig.4 .

The different comparisons with the numerical results (fig. 1,2,4) show a good agreement up to t*=4 (time limit for the calculations) except near the plate for which a certain discrepancy remains.

The study has been completed by an analysis of the modifications of these endplate effects by an increase of the aspect ratio of the cylinder and of the Reynolds number.

References :

1. SLAOUTI & GERRARD : "An experimental investigation of the end effects on the wake of circular cylinder towed through water at low Reynolds numbers." - J. Fluid Mech., vol. 112, 297-314, 1981
2. GERICH & ECKELMANN : " Influence of end plates and free ends on the shedding frequency of circular cylinders" - J. Fluid Mech., vol. 122, 109-121, 1982
3. WILLIAMSON : "Oblique and parallel models of vortex shedding in the wake of a circular cylinder at low Reynols numbers." - J. Fluid Mech., vol. 206, 579-627, 1989
4. SZEPESSY & BEARMAN : "Aspect ratio and end plate effects on vortex shedding from a circular cylinder" - J. Fluid Mech., vol.234, 191-217, 1992
5. POLIDORI, PINEAU, ABED MERAIM & COUTANCEAU : "Shedding process of the initial vortices from impulsively started cylinders at Re=1000: end and geometry effects". I.U.T.A.M. Symposium on Bluff BodyWakes, Göttingen, 1992, to be published.
6. LABIDI & TA PHUOC LOC : "Numerical solution of Navier-Stokes equations in velocity-vorticity formulation: application to the circular cylinder" - Notes on Numerical Fluid Mechanics, vol. 20, 159-166, 1988

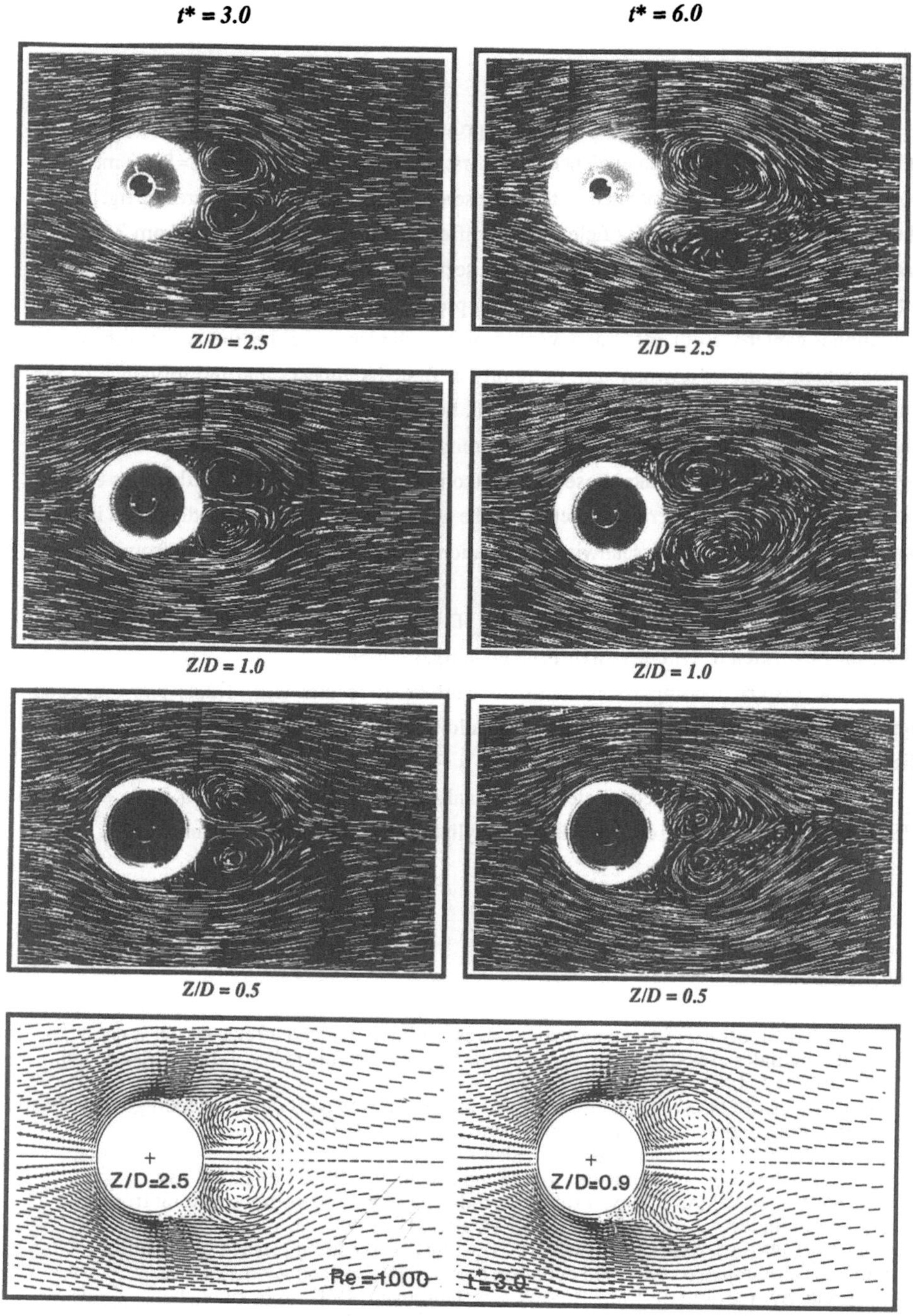

Fig. 1: Flow structure around a short cylinder for Re = 1000 and AL = 5.

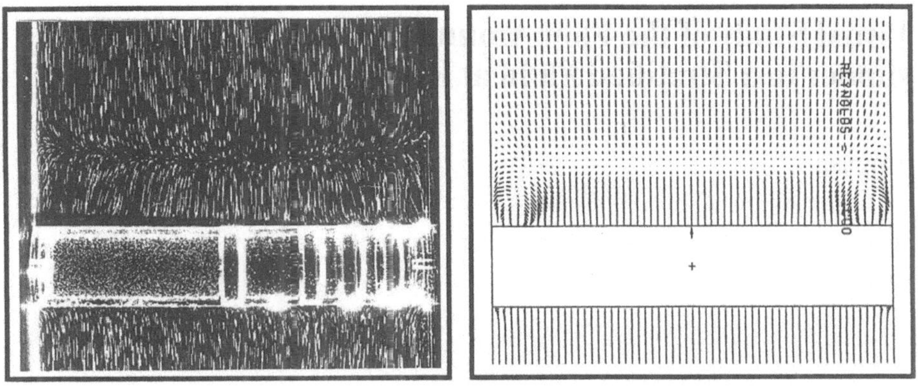

Fig. 2 : Flow stucture in spanwise direction around a short cylinder for Re=1000 and AL=5 - t=3.0 -*

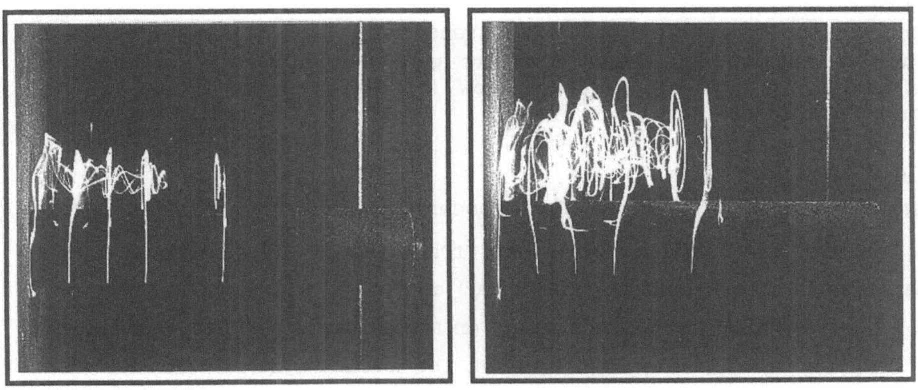

Fig. 3 : Spanwise visualizations (dye filaments) for Re=1000 and AL=5 - t=3.0 and t*=6.0 -*

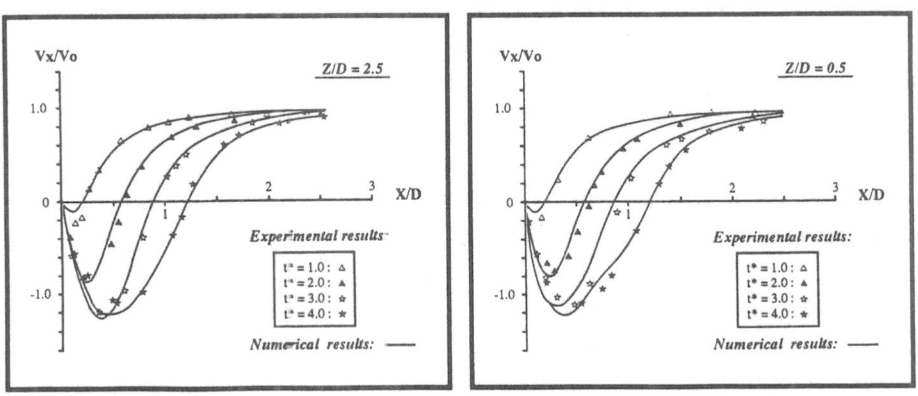

Fig. 4 : Comparison between experimental and numerical results for the velocity distribution along the downstream flow axis for Re=1000 and Al=5

A Visual Study of the Flow Around an Oscillating Cylinder in a Stratified Fluid at Rest

M. TATSUNO

Research Institute for Applied Mechanics, kyushu University
Kasuga, 816 Japan

Summary
When a circular cylinder oscillates transversely in a linearly stratified fluid at rest, the streaked flows are generated alternately along the cylinder axis and each streaked flow follows up a wavy path in the horizontal plane.

1.Introduction

Many investigations have been done concerning the secondary flow induced by a circular cylinder oscillating transversely at relatively large amplitude in a homogeneous fluid (water) at rest. The structure of the flow in a homogeneous fluid depends mainly on two parameters ; $KC=2\pi a/d$ and $\beta = f d^2/\nu$, where a is the amplitude of the oscillations, d the cylinder diameter, f the frequency and ν the kinematic viscosity of the fluid. Honji [1] has shown that a three-dimensional instability of the flow occurs in a certain range of KC and β and streaks arrange alternately in a double row along the cylinder axis. Tatsuno and Bearman [2] have confirmed the appearance of the streaked flow arranging alternately along the cylinder axis in the range $2 < KC < 4$ and $\beta > 50$.

The purpose of this study is to investigate the structure of the flows around a vertical circular cylinder oscillating transversely in a linearly stratified fluid at rest in the range of KC and β at which the streaked flows are produced alternately along the cylinder axis in a homogeneous fluid.

2.Apparatus and experimental methods

Experiments were performed using a tank filled with stably stratified salt solution, in which a vertical circular cylinder was oscillated transversely by means of a scotch yoke mechanism. Two circular cylinders were used with diameters of 1.5cm and 2.0cm. The tank has internal dimensions of 64cm long and 38cm deep by 32cm wide. The non-dimensional density gradient is defined as $\rho^* = H_0 \cdot |\partial \rho / \partial H| / \rho_0$, where ρ and ρ_0 are the local and standard fluid density respectively, and H the reference vertical coordinate and H_0 the depth of the fluid. The electrolytic precipitation method and the aluminium dust method were used for flow visualization. The coordinate system used is shown in Fig.1.

3.Results

When a vertical cylinder is oscillated transversely at small values of KC and/or β, two-dimensional streaming flow is induced around it. Figure 2 shows a view of the flow in the (X,Z)-plane. It is clear, from the dye patterns running parallel to the cylinder, that the flow is two-dimensional and proceeds along the (X,Z)-plane away from the cylinder.

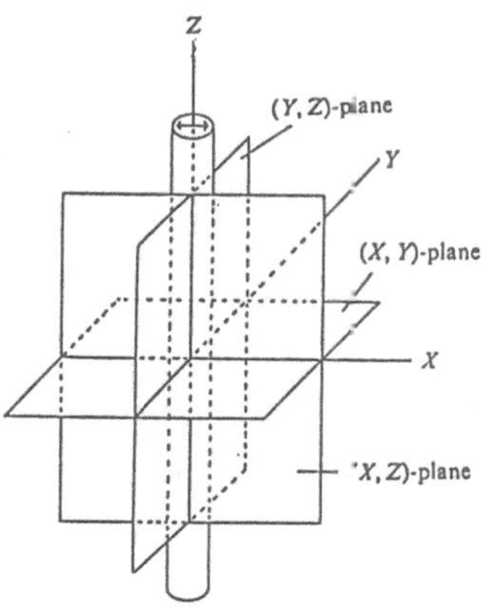

Fig.1. Coordinate system.

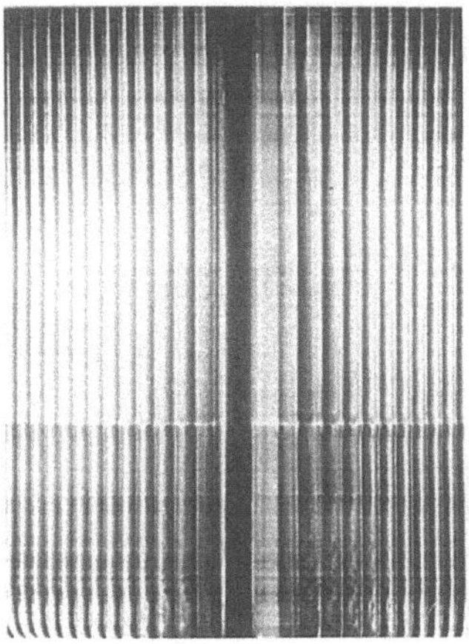

Fig.2. Two-dimensional flow in the (X,Z)-plane, d=1.5cm, a=0.8cm, f=0.313Hz, KC=3.35, β=57.7, ρ^*=1.71$\times 10^{-2}$.

When the values of KC and/or β are increased further, a three-dimensional structure appears near the cylinder surface. The streaked flows are generated alternately along the cylinder axis and each streaked flow follows up a wavy path in a horizontal plane; hence this flow mode may be called the wavy mode'. The wavy trajectories of the streaked flows are shown in Fig.3. In Fig.3, the lighter streak differs in spanwise position from that of the more intensely white streak. The periodic motion of a streaked flow is 180 deg. out of phase with adjacent streaked flows. Vertical views of the flow structure in the (X,Z)-plane are shown

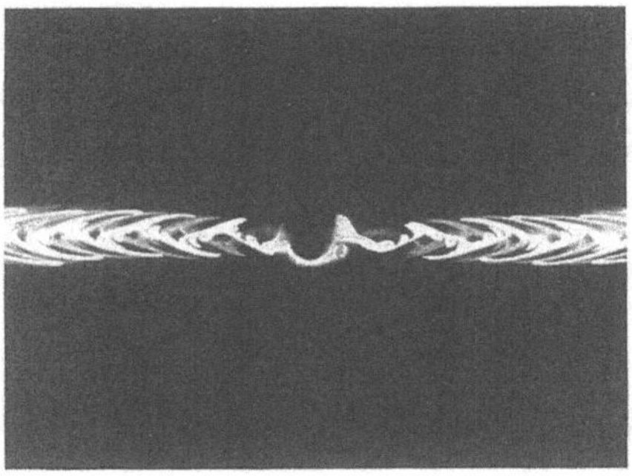

Fig.3. Plane view of flow in the (X,Y)-plane, d=1.5cm, a=0.6cm, f=0.543Hz, KC=2.51, β=139.2, $\rho^*=5.69\times10^{-2}$.

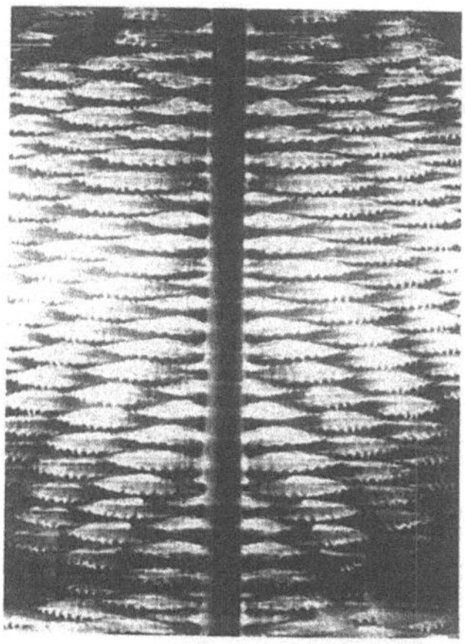

Fig.4. Plane view of flow in the (X,Z)-plane, d=1.5cm, a=0.5cm, f=0.903Hz, KC=2.09, β=199.2, $\rho^*=5.01\times10^{-2}$, without end plates.

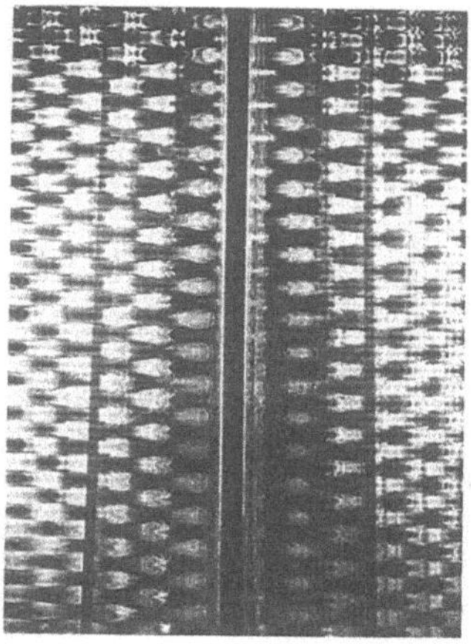

Fig.5. Plane view of flow in the (X,Z)-plane, d=1.5cm, a=0.6cm, f=0.777Hz, KC=2.51, β=129.5, $\rho^*=6.86\times10^{-2}$, with end plates.

in Figs. 4 and 5. Only inphase parts of the periodic trajectory of the streaked flow can be seen, because a vertical thin plane parallel to the direction of oscillations is illuminated through slit in these pictures. It is clear that each streaked flow is 180 deg. out of phase with adjacent streaked flows. Figure 6 shows a cross-section through the streaked flow in the (Y,Z)-plane at X≒2cm. It will be seen from this figure that streaked flows are generated alternately along the cylinder axis.

Figure 7 shows the streaked flow pattern when β is increased beyond the critical value. The streaked flows generated alternately along the cylinder axis do not meander but extend straight in the direction of oscillation in the same manner as in the homogeneous fluid.

In Fig.8, the closed circles show the values of ρ^* and β when the wavy mode of the streaked flows appeared in the horizontal plane at KC=2.51. The open squares and the open circles in Fig.8 show the data for the two-dimensional flow patterns and the alternate streaked flows extending straight, respectively. When the values of ρ^* is small, the wavy mode of

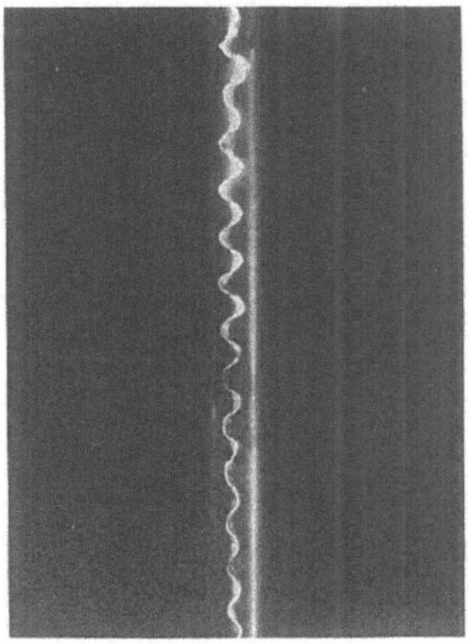

Fig.6. Plane view of flow in the (Y,Z)-plane at X≒2cm, d=1.5cm, a=0.6cm, f=0.548Hz, KC=2.51, β=141.7, $\rho^*=5.69\times10^{-2}$.

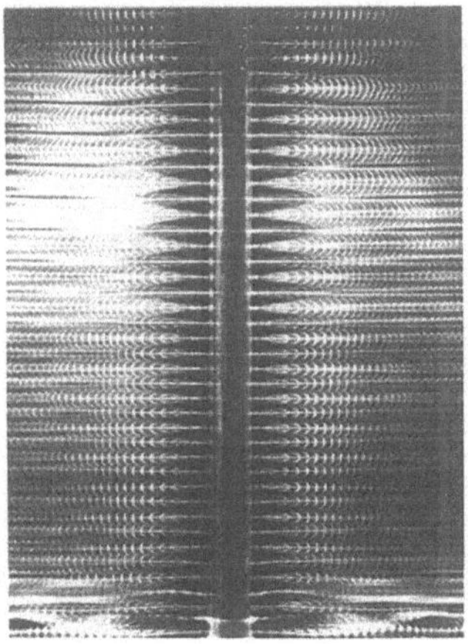

Fig.7. Plane view of flow in the (X,Z)-plane, d=1.5cm, a=0.6cm, f=0.667Hz, KC=2.51, β=123.0, $\rho^*=1.71\times10^{-2}$.

the streaked flows was observed at small values of β. Figure 9 shows the values of *KC* and β at which the wavy mode of the streaked flows was observed for several values of ρ^+. The wavy mode of the flows was observed at small values of β as the value of *KC* was increased.

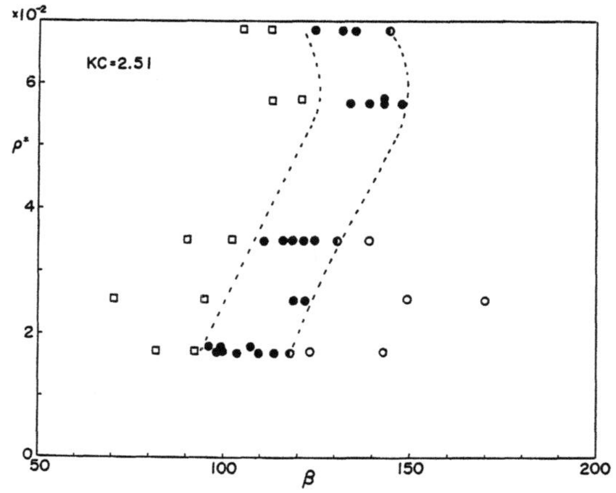

Fig.8. Formation region of the wavy mode.

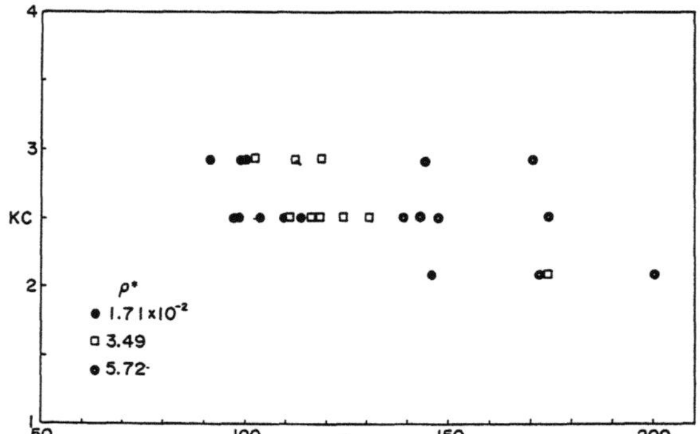

Fig.9. Relationship between *KC* and β for each ρ^+ when the wavy mode appeared.

References

1. Honji, H. : J. Fluid Mech. 107 (1981) 509-520.

2. Tatsuno, M. & Bearman, P.W. : J. Fluid Mech. 211 (1990) 157-182.

Application of an Electrolysis Method for Flows Around Ellipsoids

N. NISHIKAWA, Y.NIIMURA, and S.KIDA

Faculty of Engineering,Chiba University
Yayoi 1-33 Chiba,Japan

Summary

A simple visualization technique,which was proposed by present authors, is applied to ellipsoid,where the electrolysis method is extended with painting the phenolphtalein . The technique provides streamline directions even at unsteadiness in liquid ,especially shows direction of wall shear stress. The vortical structures normal to or lateral to the ellipsoid surface are obtained. The resultis discussed for the range of incidence angle 20 deg to 65 deg. The results as for the critical angles for the onset of closed separation as high as 60 degrees with existing experiment.

Introduction to 3-D Separated Flow over Ellipsoid

Recently,a number of visualization techniques for three dimensional flow have been attempted for a wide range of incidence angle[1,2,3]

Although the separation phenomena have not been visualized in detail and not been interpreted enough clearly.The oilflow method has been used for various flows, however it seems to be appropriate for the flow with large wall shear and the method is rare to give enough clear pattern for the flow in towing tank. An application of visualization technique[4] by electrolysis method is presented here.

The principle of the present technique consist in that coloring with phenolphthalein by the electrolysis is limited only on uncovered 'spots'on the model surface as described later.We can refer the color visualization method using ink dots for a wind tunnel by Langston and Boyl[1]. In the present study clear images of separation line and limiting stream line were obtained ina wide range of angle of attack of the ellipsoid.

Electrolysis method

Usually in this sort of method the cathode or the source of tracer are a metal rod. While in the present study several dots on model surface are used as numerous cathodes. The dots are uncovered spots surrounded by painted and insulated surfaces. The present technique also has the following two principal advantages. First the distribution of dots can be shifted by repainting the model.Thus,in the present method the control of the positions of tracer source is easier than the oil-dot-method and the reproducibility is expected to be larger.Second, if we apply high voltage, the hydrogen bubble can be generated enough for them to accumulate and make a series of tracer and can be used for visualization of vortical flow around the model,if large bubbles do not appear.

The model made of brass is the axisymmetric ellipsoid with 160 mm length and with 1/4 semi-axis ratio. The model is fixed by a 1.5mm radius rod to the bottom of channel. In this study ,when the supplied voltage is lower than 30 Volts or so,hydrogen bubble is rarely seen on the body surface.Magnitude of electric current through the present model is in the range of 0.03 Ampere to 0.15A for the voltage variation from 10 Volts to 50 Volts.

Experimental apparatus and procedure

In the channel with 0.2m width and 0.3m depth, in addition to the model of ellipsoid as the cathode we need the anode of cupper plate of 25mm by 100mm rectangle with 2mm thickness. The plate is located near the tunnel side-wall on the bottom of channel and 100mm downstream of the model. To distribute 'cathode dots',whole surface of the model is painted and dried,and then the small parts of the 'skin' of the paint is removed carefully by needle so that the size of dots may not be larger than 1mm. At these uncovered dots the brass of body can contact with the flowing water and electrolysis can occur.After turn-on of power for electrode the production of the coloured liquid by phenolphthalein is continued during 1 minutes or so enough long for us to take photographs from several directions. At ordinary voltage as 20 or 30 Volts the procedure can be repeated four or five times until the phenolphthalein crystal are fully con-sumed through the repeated tests.

Results and discussion

Here,the results for the following distribution of dots is mainly shown:that is, the axial spacing of the dots is 0.5cm and circumferential pitch is 15 or 20deg.In Fig.1 side view of surface flow pattern is shown for α =20 deg.,where we can see an converging point of streamlines near the 1/3 length,i.e.at (x,θ)=(0.68,105 deg.) ,

where x measured from upstream vertex.Although the starting point seems to be more easily observed at $\alpha > 30$ deg even at $\alpha = 20$ deg streamlines from leeside symmetry line and those from windward side are converging and starting run-together. As we examined in the previous study[4] for paraboloid this 'co-running' line can be identified as primary separation line S_1 ,although converged point is very ambiguous if we only see Fig.1 and whose sketch in Fig.2. At incidence $\alpha = 30$deg the S_1 line starts at $(x, \theta) = (0.5, 110$deg,) as shown in Fig.3 and Table 1, which summarizes the present results Although in the sketch in Fig.2 we can already find another separation line S_2,the secondary separation line at $\alpha = 30$ deg starts at $(x, \theta) = (0.5, 150$deg), whose starting point shifts upstream as α increases. If we compare the side views for $\alpha = 20$ deg and $\alpha = 30$ deg ,S_1 for $\alpha = 20$ deg S_1 has larger curvature until it approaches side of body around $\theta = 90$. and then gradually changes cicumferemtial location toward rearend . In other words the region with large curvature-streamlines at forebody shrinks as α increases and S_1 on side view has straightline as in Fig.4.for $\alpha = 60$ where the starting point of S_1 locate nearer to the leeside symmetry line.
This fact seem to be attributable to the weak reversal flow from leeside at smaller α ,while at larger α ,a streamline is pushed by S_2 or is not accessible to leeside region. It is known that at $\alpha = 0$ the separation line runs orthogonally to the generators at rear end of body . Therefore present results can be interpreted as follows,i.e.the starting point of S_1 shifts away from rear-leeside symetry line as α increases,and reaches nose-leeside-symmetry line at $\alpha = 65$ as the sketch in Fig.5 from our photograph.

The present results are compared with the visualization for ellipsoid by Han & Patel[2] for 4:1 axis ratio and by Bippes[3] for 6:1 axis ratio ellipsoid at Re=3×10^6 for water tunnel. Reynolds number in the present study is so low as 7000 referred to the ellipsoid length at the oncoming velocity 5.2 cm/sec. However the essential flow patterns are not different from results by abovementioned authors[2,3],except the critical values of incidence angle agrees only with Bippes[3],where closed type separation appears.
The behaviors for the ellipsoid contrast to the patterns on paraboloids[3] is as follows,which induced by the decreasing crosssection at aftbody of ellipsoid.
In the narrow region between the primary and secondary separation line the strong axial flow appears away from the rearvertex.This large velocity region consist the upper part of the elongated and

clockwise vortical pattern between S_1 and S_2 At a range of incidence angle the counterclock-wise vortex attaches to these vortical pattern near the rear end of body. This pattern can be observed at incidence angles $\alpha < 50$ and at larger α ,where a strong axial flow occuppies whole region between S_1 and S_2.In Fig.6 the crossesction of leeside-vortex is shown by a light sheet at $\alpha = 30$ deg.

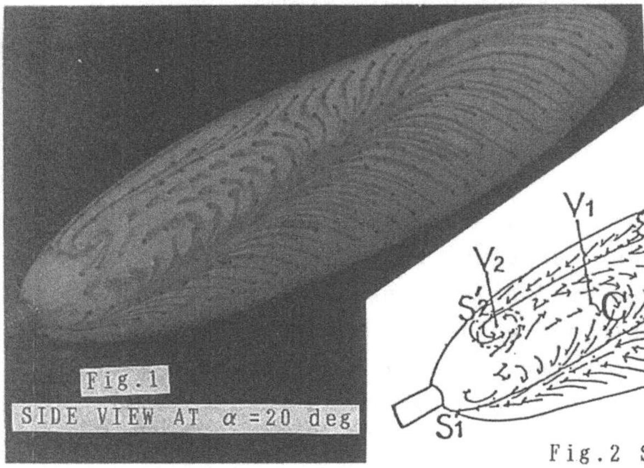

Fig.1
SIDE VIEW AT $\alpha = 20$ deg

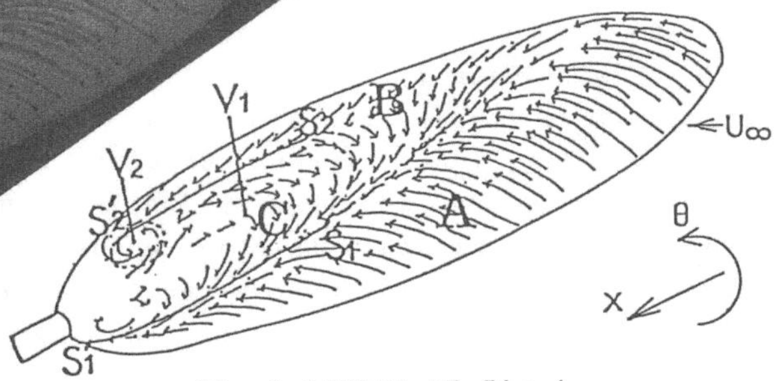

Fig.2 SKETCH OF Fig.1.

A:Windward Unseparated Region
B:Crossflow Reversal
C:Vortex between Two Sep.LIne
V_2:Small Vortex attatced C area

α	S_1 θ	X	S_2 θ	X
20°	105°	0.68	150°	0.71
30°	110°	0.50	150°	0.50
40°	120°	0.29	150°	0.41
50°	120°	0.18	150°	0.29

TABLE 1
STARTING POINT S_1, S_2

Fig.3 SIDE VIEW AT $\alpha = 30$ deg

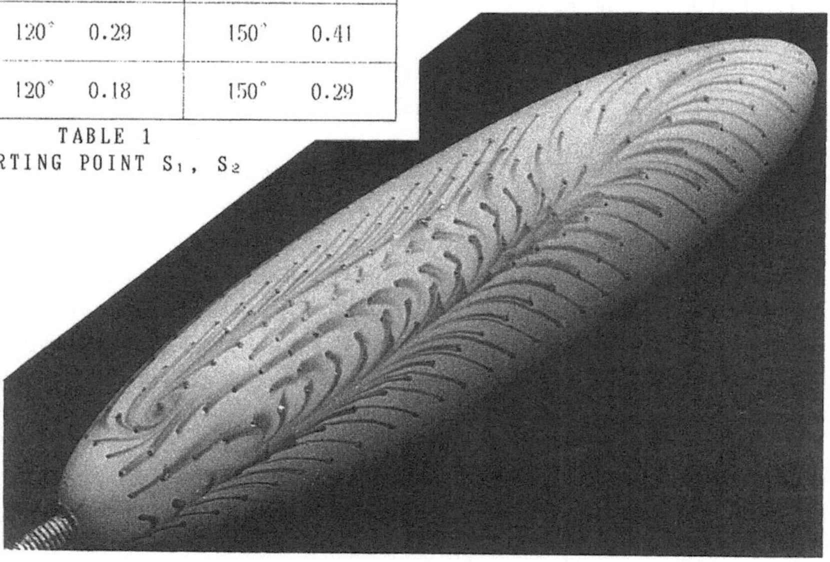

Conclusion

1. The clear surface streamlines can be visualized by the present visualization technique for the ellipsoid at the range of incidence from 20 deg to 65 deg.
2. Although the Reynolds number is as low as 7000 in the present study, the onset of the closed separation line, which is identified here at 60 deg, is analogous to the visualization by Bippes[3].

Fig.4 SIDE VIEW AT α =55 deg

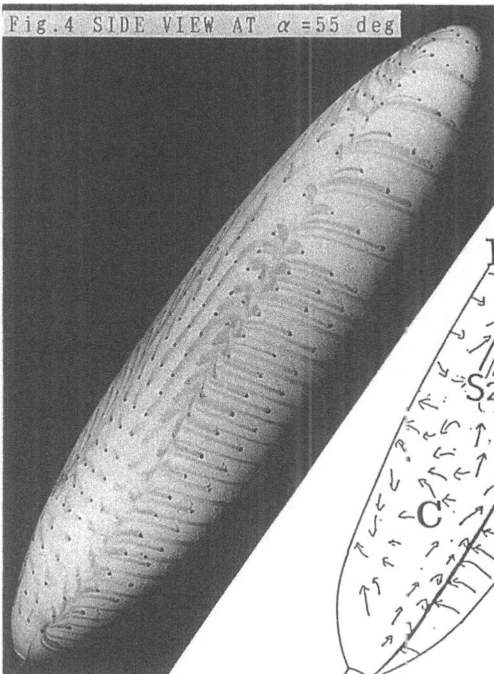

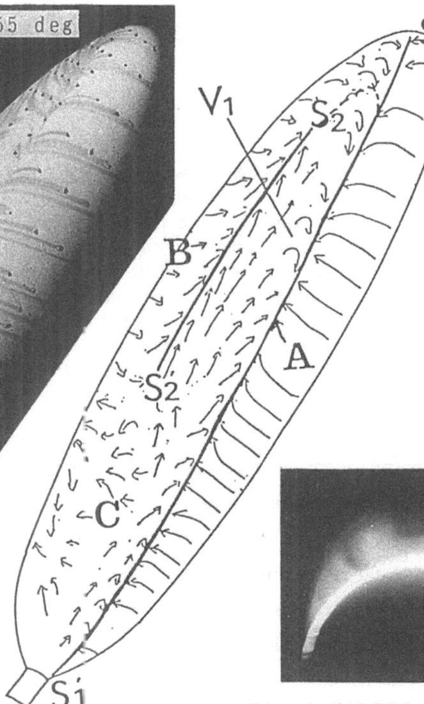

Fig.5 SKETCH for α =65 deg

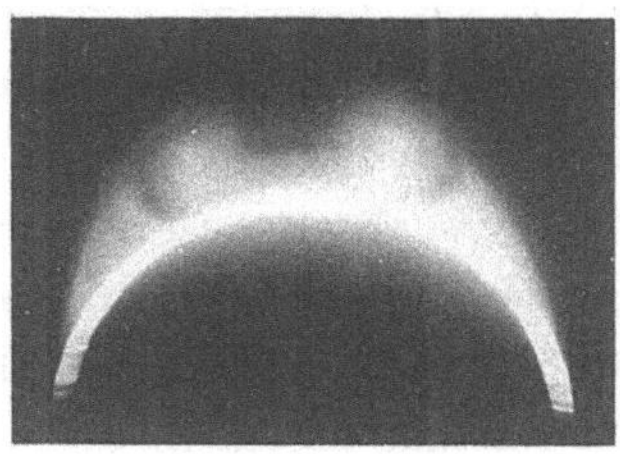

Fig.6 VORTEX CROSS SECTION AT α =30 deg, x=8cm

References

1. Langston,L,S, and Boyle,M.T,A new surface-streamline flow visualization technique,J.Fluid Mech.,Vol 125,pp53-57,1982.
2. Han,T.and Patel,V.C. Flow Separation on a Spheroid at Incidence,Journal of Fluid Mech.,Vol.92 p643,1979
3. Bippes.H,Visualization of the Flow along a Prolate Spheroid at High Incidence in the Water Towing Tank of The DFVLR, --DFVLR Rep.-1B-251-79-a18,1979.
4. Nishikawa,N.and Suzuki,T. Trans.,Japan Soc.Aeronautical and Space Sciences,Vol 31,No.91 pp61-66(1988).
5. Ishii,Y., and Oshima,Y.,Three Dimensional Separation on a Spheroid at Incidence,Inst.-Space-A-Sci.-Report SP 4, pp43-51,1986.

Flow Visualization of Karman Vortex in Shear and Turbulent Flow

S.Saito, M.Hashimoto Y.Higuchi

Dept. of Mechanical Engineering,
Science University of Tokyo

1. Introduction

Karman vortex velocimeter and flowmeter by use of laser diode have been developed in our laboratory [1),2)]. However the effects from such as shear flow, turbulent flow, channel wall and etc. are little known. In fact, the frequency of the Karman vortex in the shear flow which was measured by the method developed in our study was lower than that of the uniform flow. As a result, the value of velocity which was measured by Karman vortex velocimeter did not accord with the value of that which was measured by Pitot tube. This phenomenon have to be removed for the development of the velocimeter. To investigate these phenomena and to improve them, flow visualization was carried out.

2. Experimental apparatus and methods

Water was used as a working fluid. The test section of the channel was rectangular shape and had dimensions of 0.1m in width, 0.1m in depth and 1.8m in length. The columns (diameter; 3mm and 4mm) were used as the vortex generator. The flow visualization were carried out by streak line method and time line method. The former injected the dyestuff from the upper stream and the latter oozed the dyestuff from the generator which was drilled small holes along its axis.

The Karman vortex in the shear flow was visualized, and flow pattern of the Karman vortex and the frequency was compared with those of the uniform flow. The experiment of the vortex generator putting two plates with small distance was carried out for the aim of the improvement of the vortex shedding in the shear flow (Fig.1).

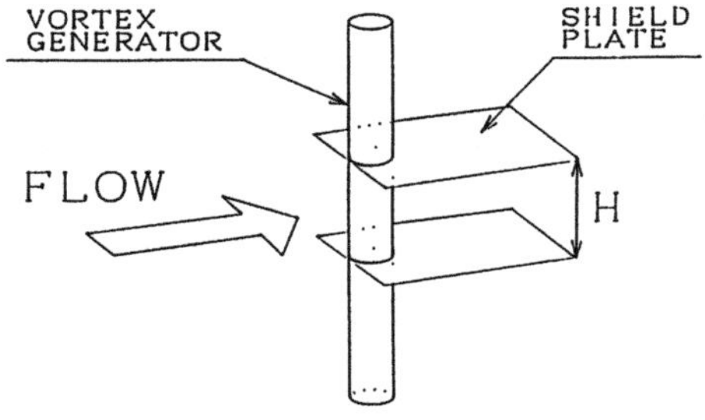

Fig.1 Vortex generator with plates

3. Results and discussion

3-1 Uniform flow

The side view of the Karman vortex for the uniform flow is shown in Photo.1 which is taken by time line method. The vortex occurred with same phase along the generator, and the vortex filament was perpendicular to the flow. The pictures for the uniform flow are shown in Photo.2(a) and 2(b), which are taken by streak line method. The vortex shedding is very stable in the uniform flow.

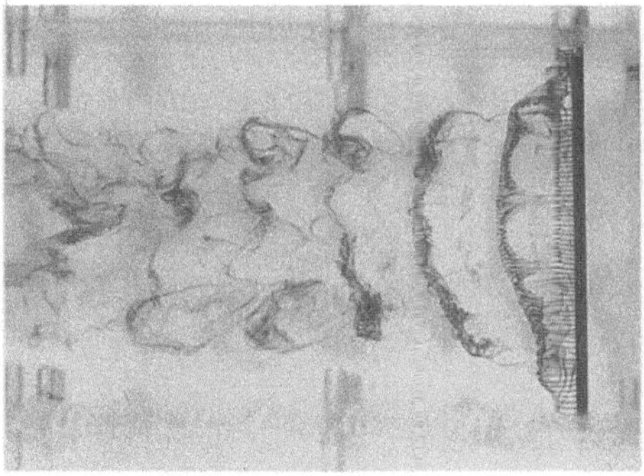

Photo.1 Uniform flow (D=4mm,Re=106)

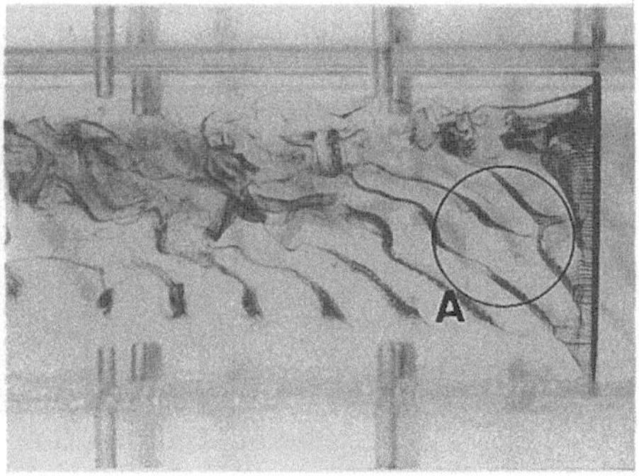

Photo.3 Shear flow (D=3mm,Re=106)

3-2 Shear flow

Photo.3 shows the side view of the Karman vortex for the shear flow (Constant velocity gradient 0.176 (1/s)). The velocity at the center of the vortex generator is almost same one with Photo.1. This picture shows that the conditions of vortex shedding are completely different from Photo.1. The vortex filament is tilted, because of the phase difference of the vortex along the generator as shown in circle A of Photo.3. As this phase difference becomes larger, the vortex filament is cut off periodically. Photo.4(a) and 4(b) show the photographs which are equivalent one to Photo.2(a) and 2(b). The secondary flow is recognized (Photo.4(b)) and therefore the vortex pattern is obscure (Photo.4(a)).

Fig.2 shows the frequency of the Karman vortex for the uniform flow and the

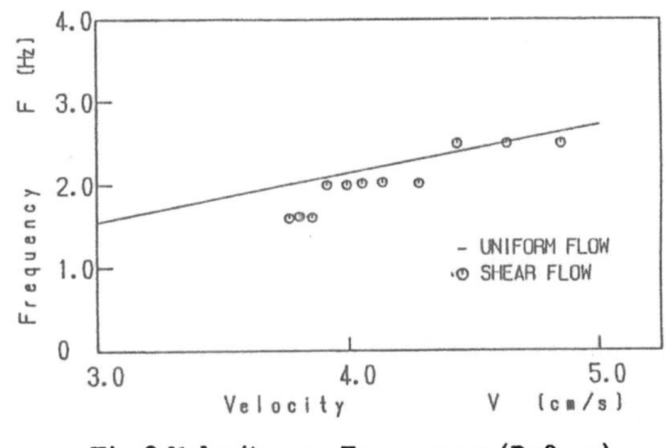

Fig.2 Velocity vs. Frequency (D=3mm)

shear flow. In this figure it is seen that the vortex shedding in the shear flow is discrete and synchronizes at a certain range. This phenomenon accords with the above result which the vortex filament is cut off periodically. It is also revealed that the frequency in the shear flow is lower than the value in the uniform flow as the result of the secondary flow.

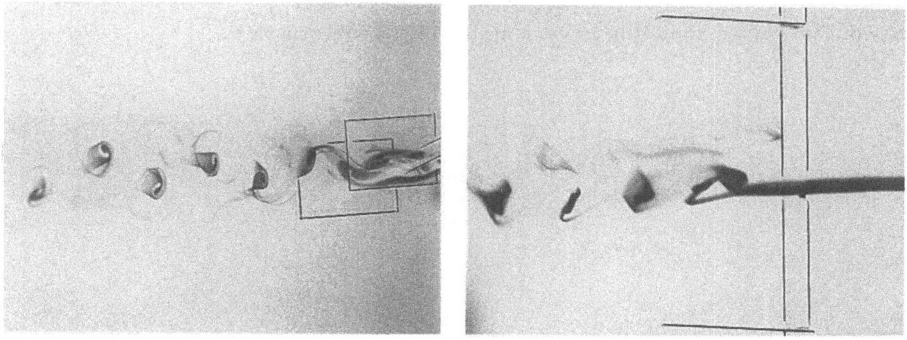

(a) (b)

Photo.2 Uniform flow (D=3mm,Re=99.8)

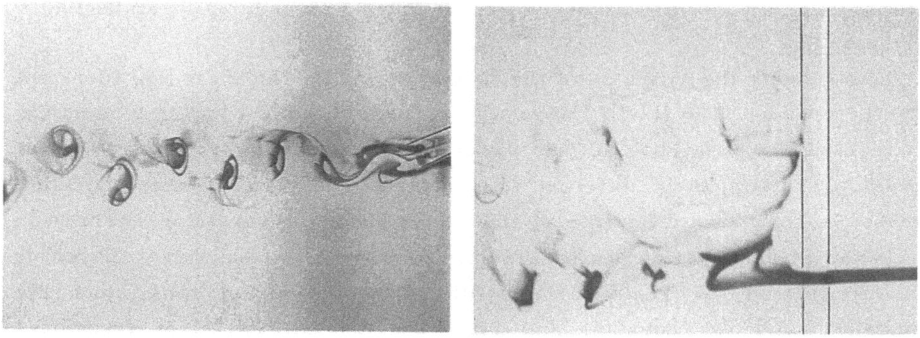

(a) (b)

Photo.4 Shear flow (D=3mm,Re=99.2)

3-3 Vortex generator with plates in the shear flow

In this section, to shield the secondary flow which influences unfavorable effect on the velocimeter, two parallel plates are put normal to the generator with small distance. Photo.5(a) and 5(b) show the photographs that the distance between two plates is 40 mm. As compared with Photo.4(a) and 4(b), the secondary flow is restrained a little (photo.5(a)). But the conditions of vortex shedding dose not change significantly, and the vortex pattern is still obscure.

Photo.6(a) and 6(b) are the photographs that the distance between two plates is 30 mm. From these pictures, it is recognized that the secondary flow was able to

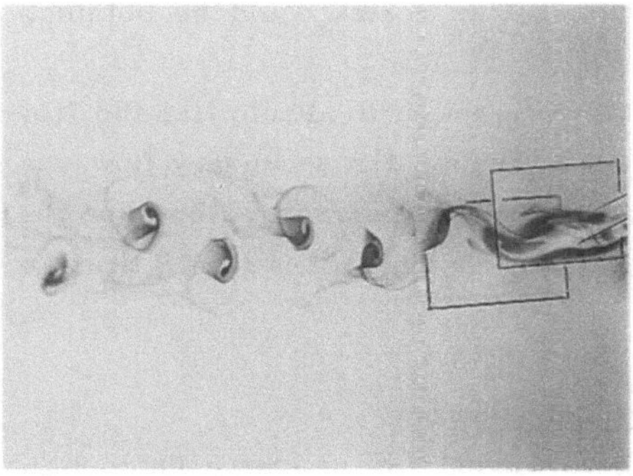

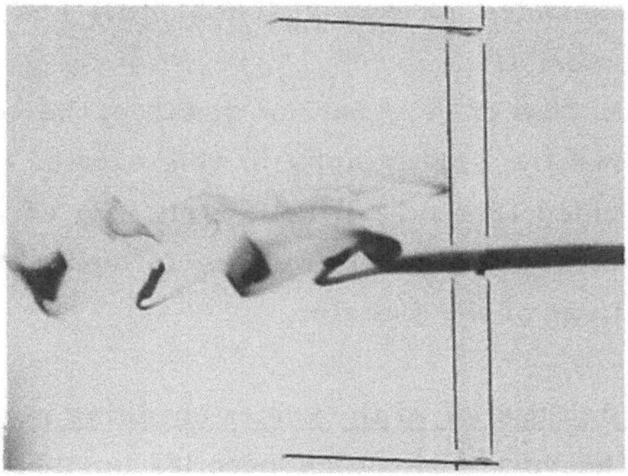

(a) (b)

Photo.5 Shear flow with plates (D=3mm,H=40mm,Re=99.8)

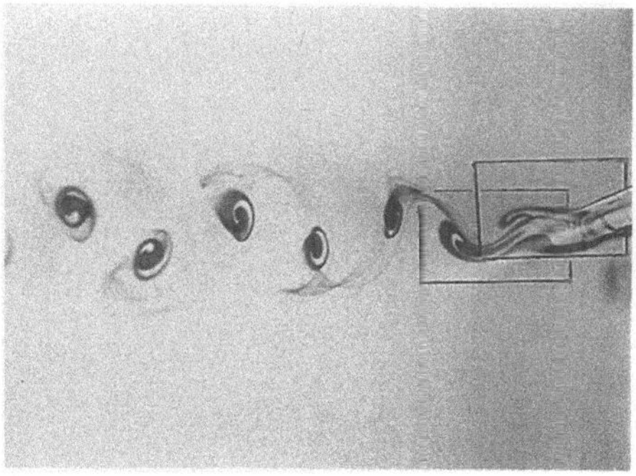

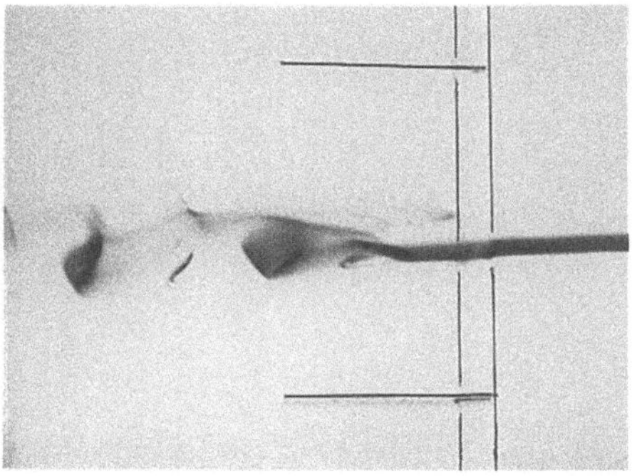

(a) (b)

Photo.6 Shear flow with plates (D=3mm,H=30mm,Re=99.2)

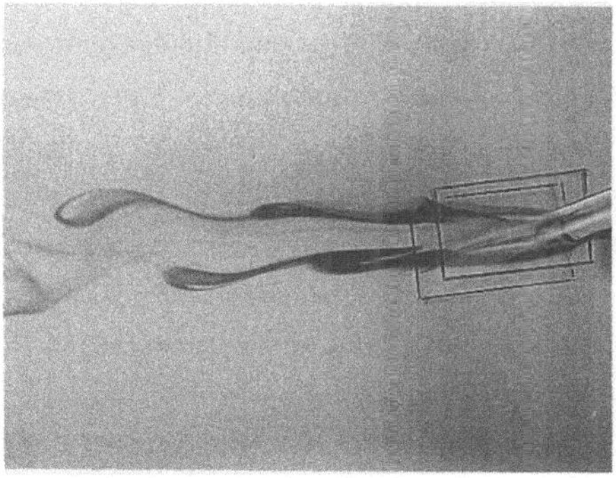

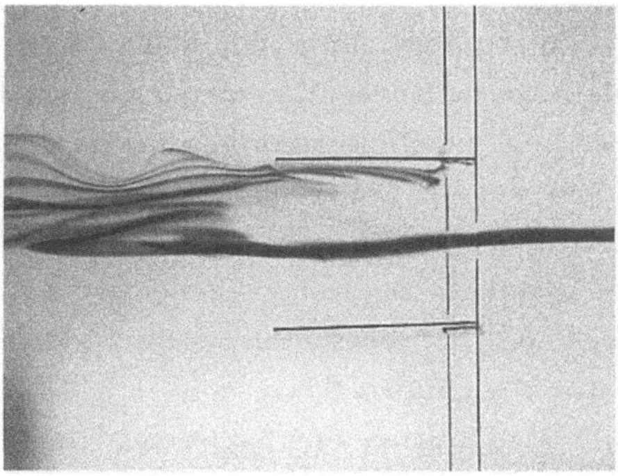

(a) (b)

Photo.7 Shear flow with plates (D=3mm,H=15mm,Re=99.2)

be shielded obviously (Photo.6(b)) and much better vortex could be obtained (Photo.6(a)).

In case of very narrow distance, the vortex dose not occur. Photo.7(a) and 7(b) shows the photographs in this case of H=15mm. Though the secondary flow was shielded (Photo.7(b)) completely, the vortex was greatly disarranged because the boundary layer of two plates influenced. The optimum distance is from 8 times to 10 times of the diameter.

4. Stabilization of the vortex shedding in turbulent flow

The vortex shedding becomes unstable in turbulent flow as shown Photo.8(a). Photo.8 (b) shows one the example of the stabilization of the shedding which was carried out by setting two vortex generators in series and the aluminium honeycomb between the two generators.

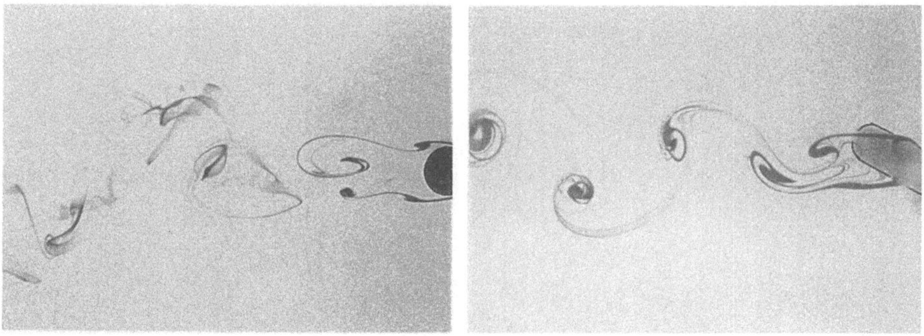

(a) Single generator (Re=338) (b) Two generator (Re=355)

Photo.8 Stabilization of the vortex shedding in turbulent flow (D=3mm)

5. Conclusions

(i) In the shear flow the vortex filament is tilted with the phase difference. As this becomes larger, the vortex filament is cut off. By this phenomenon the vortex shedding synchronizes at a certain range.

(ii) Because of the secondary flow just behind a vortex generator, the frequency in the shear flow becomes lower than that in the uniform flow. However, by putting two plates to the vortex generator with small distance, the secondary flow is shielded. As a result, the vortex becomes stable and its frequency approaches to that in the uniform flow.

(iii) Stabilization of the vortex shedding in turbulent flow was undertaken successfully.

References

1). Saito,s. et al., Tras. of JSME(Japanese) Vol.53, No.496, p3667 (1987)

2). Saito,s. et al., FLUCOME'91 ASME p571 (1991)

Study of Holography Applied to the Piles Structure for Wave Damping in Sea Shore

Kazuhide Dan
Department of Civil Engineering
Akashi College of Technology
Akashi , Japan

and

Motoaki Yano
Department of Civil Engineering
Kobe Univ.
Kobe , Japan

Summary

The purpose of this study is to design the piles structure by using the holographic theory for wave damping in the desired sea shore region.We shall consider holography as the photographic recording of an interference pattern between two sets of light waves,an information wave and a reference wave. These waves are not always light waves in holography. The real sea wave makes also the interference pattern and we can also apply the holographic theory in sea wave.We can obtain the real and virtual image in front of the hologram and behind the hologram respectively in the optical holography when we illuminate the reconstructive wave to the hologram.When we view these images we consider as if there were the object in the region of these images.We can do the same experience in sea wave and we think as if there was the object in the regions of images and it is comparatively still in these regions.Thus we apply the holographic theory to the wave damping in the desired sea shore region.

Introduction

This study is based on the inline holography.The real sea wave plays a role of the incident wave in this holography.An example of the object is an island.The diffractive wave from the island is the information wave in this holography.We have three steps when we apply the holographic theory in order to control the wave height in the desired sea shore region(Fig.1).The first step is to compute the wave height distribution at the hologram position.The second step is to transform this wave height distribution to the piles structure.The third step is to apply the reconstructive wave equivalent with the incident wave to this structure in order to reconstruct the real image.This reconstructed real image of the island can be obtained behind the piles structure.We consider that this reconstructed real image becomes the desired wave damping region.

Diffractive waves from the piles structure

The wave length,the wave period and the water depth used in this study are the data measured in the front zone of Kobe harbor.

When we apply the incident wave to the piles structure the diffractive waves occur from every pile. When these diffractive waves travel to the neighbor piles every new diffractive wave occurs from every neighbor pile. Thus these diffractive waves occur infinitely from every pile. This diffractive process is shown in Fig.3. The total sum of these diffractive waves and the incident wave becomes the last state of this case. We introduce the velocity potential. The wave height distribution is expressed by the summation of these potential data.

The potential function $\phi_0{}^m(r,\theta)$ for the incident wave is expressed as

$$\phi_0{}^m(r,\theta) = J_0(kr_m) + 2\sum_{n=1}^{\infty} i^n \cdot J_n(kr_m) \cdot \cos(n\theta_m) \tag{1}$$

where r_m is the distance from the origin that is the center of pile m and θ_m is the argument in polar coordinates system(Fig.2) and Jn() is Bessel function. m=1,2,...,N. N is the total number of piles. $i=\sqrt{-1}$.

The potential function $\phi_p(j,m)$ for the j-th diffractive wave which occurs when the incident wave travels first to the pile p is expressed as

$$\phi_p(j,m) = \sum_{n=0}^{\infty} \alpha_b(j,n) \cdot H_n{}^{(2)}(kr_b) \cdot \cos(n\theta_b) \tag{2}$$

where $H_n{}^{(2)}()$ is Hankel function. We decide the coefficient $\alpha_b(j,n)$ by using the boundary condition $\partial\phi/\partial n=0$ where n is the normal. The coefficient $\alpha_b(j,n)$ is expressed as

$$\alpha_b(j,n) = -\alpha_s(j-1,n)\left.\frac{-H_n{}^{(2)}(kr_s)n\sin(n\theta_s)\dfrac{\partial\theta_s}{\partial r_b} + \cos(n\theta_s)k\left\{\dfrac{n}{kr_s}H_n{}^{(2)}(kr_s) - H_{n+1}{}^{(2)}(kr_s)\right\}\dfrac{\partial r_s}{\partial r_b}}{k\cos(n\theta_b)\left\{\dfrac{n}{kr_b}H_n{}^{(2)}(kr_b) - H_{n+1}{}^{(2)}(kr_b)\right\}}\right|_{r_b=R_b} \tag{3}$$

$$\theta_s\Big|_{r_b=R_b} = \begin{cases} \tan^{-1}\dfrac{R_b\sin\theta_b - P_{sb}}{R_b\cos\theta_b} & s>b \\[2ex] \tan^{-1}\dfrac{R_b\sin\theta_b + P_{sb}}{R_b\cos\theta_b} & s<b \end{cases}$$

$$
\left.\begin{array}{lll}
r_s^2 = \dfrac{r_b \pm P_{sb}\sin\theta_b}{\sqrt{(r_b\cos\theta_b)^2+(r_b\sin\theta_b)^2}} & \begin{array}{l} s>b \\ s<b \end{array} & \begin{array}{l} - \\ + \end{array} \\[2ex]
\dfrac{\partial r_s}{\partial r_b} = \dfrac{r_b \pm P_{sb}\sin\theta_b}{\sqrt{(r_b\cos\theta_b)^2+(r_b\sin\theta_b \pm P_{sb})^2}} & \begin{array}{l} s>b \\ s<b \end{array} & \begin{array}{l} - \\ + \end{array} \\[2ex]
\dfrac{\partial \theta_s}{\partial r_b} = \dfrac{\mp P_{sb}\cos\theta_b}{(r_b\cos\theta_b)^2+(r_b\sin\theta_b \pm P_{sb})^2} & \begin{array}{l} \text{upper sign} \\ \text{lower sign} \end{array} & \begin{array}{l} s>b \\ s<b \end{array}
\end{array}\right\} \quad (4)
$$

where the symbol $|r_b=R_b$ implys the value at the condition of $r_b=R_b$.

The wave height distribution H will be written by the summation of these potential data. H is expressed as

$$
H = \frac{2\phi_0\,\sigma}{g}\,|\,\phi\,|\cosh(kh)
$$

$$
\phi = \phi_0^1 + \sum_{p=1}^{N}\sum_{j=1}^{\infty}\sum_{m=1}^{m(j)}\phi_p(j,m) \qquad (5)
$$

Where ϕ_0 is the constant, k is the wave number, σ is the angular frequency, h is the water depth, g is the gravity acceleration and $m(j)=2^{j-1}$.

Making piles structure

The diameter of piles is 4 m. The distance l between the hologram position and y-axis is 103.5 m(Fig.1). We introduce the following rule when we transform this wave height distribution to the piles structure. The rule is that the density of piles arranged in straight line is high in the low wave height intensity or low in the high wave height intensity.

An example of the desired wave damping region is one circular island whose diameter is 100 m. We calculate the wave height distribution at the hologram position in case of one circular island. We use the rule explained above for arranging the piles and show an example of piles arrangement in Fig.4.

Reconstruction

When we apply the incident wave to these piles we have the following way to calculate the wave height distribution behind the piles structure. By summing up the velocity potential data of all diffractive waves and the incident wave we can calculate the wave height distribution(Fig.5).

Experiment in model

The scale is 1/250.Tank in model is 80 cm×180 cm large.The water depth in model is 4 cm.Froude's similarity requires that model's period is 0.25 sec.The diameter of piles is 1.7 cm.There is the wave generator 49.5 cm far from the piles structure.We measured the wave heights by using the wave height meter.The maximum wave height distribution from the first wave to 20-th wave is shown in Fig.6.We can consider that there is the reconstructed region in Fig.6.The object zone(a circular island) is also shown in Fig.6 by a dotted line.

Conclusion

We can decide easily the arrangement of piles structure for wave damping in the desired sea shore region by using the holographic theory.

References

1.K.Dan and M.Yano:Computer Holography in Coastal Engineering,pp.65-70,Journal of the Visualization Society of Japan,Vol.9,1989.
2.K.Dan and M.Yano:Application of Holographic Theory to the Wave Height Control in Coast,pp.137-142,Journal of the Visualization Society of Japan,Vol.10,1990.
3.K.Dan and M.Yano:Wave Height Control in Coast using Holographic Theory,pp.109-114,Journal of the Visualization of Japan,Vol.11,1991.

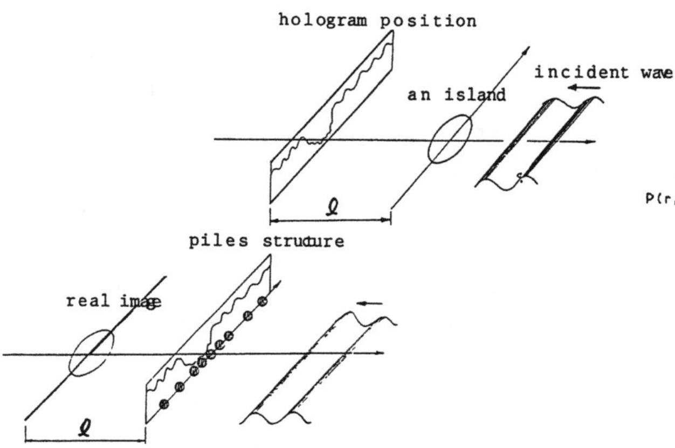

Fig.1 Sketch of Water wave holography

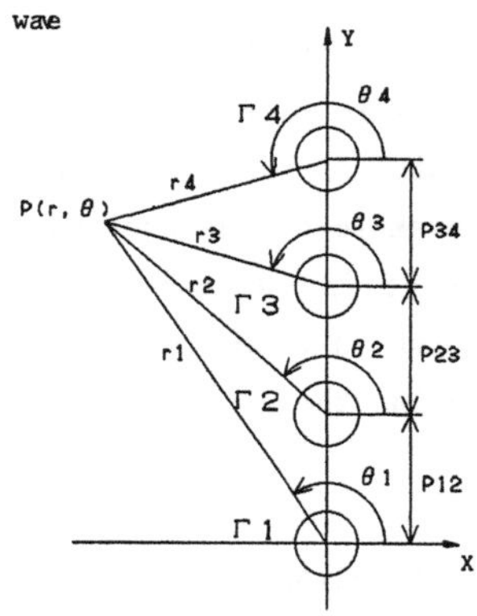

Fig.2 Coordinates

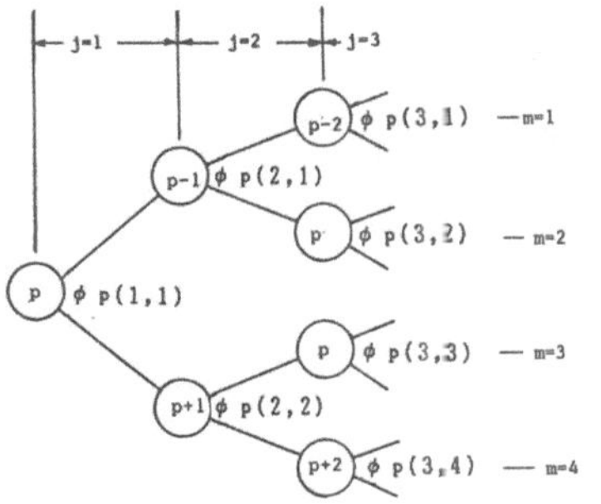

Fig.3 Diffractive process in many piles.
The first diffractive wave $\phi_p(1,1)$ occurs when the incident wave travels to the pile p.The second diffractive waves $\phi p(2,1)$ and $\phi p(2,2)$ occur when the first diffractive wave travels to the neighbor piles p+1 and p-1.....

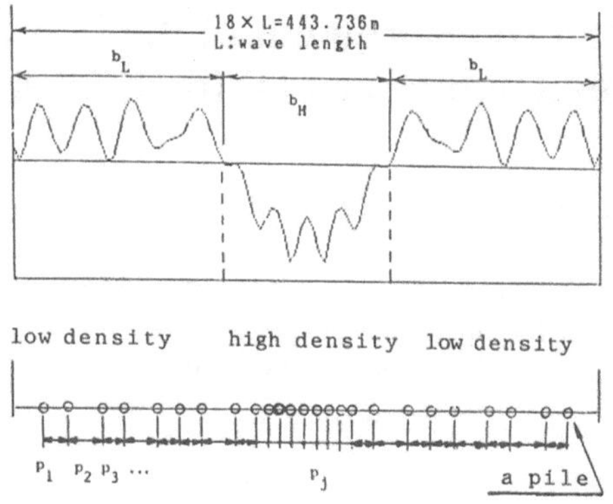

Fig.4 Wave height distribution at the hologram position and an example of piles arrangement

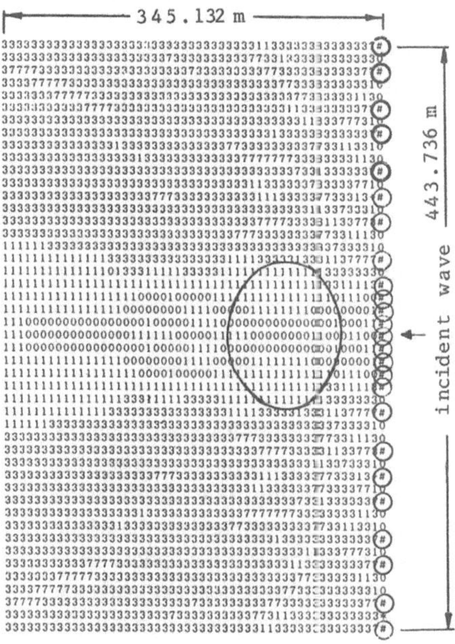

Symbol	Wave height
0	0.713 ~ 0.827
1	0.827 ~ 0.941
3	0.941 ~ 1.054
7	1.054 ~ 1.168

○ — a pile

Fig.5 Strict solution

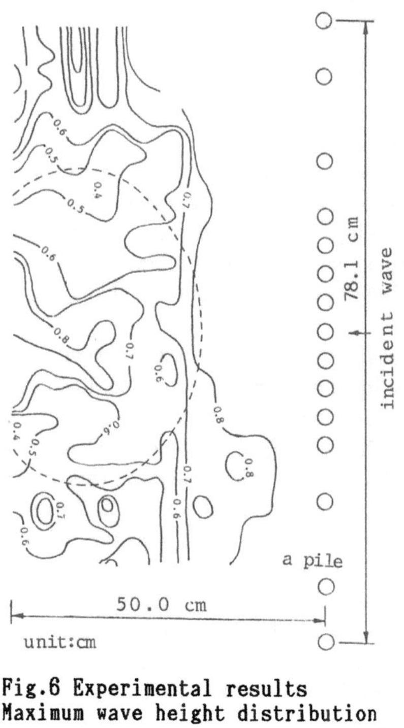

Fig.6 Experimental results
Maximum wave height distribution

Direct Measurement of the Density Field Using High Speed Differential Interferometry

J.M. DESSE and J.C. PEGNEAUX

Office National d'Etudes et de Recherches Aérospatiales
Institut de Mécanique des Fluides de Lille
5,Boulevard Paul Painlevé, 59000 Lille, France.

Abstract
The optical method presented here is based on Wollaston biprism differential interferometry using a white light source. It yields high-speed, instantaneous interferograms which directly display the isochoric lines of an instationary, two-dimensional flow. The setup is equivalent to conventional separate-reference interferometric systems, but remains differential since one of the beams - the one used as reference - lies inside the undisturbed upstream flow. The method was used to get instantaneous, high speed displays of the instationary flow around a cylinder placed across the test section.

1 Introduction

Interferometry has proven to be an excellent tool to analyse instationary aerodynamic phenomena since it provides a method for getting quantitative data on the phenomena to be studied such as density or temperature. If the variations in the index of refraction are small, or if the index itself is small, differential interferometry using a Wollaston biprism seems to be a well-suited display technique for studying two-dimensional or even axisymetric flows.

The works carried out by Desse [1] have shown that instantaneous interferograms could be obtained in white light at high speed. Each interferogram was analysed from the calibration curve linking the various colors observed in the test section and the position of Wollaston biprism. This technique has some drawbacks: the presence of the mock-up prevents the whole observed field from being analysed. This problem can be solved if one uses the optical setup proposed by Desse and Pegneaux [2]. This setup includes two interferometers and makes it possible to analyse the whole density field around the obstacle. The two interferograms are obtained at the same time and the two interference fringes are not positioned in the same way on each device : one of the interferograms is recorded with the vertical bands and the other with the horizontal bands. Since the optical method yields the first derivative of the density, analysing the interferograms requires integrating along the direction perpendicular to the interference fringes, and knowing the value of the density at the start of integration. Using the two interferograms thus makes it possible to get the whole field around a cylinder. One can also show that measurement accuracy significantly decreases as integration distance increases.

The purpose of this paper is to present a differential inteferometry technique which no longer requires integration, which offers the advantage of displaying directly the lines of same density and of increasing measurement accuracy. To this effect, the differential interferometer setup includes a Wollaston biprism which highly separates the interfering beams. This type of setup is equivalent to conventional separate-reference interferometry systems as interferograms can be interpreted directly, but remains differential since one of the beams - the one used as reference - lies inside the undisturbed upstream flow.

Some articles in the literature deal with highly-separated beam differential interferometry. In Smeets' works [3], the displayed field is on the order of 13 mm wide and the birefringent system used is a quartz with a large α angle (twenty degrees or so). The method of analysis presented by Smeets is not a simple one, but it gives the distribution of the medium's index of refraction. Also worth mentioning is Carlomagno's work [4] where the partial beams are highly separated by going through Wollaston prism five times. Carlomagno obtains a 20 mm field with a birefringence angle of 0.14 10^{-2} rd, and he plans to use a calcite biprism in future tests in order to reduce the optical defect problem encountered as the beams go through the biprism several times.

2 Description of the Mock-up

We chose to display the isochoric lines of the instationary flow around a cylinder 20 mm in diameter placed across the test section of the IMFL's transonic wind tunnel. This facility is a continuously operating return flow wind tunnel. The two-dimensional test section is 200 mm high and 42 mm wide. The permeability of the high and low walls is adjustable. The cylinder is equipped with an instationary pressure sensor with a flush membrane area on the order of 1 mm^2 area. A pressure tap is located along the same cylinder generating line as the sensor in order to get comprehensive information on pressure. The tap and the sensor are at a 90° azimuth with respect to the direction of the upstream flow.

3 Optical setup

The schlieren interferometer whose operation is described in detail in Gontier [5] has the optical rays which cross twice the test section and offers the advantage of compact size while doubling the sensitivity. Adjustment is quick and easy as no compensation is needed for the optical thickness of the test section windows since both beams pass through them. In the schlieren interferometer the biprism is placed at a right angle with respect to the axis of the spherical mirror, near the latter's center of curvature (the biprism is said to be centered). Before passing through the biprism, the beams go through a polarizer; when they leave the biprism, they pass through an analyser which intersects or is parallel to the polarizer. As the optical beams go through the biprism twice, the system provides for compensation since the initial path difference can be compensated by translating the biprism. It is also self-compensating in that, the prism being fixed, any deviation in the light beams generates variations of optical thickness in the air and the biprism that cancel each other: the observed path difference is directly related to the properties of the studied medium (Fig.1). As we work in white light, we observe a system of colored fringes whose tints are arranged in series roughly equivalent to the Newton scale. Observing the interference fringes helps to measure the optical path difference resulting from beams going through the crystal, the test section and the ambient air: we draw from that the component of the gas density gradient along the direction perpendicular to the fringes. We then get the density itself by integration.

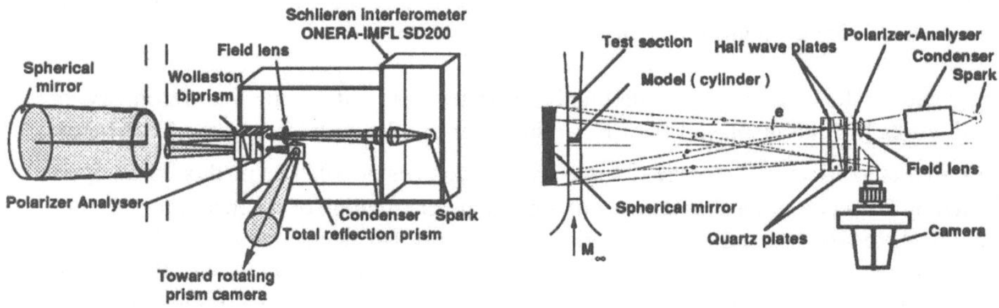

Fig.1. Differential interferometry setup and optical beam path

The following changes were made to adapt the above setup to the new optical method:

- The birefringence angle ε needs to be significantly higher to enable one of the beams to go through the undisturbed upstream flow and the other to go through the flow disturbed by the cylinder; moreover, we want to display the biggest posssible field and get a uniform background tint when the biprism is centered.

-Since the spherical mirror located behind the test section is 250 mm in diameter, the distance X between the two beams that interfere must be not more than half the diameter of the mirror. We want to increase as much as possible the birefringence angle while realizing that the prisms angle cannot be increased indefinitely, otherwise the biprism would be very large resulting in a very small interfringe. We therefore had to find a material with very different ordinary and extraordinary indexes. Calcite meets this requirement. In order to able to observe a uniform background tint on the mirror, we need to select a g/R ratio small enough so that the interfringe i given by :

$$i = \frac{1}{2\varepsilon}\frac{3R}{g}\lambda$$

is comparatively large with reference to the diameter of the spherical mirror. If these various criteria are met, the radius of curvature of the spherical mirror is 8 meter and the birefringence angle is 0.9 degree. Since the radius of curvature of the spherical mirror is large (8 m) and the laboratory is too small to enable us to align the light source, the biprism and the spherical mirror, we hed to use flat mirrors to reflect the light beam. In this configuration, the table of the schlieren interferometer is off-center. Rigidifying the setup helped us to highly reduce outside vibrations (on the order of the micron).

The flashlight source adapted to the interferometer was specially designed to get instantaneous interferograms. We chose an HL 20/50 xenon light electronic flash unit which can deliver 500 μs flash corresponding to an energy of 1,040 joules. The interferograms are recorded during this time by a high-speed camera. The camera used is a CORDIN model 350 (DYNAFAX); its recording speed can vary from 200 to 35,000 frames per second. The peak of number of recordings is 224 frames; in this study, it is the duration of the flash which sets the number of frames (10). The size of each recording is 10 x 8 mm^2 and the pictures are taken in a staggered pattern on a 35 mm film. Here, the exposure time for each interferogram is 3 μs, and the time interval between each interferogram is 35 μs. The setup presented here is about 20 times more sensitive to the index gradient than the interferometry setups using a quartz biprism. Indeed, with a quartz biprism (α = 4°), we get two or three fringes per mm in the localization plane of the interference fringes (near the prisms angle plane), while in the case of calcite (α = 4°), we get almost 60 fringes per mm. As a result the interference fringes had to be calibrated in slightly special environment conditions. We had to insulate all the optical paths from the thermal gradients which are naturally present inside the laboratory, and calibrate the fringe system in the absence of vibrations from nearby installations. We used a 600 mm lens with an aperture of 4 to take the pictures. In our optical setup, the aperture is the same as the angle formed by the light beam as it leaves the biprism, i.e. 1/32.

4 Interferogram recording

Figure 2 shows the interferograms made at very small Mach number (0.2). We also plotted the signal recorded by the instationary pressure sensor, also indicating the times at which the interferograms were recorded. Note that this type of interferogram directly shows the isochoric curves since the reference is located inside the undisturbed upstream flow. We see that the vortices are represented by concentric rings of same color.

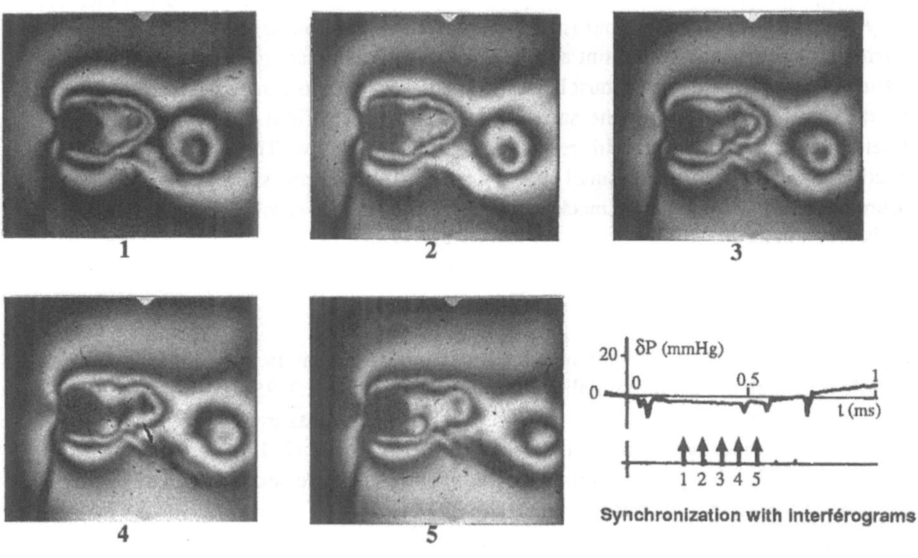

Fig.2. Interferograms recorded with uniform background color $M_\infty = 0.2$ $\Delta t = 70\ \mu s$

The analysis of interferogram 1 is plotted in Figure 3 for both values of M_∞. This analysis reveals that the density is lowest at the center of the vortices, which was already shown by previous interferometric studies around the same cylinder [2].

Interferograms were also made for higher Mach numbers (0.4 and 0.5). The pictures are not so longer of good quality as the gradients become too high to be measured correctly.

If tests are carried out when the biprism is no longer located at the center of curvature of the spherical mirror, the interference fringes appear inside the field of the display. Moving the biprism toward the sperical mirror or toward the schlieren interferometer constricts the interference fringes more or less, thus making it possible to displace the fringes in one direction or another. In addition, the displacement ment of this fringe with respect to the reference interferogram gives the value of the Mach number outside the wake. This type of interferogram is analysed in the conventional way, which boilds down to measuring the displacement of the fringes at all points in the field.

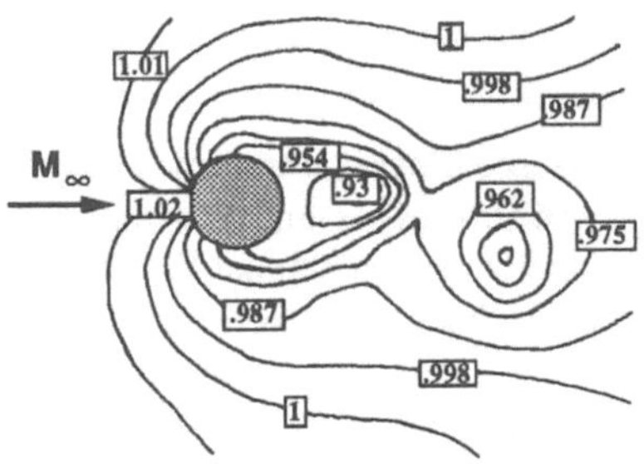

Fig.3. Isochoric curves ρ/ρ_∞ of interferogram n°1 from fig. 2 $M_\infty = 0.2$

5 Conclusion

We have shown that if we want to greatly increase the birefringence angle in order to measure density directly, using a calcite biprism is an interesting solution. If the criteria relating to interfringe, uniform background tint and field measuring range are to be met, the radius of curvature of the spherical mirror must be large. The interferograms can be interpreted directly since they display the lines of the same density, but density field measurement remains differential as it is carried out with respect to the upstream flow. The optical technique is perfectly suited for measuring small index variations in compressible flows. Moreover, measurements can be made on the mock-up itself, which is not the case when the birefringence angle is very small.

This work was supported by the french Ministry (DRET G6).

Références

[1] Desse, J.M. 1990 : Instantaneous density measurement in two dimensional gas flow by high speed differential interferometry. Exp. Fluids 9, 85-91.

[2] Desse, J.M.; Pegneaux, J.C. 1990 : Whole-field measurement of gas density from simultaneously recorded interferograms. Exp. Fluids 12, 1-9.

[3] Smeets, G. 1968 : Interféromètre différentiel à faisceaux fortement séparés. Dépouillement des interférogrammes. ISL TN, 41/68.

[4] Carlomagno, G. M. 1986 : A wollaston prism interferometer used as a reference beam interferometer. Flow visualization IV (Ed. Veret, C.) pp 105-110.

[5] Gontier, G. 1959 : Description et réglage du strioscope I.M.F.L. IMFL TN Gr 89.

Visualization of Flow Around Rectangular Cylinders with Phase-Averaging Technique

Atsushi OKAJIMA, Hisanori UENO, and Toru NAKAMURA

Department of Mechanical Systems Engineering, Kanazawa University
2-40-20, Kodatsuno, Kanazawa 920, Japan

ABSTRACT
Flow around two-dimensional rectangular cylinders at high Reynolds number over 10^4 was visualized by a phase-averaging technique using a tandem type of hot-wire probe. Flow visualization with the aid of a computer was carried out for the cylinders with rectangular section of two different width-to-height ratios. One is a cylinder whose width-to-height ratio is 2.5, with a mixture of two flow patterns; i.e. a fully separated and a reattached flows around the cylinder, and the other has the ratio of 3.0 with a periodically (alternately) reattached flow. For the former cylinder, two different flows can be easily separated with a phase-averaging technique. Finally, experimental results distinguish the flows into three modes, and show conspicuous differences among their flow configurations.

INTRODUCTION

Detailed information regarding flows around rectangular cylinders in a uniform flow is both of special interest and of great importance in the study of aerodynamics and aeroelastic instability. There has been reported that flow around rectangular cylinders greatly depends upon Reynolds number and the shape of cylinders i.e. the value of a width-to-height ratio (Okajima (1982)). Especially at high Reynolds numbers over 10^4, both a fully separated flow and a reattached one were found to exist for cylinders with a width-to-height ratio between 2.0 and 2.8. In addition, it has been well-known that just a periodically reattached flow can be seen for a width-to-height ratio of 2.8 to 6.0 (Okajima et al. (1985)). In the former case, however, it is very difficult to visualize selectively each of flows by means of a familiar visualization technique because of a mixture of two different flow patterns. So, the present study is a computer aided visualization to distinguish the two flow patterns by using a phase-averaging technique for data measured with a tandem type of hot-wire probe.

EXPERIMENTAL APPARATUS AND METHOD

The test apparatus is an open jet wind tunnel with a working section of 165mm wide and 330mm high. The size of models made of acrylic resin is H=14mm in height and B=14 to 56mm in width, and then the value of the width-to-height ratio, B/H is 1.0 to 4.0. The tested wind velocity U_0 is 16m/s, so that the Reynolds number Re is 1.4x10^4, and the intensity of free-stream turbulence in the working section is less than 0.5%. Our phase-averaging technique is applied as follows. A tandem type of hot-wire probe is employed as a measurement signal for educing velocity vector components, as the probe is available even in the reversing flow region about the bluff cylinders. At the same time a single hot-wire probe is fixed at a certain point to get a phase (i.e. time) reference signal, and then the reference signal is bandpass-filtered around a frequency of vortex-shedding. Finally, the phase-averaged velocity data for individual flow patterns can be obtained by using the measurement signal and the filtered reference one in the phase-averaging process.

RESULTS

FLOW CHARACTERISTICS

(a) Three kinds of coherent structure patterns : mode I, mode II, and mode III

Figure 1 shows how the value of Strouhal number St varies with the B/H ratio at $Re=1.4\times10^4$. Obviously, a value of Strouhal number changes with a sudden discontinuity at the B/H ratio of 2.8. And two values of Strouhal number appear for the rectangular cylinders with the B/H ratio of between 2.2 and 2.8, therefore it is evident that two different coherent structures exist for their cylinders. Then the coherent structure due to the lower Strouhal number is referred to as 'mode I' and the higher Strouhal number is as 'mode II'. On the other hand, for the cylinders with the B/H ratio of 3.0 to 4.0 there is found to exist only one coherent structure in the flow field, so that it is referred to as 'mode III'. Here both mode II and mode III are very similar in point of a phenomenon of flow reattachment, while we clearly distinguish mode II from mode III because of the discontinuity of Strouhal number between mode II and mode III in figure 1 and the periodicity of the reattachment seen in mode III.

(b) Contours of power spectra of velocity for three modes

Figures 2(a,b) exhibit the contours of power spectra of velocity in the flows for the rectangular cylinders with B/H=2.5 and 3.0, respectively. For B/H=2.5, the power spectra for mode I and mode II are separately exhibited in the upper and lower sides of the wake centerline. The distribution of mode I shows a gradual increase of spectral values from the leading edges of the cylinder and the configuration is flatter in the streamwise direction than mode II, which indicates the mode-I flow is a fully separated flow. And for mode II, only the components of the low values of the power spectra can be seen near the side surfaces of the cylinder. Therefore it can be concluded that the vortex street of mode II is shedded behind the cylinder. For mode III in figure 2(b), the components of the first and the second peaks of the power spectra exist near the side surfaces and behind the cylinder, respectively. The location of the second peak is further upstream and the wake-width becomes more narrow in comparison with mode I and mode II. Furthermore, it is very interesting that the streamwise distance between the locations of the first and the second peaks equals to about the width of the cylinder $B(=3.0H)$, which may be characteristic of a periodically reattached flow.

(c) Mixture of mode I and mode II

Figure 3(a) shows a typical example of recorded signal traces and the power spectrum of velocity in a near-field of the cylinder with B/H=2.5. In the signal traces there is seen to be three kinds of signal. Individual signals can be separately educed by data-sampling in a shorter period and the results are shown in figures 3(b-d). Figures 3(b, c, d) represent the flow where mode I is predominant, the flow mixed by two components of the same levels, and the flow where mode II is predominant, respectively. In our experiment, we have experienced the phenomenon that the flows alternately switch at short intervals of about 80ms.

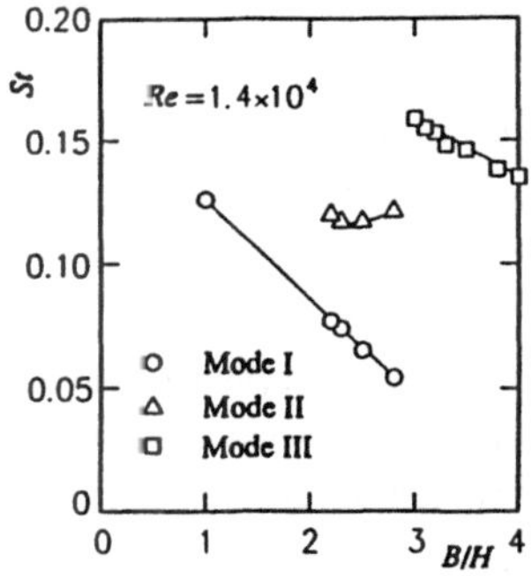

Figure 1. Strouhal number against B/H ratio

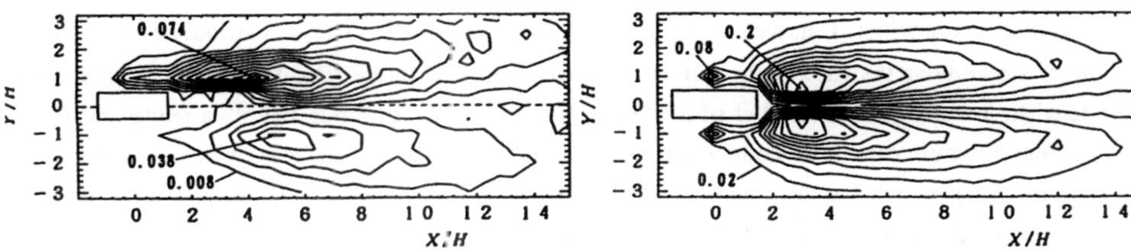

(a) Mode I (upper) and mode II (lower), B/H=2.5 (b) Mode III, B/H=3.0

Figure 2. Contours of the power spectra of a fluctuating velocity in the wake for three modes

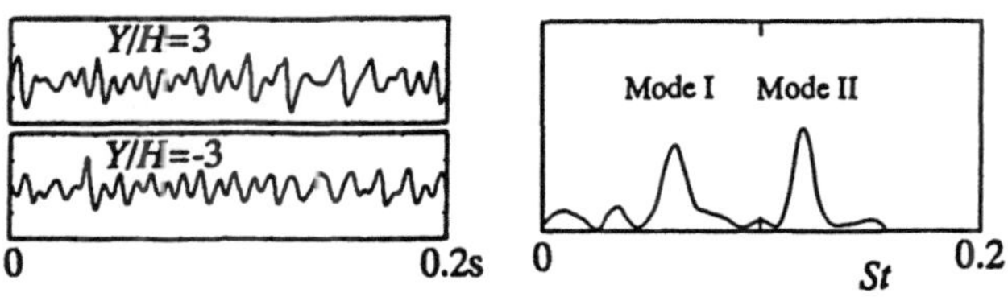

(a) Signal traces and the power spectrum of a fluctuating velocity in the wake

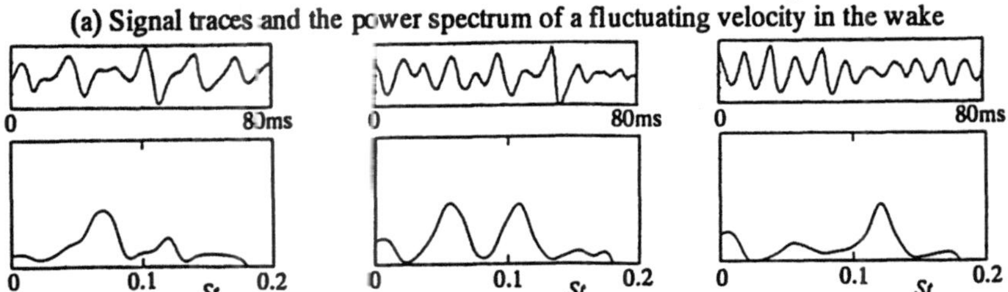

(b) The flow where mode I is predominant (c) The mixed flow at the same levels (d) The flow where mode II is predominant

Figure 3. Mixed flow of mode I and mode II, B/H=2.5

FLOW VISUALIZATION

(a) Coherent vorticity

Figure 4 shows contours of instantaneous coherent vorticity at a certain phase for several modes obtained by a phase-averaging technique. In this study, the Karman vortex of mode I collapses rapidly in the wake, while that of mode II survives with a strong coherence even in a far wake. Figure 4(c) presents the superposed flow of mode I and mode II given in figures 4(a,b) that is, the mixed flow at the same levels shown in figure 3(c). In figure 4(c) a mixture of mode I and mode II results in stretching in the streamwise direction and obscurity of the vortex street. Finally for mode III, the high value of vorticity exists behind the rectangular cylinder because of a strong rolling-up of the reattached flow near the base surface of the cylinder. And the Karman vortex of mode III moves downstream almost in the streamwise direction without spreading in the transverse direction. This may result from no interference between different modes such as mode I and mode II, that is, existence of just a single mode flow.

(b) Coherent Reynolds stress

Figure 5 exhibits contours of instantaneous coherent streamwise Reynolds normal stresses of *the periodic motion* which Cantwell & Coles (1983) described. The stress distribution of mode I has longer streamwise distance of the vortex street than those of mode II and mode III, which is deeply related to lower Strouhal number of mode I in figure 1. And for mode II with a reattached flow, the coherent structure cannot be conspicuously observed near the side surfaces of the cylinder as shown in figure 5(b), which implies that the mode-II flow is not a *periodically* reattached flow. Finally, the distribution of mode III has fairly shorter streamwise distance of the vortex street than mode II. This difference corresponds to the discontinuity of Strouhal number between mode II and mode III in

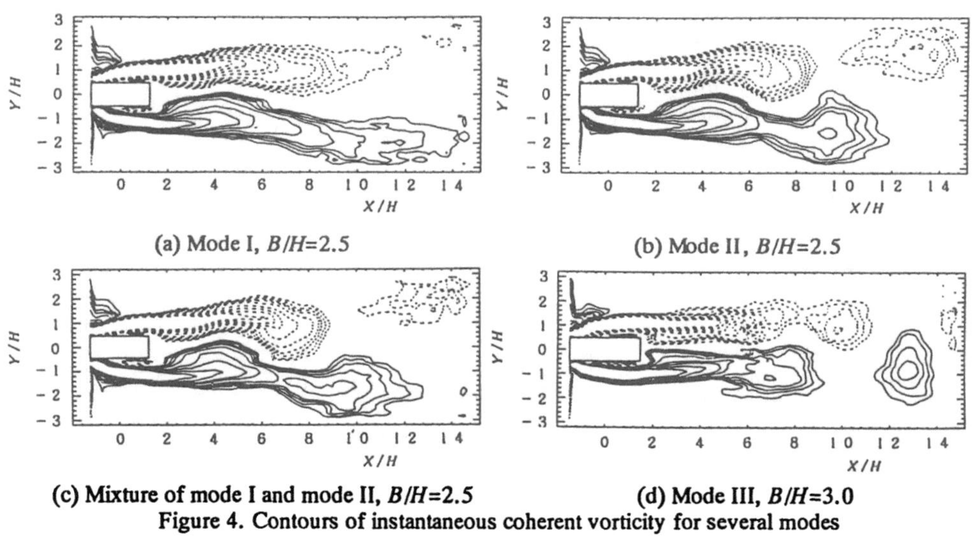

(a) Mode I, B/H=2.5 (b) Mode II, B/H=2.5

(c) Mixture of mode I and mode II, B/H=2.5 (d) Mode III, B/H=3.0

Figure 4. Contours of instantaneous coherent vorticity for several modes figure 1. Consequently, strong coherence near the side surfaces suggests that the mode-III flow is a periodically reattached flow.

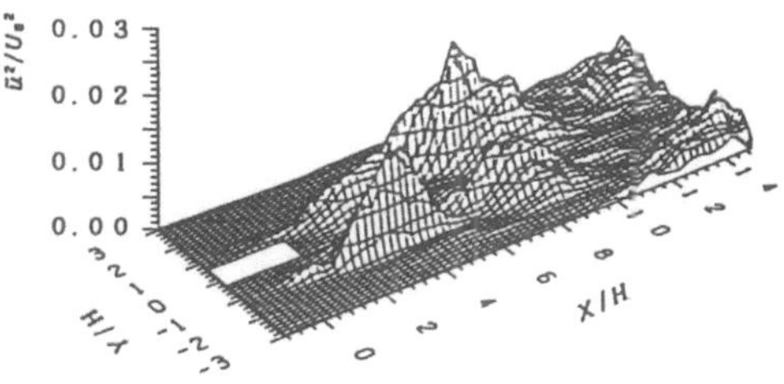

(a) Mode I, B/H=2.5

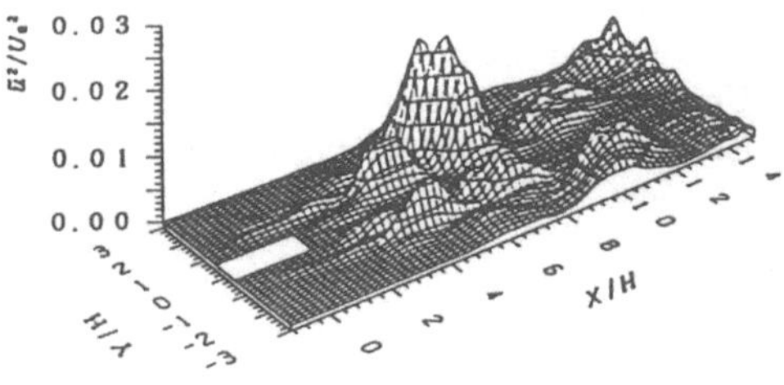

(b) Mode II, B/H=2.5

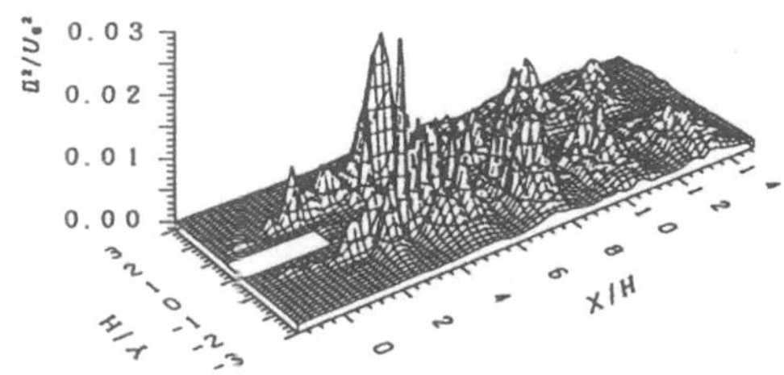

(c) Mode III, B/H=3.0

Figure 5. Contours of instantaneous coherent streamwise Reynolds normal stresses for three modes

CONCLUSIONS

The complex flows around two kinds of rectangular cylinders are visualized by means of a phase-averaging technique with a tandem type of hot-wire probe. Consequently, we find three different modes and discriminate the differences of the flow characteristics among them.

REFERENCES

1. Cantwell, B. & Coles, D., An experimental study of entrainment and transport in the turbulent near wake of a circular cylinder, J. Fluid Mech., vol. 136, pp. 321-374, 1983.

2. Hussain, A, K, M, F., Coherent structures and turbulence, J. Fluid Mech., vol. 173, pp. 301-356, 1986.

3. Okajima, A., Strouhal numbers of rectangular cylinders, J. Fluid Mech., vol. 123, pp. 379-398, 1982.

4. Okajima, A., Mizota, T. & Tanida, Y., Observation of flow around rectangular cylinder, Flow visualization III, Ed. W. J. Yang, pp. 381-386, Hemisphere, Washington, D.C., 1985.

Visualization of the Flow Around Surface-Mounted Prismatic Obstacles

R. Martinuzzi and A. Melling
Lehrstuhl für Strömungsmechanik
Universität Erlangen - Nürnberg
Cauerstr. 4, D-8520 Erlangen, FRG

Abstract

The flow field around surface-mounted, square cross-section obstacles of half-channel height placed in a fully-developed, turbulent channel flow was investigated by means of the crystal-violet, oil film and laser-sheet visualization techniques. The results were analyzed for a qualitative description of the flow and for the dependence of various macroscopic parameters (*eg.* separation and reattachment lengths) on the aspect ratio (*i.e.* ratio of obstacle spanwidth, W, to height, H). The flow geometries can be classified in two categories, corresponding to aspect ratios greater or less than 6.

Introduction

The flow patterns around sharp-edged obstacles placed in a channel provide an insight into the underlying flow mechanisms for complex two- and three-dimensional geometries. Past experimental studies have mainly concentrated on the two-dimensional case, either fences [1] or ribs [2]. With the exception of the work of Larousse *et al.* [3], three-dimensional geometries have been typically studied as perturbations to the boundary layer [4,5], although a few studies have dealt explicitly with the flow pattern around such obstacles [6]. Furthermore, the flow pattern transition between two- and three-dimensional geometries remains poorly documented. The present work aims to provide information on such flow patterns.

Experimental Apparatus and Methods

The on-coming flow conditions were those of a fully-developed, turbulent channel flow and the Reynolds number, based on the bulk velocity and the channel height ranged from 80 000 to 120 000. The experiments were performed in a water channel (550 mm×50 mm) and in a blower-type wind tunnel (600 mm×50 mm). The crystal-violet and the oil film techniques were used to obtain the time-averaged flow patterns adjacent to the channel and obstacle surfaces in the water and wind tunnels repectively. The results from these experiments were recorded on black-and-white photographic film. The laser-sheet visualization experiments provided an instantaneous view of the flow field in a given observation plane. They were performed using smoke or TiO_2 tracer particles and were recorded with a high-resolution video-camera.

Results

Oil film results for the full-span rib (W/H = 24, Fig. 1) corroborated the view that the flow is nominally two-dimensional over most of the flow field except for the region close to the channel side walls. However, a pattern of almost regularly spaced, alternating node and saddle points (Fig. 2a) was observed with the crystal-violet technique on the windward face of the obstacle and on the channel floor. This pattern appears to be related to the cellular structure seen in Fig. 2b, which indicates that the time-averaged flow there is locally three-dimensional. This pattern could be identified for obstacles with aspect ratios $W/H \geq 10$.

Stationary surface-flow patterns upstream of the low aspect ratio obstacles ($W/H < 6$), obtained using the oil film technique, show a complex, multiple separation zone (Fig. 3). A time-resolved view, obtained using the laser-sheet technique, suggests that the flow in the zone between the primary (**A**) and secondary (**B**) separation lines is unsteady and highly complex. Figure 4a shows that, at certain instances, up to four vortices could be identified in this region. These vortices were seen either to undergo vortex-stretching in the direction normal to the plane of observation and eventually disappear, or to pair-up with the horseshoe vortex (Fig. 4b-c). This latter process was followed by an ejection adjacent to the channel floor of reverse flowing fluid (Fig. 4d) which gave rise to a zone of intermittent backflow. This bimodal behaviour was found to be typical only for low-aspect ratio obstacles.

Other pictures from the sequence of Fig. 4 (not included here) show that the recirculating flow on the top of the obstacle does not always extend to the leading edge, but rather that it is occasionally "blown" back by the undeflected, on-coming flow. This process may explain the bimodal velocity distribution observed by a number of investigators. Furthermore, the corner vortices aft of the obstacle, shown in Fig. 3 and Fig. 5a, are shed as can be seen in Fig. 5b, at an almost regular frequency.

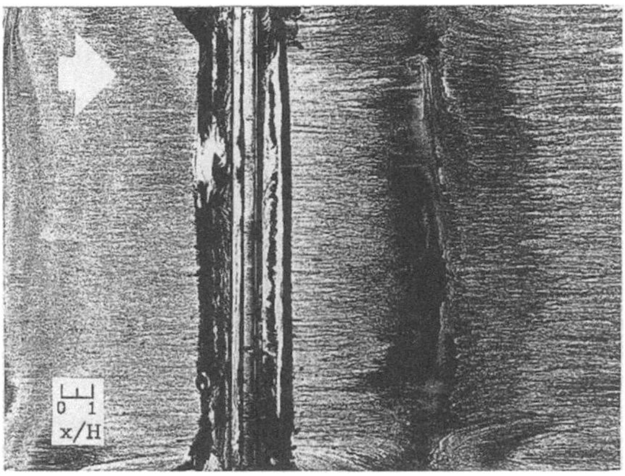

Figure 1: Oil film visualization results for a rib (W/H=24).

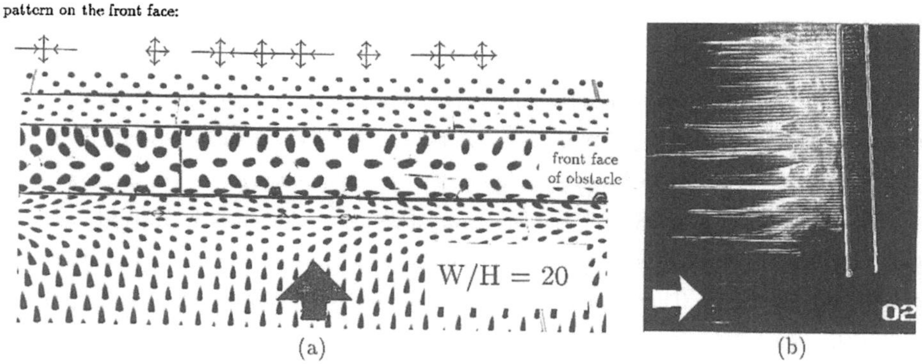

Figure 2: Crystal-violet visualization results on the front face of a W/H=20 obstacle.

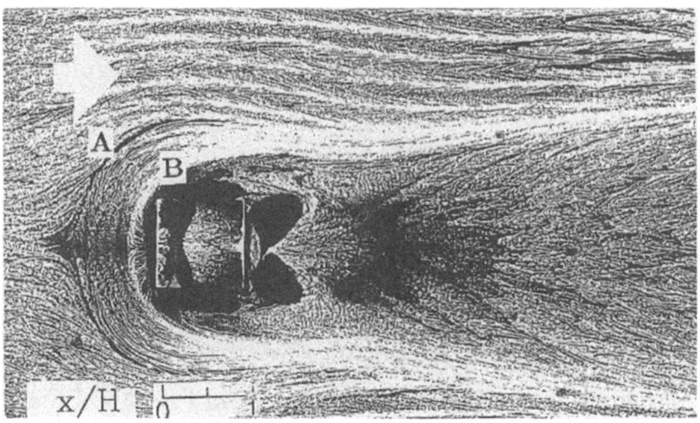

Figure 3: Oil film visualization results for a cube.

The positions of the separation , x_f, and the reattachment , x_R, along the axis of symmetry on the channel floor as well as those of the stagnation point on the windward face, h_f, and of the reattachment points on the leeward face, H_h, of the obstacles are shown as a function of the aspect ratio in Fig. 6. Clearly, the values of h_f, H_h and x_f are independent of the aspect ratio for W/H>6. The reattachment length, x_R, increases linearly up to W/H≈6 and, thereafter, approaches asymptotically the value for the two-dimensional case of x_R/H≈7.2.

Conclusion

Surface flow patterns show a nominally two-dimensional region in the wake of obstacles with aspect ratios greater than 6. For these geometries, the flow in the upstream separation region exhibits a characteristic cellular structure. For shorter obstacles, the interaction between the two-ends of the horseshoe vortex and the shear-layer originating over the top of the obstacle dominates the structure of the three-dimensional wake. Upstream of the obstacle, bimodal flow behaviour is found to be typical for low aspect ratio geometries.

References:

[1] Durst, F. and Rastogi, A.K. (1979) *Turbulent Flow Over Two-Dimensional Fences*, Proc. 2nd Symp. on Turbulent Shear Flows, London, 16–30.

[2] Dimaczek, G., Kessler, R., Martinuzzi, R. and Tropea, C. (1989) *The Flow Over Two-Dimensional, Surface-Mounted Obstacles at High Reynolds Numbers*, Proc. 7th Symp. on Turbulent Shear Flows, Stanford, CA.

[3] Larousse, A., Martinuzzi, R. and Tropea, C. (1991) *Flow Around Three-Dimensional, Surface-Mounted Obstacles*, Proc. 8th Symp. on Turbulent Shear Flows, Munich, FRG.

[4] Castro, I.P. (1979) *Relaxing Wakes Behind Surface-Mounted Obstacles in Rough Wall Boundary Layers*, J. Fluid Mech., **93**, 631–659.

[5] Dianat, M. and Castro, I.P. (1984) *Fluctuating Surface Shear Stresses on Bluff Bodies*, J. of Wind Engineering and Industrial Aerodynamics, **17**, 133–146.

[6] Hunt, J.C.R., Abell, C.J., Peterka, J.A. and Woo, H. (1978) *Kinematical Studies of the Flows Around Free or Surface-Mounted Obstacles; Applying Topology to Flow Visualization*, J. Fluid Mech., **86**, 179–200.

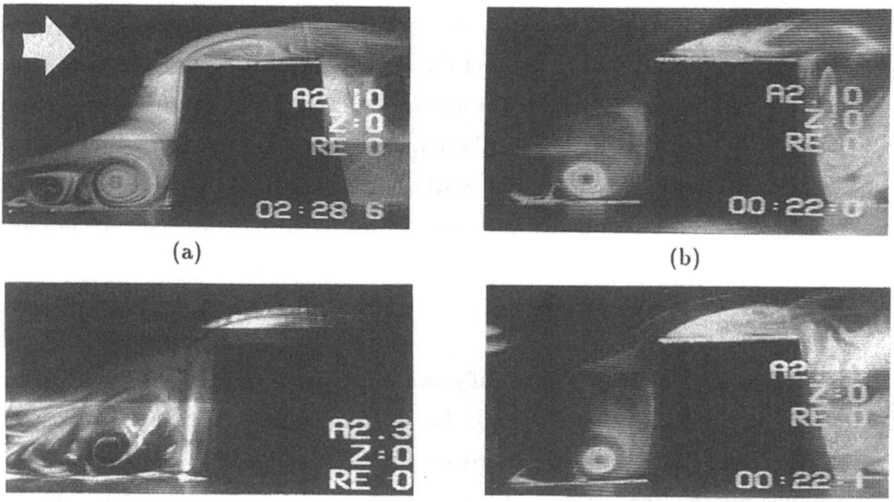

(a) (b)

(c) (d)

Figure 4: Laser–sheet visualization of the separation region upstream of a cube.

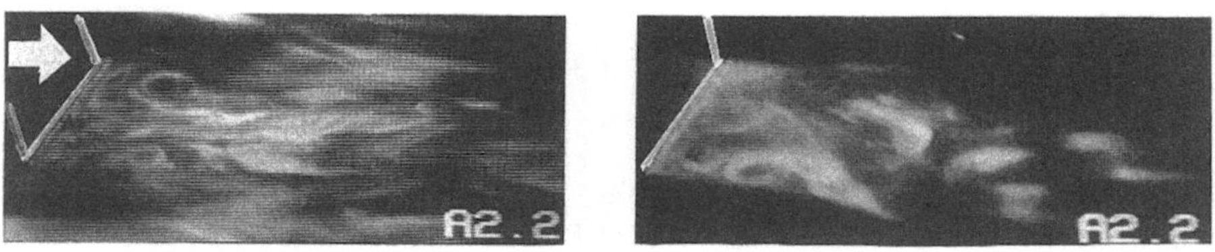

(a) (b)

Figure 5: Laser–sheet visualization results in separation region in the cube wake.

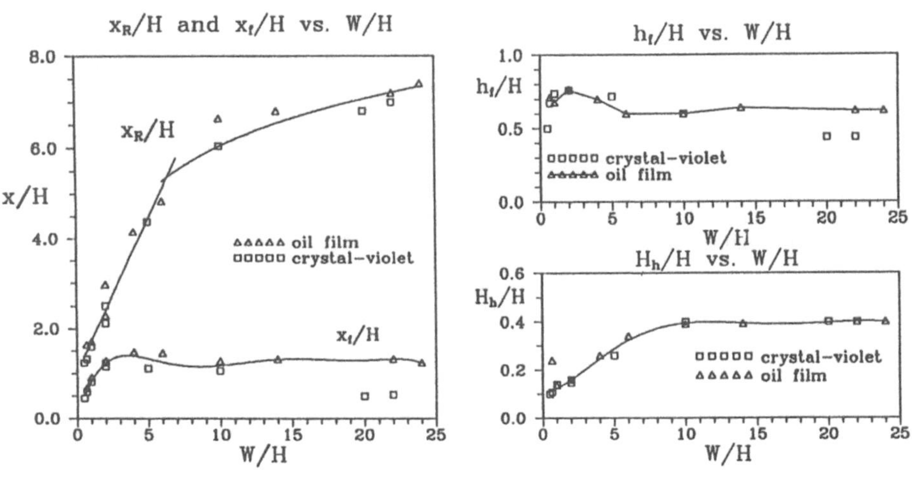

Figure 6: Positions of the separation , x_f, and the reattachment , x_R, along the axis of symmetry on the channel floor as well as those of the stagnation point on the windward face, h_f, and the reattachment point on the leeward face, H_h.

An Interferometric Study on Vortex Shedding from Trapezoidal Prisms in Transonic Flow

Takeo NAKAGAWA

Kanazawa Institute of Technology, Kanazawa, Ishikawa 921 Japan

Summary

An interferometric study has been performed to examine the influence of shape on vortex shedding from quasi-twodimensional prisms at high Reynolds number and high subsonic Mach number. It is found that overall scale of the vortex formation region, which is the space just behind a prism enclosed by the upper and lower separating shear layers, governs the vortex shedding phenomena: As scale of the vortex formation region increases, amplitude and way of the pressure variation with time become larger and more regular respectively, and the pressure spectrum at the fundamental frequency becomes larger, but the fundamental frequency itself decreases.

Introduction

The phenomena of periodic vortex shedding from a symmetric bluff body and formation of vortices in a street have been a subject of research for more than several century. On sharp-edged bluff bodies, point of flow separation is fixed at salient-edges, whereas on smooth bluff bodies location of the separation varies depending both on body shape and state of the boundary layer.

At low Reynolds numbers flow around a bluff body remains stable, but there is critical Reynolds number beyond which flow instability develops. These instabilities normally lead to organized unsteady wake motion, disorganized motion, or combination of both. Regular vortex shedding is a dominant feature of bluff-body wakes and appears irrespective of whether separating boundary layers are laminar or turbulent.

It may be considered that any adequate model of vortex shedding should take into account following facts.

(1) Bluffness of a prism is related to ratio of the wake to the prism dimension. It may be obvious that bluff body tends to diverge flow more than streamlined body, to create wider wake and induce greater drag.

(2) Vortex shedding frequency depends on width of the wake. The relation is roughly inverse, so that bluff body shows lower Strouhal number than streamlined body.

(3) For a given prism, the vortex shedding frequency is well correlated with the base pressure. In general, an increase in base pressure is accompanied by a decrease in vortex shedding frequency.

Kronauer[1] has proposed a theory that the vortex shedding frquency is mainly determined by the feedback of velocity fluctuations to the point of flow separation from the wake. Recently, Koch[2] has suggested that the discrete frequency, self-excited vortex shedding process is governed by a resonance-like feedback mechanism within the vortex formation region. For the more detailed reviews on vortex shedding phenomena, see Rosenhead[3], and Nakagawa[4].

Irrespective of these fairly extensive researches on vortex shedding, the existing knowledge is still insufficient to control it. For example, at present it is impossible to suppress the vortex shedding from a body in the flow, or to disign such a vortex generator that produces vortices having the wanted strength, geometry or frequency. This prompts the present research, which examines the influence of shape on vortex shedding from quasi-twodimensional prisms at high Reynolds number and high subsonic Mach number.

Experiment

Fig.1 shows a schematic diagram of the experiment. The experiment has been conducted with a transonic wind tunnel at Max-Planck-Institut für Strömungsforschung, Göttingen, which has the test section of 620 mm long, 330 mm high and 100 mm wide. The air is drawn from the laboratory and then flows into the vacuum tank of 132 m^3 through the test section. The velocity in the test section is controlled by adjusting the openings of the diffusor.

A trapezoidal prism being symmetrical with respect to the horizontal axis is installed between two transparent glass windows, which have the diameter of 230 mm and are of interferometric quality. A single trapezoidal prism (upper base 10 mm, lower base 20 mm, and height 10 mm) is fabricated in great accuracy, but two different configurations of the prism are investigated: One(prism 1) is that the upper base is facing to the upstream, while the other(prism 2) is that the lower base to the upstream. These prisms

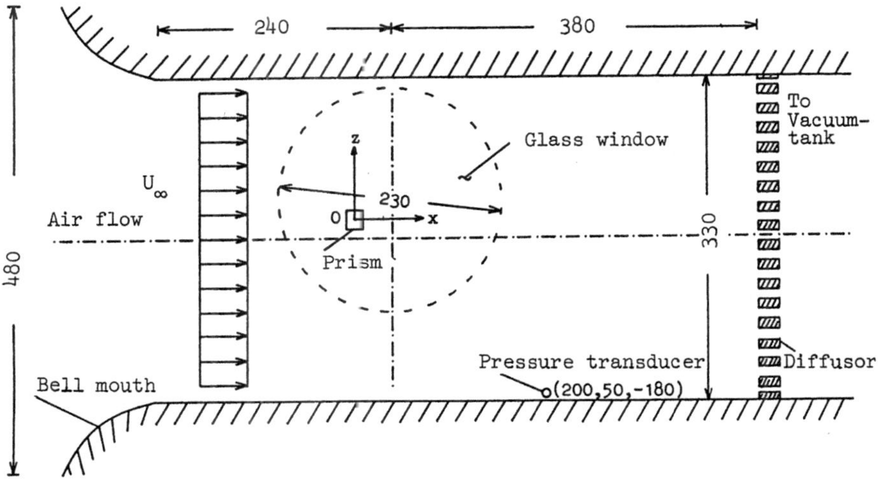

Fig.1 Schematic diagram of the experiment(units:mm)

Fig.2 Geometry of the prism(units:mm)

are delineated in fig.2.

Mach-Zehnder interferometer is used for the optical investigation. When the test section is vacuum, the interferogram shows a total of 96 fringes. The light source is a spark chamber(Impulsphysik). High speed recordings of the interferograms are made at a frame rate of about 7 kHz with a rotating drum camera.

The computer used in this study is a PDP-11/34(Digital Equipment) with a connected array processor AP-120B(Floating Point Systems). The controller CA11-FP(Digital Equipment) is the interface for a 16 channel transient recorder(Le Croy), the real time clock and digital Input/Output port. The transient recorder channels have been used for recording 8 k data at a sampling frequency of 20 kHz. A high sensitivity pressure transducer YCS-093 (Kulite) is used for pressure measurement in the test section. The transducer is fixed to the sidewall, and its sensor has circular form of 6 mm in diameter and the natural frequency is 100 kHz. The data analyses are done with the array processor, being capable of performing the fast Fourier transformation of 1024 data within 3 ms.

Result

Fig.3(a,b) show how the pressure spectrum varies with frequency. Figs.4 and 5 show the high speed flow visualization of periodic vortices from prisms 1 and 2, respectively. It is realized that the amplitude and way of variation of pressure around prism 1 are smaller and more irregular than those around prism 2. The amplitude of pressure spectrum at the fundamental frequency around prism 1 is smaller and spread over a wider frequency range than that around prism 2, while the fundamental frequency around prism 1 is higher than that around prism 2, as shown in fig.3. Figs. 4 and 5 indicate that the vortex formation region behind prism 1 is smaller than that behind prism 2, and the vortices behind prism 1 are more well-defined but smaller than those behind prism 2.

Discussion and Conclusion

It may be certain that if the upper and lower separating shear layers are brought closer together by shrinkage of the vortex formation region, their interactions are facilitated and thus the shedding period is shortened. On the other hand, if the space between the two shear layers is wider, it will

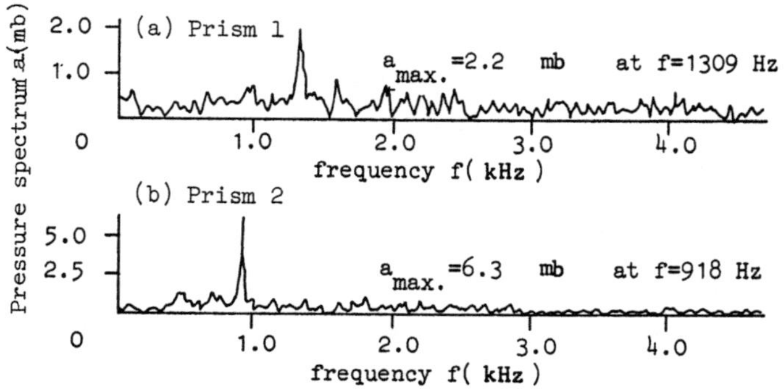

Fig.3 Pressure spectrum a against frequency f

take longer time for the sufficient concentration of vorticity to be carried across the wake. It is, therefore, concluded that overall scale of the vortex formation region, but neither merely its length nor width of the wake, are of more relevance to the vortex shedding phenomena: The vortex shedding frequency increases with decreasing the overall scale of the vortex formation region, and *vice versa*.

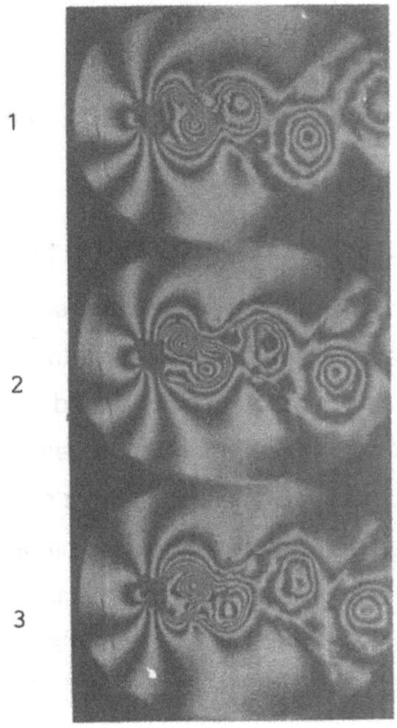

Fig.4 High speed flow visualization around prism 1.
Δt = 147 μs, U_∞= 129 m/s, f = 1309 Hz, St = 0.2029, Re = 1.720×10^5, M = 0.3762.

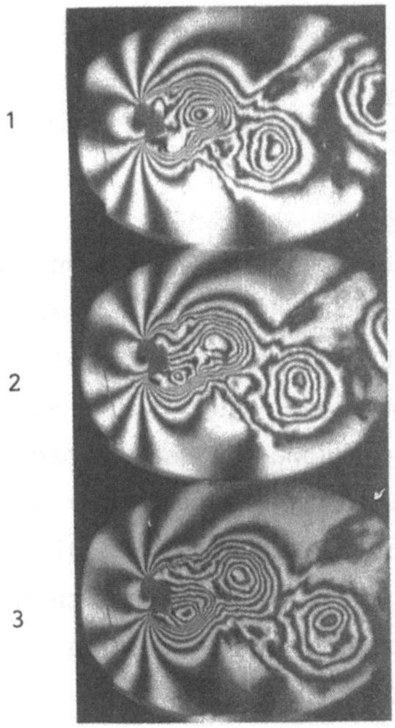

Fig.5 High speed flow visualization around prism 2.
Δt = 146 μs, U_∞= 129 m/s, f = 918 Hz, St = 0.1423, Re = 1.720×10^5, M = 0.3762.

References

1. Kronauer, R.E.; A paper presented at IUTAM Conference on concentrated vortex motions. Ann Arbor(1964)101-115.

2. Koch, W.; Local instability characteristics and frequency determination of self-excited wake flows. J. Sound & Vibration, 99(1985)1-25.

3. Rosenhead, L; Vortex systems in wakes. Advances in Applied Mechanics, 3(1953) 183-195.

4. Nakagawa, T.; Vortex shedding behind a square cylinder in transonic flows. J. Fluid Mech. 178(1987) 303-323.

Flow Visualization Study of Flow Past an Oscillating Square Cylinder

S.C.LUO

Department of Mechanical & Production Engineering, National University of Singapore, 10 Kent Ridge Crescent, Singapore 0511

Key Words
Galloping, square cylinder, flow induced oscillation, flow visualization study, wake structure.

Introduction

A square cylinder is one of the most commonly employed shape in the study of the galloping type of flow induced oscillation because most buildings have either square or rectangular cross sections and because the square cylinder is susceptible to both galloping and vortex resonance, thus making the problem complicated and challenging. Many researchers had already carried out measurement on forces and pressure acting on a transversely oscillating square cylinder. Some of the author's own measurement had been reported in Bearman and Luo [1]. The objective of the present paper is to carry out flow visualization on a transversely oscillating square cylinder to gain more understanding of the flow and make comparison with the author's own measurement.

Experimental Arrangement

A horizontally installed square cylinder with a dimension of d = 50.8mm was forced by a Scotch-Yoke mechanism to oscillate in a up and down manner, transverse to the free stream in the 0.91m x 0.91m x 5.49m working section of a wind tunnel. End plates were installed at the two ends of the cylinder, at a distance of 17d apart. The visualization of the flow was achieved by employing the smoke wire technique which involved the electrical heating of an oil coated high resistance fine wire that was transverse (hence vertical in the present case) to the cylinder and upstream to it. All the visualization were conducted at a constant wind speed of U = 2.72m/s. Different reduced velocity U_r (= $U/f_N d$) was obtained by varying the cylinder's oscillation frequency f_N. The amplitude of oscillation was also kept constant at A/d = 0.675.

<u>Experimental Results</u>

In the present investigation four different flow structures were identified from flow visualization. They are:

<u>Lock-On Type A</u>

At f_N = 7Hz, U_r = 7.65. At this reduced velocity the vortex shedding frequency f_s locks on to the oscillation frequency. Each shear layer sheds one vortex per cycle of oscillation. For the present flow structure the top shear layer sheds a vortex when the cylinder is near its highest position (from now on referred to as the top dead centre, abbreviated as TDC) and the bottom shear sheds a vortex half a cycle later, when the cylinder is near its lowest position (from now on referred as the bottom dead centre, abbreviated as BDC). A photograph of this flow structure is shown in Plate 1. By projecting many frames of the film of the flow onto a screen, the longitudinal and lateral vortex spacing (a/d and b/d respectively) were estimated to be 2.74 and 2.39 respectively. The vortex spacing ratio b/a is thus 0.87. By using the inviscid flow result of $\Gamma/(Ud) = 2(a/d)(1-S(a/d))/\tanh(\pi b/a)$ where S ($=f_s d/U$) is the Strouhal number that had been measured and reported in Bearman and Luo [1], the dimensionless circulation of each vortex was estimated to be $\Gamma/(Ud)$ = 3.71.

<u>Lock-On Type B</u>

At f_N = 4Hz, U_r = 13.39. At this reduced velocity the vortex shedding frequency is still locked-on to the oscillation frequency. Because the cylinder is oscillating slower than in Lock-On Type A, the period of one shedding cycle is longer and the vortex appears to have grown much larger than in Type A. However, unlike Type A, in the present flow structure the vortex fed by the top shear layer is shed near the BDC and the vortex fed by the bottom shear is shed near the TDC. As a result, an unusual "thrust" type of vortex street is seen. That is, each vortex ends up on the other side of the wake centre line as its feeding shear layer. In ordinary flow a "drag" type of vortex street where each vortex is on the same side of the wake centre line as its feeding shear layer is usually seen. In the present flow structure a/d and b/d were estimated to be 6.84 and 1.16 respectively. b/a is thus 0.17. $\Gamma/(Ud)$ was estimated to have a magnitude of 14.36. A photograph of this flow structure is shown in Plate 2.

<u>Triple Lock-On</u>

At f_N = 2.5Hz, U_r = 21.4. At this reduced velocity a triple lock-on situation is observed. That is, each shear layer sheds three vortices per cycle of oscillation, two when the cylinder is moving away from the shear layer and a third one when the cylinder is moving towards the same shear layer. The same situation applies to both shear layers but they are 180° out of phase. In the present flow structure a/d and b/d were estimated to be 3.49 and 1.73 respectively. b/a is thus approximately 0.50. $\Gamma/(Ud)$ was estimated to be 3.93. A photograph of this flow structure is shown in Plate 3.

Quasi-Steady

At f_N = 1Hz and 0.55Hz, U_r = 53.56 and 97.38 respectively. At such high reduced velocity the vortex shedding frequency is decoupled from the oscillation frequency and the Strouhal number has a magnitude that is very close to its stationary cylinder counter part. For U_r = 97.38, a/d and b/d were estimated to be 6.36 and 0.87 respectively. b/a is thus approximately 0.14. Γ/Ud was estimated to be approximately 5.47. A photograph of this flow structure at U_r = 85.02 is shown in Plate 4.

Flow Visualization Findings And Conclusions

For a square cylinder oscillating at a constant amplitude of A/d = 0.675 in a constant free stream of 2.72m/s, four flow structures were observed at different reduced velocities. By studying the films and the photographs of the flow carefully, the following are observed.

1. The vortex shedding mechanism proposed by Gerrard [2] is applicable only to Lock-On Type B and Quasi-Steady but not to the other two flow structures.
2. The vortex spacing ratio b/a of 0.281 predicted by Von Karman does not apply to the present flow situation.
3. During lock-on, when f_N decreases, both a/d and Γ increase.
4. Under certain flow conditions the unconventional thrust type of vortex street exists.
5. The position of the cylinder within its cycle of oscillation when vortices are shed affects the magnitude of b/d and whether the flow structure is an unstable (ie. galloping will occur) one.

References

1. BEARMAN, P.W. and LUO, S.C., Investigation of the aerodynamic instability of a square-section cylinder by forced oscillation. Journal of Fluids and Structures, 2 (1988), 161-176.

2. GERRARD, J.H., The mechanics of the formation region of vortices behind bluff bodies. Journal of Fluid Mechanics, 25 (1966), 401-413.

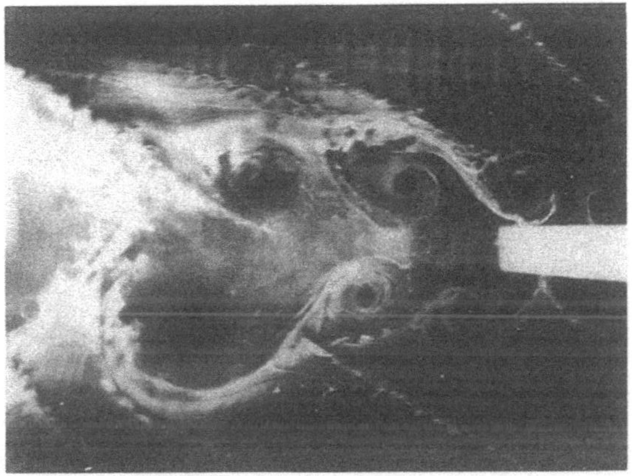

Plate 1 : Lock-On Type A

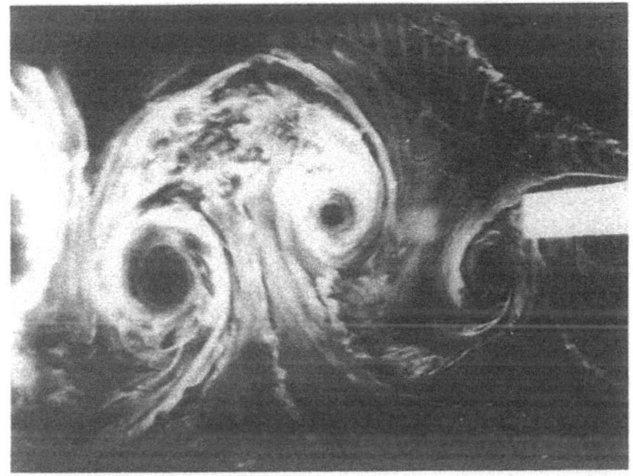

Plate 2 : Lock-On Type B

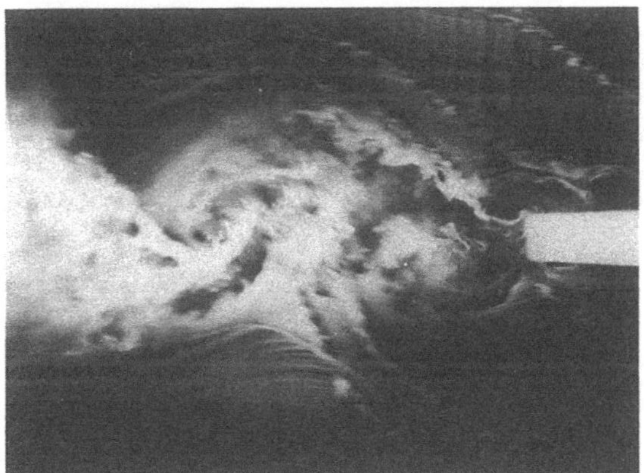

Plate 3 : Triple Lock-On

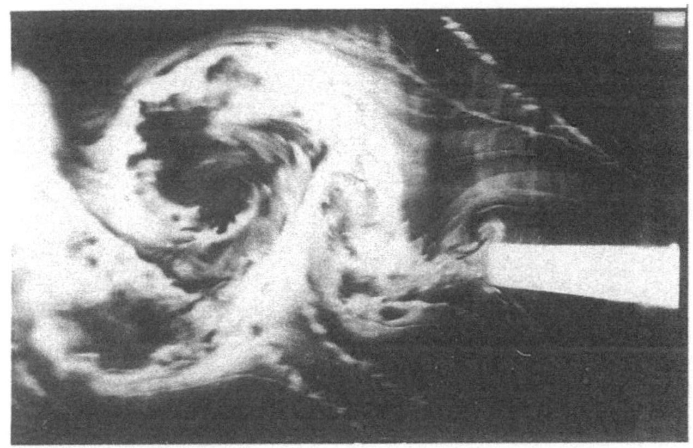

Plate 4 : Quasi-Steady

Multiphase Flows

Visualization of Particulate Two-Phase Flow

M.C. ROCO

National Science Foundation, Washington, D.C., and
University of Kentucky, Lexington

Summary

A brief survey of the experimental visualization techniques for particulate two-phase flow is presented. Various non-intrusive and computer-assisted methods are considered for the identification of the flow microstructure, including measurements of particle concentration and velocity distributions. The results provide new insights into the physics of two-phase flow, and constitute a data base for modeling.

Two-phase Flow Issues

Particulate flows, with or without an interstitial fluid, are encountered in nature and a large variety of industrial processes. Experimental visualization of the flow microstructure and overall flow pattern plays an important role in understanding the basic flow phenomena and phase interaction mechanisms [1] - [3]. In technical literature, there are more models on particulate flows than direct measurements, but this balance is slowly changing towards experimental verification. Too often models have been built like towels on clay foundations. In the last few years, significant advancements have been made on both the nonintrusive measuring instrumentation, and digital image processing and computer-generated graphics. Visualization has been used to identify spatial and temporal microstructures, as well as particle size, velocity and concentration distributions. This paper will outline several research directions for experimental flow visualization of particulate systems. The inroads made by the recently developed instrumentation (nuclear magnetic resonance, chemical and radioactive tracers, etc.) and related computational methods will be highlighted with examples. Several issues should be considered when selecting the measuring method for two-phase flow visualization studies:

a. A first issue is the possibility of performing simultaneous velocity measurements for the carrier fluid and solids. Fluid and particles with same optical, magnetic or electric index may be considered for this purpose as a function of the measuring method.

b. Three-dimensional flow measurements are needed because the flow structure can not be fully identified from two-dimensional images. The opacity of the particulate medium makes it more difficult than in single-fluid flow.

c. Another issue is the effect of particle size on the reliability and accuracy of the measurement, particularly for small scale flows and thin layer flows. Special image enhancement techniques (special cameras, microscopes, ultraviolet illumination, fluorescent particles, etc.) have to be used for finer tracer particles.

d. If the goal of visualization is the description of the flow microstructure, the spatial and temporal resolutions should be related to the particle size and particle response time. In complex flows, where the primary goal is the description of the general flow pattern, the measurement requirements are related to the flow domain.

e. It is desirable to measure quantities which are directly related to the parameters and functions needed in modeling, for instance the pair distribution functions and velocity distribution functions.

f. Methods which provide real-time data representation, even with a lower accuracy, may be useful.

g. If the data collected is too large for understanding, it may be useful to adopt an intermediary data reduction step in order to synthesize the essential flow characteristics.

h. Experiments in simple geometries, such as Couette flow or two-dimensional single particle layer flow, are to be considered at the beginning of a comprehensive study in two-phase flow.

i. Similar flow visualization techniques for particulate systems are used in various research areas, such as physics, medical field and engineering. Many methods are extensions of single-fluid flow visualization methods [4], [5].

Experimental Visualization Techniques

Visualization is based either on measurements or numerical simulations. Experimental flow visualization is more important for understanding, and is the object of this paper. The most relevant nonintrusive methods of measurement used for particulate flow visualization are:

- Direct optical methods, including two- and three-dimensional particle tracking velocimetry and holography. Holocinematography [6] provides three-dimensional time-dependent results which are not yet repeatable by direct numerical simulation.
- Particle image velocimetry (PIV) and laser speckle velocity (LSV) (Liu and Adrian, Chapter 2 [2]; and [5]).
- Interferometry [7] and spectrometry [8].
- Chemical/radioactive/magnetic tracers. Besides the more traditional methods as fluorescent tracers [4], a series of techniques have been used: nuclear magnetic resonance imaging (NMR; [9], [10]), laser induced photochemical anemometry (LIPA; R. Falco and D. Nocera, Chapter 3 [2]), computer-automated radioactive particle tracking (CARPT; [11]), to name a few. The main issue is the averaging time and averaging volume in the flow domain. Besides velocity and concentration distributions, NMR can be applied to measure acceleration, fluctuations, and bulk diffusion coefficients.
- Positron emission tomography (PET).
- X-ray, γ - ray and neutron-ray attenuation methods, commonly used in conjunction with the computer tomography scanning method.

- Electrical probes (autocorelation techniques for concentration fluctuations, electrical impedance technique).
- Acoustic diagnostics [12], [13] (attenuation methods, back-scattered methods, Doppler velocimetry, acoustic microscopy, and acoustic tomography).
- Laser Doppler velocimetry with various discrimination techniques between phases, phase Doppler velocimetry [14], and Doppler global velocimetry [15].
- In order to visualize the flow and temperature surrounding the particles in suspension, other methods currently used for single-fluids can be applied (liquid and gas tracers, hydrogen bubbles, electric sparks, infrared imaging, liquid crystals, etc.).

The computer-assisted visualizations methods provide both quantitative and qualitative pictures for data interpretation. The commonly used methods are:
- Digital image processing, including image enhancement and image synthesis.
- Pattern recognition, including use of fractal techniques, Fuzzy logic algorithms, and graf theory.
- Multi-particle tracking techniques [16].
- Computer tomography.
- Optical tomography, including image reconstruction from individual point measurements or from projections (for instance [17]).
- Video animations [3].

Phenomena Identified by Visualization

The main benefit from flow visualization is the insight gained on two-phase flow phenomena. Several recent results, which have changed our understanding of particulate flows, are exemplified below.

a. Illustrations for two-phase flow mechanisms

- The particle-particle interaction caused by the wakes formed behind particles has been visualized in a domain bounded by two parallel transparent walls (D. Joseph et al., Chapter 2 [2]). Spherical particles lock together in a stable "nested wake structure" when moving through a fluid if stationary eddies appear in the wake behind particles (22 < particle Reynolds number < 43). Otherwise, neighboring particles draft, touch each other for a short time, and then tumble.
- The pair distribution functions of spherical particles in mixture with a liquid in a Couette flow have been measured by D. Leighton and I. Rampall (Chapter 6 [2]). These functions play a determinant role in the governing equations, and their expressions were previously assumed.
- The particle spin distribution was measured in turbulent Couette flow [18]. The experimental particle angular velocity is larger that the strain rate of velocity field, and increases near a containing wall. Particle rotation in turbulent flow was neglected in previous studies, in part because of the lack of reference data.

b. Particle-wall interaction

- By using a scanning tunnel microscope, the plastic interaction between 5 nm diameter gold clusters and a smooth surface was visualized [19]. It was found that the cluster surface tension determines a cluster behavior similar to that of a fluid droplet on a flat surface.
- NMR imaging was used to document the flow induced structure in a Couette flow characterized by radial particle migration (J. Abbott et al., Chapter 1 [2]). After a large number of rotations, the particle concentration near the outer wall reached its maximum packing value, and particles were arranged in layers.
- Spherical particles have the tendency to be stratified in the proximity of a moving rigid wall. The first layer of spherical particles, which is in direct contact with the wall, has about 50% of the velocity of the wall. More than half of this velocity is due to the particle rotation in this layer. Macroscopically, this large local rate of deformation has been interpreted as a "slip" velocity at the wall. The measurement was performed in a two-dimensional single-particle layer flow, between a fixed wall and a cylinder [20].
- By visualizing the interaction of solid particles with the coherent structures near a solid wall, it was found that the particles of diameter over 5 wall units enhance the fluid turbulence (G. Hetsroni, Chapter 8 [2]). These particles increase the frequency of ejections of coherent structures, or provide those structures with solid surfaces to which they can attach. Smaller particles suppress the turbulence.

c. Velocities and concentrations distributions

- Visualizations of particle concentration distribution in the wake of a blunt body showed the importance of the particle Stokes number (C.T. Crowe et al., Chapter 18 [2]). At Stokes numbers less than 1, the particles are distributed in the whole wake. At numbers about 1 the particles are centrifugated at the extremities of the wake eddies, and at numbers much larger than 1 the particles traverse the region covered by eddies.
- By using LIPA (R. Falco and D. Nocera, Chapter 3 [2]), the velocity and velocity fluctuations were measured in a 27% by volume solids-liquid mixture flow in a downward vertical pipe. It was found that a core has higher solids concentration, and slides down the center of the pipe on a surrounding film of lower concentration. The largest fluctuations are in the pipe core, and not near boundary region as it is predicted by the traditional models.
- Velocity and concentration distributions were visualized by NMR imaging in a circular pipe with slurry flow (Altobelli et al. [9]). The particles had the tendency to stratify in the pipe at lower mixture velocities.
- The slip velocity between glass beads and water was measured in a centrifugal slurry pump by using an LDV amplitude discrimination technique (Roco et al., Chapter 13 [2]). The reconstructed particle slip velocity distributions in space and time describe a periodical macrostructure dominated by stationary waves, which affect the overall pump performance.

- The relative size between the solid particles and bubbles in a vertical liquid column, and the flow induced by the wakes developed behind larger bubbles, determine a variety of flow patterns in fluidized beds and pipes (L.-S. Fan and K. Tsuchiya, Chapter 23 [2]; and T. Sakaguchi [21]). The degree of segregation of a three-phase system in a three-dimensional cyclone flow was visualized by using X-ray tomography [22].

These examples illustrate the relevance and variety of applications of flow visualization.

Concluding Remarks

Experimental flow visualization has brought new insights in the physics of particulate two-phase flow, including identification of various two-phase flow microstructures, measurement of parameters and functions used in governing equations, as well as field distributions for velocity and concentration distributions. The new nonintrusive instrumentation and the measurements already performed, have changed qualitatively the perspective in multi-phase flow research, and have improved our understanding in numerous flow processes.

The new experimental techniques together with the capabilities of parallel processing are moving the methods of investigation and the research focus from the macroscopic to the microscopic.

References

1. D.D. Joseph and W.-J. Yang (eds.): Workshop on particulate two-phase flow visualization. NSF, Washington, D.C., 1991.

2. M.C. Roco (ed.): Particulate two-phase flow. Boston: Butterworth-Heinemann, 1992 (in press).

3. International Video Journal of Engineering Research: Special Issue on Particulate Two-phase Flows. John Wiley & Sons, Vol. 2, 1992.

4. Yang, W.-J. (ed.): Handbook of Flow Visualization. Washington, D.C.: Hemisphere, 1989.

5. Merzkirch, W.: Flow visualization. New York: Academic Press, 2nd ed., 1987.

6. Ruff, G.A., Bernal, L.P. and Faeth, G.M.: High-speed in-line holocinematography for disperses-phase dynamics. J. Appl. Optics, Vol. 29, No. 1, 1990.

7. Trolinger, J.D. and Havaner, G.: A renaissance in holographic flow field diagnostics. AIAA Paper 91-0566, Washington, D.C., 15 p., 1991.

8. Fuller, G.G.: Optical rheometry. Ann. Rev. Fluid Mech., Vol. 22, 387-417, 1989.

9. Altobelli, A., Givler, R.C. and Fukushima, E.: Velocity and

concentration measurements of suspensions by NMR imaging, J. of Rheology, Vol. 35, 721-740, 1991.

10. Kose, K.: Flow visualization by NMR imaging. Proc. Int. Conf. on Multiphase Flows, Tsukuba (Japan), Sept. 24-27, Vol. 3, 25-29, 1991.

11. Devanathan, N., Moslemian, D. and Dudukovic, M.P.: Flow mapping in bubble columns using computer-automated radioactive particle tracking. Chemical Engineering Science, Vol. 45, No. 8, 2285-2291, 1990.

12. Aas, M., Bacri, J.-C., Frenois, C., Salin, D. and Woumeni, R.: 3-D acoustic scanner. SPE 20599, Society of Petroleum Engineers, 725-732, 1990.

13. Atkinson, C.M. and Kytomaa, H.K.:Acoustic properties of solid-liquid mixtures and the limits of ultrasound diagnostics - Experiments. J. Fluid Eng., 1991.

14. Bachalo, W.D.: Method for measuring the size and velocity for dual-beam light scattering interferometry. Applied Optics, Vol. 19, No. 3, 363-370, 1980.

15. Komine, H. and Brosnan, S.J.: Instantaneous, three-component, Doppler global velocimetry. In Laser Anemometry, ASME, Vol. 1 (A. Dybbs and B. Ghorashi, eds.), New York, 273-277, 1991.

16. Uhlmann, J.K.: Algorithms for multiple-target tracking. American Scientist, Vol. 80, No. 2, 128-141, 1992.

17. Herman, G.T., Lewitt,R.M., Odhner, D. and Rowland, S.W.: SNARK89, a programming system for the image reconstruction from projections. Medical Image Processing Group, Dept. of Radiology, Univ. of Pennsylvania, 1989.

18. Ye, J. and Roco, M.C.: Particle rotation in Couette flow. Phys. of Fluids - A, Vol. 4, No. 2, 220-224, 1992.

19. Castro, T., Li, Y.Z., Reifenberger, R., Choi, E., Park, S.B. and Andres, R.P.: Studies of individual nanometer-sized metallic clusters using scanning tunneling microscopy, field emission, and field ion microscopy. J. Vac. Sci. Technol. A, Vol. 7, No. 4, 2845-2849, 1989.

20. Ye, J., Roco, M.C. and Stafford, D.: Particulate flow between a fixed wall and a rotating cylinder. Proc. 2nd World Congress on Particle Technology, Kyoto, Vol. 1, 79-86, 1990.

21. Sakaguchi, T.: Flow characteristics of gas-liquid-solid three-phase flow in pipes. Conf. on Multiphase Flows, Tsukuba (Japan), Sept. 24-27, Vol. 3, 61-75, 1991.

22. Miller, J.E., Lin, C.L. and Cortes, A.B.: Review of X-ray computed tomography and its application. Mineral Processing and Extractive Metallurgy Review, Vol. 7, 1-18, 1990.

Rising Bubble Behavior in Air-Particle Injection by Means of Image Processing

Shuichi Torii** and Wen-Jei Yang
Department of Mechanical Engineering and Applied Mechanics
University of Michigan
Ann Arbor, Michigan, 48109, U.S.A.

ABSTRACT

An experimental study is performed to observe and record the dynamic behavior of rising bubbles in the injection of air-particle mixture into a slender ladle. Real-time image processing on slow-motion video recordings is employed to analyze both the movement of rising bubbles and the change in their morphology. The variation in the transverse distributions of gas fraction (defined as the ratio of bubble lifetime to total measurement time) is determined. It is disclosed that (1) air bubbles continue to spread within a plume, and (2) the profile of gas fraction distribution becomes flatter and spreads wider with ascension of bubbles. This trend is intensified when an injected air stream is mixed with fine solid particles. The shape of rising bubbles is substantially distorted in the presence of solid particles. The results provide vital information of the theoretical modeling of recirculating flow resulting from gas-particle injection into a ladle with a througflow of liquid.

INTRODUCTION

Flow visualization is one of the most useful methods of flow analysis and provides a clue in understanding flow mechanics. Recently, with the advent of image processing, it can extract quantitative information contained within flow visualization images. Image analysis is stimulated by the remarkable development in the hardware of the computer. Results can be compared with those obtained by other experimental methods (for example, Balint et al. [1], Hernan et al. [2]). In addition, calculations based on measurements produced by this technique have the possibility of providing a valid approximation for theoretical investigations.

Injection metallurgy in which powders are injected into molten metals, is of great importance in metallurgical refining operations such as desulphurization, decarburization and deoxidization. In such industrial injection processes, powders are pneumatically transported, and then injected into molten metal through a submerged lance. The carrier gas, while rising as a plume to the free surface, induces circulatory flows in the fluid within the vessel, thereby

---

* The work reported was supported by the National Science Foundation under the Grant No. CTS-8921584.

** Visiting scholar on leave from the Mechanical Engineering Department, Kagoshima University, Kagoshima, Japan.

helping to promote metal homogeneity. The hydrodynamics of injection metallurgy plays an important role in the kinetic efficiency of chemical reactions such as desulphurization. In understanding the mechanisms of gas-stirred solid-liquid mixing phenomena in a ladle, numerical simulation may serve as an important tool to provide some insight into the flow structure. Numerical simulation requires accurate and detailed information on: (i), growth and departure of bubbles injected through a submerged nozzle as well as the manner of particles entering the ladle; (ii), the geometrical extension of the bubble jet; (iii), the local gas fraction in the ladle; (iv), liquid and gas velocities; and (v), the effect of a particle injection on the bubble morphology.

Tacke et al. [3] measured the radial and vertical distributions of gas fraction in gas bubble plumes using an electroresistivity probe. It was found that the radial profiles are close to a Gaussian function. Torii and Yang [4] investigated the bubble and particle behavior at the outlet of an air injection nozzle with the aid of the direct photographic method. The following results were obtained: (i) A larger deformation in the bubble shape occurs at a higher particle flow rate, particularly with the horizontal injection nozzle; (ii) In addition to the effects of the buoyancy and viscous forces, the distortion of bubble shape can be attributed to contamination of bubble surface in the presence of particles, which results in a reduction of surface tension; and (iii) The inertial force of traveling particles causes the bubble to burst.

This paper investigates rising bubbles in which a stream of gas mixed with solid particles is injected through a vertical nozzle submerged in a ladle. Emphasis is placed on the evaluation of the transverse distribution of gas fraction and the influence of particles on its distribution. Image processing is employed to aid in disclosing the movement of rising bubbles and the change in their morphology. Results will contribute to the development of a numerical model for the flows of a gas-particle mixture through an injection nozzle installed in the ladle.

EXPERIMENTAL APPARATUS

The experimental apparatus employed in the present study is depicted in Fig. 1. It consists of a slender rectangular tank (simulating a ladle with a 2-D flow field in the x and y directions) with two large side walls which were made of acrylic windows for illumination and observation. The tank, used as a ladle, had the dimensions of B=0.20 m and H=0.40 m, so that fluid flow was treated as two-dimensional. Four light sources (4x200 W bulbs) were used for illumination and were placed on both sides of the tank, as shown in the figure. A stream of air (simulating the carrier gas), Q_g, mixed with solid particles (simulating desulphurization or deoxidizing agent), Q_s, was injected into the tank through a submerged nozzle of 0.005-m inner diameter. The air nozzle was vertically installed at the center of the tank bottom, B_n/B=0.50. The solid particles used were polypropylene spheres of 100 μm in diameter.

The change in the rising bubble morphology was recorded by a video camera with 640x480 resolution, which can record a fast-moving object at 1/4000 sec. shutter speed. The

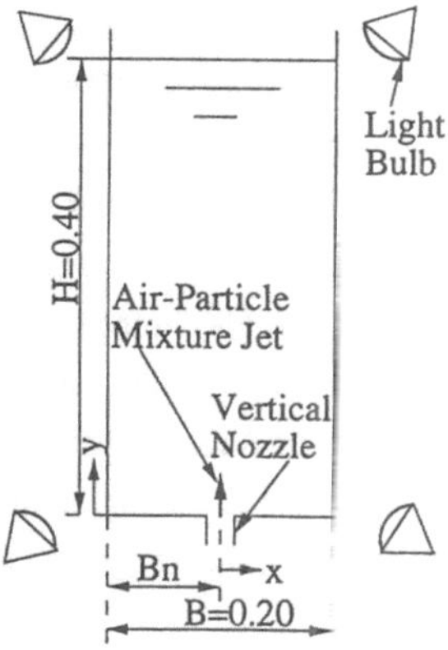

Fig. 1 A schematic of flow visualization apparatus

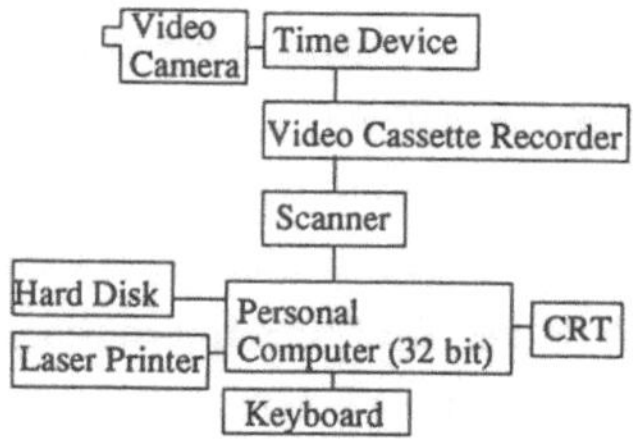

Fig. 2 A real-time image processing system

image data was supplied to an automatic data collection system, as illustrated in Fig. 2. The system consists of a video cassette recorder, a time device (being read to 1/60 sec), a scanner (for converting an optical image into a digitizing image) with 640x480 pixels, a monitor, a 32-bit personal computer, and a laser printer (printing dots at 300 DPI).

RESULTS AND DISCUSSION

Figure 3, for Q_g=0.20x10^{-3} m^3/s, is a photograph of rising bubbles in the tank with a vertical nozzle installed at the center of bottom wall, B_n/B=0.50. Corresponding still-image by means of image processing techniques is shown in Fig. 4. Based on the results of image processing, the bubble morphology became clearer, but information on its upward continuous motion was not extracted. It is desirable to have real-time imaging of the bubble movement along the upward stream. To achieve this objective, a slow-motion function of the video cassette recorder was employed to observe the phenomena of rising bubbles.

Fig. 3 Photograph of rising bubble for Q_g=0.20x10^{-3} m^3/s with the vertical injection nozzle at the center of bottom wall, B_n/B=0.50

Fig. 4 Still-image of the vertical movement of bubbles for Q_g=0.20x10^{-3} m^3/s with the vertical air injection nozzle at the center of bottom wall, B_n/B=0.50

Figure 5 depicts the rising bubble movement under the same operating conditions as those of Fig. 3. Each figure corresponds to the motion at 0.017-second intervals. Bubbles are shown as white in color. When the air was injected through a vertical nozzle at t=0 sec, the bubble at its outlet was mushroom-shaped and almost symmetrical in shape. After that, the bubble became larger and left the nozzle at t=0.050 sec. and at the same time new bubble was formed at its outlet, as shown in Fig. 5(d). The same bubble size as Fig. 5(a) in the vicinity of the air injection nozzle was seen at t=0.083. It was seen that the rising bubbles were rather deformed along the upward stream and simultaneously spread in the transverse direction. This situation becomes clearer if attention is focused on the bubble movement in the region marked by A in each figure. Real-time image processing on slow-motion video recordings was found to disclose both the rising bubble movement and its shape in the ladle.

An attempt was made to obtain the gas volume fraction in the air-water plumes by means of image processing on slow-motion video recordings. There is no widely accepted definition of the gas fraction, when the technique is employed. In the present study, this was defined as the ratio of bubble lifetime to total measurement time at an arbitrary location in the ladle. The results of the time-averaged gas fraction with and without particle injection are illustrated in Figs. 6 and 7, respectively. The measurements reveal the continuous spread of the gas bubbles within the plume, and the profiles become flatter and spread with ascension of bubbles. This trend was intensified when a stream of air mixed with solid particles was injected into the ladle, as shown in Fig. 7. It is postulated that the mechanisms which lead to a wider spread of the gas fraction is as follows: A reduction in the surface tension resulting from the contamination of the bubble surface by particle makes the break-up of the bubble surface easier, thereby promoting rapid dispersion.

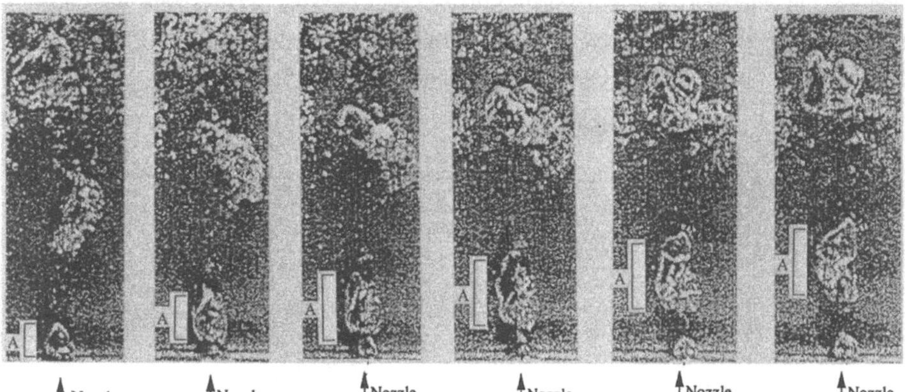

(a)t=0 Sec. (b)t=0.017 Sec. (c)t=0.033 Sec. (d)t=0.050 Sec. (e)t=0.067 Sec. (f)t=0.083 Sec.

Fig. 5 Image processing of rising bubble behavior for Q_g=0.20x10^{-3} m^3/s from the vertical air nozzle at the center of bottom wall, B_n/B=0.50

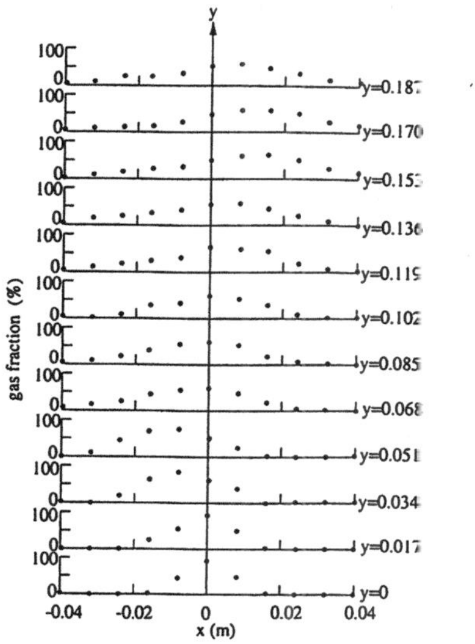

Fig. 6 Gas fraction distribution for Q_g=0.20x10^{-3} m^3/s with the vertical air nozzle at the center of bottom wall, B_n/B=0.50

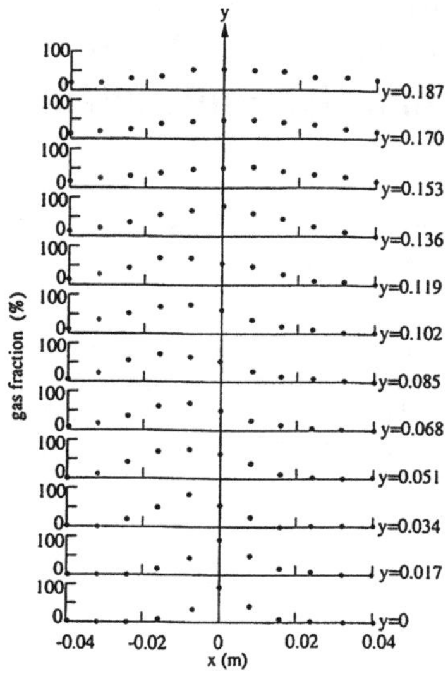

Fig. 7 Gas fraction distribution for Q_g=0.20x10^{-3} m^3/s and Q_s=3.50x10^{-6} m^3/s with the vertical air nozzle at the center of bottom wall, B_n/B=0.50

SUMMARY

Image processing has been employed to investigate the rising bubble morphology and the effect of particles on the air volumetric fraction with air-particle injection into a slender tank. The results obtained are summarized as follows: Real-time image processing with the aid of slow-motion video recording discloses the bubble and particle behavior in the upward direction. Particles cause a reduction in the surface tension of bubbles so that the rising bubble morphology is substantially distorted. Correspondingly, the profiles of the transverse distributions of air fraction become flatter and wider with ascension of bubbles than those of the no-particle injection.

REFERENCES

1. Balint, J. L.; Ayrault, M.; Schon, J. P.: Quantitative investigation of the velocity and concentration fields of turbulent flows combining visualization and image processing. In: Flow visualization III (ed. Yang, W. J.), 254-258. Washington: Hemisphere 1985.

2. Hernan, M. A.; Parikh, P.; Sarohia, V.: Digital image processing (Application to spray and flammability studies). In: Flow visualization III (ed. Yang, W. J.), 269-273. Washington: Hemisphere 1985.

3. Tacke, K. H.; Schubert, H. G.; Weber, D. J.; Schwerdtfeger, K. 1985: Characteristics of round vertical gas bubble jets. Metallurgical Transactions B. 16B, 263-275.

4. Torii, S.; Yang, W. J. 1991: Bubble and particle behavior in air-solid injection, under review process.

Application of Real-Time Neutron Radiography to Visualization and Measurement of Thermal Hydraulic Phenomena

N.Takenaka [1], T.Fujii [1], A.Ono [2], K.Sonoda [3], S.Tazawa [4] and T.Nakanii [4]

1) Dept. of Mechanical Engng. 2) Col. Lib. Arts, Kobe Univ., Nada, Kobe 657 Japan
3) Col. Lib. Arts, Himeji-Dokkyo Univ., Himeji 670 Japan
4) Sumiju Exam. & Insp. Co., Ltd., Toyo, Ehime 799-13 Japan

Summary

Various flow patterns of multiphase flows in a metallic tube and container and a liquid metal flow are visualized by real-time neutron radiography. Visualized images are image-processed for the measurement. Void fraction distribution of a water dispersed flow in a nozzle were measured. The velocity field in a liquid metal flow was obtained. It is shown that the neutron radiography is efficiently applied to flow visualization and measurement which are difficult by the other methods.

Introduction

Visualizations of a flow pattern and streak lines are very important to study thermal hydraulic phenomena. Experimental systems can be classified into three cases concerning to the visualization techniques;

1. systems in which both the working fluid and the wall are transparent for visible rays,
2. those which can be simulated by the transparent systems, and
3. those in which either the working fluids or the wall is opaque for visible rays and are difficult to be simulated by the transparent systems.

New techniques have been developed for the visualization of the case 3 and some of them are radiography techniques. Applications of neutron radiography have been proposed for nondestructive examinations since the attenuation characteristics of neutron ray are quite different from those of X-ray and γ-ray. Many applications have been reported [1]-[4] and its techniques were summarized in a handbook [5]. Some of them are methods to obtain still images on films and others are real-time methods by a video camera. The real-time neutron radiography methods applicable to observe the dynamic behaviors of the thermal hydraulic phenomena which are difficult to be visualized by the other methods.

The purpose of this study is to demonstrate the applicability of the neutron radiography to visualization and measurement of various thermal hydraulic phenomena. Cyclotron based real-time neutron radiography system at Sumitomo Heavy Ind. [6] was used.

Principle of Neutron Radiography

Radiography is a technique to visualize the structure of an object by the difference of the attenuation rate of radio rays to the materials of the object. Roentgen is a popular method of X-ray radiography.

A radio ray is attenuated exponentially in an object depending on the characteristics of the ray and the materials of the object. The ray intensity in a one-dimensional system is expressed as

$$N = N_0 \exp(-\mu_m \rho x) \qquad (1)$$

where N_0 is the initial intensity of the ray, N is the intensity in the object at the distance x, ρ is the density of the material and μ_m is the mass attenuation coefficient. The material is more transparent for smaller values of μ_m and is more opaque for larger values.

The mass attenuation coefficients of thermal neutron and X-ray of 125 keV for various elements are shown in Fig.1 [5]. The mass attenuation coefficients of X-ray increases with increasing atomic number. On the other hand, those of neutron ray depend on each element. They are large for small atomic number elements of H, B, Li and for some special elements such as Cd, Gd, etc.. They are small for most of metals. Therefore, the neutron radiography is suitable for visualizing the behaviors of light materials like water inside of a metallic wall and a liquid metal flow.

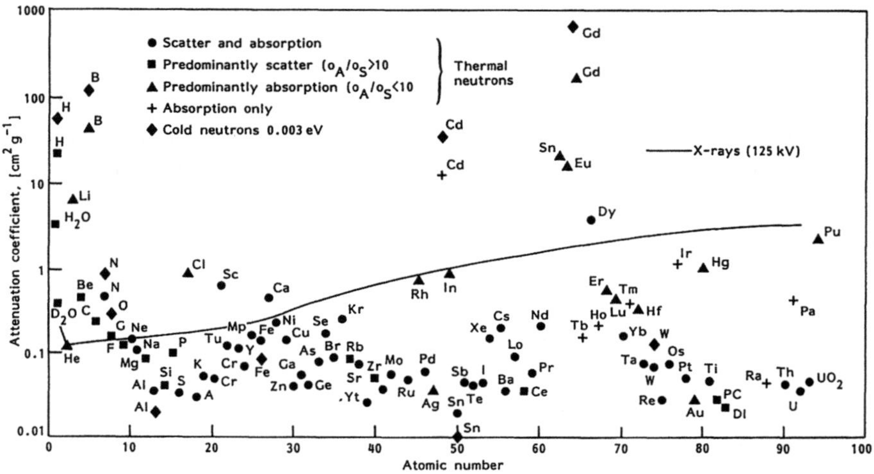

Fig.1 Mass attenuation coefficients of neutron and X-ray.

Visualized and Processed Images

Flow patterns of various multi-phase flows and liquid metal flows were visualized. Examples of the visualized and processed images are shown as follows:

Figs.2 (a)-(f) show various flow patterns of nitrogen gas-water two-phase flow in a stainless steel tubes of 25 mm in I.D. and 1 mm in thickness, water inverted annulars flow with and without a twist tape in a stainless steel tube of 20 mm in I.D. and 1 mm in thickness and fluidized beds of boron oxide powder in an aluminum pipe of 23 mm in I.D. and 1 mm in thickness.

Fig.3 shows an example of image process results on the two-dimensional void fraction distribution of water flushing dispersed flow in an aluminum divergent-convergent nozzle.

Fig.4(a)shows the flow visualization of Leed-Bismuth eutectics natural convection by a dye method. The dye of Leed-Bismuth-Cadomium alloy was filled in pipe and the dye followed the flow penetrating through holes in the pipe. Fig.4(b) shows the tracer in a Leed-Bismuth forced convection flow. Particles of intermetallic compound of Gold and Cadomium, $AuCd_3$, were employed as the tracer since its density is close to the eutectics and the eutectic wets well the compound with using solder paste. The flow vector field is shown in Fig.6.

Conclusion

Various flow patterns of multiphase flows in metallic tubes and nozzle and liquid metal flows were visualized by neutron radiography. Void fraction distributions were obtained by image processing methods. It is shown that neutron radiography can be effectively applied to thermal hydraulics researches.

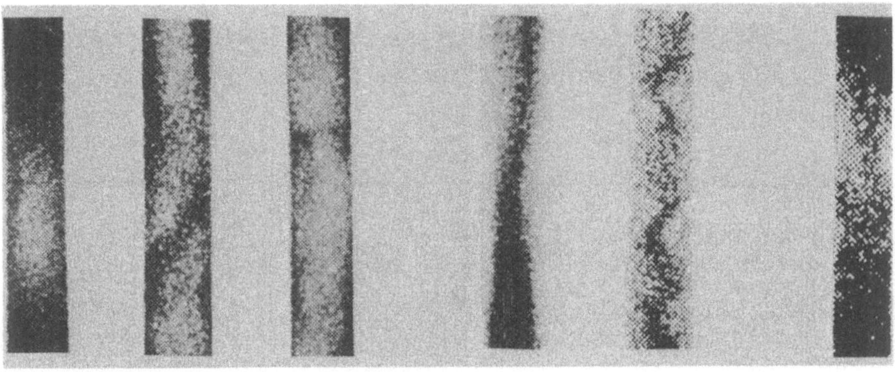

(a) (b) (c) (d) (e) (f)

Figs.2 Flow pattern of multiphase flows
(a)-(c) nitrogen gas-water two-phase flow,
(a) slug flow, (b) froth flow, (c) annular flow,
(d),(e) inverted annular flow,
(d) without a twist tape, (e) with a twist tape
(f) fluidized bed in tube.

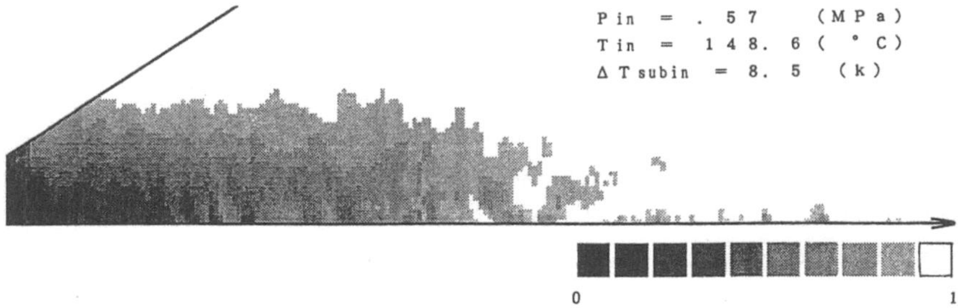

Fig.3 Two-dimensional void fraction distribution in a nozzle

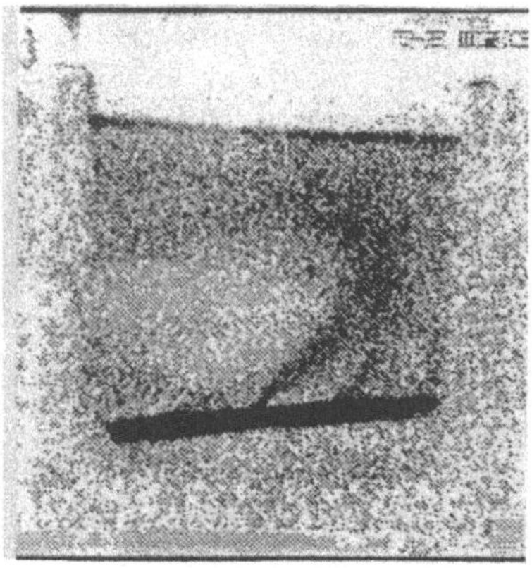

(a) by dye method

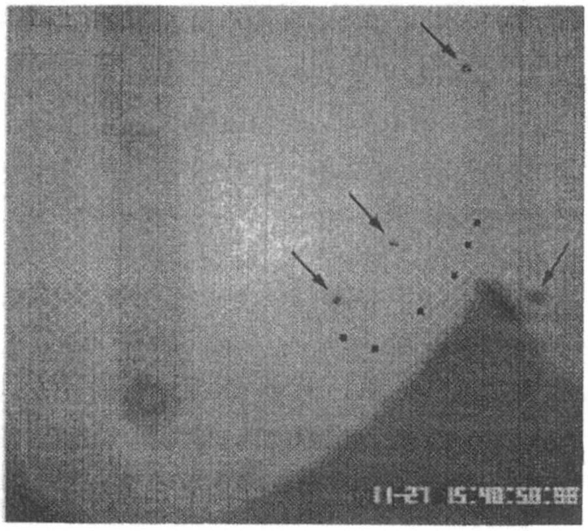

(b) by tracer method

Figs.4 Visualization of liquid metal flow

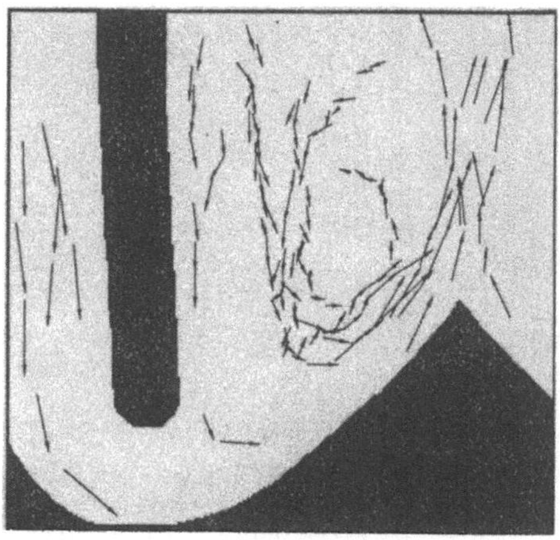

Fig.5 Flow vector field obtained by tracer method

References

1,2."Neutron Radiography ", Reidel Pub. Comp., 1981, 1986.
3,4."Neutron Radiography (3),(4)", Kluwer Acd. Pub., 1989., to be published.
5."Neutron Radiography Handbook", Reidel Pub. Comp.,1981.
6. Tazawa,S. et al., "Cyclotron-Based Neutron Radiography",in "Neutron Radiography 1986", pp.231-238.

Application of Digital Image Processing in Two-Phase Flow

Donevski, B.[1], Saga, T.[2], Kobayashi, T.[2] and Shoukri, M.[3]

1. Faculty of Engineering, University of Bitola, 97000 Bitola, Macedonia.
2. Institute of Industrial Science, University of Tokyo, 7-22-1 Roppongi Minato-ku, Tokyo 106, Japan.
3. Department of Mechanical Engineering, McMaster University, Hamilton, Ontario, Canada, L8S 4L7.

ABSTRACT

A method for digital image processing and analysis of two-phase bubbly flow is presented. The procedure determines : bubble size distribution, shape and trajectory. The digital image processing and analysis of the dynamics behaviour of vapor bubbles in subcooled flow boiling in vertical annular tube are described. The procedure can be employed to determine : an overall average bubble diameter, bubble life time, interfacial area, and bubble trajectory. These parameters are of fundamental importance in two-fluid model formulation, especially in relation to the rational design and effective operation of facilities in the energy power systems and in various industrial processes.

INTRODUCTION

In this work, a study of the dynamic behaviour of vapor bubbles in order to provide qualitative data about the physical mechanisms that have previously been assumed in subcooled flow boiling as one of the most important non-thermal equilibrium two-phase flow system. Present research involves a quantitative assessment of some basic parameters such as local bubble size and size distribution in the range of void fraction between $0.03 < a < 0.07$, recognized in the present state-of-the art of vapor generation in channels as undeveloped subcooled boiling region or " wall voidage region " [1]. The measurements include bubble life time, average distance traveled, and bubble growth and collapse as a function of the test parameters. The visualization experiments were conducted in a low pressure test loop located in the Department of Mechanical Engineering at McMaster University. Details of the experimental program, which also included measurements of void generation and condensation are given in refes [1,2]. However, the results indicate, that an understanding of the effects of various hydrodynamics and geometric parameters on the distribution of local conditions of the two-phase flow systems is of fundamental importance in model formulation and in developing constitutive equations for interfacial transfer terms which control transport of mass, momentum and energy in steady and transient subcooled flow boiling.

The subject of this paper is to describe the development of a digital image processing methodology based upon image processing of high speed motion pictures obtained of the flow field in the region of undeveloped subcooled flow boiling for a variety of experimental conditions, which appears to hold promise for determining the governing mechanisms in subcooled flow boiling, particularly near the point of net vapour generation.

DIGITAL IMAGE PROCESSING SYSTEM

The two-phase fluid motion in the vertical annular tube was photographed by a high speed cine camera. The 16 mm high speed motion test pictures (2000 frame / sec) were first converted to NTSC video image by television cine converter on a S-VHS video tape. The TV frames have a 0.0005 sec frame interval. Once the film was converted, digitization of the filmed images were undertaken on the Institute of Industrial Science at the University of Tokyo's Digital Image Analysis system, as shown in fig. 1. This facility is composed of an optical video disk recorder system (SONY LVS-5000, LVR-5000), an image processor (NEXUS 6410) with external image memory (5MB, 512×480 pixel, 8 bit) and a 16 bit micro computer (NEC PC9801VX). The video images converted from high speed film were recorded on an optical disk through a time base corrector. The consecutive TV frames can be selected arbitrary from the optical disk to the image processor. The selected images were digitized into the frame memory in real time by the image processor, and then are processed and analyzed. All of procedures of data acquisition and reduction are executed by the original computer program. The software was developed in Turbo Pascal.

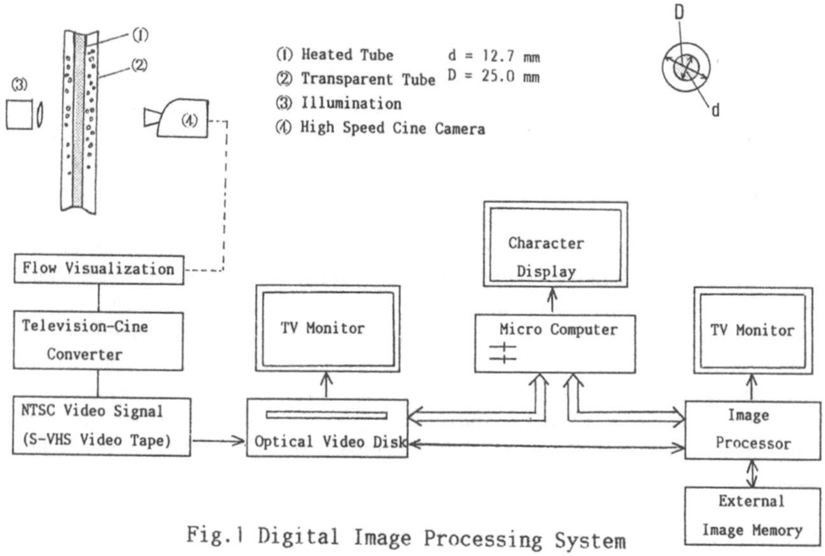

Fig.1 Digital Image Processing System

IMAGE PROCESSING AND DETERMINING THE LOCAL CONDITION

The software designed to analyze the films has two main functions : (1) - to perform statistical analysis on bubble size for a particular void fraction, and (2) - to collect a time of history of bubble growth and trajectory. A flow chart of the structure of the developed program software is presented in fig. 2.

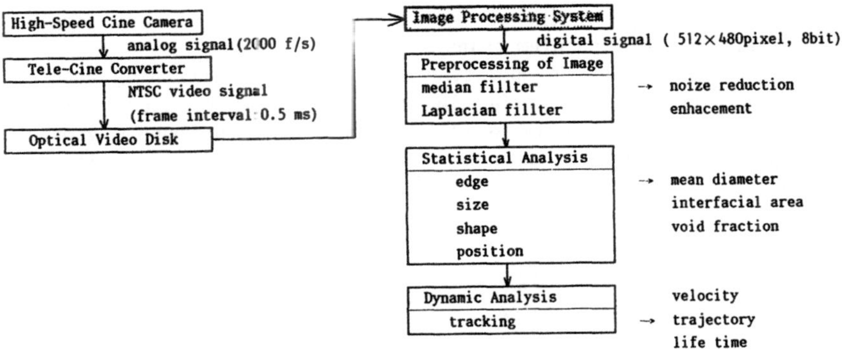

Fig. 2 Flow Chart of Image Processing for Bubble Size Measuring and Bubble Tracking

A typical 512 × 480 pixel, 256 gray-level digital image is presented on fig. 3(a), (bubbles moving upward downstream). The images had to be enhanced to remove the noise while still maintained the integrity of the image. The filters used for noise reduction and enhancement of input image, fig.3 (b) shows enhanced image using Laplasian filter.

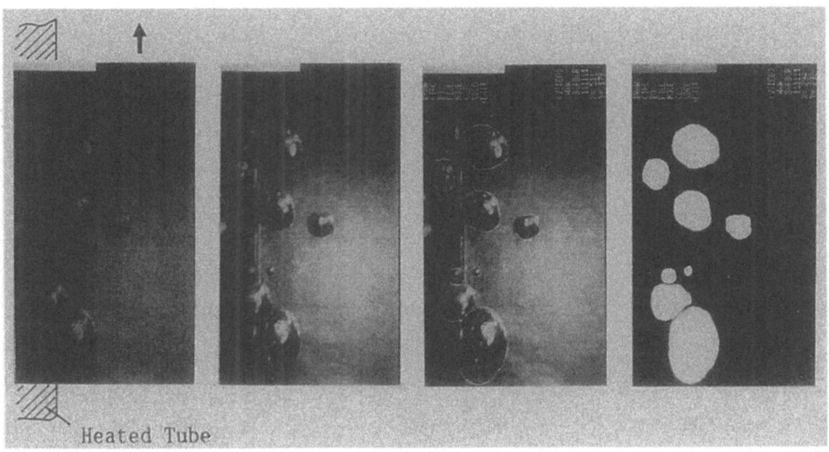

(a) Input Image (b) Enhancement (c) Edge Detection (d) Binarization
Fig. 3 An Example of Bubble Image Processing

The edges of bubbles were extracted by providing a thin homogenous gray-level strip along their surface, as shown in fig. 3(c). After a successful extraction of coding, the closed chain of each bubble was labeled and was stored with its interior pixels removed from the gradient image. Figure 3(d) shows binary image of original image. The description of coding a chain, is sufficient to evaluate the position of the bubble gravitational center, an average equivalent diameter, interfacial area and bubble trajectory.

At the present stage of the project it was decided that only single bubbles are suitable for inoculum transfer.

The " tracking mode " of operation collects data concerning bubble growth departure time, bubble life time and could also be used to find bubble velocities. Typical example of process during the bubble growth while attached to the surface, detachment and rising period until collapse is shown on fig. 4.

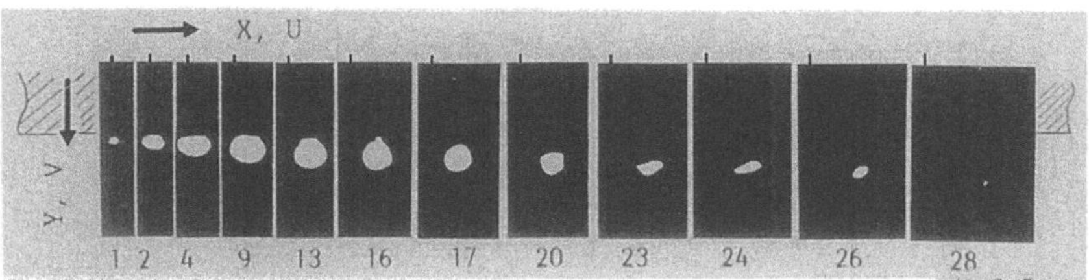

Fig. 4 Typical Process of Bubble Growth and Collapse

The bubble trajectory and the bubble size change which evaluated the bubble projection area S are also presented on figs. 5 (a) and (b). The velocity components in U and V directions for the bubble are shown as function of time on fig. 6. All these data, along with information about the level, heat flux, void fraction and degree of subcooling are stored in data file.

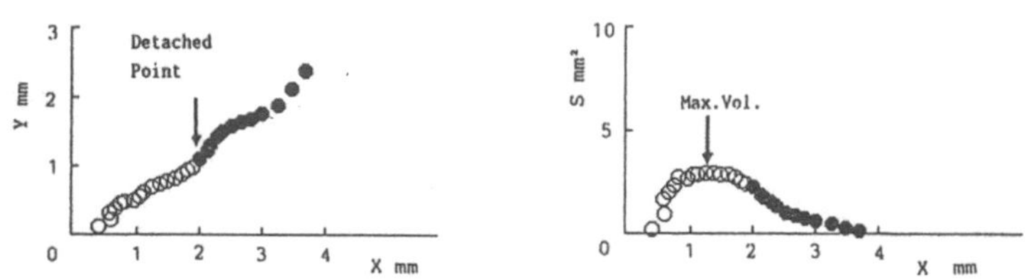

(a) (b)

Fig. 5 Bubble Trajectory and Bubble Shape Deformation

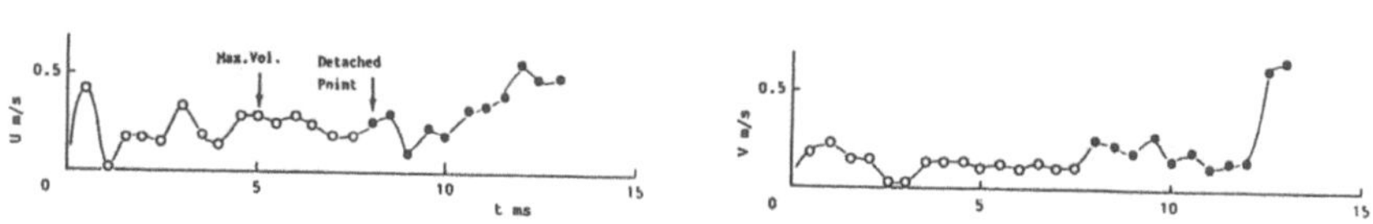

Fig. 6 U and V Components Distribution

SUMMARY AND CONCLUDING REMARKS

High speed photography, complemented with digital image processing technology was used to examine the two-phase bubbly flow characteristics. The results, accordingly, should be accurate, since the scale used was usually 0.024 mm / pixel, the measurements should be accurate within ± 0.048 mm.

It was found from the image processing that, when growing bubbles on the heated surface reach a certain diameter determined by hydrodynamics and thermal force balance, before detachment from the surface by rapid ejection and continue to flow downstream until they collapse.

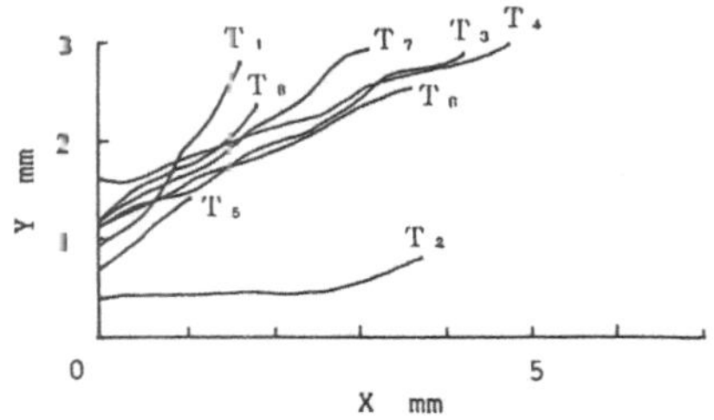

Fig.7 Typical Bubble Trajectories

However, some bubbles detach from the surface and sweep downstream moving parallel to the wall recondensing slowly in the bulk flow, as shown the pattern of trajectory of T2 on fig. 7. Whether bubbles slide along the wall or detach by rapid ejection, they will contribute to axial volumetric flux of the mixture. Because of the bubble migration into the bulk stream, the void concentration profile in the radial direction exhibits maximum and minimum point.

The estimated parameters are of fundamental importance in flow modeling and development of constitutive equations for interfacial terms in low pressure subcooled flow boiling. The data collected by the image analysis software can be incorporated into new models currently under development, which are aimed at incorporating the effect of vapour generation and condensation separately.

REFERENCES

1. Donevski, B., et. al., Mech. Eng. Dep., MacMaster University, Report No.ME/89/TF/RI,1989.
2. Shoukri, M., et.al., Phase-Interface Phenomena in Multiphase Flow, Eds. Hewitt, G.F., et.al., Hemisphere Pub. Co., New York, 1991, pp.413-421.

AKKNOWLEGEMENTS

This work was supported by National Science and Engineering Research Council of Canada, and also by Institute of Industrial Science, the Univ. of Tokyo. The authors are grateful to Mr. Segawa, S. of I.I.S. the Univ. of Tokyo for his technical assistance during the course of this work.

A Transmission-Mode Ultrasound Computerised Tomographic System for Gas/Liquid Two-Phase Flow Visualization

Lingan Xu*, Zhenxiang Chen**

*Department of Electrical Engineering and Automation, Tianjin University, Tianjin, China.
**Department of Automation, Tianjin Institute of Light Industry, Tianjin, China.

Abstract

The authors have developed a ultrasound transmission-mode tomography system for monitoring the gas/liquid two-phase flow developed by the authors. A fast and effective simple back-projection algorithm based on the time of flight measurement is used.

The experimental results for imaging a static bubble model shown that this system can run at a fast speed and the quality of reconstructed image is acceptable.

1. INTRODUCTION

Since the end the of 1970s', in order to obtain distribution pattern of dispersed phase, scientist and engineers have been trying to "transplant" the medical computerized tomography (CT) technique to two phase flow measurement. This technique (called as the Process Tomography(PT)) might offer a image of distribution of the interested physical parameters over the cross section of a pipe. In this paper, a transmission-mode ultrasound computerized tomographic system developed by the authors is described, in which a simplified backprojection algorithm is adopted. The simulation and experiment results shown that the quality of the obtained image is reasonable at a reconstruction speed of 10 frames per second.

2. Principle

The developed UCT system is based on the measurement of the flight time of ultrasound pulse in between receivers and transmitter.For the bubbly air/water two-phase flow, the following assumpticns are adopted . (1) The ultrasound velocity is constant in the water. (2) The bubble is opaque. (3) the diameter of the bubble is much larger than the wavelength. The usual Radon transform equation can be re-written as

$$[\mathcal{R}f](\ell,\theta)=\begin{cases} T, & \text{If there is no bubble in the line} \\ \text{Not } T, & \text{If there is bubble in the line} \end{cases} \quad (1)$$

Where T is the time-of-flight taken for the transmitted pulse traveling along the line in water. which is determined by the integral path uniquely, i.e. by the distance between the transmitter and corresponding receiver. In our system the "T" is defined as "true; 1 " and the "NOT T" as "false; 0". then the equation (1) become

$$[\mathcal{R}f](\ell,\theta)=\begin{cases} 1, & \text{If there is no bubble in the line} \\ 0, & \text{If there is bubble in the line} \end{cases} \quad (2)$$

After the revised Radon transform above takes place at many different angles, a set of projection is obtained. The projections obtained can be used to reconstruct the image.

3. The experimental system

As shown in Fig. 1, the UCT system consists of transducer sub-system, pulser, data acquisition system and a image reconstruction computér. The scheme of transducer sub-system is shown in Fig. 2. Thirty six transducers made by PZT ceramic are mounted in the wall of the pipe symmetrically. Six of them are half-cylindrical shape and others rectangular. The former can fire a fan-shape beam and act as receiver too. The later act only as receivers.In a complete data acquisition cycle, every transmitter is excited by the pulser one times alternatively. On the wall of the pipe opposite to the excited transducer, 25 transducers are selected to be listener and 25 measurements are made. The data acquisition cycle repeats automatically under the control of a electrical complexer board and a IBM PC/XT, thus lead to faster scan. The time-of-flight is measured by the "threshold method". The acquired data are fed into the computer through a buffer and a parallel interface.

4. Image reconstruction algorithm

The used backprojection algorithm is implemented in a rather different way from the traditional one. It backprojects the data onto the corresponding region, not only onto the line which the data are made by Radon transform. As a example, the regions for transmitter 1 and corresponding receivers are plotted on the cross section of the pipe as shown in Fig. 3. The backprojection can be implemented as following

$$[G]=\sum_{k=1}^{m}[R_k]\ M_k \qquad (3)$$

Where [G] is the image matrix. M_k is Kth measurement and R_k describes the relationship between each pixel and the Kth measurement.

In order to improve the speed of image reconstruction, a simplified backprojection algorithm is designed by the authors, which is called as the logical backprojection algorithm. It can be described as follows,

If $A=[a_{ij}]$ $B=[b_{ij}]$

defining

$$A\odot B=C=[c_{ij}]\ ;\quad c_{ij}=\begin{cases}0, & \text{if } a_{ij}=b_{ij} \text{ and } M=1\\ a_{ij}, & \text{if } a_{ij}\neq b_{ij} \text{ and } M=1\\ a_{ij}, & \text{if } M=0\end{cases} \qquad (4)$$

Then, the reconstructed image [G] can be obtained by

$$[G]=E\odot R_1\odot R_2\odot\cdots\odot R_m \qquad (5)$$

Where E is a Unit Matrix.

5. Simulation and Experimental results

A simulation results for imaging a three bubbles model by using different transducer arrangements are shown in Fig.3 . It shows that for a tread-off between the quality of the image and scanning time. 8-10 transmitters system is suggested.

For a static physical model of three air bubbles in water. The reconstructed image obtained by the developed system is shown in Fig. 4. It is estimated that the system can produce one frame in 100ms by using an IBM PC/XT(4.77MHz).

7. Discussion

The results confirms that the developed UCT system based on the time of flight principle and simplified backprojection algorithm is reasonable. To improve the quality of the reconstructed image, besides the number of observing angle should be increased, the geometrical accuracy and identity of the transducer sub-system should be guaranteed.

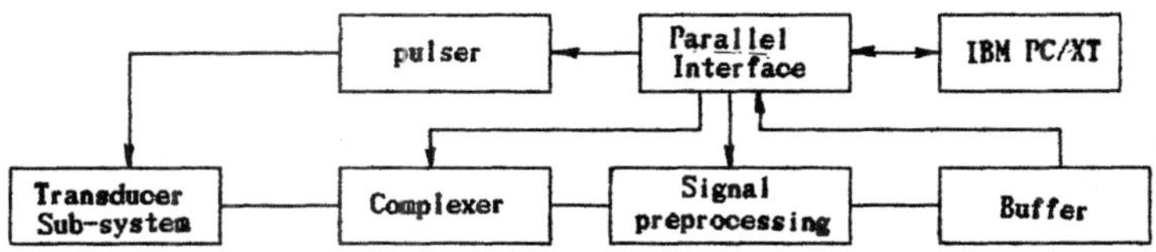

Figure 1.

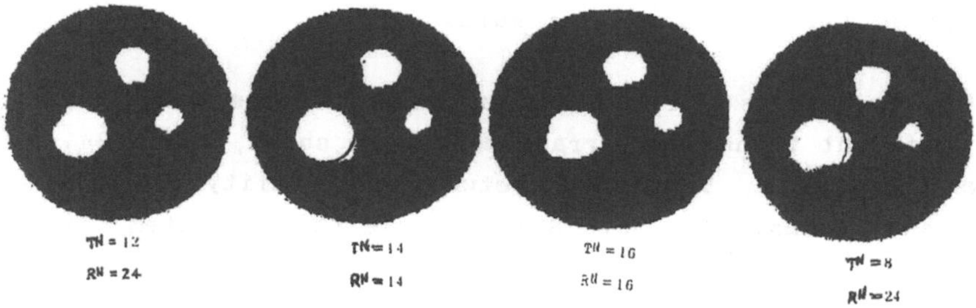

Figure 3.

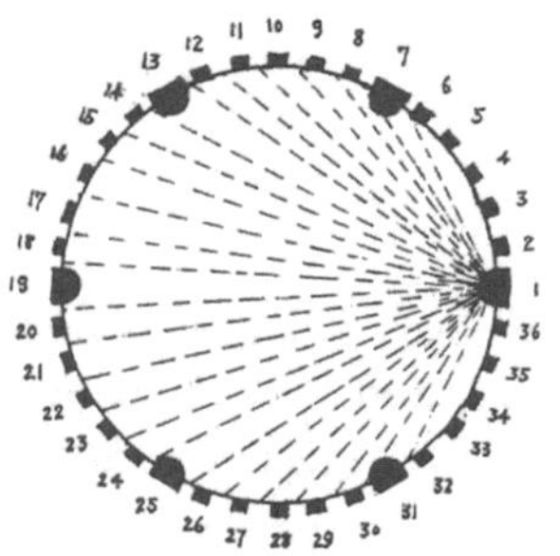

Figure 2

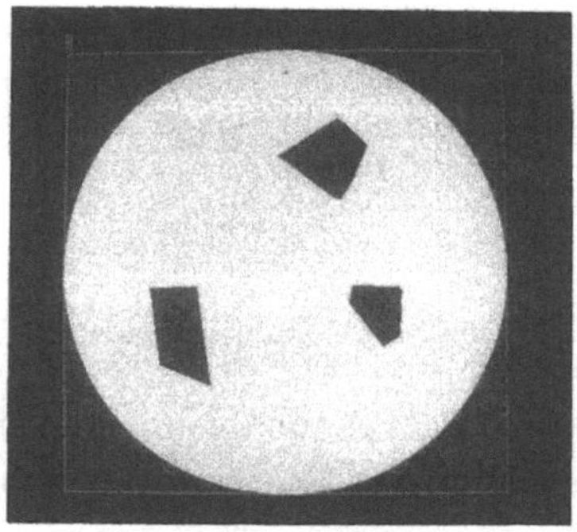

Figure 4.

REFERENCES:

1. Gai,H, Y.C. Li, A, Plaskowki and M. S. Beck 'Flow Imaging Using Ultrasonic Time-Resolved Transmission Mode Tomography' IEE 3rd Int. Conf. on Image Processing and Its Appl., University of Warwick.

2. Wolf, J., 'Investigation of bubbly flow by ultrasonic tomography', Part. & Part., Vol. 5, No.4, pp.170-173.

3. Zhenxiang Chen, Lingan Xu etc. 'Investigation of ultrasonic process tomography system for gas/liquid two-phase flow', The Third National Conference on Multi-phase Flow Measurement, Shanghai, Nov. 1990.

4. Herman,G.T. 'Image reconstruction from projections', Academic Press, London 1980.

High-Speed Photography, High-Speed Cinematography and High-Speed Holography as Tools to Investigate fast Flows

Fritz Eisfeld University Kaiserslautern Germany

Summary

First the charateristic values to estimate the quality of high speed images are described. Then the problems of exposure time and frequency are discussed. Some systems of high speed photography, high speed cinematography and high speed holography are demonstrated. Therefore also examples of high speed frames of diesel injection jets are shown. Finally, the possibilities of the analysis of high speed frames are regarded with.

Introduction

Already in the last century high speed photography was used to investigate fast motions and nowadays the high speed photography, the high speed cinematography, and the high speed holography are suitable especially to investigate fast flows. The advantage of these methods is that in such images all information of the events of the whole image field are included. Other optical methods, i.e. the Laser-Doppler-Anemometry, supplies values of the flow only in one point. An other advantage is the very good resolution of the photographic material, which is better than the resolution of a screen using fast video recording systems. Therefore, also the motion of very small particles with length of some μm could be made visible on a photographic film.

The frequency on high speed cinematography is up to 20.000.000 f/s and above. In combination with the shadow, schlieren or interferometric methods fields of concentration or temperature can be made visible. A great advantage is that the information about fast events are stored so that it is possible to analyse such images later. When we take an image series then we can give declarations on local and temporal changes of the flow.

Characteristic values to estimate the quality of high speed photo-graphic systems

A problem is the choice of the optimal high speed photographic system. Therefore Schardin has developed characteristic values for the time resolution and the optical resolution [1]. The characteristic value for the time resolution is

$$K_1 = f_w * ^2/3 \qquad (1)$$

$$g = t_w/t_p \qquad (2)$$

and the optical resolution

$$K_2 = A/\lambda^2 = A*n^2 \qquad (3)$$

f_w : frame frequency
t_w : temporal distance between 2 exposures
t_p : exposure time at one frame element
A : frame area
n : optical resolution lines/mm $\lambda = 1/n$

Table 1 shows some examples for K_1 and K_2 using a multispark camera , a rotating prism camera and a rotating mirror camera.

Table 1 Characteristic values for several high speed cameras

System	number of frames	n lines/mm	A mm^2	f_w f/s	t_p s	t_w s	K_1 s^{-1}	K_2 points
multi spark	24	80	8x11	$2*10^5$	$5*10^{-7}$	$5*10^{-6}$	$9,3*10^5$	$5,6*10^5$
rotating prism	2000	60	7,5x10	$5*10^3$	$4*10^{-5}$	$2*10^{-4}$	$1,5*10^4$	$2,7*10^5$
rotating mirror	200	80	8x24	10^6	10^{-8}	10^{-6}	$2,2*10^7$	$1,2*10^6$

The possible resolution is a function of the photographic material but also of the optical system. If the resolution is limited by diffraction the difference between two points is then

$$\Delta x = 1,22*\lambda*z/D_{eff} \qquad (4)$$

with λ : wave length of the light z : distance of the object

D_{eff} : effective diameter of the aperture

Hence it is difficult to take smallest particles. If the particles are too small, it is necessary to reproduce the object with a scale larger than 1 of the film.

The necessary exposure time and frame frequency

To receive optimal high speed frames it is necessary to optimize the exposure time and the frame frequency. In both cases the characteristic length and the velocity of the object are the important parameters for the relative path of the object e. g. of a particle, especially if the object is very fast and small. The relative path of the object during the exposure time is given by

$$F_{ft} = \frac{t_e * v_p}{L_p} \qquad (5)$$

with F_{te} : dimensionless path of the object during the exsosure time
t_e : exposure time
v_p : velocity of the object
L_p : characteristic length of the object

The relative path between two exposures is given by

$$F_t = \frac{v_p}{L_p * f} \qquad (6)$$

with F_t : dimensionless path of the object between two exposures
f : framing frequency

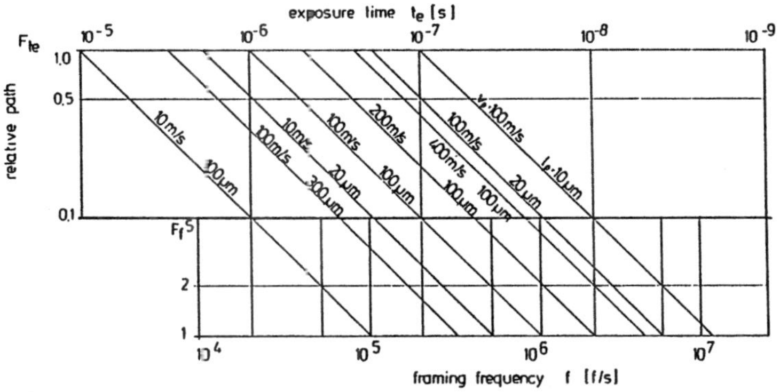

Fig.1 Diagram to estimate the exposure time and frequency

Fig. 1 shows F_{te} as function of the exposure time and F_f as function of the framing frequency for very small objects e. g. fuel droplets [2]. The limit is given for F_{te} by 1,0 and for F_f by 5,0. This demonstrates that by investigations of fast fuel droplets exposure times of 10^{-8} s and frequencies of 1.000.000 f/s are necessary. The range of the rotating prism cameras reaches up to $2*10^3$ f/s, the range of the drum cameras up to $2*10^5$ f/s and the range of the rotating mirror cameras up to $2*10^7$ f/s.

<u>The high speed photography.</u>

Only in a few cases a single high speed picture is sufficient for the explanation of a fast event. An example is the estimation of the number of droplets of an atomized injection jet, but to estimate the velocity of a droplet, 2 pictures are necessary. This double pulse method is often using a pulsed solid state laser. To investigate the motion of objects more images are needed.

A method which gives more images is the Cranz-Schardin-Method. Here a multi spark light source is used which gives the same number of images like the number of sparks. Here the photographic film is not moved. Fig. 2 shows the scheme of a Cranz-Schardin-Camera [3]. This system works with shortest exposure times and very high frequencies. However the disadvantage is the small number of images.

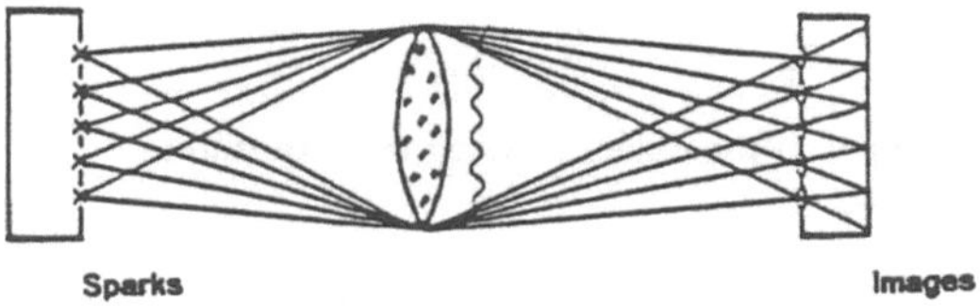

Fig. 2 Scheme of a Cranz-Schardin-Camera

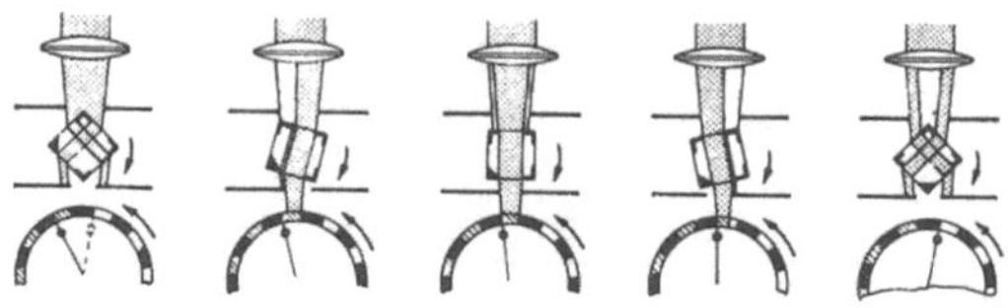

Fig. 3 Scheme of a rotating prism camera

The high speed cinematography.

For high speed cinematography several types of cameras are used. The most important ones are rotating prism cameras, drum cameras and rotating mirror cameras.

The rotating prism cameras are appropriate for frequencies up to 30.000 f/s. To reach the higher frequencies it is necessary to use a half or a quarter of the images format. The length of a frame series is 2000 and above. The film gives moving images. These cameras have an optical image separation [4]. This system is demonstrated in fig. 3.

For faster events, drum cameras in combination with pulsed light sources, i.e. sparks or pulsed lasers are used. The frame separation is the result of the light pulses. Such cameras are suited for frequencies up to 200.000 f/s. There are no corrections of image shifting. The height of the images is a function of the frequency.

A new drum camera using a polygon mirror has an optical image separation and no image shifting. The exposure time can be varied by a variable slot . The scheme of this system is shown in fig. 4. This camera works with 35mm or 70mm film. The largest image format is 15x24 mm or 15x58mm and the attainable frequency goes up to 200.000 f/s.

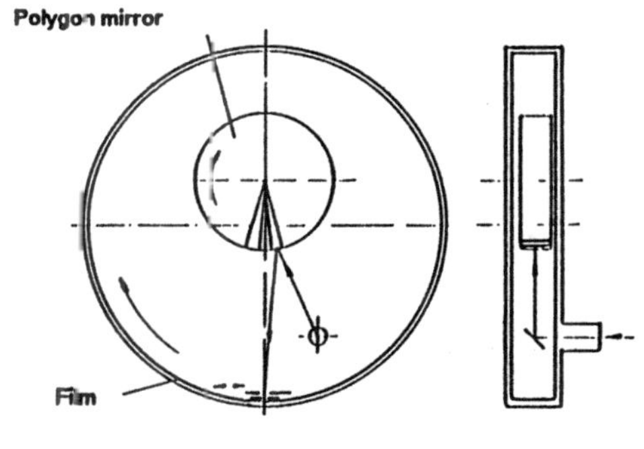

Fig. 4 Scheme of a new drum camera

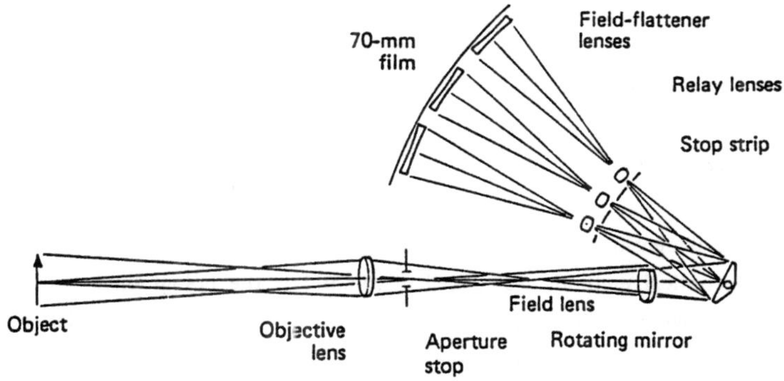

Fig. 5. Scheme of a Framing camera

For very fast events, rotating mirror cameras are suitable. There exists two types, either with a relay lens bank for image separation or the streak camera. Fig. 5 shows the type with a relay lense bank [5]. The problems of this system are the small optical resolution and the not variable image format. The frequencies are up to 20.000.000 f/s and above. Because of these reasons, a new rotating mirror camera was developed. This camera works with an argon laser as light source using a cavity dumper as optical shutter and an external image reproducing system, which allows a reproducing scale up to 2o. The frame frequency is up to 1.000.000 f/s and above with exposure times of 10^{-8} s. This camera system has an optical resolution of 16 µm limited by diffraction and is used to investigate the spreading and atomization of diesel injection jets [6]. Fig. 6 shows the scheme of this camera system which can be used with one and two 35 mm films or one 70 mm film [7]. Herewith simultaneous events in two planes can be taken. Fig. 7 demonstrates some images of an injection jet taken with 480.000 f/s .

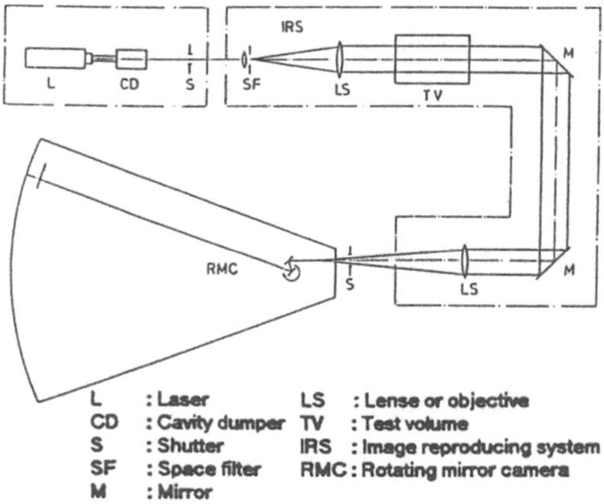

Fig. 6 Scheme of a rotating mirror camera

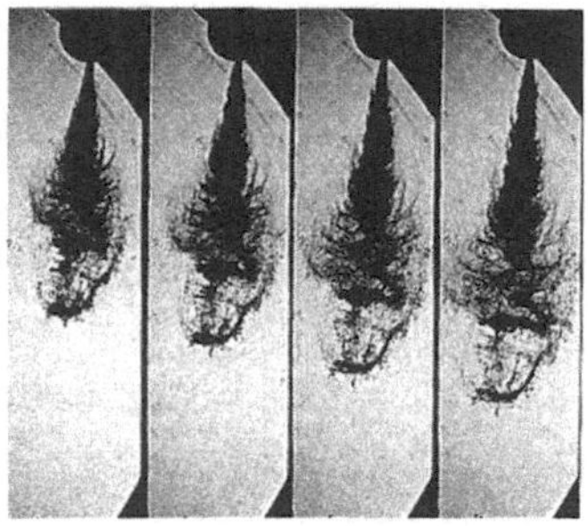

Fig.7 Frames of injection jet

The high speed holography

Many events are 3-dimensional e.g. the spreading of an injection jet. Now the question is, if there are suitable optical methods. An excellent method is the holography, but the small sensitivity of the holographic film material causes problems. A light source which produces coherent light of highest intensity is necessary. A solid state laser is not suitable, if a frame frequency larger then 10.000 f/s is required [2]. For these cases a new drum camera for frequencies up to 200.000 f/s was developed. An argon laser as light source is sufficient for using this camera. The holography allows only an image shifting of a tenth of the wave length of the light. This camera works with a compensation of the image shifting and with an optical image separation. In this new drum camera, which is a modification of the camera in fig. 4 the light beam moves synchronized with the film, i.e the velocity of the beam in the film plane is equal to the velocity of the film. There are two polygon mirrors and two objectives. This optical system is shown by fig.8.

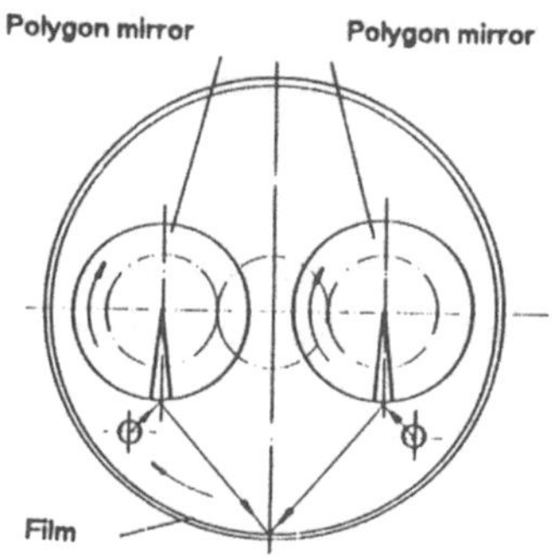

Fig. 8 Scheme of a drum camera for high speed Holography

The analysis of high speed frames.

A high speed image of a flow field contains a very large number of information. The problem is to analyse all this information. The best method is to use an electronic image analysing system consisting of a CCD camera and a computer. The advantage is that the CCD camera differs 256 degrees of grey values but the human eye only about 60. Hence it is possible with such a system to analyse images from injection jets, treating lots of aspects e.g. the velocity of the jet, droplets and fuel files, the shape of droplets, sources, sinks, and vortices in the flow field, and beyond that the number of fuel particles [8].

Conclusions

To investigate fast events especially flows of gases and liquids high speed photography, high speed cinematography and high speed holography are suitable methods. For this purpose several types of high speed cameras are available. A new type of drum camera was developed especially for high speed holography. With this camera it is possible to receive 3-dimensional information. The use of electronic analysis systems for an optimum of information is essential.

References

1. Schardin, H.; Grundlagen für meßtechnische Anwendung der Kinematography, insbesondere zur Untersuchung schnellveränderlicher Vorgänge. Schweiz. Photo-Rundschau 1951, S.294-303.
2. Eisfeld, F.; Investigation of Diesel injection jets using high speed photography and high speed holography. Proc. 19th ICHSPP 1990 SPIE Vol. 1358 p. 660-671.
3. Vollrath, K.; Thomer, G.; Kurzzeitphysik. Springer, Wien, 1967.
4. Patzke, H.G.; Früngel, F.; 3-D HIGH-SPEED PHOTOGRAPHY. Proc. 6th ICHSPP 1962, H.D. WILLINK & ZOON N.V. - HARLEM 1963 p. 250-257.
5. Shaw, L.L.; Sonderman, J.B.; Seppala, L.G.; LARGE FORMAT 70-mm HIGH -SPEED FRAMING CAMERA. Proc. 13th ICHSPP; Jap. Soc. of Precision Engn. 1979, p. 218-225.
6. Eifler, W.; Untersuchungen zur Struktur des instationären Dieselöleinspritzstrahles im Düsennahbereich mit der Methode der Hochfrequenz-Kinematografie. Diss. ; Kaiserslautern 1990.
7. Eisfeld, F.; Investigation of the Penetration and Spreading of an Instationary Injection Jet Using High Speed Cinematography with Frequencies up to 1.000.000 f/s. Proc. 18th ICHSPP 1989 p. 902-909.
8. Wagener, J.; Ein digitales Bildverarbeitungssystem zur Auswertung hochfrequenter Bildfolgen von Dieseleinspritzstrahlen. Diss. Kaiserslautern 1992.

Laser Light-Sheet Visualization of the Spatial Evolution and Mixing of Coherent Droplet Structures

Jeffrey M. Grace[1] and Patrick F. Dunn[2]
Particle Dynamics Laboratory
Hessert Center for Aerospace Research
Department of Aerospace & Mechanical Engineering
University of Notre Dame
Notre Dame, IN, USA 46556

Summary

This paper presents the results of using the laser light-sheet technique to examine the spatial organization of droplets within an electrohydrodynamic spray. This technique is well suited for such sprays in that it can clearly delineate the boundaries of the spray and identify spatially segregated groups of droplets that occur within the spray. The technique further provides a means to quantify the extent of mixing between groups of droplets.

An electrohydrodynamic spray of typically 1 to 40 μm ethanol droplets was examined using this technique. Two electric field configurations were studied: normal and perturbed. The EHD fine spray generated using a standard capillary-plate configuration defined the normal case. This spray then was modified by introducing a small probe with an electrically-grounded tip into the spray. This modification defined the perturbed case. Video images of these sprays were acquired at a rate of 30 frames/s at two viewing angles: a side view and a top view. These were digitized into a personal computer using a frame grabber board and then displayed and enhanced using imaging software. The resulting images supported that the laser light-sheet visualization method can be used to qualify and quantify droplet mixing and the lateral spread of an electrohydrodynamic spray.

Introduction

Electrohydrodynamic (EHD) atomization generates micrometer sized droplets through the action of an applied electric field on a working fluid. The technique offers the additional control parameter of droplet charge over the traditional atomization techniques, e.g., ultrasonic, airblast or pressure atomization. In the normal EHD capillary-plate configuration (see Fig. 1), the capillary tube directs the working fluid and concentrates the electric field to a sufficiently high intensity. The resulting EHD spray exhibits distinctly different modes as a function of applied potential difference between the capillary and the tip (Drozin[1]). The experiments discussed are restricted to investigating only one particular mode, the fine spray mode. This mode is characterized by a polydisperse size distribution, a source-like spread of the droplets and a rapid decay in both droplet number density and velocity away from the point of droplet generation. The source-like dispersion of droplets is a direct consequence of the divergence of

---

[1]Doctoral Candidate
[2]Associate Professor: To whom correspondence should be addressed.

the electric field lines. Analytical calculations show that these lines rapidly diverge in the vicinity of the capillary tip and become parallel in the vicinity of the ground plate. An analysis of the momentum equation for a single droplet reveals that only the electrical and drag forces are dominant; gravitational and inertial forces are negligible everywhere (Snarski and Dunn[2]). The droplets must, therefore, follow the electric field lines and, as these lines cannot cross one another, there can be no mixing of droplets beyond the generation region. In this paper, mixing is defined as the convective transport of droplets across diffusional paths. Thus, monospecific droplet diffusion is not considered mixing, whereas droplet diffusion combined with convection is (Ottino[3]).

Some applications of EHD sprays can be improved through induced droplet mixing and control of the lateral extent of the spray. The focus of this study is not to describe mechanisms that control droplet mixing and the lateral extent of the spray. Rather, it is to determine if, in fact, mixing can occur and the lateral spread can be controlled by shaping the electric field within the spray. The laser light-sheet visualization technique provides an excellent method to understand mixing within modified EHD sprays.

Experimental Apparatus

The experimental apparatus used in this investigation consisted of an EHD spray system, a laser light-sheet and an imaging system. The EHD spray system consisted simply of a positive DC high voltage supply connected between a metal capillary (406 μm outer diameter), through which ethanol was pumped at 5 mg/s, and a distant, electrically-grounded plate. Typical electric field intensities at the capillary tip were on the order of 10^7 V/m. The laser light-sheet was produced by directing the beam of a 5 W Ar-ion laser through a cylindrical lens. The light scattered from the droplets crossing the laser light-sheet was recorded with a video camera at 30 frames/s. The orientation of the light sheet defined two views: horizontal and vertical. The horizontal view had a light sheet perpendicular to the capillary axis. The viewing angle for this configuration was at 60° (back-scatter) and 45° elevation from the ground plate plane as shown in Fig. 1. The vertical view had a light sheet parallel to the capillary axis and a viewing angle of 90° (side-scatter) and 0° elevation from the ground plate plane. The video images were digitized into a personal computer using a frame grabber board and then analyzed using imaging software.

Visualization

The horizontal and vertical images obtained at different spray locations are displayed in the following figures. The vertical images were obtained along the spray centerline while the horizontal images were obtained at two axial, z, locations. The axial locations were nondimensionalized by a characteristic distance, z', of the investigation equal to 0.2 L, where L = 0.2 m defined the capillary tip-to-ground plate distance. Two different electrical configurations were studied. The first configuration, the normal case, was the reference case from which the degree of mixing and spread control was assessed. This was the standard EHD spray configuration as described above. The second case involved the presence of an isolated grounded surface (~2 mm^2), an electrically insulated probe with an electrically grounded tip located z below and 0.5 z radially off-axis from the capillary tip. This probe essentially modified the boundary conditions of the electrical potential (Laplace) equation so as to introduce a large curvature in the electric field lines within a region of the spray.

The normal case is considered first. Figs. 2a-c show the horizontal images obtained at z^* (=z/z') = 0.125, z^* = 0.5 and a vertical image at the spray centerline, respectively. These images reveal several small coherent structures very near the point of droplet origin and a relatively even distribution of droplets outside them. The digitized images of Fig. 2a show four slightly preferred droplet locations. The spatial evolution of the spread of the droplets was followed by scanning the light sheet axially along the spray. The droplet locations seen in Fig. 2a do not exist in Fig. 2b, indicating rapid diffusion. These horizontal views show the rapid decay in number density with axial position that is characteristic of the source-like spread of the droplets within the normal EHD spray. The vertical view, Fig. 2c, shows droplet streaklines which coincide with calculated electric field lines. This view clearly displays the lack of mixing, as predicted by the momentum equation, for the normal case. This view also defines the boundary of the spray and sets a reference for comparison with other configurations.

Digitized video images of the perturbed case, Figs. 3a-c, were used to assess the degree of mixing and the effect of the perturbation on the lateral extent of the spray. Figs. 3a-c show identical viewing angles and laser light-sheet orientations as in Figs. 2a-c. The horizontal view at z^* = 0.125, Fig. 3a, shows the existence of seven coherent structures. These structures do not exist with this degree of spatial resolution in the normal case. Tracking these structures axial motion by scanning the laser light sheet showed that there was not only convective mixing present in the form of large-scale structure interaction, but also diffusional mixing in the form of structure expansion. These views show significant modification of the source-like lateral extent of the normal case. The vertical view, Fig. 3c, provides the most compelling evidence for droplet mixing and control of the lateral spread of the spray. This pattern of mixing can be further understood by examining the analytical solutions to the electric potential (Laplace) equation for these cases. The solution shows that the field lines for the perturbed case converge near the probe tip. This creates an unstable flow situation, where the local number density of the converging droplets varies in time. Fig. 3c shows that these droplets do not cross over the centerline axis of the probe, but are displaced redirected outward, creating a turbulent-like motion and convective mixing. In this situation, droplets from one location interact with droplets from previously inaccessible locations, which constitutes mixing.

Finally, the sequential digitized images of the vertical views show the droplet streaklines. These can be used to determine convective length scales, which are another, more rigorous measure of mixing. This is the subject of future work. Also, the laser light-sheet technique has the ability to globally map a spray by determining relative positions of characteristic structures in real time. This ability, when coupled with laser-based, single-particle counters, can form a dependent technique capable of measuring droplet parameters in the time-dependent structures of complex sprays.

Conclusions

Laser light-sheet methods can track individual particles and global trends for analysis of the lateral extent, convective length scales and diffusional scales of an EHD fine spray. Laser light-sheet techniques coupled with other measurement techniques, such as laser-based, single-particle counters, can enhance the droplet mixing measurement capabilities to a level greater than the individual contributions. The laser light-sheet technique presented is useful for a qualitative study and limited independent quantitative study of droplet mixing as well as for a quantitative study of the control of lateral spray extent of an EHD fine spray.

References

1. Drozin, V.G., "The Electrical Dispersion of Liquids as Aerosols," Dept. of Chemistry, Columbia University, New York, pp. 158-164, 1954.
2. Snarski, S.R. and Dunn, P.F., "Experiments Characterizing the Interaction Between Two Sprays of Electrically Charged Droplets," Exp. in Fluids, Vol. 11, pp. 268-278, 1991.
3. Ottino, J.M., The Kinematics of Mixing: Stretching, Chaos, and Transport, Cambridge Univ. Press, 1989.

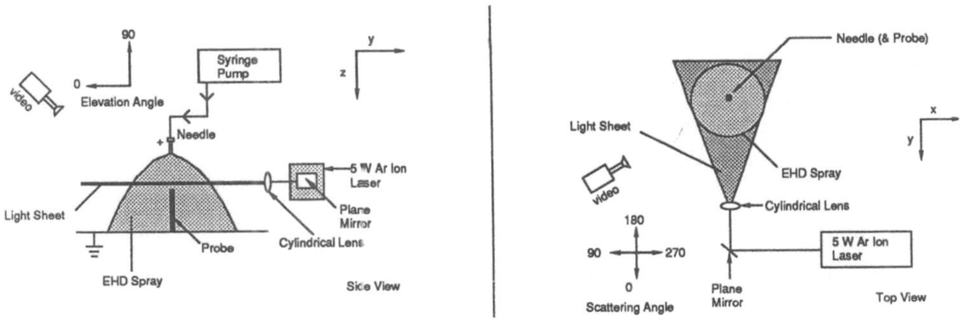

Figure 1. Experimental Apparatus

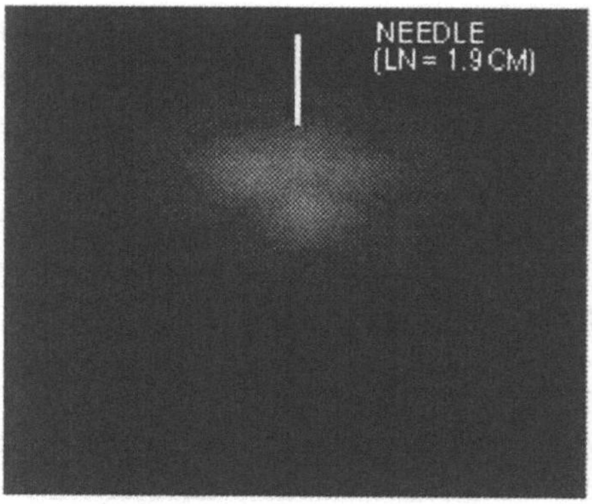

Fig. 2a: Case 1, z* = 0.125, 15 kV

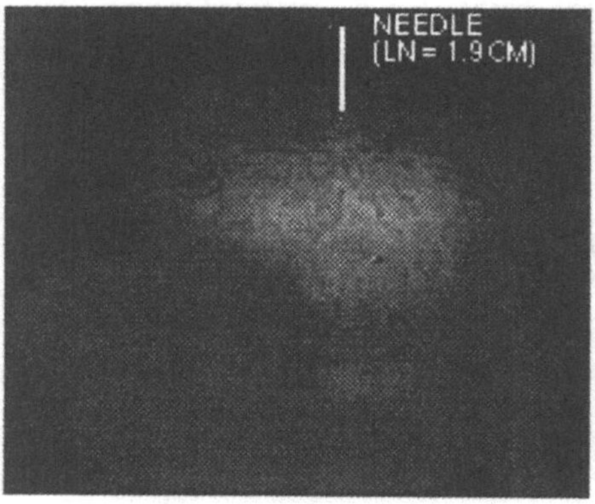

Fig. 2b: Case 1, z* = 0.5, 15 kV

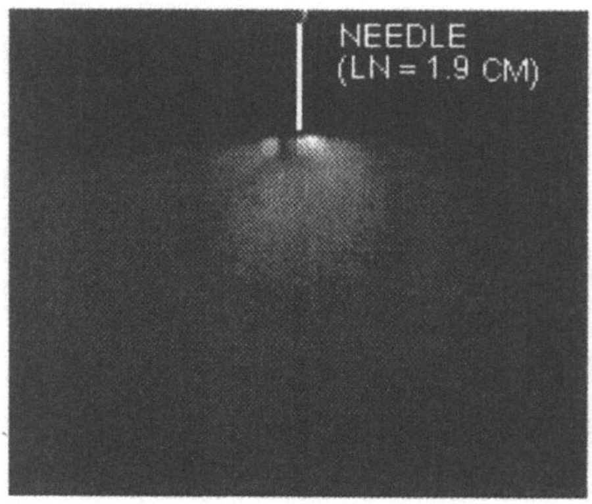

Fig. 2c: Case 1, centerline, 15 kV

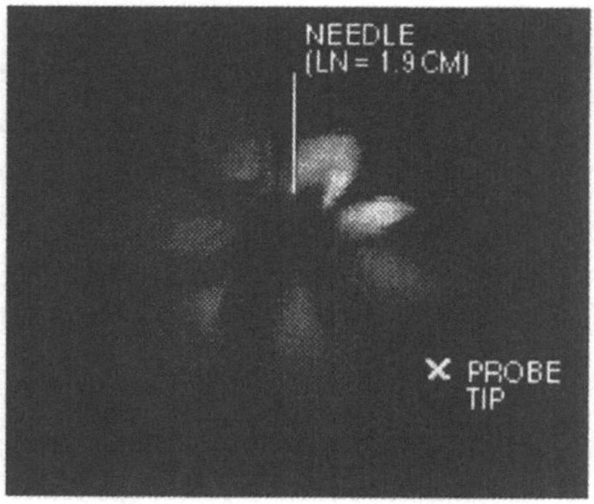

Fig. 3a: Case 2, z* = 0.125, 15 kV

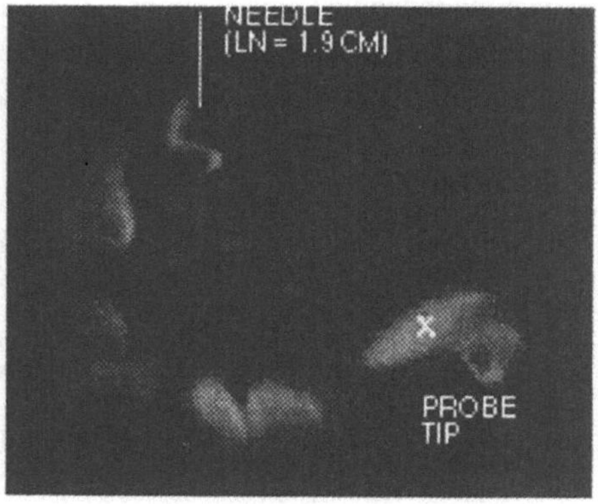

Fig. 3b: Case2, z* = 0.5, 15 kV

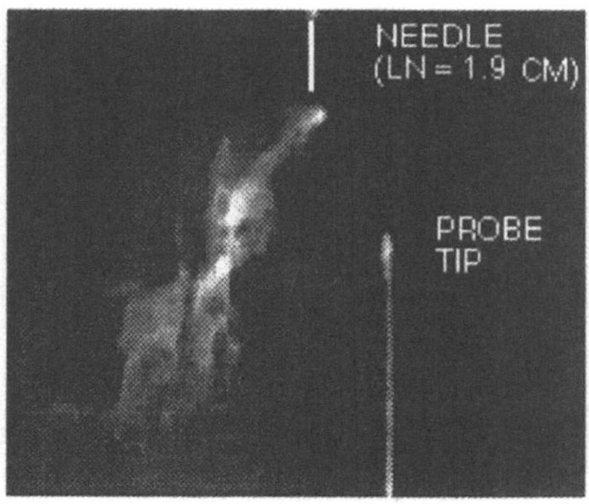

Fig. 3c: Case 2, centerline, 15 kV

Figures 2 and 3. Light-sheet Visualizations of the Reference and Perturbed Cases

Automatic Assessment of Aerosol Holograms for Granulometry and Velocimetry

C. Geiler, M. Stanislas†, H. Royer‡, T.Fournel*.

IMFL-ONERA, 5 Bvd Paul Painlevé 59000 Lille FRANCE.

†also LML, URA CNRS 1441, Cité Scientifique 59650 Villeneuve d'Ascq FRANCE.

‡ISL Rue de l'Industrie, BP 43, 68301 St Louis FRANCE.

* TSI AV L. Jouaux 42023 St Etienne FRANCE

Summary

A digital image analysis system is used for a fully automatic assessment of in line holograms of small particles. The software developed is tested on a reference hologram for granulometry. First results of its application to a multi-exposure hologram are presented.

Introduction

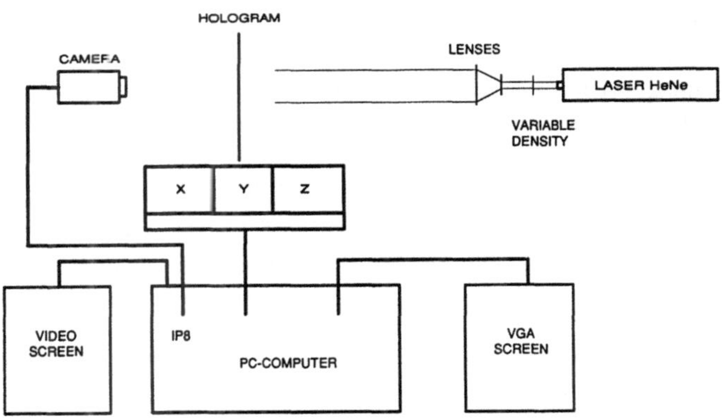

Figure 1 : Analysis bench

Micro-holography is an original method of analysis which has applications in granulometry and velocimetry and also in the study of multiphase flows. This method allows the measurement of the size, position in space and velocity of particles which equivalent diameter ranges from 5 μm to several millimeters. This can be done in a field ranging from a few centimeters cube to several meters cube using simple in line holography [1]. Unfortunately, this method is actually limited in its applications, essentially by the time necessary to analyse the hologram. This analysis is done manually, using an optical system with a high magnification ratio to see the particles images. To reduce the time and cost of the analysis and to increase the accuracy, an automation of the processing is thus necessary. Various solutions have been tried for in line holography [4,5] with apparently partial success. A solution is working for separated reference holograms of relatively large cavitation bubbles (300 to 500 μm) [6]. We present here the recent progress done in the set-up of a completely automatic analysis system [2].

Analysis system

Figure 1 presents a scheme of the analysis bench used. The hologram is fixed on a 3D motorized table with increments of $1\mu m$. The holographic image is reconstructed with a plane wave issued from an He-Ne 15 mW laser through a telescope. It is observed through a f= 50 mm lens with a PULNIX 750 CCD camera. This camera is connected to a MATROX IP8 board settled in a PC386-25 type micro-computer. The computer also pilots the displacements of the table carrying the hologram.

Analysis algorithm

As a result of previous studies [2], a software working in two steps has been developed. A first step allows a fast detection and approximate location and a second step the precise location and size measurement. This choice is essentially due to the observed fact that the volume of the holographic image which effectively contains particles represents a small part of the total volume (of the order of 10% for a velocimetry hologram).

The first step is done in a user defined parallelepiped, which is divided in tubes parallel to the light beam (Oz). The tube section is that of the video image. It can be varied by changing the magnification ratio of the optics. An overlapping is defined as a function of the size of largest particles to be detected. At the beginning of each tube, the gain and offset of the IP8 board are automatically adjusted on the basis of an histogram analysis and a detection threshold is selected. It is then proceeded to a scanning in depth of the holographic image with a step which is a function of the size of the smallest particles to be detected (typically 200 to 500 μm). At each step, the image is filtered, thresholded and analysed. If particles are detected, boxes are created containing them. The position and size of the boxes are stored on disk for the second step of the analysis.

In the second step, all the boxes registered in the detection scanning are analysed again with a smaller step in depth (typically 20 to 50 μm). The analysis is done only on the window containing the particle image, which size ranges generally between 32x32 and 128x128 depending on the size of the particle. The gain and offset are once again optimized and various thresholds are determined. For each step in depth, the image is filtered and different parameters are measured to try to determine the best focus position of the image. When this position is found, the three coordinates and diameter of the particle image are recorded. A segmentation is also realized to take into account the possibility of more than one particle in one box.

Results for granulometry

For granulometry, a reference hologram has been registered at ISL by Dr H. Royer. It is a single pulse hologram of a mist of spherical water droplets which sizes ranges from 10 to 150 μm. A field of 2.68 cm^3, containing about 250 particles of various sizes, has been analysed manually by Dr Royer with a magnification ratio of 12.5. To minimize the assessment time, this ratio was first set to 7 for the automatic analysis.

With a detection threshold set at the mean value of the histogram plus six standart deviation σ and a step in depth of 250 μm, all the particles larger than 20 μm were detected at the first step. This analysis takes 15 hours for the whole volume. Figure

2 presents the depth of the detection boxes as a function of the particle diameter. The correlation is quite correct and can be used to choose the step in depth for the detection. It can also be used to minimize the depth of analysis for the second step of precise location.

To detect particles smaller than 20 μm, the magnification ratio has then been set to 12 and the step in depth to 100 μm. A smaller volume has been analysed, in which all the particles detected manually, down to 5 μm, has been located automatically.

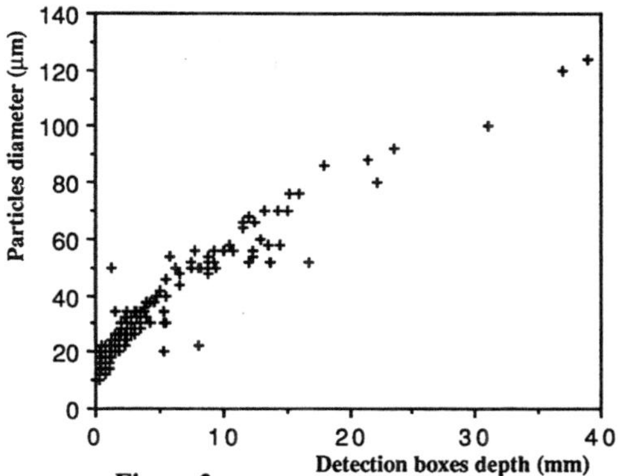

Figure 2

For the precise location of the particles, it is necessary to determine the best focussed image as a function of the position in depth of the hologram. This corresponds generally to the maximum gradient of the edge of the particle image.

Many solutions are proposed in digital image analysis books for such a detection, but most of them are based on convolution filtering which is time consuming. Here, due to the number of images to be created to analyse one hologram ($\simeq 500,000.$), we have limited ourself to simple criteria based on thresholding. Starting from the criterion proposed by Bexon & al [4] several solutions have been tested [2]. It appears that the detection is quite difficult due to the coherent nature of the light and to the existence of very bright focalization points on both sides of the particle images. No universal criterion has been found and it is a logical combination of several of them which allows the correct location in depth of all the particles.

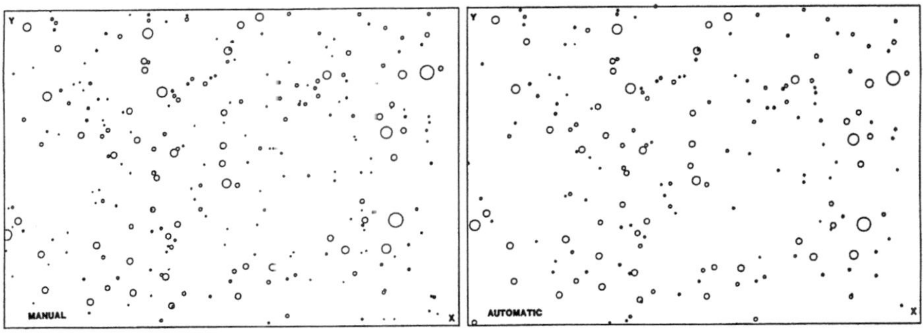

Figure 3

When the particle is located in depth, its size and position in the plane parallel to the hologram are determined by simple thresholding. Figure 3 presents a comparison of the size and location obtained manually at ISL and automatically at IMFL with a step in depth of 20 μm. All the particles larger than 20 μm has been correctly located and the difference in diameter with the manual analysis is always smaller than 5 μm. Figure 4 presents a comparison of the diameter histogram obtained manually and automatically.

The agreement is quite good for the particles larger than 20 μm. Concerning the small particles, the same algorithm applied to those detected in a smaller volume with a magnification ratio of 12 allowed the correct location and measurement of all of them.

Results for velocimetry

For velocimetry, a triple exposure hologram registered at IMFL in the water flow around a circular cylinder at Re = 500 [3] has been used. This hologram had been analysed by hand and an instantaneous velocity field was available (Figure 5).

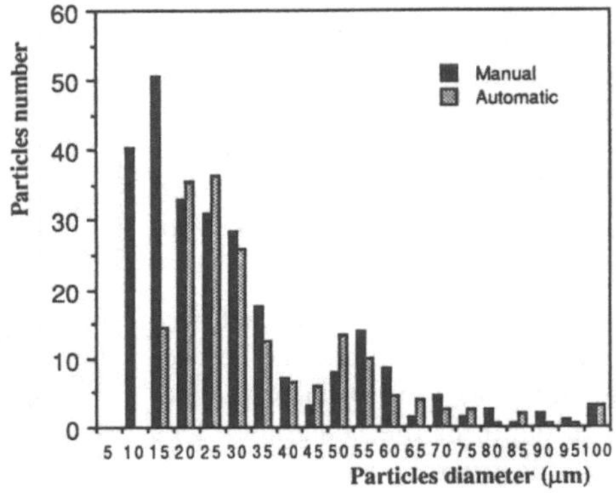

Figure 4 : Diameter histogram

Using the algorithm developed for granulometry, this hologram was analysed in the window indicated in Figure 5. Due to the triple exposure, the contrast of the particles images is reduced (Figure 6) and the thresholds had to be readjusted. For the detection step for example, it was lowered to 5σ.

When the size and position of the particles images are determined, it is necessary to compute the real position of the object particles, taking into account the wavelength change and the eventual beam divergence. An algorithm was developed for this purpose [3]. It is then necessary to associate the different images of a same particle, which is done by a sorting algorithm.

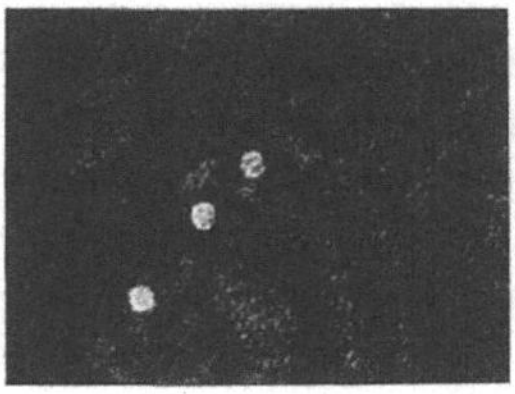

Figure 6: triple exposure image

Figure 7 presents the results obtained by automatic analysis in the window indicated in Figure 5, compared to the previous manual one. A detailed examination of the results has led to the following observations: in most of the cases, the third image is not detected, due to a lower contrast caused by a lower pulse energy at registration of the hologram. The two first images, which are of comparable contrast are correctly detected, but more difficult to locate due to the background noise superimposed on them. In the domain analysed, 970 images has been detected, 64 couples and 39 triplets has been found. As appears in Figure 7, the number of velocity vector obtained automatically is already comparable to that obtained manually (in the manual analysis, not all the particles had been measured, due to their number). For all the particles which are near enough to be compared, the velocity difference between the two analysis is of the order of 1%.

Conclusion

An analysis system to automatically reduce the data of granulometry and velocimetry holograms has been developed. For granulometry, the results obtained on a reference hologram are satisfactory. The particles are located correctly and the diameter measured is within the accuracy of the method. Concerning velocimetry, the number of particles detected can be increased by adequate filtering of the images. This may not cost too much in time because, in the second step where it is necessary, the images analysed are small (typically 64x64). Constraints have also to be put on the recording procedure to obtain images of comparable contrast.

We are thankfull to the DRET G6 for the financial support to this study.

References

[1] H. Royer. Particle velocity measurement via high speed holography. *Proc of the 11th Int. Cong. on high speed phot., London 1974.*

[2] M. Stanislas, C. Geiler, H. Royer, M. Courbon, J. Azema, J. Danieres. Particle detection criteria for automatic data reduction from micro-holograms. *Euromech Colloquium n⁰279: Image Analysis as measuring technique in flows, Delft, July 1991.*

[3] F. Beluche. Etude du sillage instationnaire d'un cylindre à nombre de Reynolds modéré par des techniques de visualisation et de vélocimétrie optiques complémentaires. *Thèse de Doctorat USTL n⁰1370, 1986.*

[4] R. Bexon, J. Gibbs, G.D. Bishop. Automatic assessment of aerosol hologram. *Journal of Aerosol Sciences, vol 7, p 397, 1976.*

[5] R.W. Menzel, L.D. Vandergriff. Program for automated holographic data reduction. *AEDC report tr 80-39, 1980.*

[6] G. Haussmann, W. Lauterborn. Determination of size and position of fast moving gas bubbles in liquid by digital 3-D image processing of hologram reconstructions. *Applied optics, vol 19 n 20, 1980.*

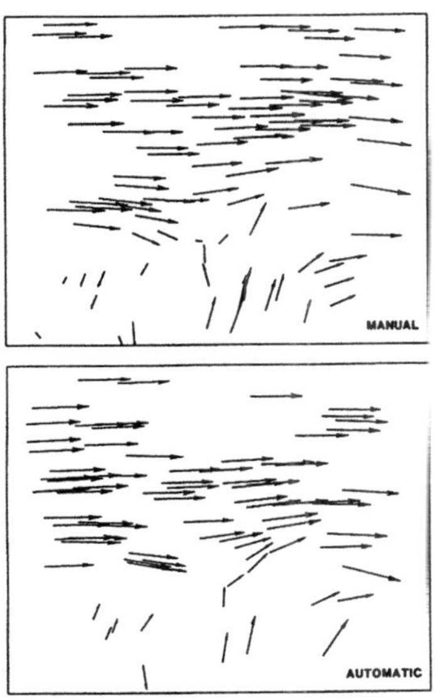

Figure 7

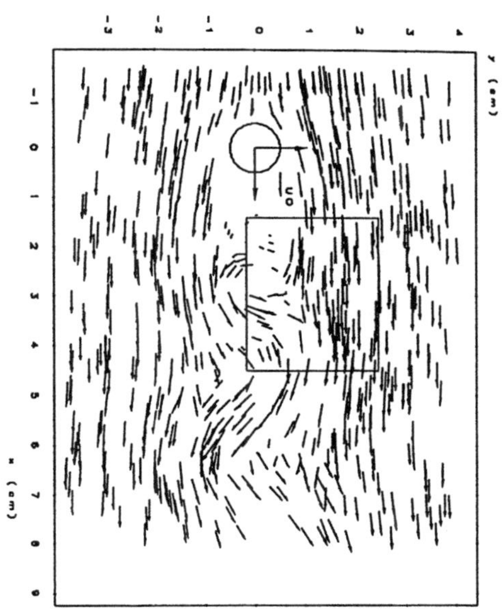

Figure 5

Thermal Problems

Liquid Crystal

Liquid Crystal Thermography and Image Processing in Heat and Fluid Flow Experiments

J. STASIEK* and M. W. COLLINS

Thermo-Fluids Engineering Research Centre,
City University, London EC1V OHB, U.K.

Summary

This paper describes new methods which can determine quantitatively two-dimensional temperature distributions on a surface and in a fluid from colour records obtained using a thermosensitive liquid crystal material combined with image processing. Application-type experiments have been carried out both to visualise the complex temperature distribution over a cooled surface disturbed by different solid obstacles, and also to investigate temperature and flow patterns in a rectangular cavity for natural convection.

Introduction

The purpose of this paper is to report a new technique that allows visual determination of both qualitative and quantitative heat transfer and fluid flow in thermofluid investigations. Moffat [8] and Jones et al [7] have recently reviewed available methods of heat transfer/temperature measurement and their applications. Thermochromic liquid crystals (TLC) are identified as a convenient measurement method offering satisfactory accuracy and resolution. Also, during the past several years, there has been an increasing use of computers and automatic recording and fringe and data processing software in heat transfer laboratories, first to facilitate conventional experiments and then to expand the range of possible work. Today, due to a synergestic combination of thermochromic liquid crystal and fast small computers, we are seeing what amounts to a revolution in heat transfer and fluid flow research. With the added advantage of automatic interpretation using image processing techniques, liquid crystal thermography is a very powerful, yet simple, measurement tool for the determination of quantitative heat transfer data. It is one of a range of new non-invasive, whole-field optical measurement methods.

* On leave of absence from Technical University of Gdansk, Poland.

In this paper we carefully review the above issues, and use illustrative examples from literature and our own work in applying TLC to the study of forced and natural convective heat transfer, on a cooled surface heated by air flow disturbed by a number of complex geometrical configurations, and for temperature and flow visualisation in rectangular cavity respectively.

Liquid Crystals: A Background Survey

In 1888 an Austrian botanist, Fredich Reinitzer [10], observed that certain organic compounds appeared to possess two melting points, an initial m.p. at which the solid phase turned to a cloudy liquid and a second m.p. at which the cloudy liquid turned clear. Further research revealed that an intermediate phase or *mesophase* did, indeed, exist between the pure solid and pure liquid phases of some organic compounds. Reinitzer termed this phase *liquid crystal*, an appropriate designation when one considers that the material exhibits the fluidity of a liquid while maintaining the characteristic of the anisotropic, ordered structure of a crystalline solid. The mesophases have been classified as nematic, cholesteric, smectic and blue [7][9]. However, from a structural viewpoint, it can be argued that there are only two basic types of liquid crystal; chiral-nematic and smectic, with cholesteric being regarded as a special kind of nematic. Their structures are shown schematically in Fig. 1.
A question of great relevance is the texture taken up by the cholesteric phase which can exhibit two forms: the focal-conic and the grandjean textures. In the focal-conic texture the cholesteric helices pack around elliptical and hyperbolic paths whereas in the grandjean form the helical axes are in the same direction and large regions of aligned molecules are present.

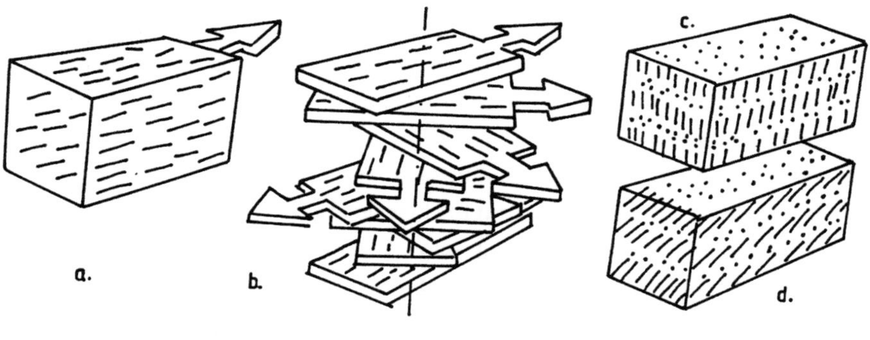

Fig. 1. Mesophase structures: a) Nematic b) Cholesteric c) Smectic A d) Smectic C (Jones et al 1992)

This texture is birefringent but optically inactive. In the grandjean texture, which forms from the focal-conic by mechanical shear, the helices are more or less all aligned with their axis parallel to incident light. This is the texture that exhibits the unique optical effects of the mesophase-iridescence, optical activity, circular dichroism and selective reflection of white light to show brillant colour. Over a known, reproducible range of temperature, the *event temperature range*, the cholesteric liquid crystal will progressively exhibit all colours of the visible spectrum as it is heated through the event temperature range. The phenomenon is reversible, repeatable and, with proper care, colour can be accurately calibrated with temperature. Hence thermal transients are measurable.

Both the colour play interval (that is, the range of temperature over which a certain colour is visible) and the event temperature range of a liquid crystal can be selected by adjusting its composition, and materials are available with event temperatures from -30°C to 120°C and with colour play bands from 0·5°C to 20°C [7][8] although not all combinations exist of event temperature and colour play band widths. Widths of 1°C or less will be called *narrow band* materials , while those whose band width exceeds 5°C will be called *wide-band*. The types of material to be specified for a given task depend on the type of image interpretation technique to be used.

Pure liquid crystal materials are thick, viscous liquids, greasy and difficult to deal with under most heat transfer laboratory conditions. Two approaches have been used to make them more practical to use: (i) encapsulation of the liquid crystal in a gelatine-like material forming nearly spherical particles from 10-30 microns in diameter and then making a paint using those particles as the pigment, or (ii) application of the unencapsulated material (unsealed liquids) to a clear plastic sheet and sealing it with a black backing coat to form a pre-packaged assembly. This is because in their (pure) exposed state TLC quickly degrade.

Calibration of Liquid Crystals and Image Processing

Before the execution of a thermal or flow visualisation experiment, we should recognise the characteristics of the overall combination of the TLC, the light source, the optical and camera system (ordinary, CCD black-and-white or RGB colour camera), and make a rational plan for the total measurement system. Figures 2 and 3 show the apparatus of a calibration experiment. A known

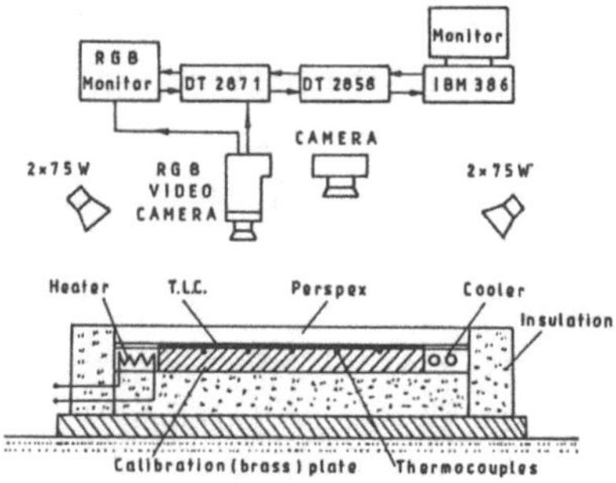

Fig. 2. Experimental apparatus for the calibration experiment

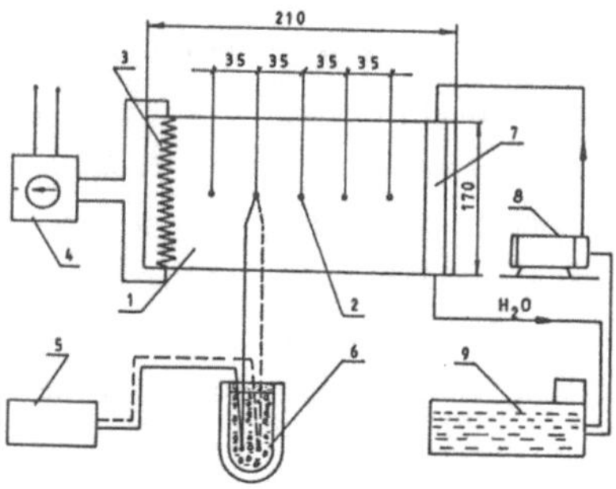

Fig. 3. Schematic drawing of circuit diagram for test plate

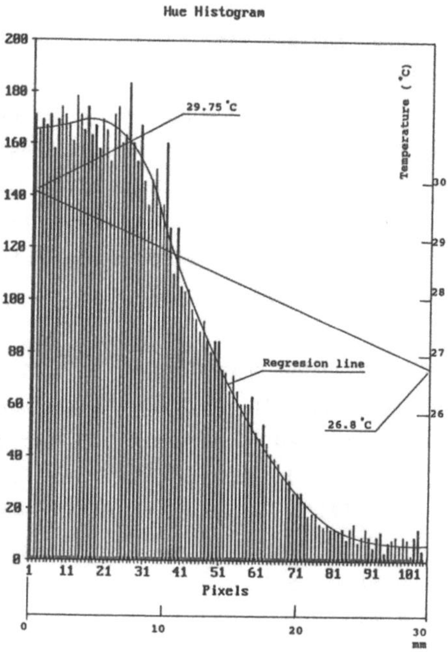

Fig. 4. Hue histogram and temperature distribution along test plate

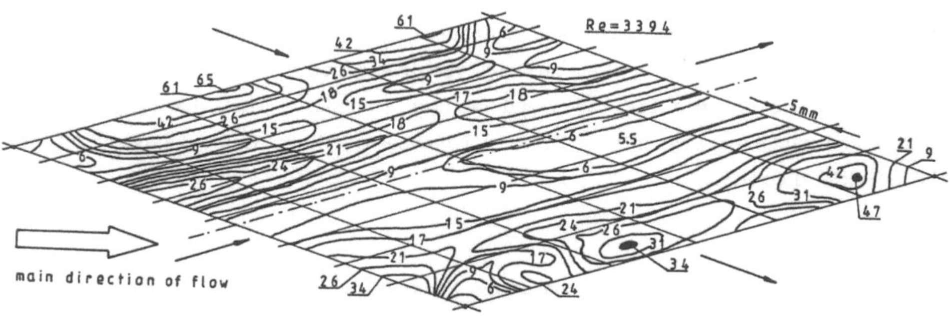

Fig. 5. Local Nusselt number distribution on the bottom wall of the measurement diamond for the corrugated geometry $\varphi=36^\circ$, Re=3394, $Nu_{av} = 20.44$, $h_{av} = 37.0\ W/m^2K$

temperature distribution exists on a *calibration plate* to which is attached the liquid-crystal layer. This experiment was also designed to investigate viewing angle effects. In order to maintain a linear temperature distribution with desired temperature gradients, one end of a brass plate was cooled by stabilised water and the other end controlled electrically to give a constant temperature. The distribution of the colour component pattern on the liquid-crystal layer was measured using a photo-camera with attached optical polariser and also a RGB colour camera. In the present study several temperature gradients were tested. Figure 4 shows the relationship between Hue (one of the chromaticity characteristics explained later) and temperature ($\Delta t/\Delta X$ = 94 K/m) on a RGB colour co-ordinate - there is a tendency for larger Hue values to occur for larger temperature gradients. In this particular experiment uncertainty was estimated as about $\pm 0{\cdot}075^{\circ}$C because the specimen was calibrated in place with the lighting and RGB camera system arranged just as it was during the application.

In the early days of image processing, only monochrome systems were available. Soon after, colour image processing became available, but monochrome continued to dominate because it was easier and less costly to manufacture and less complicated to use. However, in the last four years or so, colour image processing has gone from being used mainly in highly technical applications involving expensive image-processing systems to being available to virtually anyone who can use a desktop computer.

There are several commercial packages of hardware and software available which will digitize a video image from either black-and-white or colour cameras and store individual frames pixel by pixel in arrays. Three separate arrays are used to store a colour image, one each for the red, green and blue signal components. Also, three styles of camera have been used; NTSC composite (the usual home video camera), CCD (change-coupled device) and CID (charge injected device) [7].

According to Akino et al [2], Moffat [8] and Jones et al [7] there are four broad classes of image interpretation techniques available.

- human observers,
- multiple narrow-band spectral intensity image processing system,
- multiple linear regression method, and
- true-colour image processing systems.

Human observers can interpret liquid crystal images by direct visual inspection of colour photographs or tape recorded video images, usuallly using narrow-band paints. Calibrations for such use are generally limited to identifying the temperature associated with a particular colour often the red or the yellow-green colour near the centre of the colour-play band. The uncertainty associated with direct visual inspection is about 1/3 the colour-play band-width, given an observer with normal colour vision and a little experience [8] or about $\pm 0{\cdot}2^{\circ}C$ to $0{\cdot}5^{\circ}C$, depending on the band-width.

The second and third methods are based on the interferential optical filters with narrow band-play characteristics of transmittance, through which equally coloured regions can be extracted. Akino et al [1] developed a multiple narrow-band spectral intensity interpreter using a set of eighteen narrow band-pass filters, whose central wave-length were from 400 to 750 nm, and the full width of the half maximum (FWHM) of the filters were less than 10 nm. The behaviour of liquid crystals can be interpreted in terms of the variation of intensity with wavelength, at a given temperature instead of the variation of dominant with Hue temperature. This approach gives rise to calibrations based on spectral intensity, as reported in [1] - the resolution of this method is better than $0{\cdot}1^{\circ}C$.

In the fourth method, colour video information is displayed as a unique brightness combination of red, green and blue (RGB) light. A displayed RGB image is immediately perceived by a user as exhibiting three distinct attributes:

(1) Hue - the spectral colours present.

(2) Saturation - how deep or faded the colours appear to be.

(3) Intensity - the edge information or what would be seen if a black-and-white version of the image were displayed. Hence, only intensity is available from a monochrome system.

Many image processing operations developed for processing grey-level (intensity) images can be readily applied to HSI colour images. These include image transformations, enhancement, analysis, compression, transformations and restorations. Building on gray-scale image processing hardware and software technology, Data Translation has produced a 512 x 512 pixels x 8 bit (256 grey levels) colour frame-grabber board for PC/ATs.

A schematic view of such an image processing system (developed by the authors) is shown in Figure 2. The two-dimensional temperature distribution is determined using RGB video-camera, IBM 386 Personal Computer AT, HSI Colour Frame Grabber DT 2871 and Auxiliary Frame Processor DT 2858.

Experimental Technique

Two main methods of measurement are performed involving steady state and transient techniques. A brief history of these is given, Baughn et al [5]. Recent reviews of heat transfer measurements have also been produced by Moffat [8], Jones et al [7] and Ashforth-Frost et al [3].

** Steady state analyses - constant flux method*

The steady state techniques employ a heated model and the TLC is used to monitor the surface temperature. Usually, a surface electric heater is employed such that the local heat flux, $\dot{q}$ is known and this, together with the local surface temperature, T_w, (found from the TLC), gives the local heat transfer coefficient, h,

$$q = I^2 r \quad \text{and} \quad h = \frac{\dot{q}}{T_a - T_w}$$

T_a is a convenient driving gas temperature, I is the current and r is the electrical resistance per square of the heater.

** Steady state analyses - uniform temperature method*

The TLC-coated test specimen forms one side of a constant temperature water bath and is exposed to a cool/hot air flow. The resulting thermograph is recorded on film or video and further measurement positions are obtained by adjusting the water bath temperature. This method is more time consuming due to the large volume of water that needs to be heated. In this case, the heat transfer coefficient is determined by equating convection to the conduction at the surface,

$$h(T_a - T_w) = \frac{k}{x}(T_w - T_b)$$

where, T_b is a water temperature, x the wall thickness and k the thermal conductivity.

** Transient method*

This technique requires measurement of the elapsed time to increase the surface temperature of the TLC-coated test specimen from a known initial temperature to a predetermined value. The rate of heating is recorded by monitoring the colour change patterns of the TLC with respect to time. If the specimen is made from a material of low thermal diffusity and chosen to be sufficiently thick, then the heat transfer process can be considered to be one-dimensional (1-D) in a semi-infinite block. Numerical and analytical techniques can be used to solve the 1-D transient conduction equation. The relationship between wall surface temperature, T, and heat transfer coefficient, h, for the semi-infinite case is,

$$\frac{T - T_i}{T_a - T_i} = 1 - e^{\beta^2} \operatorname{erfc}(\beta); \quad \beta = h\left(\frac{t}{\rho c k}\right)^{0 \cdot 5}$$

where, ρ, c and k are the model density, specific heat and thermal conductivity. T_i and T_a are the initial wall and gas temperatures and t is time from initiation of the flow [5][7].

** Shear stress measurements*

TLC's can also indicate shear stress and their use in neat (unencapsulated) form in this mode and gives transition data. It was first investigated at NASA [8] and a quantitative measurement method for skin fraction was invented later by Bonnett et al [4]. They observed that if a material in the focal-conic texture on a given surface is suddenly subjected to a steady air flow, then the time for the material to align into grandjean texture is a unique function of the shear stress. This time can be easily measured as it corresponds to the time for the surface to change from a colourless state. Thus, a robust technique independent of viewing angle or illumination became available. In this last case transition was readily detected [4].

** Temperature and flow visualisation in fluids*

Liquid crystals have been used to make visible the temperature and velocity fields in fluids as well as air. Temperatures have been visualised using the materials in the unsealed (neat) and microencapsulated forms as both tracer particles in flow field studies, and as surface coatings [6][8][9].

Application Experiments

In order to demonstrate the feasibility of TLC techniques in practical heat transfer contexts the authors have performed several experiments. The first set was carried out to investigate temperature and heat transfer coefficient distributions on a cooled surface heated by an air flow and disturbed by a number of complex geometrical configurations, namely,

(i) Crossed-corrugated geometrical elements as used in rotary heat exchangers (regenerators) for fossil-fuelled power stations.

(ii) Cylinders (both single and double) and a square section column.

(iii) Square roughness elements

} relevant to local heat exchanger performance or enhancement

Figure 5 shows the distribution of local Nusselt number for uniform wall temperature on a corrugated-geometry. The contours of constant Nusselt number were obtained using the red colour to define a specific isotherm (27·7 for these experiments). Eight to sixteen isotherms (each corresponding to a different heat flux) are photographed and then superimposed to record local contours under an oblique viewing direction, and then to use human observation.

Figure 6 shows the contours of constant Nusselt number distribution around a roughness plate at Re = 10000; $q = 228\ W/m^2$ and $D_h = 0{\cdot}009$ m. These isochrome lines were determined by the true-colour image processing method. In the actual measurements, only the green colour band, corresponding to $t = 28{\cdot}3 \pm 0{\cdot}075^{\circ}C$, was used (Hue value range from 55 to 70).

In the second set of experiments flow structures and temperature distribution in rectangular cavity were visulaized using photographic records of the motion of tracer particles illuminated by a sheet of white light. A thermal (natural convective) flow was generated in a (18cm long x 6cm wide x 3cm high) cavity. Unsealed TLC's were used to make visible the temperature and velocity fields in a glycerol-filled cavity, which could be angled from *horizontal* through *inclined* to *vertical*.

Figure 7 shows the visual image obtained for the *horizontal* case. The right-hand edge is, in fact, centre-line of complete TLC cavity record and the frame shows the area used for automatic processing (segmentation). The temperature difference between the heated (lower plate) and the cooled-upper plate was 10K

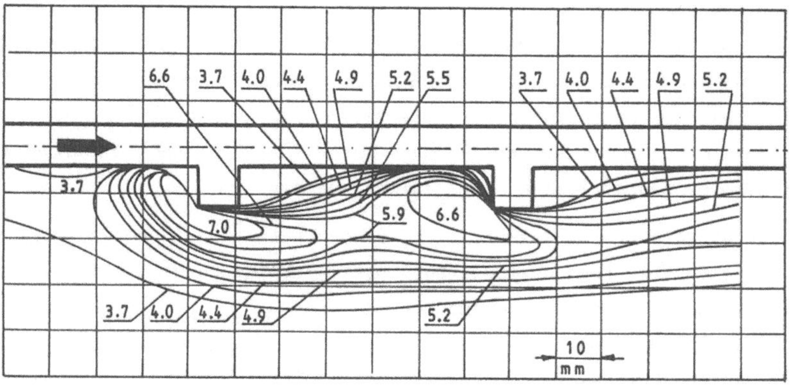

Fig. 6. Contours of constant Nusselt number distribution around a roughness plate at Re = 10000, $q = 228\ W/m^2$, $D_h = 0.009$ m

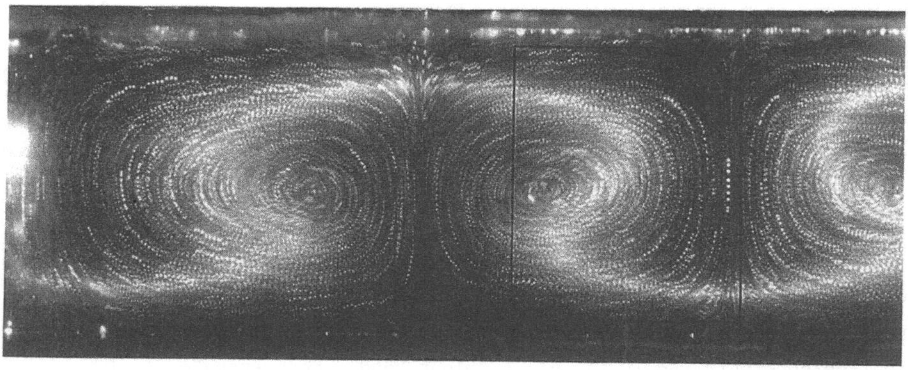

Fig. 7. Liquid crystal visualisation of free convection in a horizontal glycerol-filled cavity

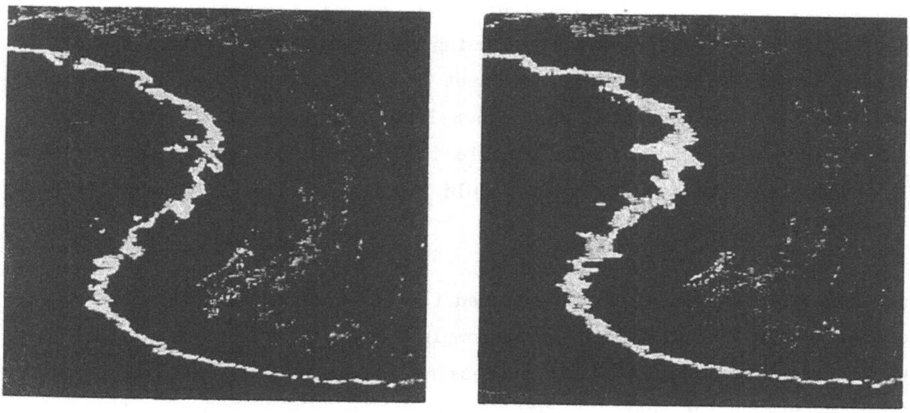

Fig. 8. Segmentation transform of the cavity image shown in framed region of Figure 7. a) Hue: 80-100 b) Hue: 100-120

(in these particular experiments the temperature range was between 14·5°C and 24·5°C). The time interval between traces was 6 sec. The TLC's which we used here were chiral nematics type TM256 (BDH Chemicals Ltd.). The red colour start temperature given by the manufacturers is 19·6°C, with a 3·8°C colour play temperature range. The isotherms developed from the image in Figure 7 (using true-colour image processing system) - are shown in Figures 8 and 9.

Conclusions

In this paper, we describe briefly the mesophase structure which gives rise to the thermochromic liquid crystal phenomenon, and the application of TLC's to temperature, and also shear stress, measurement. The former is illustrated with typical results from various research investigations of the authors involving forced convection. Stress is laid on the ability of the TLC's to give quantitative data of practical heat transfer significance. Finally, we report the use of TLC's to give both velocity and temperature data in a natural convection cell.

Throughout the paper, we emphasise the desirability and practicability of using automatic processing methods; we have used a *true-colour* approach.

Acknowledgement

We are very grateful to PowerGen UK for their sponsorship of the cross-corrugated heat transfer project.

References

1. Akino, N.; Kunugi, T.; Ichimiya, K.; Mitsushiro, K.; Ueda, M.: Improved liquid-crystal thermometry excluding human color sensation. J. Heat Transfer, 111, (1989) 558-565.

2. Akino, N.; Kunugi, T.; Shiina, Y.; Ichimiya, K.; Kurosawa, A: Fundamental study on visualization of temperature fields using thermosensitive liquid-crystals. Flow Visualization V. (ed. R. Reznicek), Washington: Hemisphere (1990) 87-92.

3. Ashforth-Frost, S.; Wang, L. S.; Jambunathan, K.; Graham, D. P.; Rhine, J. M.: Application of image processing to liquid crystal thermography. Procs. Optical Methods and Data Processing in Heat and Fluid Flow, City University, London (1992) 121-126.

4. Bonnett, P.; Jones, T. V.; McDonnell, D. G.: Shear-stress measurement in aerodynamic testing using cholesteric liquid crystals. Liquid Crystal, 6, (1989) 271-280.

5. Baughn, J. W.; Ireland, P. T.; Jones, T. V.; Saniei, N.: A comparison of the transient and heated-coating methods for the measurements of local heat transfer coefficients on a pin fin. J. Heat Transfer, 111, (1989) 877-881.

6. Hiller, W. J.; Kowalewski, T. A.: Simultaneous measurement of temperature and velocity fields in thermal convective flows. Flow Visualization IV (ed. C. Veret), Washington: Hemisphere (1986) 617-622.

7. Jones, T. V.; Wang, Z.; Ireland, P. T.: The use of liquid crystals in aerodynamic and heat transfer experiments. Procs. Optical Methods and Data Processing in Heat and Fluid Flow, City University, London, (1992) 51-65.

8. Moffatt, R.J.: Experimental heat transfer. Procs. 9th Intl. Heat Transfer Conf., Jerusalem, Vol.1, (1991) 308-310.

9. Parsley, M.: The use of thermochromic crystals in heat transfer and flow visualisation. Research. FLUCOME'88, Sheffield University, England, (1988) 216-220.

10. Reinitzer, F.: Beiträge zur Kenntniss des Cholestrins, Monatschr. Chem. Wein, 9, (1888) 50-90.

11. Stasiek, J.; Collins, M. W.; Chew, P. E.: Liquid crystal mapping of local heat transfer in crossed-corrugated geometrical elements for air heat exchangers. EUROTECH DIRECT '91 Congress, Birmingham, England, (1991). Paper C413/040.

12. Stasiek, J.; Collins, M. W.: Internal Reports, City University, London, (1989-92).

Opto-Thermal Properties of Water Suspensions of Thermo-Sensitive Liquid-Crystal Particles

N. Akino

Japan Atomic Energy Research Institute
Tokai-mura, Ibaragi-ken, 319-11 JAPAN

S. Sagiya and M. Nakanishi

Japan Capsular Products Inc.
1-3, 2-chore, Iwamoto-cho, Chiyoda-ku, Tokyo, 101 JAPAN

Summary

This paper describes the relation between color and temperature of suspensions of thermo-sensitive liquid-crystal particles into water carrier which disperse colored light according to temperature. This suspension becomes often to be used to visualize a temperature field in a liquid qualitatively or quantitatively. Being observed by naked eye, a suspension shows colors of dark-reddish, green, blue, purple and again dark-reddish as temperature increases. These color-plays were quantitatively evaluated using a color-meter, and it was concluded that two dark-reddish colors could be distinguished by the saturation value. Spectra of the colored lights were analyzed using monochrometer and it was concluded that the spectra have no predominant component at the temperatures of both dark-reddish colors within wavelength band of visible light.

Liquid-Crystal Suspension Method

A choresteric liquid-crystal material shows thermo-chromatism of rainbow according to temperatures. They can convert a temperature distribution into a beautifully colored map and are often applied to visualize temperature fields on a surface and in a liquid, in many recent investigations on thermo-fluid phenomena[1]. For liquid application, drops or stabilized particles of a liquid-crystal material are dispersed and suspended into a working fluid. In case of studies using silicone oils as working fluid, even quantitative mapping of temperature distributions have been carried out successfully [2, 3]. On the other hand, no paper has yet been presented, in which a temperature was quantitatively measured using water suspension of liquid-crystal.

The object of this study is to clarify the thermo-chromatic properties of suspensions of micro-encapusulated liquid-crystal particles in water.

Experimental apparatus and procedure

Four components materials of kiral-nematic liquid-crystal, made by BDH Chemicals Ltd., were mixed to give five formulations with different beginning temperatures and ranges of color-plays as shown in Fig.2. For the purpose of chemical stabilization, each mixture was micro-encapusulated by urea resin into small spherical particles with diameter of about 20 micron-meter. The particles was dispersed into water to prepare suspensions with concentrations of 0.01, 0.02 and 0.04 percent.

A suspension was poured into a small transparent cell (size: 10x10x50mm) for spectro-photometry. As shown in Fig.1, the cell was inserted into a chamber of which temperature was finely controlled by flowing water from a thermobath. The suspension in the cell was agitated by a small mixer and its temperature was measured by a calibrated thermocouple with accuracy of 0.1 deg.C. Colors of each suspension at each temperature were observed and recorded using a 35mm camera with a color positive film through a window on one side of the chamber being illuminated by light from a halogen lamp from the other side perpendicular to the observation window.

Quantitative evaluation of the colors was carried out using a color meter and a monochrometer. Output of the color-meter was lightness Y and chromaticity coordinate values of x and y. The monochrometer measures spectra with a resolution of 0.2nm in wavelength ranges of 300 to 900nm.

Experimental results and discussions

(1)Observation by naked-eye By the naked-eye observation, a suspension shows colors of Dark-reddish(orange), green, blue, purple and again dark-reddish as temperature increases. This relation between color and temperature are fundamentally same to all liquid-crystals with the five formulations. The observation results are summarized in Fig.2. It, however, is a serious problem of temperature determination, because two colors of same kind, that is, dark-reddish or orange were observed at different temperatures.

(2)Result of color-meter By the color-meter, the Y-x-y values of colors of suspensions were measured changing liquid-crystal formulation, concentration and temperature. Fig.3 shows data of a liquid-crystal with a formulation under three concentrations. The lightness Y depends clearly on the concentration, which is not adequate as a temperature indicator. And the chromaticity values (x,y) do not depend on concentrations, but it is difficult to recognize the relation between temperature.

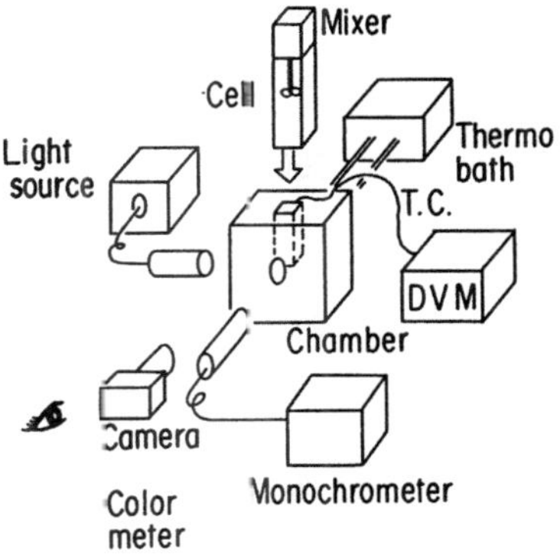

Fig. 1

Experimental apparatus

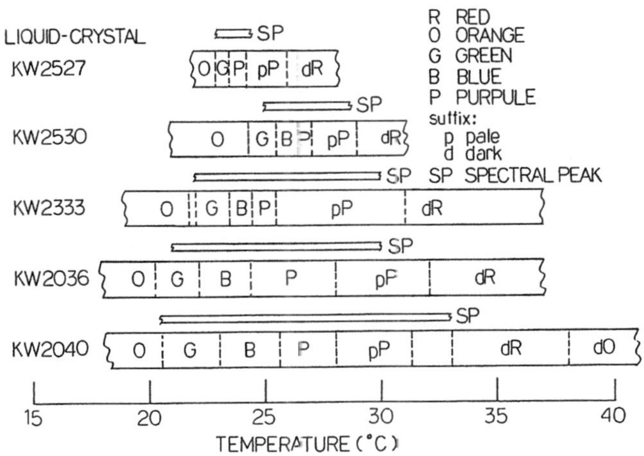

Fig. 2
Results of observation by naked-eye

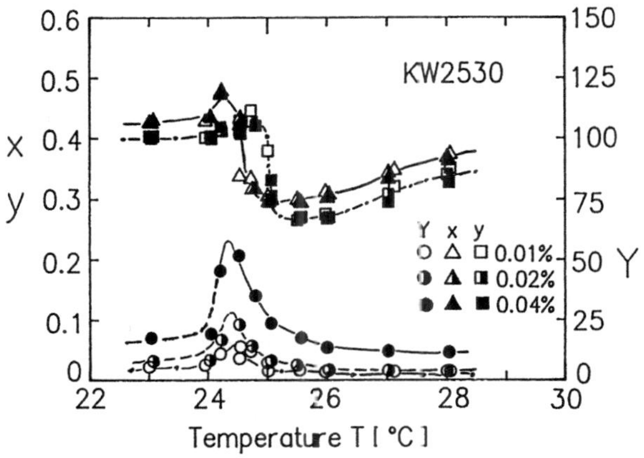

Fig. 3
Color-meter data (Y-x-y) and temperature

Y : lightness
x, y: chromaticity coordinate

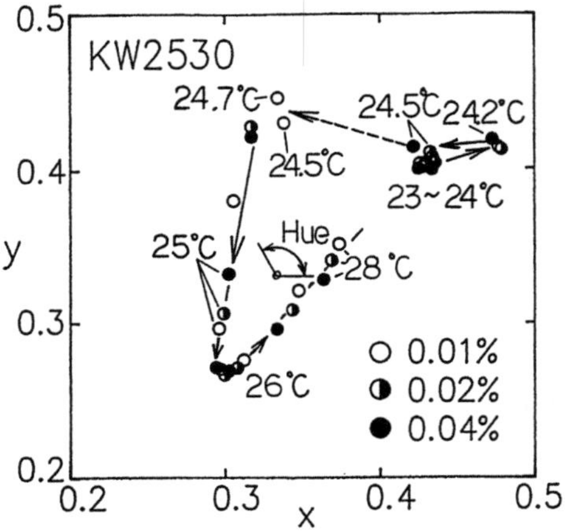

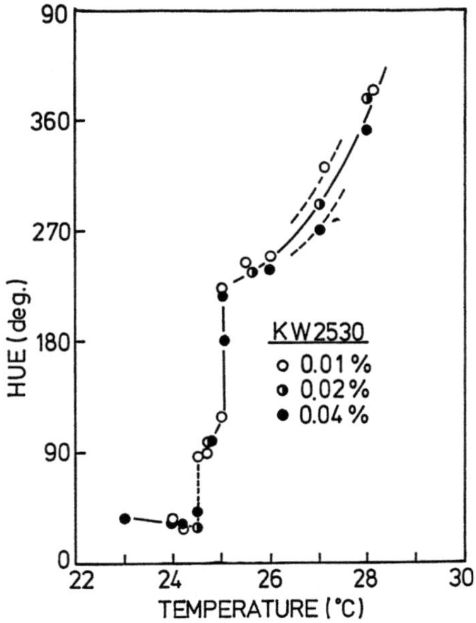

Fig.4 Chromaticity coordinate (x,y)

Fig.5 Hue and temperature evaluated by angle between white point and data points in Fig.4

Then, data were plotted on a chromaticity diagram, as shown in fig.4. Hue is one of the three attributes of color which corresponds to the perceived color, such as red, green or blue. The hue can be evaluated by a angle between a central white point and each data, and can uniquely correlate to almost temperatures. Below 24 deg.C and above 28 deg.C, the observed colors were same dark-reddish (or orange) and the hue angles are nearly same. But the distance between the white point and the data points are clearly different and the two dark-reddish colors can be distinguished. In other word, we can distinguish two red or orange colors by the saturation values.

Fig.5 shows the relation between hue and temperature. Unfortunately, there are abrupt changes of hue at about 24.5 deg.C and about 25 deg.C. The relation is unique function with no effect of concentration, but neither linear nor continuous. These make it difficult to determine temperature from hue value quantitatively.

(3)Measurement by monochrometer

Fig.6 shows a series of spectra of the KW2530 liquid-crystal under a concentration of 0.02%, which were measured changing temperature. Below 24 deg.C, a spectrum shows broad convexity with maximum at around 700 nm, and

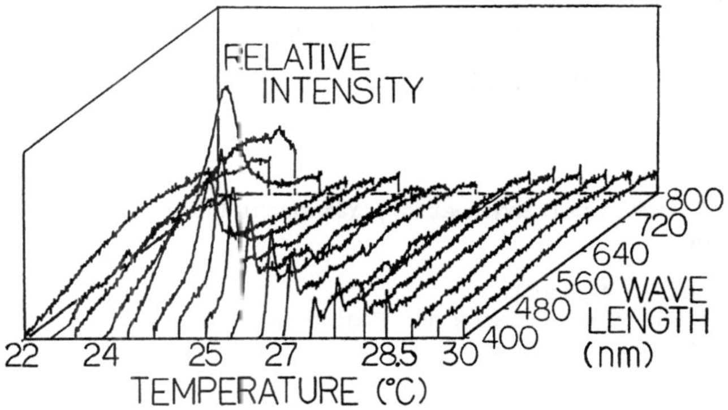

Fig. 6
Spectra of KW2530 liquid-crystal at various temperature measured by monochrometer

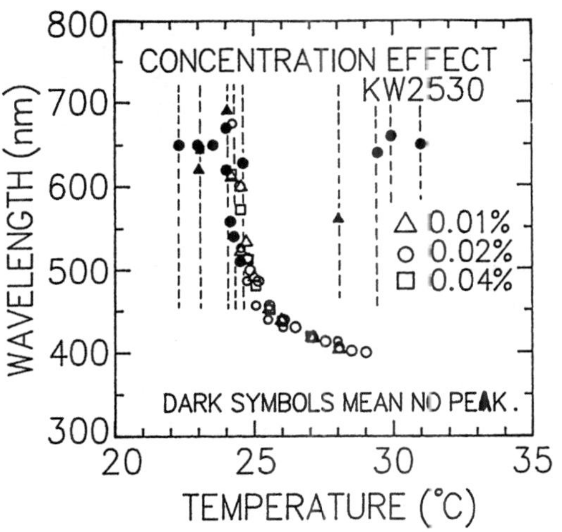

Fig. 7
Relation between peak wave-length and temperature

the color may look dark-reddish or orange. As temperature increases, a large and clear peak appears suddenly at about 24.5 deg.C and it looks green. For further temperature increase, the peak gradually decreases height and moves towards shorter wavelength. At about 28 deg.C, the wavelength of the peak becomes shorter than 400nm and, within wavelength region of visible light, there remains a broad convex spectrum having maximum of low intensity at around 700 nm, which maybe look dark-reddish.

Reference

1. Moffat, R. J., Experimental heat transfer, Heat transfer 1990, Vol. 1, p. 187
2. Kimura, I, et al., Quantitative thermal flow visualization using color image (application to a natural convection visualized by liquid crystal), ASME FED-Vol. 85, p. 69 (1989)
3. Akino, N., et al., Thermo-chromatic characteristics of liquid-crystal suspensions, ASME FED-Vol. 128, p. 247 (1991)

Visualization of Temperature and Flow Fields Using Liquid Crystals in the Fluid Region During Various Solidification Processes

T. Nishimura
Department of Mechanical Engineering, Yamaguchi University
Ube, 755 Japan

M. Fujiwara and H. Miyashita
Department of Materials Science and Engineering, Toyama University, Toyama, 930 Japan

This article presents a simple technique for temperature visualization using liquid crystals in the liquid region during the process of cooling and solidification. This method shows interesting timewise evolution of flow and temperature fields, i.e. the process of density inversion for freezing of water and the double-diffusive process for solidification of aqueous solutions.

Introduction

We often encounter a situation in which we want to know instantaneous temperature fields with thermal convection in transient transport processes. Thermocouples have been used conventionally for temperature measurements. However, in the above situation many thermocouples are necessary and the data deduction is not easy. Therefore, the usual methods of temperature visualization, such as schlieren, Mach-Zehnder and holography, provide information about temperature variations. However, since these methods are based on the variations in density and refractive index of fluid with temperature, temperature visualizations by these methods are difficult for the systems of thermal convection with density inversion and double-diffusive convection produced by interaction of heat and mass transfer.

The present study is demonstrated that encapsulated thermochromic liquid crystals are very useful for temperature and flow visualizations even in the above systems. We first attempted to observe the isotherms near 4 °C during freezing of water with density inversion in a small enclosure. Secondly, we investigated natural convection patterns occurring in aqueous solutions during several solidification processes. Although in recent years, thermochromic liquid crystals have been adopted as a temperature-measuring probe to indicate temperature variations in several systems [1,

2], to the knowledge of the authors, no one has performed any work which uses liquid crystals to visualize flow and temperature fields during a solidification process.

Experimental Apparatus and Procedure

Figure 1 shows a schematic diagram of the experimental apparatus. Experiments were performed in a rectangular test cell with the following inner dimensions: 40 mm long, 50 mm high and 30 mm wide. The test cell was made of acrylic resin, except for the cold wall, which consisted of a copper plate. A cooling chamber was attached to the rear side of the cold wall, wall temperature T_c was maintained below 0 °C by refrigeration unit 1. Refrigeration unit 2 was used to control the initial temperature of fluid T_i. To minimize heat losses, the test cell was insulated with 60 mm of styrofoam on all sides, and the experimental apparatus was used to controlled room. Two working fluids were used in this experiment; one is water, and the other is a super-eutectic aqueous sodium carbonate (Na_2CO_3) solution.

Liquid crystals change color according to the environmental temperature, the change being reversible and repeatable as long as the crystals are not physically damaged. The liquid crystals used here were of the chiral nematics type manufactured by BDH Chemical Ltd. and which were found in preliminary experiments to be stable in aqueous solutions of a number of salts. To represent two different event temperatures, we used two-event narrow band liquid crystals, which can be made simply by physically mixing liquid crystals of the desired event temperatures. Each liquid crystal covered a temperature range of 2.5 °C, and the colors in the range from the highest to the lowest temperature included blue, green, yellow and red. The density of the encapsulated liquid crystals was about 1.02×10^3 kg/m^3, and their diameters ranged from 10 to 15 μm. A small amount of liquid crystals (0.004% by volume) was suspended in the working fluid for good visualization just before the experiment. A light bean was in-

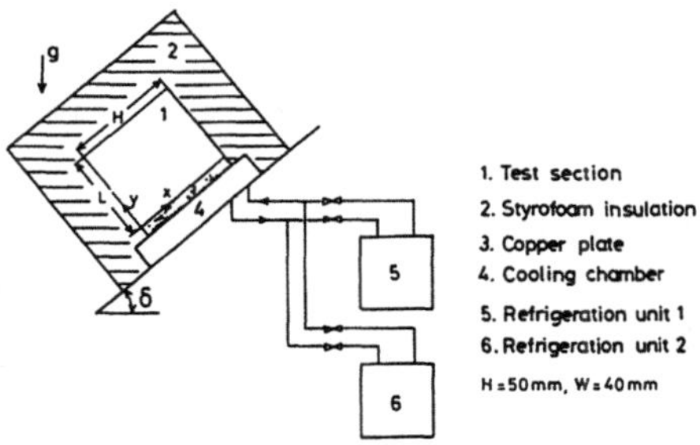

Fig. 1. Experimental apparatus

troduced through a slit in front of a projector lamp. Color photographs of time-dependent temperature fields were taken with an exposure time of 1-2 s at an angle of 90° to the plane of the incident light beam. The temperatures were also measured with thermocouples.

Results and Discussion

In the case of natural convection in water, the flow pattern can seriously be affected by the occurrence of the density maximum associated with the temperature of 4 °C. Figure 2 shows the temperature visualization for freezing of water in an enclosure with lateral cooling. The isotherms near 4 °C represent the interface between two-dimensional counter-rotating vortices produced by density inversion. A downward flow is formed at the interface and the counter-clockwise vortex exists along the ice surface. Density inversion is found to play an important role in natural convection patterns in the fluid region.

Figure 3 shows the result for upward cooling. Three-dimensional mushroom-like vortices appear on the ice surface and grow through an unstably straified layer less than 4 °C. The fluid below 4 °C undergoes convective motions when the unstably stratified layer reaches a critical value of the layer depth. We determined the critical Rayleigh number from the isotherms indicated by liquid crystals for various parameters. The Rayleigh number is defined as

$$Ra_c = g\beta(T_m - T_f)d^3/(\alpha\nu)$$

where d is the depth of unstably stratified layer at temperatures between T_m and T_f which are 4 and 0 °C, respectively. g, β, α and ν are gravitational acceleration, volumetiric expansion coefficient, thermal diffusivity and kinematic viscosity of water, respectively. Figure 4 shows that the critical Rayleigh number is constant in spite of the cold wall temperatures and initial temperatures and is about 500. This value is almost identical to that deterimined from Mach-Zehnder interferometer by Tankin and Farhadieh [3] and also agrees with the

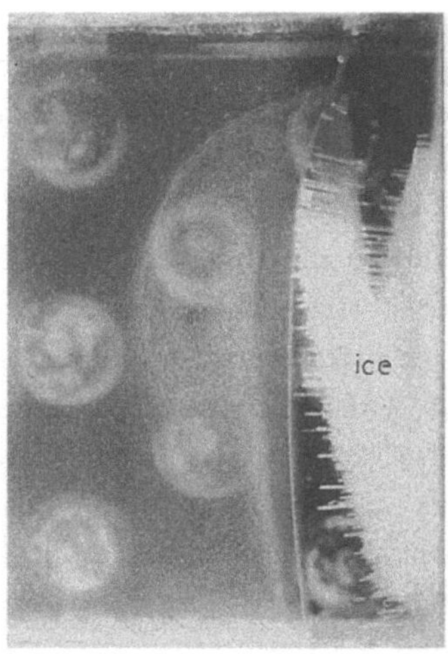

$T_i = 25$ °C
$T_c = -15$ °C

Fig. 2. Lateral freezing

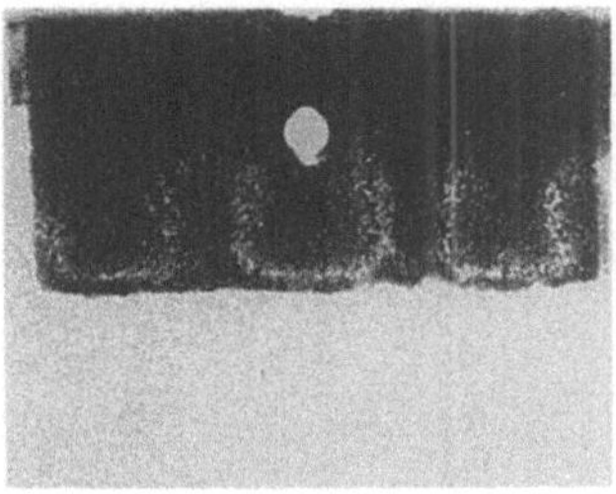

Fig. 3. Upward freezing

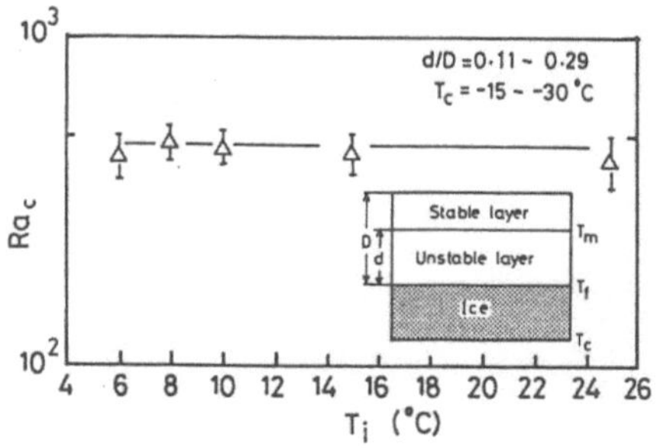

Fig. 4. Critical Rayleigh number

Rayleigh number for a water layer without freezing determined from linear stability analysis by Katto and Iwanaga [4]. Thus it is confirmed that quasi-steady assumption is held for the onset of convection during the ice growth process.

Solidification of a super-eutectic aqueous solution of Na_2CO_3 to form Na_2CO_3-$10H_2O$ preferentially extracts the heavy solution and leaves behind depleted fluid that is relatively less dense. Temperature visualization photographs for lateral cooling is shown in Fig. 5. The isotherms meander in the vertical direction with progressing solidification, showing S-shaped temperature profiles. The reason for this is that the development of the vertical density stratification from the top of the enclosure effectively damps thermal convection during solidification and leads to establishment

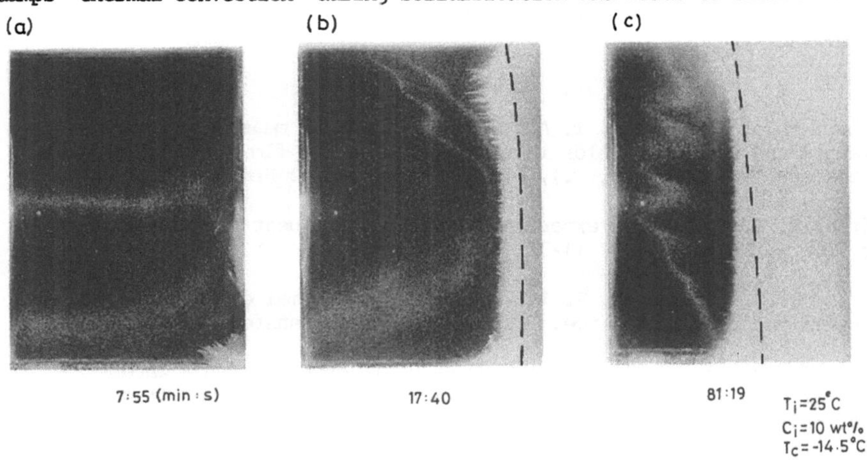

Fig. 5. Solidification of Na_2CO_3-H_2O system (lateral cooling δ=90°)

(a) (b)

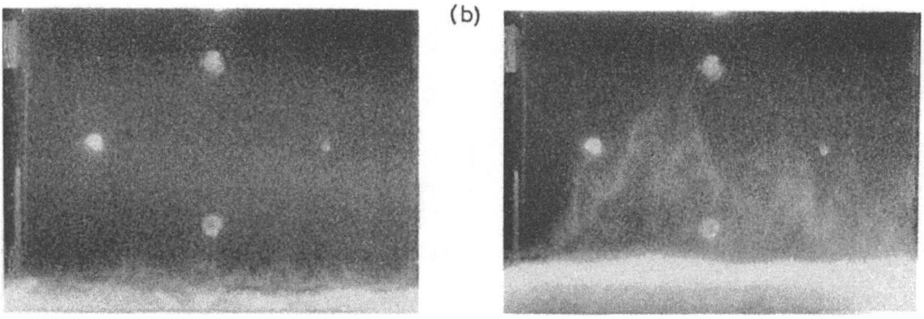

Fig. 6. Solidification of Na_2CO_3-H_2O system (upward cooling $\delta = 0°$)

of the horizontally stacked clockwise circulations which form double-diffusive layers. The clockwise circulations are easily identified by the motion of liquid crystal particles in the magnified photographs, but are not shown here. Figure 6 shows the result for upward cooling. Since thin plumes of light residual fluid are free to rise upward from the spiky crystals, which is referred to as finger convection, the isotherms in the fluid region exhibit a sawteeth-like profile in the horizontal direction, in sharp contrast to the case of thermal convection as shown in Fig. 3.

Chen and Turner [5] was the first to visualize double-diffusive layers and finger convection using a shadowgraph for a similar system, but they provided no information about temperature fields. A simple technique for temperature measurement by use of liquid crystals is much more helpful in interpreting the double-diffusive process during solidification than such conventional techniques as thermocouples.

References

1. Hiller, W.J.; Kowalewski, T. A. 1987: Simultaneous measurement of temperature and velocity fields in thermal convective flows. In: Flow Visualization IV (ed. Veret, C.), 617-622. Washington: Hemisphere

2. Moffat, R. J. 1990: Some experimental methods for heat transfer studies. Exp. Therm. Fluid Sci. 3, 14-32

3. Tankin, R. S.; Farhadieh, R. 1971: Effects of thermal convection currents on formation of ice. Int. J. Heat Mass Transfer 14, 953-961

4. Katto, Y; Iwanaga, Y. 1975: Onset of natural convection in a water layer with a upper stable layer. Preprints of 12th Jpn. Heat Transfer Symp. A331, 178-180

5. Chen, C. F.; Turner, J. S. 1980: Crystallization in a double-diffusive system. J. Geophys. Res. 85, 2573-2593

Heat Transfer

Numerical Analysis and Visualization of Air Ventilation Systems

M.AKIYAMA, H.SUGIYAMA, N.NINOMIYA and A.LEONI–SCHMID

Faculty of Engineering, Utsunomiya University, Utsunomiya City 321, Japan

Summary

Recent literature on numerical and experimental analysis of air ventilation systems of different geometrical configurations is reviewed. Application to passive solar systems is emphasized. A modified configuration of the conventional Trombe wall system is presented and comparative simulations are performed. Visualization of the calculation results gives insight into the behavior of the new heating system, pointing out some aspects worthy of further investigation.

Literature review

Recently, advances in the analysis of ventilation systems have been often reported in the literature. In these studies, visualization of the results plays a fundamental role, providing a qualitative comparison between calculations and experiments and the validation of the computational analysis techniques. Thus, a faster engineering development is attained than when relying solely on experiments. Especially in the field of passive solar energy, where geometrical factors affect significantly the thermal efficiency of the buildings, visualization techniques provide useful information which accelerates the process of decision making and encourages an extensive investigation of new configurations. Reviews on the subject were presented by Bauman et al.[1], Tosaka et al.[2], and Murakami et al.[3]. Nevertheless, there are many other works more closely related to the present research. Some remarkable examples among them are mentioned below.

Yoshikawa and Yamaguchi reported a series of investigations, comprehending a description of the numerical method and results for laminar flow[4], results for turbulent flow[5], an extension for the three–dimensional case[6], a non–isothermal calculation[7], a comparison between different air supply and exhaust locations[8,9] and an unsteady analysis[10], adding several comparisons between computer plots and visualization photographs. With the same room geometry, Akiyama et al.[11] investigated numerically the two–dimensional laminar mixed convection by finite element method, Yokoyama[12] performed a three–dimensional turbulent analysis with consideration of radiative exchanges, and Masuda et al.[13] performed a two–dimensional time–dependent analysis and comparison with experimental results for a lower range of Rayleigh numbers. Akiyama et al.[14] investigated the effects of mesh number and temperature boundary conditions for laminar natural convection within a square enclosure, comparing the results with experimen–

tal data. Li et al.[15] analyzed the effects of radiative heat exchanges between the wall surfaces, and Akiyama et al.[16] studied the influence of heat conduction through the solid walls, comparing numerical and experimental results. Murakami et al. and Kato et al. presented numerous works on flow simulation and visualization in buildings of various geometries, with the sizes of the spaces ranging from a room[17,18] to an indoor sports gymnasium[19,20]. Flow in an atrium and in an arena of complicated geometry were analyzed both numerically and through full scale measurements by the Kajima Corporation[21]. Tanguy et al. and Dupuis et al. performed various comparisons between visualization photographs and numerical calculations through different techniques for an industrial building[22-25]. Barozzi et al.[26] analyzed both experimentally and numerically the flow in a passive solar cooled house. There are many studies which analyzed theoretically and numerically different passive solar devices, mainly variations of the Trombe wall for single-story buildings. Ormiston et al.[27] and Bilgen and his co-workers[28] analyzed numerically Trombe wall models. In contrast to these studies, the purpose of present research is to extend the Trombe wall concept to a multistory building and to verify the efficiency of an alternative system presently suggested. Based on these results, some modifications in the alternative configuration were done and their effect is also reported.

Description of the model

A Trombe wall system consists of a massive wall placed some centimeters inwards from an insulated window of a building, forming an air gap or channel. Having a black surface outside, the wall absorbs the solar radiation and transmits part of it through convection to the air. Vents connecting the room and the channel provide ventilative heating of the room. The remaining energy is conducted through the wall and reaches its inner surface with a time lag that, when the wall thickness is well tuned, allows the heat to be released in nighttime. Considering a unique story, most of previous numerical analyses assumed the floor and ceiling surfaces to be thermally insulated. Whereas this can be a somehow realistic assumption for single-story buildings, it is not the case in multistory buildings, where heat exchanges take place across the slabs, i.e., the floors. Therefore, the numerical analysis should also include the effect of heat conduction through the slabs, allowing the prediction of interactions between the building floors.

As the warm air reaches the room space from the top of the Trombe wall, the ventilative heating of the room will be slow because of thermal stratification. In a multistory building equipped with air heaters on each floor, an alternative device, as shown in Fig.1, can be conceived in order to avoid the stratification and obtain a more uniform temperature distribution in the vertical direction. In a generic i^{th} floor, air taken from outside at floor level is heated along the channel and reaches the $i+1^{th}$ floor at the floor level. Except the first floor, a more convenient ventilation system than the conventional Trombe wall system is obtained.

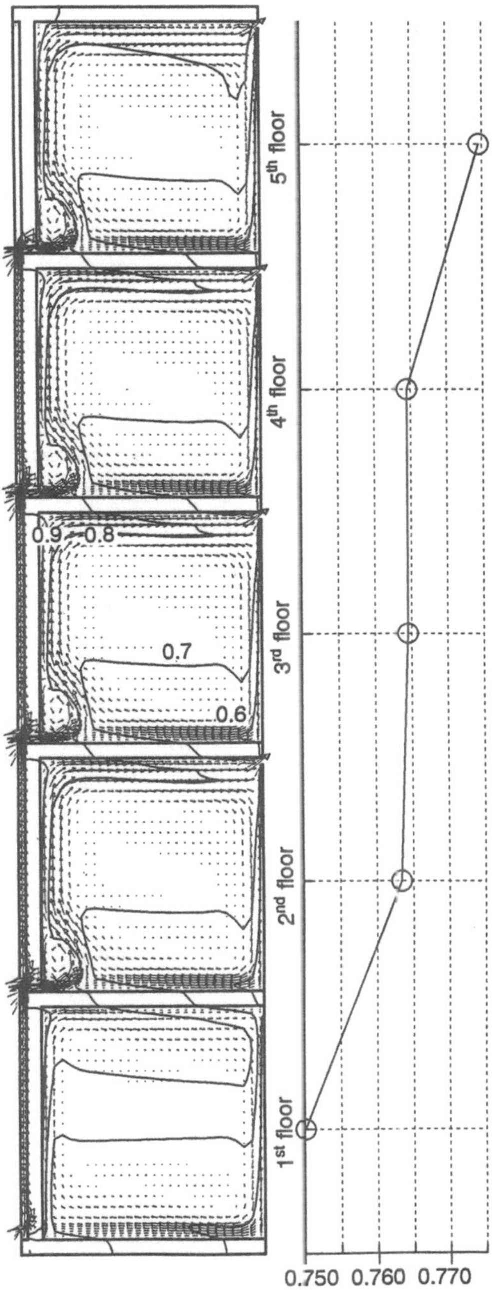

Fig.1 Analysis of a 5-story building, Ra=10^7: a)velocity vectors and temperature isolines b)floor average temperatures

Description of the calculations

Continuity equation, energy advection equation and the two-dimensional (laminar) steady-state Navier-Stokes equations were numerically solved for both cases. Buoyancy effects were accounted for through the Oberbeck-Boussinesq approximation. A finite control volume approach is used in the discretization, with a 50x50 grid refined over the channel and vents. As an analysis of the whole building would involve a repetitive pattern (Fig.1), computer time and storage can be saved through consideration of a unique, representative floor. Boundary conditions for the floor and slab surfaces are obtained by applying a cyclic condition, which connects the variables at the ceiling surface of ith floor to the floor surface of the $i+1^{th}$ floor. In order to compare the performances of each configuration, a wide range of parameters was covered: Rayleigh number from 10^4 to 10^7, non-dimensional glass temperatures T_{glass} from 0.00 to 1.00 and relative heat conductivity between walls and air from 1 to 100. Dimensionless values are defined as shown in Eq.(1) and (2). The solar heat flux was determined from a fixed temperature of T_{hot}=1.00 at the outer surface of the massive wall, and the opposite wall was given the temperature T_{cold}=0.50. For the alternative system, an outside temperature of T_{out}=0.00 was adopted and this was also used as a reference temperature in the buoyancy term of the momentum equations.

$$T = \frac{\theta - \theta_{out}}{\theta_{hot} - \theta_{out}} \quad (1) \qquad Ra = \frac{g\,\beta\,(\theta_{hot} - \theta_{out})L^3}{\nu\,\kappa} \quad (2)$$

(θ: temperature, g:gravity acceleration, L: room's width, β: volumetric expansion coefficient, κ: thermal diffusivity, and ν : dynamic viscosity).

Discussion of the results

Typical results for the conventional Trombe wall and the alternative model show differences in the temperature isolines close to the horizontal surfaces. Comparing the calculations which apply a cyclic condition with a case considering adiabatic surfaces, the latter shows a stronger thermal stratification. At Ra = 10^4, the flow through the channel is slow and has little significance when compared to the circulation which takes place in the room between the heating wall and the opposite wall. At Ra = 10^6, which means comparatively stronger buoyant forces, the same recirculation pattern takes place in the room, but a more intense flow is found in the channel.

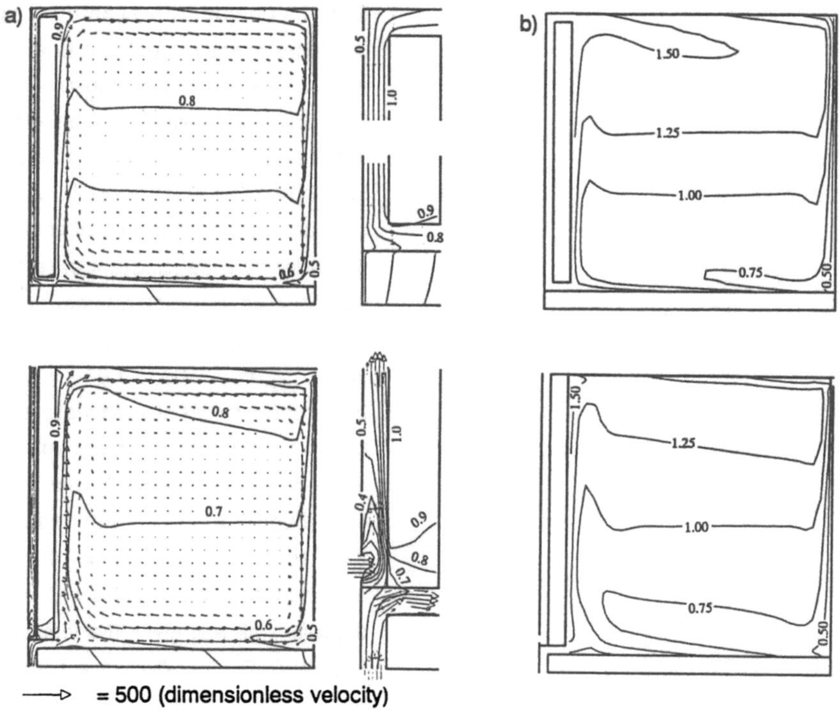

Fig.2 Results at Ra=10^7, T_{glass}=0.5, $\lambda_{wall}/\lambda_{air}$=10 (top: conventional Trombe wall, bottom: alternative model). a) temperature isolines and velocity vectors; b) PMV comfortness factor isolines

For Rayleigh number varying from 10^6 to 10^7, an even higher relative increase of channel air flow is observed. Nevertheless, even for Ra=10^7, the relative influence of the flow from the channel seems to be little. Warm air from the channel enters the room and goes up to the ceiling along the heating wall, without being sufficiently mixed with the room air.

When comparing the conventional and the alternative models in Fig.2 a), it should be remembered that the first is a closed space, and the second has a cold air influx from outside. Accordingly, a lower average temperature is expected for the alternative model. Nevertheless, as persons occupy the lower half of the room space, this does not mean a cooler room. The indoor comfortness can be quantified by the index PMV (Predicted Mean Vote, see Fanger[29]).The PMV is based on many variables and varies from −3 to +3, which represent the sensations of cold and hot, respectively; a null value represents a thermally neutral environment. Fig.2 b) shows the PMV isolines as obtained from the Fanger tables for a typical case of clothing insulation and metabolism level.

As the air circulation in the room proved to be the dominant flow pattern in most cases, an attempt was done to reduce its relative influence by increasing the temperature of the cold wall from 0.5 to 0.75. A slight effect could be noticed, the air current from the channel reaching a little deeper in the room before going up. Same effect was observed when enlarging the channel to two and three times its original width.

A recent tendency in air conditioning is the air supply through a gap under the floor. This means, a false floor is placed some centimeters over the slab, permitting air flow between the two surfaces. This allows a personalized ventilation, desirable, for example, in office buildings. In addition, air being supplied from the floor and exhausted at the ceiling provides a better air quality inside the room with prompt elimination of contaminants like tobacco smoke. Presently two cases were investigated, one with air inlet to the room at the center of the floor, the other by the cold wall. A visual inspection of the temperature distribution patterns shows how the underfloor heating can interrupt the thermal stratification in the room, providing simultaneously a warmer floor, and a warm air current across the lower half of the room.

Conclusions

A review on numerical and experimental analysis of ventilation systems was presented, giving a glance at the development in the field. As a further step, the case of a multistory building was considered with analysis of different type of passive heating systems. Despite many simplifying assumptions done in the calculations, results for the different cases provided enough qualitative information as to recognize the causes of the flow patterns obtained and suggesting modifications to be added in order to improve the heating efficiency. This confirms the importance of numerical analysis and flow visualization techniques in the project of air ventilation systems.

References

1. F.Bauman, A.Gadgil, R.Kammerud, E.Altmayer, and M.Nansteel, Trans. of ASHRAE, pp.215–233, (1983).
2. N. Tosaka and M. Kaizuka, Kenchiku Zasshi 100, No.1232, pp. 37–41, (in Japanese), (1985).
3. S.Murakami and S.Akabayashi, J.Flow Vis.Soc.Japan, 10 (36), pp.2–9, (in Japanese), (1990).
4. A.Yoshikawa and K.Yamaguchi, Air Cond.and Sanit.Eng., 48 (1), pp.5–17, (1974).
5. A.Yoshikawa and K.Yamaguchi, Air Cond.and Sanit.Eng., 48 (10),pp.11–25, (1974).
6. A.Yoshikawa and K.Yamaguchi, Air Cond. and Sanit.Eng., 49 (12), pp.7–18, (1975).
7. A.Yoshikawa and K.Yamaguchi, J.Air Cond. and Sanit.Eng., 5 (10), pp.55–64, (1977).
8. K.Yamaguchi, Y.Inoue, K.Yoneyama, and A.Yoshikawa, J.Air Cond.and Sanit.Eng., 6 (2), pp.75–85, (1978).
9. K.Yamaguchi, Y.Inoue, K.Yoneyama, and A.Yoshikawa, J.Air Cond.and Sanit.Eng., 6 (2), pp.85–96, (1978).
10. K.Yamaguchi and A.Yoshikawa, Air Cond.and Sanit.Eng.,7 (6), pp.71–80, (1978).
11. M.Akiyama, Y.Odaka, S.Yazaki, and K.Masuda. J.Flow Vis.Soc.Japan, 3 (9), pp.43–50 (in Japanese), (1983).
12. H.Yokoyama, MSc Thesis, Utsunomiya University, Japan, (in Japanese), (1981).
13. K.Masuda, M.Akiyama, S.Yamazaki, and M.Suzuki, Kuki Chowa Eisei Gakkai, 13 (6), pp.107–115, (in Japanese), (1980).
14. M.Akiyama, B.C.Huang, H.Sugiyama, N.Ninomiya, I.Urai, and K.Huang, 2nd China–Japan Symp.on Flow Vis., (1992).
15. S.Li, M.Akiyama, H.Sugiyama, and I.Urai, J.Vis.Soc.Japan, 11 (1), pp.51–54, (in Japanese), (1991).
16. M.Akiyama, H.Sugiyama, N.Ninomiya, and Y.Okada, to appear in the J.Vis.Soc.Japan, (1992).
17. S.Kato and S.Murakami, Trans.of ASHRAE, pp.309–330, (1988).
18. S.Murakami, S.Kato, and S.Suyama, Trans. of ASHRAE, pp.621–642, (1987).
19. S.Murakami, S.Kato, and Y.Ishida, Trans. of ASHRAE, pp.30–57,(1990).
20. S.Kato, 4th Symp. on CFD, Ryutai Kaiseki System Kenkyukai, pp.83–97, (1992).
21. S.Hayakawa, K.Horikoshi, Y.Korenaga,S.Kasahara, and M.Sato, J.Vis.Soc.Japan, 11 (41), pp.35–40, (in Japanese), (1991).
22. M.Dupuis, J.C.Methot, and E.Dernedde, Can.J.Chem.Eng., 63, pp.155–161, (1985).
23. P.A.Tanguy and M.Dupuis, Can.J.Chem. Eng., 64, pp.1014–1019, (1986).
24. P.A.Tanguy and M.Dupuis, Can.J.Chem. Eng., 66, pp.324–329, (1988).
25. M.Dupuis, E.Dernedde, and J.C.Methot, Can.J.Chem.Eng.,67 (10), pp.713–721,(1989).
26. G.S.Barozzi, M.S.Imbabi, E.Nobile, and A.C.M.Sousa, Proc. ISES Solar World Congress 1991, pp.2997–3002, (1991).
27. S. J. Ormiston, G. D. Orthby, and K. G. T. Hollands, Int.J.Heat Mass Transfer, 29 (6), pp.869–877, (1986).
28. R.B.Yedder and E.Bilgen, Int.J.Heat Mass Transfer, 34 (4/5), pp.1237–1248, (1991), etc.
29. P.O.Fanger,Thermal Comfort, R.E.Krieger Publish.Comp., Malabar, Florida,(1982).

Three-Dimensional Concentration Measurement in a Ventilated Room Model Through Digital Image Processing

Akikazu KAGA, Yoshio INOUE and Katsuhito YAMAGUCHI

Department of Environmental Engineering,
Osaka University, Osaka

Summary
A method to measure a 3-D concentration distribution of tracer particles in the airflow through visualization and digital image processing is presented. The tracer particle images obtained in different depths illuminated with a laser light sheet are analyzed synthetically including the effect of ray extinction due to light scattering.

Introduction

The experiment of mass transfer in a turbulent flow, such as the dispersion of pollutant in a ventilated room model, has implied the measurement of 3-D distribution of tracer concentration. However, a measurement with probes and sensors requires much of labor to obtain large amount of data for 3-D distribution; moreover probes can disturb the airflow. The work presented here describes the approach to measure 3-D concentration distribution of tracer particles through digital image processing. Similar approaches to 2-D measurement have been reported [1,2], where the effect of ray scattering has not been considered.

Ray Extinction due to Scattering

In this work, oil mist particles were used as visible tracers, a model chamber was illuminated with a laser light sheet and visualized images were recorded in a video recorder as shown in Fig.1. Relatively high concentration tracer particles are necessary in a measurement with a low power light source and a video camera of normal sensitivity, and the ray extinction due to light scattering can not be neglected.

The turbidity of tracer particles is defined by Eq.(1) and the ray extinction rate between point P and P' in Fig.2 is given by Eq.(2).

$$\tau = (\pi / 4)\ \delta^2 c K_{scat} \tag{1}$$

$$I_P / I_{P'} = \exp\{-\int_0^{\xi} \tau(\xi)\, d\xi\} \equiv E(\overline{PP'}) \tag{2}$$

in which δ is the average diameter of tracer particles, c is the number concentration, K_{scat} is the scattering efficiency and I is the ray intensity. The luminous intensity at point p (which is the image of point P as shown in Fig.3) by the first order scattering ray, $\overline{OPRp}$, is given by

$$L \propto I_Q\ (r_Q / r_P)\ E(\overline{QP})\, E(\overline{PR})\, \tau(P)\, f(\angle QPR)\, dV \tag{3}$$

in which I_Q is the intensity of incident ray at point Q, $\tau(P)$ is the turbidity at point P, and $f(\angle QPR)$ is the function to represent the direction dependence of relative scattering intensity. Also the second order scattering ray such as $\overline{OP'P''Rp}$ will influence the luminous intensity at point p.

Transformation of Luminous Intensity to Concentration

The luminous intensity corrected for the effect of $f(\angle QPR)$ is given by

$$\begin{aligned} L^* &\equiv \{f(\pi/2) / f(\angle QPR)\} \\ &= K_Q\ (r_Q / r_P)\ E(\overline{QP})\, E(\overline{PR})\, \tau(P) \end{aligned} \tag{4}$$

in which K_Q is a proportional constant. Since I_Q is usually not uniform in a light sheet, the value of K_Q is different for each incident ray.

Before measuring the concentration distribution, the luminous intensity distribution $\overline{L^*}$ in the chamber filled with tracer particles uniformly $(c = \overline{c})$ was measured. On the surface of $\overline{PR} \fallingdotseq 0$, Eq.(4) gives

$$\overline{L^*} = K_Q (r_Q / r_P) \exp\{-\overline{\tau}(r_P - r_Q)\}\overline{\tau} \tag{5}$$

in which $\overline{\tau}$ is the turbidity for $c = \overline{c}$ and estimated from the ratio of $\overline{L^*}$s at two different points on the same ray by Eq.(5). On the line $\overline{QP} \fallingdotseq 0$ and $\overline{PR} \fallingdotseq 0$ (point T in Fig.3), substitution of $r_P = r_Q$ into Eq.(5) leads to Eq.(6) and the value of K_Q for each incident ray can be estimated.

$$\overline{L^*} = K_Q \overline{\tau} \tag{6}$$

Therefore, by estimating the value of $E(\overline{QP})$ and $E(\overline{PR})$ from numerical integrations of Eq.(2), the value of L^* is calculated by Eq.(4), and the concentration distribution is determined as the relative value to $\overline{c}$. Since the values of τ already determined at the points closer to the video camera and light source are necessary in the integrations of Eq.(2), the calculation of Eq.(4) has to start from point T.

Transformation of Gray level to Corrected Luminous Intensity

The spatial sensitivity distribution of the hardware (Fig.4), the relationship between average gray level and luminous intensity (Fig.5) and the direction dependence of relative scattering intensity (Fig.6) were used on the process of the transformation of gray level to corrected luminous intensity,

Model Experiment of Ventilation

The model chamber used in the experiment is shown in Fig.7. The ventilated room in which the fresh air is supplied from an inlet on the ceiling and the contaminated air is generated on the floor was assumed. DOP mist particles of about one micrometer in diameter generated by an atomizer were used as tracer particles. An example of the concentration distribution is shown in Fig.8.

References

1. Sanders,D.D.;L.D.Albrigt: Building Systems: Room Air and Air Contaminant Distribution,87-85,ASHRAE 1989
2. Ohba,M.: Application of Video Image Processing Techniques to Concentration Measurement of Particles Emitted from Stack. J. of Visualization Society of Japan,10,Suppl.1(1990)127-130

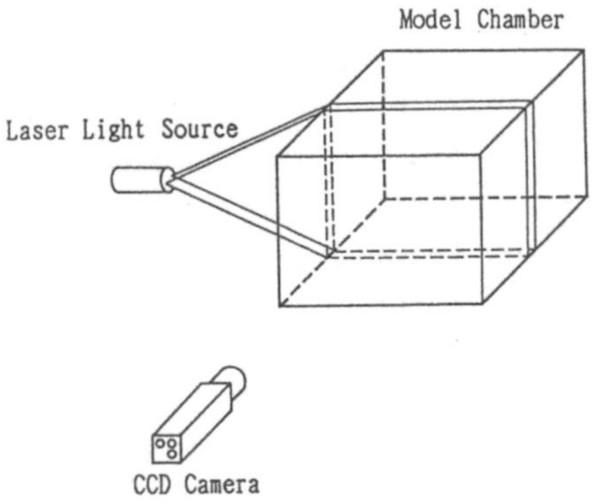

Fig.1. Schematic of test equipment

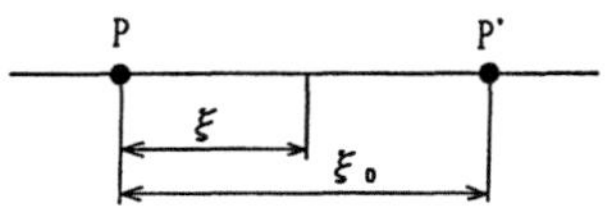

Fig.2. Coordinate in Eq.(2)

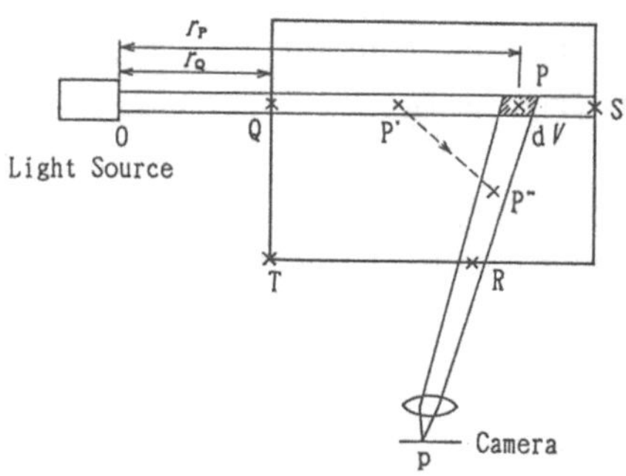

Fig.3. Top view of light paths

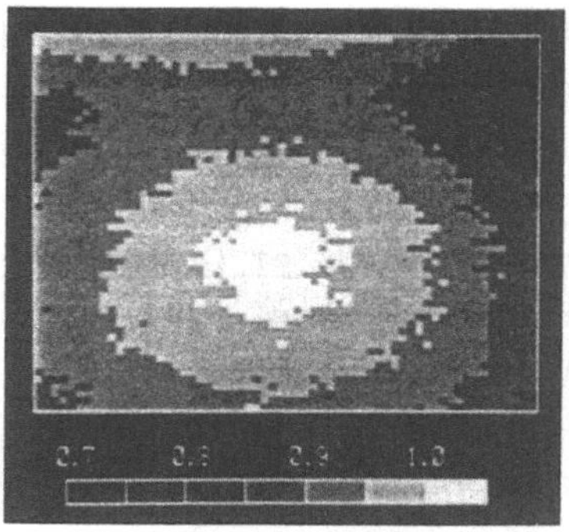

Fig.4. Spatial gray level distribution of uniformly illuminated plain

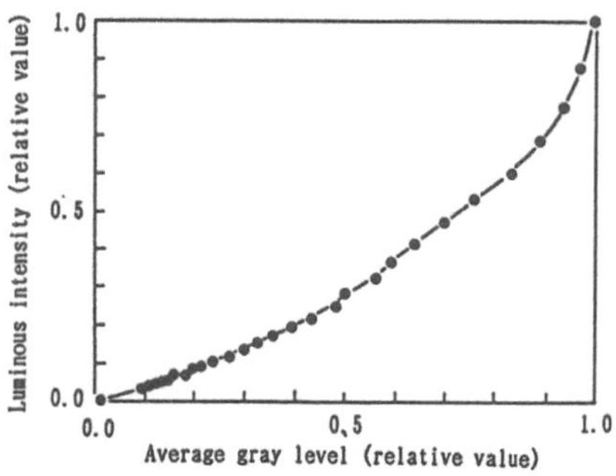

Fig.5. Relation ship between average gray level and luminous intensity

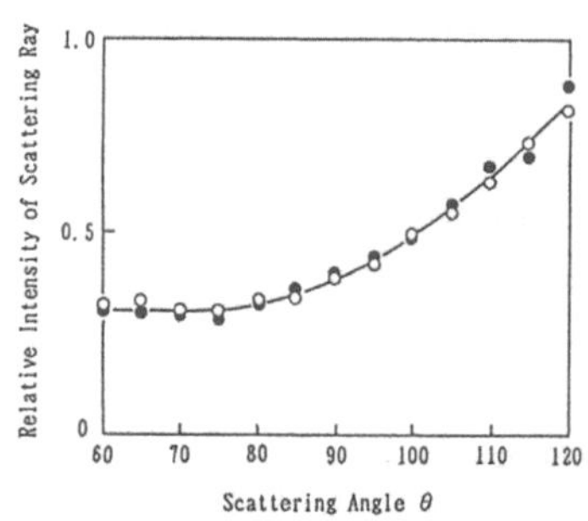

Fig.6. Direction dependence of relative scattering intensity

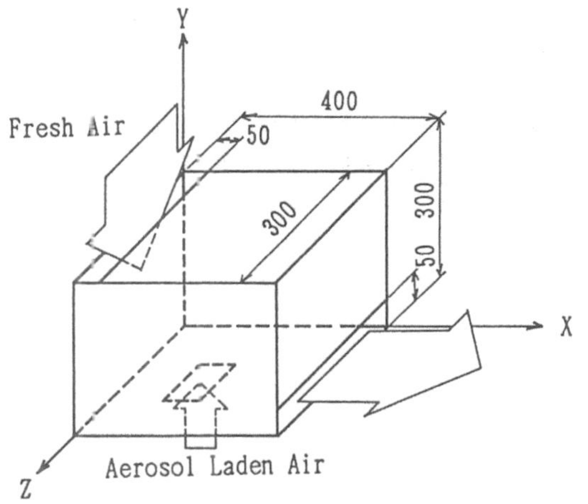

Fig.7. Model chamber of a ventilated room

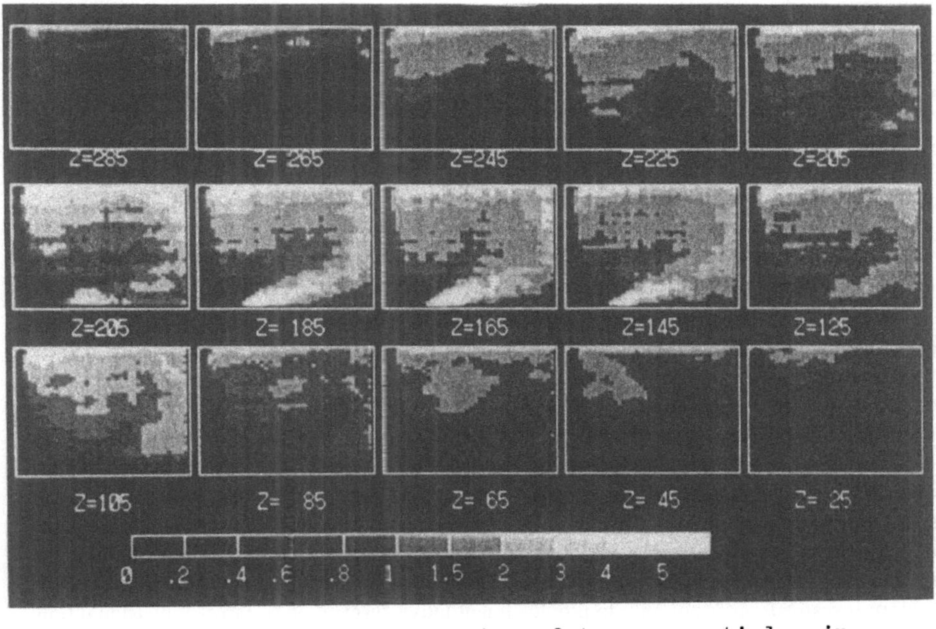

Fig.8. Concentration distribution of tracer particles in various cross-sections.

Visualization of 3-D Natural Convection

W.J. Hiller, St. Koch, T.A. Kowalewski, P. Mitgau & K. Range
Max-Planck-Institut für Strömungsforschung, 3400 Göttingen, Germany

SUMMARY

To provide a better understanding of three-dimensional convective flow and for the purpose of a comparison with existing numerical models, two computer aided visualization techniques were developed: analysis of the light colour (hue) which is reflected by liquid crystal tracers suspended in the flow, and stereoscopic observation of tracer paths. They allow non-intrusive measurement of temperature fields and 3-D registration of tracer tracks conveyed with the flow, respectively.

1. INTRODUCTION

Convective flow is one of the most effective mechanisms for heat and mass transport, both in technical applications and in nature. However, experimental knowledge of natural convection within enclosures is still far from being complete. Most of the investigations are performed using two-dimensional methods, like interferometry to measure temperature, or the light sheet method to visualize velocity fields. The real 3-D structures, caused by the unavoidable bounding walls, are usually neglected. Hence, 3-D numerical calculations, which are nowadays performed extensively, do not match corresponding experimental data to compare with.

In the present paper we describe two visualization techniques, which allow to collect data about temperature field and flow structure in a true three dimensional flow. Both methods are used to investigate natural convection in a simple cube-shape cavity. The use of the cubic cavity emphasis the aim to concentrate our interest on 3-dimensional flow characteristics. Two vertical walls of the cavity are isothermal, whereas the thermal properties of the four other walls are not ideal, i.e. neither adiabatic conditions nor ideal conductivity can be assumed. In the simplest case, at low Rayleigh number ($Ra < 40000$), one convection roll develops, with a small inner spiral transporting liquid from the side walls, and a large outward spiraling structure in the plane of symmetry. It was previously observed [1], that for higher Ra the single inner spiral splits up into two, while close to the front and back walls still only one roll exists. The purpose of the present investigations is not only to provide quantitative description of three-dimensional flow and temperature structures in the cavity, but also to get real temperature boundary conditions at the non-isothermal walls. This information supplied to the numerical model evidently moderates existing discrepancies between observed and simulated flow structures.

2. EXPERIMENTAL METHODS

2.1. Temperature and velocity fields

The experimental set-up used for measuring temperature and velocity fields is shown schematically in Fig. 1. It is essentially similar to that previously used [1, 2]. The convection cavity, the xenon flash tube and the camera are mounted on a rectangular frame, which can be tilted to observe vertical and horizontal cross-sections without re-adjusting the optics. The box with an inner length of 50 mm consists of two vertical opposite isothermal walls made of black anodized metal and four 8 mm thick plexiglas walls. The cavity is filled with a very dilute suspension of uncoated thermochromic liquid crystal (LC) droplets in glycerol or a glycerol-water mixture. The LC-droplets of about $50\mu m$ are used as tracers and temperature sensors.

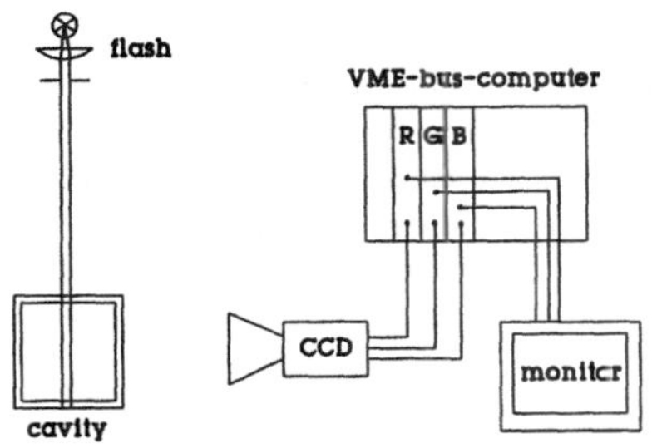

Fig.1. *Experimental set-up.*

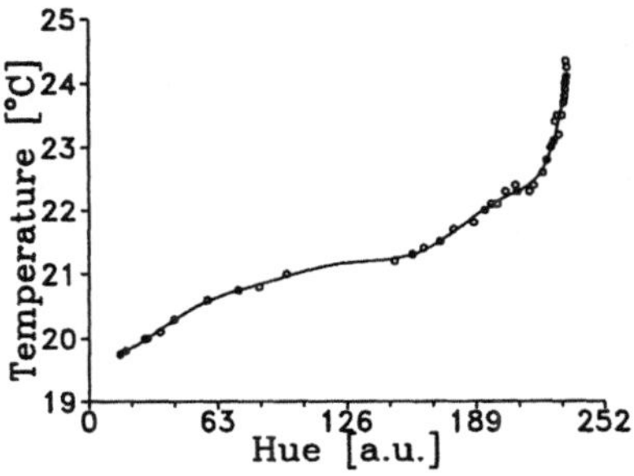

Fig.2. *Typical calibration curve. The hue values of RGB colours are (0,126,252), respectively.*

The xenon flash tube generates a 2 mm thick white light sheet, which illuminates a cross-section of the flow. The wavelength of the reflected light depends on the temperature of LC-tracers. A three chip CCD color camera (Sony DXC-750P), which gives a RGB-signal for the red, green and blue portion of the incoming light, is used to observe the flow. The images are acquired by three identical frame grabbers (PPI- Eltec) in a VME-bus computer. Besides the further image processing, the computer entirely controls the experimental conditions, i.e. the wall temperatures, position of the cavity and registration sequences. Moving the convection cavity perpendicular to the light sheet the temperature and velocity fields of the entire flow are scanned.

For evaluating the temperature the following, generally applied [3] method is used. The incoming RGB signals are transformed pixel by pixel into intensity, saturation and hue. Relating the hue to a temperature by a calibration function, the value of temperature is estimated. The values of intensity, saturation and hue are computed with a quite simple and fast algorithm, which nevertheless gives the same accuracy as other more complex formulas. For the given three values of R, G and B, the intensity is defined as $I = \sqrt{(R^2 + G^2 + B^2)}$ and the saturation as $S = 1 - min(R, G, B)/I$. The hue is the ratio of those two values of (R, G, B), which, after subtracting $min(R, G, B)$ are non-zero. As liquid crystals reflect pure spectral colours, there are no hue values registered simultaneously by the blue and red sensitive chip of the camera. Hence, only the RG and GB hue regions of the chromaticity diagram have to be considered.

The calibration is performed at the experimental conditions, increasing gradually (in steps of 0.1 K) the temperature in the entire convection cavity, and evaluating the hue of light reflected from tracers close to the isothermal walls. The non-linear relation between temperature and hue (predetermined by the camera construction), is described by a 9th order polynomial fitted to the points (Fig. 2), measured for LC type TM317 (BDH). The accuracy of the measured temperature depends on the hue value, and varies in the temperature range $20 - 24°C$ from $\pm 0.1K$ to $\pm 0.3K$. The results of the computer evaluation are displayed as a false colour image, using a median filter to smooth picture noise, and to paste area without colour information.

The displacement of the tracers captured on the two subsequent images are used to evaluate two-dimensional velocity components of the flow. This is done with help of a specially developed software [4] for particle imaging velocimetry (PIV), which runs on a personal computer equipped with a FFT processor.

2.2. Particle Tracks

To observe three dimensional flow structure, a special procedure has been developed [5], which allows detection and full automatic tracking of particles suspended in the flow medium. For this purpose a few, nearly neutrally buoyant, and about $150\mu m$ in diameter tracers, are observed against the bright background from two perpendicular directions. The images are captured by two CCD cameras (Sony XC77CE) integrated with two identical frame grabbers (VS100 Imaging Inc.) mounted in a personal computer. The evaluation software allocates particles and evaluates its position within about 10s. It allows to repeat data acquisition with storing solely a small data file containing informations essential for the further evaluation. This saves a lot of disk space, comparing to storage of the acquired pictures, thus allowing registration of particle paths over periods as long as several days.

3. APPLICATIONS

3.1. Temperature measurements

The temperature visualization technique has been applied for the investigation of the onset of convection in the cavity, and to measure the temperature distribution of the non-isothermal walls.

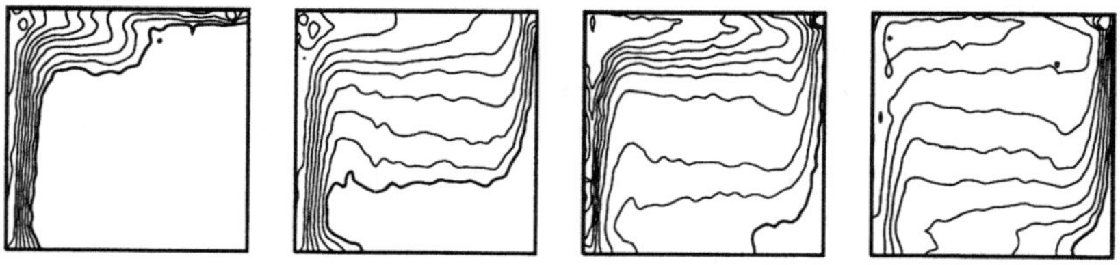

Fig.3. *Isotherms measured in the centre-plane of the cavity: 2min, 10min, 20min and 60min after the temperature gradient is applied; Ra=180000, Pr=980.*

Onset of convection

The observations of the transient flow development are limited to the centre-vertical plane of the cavity, because the simultaneous observation of its different cross-sections is presently not possible. The experiment commences when in the well insulated cavity, heated to an overall uniform temperature, a sudden temperature gradient is applied to the two opposite vertical walls. A sequence of 40-60 images of the flow taken subsequently shows that during the first 10-20 minutes primary vertical isotherms continuously transform its shape into the final characteristic S-form. A few isotherms captured at different instants, scaled to the applied temperature difference $(20 - 23°C)$, are displayed in Fig. 3. The isotherms show how the heat flux from the isothermal walls into the adjacent fluid layer gradually generates convective flow in the whole cavity. The flow achieves its final steady state after more than one hour.

Measurements of the wall temperature.

The temperature of non-isothermal walls is measured to explain some observed discrepancies [2, 6] between experiment and numerical modelling. It was previously found, that neither the adiabatic boundary condition nor the assumption of constant heat flux through the side walls are suitable to model observed particle tracks. Satisfactory results obtained by a direct application of the measured thermal conditions to the numerical code, implies directions for further improvements of the modeling procedure.

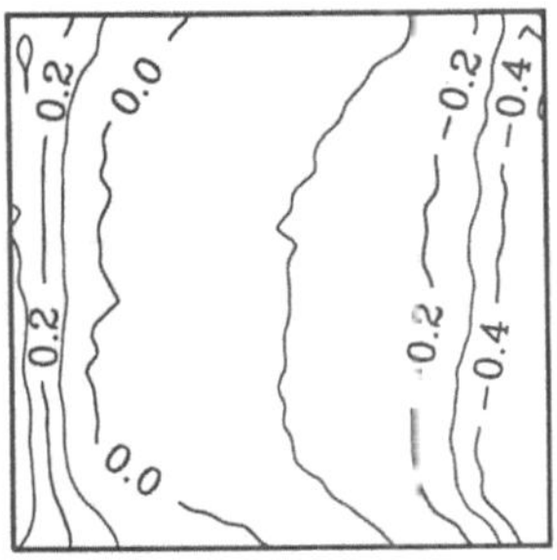

Fig.4a. *Isotherms measured on the bottom wall.* $T_H = 23°C$, $T_C = 20°C$.

Fig.4b. *Isotherms measured on the side wall.*

As a direct observation of the side walls of the cavity has several disadvantages, the temperature measurements are performed in several vertical cross-sections gradually approaching the investigated wall. In such a way, a 3-dimensional mesh of temperature values, with a mesh size less than 1 mm, is obtained. Extrapolation of the mesh data allows to evaluate accurate values of the wall temperature. As the position of the light sheet is controlled by the computer, the whole experiment runs automatically. Examples of the isotherms measured for the fully developed convection in the cavity at $Ra = 180,000$ and $Pr = 980$ are shown in Fig. 4.

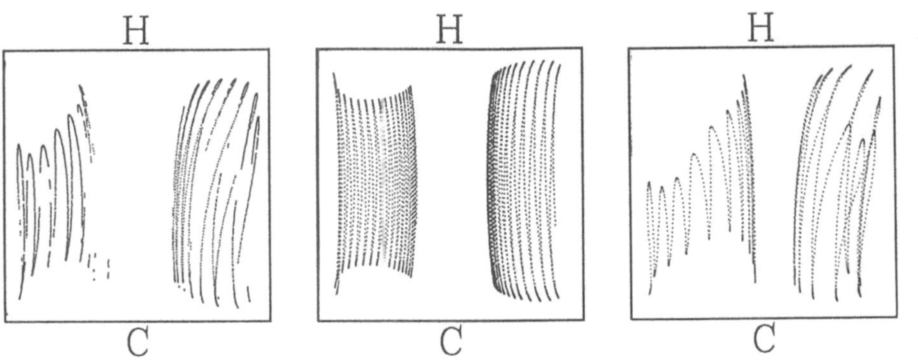

Fig.5. *Particle tracks measured (left), calculated with adiabatic boundary conditions (center), and calculated with measured temperature imposed as the thermal boundary conditions (right).*

3.2. Particle tracks

The particle tracks with adequately chosen starting points are a good tool to investigate the flow topology. They are also useful for the verification of the numerical results. The tracks presented in Fig. 5 in a 2-dimensional projection are obtained experimentally for the flow in the cavity at Ra=40,000 and Pr=1,900. They are compared with the tracks generated with help of a numerical code FRECON3D [7]. It can be seen, that the measured wall temperature, imposed as thermal boundary condition in the numerical simulation, evidently improves the overall agreement between both tracks.

CONCLUDING REMARKS

The computer-aided analysis of colour images of uncoated LC-tracers is a promising method of visualizing both velocity and temperature in a convective flow. Together with the the particle tracking, performed in parallel experiments, it allows complex analysis of the investigated flow, and provides an exceptional tool to verify the accuracy of temperature and velocity fields obtained from numerical models.

References

[1] Hiller W.J., St. Koch, Kowalewski T.A.; Three-dimensional structures in laminar natural convection in a cube enclosure. Exp. Therm. and Fluid Sci., **2**, (1989), pp.34-44.

[2] Hiller W.J., St. Koch, Kowalewski T.A., de Vahl Davis G., Behnia M; Experimental and numerical investigation of natural convection in a cube with two heated side walls. Proc. of the IUTAM Symposium Cambridge U.K., Aug. 13-18, 1989, pp. 717 - 726, CUP 1990.

[3] Hirsch C., Platzer K.H., Metzger D.E., Witting S.; Instationäre Wärmeübergangsmessung unter Nutzung von thermochromen Flüssigkristallen als Temperaturindikatoren, in 2D-Meßtechnik, DGLR-Workshop 18-19 Oct. 1988, Markdorf. Bericht 88-04, pp. 211-221. Edt. DGLR Bonn.

[4] Paul R.; Die Erfassung von Geschwindigkeitsfeldern durch automatische Bewgungsanalyse am Beispiel einer rotierenden Flüssigkeit. Mitteilungen aus dem MPI Göttingen FRG, No. 102, 1991.

[5] Bartels-Lehnhoff, H.-H; Erfassung dreidimensionaler Bahnlinien und Geschwindigkeitsfelder mittels automatischer Bildverarbeitung. Mitteilungen aus dem MPI Göttingen FRG, No. 101, 1991.

[6] Hiller W.J., Koch St., Kowalewski T.A., Range K., Behnia M., de Vahl Davis G.; Visualization of 3-D natural convection - comparison with numerical results in: Proc. of 11th ABCM Mech. Eng. Conf. São Paulo (Brasil), Dec. 1991, vol.2, pp. 21-24, Ed. Esp. da Revista Brasileira de Ciências Mecânicas, São Paulo 1991.

[7] Goh L.P, Leonardi E., de Vahl Davis G.; FRECON3D - Users Manual. A program for the numerical solution of mixed convection in a three- dimensional rectangular cavity. The University of New South Wales, School of Mech. and Ind. Eng., Rep 1988/FMT/7, 1988.

Corresp. to: St. Koch, MPI für Strömungsforschung, Bunsenstr. 10, D-3400 Göttingen, FRG.

Flow Visualization of Natural Convection in a Vertical Fluid Layer with Internal Heating

Z. KAWARA, I. KISHIGUCHI, M. AOKI and I. MICHIYOSHI

Department of Nuclear Engineering, Kyoto University
Yoshida, Sakyo-ku, Kyoto 606-01, JAPAN

Abstract

Flow visualization with tracer method has been carried out on laminar natural convection in a vertical fluid layer of water with uniform internal heat generation. A pair of opposite vertical walls was cooled and maintained at constant temperatures respectively, and other boundaries were adiabatic. When the fluid layer was heated both externally and internally, complicated flow pattern was observed. Flow patterns were classified into three types with the internal and external Rayleigh numbers. Similar flow pattern was observed for different aspect ratio.

Introduction

Natural convection in a confined regime with internal heat generation is important phenomena in the area of nuclear engineering, chemical engineering, geophysics and so on. It has been investigated both experimentally and theoretically, but almost of all are for horizontal fluid layer, which is heated from below and cooled above. The studies for vertical fluid layer, which has a pair of cooled side wall, are a few. Especially study for the fluid layer heated both externally and internally had been starting recently (for example, numerical study of Acharya and Goldstein [1], and experimental and numerical study of Wang and Wang[2].)

In this work, flow field of laminar natural convection in a vertical rectangular fluid layer heated internally and from side wall was visualized with tracer method.

Experimental Apparatus

Figure 1 shows schematic of experimental conditions. A pair of vertical opposite walls were made of copper blocks, which were the constant temperature boundaries. Other walls were adiabatic boundaries made of acrylic resin plates. The height of the layer ranged from 30mm to 100mm, and distance between copper blocks from 20mm to 100mm. The width of copper blocks was 50mm.

Test fluid was dilute potassium chrolide aqueous solution, and was heated internally by joule heating. Alternating electric current was passed horizontally between two copper blocks, and all region of the fluid layer is heated uniformly. The fluid layer was also cooled or heated externally by circulating water through the channel in the copper blocks. Temperatures of the copper blocks were measured by embedded thermocouples, and were maintained at a constant temperatures respectively.

Flow pattern was visualized by using tracer method. Powder of lycopodium was mixed into the fluid layer as tracer with the concentration of 0.2g/l. Lycopodium is spore of plant whose diameter is about 20µm, and its density is almost equal to water. Flow was very slow (~ 0.1mm/s) especially in the core region of the fluid layer, so lycopodium powder was best tracer for present experiment. Sheet-like light of 5mm in thickness passed through the fluid layer, and two-dimensional cross sectional view of flow field was observed and photographed. All photographs in the present report were taken in steady states, and exposure time was 20 seconds.

Results and Discussion

Figure 2 shows photographs of the typical flow patterns at various heating conditions. Overall dimensions of the fluid layer is 100mm hight ×100mm width. The vertical wall at the right side is cold wall, and the left side wall is hot wall. When two vertical walls are cooled at the same temperature ($\Delta T=0$) and fluid layer heated internally (as shown in Fig.2(A)), two symmetrical flows develop. Fluid close to the side walls is cooled and flows downward quickly to the bottom of the fluid layer. In central region of the layer, fluid flows upward slowly. On the other hand, the case of external heating only (Fig.2(F)), one large scale flow develops. Flow is downward near the cold wall, and upward near the hot wall. Central region is stagnant.

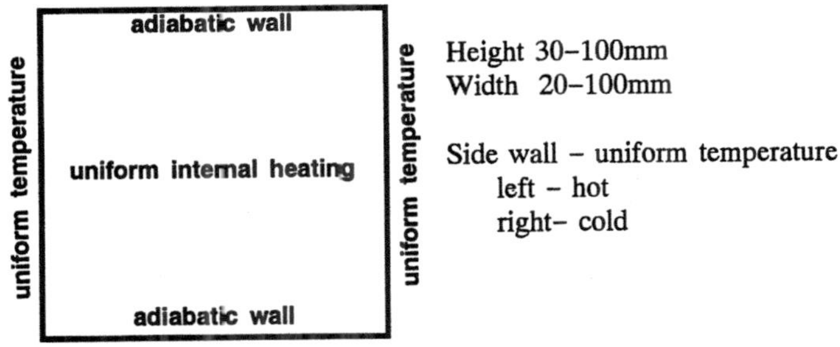

Fig.1. Schematic of Experimental Condition

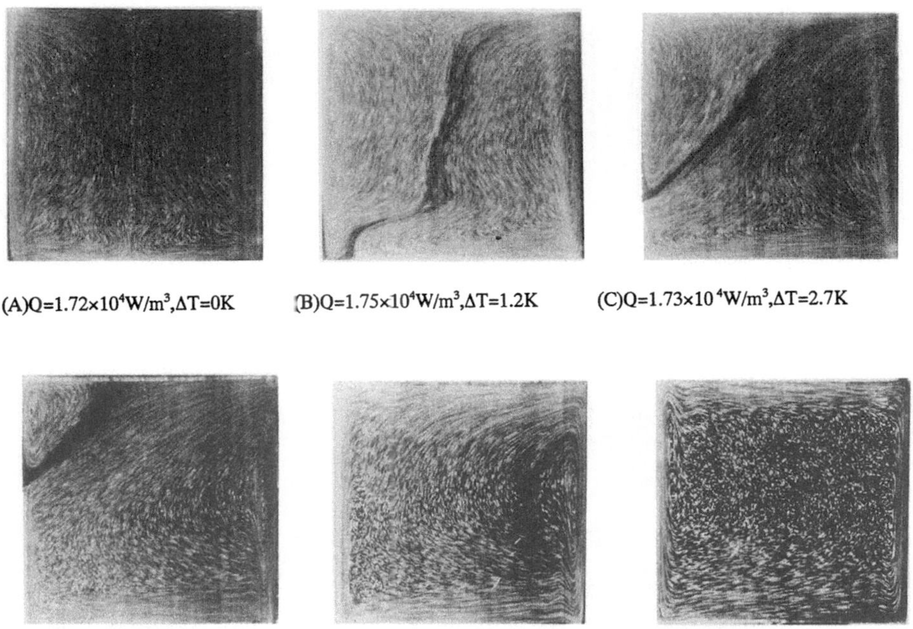

(A)Q=1.72×10^{4}W/m^3,ΔT=0K (B)Q=1.75×10^{4}W/m^3,ΔT=1.2K (C)Q=1.73×10^{4}W/m^3,ΔT=2.7K

(D)Q=9.47×10^{3}W/m^3,ΔT=2.7K (E)Q=5.50×10^{3}W/m^3,ΔT=2.5K (F)Q=0 W/m^3,ΔT=2.6K

Fig.2. Photograph of flow patterns (100mm × 100mm)

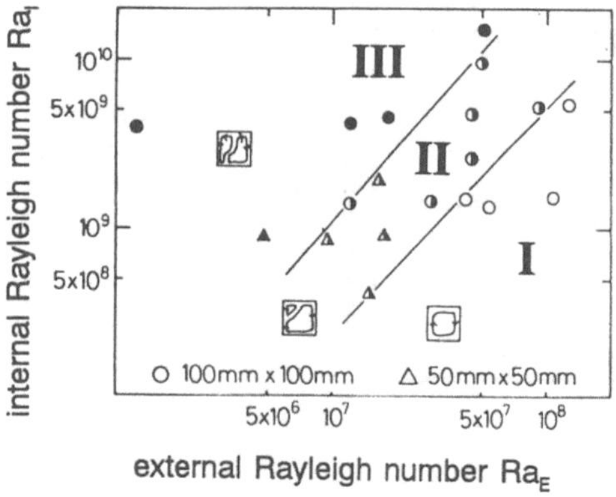

Fig.3. Flow pattern map (100mm × 100mm, 50mm × 50mm)

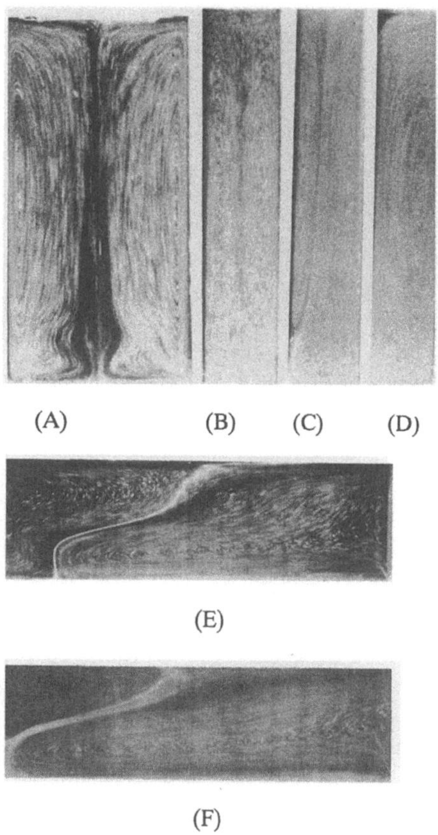

(A)100mm × 50mm
$Q = 8.5\times10^4 W/m^3$
$\Delta T = 0$ K

(B)100mm × 20mm
$Q = 1.43\times10^4 W/m^3$
$\Delta T = 0$ K

(C)100mm × 20mm
$Q = 1.40\times10^5 W/m^3$
$\Delta T = 2.5$ K

(D)100mm × 20mm
$Q = 1.68\times10^4 W/m^3$
$\Delta T = 2.5$ K

(E)30mm × 100mm
$Q = 5.11\times10^4 W/m^3$
$\Delta T = 3.3$ K

(F)30mm × 100mm
$Q = 2.67\times10^4 W/m^3$
$\Delta T = 3.3$ K

Fig.4. Photograph of flow patterns at various aspect ratio

When the layer heated both internally and externally, complicated flow patterns were observed, and they are classified into three patterns. The first pattern (I) is the case of one vortice (Fig.2(E)), and in this case fluid is heated on the left side wall. The second pattern (II) is the case that two vortices develop and they meet on the hot wall (Fig.2(C),(D)). The third pattern (III) is the case that two vortices meet on the bottom of the enclosure, and fluid is cooled on both side walls (Fig.2 (B)). These flow patterns (I, II, III) were arranged and correlated with internal heating rate and temperature difference between two side walls. Figure 3 is flow pattern map classified with nondimensional parameters, external Rayleigh number Ra_E ($=g\beta\Delta TL^3/a\nu$) and internal Rayleigh number Ra_I ($=g\beta QL^5/\lambda a\nu$).

Flow patterns at various aspect ratio were also obtained as shown in figure 4. Flow patterns deformed according to the variance of aspect ratio, but they were similar to the one of aspect ratio 1.

Concluding remarks

Varoius flow patterns are observed in natural convection of vertical fluid layer with internal heat generation, and they are correlated well with non-dimensional parameters. Flow pattern changes according to the parameters sensitively and systematically. Therefore, the present results can be used as good example for bench mark test of computer code on thermal problem.

Nomenclature

a:thermal diffusivity, g:gravitational acceleration,
L:distance between side walls,
Q:volumetric rate of energy generation,
β:isobaric coefficient of volumetric expansion
λ:thermal conductivity ν:kinematic viscosity

References

1.Acharya, S. and Goldstein, R.J., Natural convection in an externally heated vertical or inclined square box containing energy sources, ASME J. of Heat Transfer, Vol.107, pp.855, 1988.
2.Wang, H.S. and Wang, Q.J., Natural convection in an externally heated vertical rectangular enclosure containing internal energy sources, Proc. of 1st Conf. Experimental Thermal and Fluid Science, pp.709, 1988.

Numerical Analysis on Two-Dimensional Flow and Heat Transfer in Compact Heat Exchangers of Louvered Fins

C.S.Kang
Department of Mechanical Engineering, College of Engineering
Kyungbook National University, Taegu, Korea

J.L.Sohn and T.M.Choi
Fluid Engineering Laboratory
Korea Institute of Machinery and Metals, Changwon, Korea

J.H.Lee
Department of Mechanical Engineering, Graduate School
Kyungbook National University, Taegu, Korea

ABSTRACT

A finite difference method with a SIMPLE algorithm was applied to numerically handle a two dimensional problem in a compact heat exchanger with louvered fins. The characteristics of physical phenomena for the fluid flow and heat transfer inside a heat exchanger are investigated by changing geometric and flow parameters such as louver angle of the fin(θ), the ratio of louver pitch to fin pitch(F_p/L_p) and Reynolds number($Re=UL_p/\nu$). Computational results are graphically visualized and many important physical characteristics could be interpreted from those computational visualization.

INTRODUCTION

Heat exchangers for automobiles, such as radiators and/or condensers, must be compact because of requirements of high performance with light weight in small sizes. To improve the performance of heat exchangers, louvered fins are widely employed in various types of compact heat exchangers for automobiles[1,2]. Figure 1(a) shows array of louvered fins in compact heat exchanger and flow characteristics around fins. Dimensions governing the chracteristics of heat exchangers are fin pitch(F_p), louver pitch(L_p) and louver angle of fins(θ). Important parameters for the analysis of fluid flow and heat transfer are Reynolds number($Re=UL_p/\nu$) and the ratio of louver pitch to fin pitch(F_p/L_p). Because the speed of air running through heat exchanger is in the range of 1-8 m/s, the Reynolds number based on this range of speed is less than 500 which means that the flow can be reasonably assumed as the laminar flow. In this study, the characteristics of fluid flow and heat transfer in compact heat exchanger with louvered fins are investigated by solving governing equations using finite difference method.

COMPUTATIONAL DOMAIN AND BOUNDARY CONDITIONS

Computational domain and related boundary conditions are shown in Figure 1(b) and 2. Because of its periodic characteristics of the array of louver fins, it is reasonable to choose one array of them with the appropriate cyclic boundary conditions. Shapes and louver angle of fins inside computational domain could be changed by the number of finite difference meshes and their densities inside the domain. Effect of the location of fins could be treated by the blockage thechnique.

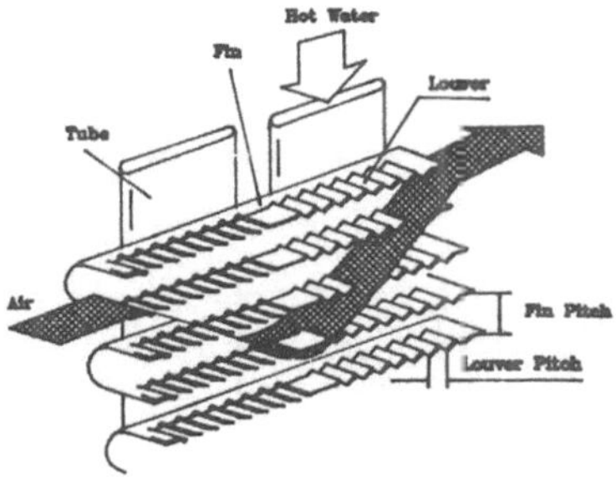

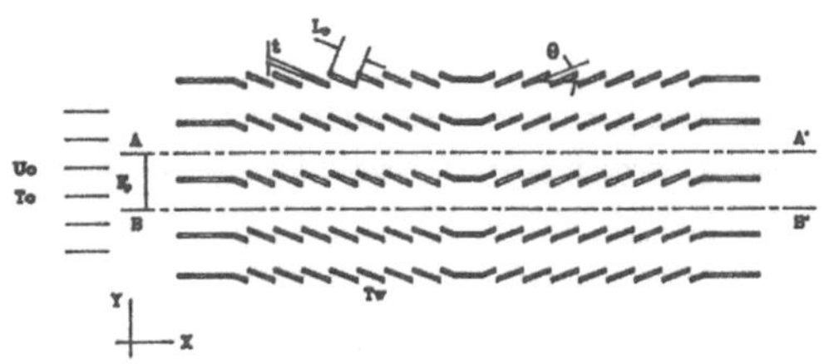

(a) Three-D. view of heat exchanger (b) Calculation domain

Fig. 1 Louvered-fin array in compact heat exchanger

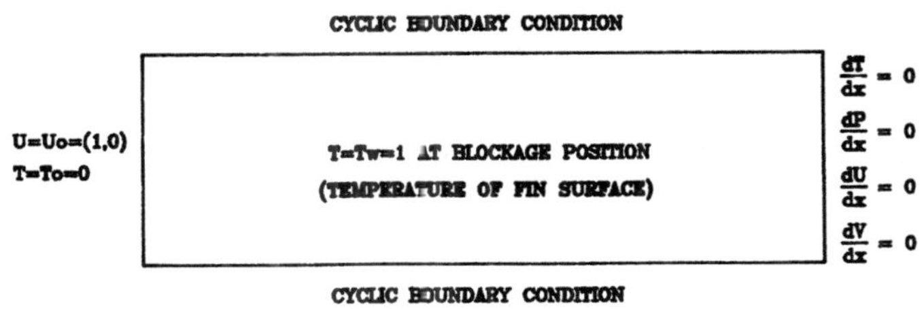

Fig. 2 Boundary condition

MATHEMATICAL FORMULATION

Basic assumption for this study are as follows:

(1) Two dimensional steady state
(2) Incompressible laminar flow
(3) Forced convection without considering radiation effects

Governing equations based upon the above assumptions are as fllows:

Continuity Equation

$$\frac{\partial U}{\partial X} + \frac{\partial V}{\partial Y} = 0 \tag{1}$$

Momentum equation

$$U\frac{\partial U}{\partial X} + V\frac{\partial U}{\partial Y} = -\frac{\partial P}{\partial X} + \frac{1}{Re}\left(\frac{\partial^2 U}{\partial X^2} + \frac{\partial^2 U}{\partial Y^2}\right) \tag{2}$$

$$U\frac{\partial V}{\partial X} + V\frac{\partial V}{\partial Y} = -\frac{\partial P}{\partial Y} + \frac{1}{Re}\left(\frac{\partial^2 V}{\partial X^2} + \frac{\partial^2 V}{\partial Y^2}\right) \tag{3}$$

Energy equation

$$U\frac{\partial T}{\partial X} + V\frac{\partial T}{\partial Y} = \frac{1}{RePr}\left(\frac{\partial^2 T}{\partial X^2} + \frac{\partial^2 T}{\partial Y^2}\right) \tag{4}$$

In the above equation, all variables are non-dimensionalized based upon the speed of air running through heat exchanger(U) and louver pitch(L_p).

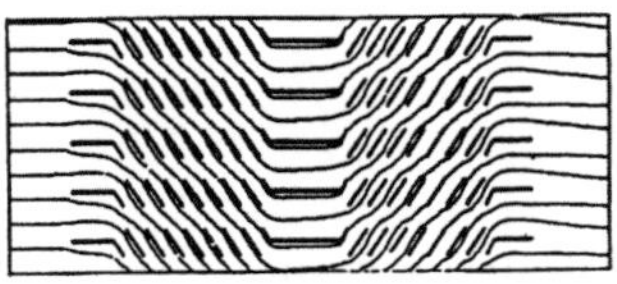

(a) Re=100, Fp/Lp=1.0, θ=28°

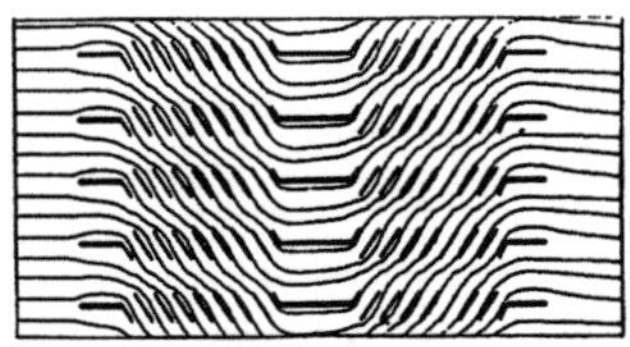

(b) Re=100, Fp/Lp=1.23, θ=28°

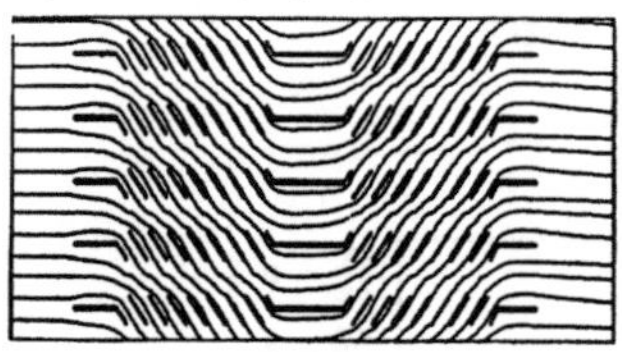

(c) Re=300, Fp/Lp=1.23, θ=28°

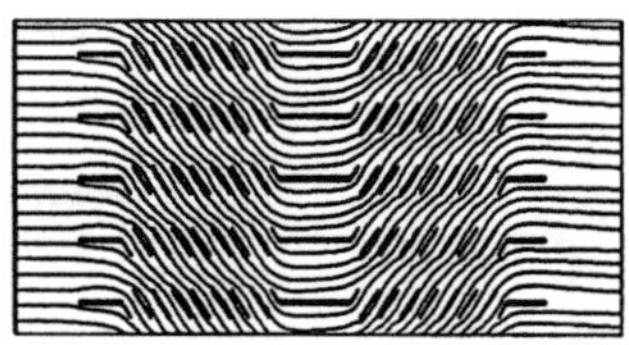

(d) Re=500, Fp/Lp=1.23, θ=28°

Fig.3 Streamline contours

RESULTS AND DISCUSSIONS

As well-known, the above equations are elliptic partial differential equations. To numerically handle this kind of equations, the EL2D code , which was developed based on the SIMPLE algorithm, is used. Some modifications are made to handle cyclic boundary conditions effectively. Sun SPARC station 1 was used to solve this problem. KIMMPOST program was used to graphically visualize computational results. KIMMPOST was dveloped for the postprocessing of the results of finite difference computation based on the SunPhigs graphic routine on the sun workstation.

To investigate the characteristics of fluid flow and heat transfer in compact heat exchanger with louvered fin, effects of Reynolds number(Re=100, 200, 300), the louver angle of fins(θ=22deg, 24deg, 28deg) and the ratio of louver pitch to fin pitch(F_p/L_p=1.0, 1.23, 1.5) are considered. Streamline patterns around louvered fins are shown in Figure 3. In the entrance region, it is observed that streamlines are passing close to the surface of fins in case of low F_p/L_p. As shown in Figure 4, the region influenced by the effect of convective heat transfer becomes larger and the thickness of thermal boundary layer is thinner as the Reynolds number increases. Also, it is observed that the effect of convective heat transfer becomes wider as the value of F_p/L_p decreases. To consider the effect of louver geometry to drag on the surface of louvered fin, the local pressure coefficient on the surface of louver is defined as,

$$C_f = \frac{\Delta P}{1/2\ \rho U_0^2} \cdot \frac{D_h}{4L} \qquad (5)$$

From the computational results, it could be concluded that the local pressure coefficient on the surface of louver decreased with the increse of F_p/L_p.

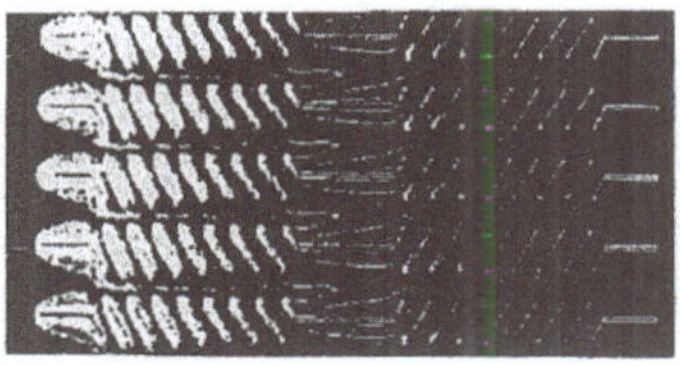

(a) Re=100, Fp/Lp=1.0, θ=28°

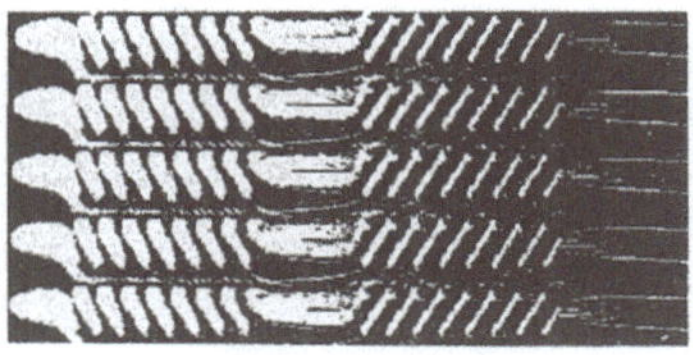

(b) Re=500, Fp/Lp=1.0, θ=28°

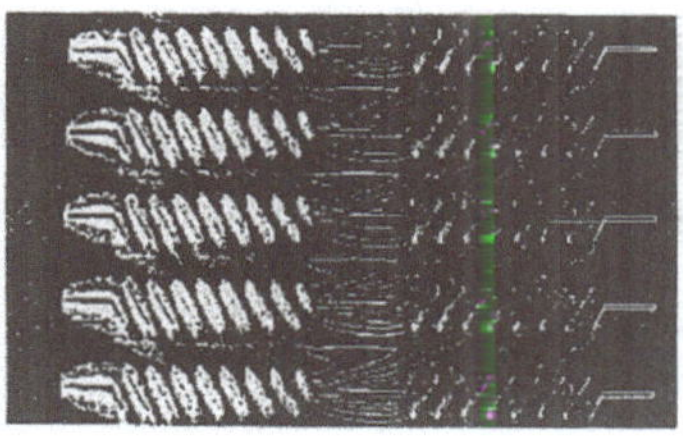

(c) Re=100, Fp/Lp=1.23, θ=28°

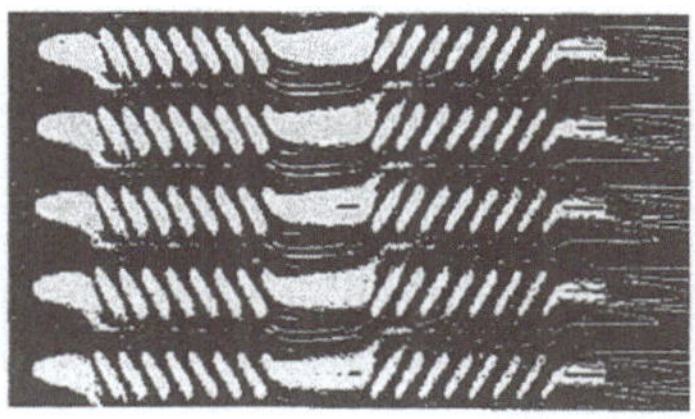

(d) Re=500, Fp/Lp=1.23, θ=28°

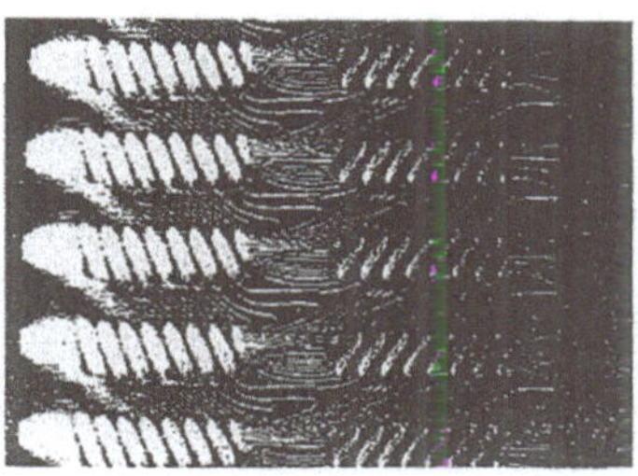

(e) Re=100, Fp/Lp=1.5, θ=28°

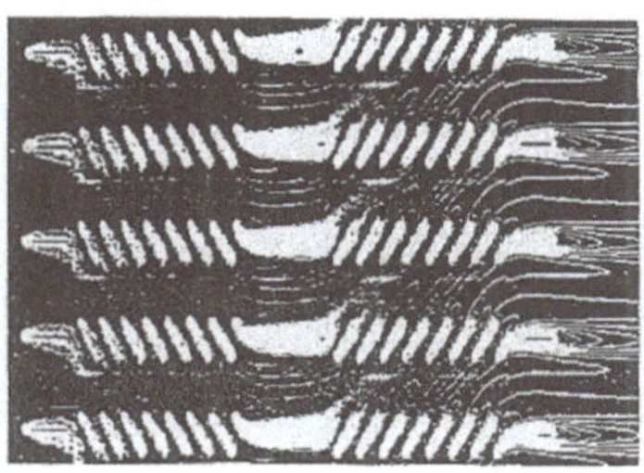

(f) Re=500, Fp/Lp=1.5, θ=28°

Fig. 4 Temperature distribution

Also, the local Nusselt number on the surface of louver is defined to consider heat transfer characteristics on the surface of louvered fin,

$$Nu_x = \frac{|q_w| L_p}{k(T_w - T_b)} \quad , \text{ where } \quad q_w = -k\frac{\partial T}{\partial Y} \quad , \quad T_b = \frac{\int UTdY}{\int UdY} \qquad (6)$$

The local Nusselt number on the surface of louvered fin is increased as the louver angle and F_p/L_p decreased.

REFERENCES

1. HIRAMATSU,M. and KASINO,M. : Research and Improvement of Automotive Radiators, Trans. UDC, 621.43.036/.01
2. HIRAMATSU,M. , ISHIMARU,T. , MATSUZAKI,K. : Research on Fins for Air Conditioning Heat Exchanger, Trans. J.S.M.E., 55-519, B(1989-11), 3449.

Automatic Analysis of Thermocolor Paints for Heat Flow Measurements in Wind Tunnel Facilities

X. BRIOTTET, M. DINGUIRARD

ONERA CERT DERO
2 Avenue Edouard Belin
31055 TOULOUSE CEDEX

SUMMARY

An automatic analysis of thermocolor paints is done. All the images have been masked, corrected from non uniform illumination and registered. Then, after having compared different colorimetric representation systems (R-G-B, I-H-S and K-L bases), we use the bidimensional histogram to extract the different colors and so detect the iso-flux lines on each image. A relative thermic exchange map is obtained by superimposing a sampling of processed images.

1- INTRODUCTION

Heat flow calculation over any surface is difficult specially with complex geometric shapes. In that case it is necessary to perform experiments in wind tunnels. The use of models and thermocouples allowed precise results but the detail study of thermic phenomenons is limited by the number of sensors and the fact that exact location of the best acquisition points is unknown [1]. The use of thermocolor paints avoid these two limitations. As they have the ability to change of color with temperature, their different hues during the run are directly related to the local heat flow level.

The extraction of the transition lines between the colors is mainly done by hand, marking the lines on each image acquired during the run and in superimposing them on a same graph. Using calibrated models, the iso-flux lines can be quantified to give a thermic exchange map. The work presented here is an automatic procedure to analyze the images, recorded by a tri-CCD digital camera, of an experiment performed during a run in the ONERA-MEUDON wind tunnel facility. It is a complement of previous studies [2].

2- PRE-PROCESSING

The details of the different pre-processings are given in Briottet and al [2]. It contains different steps : the mask extraction which is easily extracted as the background has levels different of those of the model, the registration of the model in the images flow by putting in coincidence the partial edges, the correction of the illumination non-uniformities by dividing pixel to pixel a color plan at the instant t by the same color plan at to (image recorded before the run) assuming that the illuminating source is stable during the sequence.

3- COLORIMETRIC SEGMENTATION

The colorimetric segmentation could be done in many different bases, the choice of the base depends on the image type. In this paragraph, we describe and compare the performances of the principal systems of colorimetric representation (R-G-B, I-H-S and K-L transform).

The most common is the R-G-B base which is directly obtained by a tri-CCD camera. Several bases are deduced by a linear combination of the R-G-B plans. But this base is far from the human perception of the color which is better described by the I-H-S base introduced by JR Kender [3] where the intensity is proportional to the collected energy, the hue is the color and the saturation is the color purity. This base takes a long computing time due to the complexity of the axis transform and these calculations introduce artefacts [3]. The colors often are weakly saturated so the repartition of the cloud of points in the 3D histogram is centered along the achromatic axis. The computation of the eigen vectors of the covariance matrix defines the base where the colors are decorrelated. This transform, named Karhunen-Loeve or Principal Component Analysis (PCA), is largely described in the literature [4], [5]. In the case of multispectral images with 3 color plans, only the first three eigen values of the K-L base are used corresponding to the larger eigen values of the decomposition.

The choice of the best base is done by comparing the monodimensional histogram of each image plan in the different bases. We used an image recorded at the ONERA-Meudon wind tunnel facility by H. Consigny and al representing a NASA shuttle model where four colors must be found (green, blue, yellow and brown) or, which is equivalent, three colorimetric transitions. The image plans have been masked, put in coincidence and normalized. The normalization introduces a spike at level 100 representing the saturated points in the plan.

In the R-G-B base (figure 1-a), the histograms of the R and G plans are quite similar, the valley between the 2 principal modes giving the brown-yellow transition. The histogram of the B plan shows two principal modes of same shape separated by the yellow-blue transition.

In the I-H-S base (figure 1-b), the histogram of the intensity plan shows 2 modes separated by the brown-yellow transition, the weaker one is the brown color. In the histogram of the hue plan, only the modes centered at 200° and 300° are significant : the first one is the yellow color and the other one all the resulting colors; the valley is the yellow-blue transition. The histogram of the saturation plan brings no information.

In the K-L base (figure 1-c), only the 2 first plans are significant for colorimetric segmentation purpose, as shown by YU-OTHA and al [6]. In the first plan, there are 3 modes representing the blue, yellow and brown colors. From the second plan, we can also extract the yellow-blue transition.

In conclusion, the best results are obtained with the green plan in the R-G-B base and with the first plan of the K-L transform. These 2 plans have complementary advantages such as region homogeneity, sharpness of the transitions... But at any time, we are not able to extract the blue-green transition because the two colors are too near. The use of bidimensional histogram is an original mean to separate the 2 hues.

For bidimensional histograms, we have to consider two color planes of a same image. The histogram is then a surface in a tridimension coordinate system composed of two "grey levels" axis and one " number of pixels" axis.

The bidimensional histogram is calculated with G and B plans and is given in pseudo color in the figure 2 (the 4 modes are presented). If we project this histogram along one of the axis, we obtain a monodimensional histogram where the 2 middle modes are confounded. The colorimetric segmentation is then solved in two steps: a peak detection followed by a geodesic squiz to affect a color to the different points of the neighborhood of each peak. The segmentation has to be done only once on an image chosen for having all the wanted colors.

Each pixel is then labelled so we obtain the segmented image on which an edge detection is done to extract the line transitions in each image.

The superimposition of the different lines is done by a simple addition of the labelled images.

4- RESULTS AND INTERPRETATIONS

We used a sequence of 28 images recorded, in the ONERA-Meudon wind tunnel facility, by a tri-CCD camera representing a thermocolor painted sphere. The first image, recorded before the run, is the reference image.

After the user gives the maximum number of modes to detect, all the following operations are done automatically : masking, normalization, registration, colorimetric segmentation (done on the 25th image).

The figure 3 gives different images recorded during the run after the segmentation. The color coding is mauve color for green region, green color for blue region, clear blue color for yellow region and hard blue color for brown region. Before segmenting the images, we use morphologic median filter which introduces artefact on the top of the sphere. The images which have been hardly translated have their upper border truncated so during the registration operation, which translates the sphere in the low part of the image, the free levels are putting at background levels. Artefacts are also introduced due to hand writing along equator plan of the sphere.

The figure 4 gives the final result obtained from a sampling of superimposed images. Superimposing all the images makes the final result difficult to understand so the user can choose the principal images to calculate the relative thermic exchange map.

The automatic analysis takes about 2 hours to process the 28 images on a SPARC station without optimizing the processing time.

5. CONCLUSION

This study is concerned by the automatic detection of the different hues which appear on a thermocolor painted model during a wind tunnel run, in order to derive the iso-flux lines on the model.

After comparing different colorimetric representations, the best results are obtained with the green plan and the first K-L plan. But with a monodimensional histogram, it is not possible to extract all the transitions specially the blue-green transition. This transition is detected with a bidimensional histogram representation. We used the histograms of the sequence last image planes to perform the color detection. An other mean could be to detect the histogram modes image by image as they appear. We have to estimate the performance of our algorithm in the case of an important number of colors to detect.

Finally, the detected transition lines are noisy so we will smooth them in a near future.

To improve the absolute measurement, a correction of the bidirectional reflectance effect can easily be introduced in our algorithm but this need better characterization of these effects. A characterization campaign is done to evaluate the sensitivity of thermocolor paints to the different parameters such as the geometrical conditions of viewing and illumination, the environment state, the spectral and temperature conditions.

This algorithm gives good results and could be used for liquid crystals.

We thank H. CONSIGNI team (ONERA Meudon/OA division) for his cooperation.

6. REFERENCES

[1] J. Ponteziere, A. Betremieux; Méthodes de mesures des flux thermiques pariétaux par peintures thermosensibles, Communication présentée à l'association Maritime et Aéronautique, Paris 24-27 Avril 1967

[2] X. Briottet, C. Henrion, M. Dinguirard; Automatic image processing of thermocolor paints for heat flow measurements in wind tunnel facilities; Optical methods and data processing in heat and fluid flow; City University; London; 2-3 April 1992; pp145-149

[3] J.R. Kender; Saturation, Hue and normalized Color : calculation, digitization effects, and use; Master's thesis; November 1976

[4] R.C. Gonzalez, P Wintz; Digital image processing; Addison-Wesley Publishing Company; 1977; pp 103-107

[5] W.K. Pratt, Digital image processing; John Wiley & Sons; 1978

[6] Y.I. Ohta, T. Kanade, T. Sakai; Color information for region segmentation; Computer Graphics and Image Processing 13; 1980; pp 222-241

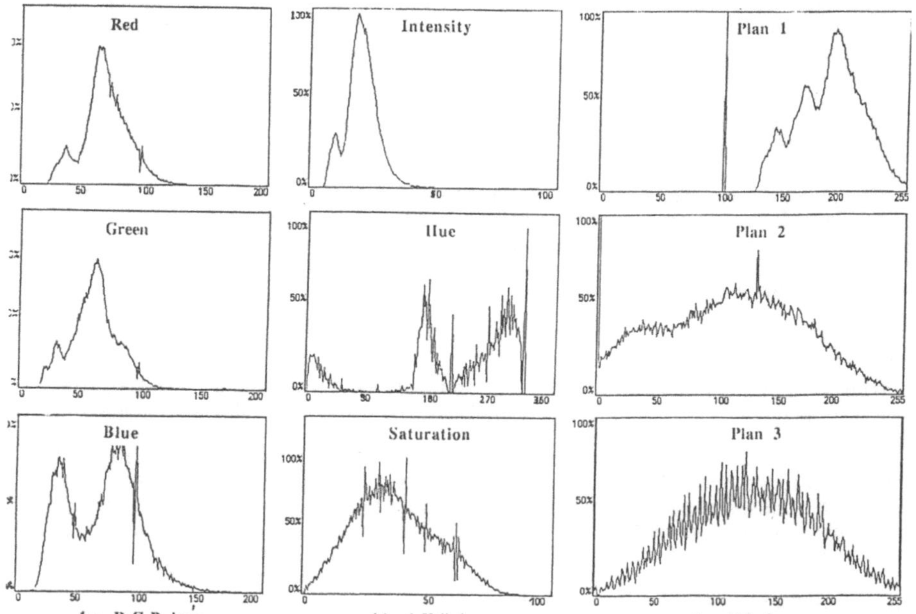

1.a. R-G-B base   1.b. I-H-S base   1.c. K-L base

Figure 1 : Monodimensional histograms in different colorimetric bases

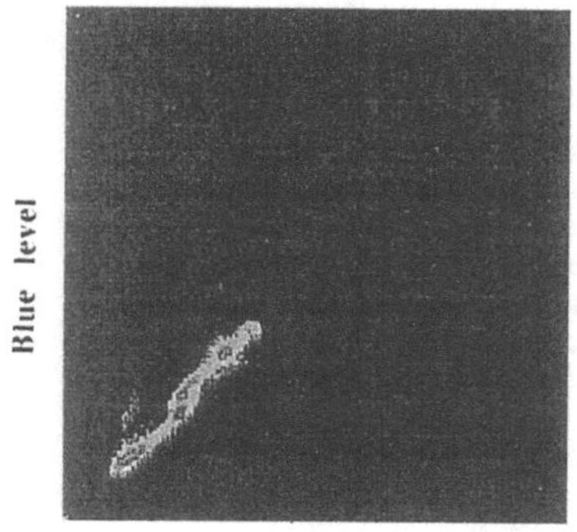

2.a. In pseudo-color

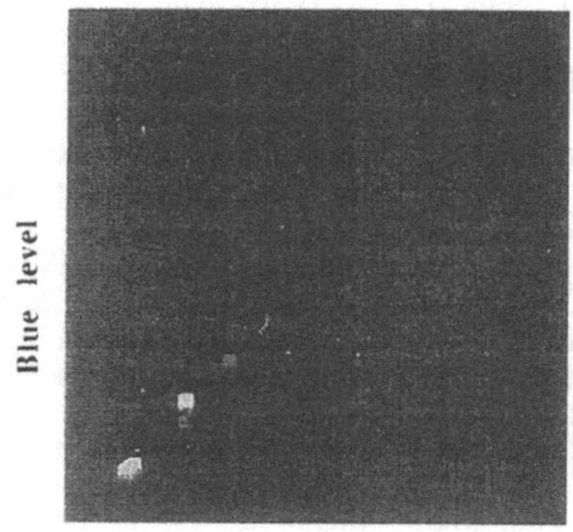

2.b. Peak extraction

Figure 2 : Bidimensional histogram of sphere 27

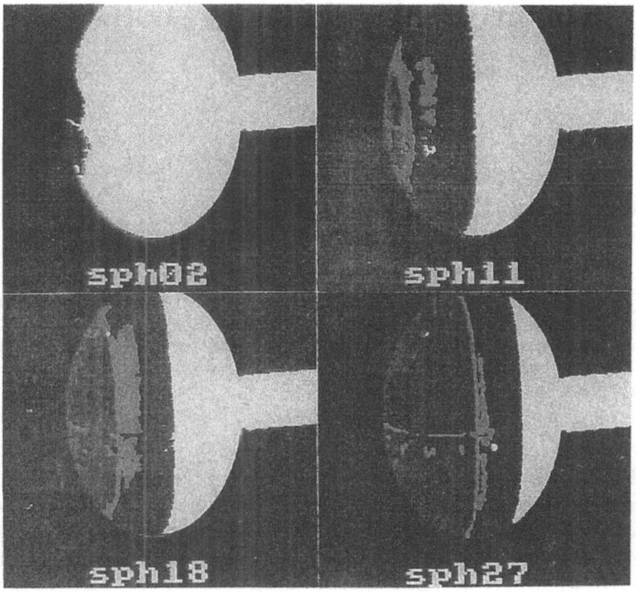

Figure 3 : Colorimetric segmentation of some images acquired during the run

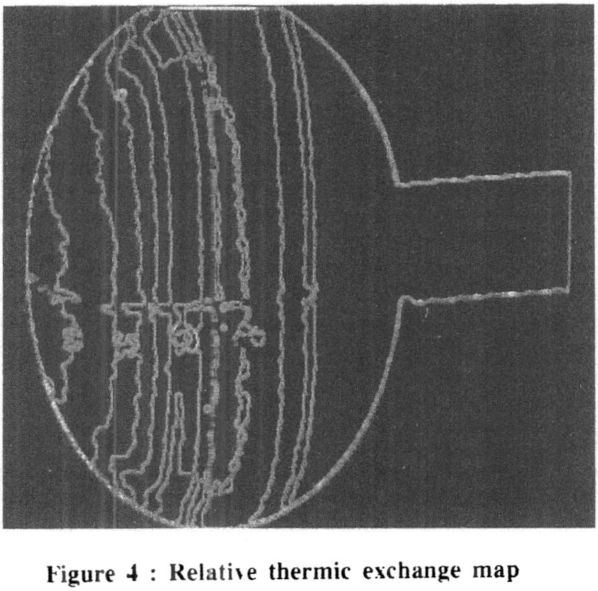

Figure 4 : Relative thermic exchange map

Transport Phenomena in Fluid Flow of Finned Tube Heat Exchangers

P. Geiser and V. Kottke

Institute of Food Technology, Food Process Engineering

Hohenheim University, Stuttgart, F.R.G.

Summary

The influence of geometric and fluiddynamic parameters on local heat and mass transfer is analyzed for different finned tube configurations. A visualization method is used to understand the near wall transport phenomena. The local heat transfer coefficients are quantified by a measuring technique which allows the calculation of the heat flux distributions to optimize heat exchangers.

Introduction

Externally finned tubes are widely used for heating, cooling, refrigeration and air conditioning. Their growing applications explain the extensive investigations to produce more effective heat exchangers. Advanced design of tube heat exchangers relies on highly accurate evaluation of the performance as these sometime will be the most critical components for output power and efficiency. The flow phenomena strongly influence the efficiency of these heat exchangers. With respect to the optimization of finned tubes accurate information on the local transport phenomena is required. Experimental investigations have been conducted to clarify transport phenomena and flow behaviour around finned tubes in order to increase heat transfer coefficients. The present work demonstrates the influence of the variation of fin spacing, fin height, tube shape and main velocity on flow phenomena and the local heat and mass transfer coefficients. Moreover, it shows the feasibility and the suitability of the used experimental tool.

Measuring technique

A physico-chemical method is used, based on convective molecular mass transfer with absorption and chemical colour reaction /1/. The surface of the fins is coated with filter paper, soaked with a reactive solution and mounted in the test section of the wind tunnel. A defined amount of tracer gas (ammonia) is pulsed into the air flow. The reaction gas is absorbed on the wet surface and a colour reaction takes place. The visible colour intensity corresponds directly to the amount of the transferred mass. Dark colour indicates high, light colour low mass transfer. The local mass transfer coefficients are evaluated by means of a new stepping motor driven remission-spectral-photometer based on on-line processing and a calibration curve. The analogy between heat and mass transfer allows to quantify the heat transfer coefficients. Furthermore, this method enables to visualize the near wall transport characteristics.

Effects of fin height

In the following survey, illustrated in figure 1, a duct height - the so called fin spacing - of $S = 8$ mm, a duct width of $B = 150$ mm and a cylinder diameter of $D = 75$ mm is chosen. The main flow velocity w is varied from $w = 0.28\ \frac{m}{s}$ to $8.8\ \frac{m}{s}$, the fin height - the distance between the leading edge of the fin and the cylinder - is adjusted for $H = 18$ mm, $H = 138$ mm and H = 2200 mm. The interaction between fin and cylinder boundary layers produces counterrotating horseshoe vortices in the circumference of the tube which affect extremely high heat transfer coefficients in the vicinity of the flat plate-tube junction /2/. The size and the shape of these three-dimensional secondary flow phenomena are strongly determined by the thickness of the flat plate or duct boundary layer respectively. Thick boundary layers cause the generation of large horseshoe vortex systems. Concerning the examined configuration, the "thickest" boundary layer occurs for H = 2200 mm. In this case the fluid flow can be considered as fully developed. Figure 1 documents that the magnitude of the local heat transfer distribution strongly depends on the Re_D which represents the Reynolds number referred to the diameter of the cylinder. The left column of figure 1 shows that up to w = 2.2 $\frac{m}{s}$ a large separation zone exists downstream of the cylinder accompanied by very low local heat transfer coefficients in the rear of the cylinder. The comparison between the left and the middle column in figure 1 makes plausible that an increase of H is combined with a significant augmentation of the total heat transfer for w $\geq$ 1.1 $\frac{m}{s}$, because the separation zone is reduced and the size of the horseshoe vortex system considerably enlarged. An additional increase of H, represented by the right column of figure 1, leads only to a slight improvement of heat and mass transfer for low and moderate Re_D. In the case of higher Re_D - w $\geq$ 4.4 $\frac{m}{s}$ - an additional augmentation of H from H = 138 mm to H = 2200 mm even leads to a reduction of the horseshoe vortex system width. These high main velocities induce turbulent flow upstream of the cylinder which leads to thin horseshoe vortex systems in front of the cylinder and high heat and mass transfer at the rear part in the separated flow. Obviously, the limiting parameter to achieve large horseshoe vortices and small separation zones in case of sufficient fin height - H = 138 mm - and moderate Re_D, is not represented by H, but by Re_D.

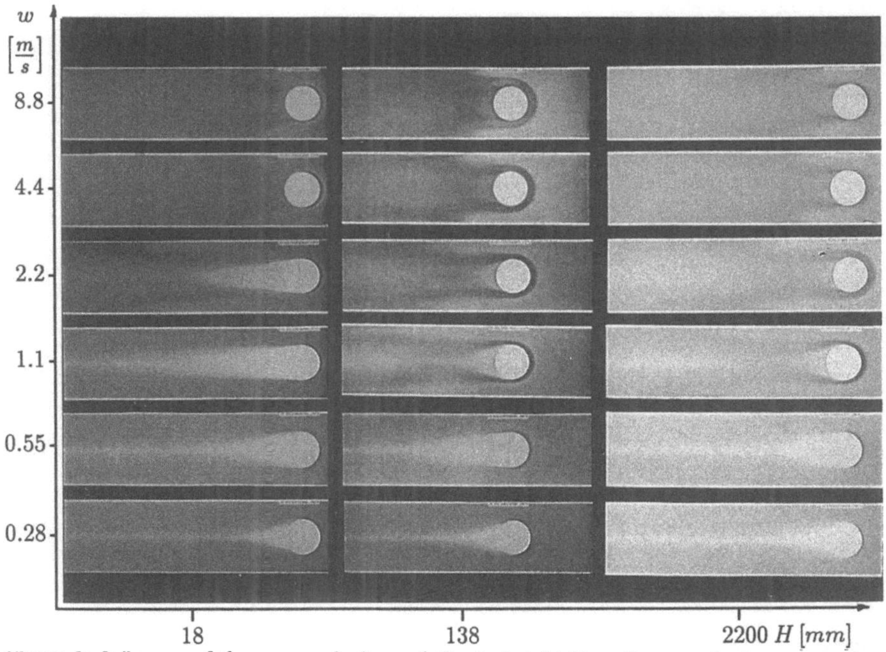

Figure 1: Influence of the mean velocity and the fin height H on the near fin transport phenomena at finned cylindrical tubes.

Effects of fin spacing in fully developed flow

The variation of the S/D relation in the following permits a survey of the flow phenomena and the corresponding heat and mass transfer of circular cylinders in fully developed laminar and turbulent duct flow (fin height H = 2200 mm).
The visualized heat and mass transfer distributions of fully developed fluid flow, illustrated in figure 2, are based on a fin spacing of S = 32 mm. For example in the case of w = 1 $\frac{m}{s}$ a doubling of S/D = 0.43 is joined with the creation of a light zone in the rear of the tube which is responsible for low heat transfer coefficients, because Re_D is halved.
The results, obtained by varying the tube diameter lead to the conclusion that for the high fin spacing, a change in relation S/D, realised by reduction of the cylinder diameter, is not appreciably affected by the relation of horseshoe vortex system width to cylinder diameter. That can be put down to the fact that in the case of fully developed flow, the horseshoe vortices reach their maximum of extension for high fin spacings. In this case, the fin spacing only obstructs the generation of the horseshoe vortices negligibly.
Nearly no essential change in local heat and mass transfer occurs if the S/D relation is modified for Re_D = const., because the fin spacing is sufficient large. Thus, under the mentioned circumstances the flow phenomena are basically influenced by Re_D.

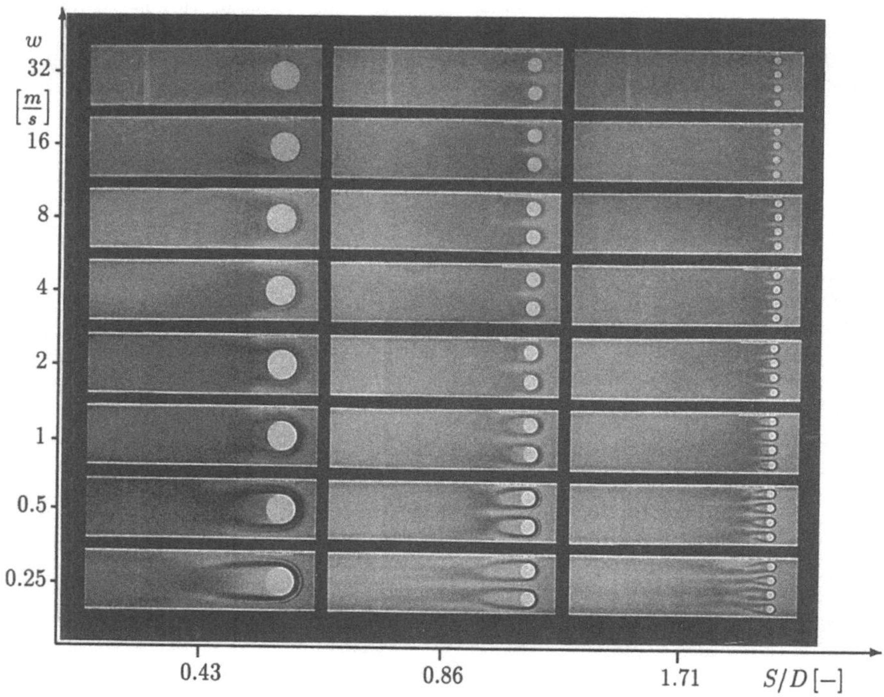

Figure 2: Influence of main flow velocity w and the relation S/D on the transport phenomena in fully developed fluid flow.

Non circular tubes

The fluid flow around circular tubes is combined with high pressure losses.

Thus, finned tubes with fluiddynamic more favorable shapes are of interest.
The outstanding features of the long elliptical tube geometry are its great heat transferring surface and its small separation zone.
Figure 3 shows flow visualizations in dependence on main flow velocity of three elliptical tubes with different relations between principal and secondary tube axis A/B. These results are based on fully developed fluid flow (H = 2200 mm).
The samples demonstrate that the difference in the horseshoe vortex dimension, associated with the modification of the tube shape by an augmentation of the axis relation A/B, can be explained by a decrease of the nose section diameter. At low flow velocities, the influence of the tube geometry on the horseshoe vortex extension is of lower importance because of the small contribution of the horseshoe vortices on the heat and mass transfer.
Moreover, figure 3 illustrates that long tube models, accompanied by regions of smooth expansion, lead to lower pressure losses and cause smaller separation zones in comparison with cylindrical tubes.

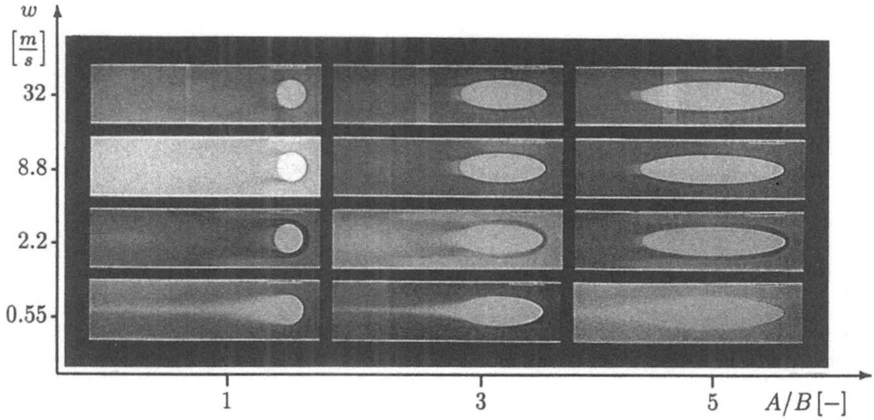

Figure 3: Influence of the mean velocity and the tube shape on the transport phenomena, S = 8 mm and H = 2200 mm.

Heat transfer at a finned elliptical tube

Exemplary, a local heat transfer distribution is plotted in figure 4, achieved with a finned elliptical tube in air flow. Special attention is given to this case in which w = 2 $\frac{m}{s}$.
The high heat transfer coefficients h near the leading edge of the fin and at the base of the tube are remarkable. The heat transfer coefficients in the thin horseshoe vortex region, adjacent to the tube, decrease strongly in flow direction.
It is deducible that the small nose section diameter leads to a small extension of the main horseshoe vortex. In the main flow direction, the local heat transfer coefficients decay in dependence on the growing duct boundary layer thicknesses.
In the rear of the elliptical tube relative high local heat transfer coefficients appear, caused by separation and backward flow phenomena.
It will be noticed that the considerable resolution in location and intensity of the measuring techniques delivers local heat and mass transfer coefficients with high accuracy.
These obtained quantities can be used to calculate temperature distributions of the fin, tube and fluid, which enable the computation of the local and total heat flux transferred by the fin and the tube /3-4/.

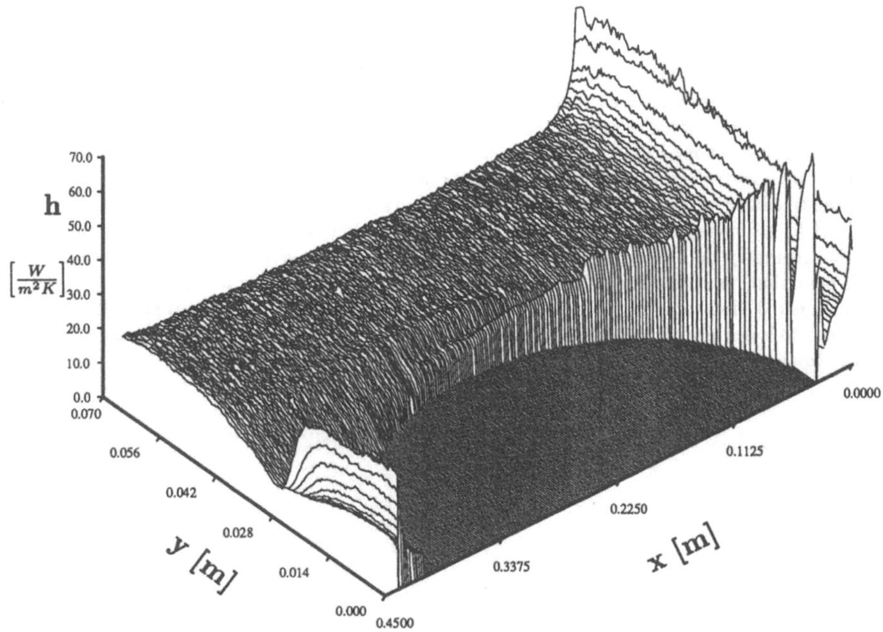

Figure 4: Local heat transfer coefficient distribution of a finned elliptical tube, A/B = 5, S = 8 mm, H = 36 mm and w = 2 $\frac{m}{s}$.

Results

The presented local heat transfer distribution has indicated that the used new measurement equipment provides results of high resolution in location and intensity. The investigations of flow, heat and mass transfer visualization have shown that the local heat and mass transfer coefficients of the fin vary strongly in dependence on the location. The size of the region affected by the horseshoe vortex system is determined by the Reynolds number, the tube shape and the fin spacing. The enlargement of the fin height for moderate Re_D leads to an intense augmentation of heat and mass transfer. In the case of finned cylindrical tubes with high fin spacing, associated with fully developed flow, a recognizable improvement of heat and mass transfer can be achieved by the increase of Re_D at low Re_D numbers. The width of the horseshoe vortex is also governed by the nose section diameter of the tube. Long elliptical tubes cause small separation zones and thin horseshoe vortices.

Acknowledgement

This work has been supported by the Federal Ministry of Research and Technology (BMFT) of Germany and GEA Luftkühlergesellschaft GmbH, Bochum, Germany.

References

1. Kottke, V.; Blenke, H.: Verfahren zur Bestimmung örtlicher Stoff- und Wärmeübertragung an beliebig geformten Oberflächen. verfahrenstechnik 16, pp 504-509, 1982.
2. Schüz, G.; Kottke, V.: Enhancement of heat transfer by horseshoe vortices in finned-tube heat exchangers. Proc. Eurotherm No.9., Entropie, Bochum, pp.9-12, 1990.
3. Schüz, G.; Kottke, V.: Strömungs- und Transportvorgänge in Rippenrohr-Wärmeübertragern. Chem. Eng. Technol., in press.
4. Kottke, V.; Schüz, G.; Geiser P.: Combination of local heat transfer and thermal contact resistance between fins and tubes. Proc. Europ. Conf. on Finned Tube Heat Exchangers, Stuttgart, 1991.

Visualization of Flow Phenomena Behind Rows of Vortex-Generators in Internal Flow

E. Hochdorfer*, V. Kottke**

* Institute of Chemical Engineering, University of Stuttgart, FRG

** Institute of Food Technology, Food Process Engineering, Hohenheim University, Stuttgart, FRG

SUMMARY

Different rows of vortex generators are investigated in an open wind tunnel. By means of an absorption method the mass transfer is visualized for different geometries. For specific rows of vortex generators the local mass transfer as well as the pressure drop for different Reynolds numbers are determined.

INTRODUCTION

In chemical engineering vortex generators are successfully employed to optimize mass and heat transfer in heat exchangers, catalyzers or recuperators. It is possible to achieve a distinct increase of mass or heat transfer with a simultaneous modest increase of pressure drop.

EXPERIMENTAL APPARATUS AND PROCEDURE

The experiments are carried out in an open wind tunnel as shown in Fig. 1. The flow velocity is measured with a bellmouth intake and a Betz pressure gauge. The test section has a length of L=1m and consists of 6 rectangular ducts, each h x b = 8x150mm in cross-section. The ducts were shaped by 1mm aluminum plates.

The rows of vortex generators are cut from a thin aluminum sheet and glued onto the plates. The mean pressure loss of the vortex generators is measured between two points 30mm before and behind the ducts. As a reference measurement the pressure loss of the ducts without vortex generators is measured. For visualization and mass transfer measurements an absorption method /1/ is employed: The top and the bottom of one duct are covered with sheets of filter paper, soaked with reactive solution. After closing the channel, switching on the blower and adjusting the airflow, a short pulse of reaction gas is blown into the bellmouth.

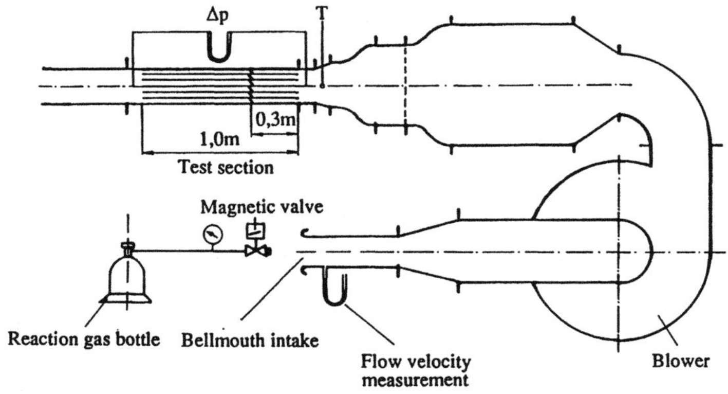

FIG. 1: Open wind tunnel (no scale)

The flow is mixed in the blower so that the reaction gas reaches the ducts in a homogenous concentration. Depending on the local mass transfer coefficients, the gas reacts with the solution to a coloured product leading to a colour density distribution, which first of all shows the quantity of the reaction product by the colour intensity. The colour density distribution and thus the transferred mass can be evaluated by means of a photometric measuring method. A computer program provides the local mass transfer coefficient as well as the mean local Sherwood numbers (related to the hydraulic diameter) at each position L downstream of the vortex generators.

Of particular interest are the types of vortex generators shown in Fig. 2. All vortex generators reach from the bottom to the top of the duct. There are deltawings with a distance of 16mm between the tips, a basical width of 8mm and an angle of attack of 45°, double deltawinglets with a distance of 16mm between the tips, a basical length of 20mm and an angle of attack of ±15° and single deltawinglets with a distance of 16mm between the tips, a basical length of 20mm and an angle of attack of 15°. Each row consists of 9 elements corresponding to the duct width. The row of single deltawinglets induces secondary vortices further downstream which also lead to a regular vortex field, but at a lower pressure loss as the double deltawinglets. All presented results have been achieved at a Reynolds number of 1500 based on the hydraulic diameter of the duct: $Re = u \cdot D_{hyd} / \nu$, with $D_{hyd} = 4 \cdot b \cdot h / (2 \cdot (b + h))$.

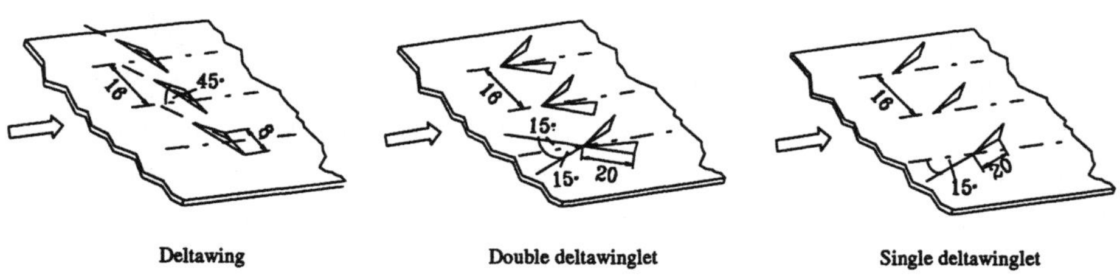

FIG. 2: Geometry of the investigated vortex generator rows

RESULTS AND DISCUSSION

Fig. 3-5 show the photographs of two mass transfer distributions, one from the upper, the other from the bottom wall of the ducts. The colour density distribution indicates the amount of the locally transferred mass. The darker the colour, the more mass has been transferred to the surface. A mass transfer maximum on one side corresponds to a minimum on the other side. There is a great difference visible between the three geometries in the development of the vortices behind the vortex generators. Both deltawings and double deltawinglets show regions of intensive dark colour directly behind the vortex generators.

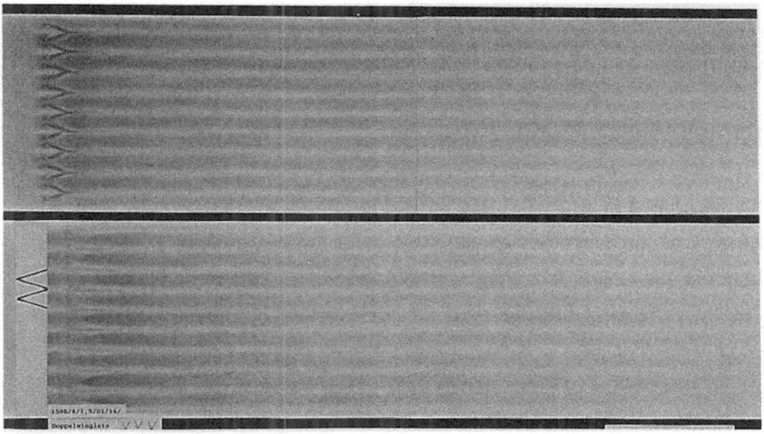

FIG. 3: Mass transfer distribution behind double deltawinglet row

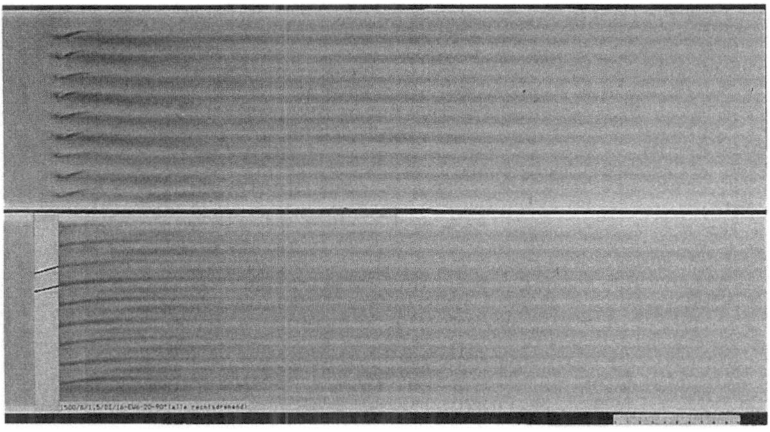

FIG. 4: Mass transfer distribution behind single deltawinglet row

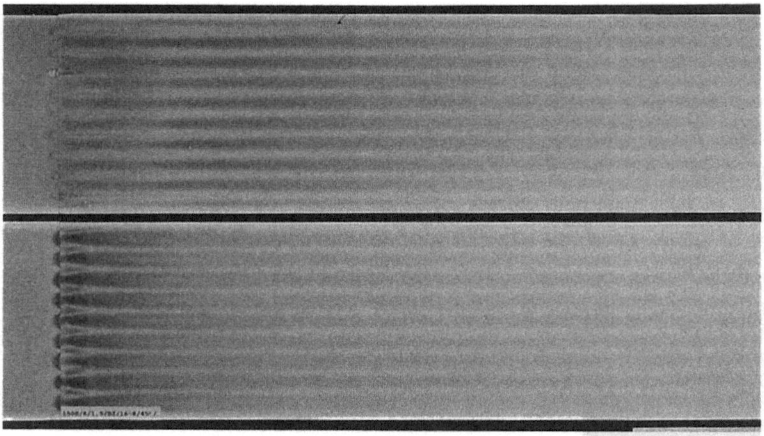

FIG. 5: Mass transfer distribution behind deltawing row

This regions cover a wider area than the dark regions behind the single deltawinglet type. But about 0.4m downstream all three geometries show the same pattern of lines even with the same intensity. Therefore the single deltawinglet row provides a lower mass transfer in the region behind the vortex generators than the other two geometries, but also a regular vortex field further downstream.

Fig. 6 shows the mean local Sherwood number $Sh = \beta \cdot D_{hyd} / D$ at a Reynolds number of 1500, where β is the local mass transfer coefficient and D the diffusivity. In all cases the rows of vortex generators are in fully developed velocity (/2/) and concentration profiles. In the distance L = 0.3m downstream of the leading edge, for $Sc = \nu / D = 0.616 < 1$ the concentration profile has developed before the velocity profile.

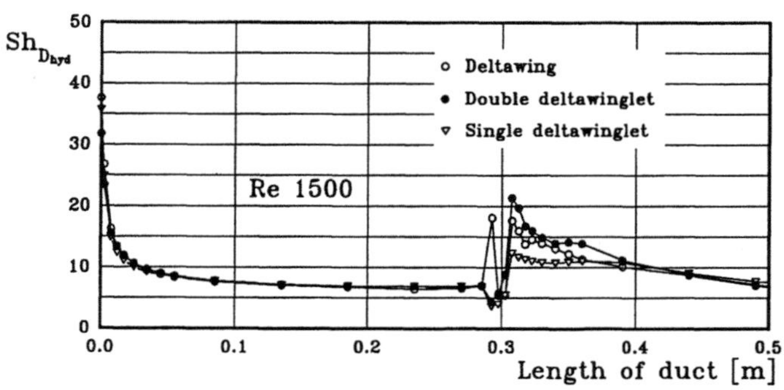

FIG. 6: Mean local Sherwood number versus duct length of vortex generators at Re 1500

Thus the Sherwood number of the fully developed flow is Sh = 6.88 = const. for L > 0.26m (/3/). The measured mass transfer for the developing undisturbed duct flow (0 < L < 0.26m) shows good correspondence for all three geometries.

Fig. 7 shows the results of the pressure loss measurements. From the measured pressure loss Δp a pressure loss coefficient $\zeta = \Delta p / (½\rho u^2 \cdot L / D_{hyd})$ is calculated and related to the pressure loss coefficient of the rectangular duct without any vortex generators. As Fig.7 shows, the single deltawinglet row provides the lowest, the deltawing the highest pressure loss ζ at all Reynolds numbers.

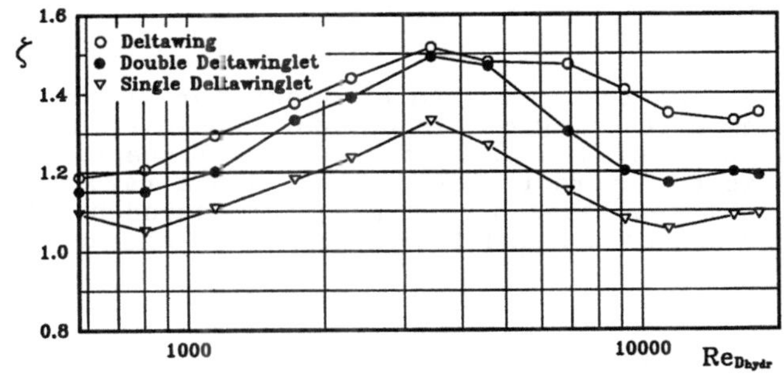

FIG. 7: Pressure loss coefficient ζ versus Reynolds number for different vortex generators

CONCLUSIONS

As the Figures 3 - 6 prove, the single deltawinglet type vortex generator row leads further downstream to the same field of vortices although only half of their number is directly induced by the generator elements. The pressure loss of the single deltawinglet row is clearly lower than the pressure loss of the other two geometries, but the downstream mean mass transfer is also the lowest. The Figures also prove, that the double deltawinglet row provides a higher mass transfer at a lower pressure loss than the deltawing row. In all cases the integral mass transfer and the pressure losses are in a highly interesting relation for practical applications.

REFERENCES

/1/ Kottke, V.; Blenke, H.; Schmidt, K.G.: Wärme- und Stoffübertragung, 10 (1977) pp. 9-21.

/2/ Merker, Günter P.: Konvektive Wärmeübertragung. Berlin [u.a.]: Springer, 1987.

/3/ Klein, G.: Bestimmung lokaler Stoffübergangskoeffizienten in monolithischen Katalysatoren mit geraden Strömungskanälen. Studienarbeit, Institut für Chemische Verfahrenstechnik der Universität Stuttgart, 1990.

Visualization of Flow Mixing Mechanisms in Pin-Fin Arrays

K.MINAKAMI, S.MOCHIZUKI, and A.MURATA
Tokyo University of Agriculture and Technology
2-24-16 Naka-cho, Koganei-shi, Tokyo, 184, Japan

Y.YAGI
Tokyo Polytechnic College
2-32-1 Ogawanishi-cho,
Kodaira-shi, Tokyo, 187

H.IWASAKI
Mechanical engineering Lab.
R & D Center, Toshiba Corporation
4-1, Ukishima-cho, Kawasaki-ku,
Kawasaki-shi, 210

Summary
The flow visualization tests were carried out for enlarged pin-array models using a reacting-dye injection technique in order to evaluate the effects of the pin cross-sectional shape. Measurements on heat transfer characteristic of pin-fin arrays were also conducted. The test results showed a good correspondence between the heat transfer phenomenon and the flow pattern characteristic.

Introduction

Recently, active research on high performance compact heat exchangers has been conducted. The pin-fin arrays used in the present study were made of closely arranged very thin wires. This configuration can provide very high area density of minute structure heat transfer surfaces, and can produce noticeably high performance and compact heat exchanger cores. In references [1]-[5], the evaluation of the heat transfer performances for pin-fin arrays are presented. In this report, the effect of pin cross-sectional shape is investigated in detail on the basis of visualization of flow mixing zones.

Experimental Apparatus

Figure 1 shows the pin-fin array for the heat transfer performance test. Table 1 shows the geometrical properties of the individual pin-fin arrays tested. In the present study, two types of pin-fin arrays, circular pins and square pins, arranged in a rectangular form were used. These arrays were made of copper. The pin pitches in the flow direction and the flow normal direction were $b/dp=4$ and $a/dp=4$, respectively. The number of pin rows in the flow direction was fifteen.

The modified single blow transient testing method [1] was applied to evaluate the heat transfer performance of the pin-fin arrays.

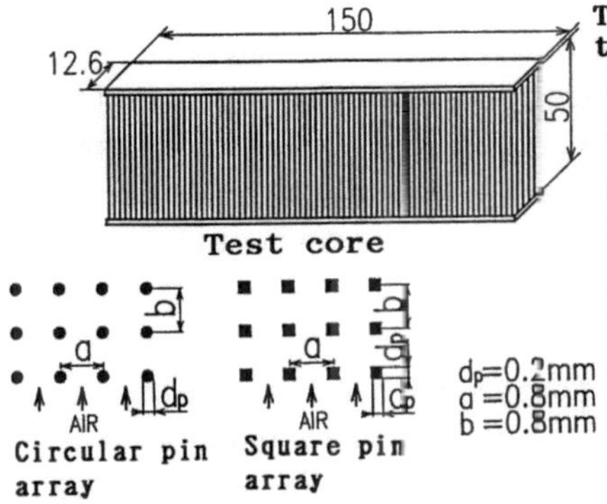

Figure 1 Pin-fin array for heat transfer performance test.

Table 1 Geometrical properties of the individual pin-fin arrays tested.

Pin cross-sectional shape	Circular Pin	Square Pin
a : pin pitch in flow normal direction (mm)	0.8	0.8
b : pin pitch in flow direction (mm)	0.8	0.8
dp : diameter of circular pin or side length of square pin (mm)	0.2	0.2
a／dp	4	4
b／dp	4	4
$A \times 10^2$: heat transfer surface area (m^2)	8.86	10.7
$V \times 10^6$: volume of pin-fin array (m^3)	9.45	9.45
A／V (m^2/m^3)	937	1130

This method is one of the transient techniques and has the following advantages as compared to the steady-state method: (1)The time needed for measurement is quite short. (2)The heat transfer performance can be directly evaluated, not being affected by the fin/base plate contact resistance and fin efficiency. Figure 2 shows a schematic view of the heat transfer test apparatus. It consists of a suction-type wind tunnel and an instrumentation system. The test section has a cross section of 50 mm by 150 mm. The test procedure is described in detail in the reference [1].

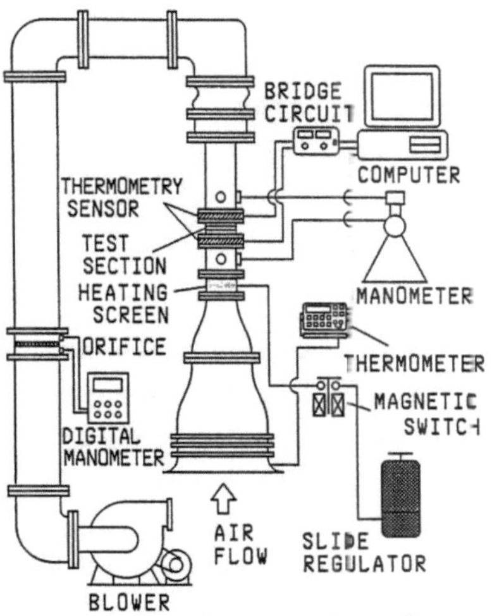

Figure 2 Schematic view of heat transfer test apparatus.

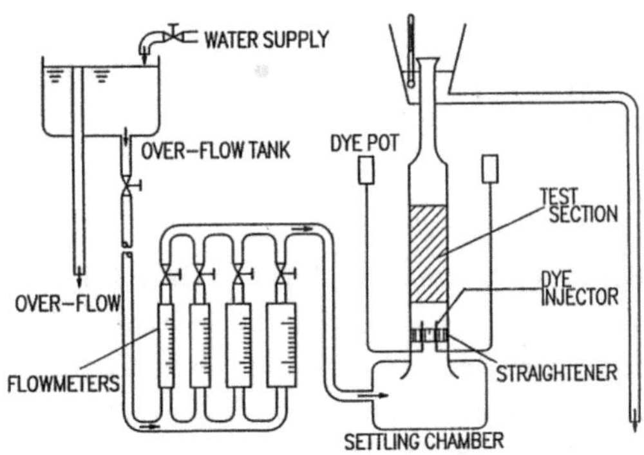

Figure 3 Schematic view of flow visualization test apparatus.

Figure 3 shows a schematic view of the flow visualization test apparatus. The model used was fifteen times larger than the actual pin-fin heat exchanger core. Water was used as the testing fluid. The test section has a cross section of 50 mm by 60 mm. Dyes were injected by using thin stainless steal tubes, 0.8 mm in outer diameter. To observe the overall flow pattern, black ink was injected from several points in the upstream region of the pin array. In order to investigate the flow mixing phenomena, reacting dyes were injected from two locations in the upstream. Phenolphthalein and alkaline solution, NaOH, were used as the reacting dyes. Both dyes are originally colorless. However, the mixing of these two liquids results in red color. Therefore, the flow mixing area in the pin array could be recognized clearly. The specific weights of all the dyes were adjusted to 1.0.

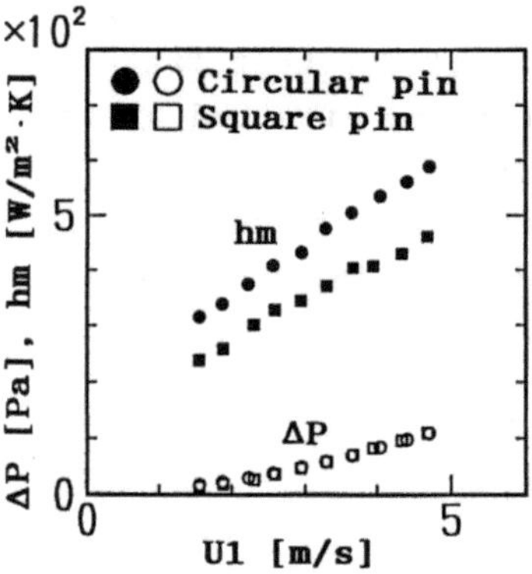

Figure 4
Heat transfer coefficient and pressure loss versus upstream velocity.

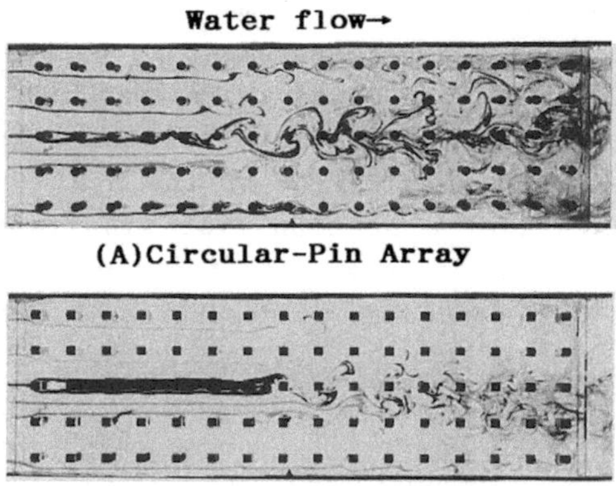

Figure 5 Overall flow pattern. [Re=60]

Test Results

Figure 4 shows the heat transfer coefficient, hm, and pressure loss, △P, versus upstream velocity, U1, for the pin-fin arrays. Within the range of the parameters tested, pressure losses were almost the same for both cases. However, the circular pin-fin had higher values as compared to those of the square pin-fin in terms of the heat transfer coefficient. Figure 5 shows typical flow visualization results when observing the overall flow pattern in the pin array. The Reynolds number, Re, in both cases were about

60. The characteristic lengths used in the definition of the Reynolds number are the diameter (3 mm) for the circular pin and the side length (3 mm) for the square pin. These figures show that flow perturbation occurs from the fifth row in the circular pin array. On the other hand, perturbation starts at the seventh row in the square pin array. Figure 6 shows the flow mixing visualization results for the same Reynolds number. These figures show clearly a wider flow mixing area and an earlier start of flow mixing in the case of circular-pin array as compared to the case of square-pin array. Thus, it has been disclosed that the reason behind obtaining a higher heat transfer coefficient in the case of the circular-pin array as compared to the square-pin array is due to the difference in the flow phenomena described above. Further basic research is necessary in order to get a more detailed understanding of the flow mechanisms.

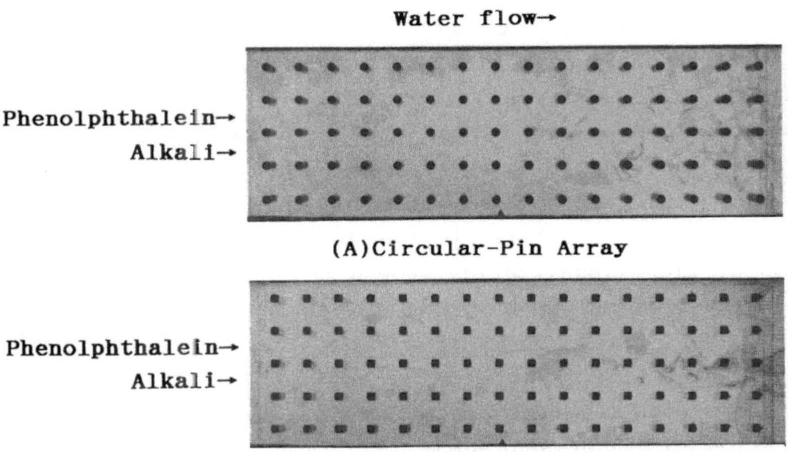

(A)Circular-Pin Array

(B)Square-Pin Array

Figure 6 Visualization of flow mixing area. [Re=60]

Figure 7 shows the heat transfer per unit core volume and unit temperature difference, hm·A/V, versus the friction loss power per unit core volume, P. A representative performance of a modern high performance louvered-fin is also indicated in the figure by a solid line. As shown in the figure, plots for pin-fin arrays fell in the region above the louvered-fin performance. Within the range of parameters tested, a square pin-fin indicated almost the same performance as a circular pin-fin. This is due to the compensation of lower heat transfer coefficient by a larger heat transfer surface area in square pin-fin.

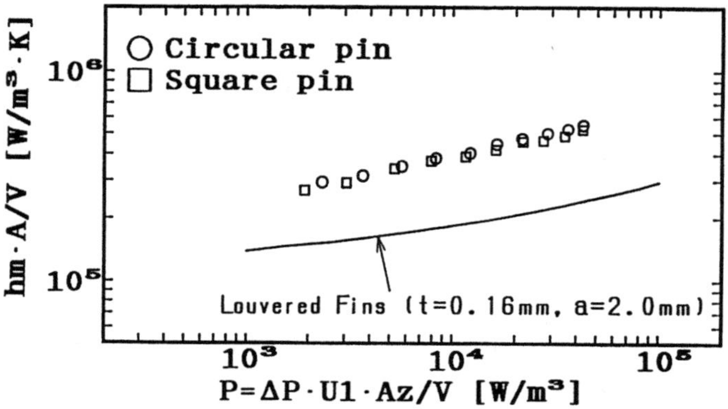

Figure 7 heat transfer per unit core volume and unit temperature difference versus the friction loss power per unit core volume. (Az:frontal area of pin-fin heat exchanger core)

Conclusion

The effects of the pin cross-sectional shape on pin-fin arrays was investigated. Test results showed that the circular pin array had almost the same overall performance as that for the square pin array. However, so far as the heat transfer coefficient is concerned, the circular pin array indicated a higher value than the square pin array. The reason for this difference has been found to be closely related to the flow phenomenon in the array. Results of flow visualization showed that the circular-pin array has a wider flow mixing area than the square pin array.

Acknowledgment

The authors wish to thank Messrs. M.Sato and T.Minato for their cooperation in the present work.

[REFERENCES]

(1)S. Mochizuki, Y. Yagi, The 9th IHTC G.P., Vol.3, 217, (1990)

(2)T.Hara, S.Mochizuki et al., Proc. of 28th National Heat Transfer Symposium of Japan, Vol. I, D111, 217, (1991)

(3)K. Minakami et al., Proc. of 28th National Heat Transfer Symposium of Japan, Vol. I, D112, 220, (1991)

(4)M.Sato, S.Mochizuki et al., Proc. JSME,No.910-84,B32, 171, (1991)

(5)K.Minakami et al., Proc. JSME,No.920-17,Vol.B,242,(1992)

Thermal Design Approach for a Copy Machine by Use of Flow Visualization and a Personal Computer

M. ISHIZUKA AND T. SASAKI
Research & Development Center
Toshiba Corporation, Kawasaki 210 ,Japan

Summary

This paper reports on an approach based on lumped models combined with experimental data of the flow resistance coefficient and heat transfer coefficients. Simulations were carried out for the thermal design of a copy machine by use of flow visualization and a personal computer. The calculated temperature values were compared with measured values for several cases. The compared temperatures were for the drum, power supply, and drum ambience. The selected design parameters were the fan casing size, the presence or absence of holes on the base frame,the rate of heat dissipation from the power supply, and so on. The proposed method is proven to serve thermal designers satisfactorily.

Introduction

It is very important to cut down the lead time required for a new model to reach its production stage to keep up with the rapid progress being made in its technology. Accurate thermal design for copy machines is very important, because the machine has heat-sensitive components such as the drum located very near a heat source. Also, the copy table glass, on which the original document is placed, must also be protected from thermal stress resulting from temperature gradients in the glass. Therefore, it has now become necessary to apply numerical simulation using a computer to the thermal design of a copy machine to save time and speed up the work.

In this paper, an approach to the thermal design of a copy machine is presented by use of a personal computer. This is a simple analysis based on flow visualization and lumped models combined with experimental data of flow resistance coefficients and heat transfer coefficients .

Copy Machine Model Structure

The analysis and experiments were carried out using a model copy machine as shown in Fig.1, due to the complicated internal structure of an actual copy machine. Some important components for fluid and heat flow are incorporated in this model machine. The flow rate was matched with that for an actual machine. The heat roller temperature is controlled to maintain 180 degrees centigrade, the same temperature as in an actual machine.

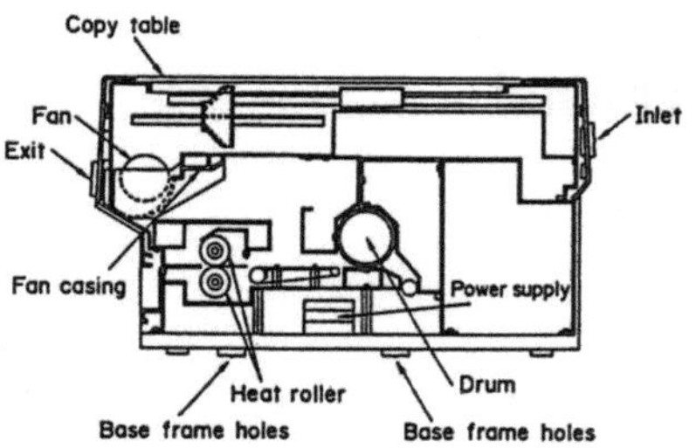

Fig .1 Copy Machine Experimental Model

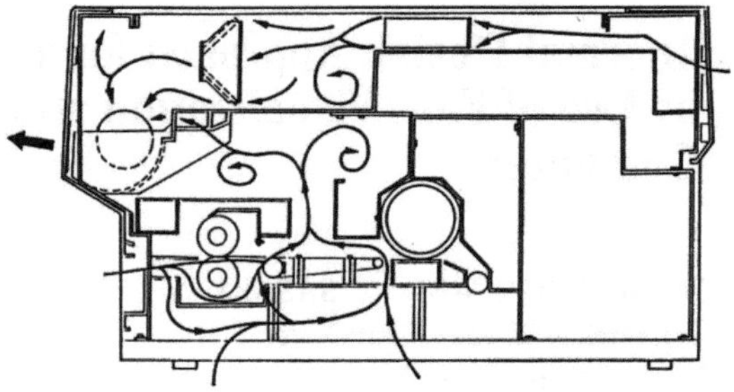

Fig. 2 Flow Visualization Results

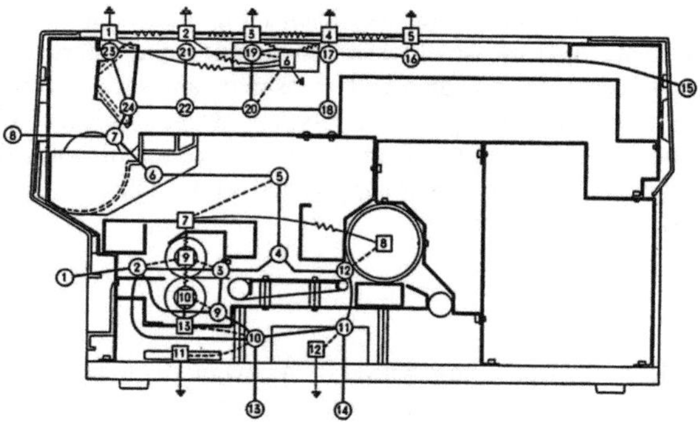

Fig. 3 Model for Analysis

First, the authors carried out flow visualization tests using paraffin smoke. The results are shown in Fig.2. It was found that two main flow streams existed under the glass and between the heat roller and the drum. Next, on the basis of these results, a network was set up for analysis using a lumped model, as shown in Fig.3. This model included the fluid flow field nodes indicated by the circles, the heat flow field nodes indicated by the squares, and links between the flow and heat nodes by broken lines.

Flow Field Calculation

The pressure field network shown in Fig.4 was considered.The circle symbols represent the pressure field. This is one simple example of the network. Each point corresponds to a different pressure value. There is a certain amount of air flow between any two points. The following four equations are obtained, From mass conservation and from the definition of the flow resistance coefficient, the following four equations are obtained with the pressure difference, ΔP, air velocity in the flow path, u, the air density, ρ, and the fluid resistance coefficient, K.

$$V_1=V_2+V_3 \quad (1)$$

$$P_1-P_2 = K_1 \rho u_1^2/2$$
$$P_2-P_3 = K_2 \rho u_2^2/2 \quad (2)$$
$$P_2-P_4 = K_3 \rho u_3^2/2$$

where air density changes are neglected. At the same time, the volume flow rate, V, can be related to the flow velocity, u, using the flow path area, A, as follows,

$$u=V/A \quad (3)$$

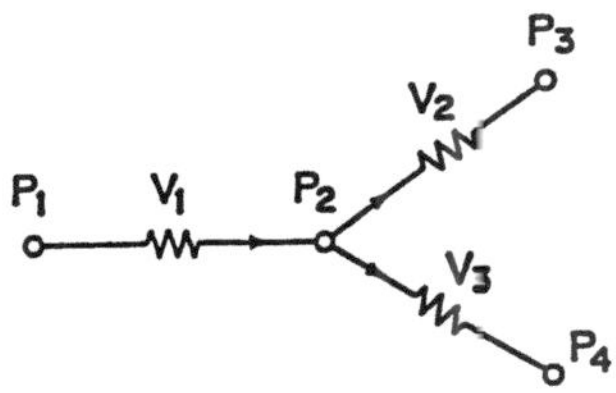

Fig.4 Flow Resistance Network

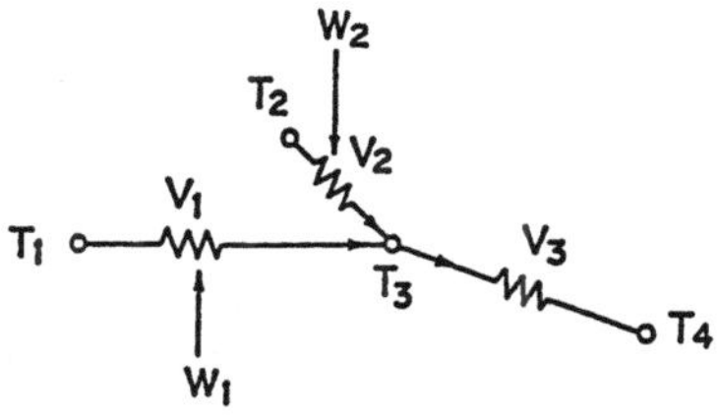

Fig .5 Thermal Network

The results are a set of non-linear equations for pressure and air velocity. These equations can be solved to determine the flow field if the K values are given. When the equations are solved, the flow distribution and pressure drop are obtained.

Temperature Calculation

Next, the temperature field of cooling air will be considered using Fig.5. Considering the enthalpy in and out of the nodal point T_3, the enthalpy equation is written as follows;

$$\rho Cp\ (V_1T_1+V_2T_2)+W_1+W_2=\rho Cp\ (V_1+V_2)T_3 \quad (4)$$

where Cp is the specific heat and W_1 and W_2 are the dissipation values along the flow path between the nodal points T_1 and T_2 and between the nodal points T_2 and T_3, respectively. The air temperature values for the cooling air are obtained from these results. Then, the surface temperatures are obtained using the thermal network method (Fukuoka and Ishizuka[1]), shown in Fig.6. The heat flow Q_1 among the total heat generated by the component is dissipated to air. The heat transfer coefficient, α, can be estimated using the u values from Equation.(3). The following expression is obtained, where ΔT is the temperature rise, W is the heat generated by the component, and the surface area is A:

$$\Delta T_S = Q_1/(\alpha S) \tag{5}$$

Therefore, the surface temperature T_S is obtained as follows.

$$T_s = T + \Delta T_s \tag{6}$$

Thus, equation formulation is very simple in the thermal network method. In Fig.6, the heat flow conservation equation at\point T_j and the Ohm's law for the electric field between the temperature points are as follows;

$$W = Q_1 + Q_2 \tag{7}$$

$$\begin{aligned} T_s - T &= R_1 Q_1 \\ T_j - T_s &= R_2 Q_1 \\ T_j - T_1 &= R_3 Q_2 \end{aligned} \tag{8}$$

where R indicates thermal resistance, for example, $R_1 = 1/(\alpha A)$.

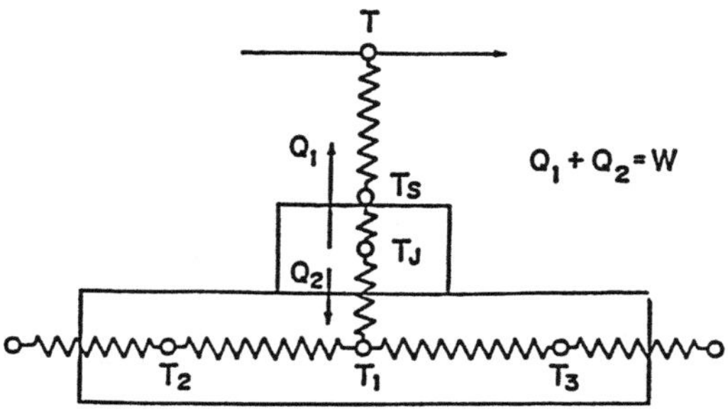

Fig.6 Thermal Network for the Assembly of Components

Parameter Survey Using an Analytical Model

Not all the necessary experimental data for the calculation can be previously specified. Therefore, unspecified data, such as some of the fluid resistance coefficient values ,were treated in the code as parameters so that previously measured velocity and temperature values and the calculated results might be matched by adjusting the parameters.After that, another parameter survey using an analytical model was carried out.

Figure 6 shows the results compared with some measured values for six cases. The selected design parameters were the fan casing size, base frame with or without holes, dissipation from the power supply, and so on. The compared temperatures were for drum, power supply and drum ambience. Considerably good agreement was reached. More experimental data are needed to develop this network approach. Flow resistance values for a low Reynolds number, such as natural air convection, are needed, for thermal design, because a very slow flow exists even in a forced air convection cooling cabinet. In this low Reynolds number region, some flow resistance values for natural air convection were given by Ishizuka,et al.[2] and Ishizuka[3].

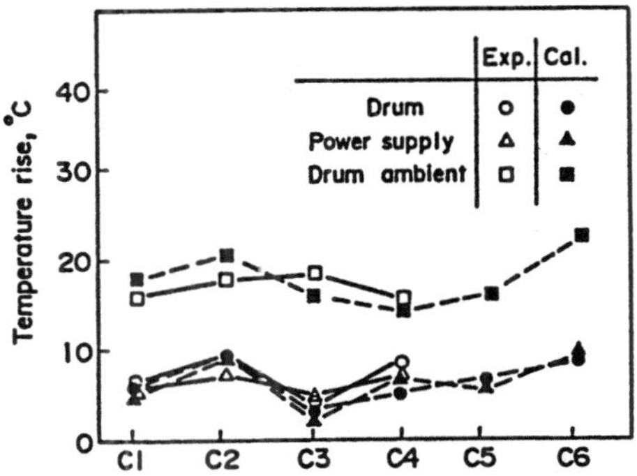

C 1	Before parameter survey experiment
C 2	Without holes in base frames
C 3	Without exit for papers
C 4	Blocking heat flow from heat roller
C 5	With double-sized fan casing
C 6	With half-sized fan casing

Fig.7 Parameter Survey Examples

Conclusion

A simple approach is proposed to offer the equipment designer a comvenient analytical tool. tr. This approach is based on flow visualization and lumped models combined with experimental data of the flow resistance coefficient and heat transfer coefficients. The thermal design of a copy machine is studied. To use the proposed method what the designer needs is a small-sized program that can be run on a personal computer.

References

1. Fukuoka,Y. and Ishizuka,M.,1989,"An Application of the Thermal Network Method to the Thermal Analysis of Multi-Chip Packages,Japanese J.of Applied Physics, Vol.28, No.9, pp.1578-1585.

2. Ishizuka, M.,Miyazaki,Y. and Sasaki,T.,1987,Air Resistance Coefficients for Perforated Plates in Free Convection,ASME J.of Heat Transfer,Vol.109,pp.540-543.

3. Ishizuka, M.,1987,Air flow Resistance of Wire,Nettings in Natural Convection,ASME J.of Fluids Engineering, Vol.109 ,pp.389-393.

Flow Visualization and Heat Transfer Measurements by Infrared Thermography

Giovanni Maria CARLOMAGNO

University of Naples, Faculty of Engineering, DETEC
P.le Tecchio 80, 80125 Naples, Italy

Introduction

In the last decade the good results attained by means of a widespread use of the *Infrared Scanning Radiometer* (*IRSR*) in the experimental study of convective heat transfer problems have proved *IRSR* itself to be an effective tool to overcome several limitations of the standard sensors, both from the measurement and the visualization points of view [1].

The exploitation of quantitative infrared (*IR*) thermography needs the solution of several problems mainly concerned with: an accurate characterization of the *IR* system performance and its calibration; the use of external additional optics and/or mirrors, e.g. to increase the spatial resolution; the choice of the most appropriate heat flux sensor and its characterization especially with regard to the lateral thermal conduction effects and to radiation losses; the determination of the body surface emissivity; the correct geometrical identification of the measured points; the design of the optical access window including the choice of the most appropriate *IR* material. An exhaustive treatment of these general aspects is made by Carlomagno and de Luca [1] and Balageas et al. [2].

This paper intends to discuss the use of the *IR* imaging system in various heat transfer and fluid dynamics problems. The applications herein discussed are selected to the aim of emphasizing some relevant questions that arise in the use of the *IR* technique.

Measuring Convective Heat Fluxes and Visualizing Flow Fields

When studying the problem at issue, a temperature difference is required between the flow and the tested body; it is therefore useful to make at first some remarks about the different experimental procedures to be followed depending on the flow Mach number M range.

In the case of iposonic flow regime ($M \ll 1$), since the aerodynamic heating is not adequate to the sensitivity of *IRSR*, it is necessary to artificially produce a temperature difference between the model surface and the flow (*active* mode). This may be achieved either by varying the flow temperature or by heating the model in steady state or transient ways. A very convenient method to steadily heat models, having a cylindrical geometry of their surface, is the so-called *heated-thin-foil* technique. This consists of realizing (or coating) the model surface with a very thin metallic foil and heating it by Joule effect. By measuring the surface temperature, it is possible to compute the heat transfer coefficient from the foil to the flowing stream.

At high Mach numbers, because of the stream high kinetic energy content, the detection of the thermal image by *IR* thermography may be relatively simpler. In fact, the strong aerodynamic heating makes possible the use of the so-called *passive* mode.

In general the model, which is initially at uniform ambient temperature, is suddenly exposed to the flowing stream. The *thin-skin* (e.g. Carlomagno et al. [3]) or the *thin-film* (e.g. de Luca et al. [4]) heat flux sensors are generally used to obtain the convective heat transfer coefficients and/or to make diagnostics of the boundary layer. The possibility of evaluating the behaviour of the boundary layer over the surface of a body immersed in a flow by means of *IR* thermography derives from the close relation existing between skin friction and heat flux at the wall (Reynolds analogy).

When visualizing flow fields and/or measuring heat fluxes, the problem of restoring (in a general sense) the recorded thermal image (*thermogram*), may be very important both from the *imaging* and the *measurement* points of view. The thermogram, in fact, may be degraded both because of the limited spatial resolution of *IRSR* and due to the thermal conduction inside the heat flux sensor in the direction which is parallel to its surface. The first problem is addressed by Arconada et al. [5] and the more general one by de Luca and Cardone [6]. Thermogram restoration can be performed by means of *FFT* inverse filter algorithm:

$$f(x,y) = \mathfrak{F}^{-1}\left\{G(\nu,\mu)/H(\nu,\mu)\right\} \qquad (1)$$

where $f(x,y)$ is the restored thermogram as a function of the spatial coordinates x and y, G is the Fourier Transform of the recorded (degraded) image, and H is referred to as the generalized *System Transfer Function*. Both G and H are functions of the spatial frequencies ν and μ. Imaging and sampling as well as lateral thermal conduction effects may be accounted for by means of the *Average Sampled System Modulation Transfer Function* (*ASSMTF*) and of the *Temperature Amplitude Transfer Function* (*TATF*), respectively [6]. By assuming cascaded the above effects, the relationship $H = ASSMTF \times TATF$ may be employed to solve eq.(1).

Natural Convection Over a Vertical Plate

Some meaningful results concerned with the use of *IRSR* in natural convection studies are discussed. They deal with the transient natural convection over a heated vertical plate immersed in quiescent air which is subjected to a boundary condition of a step varying uniform heat flux. The *heated-thin-foil* technique is applied. Following Cardone and Carlomagno [7], the experimental set-up consists of a vertical thin stainless steel foil (*245mm* high, *960mm* wide, *40µm* thick) electrically heated. *IR* camera records the thermal image of the plate at a distance of about *1m* through a lens of *20°*, which gives a spatial resolution of about *0.5 pixels/mm*.

The measured total emissivity of the foil back surface is relatively low ($\simeq 0.15$, typical of polished stainless steel) but, since the one of the viewed surface is very high ($\simeq 0.95$; in fact for measuring purposes this surface is coated with a thin layer of black paint), the radiation heat flux is about one third of the Joule heating, i.e. of the same order of magnitude of the convective one. Therefore the former one (from both surfaces) must be subtracted from the input electric power flux in order to evaluate the latter one. The radiation flux is computed by means of the same thermographic technique which proves to be very accurate in this case.

Tests are carried out for several values of the heat flux step Q mostly starting from zero Joule heating. The transient natural convection heat transfer from a suddenly heated wall depends on the value of Q, on the wall thermal capacity and on the fluid properties. The natural convection regime, however, is always preceded by a regime of thermal conduction alone, whose duration increases as Q decreases [7].

The plot of Fig. 1 reports the time trend of the temperature difference $\Delta T = T_w(t) - T_{wi}$ measured on the foil centerline at two vertical locations (from the foil leading edge), y = *61mm* and *169mm* for the power input step *0-130* W/m^2. T_w is the local wall temperature measured at time t; T_{wi} is the initial foil temperature that, in the present case, coincides with the ambient one. Such trends are compared with the analytical prediction obtained by Goldstein and Briggs [8] for a pure conduction regime and by Cardone and Carlomagno [7] which include the effects of the radiative heat exchange. As shown by Fig. 1, the presence of the radiative term (that is of the same order of magnitude of the conductive one) protracts of one order of magnitude the duration of the pure conduction regime; this latter appears longer at the highest vertical location. During the period in which the conductive mode alone is present, data well fit the analytical prediction of [7].

In the present natural convection studies, apart from the ones due to the strong radiation flux, no corrections were practically needed to take into account either the spatial resolution of the *IR* system or the effects of the lateral thermal conduction within the foil.

Heat Transfer from a Plate to Impinging Jets

Also the convective heat transfer to impinging jets is studied by means of the *heated-thin-foil* technique. Carlomagno and de Luca [9] describe in detail the experimental apparatus and procedure. Being the Biot number (based on the foil thickness s) very small as compared to unity, the surface temperature distribution can be measured by viewing with the *IR* camera the rear face (opposite to the one where jets are impinging) of the foil. In order to increase the surface emissivity of the foil, this latter is covered by a thin film of black paint. Adiabatic wall temperature is assumed to coincide with the temperature exhibited by the foil when Joule heating is suppressed and it is measured by the very same thermographic technique.

Tests herein reported are performed for the case of a single jet. Truncated cone shaped nozzles, with rounded inlet sections and different nozzle exit diameters D are generally used. The dimensionless nozzle-to-plate distance Z/D and the Reynolds number Re (based on D and nozzle exit velocity V) are varied during the tests. Stagnation temperature of the jets is kept equal to ambient temperature by means of a heat exchanger.

The relief map of the recovery factor distribution $r = (T_{aw} - T_j)/(T_o - T_j)$ is reported in Fig. 2 for Re = *115,000*; M = *0.52*; D = *10mm*; Z/D = *8*. T_{aw} is the adiabatic wall temperature and T_j and T_o are respectively the static and the total temperatures of the jet at nozzle exit. The recovery factor presents a local minimum in the annular region having a radius $R/D \cong 1$. This minimum is attributed to a vortex ring in the shear layer surrounding the jet and, as shown in Fig. 3, it slightly moves outward as the distance Z/D decreases. Data of Fig. 3 refer to the case of D = *5mm*, Re = *65,000* and M = *0.59*; they show, for different Z/D values, the r radial profile which is symmetric with respect to the jet axis since it is obtained by performing averages in the azimuthal direction.

Data relative to the performed experimental studies on jets do not generally need imaging restoration. Furthermore, because of the small temperature differences that are encountered in the measurements of the adiabatic wall temperature, these latter do not practically need any correction for radiation heat exchange and for radial thermal conduction. The evaluation of the convective heat transfer coefficient h, instead, must be done by taking into account both radiation and conduction corrections:

$$h = \left\{Q - \left[-\lambda f s \left(\frac{\partial^2 T_w}{\partial R^2} + \frac{1}{R}\frac{\partial T_w}{\partial R}\right) + \varepsilon\sigma\left(T_w^4 - T_o^4\right)\right]\right\} \Big/ \left(T_w - T_{aw}\right) \quad (2)$$

where λ_f is the foil thermal conductivity, σ the Stefan-Boltzmann constant.

The Nusselt number, defined as $Nu = hD/\lambda$ where λ is the air thermal conductivity coefficient evaluated at film temperature, gives the convective heat transfer coefficient in a dimensionless form. In Fig. 4 the radial distribution of Nusselt number is shown for $D = 10mm$, $Z/D = 2$, $M = 0.13$ and $Re = 28{,}000$. Nu profiles are obtained by averaging data over circular rings. For such a short nozzle-to-plate distance there is an inner peak which occurs at $R/D = 0.5$; the outer peak, located at $R/D \simeq 2.5$, has the form of an annular hump. Also the minimum in between these two peaks can be attributed to the vortex ring in the shear layer surrounding the jet. The four profiles presented in Fig. 4 refer: to the coarse data (Q), to those corrected for radiation alone ($Q-Q_r$) or conduction alone ($Q-Q_c$) and to those including both corrections ($Q-Q_r-Q_c$). Differences between corrected and uncorrected data are small but not negligible.

Goertler Vortices in Hypersonic Flow

Within the European Community Space program *HERMES* to develop the first european Space Shuttle, a series of experimental tests have been performed in a hypersonic wind tunnel (CEAT, Poitiers) on delta wing with ramp and double ellipsoid models. The operating fluid of tunnel is dry air which has typically a stagnation temperature $T_o = 800°K$ and a stagnation pressure $4.5\ MPa < p_o < 10MPa$. Unit free stream Reynolds number Re_m is changed mainly by varying the stagnation pressure and, in present tests, ranges from $.89 \times 10^7/m$ to $2.1 \times 10^7/m$. The nominal free stream Mach number is equal to 8.15.

Particular attention is paid to visualize the presence of a transversal instability of the boundary layer over the ramp downstream of the delta wing, that leads to the formation of counter-rotating vortices in the reattaching flow, referred to as Goertler vortices [6]. The model, made of solid *NORCOAT 4000* to measure heat fluxes by means of the *thin-film* technique, has a swept angle of 70°, a ramp angle of 15° and a bottom wedge angle of 11°. At the beginning of each test run the model (which is initially in a remote position at room temperature) is vertically injected into the stream. A window, made of either germanium or zinc-selenide, AR coated is mounted on one side of the wind tunnel test section in order to provide an *IR* optical access to the tested model. A time sequence of thermograms (average time step is *0.24s*) is generally recorded during the test runs. Injection time is about *0.15s* and testing duration of a few seconds.

The spanwise periodic variation of the convective heat transfer coefficient produces a variation of the model surface temperature which allows the detection of the vortices by means of the *IR* technique. Thermograms may also be used to visualize the overall surface flow. It is stressed that, owing to the relatively high spatial frequencies contained in the recorded thermal images, a thermogram restoration is strongly needed in this case. The recorded image, in fact, is degraded both because of the limited spatial resolution of the *IR* imaging system and due to the tangential thermal conduction inside the *thin-film* heat flux sensor.

The thermogram of Fig. 5 shows the restored temperature map relative to a square small frame ($21x21mm^2$) viewed on the ramp downstream of the delta wing model. Testing conditions are: $M = 8.15$, $Re = 1.6x10E6$ ($Re_m = 1.3x10E7/m$), zero angle of attack. Since the flow comes from the bottom of the thermogram, the presence of the spanwise vortices is made evident by the horizontal periodic variation of the temperature. The average wavelength of the vortices results to be about *2mm*. The amount of the corrections due to the image restoration is shown in Fig. 6 where the Stanton number spanwise profile over the ramp (in particular along the line drawn on the thermogram of Fig. 5) is reported. The dotted line represents the coarse data, the dashed line the ones restored from imaging and sampling degradation effects, the continuous line the overall restoration which takes into account also the lateral conduction degradation. As may be seen, in this specific case thermogram restoration is very important, particulary that involving imaging.

Conclusions

The capability of the *IR* thermography to measure convective heat fluxes and to visualize flow fields has been analysed. The different experimental procedures to be followed, according to if the flow Mach number is relatively low or not, have been addressed. The needs of *restoring* the recorded thermal images (i.e., of removing the degradation effects), as well as of taking into account radiation losses of the sensor, have been also evidenced.

Applications to natural and forced convection problems have been reported. In the first one (study of free convection on a vertical plate in a transient state) it has been found that the presence of the radiative flux (which is also computed on the basis of *IR* measurement) determines a pure conduction heat transfer regime much longer than that for the case where this flux is negligible. The results concerning the application of IRSR to forced convection studies dealt with the mapping and the measurement of the cooling of a plate by means of impinging jets, as well as the aerodynamic heating in hypersonic flow. In particular, the detection of Goertler vortices in the reattaching flow region on a delta wing/ramp configuration furnished the opportunity of successfully applying a restoration technique which included imaging and measurement effects.

References

1. Carlomagno G.M. and de Luca L., Infrared Thermography in Heat Transfer, in Handbook of Flow Visualization (Ed. Yang W.J.), Ch. 32, pp. 531-533, Hemisphere, 1989.
2. Balageas D.L., Boscher D.M., Déom A.A. et Fournier J., Application de la thermographie infrarouge passive et stimulée à la measure des flux thermiques en soufflerie, La Recherche Aérospatiale, 1991-4, pp. 51-72, 1991.
3. Carlomagno G.M., de Luca L. and Alziary de Roquefort T., Mapping and Measurement of Aerodynamic Heating and Surface Flow Visualization by Means of IR Thermography, in Multiphase Flow and Heat Transfer (Eds. Chen, X.J. et al.), Vol. 2, pp. 1316-1324, Hemisphere, 1991.
4. de Luca L., Cardone G., Carlomagno G.M., Aymer de la Chevalerie D. and Alziary de Roquefort T., Flow Visualization and Heat Transfer Measurement in Hypersonic Wind Tunnel, Experimental Heat Transfer, Vol. 5, pp. 65-79, 1992.
5. Arconada A., Argiriou A., Papini F. et Pasquetti R., La mesure en thermographie infrarouge: calibration et traitment du signal, J. Modern Optics, Vol. 34, pp. 1327-1335, 1987.
6. de Luca L. and Cardone G., Experimental Analysis of Goertler vortices in Hypersonic Wedge Flow, in Thermosense XIV (Ed. J.K. Eklund), SPIE Vol. 1628, pp. 271-281, 1992.
7. Cardone G. and Carlomagno G.M., Convenzione naturale su lastra piana nei regimi stazionario ed instazionario, Atti IX Congresso Nazionale UIT, pp. 397-410, Pisa, 1991.
8. Goldstein, R.J. and Briggs, D.G., Transient Free Convection about Vertical Plates and Cylinders, Trans. ASME C, J. Heat Transfer, Vol. 86, pp. 490-500, 1964.
9. Carlomagno G.M. and de Luca L., Heat Transfer to Impinging Jets Measured by Infrared Thermography, Próc. COBEM 89 (Eds. Hirata, M.H. et al.), pp. 209-212, Rio de Janeiro, 1989.

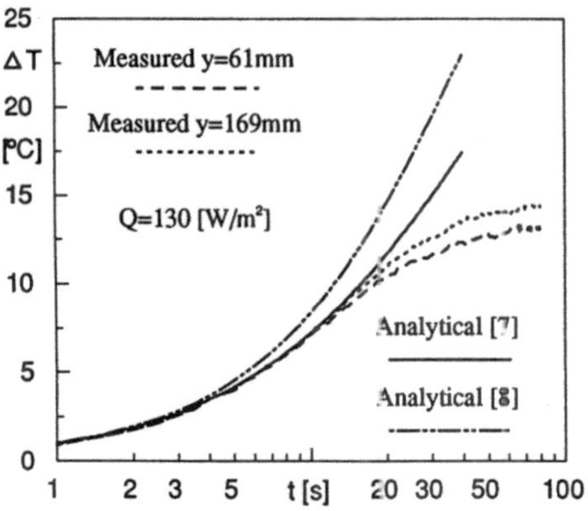

Fig.1 Temperature difference ΔT measured on the foil centerline

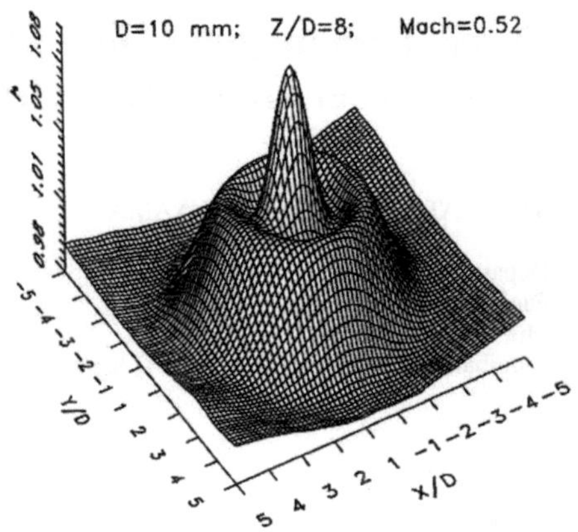

Fig.2 Relief map of the recovery factor

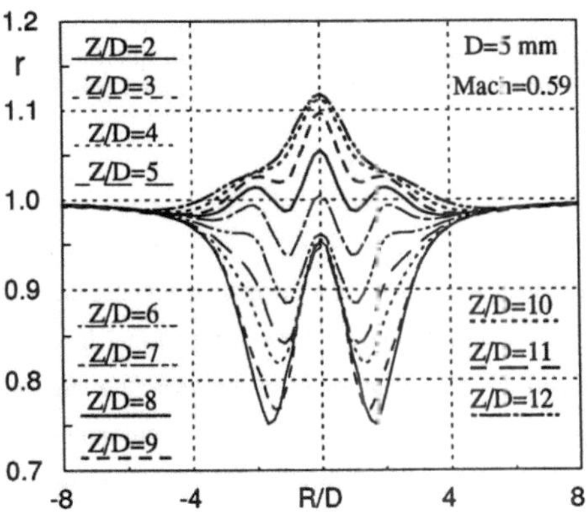

Fig.3 Recovery factor radial profiles

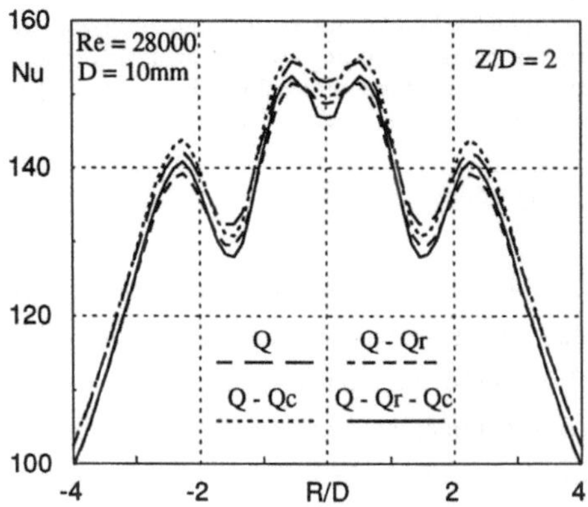

Fig.4 Nusselt number radial profiles

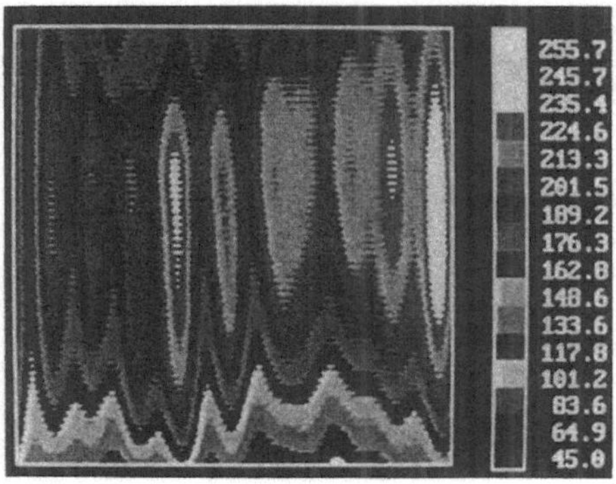

Fig.5 Restored thermogram of a small frame over the delta wing ramp

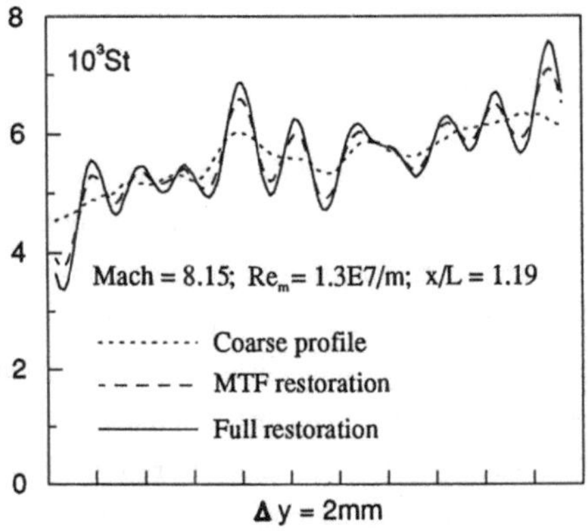

Fig.6 Stanton number spanwise profiles over the delta wing ramp

Three-Dimensional Heat Flow Visualization with a Laser Holographic Interferometry

Seizo KATO and Naoki MARUYAMA

Department of Mechanical Engineering
Faculty of Engineering
Mie University
Kamihama-cho 1515, Tsu, Mie 514, Japan

Abstract

Three-dimensional heat flows generally involve difficulties in being visualized with nonintuitive optical methods. In this paper, however, an attempt is made to visualize such flows using a laser holographic interferometry of which measuring light rays are set to be propagated in the stream direction of concerned heat flows. An analysis of the taken interferogram is performed to obtain quantitative values of three-dimensional local temperatures with good accuracy, and a computer aided processing is made to visualize the heat flows.

Introduction

Combined free- and forced-convective heat flows in the entrance region of a horizontal rectangular duct with a constant wall temperature is chosen here as a typical instance of three-dimensional heat flows. The three-dimensional temperature distribution and the Nusselt number in the heat flow are estimated from the interferogram, and the three-dimensional heat flow is visualized by the computer-aided-processing. In the interferogram estimation, light deflection and deformation of the measuring beam are compensated as mentioned below, and this procedure yields a good accuracy of measurement.

Interferogram analysis

Figure 1 shows schematically the thermal boundary layer developing along an isothermally heated surface. The direction of the measuring beam light rays are adjusted to be parallel to the wall surface[1]. The thermal boundary layer is divided into sublayers so thin that each refractive index is assumed to be constant. Here the thermal boundary layer thickness δ_T developing in this duct is assumed to be expressed

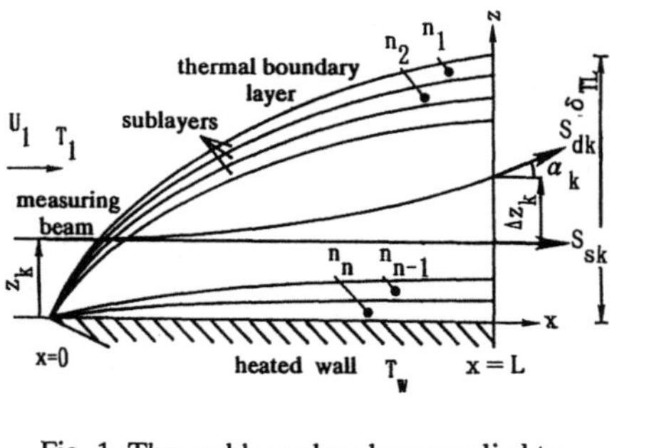

Fig. 1 Thermal boundary layer applied to interferogram analysis.

$$\delta_T/x = C \cdot Pr^a (Re_{dh} \cdot x/d_h)^b \qquad (1)$$

(C, a, b; constants, Pr; Prandtl number, $Re_{dh} = U_1 d_h/\nu$; Reynolds number based on the hydraulic diameter d_h, ν; dynamic viscosity of fluid). The constants C for the respective sublayers vary corresponding to the respective thermal boundary thicknesses at $x = L$, and a and b are assumed to $a = -1/3$, $b = -1/2$ respectively which are reasonable for laminar flow field[2]. The total interferometric fringe order shift of the measuring beam of which height is z_k from the heated wall, S_k, is expressed by the following equation;

$$S_k = \sum_{i=2}^{n} (n_\infty - n_i)\, \Delta l_{i,k} / \lambda \qquad (2)$$

(λ; wave length of light source, $\Delta n = n_\infty - n_k$; refractive index difference during the recording between the null hologram and taken interferogram). Then the refractive indices for the respective sublayers n_k ($k = 2, ..., n$) are estimated and be converted into the density ρ by the Gladstone–Dale relation and the temperature by the equation of state.

Since a real interferogram includes additional fringe order shifts which are caused by light deflection of the measuring beam rays and temperature change of the experimental surroundings, the fringe order shifts and their fringe positions of the taken interferogram are compensated according to the light path simulation mentioned below. The measuring beam rays are deflected during travel through the thermal boundary layers, and now their path deflection behaviors are numerically simulated with the aid of the Snell law. Then the total fringe order shifts for the respective measuring beam rays, S_{dk}, are calculated and compared with the ideal fringe order shifts, S_{sk}, which arise in the case of the ideal light rays that do not deflect but travel straight. The additional order shifts due to light deflection are estimated from the difference of $\Delta S_{dk} = S_{dk} - S_{sk}$. The fringe positions are also deformed due to deflection, and the deformations are calculated from the deflection angles at the test section exit and the optical geometry of the TV–camera. Therefore the deformed fringe positions are revised to the correct ones. The other additional order shift due to the temperature change of the experimental surroundings can be also compensated, and for further details, please refer to the previous paper[3].

Experimental setup and procedure

A schematic layout of the experimental apparatus is shown in Fig. 2. This apparatus consists of a holographic interferometer, a test section and an image processing system. The measuring beam rays are adjusted to be parallel to the direction that the thermal boundary layer develops. The test section is a rectangular duct of 20.0X20.0 mm in cross section(hydraulic diameter d_h = 20.0 mm) and the test section length L is fixed to L = 300 mm. The duct walls are uniformly heated, and the experiments are performed for the temperature differences of $\Delta T = T_w - T_1$ = 35 and 55 K. Air flow is generated by a variable speed suction pump along the bell mouth with minimum disturbance, and the Reynolds number is fixed in this experiment at about $Re_{dh} = 1000$. For $Re_{dh} \approx 1000$, the thermal boundary layers developing along the respective isothermal wall surfaces do not intersect at the test section outlet and the real time interferogram is kept relatively stable. After the steady state condition is reached, the real time interferograms are taken by a CCD camera within 1/30 second.

Before the experiment, the optically exact wall position is determined by a dimension standard inserted into the test section. The reference null hologram is recorded under the unheated condition at an infinite fringe adjustment. The video signals are recorded and processed by the micro computer with an image processing system.

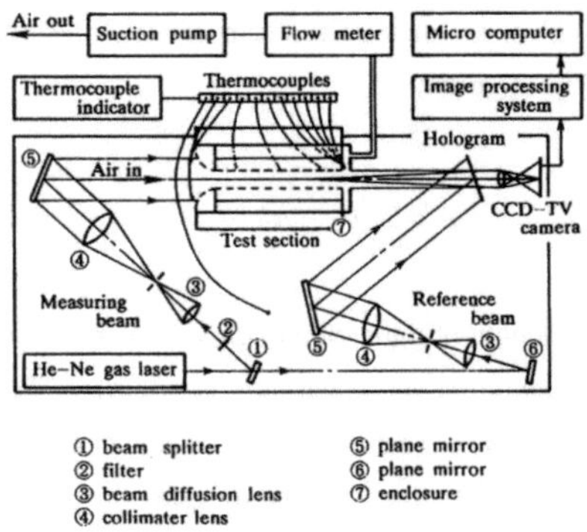

Fig. 2 Schematic layout of the experimental apparatus.

Results and discussion

The typical interferogram taken at the test section of interest is shown in Fig. 3. The fringes approximately show the isothermal contour lines. However, this interferogram includes the additional fringe order shifts as mentioned before, and the dimensionless temperature distribution is the corrected one using the compensation procedure described before. The core fringe represents the minimum temperature zone where the temperature is equal to the free stream inlet temperature T_1. This core region represents the regime of the thermal entrance region. In addition, the fringe density, namely temperature gradient, in the upper region is much lower than that in the lower region, and therefore the secondary flows caused by buoyancy can be recognized schematically as shown in Fig. 3(b). The isotherm contours represent the longitudinally averaged temperatures.

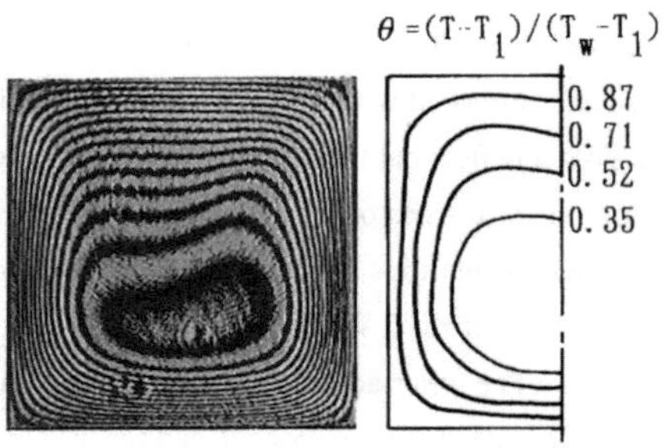

(a) (b)

$Re_{dh} = 992$ $L = 0.30$ m

$Pr = 0.7$ $d_h = 0.020$ m

$T_w = 332.7$ K

$T_1 = 297.0$ K

Fig. 3 Typical interferogram observed by the real-time method. (a) Taken interferogram. (b) Temperature distribution and secondary flow

Figure 4(a) shows the deflected ray paths traveling through the thermal boundary layer. The ray paths are deformed by the temperature gradient, namely refractive index differences, and the deflection effect becomes remarkable as the temperature difference $\Delta T = T_w - T_1$ increases. The total deflection angles α_t at the test section exit are also shown in Fig. 4(b). α_t increases as z_k becomes zero, namely approaches to the heated wall neighbor.

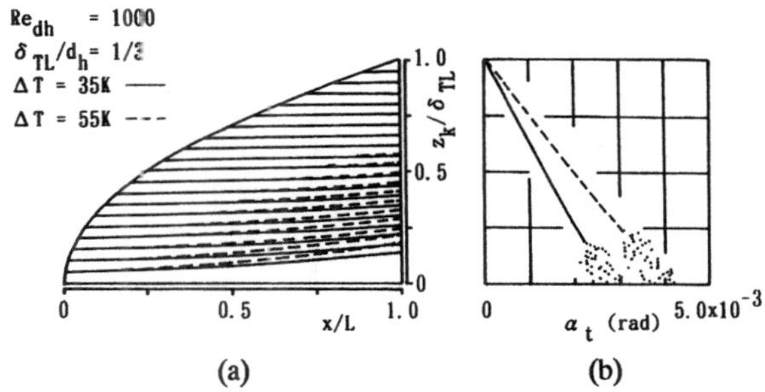

Fig. 4 Light deflection simulation of measuring beam rays passing through thermal boundary layer(a) and deflection angle at test section exit α_t(b).

Figure 5 shows the fringe intensity numerically estimated from the ray path simulation. Fig. 5(a) indicates the intensity in the case of the ideal light rays, in which the ideal fringe order shift and its positions are denoted by S_{sk} and z_{sk} respectively, and Fig.5(b) shows the fringe intensity in the case of the deflected ray paths, in which are denoted by S_{dk} and z_{dk} respectively. Thus the fringe positions can be numerically analyzed for the respective rays, and consequently the required compensation should be made based on such results as Figs. 4 and 5 which characterize quantitatively the difference between the optically ideal and real situations.

The three-dimensional temperature distribution and the Nusselt number are calculated taken account of the deflection compensation and are shown in Fig. 6. The thermal boundary devel-opment in the upper region is much rapidly than that in the lower region because of the secondary flow caused by buoyancy. The averaged Nus-selt number is evaluated from the obtained temperature distribution. The Nusselt number is well agree within 2% to the Nusselt number evaluated from the heat balance, when the addi-tional fringe order shifts and fringe deformations are taken into account to evaluate the tempera-ture from the interferogram.

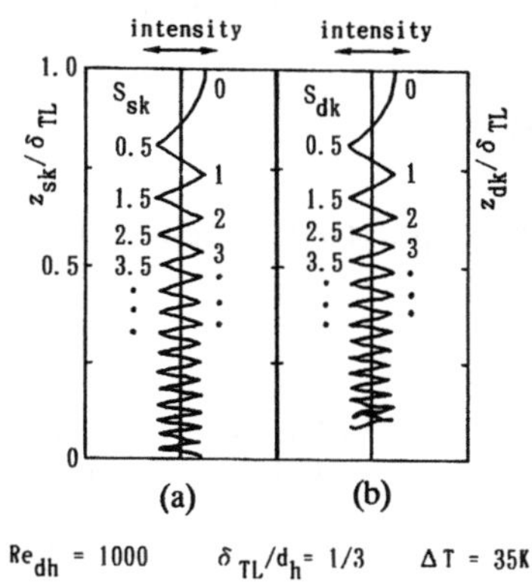

Fig. 5 Fringe intensity estimated from theo-retical ray path simulation. (a) Ideal measur-ing ray path (measuring beams do not de-flect). (b) Real measuring ray path (measur-ing beams deflect).

The Nusselt number at near the inlet region becomes so remarkably large that it is beyond imagination from the temperature distribution as shown in Fig. 6(b). As a result, the three-dimensional temperature and Nusselt number distributions can be visualized well with the aid of two-dimensionally obtained interferogram.

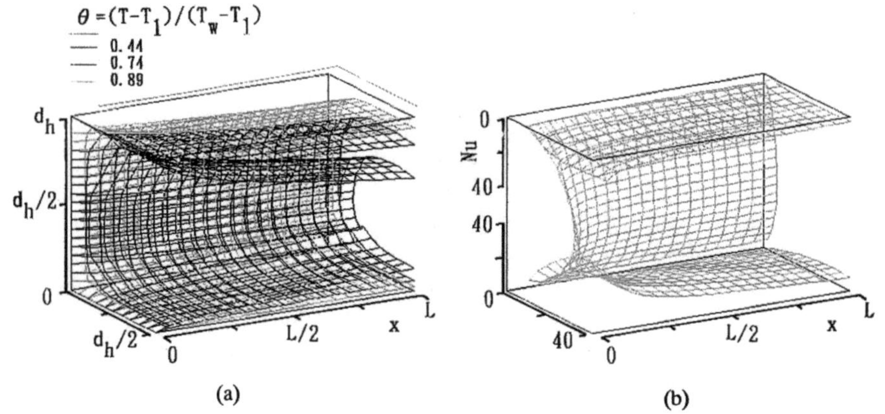

Fig. 6 Three-dimensional distribution for temperature(a) and Nusselt number(b).

Conclusion

The holographic interferometry is applied to visualize the three-dimensional temperature field of the thermal entrance region in the rectangular duct. The three-dimensional heat flow can be well visualized with the aid of the present interferogram analysis with compensation for light deflection.

Acknowledgement

This work was supported by the Research Foundation for the Electrotechnology of Chubu, Japan.

References

1. KATO, S. and MARUYAMA, N., Holographic Interferometric Measurements of the Three-Dimensional Temperature Field with Thermally Developing Flow in the Measuring Beam Direction, *Experimental Thermal and Fluid Science*, **2**-3, 333–340, Elsevier, 1989.
2. Eckert, E.R.G. and Drake, R. M. Jr., Analysis of Heat and Mass Transfer, 281–327, McGraw-Hill, 1972.
3. KATO, S. and MARUYAMA, N., A Theoretical Analysis of Light Deflection Effect on Quantitative Evaluation in Holographic Interferometry, *Laser Interferometry IV: Computer-Aided Interferometry*, SPIE, **1553**, 298–305, 1991.
4. Hauf, W. and Grigull, U., Optical Methods in Heat and Mass Transfer, *in Advances in Heat Transfer,* **6**, *T. F. Irvine, Jr., and J. P. Hartnett, eds.*, 133–366, Academic, New York, 1970.

Temperature Measurement of Transonic Nozzle Flow by Laser-Induced Fluorescence Method

M. INOUE, M. MASUDA, T. MURAISHI and Y. HYAKUTAKE

Kyushu University, Hakozaki, Fukuoka City, Japan

Summary

To measure the spatially- and temporally-resolved temperature and velocity in a transonic turbine and compressor, the laser-induced fluorescence (LIF) method was developed for the flow field with atmospheric pressure. The temperature distribution was obtained from the fluorescence intensity of the seeded iodine excited with an argon-ion laser, and the velocity was measured by the Doppler shift of the spectral peak of the fluorescence.

Introduction

In the large-scale gas or steam turbine, the flow around the blade tip becomes transonic, and the shock waves are generated in the blade passage. These shock waves induce the boundary layer separation and the vibration of blades, which not only degrade the efficiency but also induce the large vibrational stress in the blade. In addition, the secondary flow is induced by the centrifugal force associated with the large changes in the flow direction in the cascade. Our research on these phenomena suggested that this secondary flow was related to the boundary-layer separation and the vibration of shock waves [1].

The flow fields inside the transonic cascades have been measured by Pitot tubes and hot-wire anemometers; the Schlieren method has also been employed as an optical diagnostic technique. However, the transonic flow field is disturbed severely by the insertion of Pitot tubes and hot-wire anemometers, and the Schlieren method measures the flow pattern averaged along the line-of-sight. This means that the local transonic-flow parameters can not be measured accurately by these methods.

To clarify these phenomena, the LIF method [2] was developed and applied to the flow field in the transonic nozzle with rectangular cross-section. The air seeded with iodine was used as a working gas. The stagnation pressure was taken as an atmospheric pressure. The fluorescence was obtained by exciting iodine with the argon-ion laser. When the pressure of the flow-field is nearly atmospheric, the fluorescence intensity obtained by the broad-band lasing of the argon-ion laser depends only on temperature. By using this, the temperature distribution in the nozzle was obtained. The velocity along the nozzle center line was also measured by scanning the laser wavelength and observing the Doppler shift of the spectral peak of the fluorescence. The measured temperature and velocity were found to compare very well with those calculated by the measured static pressure with an isentropic assumption.

Experimental Setup

Figure 1 shows the experimental setup for the temperature measurement. The wind tunnel was a blow-down type with an atmospheric air as a working gas. Air flowed into the iodine cell in which iodine was sublimated into air (4wt%), and then expanded through the nozzle into the expansion chamber. The nozzle was a two-dimensional nozzle with a rectangular cross-section. The depth of this was 20mm, the throat height was 8mm and the exit height was 10mm. The stagnation pressure was measured and found to be nearly atmospheric. In the temperature measurements, the laser beam was shaped to a 0.5mm thick sheet and injected to the nozzle test section. The fluorescence was collected by lenses, amplified by the image intensifier and detected by the CCD camera.

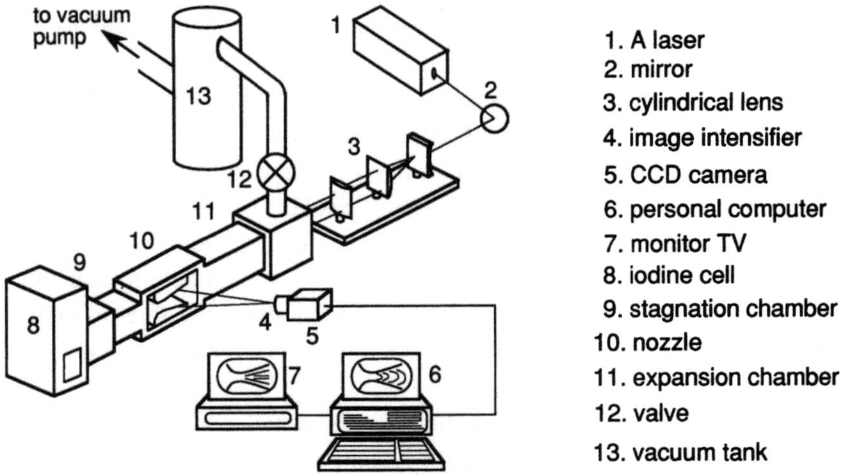

Fig. 1 Experimental setup for temperature measurement

In the temperature measurement, following conditions were assumed. (i) the laser was operated with a broad-band mode, (ii) the laser intensity was much less than the saturation intensity and (iii) the gas pressure was nearly atmospheric so that the collisional quenching dominated the natural emission rate. Under these assumptions, the detected fluorescence intensity was a function only of temperature. This meant that the temperature of the flow field could be calculated from the fluorescence intensity, if the temperature of the one specified point in the flow field was known [3,4,5]. The impact and static pressures at the center of the nozzle exit plane were measured in the present work with the impact tube and the static pressure hole of the nozzle side-wall. The temperature of this point was calculated with these two measured pressures by using an adiabatic assumption. With this calculated temperature, the temperature distribution over the whole flow field was obtained from the measured fluorescence intensity. Since the pressure was nearly atmospheric and the laser power was 0.5W that was much less than the saturation intensity, the conditions described above were satisfied.

To measure the velocity, a tunable etalon was inserted into the laser cavity and scanned by the stepping motor. By this modification, the laser wavelength was changed to get the spectral profile of the fluorescence. In this experiment, the laser beam with 0.5mm in diameter was injected along the nozzle center line and the photomultiplier was used to detect the fluorescence. The Doppler shift of the fluorescence spectrum was measured, and the velocity was obtained by comparing the spectral peak of the fluorescence with that of the standard iodine cell.

Experimental Results

Figure 2 shows the temperature contour of the central plane of the nozzle, where the abscissa is the distance from the geometrical throat of the nozzle. This clearly shows the characteristic feature of the transonic nozzle flow. The critical temperature is 250K, and the position of the critical state is seen to be located 1.6mm downstream of the geometrical throat of the nozzle. As shown in the figure, the temperature decreased from the stagnation chamber, through the critical state, to the 7mm downstream of the geometrical throat, but then it increased, took maximum value and decreased again. This experiment used an atmospheric air with the temperature of 295K and the relative humidity of 75%. When the air with this parameter was accelerated to a supersonic speed, the shock wave was reported [6] to appear due to the condensation of water vapor.

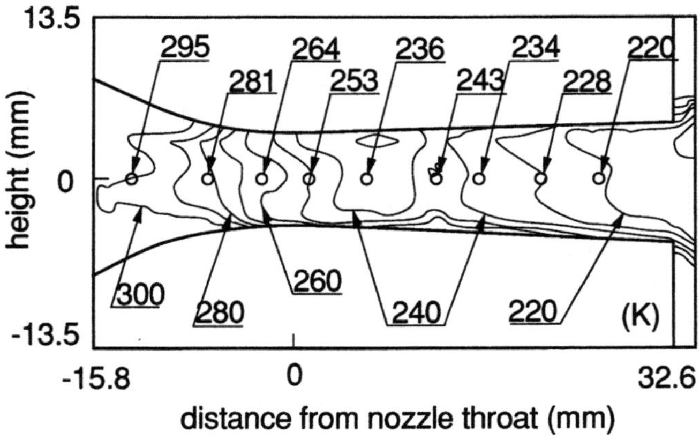

Fig. 2 Temperature distribution

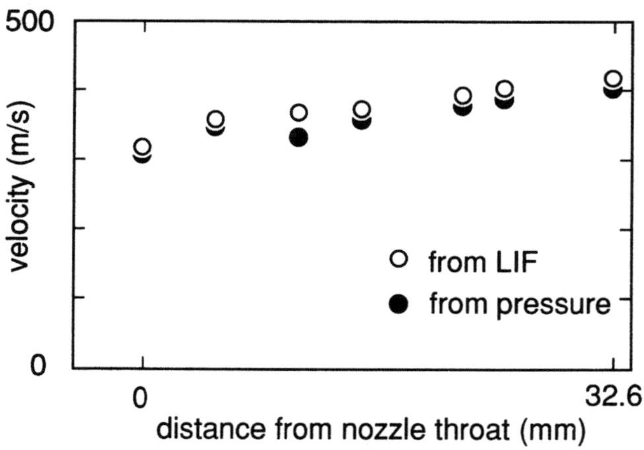

Fig.3 Velocity along nozzle center line

The temperature increase observed in Fig.2 was caused by this condensation effect. Open circles show the location of the static pressure holes, and the temperatures denoted on the upper part of this figure are those calculated from the measured stagnation and static pressures using an isentropic assumption. As shown in the figure, good agreement was obtained for the temperatures measured with these two methods, and the LIF diagnostic technique used in the present research was proved to be very accurate.

Figure 3 is the velocity along the nozzle center line. The open circles indicate the velocities measured by the LIF method and the solid circles are those calculated from the measured static pressures with an isentropic assumption. Since the fluorescence intensity was not high in the pressure regime of the present experiments, the velocity data taken by the LIF method showed a scatter of about ±20%. Although the results shown in Fig.3 were the average of five data, these compared well with the velocity calculated from the static pressure.

Conclusion

The diagnostic system based on the laser-induced fluorescence method was developed to measure the transonic flow field. This system used the argon-ion laser as a light source and the target gas was an atmospheric air seeded with iodine. By operating the laser with a broad-band mode, the temperature distribution in the transonic nozzle was measured. Also by scanning the laser wavelength and observing the Doppler shift of the fluorescence spectra, the velocity along the nozzle center line was obtained. These were shown to agree very well with those calculated from the measured stagnation and static pressures. The present work proved that the temperature and velocity in the transonic flow-field near an atmospheric pressure could be measured with high accuracy by the laser-induced fluorescence method.

References

1. Inoue,M.; Furukawa,M.; Muraishi,T.: Shock wave oscillation with inherent frequency in transonic turbine cascades. Proc. 2nd KSME-JSME Fluid Eng. Conf. 2 (1991) 111-116.
2. McDaniel,J.C.: Investigation of laser-induced iodine fluorescence for the measurement of density in compressible flows. SUDAAR No.532, Stanford University (1982).
3. Inoue,M.; Muraishi,T.; Masuda,M.; Natsunari,K.: Flow visualization of curved transonic nozzle by LIF method. J. Visualization Soc. Japan 10-2 (1990) 103-106 (in Japanese).
4. Hartfield,R.J.; Hollo,S.D ; McDaniel,J.C.: Planar temperature measurement in compressible flows using laser-induced iodine fluorescence. Optics Lett. 16-2 (1991) 106-108.
5. Inoue,M.; Masuda,M.; Muraishi,T.: Flow visualization of transonic nozzle by laser-induced fluorescence method. J. Visualization Soc. Japan 11-2 (1990) 35-38 (in Japanese).
6. Matsuo,K.; Kawagoe,S.; Sonoda,K.; Kwon,S.B.; Yamamoto,H.; Sugiyama,E.: Studies of condensation shock waves. Bull. JSME, 29-248 (1986) 439-443.

Heat Transfer Visualization of Light Emitting Diode by Means of Infrared Radiometer

N.TSUYUZAKI

Corporate Planning Room, Okumura-uki Co. Ltd
Porshe Platz Bild., 8-6-18 Ginza, Chuo-ku, Tokyo 104, Japan.

S.Sekiya, F.Kaminaga, Y.Okamoto

Dep. of Mechanical Engineering, Div. of Engi., Ibaraki Univ.
4-12-1 Narusawa-cho, Hitachi-city, Ibaraki 316, Japan

Abstract
Temperature distribution, displayed to the two dimensional CRT, of Infrared Light Emitting Diode was measured by Infrared radiometer to visualize the thermal emission mechanism and the temperature under real time condition. The temperature deviation of the LED element and the temperature response time were evaluated numerically by means of the thermal network method. The compare with the measured data is not only useful for estimation of the LED temperature but also applied to the final inspection of the LED after construction under production process.

Introduction

In these day, the infrared LEDs attached to the remote controller and the touch sensor in front of the display are increased in the field. the input current of the LEDs are used a pulse type current which is stronger than normal direct current in order to pick the emitting light up in the ambient. The duty ratio of the input pulse current is nearly same as normal direct current. Some of the LEDs under the above conditions failed at the gold wire bonding area and chip bonding part. Then, we estimate that failure of disconnection is caused by difference of the thermal expansion. Therefore, we measured the maximum temperature and its distribution of the inside and outside of the LED using a infrared radiometer and thermocouples. The cooling efficiency of the LED surface was evaluated by the calculation from measured temperature distribution.

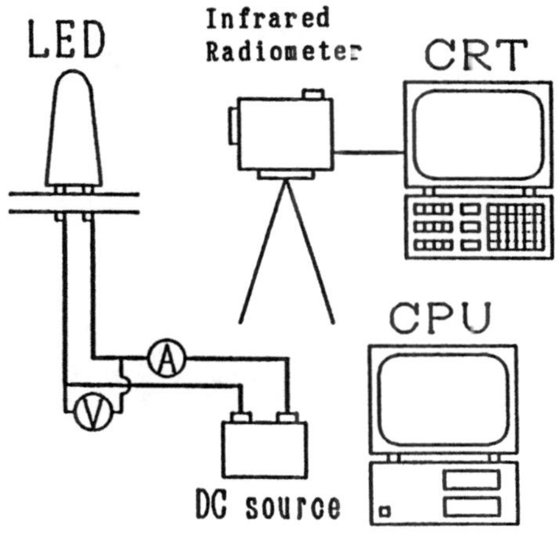

Fig.1 Measurement Unit

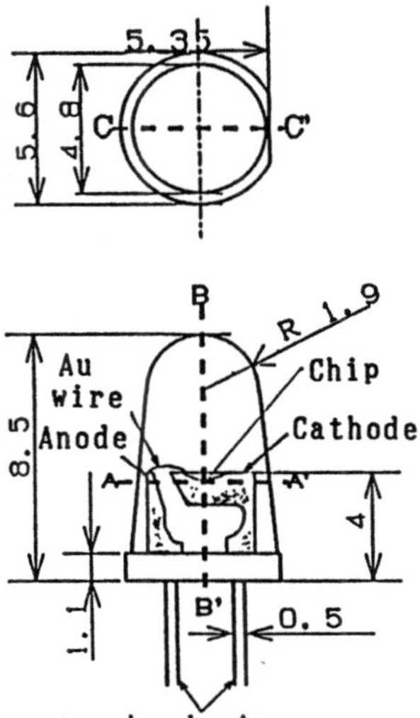

Fig.2 Detail of LED

Measurement System

The lead frame of the LEDs are connected to a DC source with ampere and voltage meter. The maximum constant direct current of LED is about 50mA. A infrared radiometer, 0.5° C deviation, is settled front of the testing LED and connected to the data analyzing computer. These measurement system is shown in the Fig.1. The figure 2 is the detail of outer dimensions of the LED.

Measurement

The temperature distribution of the LED mold is measured by the infrared radiometer to check the maximum temperature position. The measurement condition of the LEDs is shown in the Table 1. The measured temperature distribution of the LED surface is shown in Fig.3. Its input conduction is 50mA and 1 minute. The Fig.4 shows detail of the lead frame construction in the LED epoxy mold. Figure 5 is temperature distribution of the lead frame before molded of the epoxy resin. The C.A thermocouples

Table. 1 Measurement Condition

Item	Parameter
Current (mA)	50,100,150,200
Input Time Length (min)	1, 20, 30

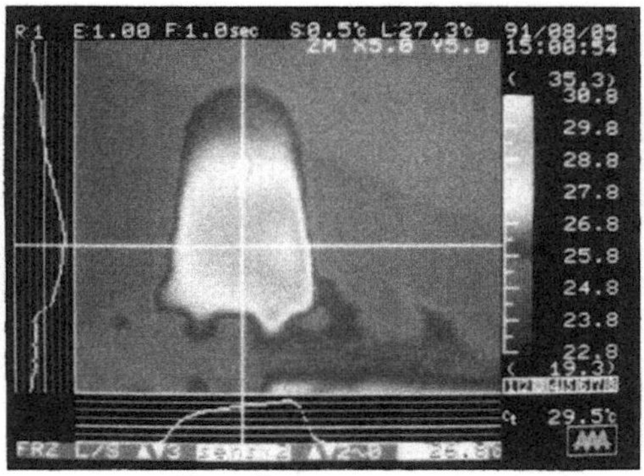

I=50mA, E=1.28V, W=0.064W, τ=1min
Fig.3 Termography of LED

welded to the lead frame are measured during check the temperature by the infrared radiometer. After the measurement of bare lead frame shown in Fig.4, the lead frame with thermocouples are molded by the epoxy resin. The molded frame with thermocouples is measured again to compare with the temperature data of Fig.3. The temperature distribution after the cut of the thermal conduction to the lead frame during on-power and cut off power is shown in Fig.6. The material of thermal insulation ia quart wall. The input power is about 50mA D.C..

Discussion and Conclusion

The maximum temperature point of the inside of the LED based on the Fig.3 and Fig.5 is the chip bonding area of the semi-conductor. The C.A thermocouple also indicated that the highest point is a part of the reflection miller. The Fig.6 shows that the rise up time from room temperature is few minutes. Then, it is clear that the maximum temperature area is chip bonding part and its point is 20°C higher than room temperature under 50mA DC. We estimate that this method is very useful for the final check of the LED because the temperature would be unstable if

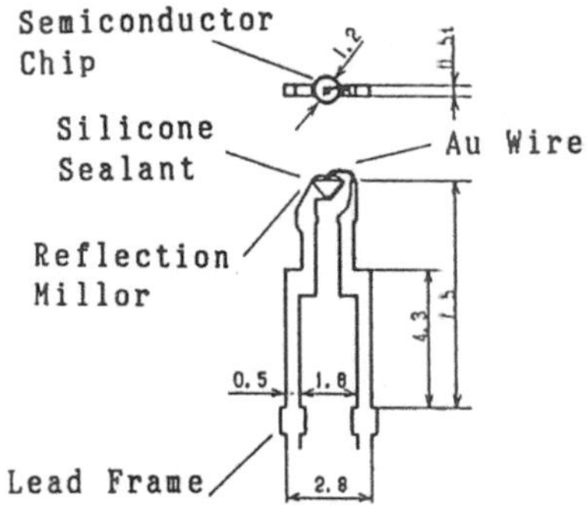

Fig.4 Detail of LED Lead Frame

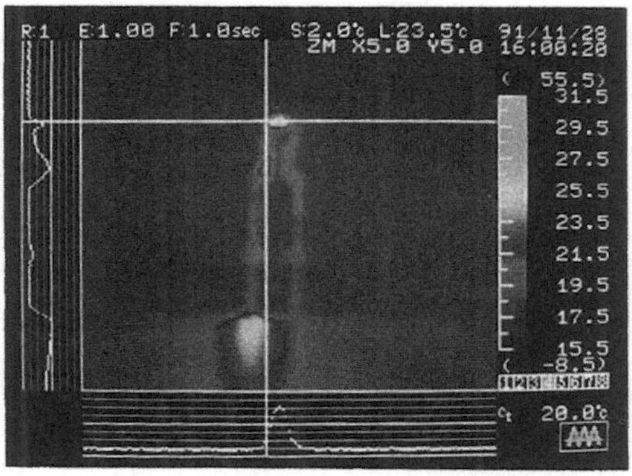

Fig.5 Thermoview of Lead Frame

the chip bonding conditions also unstable. The heat transfer ratio is calculated from the temperature data comparison of Fig.4 and measured temperature using the LED surrounding by the thermal insulation. Figure 7 shows the heat generation rate and surface temperature. It is clear that heat conduction of the lead frame is higher ratio than surface convection of the LED mold when the thermal generation of the LED is about 100mW and its ratio is 2.9.

Acknowledgment

These experiment and examinations are held by Mr. Wakashima(Ibaraki Univ.). Infrared LEDs are given from Mr. Amigo Huang(Taiwan LITON, now JAMES & AMY CORP. in Taiwan) and Nippon LITON. We authors express the thankfulness to these members.

References

1. Taiwan LITON Electonic Co. LTD.: Optoelectronics Data Book 1991-1992. Second Edition. Dec, 1990.
2. Kotake, S.; Hijikata, K.: Thermal and Flow. 6th Edition. Maruzen, Tokyo, Japan. June, 1990.

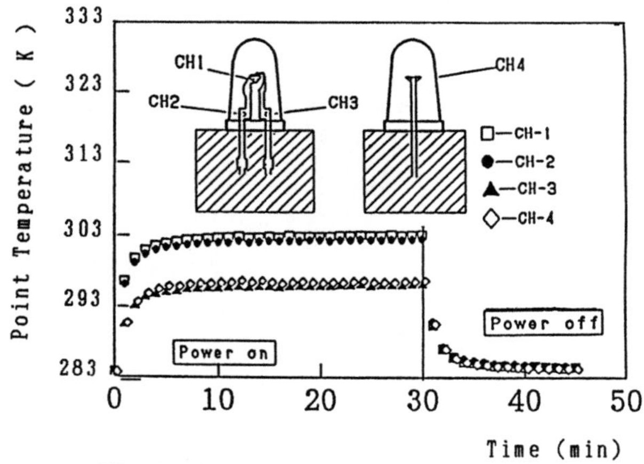

Fig.6 Time Dependant Temperature of Lead Frame

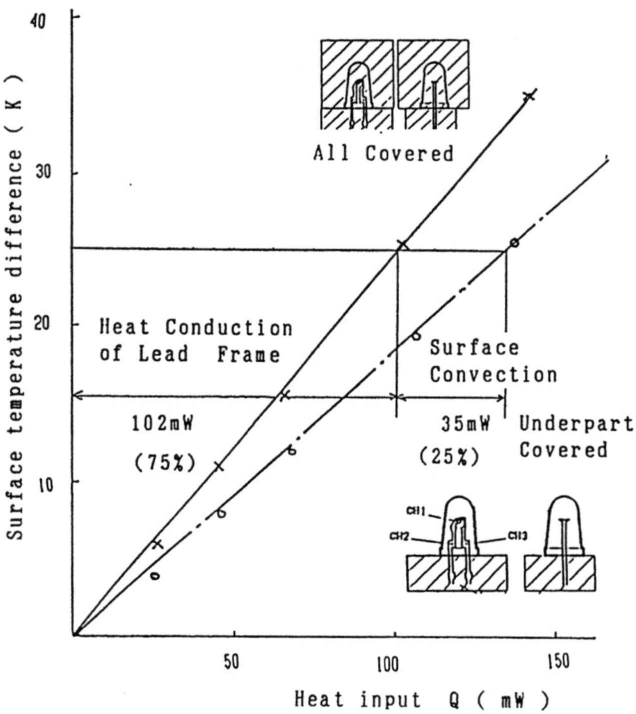

Fig.7 Heat Generation Rate and Surface Temperature

A New Technique for Visualization and Determination of Local Mass Transfer at Solid Walls in Liquid Flow

V. Kottke*, W. Kühnel*, S. Becker**

* Institute of Food Technology, Food Process Engineering
Hohenheim University, Stuttgart, FRG

** Institute for Chemical Engineering
University of Stuttgart, FRG

SUMMARY

A new technique for visualization and quantification of local heat and mass transfer is presented. The technique is based on chemisorption of dyes at solid surfaces. The intensity and the distribution of the colour is directly correlated to the transferred mass. With a remission photometer the test samples are evaluated. First results with this new technique are presented for mass transfer in multiple-jet-stagnation flow.

1. INTRODUCTION

Methods for visualization and quantification of local heat and mass transfer are very important for the effective use of energy and material. With such methods it is possible to optimize complex surfaces e.g. of plate heat exchangers, regenerators, membrane systems, catalysts, adsorbers, column inserts and static mixers.

For simple geometries it is possible to predict the local heat and mass transfer by numerical calculation. But for complex geometries in industrial scale the prediction is often not possible up to now. In this case it is necessary to have a powerful method for the visualization of the heat and mass transfer which allows as well the quantification of the local and integral heat and mass transfer in a short time.

2. THE NEW TECHNIQUE

There are several methods to visualize flow pattern on surfaces in gas and liquid flow /1,2/. The methods for the determination of local heat and mass transfer in liquid flow can be divided principally in four basic types : electrochemical methods, solution techniques, swollen gel technique and adsorption methods. Up to now the use of all techniques is limited to simple geometries. Especially the adsorption method allows the visualization of the local heat and mass transfer and is therefore here of special interest.
In the past there have been especially two publications using adsorption methods. Koncar-Djurdjevic /3/ coated the surface of aluminium plates with a thin layer of silica gel and mounted them in a water channel. A dilute aqueous solution of methylene blue was used as adsorbate. Varol and Beg /4/ made their investigations with a system consisting of a transparent methoxymethyl nylon-66 polymer film and an acid dye. The amount of mass transfer was measured with a microdensitometer. This assumed a transparent specimen.

The new technique presented here for the visualization and determination of local heat and mass transfer in liquid flow is based on chemisorption of dyes at solid surfaces. Fixation by chemisorption is essential because the visualized local mass transfer distribution should not be changed by local diffusion processes at or in the wall. The use of molecular mass transfer allows a high local resolution which is only limited by the structure of the surface.

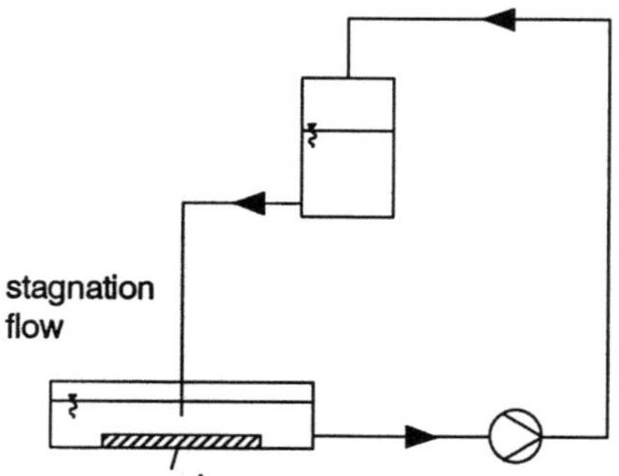

Fig.1 : test apparatus for impinging jet flow

The surface of the specimen is coated with a smooth and cheap sheet. The sheet has to be resistant to water and has to have a high affinity for the dye used. On the other side there are several requirements on the dye : good affinity to the sorption medium of the surface sheet, fixation of the transferred dye at the surface by chemisorption, water-soluble, non-toxic and cheap. To find a dye which fits best to these demands many different types of dyes with different chemical structures and different colours has been investigated. These first investigations were made in a very simple test apparatus with an impinging jet flow (fig. 1).

Many of the tested combinations of dyes and foils did not produce a good visualization of the stagnation flow. Mostly the foil was evenly coloured because the dye was not fixed onto the surface. It diffused in the wet foil.

The sheets are normal chromatographic sheets coated with a very thin layer of polyamide as sorption medium. These thin foils are mounted onto the model surface. The blue anionic dye which is chosen shows the mass transfer distribution of the stagnation point (Fig. 2) with a high local resolution. By salt bond the dye is absolutly fixed to the sorption medium on the foil. Allthough the dye is solved in the liquid in a very low concentration the solution has a very intensive blue colour.

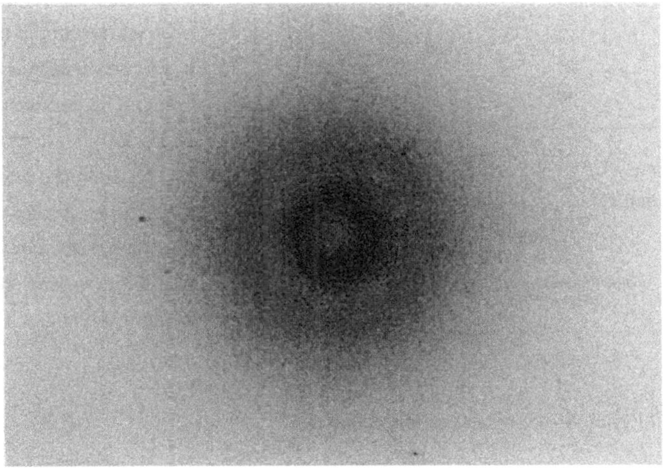

Fig. 2 : stagnation point visualized with the new visualization technique

After chemisorption the dye is locally fixed onto the surface. The colour intensity corresponds directly to the amount of the transferred mass. Dark zones indicate high mass transfer and a thin boundary layer. Light-coloured zones indicate low mass transfer and an area with a thick boundary layer. The amount of the transferred and adsorbed dye is evaluated by photometric measurements.

For the evaluation a simultaneous spectral photometer with a line of diodes is used which is measuring the entire spectrum within the wavelength range from $\lambda = 290$ nm to $\lambda = 1010$ nm in a few milliseconds. The dry coloured foil is fixed onto a cross table. The measuring head of the photometer is stepper motor driven and can be moved to any position of the cross table. The whole measuring equipment - stepper motor and data acquisition by the photometer - is controlled by a personal computer (Fig. 3).

The light of a xenon light source is conducted by a three legged fiber-optic bundle to the measuring head. Each fiber-optic is terminated with a special optic which focuse the light beam to the measuring area under an angle of 45°. The remitted light is measured by a two-legged fiber-optic bundle which conducts the light to the two modules of the photometer. Each module works in a different range of wavelength thus both module have together a range from λ = 290 nm to λ = 1010 nm. After a very short time (5.6 ms) the acquisition of the whole spectrum is finished.

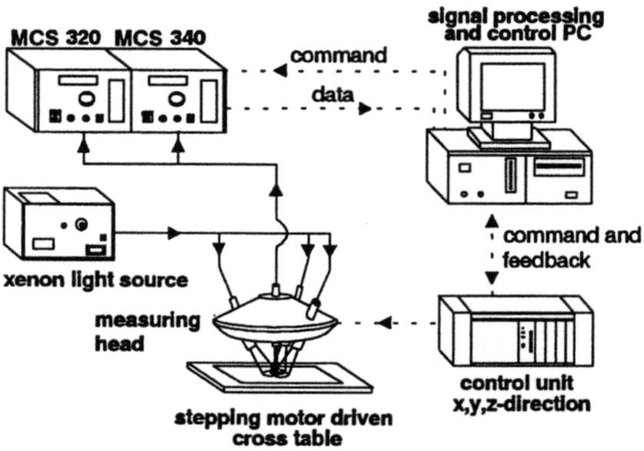

At the beginning of the measuring the remission of a pure white sample of the foil has to be measured as a reference. The remission of this white foil is determined as 100%. Then the dried foil of the sample is measured and the data are converted to remission values in relation to the reference.

Fig. 3 : diagram of the remission photometer

3. APPLICATION OF THE VISUALIZATION TECHNIQUE

For the investigation of multiple-jet-cleaning systems it is necessary to have a technique to compare the effect of different cleaning devices. One way is to coat very carefully a sample with a test layer and to evaluate the removal. The other way is to use a visualization technique like those presented here. There all the problems of coating (different layer thicknesses) do not exist. The effect of cleaning is depending on the momentum of the jets. The visualization technique shows the mass transfer from the liquid to the surface of the sample. Therefore the results of the experiments with a test layer and with the visualization technique has to be compared. The comparison has shown that it is possible to investigate jet-cleaning devices by using a visualization technique. Fig.4 shows an example of three impinging jets.

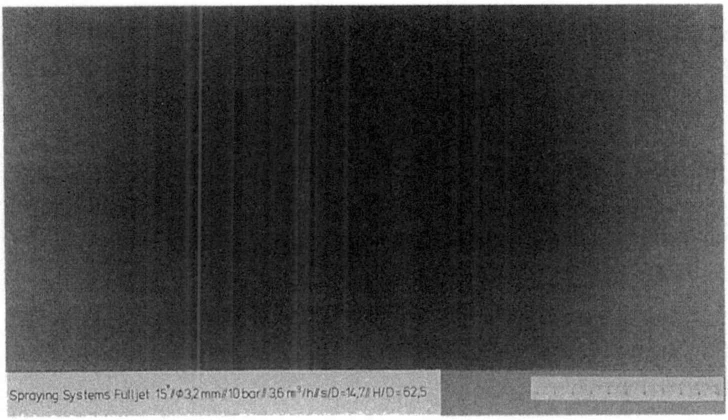

Fig.4 : Visualization of three impinging jets

CONCLUSION

The technique for visualization of heat and mass transfer in liquid flow presented here is based on chemisorption. A special sheet coated with a thin layer of a sorption medium is fixed onto the surface of the sample. The dye solved in the liquid is adsorbed and fixed by chemisorption on the surface of the sheet. Depending on the flow pattern there is in the region of high mass transfer a high intensity of colour and vice versa. The colour distribution and intensity is evaluated and quantified with a spectral remission photometer. First investigation of multiple-jet-cleaning systems were made by using this new visualization technique. They have shown the suitability of this technique.

REFERENCES

1. Reznicek, R. : Two Examples of Using Surface Tracing Flow Visualization Methods
   Proceedings of the 5. International Symposium on Flow Visualization, Prague 1989,
   pp. 93-108
2. Kottke, V. : Verfahren zur Sichtbarmachung von Gasströmungen
   Chem.-Ing.-Tech. 52 (1980) Nr.9, pp.687-695
3. Koncar-Djurdjevic, S. : Eine neue allgemeine Methode zur Untersuchung der Wirkungsweise chemischer Apparaturen
   DECHEMA Monographien 26(1956), pp.139-153
4. Varol, K. ; Beg, S.A. : A Novel Technique for Mass Transfer Studies
   Wärme- u. Stoffübertragung 4(1971), pp.82-86

Taylor-Goertler Instability in Sinusoidal Wavy Channels

V. Kottke*, G. Gaiser and P. Gschwind**

* Institute of Food Technology, Food Process Engineering
Hohenheim University, Stuttgart, FRG

** Institute for Chemical Engineering
University of Stuttgart, FRG

SUMMARY

Wavy channels are especially used for the enhancement of heat and mass transfer in heat exchangers, regenerators, catalyzers and membrane moduls. In the concave part of the ducts a flow instability can be observed leading to longitudinal vortices comparable to the Goertler-vortices at concave walls or the Dean-vortices in tube flows. The concave-convex walls in sinusoidal wavy channels alternatively enhances and destructs the longitudinal vortices resulting in highly complex flow and transport phenomena.

INTRODUCTION

The application of channels with wavy walls in technical apparatus allows a considerable enhancement of heat and mass transfer at low pressure losses. Examples of their application are plate heat exchangers, finned tube heat exchangers, regenerators and membrane systems i.e. electrodialyzers and blood oxygenerators. Previous analysis of sinusoidal wavy channels in plate heat exchangers /1-3/ revealed the existence of longitudinal vortices. The local heat and mass transfer in this case is drastically changed to a highly inhomogeneous distribution. Therefore a systematic analysis on flow instability and transport phenomena, especially on the local heat and mass transfer is required.

EXPERIMENTAL PROCEDURE AND APPARATUS

For the experimental analysis of sinusoidal wavy walls at constant duct width $B = 150$ mm and constant duct length $L = 400$ mm four parameters are especially of importance. Fig. 1 shows the sinusoidal wavy duct with constant wavelength $\lambda = 26$ mm and amplitude

$a = 1.825$ mm and the dimensionless wavelength $\lambda/a = 14.25$. The duct height s is varied in the range $1.8 \text{ mm} \le s \le 6.4$ mm with the dimensionless parameter $s/a = 1$; 1.5; 2; 2.5; 2.9; 3.2 and 3.5. The Reynolds-number related to the duct height s is varied in a wide range $50 \le Re \le 10.000$ with air as working fluid. For the longitudinal vortices the low Reynolds-number range is of special interest in the transitional regime. Note that the critical Re-number in wavy ducts can be very low, depending on λ/a and s/a.

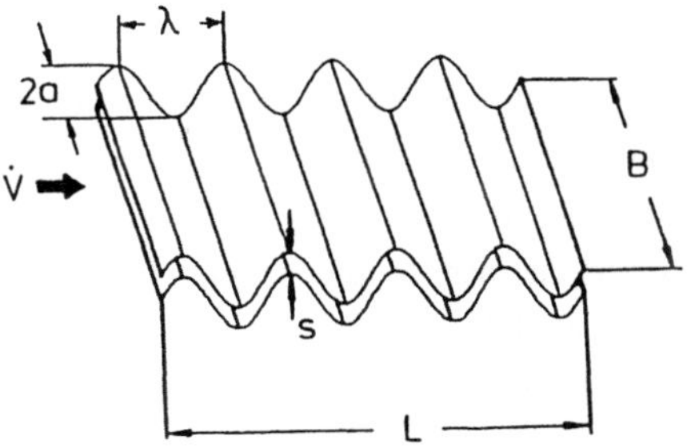

Fig. 1 Geometrical parameters of the sinusoidal wavy duct

Fig. 2 shows the wind tunnel. The flow processes strongly depends on the distance x from the entrance in the wavy duct. The analysis of this parameter is limited by the duct length L = 400 mm.

MEASURING TECHNIQUE

For the visualization and determination of the local mass transfer a physico-chemical method is used based on absorption, chemical reaction and coupled colour reaction. The surface is coated with a wet filter paper. The mass to be transferred (ammonia) is added as a short gas pulse. The locally transferred mass is visible as colour density distribution and the colour intensity corresponds directly to the locally transferred mass. The wall tracing by mass transfer allows also direct information on the local flow processes, especially on the vortex formation and flow separation.

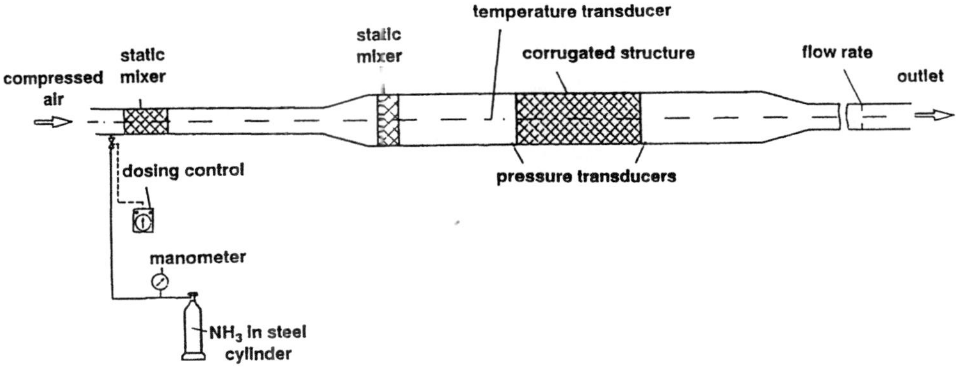

Fig. 2 Wind tunnel with mass dosing unit.

MASS TRANSFER DISTRIBUTION

Fig. 3 shows three mass transfer distributions for Re = 100, 250 and 1000 from top to bottom. In all cases, only one side of the wavy duct is coated with filter paper. At the lowest Re-number Re = 100 (top) the flow is two-dimensional. At the crests of the duct in the convex part the mass transfer is high, in the valleys - in the concave part - the mass transfer is poor caused by the two-dimensional laminar separated flow. At the higher Re-number Re = 250 (Fig. 3, central) the centrifugal instability along the concave part of the duct leads to longitudinal vortices over the whole duct length. One vortex pair leads to one dark streak with high mass transfer. Then the distance between two minima corresponds to the vortex wavelength λ_V. The vortices can be observed behind the concave part and end behind the crests of the wavy duct. Here the vortices are amplified at the opposite side of the duct

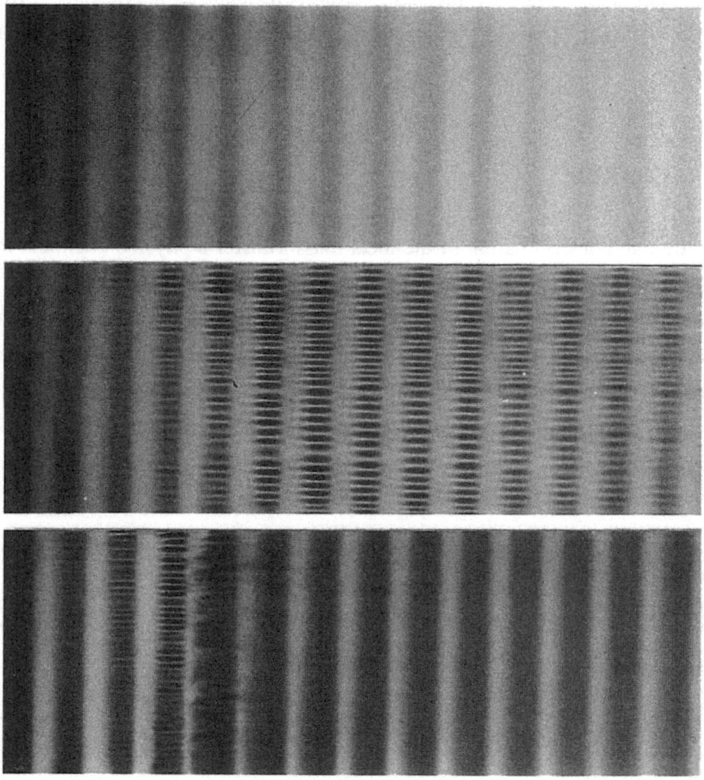

Fig. 3 Mass transfer distribution in sinusoidal wavy ducts ; $\lambda/a = 14.25$; $s/a = 2$; Re = 100, 250 and 1000 from top to bottom.

(not shown here). Following the flow direction, the vortex pairs have a phase shift of 1/2 λ_V. Thereby the vortex diameter $\lambda_V/2$ is in the order of one half of the duct heigth s. If the flow velocity is further enhanced (Fig. 3 bottom), the longitudinal vortices are amplified in a shorter distance from the duct entrance. But the existing range downstream now strongly depends on the Reynolds-number. The higher the Re-number, the shorter the existing range of the vortices. At higher Re-numbers the effects of flow separation suppress the formation of longitudinal vortices. The finite duct length of L = 400 allows no information at Re = 250 on the existing range of the longitudinal vortices further downstream (absolute instability?)

STABILITY DIAGRAM

In a stability diagram (Fig. 4) corresponding to the Taylor, Goertler and Dean problem the centrifugal forces are represented by an analogous Goertler-parameter Re $\sqrt{s/R}$, where R is the smallest radius R_{min} in the valley, as also used by Nishimura et.al. /4,5/. The wave number parameter is $2 \pi s / \lambda_V$.

Normally the wavelength is almost not changed over the whole duct length. Then the wave number parameter is constant and the longitudinal vortex flow in Fig. 4 is always represented by one point. Therefore all results - independent on the location in the duct - are shown in Fig. 4 for different duct heights s/a and Re-numbers. At small duct heights s/a = 1 the wavelength and therefore the wave number parameter is constant. At larger duct heights the dimensionless wavelength λ_V /s decreases with increasing Re-number and duct height s/a. At low Goertler numbers the neutral curve corresponds to the critical Re-number for the transition from two to three-dimensional flow (Fig. 3 top to central). The results of Nishimura et.al. /4/ measured at the end of the test section at a fixed point in the duct are in good agreement with the new experiments.
After the transition from laminar to turbulent flow, longitudinal vortices can also be observed in turbulent flow. The wavelength is then larger as in the laminar and transitional flow. This case is not included in the stability diagram (Fig. 4).

At small duct heights and high Re-numbers an other kind of vortex formation appears. Here the phase shift disappears and the vortices can be observed for a longer distance downstream over several crests and valleys.

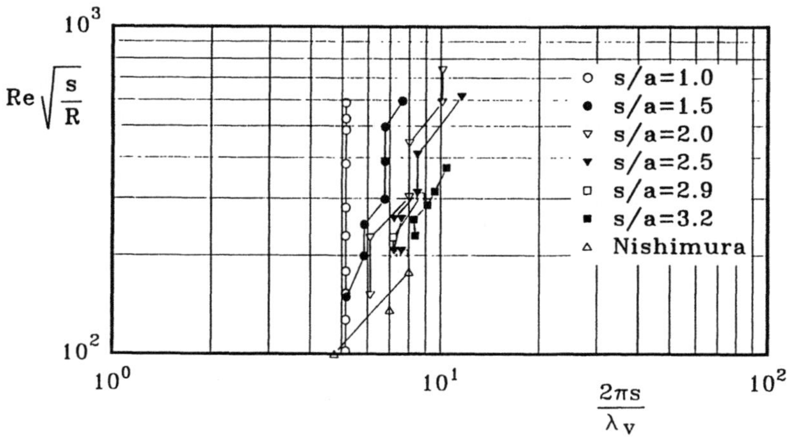

Fig. 4 Stability diagram for sinusoidal wavy ducts ; $\lambda/a = 14.25$. Comparison with /4/.

REFERENCES

1. Gaiser, G.; Schüz, G.; Kottke, V.; Blenke, H. : Zur Verteilung des Wärmeübergangs in Plattenwärmeübertragern. Forschungsbericht Institut für Chemische Verfahrenstechnik, Universität Stuttgart, Nov. 1986.

2. Gaiser, G.; Kottke, V.; Blenke, H. : Strömungsvorgänge, Druckverlust und Wärmeübergang in Plattenwärmeübertragern unterschiedlicher Abmessungen und Oberflächenstrukturen. Forschungsbericht Institut für Chemische Verfahrenstechnik, Universität Stuttgart, März 1988.

3. Kottke, V. : Effects of (quasi-) steady velocity distribution behind screens and grids on local heat and mass transfer from flat and concave plates. Proc. First Symposium on Goertler vortex flow, Euromech 261, Nantes 1990.

4. Nishimura, T.; Yano, K.; Yoshino,T.; Kawamura; Y. : Occurence and structure of Taylor-Goertler vortices induced in two-dimensional wavy channels. J. Chem.-Eng. of Japan, 23 (1990) 697-703.

5. Nishimura, T.; Yoshino,T.; Kawamura : Instability of flow in a sinusoidal wavy channel with narrow spacing. J. Chem.-Eng. of Japan, 20 (1987) 102-103.

Determination of Temperature Fields Above Plates with Different Geometry Heated by Moist Air

S. Urgela - O. Lupták

Technical University in Zvolen, Department of Physics and Applied Mechanics, T.G.Masaryka 24, 960 53 Zvolen, Czecho-Slovakia

Summary

The heat transfer coefficients at the different profiles (plane, saw-toothed, rippled, rectangular) of the aluminium plate are obtained by means of holographic interferometry. The plate is heated by the moist air. The most convenient profiles have been selected to be used in the recuperator.

Introduction

Recuperation of low temperature heat energy and exploitation of new sources of energy require the enhancing of our knowledge about heat transfer through the walls of varied profiles. The intensity vector of the heat transfer depends on their geometry. Moist air frequently represents a heat transfer medium and in this case water condensation originates on the wall. The solution of the Fourier equation of heat transfer is complicated in this case and that is why we decided to solve this problem by visualization of the temperature field in the surroundings of the wall - outside the chamber - by means of the holographic interferometry.

Experiment and discussion

The scheme of the experimental equipment is shown in Fig.1. The heat plate and the water reservoir are placed into the chamber with vertical thermal insulating boards. The upper horizontal wall is constructed from a thin aluminium sheet. The thickness of the sheet is 0.099 mm. Heat has an effect on water evaporation. There is moist air in the volume surrounded by water, the aluminium sheet and the insulting boards. The experimental model creates quasi - steady state with defined temperature and dampness inside the chamber The heat flow from the wall takes place by free convection. The experimental model is placed into object beam of the holographic interferometer. Interferograms of warm air above the aluminium plate are interpreted as two-dimensional ones [1].

The profiles of the sheets can be different, e.g. plane, saw-toothed, rippled or rectangular. Fig.2a,b,c,d illustrate four profiles which were investigated. The tempera-

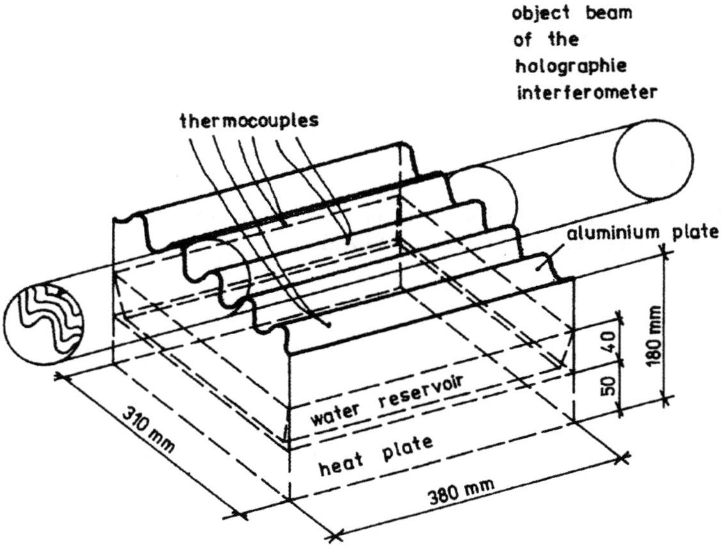

Fig. 1. The scheme of the experimental equipment

ture fields in the surroundings of the profiles were measured under conditions inside the chamber which are given in Table 1. These temperatures correspond with the real recuperation values. The visualized temperature fields are shown in Fig.3a,b,c,d (measurement number 2,4,6,8). The interference fringes correspond to the isoterms of temperature field.

In the steady state the convection heat transfer is equal to the conduction heat transfer through the thermal boundary layer. It may be quantified by the expression

$$\alpha = \lambda \frac{\text{grad } T}{T_\infty - T_w}, \qquad (1)$$

where λ is the thermal conductivity factor, T_∞ is the temperature of the surrounding medium and T_w is the temperature on the surface. The local values of grad T and T_w are determined from the interferograms in the elementary lines designated by Roman numerals in Fig.2a,b,c,d. The mean value of the heat transfer coefficient can be expressed by the relation

$$\alpha_{av} = \frac{\Sigma \alpha_i \, l_i}{\Sigma l_i}, \qquad (2)$$

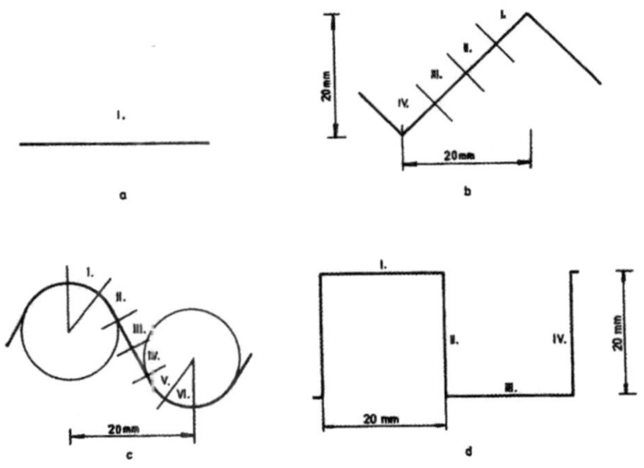

Fig.2a,b,c,d. The profiles of the aluminium plate

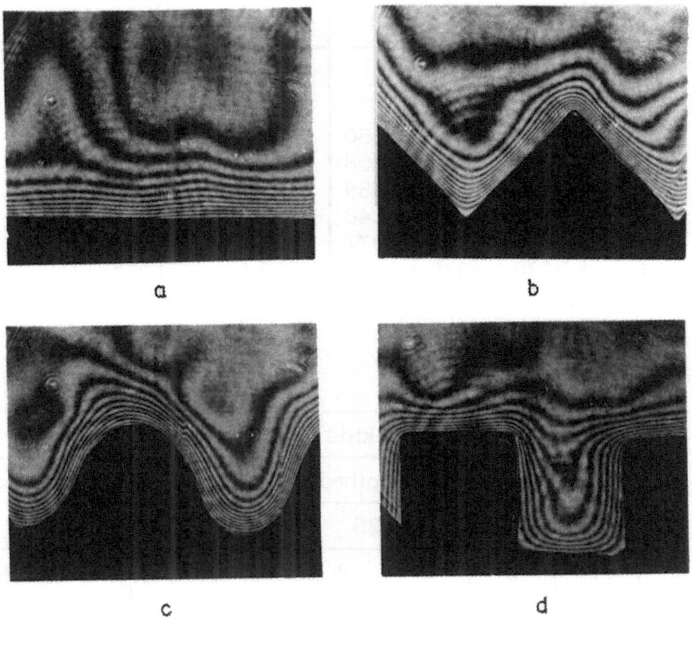

Fig.3a,b,c,d. Interferograms of temperature fields above plates drawn in Fig.2a,b,c,d where α_i and l_i represent the heat transfer coefficient and length of the i-th elementary line designated by Roman numerals. The temperatures on the surface and the resulting temperature gradients are given in Table 1 and 2. The mean values of the heat transfer coefficient are given in Table 3. λ is taken from [2]. These values are

Table 1. Condition of the measurements

measure-ment number	temperature outside the chamber T_∞	temperature inside the chamber	temperature on the Al plate T_w	temperature of water in reservoir
1,3,5,7	297.7 K	313.5 K	309.0 K - 313.0 K	315.0 K
2,4,6,8	297.7 K	325.3 K	320.0 K - 324.0 K	330.0 K

Table 2. Local temperature gradients

measu-rement number	grad T (K m^{-1}) in the area					
	I	II	III	IV	V	VI
1	1649	-	-	-	-	-
2	2949	-	-	-	-	-
3	1539	1631	1560	1320	-	-
4	3442	3387	3208	2559	-	-
5	1424	1549	1589	1506	1404	1365
6	2796	3081	3246	3136	2965	2913
7	1181	821	977	821	-	-
8	2728	1845	2510	1845	-	-

Table 3. Heat transfer coefficients

	kind of profile							
	plane		saw-toothed		rippled		rectangular	
S	1		1.425		1.475		2	
measur. num.	1	2	3	4	5	6	7	8
α_{av} ($Wm^{-2}K^{-1}$)	3.08	3.18	2.83	3.09	2.65	3.23	2.00	2.44
α_{res} ($Wm^{-2}K^{-1}$)	3.1	3.2	4.0	4.4	3.9	4.8	4.0	4.9

multiplied by the coefficient S (proportion between the plane surface and the surfaces of the other three profiles). The resulting mean values of the heat transfer coefficient α_{res} are given in the bottom line of the Table 3. Heat transfer, in this case, is influenced by the layer of the condensate. Thickness of this film (average thickness is approximately 0,1 mm per hour) is uneven and it is a function of the profil kind, of the space coordinates and of time (in quasi-steady state too). This fact and the rise of the turbulences in air 1 - 2 cm above the aluminium plate increase the dispersion of the measured values. Therefore, to obtain the values (grad T and T_w) 5 interferograms have been used for every measurement (1-8).

The radiant energy from the aluminium plate is more than 12 times smaller when compared to the free convection heat transfer. It is calculated by means of the data given in [3]. The termal resistance of the layer of the condensate is 1000 times smaller than the quantity $1/\alpha$. The heat transfer coefficient inside the chamber α_1 and the heat transfer coefficient outside the chamber α_2 can be expressed as [3]

$$\alpha_1(T_{in} - T_w) = \alpha_2(T_w - T_\infty) , \tag{3}$$

where T_{in} is temperature inside the chamber.

Conclusions

When considering the conditions presented we can state that the most convenient profiles of the heat exchange surface of the recuperator are the saw-toothed or the rippled ones. Although the differences between the rectangular, the saw-toothed and the rippled profile are undistinguishable the rectangular profile is made of a larger amount of material. To improve the properties of the rectangular profile it would be necessary to change the proportions of its sizes.

References

1. Vest, Ch.M.: Holographic Interferometry, John Wiley & Sons, New York, 1979

2. Gray, E.: American Institute of Physics Handbook, McGraw-Hill Book Company, New York 1972

3. Bird, R.B.; Stewart, W.E.; Lightfoot, E.N.: Transport Phenomena, John Wiley&Sons, New York, 1965

Photochromic Flow Visualization in Non-Isothermal Liquid-Liquid Two-Phase Flow

C. MANSSON, J. MADDREN, and E. P. MARSCHALL

Department of Mechanical and Environmental Engineering
University of California
Santa Barbara, CA 93106

Summary

The photochromic flow visualization technique was used to study non-isothermal liquid-liquid two-phase flow.

Introduction

Drop formation and release as well as the movements of drops or drop assemblies in liquid-liquid two-phase systems has been studied extensively during the past decades. Summaries of many of these studies can be found in review articles or in books, for instance in Ref. [1]. Almost without exception, flow characteristics of the liquid-liquid two-phase systems are described assuming that temperature variations within the system have no influence or only a negligibly small influence on the flow fields. This assumption has been recently challenged by Hutchins et al [2,3] and Maddren et al [4]. Hutchins carried out a numerical analysis of heat transport to a liquid drop which is rising in a second immiscible liquid. He assumed that the liquid-liquid interface is rigid so that movements inside the drop must be attributed to either the drop formation and release process or to free convection. His numerical predictions were compared with experimental heat transfer results and excellent agreement was found between predicted and experimentally obtained Nusselt numbers.

The assumption of a rigid liquid-liquid interface is somewhat controversial, however, not more so than the popular assumption of a fully mobile interface. Many studies have shown that surface-active substances have a marked influence on the motion of liquid-liquid interfaces. Clift et al [1] provided an extensive review of the interface behavior and of the effects of traces of surface-active contaminants on the characteristics of drops and bubbles. Evidently, purifying liquids and maintaining them free of contaminants is so difficult that in most situations of industrial importance the presence of some contaminants must be expected. Therefore, a fully mobile interface constitutes a limiting or ideal situation, which is rarely observed to exist. The other limiting case, the rigid interface, has shown to exist for small drops and for pairs of immiscible liquids with a high interfacial tension. The following study is concerned with the influence of free convection on the drop formation and release process in a non-isothermal environment. The study is, at the present time, of qualitative nature and it is based on flow visualization experiments.

Experimental Technique

Radiachromic flow visualiation techniques have established themselves as reliable and useful experimental techniques. Especially the photochromic flow visualization methods are useful tools for determining qualitatively and quantitatively flow fields in small test volumes. Detailed descriptions of the photochromic techniques may be found in Ref. [5-8]. The technique uses soluble radiachromic, mostly photochromic, dyes that, in the absence of short wavelength light, are colorless when in solution in many liquids. The dye may be selectively activated by an ultraviolet light source to produce a bright color. When the light source is a sharply focused beam of sufficient power, a narrow and well defined colored line may be produced in the dye and solvent solution. The subsequent motion and deformation of the line may be either simply observed or they may be recorded with still photography or high speed photography for later analysis.

Even though thousands of radiachromic substances are known, only a few have proven to be suitable for flow visualization experiments. The dye used in the experiments described here was 1′, 3′, 3-trimethyl-6′-nitroindoline-2-spiro-2-2′-benzopyron, also known as TNSBP. TNSBP is soluble in most aliphatic and aromatic solvents, and many other nonpolar liquids, it activates in less than 3 microseconds, and it will remain in the colored state up to and in excess of thirty second depending on temperatures. It is a true photochromic dye, that is, once it has been activated it will deactivate after a certain period of time and return to the colorless state.

Drop formation experiments were all carried out in a single-nozzle vertical spray column. The continuous phase was either pure water with AMSCO petroleum solvent serving as the dispersed phase, or it was a solution of ethanol and water with mineral oil serving as the dispersed phase. The spray column was fitted with windows so as to allow one the use of a nitrogen pulse laser as the light source. Because the emission of the nitrogen pulsed laser is at a wavelength of 0.3371 μm, which is close to the visible range of wavelengths, many of the optical problems normally associated with ultraviolet absorption in optical components are not present. As a result lenses and windows of crown or flint glass work sufficiently well. Windows were also provided for visual observation and for producing still photos or high speed movies.

Isothermal Drop Formation

Isothermal drop formation has been studied extensively. Even so, prediction of drop diameter, drop volumes, or release frequency range only from good to marginal. The reasons for this include uncertainties about the liquid-liquid interface conditions, sensitivity to nozzle design and nozzle material, and difficulties in describing correctly the drop dynamics. The flow fields inside forming drops depend strongly on the drop formation and release rate. Figure 1 shows a typical sequence of AMSCO-solvent drops forming slowly in water. In slow drop formation, the drop rise velocity during detachment is much larger than the mean velocity in the dispersed phase nozzle. As can be seen from the distortion of the initially straight dye lines, very little recirculation exists in the developing drop. However, after drop release, the lower end of the drop is accelerated towards the drop center thereby enhancing strongly the initially weak circular motion within the drop. The situation is somewhat different when drops form at a high rate. In this case the drop rise velocity during detachment is not much larger than the mean velocity in the dispersed phase nozzle. In contrast to the slow drop formation process, strong recirculation patterns exist while drops form rapidly. Countless variations of the basic flowfield shown in Fig. 1 have been studied. They include assymmetric flow fields, drop formation in cross-flow, drop formation at nozzles which were constructed from wetting materials and drop formation from nozzles containing obstructions and drop formation at short length nozzles.

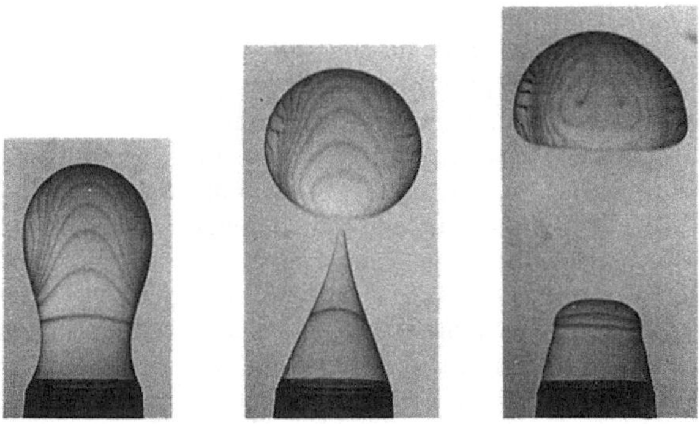

Fig. 1. Isothermal Drop Formation

Most detached drops in these flow visualization experiments had equivalent diameters in the range of 5 to 8 mm. Common to all these experiments was that the drops had initially a strong recirculation pattern right after detachment which was either caused by the release process or by a high flow rate in the dispersed phase nozzle.

Non-Isothermal Drop Formation

Little attention has been given to drop formation in a continuous phase with a temperature different from the temperature of the dispersed phase. Figure 2 shows the slow formation of AMSCO-solvent drops in water. The water temperature was about 20°C higher than the solvent temperature at the nozzle exit. The nozzle was well insulated so as to minimize heat conduction across the nozzle walls. The figures show that strong free convection flows exist in the developing drops. As the fluid at the liquid-liquid interface within the drops is heated up, it flows upward along the interface, thereby creating a flow field which is entirely different from the flow fields usually observed in isothermal flow. While in isothermal drop formation there is a strong flow upwards along the nozzle axis and, depending on the flow rate in the nozzle, a flow downwards along the liquid-liquid interface, in non-isothermal slow drop formation, this flow pattern is reversed. At medium rates of drop formation free convection currents and forced convection currents balance each other in such a manner that the flow field inside the developing drop has the appearance of slug flow. At high drop formation rates, approaching jetting, the forced convection flow dominates again and flow fields within forming drops are similar to the flow fields in isothermal drop formation.

Fig. 2. Non-isothermal Drop Formation

Figure 3 presents mineral oil drops at various states during formation and after detachment. The locations of the detached drops are approximately 2, 15, and 30 cm above the nozzle. Even though strong free-convection flow can be observed during drop formation, forced convection recirculation patterns caused by the release process dominate after the detachment. These recirculation patterns are very similar to the ones observed in isothermal flow. Therefore, a reasonable assumption in any numerical analysis of heat transfer right after drop detachment is the assumption of the observed recirculation, even though free convection flow may have dominated during the formation process.

Conclusion

The photochromic flow visualization technique was used to study qualitatively drop formation and release in liquid-liquid two-phase flow. While free-convection effects clearly affect the flow inside developing drops, immediately after release, the flow pattern inside drops are caused mainly by the formation and release process.

References

[1] Clift, R., Grace, J.R., and Weber, M.E., Bubbles, Drops, and Particles, Academic Press, New York, 1978.

[2] Hutchins, J., Moresco, L., Pickens, K., and Marschall, E., "Prediction of Heat Transfer Rates in a Liquid-Liquid Direct-Contact Heat Exchanger," in *Design and Operation of Heat Exchangers, Proceedings of the Eurotherm Seminar*, No. 18, pp. 420-439, March 1991.

[3] Pickens, K., Hutchins, J., and Marschall, E., "A New Model of Liquid-Liquid Direct-Contact Heat Transfer," *ASME/JSME Thermal Engineering Proceedings*, Vol. 2, pp. 235-240, ASME 1991.

[4] Maddren, J., Mansson, C., and Marschall, E., "Photochromic Flow Visualization in Isothermal and Non-Isothermal Liquid-Liquid Two-Phase Flow," *FED*, Vol. 125, Experimental Techniques in Multiphase Flows, pp. 27-31, ASME 1991.

[5] Popovich, A.T. and Hummel, R.L., "A New Method for Non-Disturbing Turbulent Flow Measurements Very Close to a Wall," *Chem. Engng. Sci.*, Vol. 22, pp. 21-25, 1967.
[6] Johnson, G., Marschall, E., and Esdorn, J.H., "Improved Flow Visualization Technique for Quantitative Velocity Measurements in Small Test Volumes," *Rev. Sci. Instrum.*, Vol. 56, pp. 264-267, 1985.
[7] McWaid, T. and Marschall, E., "An Improved Photochromic Flow Visualization Technique," *Experimental Thermal and Fluid Sciences*, Vol. 3, No. 2, pp. 232-241, 1990.
[8] Yourechko, V.N., Ryazanlzev, Y.S., and Ilminsky, V.N., "The Method of Photochromic Visualization in Liquid Flows," in Flow Visualization V, Ed. R. Reznicek, pp. 65-71, Hemisphere, New York, 1990.

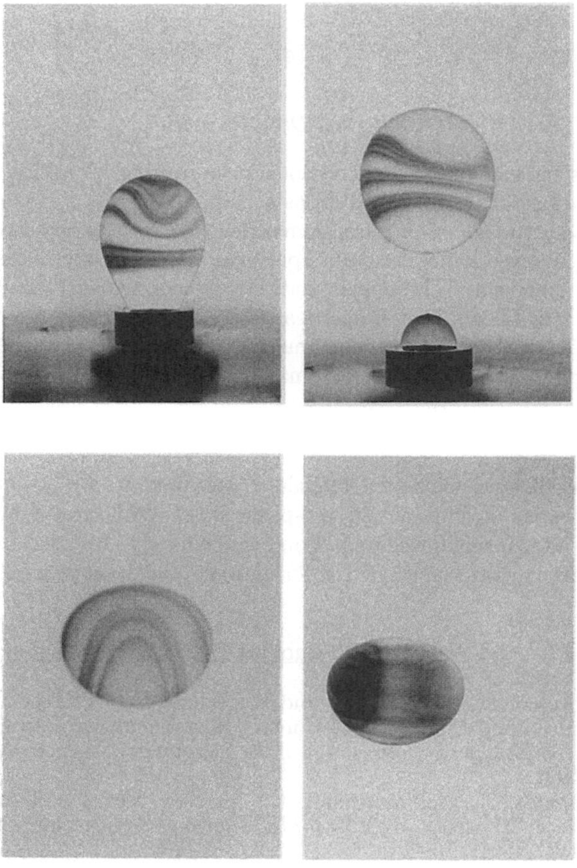

Fig. 3. Non-isothermal Drop Formation

Droplet Dispersion in Turbulent Flow

P. HEBRARD, Y. BISCOS, P. BERTHOUMIEU, P. TRICHET
CERT - ONERA / DERMES
2, Avenue E. Belin - BP. 4025 - 31055 TOULOUSE CEDEX (FRANCE)

SUMMARY

A new experimental method using a high speed video camera associated with an image processing system is presented. Some examples dealing with the use of this system to study the dynamic behaviour of liquid droplets during their transport (turbulent dispersion, influence of coherent structures, ...) or wall interaction (rebound, transformation droplet -> film, vaporization, ...) are given.

INTRODUCTION

In gas turbine combustors, afterburner, ramjet or ramrockets, the fuel injection process plays an important role on the performances : stability range, combustion efficiency, NOx production ... The actual 3D numerical codes which are developed to predict these performances need to use accurate boundary conditions and data base to validate the modeling of the liquid phase. To answer this question, CERT-DERMES has undertaken a research program to study some of the elementary phenomena occuring in combustion chambers for the liquid phase : liquid sheet disintegration, spray characteristics, turbulent and unsteady dispersion, vaporizing, ...[1, 2, 3].

In this paper, some examples using a high speed video camera and associated image processing to obtain 2D individual droplets trajectories are presented. The application deals with droplet turbulent dispersion and hot wall - droplets interaction.

FACILITIES AND INSTRUMENTATION

. Droplet generator

In order to get precise and well known initial conditions for the liquid phase (drop size and velocity), monosized droplets are injected with a TSI Vibrating Orifice Aerosol Generator. This system, which was calibrated at CERT with a two components Aerometrics PDPA, provides a regular stream of given droplets velocity and size (Fig. 1).

. Facility for turbulent dispersion study (Fig. 2)

A vertical wind tunnel (150 x 150 x 1500 mm^3) in which variable air velocity ($V \leq 10$ m/s) and turbulence level and scale are generated by various calibrated grids is used. Calibrated droplets can be injected laterally in this air flow and visualized.

. Hot plate (Fig. 3)

An horizontal hot plate (200 x 150 x 40 mm^3) made of copper, in which five heating cartridges are imbeded to produce a uniform wall temperature (Tw $\leq$ 600° C) is placed downstream an horizontal wind tunnel (V $\leq$ 5 m/s). The wall temperature uniformity is checked by an infrared camera and its value stabilized by a loop regulation using three thermocouples inside the plate.

. Droplet visualization

To observe the droplets behaviour, a high speed video camera (resolution 192 x 240) KODAK EKTAPRO 1000 ($\leq$ 1000 frames/s full frame or 6000 frames/s for 1/6 full frame) is used with a shadowgraphy mode with a continuous or synchronized pulsed light source. Following the lighting conditions and droplet velocity, an intensified imager and one single shot camera can be used. Images, stored on a special video recorder, can be viewed frame after frame and processed afterwards.

. Image processing

From the video record, simplified digitized images are stored and processed by the image analysis system [4]. This system, which uses various DATACUBE cards and a VME computer (Fig. 4), can provide at each frame the 2D coordinates of each elementary droplet (images subtraction, thresholding, ...). From these elementary coordinates, various programs allow to compute the 2D trajectories, drop number density, droplet velocities, ...

EXPERIMENTAL RESULTS

. Turbulent dispersion

An example of spray visualization corresponding to the following conditions : air velocity 3,6 m/s, turbulence level : 10 %, drop size : 250 µm, drop velocity : 4,5 m/s is given on figure 5. From image analysis (Figure 6) of a sufficient number of elementary frames, the individual trajectories of droplets can be obtained. These results are also completed by various statistics such as : mean trajectory, standard deviation and local density rate in a downstream section (Fig. 7 and 8).

This kind of experiment is particularly useful to study the effect of the air flow turbulence on the dispersion of elementary droplets. It can also be used to validate various numerical simulation or stochastic modeling approaches (Ref. 5) in order to choose an optimized model to be introduced in the numerical codes for combustion chambers.

. Droplet - Wall interaction

From the analysis of droplets visualization for various impinging conditions (droplet size and normal incident velocity) and wall temperature, the dynamic behaviour of the impinging droplets can be studied (Ref. 6) : rebound, deformation, Leidenfrost phenomena, vaporizing.

A particularly useful result to be used as boundary conditions in the computation codes concerns the droplet rebound. From the computed trajectories of individual droplets before and after the rebound (Fig. 9). It is possible to compute, for example, the rebound angle as function of the incident one (Fig. 10) or the rebound velocity (or Weber number $We = \rho d\, V^2/\sigma$) as function of the incident one) (Fig. 11).

CONCLUSION

From this new method which combines a high speed video and an image processing, the possibility to obtain various experimental results on the behaviour of liquid droplets is demonstrated. This method, provides useful results to initiate or validate numerical codes on combustion chambers.

ACKNOWLEDGMENTS

Part of this research was supported by the french Ministry of Defence and Turbomeca.

The authors appreciate the helpfull discussions provided by G. LAVERGNE and gratefully acknowledge Mr B. PLATET and F. BISMES for development of test rigs and for data acquisition.

REFERENCES

1 P. HEBRARD, G. LAVERGNE, P. TRICHET
Experimental study of boundary conditions for spray systems in combustion chambers
6th ILASS - Europe, Pise - July 1990

2 P. HEBRARD, P. TRICHET, X. BARDEY
Experimental investigation of two-phase flow in the near field of an airblast atomizer
5th ICLASS - Gaithersburg - July 1991

3 G. LAVERGNE, P. TRICHET, P. HEBRARD, Y. BISCOS
Liquid sheet disintegration and atomization process on a simplified airblast atomizer
IGTI conference Köhln - June 1992

4 Y. BISCOS, P. BERTHOUMIEU, P. HEBRARD, G. LAVERGNE
Développement d'une méthode de trajectographie 2D de gouttes par traitement d'images obtenues en vidéo rapide.
14ème Colloque national ANRT - Paris - Mars 1992

5 Ph. DONADILLE, A. ROMPTEAUX, G. LAVERGNE, P. HEBRARD
Turbulent dispersion of two-phase flows : experimental and computational approaches
International conference on multiphase flow 91 - Tsukuba - September 1991

6 G. LAVERGNE, P. HEBRARD, Y. BISCOS, P. TRICHET
Experimental study of boundary conditions for two-phase flow in combustion chambers - Spray and aerosols - ILASS Europe - Guildford - October 1991

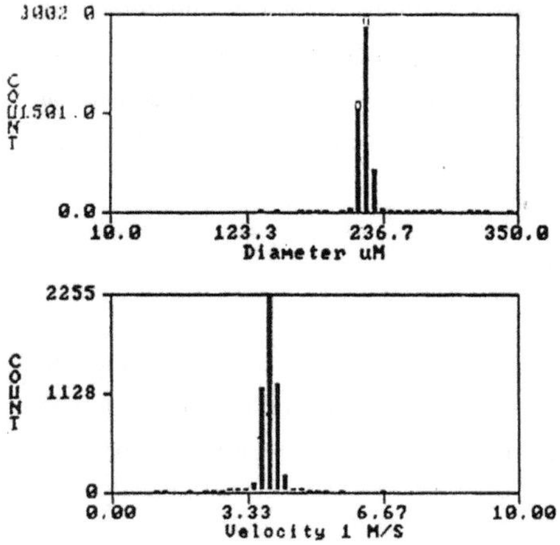

Figure 1 :

Droplet generator size and velocity histogram.

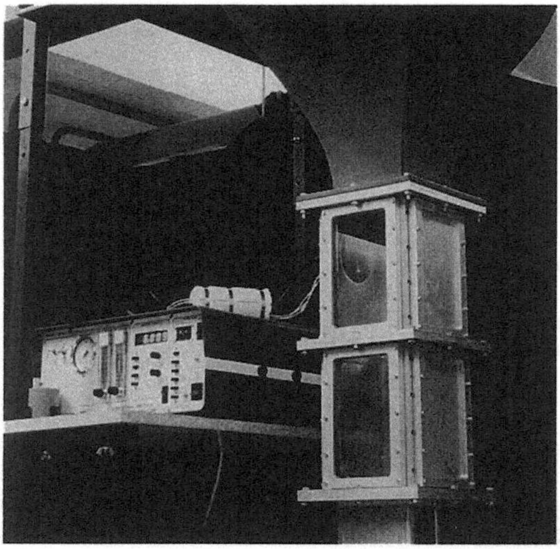

Figure 2 :

Test rig for turbulent dispersion study.

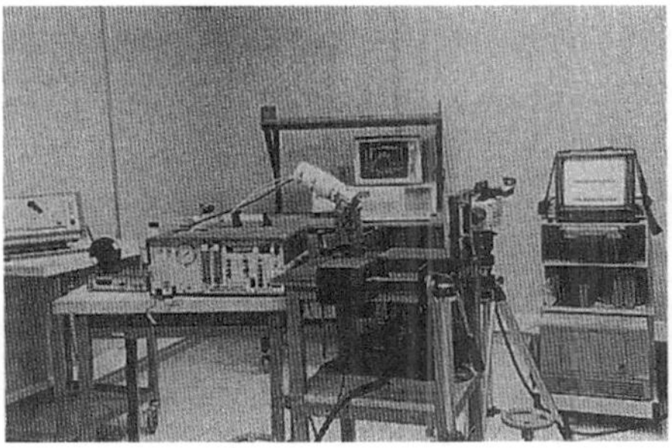

Figure 3 :

Experimental set up for drop-wall interaction.

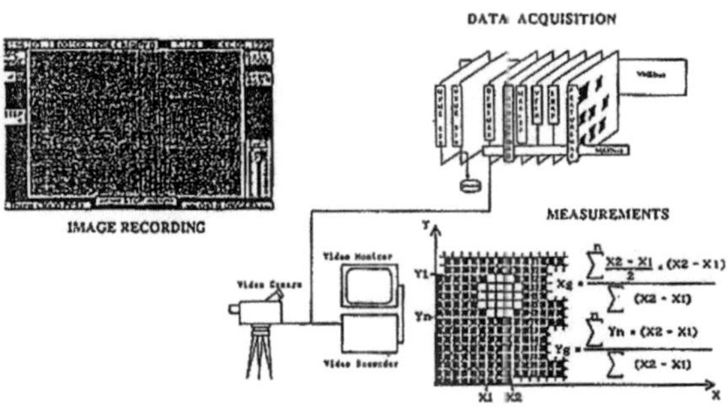

Figure 4 :

Image acquisition and processing system.

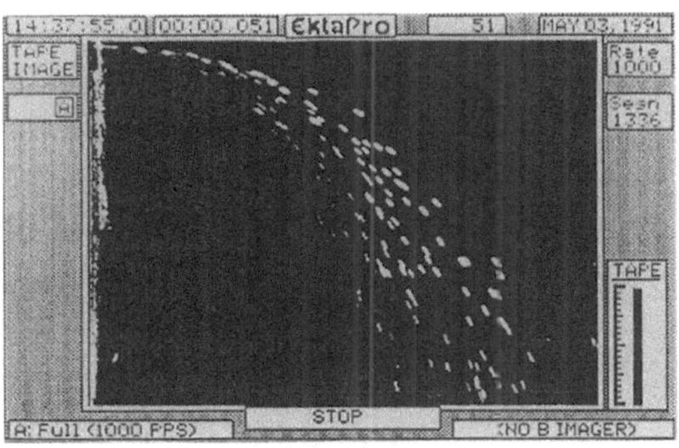

Figure 5 :

Droplet dispersion visualization (Video recording).

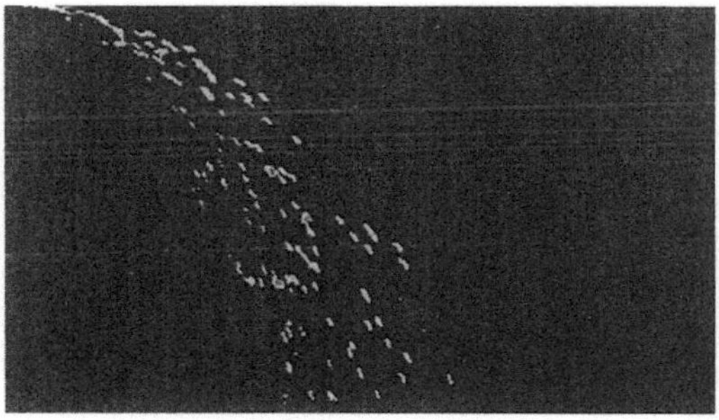

Figure 6 :

Droplet dispersion visualization (Binary image).

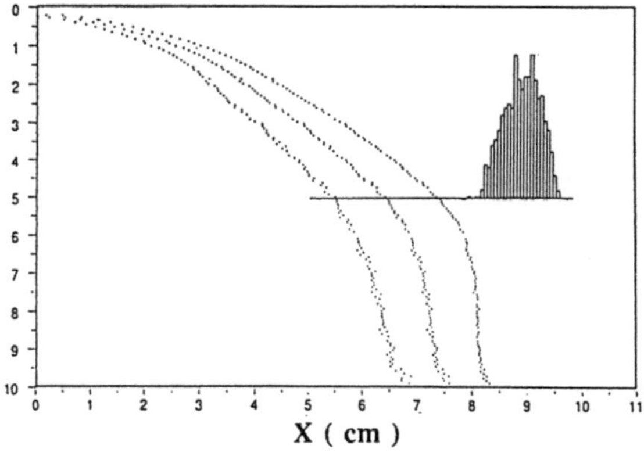

Figure 7 :

Droplet mean trajectory, droplet number density (Large grid turbulence)

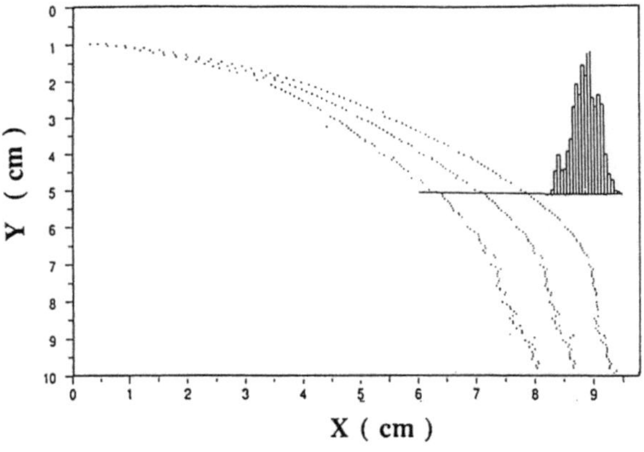

Figure 8 :

Droplet mean trajectory, droplet number density (Small grid turbulence)

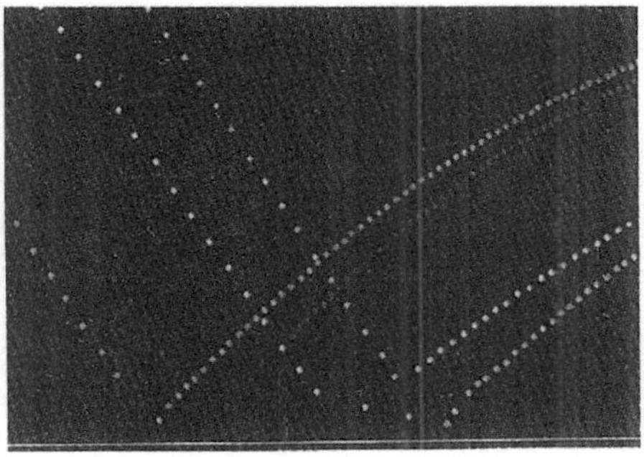

Figure 9 :

Droplet rebound.

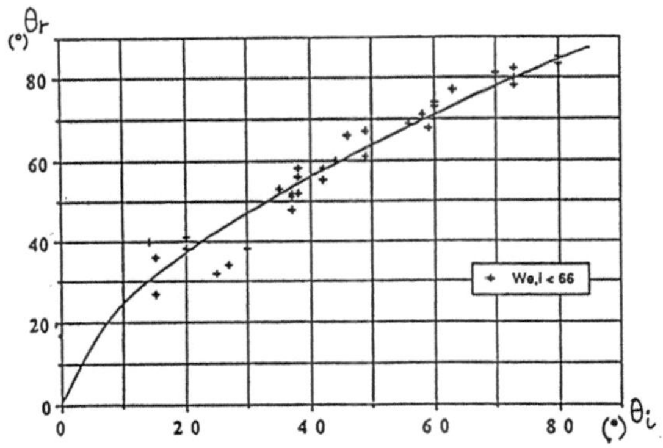

Figur 10 : -

Variation of rebound angle with incident angle.
(Water droplets, Tw=400°C)

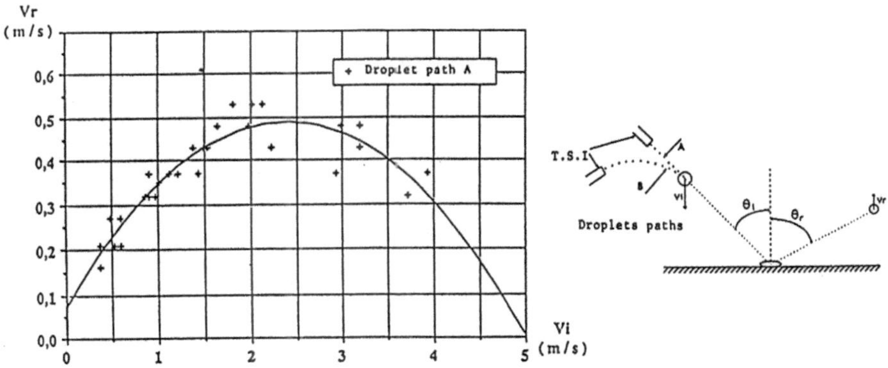

Figure 11 :

Variation of rebound velocity with incident velocity.
(Water droplets ,Tw=400°C)

Combustion

Test Facility for the Flow Structure Study in the Supersonic Combustion Chamber

Baranovsky S.I., Davidenko D.M., Levin V.M.

Moscow Aviation Institute, Moscow, Russia

Summary

A facility for the investigation of flow structure in a supersonic combustion chamber has been constructed and tested. The side walls of the rectangular combustion chamber model can be moved during the experiment to the alternately expose water cooled stainless steel or quartz windows. Window exposure time is limited by thermal loading restrictions on the windows. A Schlieren system is used for visualization. Results are presented for tests using both hydrogen fuel and with air being injected through the fuel pylons.

Introduction

Very little research has been conducted to investigate in detail the flow structure inside the combustion chambers.

The research work of Northam, et al. [1] detailed the flow structure around fuel injectors. For flow visualization in that work, the side walls were removed, altering the basic flow structure and preventing the formation of the system of shocks that result from thermal throttling.

Research on the flow structure in a combustor of the dual-mode scramjet was conducted by Bayev, et al. [2]. This work studied the flow in a closed combustion chamber with a step for the coflow fuel injection and flameholding. The model side walls had quartz windows for visualization of the flow structure. Shadowgraphs and pictures of flame were taken ant parameters inside the chamber were measured. Material limits restricted inlet temperatures to less than 1200 K.

To research the flow structure in a supersonic combustion chamber with pylon fuel injectors and the interaction between the fundamental processes of supersonic combustion, a special experimental facility had to be designed and built.

Description of test facility

The experimental setup consists of a heater, a supersonic nozzle and a combustion chamber model shown in Fig.1.

Fig.1. Combustion chamber model (side walls in central position).

All three parts are water-cooled. The gasoline-fueled heater is used to heat the incoming compressed air. Fresh oxygen is fed into the heater to model the composition of atmospheric air. At the heater's exit the flow has the following stagnation properties: $P_o \leq 30$ atm, $T_o \leq 2000$ K. Interchangeable supersonic nozzles allow the Mach number at the model entrance to be 2.0 or 2.5.

The combustion chamber is 610 mm long and has a rectangular cross section 50x100mm at the chamber's entrance. By inclining the upper and lower walls, the height of the exit section can be varied from 50 to 70mm.

Fuel is injected 240mm downstream from the entrance of the chamber. Gaseous hydrogen is injected perpendicular to the axis of the flow through blunt cantilever pylons with 1mm openings. Fuel is simultaneously injected through six pylons, three on the upper wall and three on the lower wall. The outer diameter of the pylons is 4mm and their heights can be varied from 0 to 20mm.

For observation of the flow structure, 100x400mm quartz windows are at the center of the two side walls of 410x480mm (see Fig.1). A hydraulic system is used to move the walls up or down during the experiment, alternately exposing the windows or water-cooled stainless steel. Clearances between mobile and immobile part of the setup are closed by a system of inflatable rubber tubes.

During the startup of the heater, the side walls are situated such that the windows are not exposed to the chamber. After startup, the walls are moved to expose the windows to the chamber, allowing the observation of the flow. The time the windows are left in this central position is set by the experimental program. The windows are then moved out of the section again. This series of moving the windows in and out of the section significantly reduces thermal loads and soot depositions (from the startup of the heater on the windows).

The setup is equipped with a system for measuring and controlling heater parameters and the pressure distribution on the walls of the chamber. A Schlieren system is used to observe the flow structure. Non-intrusive methods and devices for measuring fields of flow parameters and exit gas compositions will be used in future research.

Results of first test series

The first series of tests was conducted to check out the capabilities of the experimental setup and to obtain initial data on the internal flow structure. For the first series of tests, heater exit parameters were P_o = 7.1 atm and T_o = 1530 K, which correspond to mean static values of P_1 = 1 atm, T_1 = 964 K and M_1 = 1.88 at the nozzle exit. Shadowgraphs and pictures of the flow were taken in tests using both air and hydrogen fuel injection through the fuel pylons. Mass flows of air and hydrogen through the fuel pylons were $\dot{m}$ = 0.0042 and 0.0097 kg/sec respectively, which correspond to injection total pressure P_{jo} = 36.8 and 30.6 atm. Hydrogen mass flow of 0.0097 kg/sec gives an equivalence ratio of ER = 0.16.

A photograph of flow inside the combustion chamber and a Schlieren photograph for air injection are given in Fig. 2.

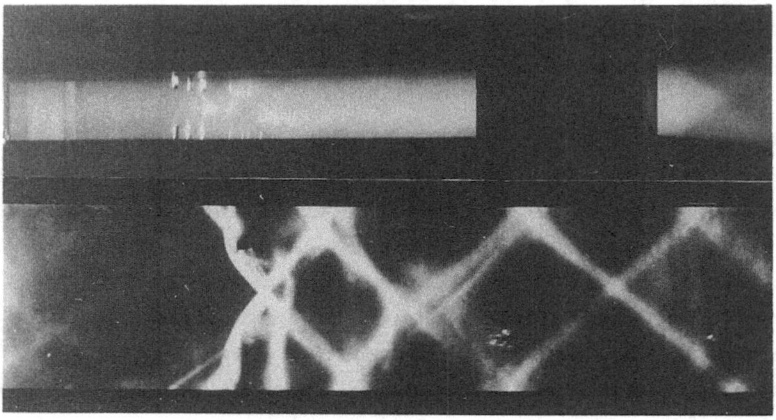

Fig. 2. Air injection.

The upper photograph shows a side view through the entire window into the combustion chamber duct and a part of the exhaust jet at the chamber's exit. All the fuel pylons can be seen. The other photograph shows a shock pattern, caused by pylons and fuel injection. The left end of the view is 70 mm upstream from the pylons and the right one is 160 mm downstream from the pylons. Intense shocks in front of the pylons and jets lead to appearance of separated flow zones on the walls and a reflected shock system. Wall pressure grows slowly to 1.3 atm at the chamber's exit.

A side view of the combustion chamber and a Schlieren photograph for hydrogen combustion are given in Fig. 3.

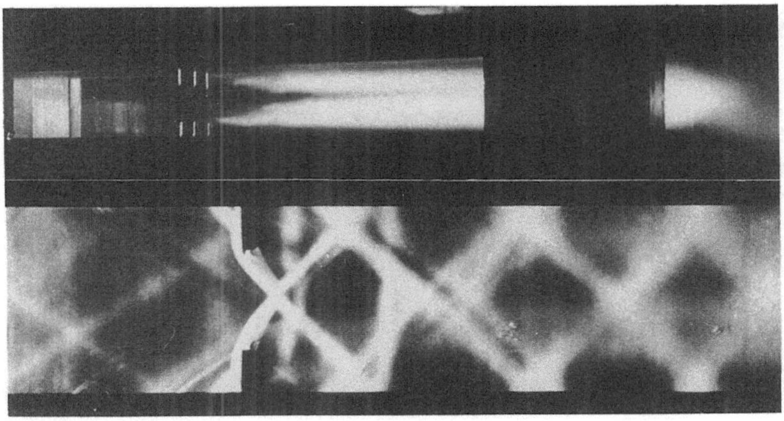

Fig.3. Hydrogen combustion.

The side view shows a bright hydrogen flame inside the combustion chamber. Separated flames from each pylon merge in the common flame occupying the whole chamber's cross section. Fuel ignition occurs spontaneously in the core flow, without any flameholding on the walls. It is important to note, that the heat release does not considerably change shock pattern as far as the right end of the view in spite of the twofold pressure rise toward the chamber's exit. Wall pressure rise is monotonic from the pylons to the exit. Hydrogen injection leads to the contraction of the shock system and to the shocks diffusion. Fuel penetration and spreading can be seen in the Schlieren photograph. The use of the fuel pylons provides the fuel penetration sufficient for rapid mixing and combustion.

References

1. Northam, G.B.; Greenberg, I.;Byington C.S.: Evaluation of Parallel Injector Configurations for Supersonic Combustion. AIAA-89-2525.

3. Baev, V.K.; Golovichev, V.I.; Tretiakov, P.K.; et al.: Gorenie v Sverkhzvukovom Potoke. Novosibirsk: Nauka 1984.

The Structure of Conical Methane-Air Flames

A. A. KONNOV, I. V. DYAKOV, and G. I. KSANDOPULO

Combustion Problems Institute
Alma-Ata, Kazakhstan, CIS

Summary

Relative OH concentration profiles in the cone section of an atmospheric pressure methane-air flames stabilized over Bunsen burner have been obtained on the laser induced fluorescence arrangement. It has been established that the profile of the OH radical concentration in the pre-flame cool zone has a flattened section and deviates from the calculated dependence expectable on the assumption of purely diffusion process. It has been concluded that the structure of the conical flames in the gravity field differs from those of flat flames. These differences are attributed to the occurrence of flow instability in the pre-flame zone of Bunsen flames.

Introduction

Experimental studies of the structure of conical hydrocarbon flames under atmospheric pressure established that in the pre-flame zone the fuel conversion proceeds with an insignificant heating-up of the mixture. For instance, conversion of almost 50 % of propane and other hydrocarbons into intermediates, mainly CO, CO_2, and H_2, is observed when the temperature rise does not exceed 100 K [1,2]. This difference in structure of conical flames from flat ones may be attributed to the difference in chemical processes due to possible entrainment and diffusion of secondary air in the inner cone [3] as well as to hydrodynamic features of Bunsen flames and the influence of gravity on the flame structure. The vertical pumping frequency of Bunsen flame was detected [4] by measuring oscillations of the velocity of cool gases upstream of the flame front. The effect vanishes completely under microgravity conditions. Our previous investigations of propane-air flame [5] now have been extended to get a picture of the effect of environment conditions and gravity field on the structure of pre-flame zone.

Experimental

Hydroxyl concentrations were measured in the 10 mm burner by laser-induced fluorescence. The schematic of the experimental rig is shown in Fig.1. The probing source was a frequency-doubled computer-controlled home-built tunable dye laser which was pumped by a frequency-doubled Nd:YAG laser. The output parameters of the probe laser were as follows: the spectral tuning range from 295 to 330 nm, bandwidth of 0.01 nm, pulse energy 0.03 mJ, pulse duration 10 ns, repetition frequency 12.5 Hz. The probing wavelength could be checked by a grating spectrograph with an OMA with an accuracy of 0.01 nm in the red. The relative intensity of each laser pulse was measured by vacuum photoelements and digitized by ADC. Laser radiation was focused by a lens which made a 100-μm diameter spot in the burner flame. OH fluorescence was collected at a right angle to the exciting beam by a quartz lens and focused onto a slit of monochromator outfitted with an photomultiplier (PM). The PM output was sampled by stroboscopic integrator and digitized by ADC. All measured data were stored in a Schneider EURO PC computer after each laser shot. Besides, the flame luminescence proper was measured with a 2 ms delay after the laser pulse.

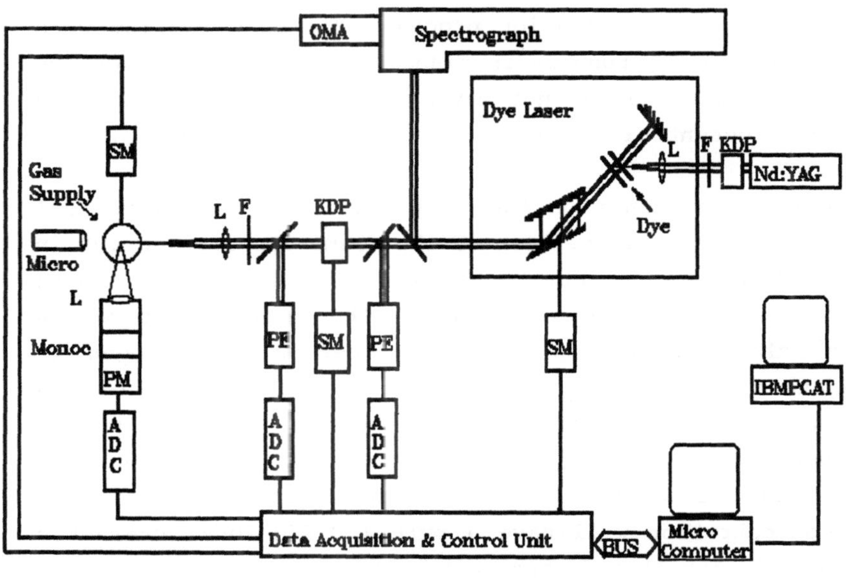

Fig.1. The schematic of the experimental rig.

The fluorescence signal was estimated as a difference between these two measurements normalized to the intensity of the laser pulse. The Bunsen burner was moved with a predetermined step in horizontal and vertical directions by step motors. Thus, the apparatus enables to measure the excitation spectra of OH radicals and scan the flames along the cross section in different directions.

Results and Discussion

It has been established that the profile of OH radical concentration in the pre-flame cool zone has a flattened section. Similar flattened sections are observed for Bunsen flames with different front curvature, and it enables us to rule out the possibility of the signal registration from the neighboring flame zone. The OH rotational temperatures in the pre-flame zone were also measured. These temperatures agree with thermocouple measurements within the limits of a measurement error.

Figure 2 shows the profile of the fluorescence signal obtained with a vertical Bunsen burner (right curve). The drop of the OH radical concentration in the cold pre-flame zone clearly deviated from the calculated dependence expectable on the assumption of purely diffusion process. The same results were obtained for Bunsen propane-air flames [5]. Chemical reactions in the pre-flame zone can not explain of the observed effects due to their essential difference for methane and propane.

A number of experiments on the measurement of the OH radical profiles at horizontal position of the flame was carried out in order to study the potential hydrodynamic features of Bunsen flames and the influence of gravity on the flame structure. In addition, the flame was placed in a co-flow of nitrogen to investigate the effect of surrounding air. The profile of the OH radical concentration in the pre-flame zone practically does not change in this case. On the other hand, if methane or propane flame has the horizontal luminous front, the profile of the OH concentration will have a purely diffusional character (left curve in Fig.2) and agree with literature data [6].

These differences of structure of conical flames from the flames of other shape or from numerical calculations enable us to make conclusion about the effect of stabilization mode on its aerodynamics. In particular, in [6] the front of methane flame is arranged horizontally in gravity field and differs from Bunsen flames. An analysis of two-dimensional Navier-Stocks equation results in that in the field of gravity for vertical or inclined flame front the horizontal pressure gradient, i.e. an added convection arise in the region of significant mixture density gradient. Thus, such a gradient together with the buoyancy of the hot layers can give the little scale vortex and therefore some mixing in the cold region near the flame front.

OH radicals were observed in cold region far from the flame front of methane [7], ethylene [8], propane [5]. This process may be the reason for appearance of OH radicals in methane flame which have non-Boltzmann rotational distribution [7]. It

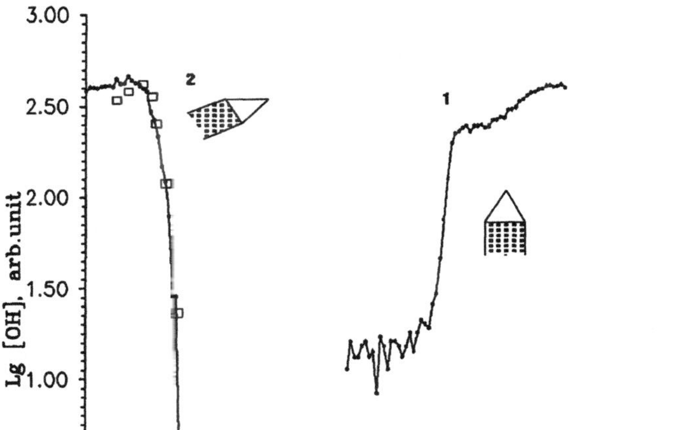

Fig.2. Profiles of the fluorescence signal for methane -air flames (ϕ=1.3). 1. Vertical position of Bunsen burner.
2. Horizontal position of luminous front. Points - OH concentration for flame with ϕ = 1.25 [6].

should be noted that such a flow instability in cold region together with heating from the curved surface [9] may be one of the reason of the increase of burning velocity near the top of the cone.

Conclusion

On the basis of obtained experimental results we can suggest that a local flow instability appears in vertical Bunsen flames located in gravitational field, in schlieren cone region. Such an instability can cause a partial agitation of the mixture in low-temperature region and hence may be a reason of the observed difference between the experimental profiles of the product accumulation for conical and flat flames. Apparently these perturbations are small enough and thereby can not induce unstable propagation of the flame front itself.

References

1. Ksandopulo, G.I. Flame chemistry. Moscow: Khimiya 1980.

2. Manzhos, V.K.; Kolesnikov, B.Ya.; Ksandopulo, G.I.: The structure of the low temperature zone of flames and its dependence on the nature of the combustible mixtures. Archivum Combustionis 8 (1988) 149-165.

3. Claesson, O.; Noda, P.; Yoshida, P.: ESR studies of Bunsen-type methane-air flames.I.Effects of the entrainment and diffusion of secondary air on the chemical reactions in the flame. Bull.Chem.Soc.Jpn. 56 (1983) 2559-2561.

4. Durox, D.; Baillot, F.; Scouflaire, P.; Prud'homme, R.: Some effects of gravity on the behavior of premixed flames. Combust. & Flame 82 (1990) 66-74.

5. Konnov, A.A.; Ksandopulo, G.I.: 23rd Symp. (Int.) on Combustion Abstr. Symp. Pap. The Combustion Institute, (1990) 299.

6. Bechtel, J.H.; Teets, R.E.: Hydroxyl and its concentration profile in methane-air flames.Appl.Opt. 18 (1979) 4138-4144.

7. Wang, C.C.; Davis, L.I.: Ground-state population distribution of OH determined with a tunable UV laser. Appl. Phys. Lett. 25 (1974) 34-35.

8. Schafer, M.; Ketterle, W.; Wolfrum, J.: Saturated 2D-LIF of OH and 2D determination of effective collisional lifetimes in atmospheric pressure flames. Appl. Phys. B52 (1991) 341-346.

9. Lewis B., G. von Elbe: Combustion, Flames and Explosions of Gases. Third ed. Academic Press Inc. 1987.

Analysis of Swirl Formation in the Combustion Chamber of a Direct-Injection Diesel Engine

W. Hentschel, K.-P. Schindler,

Volkswagen Research, Wolfsburg, Germany

Summary
The purpose of this paper is to demonstrate, how in-cylinder flow analysis by means of laser light-sheet techniques and laser doppler anemometry (LDA) can help to understand the interaction of swirl flow and combustion in a direct-injection diesel engine and thus supports the development of engines with lower fuel consumption and emissions.

Introduction

In spite of the high efficiency of the diesel engine there is still a potential for further reduction of fuel consumption and emissions. Therefore, there is a need to improve our fundamental understanding of the very complex processes which take place throughout diesel combustion. This is the aim of the research programme association known as **IDEA** (Integrated Diesel European Action), which is supported by the automobile manufacturers Fiat, Peugeot, Renault, Volvo and Volkswagen, who are members of the Joint Research Committee (**JRC**), by the Commission of the European Communities, and by the National Swedish Board for Industrial and Technical Development.

The combustion in a modern fast running direct-injetion (DI) diesel engine for passenger cars is a non-stationary diffusion controlled process in a non-premixed turbulent flow field. To achieve a good mixture formation the strength of the swirl flow has to be optimized for an engine operation in a large band of speed and load. Therefore, a precise knowledge of the swirl build-up process and of the flow velocity distribution is important, especially at the time of injection.

Experimental set-up for flow investigations

Direct optical access to the combustion chamber is a prerequisite for the application of optical measuring techniques at internal combustion engines. Volkswagen has developed a transparent version of the 1.9 l DI diesel engine which is equiped with elongated pistons to provide a large scale visual access to the piston bowl, both through a quartz glass window in the bowl bottom and quartz glass plates in the upper bowl area; see Fig 1 a and b (for more details see literature [1-3]).

Characteristic features of the DI diesel engine (Fig. 1 c) are the swirl port in the cylinder head, the bowl-shaped combustion chamber in the

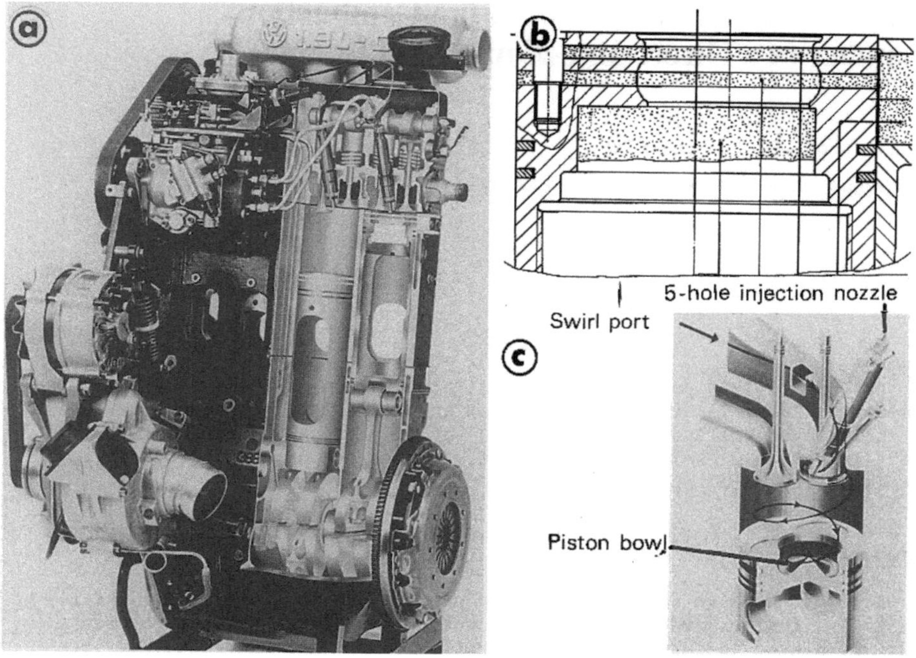

Fig. 1: Transparent 1.9 l DI diesel engine (a) equiped with the sandwich piston for flow investigations in the piston bowl (b) and characteristic features of the design (c).

piston, and the five-hole injection nozzle which is centrally located with respect to the bowl. Measurements were performed at engine speeds of 1000 and 2000 rpm, motored and fired at low load.

Flow field visualization: To observe the development of the flow field inside the cylinder and the piston bowl, laser light-sheet studies were performed with lycopod spores as tracer particles and by using a 20 W argon-ion laser for illumination. The movement (path lines) of the tracer particles were filmed with a high-speed camera (16 mm, 4000 frames/sec) through the window in the piston bowl bottom via a fixed mirror inserted from the side (see Fig. 2). Quantitative information were achieved from the films by particle tracking.

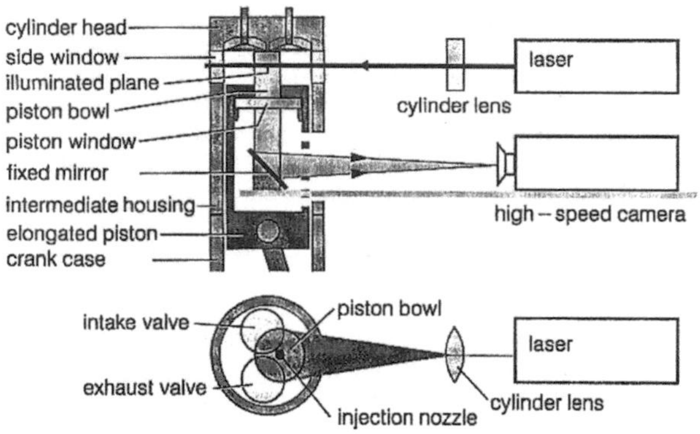

Fig. 2:
Set-up of the laser light-sheet technique for the transparent engine.

Crank-angle resolved flow velocity measurement: To measure the temporal development of the flow field (velocity and turbulence) LDA-measurements were done through the piston bowl window with MgO as a seeding material. A two-component backscatter LDA is used with a probe size of 0.08 mm (diameter) x 1.9 mm (length). A set of measuring locations in a plane 30 mm below the flat cylinder head were scanned automatically by a traversing system. Ensemble-averaged mean bulk velocities and RMS-values are calculated depending on the crank-angle.

Experimental results and discussion

The development of the in-cylinder and in-bowl flow 5 mm below the cylinder head during intake and compression stroke is shown in Fig. 3. The sketches show typical situations at selected crank angles. As long as the intake valve (IV) is open, the flow passes the valve gap with a tangential component - a consequence of the swirl port. Around BDC (low piston position) no significant swirl is visible. At the end of the compression stroke a well-defined swirl flow is developed. At the time of injection (around TDC) the swirl is similar to a solid-body-rotation.

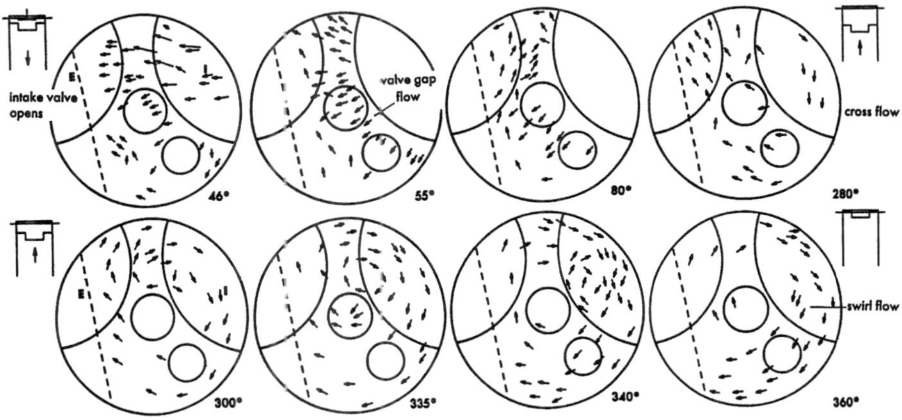

Fig. 3: Sketches of the in-cylinder and in-bowl flow development during intake and compression stroke at 1000 rpm.

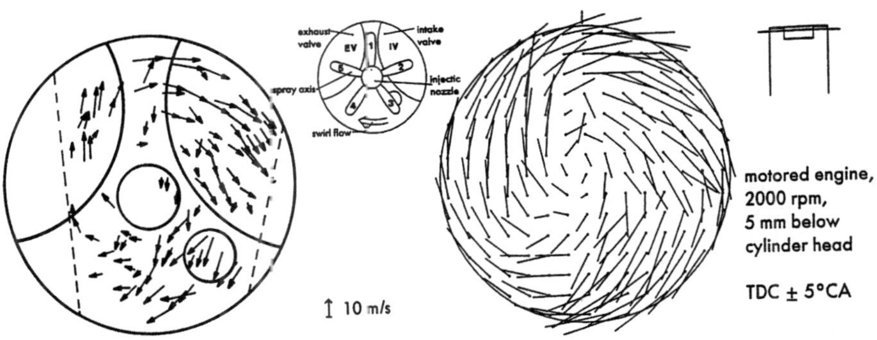

Fig. 4: 2D cycle-averaged quantitative flow field at the time of injection; measured flow vectors (left) and interpolated flow field (right).

A frame-by-frame analysis of a large number of cycles leads to a quantitative cycle-averaged flow field determination at the time of injection. In Fig. 4 the evaluated flow vectors (left) are presented together with the interpolated flow field (right); for the interpolation procedure see [4]. An averaged value of the tangential component of the flow velocity 10 mm from the bowl centre is 9 m/s.

The LDA flow measurements in a plane 30 mm below the cylinder head show a different situation (Fig. 5). The swirl flow is initiated by the intake flow and remains during intake and compression stroke until TDC. The swirl centre moves around and finally centres in the piston bowl. The swirl development can be studied best in a synthetic film constructed from time-dependent LDA data.

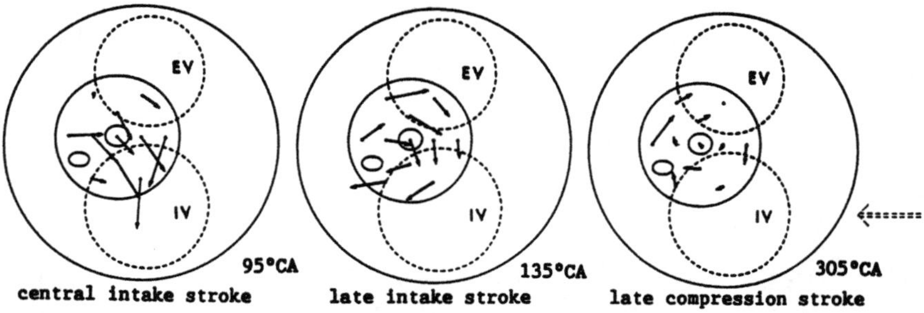

Fig. 5: LDA flow measurements 30 mm below the cylinder head at 1000 rpm.

The swirl has a strong influence on spray formation, vaporization and auto-ignition. In Fig. 6 four frames of a cycle-averaged and time resolved synthetic representation of the spray development are shown. They are achieved by 2-D averaging of the spray contour lines of 280 single frames of 60 cycles taken from laser shadowgraphy high-speed films. Due to the swirl flow the spray is shifted in the outer regions of the piston bowl.

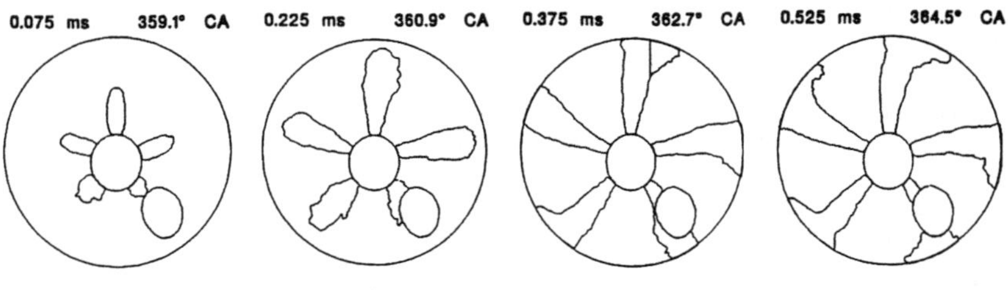

Fig. 6: Cycle-averaged time-resolved synthetic spray formation (2000 rpm).

Auto-ignition is observed on the films by the bright emission of soot particles. The spatial distribution of the auto-ignition locations averaged over 63 cycles is shown in Fig. 7. The fuel vapour is shifted by the swirl flow and ignites on the lee side of the sprays.

Conclusion

In summary, the results presented have demonstrated the application of optical measuring techniques for flow analysis in a production-line DI diesel engine. Laser light-sheet experiments combined with high-speed

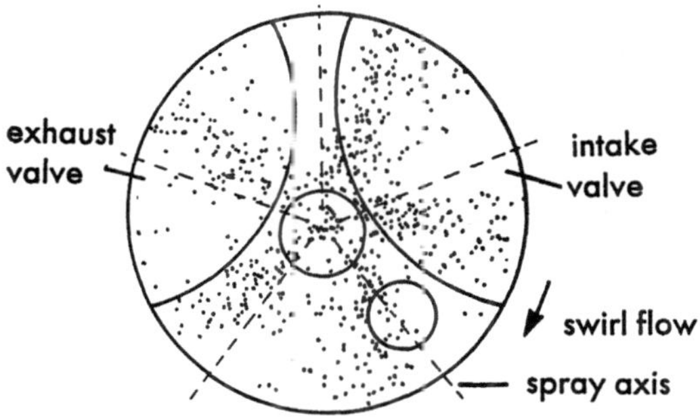

Fig. 7:
Spatial distribution of the auto-ignition locations in the piston bowl (2000 rpm, low load).

filming and LDA measurements complement one another to get both, qualitative and quantitative information of the swirl development. At the time of injection a well-defined nearly solid-body swirl flow is formed. The fuel is injected into the swirl flow, vaporizes, and ignites on the lee side of the spray due to the high compression temperature and afterwards burns as a diffusion flame. Soot and NO_x formation depend on the local air-fuel ratio and on the combustion temperature and can be controlled by the mixture preparation, which strongly depends on the strength of the swirl flow. Additional laser optical techniques as Mie scattering and laser induced fluorescence have been applied to provide more knowledge of the processes occurring during mixture preparation and combustion [5].

Acknowledgement

The authors thank E. Gaschler, W. Oppermann and E. Schulz for design and construction of the transparent DI-diesel engine, N. Dukiewitz for engine set-up and operation, G. Ohmstede and P. Tangni for light-sheet flow investigations, M. Haug and B. Stephan for LDA measurements, and E. Francksen for spray image processing.

Research funded by the Commission of the European Communities within the frame of the JOULE Programme, by the National Swedish Board for Industrial and Technical Development (NUTEK), and by the Joint Research Committee of European automobile manufactures (Fiat, Peugeot S. A. Renault, Volkswagen and Volvo) within the IDEA Programme.

References

[1] Hentschel, W.; Schäfer, H.-J.: Visualization and measurement of charge motion in lean burn spark-ignition engines. Proc. of 5th Int. Symp. on Flow Visualization, Prag 1989

[2] Hentschel, W.; Application of lasers for in-cylinder studies and flow visualization. in: Frankowski, G. et. al. (eds): Application of metrological laser methods in machines and systems. Physical Research Vol 15, Adademie-Verlag, Berlin 1991.

[3] Hentschel, W; Hesse, A.; Schindler, K.-P.: Experimental investigation of spray formation and combustion in a real diesel engine. Auto-Tech. Seminar C399/19 (I Mech E) 1989

[4] Hentschel, W.; Hehn, S.: Flow analysis inside a passenger car silencer by laser light-sheet and coded particle trace techniques. Proc. of 6th Int. Symp. on Flow Visualization, Yokohama 1992 (in this volume)

[5] Arnold, A.; Dinkelacker, F.; Heitzmann, T.; Monkhouse, P.; Schäfer, M.; Sick, V. Wolfrum, J.; Hentschel, W.; Schindler, K.-P.: DI diesel engine combustion visualized by combined laser techniques. Proc. 24th Symp. on Combustion, Sydney 1992 (in press)

Visualization of Counterflow Flames

E. Mastorakos & A. M. K. P. Taylor
Mechanical Engineering Dept., Imperial College of Science, Technology and Medicine, London SW7 2BX, U. K.

1. Introduction

Laminar counterflow flames have been used to study a variety of combustion problems, for example extinction conditions and chemical kinetic mechanisms with a review in Ref. [1], and to provide information for flamelet models of turbulent flames [2]. Recently, this flow configuration for *turbulent* premixed and non–premixed flames has also attracted interest [3,4] because the velocity field provides a uniform strain rate, a quantity important for flame extinction, and because the simple geometry facilitates theoretical analyses [5,6]. Mastorakos *et al.* [4] found that partial premixing improved flame stability and that a spectrum of flames from "pure" diffusion to premixed flames could be established in the counterflow geometry. The present paper is concerned with the *appearance* of the range of flames that can be stabilised in turbulent counterflowing streams and gives additional information on the consequences of stretching on laminar and turbulent counterflow flames with partial premixing and on the mixing between two turbulent isothermal opposed jets, information relevant to opposed jet combustion. In the next Section, the geometies and the methods used to visualise the flames are described, with the results presented and discussed in Section 3. Finally, Section 4 summarises the more important conclusions.

2. Experimental apparatus

The burner used to stabilise *laminar* counterflow diffusion flames, Fig. 1, consisted of two vertically opposed assemblies separated by a distance H, each feeding reactants through a contoured contraction of area ratio 9 ending in a diameter (D_i) of 30mm, a straight section after the contraction 50mm long, surrounded by a nitrogen–carrying annulus with outer diameter (D_o) 50mm. The *turbulent* flames presented here were established in the burner described in Ref. [4], the only difference from Fig. 1 being that straight pipes were used instead of contractions with D_i=25.4mm and D_o=51.8mm and, in order to promote turbulence, perforated plates were fitted inside the pipes. The turbulence intensity at the exit plane of the burner was 0.085 times the bulk velocity [4].

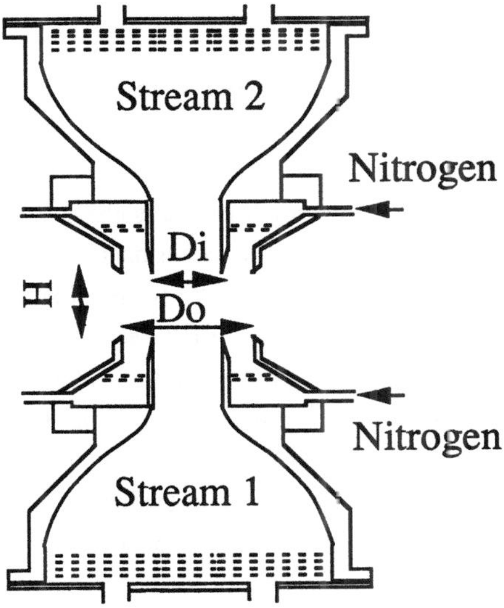

Figure 1. The burner arrangement for laminar flames. For turbulent flames, straight pipes were used instead of contractions and turbulence generators were placed before the nozzle exits.

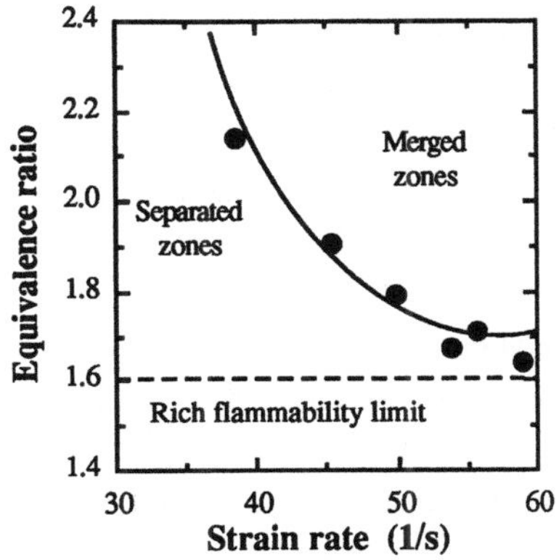

Figure 5. The equivalence ratio when the two reaction zones in a partially premixed diffusion flame merge as a function of strain rate, the latter taken equal to 2 U / H with H = 15mm.

For non–premixed laminar or turbulent flames, the upper, or oxidant, stream was air and the lower, or fuel, stream was natural gas (methane 94% by volume), either pure or partially premixed with air. When the degree of partial premixing, described by the air volume fraction X in the fuel stream, fell within the flammability limits for premixed flame propagation, a single *premixed* flame was established. *Two* premixed flames were stabilised by impinging identical flammable reactant streams. To limit reaction to the stagnation region between the two jets, entrainment of ambient air was prevented by the nitrogen "shroud" streams. All flow rates were metered by calibrated rotameters and the flow conditions are described by the bulk velocities of the lower and upper streams, U_1 and U_2 respectively, the nozzle separation, and by the extent of partial premixing given by X or the equivalence ratio ϕ, the latter used for premixed flames (for methane–air combustion, ϕ=9.524 (1/X – 1)).

Colour photographs of the flames were taken with a 35mm SLR camera using 1000 ASA slide film; two–dimensional images were made by densely seeding one of the streams with Al_2O_3 particles from a reverse cyclone feeder and illuminating with a laser sheet formed from an 130mW Ar–ion laser. The alumina particles also emitted intense red light due to thermal radiation in regions of high temperatures and this was used to visualise hot regions of the flow.

3. Results and Discussion

The mixing between two *isothermal* turbulent air jets was visualised by seeding the upper stream and illuminating with a laser sheet; Fig. 2a shows that the interface between the two jets is located approximately mid–way between the two nozzles and the lengthscale of the "wrinkles" is larger in the radial than in the axial direction. The lengthscale anisotropy is consistent with the bulk flow strain rate elongating turbulent eddies along the extensional strain axis and compressing these along the compressional axis. Figure 2a shows that the radial lengthscale is of the order of the jet radius, i.e. about 13 mm, which is larger than the estimated value of the turbulent integral lengthscale of the turbulence in the jet exit plane of 4mm, based on the size of the turbulence–producing grid [4]. Non–premixed combustion would occur along the interface shown in Fig. 2a, while premixed combustion, which does not require any additional reactant from the opposed stream, occured towards the reactant side of the interface, as shown by the "black" region between the blue reaction zone and the green air stream in Fig. 2b.

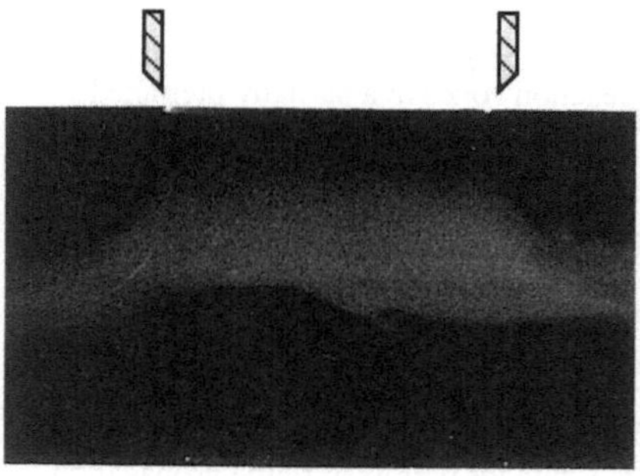

(a)

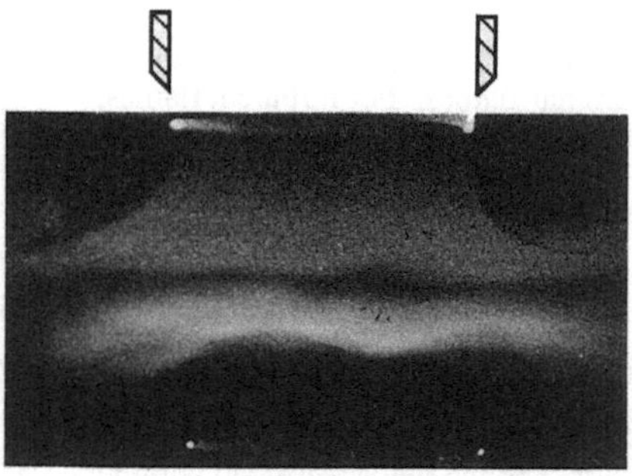

(b)

Figure 2. (a) Turbulent isothermal opposed jet flow with $U_1=U_2=3.0$m/s and H=25.4mm. (b) Single premixed flame against an opposed seeded air stream. $U_1=U_2=2$m/s, $\phi_1=0.8$, and H= 25.4mm.

Changes in partial premixing and strain rate in *turbulent* counterflow flames do not modify the appearance with the flames always being flat, blue, wrinkled, and located approximately mid-way between the two nozzles, Fig. 3. In contrast, *laminar* partially premixed flames may have *two* reaction zones instead of one, Fig. 4, with the lower reaction zone having the characteristics of a premixed flame and the other those of a diffusion flame

[7,8]. The presence of these "double structures" is a function of the equivalence ratio and the strain rate and Fig. 5 (next to Fig. 1) shows that the equivalence ratio at which the two flames merge decreases with an increase in the strain rate, consistent with Ref. [7]. This result suggests that single reaction zones should be expected in *turbulent* partially premixed flames, where the strain rates are high because of the small scale motions.

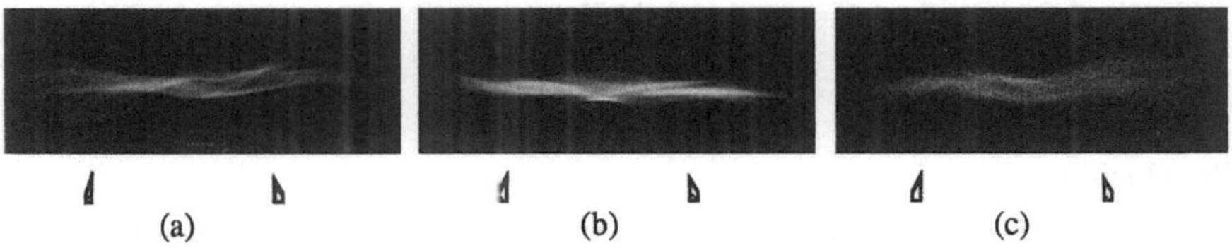

Figure 3. Turbulent counterflow partially premixed flames with (a) U_1=1.34m/s, U_2=1.0m/s, X_1=0.0, (b) U_1=1.1m/s, U_2=1.0m/s, X_1=0.5, and (c) single premixed flame with U_1=U_2=2.2m/s, ϕ_1=0.8, ϕ_2=0.0. For all cases, H=20mm.

Figure 4. (a) Laminar counterflow non–premixed flames at U_1=U_2=0.45m/s, H=15mm and ϕ_1=1.75 (or X_1=0.84), (b) ϕ_1=1.45, and (c) ϕ_1=1.25. The "double structure" is visible in (a) which shows a flame still in the diffusion flame regime.

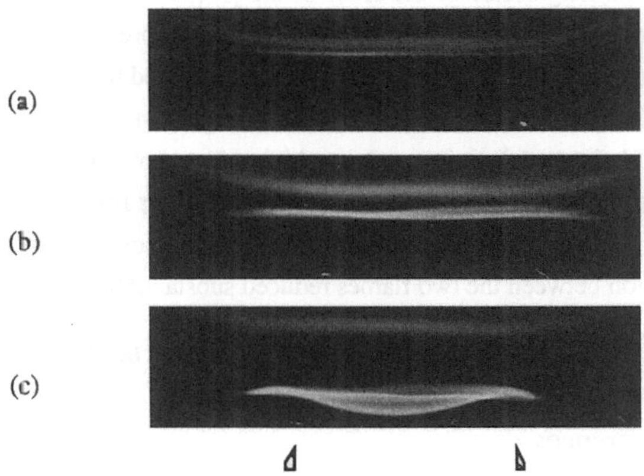

When two identical reactant mixtures impinge on each other, *two premixed* flames are stabilised, one on each side of the stagnation plane, with the region between the two flames containing hot products [4]. Figure 6 shows that as the strain rate increases, the hot product region, visualised by the thermal radiation of Al_2O_3 particles, becomes thinner. Extinction of these flames is promoted by decreasing the mixture strength or by increasing the strain rate and Bray *et al.* [6] predicted that extinction occurs when the mean concentration of products in the post–flame region is smaller than a critical value; this prediction is qualitatively confirmed by the present visual evidence of almost lack of products in the stagnation region as the flames approach extinction.

(a) (b)

Figure 6. Two identical turbulent premixed flames with $\phi_1=\phi_2=0.7$, $U_1=U_2=2.0$m/s, and (a) H=20mm and (b) H=10mm. The flame shown in (b) is close to extinction.

4. Conclusions

Colour photographs have been presented of turbulent opposed isothermal jets and of laminar and turbulent diffusion, partially premixed, and premixed flames in order to show the types of flames that can be stabilised in counterflow burners by varying the compositions of the reactant streams. The interface between the two isothermal opposed jets exhibited a waviness with a lengthscale of the order of the jet radius in the radial direction but smaller in the axial direction. The progression from diffusion to premixed *turbulent* flames was smooth with the flames always being blue, flat, wrinkled, located near the symmetry plane and with appearance independent of strain rate. In contrast, *laminar* counterflow partially premixed flames could result in two reaction zones which merged at high strain rate or high equivalence ratio. Impinging identical *turbulent* reactant streams resulted in two premixed flames which came closer with increasing strain rate and, close to extinction, the concentration of products in the region between the two flames reduced substantially, in agreement with theoretical predictions.

This work has been partially supported by CEC SCIENCE Contract No. SC1–0459.

References

1. Tsuji, H. (1982) Counterflow diffusion flames. *Prog. Energy Combust. Sci.* **8**, 93.
2. Peters, N. (1984) Laminar diffusion flamelet models in non–premixed turbulent combustion. *Prog. Energy Combust. Sci.* **10,** 319.
3. Kostiuk, L. W., Bray, K. N. C. & Chew, T. C. (1989) Premixed turbulent combustion in counterflowing streams. *Combust. Sci. and Tech.* **64**, 233.
4. Mastorakos, E., Taylor, A. M. K. P. & Whitelaw, J. H. (1992) Extinction and temperature characteristics of turbulent counterflow diffusion flames with partial premixing. To appear in *Comb Flame.*
5. Mastorakos, E., Taylor, A. M. K. P. & Witelaw, J. H. (1992) Scalar dissipation rate at the extinction of turbulent counterflow non–premixed flames. To appear in *Comb Flame.*
6. Bray, K. N. C., Champion, M. & Libby, P. A. (1991) Premixed flames in stagnating turbulence: Part I. The general formulation for counterflowing streams and gradient models for turbulent transport. *Comb. Flame* **84**, 391.
7. Seshadri, K., Puri, I. & Peters, N. (1985) Experimental investigation of partially premixed diffusion flames at extinction. *Comb. Flame* **61**, 237.
8. Yamaoka, I. & Tsuji, H. (1976) Structure analysis of rich fuel–air flames in the forward stagnation region of a porous cylinder. 16th Symposium (Int) on Combustion, 1145.

Oscillation of Impingement Flow in a Combustor with Dual Side-Inlets

T.M. Liou, Y.Y. Wu and S.M. Wu

Department of Power Mechanical Engineering
National Tsing Hua University, Hsinchu, Taiwan 30043

Abstract

The oscillation of streamwise vortices resulting from a jet-on-jet impingement in a round duct with dual rectangular side-inlets was investigated using the particle-tracer visualization technique for Reynolds numbers of 2000 to 26000. The results showed that the vortex patterns could be classified into three oscillation modes: clockwise, impingement, and counter-clockwise modes, and the frequency of the mode shift increased as the Reynolds number was risen.

Key Words: impingement flow, side-dump combustor, particle-tracer technique, oscillatory vortex, flow visualization

Introduction

In general, the jet-on-jet impinging flows are intrinsically unstable and lead to an oscillatory and flapping flowfield in which strong streamwise vortices are generated in the impinging region. Such a flowfield appears in several practical applications such as reaction injection moldings, utility boiler furnaces and side-dump combustors. A better understanding of the relevant flow structures is, therefore, highly desirable. Several researchers have studied the turbulent flows in side-dump ramjet combustors (Fig. 1). Nosseir and Behar [1] and Miau et al. [2] individually conducted the water channel tests to investigate the impinging flows in two-dimensional combustors with rectangular and square cross-section, respectively. They showed that the impingement of flows leads to a flapping motion in the stagnation region. Because the cross section of a practical ramjet combustor is circular rather than rectangular or square, Liou and Wu [3] employed the laser-Doppler velocimetry (LDV) to study the flowfield in a cylindrical combustor with two 60-deg rectangular side-inlet ducts and found the existence of spiral vortices which are distinct from those previously reported in the two-dimensional models. The purpose of the present work is to complement the previous LDV measurements made by Liou and Wu [3].

In this paper the visualization results of the instantaneous structures of streamwise vortices in the impinging region of a cylindrical combustor are presented and the effect of the Reynolds number on the oscillation of streamwise vortices is addressed.

Experimental Apparatus and Method

The configuration, coordinate system, and dimensions of the dual-inlet side-dump combustor model are depicted in Fig. 1. The test model was made of 5-mm thick Plexiglas. The dome height (L_d in Fig. 1) was fixed at 50 mm and the Reynolds number, based on the combustor diameter (D) and bulk mean velocity, was varied from 2000 to 26000.

Particle-tracer visualization technique was used for visualizing the impinging flow structures. The saline seeding particles, nominally 1 μm in size, were generated by atomizers and introduced into the combustor at upstream settling chambers. A 5-W argon-ion laser operated at 514.5 nm was employed as the light source. Through a module consisting of a glass rod and a plano-cylindrical lens, the green beam was transformed into a 1.5-mm thick light sheet to illuminate the examined cross-sectional planes. Pictures were recorded by a video camera located in front of the dome plate (Fig. 2) and aligned along the combustor axis.

Results and Discussion

Figure 3 shows the three typical modes of streamwise vortices formed by flow impingement at $X^*(\equiv X_c/D)=0.25$ in the combustor for Re=26000. The oscillation patterns were categorized into three modes: clockwise (CW) mode, impinging (IM) mode and counter-clockwise (CCW) mode, and the flow shifted back and forth among these modes. It is worth pointing out that the bimodal shape in the velocity probability density function measured by Liou et al. [4] was associated with these oscillation phenomena.

In the CCW mode each side-inlet jet bifurcated at the combustor wall to form one clockwise vortex with respect to the combustor axis. Meanwhile, the axes of two side-inlet jets swung transversely as shown in Fig. 4. The situation under the CW mode was similar except the direction of the vortex.

During the alternation between the CCW and the CW modes, the axes of two opposing jets were occasionally aligned at the vertical symmetry plane (θ=90 deg), and the IM mode was formed. In this mode there were four vortices formed in the cross-sectional plane (Fig. 3b) and the impinging plane moved up and down and sometimes curved (Fig. 5) due to the temporal and spatial phase differences of velocities between the two side-inlet jets. The duration ratio of CW, IM and CCW modes was about 2:1:2, and the frequency of oscillation was about 0.5 Hz. It is clear now that the mirror image concept (Powell [5]) is not suitable in such flows because the movement and distortion of the impingement plane make it plausible to place a flat plate to model the impinging plane.

According to the resonant feedback model [6], the disturbances generated at the downstream impingement region are propagated upstream to influence the initial jet flows. In order to confirm this model, a short rectangular duct was attached at each side-inlet port to protect the roots of the jet flow from the feedback disturbances. It was found that the impinging flow stayed longer at the IM mode. This observation suggests a possible method to control the oscillatory impingement flows in the side-dump combustor.

As the Reynolds number decreased to 9500, the oscillation patterns were similar to those at Re=26000. The duration ratio and the oscillatory frequency, however, changed to 10:1:10 and 0.03 Hz, respectively; this trend showed that the probability of the IM mode descended with decreasing Reynolds number and implied that the IM mode become more difficult to persist and more unstable at a lower Reynolds number. As the Reynolds was further decreased to 2000, the flow pattern was always fixed at CCW mode or CW mode depending on the initial condition and the geometrical errors.

Concluding Remarks

In this paper the instantaneous streamwise vortex structures in a three dimensional side-dump combustor model are described. Three modes of streamwise vortices, clockwise, impingement and counter-clockwise modes, are presented and the frequency of the mode shift becomes lower with decreasing Reynolds number. By understanding the mechanism of the instability of jet-on-jet impingement, it is possible to control the flow patterns in the combustor by use of a rectangular duct to protect the roots of the side-inlet flows from downstream perturbations. The present flow-visualization results also indicate the inadequacy of the mirror image concept for the flowfields investigated.

References

1. Nosseir, N.S. and Behar, S., 1986, "Characteristics of Jet Impingement in a Side-Dump Combustor," *AIAA Journal*, Vol. 24, No. 11, pp. 1752-1757.
2. Miau, J.J., Mong, S.M., Sun, D.J., and Hong, Z.C., 1988, "Visualization of Flow in a Side-Inlet Square Duct," *AIAA* Paper no. 88-0502.
3. Liou, T.-M. and Wu, S.-M, 1988, "Flowfield in a Dual-Inlet Side-Dump Combustor," *Journal of Propulsion and Power*, Vol. 4, No. 1, pp. 53-60.
4. Liou, T.M, Wu, S.M., and Hwang, Y.H., 1990, "Experimental and Theoretical Investigations of Turbulent Flow in a Side-Inlet Rectangular Combustor", *Journal of Propulsion and Power*, Vol. 6, No. 2, pp131-138.
5. Powell, A., 1960, "Aerodynamic Noise and the Plane Boundary," *Journal of the Acoustical Society of America*, Vol. 32, pp.982-990.
6. Rockwell, D., 1983, "Oscillations of Impinging Shear Layers", *AIAA Journal* Vol. 21, No. 5, pp.645-664.

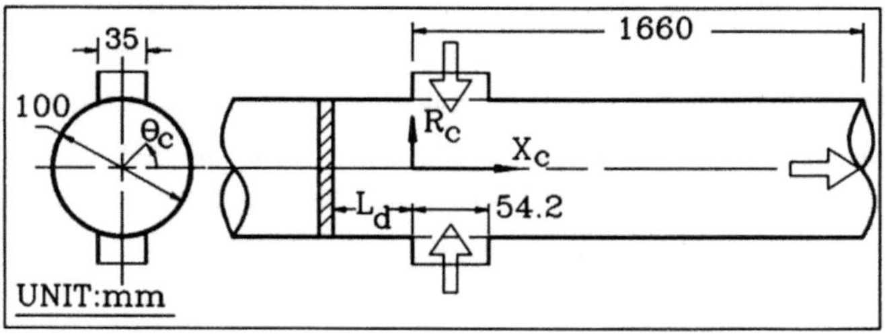

Fig. 1 Sketch of the simulated combustor model

Fig. 2 Set-up of the flow visualization experiment

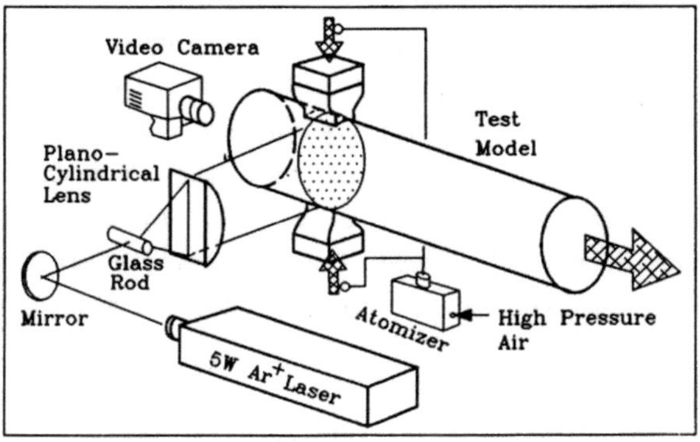

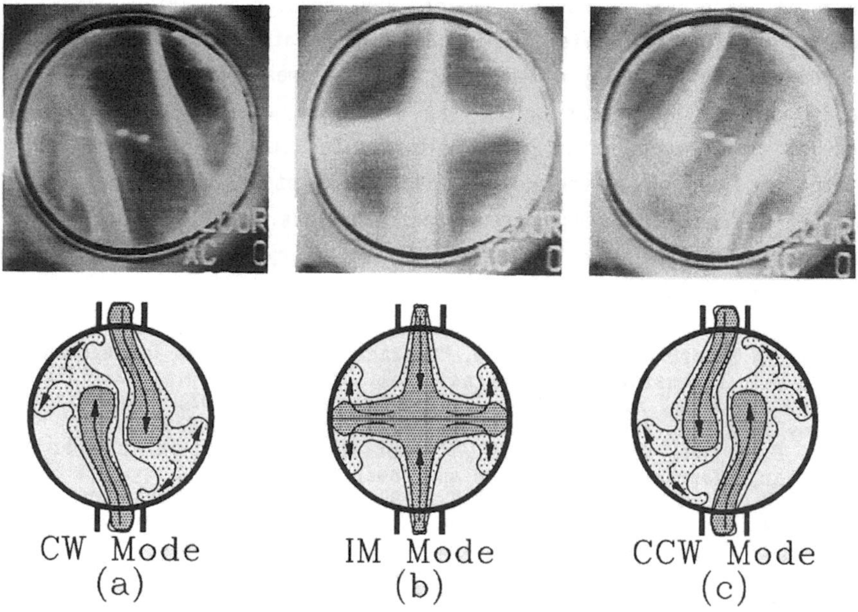

Fig. 3 Photographs and schematic drawings of three oscillation modes in the impingement region

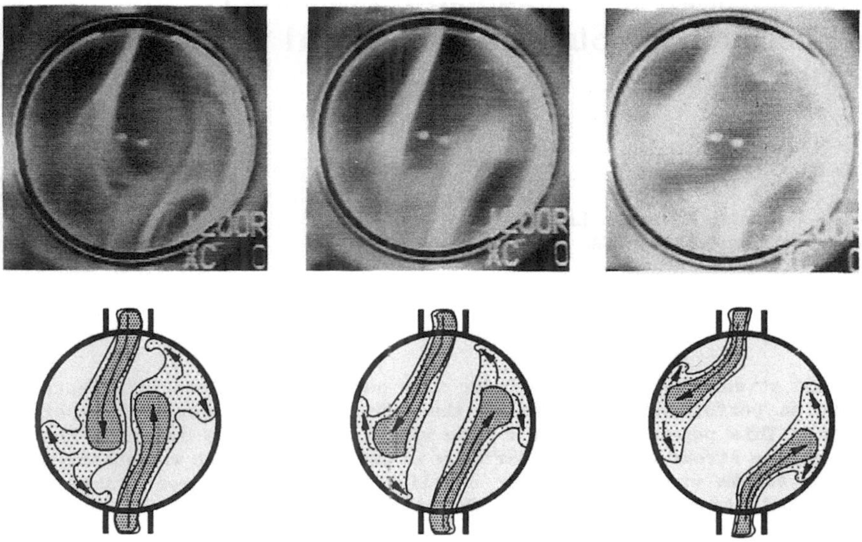

(a) (b) (c)

Fig. 4 Photographs and schematic drawings of the CCW mode in the impingement region

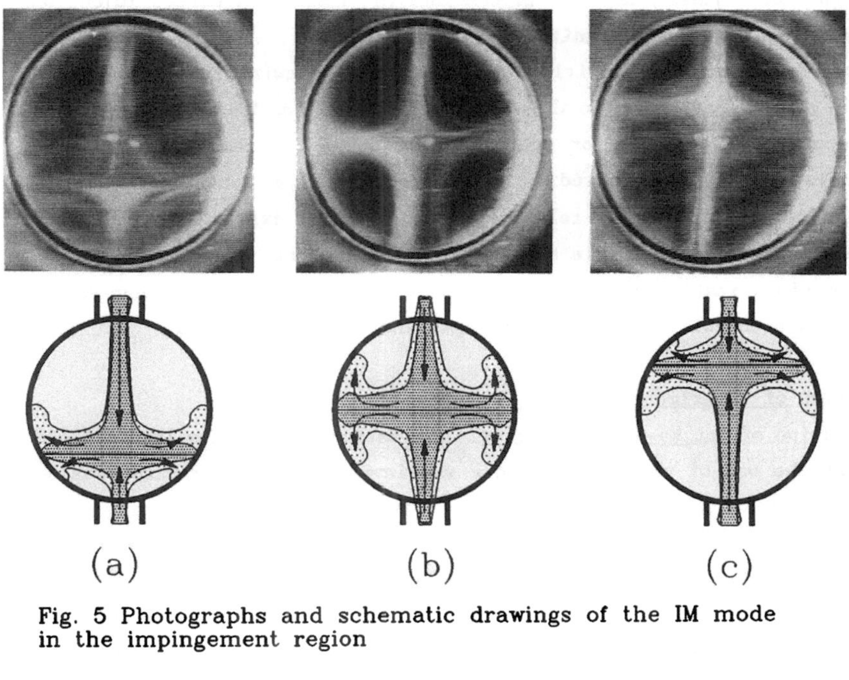

(a) (b) (c)

Fig. 5 Photographs and schematic drawings of the IM mode in the impingement region

Flow Visualization and Quantitative Evaluation of the Effect of Bore/Stroke Ratio on Swirl in a Piston Engine

A. MIKULEC, J.C. KENT

Ford Motor Company, Research Laboratory
Dearborn, Michigan 48124, USA.

Summary

The type and strength of the in-cylinder flow pattern significantly affects piston engine performance in terms of the rate and cyclic variability of combustion. This paper describes a flow visualization study in which the effect of engine stroke on in-cylinder flow is investigated. It was observed that longer stroke produces stronger swirling motion and in most cases a larger vertical (tumble) component of the rotational motion.

Introduction

It is now well established that formation of swirl during the induction process is important for fast, repeatable combustion in homogeneous charge spark ignition engines, as well as for fuel/air mixing in direct injection engines (diesel or spark ignited stratified charge). Swirl strength is expressed in terms of the swirl ratio, which is the swirl rotational speed divided by crankshaft rotational speed. While swirl has been conceptualized as a rotation of the air, or fuel/air mixture, about the cylinder axis, recent flow visualization studies (1) showed that the rotational axis is generally tipped at an angle relative to the cylinder axis. This indicates that components of both tumble and swirl can be present. Reported here are new flow visualization studies that examine the influence of changes in engine stroke on swirl rotational strength and axis orientation.

Description of Experiment

Visualization of Intake-Generated Flow. The nature of the in-cylinder fluid motion at the end of the intake stroke was investigated using water analog simulation and particle image velocimetry (PIV) using video-based techniques described in references (1-7). In each test case, the cylinder head and valve train of the respective engines were mounted on the water analog, which was equipped with the appropriate bore and stroke. The water analog was operated at 25 cycles per minute, which simulates an engine idle speed of 850 rpm.

<u>Hardware tested</u>. Three different cylinder heads were utilized in the experiments. One was a 4-valve head (two intake and two exhaust valves) that was operated with only one intake port active to produce swirl. The others had one intake and one exhaust valve. All utilized intake valve "masking" -- a blockage of a portion of the valve flow area -- to enhance swirl strength. The combinations of bore/stroke ratios tested are as follows:

Engine Type	I4, 4-valve		I6, 2-valve		V8, 2-valve		
Case	1a	1b	2a	2b	3a	3b	3c
Bore/Stroke	1.16	0.93	1.18	0.93	1.29	1.00	0.85

Results and Discussion

<u>Case 1</u>. The swirling motion is seen in Fig. 1 via the lower portion (bottom view through the piston), and the inclined swirl axis is seen in the view through the cylinder. Table 1 indicates that the swirl inclination angle increased as the stroke was increased (bore/stroke ratio decreased). It also shows that the swirl ratio, obtained from 3-D PIV, increased as the stroke was increased. The latter confirms the trend predicted by a simplified simulation model for engine swirl (8).

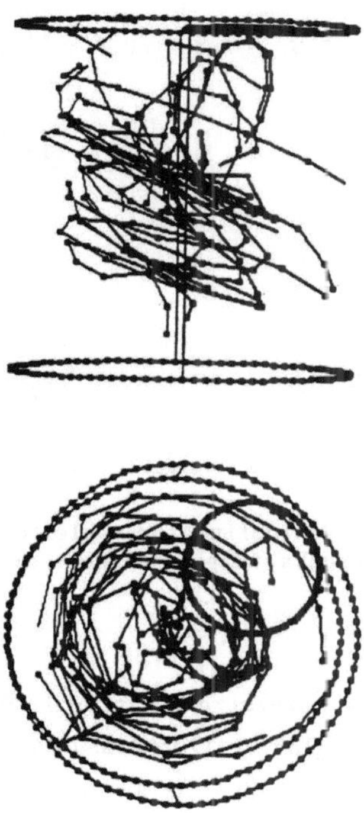

Figure 1. Velocity fields obtained from particle image velocimetry for Case 1b, Bore/Stroke = 0.93.

Table 1.

EFFECT OF STROKE ON SWIRL RATIO AND INCLINATION ANGLE 4-valve head, 1 intake port active		
BORE / STROKE RATIO	0.93	1.16
SWIRL INCLINATION ANGLE,deg.	42.10	29.00
SWIRL RATIO	3.72	2.70

Case 2. Video images of the flow tracer particles are shown in Figure 2 at BDC of intake. Analysis of individual frames on a video tape revealed that the swirl inclination angle increased from about 20 to 45 deg. as the bore/stroke ratio was decreased from 1.18 to 0.93.

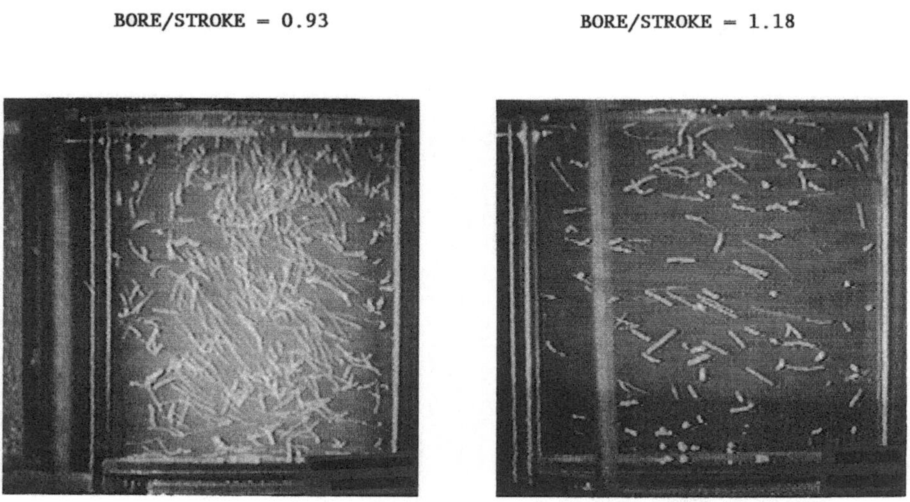

Figure 2. Effect of stroke change on swirl inclination angle for Case 2 as indicated by flow tracer streaks imaged through the cylinder.

Case 3. For this cylinder head, the following bore/stroke ratios 1.29, 1.00 and 0.85 resulted in swirl axis inclination angles of 12, 24 and 34 degrees respectively and swirl ratios of 1.95, 2.21 and 2.60. As in Case 2, these results were obtained from analysis of video recordings.

Compared to Case 2, the flow patterns are similar, but the swirl inclination angle increased less, even though the bore/stroke ratio change was significantly higher (from 1.29 to 0.85 versus 1.18 to 0.93 in Case 2). It is speculated that this may be caused by variations in directed and helical components of flow for different port designs.

Conclusions

For the cylinder head configurations tested, flow visualization and PIV showed that both the swirl inclination angle and rotational speed increase with increasing piston stroke. These observations, along with unpublished engine combustion data for different bore/stroke ratios, provide additional evidence of correlation between intake-generated fluid motion and combustion.

References

1. Kent, J. C., Mikulec, A., Rimai, L., Adamczyk, A. A., Mueller, S. R., Stein, R. A., and Warren, C. C., "Observations on the Effect of Intake-Generated Swirl and Tumble on Combustion Duration", SAE 892096, 1989.

2. Ekchian, A. and Hoult, D. P., "Flow Visualization of the Intake Process of an Internal Combustion Engine," SAE Paper 790095, 1979.

3. Kent, J. C. and Mikulec, A., "Visualization of Induction-Generated Flow in the Cylinder of a Piston Engine Via Water Analog Simulation," Flow Visualization IV, Proceedings of the Fourth (International) Symposium on Flow Visualization, Paris, France, August 26-29, 1986, 735-740, Hemisphere Publication Co., Washington, D.C., 1987.

4. Khalighi, B. and Heubler, M. S., "A Transient Water Analog of a Dual-Intake-Valve Engine for Intake Flow Visualization and Full-Field Velocity Measurements," SAE Paper No. 880519, 1988.

5. Mikulec, A., Kent, J. C., Adamczyk, A. A., and Rimai, L., "Dynamic Measurements of Swirl in Piston Engines using Water Analog Simulation and Quantitative Flow Visualization," 5th International Symposium on Flow Visualization, Prague, Czechoslovakia, August 21-25, 1989.

6. Rimai, L., Adamczyk, A.A., Mikulec, A., and Kent, J. C., "Quantitative 3D Reconstruction of Flow Fields from Orthogonal views of Particle Tracks: Application to Water Analog of a Piston Engine," 5th International Symposium on Flow Visualization, Prague, Czechoslovakia, August 21-25, 1989.

7. Adamczyk, A. A. and Rimai, L., "Reconstruction of a 3-Dimensional Flow Field from Orthogonal Views of Seed Track Video Images," Exp. Fluids, 6, 380-386, 1988.

8. Davis, G. C. and Kent, J. C., "Comparison of Model Calculations and Experimental Measurements of the Bulk Cylinder flow Processes in a Motored PROCO Engine," SAE Paper 790290, 1979.

Roll-up Vortex on the Reciprocating Piston in a Cylinder

Tomio Obokata
Gunma University, 1-5-1 Tenjin, Kiryu
and
Atsushi Okajima
Kanazawa University, 2-40-20 Kodatsuno, Kanazawa

Summary

The behavior of a roll-up vortex in a cylinder is studied experimentally and numerically. The vortex is visualized by smoke under illumination of a laser sheet and the unsteady velocity field is investigated by means of laser Doppler anemometer (LDA). A numerical solution of the Navier-Stokes equations and a continuity equation by a finite difference method is also presented. The path line of calculated vortex on the piston at Reynolds number (based on a cylinder diameter and the maximum piston speed) of 5,000 corresponds well with that of observed vortex in an engine at a rotation speed of 756 rpm.

Introduction

During the exhaust stroke of a reciprocating engine a viscous boundary layer on the cylindrical wall is scraped up by the moving piston and forms into a roll-up ring vortex. It is important to analyze the behavior of the vortex for understanding the characteristics of exhaust emission of the engine because the vortex contains the greater part of unburned hydrocarbon in the cylinder [1]. Although some experimental [2, 3] and numerical [4] works on this subject had been reported,

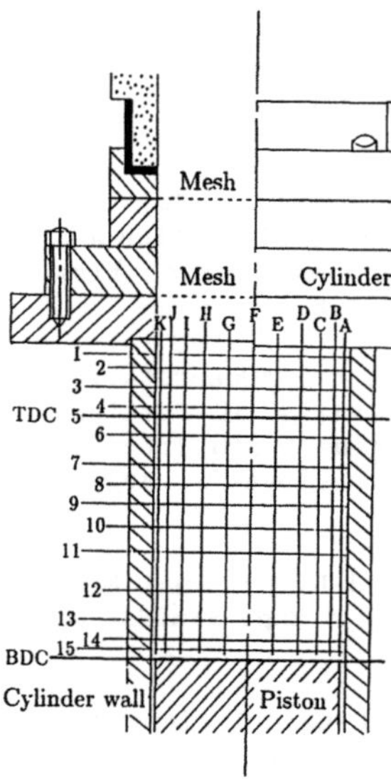

Z	mm	r	mm
1	-13	A	-19
2	-10	B	-17
3	-6	C	-14
4	-2	D	-10
5	0	E	-5
6	4	F	0
7	10	G	5
8	14	H	10
9	18	I	14
10	23	J	17
11	28	K	19
12	36		
13	42		
14	46		
15	48		

Fig. 1 Cross section and measuring points of the engine.

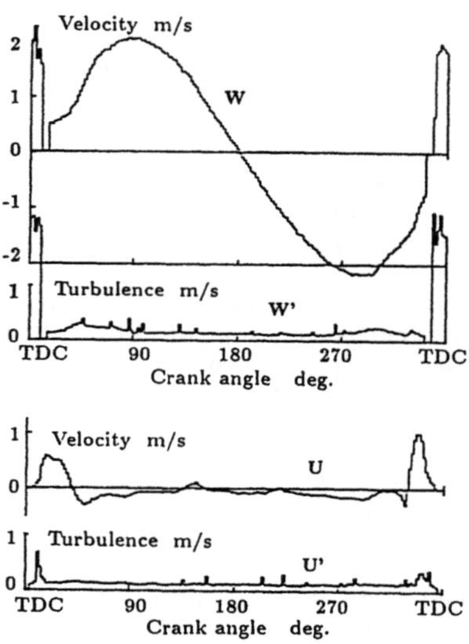

Fig. 2 Examples of velocities measured at Z=0mm and R=14mm (756 rpm).

more detailed experimental investigations and comparison with numerical computation were expected. As the first step of the study, roll-up vortices on the piston in the open ended cylinder of an internal combustion engine are visualized by an injected smoke, and the velocity vectors are measured by an LDA. Further more, the flow in a cylinder with a moving piston is simulated by a finite difference method.

Velocity Measurement

Through the transparent cylinder of a test engine (Megatech Mark III, Bore × Stroke = 40.6mm × 50.8mm) without cylinder head as shown in Fig. 1, gas flow velocities at 165 points in the engine are measured by a fiber optic laser Doppler anemometer (LDA). Fig. 2 shows examples of ensemble averaged mean velocities (W: axial, U: radial) and their fluctuations (W', U'). Almost similar flow velocity to the piston speed is measured in axial direction and outwards flow from the center of the cylinder is observed near the piston surface in radial direction. At the LDA measurement of radial velocity, launching angle and position of the incident beams are adjusted to compensate the lens effects of transparent cylinder [5]. Silicone oil droplets (about 4μm in diameter) are used as the scattering particles and a frequency counter (Dantec 55L90a) converts the Doppler signal to the flow velocity.

Flow Visualization

A smoke injection method was applied to the visualization of fluid flow produced by the motion from rest of a piston in the cylinder. Before starting the engine, an oil mist (Moleffect Fogmaker 1963) was injected on the piston at bottom dead center (BDC). A video camera with shutter speed of 1 ms and frame speed of 60 f/s has been used to record the roll-up vortex, illuminated by a laser sheet of 1 mm in depth and a power of 1 W. Fig. 3 shows the vortex pattern of the second rotation cycle of the engine, which the rotation speed of engine is almost constant. The first vortex formed at the first rotation cycle is shown at the BDC (θ = 180 deg.). During the second rotation cycle, we can find more growth of the first vortex (θ = 227 deg.), development of the second vortex (θ = 287 deg.) and merging of both the vortices (θ = 62 deg.).

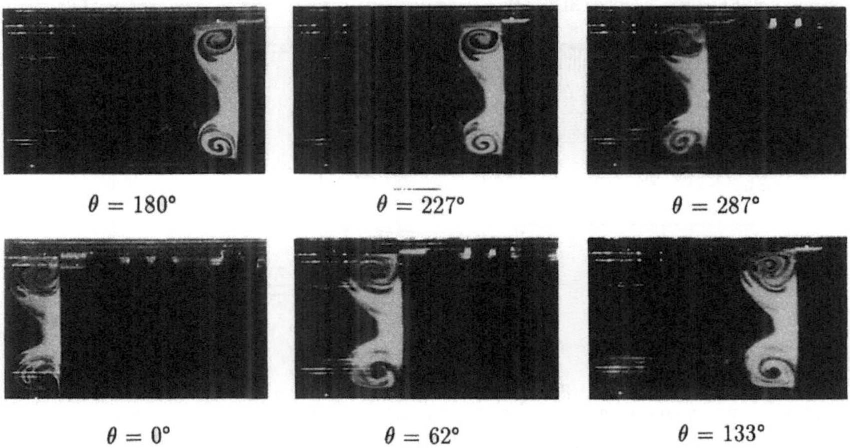

Fig. 3 Vortex patterns visualized by smoke and laser sheet (756rpm, 2nd rotation cycle).

Numerical Simulation

The flow is assumed to be axisymmetric unsteady, incompressible and viscous, and has been computed about the primitive variables, i.e., velocity **u** and pressure P. Fundamental equations are the continuity equation and the Navier-Stokes equations as shown here.

$$\nabla \cdot \mathbf{u} = 0 \tag{1}$$

$$\partial \mathbf{u}/\partial t + (\mathbf{u} \cdot \nabla)\mathbf{u} = -\nabla P + (1/Re)\nabla^2 \mathbf{u} \tag{2}$$

Where Reynolds number is $Re = U_w D/\nu$ based on the cylinder diameter (D), the maximum piston speed (U_w) and kinematic viscosity of air (ν). A finite difference scheme has been applied to this flow field divided into 44 × 87 points [6]. Boundary condition of no slip ($\mathbf{u} = 0$) are imposed on the surface of the piston, and the surface of the cylinder is moved with -V(t), where V(t) is the velocity of the piston. The conditions at the upper region are given by the use of zero-gradient boundary conditions. Namely, a moving coordinate system is adopted.

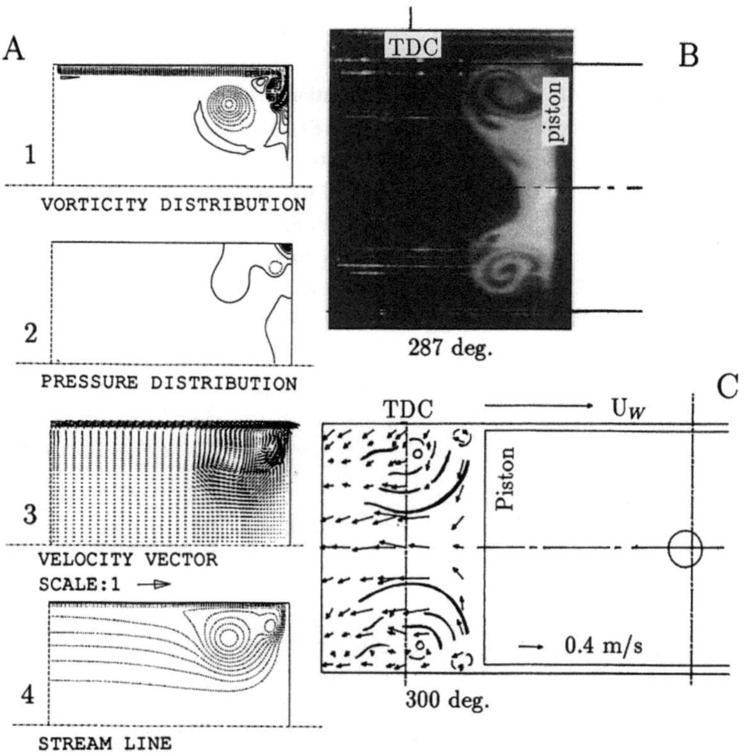

Fig. 4 Comparison of calculated flow (A) with visualized pattern (B) and measured flow (C). A: Calculation (Re=5,000, θ=300 deg.), B: Visualization (N=756 rpm, Re=5,440, θ=287 deg.), C: Measured velocity vector relative to the piston speed (N=756 rpm, θ=300 deg.).

Discussion

An example of comparison between experimental results and numerical simulation is shown in Fig. 4. Numerical results (A) and photograph (B) are taken during the second rotation cycle from the start, however velocity distribution (C) is obtained at continuous rotation mode. We make comparisons between vorticity distribution (A-1) and streak line of vortex (B), velocity vectors (A-3) and (C), and stream line (A-4) and the estimated stream line (C). There are good agreements between them, respectively.

The radial positions from the cylindrical wall (Ro−r) and the axial positions from the piston (Z) of the vortex center are determined from these three results, and are plotted in Fig. 5. Top dead center (TDC) positions of the piston are denoted by the crank angle of $\theta = 0$ and 360 degrees. Reynolds numbers are Re=5,000 and 10,000 for this simulations, and are 5,440 (rotation speed of the engine N=756 rpm) for experiments. The similar development patterns of the vortices are shown in this studies. The path line of the vortex center by the simulation also corresponds well with that by the experiments as shown in Fig. 6.

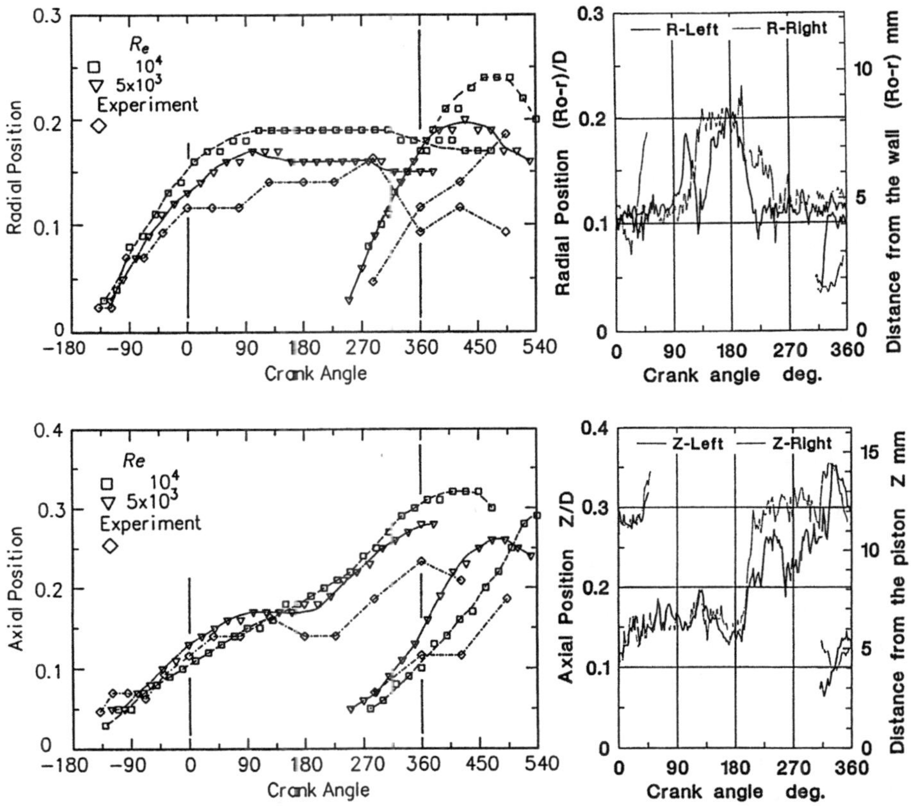

a) Calculated and visualized vortices.

b) Measured vortex by LDA.

Fig. 5 Axial and radial positions of the vortex center.

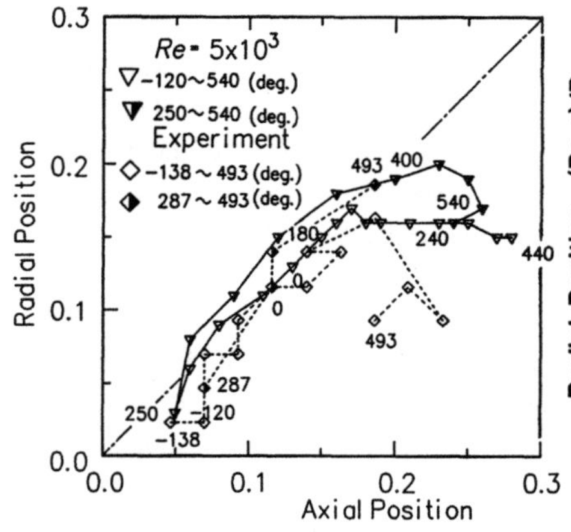

a) Calculated and visualized vortices.

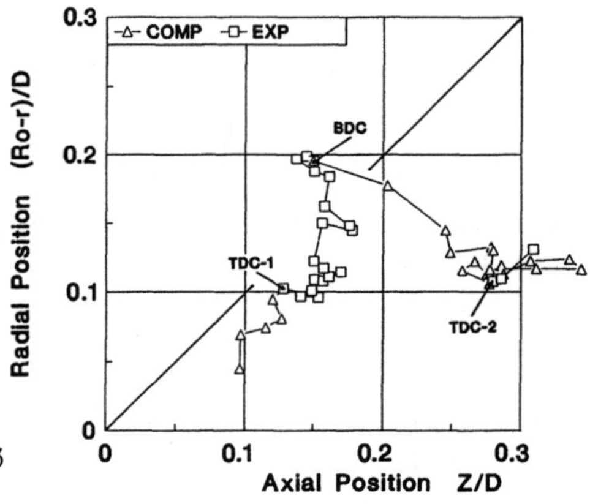

b) Measured vortex by LDA.

Fig. 6 Traveling of the vortex.

Conclusions

A roll-up vortex in an engine cylinder is visualized by smoke injection, and the vectors of flow velocity are measured by LDA. Furthermore, flow in the cylinder are numerically simulated by the finite difference method. The development patterns of vortex and traveling paths of vortex center are observed in these three experiments. There are good agreements between the results by two kinds of experiments and numerical simulation for the patterns of vortex-development and the traveling paths of vortex center.

Acknowledgments

The authers wish to express their appliciations for financial support from Honda R & D. Thanks are also due to Mr. Y. Hyuga and Mr. T. Koyama of Gunma University and Mr. H. Endoh of Kanazawa University for their cooperation.

References

1. Lavoie, G.A.; Lorusso, J.A.; Adamczyk, A.A.: Hydrocarbon emissions modeling for spark ignition engines. (eds., Mattavi, J.N.; Amann, C.A.: Combustion modeling in reciprocating engines). New York: Plenum Press 1980. 409-445.
2. Tabaczynski, R.J.; Hoult, D.P.; Keck, J.C.: High Reynolds numder flow in a moving corner. J. Fluid Mech. 42-2 (1970) 249-255.
3. Hughes, M.D.; Gerrard, J.H.: The stability of unsteady axisymmetric incompressible pipe flow close to a piston. Part 2.: J. Fluid Mech. 50-4 (1971) 645-655.
4. Gerrard, J.H.: Part 1. ibid. 625-644.
5. Obokata T.; Hashimoto T.: LDA measurement of roll-up voltex in a cylinder. Prepr. of JSME No. 910-62. B. (1991) 717-719 (in Japanese).
6. Okajima, A.; Nagahisa, T.; Rokugoh, A.: Numerical analysis of flow around rectangulsr cylinders.: JSME Int. J. II. 33-4 (1990) 702-711.

Shock Waves

Visualization of Shock Wave Propagation Through Curved Channels

N. P. MENDE

A.F.Ioffe Phvsico-Technical Institute
Russian Academy of Sciences
St.Petersburg. 194021
Russian Federation

Summary

Shock wave propagation through channels with a rectangular elbow and a circular bend is considered. An original construction of a shock tube allows one to change easily the channel configuration in the working section. To visualize shock waves and the induced flows a grating interferometer is employed. The interferometer is equipped with two ruby lasers to produce high speed two-frame registration of the process. Large-scale recording of the interferogram provides their high quality and resolution. At late stages of the propagation complicated patterns of the diffraction-reflection interaction are observed.

Introduction

For a long time the processes accompanying shock wave interactions with rigid surfaces are studied separately to elucidate their nature without outside influences. During a half of century the shock wave diffraction and reflection have attracted researchers attention throughout the world. However. in reality these two processes often occur simultaneously and affect each other. The complexity of such interactions practically rules out a description of them by analytical methods and restricts a role of experiments to data accumulation without profound general conclusions. The leading role in investigations of such processes shifts to the computational simulation. In such a situation the high-quality experiment can be helpful for the aim of testing numerical schemes and checking the applicability of gas flow models used. Certainly. the experiment must supply accurate quantitative data.

Experimental equipment

To observe shock wave propagation through a curved channel we designed a shock tube of an explosive type with a working section of changeable geometry. The distinctive feature of the shock tube consists in combining the gas receiver and the working section into a common unit. The rectilinear channel of 1.5 *m* in length with cross-section of 30x30 mm^2 enters into the receiver and ends in the working section formed by two glass windows of 230 *mm* in diameter with interchangeable steel details between them. This does not allow to take care of a hermetic conjunction of the changeable working section outlet with the receiver. Naturally, the receiver has windows too which are co-axial with the test section windows. Besides, such a working section construction allows for distinctive features of a grating interferometer. It is known that a shearing interferometer can be successfully applied if in the working field there is a portion undisturbed by a flow under study [1]. The space outside the channel within the window area can be used as a referring surface because it always remains undisturbed. The same distinctive features of the interferometer allow us to glue onto the working section window near the interchangeable details some glass scales in such a way that their images in one of the coherent beams being shorn by the action of the interferometer would be superposed onto the shadows of the non-transparent channel walls in the other beam. The presence of the scale in the photo allows us to detect the image magnification and measure the path of shock waves with the accuracy comparable with the spatial resolution of our optical device (about 0.1 *mm*). We equipped the grating interferometer with two ruby lasers with the polarization planes of their beams to be orthogonal to each other [2]. This allows one to select the laser beams at the output of the interferometer using a birefringent prism and obtain two images of an object under study in one photo separately. So, if the lasers operate in different instants we have two-frame registration with a time interval between them which is measured with the accuracy not worse than 0.1 μs. Thus, we can determine the average shock wave velocity with the error about 2% at the velocity of 1 *km/s* along the path about 10 *mm*. It is quite enough to identify reliably fringes in the interferogram using the gasdynamic relation for the gas density ratio across the shock. (The isopycnic step in terms of non-dimensional density in our interferograms varied from 0.085 to 0.15 depending on the initial pressure in the channel, the gas density ratio across M=3 shock is 3.86 in air). In Fig.1 one can see a slightly reduced copy of the working field image as seen by the experimenter through the interferometer. Due to high power of the ruby lasers we take the photos even with a small magnification what results favorably in their resolution.

The shock wave intensity varied depending on the chemical charge value and the gas pressure inside the channel and the receiver. There is no diaphragm between the blast chamber and the channel. This circumstance along with small time dispersion of the explosion development radically facilitates

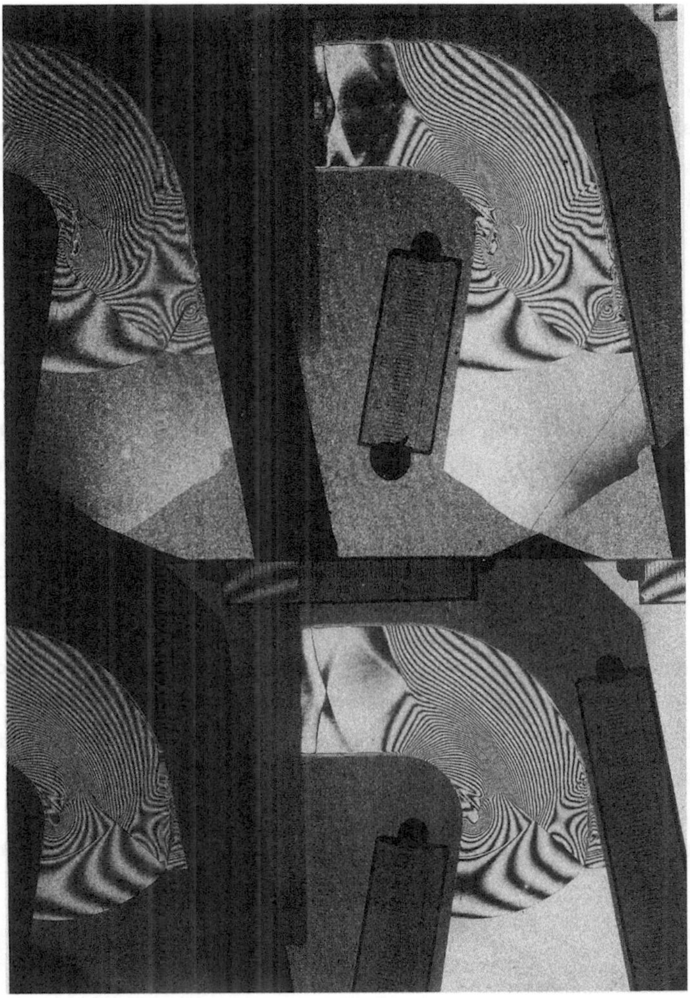

Fig.1. A fragment of the interferometer field of view.

the synchronization of the recording system operation with the process. The gas density behind a blast wave is known to vary with distance from its front. The isopycnics on the background of the induced flow behind the M=3 incident shock wave show that the non-dimensional gas density varies by 0.15 along the distance comparable with the channel width. Such a change is considered to be negligible when numerical simulation of the same regimes (see Ref.[3]).

Experimental results

We studied the shock wave propagation through curved channels of two types: with a rectangular elbow and a circular bend. The shapes of the channels can be seen in the Figures below. The radii of the wall roundings of the circular bend are 15 and 45 *mm*, the plane nozzle full angle behind the circular bend is 15°. The incident shock wave propagates through the horizontal channel section from the left to the right. Its intensity varied from Mach Number 1.2 to 3.0 in air. The isopycnic step at Mach number 3 is 0.15, at the other Mach number - 0.085.

Figs.2,3 show various stages of the interaction of the shock wave diffraction and reflection processes at various Mach numbers. At initial stages of the shock wave propagation the diffraction - reflection interaction has a one-sided character - the main features of the diffraction remain undisturbed while the reflection is distorted by the rarefaction wave and the slipstream. The regime in Fig.2 for the circular

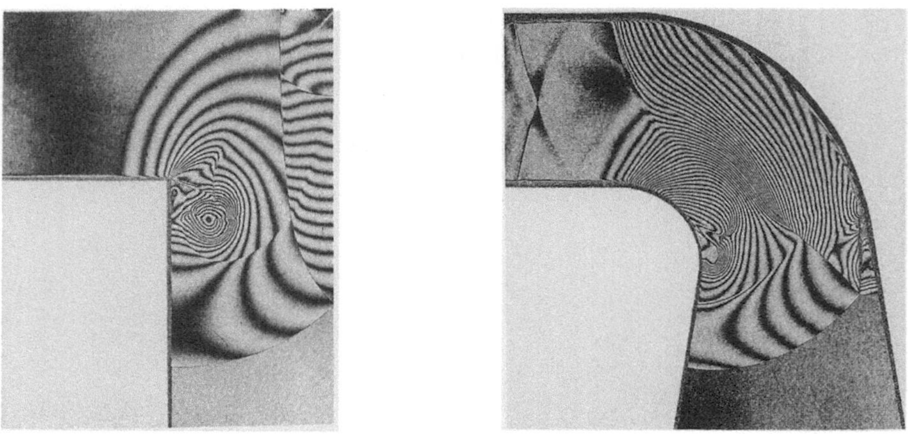

Fig.2. Initial stages of shock wave propagation through channels with rectangular and circular bends at Mach numbers 1.7 and 2.1 respectively.

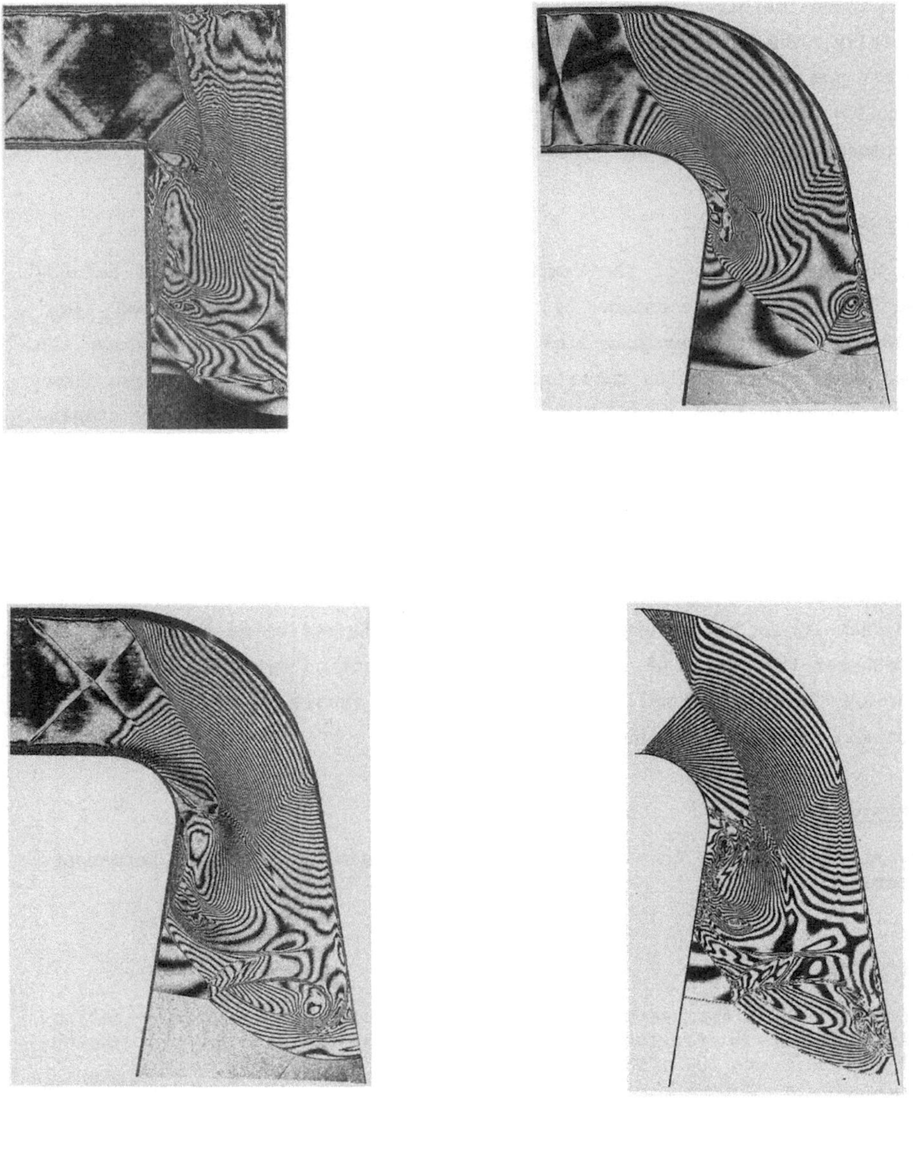

Fig.3. Later stages of shock wave propagation through channels with rectangular and circular bends at Mach number 3. In the second row experimental and calculated interferograms [3] of the same stage are shown.

bend approximately corresponds to the sonic induced flow behind the initially plane incident shock wave. As the induced flow becomes supersonic one can see weak shock waves emanating from prominences in the place of the channel - working section conjunction. The developed interaction of the processes stimulated by the opposite walls one can see in Fig.3.

Conclusive remarks

For the presentation of the computational results the authors of Ref.[3] used the original technique [4] allowing to visualize calculated flow fields in the interferogram-like form. It is interesting to compare the experimental results with those obtained in the frame of the Eulerian flow model. The comparison shows their acceptable agreement. At the early stages of the shock wave propagation the numerical and physical interferograms are actually identical. At the late stages characterized by the interaction between the diffraction and reflection the numerical simulation provides as a whole good agreement with the experimental data. Distinctions appear in the regions where viscous effects in the flow starts to play an essential role, namely, in the domains of interaction of the reflected shock with the vortex near the convex wall as well as of interaction between the boundary layer on the concave wall and the slipstream emanating from the triple point and rolling up in the vortex near the wall.

Acknowledgements

The author wishes to thank Dr. V.A.Komissaruk for the permanent cooperation.

References

1. Kovalev, P.I.; Komissaruk, V.A; Martynov, V.P.; Mende N.P.: Polarization and grating interferometers in ballistic experiment. Opt. and Laser Tech. 11, (June 1979) 137-142.

2. Dukhovskii, I.A.; Komissaruk, V.A.; Kovalev, P.I.; Mende, N.P.: High speed photography of the interaction of a water drop with a supersonic sphere. Opt. and Laser Tech. 17, (June 1985) 148-150.

3. Fursenko, A.A.; Oshima, K.; Sharov, D.M.; Timofeev, E.V.; Voinovich, P.A.: Numerical simulation of propagation of shock waves through channel bends using TVD and ENO schemes. Collection of Technical Papers of the Fourth Int. Symp. on Computational Fluid Dynamics, (Davis, 1991), vol.1, pp.371-376.

4. Tamura, Y.; Fujii, K. Visualization for computational fluid dynamics and the comparison with experiments. AIAA 8th Applied Aerodynamics Conference. (Portland, August 1990) AIAA-90-3031.

Flow Visualization of Secondary Separation and Oscillating Shock Waves in Three-Dimensional Shock Waves-Turbulent Boundary Layer Interaction Region

Shigeru ASO*, Keiichi KARASHIMA**, Kiyoshi SATO**,
Satoshi OKUYAMA* and Shozo MAEKAWA*

* Dept. Aeronautics and Astronautics, Kyushu University, 6-10-1 Hakozaki, Higashi-Ku, Fukuoka 812, JAPAN
** Institute of Space and Astronautical Sciences, Sagamihara, JAPAN

Summary

Secondary separation phenomena in three-dimensional shock wave/turbulent boundary layer interaction regions induced by sharp fin and blunt fins have been investigated. A color oil source method, in which various color oil are supplied through many holes on the interaction surface, are applied for detailed visualization of separated region as well as conventional oil flow technique. Experiments are conducted under the testing conditions of freestream Mach number of 2-4, total pressure of 0.3-0.62 MPa and unit Reynolds number of $2.6\text{-}3.9 \times 10^7$. The results show the new technique is quite useful for detecting the secondary separation. Also quite interesting, complicated features in the separated regions are revealed.

Also oscillating shock waves are visualized by the Schlieren technique using high speed video camera, whose maximum frame speed is 4000 frames per second. The results show quite unsteady properties of the interaction flow fields between shock wave and turbulent boundary layer.

Introduction

Recently, the projects of space planes and winged vehicles are in progress. In those projects, shock wave/turbulent boundary layer interaction phenomena become one of the most important subject for the design of the spaceplanes and winged vehicles. It is hoped to reveal the detailed structure of shock waves/turbulent boundary layer interaction which causes local peak pressure and high aerodynamic heating load.

In those interaction region primary and secondary separations might be observed. Though primary separation is observed clearly by oil flow technique, secondary separation is quite difficult to detect in the interaction region[1,2]. In the present study an intensive study for capturing secondary separation in three-dimensional shock wave/turbulent boundary layer interactions induced by sharp and blunt fins has been conducted intensively by using the supersonic wind tunnel with wider testing section.

Also oscillating shock waves are visualized by the Schlieren technique using high speed video set, whose maximum frame speed is 4000 frames per second. The results show quite unsteady properties of the interaction flow fields between shock wave and turbulent boundary layer.

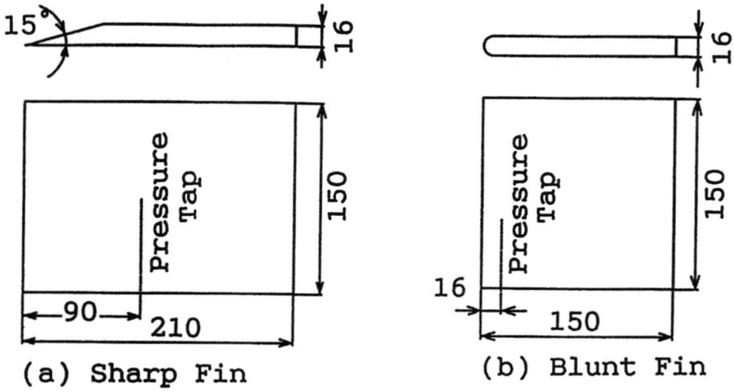

Fig.1 The schematic diagram of the fin models

	α [deg]	M_∞	P_0 [MPa]	T_0 [K]	Re [1/m]
Sharp Fin	12 15 22	4.16 4.02 4.13	0.590 0.618 0.623	289 292 296	2.68×10^7 2.97×10^7 2.93×10^7
Blunt Fin	0 0 0	2.01 3.04 4.18	0.295 0.491 0.590	287 290 285	3.86×10^7 3.82×10^7 2.71×10^7

Table-1 Testing conditions

Experimental Apparatus and Procedures

In the present study the variable Mach number supersonic wind tunnel which is newly installed in the Institute of Space and Astronautical Science. The testing conditions are shown in Table-1. The schematic diagram of the fin models are shown in Fig.1. Adding to the conventional technique a new color oil flow technique is used. Which uses a mixture of Color powder and Silicon oil and various color oil ooze through small holes prepared on the wall and the oil flows are recorded by video camera system during experiments.

Results and Discussions

For sharp fin separated region increases as attack angle is increased as shown in Fig.2. Also at higher attack angle a separated region is divided into two parts. The outer part is located next to the primary separation line and oil flow lines are almost parallel to the primary separation line. The inner part is located between the outer part and fin and 'S' type oil flow lines are quite significant.

For blunt fin separated regions at freestream Mach number of 2, 3 and 4 are visualized as shown in Fig. 3. The results show separated flow patterns changes between Mach number of 3 and 4. In order to reveal the detailed flow structure, a new color oil flow technique is applied to the flow visualization at Mach number of 4. Various color of oil are supplied through small holes located at specified location on the wall during the test as shown in Fig.4. The results show the primary and secondary separation ahead of the blunt fin clearly.

Also oscillating shock waves are visualized by the Schlieren technique using high speed video set, whose maximum frame speed is 4000 frames per second. The instantaneous Schlieren pictures are shown in Fig. 5. The results show quite unsteady properties of the interaction flow fields between shock wave and turbulent boundary layer.

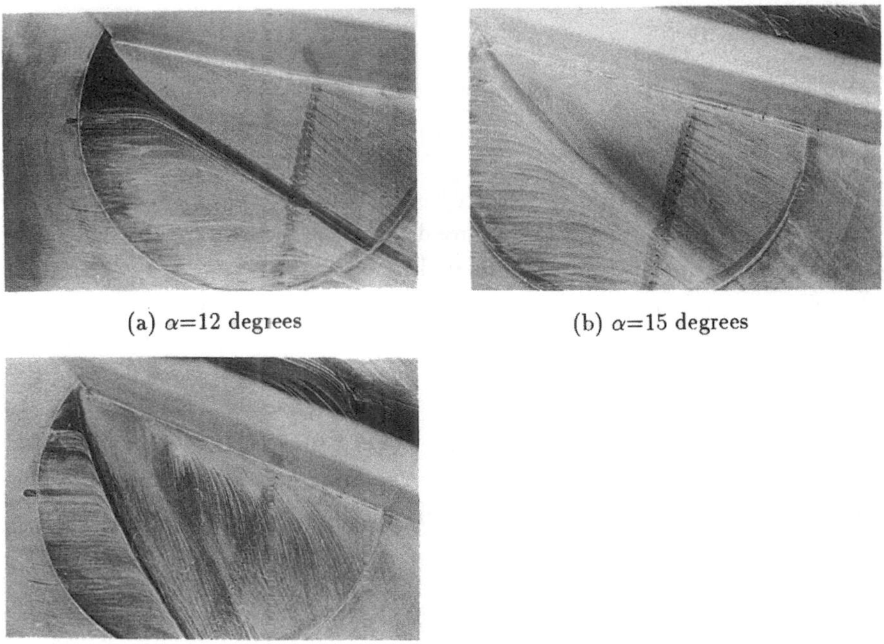

(a) α=12 degrees

(b) α=15 degrees

(c) α=22 degrees

Fig.2 Oil flow pictures(Sharp fin)

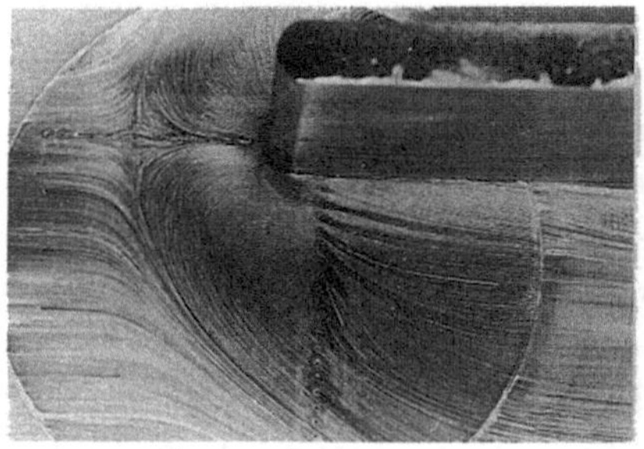

(a) $M_\infty=2$

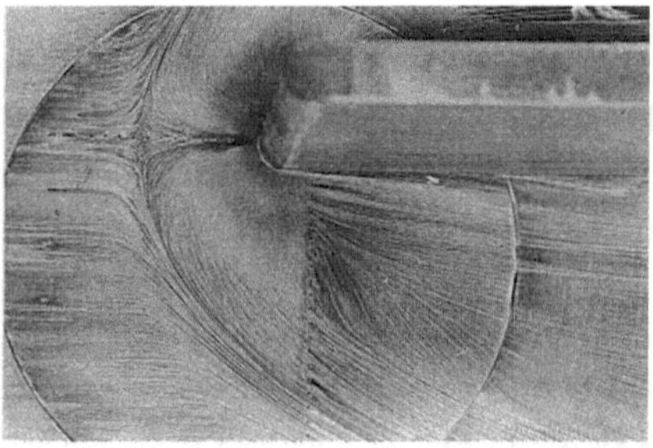

(b) $M_\infty=3$

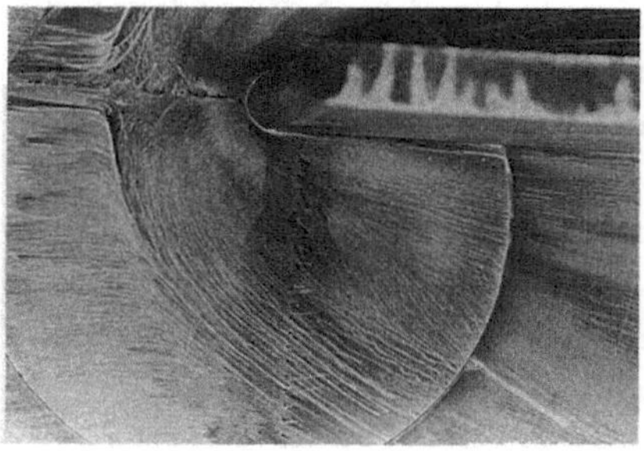

(c) $M_\infty=4$

Fig.3 Oil flow pictures(Blunt fin)

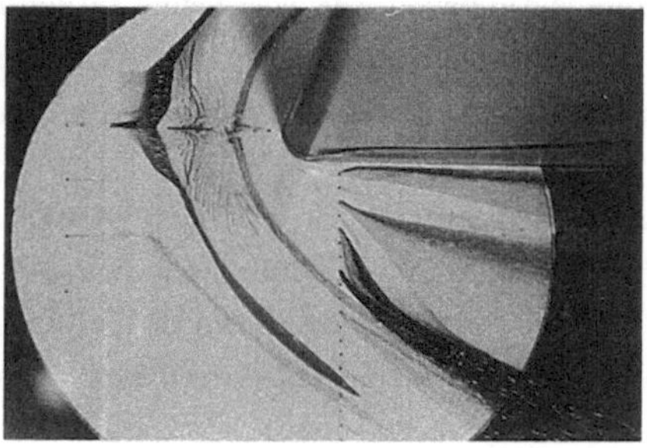

Fig.4 Color oil flow picture (Blunt fin, $M_\infty=4$)

Conclusions

Secondary separation phenomena in three-dimensional shock wave/turbulent boundary layer interaction regions induced by sharp fin and blunt fins have been investigated. A color oil source method are applied for detailed visualization of separated region as well as conventional oil flow technique. The results show the new technique is quite useful for detecting the secondary separation. Also quite interesting, complicated features in the separated regions are revealed.

Also oscillating shock waves are visualized by the Schlieren technique coupled with high speed video camera. The results show quite unsteady properties of the interaction flow fields between shock wave and turbulent boundary layer.

References

1) Hung, C.M.: NASA Tech. Memo. 102266, 1989.
2) Sedney, R. and Kitchens, C.W.: AGRAD CP-168, Paper 37, 1975.

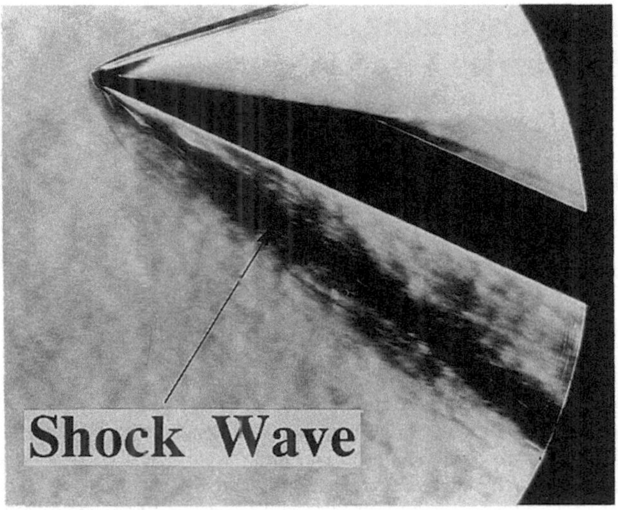

(a) Sharp fin(α=22 degrees)

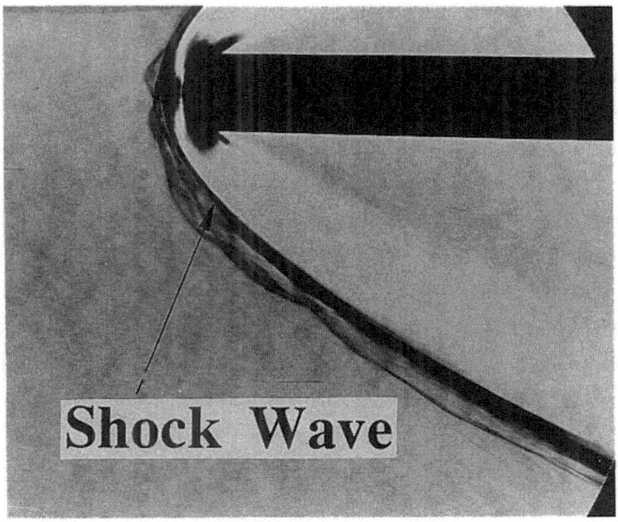

(b) Blunt fin

Fig.5 Visualization of oscillating shock waves(M_∞=4)

Measurement of Oscillating Shock Wave in Supersonic Nozzles

Kazuyasu MATSUO and Heuy-Dong KIM
Graduate School of Engineering Sciences, Kyushu University,
Kasuga, Fukuoka 816, Japan

Summary

The paper describes experimental investigations for the nonintrusive detection of oscillating shock positions in supersonic nozzles. An array of wall-mounted transducers and especially line image sensor were selected for detailed study. The line image sensor was collimated with a conventional schlieren optical system and was a good indicative of capturing the shock flapping motions in the present configuration. This study shows that amplitude of the shock motions increases with approaching flow Mach number. In addition, this system can be effectively applied to a variety of practical problems for determining the shock-wave positions.

Introduction

Flows through supersonic nozzles possess several characteristic features that make their experimental investigation or theoretical prediction exceptionally difficult. The most important one of these features is the shock wave/turbulent boundary layer interaction that is ubiquitous in supersonic-bounded flows. Very frequently such an interaction exhibits oscillations of shock wave [1]. Many past studies of normal shock wave/boundary layer interaction are available from those of unsteady transonic flowfields over airfoils [2]. The wall pressure distribution on the suction side of airfoils is often similar to that on nozzle or diffuser walls, and many of the dynamic features are also similar. In particular, upstream propagation of pressure perturbations with significant amplitude has been established in both situations. However, little is known about how such flows behave in real supersonic nozzles, and the mechanism causing oscillation

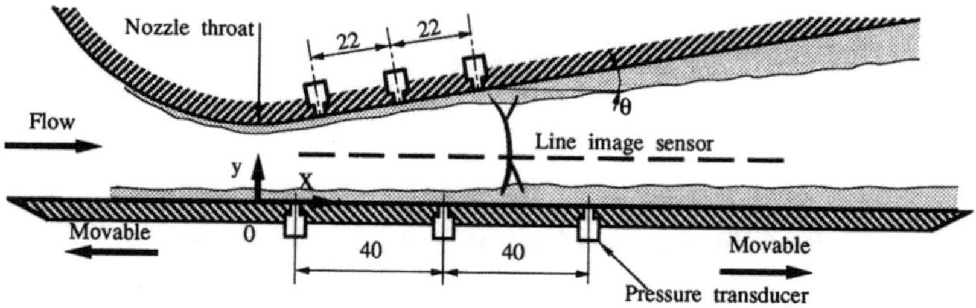

Fig.1. Test nozzle

of the interacting shock wave is poorly understood. On the other hand, recently many interests [3-4] are concentrated on the performance and stability of high-speed, mixed-compression inlets which are strongly affected by the location of the terminal shock wave. The propulsion system is very desirable to include means for determining and controlling the shock position [5]. In order to effectively detect the shock positions and ultimately characterize the mechanism responsible for the unsteady behavior of such shocks, the detailed experimental explorations of two-dimensional supersonic nozzle flows were carried out, as shown in Fig.1. Experiments were carried out in a nozzle with the divergence angle of 9 degrees. Time-mean and fluctuating surface pressures were measured by means of multiple pressure transducers, and visualization of the flowfields was made by means of schlieren optical system which was also employed for a line image sensor. The system was incorporated a 1024-element, line-scan imaging camera, whose output was statistically processed to yield a real-time, analog signals proportional to shock positions. For the line image sensor, the data of about 2 seconds with the 2 kHz sampling rate were adequate to contain all significant information about the shock motions. The resolution of streamwise shock motion was less than 0.04mm. Surface oil-flow tracers were made to measure time-mean separation induced by the interacting normal shock, and the pitot probe for the nozzle wall boundary layer measurements was also employed, in which the boundary layer in approaching flows obeyed both the law of the wall and law of the wake. The wall Mach numbers immediately upstream of the normal shock wave in a range of 1.10 to 1.50, were mainly presented to characterize the interacting shock motions.

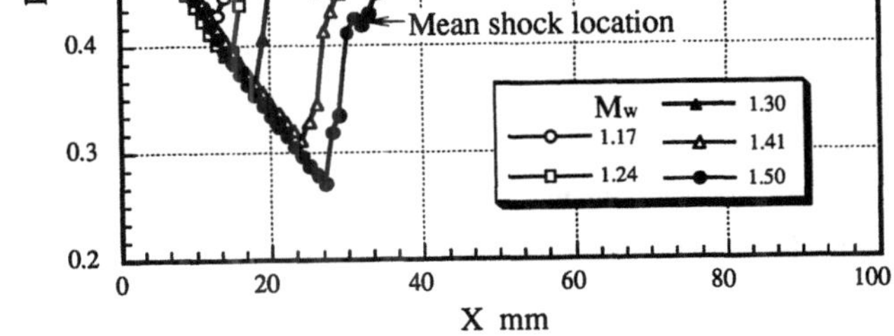

Fig.2. Streamwise distributions of time-averaged wall pressures

Results and Discussion

Typical time-mean wall pressure distributions in supersonic nozzle flow are illustrated in Fig.2 for a range of wall Mach numbers (M_w) immediately upstream of the shock wave. The abscissa indicates the streamwise distance from the nozzle throat. In carrying out experiments, the time-mean shock positions are controlled by a wedge located downstream the test section. The wall pressures (p_w) are normalized by the stagnation pressure(P_0) of the upstream reservoir chamber. When normalized in this fashion, supersonic portions of the wall pressure distributions collapse into a single line, but the subsonic portions do not. The data show that as approaching flow Mach number (M_w) increases, the time-mean shock position moves downstream, and as the Mach number goes beyond 1.30, the boundary layer at point meeting with the front leg of lambda shock separates from wall surface. Associated with such a shock-induced separation, the wall pressures for the Mach number above this value show re-expansion of flow downstream the pressure peak due to compression shock [6], that is, the post-shock expansion. Especially, common characteristic of the interacting shock wave of all strengths is unsteadiness; the shock tends to oscillate at relatively low frequencies. During such oscillations the shock structure itself varies.

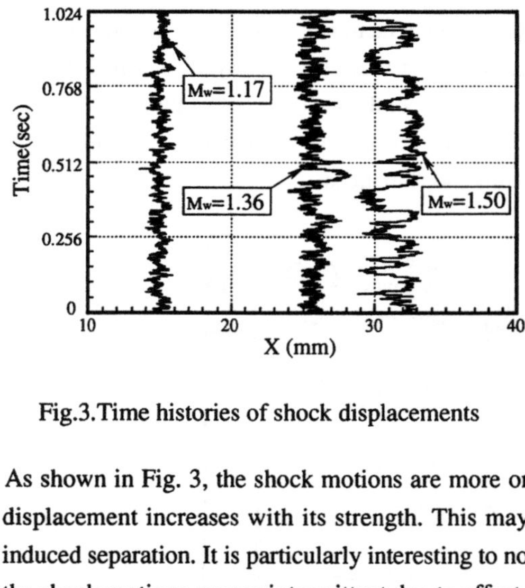

Fig.3.Time histories of shock displacements

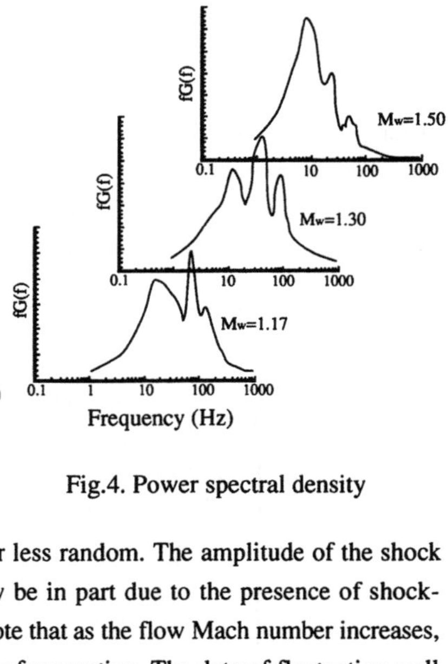

Fig.4. Power spectral density

As shown in Fig. 3, the shock motions are more or less random. The amplitude of the shock displacement increases with its strength. This may be in part due to the presence of shock-induced separation. It is particularly interesting to note that as the flow Mach number increases, the shock motions appear intermittent due to effect of separation. The data of fluctuating wall pressures showed that the location of maximum rms pressure corresponded with the time-mean shock position. Thus the fluctuating wall pressures with large scales can be regarded as being due to the flapping motions of shock. The time scales of the shock motions are presented in terms of power spectral density of displacement fluctuations, as shown in Fig.4. Most of the frequency components of the shock motions shown were below 200 Hz and seemed to have triple peaks related with the characteristic of shock motions. These dominant frequency range may be related with characteristic size of the channel. Fig. 5 indicates the probability density functions(PDFs) for the shock displacements. Inserted sign (↓) indicates ensemble-mean shock position. For the cases of relatively low Mach numbers, the shock motions are likely to nearly follow Gaussian with only small skewness. However, as the Mach number increases over 1.30, the distributions appear significantly skewed. This means the intermittent excursions of shock wave upstream. These appear as double peaks in the distributions.

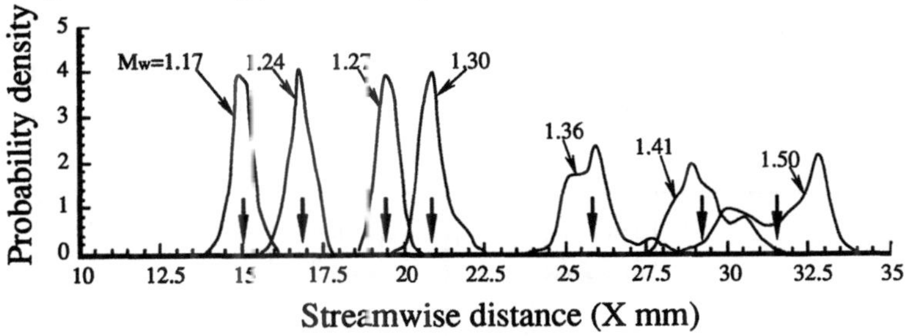

Fig.5. Probability density functions for shock displacements

Furthermore, the spatial extent of shock motions is significantly increased with flow Mach number, comparable with an order of 2-4 times turbulent boundary thickness at beginning of the interaction between normal shock and turbulent boundary layer. Variation of amplitude of displacement fluctuations with the Mach number is presented in Fig.6. Surface oil flow showed that for the wall Mach number over about 1.27 the shock-induced separation occurred. For the flow with shock strength not sufficient to cause the separation, variation of the displacement fluctuations with the Mach number are not so great. However the fluctuation is increased by a factor of four as Mach number increases from 1.27 to 1.50. Thus the flow separation may not be essential to cause the shock oscillations, but rather play an important role to amplify them by any unknown mechanism(s). From time derivatives of the time-displacement diagram for the shock motions, presented in Fig.3, the velocity of interacting shock can be approximately calculated if a linear motion should be assumed between two points.

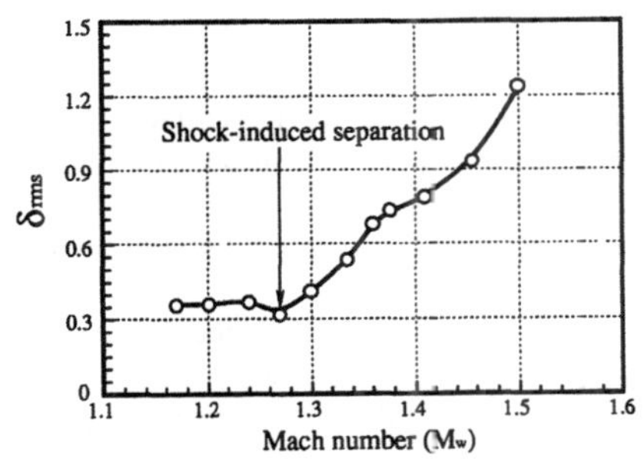

Fig.6. Variation of rms values with Mach number

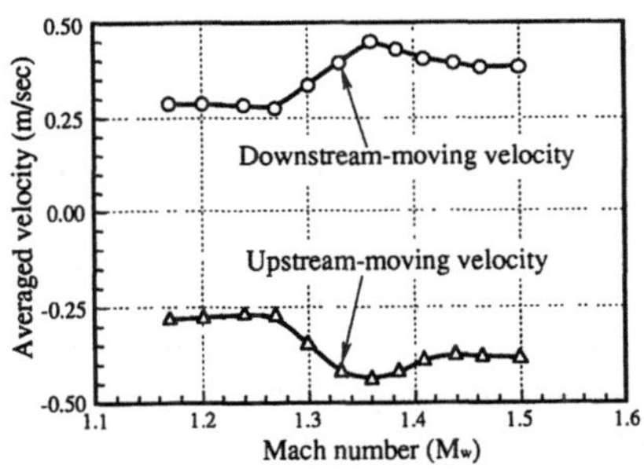

Fig.7. Ensemble-averaged velocity

Fig.7 indicates ensemble-averaged velocities for the shock motions obtained in this fashion. The shock velocities moving up-and downstream is nearly symmetric. As the Mach number is relatively low below 1.27, the velocity are nearly constant. For a larger Mach number, it increases until reaching a peak at about M_w=1.36 and then approaches a nearly constant level. These data show that the velocity of shock is very low, only a small fraction of sonic velocity in approaching flow. The velocities of oscillating shock are also presented in terms of PDF, as shown in Fig.8. The distributions of both up-and downstream-moving velocities are widely spread with a peak at about 0.2. This implies that the shock displacements should be appeared in such a way that fluctuations with relatively large scales are superimposed on small-scaled ones.

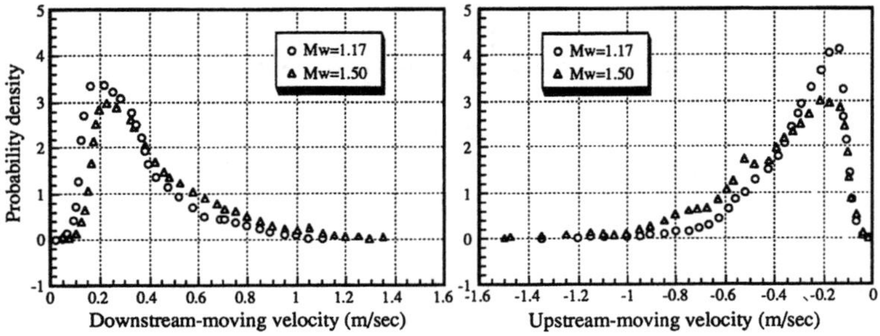

Fig.8. Probability densities of velocity components

It thus seems reasonable to believe that the cause triggering the shock motions should be not in the only one. Further study needs to solve this problem. Finally the data obtained in this configuration are considered representative of the data obtained for all other configurations in the present tests. The line image sensor designed for the detection of normal shock motions was applied to supersonic nozzle flows. By using this system, the displacement fluctuations of interacting shock wave could be determined simply and reliably. Thus this system can be widely used in a variety of flow problems to determine the shock locations.

References

1.Kim,H.D.; Kondoh,H.; Kawagoe,S.; Matsuo,K.: Detection of shock wave oscillation in a supersonic nozzle using line image sensor. Journal of the Flow Visualization Soc.Japan (1991) 11-2 (In Japanese)

2.Levy,L.L.Jr.: Experimental and computational steady and unsteady transonic flowfields about a thick airfoil. AIAA Journal (1978) 16-6.

3.Vinogradov,V.A.; Stepanov,V.A.; Alexandrovich,E.V.: Numerical and experimental investigation of airframe-integrated inlet for high velocities. J. Propulsion and Power (1992) 8-1.

4.Sajben,M.; Bogar,T.J.; Kroutil,J.C.: Response of a supersonic inlet to downstream perturbations. J. Propulsion and Power (1985) 1-1

5.Sajben,M.; Donovan,J.F.; Morris,M.J.: Experimental investigation of terminal shock sensors for mixed-compression inlets. J. Propulsion and Power (1992) 8-1.

6.Kim,H.D.; Matsuo,K.; Kawagoe,S.; Kinoshita,T.: Flow unsteadiness by normal shock wave/turbulent boundary layer interaction in internal flow. International Journal, JSME (1991) 34-317.

Pressure Visualization on Aerodynamic Surface by the Method of Luminescent Coating

G.M.Zharkova, A.I.Maksimov, A.A.Pavlov, V.M.Khachaturyan

Institute of Theoretical and Applied Mechanics, Siberian Division of the Russian Academy of Sciences, Novosibirsk, 630090, RUSSIA

Key words: *aerodynamics, liminescent coatings, pressure, oxygen*

Introduction

The development of drain-free methods for registering gas dynamic parameters on the surface of models is one of the most important tasks of modern aerodynamics. In the recent years under intensive development are the non-contact optical methods of gas flows diagnostics near surfaces. The ideal method for the registration of pressure distribution on a model could be the application of a thin coating on the surface, wherein the optical properties of the coating depend on the static pressure of air in the flow.

One of the promising physical principles underlying the formation of such a coating could be the phenomenon of a luminescent quenching of organic dyes, encapsulated in a polymer matrix, by oxygen molecules contained in the air [1,2]. The most simple method for the visualization of pressure using a luminescent coating is the registration of a stationary value of the intensity of fluorescence or phosphorescence during the continuous excitation of coating. A disadvantage of the method is is the dependence of measured intensity on the parameters of exciting radiation and also on the properties of the luminescent coating used (the thickness of coating, concentration of luminophore molecules). Another nethod for the pressure measurement is the registration of pressure by the neasurement of luminescence attenuation time. If the attenuation law is exponential then the signal being registered in this case will not depend on the intensity of exciting radiation and on the number of excited luminophore molecules.

To visualize the pressure, the organic dyes like tripaphlavin, acrydine orange are used as the luminophores. A demerit of these compounds is the dependence of luminescence quenching on temperature. A more appropriate luminophore in this case could be pyron featuring in lesser temperature dependence. But this compound under the conditions of aerodynamic flow is volatile and that complicates the acquisition of stable results.

The paper is aimed mainly at the search and investigation of luminophores meeting the requirements for an aerodynamic experiment.

Investigation of kinetics of luminescence quenching in air oxygen.

As an object of investigation the complexes of rare earth elements and ceraidonine derivatives were taken [3.4]. The luminophores were dissolved in the polysiloxane rubber featuring in a high gas permeability and in a small time for the oxygen equilibrium concentration in a film to reach the steady state (1 s with the thickness of 20 mkm). The luminophore concentration in the polymer was varied within $5x10^{-3}$ to $1x10^{-3}$ mol/kg. For all the luminophores under study the absorption spectra were meausured.The absorption maximum for the ruthenium complex lies within the spectrum area 450 to 500 nm. The absorption maxima for the europium and palladium complexes lie within the area 500 to 525 nm. The ceramidonine derivatives absorb in the area 600 to 650 nm.

The kinetics of luminescence quenching for luminophores encapsulated in polymer was studied in the vacuum chamber at pressure variation from 1 atm to 0.004 atm. The exitation of luminescence was performed by the laser OGM-40 in the single pulse mode with the wavelength 530 nm. The luminescence attenuation curves were registered by an oscilloscope and a photocamera. The life time is read at the instant when the luminescence intensity falls by *e* times.

The life time of excited molecules of the europium and palladium complexes and the ceramidonine derivatives amounted to 50 ns. The air pressure variation in the chamber within 1 to $2x10^{-3}$ atm did not result in a temporary change in the luminescence attenuation curve. The life time of luminescence of the ruthenium complexes is much greater and depends on pressure. The luminescence attenuation curve of the ruthenium complex in a polymer is described by an exponent. The sample thickness within 5 to 25 mk does not affect the attenuation kinetics.The relaxation times for a sample of 20 mk in thickness were not more than 1 s.

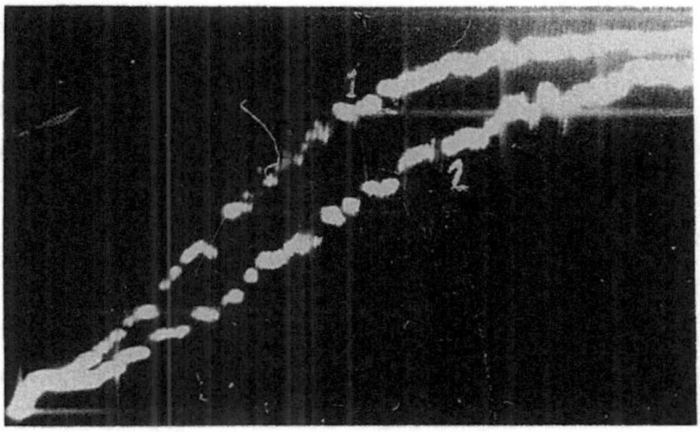

Fig.1. Kinetics of luminescence attenuation of ruthenium complex in a polymer (10^{-3} *mol/l*) at various pressures:
1 - 1 atm, 2 - $2x10^{-3}$ atm; one division = 1 mks

Under atmospheric pressure the luminescence life time 2±0.1 ms for the ruthenium complex increases with the decrease in pressure within the barometric chamber and at the pressure $2x10^{-3}$ atm it amounts to 3 ± 0.1 mks for a newly prepared sample. In the course of time the pressure variation sensitivity drops and in approximately 15 days the coating based on the ruthenium complex becomes inreactive. Such dynamics of time characteristics of luminescence in the ruthenium complex may be explained by its instability. Under the effect of laser radiation and air oxygen there the irreversible changes in the ruthenium complex occur gradually that result in the insensitivity of its excited state to the presence of oxygen.

<u>The tests of luminophores in a supersonic wind tunnel</u>.

The luminescent coatings on the basis of ruthenium complex were tested in a supersonic wind tunnel at the Mach numbers 2.27 - 3.4. The general experimental setup comprises:

1) a system for luminophore excitation; 2) cinema-camera "Kanvas-avtomat;
3) a PC;

The PC serves to fix the time for the development of film frames at the instant of flash sent from the excitation source.

As the test model an angular configuration composed of two plane slabs with sharp front edges was used. The lateral side of the external right angle of

50 mm in width was drained in 6 cross sections so as to measure the pressure distribution. The tests were performed at the angles of attack of the model $\alpha = 16^\circ$. Herewith, a plane sliding shock wave, generated by a special generator being placed at the same angle of attack, was incident upon the lateral side. For in the streamlining about the external right angle an increased pressure arises at the upper side (due to the slanted shock wave directed from the frontal edge of the side), a fairly complex interference diffraction flow is observed.

The nature of flow is fairly well illustrated by a photo of oil-film visualization of the limiting stream lines presented in Fig.4 ($M_\infty = 2.27$). In the free diffraction zone in the frontal part of model due to the pressure drop between the upper and lateral sides a rather intensive stall vortex is formed. As the pressure distribution measurements show, under the nucleus of the stall vortex a considerable rarefaction of flow is observed, and near the external boundary a slight pressure rise occurs.The shock wave falling from the generator interacts with the boundary layer of the lateral side and under its given intensity ($M_\infty = 2.27$; $\xi = P_2 / P_\infty = 2.45$) causes its separation with the formation of explicit outflow line of the limiting stream lines. The outflow line is located much higher than than the trace of the shock wave, as is shown by the shaded line in the photos.

It is known that during the separative interaction of shock wave with the boundary layer a specific pressure distribution occurs in the form of initial rise up to the "plateau" value and subsequent decrease with further abrupt rise up to the value determined by the intensity of the incident shock wave ξ. With the increase in Mach numbers under the same angle of attack $\alpha = 16^\circ$ the intensity of the shock wave rises considerably (at $M_\infty = 2.27$, $\xi = 2.45$, at $M_\infty = 3.0$, $\xi \approx 3$, and under $M_\infty = 4$ $\xi = 4$) resulting in the increase of pressure drops both in the zone of interaction of the shock wave with the boundary layer and in the zone of stall vortex near the rib of the external right angle.

It should be noted that a more complex flow pattern is observed in the region of interaction between the shock wave and the separated flow with the stall vortex caused by the shock wave. In this area abrupt pressure drops are observed . Next downstream the flow pattern becomes more simple essentially and the pressure field becomes almost smooth.

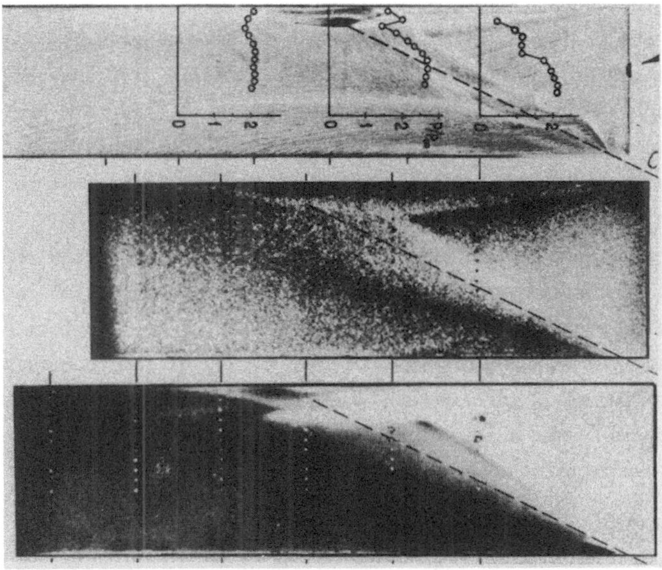

a)

b)

c)

Fig.2. Streamlining about external right angle with the shock wave generator.

a) oil-film visualization of limiting stream lines, $M_{\infty} = 2.27$, $\alpha = 16^{c}$

b,c) visualization of pressure distribution by the method of luminescence coating; b) - Mach = 2.27; c) Mach = 3; $\alpha = 16^{\circ}$;

All the described features of supersonic stream- lining about the model are distinctly registered also my means of luminescent coating. The luminescent coating contains the ruthenium complexes in a polymer matrix (polyvinilpirroli- don). The alcohol solution of the polymer with the luminophore is applied on the surface of model by means of a sprayer. After the solvent has evaporated a thin film of the coating (30 - 40 mk) is formed on the surface of model. The coating adheres readily to the metal surface, has a small roughness (up to 2 mkm). The time necessary for the oxygen equilibrium concentra- tion to reach the steady state is not more than 2 min (within the 2% accuracy).After the wind tunnel reaches the operation mode the light exposure of model is performed and in 2 min the film shooting of the model starts. The pattern of pressure distribution over the model obtained by means of the luminescent coating under various Mach numbers is presented in Fig.4.

With the increase in the Mach numbers the angle of deflection of the shock wave trace reduces and the zone of interaction with the boundary layer shifts downstream (under the fixed position of shock wave generator). Simultaneously there increase the pressure drops and the boundaries of interaction zones. Very distinct is the separation of plateau pressure (moderate pressure rise) from the region of further abrupt pressure rise (dark area in the lower part of the model). At M_∞ 3 the details are distincltly seen in the region of interaction of shock wave with the stall vortex.

Thus, the experiments conducted with a series of luminophores in the wind tunnel T-313 demonstrated the possiblity for their application in visualizing the panoramic pressure distribution over the model surface during its streamlining by a supersonic flow. (M_α = 2.27, 3, 4). The preliminary results on visualization of flow pattern and on the qualitative estimation of pressure levels observed on the model surface testify to that the luminescent coatings prove to be operative.

REFERENCES.

1. Peterson J., Fitzgerald P.V., New technique of surface flow visualisation based on oxygen quenching of fluorescence. *Rev. Sci. Instrum.,* 51 (1980), 670 - 671.
2. Ardasheva M.M., Nevsky L.B., Pervushin G.E., Techniques for measurement of distributed pressures by means of indicator films, *PMTF,* 4 (1985), 24-30, (in Russian).
3. Savinska M., Launay J.P., Mugnier J., Pouget J., Valeun B., Photoluminescence of ruthenium complexes with molecules exhibiting twisted interval change transfer states. *J. Photochem.,* 37 (1987) 69-79.
4. Fokin E.P., Shishkina R.P., Amino groups orientation variations in the direct amination of ceramidonines with alkali, *Izvestiya SO AN SSSR,* 12 (1969), 104-110, (in Russian).

Visualization of a Pseudo-Shock Wave in a Rectangular Duct

H. SUGIYAMA, T. ARAI, N. UNO, T. MATSUMOTO and M. HOMAREDA

Mechanical Systems Engineering, Muroran Institute of Technology
27-1 Mizumoto, Muroran, Hokkaido, 050, Japan

Summary

The internal structure of a pseudo-shock wave in a straight square duct was visualized using a color schlieren and a laser holographic methods. The Mach number just upstream of the pseudo-shock wave was 1.8, and the Reynolds number was 6.8×10^4. The velocity distributions in the pseudo-shock wave were also measured in detail using a two-dimensional LDV system. The density distribution in the pseudo-shock wave and the effect of top and side wall boundary layers on the structure of the pseudo-shock wave were clarified.

Introduction

When a supersonic flow is decelerated to subsonic flow in a duct, a series of plane or lambda shock waves, known as a pseudo-shock wave [1], is produced as the result of shock wave and wall boundary layer interaction. The study of pseudo-shock wave has important implications for the design and operation of scramjet/combuster isolators, ejectors and wind tunnel supersonic diffusers, and so on.

The macroscopic structure and oscillation phenomena of pseudo-shock waves in straight ducts are known well [2]. However, the internal structure of pseudo-shock wave has not been fully clarified[3]. This paper describes visual and LDV investigations of internal structure of pseudo-shock wave in a square duct.

Experimental Apparatus and Visualization Methods

Figure 1 shows the schematic diagram of the experimental apparatus and the laser holographic interferogram system. The test section is a square duct of width D=50mm and length L=1030mm.

The supersonic nozzle was mounted on the top and bottom walls and side-walls, as shown in Fig. 2. The light source of the color schlieren system is a Xe-flash lamp of 1 μsec duration. A pulse ruby laser was used for the holographic system. The velocity profiles of the pseudo-shock wave were measured using a fiber optic dual-beam LDV (TSI Co., SYSTEM 9100). The DOP particles were used as tracer particles.

Color Schlieren Photographs of a Pseudo-Shock Wave

Figure 3(a) shows the black and white prints of colored schlieren photographs of the pseudo-shock wave in the x-y plane (Case A). The location of the pseudo-shock wave was X_f=2.8D. The flow Mach number, boundary layer thickness and Reynolds number just upstream of the pseudo-shock wave were M_1=1.8, δ_1=0.2h and $Re_{\delta 1}=6.8\times10^4$ respectively, where h=25mm. From the top photograph taken with a horizontal color filter, it can be seen how the top and bottom wall boundary layers interact with λ-shape first, second and third shocks. From the bottom photograph with a vertical color filter, expansion regions (dark shaded regions) can be seen qualitatively just downstream of each shock wave.

Figure 3(b) shows the schlieren photographs of the pseudo-shock wave in the x-z plane (Case B). From the top photograph (horizontal color filter), it can be seen that, in this case, the side wall boundary layers interact more intensively with the first shock, resulting in an X-shape first shock wave. From the bottom photograph (vertical color filter), a dark shaded diamond region which is the expansion region is clearly observed.

Holographic Interferograms of a Pseudo-Shock Wave

Figures 4(a) and (b) show the laser holographic interferograms of the pseudo-shock wave corresponding to Figs.3(a) and (b), respectively. The interferometric fringes mean the density contours, so we can observe clearly the density distributions in the pseudo-shock wave. Figure 5 shows the nondimensionalized density distribution along the center line of the duct, which was obtained from the Figure 4(b), where ρ_1 is the air density just upstream of the pseudo-shock wave. From this figure, we can see how the density changes in the pseudo-shock wave and that compression and expansion regions appear alternately in the pseudo-shock wave.

Velocity Profiles in a Pseudo-Shock Wave

To investigate the structure of the pseudo-shock wave and the mechanism of formation of the second shock, velocity profiles between the first and third shock were measured in detail using the LDV system. Figure 6 shows the streamwise velocity profiles measured in the x-y plane (Case A), where U_∞ is the velocity just upstream of the pseudo-shock wave. At location 4, the wall boundary layer thickness increases abruptly. At locations 5, 6 and 7 following the first shock, the peak velocity occurs at the edge of the wall boundary layer. The core flow behind the first shock re-accelerates, and at locations 8 and 9 the core flow becomes a nearly uniform flow outside the boundary layer. The streamwise velocity profiles behind the second shock wave are similar to those behind the first shock wave.

Concluding Remarks

The internal structure of a pseudo-shock wave in a straight square duct was investigated by a color schlieren and a laser holographic interferogram methods, and a LDV system. The effects of the top and bottom wall boundary layers and the side wall boundary layers on the internal structure of the pseudo-shock wave were clarified.

References

1. Crocco, L.: High Speed Aerodynamics and Jet Propulsion, Vol. 3, Ser. B, p. 110, Princeton Univ Press 1958.
2. Sugiyama, H.; Takeda, H.; Zhang J.; Okuda, K.; Yamagishi, H.: Locations and Oscillation Phenomena of Pseudo-Shock Waves in a Straight Rectangular Duct. JSME Int. J, Ser. II, 31-1(1988), 9-15.
3. Sugiyama, H.; Arai, T.; Abe, F.; Takahashi, T.; Takayama, K.: Flow Mechanism of λ-type Pseudo-Shock Waves in a Straight Square Duct. Trans. JSME, 56-522 (1990), 330-335.

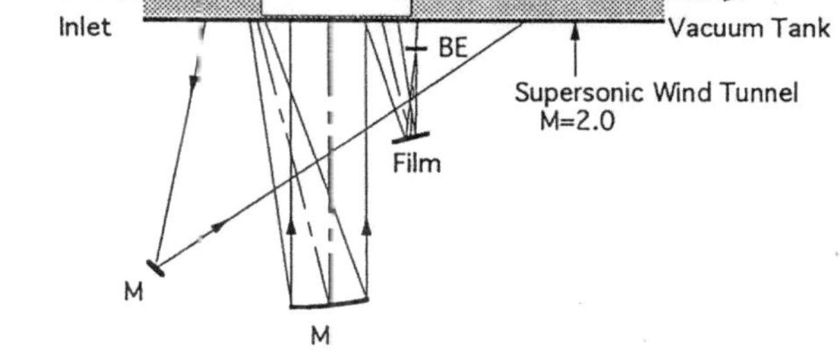

Fig. 1. Schematic diagram of experimental apparatus and holographic interferogram system

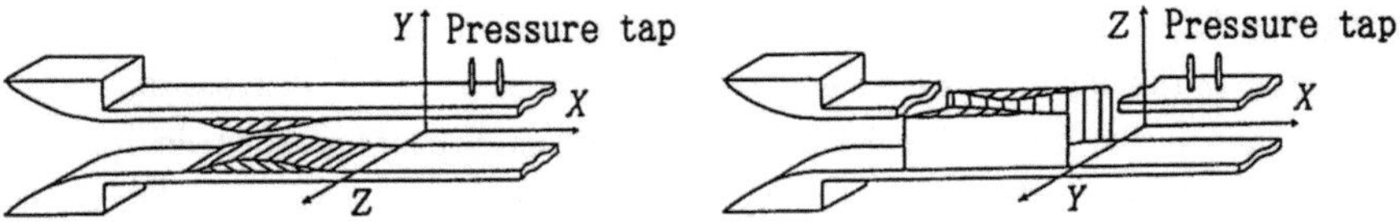

(a) Case A

(b) Case B

Fig. 2. Supersonic nozzle and coordinate system

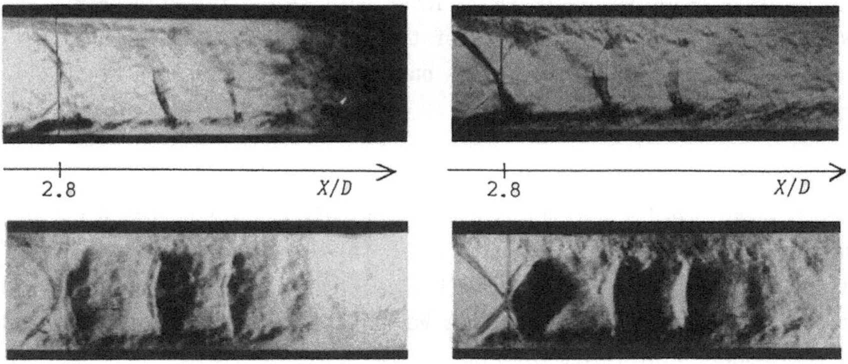

(a) Case A (x-y plane)

(b) Case B (x-z plane)

Fig. 3. Black and white prints of colored schlieren photographs of the pseudo-shock wave in a square duct (M_1=1.8, $Re_{\delta 1}$=6.8×10^4)

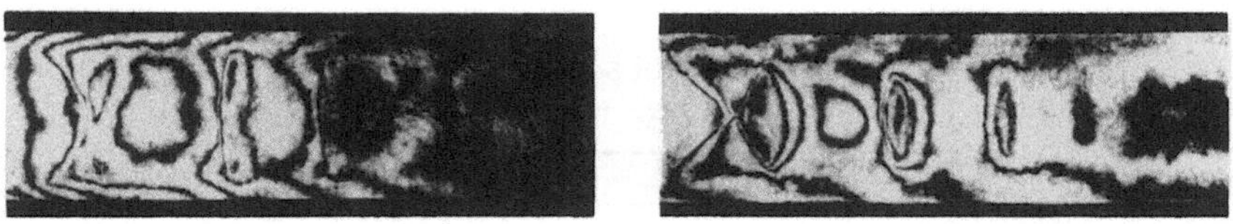

(a) Case A (x-y plane)

(b) Case B (x-z plane)

Fig. 4. Holographic interferograms of the pseudo-shock wave (M_1=1.8, $Re_{\delta 1}$=6.8×10^4)

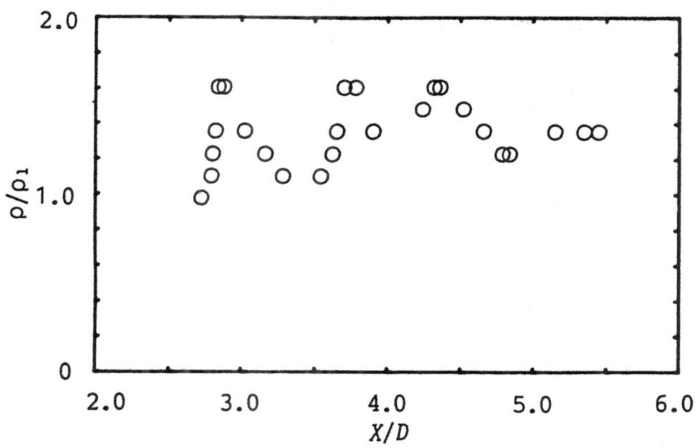

Fig. 5. Density distribution in the pseudo-shock wave along center line of duct

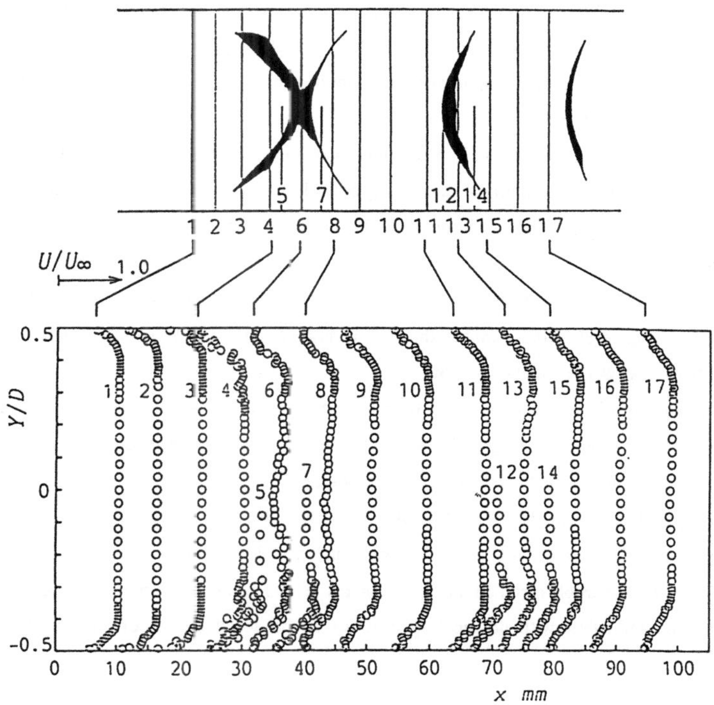

Fig. 6. Velocity profiles in the pseudo-shock wave (Case A, x-y plane)

An Overview of Planar Laser Scattering for the Visualization of High-Speed Airflows

G. S. Settles
Gas Dynamics Laboratory, Pennsylvania State University
University Park, PA 16823 USA

Introduction

The use of planar illumination to observe the scattering from minute particles in high-speed gas flows dates back about 3 decades [1]. Initially, poorly-collimated non-coherent illumination was used to form the light sheet or screen, while the scattering medium usually consisted of an aerosol of moisture particles. This "vapour-screen" technique was generally applied in a purely qualitative fashion. Its name was a misnomer in that the scattering centers were clearly much larger than molecules of water vapor. Subsequent investigators have called this approach Mie scattering, which is also a misnomer in that Mie theory strictly applies only to spherical particles. Some current investigators use the term Rayleigh scattering under the unproven assumption that their scattering particles approach molecular dimensions. We call this method simply Planar Laser Scattering, or PLS for abbreviation.

Whatever one wishes to call it, the PLS technique has some impressive advantages to offer in high-speed flow visualization. Unlike traditional integrating optical methods, it is truly tomographic in nature, allowing one straightforwardly to "slice" a complex 3-D flowfield into a series of 2-D images, whence the 3-D structure may be reconstructed. In that 3-D holography never fulfilled its potential in this regard, PLS stands alone as the technique of choice for the visualization of high-speed 3-D flowfields. Moreover, the advent of modern lasers and imaging electronics makes the PLS technique amenable to careful control and to the capture of quantitative as well as qualitative data.

Many publications have considered PLS imaging in high-speed flows, and it is well beyond current space limitations to review them all. Instead, this brief paper gives an overview solely of recent uses and advances of the PLS technique conducted by the present author and graduate students F. S. Alvi [9, 10], L. N. Cattafesta [7, 8], T. J. Garrison [11], S. S. McIntyre [5,6], and J. W. Naughton [2-4] at the Penn State Gas Dynamics Laboratory. The subject matter of this work includes viscous/inviscid flow interactions in the supersonic range and compressible turbulent mixing problems. These examples stem from research supported by NASA, AFOSR, and ONR with practical applications in the aerodynamics of high-speed flight and airbreathing propulsion.

Examples

The first example is the use of PLS to image a supersonic streamwise vortex and its turbulent mixing with a supersonic external stream [2-4]. Vortices were produced by the vortex injector shown in Fig. 1 so as to flow coaxially with the freestream of a supersonic wind tunnel. The 10-Watt Copper-vapor laser light-screen was oriented perpendicular to the vortex axis and the freestream (air) was seeded with submicron ice crystals, while the vortex (consisting of a Helium-Nitrogen mixture) was unseeded. Figs. 2a and 2b show mean and instantaneous images of the scattered light from a cross-section of the vortex. The intensity distributions of a large array of such images were used to deduce the quantitative mixing rate of the vortex with the freestream [2]. A mixing enhancement of up to 60% was found. Further, ensembles of instantaneous images like Fig. 2b were found to yield the same mixing rate as that of long-exposure images such as Fig. 2a.

Fig. 2c reveals the scattered light from the condensation of Nitrogen in the core of a streamwise Mach 3 Helium-Nitrogen vortex in a supersonic freestream. Similar effects are shown in Fig. 3, where the injected stream is comprised of CO_2. In Figs. 3a and 3c a pure jet with no swirl is imaged. The addition of swirl (Figs. 3b and 3d) leads to low temperatures in the vortex core which cause visible condensation of the CO_2. The "notch" at the top of the jet and vortex cross-sections is due to interference from the wake of the vortex-generator support strut (see Fig. 1).

Figs. 4a and 4b are instantaneous PLS images of an axisymmetric Mach 3 airjet exhausted into room air, taken using a frequency-doubled Nd-YAG laser. The scattering in the turbulent mixing region is due to entrainment of moist atmospheric air by the dry jet and subsequent moisture condensation due to the low static temperature of the jet. Much larger individual scattering centers are also seen in the center of the jet (probably due to rust particles from a new piping installation upstream). The inclined large-scale turbulent structures of compressible mixing are clearly revealed in Fig. 4 [5, 6]. PLS flow visualization has been helpful in gaining a better understanding of compressible turbulent mixing, upon which supersonic combustion ramjet (SCRAMJET) engines for hypersonic propulsion depend.

Fig. 5 [7, 8] illustrates shock wave/vortex interactions created by passing an air vortex seeded with an ethanol aerosol through a Mach disk. The Mach disk, or normal shock, is produced in the exit flow of an axisymmetric supersonic nozzle by overexpansion leading to an irregular shock crossing. In Fig. 5a the combined shock and vortex strengths are only sufficient to cause a weak shock/vortex interaction, but in Fig. 5b a strong interaction with vortex breakdown is produced. A conceptual model of this breakdown, based on PLS and other flow visualization, is shown in Fig. 5c. We believe that the structure of such a shock/vortex interaction has never before been visualized. Moreover, the PLS technique is probably the only available means of revealing the internal structure of this complex flow. The images of Figs. 5a and 5b were obtained from single 20 ns pulses of a frequency-doubled Nd-YAG laser.

Next, PLS has been used to study the structure of fin-generated, swept shock wave boundary layer interactions [9, 10]. Fig. 6a shows the experimental setup, Fig. 6b gives a typical PLS image, and Fig. 6c is a scaled flowfield map derived from the PLS images and other flow imaging data. Such images led to the discovery of the impinging-jet structure which is responsible for high local heating, shear, and pressure levels on flight vehicles. Fig. 6c also shows that the entire boundary layer is scooped up into the helical vortex generated by such an interaction. Since the interaction is found to be very nearly conical, a single PLS cut normal to the conical interaction rays is sufficient to reveal the structure of the flow.

Fig. 7 consists of a diagram and 6 individual PLS slices of the complex 3-D interaction of an oblique shock wave and a supersonic streamwise vortex [8]. The vortex is generated by the apparatus of Fig. 1, while the oblique shock arises from a wedge mounted in the wind tunnel freestream. The curved oblique shock is seen to be just touching the bottom of the vortex in frame 1. By frame 6 a complex 3-D shock/vortex interaction has occurred. Ordinary schlieren images of this flow (not shown here) give almost no clue to its complex structure.

The final PLS flow visualization example is the interaction of crossing oblique shock waves with a turbulent boundary layer at Mach 4 [11]. We have recently obtained the world's first flowfield visualizations of this interaction using the arrangement illustrated in Fig 8a and seeding of the wind tunnel freestream by submicron ice crystals. The 10-Watt Copper-vapor laser screen is scanned from the front to the rear of the interaction, revealing the flow structure to a miniature CCD camera mounted in the flow downstream and moving in tandem with the laser screen. Frames B-I constitute a sequence of individual images taken from the resulting videotape and showing the merging and crossing of the two single-fin interactions which serve as input to this complex flowfield. A tomographic analysis of such images has allowed a 3-D model of the crossing-shock flowfield structure (Fig. 8j) to be drawn. Among other conclusions, it was found that a significant bell-shaped region (see Fig. 8I) is formed in which the fluid has low momentum. This region tends to defeat the purpose of sidewall compression in the inlet of a high-speed airbreathing engine.

References

1. McGregor, I., "Development of the Vapour Screen Method of Flow Visualization in a 3ft x 3ft Supersonic Tunnel," in AGARDograph 70, "Flow Visualization in Wind Tunnels Using Indicators," compiled by R. L. Maltby, April 1962.

2. Naughton, J. W., and Settles, G. S., "Experiments on the Enhancement of Compressible Mixing via Streamwise Vorticity, Part I - Optical Measurements," AIAA Paper 92-3459, July 1992.

3. Naughton, J. W., Cattafesta, L. N., and Settles, G. S., "An Experimental Study of the Effect of Streamwise Vorticity on Supersonic Mixing Enhancement," AIAA Paper 89-2456, July 1989.

4. Naughton, J. W., Ph.D. Dissertation, M.E. Dept., Penn State University, 1992.

5. McIntyre, S. S., Ph.D. Dissertation, M.E. Dept., Penn State University, 1992.

6. McIntyre, S. S., and Settles, G. S., "Optical Experiments on Compressible Turbulent Mixing Layers," AIAA Paper 91-0623, January 1991.

7. Cattafesta, L. N., and Settles, G. S., "Experiments in Shock/Vortex Interaction," AIAA Paper 92-0315, January 1992.

8. Cattafesta, L. N., Ph.D. Dissertation, M.E. Dept., Penn State University, 1992.

9. Alvi, F. S., and Settles, G. S., "A Physical Model of the Swept Shock Wave/Turbulent Boundary-Layer Interaction Flowfield," AIAA Paper 91-1768, June 1991, (to be published in *AIAA Journal*).

10. Alvi, F. S., Ph.D. Dissertation, M.E. Dept., Penn State University, 1992.

11. Garrison, T. J., and Settles, G. S., "Flowfield Visualization of Crossing Shock-Wave/Boundary Layer Interactions," AIAA Paper 92-0750, Jan. 1992.

Keywords: Interacting Flows, Shock Wave Interaction, Lasers, Particle Tracers, Gas Dynamics

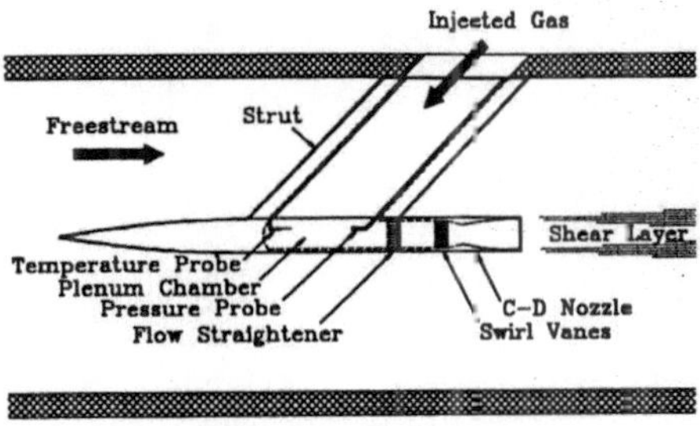

Fig. 1 - Vortex Injector in Wind Tunnel.

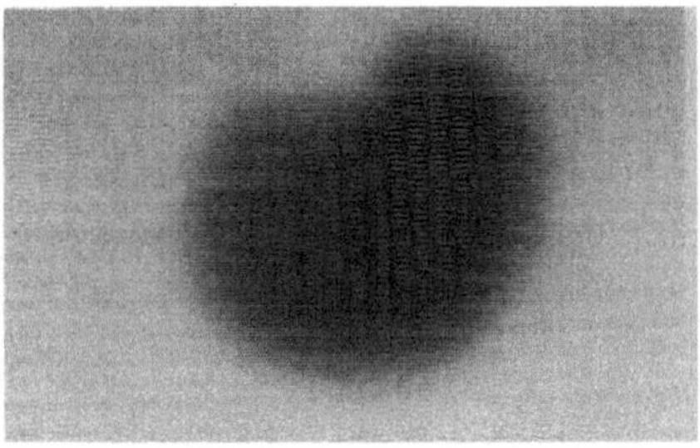

Fig. 2a - Mean Image of M=3 He-N_2 Vortex.

Fig. 2b - Instantaneous Image of M=3 He-N_2 Vortex.

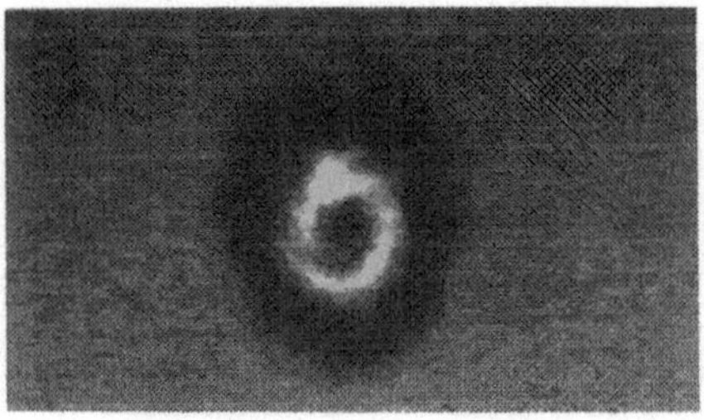

Fig. 2c - N_2 Condensation in He-N_2 Vortex Core.

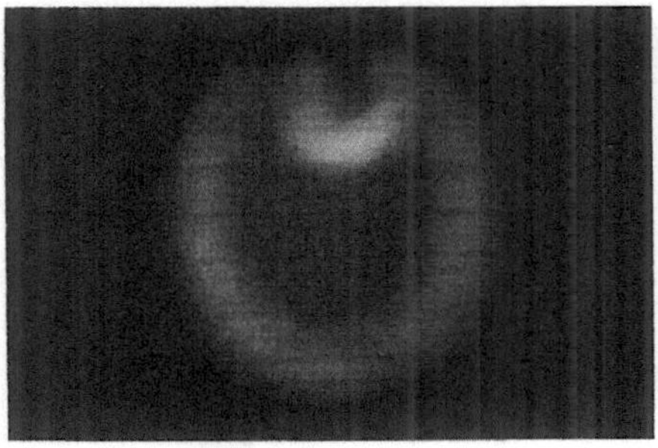

Fig. 3a - Cross-Section of CO_2 Jet.

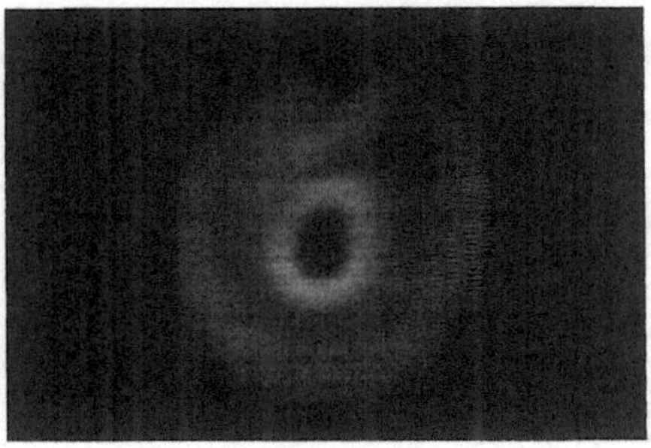

Fig. 3b - Cross-Section of CO_2 Vortex.

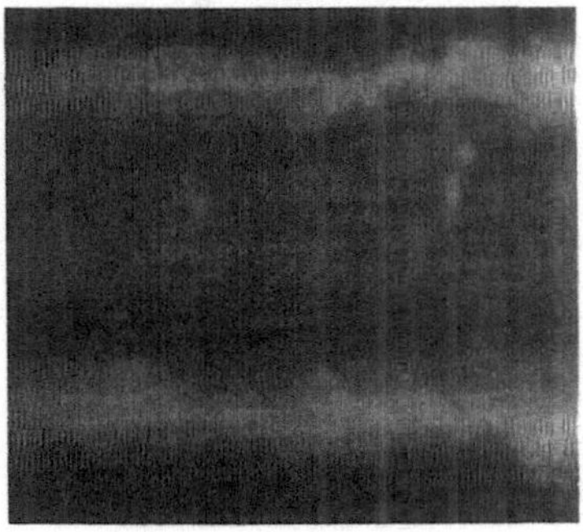

Fig. 3c - Streamwise Cut of CO_2 Jet.

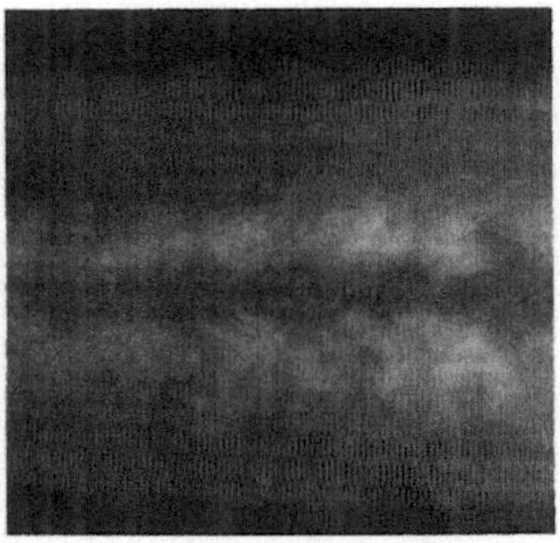

Fig. 3d - Streamwise Cut of CO_2 Vortex.

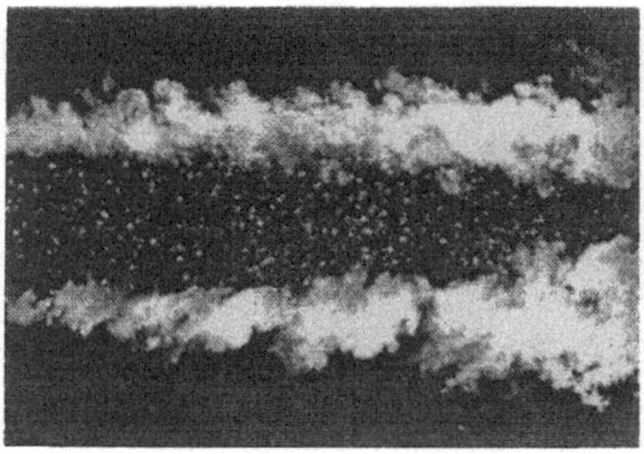

Fig. 4a - Streamwise Cut of Mach 3 Airjet.

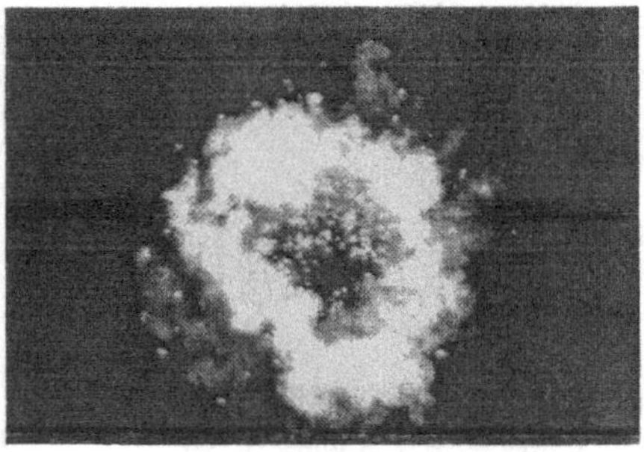

Fig. 4b - Cross-Section of Mach 3 Airjet.

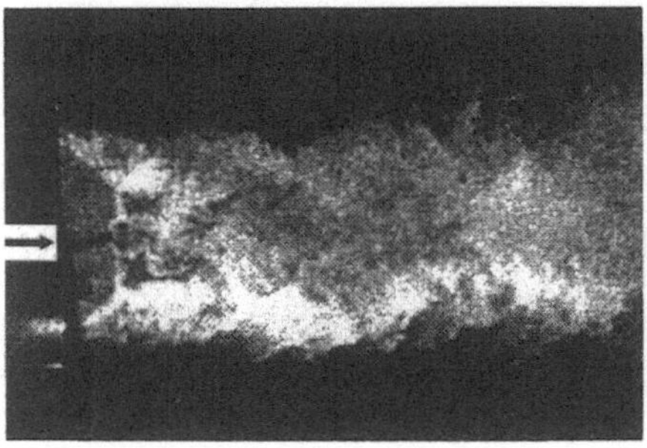

Fig. 5a - Weak Shock/Vortex Interaction.

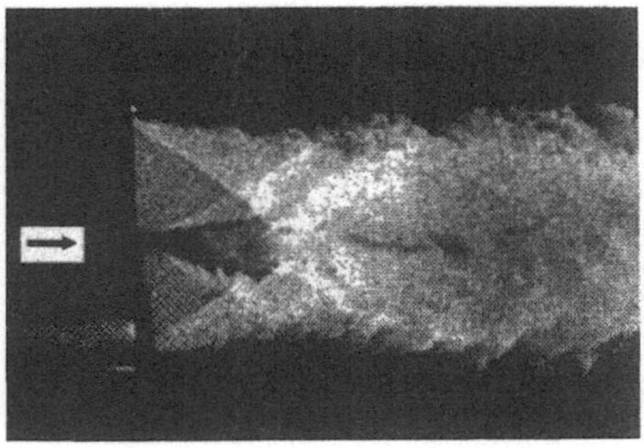

Fig. 5b - Strong Shock/Vortex Interaction.

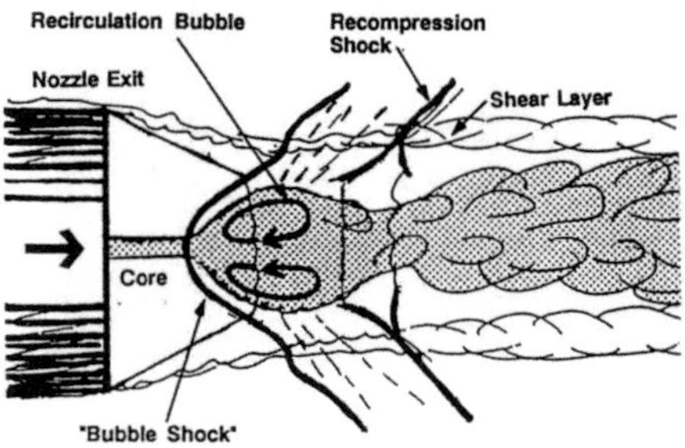

Fig. 5c - SWVI Conceptual Flow Model.

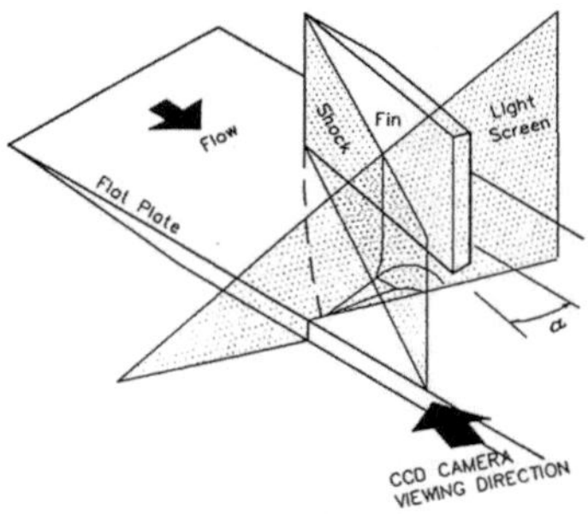

Fig. 6a - SWBLI Model Diagram.

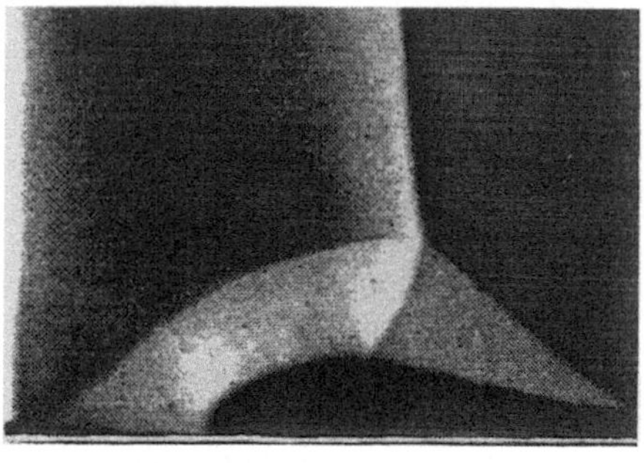

Fig. 6b - PLS Image, M=3, $\alpha=16°$

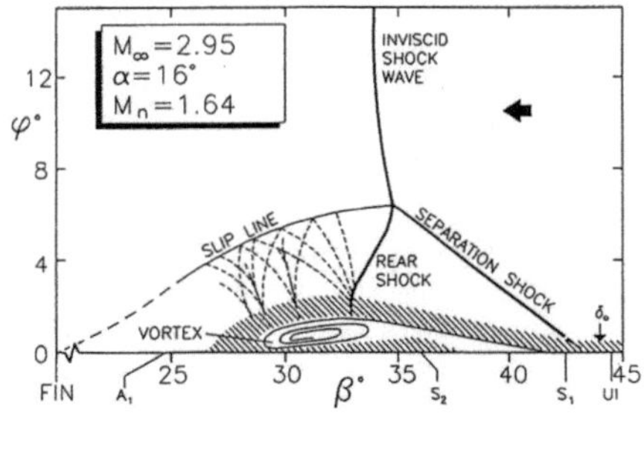

Fig. 6c - Physical Model of Flowfield.

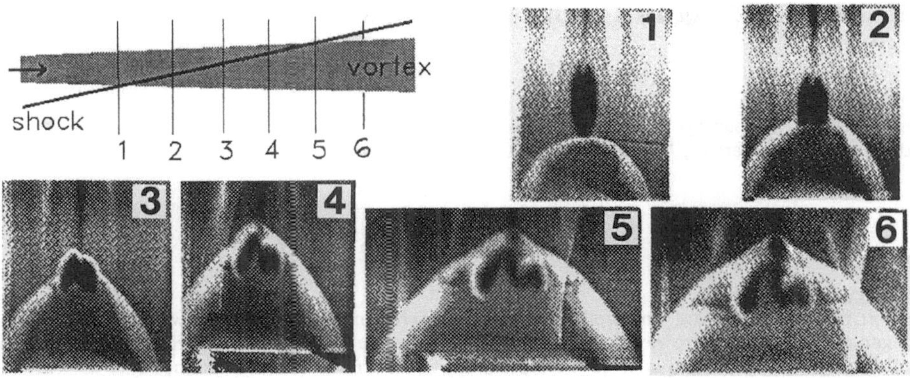

Fig. 7 - Oblique Shock Wave/Vortex Interaction: Series of Transverse PLS Cuts Through Flowfield.

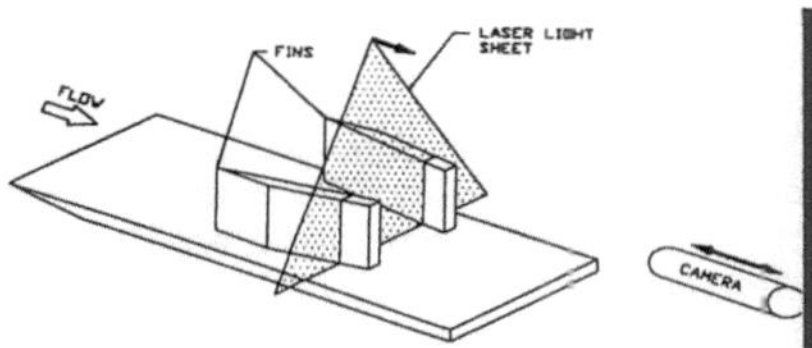

Fig. 8a - X-Shock Interaction Test Model.

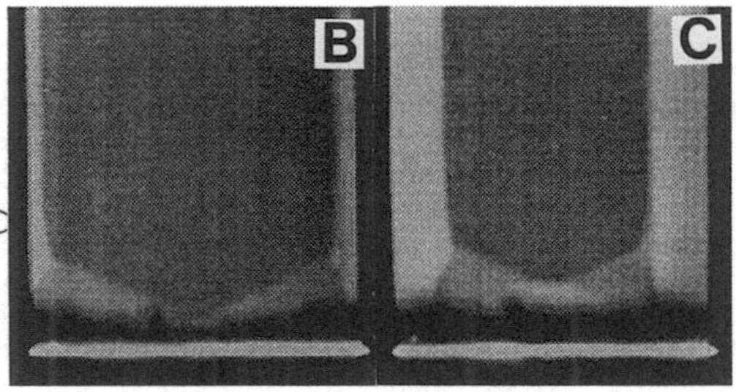

Fig. 8B-I - PLS Flowfield Cuts (Front-to-Back).

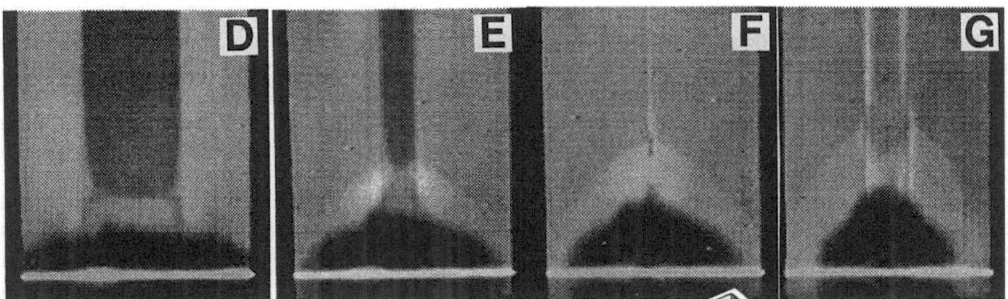

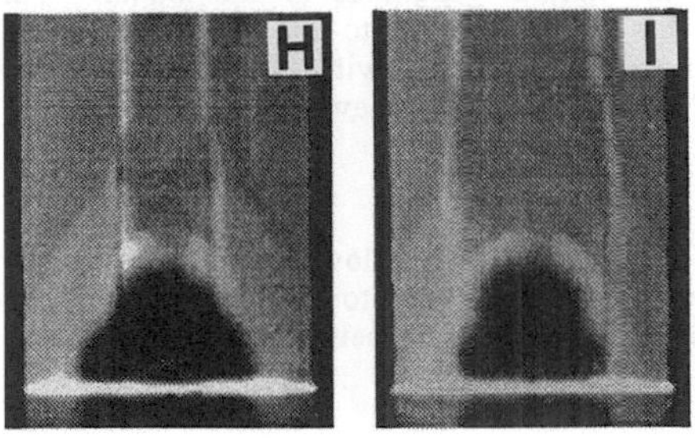

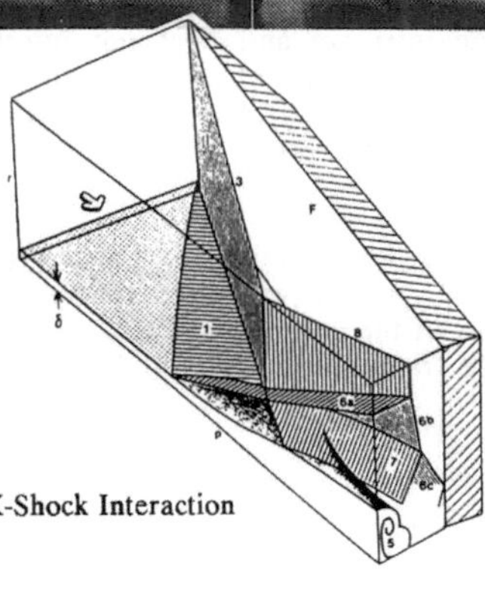

Fig. 8j - Physical Model of X-Shock Interaction Flowfield.

Infrared Imaging in Cryogenic Wind Tunnels

Ehud Gartenberg
Department of Mechanical Engineering and Mechanics
Old Dominion University, Norfolk, Virginia 23529-0247, USA

Summary
The principles governing infrared imaging for aerodynamic surface-flow visualization in cryogenic wind tunnels are reviewed, with emphasis on boundary-layer transition detection. The applicability of the technique to low temperatures is illustrated through examples of transition, shock and vortical burst detection.

Introduction

Experimental verification of drag predictions for airplanes on wind-tunnel models depend upon proper reproduction of the boundary layer, as it would occur in flight. The boundary-layer development, its transition to turbulence, shock interaction and possible separation, are Reynolds and Mach numbers dependent, and hence the need to test models at high Reynolds numbers in subsonic and transonic testing. Subsonic and transonic cryogenic wind tunnels are the only facilities with proven capability of testing models of large transport airplanes at Mach and Reynolds numbers combinations occurring in actual flight. This capability is made possible by testing in gaseous nitrogen at low temperatures, down to 100 K and moderate pressures, up to 10 atmospheres. NASA Langley Research Center, where the results reported herein were obtained, has two cryogenic wind tunnels, the National Transonic Facility (NTF) and the 0.3-m Transonic Cryogenic Tunnel (0.3-m TCT). The most capable installation is the NTF which has a test section of 2.5 X 2.5 m^2 and can generate Reynolds numbers up to $490X10^6$ per meter, at Mach 0.85, total temperature of 110 K and total pressure of 8.5 ata. For comparison, large commercial airplanes fly at mean chord Reynolds numbers up to $60X10^6$. The 0.3-m TCT is smaller and it is used mainly for 2D airfoil testing. It has a test section of 0.3 X 0.3 m^2, identical minimum temperature capability as the NTF, but is limited to total pressures of 6 ata, and its capability to generate Reynolds numbers is limited accordingly.

Recently renewed interest in laminar flow technology for large transport airplanes raised the need to confirm experimentally the location of laminar boundary-layer transition to turbulence in cryogenic, high Reynolds-numbers flows. For transition detection in wind-tunnel testing, infrared (IR) imaging is a well established technique. It is attractive because it is global, fast, clean, non-intrusive and does not interfere with the natural roughness of the model. For cryogenic wind-tunnel application, this technique has to be adapted to the low temperature environment. In what follows, the principles governing this application will be enumerated,[1] and a few experimental results obtained in the 0.3-m TCT will be briefly reviewed.[2]

Thermal Signature of Boundary-Layer Transition

The thermal signature of boundary-layer transition in thermally stable flows is characterized by an increase in the adiabatic wall temperature between the areas exposed to laminar and turbulent regime. At transonic speeds and ambient temperatures, the difference between the laminar and the turbulent adiabatic wall temperature is a few degrees K. At typical operational temperatures in cryogenic wind tunnels, around 110 K, this difference drops below 1 K. Thermal enhancement of the transition and turbulent regions can be obtained by convectively heating the model, taking advantage of the increased value of the heat transfer coefficient in the turbulent versus the laminar regime. Convective heating of the model is conveniently done, through the wind-tunnel control sytem, by assigning the flow a higher temperature relative to the current running condition.

IR Radiation Emission and Detection at Low Temperatures

Analysis of the Plankian law describing black-body radiation indicates the following effects of decreasing surface temperature on the infrared energy power flux:

1. The overall radiated power decreases proportionally with T^4.
2. The peak of the spectral radiation moves to increasingly longer wavelengths. For example, at 300 K, the radiation power peaks at 10 microns; at 100 K, the radiation power peaks around 30 microns and the photon emission peaks around 36 microns.
3. 75 percent of the total energy at any surface temperature is radiated in the waveband between the peak-power wavelength and infinity.

These facts combined with the sharp cut-off sensitivity of IR detectors at their upper limit of wavelength sensitivity, causes an ever decreasing sensitivity of IR imaging systems at temperatures below the nominal operating conditions.
Thus, it is desirable to scan low temperatures targets with long wavelength sensing IR imaging systems. Current commercial "long wavelength" IR imaging systems operate in the 8- to 12-micron waveband. This waveband is not ideal for scanning targets in the range of temperatures under consideration, but it is the better choice when compared with "short wavelengths" systems that operate in the 2- to 5-microns waveband.
Models for cryogenic tunnels are usually metallic and highly polished, to prevent boundary-layer instability. These characteristics have two consequences:

1. The thermal signature of transition on the model surface is leveled-out through longitudinal heat conduction.
2. The radiation is attenuated by low surface emittance (characteristically below 0.1), and the resulting high reflectance, that decrease the signal to noise ratio.

These features are true for metallic models in general, and are not particular to cryogenic models. The generally accepted solution to these problems is to use models with composite skins, e.g., fiberglass, or with epoxy inserts or coatings.
Installation of the camera outside the tunnel requires IR transparent windows cut of materials that should be compatible with the cryogenic wind-tunnel environment, i.e., capable of withstanding the combined pressure and thermal loads. Furthermore, the windows should be anti-reflective coated to enhance their transmittance in the camera sensing waveband. Germanium and Zinc-Selenide, for example, proved to be useful for this application.

Test Procedure and Results

The feasibility of transition detection was investigated at NASA Langley Research Center in the 0.3-m Transonic Cryogenic Tunnel using a commercial 8- to 12-micron IR imaging system (Fig. 1).
The target was a 9-inch chord, supercritical SC(3)-0712 airfoil, with an epoxy-covered fiberglass skin (Fig. 2).

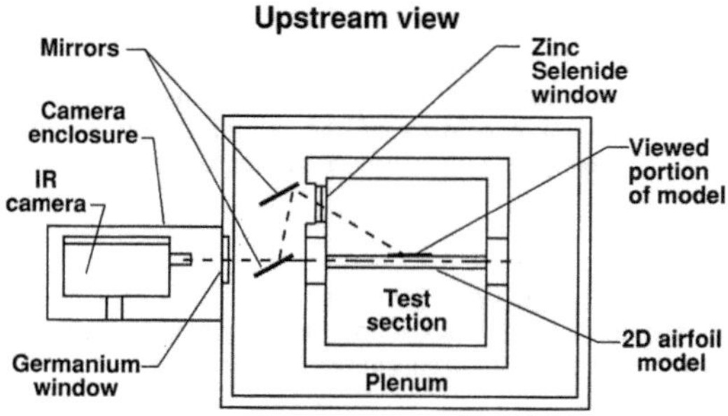

Fig. 1. Optical path of IR radiation from airfoil to IR imaging system in the 0.3-m TCT.

Fig. 2. Airfoil area scanned by IR imaging system.

Since previous experience indicated that a direct imaging approach may not produce the desired results, the thermal signature of boundary-layer transition was enhanced by a combination of means. The adopted procedure was to superimpose the effect of a low and controlled transient heat transfer on the steady, adiabatic wall temperature distribution; to subtract the "steady state" thermogram from the "transient" thermogram; and to use binary shading on the final thermograms (light for laminar areas and dark for turbulent ones) to intensify the visual contrast between the two boundary-layer regimes. Using this method, transition was detected on a composite model at temperatures down to 170 K (Figs. 3 & 4).

Other flow phenomena of interest that were also detected included shock waves with boundary-layer separation and reattachment (Fig. 5), and laminar vortical wedges undergoing transition to turbulence (Fig. 6).

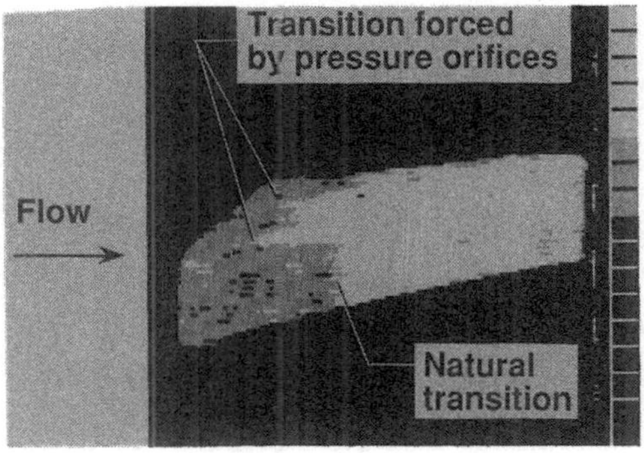

Fig. 3. IR imaging transition detection at 220 K.
T_t = 220 K, ΔT = 1.0 K, α = -1.5°, M = 0.56, R_c = 4.7 X 10^6.

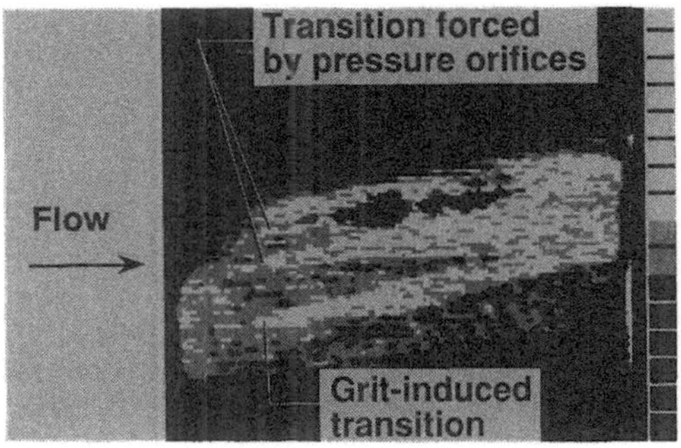

Fig. 4. IR imaging transition detection at 170 K.
T_t = 170 K, ΔT = 1.5 K, α = -5°, M = 0.66, R_c = 8.5 X 10^6.

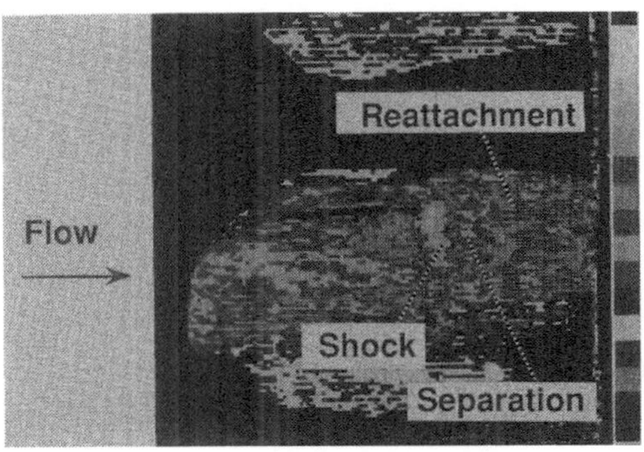

Fig. 5. IR imaging shock detection at 200 K. Notice boundary-layer separation and reattachment.
T_t = 200 K, ΔT = 1.5 K, α = 3°, M = 0.69, R_c = 7.8 X 10^6.

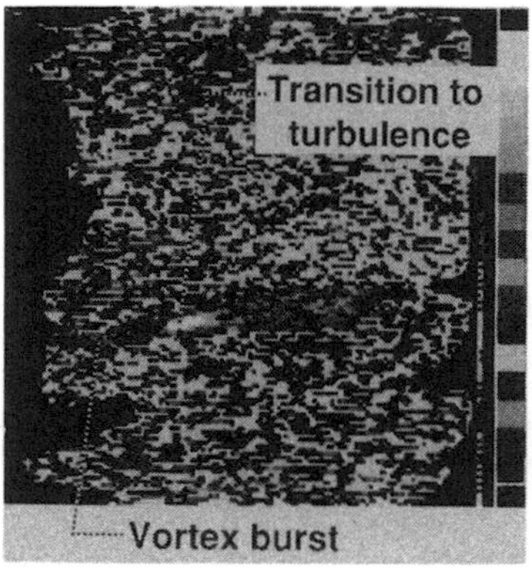

Fig. 6. IR imaging detection of laminar vortex burst at 230 K. Notice transition to turbulence . Vortex duration about 6 sec.

Acknowledgment
This work was supported under NASA grant NAS 1-18584-58.

References

1. Gartenberg, E. and Wright, R.E. Jr., "Principles of Detecting Boundary-Layer Transition in Cryogenic Wind Tunnels by Infrared Imaging," Proceedings of the International Congress on Instrumentation in Aerospace Simulation Facilities (ICIASF '91), Rockville, Maryland, U.S.A., IEEE, October 1991, pp. 156-168.

2. Gartenberg, E., Johnson, W.G. Jr., Johnson, C.B., Carraway, D.L. and Wright, R.E., "Transition Detection Studies in Cryogenic Environment," Proceedings of the 8th Applied Aerodynamics Conference, Portland, Oregon, U.S.A., AIAA, August 1990, pp. 234-244.

Gas Flows Imaging by Nanosecond Gas Discharge

GAS FLOWS IMAGING BY NANOSECOND GAS DISCHARGE

A.A.ALYAPIN, I.V.FILIOUGINE, S.V.KOSTIOUCHENKO, N.N.KOUDRIAVTSEV

Moscow Institute of Physics and Technology.
Moscow Region, Dolgoprudny, Russia, 141700

Summary

A method of separate species concentration flow-field imaging in the high enthalpy gas flows have been developed. The method is based on instantaneous volume uniform excitation of the gas by nanosecond gas discharge with subsequent radiation spatial distribution recording. An experimental facilities consist of a shock tube with a discharge section for flow area 80x1200 cm×cm imaging. The investigations of nonequilibrium chemical kinetics behind incident shock wave in nitrogen, oxygen and air and in associated flow field of a solid body in hypersonic flow of these gases have been performed.

Introduction

Several classical techniques such as *interferometry* [1], *speckle* [2], *laser-doppler* [3], etc. are widely used in modern gasdynamic investigations for flow fields imagination [4]. These methods permit somebody to obtain gas density and velocity spatial distributions pertaining for various gasdynamics phenomena. For investigations in modern hypersonic gasdynamics where gasdynamics phenomena are accompanied by chemical reactions, it's important to have, a time-resolved monitoring of the spatial distribution of the gas mixtures composition. Some techniques using high-power pulsed lasers are now under intensive developing. One of the most attractive for time-resolved gasdynamic investigation is *Laser-Induced Fluorescence* method (LIF) [5], enabling point-to-point concentrations of some species (NO, etc.) measurements, or under modification, obtaining the pictures of their spatial distribution. But LIF may be used only for the limited numbers of species which are able to absorb laser radiation and lets a single species to be monitored during the experiment.

Mixture composition imaging

In this work an universal method for simultaneous, time-resolved ($\tau \leq 10^{-8} \div 10^{-6}$s) and separate visualization of the several species in the high-enthalpy gas flows have been described. The background of this method is an instantaneous in the gasdynamic time scale and volume uniform excitation of atoms and molecules by *Nanosecond Gas Discharge* (NGD) [6] with subsequent recording of radiation arising in spontaneous transitions from short living quantum states of the particles. This technique have been realized on the *Shock Tube* (ST) facilities for investigation of the modern gasdynamics problems: modeling big Space Shuttle type spacecrafts reentering, hypersonic flights of aircrafts in the rarefied atmosphere, and *Shock Wave* (SW) propagation through the reactive gas mixtures and plasma.

Nanosecond gas discharge

It seems only this type of discharge techniques gives an unique opportunity for uniform, instantaneous gas excitation in large volume. NGD arises after supplying of the high voltage pulse with a very short rise time about several nanosecond and amplitude 10^4-10^6 V to the gas gap. A discharge develops from electrodes as uniform *Breakdown Ionization Wave* (BIW), having sublight velocity and so it provides instantaneous gas excitation on time scale of gasdynamics processes. For density ranges 10^{15}-10^{19}cm^{-3}, which is of gasdynamic standpoint interest, this excitation may be done also volume uniform. The main portion of the discharge energy ($\geq$98%) is converted into excitation of electronics and vibrations of the gas. By recording of a radiation intensity picture in appropriate spectral bands it's possible to obtain concentration distribution of the separate gas species. The resolution time determine only on decay of exited particles (both radiative and nonradiative) and have typical values 10 ns - 1 μs), which corresponds to spatial resolution in hypersonic gas flows 0.01-10 mm.

Experimental facilities

The principal set-up's layout is shown in Fig. Hypersonic gas flow is created behind *Incident Shock Wave* (ISW), generated in the ST channel. Experiments have been carried out on a ST facility with inner diameter 80 mm and length 8.5 m. The low

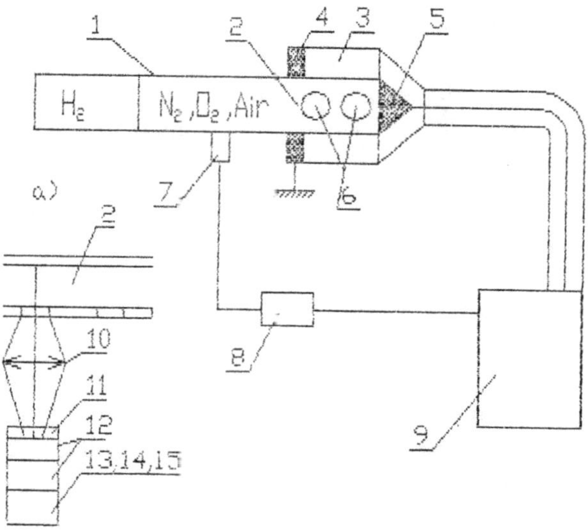

Fig. Experimental set-up layout

pressure section (1) is connected with a discharge chamber (2) - dielectric quartz tube surrounded by metallic screen (3), and having the same as ST channel inner diameter and 1.2 m length. The space between quartz tube and a metallic screen is filled with organic glass and pumped by nitrogen at a pressure up to 20 bar permitting a voltage impulses of up to 10^6 V amplitude to be applied to the ring (4) and conical (5) high voltage electrodes. Special quartz windows (6) are provided for passive and active optical diagnostics of the gas in the discharge chamber. The ST facilities, described in detail in [7], enables an intensive SW with the Mach number up to 15 and associated gas flows with Mach number up to 10 to be created . When SW comes into a discharge chamber the signal from piezoelectric transducer (7) is passes to synchronization unit (8). This unit turns on a nanosecond voltage generator (9) with demanding delay time and a high voltage impulse is applied to the electrodes to make BIW, which then propagates SW with the speed 10^9 -10^{10}cm/s [6], exciting gap volume in the short time 10-20 ns. The Marx electric generator is employed to make an impulse discharge [8]. It produces impulses of 100-400 kV amplitude, 30-40 ns duration, and a rise time less than ns , allowing BIW in such gases as N_2, O_2 and similar,in the density range 10^{15}-10^{19}cm^{-3} to be created. A typical discharge current and an impulse power are 5-20 kA and 10^8-10^{10}W correspondingly.

Set-ups operation controlling and gas flows imaging technique.

The following parameters of NGD have been registered: BIW velocity and time-dependent electron concentration before and behind SW, BIW current and electrical power inserted into discharge gap. A Doppler frequency shift of microwave, reflecting form SW front is the basis for continuous SW velocity recording. The imaging optical system (insert in fig) consists of quarts windows (6), lenses (10), wide band interference filters (11), two or three steps of brightness amplifiers (12) camera (13), TV camera (14), or multichannel optical analyzer (OMA) displacing in x or y directions (15) for radiation intensity picture or its cross-section fast recording. Camera and TV camera are utilized for gualitative flow field imagination, OMA - for quantitative one-dimensional concentration distribution measurements. The radiation intensity in the 2^+ positive band system of N_2, Shuman-Runge and Hertzberg bands of O_2, NO-$\alpha,\beta,\gamma,\delta$ UV bands, as well as in several atomic lines of O and N were monitored. After discharge afterglow has fully quenched the next voltage pulse is applied and so flow field be continuously visualized with desirable time steps 1-100 μs.

Investigations

Several series of experiments for testing of the developed imaging system have been performed. In the first ones we explored a plane SW with M=5-10 propagating through nitrogen, oxygen or air at initial pressure 0.005-0.1 bars. Atomic oxygen, nitrogen and nitrogen oxide molecules (in air) longitudinal concentration distribution have been monitored behind SW. The second experimental series was devoted to investigation of concentrations flow-fields near the solid bodies (blunt body or flat plate with sharp leading edge) in the hypersonic high enthalpy gas flows. Utilizing OMA we obtained a quantitative results on concentration variations behind leading SW bow . Only qualitative results have been so far obtained for boundary layer near the solid body, because of its thickness - several tens of millimeters.

References

1. Beams J.W. Physical measurements in gas dynamics and combustion.New-York,London: Oxford Univ. & Princeton Univ. Press 1955.
2. Laser speckle and related phenomena.Ed. Dainty J.C. Berlin: 1975.
3. Combustion measurements. Modern techniques and instrumentation. Ed. Goulard R. New-York: 1976.
4. Optical methods of the gas-flow investigations.(in Russian) Proc. II All-Union Conference on Methods of Aerophysics Investigations. Ed. Soloukhin R.I. Minsk: 1979.
5. Laser analytical spectroscopy. (in Russian) Ed. Letochov V.S. Moscow: 1986.
6. Asinovsky E.I., Vasilyak L.M.,Markovetc V.V. Wave breakdown of the gas gaps. (in Russian) Thermophysics of High Temperatures N2 (1983) 371-381; N3 (1983) 577-590.
7. Doroshenko V.M., Koudriavtsev N.N., Novikov S.S. Determination of the time-resolved probabilities of heterogeneous recombination of atoms in shock tube experiments. Int. J. Heat & Mass Transfer 29 (1986) 1-19.
8. Kovalchuk B.M., Mesyats G.A., Semin B.N, Shpak V.G. High-current nanosecond accelerator for investigations of the high-speed processes. (in Russian) J. of Scientific Device and Experimental Technique N 4 (1981) 15-18.

The First Electric Discharge Method for Visualizing Three Dimensional Shock Shapes Around Hypersonic Vehicles

Masatomi NISHIO* and Mutsuo KOTAKE**

* Department of Mechanical Engineering,
Fukuyama University,
Fukuyama 729, Japan.

** Department of Aerospace Engineering,
National Defense Academy,
Hashirimizu, Yokosuka 239, Japan.

Summary

A qualitative theory of the first electric discharge method for visualizing three dimensional shock shapes around hypersonic models was established by considering the relationship among the radiation intensity from the electric discharge, the excitation function vs. electron energies, and the gas molecular number density. By establishing the qualitative theory of this method, it was found there exists the most suitable experimental condition for visualizing shock shapes. Utilizing this knowledge, three dimensional shock shapes have been visualized successfully. As an example of this method, the visualization of a cross sectional shock shape around a delta wing with a body traveling at a hypersonic speed has been demonstrated.

1. Introduction

To visualize three dimensional shock shapes around hypersonic vehicles is very important for understanding the flowfield. However, there are few methods available for visualizing three-dimensional shock shapes.

The schlieren method, Mach-Zehnder interferometers, shadowgraphs, etc., are useful only for visualizations of two-dimensional shock shapes and shock shapes around bodies of revolution.

Visualizations of shock waves using an electric discharge had been performed by MaCroskey, Bogdonoff, Mcdougall, Horstman, Kussoy,[1,2] etc. They thought that when an electric discharge is generated across a shock wave, the radiation intensities from the two regions, one in the freestream and the other in the shock layer, become different from each other. Therefore, the shock shape can be visualized by taking the discharge photograph. However, they did not have reasonable theoretical ideas why the radiation intensities become different from each other. Therefore, it was very difficult to select the suitable experimental conditions. Consequently, the visualizations of shock shapes were very difficult, and shock shapes were not necessarily visualized successfully, as mentioned by Nishio[3].

In this paper, a qualitative theory of the first electric discharge method for visualizing three dimensional shock shapes is described. By establishing the qualitative theory of this method, it was found there exists the most suitable experimental condition for visualizing shock shapes. Utilizing this knowledge, three dimensional shock shapes have been visualized successfully.

2. Visualizing Principle

Radiation intensity I from an electric discharge is proportional to the product of the exciation function F and the gas molecular number density N. In general, excitation functions rise according to the increase of electron energies and once they have reached their maximums, decline according to the increase of electron energies.[4] Therefore, if the gas molecule type is same, the radiation intensity curve can be drawn as shown in Fig.1, with the molecular number density N as the parameter. In the case of the figure, k is larger than unit. When we express the horizontal scale with the electric field E instead of the electron energy W, Fig.2 may be drawn since $W \propto E/N$.[5]

Now we consider the case shown in Fig.3. If the gas molecular number density $N_\infty = N1$ and $Ns = kN1$ by selectin the suitable value of k, Fig.4 may be drawn. In Fig.4, I_∞ is the radiation intensity from the electric discharge column in the freestreem and Is is the one in the shock layer. Judging from Fig.4, we can draw the curve of the radiation intensity ratio I_∞/Is as shown in Fig.5.

When the elecric field is small enough, it seems to be suitable for the visualization of shock shapes since $I_\infty/Is \gg 1$ is realized.

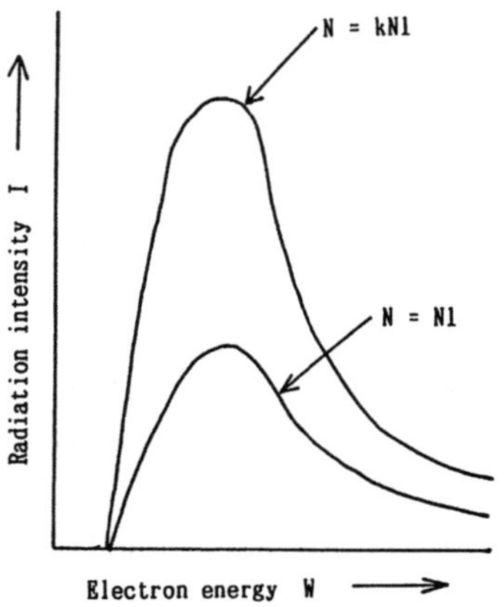

Fig. 1 Radiation intensity curves in arbitrary units vs. electron energy in case that molecular number density is parameter.

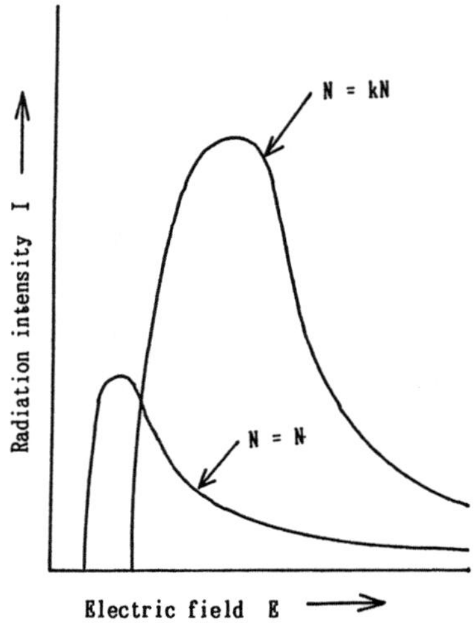

Fig. 2 Radiation intensity I vs. electric field E.

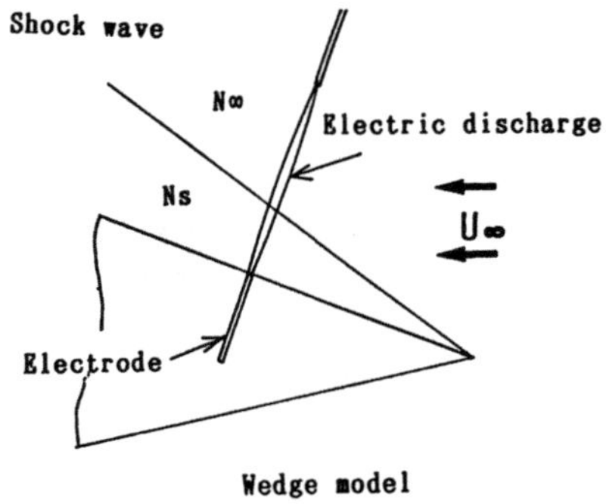

Fig. 3 Illustration of shock wave over wedge and electric discharge.

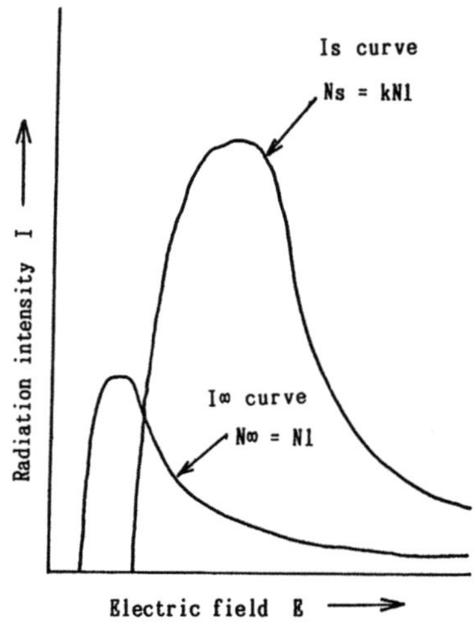

Fig. 4 Radiation intensity curves I∞ and Is.

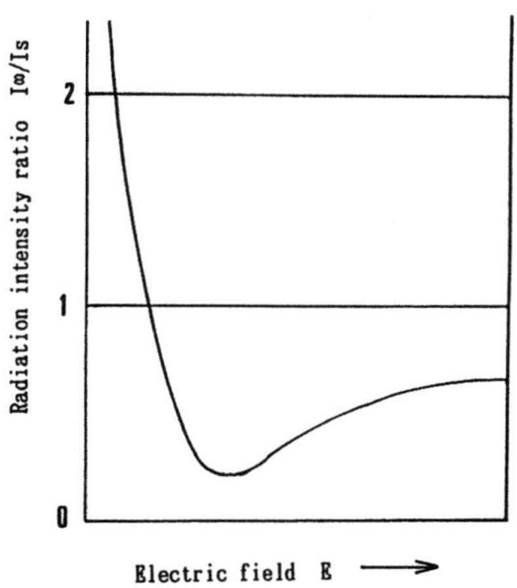

Fig. 5 Relation between radiation intensity ratio I∞/Is and electric field E obtained qualitatively.

However, it is very difficult to generate an electric discharge under the condition of a small electric field without flowing a great amount of electric current. In the case of $I\infty/Is=1$, the shock shape cannot be visualized since there exists no radiation contrast at the shock position. Therefore, it is considered to be best to select the electric field value at which the value of $I\infty/Is$ is minimun.

3. Experiments

The characteristics of the hypersonic tunnel used in these experiments were: Mach number is 10. The type of gas is air. The molecular number density in the freestream is about $10^{17}/cm^3$.

First, the lateral shock shape over a wedge is visualized. The arrangement of the model and the pair of point-line electrodes[3] is shown in Fig. 6. The visualized shock shape is shown in Fig. 7. The shock shape is compared with the result obtained by the schlieren method. From this comparison, it was found that the shock shape obtained by the first electric discharge method was considerably exact.

Second, the cross-sectional shock shape around a delta wing with a body is visualized. The arrangement of the model and the pair of electrrodes is shown in Fig. 8. The visualized shock shape is indicated in Fig. 9.

References

(1) McCroskey, W. J., Bogdonoff, S. M. and McDougall, J. G., "An Experimental Model for the Sharp Flat Plate in Rarefied Hypersonic Flow," AIAA Journal, Vol. 4, No. 9, 1966, pp. 1580-1587.

(2) Horstman, C. C. and Kussoy, M. I., "Hypersonic Viscous Interaction on Slender Cones," AIAA Journal, Vol. 6, No. 12, 1968, pp. 2364-2371.

(3) Nishio, M., "New Method for Visualizing Three-dimensional Shock Shapes Around Hypersonic Vehicles Using an Electrical Discharge," AIAA Journal, Vol. 28, No. 12, 1990, pp. 2085-2091.

(4) Nasser, E., "Fundamentals of Gaseous Ionization and Plasma Electronics," Wiley Interscience, (1970), pp. 74-75.

(5) Engel, V., "Ionized Gases, Oxford University Press," (1965), pp. 122-123.

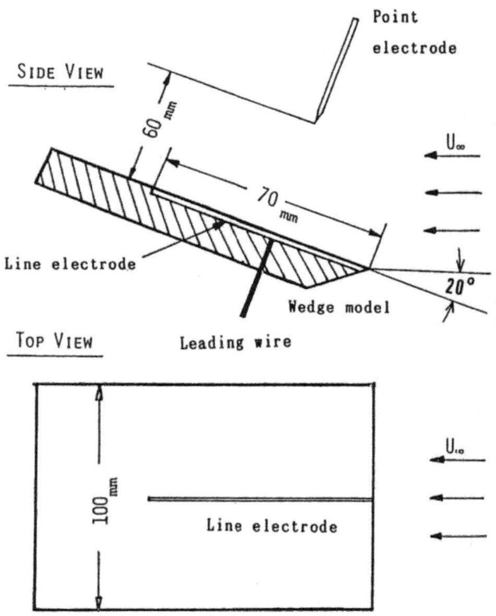

Fig. 6 Arrangement of wedge model and pair of electrodes.

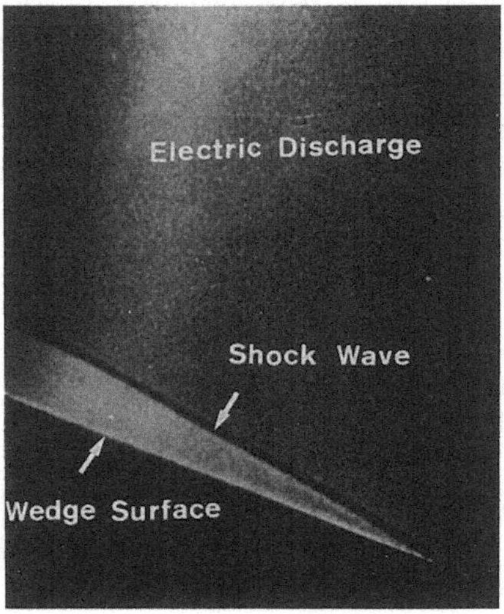

Fig. 7 Visualization of shock shape over wedge.

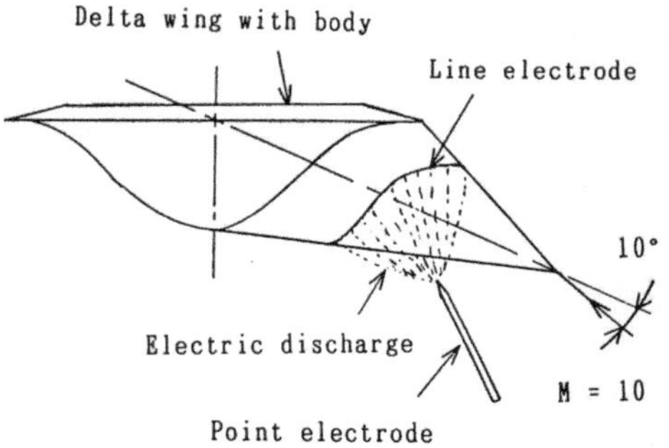

Fig. 8 Arrangement of delta wing with body and pair of electrodes.

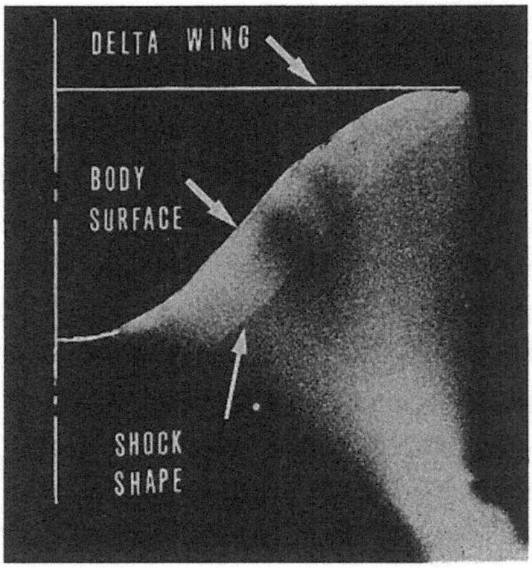

Fig. 9 Visualization of three-dimensional (cross-sectional)shock shape over delta wing with body.

Schlieren Device and Holographic Interferometer for Hypersonic Flow Visualization

J. SURGET, M. PHILBERT, D. BIZE

Office National d'Etudes et de Recherches Aérospatiales
BP 72 - 92322 Châtillon CEDEX

Summary

A new high enthalpy arc heated hypersonic facility, called F4, has been recently built at Le Fauga-Mauzac ONERA Center. The initial visualization equipment of this wind-tunnel is composed of two optical devices: a schlieren apparatus (presently in exploitation), and a multipass holographic interferometer (under construction). This paper presents these apparatus, and some visualization pictures demonstrating their respective performances.

Introduction

Visualization of gas flows generated into the test chamber of the new high enthalpy arc heated hypersonic facility F4 of Le Fauga-Mauzac ONERA Center is rather difficult due to the following main characteristics: run duration is short (20 to 100 ms) and density is very low (10^{-3} down 6.10^{-5} $kg.m^{-3}$).

The initial equipment of F4 is composed of two complementary apparatus:

- a large optical field schlieren device, to visualize the whole aerodynamic phenomena surrounding any model set in the test section;
- a multipass holographic interferometer, which allows to establish density maps around two dimensional or axisymetric small models and also to characterize the flow upstream of these models.

Schlieren Device

Description

The optical scheme of the schlieren device corresponds to the Z type (fig. 1). The flow field being 500 mm wide, the mirrors M1 and M2 have to be set at the same diameter. To limit the light wavefront distorsion, the spherical aberration is reduced using a focal length of 5 meters, and the mirror surfaces are polished with 0.1" precision.

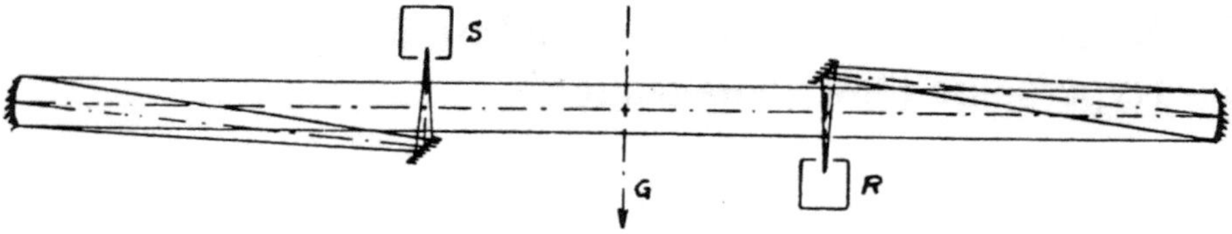

Fig.1: Optical scheme of the schlieren device.

The emission set-up S (fig. 1) contains three commutable sources:
- a continuous Hg vapor source (HBO,200W),
- a snapshot source (FX 12),
- a continuous halogen source.

The reception set-up R also offers several options:
- a high speed video camera up to 10,000 frames/s (NAC E10),
- a CCD video camera,
- a HASSELBLAD camera (2000FCW 6x6 500mm),
- a NIKON camera (F3 24x36 85mm).

The mirrors and the schlieren knives are monitored from the wind tunnel control room.

<u>Results</u>

Before building the F4 schlieren apparatus, the capability of the method for low density gas visualization was tested in the R5 Ch wind tunnel of Chalais-Meudon ONERA Center (minimal density: 4.10^{-4} $kg.m^{-3}$) [1], using a Z schlieren device 200 mm field of view.

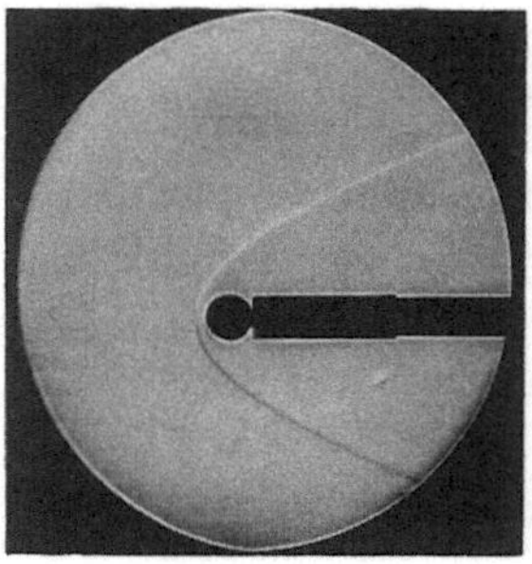

Fig. 2: Schlieren visualization at R5Ch.

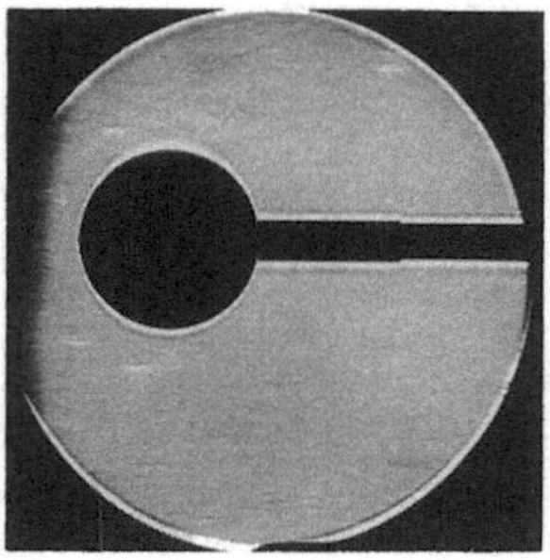

Fig. 3: Phase constrat picture at R5Ch.

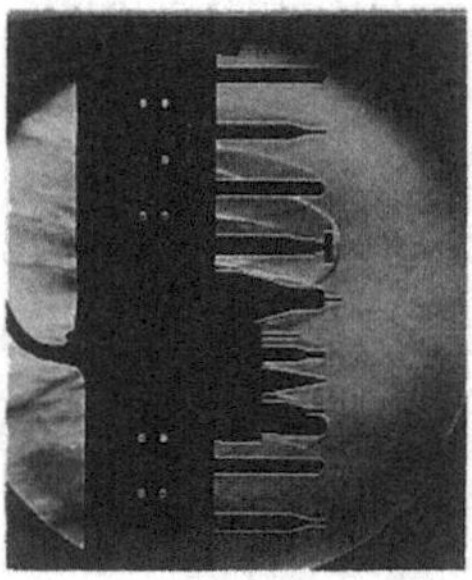

Fig. 4: Schlieren visualization at F4.

Fig. 2 shows a schlieren picture obtained with knives parallel to the gas flow. The model is a cylinder (diameter: 15 mm, length: 150 mm). For this run the density was: $4.3.10^{-4}$ $kg.m^{-3}$.

Fig. 3 was obtained using phase contrast knives, in order to increase the sensitivity, because of the axisymetric configuration of the sphere model, 60 mm in diameter. In spite of the low density during the run (5.10^{-4} kg.m^{-3}), the shock wave is well visible.

Fig. 4 is one of the earliest schlieren pictures recorded at F4 wind tunnel; it shows a nitrogen flow with shock wave around the probes for the aerodynamic calibration (density: $1.3.10^{-3}$ kg m^{-3}).

Multipass Holographic Interferometer

Single pass holographic interferometry, currently used for supersonic gas flow studies, can also be applied to visualize hypersonic flows generated in shock tunnels [2]. However its sensitivity is not high enough to enable accurate measurements of the lowest densities existing in the flows generated by the F4 nozzle, ranging from 10^{-3} down to 6.10^{-5} kg.m^{-3}. Consequently the apparatus is designed to achieve improved sensitivity thanks to multipassages of the test beam through the aerodynamic phenomenon: the distorsion of the test light wavefront, due to gas index gradients, increases each time the beam crosses the flow [3] [4].

First Multipass Holographic Interferometer

A holographic set-up has been built in order to test the practical possibilities of application of the multipass process [5].

The device was installed at the R5Ch wind tunnel of ONERA Chalais-Meudon Center [1] to visualize a low density gas flow (density about 4.10^{-4} kg.m^{-3} and 10 Mach number).

Fig. 5 is a set of three interferograms of the R5Ch flow, with a wedge as a model (15° angle, 150 mm spread), recorded respectively for one, two and three double crossings . The sensitivity of a single pass system is multiplied respectively by 2, 4 and 6. The distorsion of the initial fringe pattern, due to the shock wave, increases with the number of passages: just visible on 5a, it clearly appears on 5c and becomes easily measurable.

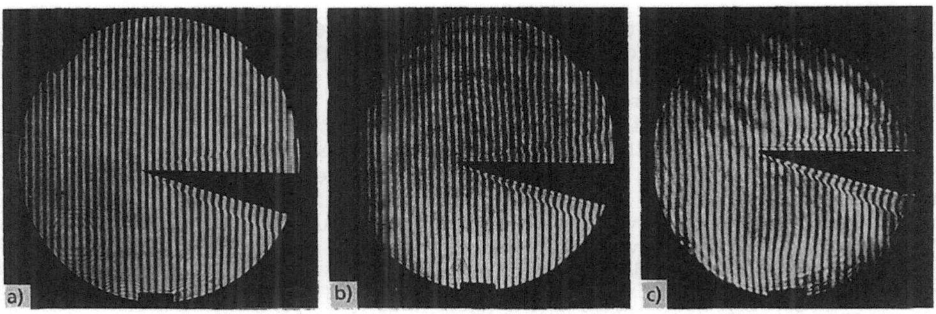

Fig. 5: Finite fringe multipass interferograms recorded at R5Ch: a)N=1, b)N=2, c)N=3.

F4 Multipass Holographic Interferometer

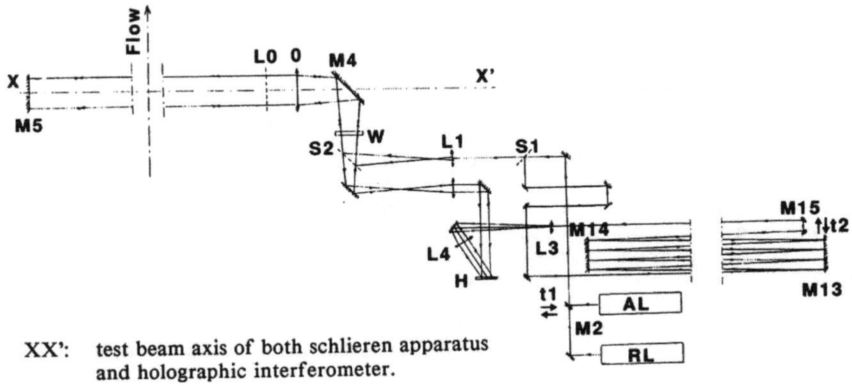

Fig. 6: Optical diagram of the multipass holographic interferometer designed for F4 facility.

Fig. 6 shows the optical scheme of the apparatus. Both the ruby laser RL or the argon laser AL can be used as light source, thanks to the translation t1 of the movable mirror M2.

On the test arm of the holographic device, the plane mirror M5 and the semi-transparent glass window LO are disposed on both side of the gas flow G. They constitute the optical cavity necessary for to create multicrossings of G by the test beam.

On the reference arm, the distance between the mirrors M13 and M14 being equal to the optical cavity gap M5-LO, the translation t2 of the mirror M15 makes it easy to equalize the reference pathlength to the test pathlength corresponding to the chosen number N of double crossings.

The optical configuration represented fig. 6 is suitable for the number pass selection method only founded on the limited coherence length of the laser light. Three double crossings are pratically the upper limit for this method application (sensitivity multiplied by 6), because of the unwanted light falling on H due to the uneselected test beams [5]. To multiply the sensitivity by 8, 9 etc... using N>3, the unuseful test light must be eliminated. A solution consists in a time filtering method which takes advantage of the short duration t of the ruby laser pulse [5]. However the optical cavity must be enlarged for to obtain a transit time T of the light into the cavity (path LO-M5-LO) superior to t. So, it is possible to select the test beam corresponding to the chosen number N of double crossings, using a Pockels cell associated to H.

The pathlengths corresponding respectively to the two F4 configurations, and to the R5 set-up (by comparison) are indicated in the table of fig. 7.

N	R5 config.		F4 1st config		F4 2nd config	
	LC	LT	LC	LT	LC	LT
1	4.2	8.5	3.6	11.2	10	17.5
2	8.4	12.7	7.2	14.8	20	27.6
3	12.6	16.9	10.8	18.4	30	37.6
4					40	47.6
5					50	57.6
7					60	67.6
8					70	77.6

N: number of double crossings.

LC: pathlength into the optical cavity (m).

LT: total pathlength (m).

Fig. 7: Optical pathlengths for R5 and F4 configurations.

This interferometer (150 mm optical field diameter) has been designed as a complementary apparatus to the schlieren device. Only a few optical components of the schlieren device must be disassembled for allowing the interferometric installation. Moreover, the schlieren window W is also used to introduce laser light into the test chamber, and to return it towards the photo plate H used for hologram recording. Of course, such an arrangement prevents to simultaneously use the schlieren apparatus and the interferometer. However, visualization method exchange will require only one or two hours.

Conclusion

A schlieren apparatus is presently in exploitation at the F4 high enthalpy hypersonic wind tunnel of Le Fauga-Mauzac ONERA Center, and a multipass holographic interferometer, under construction, is intended to be operational at the end of 1992.

References

1. Chanetz, B.; Coet, M-C.; Nicout, D.; Pot, T.; Broussaud, P.; François, G.; Masson, A.; Venneman, D. : Nouveaux moyens d'essais hypersoniques développés à l'ONERA: les souffleries R5 et F4. AGARD Symposium. "Theoretical and experimental methods in hypersonic flows". 4-8 May 1992. Torino (Italy).
2. Fleck, B.; Mc Intyre, T.J.; Eitelberg, G. Holographic; interferometry on the new high enthalpy shock tunnel in Gottingen (HEG), 3rd French-german congress on applications of holography, Saint Louis, France, (1991).
3. Ostrosky, Yu I.; Bustov, M.M.; Ostrovskaya, G.V. Interferometry by holography. Spring-Verlag, Berlin, Heidelberg, New-York, 178, (1980).
4. Trolinger, J.D. AGARDograph 296, Laser applications in flow diagnostics, 93, (1988).
5. Surget, J.; Dunet, G. Multipass holographic interferometry for low density gas flow analysis. 19th International congress on high speed photography and photonics, Cambridge, Great Britain, (1990).

A Semiconductor Laser with Unusual Characteristics for Flow Visualization

A. N. GOLCHENKO, M. V. GUSHCHIN I. M. OLIKHOV

GAMMA Scientific Production Corp. Ltd.
Shcholkovo, Moscow region, 141100

I. M. DEMENTJEV, N. P. MENDE

A.F.Ioffe Physico-Technical Institute
Russian Academy of Sciences
St. Petersburg, 194021
Russian Federation

Summary

Along with the brilliant advantages of lasers from the standpoint of their application in interferometric studies of high-speed gasdynamic processes they possess a certain demerit. High spatial coherency of the ruby laser beam leads to appearance of a speckle structure in an image obtained with the help of interferometers or schlieren devices. A new semiconductor laser with electron-beam pumping is free from such a demerit.

A semiconductor laser with electron-beam pumping

Application of optical methods for studying high-speed non-stationary processes in gasdynamics imposes rigid requirements on light source characteristics which are contradictory.

The emergence of pulse lasers allowed one to overcome these difficulties. However, the laser radiation features beget a specific structure of the image. High coherency of a laser beam leads to the speckle structure of the image which reduces clearness of fringes in interferograms and their resolution. As to schlieren method, employment of laser illumination with a narrow slit turns out to be impossible at all because of diffraction phenomena at the slit edges which give rise to an image striped structure.

A new semiconductor laser is void of this "imperfection". Its radiation has a relatively wide spectral band in contrast to, for example, the ruby laser widely used for interferometric measurements [1]. Operation principles of semiconductor lasers are described in monograph [2]. They are based on transformation of electron beam energy into optical radiation within a semiconductor mono crystal which serves as an active medium. Our semiconductor laser is a high-voltage electrovacuum device of diode type in which a cold cathode serves as a source of electrons with high uniformity of the beam density; a multi-element semiconductor target is used as an anode. Radiation wave length depends on both the semiconductor material type and the target temperature. The target is fabricated from a semiconductor material representing a solid solution of CdS_xSe_{1-x} type. The unit containing the semicondactor target is shown in Fig.1.

Fig.1. Optical head of the semiconductor laser.

Depending on a relation between the components of the target material, the semiconductor laser has the following characteristics:

Light pulse energy , *mJ*	$20 \div 40$
Pulse duration at a level of 0.1, *ns*	5
Light pulse power, *Watt*	$(4 \div 8)\cdot 10^6$
Beam divergence, *deg.*	$20 \div 30$
Radiation wave length range, *μm*	$0.51 \div 0.66$

The number of wave lengths, radiating simultaneously	1 ÷ 4
Spectral band width, Å	20 ÷ 50
Emitting surface diameter, *mm*	15 ÷ 20
Light pulse frequency, *Hz*	up to 3
Light pulse time delay relative to input control pulse, *μs*	5 ± 0.1
Reliability, the number of operation cycles until failure	10^5
Operation temperature range, *deg.*	-20 ÷ +30
Power consumption, *Watt*	50
Power requirements	220 *Volts*, 50 *Hz*
Dimensions, *mm*	
the power source	310x310x150
the laser unit	Ø160x500
Total weight, *kg*	22

Due to the wide spectral band the semiconductor laser provides uniform illumination of the working field without the speckle structure. At the same time the spectral band width is not too large (about 20-50 Å), therefore the resolution of fringes remains high enough. Elimination of the speckle structure is also provided due to a large size of the emitting crystal surface because the diffraction phenomena at light rays emanating from different points of the emitting surface fail to coincide with each other and smear the speckle picture. The laser pulse energy is quite enough for recording interferograms in a large scale (on a film of 24x30 cm^2).

In contrast to the ruby laser, the semiconductor laser does not demand a large time delay between the control pulse (igniting pumping lamps in a ruby laser) and the generation beginning. The interval between them is constant and does not exceed 5 *μs*. This facilitates the synchronization of the laser operation with a process under study very much.

The emitting crystal surface can be of an arbitrary shape. So, one can make a light source of a desirable shape and, in particular, in the form of a narrow slit for using such a source in schlieren devices (slit-knife method). Due to the high beam divergency we did not use a condenser but simply projected the laser divergent radiation on the slit filling up completely the angular aperture of our schlieren device (focal ratio 1:8).

The emitting surface may be constructed from several crystal materials with different properties. So, one can make a light source radiating in several spectral regions simultaneously. Such a source may be used for color schlieren method or for multi-wave-length interferometric study of plasma.

Large size of the emitting surface allows one to use light guides in order to part the radiation in desirable directions, for example, for illuminating an object of interest on different sides for the objectives of tomography [3].

Experimental results

Fig.2 shows an interferogram of flow over a supersonic sphere in free flight. The sphere flies in air under normal conditions at Mach number 2.5. The sphere diameter is 30 mm. The interferogram is obtained with the help of the grating interferometer described in Ref.[4]. One can notice excellent resolution of the fringes before the sphere right up to the bow

Fig.2. An interferogram obtained with the help of a grating interferometer with the semiconductor laser as a light source.

shock wave, which is of a finite width, unfortunately. The shock wave image in the form of wide band in front of the sphere is explained by an imperfection of our optical device: strongly deflected light rays are screened by its output diaphragm.

Conclusion

In our opinion semiconductor lasers of this kind can provide conspicuous advance in optical methods in experimental gasdynamics. They can be applied for study of gas flows in wind tunnels, shock tubes and ballistic ranges.

Due to unique technical characteristics the semiconductor lasers can find many applications in other fields, namely:
- TV-observation of objects with spatial-time selection in scattering air and liquid media;
- ecological control of the atmosphere by sounding pulses of several wave lengths simultaneously;
- forming a multi-color light beam of a desirable configuration in small systems of light-house type for aerial and naval navigation.

At present the development work is done to improve technical characteristics of the laser, in particular, to increase the pulse energy up to 100-200 *mJ*.

Acknowledgements

The authors wish to thank M.V.Chichikova, V.A.Polin, A.V.Shilo, S.G.Thomson and Dr. V.A.Komissaruk for taking part in testing the semiconductor laser in a ballistic range.

References

1. Kovalev, P.I.: Komissaruk V.A.; Martynov V.P.; Mende N.P.: Polarization and grating interferometer in ballistic experiment. Opt. Laser Tech. 11 (June 1979) 137-142.
2. Bogdankevich O.V.; Darznek S.A.; Elisejev P.G. Semiconductor lasers. Moscow: Nauka Press 1976.
3. Kovalev, P.I.: Komissaruk V.A.; Mende, N.P.: A system combining several interferometers to investigate spatial gasdynamic phenomena and high-speed photography of axisymmetric processes. Opt. Laser Tech. 15 (June 1983) 141-144.
4. Dukhovski, I.A.: Komissaruk, V.A.; Kovalev, P.I.; Mende, N.P.: High-speed photography of the interaction of a water drop with a supersonic sphere. Opt. Laser Tech. 17 (June 1985) 148-150.

The Measurement of Gas Density Around a Blunt Cone Using a Differential Interferometer

Sun Qiming Le Jialing and Li Chao
China Aerodynamics Research and Development Center
P. O. Box 211, Mianyang, Sichuan, China

ABSTRACT

This paper discribes the measurement of the gas density around a blunt cone with a differential interferometer. Instruments and evaluating procedure are introduced. The measured results are given and compared with that from numerical solution.

INTRODUCTION

It is a valuable aid to the aerodynamic analysis to measure the gas density around a model in wind tunnel. It also provides a verification for numerical methods. It is difficult to measure gas density with a schlieren apparatus, but a differential interferometer, which may be modified from a schlieren apparatus, can be used for measuring gas density. Although the measurement accuracy of a differential interferometer is not so good as that of a M-Z interferometer or a holographic interferometer, the differential interferometer has some advantages, e. g. low cost and easy operation, so it could be used extensively.

Both the M-Z interferometer and the holographic interferometer are called the absolute interferometer or reference beam interferomstor, in which the reference beam does not traverse through the measured flow field, so their fringe shifts are the measure of density difference between the flow field and a reference region. However, the reference beam in a differential interferometer traverses also through the flow field and is very close to the object beam, so its fringe shift is the measure of the density difference between two neighbour points in the flow field. The evaluating procedures for two kinds of interferograms are different. Several procedures were introduced in reference [1]. In this paper another method was adopted, which had been used for evaluating the gas density around a sharp conoid[2].

EXPERIMENT

The test was conducted in a supersonic wind tunnel with a test section 300mm × 300mm at

CARDC. The test conditions are Mach number $M=2$, total pressure $P_0=0.225$MPa, total temperature $T_0=24.8$℃. The model is a blunt cone with 15° semiconic angle and 81.6mm in length. The angle of attack is equal to 0°.

The optical arrangement of the differential interferometer is shown in Fig. 1. Light source is a He-Ne laser with 20mW. A half—wave plate Wp is to change the polarized direction of the beam. Lens Lc makes laser beam expanded, and mirrors M_1 and M_2 with apertures 230mm, make the beam collimated and focused. Wollaston prism W and polarizers P_1 and P_2 compose a interferometer. The polarized axes of the polarizers are 45° with respect to the crystal axes of the Wollaston prism. The interferometer divides the beam into two parts, which are sheared by a small angle. A vertical plane in the flow field is focused on the plate P by means of lens Lo and the two parts of the beam interfere each other there.

The differential interferogram obtained in the test is shown in Fig. 2. The fringe shift distribution in the section which is 40mm from the bottom of the cone is shown in Fig. 3, in which the ordinate represents the fringe shifts, the abscissa represents the ratio between the distance of a measured point from the axis of the cone and the corresponding radius of the shock wave.

EVALUATION

From the principle of two-beam interference we get

$$D(x,y) = \frac{k}{\lambda}\left[\int_{z1}^{z2}\rho(x,y,z)dz - \int_{z1}^{z2}\rho(x,y+d,z)dz\right], \qquad (1)$$

where D is the fringe shift, k the Gladstone-Dale constant, λ the light wave length, d the distance by which the beam was sheared in the flow field.

In principle having measured the fringe shifts, the gas density could be evaluated from Eq. (1) by means of the procedures in reference [1]. In this paper another method was used, by which the fringe shifts of a differential interferogram are first transformed into that of the corresponding absolute interferogram, then the gas density of the flow field could be evaluated with the existing methods for the absolute interferogram.

In the case of the absolute interference $\rho(x,y+d,z)$ in Eq. (1) is a constant ρ_0, so the fringe shift is

$$Da(x,y) = \frac{k}{\lambda}\int_{z1}^{z2}[\rho(x,y,z) - \rho_0]dz. \qquad (2)$$

From Eq. (1) and Eq. (2) we can deduce the relation between the differential fringe shift and the corresponding absolute fringe shift[2],

$$Da(x,y) = D(x,y) + Da(x,y+d). \tag{3}$$

Let y_b be the boundary of the measured region. Outside this boundary the gas density is a known constant, so the absolute fringe shifts are

$$Da(x,y) = 0. \qquad (y \geqslant y_b) \tag{4}$$

With Eq. (3) and Eq. (4) and starting from the boundary, the fringe shifts of the absolute interference can be calculated from that of the corresponding differential interference. Then soluting the following equation

$$Da(y) = \frac{k}{\lambda}\int_y^R [\rho(r) - \rho_0] d(r^2)/(r^2 - y^2)^{\frac{1}{2}}, \tag{5}$$

the density value $\rho(r)$ can be obtained. The results evaluated from the differential fringe shift in fig. 3 are given in Fig. 4, where r is the distance of a point from the axis of the cone.

NUMERICAL SOLUTION

The flow fields around the blunt cone at M=2.0 were computed by the symmetrical dissipative difference scheme. The mixed flow field of subsonic and transonic flow around the nose of the blunt cone was computed by the time dependent method with 17×21 points, and the steady supersonic field along the cone body was computed by the marching algorithm with 120×42 points. The grids were generated to fit the body and refined in the aft-flow field. The shock waves in the flow fields were captured automatically. The computed results are also shown in Fig. 4.

DISCUSSION AND CONCLUSION

Fig. 4 demonstrates that the experiment results are coincided with the numerical solution near the shock, where $r/R = 0.75$ to 1.0. It is very natural that at the shock ($r/R = 1$) the density has a jump. But in the region where $y/R = 0.8 \sim 0.9$, the density decreases abruptly. The reason is that the flow field in this region is affected by the strong expanding flow along the body surface. At the place where $r/R=0.8$, the density just starts to increase. This indicates that the flow has entered the compressed region. When $r/R<0.75$, there is a little difference between computed and measured data. The reason should be further studied.

From this measurement, the following conclusions could be made:

1. The differential interferometer can be used to measure the gas density in a large flow field. It has the advantages of low cost and easy operation.

2. The evaluation method of transforming the fringe shift of differential interference into one of the absolute interference is valid. It can also be used for three-dimentional flow field.

REFERENCE

1. Merzkirch, W. "Flow Visualization", Academic Press, New York and London, 1974.
2. Qiming, S. "A New Method for Determination of Gas Density for Differential Interferograms", Flow Visualization IV, 1986.
3. Bradley, J. W., "Density Determination from Axisymmetric Interferogram", AIAA J., 6, 1968.

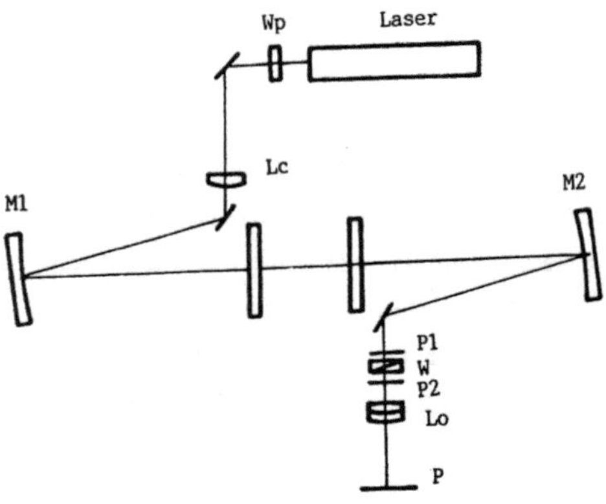

Fig. 1 The optical system of the differential interferometer

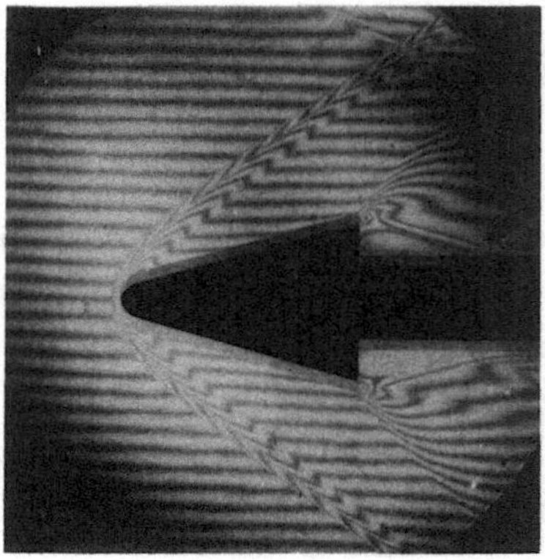

Fig. 2 The differential interferogram of the blunt cone

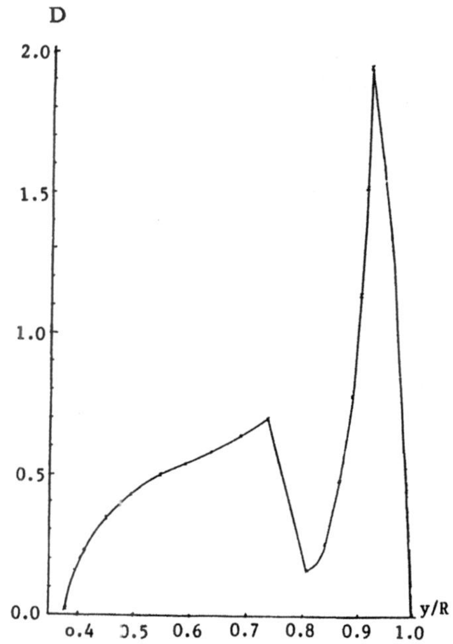

Fig. 3 The fringe shift distribution at the section of 40mm from the bottom of cone

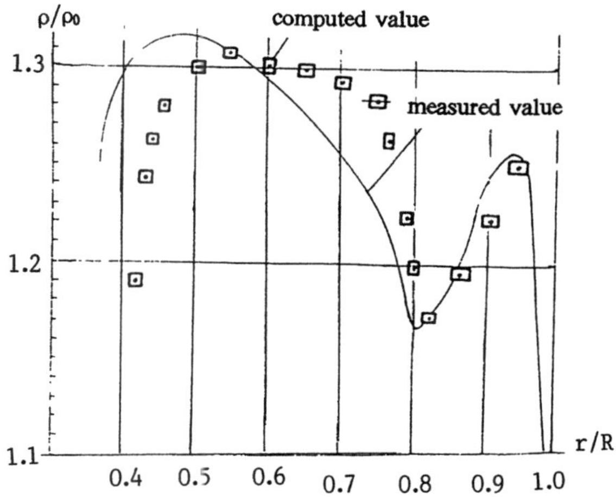

Fig. 4 The density distribution at the section of 40mm from the bottom of cone

Optical Methods

Holographic Color Schlieren Method of Flow Visualization

N. M. SPORNIK

Scientific & Technical Centre "Diacont",
Grodno, White Russia

Summary

The holographic color schlieren method of visualization of heterogeneities in transparent media are dealt with in the paper. The essence of the method is expounded, its realization chart and results of experiments on flows round the flying ball are given.

Key words: hologram, colored schlieren pattern, holographic system, ballistic route, flow visualization.

Introduction

While investigating non-stationary phase objects with the help of conventional schlieren methods observation and evaluation of the object under test are possible according to a color photograph. However, making a color photograph of satisfactory quality one comes across with a number of technological difficulties caused, in particular, by the low sensitivity of colored photomaterials, their inaccurate transmission of color, etc. Holographic schlieren method of colored images [1,2] enables to evaluate the object under test in stationary conditions directly by colored pattern without its photoregistration. In this case, one makes use of dispersive properties of the hologram illuminated by collimated beam of polychromatic light.

The colored schlieren pattern processing is performed with the help of the color tinges comparison method [3,4]. The accuracy of the method is essentially dependent on the spectral parameters of the receiving device.

The method of the colored schlieren pattern processing with the help of a spectral instrument is dealt with in the paper. The results of the schlieren pattern decoding during the visual comparison of colors and the photographs of the dispersion path being received with the help of the spectral instrument and characterizing beam angles of deviation in the heterogeneity are given in the "Experiment" section.

Color Tinges Comparison Method

Schlieren method of comparing color tinges allows to increase the accuracy of measuring beam angles of deviation in the heterogeneity under test and to simplify the analysis of experiment results [3,4]. The essence of the method lies in comparing the color of the field pure part, that is possible to change, with color tinges of various zones of the heterogeneity investigated, differing in the quantities of beam deviation angles. The accuracy of the processing increases essentially, because one directly compares field parts of equal coloration without the pattern's photoregistration.

Visual comparison of color tinges is dependent on the observer's individual capacities to percept color tinges. The use of a spectral instrument makes it possible to evaluate angles of deviation more objectively [5], what can be explained in detail with the help of the scheme shown in Fig.1.

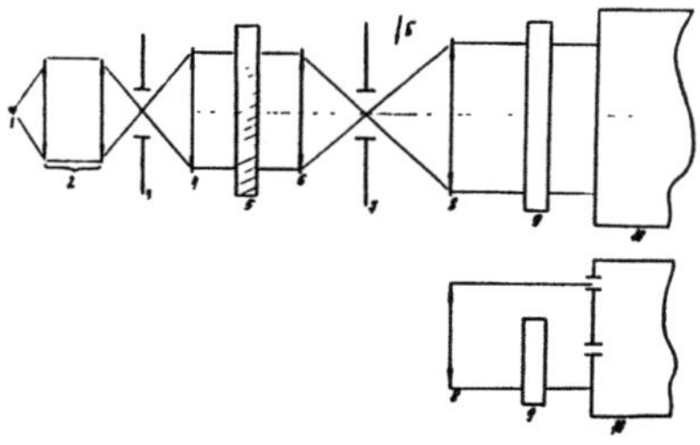

Fig.1. Instrument optical scheme.
1- light source; 2- condenser; 3- entrance aperture; 4- lens; 5- hologram; 6- receiving lens; 7- visualizing aperture; 8- lens; 9- plane parallel plate; 10- receiving and registrating device.

Hologram 5 of the image in focus is illuminated by the beam focused with the collimator and going out of the lightening aperture, illuminated by the source of light through the condenser. The wave front restored from the hologram is directed to receiving lens 6, in whose focal plane two images of different colors are received and two apertures - visualizing aperture 7 and the one, by which the light source image scanning is made - are set. The collimator forms two parallel beams. Plane parallel plate 9 is placed into one of them, formed by the beams that have gone through the heterogeneity. The scanning of the heterogeneity image in horizontal plane can be achieved by means of turning plane parallel plate 9 round the vertical axis. Moving the optical instrument aperture in horizontal plane, one cuts out the chosen part of the hologram image, the aperture singling out the color of the light source pure image.

In the exit of the spectral instrument one observes the entrance aperture image, characterizing the light dispersion path in the heterogeneity as to the pure field. The combination of maximums of quazimono-chromatic radiation for both images is achieved by scanning the light source image by the aperture. Being aware of the scanning aperture shift for every zone of the heterogeneity one finds beam deviation angles ε_x in the heterogeneity investigated

$$\varepsilon_x = \frac{S_x}{f'}$$

x being the measured co-ordinate, S_x the light source image shift, corresponding to the measured co-ordinate, f' focus distance of the receiving lens.

So as to take into account color (or wave lengths) of the patterns, received after the beams going out of the heterogeneity, it is necessary to register the initial reading line defined by the light wave length, that is used when recording the hologram. Having chosen a certain scale, one can calculate the color of any light beam deviated by the heterogeneity.

Experiment

For the wave front processing by means of the above-described methods one made use of the single-exposure hologram recorded according to the scheme given in work [6] on the ballistic path [7]. The distribution of density round the ball 20.7 mm in diameter (d=20.7 mm), flying at speed 610 metres per second (m_∞=1.8), was taken as an object. Measuring of beam deviation angles in the heterogeneity according to a colored schlieren pattern was made visually with the device, whose scheme is shown in work [4].

The results of ρ density calculation

$\bar{\rho} = \rho/\rho_\infty$ in middle section

$\bar{x} = 0{,}5;\ \bar{x} = (x-c)/d$ for different co-ordinates ξ

$\xi = (y-h)/(H-h)$ received in the experiment

are shown in Fig.2.

The density calculation was performed according to the method, described in work [8]. To illustrate the method of colored pattern analysis with the help of spectral instrument (Fig.1) one took the patterns (Fig.3a,b), characterizing the dispersion path along x axis (Fig.2) for y fixed co-ordinate.

The image (Fig.3b) was received with the use of Sabatier's effect, that allowed to increase the accuracy of the processing in evaluating the beam angles of deviation in the heterogeneity.

To go on with the field density calculation the above-mentioned method [8] is used.

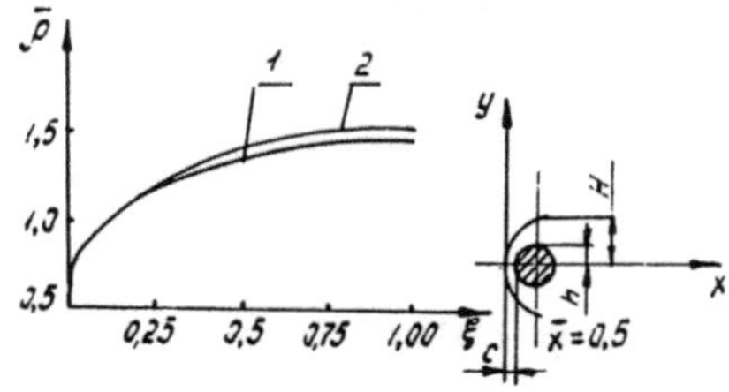

Fig.2. Density distribution round the flying ball.
1- results of the experiment; 2- theoretical results.

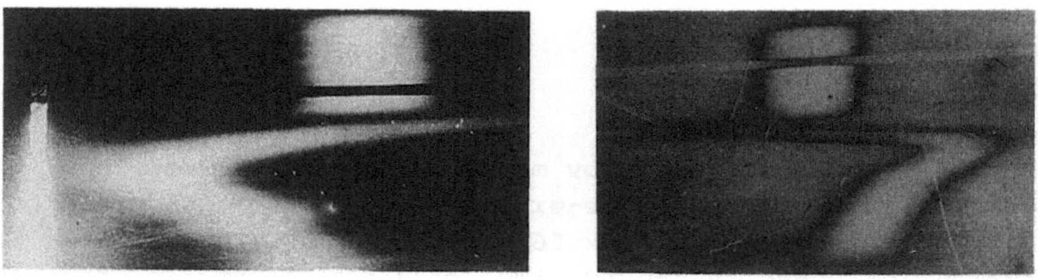

Fig.3a,b. Spectrograph entrance aperture images, characterizing light dispersion path in a colored schlieren pattern.
a- conventional pattern; b- pattern made with the use of Sabatier's effect; c- initial reading line.

Calculation of the field of beam angles of deviation was made in the following way:
1- one finds the co-efficient of proportion $K=\frac{\Delta \ell}{\Delta S}$, ΔS being the visualizing aperture shift; $\Delta \ell$ the shift of the spectrograph entrance aperture image (for the pure field); 2- one measures the shifts of aperture images in the field of heterogeneity in regard to those of the pure part; 3- one calculates ε angles of deviation in heterogeneity according to the formula:

$$\varepsilon = \Delta x / K f'$$

Δx being the shift of the aperture image in the pattern; f'- being the focus distance of the lens according to the scheme shown in Fig.1.

Conclusion

The described holographic colored schlieren method enables to carry out in practice the visualization of heterogeneities in transparent media with various sensitivity and dimension range without colored pattern registration.

The comparison of color tinges visually or with the help of the spectral instrument ensures acquiring quantitative results of high precision. In this case, the technique of experimental data obtaining and processing is considerably simplified.

References

1. Belozyorov, A.F.; Spornik, N.M. "Obtaining of schlieren patterns under wave front restoring from a hologram" (Optical and mechanical industry, 1971, No.3, p.9-11).

2. Belozyorov, A.F.; Spornik, N.M. "Obtaining of colored schlieren patterns under investigation of wave front, restored from a hologram". (Edition of higher schools.Geodesy and aerophotography. 1974, No.1, p.139-141).

3. Spornik, N.M. "Decoding of schlieren patterns, obtained from a hologram". (Letter to J.t.f., 1975, vol.1, IV edition, p.184-187).

4. Spornik, N.M. "Instrument for quantitative evaluation of heterogeneities in transparent media". (The USSR patent 748126. Bulletin of images, No.26, 1980).

5. Spornik, N.M.; Komar, V.N. "Manner of quantitative evaluation of heterogeneities in transparent media and instrument for its realization". The USSR patent 1393023.(1988).

6. Spornik, N.M. "Interferometer with laser on the basis of IAB-451 instrument". (Optical and mechanical industry, 1973, No.2, p.77-78).

7. Belozyorov, A.F.; Berezkin, A.N.; Razumovskaya, A.I.; Spornik, N.M. "Aeroballistic gas flow investigating using holographic device to schlieren system". (Prot. of the Intern. Congr. on High Speed Photography, Nica, 1972, p.401-408).

8. Beketova, A.K.; Belozyorov, A.F.; Berezkin, A.N.; Zeilikovich, I.S.; Mustafina, L.T.; Razumovskaya, A.I.; Spornik, N.M.; Chernikh, V.T. and others. "Holographic interferometry of phase objects". (L.: Science, 1979, p.232).

A Modern Focusing Schlieren for Flow Visualization

LEONARD M. WEINSTEIN

NASA Langley Research Center
Hampton, Virginia 23665-5225

Summary

A focusing schlieren system has been developed that does not have the limitations of previous versions, such as low image brightness, small field of view, and difficulty of use. Two-dimensional slices of the flow can be examined at selected depths. Good results are demonstrated, even through poor optical quality windows. A comparison with shadowgraph and conventional schlieren is also shown.

Introduction

Several problems can occur which degrade image quality when using conventional schlieren (or shadowgraph) systems. These include thermal room currents, flow patterns in the tunnel window region (including tunnel-wall boundary-layer effects), flaws in the tunnel windows, and even undesired flow features which obscure the local feature of interest. The use of a focusing schlieren system can reduce or even eliminate these problems.

Two different versions of focusing schlieren were developed in the late 1940's. One version, by Kantrowitz and Trimpi [1], required two high-quality wide-angle lenses and had a field of view smaller than the lens diameter. The other, by Ralph Burton [2], only required one lens and could cover a field of view larger than the lens diameter. A variation of the Kantrowitz, et. al. system was used by Buzzard [3] in 1968 to make holographic three-dimensional pictures of flow fields.

These techniques have not been widely used due to problems such as low image brightness, small field of view, or difficulty of use [4]. In a recent paper by Weinstein [5], a focusing schlieren system was described which overcomes the problems mentioned. The current paper will expand on that approach, and give some examples of its use.

Properties of a Focusing Schlieren System

Focusing schlieren uses multiple viewing angles to examine optical density variations in flow fields. A sketch of the focusing schlieren system in Figure 1 shows the main components. An extended light source is concentrated by a large Fresnel lens onto the face of the imaging lens. A source grid is placed a short distance in front of the Fresnel lens, and a negative photographic image of the grid is made with the imaging lens. The test section is imaged on a final screen, and the system can be focused at different locations by moving the location of the image screen. The details of the design of these systems are described in the paper by Weinstein [5]. Typical systems use lenses of 0.3 to 1.0 m focal length, and can produce sensitivities comparable to conventional schlieren systems. Designing a focusing schlieren system requires that a trade-off be made between sensitivity, depth of focus, field of view, and resolution. In general, the larger the field of view is, the greater the depth of focus.

The author has found that this type of focusing schlieren is less complex to make and adjust than most conventional schlieren systems. The cost of a large focusing schlieren system is generally far lower than conventional schlieren, since there is no longer a need for large optical-quality mirrors and windows. In addition, the design can be used to build very large systems, with possible fields of view of over 5 m.

It should be noted that focusing schlieren systems are not collimated like conventional schlieren systems, and the resulting variation in viewing angle over the flow field may be somewhat restrictive for examining two-dimensional flow fields.

Example Focusing Schlieren Systems

Results from several focusing schlieren systems are presented to demonstrate some of the capabilities of this technique. These results are described in the following discussion, with setup details given in the paper by Weinstein [5].

Two jets of Chlorodifluoromethane were injected into ambient air from 4.7 mm ID tubes. Very short duration (10 ns) photographic exposures were made with a Nd-Yag laser. The jets crossed near the middle of the field of view, with one jet located near the plane of sharp focus, and the other jet moved toward the imaging lens. Figure 2 shows the effect of the different jet spacings. Fine detail is blended out when one jet is 19 mm out of focus, and the larger detail is blended out with a separation of 38 mm. Turbulent detail in the jet at the sharp focus location is evident for both cases.

Figure 3 shows the effects of a 50 mm thick block of glass, with severe optical striations, in the optical path of a focusing schlieren system. The photographs were made with a flash lamp. The photograph on the left was made with the system sharply focused on the glass. The streaks in the glass would make such a window useless for conventional schlieren (or shadowgraph). The middle photograph shows the same setup with the optical system focused 125 mm away from the window (on the grid side). The sharp streaks are gone, and only broad brightness variations are observed. A gas jet, similar to the ones in the previous example, was then directed into the region of focus. The right photo shows the jet structure almost as well as if the poor quality window was not there.

The next example was obtained from a 25 mm diameter supersonic jet exiting into air. The photographs of Figure 4 were made with sub-microsecond exposures. The flow features that were of greatest interest were the strong shock waves within the jet. The shadowgraph and conventional schlieren pictures show considerable turbulence detail, but show the shock structure poorly. The focusing schlieren picture shows the shock structure better, even though the small size of the jet did not put the out-of-focus details, from the front and back of the jet, very far away.

Concluding Remarks

An improved large-field focusing schlieren system has been developed. Focusing schlieren systems can examine details of three-dimensional flow fields in thin planes, through out-of-focus flow structures. Focusing schlieren systems can obtain good flow visualization results through poor optical- quality windows. These systems can be constructed on a large scale and at a low cost.

References

1. Kantrowitz, A.; and Trimpi, R. L.: A Sharp-Focusing Schlieren System. Journal of Aeronautical Science, 17, May 5, 1950, pp. 311-314.
2. Burton, R. A.: A Modified Schlieren Apparatus for Large Areas of Field. J. Opt. Soc. Am., Vol. 39, November 1949, pp. 907-908.
3. Buzzard, R. D.: Description of Three-Dimensional Schlieren System. High Speed Photography. Proceedings of the 8th International Congress on High-Speed Photography, Stockholm, June 23-29, 1968. John Wiley & Sons, Inc., New York, pp. 335-340.
4. Fish, R. W.; and Parnham, K.: Focusing Schlieren Systems. C.P.No.54 (13.865), ARC Technical Report. Technical Note No. IAP 999, November 1950.
5. Weinstein, L. M.: An Improved Large-Field Focusing Schlieren System. 29th Aerospace Sciences Meeting, Reno, NV, AIAA Paper 91-0567, Jan. 7-10, 1991.

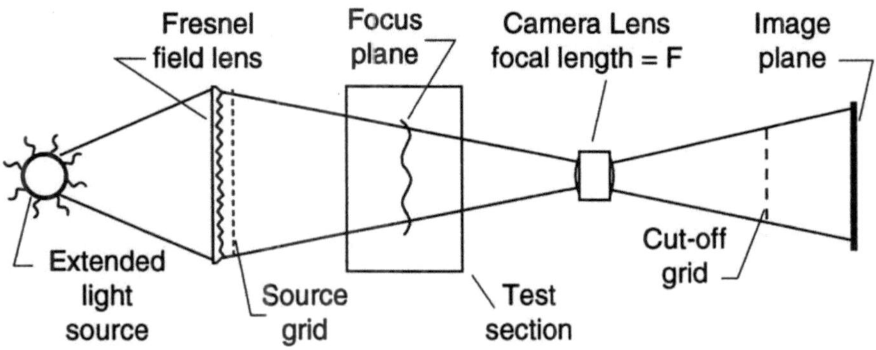

Figure 1. High-brightness large-field focusing schlieren system.

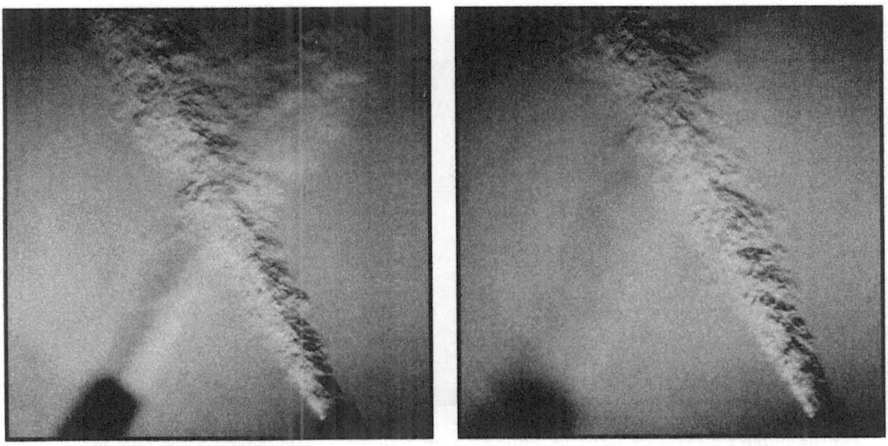

19 mm Separation

38 mm Separation

Figure 2. Focusing schlieren of crossed jets, with one jet out of focus.

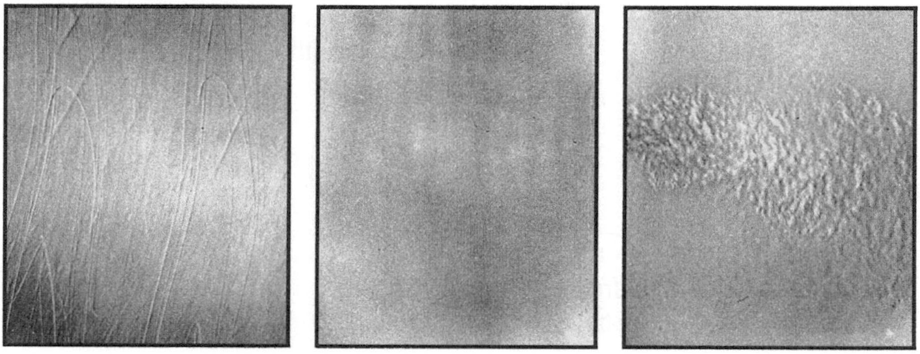

Window at Focus | Window 125mm beyond focus | Gas jet at focus seen through out-of-focus window

Figure 3. Focusing schlieren through 50 mm thick Pyrex window.

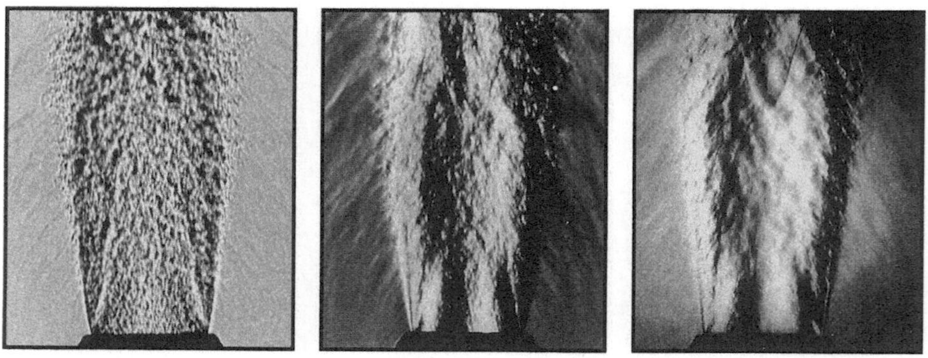

Shadowgraph | Conventional Schlieren | Focusing Schlieren

Figure 4. Comparison of conventional techniques with focusing schlieren in Mach 2 jet.

Simultaneous Measurements of Two Dimensional Velocity and Temperature Field Using Correlation Technique and LIF

J. Sakakibara, K. Hishida and M. Maeda

Department of Mechanical Engineering Keio University
3-14-1 Hiyoshi, Kohoku-ku, Yokohama 223, Japan

Summary

This study deals with the development of two dimensional image processing velocimeter and thermometer using cross-correlation technique and laser induced fluorescence(LIF). The cross-correlation technique is to estimates a displacements of particle groups in two pictures of the flow field taken a short interval time. An excited fluorescence radiation is varied as a function of temperature. These two techniques were joined together, and applied to the thermally stratified pipe flow. The flow field, which are affected by a buoyant force, was clarified by mapping of two-dimensional velocity vectors and temperature fields.

Principle of Temperature Measurement

The fluorescence solution exited by a laser radiates the light which is varied as a function of PH, concentration and temperature of the solution. Under a fixed condition except a temperature of the solution, the temperature is related to the intensity of the radiation. This technique has been applied to the temperature measurement by Nakajima.T.,et al.(1990). The fluorescence intensity(I) can be described by the following equation;

$$I=I_0 A\phi\varepsilon LC$$

where I_0 is the intensity of excitation beam, C is the concentration of fluorescence, ε is the molar absorptivity, ϕ is the quantum yield, A is the fraction of the available light collected, and L is the sampling length along the incident beam. Therefore, when the concentration and the absorptivity are constant, the fluorescence intensity has a negative correlation with temperature. A fluorescence used in this experiment is Rhodamine B which absorbs green beam and radiates orange and sensitively varies its intensity with temperature.

Figure 1 shows the relation between temperature and fluorescence intensity of Rhodamine B, which was obtained in the present experiment using 0.5mg/l solution. The detected radiation intensity was normalized by the intensity at 303K.

In an actual measurement, some difficulties are encountered to make uniform intensity distribution of laser light sheet because of fluorescence absorption. Since the fluorescence intensity is proportional to the exciting light, the ratio between measured intensity and reference intensity at a given temperature shows the relative intensity of the fluorescence. The temperature field was computed by following equations.

$$T(x,y) = T_c\left(\frac{I_c(T_{ref})}{I_{ref}(x,y)} I(x,y)\right)$$

where I_c is the normalized intensity, T_c is inverse function of I_c, I_{ref} and T_{ref} is the initial intensity and temperature, I is the measured intensity. In order to reduce the CCD camera's noise, the intensities of several pixels were evaluated to a local mean value.

Principle of Velocity Measurements

The cross–correlation technique was adopted to obtain the velocity vectors for each location. Figure 2 shows the principle of this technique. The small area at $t=t_0$, reference matrix, is searched scanning in the corresponding matrix at $t=t_0+\Delta t$. The corresponding matrix was identified by the maximum coefficient of the cross–correlation between the reference matrix and the small area in interrogation window. The velocity vector was calculated from the displacement and interval time (Δt) between the reference matrix and the corresponding matrix, as reported in the previous study (Kobayashi,K.et al.,1991).

Hardware Setup

Figure 3 shows a hardware setup. The flow marked with particles for velocity and fluorescence dye of Rhodamine B for temperature were illuminated by the laser sheet (4 W Ar ion laser) whose pulse width and interval are controlled synchronously with the two cameras which view at the same measuring region simultaneously. Making the pulse for synchronized stroboscopic illumination, an acoustic optical modulator cell (AOM cell) was switched by the vertical synchronizing signal of the CCD cameras. Output signal of CCD cameras are recorded on each VTRs with superimposing bar–code number to identify the individual planes. From individual images transferred from VTRs to the image grabber (512x512 pixels, 8 bit), the temperature distributions and the velocity vectors were evaluated using LIF and the cross–correlation technique by employing DSP(Texas Inst.,TMS320C30) in the personal computer.

Comparison between Image Processing and Point Measurements

LDV was also applied to simultaneous measurement to qualify data of the present system. The image processing results were presented for the thermally stratified pipe flow shows the time–series streamwise velocity fluctuations simultaneously measured by the present system and LDV (Fig.4,(a)). The temperature measurement compared with a point data by a thermocouple in the natural convecting flow is also in good agreement and confirmed reliable results for the evaluation of fluctuation, as shown in Fig.4.(b).

Application for a Thermally Stratified Pipe Flow

The experimental apparatus is schematically illustrated in Fig.5. The test section of a constant diameter of 60mm and a total length of 1640mm was set in the horizontal direction. The higher temperature fluid at low velocity flows into the upper settling chamber, and the lower temperature fluid at high velocity flow into the lower one. In the experiments, the bulk Reynolds number was set at 7400 and temperature difference between hot and cold waters was 28K. Upper and lower velocities were at 100mm/s and 140mm/s, respectively.

Figure 6 illustrates examples of thermally stratified flow measurements. The fluctuating velocity and temperature distributions are given. The fluctuating velocity were calculated from instantaneous velocity and average velocity which was obtained from 1000 planes of time–series velocity. The hot water flows around a cold lump and the velocity shear layer well corresponds with thermal interface. On lower figure, there are a couple of vortex, and typical eddy motions are observed.

Conclusions

Time–series image processing system using a cross–correlation technique and LIF was developed, and applied to thermally stratified pipe flow. The motion of eddies can be dynamically obtained from the series of pictures. The large eddy motion of the thermally stratified shear layer could be made clear that the fluid lump came from lower layer by turbulent mixing bounded by upper layer and came back to lower layer.

References

1. Nakajima,T., Utsunomiya,M., Ikeda,Y., and Matsumoto,R., Simultaneous Measurement of Velocity and Temperature of Water using LDV and Fluorescence Technique, 5th Int. Symp. on Appl. of LASER Tech. to Fluid Mech. Lisbon, 12.1, 1991
2. Kimura, I., and Takamori, T., Image processing of flow around a circular cylinder by using correlation technique, Proceedings, 4th. Int. Symp. Flow Visualization IV, C. Veret ed., Hemisphere, Paris, France, pp.221–226, 1986

3. Kobayashi,K., Sakakibara,J., Hishida,K., and Maeda,M., Time-series measurements of turbulent flow field using image processing system, Proceedings, Experimental and Numerical Flow Visualization-1991, Winter annual meeting ASME, Atlanta, pp.155-162, 1991

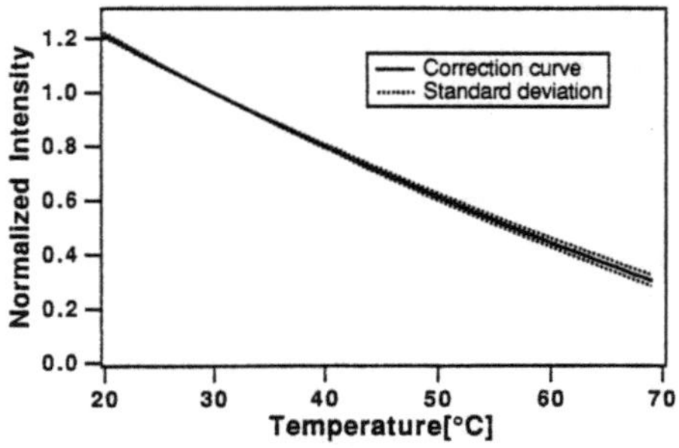

Fig.1. Relation between temperature and fluorescence intensity

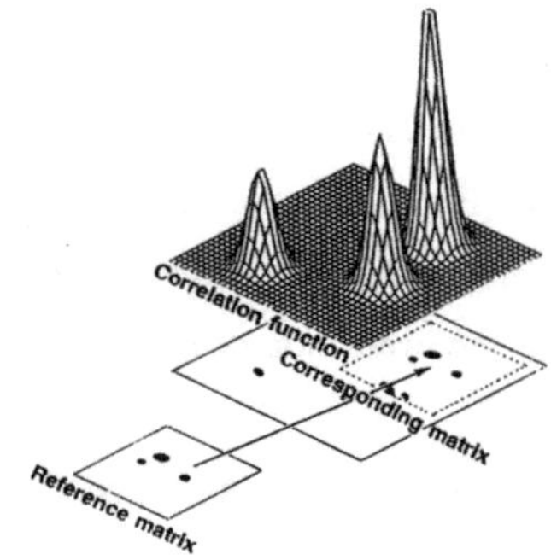

Fig.2. Principle of correlation technique

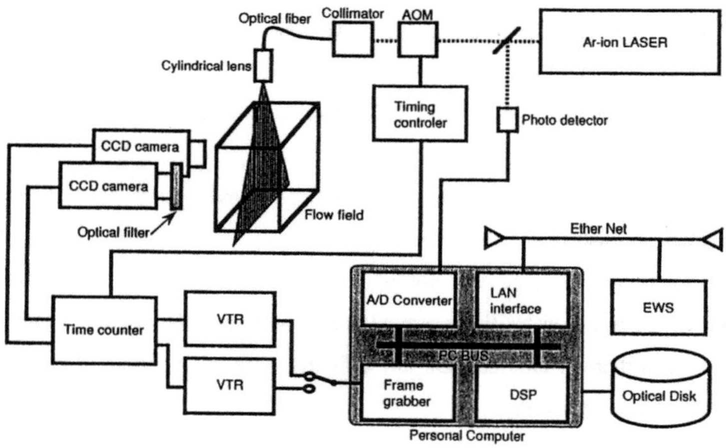

Fig.3. Hardware setup

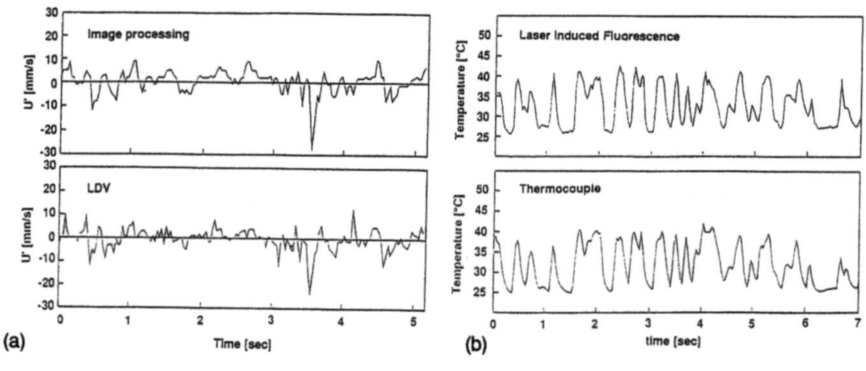

Fig.4. Comparison between image processing and point measurements

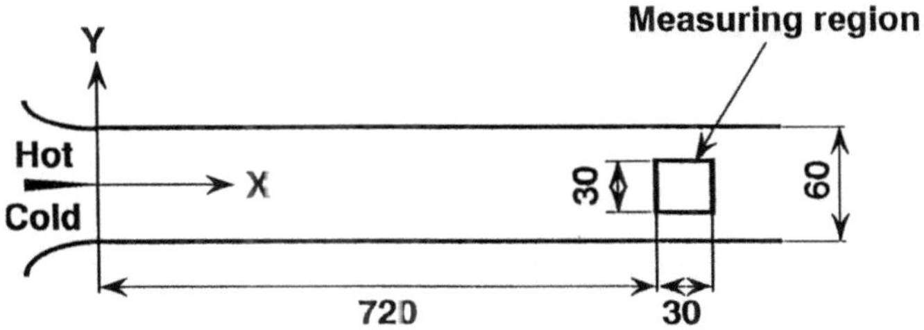

Fig.5. Experimental setup

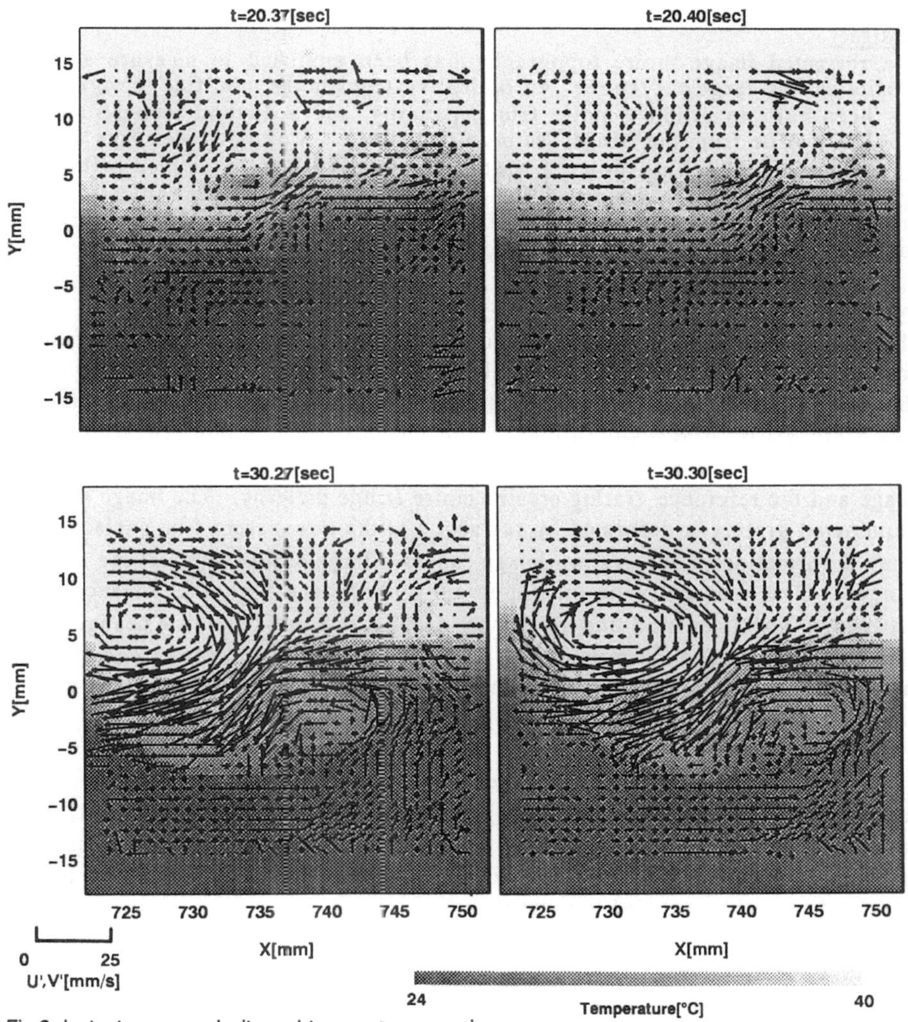

Fig.6. Instantaneous velocity and temperature mapping

The Refracted Image Moire Topography for Measuring Liquid Film Profiles

T. NOSOKO, T. NAGATA, T. SHINZATO and M. FANG

Department of Mechanical Engineering, University of the Ryukyus,

Sembaru 1, Nishihara, Okinawa 903-01, Japan.

Abstract

The refracted image moire topography has been modified to measure a film profile having large curvatures. An optical system for the modified topography has been built to produce moire fringe patterns, and a method reconstructing the film profile from the fringe has been developed. A planoconvex lens and a wavy liquid film falling down a vertical plate have been examined with the optical system and the reconstructing method. The results have indicated that the modified topography has a good accuracy and its maximum measurable curvature is 60 m^{-1} in the case of the falling liquid film.

Introduction

The moire topography is a useful method for measuring profiles of liquid films or other transparent films. The optical system of the moire method is simple as shown in figure 1. It consists of an optical lens, a screen and two gratings which are grills of straight rulings. The image of the illuminated grating is focused on the reference grating. The geometrical interference of the grating image and the reference grating creates moire fringe patterns. The image of the illuminated grating is deformed by refraction at the transparent film surface. A small deformation of the grating image causes a large deformation of the moire fringes. The sensitivity of the topography increases with the distance, L, between a specimen and the reference grating. Scott[1] and Kheshgi and Scriven[2] developed the topography where the distance, L, is long enough so that the thickness of the specimen can be neglected. Their measurements were limited to films having surfaces of small curvatures.

By decreasing the distance, L, between the reference grating and the film, the maximum measurable curvature can be increased. In the present work, we have developed the moire topography in which the distance, L, has been made as short as possible. A planoconvex lens and a falling liquid film have been measured with the modified topography.

Experimental apparatus and procedure

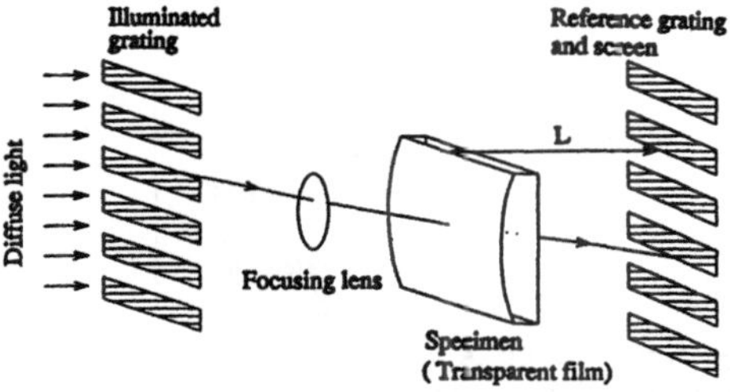

Fig. 1. Diagram of refracted image moire topography.

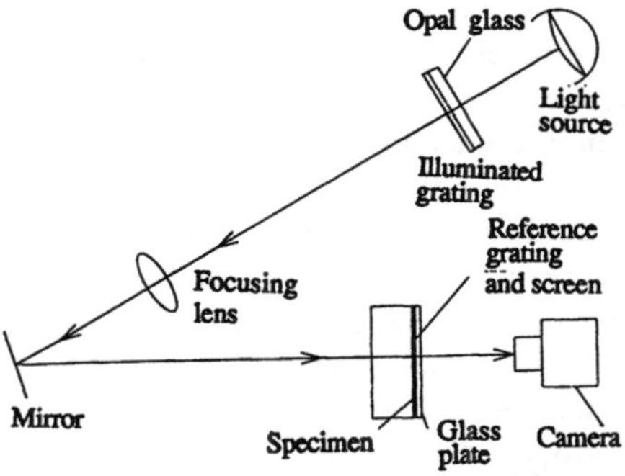

Fig. 2 Optical system for refracted image moire topography.

The schematic diagram of the experimental set up is shown in figure 2. Its heart is a couple of gratings, grills of black straight parallel rulings which are printed on glass plates. The size and the pitch of the illuminated grating are 50×50 cm^2 and 50.8 μ m respectively. Those of the reference grating are 10×10 cm^2 and 84.7 μ m. Here the pitch is defined as the distance between the center lines of the neighboring rulings. Flush light emitted from a stroboscope diffuses at an opal glass surface and illuminates one of the gratings. The image of the illuminated grating is focused on a screen with a 700 mm focal length lens. The screen is an opal paper which is sandwiched between the other grating, i.e. the reference grating, and a specimen. Moire fringes appearing on the screen are recorded with a 35 mm film camera. The distance, L, between the surface of a specimen and the screen ranges from 3 to 8 mm. The values for the distance are so short compared with the distance, 53 - 127 mm, employed by Kheshgi and Scriven[2].

Before positioning a specimen into the optical system, the optics has been aligned in order to focus to the screen the grating image whose pith is equal to that of the reference grating. The procedure of the alignment is described elsewhere[2]. Vertical parallel fringes with the approximately a 4 mm pitch have been created in the condition of the complete alignment. The rulings of both gratings must be inclined with a very small angle, approximately $\pm 2.4°$, from the horizon in order to create the vertical fringes.

Evaluation of moire fringe patterns

Consider a liquid film having a two-dimensional surface profile, which is supported by a glass plate. The surface profile varies in the vertical direction, not in the horizontal one. The rulings of the illuminated gratings can be assumed to be horizontal. Figure 3 shows that a typical light ray emitted from the illuminated rulings passes through the film and the supporting plate to the screen, so that the image of the rulings are focused on the screen . The light ray refracts twice on the film surface and on the liquid glass interface. The light ray can be assumed to be parallel to the axis of the optical system before reaching the film surface, because of the very long distance between the lens and the screen, compared with half of the screen size. The position hit by the light ray on the screen, corresponding to the position of the image of the ruling, is displaced a small distance, δx. The displacement, δx, is a function of the slope, dh/dx, of the film surface and the film thickness, h. The combining Snell's law of reflection and a geometrical relation deduces the equation.

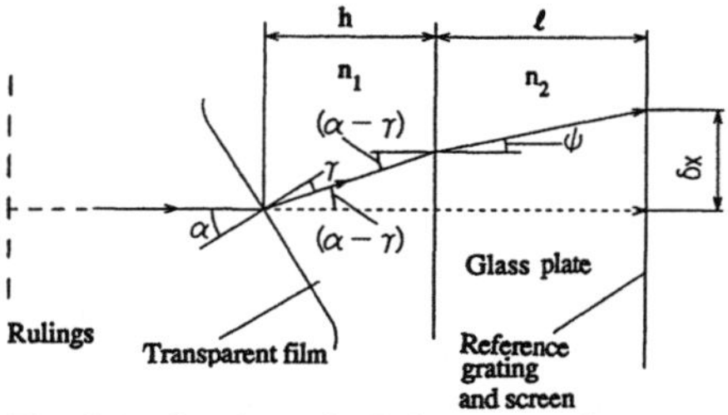

Fig. 3. Refraction of a light ray passing through a wavy file and a supporting plate

$$dh/dx = \delta x/\{(1-1/n_1)(h+n_1/n_2\cdot \ell)\} \tag{1}$$

where x is the coordinate in the vertical direction , h is the film thickness, n_1 and n_2 are the refractive indexes of the film and the supporting plate respectively and ℓ is the plate thickness. A small displacement, δx, of the ruling image in the vertical direction causes a large displacement, δy, of moire fringes in the horizontal direction. The relation between them is proportional, therefore the following equation is deduced,

$$dh/dx = \pm P_r/P_f\cdot \delta y/\{(1-1/n_1)(h+n_1/n_2\cdot \ell)\} \tag{2}$$

where P_r and P_f are the pitches of the reference grating and of the grating image respectively. The equation is more general than Scott's equation[1]. If employing the refractive index of the air for n_2 and neglecting h, which is negligibly small compared with the distance, ℓ, equation (2) is reduced to Scott's equation. The values for δy have been obtained from the distorted moire fringe patterns as a function of x. Equation (2) has been integrated using the function $\delta y(x)$ in order to reconstruct the film profile.

Measurement of a planoconvex lens

A planoconvex lens has been examined by the modified topography. The curvature and the maximum slope of the lens are 2.0 m^{-1} and 0.1 respectively. Figure 4. shows a sample of the moire fringe patterns distorted by the lens and its surface profiles reconstructed from the fringe patterns. The surface profile is along the vertical line passing through the center of the lens. The reconstructed profile is compared with the curve having the constant curvature, 2.0 m^{-1}. The measurement has been repeated four times for the lens and has found that all measurement results have fallen within a 3% error. The error is defined as the ratio of the difference between the reconstructed profile and the

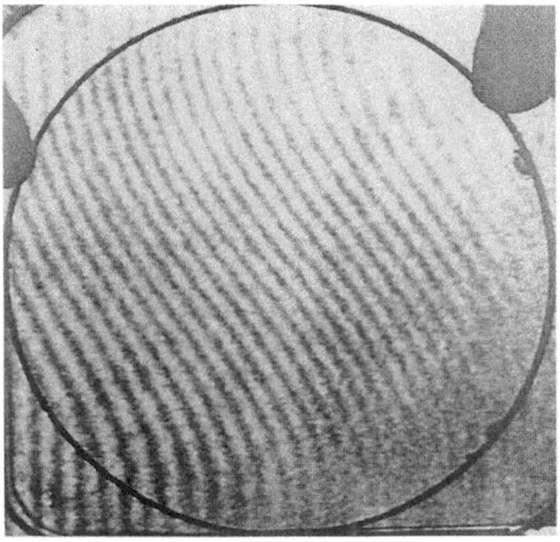

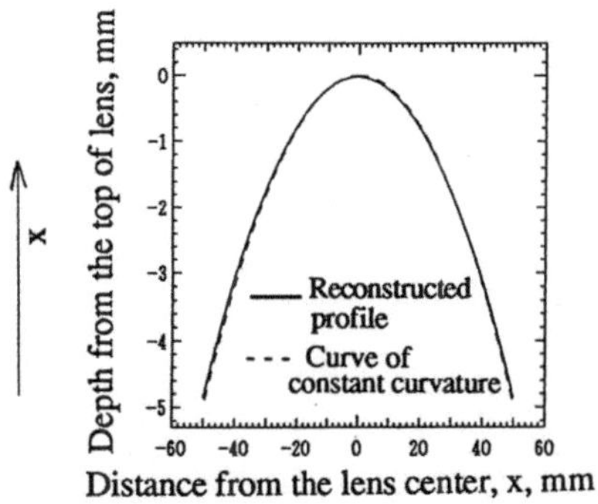

Fig. 4. Moire fringe patterns deformed by a planoconvex lens and its surface profile reconstructed from the fringe patterns.

constant curvature curve to the maximum depth of the lens surface from its top. It has been also found that the error depends significantly on how accurately the alignment of the optics has been achieved.

Wave profiles on a falling liquid film

A water films falling down a vertical plate has been measured. Figure 5. shows moire fringes distorted by a falling film and its surface profile reconstructed from the fringes. The fringe patterns demonstrate that waves appear and then grow up rapidly on the film. The waves can be assumed to be two-dimensional since the shapes of fringes vary little in the horizontal. After the figures are maintained to be straight for a distance, the deformation of the fringes develops rapidly in the flow direction. Finally, the deformation becomes so large that horizontal black strips appear, where no fringe can not be identified. Corresponding to the fringes patterns, the reconstructed waves increase in its amplitude and wave length with the distance in the flow directions. The wave shapes are nonsymmetrical with a steep front and a long, gently sloping tail. A deep trough is formed before the steep front. This tendency is enhanced with the development of a wave. Finally, a large curvature of the trough causes the appearance of the horizontal black strip since the trough works like a concave lens which leads light rays to spread out.

The present wave shape is in good agreement with those obtained with the laser interference moire topography[3,4]. The present results have found to have a good accuracy.

In the neighboring regions of the black strip, fringes are deformed substantially and the slope of the fringes approach a horizontal shape. As a result, the fringes come so close with each other that it is difficult to identify

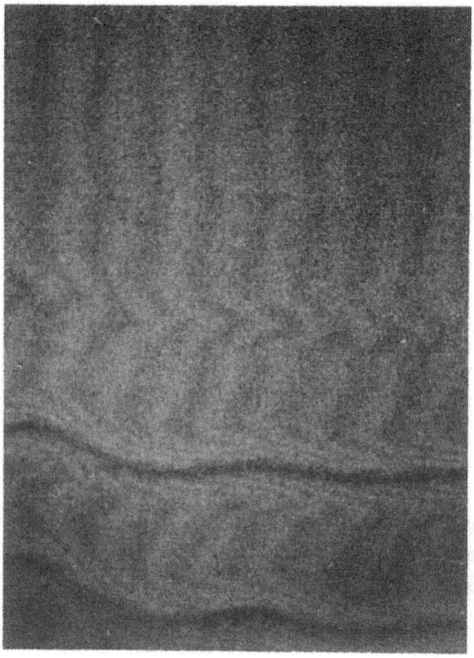

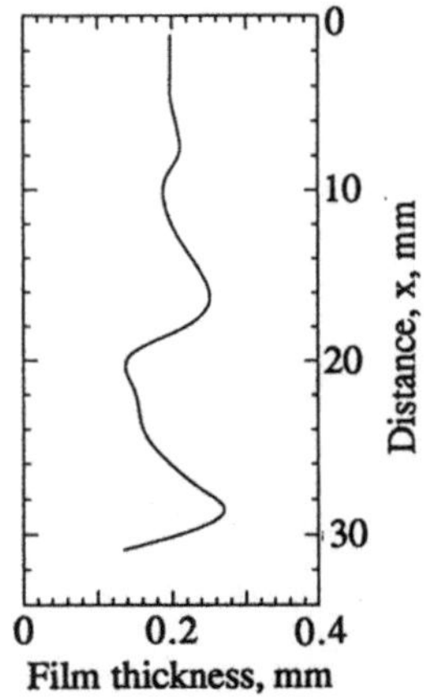

Fig. 5. Moire fringe patterns deformed by a wavy liquid film and its surface profile reconstructed from the fringe patterns; the Reynolds number is defined as Re = Γ/ν, where Γ is volumetric flow per unit wetted perimeter and ν is kinematic viscosity.

each fringe from the other ones. Either the loss of the fringes identification or the appearance of the black strip determines the maximum measurable range of the present measurements. The present results show that the maximum measurable curvature is evaluated to be 60 m^{-1}. This is a great increase compared with Kheshgi and Scriven's results in which the water surface of 4.0 m^{-1} curvature was reconstructed.

Acknowledgment -- We are indebted to T. Fukushima for his valuable advice and encouragement and to T. Shinohara for his kind advice and the helpful discussions. Particular thanks are due to M. Kamiya and S. Hatano for their assistance in the experimental work.

References

1. J.C. Scott, J. Fluid Mech., Vol. 56, pp. 591-606, 1972.
2. H.S. Kheshgi and L.E. Scriven, Chem. Eng. Science, Vol. 38, pp. 525-534, 1983.
3. M. Ishimatsu, T. Nosoko and T. Nagata, Heat Transfer-Japanese Research, Vol. 19, pp. 602-616, 1990.
4. T. Nosoko, M. Fang, T. Nagata and T. Shinzato, Proceedings of Joint Conference of 3rd International symposium on Heat Transfer, Beijing and 5th International symposium on Transport Phenomena, to be published.

Image Processing and Numerical Analysis

Image Processing

Geometric Modeling and Computer Graphics for Flow Visualization

Junta DOI

Bioinformation Engineering
Department of Biotechnology
The University of Tokyo, Tokyo

Summary

An automated shape modeling system that represents an object body by a solid model of polyhedral approximation was developed using multi-directional image inputs. Application to the non-contacting three-dimensional measurement and tracking of the object of irregular and complicated shape such as a tuft and a candle flame is reported. Parameters involved in this analysis are position, velocity, trajectory, acceleration, skeleton of the body. Tracking of a oscillating tuft, a small tracer particle and a candle flame in a draft is presented. This procedure is entirely automatic requiring no annual operation. Spatial temperature distributions in a laminar asymmetric alcohol wick flame and a turbulent propane burner flame were estimated from multi-directional holographic interferometers using computed tomography. Computer graphics of solid models made by Boolean operation on the initially reconstructed distribution models give a comprehensive understanding of the flame structure.

Principle of Three-dimensional Shape Approximation and Solid Model Generation

In the first stage, the silhouette of the object image extracted from a picture taken, for instance, with a video camera. The silhouette is represented by a polygon as an approximation. In the second stage, a semi-infinite pyramid which is constructed from a center of perspective and a silhouette polygon is determined. In this pyramid, a vertex is the center of perspective or the lens center and the cross section is the polygon which is made by the above procedure. When multi-directional image inputs are applied, a corresponding number of semi-infinite pyramids are generated in the same way. In the third stage, the intersection of all of these multi-directional pyramids or, in other words, the calculation of the common area of these pyramids generates a polyhedron. When the object has a convex shape, this polyhedron becomes a progressively better approximation to the original shape as the number of visual directions increases. If the object is not a thoroughly convex form or the number of visual directions is not large enough, this polyhedron may be used as an approximation within acceptable engineering accuracy.

Video camera parameters, such as optical axis direction, rotation, field of view, and magnification ratio, are determined prior to the solid model computation. After the multiple pyramids are generated, a wire-frame model for the intersected area is finally transferred into a so-called solid model of the form of a polyhedron. In this modeling, the spatial location of each vertex is described by a single co-ordinate system, which is defined by the co-ordinate when the camera parameters are initially calibrated. In Figure 1(a), a co-ordinate is shown illustrating the components of translation and rotation. If the approximated polyhedron is reconstructed successively by multiple pictures which are taken in a frame-to-frame manner, characteristics of the object shape and behavior are consecutively and automatically obtained.

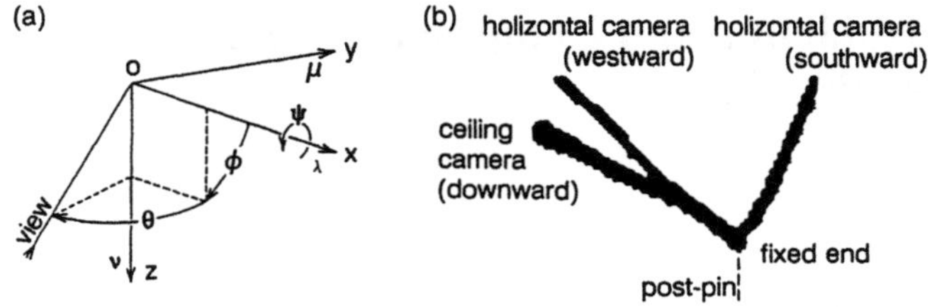

Figure 1. Co-ordinate system (a) and three-directional images of a tuft (b)

Tuft Behavior Analysis

A tuft in the air flow was observed by three CCD video cameras which were installed in the directions nearly perpendicular to each other [1]. In this observation, a tuft woolen yarn of about 1 mm in diameter and 35 mm long was attached on the top of a thin post pin. Figure 1(b) shows three simultaneous video camera images. An example of the reconstructed solid model using three simultaneous images at a moment is shown in Figure 2. As this modeling is made by only three-directional images, the resulting approximation is rather rough, however, the fundamental shape characteristics are reproduced. With this modeling system successive reconstructions of the tuft shape were made with the same co-ordinate system to show deformations through time. Transient behavior or oscillatory characteristics of the tuft can be tracked. As shown in Figure 3, typical examples of tracking the time-varying behavior are investigated by frame-to-frame imaging. It is observed that the free end of the tuft model spreads out widely as shown in Figure 3. This is due to the rapid tuft motion within a video frame time. Skeletonizing polyhedral model gives instantaneous tuft direction more comprehensively as show in Figure 4. More information on quantitative tuft behavior and flow characteristics can be obtained by applying shape analyses.

Tracer Particle Tracking

Figure 5 shows an example of the reconstruction of a tracer particle by the same modeling method [2]. Image acquisition was made simultaneously by three CCD video cameras installed nearly perpendicular to each other. A tracer of 3 mm in diameter was observed in an agitated cubic vessel of 100 mm edges. The tracer is a polystyrene sphere of a density of 1.06. The fluid is a saline solution adjusted to neutral buoyancy. In the figure, 36 consecutive polyhedra of the approximation are depicted at a constant time interval of 5 seconds started from T1. Average number of the polygonal face of each consecutive polyhedra is about 20. Each volumetric deviation to the original particle is about 4 per cent.

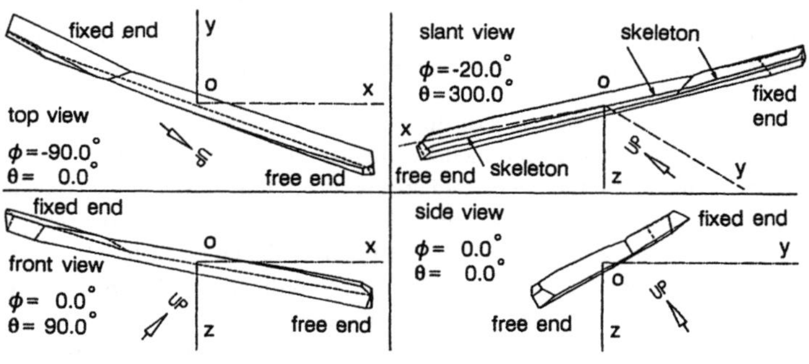

Figure 2. Solid model of a tuft at a moment (T00)

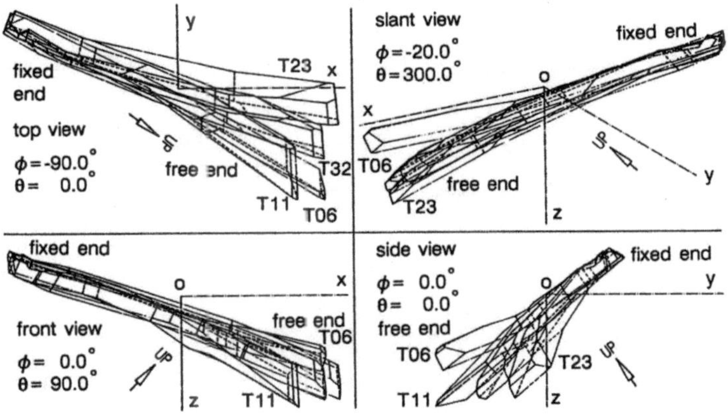

Figure 3. Solid models of a tuft at various moments (T06,T11,T23,T32)

Particle position is estimated to the gravity center of the model of the approximated polyhedron. Particle trajectory is visualized by connecting each gravity center successively. The trajectory is expected to be the path-line of the flow. Each line segment between the consecutive polyhedra is the mean velocity vector in 5 seconds. Mean velocity observed in this figure varies from 1.0 to 1.8 mm/s. Acceleration is calculated by taking the successive difference of the velocity vector in respect to time. Figures 5(c) and 5(d) are the same movement visualized from another angles of view. In the figure, acceleration vector is enlarged by three times for more distinguished presentation. This method is suitable to track the object which changes its shape and size in the course of traveling.

Shape Approximation of a Time-varying Candle Flame

A candle flame varying in time was reconstructed in the same polyhedral expression [3]. In Figure 6(a) an example at a certain moment (T1) is formed by the third angle projection method. In this figure the vertical direction is illustrated in each candle flame figure. The flame in Figure 6(a) was in still air. In Figure 6(b), a flame at T20, deformed by a light draft of about 0.2 m/s,

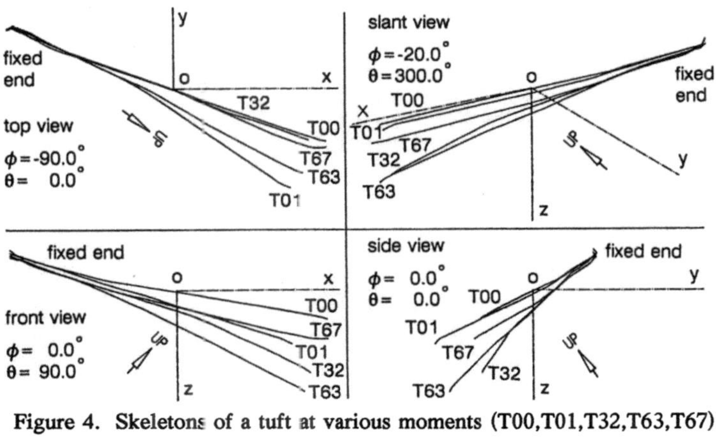

Figure 4. Skeletons of a tuft at various moments (T00,T01,T32,T63,T67)

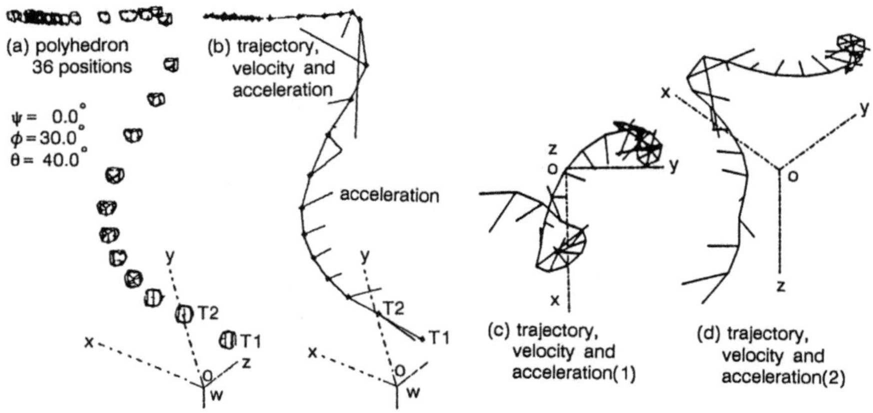

Figure 5. Solid model, trajectory, velocity and acceleration of a tracer particle

is shown. In Figure 6(b) the figures of T1 and T20 are superimposed to compare the elongated flame behavior in the light draft.

Flame Shape Modeling from Multi-directional Interferograms

It is possible to make a solid-model reconstruction using fringe patterns on interferograms [3]. This interferometric system consists of 24 object beams distributed in the total viewing angle of 168 degrees, with 4 reference beams and 4 photographic recording plates [4]. The target is an alcohol wick flame. The wick size is 24 mm by 9 mm. Figure 7(a) is a solid model of the flame reconstructed from a set of the first fringe pattern. Its height is limited to the height of the photographic recording plate. This flame is nearly rectangular near the base and becomes rounded in the upper part. It is expected that the polyhedral zone corresponding to the first fringe pattern is the external shape of the flame, comparing from the following temperature estimation.

3D Visualization of Flame Temperature Distribution using Holographic Interferometry

The three-dimensional temperature distribution in an asymmetric alcohol wick flame was estimated and displayed with computer graphics [5]. The multi-directional optical system with

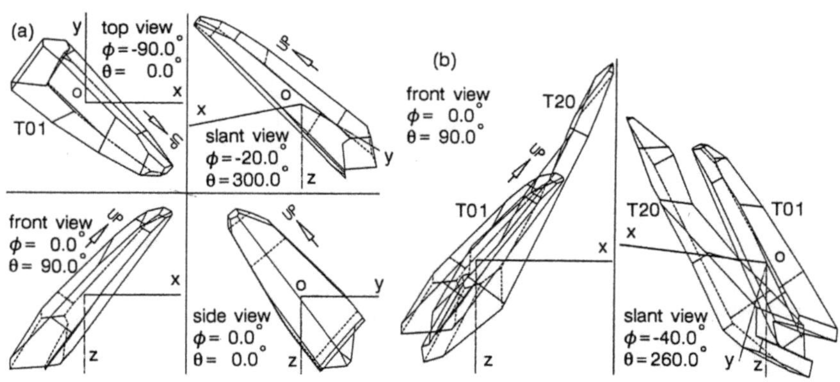

Figure 6. Solid models of a candle flame at two moments (T01,T20)

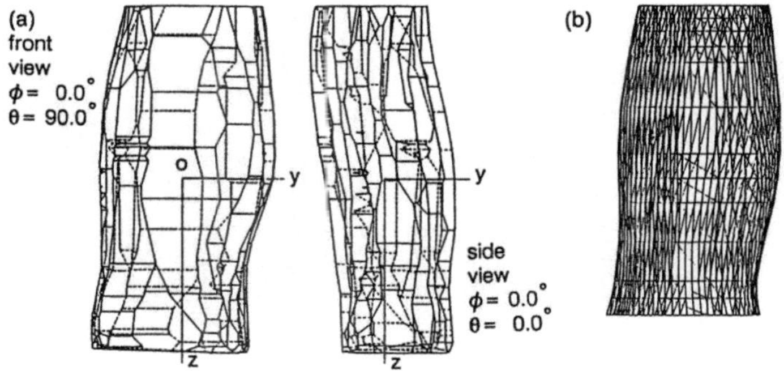

Figure 7. Solid models of an alcohol flame, first fringe solid (a) and 400K solid (b)

non-diffuse illumination is as same as the previous section [4]. The light source is the beam of an Argon-ion laser of 514.5 nm wave length and of 300 mW power. Since the alcohol flame is transparent in nature, this 24 directional arrangement covers another symmetric 24 directions, covering totally 48 equivalent directions in a circle. Interferograms were taken at a moment when the flame fluctuated because of a draft, showing an asymmetric and distorted flame.

An algorithm of the convolution method was used to reconstruct the temperature distribution. The gas composition of the flame is assumed as that of air and the combustion takes place at constant pressure, however, a temperature difference of about 5% at maximum is estimated [4]. Calculations for the temperature profile reconstruction were made separately in each of seven horizontal cross sections. The first section is located 4 mm above the base of the wick and the seventh or the highest is located 65.8 mm above. Temperature data are then summarized by making isotherms in all seven sections. Isotherms are similar to that shown in Figure 9(a). Each isotherm of the same temperature value is then stacked together successively from the bottom to the top of the section to form a three-dimensional body, of which side surface is to be isothermal. Figure 7(b) is a reconstructed temperature model in wire frame representation. Its side surface shows an isothermal surface of 400K. In this figure, the flat sections of the top and the bottom correspond to the seventh and the first sections, respectively. This figure is very similar to the figure in Figure 7(a) as mentioned before.

When a solid model has been reconstructed, computer aided visualization technique become available for more comprehensive understanding of the internal structure. For example, a solid

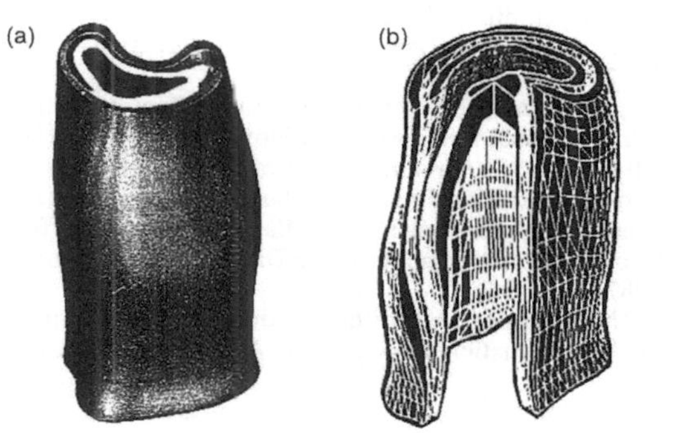

Figure 8. Temperature distribution solids with Boolean operation

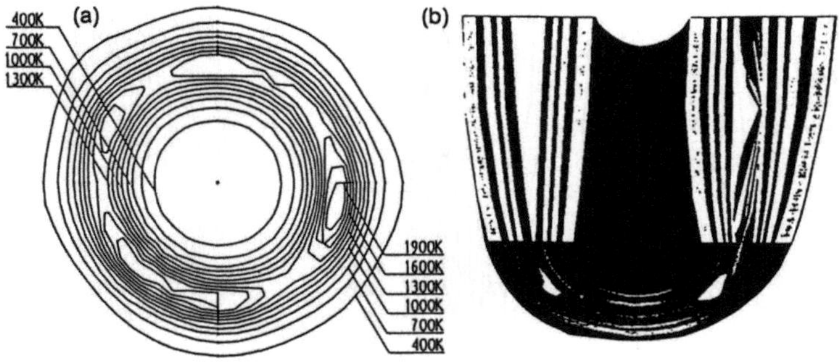

Figure 9. Isotherms (a) and plane cut model (b) of temperature distribution in turbulent flame

model, which has a temperature distribution from 400K to 550K is generated by Boolean subtraction of the solid model of 550K from that of 400K. Repeating this procedure, a nest set of a tube-like body is constructed. Figure 8(a) is an example of the solid model in a shaded image representing isothermal surfaces of 400K, 550K, 700k, and 850K. Figure 8(b) is an example of the corner-cut display to show the internal temperature distribution. Figure 9 shows a temperature distribution of a turbulent premixed propane burner. The procedure is similar to that of the laminar flame. In this case an eight-directional holographic interferometer was used [6]. Light source was the beam from a ruby laser. Isotherms are more complicated having hotspots near the middle part of the flame. In this case similar graphics technique is applied for comprehensive understanding.

Conclusion

Application of three-dimensional shape approximation and geometric solid modeling for flow and flame visualization was reported. With this modeling system successive reconstructions of a object shape are made with the same co-ordinate system to show three-dimensional motion and deformation through time. Solid models of the three-dimensional temperature distribution were reconstructed from multi-directional interferometric tomography. Combined solids made by Boolean operation were demonstrated by computer graphics to make comprehensive presentation. This development will have considerable utility in the integration of experimental data and numerical simulation.

References

1. Doi, J.; Miyake, T.: Three-dimensional analysis of time varying tuft behavior by its successive geometric shape modeling. Proc. 4th International Symposium on Flow Visualization (1986) 229-234.
2. Doi, J.; Miyake, T.; Asanuma, T.: Three-dimensional flow analysis by on-line particle tracking. Proc. 3rd International Symposium on Flow Visualization (1983) 14-18.
3. Doi, J.; Sato, S.; Miyake, T.: Three-dimensional measurement of the shape of combustion flames. Laser Diagnostics and Modeling of Combustion. Springer-Verlag (1987) 195-202.
4. Sato, S.: Measurement of three-dimensional flame temperature fields by holographic interferometry and computed tomography. Laser Diagnostics and Modeling of Combustion. Springer-Verlag (1987) 187-194.
5. Doi, J.; Sato, S.: Three-dimensional reconstruction of flame structures using multiple interferograms. Proc. 8th International Congress on Application of Lasers and Electro-Optics. (1988) 131-138.
6. Sato, S.; Kumakura, T.: Interferometric tomography measurement of temperature fields in turbulent flame. Proceedings of Joint International Conference on Combustion. (1989)

Charge-Coupled Devices in Flow Visualization

W.J. HILLER, T.A. KOWALEWSKI, V. LLORACH FORNER, B. STÜCKRAD & M. BEHNIA*

Max-Planck-Institut für Strömungsforschung, 3400 Göttingen, Germany

**University of New South Wales, Kensington, 2033 Australia*

SUMMARY

Solid state imaging devices are increasingly replacing conventional photographic techniques in the flow visualization. This paper gives the following examples of visualization and evaluation techniques, which make use of special features of the CCD's: (i) multiexposure imaging and analysis of droplet oscillations, (ii) local flow vorticity measurement with specially marked tracers, (iii) high-speed imaging during the frame transfer of the device applied to observations of liquid jet instabilities and crack propagation in a pressed coal powder.

INTRODUCTION

During the last decade, in many scientific applications CCD's have replaced the classical photographic methods. There are several reasons for this development. This is partly due to the inherent characteristics of the device itself, like the linear response of the output signal to the amount of the collected light, the division of the sensor into a well defined array of pixels, and the long term stability of its electrical and mechanical properties. On the other hand, the nearly immediate access to the information stored in the pixels by means of frame grabbers, and the on-line evaluation of the images is an important feature as this allows fast control and adjustment of the experimental conditions.

Over the last few years our group has used CCD cameras quite extensively in a number of experiments with different applications. The general features of the CCD will be elucidated by two examples. First, demonstrates evaluation of multiexposed images of oscillating droplets. This analysis allows the possibility of measuring the dynamic surface tension of liquids. In the second example, the identification and evaluation of images of the specially marked spherical tracers provides us with on-line information about the local values of the velocity and vorticity of a flow in a rotating circular duct.

The other and very interesting characteristic of the Frame-Transfer CCD is a very fast vertical transport of charges generated by incident light. This feature gave us the possibility to develop the high-speed imaging device, which captures a series of several oblong exposures of the investigated objects within periods as short as $160\mu s$. The device is used to visualize surface instabilities of liquid jets and the propagation of cracks in coal samples.

In this paper we describe some details of the aforementioned three groups of applications of CCD imaging for the flow analysis.

VISUALIZATION EXPERIMENTS

1. Oscillating Droplets

The first application is concerned with the analysis of oscillating and eventually evaporating droplets. In our experimental setup the droplets are generated by the breakup of a laminar jet discharging from a convergent nozzle to the atmosphere. The droplets are typically $0.1 - 0.3mm$ in diameter and move at a few metres per second. They are illuminated with short ($200ns$) pulses from a Light Emitting Diode (LED). The images are captured and registered under a microscope by a CCD camera (Sony XC77CE) and then recorded with the help of an imaging board (VS100 Imaging Inc.). The visualization method used is based on the principle of multi-exposuring a transient object on a single video frame (Fig.1). Due to the high dynamic response and the linearity of the CCD sensor, multi-exposed images can be easily evaluated by using a specially developed software. This allows the determination of the droplet diameter (or volume) as well as its oscillating frequency [1, 2]. From these parameters the droplet evaporation rate [3] as well as the dynamic surface tension [4] of the liquid could be determined. The multiexposure method of imaging provides high temporal and spatial resolutions for about one period of droplet oscillation. To extend the observation time, long series of droplet images are recorded using the method of "beat-frequency" stroboscopic illumination. With the help of a piezoceramic transducer mounted inside the nozzle it is possible to control the breakup process of the jet, and observe droplets stroboscopically. Then, the droplets illuminated with a frequency close to that of the jet excitation appear in a slow-motion and can be easily acquired periodically by the image processor. The long sequences of images are saved in a permanent computer storage. Figure 2 shows the variation of the oscillation amplitude measured in such a way, for the series of 270 droplet images.

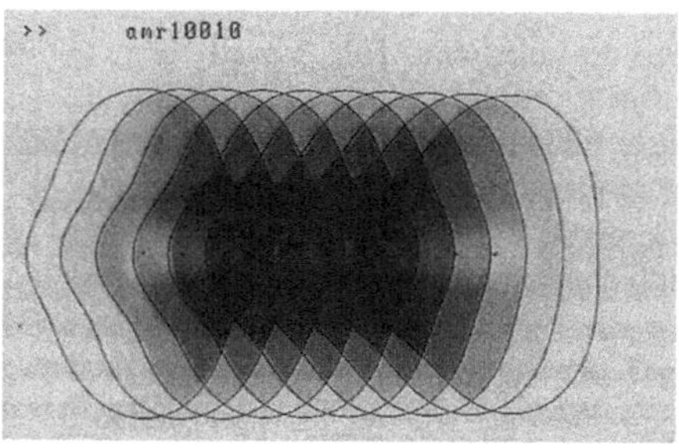

Fig.1: *Image and evaluated contour (solid line) of a multi-exposed oscillating ethanol droplet. Equilibrium radius 0.210mm. Strobe frequency 128kHz, 10 flashes of 200ns duration.*

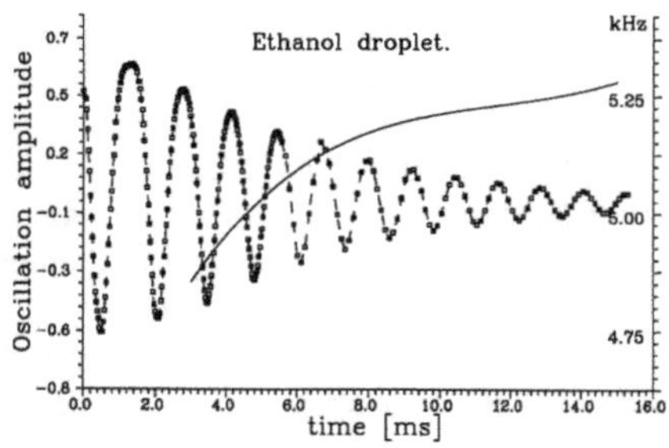

Fig.2: *Measured amplitude of the fundamental oscillation mode for an evaporating ethanol droplet. Solid line shows approximate variation of the oscillation frequency.*

The non-intrusive measurements of dynamic surface tension, which becomes possible by the oscillating droplet method, are of an intimate interest for estimation of physico-chemical properties of liquid mixtures. Figure 3 gives an example of the measured [4] variation of the surface tension using oscillating droplets of a heptanol-water mixture (0.0076mol/l bulk concentration).

For evaporating droplets besides the evaporation rate obtained from the temporal variation of the diameter, an additional analysis of the oscillation frequency is used to measure the time dependent variation of the surface tension. Making use of the previously estimated temperature dependence of the surface tension for the liquid used, the method gives the unique possibility of measuring the surface temperature of the evaporating droplets. This temperature is a primary and very important parameter in most evaporation models. Also its measurement by conventional techniques is not possible.

2. Flow Field Vorticity

With the help of a CCD camera and an imaging board, the motion and rotation of a small (3mm) solid sphere conveyed within a fluid flow field is captured. This allows the "classical" analysis of the tracer velocity as well as an evaluation of the local flow field vorticity. The rotation of the sphere is detected by observing a pattern of dots specially printed on its surface (Fig. 4). In the plane of observation, from the change in time of the position of only two of these dots, the vector of rotation can be evaluated. Making use the of standard features of the Sony XC77CE camera it is possible to capture two separate images of the sphere (two video fields) within $100 \mu s$ of one another (any longer intervals are of course possible). It allows investigation of flow fields with the maximum value of local vorticity about $10^4 s^{-1}$. At present our investigations are performed in a flow field generated in the rotating circular duct. In this flow the tracer convective velocity relative to the observer is almost nil, and practically a pure rotation movement of the sphere takes place. Future development of the experimental setup will allow us to follow the single tracer sphere conveyed by the flow field. This will further expand the application domain of the method to include a wide class of flow structures, characterized by translational velocity low enough to allow tracking of the tracer movement by a mechanically controlled camera.

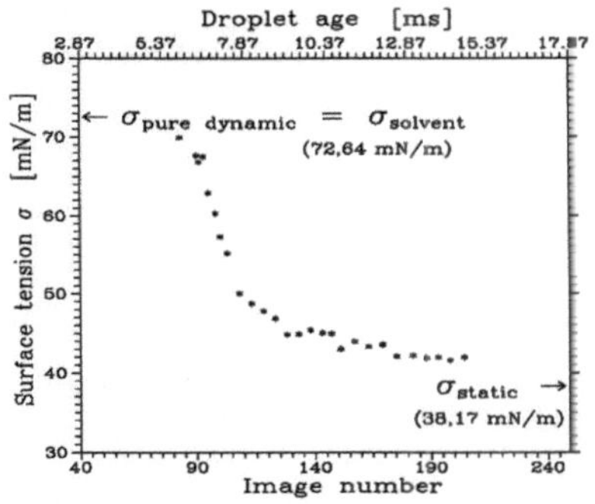

Fig.3: *Temporal variation of the surface tension of a heptanol-water mixture measured by the oscillating droplet method for the series of 215 droplet images.*

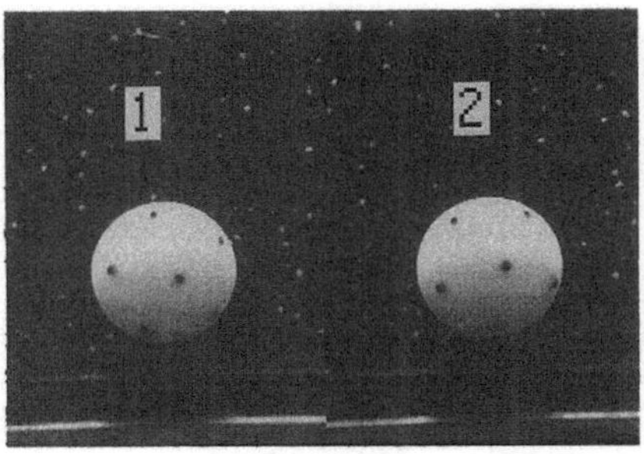

Fig.4: *Two superimposed images of a special tracer moving in the circular duct. The time interval between exposures is* $10 ms$. *Evaluated vorticity* $curl\,\vec{v} = (12.2, 13.6, 86.5)\, s^{-1}$.

3. CCD Frame Transfer Application

The frame transfer feature of the CCD can be easily adopted for high-speed imaging [5]. Due to the movement of sensor charges during the transfer cycle, observed effectively as a movement of the sensor plane, short sequences of images can be registered (like in a classical streak camera). In collaboration with *Theta System Elektronik GmbH*, a commercially available sensor (Thomson TH7863) has been adapted to construct a powerful high-speed imaging system. By modification of the frame-transfer drive-pulse sequence two fields of the video frame are shifted continuously without interruption. As a result the entire sensor area (i.e. storage and image section) are used to save one sequence of images. The minimum time necessary to execute this shift is a property of the device and for the sensor used (TH7863) was $160\mu s$. This frame-transfer time can be prolonged in steps to a maximum of $20 ms$. By doing this the image interval frequency is matched to the requirements of the transient processes. The record of an image sequence can be read out either directly in CCIR TV mode on two separate video-fields, or in the variable scan mode via a frame grabber as a single data block. The camera was successfully integrated with the VS100 board in a personal computer (IBM compatible), and with IPP-100 (Eltec) image grabber in the VME-bus computer. The facility allows the division of the sensor field (576 vertical by 384 horizontal pixels) into a sequence of separate images, with its vertical format and framing period inversely proportional to the total number of images recorded. In practice, a series of ten images, each of 57 by 384 pixel, are found to be an acceptable compromise between spatial and temporal resolution. The triggering capability of the camera (within $1\mu s$) makes it suitable for studying chaotic fast transient events such as boiling and bursts of evaporating liquid jets.

The camera has been exploited in two different applications: (i) visualization of liquid jet instabilities induced by explosive evaporation and (ii) studying the crack propagation in coal samples (saturated with gas), caused by a sudden pressure release [6]. Figures 5 and 6 show examples of the images obtained during the evaporation of ethanol and crack growth, respectively.

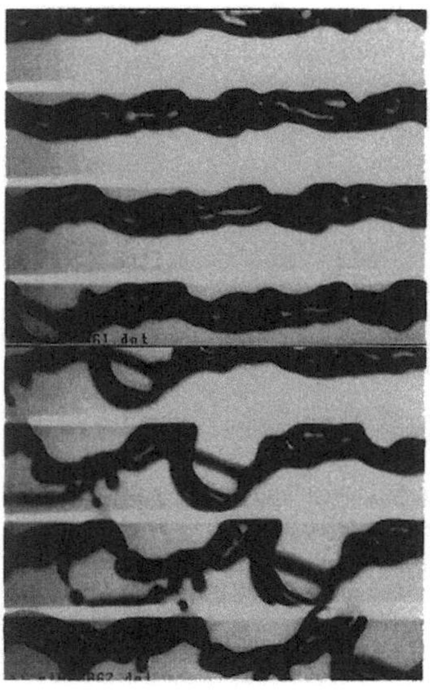

Fig.5: *8-fold frame transfer exposure of a 0.2mm ethanol jet evaporating in vacuum. Strobe frequency 51kHz, flash duration 200ns.*

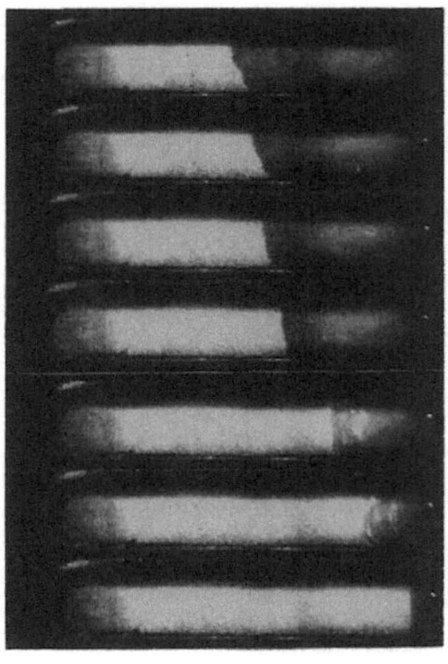

Fig.6: *7-fold frame transfer exposure of the crack propagating in a coal sample saturated with gas. Strobe frequency 1.6kHz, the crack mean velocity 4.8 m/s.*

Additional applications of the frame transfer CCD features for studying moving structures in a turbulent flow are under way. The frame transfer of the camera is synchronous with the base flow component. This allows for observation from a reference system moving with the main flow velocity. As a result the separation of important flow structures such as turbulent "bursts" or moving singular points from "uniform" background flow can be easily realized.

CONCLUDING REMARKS

The computer-aided analysis of multiexposed images gives the promiscuous method of visualizing movement of objects with well defined shape. The image noise is practically the main limit for the number of superimposed images. This can be improved by making use of special low noise devices and by calibration of the individual pixel's response sensitivity.

The presented frame-transfer based system characterizes relatively low spatial resolution. This limitation is imposed by the dimension of the commercially available CCD sensors. The fast development of CCD technology gives rise to the hope that it would be soon possible to improve this resolution.

Experimental results presented here show that CCD cameras can be successfully used for analyzing complex flows, which cannot be readily visualized by other conventional techniques.

References

[1] Hiller W.J., Kowalewski T.A.; Surface tension measurements by the oscillating droplet method. Phys. Chem. Hydrodyn., **11**(1), pp.103-112, 1989.

[2] Becker E., Hiller W.J., Kowalewski T.A; Experimental and theoretical investigation of large amplitude oscillations of liquid droplets. J. Fluid Mech., **231**, pp.189-210, 1991.

[3] Kowalewski T.A., W.J. Hiller; Unsteady droplet evaporation in: Proc. of 11th ABCM Mech. Eng. Conf. São Paulo (Brasil), Dec. 1991, vol.2, pp. 17-20, Ed. Esp. da Revista Brasileira de Ciências Mecânicas, São Paulo 1991.

[4] Stückrad B.; Die Dynamische Oberflächenspannung grenzflächenaktiver Lösungen und ihre Bestimmung mittels der Analyse von Tropfenschwingungen, Diplomarbeit, Georg-August-Universität, Göttingen 1992.

[5] Hiller W.J., Kowalewski T.A.; Application of the frame transfer charge-coupled device for high speed imaging. Optical Eng. **28**(2), pp.197-200, 1989

[6] Meier G.E.A, Gawor M., Rysz J.; Experimental study of expansion waves in mixture of pulverized coal and gas. To be published by *Strata Mech. Res. Inst., Polish Academy of Scs.*, Cracow 1992.

Corresp. to: Dr. T. Kowalewski, MPI Bunsenstr. 10, D-3400 Göttingen, FRG.
E-mail: Tkowale@ibm.gwdg.de

A Digital Image Analysis System for High-Speed Filmed Injection Jets

J. Wagener

Fachbereich Maschinenwesen, Universität Kaiserslautern, Germany

Introduction

To reduce the emission of harmful substances of Diesel engines a detailed comprehension of spray formation is essential. The investigation of injection nozzles is a very important part in this connection.

With this aim in view a high-speed camera had been presented, which is able to film injection jets with a framing rate up to 1.000.000 f/s [1,2]. A digital image processing system has been developed to carry out an automatic analysis of the obtained image sequences.

The Image Processing System

To transfer the picture information to the digital computer a positioning device is used that presses the film between two plates of glass and that can be moved horizontally controlled by the computer. A fixed light source lightens the film from below and an electronic CCD-camera transfers the brightness information to a frame grabber in a computer for further analysis.

For the processing of the up to 300 images per sequence the film is cut digitally into single, independent frames each describing the spray development at a certain time. This can be achieved by several correlation and symmetry calculations across the spray image. To remove image faults possibly caused by the high-speed camera various illumination filters based on homomorphic algorithms and contrast improving operators can be selected, which are not presented in this paper.

Determination of Spray Parameters

A quantitative analysis enables now to determine global spray parameters like the spray cone angle or the tip penetration. A special edge detector extracts the relevant contour of the spray according to the maximum gradient of the image brightness. In a postprocessing refinement all contourparts are removed which do not belong to the most important, smooth and closed contour of the spray. The geometry of the injection nozzle that can be seen on each film is available online from a database. The exact position of the nozzle inside the frame is calculated by a matching algorithm. Finally a least-square regression starts from the nozzle opening separately over the left and right half of the contour to calculate the spray cone angle. The tip position is detected automatically as well, and doing so for each image of the whole sequence, the temporal development of the spray can be judged. Fig. 1 shows from left to right a typical original spray image taken at 10 bar chamber pressure and 20°C temperature, the primary edge detector result and in the third frame the smoothed edges with the replaced idealized nozzle contour. The cone angle is sketched in the fourth frame limited by two straight lines, the tip is marked with a cross and the symmetry axis is drawn as a broken line.

Results of a Special Core Filter

The internal structure of the jet core is of special interest for the understanding of spray formation. However, in the images there are only poor brightness variations. A further edge detector has been developed which is adapted especially to this problem. Fig. 2 contains a processed image which shows an original spray, the core structure and two coarser subsets to concentrate on larger structures. The most important detail is a small jet in the second frame on the left side directly below the flat nozzle, which is normally nearly invisible on the film. The small jet is caused by the implosion of a cavitation bubble.

Analysis of Dynamical Effects

The main emphasis of the spray analysis is to determine the optical flow. Analogously to the understanding of flow in the field of fluid mechanics the term optical flow is used in image processing to describe a flowing brightness similar to a flowing fluid. The well known method of particle tracking, however, can't be accomplished, because for our film sequences no correspondence between referring image parts can be established. But the concept of Horn and Schunck [3] has been successfully adapted, an iterative algorithm based on temporal and spatial derivatives of the images in combination with a special smoothness constraint. In Fig. 3 only a few selected velocity vectors have been drawn for three separated time steps, the full information is only available on coloured diagrams or with smaller spatial resolution. Fig. 3 shows the temporal and spatial movement of turbulent zones, characterized by the spray contour with averaged vectors being faster than 30 m/s. Thus the visualized dynamical information of the injection jet is a suitable help to understand the phenomenon of spray development.

Acknowledgements

The author wishes to acknowledge the financial assistence of the "Volkswagenwerkstiftung" and of the "Deutsche Forschungsgemeinschaft".

References

1. Eisfeld, F. ; Eifler, W. : Investigation of the motion of fast injection jets with a three-dimensional optical method using high-speed cinematography, Flow Visualization V (1989) 772-778
2. Eifler, F. : Untersuchungen zur Struktur des instationären Dieselöleinspritzstrahles im Düsennahbereich mit der Methode der Hochfrequenz-Kinematografie, Dissertation, Universität Kaiserslautern, 1990
3. Horn, B.K.P. ; Schunk, B.G. : Determining optical flow, Artificial Intelligence 17 (1981) 185-203

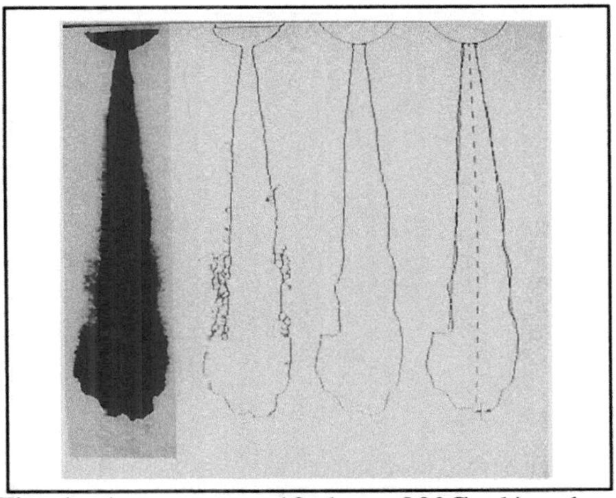

Fig. 1 a) spray at 10 bar, 20°C, b) edge detector result, c) smooth edges, d) cone angle and tip position

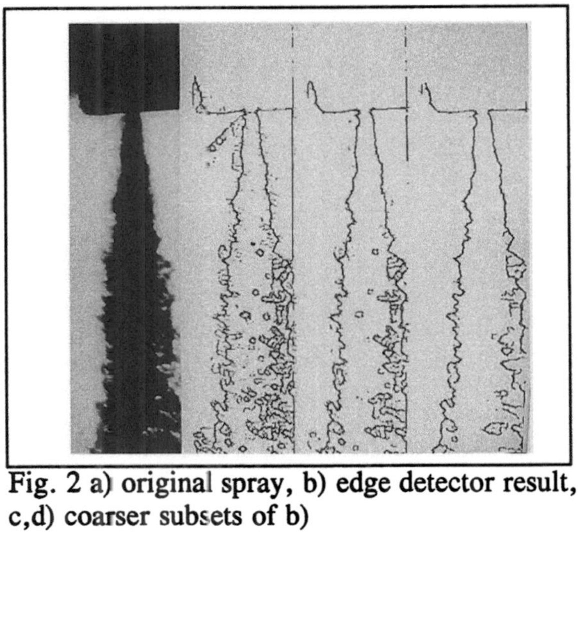

Fig. 2 a) original spray, b) edge detector result, c,d) coarser subsets of b)

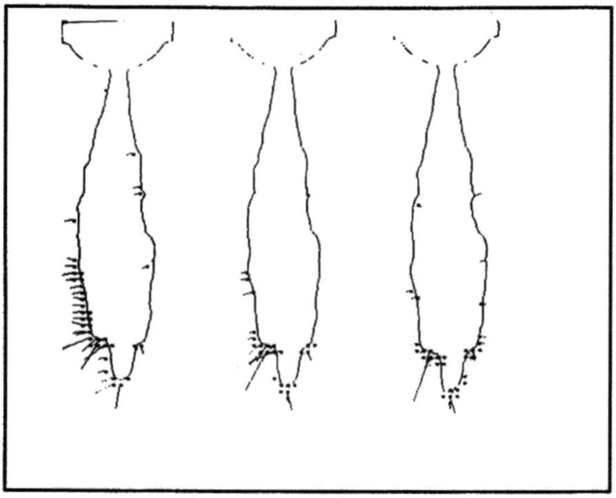

Fig. 3 Dynamical spray development for three successive time steps, max. velocity 350 m/s

CT Image Reconstruction by Matrix Method

T. YOSHIDA

National Aerospace Laboratory, Jindaiji-Higashi, Chofu, Tokyo 182

M. KITADA

Ishikawajima-Harima Heavy Industries Co.,Ltd., Tanashi, Tokyo 188

I. YAMAGUCHI

Nissan Motor Co.,Ltd., Natsushima, Yokosuka, Kanagawa 237

Summary

In the matrix method, unknown values of individual elements constituting a test object are solved by constructing a large scale matrix equation with M-dimension bigger than the number of elements. The method has possibility to provide us mathematically satisfying solutions. However this has never been performed as far as the authors are aware because it requires a big scale, high speed computer, so called super computer and an appropriate solution method for a big scale matrix equation.

The present paper deals with the description of the method itself and an example of its verification by sample data and some results from actual turbine blade data taken in a γ-ray CT apparatus.

Description of the Matrix Method

Computed tomography (CT) is an advanced computer-aided visualization technique which enables us to view the cross section of an object without cutting it. Today, CT technology is well established, and so called *filtered back projection method* (FBP) advocated by Shepp and Logan [1] is used for image reconstruction. For obtaining projection data, the same procedure is followed in the matrix method as those in FBP. **Fig.1** indicates a schematic drawing of gathering projection data. At a certain location of rotation and translation, value of projection (designated by **P** in the figure) is a ray sum resulting from attenuation of emitting electromagnetic wave due to passing through the object. Its intensity corresponds to the density of the object

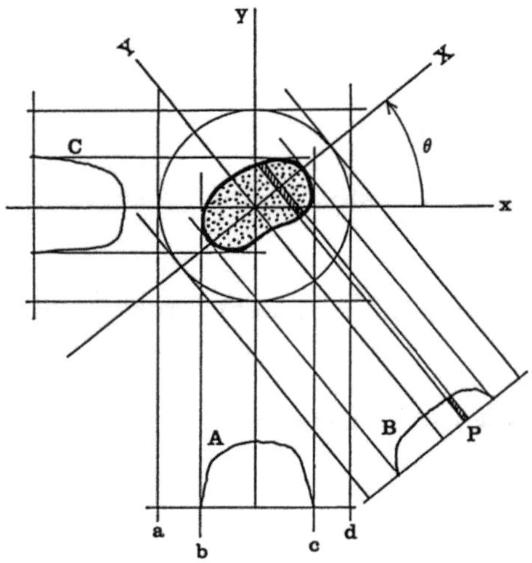

Fig.1. Projection data sampling

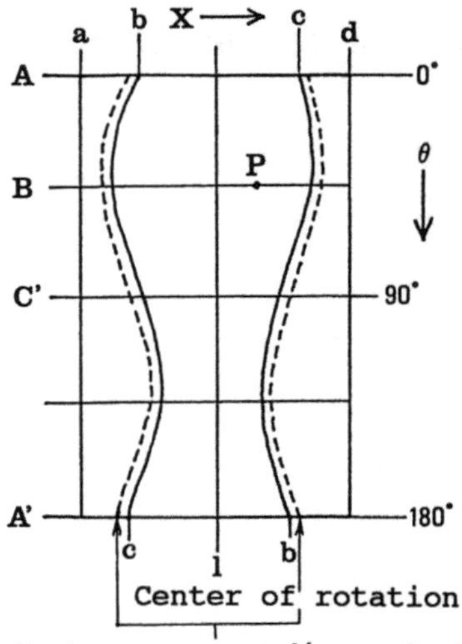

Fig.2. Geometric allocation of projection data

in the ray path. A set of projection data at an individual translation is gathered from those at 0deg.of rotation to those at 180deg.at an interval of 1deg.for example. Location of each projection can be organized geometrically as shown in **Fig.2.** Each translation is aligned downward from the top (translation **A** at 0deg.) to the bottom (**A**'at 180deg.). The relation between the two figures can easily be understood by looking at **P**. Ray sum on the border of the object is always very close to those without hitting the object. It is very difficult to distinguish the border by the value of ray sum. Thus the data situated in the center region between the two broken lines are considered substantial.

In the matrix method, unknown values (from $\mathbf{F}_1$ to $\mathbf{F}_N$) of individual elements constituting the test object are allocated as shown in **Fig.3**. The relevant matrix equation is constructed as indicated in **Fig.4**. **M** is the number equal to or less than the product of the number of times of translation and rotation. **N** is the number of elements, unknown values. M is also the number of projection data introduced in the matrix, and should be far bigger than N.

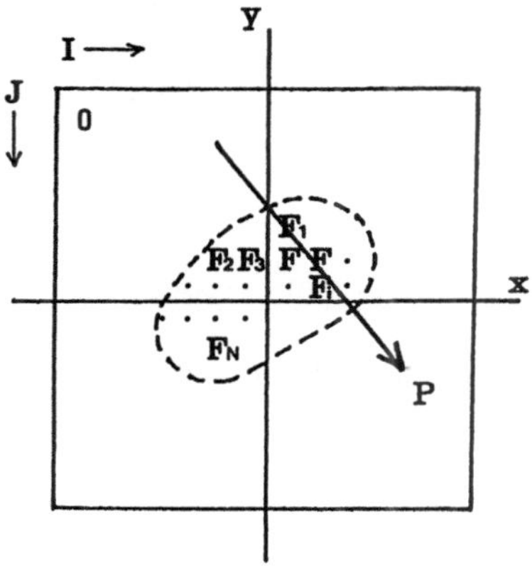

Fig.3. Allocation of unknown values

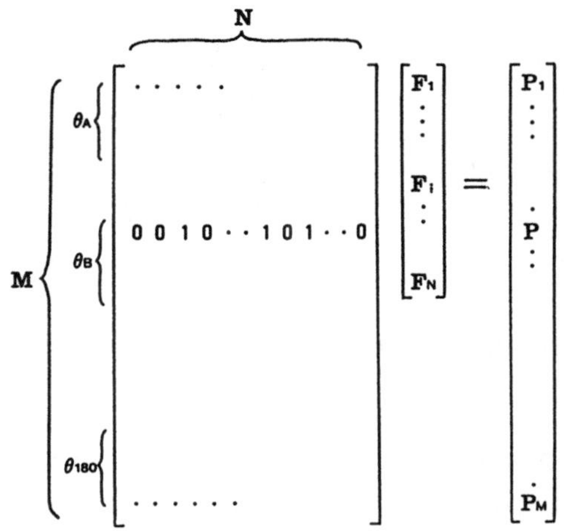

Fig.4. Matrix equation

By reference to a solid line with an arrow in Fig.3 and corresponding single line of [0010..101..0] in the matrix, the way of constructing equation may easily be realized.

Verification Test with Sample Data

Artificial projection data were created from a simple imaginary object which has similar shape to a hollow turbine blade. For solving the matrix equation, the method to obtain *the least squares and the least norms* solutions [2] was applied in the present study. This is considered suitable for solving a big scale and non-square matrix equation.

A very satisfying solution was obtained by the present method as shown in **Fig.5**. This was executed with FACOM VP-400 super computer at NAL. Reference data are as follows: number of image data; 63x63, memory size; 32MB, CPU time; 177.8sec. Conventional FBP method was also applied to the same projection data and the result is shown in **Fig.6**. In this case, the filter function advocated by Ramachandran [3] was adopted. There was no significant difference in the result by adopting the Shepp & Logan's filter function, compared to Fig.6. In the FBP method with the present sample data, some discrete artifacts are observed and resulting condition of image reconstruction is not satisfying.

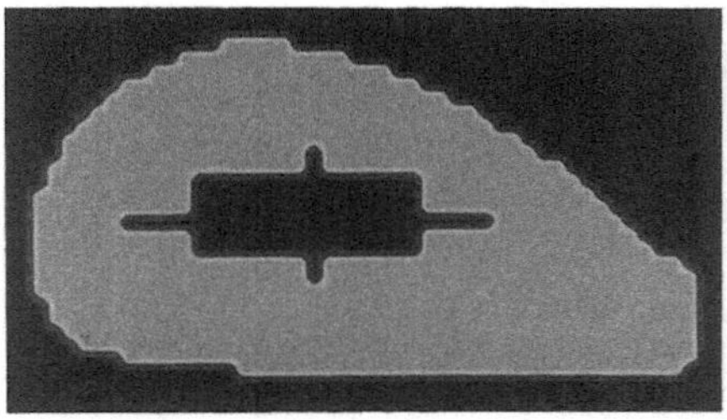

Fig.5. Verification result by matrix method

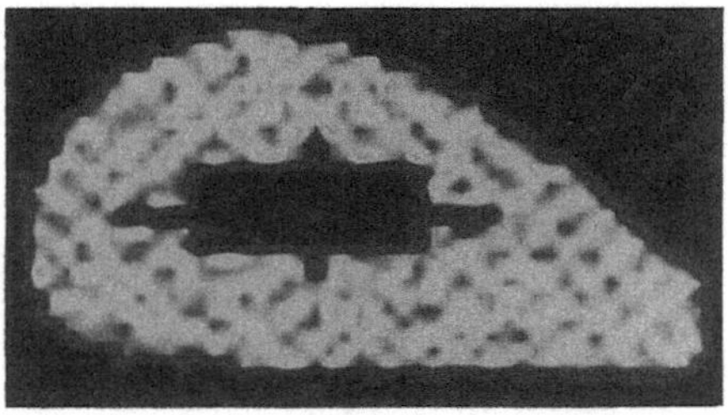

Fig.6. Verification result by FBP method

Non Destructive Inspection of a Turbine Blade

Test specimen The authors [4] have presented the work of NDI by X-ray CT applying FBP with the use of a turbine blade epoxy resin model. Here an actual turbine blade made of nickel base super-alloy, IN-100 was adopted as a test object. It was developed and installed in the turbofan engine FJR710. Schematic inner cooling construction of the test blade is shown in **Fig.7.** Its main dimensions are as follows: chord length; 19mm, shell thickness; ≈ 1.5mm, cooling hole diameter; ≈ 0.5mm. **Fig.8** shows the cross section where the projection data was gathered. This picture was taken after the CT process described below. The shape implys exact solutions that should be obtained by image reconstruction.

γ-ray CT process Very stable and intense γ-ray pencil beam of ^{192}Ir was used to obtain projection data. The test apparatus was developed at Energy Research Laboratory of Hitachi Ltd. [5] and the test was conducted there. The test conditions are as follows: projection data; four digit from 3341 to 8705, translation; 132 points with a pitch of 0.3mm, rotation; 180 times at an interval of 1deg., sampling time; 0.3sec for each point.

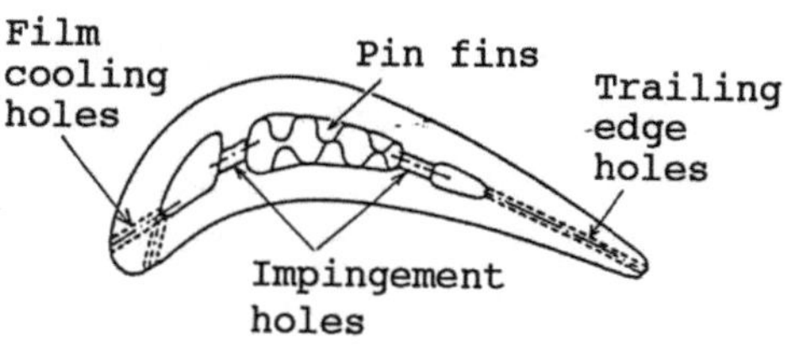

Fig.7. Cooling construction of test object

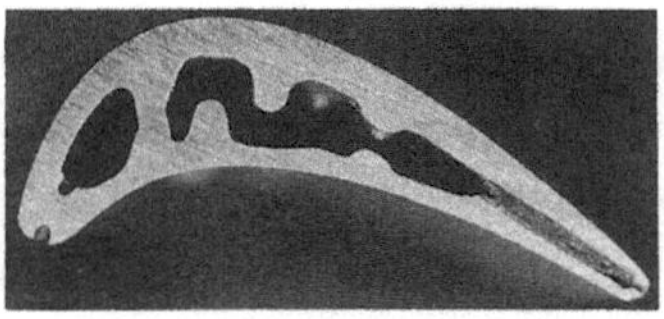

Fig.8. Actual view of test section

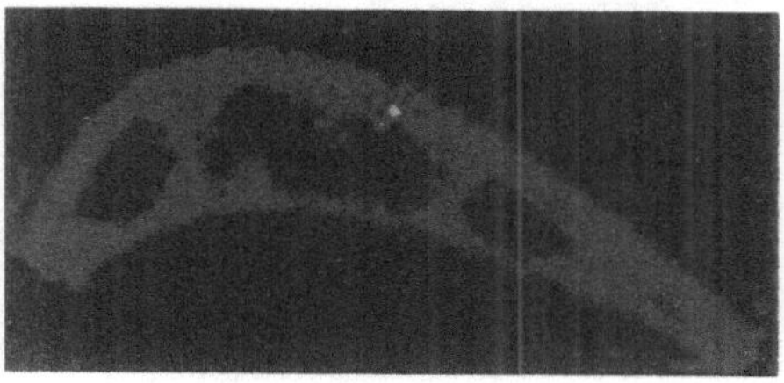

Fig.9. Image reconstruction (matrix method)

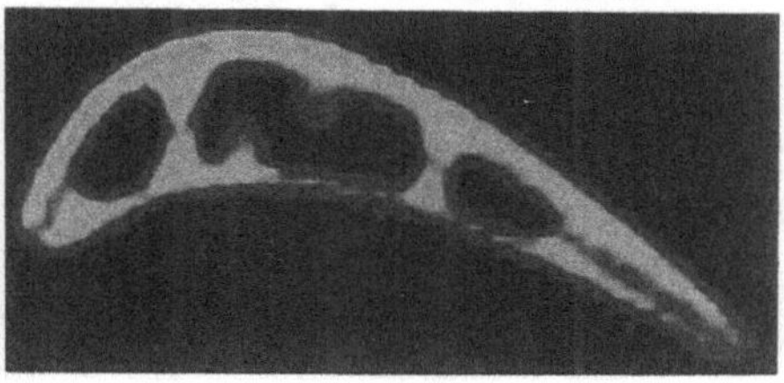

Fig.10. Image reconstruction (FBP method)

Results **Fig.9** represents a result by the present matrix method. (Data: point;76X76, memory;38MB, CPU; 237.5sec by FACOM VP-400) Smoothing process called *hysteresis method* [6] is applied to reconstructed data for reducing dot-wise noise. Although small holes and fins can be identified, inner and outer contours are still not well sharp. Conclusively, it should be said that resulting reconstruction of cross section is not satisfactory. On the other hand, **Fig.10** shows a result by the conventional FBP method. (Data: point;132x132, CPU; 2.16sec by FACOM M-780) Image processing such as smoothing and truncation of data is again applied. Although the result is not perfect, FBP provides far better image than the matrix method at least with the present projection data and a series of data processing.

Concluding Remarks

Possibility of image reconstruction by CT applying matrix method is verified by the present practical execution. Further improvements are strongly required in view of getting high quality projection data and reforming coefficient matrix consisting of 0 and 1 to that with appropriate real number depending on passing condition of ray beyond each element and others.

References

1. Shepp,L.A.; Logan,B.F.:IEEE Trans.Nucl.Sci.NS-2 (1974) 21-43.
2. Fujitsu Co.Ltd.(ed):FACOM SSL-II (1985) 341-343.
3. Ramachandran,G.N.;Lakshminarayanan,A.V.:Proc. Nat.Aca.Sci.USA 68-9 (1971) 2236-2240.
4. Yoshida,T.;Takahara,K.:Proc. GTSJ 13th Meeting (1985) 145-150.
5. Kanamori,T.:Image Technology & Information Display 17-14(1985) 55-60.
6. Onoue,M. et al.(eds):Image Processing Handbook: Shoko-do 1988.

Measurement of Unsteady Separated Flows on a Blunt Plate by a Fourier Transfrom Method

Shigeru MURATA*, Hiroshi SHIMADA**

* Mechanical and System Engineering
Kyoto Institute of Technology, JAPAN
** KUBOTA Corporation, JAPAN

Abstract
This paper provides the experimental results of unsteady separated flows on a blunt plate. The measurement was carried out for the case with the Reynolds number 900 by a Fourier transform method for visualization pictures. The Fourier transform method has been improved for the picture with velocity fluctuation by a simple procedure of normalization. Time mean flow variables measured are compared with other computational results and vorticity peak distribution is presented.

Introduction

Unsteady Separated Flows have been principal subjects of fluid engineering research, and the flow on a blunt plate kept parallel to an oncoming stream is one of the separating and reattaching flows with simple flow configuration. Many researchers have experimentally investigated the flow in detail by hot-wire anemometer. Lane & Loehrke[1], Sasaki & Kiya[2] have observed the separation bubble by flow visualization at low and moderate Reynolds numbers, and Sasaki et al. shows the three-dimensional structure in a separation bubble. Tafti & Vanka[3] have numerically approached to the three-dimensional flow mechanism by a diffence method.

The aim of this paper is to present the experimental results obtained by a particle image velocimetry for the moderate Reynolds number 900, to contribute to the investigation of the unsteady flows.

Fourier Transform Method

We have utilized a Fourier transform method to obtain two-dimensional velocity distribution from a couple of sequential visualization pictures. The Fourier transform method was originally proposed by Okuno[4], and the basic concept can be explained in the following. Let us assume that the gray level function of the second picture, $g(x,y)$, can be expressed by using that of the first one, $f(x,y)$, taken a time Δt before, and noise, $n(x,y)$, as follows:

$$g(x,y) = f(x-\Delta x, y-\Delta y) + n(x,y), \tag{1}$$

where Δx,Δy denote the average spatial shifts in the x and y- directions, respectively. The noise n means the displacement deflection of tracer particles from the average shift (Δx,Δy), which is caused by velocity fluctuation. The Fourier transform of Eq.(1) is given by

$$G(\xi,\eta) = e^{-j(\Delta x\xi+\Delta y\eta)}F(\xi,\eta) + N(\xi,\eta), \tag{2}$$

and then we may obtain the following equation,

$$D(\xi,\eta) = \frac{F(\xi,\eta)}{G(\xi,\eta)} = (1 - \frac{N(\xi,\eta)}{G(\xi,\eta)})e^{j(\Delta x\xi+\Delta y\eta)} \equiv R_d e^{j\theta}. \tag{3}$$

For the case that $n(x,y)$ is zero everywhere, Fourier transforming Eq.(3) again, we can find the average shift (Δx,Δy) by detecting the position of a large spike (Dirac delta function) as follows :

$$d(x,y) = \delta(x - \Delta x, y - \Delta y). \tag{4}$$

Thus, the velocity vector (u,v) was obtained by

$$(u,v) = (\frac{\Delta x}{\Delta t}, \frac{\Delta y}{\Delta t}). \tag{5}$$

However, $n(x,y)$ is not always equal to zero and it is obvious from Eq.(3) that the noise affects the magnitude of $D(\xi,\eta)$ and the effect may be suppressed by normalization,

$$D(\xi,\eta) = e^{j\theta}. \tag{6}$$

In this paper, instead of Eq.(3), we have utilized the following equation which is obtained by dividing the RHS of Eq.(6) by the corresponding wave number,

$$D(\xi,\eta) = \frac{e^{j\theta}}{\sqrt{\xi^2+\eta^2}}. \tag{7}$$

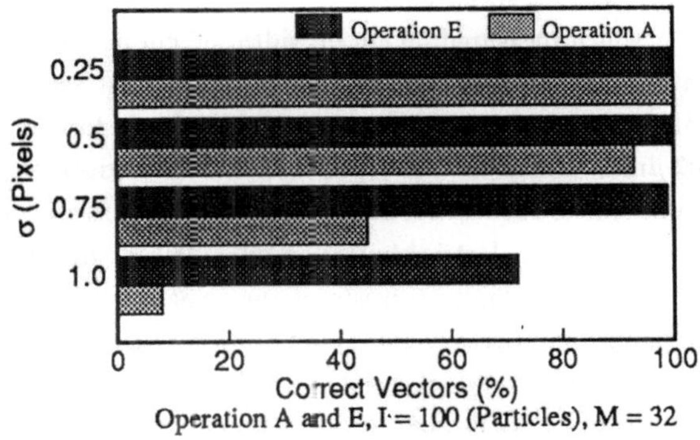

Fig.1. Effect of normalization

We have carried out the performance test for 100 models of tracer particle Image. Figure 1 shows the comparison of number of correct velocity vectors between our modified method (Operation E) and the original one (Operation A). I and M are the number of tracer particles and the size of Fourier transform subregion, respectively. σ denotes the standard deviation of particle displacement. It can be seen that the correct ratio is improved from 8% to 72% for σ=1 by using the present method and it is very robust for velocity fluctuation. Figure 2 describes the processing flow-chart of the present method. In order to obtain the smooth vorticity field, obviously erroneous velocity vectors are eliminated as follows :

$$u_{i,j} \leftarrow u_{ave} \qquad for \quad | u_{i,j} - u_{ave} | \geq u_{cr}, \tag{8}$$

where

$$u_{ave} = 0.25(u_{i+1,j} + u_{i-1,j} + u_{i,j+1} + u_{i,j-1}). \tag{9}$$

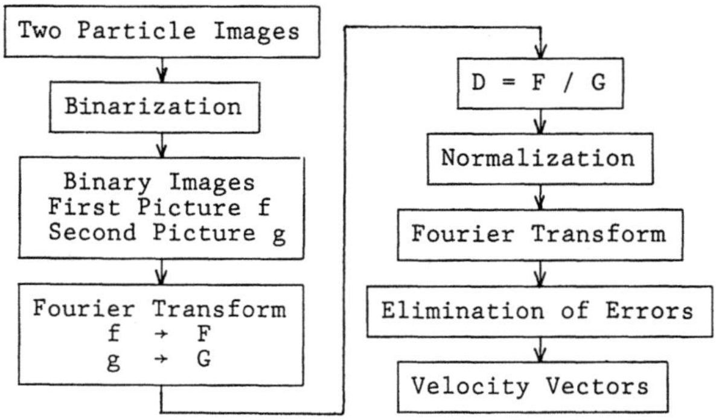

Fig.2. Image processing flow chart

Experimental Setup

Water circuit was used for the flow visualization of separation bubble on a blunt plate. The experiment was performed at a speed of 5.2 (cm/sec) measured with the LDV at an inlet of the test section, which measured 50 cm in width, 50 cm in hight and 280 cm in length. The visualization was conducted in the middle of the test section, where a blunt plate with the width 2 cm was vertically mounted, and it was performed by particle tracers, Nylon 11, 80~200 μm in diameter. The separated flow was horizontally lightened through double slit from one side of the test section and the flow pattern was recorded on videotapes from the lower view at an electric shutter speed 1/250 sec at a time interval 1/30 sec. Image processing system mainly consists of the image processor TOSHIBA Tospix-i and the host computer Sun Microsystems Sparc Station 1. Digitalized image has an active area of 512 pixel horizontally by 480 pixel vertically, and it was divided at one line interval into two pictures of 512×240 pixel which were captured at a time interval 1/60 sec. The size of a Fourier transform subregion was taken 64×64 pixel. The criterion for eliminating errors u_{cr} in Eq.(8) was assigned to 1.3(cm/sec).

Experimental Results

Figure 3 illustrates the flow configuration of separation bubble on a blunt plate and the coordinate system utilized in this study. The most important parameter is the time mean length of separation bubble, denoted by x_R, which is defined as the location of zero skin-friction measured from the leading edge of the blunt plate. The streamwise velocity distribution along the line adjacent to the surface at $y = 0.071b$, is depicted in Fig.4, where the velocity component u and the streamwise coordinate x are normalized by the uniform

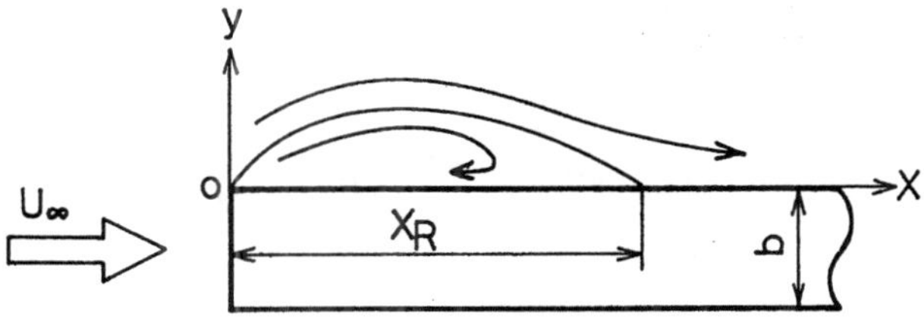

Fig.3. Flow configuration

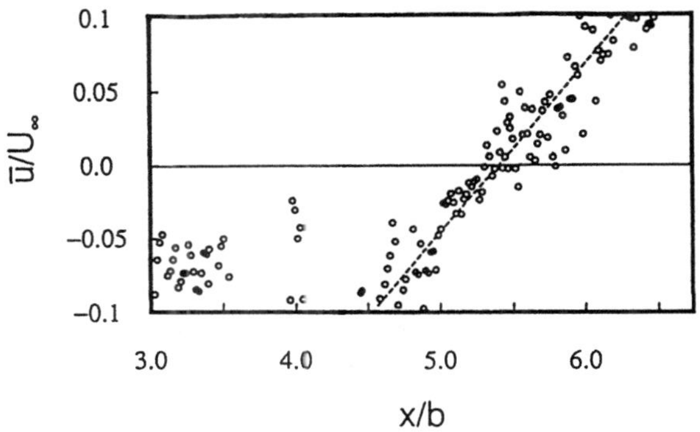

Fig.4. Velocity distribution on plate surface

velocity U_∞ and the plate thickness b, respectively. From this figure, the bubble length is found to be $5.4b$ which is a little bit larger than that of other experiments, $4.9b \sim 5.0b$, for higher Reynolds numbers than 700. Figure 5 shows the cross-stream distributions of time mean variables at six different streamwise positions, where (a),(b) and (c) indicate the mean streamwise velocity ($\bar{u}/U_\infty$), the streamwise turbulent intensity (u_{rms}/U_∞) and the cross-stream turbulent intensity (v_{rms}/U_∞), respectively. In these figures, solid line denotes the numerical solutions of three-dimensional analysis for the Reynolds number 1000, by Tafti & Vanka[3]. The experiments in this research were carried out under two different resolutions for each streamwise position. Our results of mean streamwise velocity are in good agreement with their solutions at all positions, and the agreement between the measured turbulent intensity and computed one is good for $x/x_R \leq 0.8$. At the positions 1.0 and 1.2, there are unnegligible discrepancies between them above $y/x_R = 0.2$, but it can be said that their computational analysis is valid at moderate Reynolds number. Figure 6 shows the locations of clockwise vorticity peaks. The vorticity peaks concentrate along the center of shear layer for $x/x_R \leq 0.4$ and the vertical range become wider as going downstream. Solid line denotes the center of shear layer numerically obtained by Tafti et al., and it is consistent with our vorticity peak distribution.

Conclusions

The time mean flow variables and vorticity peak distribution are presented for the unsteady flow in a leading-edge separation bubble on a blunt plate. It is shown that a Fourier

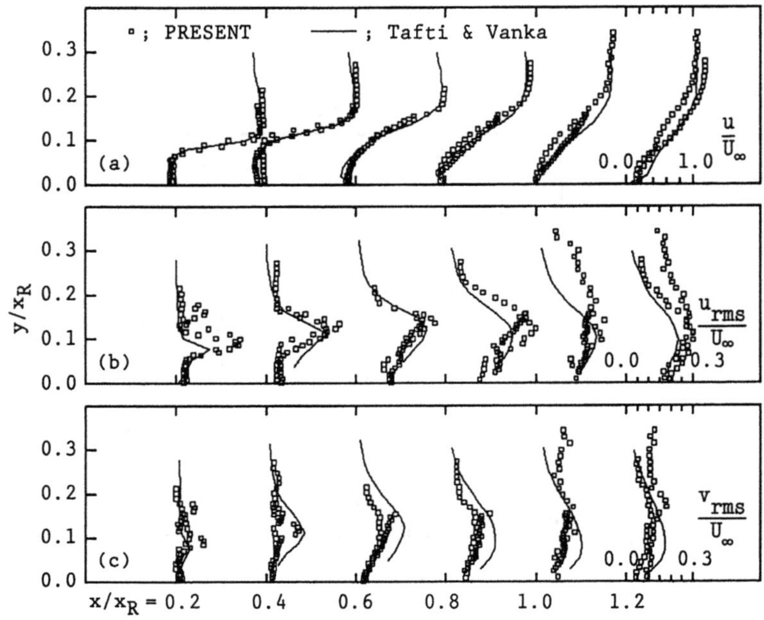

Fig.5. Time mean flow variables

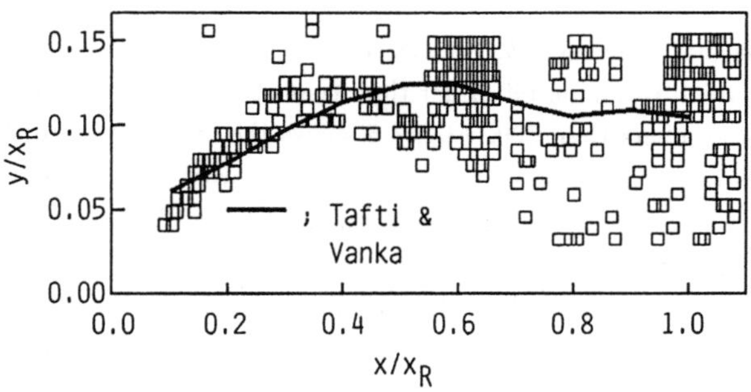

Fig.6. Locations of vorticity peaks

transform method of particle image velocimetry enables us to understand the unsteady separated flow fields well.

References

[1] Lane,J.C. and Loehrke,R.I., "Leading Edge Separation From a Blunt Plate at Low Reynolds Number," ASME *J. Fluids Eng.*, **102** (1980), 494-496.

[2] Sasaki,K. and Kiya,M., "Three-Dimensional Vortex Structure in a Leading-Edge Separation Bubble at Moderate Reynolds Numbers," ASME *J. Fluids Eng.*, **113** (1991), 405-410.

[3] Tafti,D.K. and Vanka,S.P., "A Three-Dimensional Numerical Study of Flow Separation and Reattachment on a Blunt Plate," *Phys. Fluids A*, **3** 12 (1991), 2887-2909.

[4] Okuno,T. and Sakamoto,E., " Velocity Field Measurement by Flow Images with Fourier Transform," *J. Kansai Soc. Naval Architects*, **208** (1988), 61-67 (in Japanese).

New Technique for Visualizing Three-Dimensional Flow in the Atmospheric Boundary Layer

M. P. BLEIWEISS and R. A. HOWERTON

U.S. Army Atmospheric Sciences Laboratory
White Sands Missile Range, New Mexico 88002-5501, USA

K. C. PAYNE and T. A. KING

Physical Science Laboratory
New Mexico State University
Las Cruces, New Mexico 88003, USA

Abstract

The atmospheric transmission large-area analysis system (ATLAS) was developed to characterize smokes/obscurants during U.S. Army field tests. Digital image processing of data obtained from a thermal imager allows a two-dimensional map of transmission through the cloud (in a plane perpendicular to the observer's line of sight) to be obtained. The ambient (natural) background scene in the imager's field of view is the source against which the transmission measurements are made. The spatial resolution varies depending on the geometry of the test setup; however, X-Z resolution (Y is along the line of sight) of the order of 0.5 m is easily obtained. The time resolution is 10 Hz. Conversion of these results to concentration is described and the results presented. The technique can be extended to three dimensions by using tomographic analysis.

Introduction

The use of smoke for flow visualization is not new nor is the use or proposed use of tomography to extend the two-dimensional aspects of flow visualization to three dimensions. What is new, however, is the derivation of quantitative information from the flow visualization described here and the applications of this technique in the open atmosphere. Point measurements have been made in a variety of situations, but the complete instrumentation of a chamber or tunnel would disturb the flow being measured. The ATLAS technique allows nonintrusive measurements of the aerosol concentration integrated along the line of sight of the thermal imager with high time resolution compared with most past point or integrated measurements of concentration. For observations perpendicular to the mean flow, this is referred to as cross-wind horizontal or vertical integrated concentration, CWIC or VIC, respectively. The simultaneous measurements along thousands of lines of sight have not been previously accomplished. Through the use of tomography, this technique may be extended to the three-dimensional determination of the aerosol concentration in a cloud with high time resolution for extended periods. ATLAS will be described and the extraction of aerosol concentration shown. A discussion of tomography and the most probable procedures to be used for this experiment will be presented.

ATLAS

ATLAS consists of a suite of imagers sensitive to the visible, near, mid, and far infrared portions of the electromagnetic spectrum. For the results discussed here, only the far infrared imager (an Inframetrics™ 600 thermal imager sensitive to 8 to 14 μm radiation) was used. The observations are recorded on video tape for posttest processing. A 386 based personal computer (PC) is used to control a Recognition Concepts, Incorporated, Trapix™ 5500 image processor with a real-time disc for storage of the digital images. An eight-frame running average is performed (low-pass filters the data to ensure no aliasing), and every third frame (of 30 frames per second) is grabbed for further reduction. This yields information at 10 frames per second.

Each pixel is processed by using an algorithm based on equation (1), which is a simplified equation of radiative transfer for the situation under study:

$$L_R = L_0 \, T + (1 - T) \, L_c \, , \qquad (1)$$

where the received radiance is L_R and the pretest scene radiance is L_0. The transmittance through the smoke cloud and the cloud's radiance when optically thick are given by T and L_c, respectively. The assumptions of this process are that the background radiance remains constant during the measurement period, that the background radiance is sufficiently different from the cloud radiance to enable the measurement of obscuration, and that the ambient environmental radiation scattered off the cloud into the imager field of view is negligible (or at least constant over the cloud). These assumptions and their implications are described more fully in Vonder Haar et al. [1]. The ATLAS process has been validated against more traditional techniques of making transmission measurements [2]. Also, methods to arrive at L_c continue to evolve.

Another technique similar to ATLAS would be to view a two-dimensional array of sources (flood lamps are adequate). This solves the dynamic range and constant background radiance problem, and one can also subtract out the path radiance by assuming the cloud structure is of the same order as the spacing between sources. This technique might be more appropriate for use in wind tunnels since arrays suitable for field testing become unwieldy in size and number of sources.

A color video movie has been made of the ATLAS transmission maps and a portion of it will be shown. Figure 1 is a hardcopy of one frame from the video, which is constructed from pseudocolor representations of the two-dimensional transmission maps as a function of time. The cloud is produced by a smoke generator to the left out of the field of view. The cloud moves from left to right through the field of view, driven by the mean wind. The distance of the cloud from the imager is approximately 500 m so the frame represents a field of about 70 m by 140 m at the cloud. The authors emphasize that these images are not radiance or grey-level plots that have been converted to color but are transmission maps of the cloud obtained from considerations of the equation of radiative transfer.

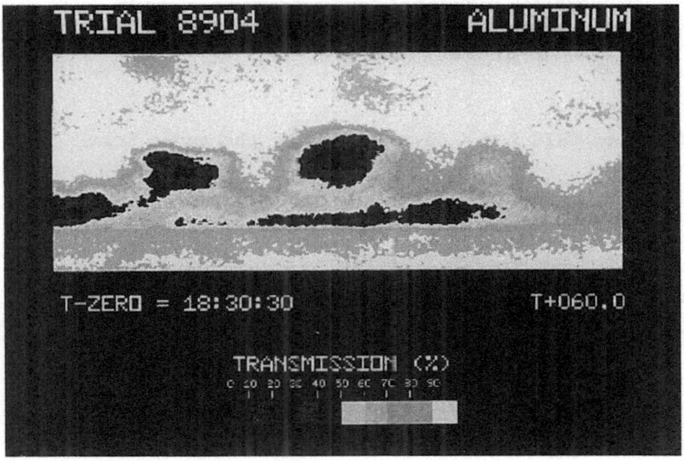

Fig. 1. Hardcopy of one frame of the ATLAS videotape that shows two-dimensional transmittance as a function of time

Diffusion and Transport

ATLAS analysis was performed on one trial from a recent field test. The output was converted to CWIC, from which comparisons with standard measurements of turbulent diffusion (probability density function, means, and variances) can be made. This has been done for a 60-s portion of the smoke trial described here, and some of the results are presented in Fig. 2. Detailed study of these results may allow comparison between reflected Gaussian and dissipated Gaussian models of dispersion and transport. In addition, the links between K-theory and Gaussian models may be drawn. The pdf is important for investigation of the phenomenon of intermittency.

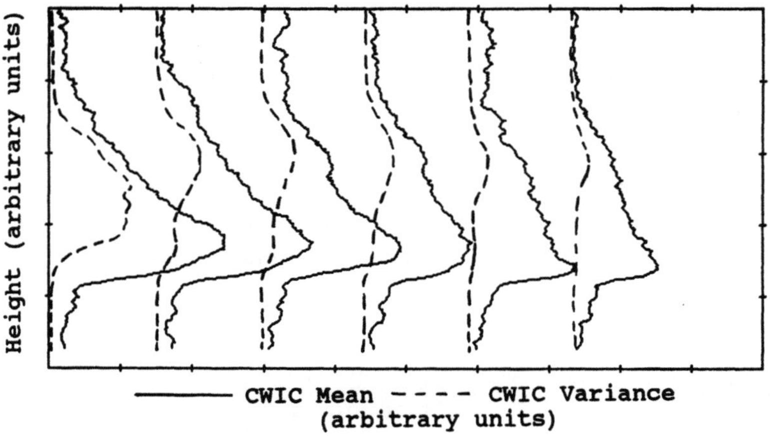

Fig. 2. Mean and variance of CWIC for a 60-s portion of the smoke trial under discussion

Tomography

Although tomography of a smoke/obscurant cloud in three dimensions in the open atmosphere has not been accomplished, a review of the literature suggests that such an experiment can be performed. We are currently reviewing the various algorithms in detail to determine which is the most applicable. Preliminary work suggests that the convolution back projection method as applied to fan beam geometry or iterative reconstruction techniques are the probable choices. We will then do a simulation to determine the specifics of the experimental setup and then perform the test.

Acknowledgments

The preparation and presentation of this paper resulted form the suggestion and encouragement of Professor G. S. Settles of Pennsylvania State University to one of the authors during the Flow Visualization Techniques Short Course of the University of Michigan in 1991. Also, many useful discussions with and the assistance of Mr. J. F. Bowers of the U.S. Army Dugway Proving Ground are gratefully acknowledged.

References

1. Vonder Haar, T.; Stephens, G.; Jones, A.; Shih, C.F.; Davis, J.: Analysis and Assessment of the Atmospheric Transmittance Large Area System (ATLAS). Metsat, Inc., 515 South Howes, Fort Collins, CO 80521, 1990.

2. Bleiweiss, M.P.; Howerton, R.; Valdez, R.; Payne, K.; King, T.; Hutchison, K.: A Comparison of the MPTR and ATLAS Transmissometers. Proceedings of the Smoke/Obscurants Symposium XV, U.S. Army Chemical Research, Development and Engineering Center, Aberdeen Proving Ground, MD 21010-5423, 1991.

On the Quantification of Cavity Extent on an Orifice in a Pipe by Use of an Image-Processing Technqiue

K.OGURA

Department of Mechanical Engineering
Tamagawa University, Tokyo, Japan

Summary

In studying the cavitation of an orifice in a pipe, attempts were made to quantify the cavity extent under different conditions. A system was developed to evaluate the cavity extent by binarizing the video-recorded cavitation patterns to obtain black-and-white fractions. With this, the cavity extent can be quantified with high reproducibility by using relatively simple software and instrumentation. This paper describes the process of this image-processing method with some examples to which this method was applied. As conclusion, it has been shown that this simple image-processing method can be a useful tool for quantifying the cavity extent at developed stages of cavitation in the downstream of orifices in a pipe.

Introduction

The author has been studying to the cavitation characteristics of orifices in pipes to show some guidelines for their design. To do this, it is necessary to quantify cavity extent over the range from incipient/desinent cavitation to supercavitation. In the case of the cavitation in the downstream of orifices, as seen from Fig.1(photographs), it is difficult to evaluate directly cavity extent, because of spatially scattered distribution of cavity and of its temporal fluctuations.
To the author's knowledge in the existing studies e.g. [1] [2] ,having been made on cavitation of orifices, quantitative evaluation of cavity extent is not seen.
So the author attempted quantifying cavity extent by image-processing with simple software and instrumentation. This paper describes the process of this image-processing method with some examples to which this method was applied.

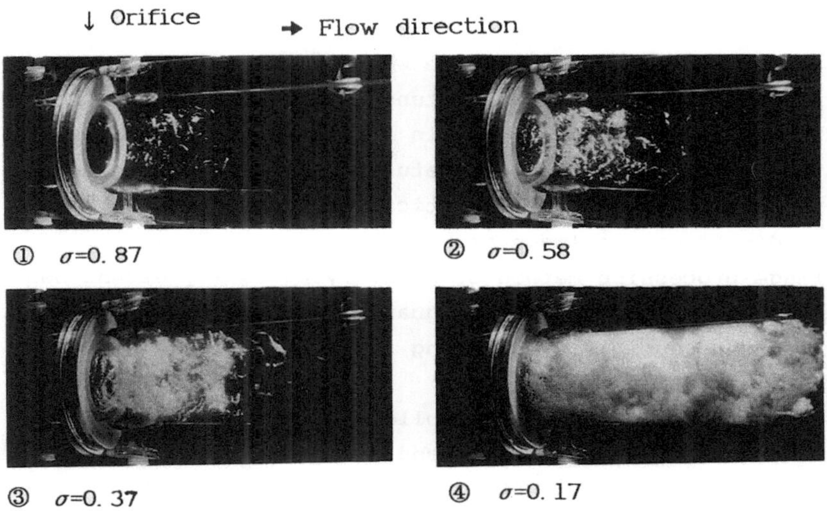

D=ϕ50mm, β=0. 7, V_0 =20m/s

Fig. 1. Photographs at each stage of Cavitation

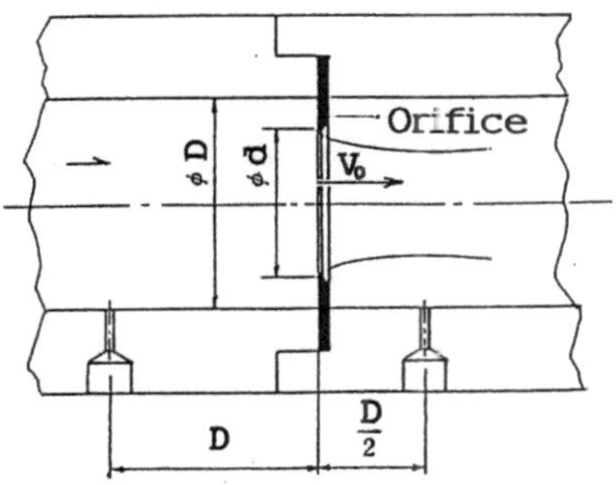

d : Inner diameter of pipe
D : Diameter of orifice
β : Diameter ratio of orifice (=d/D)
V_0 : Average velocity at orifice
P_1 : Upstream pressure
P_2 : Downstream pressure
σ : Cavitation number (= $(P_2 - P_V) / (\rho V_0^2 / 2)$)
P_V : Vapor pressure,
ρ : Density

Fig. 2. The section of an orifice

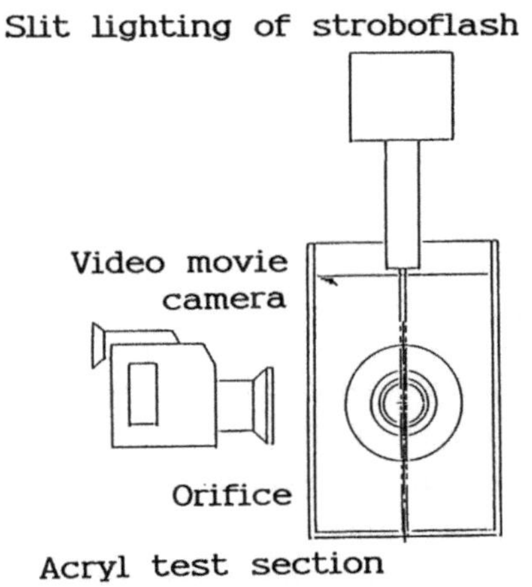

Fig. 3.
The apparatuses and their arrangements used to take pictures of the orifice and its downstream

Apparatus and Procedure

The cavitation test on an orifice was conducted in the variable pressure circulating water tunnel [3]. Fig.2 shows the measuring section of an orifice in a pipe tested in the present study. Fig.3 shows the apparatuses and their arrangements used to take pictures of the vertical section of the orifice and its downstream to image-processing.

This image-processing method of quantifying cavity extent has been developed with the view to quantifying the cavity extent with high reproducibility by using relatively simple software and instrumentation.

Procedure of this method is as follows.

(1) To take pictures of orifice and its downstream by video movie camera with slit lighting of stroboflash (duration time 22μs) under the constant photographing conditions(lighting, iris, brightness, white-balance) with all the windows blackout curtained.

(2) To file the data of pictures(62mmx80mm) in floppy disk and to binarize them by the software available,with the constant black and white level.

(3) To transfer the binarized data to VRAM in a personal computer and to compute white fractions.

(4) To repeat (2)-(3) nine times to obtain reproducible average of fluctuating cavity extent.

Results and Discussion

Fig.4 shows examples of video movie pictures and the binarized images of them. Fig.5(Ⓐ) shows the white fractions plotted to the base of cavitation numbers(see Fig.2). Since this image-processing method is suitable for developed stages of cavitation, it is advisable to quantify incipient/desinent cavitation and supercavitation by some acoustic means (Fig.5(Ⓑ)).

Fig.5(Ⓐ & Ⓒ) presents the white fractions obtained by this method for the two different air contents (32% and 55%). It is apparent that cavity extent increases as air content increases.

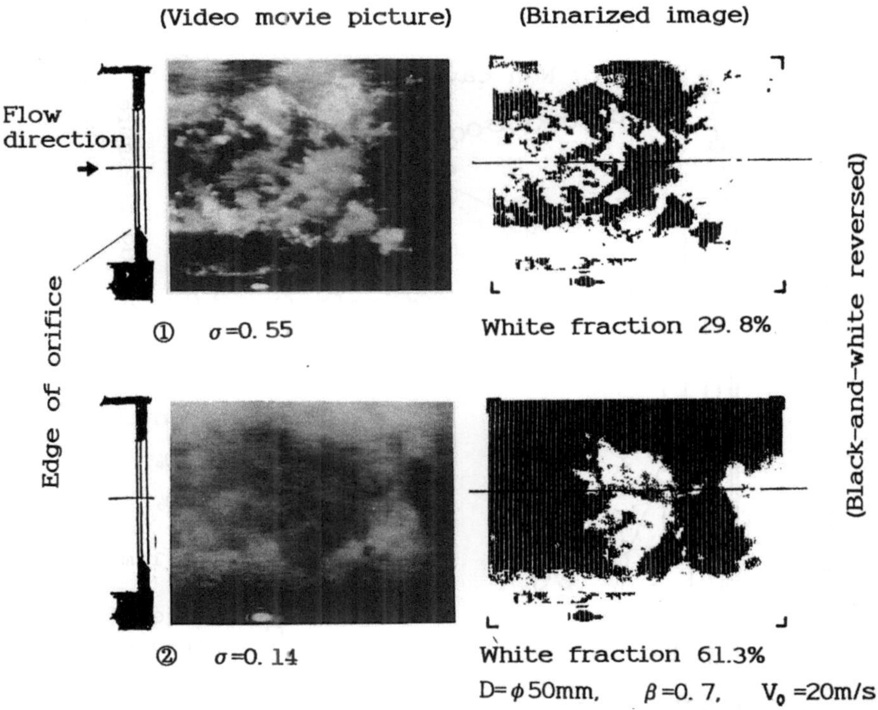

Fig. 4. Video movie pictures and the binarized images of them

Conclusion

As conclusion, it has been shown that this image-processing method with relatively simple software and instrumentation can be a useful tool for quantifying the cavity extent at developed stages of cavitation on orifices in a pipe.

Acknowledgement

The author wishes to express my thanks to Prof. H. Tanibayashi of Tamagawa University, Prof. H. Kato of The University of Tokyo and Prof. emeritus T. Inui of The University of Tokyo for their guidances, helpful suggestions and steady encouragements during the execution of this work, and to Assistant Prof. H. Obara for his instruction on image-processing system.

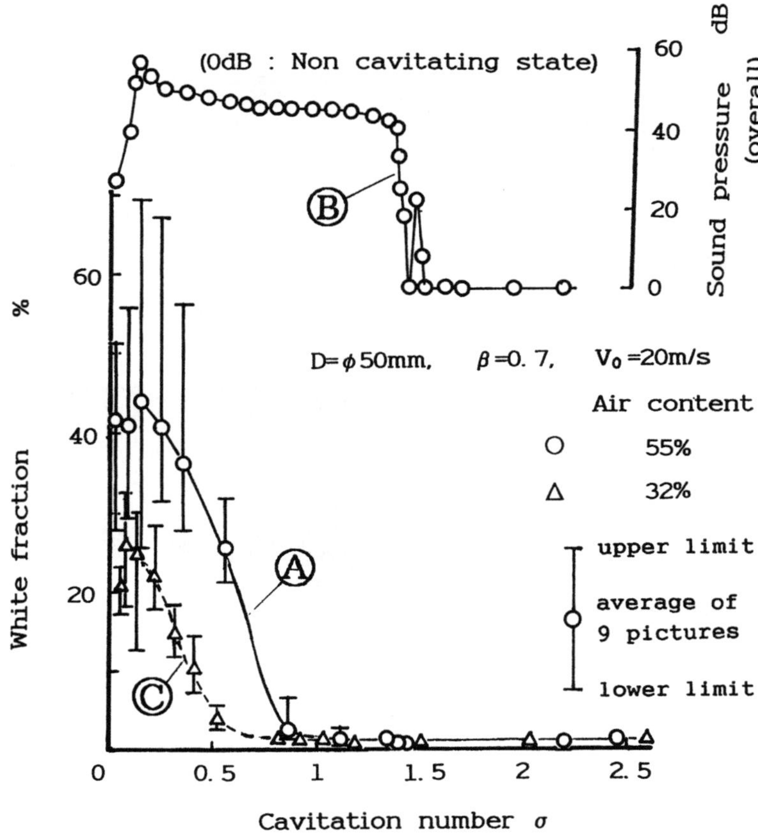

Fig. 5. White fraction and cavitation number

References

1. Oba, R. et al., Trans. Jpn.Soc.Mech.Eng.,(in Japanese),Vol.-51,No.462,B(1975),p.419.
2. Tullis, J. P. and Govindarajan, R., Proc.,ASCE,Jr.Hydr.Div., 99-HY3(1973),p.417.
3. Tanibayashi, H. and Ogura, K., Cavitation and Multiphase Flow Forum,ASME,(1991),p.199.

Higher Order Approximation for Spatio-Temporal Derivative Method

S. NISHIO

University of Osaka Prefecture
Department of Naval Architecture
Gakuen-cho 1-1, Sakai
Osaka 593, Japan

T. OKUNO

University of Osaka Prefecture
Department of Naval Architecture

S. MORIKAWA

Nissan Motor Co., Ltd.

Summary

The spatio-temporal derivative method for flow field image measurement is well known as a high spatial resolution technique. The governing equations are derived using Lagrange derivative, assuming that the visualized particle image pattern varies smoothly almost everywhere. The authors examine the formulation of governing equation and the brightness function of the flow image in detail, and try to extend this technique to the theory with higher order approximation. The obtained equation can be applied to the case that illumination intensity is not uniform in measurement area.

1. Introduction

The spatio-temporal derivative method does not always gives the results with sufficient high spatial resolution, although it has a superior ability of spatio resolution in principle. This is because the method needs both time and spatial differential coefficients to obtain velocity field: the result is very sensitive to the quality of flow field image. The another reason is also that the accuracy of the approximation in governing equation is not sufficient, and also its approximation makes worse the spatial resolution.

The authors examine the formulation of governing equation and the brightness function of the visualized flow image, and try to extend the usual spatio-temporal derivative method to the theory with higher order approximation. The higher order differential terms in the Taylor series of brightness function have been employed in the governing equation.

2. Formulation of Governing Equation

Horn et al. [1],[2] has discussed already the formulation of governing equation for the spatio-temporal derivative method. Here the procedure is examined in detail, and a more general form has been derived.

Figure 1 illustrates the flow field image obtained by a flow visualization, using brightness function $f(x,y,t)$. Assuming the tracer particle moves with the fluid completely, and it moves from point $P(x-\Delta x, y-\Delta y, t-\Delta t)$ to $P'(x+\Delta x, y+\Delta y, t+\Delta t)$ during short time $2\Delta t$, the difference of brightness between these two points is expressed as

$$f(x+\Delta x, y+\Delta y, t+\Delta t) - f(x-\Delta x, y-\Delta y, t-\Delta t) = g(x,y,t,\Delta t) \qquad (1).$$

Further more assuming that the visualized image pattern varies smoothly almost everywhere, and that the velocity can be expanded in Taylor series, the moving distance $(\Delta x, \Delta y)$ in eq.(1) are expressed as

$$\begin{aligned} \Delta x &= u\Delta t + O(\Delta t^2) \\ \Delta y &= v\Delta t + O(\Delta t^2) \end{aligned} \qquad (2),$$

where u and v are the velocities at $P_0(x,y,t)$. When the brightness function $f(x,y,t)$ also can be expanded in Taylor series in time and space,

$$\lim_{\Delta t\to 0}\frac{1}{\Delta t}\{f(x+\Delta x, y+\Delta y, t+\Delta t) - f(x-\Delta x, y-\Delta y, t-\Delta t)\} = \lim_{\Delta t\to 0}\frac{g(x,y,t,\Delta t)}{2\Delta t}$$

$$\frac{\partial f}{\partial t} + u\frac{\partial f}{\partial x} + v\frac{\partial f}{\partial y} = \lim_{\Delta t\to 0}\frac{g(x,y,t,\Delta t)}{2\Delta t} \qquad (3),$$

is obtained using Lagrange derivative. Equation (3) is the governing equation of the spatio-temporal derivative method for evaluating velocity distribution.

It is found that there remains the finite value as a function of time and space when intensity of illumination is not constant in measurement area as shown in Fig.1, although it has been assumed usually [3],[4] that $g(x,y,t,\Delta t) = o(\Delta t)$, i.e., right hand side of eq.(3) always converges to 0.

3. Velocity Field

Since the eq.(3) contains two unknowns (u,v), it can not be solved without any other relationship. One of the possibility is to use the continuous equation. It has been solved usually, however, by using the least square method applied to the small analysis area, with the assumption that the velocity field varies smoothly in space.

Assume the velocity at any point can be expanded in Taylor series in the small region,

$$u(x+\Delta x, y+\Delta y, t+\Delta t) = u(x,y,t) + \frac{\partial u}{\partial x}\Delta x + \frac{\partial u}{\partial y}\Delta y + \frac{\partial u}{\partial t}\Delta t + O(\delta^2) \qquad (4),$$

where $\Delta x, \Delta y$ and Δt are supposed to be sufficiently small. Ignore the higher order terms than second order in eq.(4) and the same form equation for v, and substitute them into eq.(3) which are set up at $(x+\Delta x, y+\Delta y, t+\Delta t)$, the following form is obtained,

$$\left.\frac{\partial f}{\partial t}\right]_{(x+\Delta x,y+\Delta y,t+\Delta t)} + \left\{u(x,y,t) + \frac{\partial u}{\partial x}\Delta x + \frac{\partial u}{\partial y}\Delta y + \frac{\partial u}{\partial t}\Delta t\right\} \left.\frac{\partial f}{\partial x}\right]_{(x+\Delta x,y+\Delta y,t+\Delta t)} + \left\{v(x,y,t) + \frac{\partial v}{\partial x}\Delta x + \frac{\partial v}{\partial y}\Delta y + \frac{\partial v}{\partial t}\Delta t\right\} \left.\frac{\partial f}{\partial y}\right]_{(x+\Delta x,y+\Delta y,t+\Delta t)} = 0 \qquad (5).$$

Equation (5) represents the relationship between the velocities (u, v) and the brightness function f at $(x+\Delta x, y+\Delta y, t+\Delta t)$. The velocities (u, v) can be obtained from the simultaneous equations which are set up in the small area around the point (x, y, t). Although the higher order terms in eq.(4) than second order are excluded in this paper, it should be considered in the case that the gradient of velocity is steep in the analysis area.

Figure 2 shows a spatio-temporal coordinate system with two flow images which are obtained at $t+\Delta t$ and $t-\Delta t$. Assuming again the brightness function $f(x+\Delta x, y+\Delta y, t+\Delta t)$ can be expanded in Taylor series at any point, and ignoring higher order terms than third order,

$$\begin{aligned} & f(x+\Delta x, y+\Delta y, t+\Delta t) \\ & = f(x,y,t) + \frac{\partial f}{\partial x}\Delta x + \frac{\partial f}{\partial y}\Delta y + \frac{\partial f}{\partial t}\Delta t + \frac{1}{2}\left\{\frac{\partial^2 f}{\partial x^2}(\Delta x)^2 + \frac{\partial^2 f}{\partial y^2}(\Delta y)^2 + \frac{\partial^2 f}{\partial t^2}(\Delta t)^2 \right. \\ & \quad \left. + \frac{\partial^2 f}{\partial x \partial y}(\Delta x \Delta y) + \frac{\partial^2 f}{\partial x \partial t}(\Delta x \Delta t) + \frac{\partial^2 f}{\partial y \partial t}(\Delta y \Delta t)\right\} + O(\delta^3) \qquad (6). \end{aligned}$$

The first order terms in eqs.(5) and (6) express the usual governing equation of the spatio-temporal method. The obtained new equations enables to criticize the reliability of the solution because these include the higher order terms than previous ones. When the velocity gradient terms are ignored as usual, the gradient velocity field (vorticity field) and the error of the solution (velocity field) cannot be separated. Since these two factors are expressed clearly here, the criterion of the error velocity becomes more reliable.

In the case that the sheet light is formed by point illumination from the center of circular cylinder as shown in Fig.3, the illumination intensity is in proportion to $1/r$, where r is the distance from the center of cylinder. Assuming the intensity of illumination is constant in time base, the right hand side of eq.(5) can be expressed as,

$$\lim_{\Delta t \to 0} \frac{g(x,y,t,\Delta t)}{2\Delta t} = -\frac{u_r}{r} f(x,y,t) \qquad (7).$$

Substitute eq.(7) into eq.(5), the governing equation can be derived as follows,

$$\frac{\partial f}{\partial t} + u\left(\frac{\partial f}{\partial x} + \frac{f}{r}\cos\theta\right) + v\left(\frac{\partial f}{\partial y} + \frac{f}{r}\sin\theta\right) = 0 \qquad (8).$$

The non-uniformity effect of illumination intensity can be considered with additional term in governing equation as shown in eq.(8). Although the effect is expressed when the illumination intensity function is known, the iteration technique would be needed to obtain the correct velocity when the background illumination is not expressed by a function.

4. Conclusion

The spatio-temporal derivative method was extended to the theory with higher order approximation. The new governing equation can be applied to the case that illumination intensity is not uniform in measurement area, and also enables to make accurate velocity distribution.

The authors would like to express their gratitude Prof. Mori, Hiroshima Univ. of his discussion.

REFERENCES

[1] E.C. Hildreth: Computation Underlying the Measurement of Visual Motion, Artificial Intelligence, 23, 309/354 (1984)

[2] B.K.P. Horn and B.G. Schunck: Determining Optical Flow, Artificial Intelligence, 17, 185/203 (1981)

[3] S. Ando: A Vector Field Measurement System Based on Spatio-Temporal Image Derivative, Trans. of the Soc. of Instrument and Control Engineers, Vol.22, No.12 (1986)

[4] T. Okuno and J. Nakaoka: Velocity Field Measurement by Spatio-Temporal Derivative Method, Jour. of the Kansai Soc. of Naval Architects, No.215 (1991)

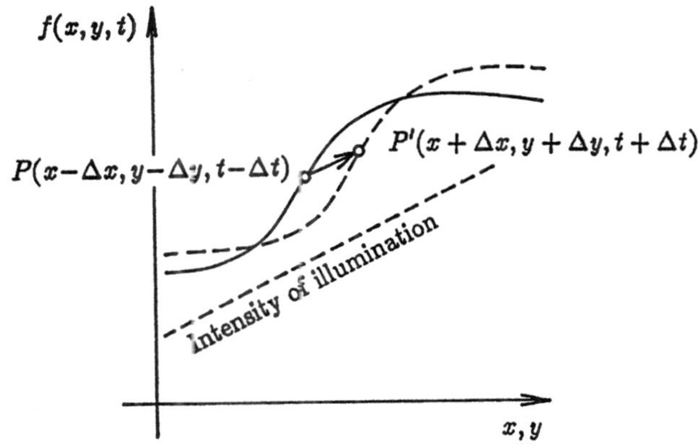

Figure 1 Principle of spatio-temporal derivative method

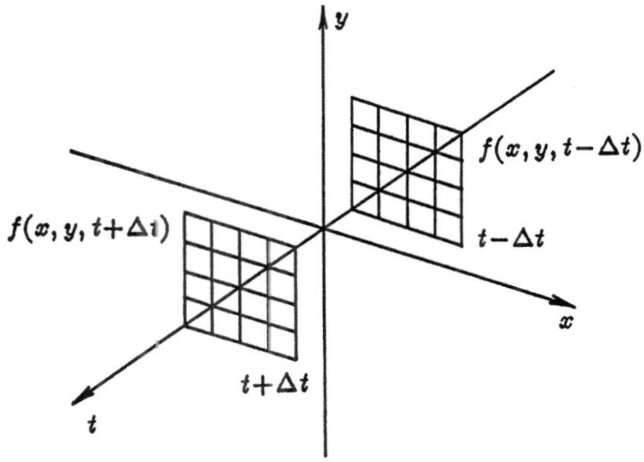

Figure 2 Spatio-temporal coordinate system

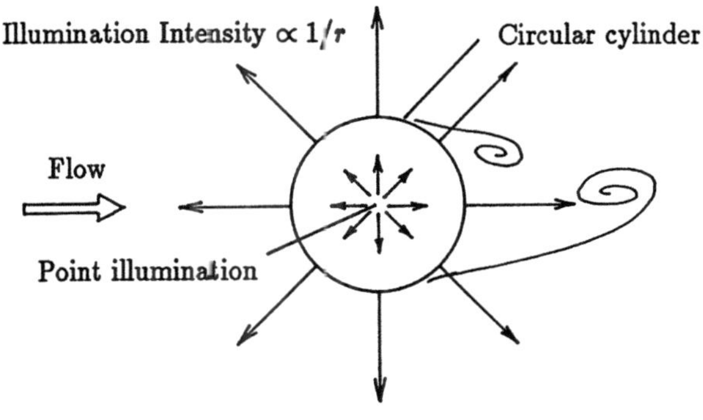

Figure 3 An example of non-uniform illumination

Photon Counting for the Measurement of Velocity and Temperature in Liquid Flow

Motoaki Yano
Kobe University, Rokkodai, Nada, Kobe 657, Japan

SUMMARY

The velocity distribution in turbulent boundary layer is measured by the data of the distribution of the average counting numbers of photon. The photons of the scattered light of the micro particles included naturally in flow are counted by the two dimensional photoelectric device. The micro particles in flow are illuminated by a laser light sheet. The temperature of flow is measured by also the photon counting of the fluorescence of the fluid included the suitable fluorescent material.

INTRODUCTION

When the fluid is illuminated by the incident laser light sheet, the intensity of the scattered light and the fluorescent light from the liquid fluid are generally weak strength. In the measurement of the intensity of these observed lights, the detector of the photoelectric device has to have the high current amplification factor over 10^6. If we use this device with the amplifier having wide frequency range and low noise level in the measurement, then we can measure the intensity of light in the photon counting mode.

The counting average numbers of photon are proportional to the average intensity of the observed light and also the counting time. When the observed light is coherent light or the counting fixed gate time is enough longer than the coherent time of the chaotic observed light, the distribution of the counting photon numbers satisfies the distribution of Poisson in the repeating measurements.

In the measurement, the using device of photon detector is a two dimensional photoelectric transfer tube that is constructed by three micro channel plates and a resistance position sensor(N1831U; HAMAMATSU).

The direction vectors concerning with the flow, the incident laser light sheet, and the scattered or fluorescent light are normal to each other. The scattered or fluorescent light is focused on the photocathode of the 2-D detector tube using a camera lens(Micro Nikkor 105mmF2.8;Nikon) with interference filter(490nm for the scattered light, 550nm for the fluorescent light;TOSHIBA).

TEMPERATURE MEASUREMENT

The temperature distribution in flow is measured by using the thermal characteristics of the intensity of fluorescence. In this study, the liquid of flow is water. The water soluble organic fluorescent materials are used in the measurement.

The fluorescence is a phenomenon of luminescence. The fluorescent molecule absorbs the energy of the incident light of short wave length and emits the fluorescent light of longer wave length. The process of fluorescent emission is explained by the quantum transition. The life time(converse number of the velocity constant) of a spontaneous emission in the process is related the absorption constant of the molecule.

The absorption constant is estimated by the two-level model[1] refined by the approximation of Born -Oppenheimer and with the factor of Franck-Condon[2] in fluorescent molecular model.

The velocity constants of the intersystem crossing k_{isc}, the internal conversion k_{ic}, etc. exist in the quantum transition of the fluorescent molecule besides the constant k_f concerning with the life time of the spontaneous emission. The averaged life time of the fluorescent molecule is $(k_f+k_{isc}+k_{ic})^{-1}$. And the quantum yield ϕ is expressed by $\phi = k_f/(k_f+k_{isc}+k_{ic})$.

When the temperature of the fluorescent molecule increases, the intensity of emitted fluorescence decreases as compared with the normal state in generally. The reasons of the decreasing intensity of fluorescent light in high temperature are considered the increasing of the loss energy of the collision of molecules, the internal conversion and the intersystem conversion.

However, there are some fluorescent material having the increasing intensity of the fluorescence in the high temperature as compared with in the normal temperature. The reason of this phenomenon is considered the existence of the metastable level between the levels of ground and excitation in the quantum transition. The quantum transition of the special fluorescent molecule passes the metastable level. The probability of the returning to the excitation level is proportional to the $\exp(-\Delta E/kT)$ (ΔE:the energy difference between the levels of the excitation and the metastable) by Boltzmann's law. The intensity of fluorescence is increasing in the rasing temperature when the materials of the thermally activated delayed fluorescence is dissolved in the fluid.

The experimental results concerning with the relation between the intensity of the fluorescence and the temperature of the fluorescent fluid and the spectrum are shown in the Fig.-1 and 2.

VELOCITY MEASUREMENT

In the preliminary experiments, few numbers of the micro particle naturally included in fluid are found in the incident laser light beam in flow. In the single photon counting mode using a photon counter, the average counting number of photons of the scattered light of the particles is proportional to the velocity of flow.

In the measurement of velocity distribution of turbulent boundary layer in vertical duct flow, the measuring section in the flow illuminates by a parallel laser light sheet. The photons of the scattered light of the micro particles in flow is received on the photocathode of the two dimensional photon counting tube using the interference filter and the camera lens. (Ref. Fig.-3)

The output signals of the tube are the x and y coordinates of each single photon in time series. When the tube receives over single photon at the same time, the window of the device rejects the signals of these photons. Therefore, the intensity of the scattered light has to keep the condition of the single photon counting mode. The neutral filters are used to adjust the intensity of the laser light sheet in flow. (Ref. Fig.-3)

The detective probability of the single photon $p(1,T)$ is obtained by the distribution of Poisson. $p(1,T)=(\alpha IT)\exp(-\alpha IT)$. Where α, I and T are the values of the photoelectric transfer factor of the tube, the intensity of the receiving light and the observing time of the photon counting. And the value αIT is equal to the average counting number of photons. The probability $p(1,T)$ under the weak intensity of the light is proportional to the average numbers of photon. $p(1,T)=\alpha IT$.

The incident direction y of the parallel laser light sheet is normal to the down stream direction x in flow and is inclined to the flat plate for the turbulent boundary layer. The averaged flow velocity is constant in the x direction with the fixed y coordinate. The accumulated average counting number of photon in x direction with the fixed coordinate y is $\int p(1,T)d(\alpha IT)=\int p(1,T)dx=(\alpha IT)^2$ in the weak intensity of the scattered light. (Ref. Fig.- 4.)

The intensity of the parallel laser light sheet in flow is not uniform. For example the intensity of the reflected light on the transparent flat plate in flow is different as shown in Fig.-5. The compensating coefficients are obtained from the photon counting data in the experiment in still fluid. The accumulated photon number in y and the same x coordinate after the calibration and the subtraction of the back ground noise photons is proportional to the average velocity in y coordinate. A result of the experiments of the photon numbers over the flat plate is shown in Fig.-6.

CONCLUSION

The method of simultaneous measurement of velocity and temperature in liquid flow is tried in this study. The both measurements in flow are based on the probability of the counting number of photon including the information. These optical information can be separated by using the interference filter. Unfortunately, only one detector can be used in the measurement. Therefore, real simultaneous measurement is not realized in this study. In the application of these methods, the calibration has to be necessary before the measurement. The relation between the concentration of the materials and temperature or velocity is rather sensitive.
In generally, it is necessary to repeat many times of the same counting job in the application of photon counting method. Then the time averaged value may be expected as the result of the photon counting.

REFERENCE

1. Simoda, K.; Yajima, T.; Ueda, Y.; Shimizu, T. Kasuya, T.: Quantum Electronics. Tokyo: Shokabo 1972: pp35.

2. Tsuchiya, S.(ed.): Laser and Chemical Reaction. Tokyo: Gakkai Shupan Center 1980: pp84.

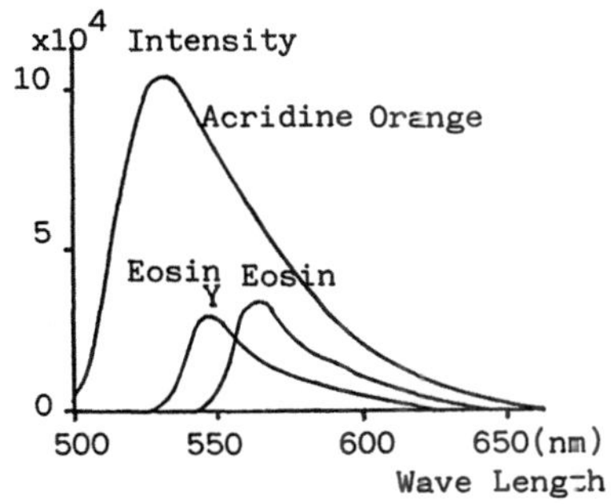

Fig.-1 Spectrum of Fluorescence Excitation with 488nm. Ar. Laser

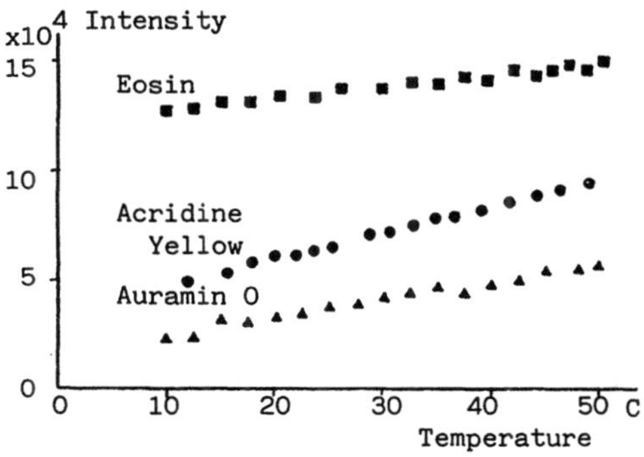

Fig.-2 Relation between the Intensity of Fluorescence and the Temperature. The intensity is the maximum value in the spectrum.

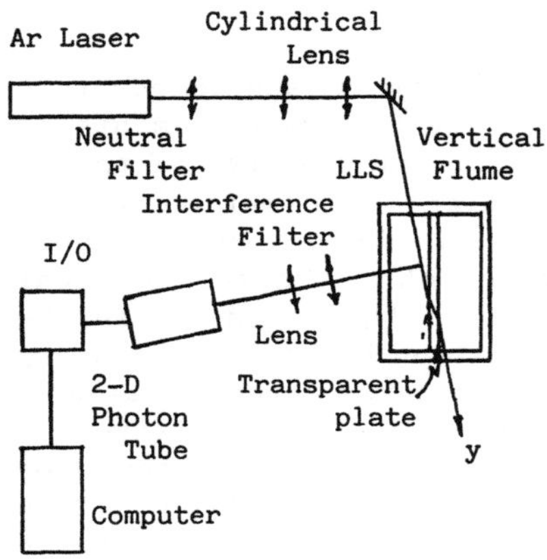

Fig.-3 Optical arrangement of the measurement of velocity and temperature in flow

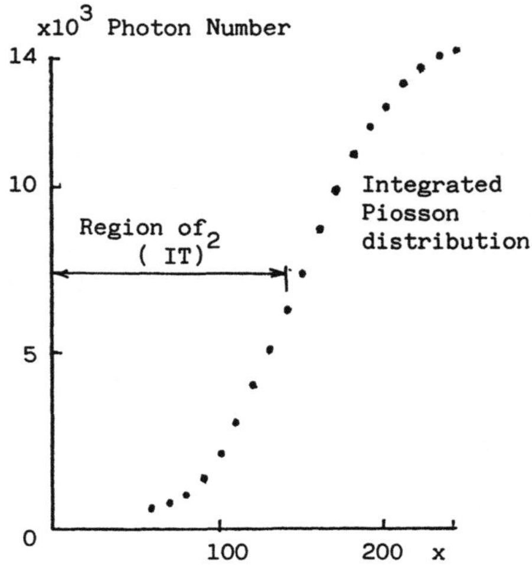

Fig.-4 Averaged integrated counting number of photon in down stream direction x at fixed y coordinate (y=132)

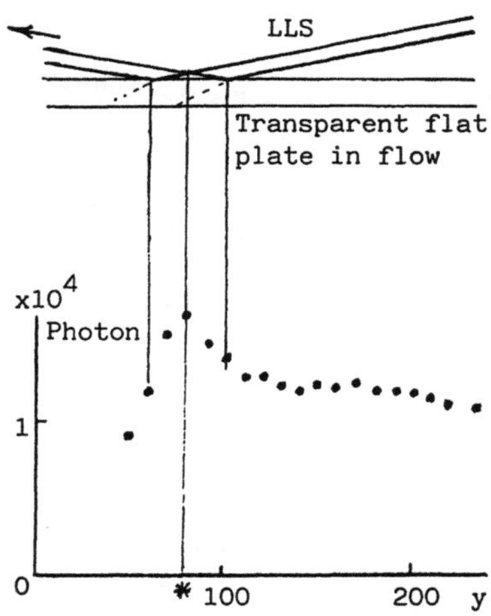

Fig.-5 Averaged counting number of photon in direction of y of incident laser light sheet at fixed x coordinate in still fluid over transparent flat plate placed in flow

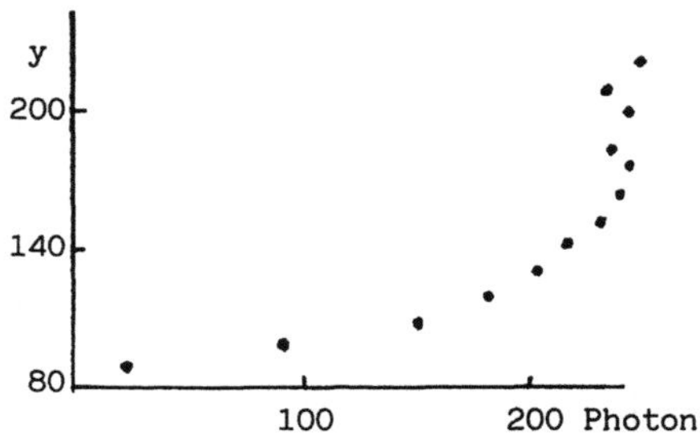

Fig.-6 Experimental result of averaged integrated number of photon in y direction at fixed x coordinate. The values are calibrated by the data of Fig.-5 and are rejected the data of background and noise. The distribution of photon number corresponds to the velocity of turbulent shear flow over the plate. The origin of y is selected at * point(y=80) marked in Fig.-5.

Investigation of a Digital Fourier Filtering Technqiue for Two-Dimensional Velocity Measurement

R.F. Larijani, G.W. Rankin and K. Sridhar
Department of Mechanical Engineering and Fluid Dynamics Research Institute
University of Windsor, Windsor, Ontario, Canada N9B 3P4

Abstract

Algorithms for performing a digital whole-field Fourier filtering technique have been developed. The effect of parameters such as filter shape and location, image resolution and particle size on the number and visibility of the whole-field isovelocity fringes was also investigated. Computer generated images were used instead of actual photographs to allow an evaluation of the method. The flow considered in this paper is that of an impinging flow.

Introduction

Particle Image Velocimetry (PIV) involves a determination of the velocity of neutrally buoyant particles in a transparent fluid by measuring the displacement between particle images. PIV involves two fundamental steps: recording and processing.

The recording process involves the use of a single beam of coherent light which is shaped into a thin sheet and passed through a fluid medium. A plane of illumination is thus generated in which the motion of small scattering particles may be photographed to reveal in-plane velocity vectors. The exposure is interrupted at a known frequency, producing what is referred to as a multiply-exposed image of the flow field.

The optical Fourier filtering setup used for processing multiply-exposed particle images is shown in Figure 1. The lens at z_3 with a focal length f_1 produces the Fourier transform of the mulvtiply-exposed image in its focal plane z_4, where an opaque mask with a small circular aperture is located. After bandpass filtering by the aperture, the lens at z_5 performs an inverse Fourier transform and a filtered image is reconstructed in its focal plane at z_6.

The filtering operation blocks all but a narrow band of spatial frequencies which correspond to a range of specific velocity vectors. If d_{oa} is the distance of the filtering aperture from the optical axis, then a filtered image displays isovelocity fringes which represent a velocity u given by [1]:

$$u=\frac{f_n \cdot f_1 \cdot \lambda}{\Delta t \cdot d_{oa} \cdot m_g} \tag{1}$$

where: d_{oa} is the distance of the filter from the optical axis, f_1 is focal length of the lens at z_3, f_n is the fringe number, Δt is the time between exposures, m_g is the magnification of the imaging system and λ is the wavelength of the coherent illumination used.

The objectives of this paper are:

(a) to present a procedure for the digital implementation of the Fourier filtering technique and

(b) to investigate the effect of input variables such as filter shape and location, particle size, and image resolution on the number and quality of the isovelocity fringes.

Digital Implementation of the Fourier Filtering Technique

Computer generated images are used in this study in order to evaluate the performance of a numerical Fourier filtering algorithm. Particles are initially placed in the field of study and their locations after a short time interval are determined using the prescribed flow equations. This represents a multiply-exposed image of the particles in the flow field being studied (Figure 2). The multiply-exposed image is then transformed into the frequency domain using a two-dimensional fast Fourier transform (FFT). A digital approximation to the optical filter is applied to the resulting frequency domain image. The filtered image is then transformed back into the spatial domain by using a two-dimensional inverse fast Fourier transform (IFFT). The resulting fringes are the isovelocity component contours within the flow field. Figure 3 shows the steps involved in implementing the above procedure.

An image containing isovelocity fringes consists of alternating bright and dark contours of constant velocity. An isovelocity fringe is the locus of points in the flow with equal velocity components. Ideal isovelocity fringes can therefore be created if the horizontal and vertical velocity components of a flow (u and v) are known. In the case of impinging flow ($\psi=3x^2y-y^3$) the horizontal and vertical velocity components are given by:

$$u=3x^2-3y^2 \qquad \text{and} \qquad v=-6xy \tag{2}$$

Figure 4 shows the ideal horizontal isovelocity fringes obtained for impinging flow.

Figure 5 shows the actual isovelocity fringes obtained by using a 2x2 square filter ($r_a=2$) located on the x-frequency axis at a distance of 47 pixels from the optical axis ($d_{oa}=47$).

In order to obtain the change in velocity represented by one full fringe, the magnitude of the velocity at one point in the flow must be known. The value of Δu is then obtained using the equation:

$$\Delta u=\frac{u(x,y)}{f_{nh}(x,y)} \tag{3}$$

where

Δu is the change in the horizontal velocity component between two adjacent bright fringes
$u(x,y)$ is the magnitude of the horizontal component of velocity at (x,y)
f_{nh} is the fringe number at point (x,y) for the horizontal velocity component.

Filter Shape

The filter used in the optical Fourier filtering setup is a small circular aperture. Both an approximation to a circular filter and a square filter were studied. The square filter produced isovelocity fringes that were smoother than the ones associated with the circular filter and therefore the square filter was used in the remainder of the work.

Filter Location

Moving the filter away from the optical axis resulted in the motion of the fringes toward the zero order fringe. This phenomenon can be used to determine the location of the zero order fringe which is of great importance in quantitative studies.

Image Resolution

Increasing image resolution is equivalent to using a camera with a higher resolution. The effect of increasing image resolution was found to be twofold: (1) it reduced the error associated with particle position and (2) shifted the frequency spectrum of the function to the low frequency region.

Particle Size

Increasing the particle size was found to reduce the visibility of the isovelocity fringes.

Randomness in Particle Distribution

The flow field considered in the previous sections had a uniform particle distribution. In an actual experiment, however, it is impossible to achieve a uniform particle distribution. Therefore it is necessary to investigate the effect of random particle distribution on the isovelocity fringes. Figure 6 shows a multiply-exposed image of impinging flow with randomly distributed particles. After filtering in the frequency domain, isovelocity fringes are obtained for the horizontal velocity component (Figure 7). These fringes are clearly inferior to the ones obtained using a uniform particle distribution (Figure 5).

Conditions for Increasing the Number of Isovelocity Fringes

In order to increase the number of isovelocity fringes, it is necessary to increase the image size while keeping the maximum particle displacement and particle size as small as possible. Once these conditions have been met, locating a filter in the high frequency region of the spectrum will produce the maximum number of fringes. Figure 8 shows the isovelocity fringes obtained by filtering a three-exposure 512x512 pixel image of impinging flow with a time between exposures of 0.000007 seconds. There are five bright fringes and four dark fringes visible in Figure 8. More details regarding the points discussed above are presented in Reference [2].

Conclusions

A digital Fourier filtering algorithm has been developed for processing multiply-exposed particle images of up to 512x512 pixels with an 8-bit spatial resolution. In order to evaluate this technique, computer generated multiply-exposed particle images were used.

A study was carried out to determine the effect of parameters such as filter shape and location, image resolution and particle size, on the resulting isovelocity fringes. Furthermore, the conditions for obtaining the largest number of isovelocity fringes are presented.

Using randomly distributed particles instead of a uniform particle distribution results in the smearing and deterioration of the isovelocity fringes. Since it is impossible to have a uniform particle distribution in an experimental setup, it would be difficult to apply this technique in an actual experiment.

References

[1] Grousson, R. and Mallick, S., Study of Flow Pattern in a Fluid by Scattered Laser Light, Applied Optics, Vol. 16, No. 9, pp. 2334-2336, Sept.1977.

[2] Larijani, R.F., Parametric Study of a Digital Whole Field Measurement Technique for Two-Dimensional Flow, M.A.Sc. Thesis, Department of Mechanical Engineering, University of Windsor, Windsor, Ontario, Canada, 1992.

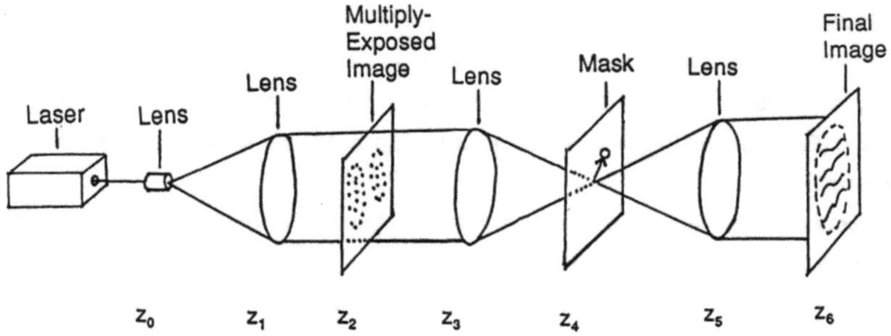

Figure 1 Schematic of Optical Setup for the Fourier Filtering Technique

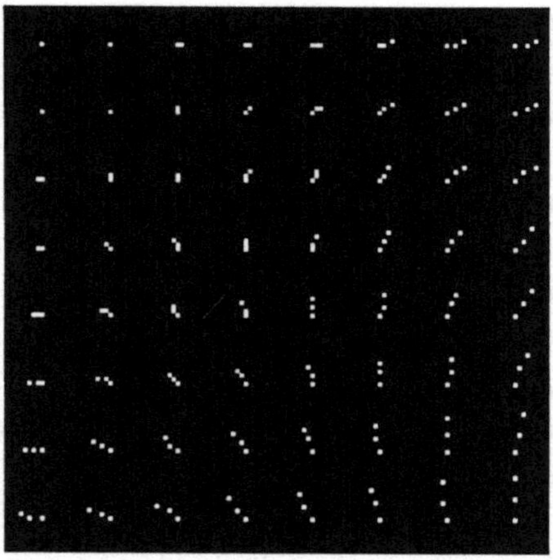

Figure 2 128x128 Pixel Computer Generated Multiply-Exposed Image of Impinging Flow with Uniformly Distributed Particles (Δt=0.00006 sec, Number of Exposures (n_x) =3)

Figure 3 Digital Implementation of the Fourier Filtering Technique

Figure 4 Ideal Contours of Constant Horizontal Velocity Component for Impinging Flow

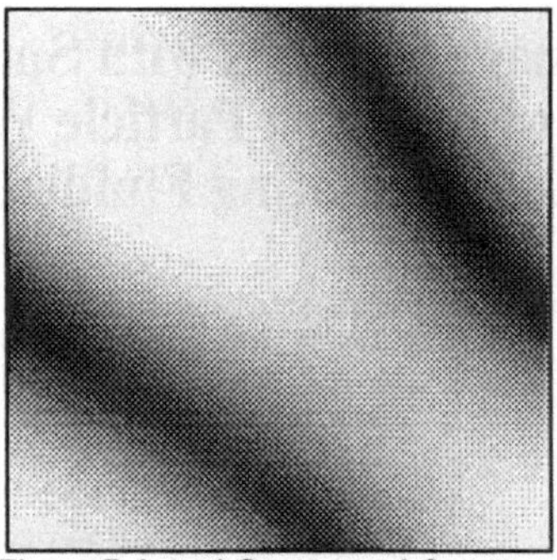

Figure 5 Actual Contours of Constant Horizontal Velocity Component of Fig. 2 (Δu=49.2 m/s, r_a=2, d_{oa}=47)

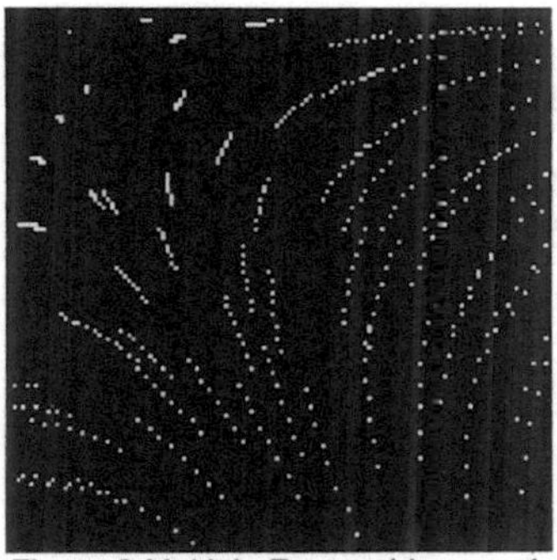

Figure 6 Multiply-Exposed Image of Impinging Flow with Randomly Distributed Particles (Δt= 0.0001 sec, n_x= 8)

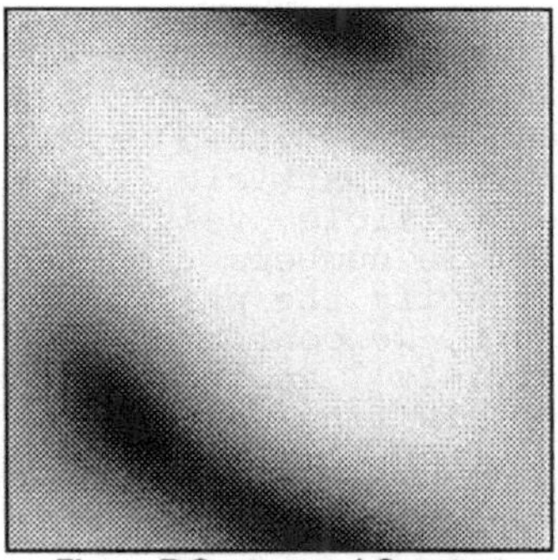

Figure 7 Contours of Constant Horizontal Velocity Component for Fig. 6 (r_a= 2, d_{oa}= 26)

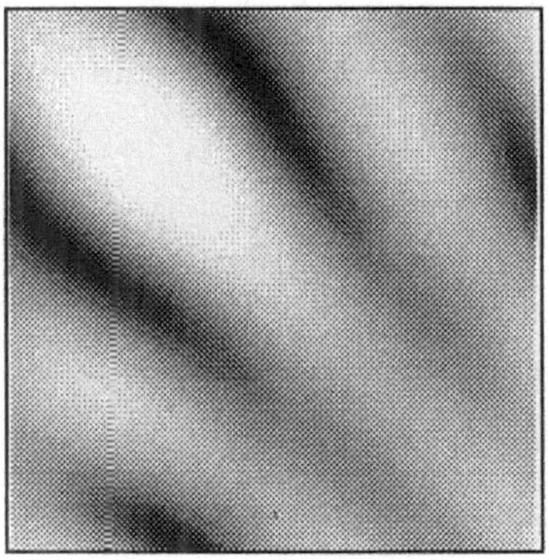

Figure 8 Contours of Constant Horizontal Velocity Component for a 512x512 Image (r_a= 4, d_{oa}= 190)

Image Analysis with Spatial Filtering Method for Fluidizing Particle Velocity in a Circulating Fluidized Bed

Yasuo KATOH, Masahide MIYAMOTO, Shingo KANEKO
Hidetoshi MIIKE, Kazutoshi KOGA

Dept. of Mechanical Engineering
Faculty of Engineering , Yamaguchi University
2557 Tokiwadai , Ube 755 , Japan

SUMMARY

This paper describes an application of a new measuring technique; the spatial filtering method based on dynamic image processing and a sequential image data obtained by a high speed video system for measuring the velocity of flowing particle in a circulating fluidized bed(CFB). The particle velocity was too high and the contained particle numbers too many. Therefore, it was very difficult to specify the peak of frequency by spectrum analysis. As a result, we confirmed that it was possible and successful to basically apply the technique for measuring the velocity of fluidizing particle in a circulating fluidized bed.

1. Introduction

There are three very interesting optical methods for the measuring of velocity, one is the laser Doppler method, one is the laser-speckle method and the other is the spatial-filtering method. Although the spatial-filtering method, proposed by Ator[1] in 1963, has some advantages such as simplicity and stability of the optical and mechanical systems, it has not been developed because of it's poor spatial resolution and measuring accuracy. In recent years, the development of opt-electronics has made it possible to study again for the development of practical instrument as well as extra both methods.

Two authors have already proposed spatial filtering velocimetry by dynamic image processing[2]. In that paper, through an analogy of spatial filter technique and laser Doppler velocimeter, a raw dynamic scene is transformed to a scene superposed with sinusoidal spatial pattern of gray-levels, that paper evaluated velocity information of moving particles by spectral analysis and confirmed the method to be valid and useful however the velocity was too low.

This paper describes a new application of the above method of measuring the relative higher velocity of flowing

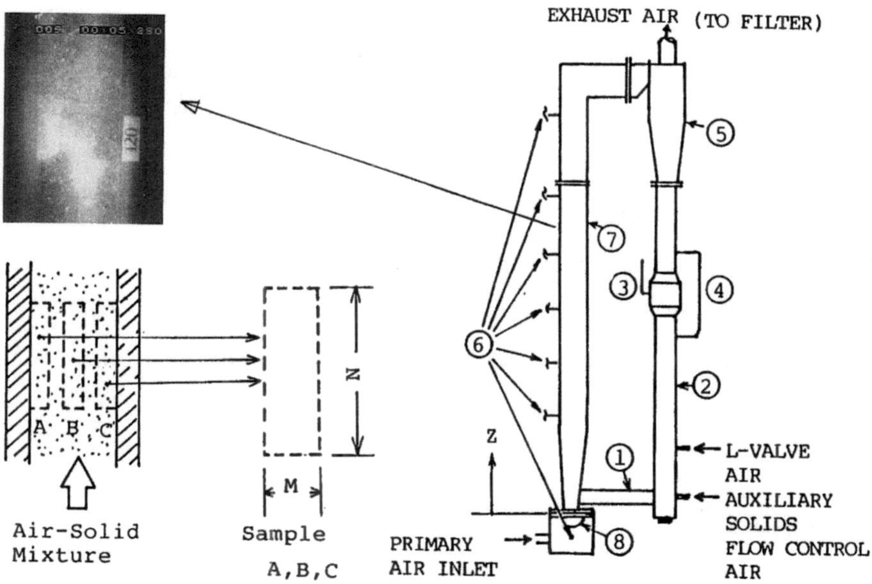

Fig.1 Experimental Apparatus and Area of Sampling Data

particle in CFB, though on the other hand there are few examples of measuring the particle velocity using the two-optical fiber probe method[3][4][5].

2. Image Analysis

There are papers[6][7] that explain the principle of the spatial filtering method precisely. Therefore, we have the following description without the minute explanation about the whole principle of this method.

2.1 Sampling data

Fig.1 shows the experimental apparatus and the area of sampling data. Image of sampling data obtained by a high speed video system(500 and 1000 frame/sec) were used for measuring the relative higher velocity of flowing particle in a circulating fluidized bed than that of previous one. The test column consists of a 33.7X33.7mm square cross sectional area. From such a experimental column as shown in Fig.1, containing a amount of glass beads in a mean diameter 200 μm, a sequential data of A, B and C regions was sampled. In this case superficial gas velocity is approximately 2 m/s and we supposed the particle velocity had slipped in gas so it's velocity is 1-1.5m/s at the most.

2.2 Data Processing

After a sequential data of A, B and C regions was sampled, a specified frequency was obtained by the spectrum analysis technique(FFT), with reduced signals when a small moving particle passes through the screen of spatial filter(λ) as shown in Fig.2.

Originally, we had to get sequential L frames on image processing. The total number of pixel were M(pixel) x N(pixel) x L(frame). But it was too large to memorise all the numbers, so in order to save the memory, these data were condensed in the one dimensional data as shown in Fig.3. We also could save the calculating time for using FFT.

3 Velocity Estimation

Once we got a specified frequency(f) by using FFT, we could calculate the velocity of particle(Vo) by the following equation,

$$Vo = f \times \lambda \qquad (1)$$

In our case, the particle velocity is too high and the contained particle numbers are too many. Therefore, it was very difficult to specify the peak of frequency as shown in Fig.4. In this case, however, we obtained an approximate particle velocity

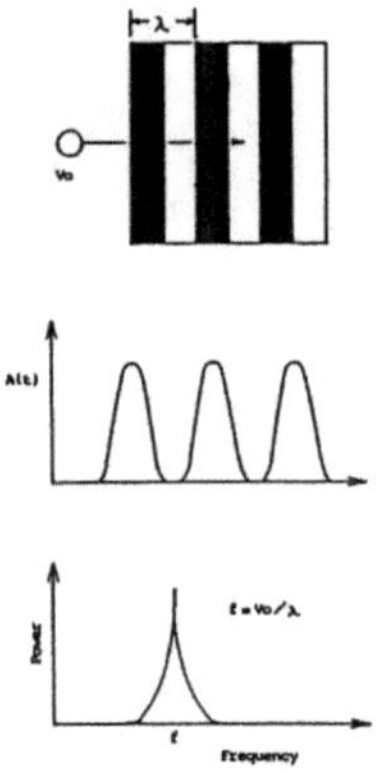

Fig.2 Principle of spatial filtering method (SFM)

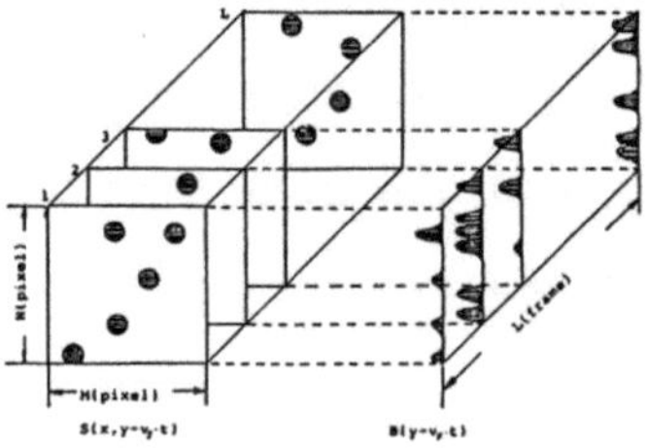

Fig.3 General construction of data transformation on SFM

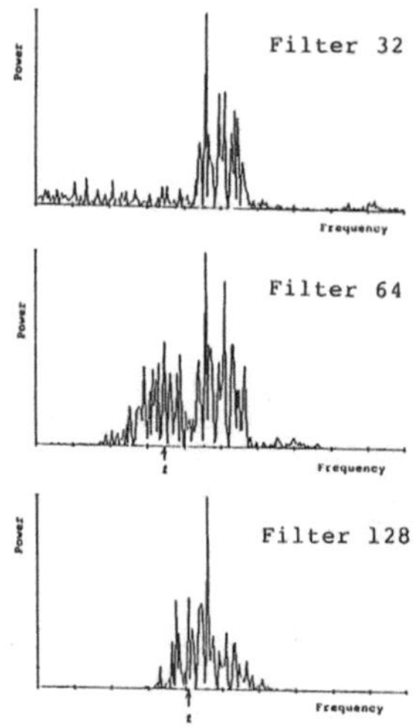

Fig.4 Spectrum analysis on B ($Vo = f \times \lambda$)

as shown in Fig.5. We also compared it with another researcher's results[5] in Fig.5. Consequently, the results were much different. However, we confirmed that it was possible and successful to basically apply the technique for measuring the velocity of fluidizing particle in a circulating fluidized bed. If we need a further precise discussion, the clearer image data and the more technical progress in image processing are necessary.

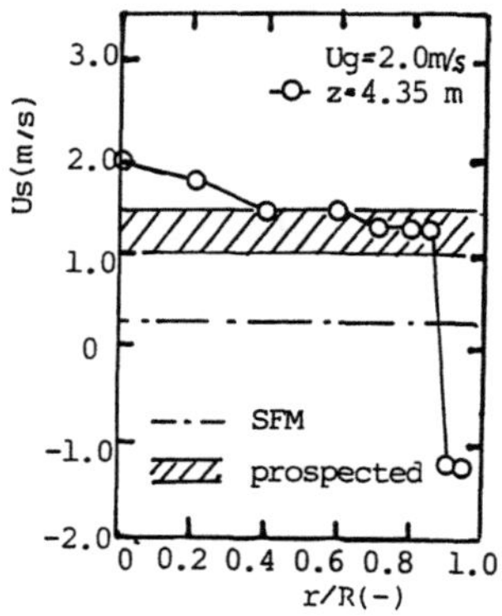

Fig.5 Distributions of Particle Velocity

4. Conclusion

We found that the spatial filtering method was a basically successful method for measuring the velocity of fluidizing particle in a circulating fluidized bed. We confirmed also that we will have more precise results in further experiments.

Reference

[1] Ator J.T., J. Opt. Soc. Am. 53(1963) 1416
[2] Miike H., Koga K., Momota M. and Hashimoto H., Japanese J. of App. Physics, Vol.26, No. 9,1987-9
[3] Horio M., Morishita K.,Tachibana O. and Murata N., C.F.B.T.,P.Basu and J.F.Large(eds.), Pergamon Press 147 (1988)
[4] Hartge E.-U.,Resner D. and Werther J., ibid.,165(1988)
[5] Mineo H., Ohtawara H., Matsumura Y., Yamazaki R. and Yoshida K., Powder Tech., in press
[6] Aizu Y. and Asakura T., Apply Physics,B43, 209-224(1987)
[7] Yamamoto H., Momota M., Koga K. and Miike H., Trans. Inst. Electron Commu. Eng., in press(Japanese)

Numerical Analysis

An Algorithm for Searching Points on Generalized Curvilinear Coordinate System and its Application to Flow Visualization

Satoshi Chiba and **Susumu Shirayama***
Institute of Computational Fluid Dynamics, Tokyo Japan
and
Kunio Kuwahara
The Institute of Space and Astronautical Science, Kanagawa Japan

* Present address; SOFTEC Systems Inc., Tokyo Japan

1.Introduction

This paper describes a useful algorithm for searching points on generalized curvilinear coordinate systems, which are used in finite difference methods. With this algorithm one can compute computational coordinates of arbitrary points, whose physical coordinates are given, very fast and precisely. Once the computational coordinates are known, physical quantities at those points can be interpolated from the data of surrounding grid points. Then, this algorithm is applicable to any stage of simulations; integration of Navier-Stokes equation and pre/post processing of the data.

Conventional algorithms for searching points inside polygons, which utilize geometric properties of polygons, are useless in view of processing speed, for practical application to 3D problems. The algorithm herein explained is completely different from those. A non-linear ordinary differential equation is solved. The reason of high-speed processing in the proposed method is the effective use of information of structured grid systems. It can be applied to both 2D and 3D grid systems and any grid systems such as O-, H- and C-type grids. It also has favorable characteristic that it is almost insensitive to an increment of grid points. The code can be parallerized as well as vectorized. Therefore, this method is appropriate for massive data processing. In the following sections, the algorithm is introduced and results of bench mark tests and some applications to flow visualization are demonstrated. A similar algorithm was introduced by Himeno and Fujitani in [1]. The difference is a derivation of the o.d.e. and the thorough investigation of present work for this kind of methods on various grids.

2.Numerical procedure

The procedure of this method is briefly described using Fig.1. Q is an arbitrary fixed point of which computational coordinates are unknown. The initial physical and computational coordinates of point P are given. Generally it is placed just on a grid point. **F** is the virtual force acted between P and Q. Its magnitude is the function of the distance between P and Q. In this paper, the linear relation is simply adopted as eq.(2). The movement of P is assumed to follow the eq.(1).

$$d\mathbf{x}_P/d\tau = \mathbf{F}, \qquad (1) \qquad\qquad \mathbf{F} = a\,(\,\mathbf{x}_Q - \mathbf{x}_P\,), \qquad (2)$$

where, $\mathbf{x}_P, \mathbf{x}_Q$ are the position vectors of P and Q. τ is the virtual time and a is the positive constant. All equations appeared hereinafter are written in non-dimensional form. The eq.(1), (2) are transformed from physical to computational space and rewritten as:

$$d\xi_{Pi}/d\tau = a\,(\,\partial\xi_i/\partial x_j\,)_P \cdot (\,x_{Qj} - x_{Pj}\,) \qquad (\,i=1,2,3\,), \qquad (3)$$

where, x_j, ξ_i are the Cartesian and computational coordinates. Summation convention is invoked here. Since initial values of $\mathbf{x}_P, \mathbf{x}_Q$ and ξ_P are given, the evolution of ξ_P value can be computed as a standard initial value problem of o.d.e.. Here, the term $(\,\partial\xi_i/\partial x_j\,)_P$ is regarded as a variable of which value changes depending on the position P. The eq.(3) is discretized as:

$$\Delta\xi_{Pi} = c\,(\,\partial\xi_i/\partial x_j\,)_P \cdot \Delta x_{Pj} \qquad (\,i=1,2,3\,), \qquad (4)$$

where, the relations; $c = a\,\Delta\tau$, $\Delta x_{Pj} = x_{Qj} - x_{Pj}$ are applied. If $c = 1$, eq.(4) is nothing but the descretized form of total differential equation of ξ_P. Therefore, it is easily understood that in case Δx_{Pj} is very small, $c > 1$ and $c < 1$ give over and under estimation of $\Delta\xi_{Pi}$, respectively. In case Δx_{Pj} is not small, that is an actual case, it is found that the stable convergence is obtained if the value of c is set properly within the range $c \leq 1$.

The computation is finalized in the following condition and ξ_P is taken as the computational coordinates of point Q.

$$|\,\mathbf{x}_Q - \mathbf{x}_P\,| < \varepsilon_1\,,$$

where, |.| denotes its absolute value. ε_1 is a given small value and set to be much less than the minimum spacing of grids. The computation is aborted in the condition:

(a) $| \mathbf{x}_P^{n+1} - \mathbf{x}_P^n | < \varepsilon_2 \quad \cap \quad | \mathbf{x}_Q - \mathbf{x}_P | > \varepsilon_1$,
(b) $n > n_{max}$.

here, n indicates nth discrete time integration steps. ε_2 is a small value which is set to be much less than ε_1. n_{max} is the limitation number of time integration step.

The $\mathbf{x}_P$ value at every virtual time step is converted from ξ_P by using appropriate interpolation formula. Here, bilinear interpolation is adopted as:

$$x_P = (1-\gamma)\{(1-\alpha)(1-\beta)\, x_{IJK} + (1-\alpha)\beta\, x_{IJ+1K} + \alpha(1-\beta)\, x_{I+1JK} + \alpha\beta\, x_{I+1J+1K}\} \\ + \gamma\{(1-\alpha)(1-\beta)\, x_{IJK+1} + (1-\alpha)\beta\, x_{IJ-1K+1} + \alpha(1-\beta)\, x_{I+1JK+1} + \alpha\beta\, x_{I+1J+1K+1}\}, \quad (5)$$

where, $I = [\xi_1]$, $J = [\xi_2]$, $K = [\xi_3]$, $\alpha = \xi_1 - I$, $\beta = \xi_2 - J$, $\gamma = \xi_3 - K$, and [.] expresses Gauss's symbol. x_{IJK} denotes a component of the Cartesian coordinates of a grid point (I,J,K). Lower suffices for x and x_P are omitted here for the sake of simplicity.

The required physical quantity at Q is calculated with the same interpolation formula. Then, it is said that the accuracy of the obtained physical quantity is determined by the degree of accuracy of the applied formula, which is used for interpolation of both physical coordinates and physical quantities, and the given value of ε_1.

The procedure at boundaries adopted in this investigation is simple and sufficient for practical use. If the point P goes outside a boundary, only the corresponding component of the computational coordinate is revised as:

$x_{P1} < 1$ then $x_{P1} = 1$,
$x_{P1} > I_{max}$ then $x_{P1} = I_{max}$,

where, I_{max} is the maximum number of I. The similar equations are applied to ξ_{P2} and ξ_{P3}. The treatment at periodic boundary is trivial and not mentioned here. In Fig.2, two loci of the integration procedure of eq.(3), which were executed on O-type grid (2D), are depicted as examples. Generally, the searching process is finalized less than a few ten steps, unless the grids are heavily distorted and deformed.

3.Bench mark test and Examples

We performed bench mark tests of this algorithm for ordinary H-H, O-H and O-O type grids on Fujitsu VP2600. An predictor-corrector method was applied to the integration of eq.(3). In order to examine effects of the number of grid points, two of 50x50x50 and 100x100x100 grids were used. The results are summarized in Table 1. We can find that it requires less than one second to search points consisting of (100x100) meshes on a given plane inside domains (see Fig.3) for all types of grid. The time differences are originated in the magnitude of deformation of those grids. As prescribed, 'c' is set to be less than unity. Concretely, c=1.0 is proper for lightly deformed grids and the smaller values for more deformed grids. The adopted values here are c=1.0 for the H-H type grid and c=0.8 for the O-H and O-O type grids. We also find that CPU times for searching of (100x100x10) mesh points are greatly reduced by using the vector processor. The reason why the left values of the vector processor are not so improved is due to recursive use of initial points. When searching of (100x100) mesh points, the computational coordinates of neighboring grid point, which are determined one procedure before, are used as initial values of point P.

It is surprisingly fast compared with other algorithms which utilize geometrical properties of polygons. An exemplar algorithm for a triangle is the utilization of area coordinates. Generally, such algorithms require very long CPU time and are useless for massive data processing in 3D problems. In addition, the CPU time of those algorithm increases in proportion to the number of grid points. On the other hand, the CPU time based on the present method is not influenced by the increment of grid points as shown in the table. It is a particular feature of this method. In addition, the present algorithm can be parallerized as well as vectorized. Therefore, it suits for massive data processing.

The most appropriate application may be a flow visualization. As last examples, two pictures are demonstrated. Fig.4 shows a result of simulation of flow around a vertical circular cylinder [3], which is a example of O-H type grid. Free surface exists and the bottom end of the cylinder locates three diameters below the mean free surface. The cylinder and the free surface are drawn as wire frames. Streamwise vorticity contours at three vertical plane behind the cylinder are computed using the present method. Fig.5 is a pressure field around a car visualized with a volume rendering technique. Volume rendering is expected to be a practical tool for analyses of 3D scalar field. The present algorithm works very effectively with such a method.

4. Acknowledgment

Authors would like to appreciate Mr. Masahiro Suzuki of iCFD for his helpful discussion.

References

[1] R.Himeno & K.Fujitani, "CG on Structured Grids", Computer Graphics for Flow Visualization, JSME, 1991 (in Japanese)

[2] S.Shirayama & K.Kuwahara, "Flow visualization In Computational Fluid Dynamics", The Journal of Supercomputer Applications, 4, 3, Summer, 1990, 66-80

[3] S.Chiba & K.Kuwahara, "A Finite Difference Formulation for Free Surface Flow Using a Free Surface Conforming Grid System", Lecture Notes in Physics 371, Springer-Verlag

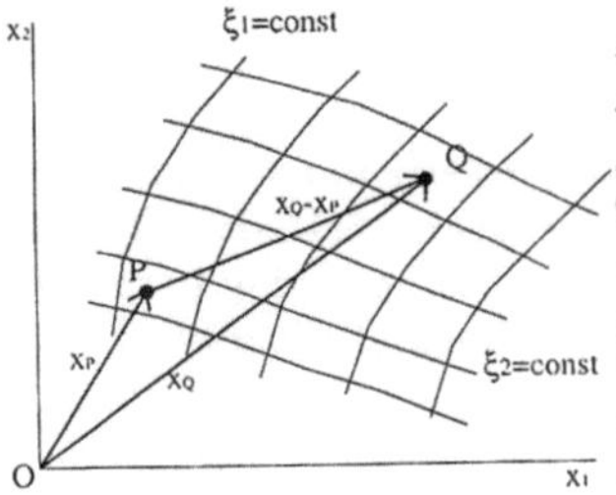

Fig.1 Schematic figure for description of the algorithm

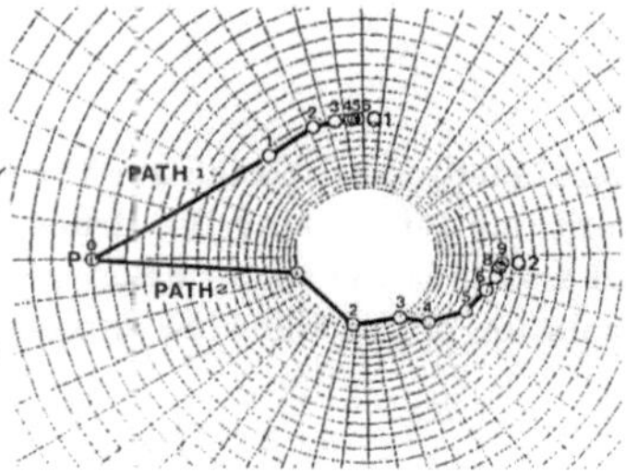

Fig.2 Examples of loci of integration process

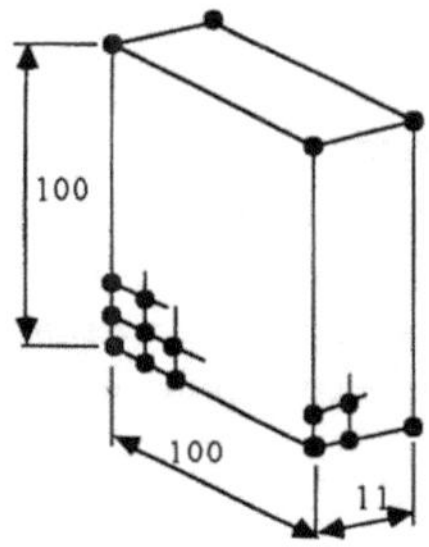

Fig.3 Searching meshes

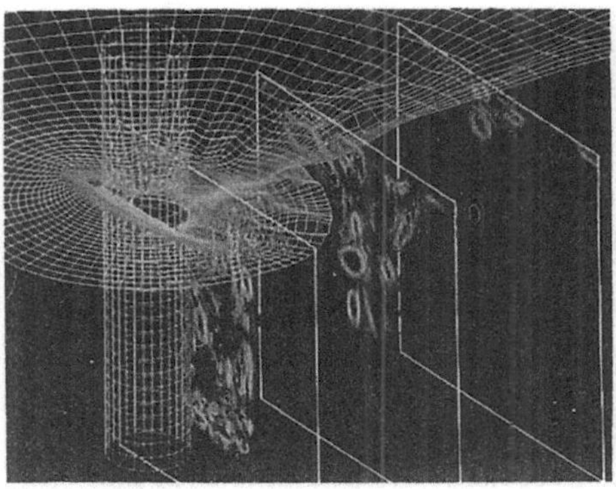

Fig.4 Contour drawing of streamwise vorticity of a wake behind a cylinder

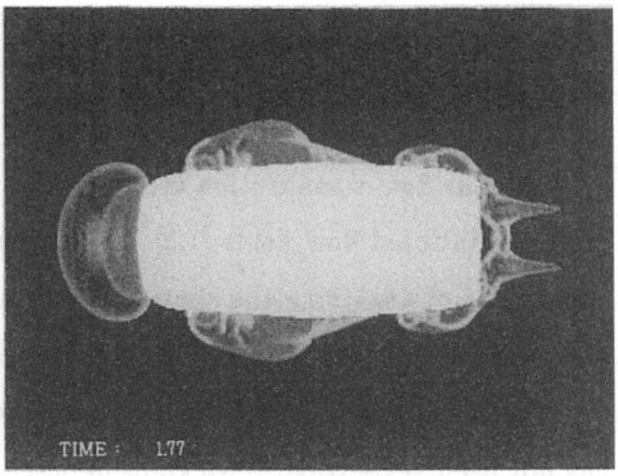

Fig.5 Pressure contour around a car (Volume rendering)

Table 1 CPU Time List

	H-H Grid	O-H Grid	O-O Grid	Remarks
Scalar Processor	0.147/1.35	0.869/8.60	0.849/8.46	50x50x50
Vector Processor	0.130/0.141	0.810/0.756	0.788/0.779	Grid
Scalar Processor	0.147/1.36	0.897/8.60	0.880/8.43	100x100x100
Vector Processor	0.130/0.100	0.831/0.642	0.810/0.661	Grid

Note) Left values; CPU time for searching points consisting of 100x100 meshes,
Right values; CPU time for 100x100x10 meshes
Performance of CPU; approx. 30MFLOPS for scalar processor
approx. 3GFLOPS for vector processor

Visual Study for Toplogical Transitions of Vector Fields

S. SHIRAYAMA

SofTek Systems, Inc.
2-34-5, 5F, Nozawa, Setagaya-ku, Tokyo 154, Japan.
phone:81-3-3412-6008, facsimile:81-3-3412-7990

Summary

This paper describes several methods of visualizing the vector fields in a flow analysis. A class of computational algorithms determining the structure of vector fields is stated. It is assumed that the visualized results in the vector fields are classified according to certain kinds of expressions for the solution of the following ordinary differential equation:

$$\frac{d\mathbf{x}}{ds} = \mathbf{u} \quad or \quad \boldsymbol{\omega}.$$

We pursue the characteristic feature of the equation by expanding around a critical point, and study the topological transitions of vector fields.

Introduction

Various flow-visualization techniques have been developed for analyzing experimental or computational flow fields [1,2]. Such techniques are divided into two groups from the standpoint of analyzing the spatial flow structure. One is the analysis of the scalar field, and the other is that of the vector field. Hesselink [1] mainly classified the flow-visualization techniques for scalar fields. In the vector field, however, the classification of approaches has not been yet settled. Two kinds of descriptions of the vector field have been proposed. One is the direct representation of vectors, i.e., the direction and magnitude of vector is presented at a certain location. The other is the Lagrangian or topological representation of the vector field; oil flow pattern, streamlines, pathlines, streaklines, etc.

In the previous symposium, we proposed the efficient techniques of particle tracking, and visualized several flow fields in order to compare them with the experimental results [3]. In this paper, we study the topological transitions of vector fields, and consider how the characteristic of the unsteady flow fields can be expressed by the singularities. This paper concentrates on a formulation of the problem.

Computational Approach

A vector field in steady or unsteady flows is described by the following dynamical system:

$$\frac{d\mathbf{x}}{ds} = \mathbf{b}. \tag{1}$$

where $\mathbf{x}$ defines a space in which the solution moves. The system is *autonomous* if the vector field $\mathbf{b}$ is independent of time, and $\mathbf{b}$ is a function of variable $\mathbf{x}$. The system is a *nonautonomous system* if the vector $\mathbf{b}$ is a function of variables $\mathbf{x}$ and time. If the transformation matrix P^{-1} operates on Eq.(1) from the left-hand side, the equation is transformed into the curvilinear coordinates:

$$\frac{d\boldsymbol{\xi}}{ds'} = \mathbf{B}, \tag{2a}$$

$$d\boldsymbol{\xi} = P^{-1} \quad d\mathbf{x}. \tag{2b}$$

The global characteristic of the solution will be examined by the linearization of Eq.(1) or Eq.(2) around a critical point. Local solutions to the Navier-Stokes equations have shown that the critical points play an important role in the investigation of vector fields [4]. The critical points are defined as follows:

$$\frac{b^i}{b^j} = \frac{0}{0} \qquad for \qquad i \neq j. \tag{3}$$

The classification of the critical point indicates the global characteristic of the solution. In order to classify the critical point, Eqs.(1) and (2) are expanded and linearized around a certain point c. The linearized equations are:

$$\frac{d\mathbf{x}}{ds} = E \quad \mathbf{x} + \mathbf{b}_c, \tag{4}$$

$$\frac{d\boldsymbol{\xi}}{ds'} = F \quad \boldsymbol{\xi} + \mathbf{B}_c. \tag{5}$$

where $E_{ij} = \frac{\partial b^i}{\partial x^j}$ and $F_{ij} = \frac{\partial B^i}{\partial \xi^j}$. If point c is a critical point, the classification of the critical point is made by inquiring into the eigenvalues and the eigenvectors of matrices E and F. Here we find the following relation:

$$F = P^{-1} \quad E \quad P. \tag{6}$$

As this relation means the similarity transformation of matrix E, the two matrices E and F have the same eigenvalues and eigenvectors. Therefore, we only investigate either E

or F. Then, we consider how the analysis of critical points is valid in the study of the detail structure of the vector field. According to the theorem of Hartman-Grobman [5], if the critical point is hyperbolic point (all eigenvalues have real parts unequal to zero), these truncated systems (Eqs.(4) and (5)) are sufficient to determine the local topology of the vector field. The detailed statements about the classification of the critical points are described in Ref.[4]. In this paper, we introduce the following two comments for the critical-point concept in the computed vector fields.

First, there is the case where first-order interpolation is selected for **b** and P^{-1}. In this case, the structure of the vector field in the reconstructed space is fully described by the critical-point concept.

The second comment is about the property of critical points in the case of two-dimensional incompressible velocity fields. At a certain critical point (x_c, y_c), Eq.(4) becomes:

$$\frac{d}{ds}\begin{pmatrix} x - x_c \\ y - y_c \end{pmatrix} = \begin{pmatrix} \frac{\partial u}{\partial x} & \frac{\partial u}{\partial y} \\ \frac{\partial v}{\partial x} & \frac{\partial v}{\partial y} \end{pmatrix}_{x_c, y_c} \begin{pmatrix} x - x_c \\ y - x_c \end{pmatrix}. \tag{7}$$

The condition for incompressibility is

$$\frac{\partial u}{\partial x} + \frac{\partial v}{\partial y} = 0. \tag{8}$$

Let d denote $\frac{\partial u}{\partial x} \ \frac{\partial v}{\partial y} - \frac{\partial u}{\partial y} \ \frac{\partial v}{\partial x}$. The eigenvalues are as follows:

If $d \leq 0$ then,

$$\lambda_1 = \sqrt{-d}, \quad \lambda_2 = -\sqrt{-d}, \tag{9a}$$

the critical point is saddle.

If $d > 0$ then,

$$\lambda_1 = i \ \sqrt{d}, \quad \lambda_2 = -i \ \sqrt{d}, \tag{9b}$$

the critical point is the center.

Therefore, we should not have node and spiral points in the velocity field. In the incompressible flow field, d is related to the pressure field:

$$d = \frac{1}{2}\left(\frac{\partial^2 p}{\partial x^2} + \frac{\partial^2 p}{\partial y^2}\right). \tag{10}$$

According to Eq.(10), the pressure value at the saddle point is higher than the mean value surrounding the point, and the value at the center is lower. These comments are useful in verifying the numerical solution.

Numerical Experiment

For example, we compute a flow past a sphere at $Re = 500$. We show the topological transitions from the standpoint of time variation of the boundary layers. Time development of the patterns of surface streamlines on the body in rear view is shown in Fig.1. It is clearly observed that the axisymmetry has been broken down, and the plane of symmetry in flow field appears near the x-y plane. The sequence of the topological changes is observed on the body surface. Initially, topological structure is characterized by the front and rear stagnation points, and these stagnation points are defined as the star-nodes. As time goes on, the boundary layer separates, but the axisymmetrical structure is preserved. The rear stagnation point changes from the stable star-node to the unstable star-node. The points on the separation line are node-saddles. Suddenly the axisymmetry is broken, then innumerable node-saddles change to one saddle point and one nodal point (indicated by Sa and N in Fig.1a) on two locations where the separation line and the symmetry plane cross each other. Then, the nodal point changes to the saddle and the foci, as shown in Figs.1b and 1c. In the developed stage, we cannot find the critical points except the rear stagnation point in the separated flow region (see Figs.1d to 1f). In this stage, two saddle points move on the separation line. It is expected that this structure is related to the periodical structure that appeared in the wake region. In these figures, we emphasize that, in all stages, the following topological rule about critical points on the body surface is satisfied:

$$\sum_{N,F} - \sum_{S_a} = 2.$$

Conclusions

We can derive many knowledge about the fluid motions from the vector field. However, the raw dataset of the vector field gives too much information to understand the structure of the flow field. Therefore, it is necessary to reduce the original vector field. A set of critical points and lines is one candidate to compress the dataset. We can obtain a rough estimate about the flow structure by this set. In this paper, we have proposed a method of computing the singularities and have inquired the characteristic nature of the critical points in the incompressible fluid flow.

References

1. L. Hesselink, Ann. Rev. Fluid Mech. **20**, 421 (1988).

2. S.E. Rogers, P.G. Buning and F.J. Merritt, International Journal of Supercomputer Applications, **1.4**, 96 (1987).

3. S. Shirayama and K. Kuwahara, in the Proceedings of 5th ISFV, also published in the International Journal of Supercomputer Applications, **4.2**, 66 (1990).

4. A.E. Perry and M.S. Chong, Ann. Rev. Fluid Mech. **19**, 125(1987)

5. J. Guckerheimer and Ph. Holmes,Non-linear Oscillations, Dynamical Systems, and Bifurcations of Vector Fields, (Springer Verlag, New York,1983)

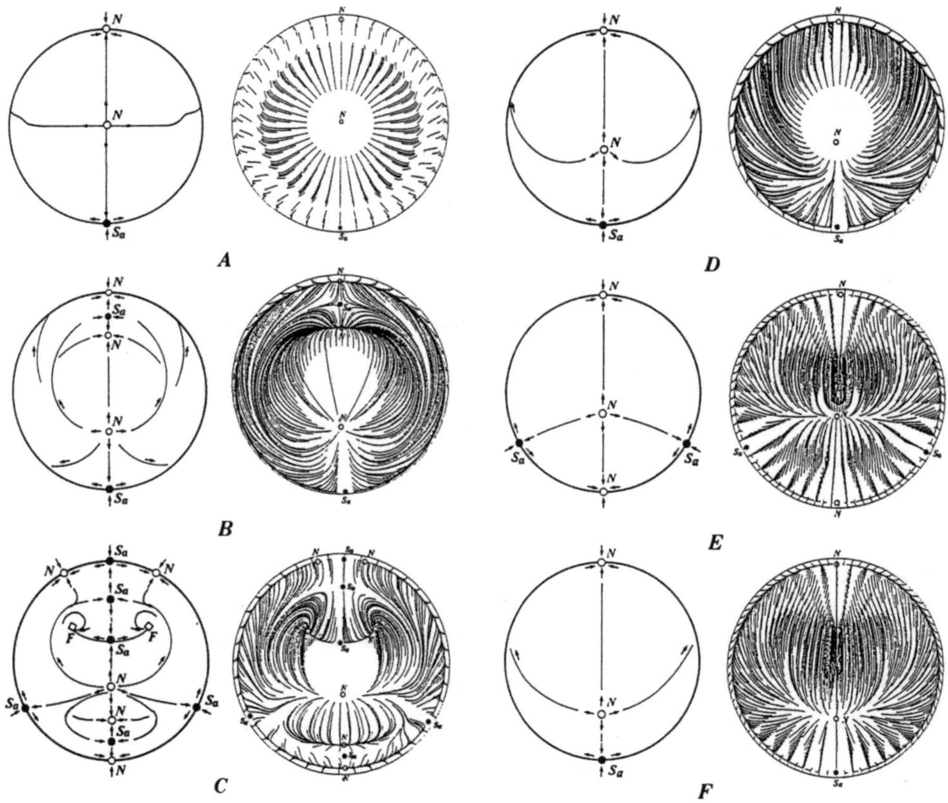

Fig.1 Time evolution of the patterns of surface streamlines on the body surface and its topological representations.

Numerical Flow Visualization of Separated Flows Around Oscillating Airfoil by Solving Incompressible Navier-Stokes Equations

S. ASO*, A. SAKAMOTO** and M. HAYASHI***

* Dept. Aeronautics and Astronautics, Kyushu University, 6-10-1 Hakozaki, Higashi-Ku, Fukuoka 812, JAPAN
** Dept. Aeronautics and Astronautics, Kyushu University, Fukuoka(at present Nissan Motor, Co. Ltd., Tokyo, JAPAN)
*** Nishinippon Institute of Technology, Fukuoka, JAPAN

Summary

Numerical flow visualization of separated flows around oscillating airfoil are conducted by solving incompressible Navier-Stokes equations by a third-order upwind scheme in order to understand the flow structure and mechanism of dynamic stall. The calculated separated region is small in pitching-up process and it becomes large in a pitching-down process. Quite different characteristics of flow patterns between in a pitching-up and pitching-down processes are obtained.

Introduction

Studies on dynamic stall have been conducted experimentally and theoretically by the many researchers[1-4]. However as the phenomena are essentially non-linear and the flow field becomes complicated as the separated region is increased, those phenomena are not understood sufficiently.

In the calculations by viscous flow calculations incompressible Navier-Stokes equations have been solved by a third-order upwind scheme in order to understand the flow structure and mechanism of dynamic stall. Especially the flow mechanics of movement of separation point on the wing surface, vortex formation from the surface of the wing and unsteady Kutta condition at the trailing edge of the wing are investigated carefully. The unsteady flow fields around a pitching airfoil are calculated by moving a grid system relative to the freestream. Remarkable characteristics of dynamic stall are obtained. Also the hysteresis curve of aerodynamic characteristics of C_L and C_M are obtained.

Numerical Methods

Unsteady flows around a oscillating airfoil are simulated by solving incompressible Navier-Stokes equations with third-order upwind scheme. The detailed numerical procedures will be given in Reference 5. The mesh system is generated as O-grid type by Steger and Sorenson's method[6]. For the calculation of separated flow around oscillating airfoil a general coordinate system with time-dependent variables[e.g.7] is introduced.

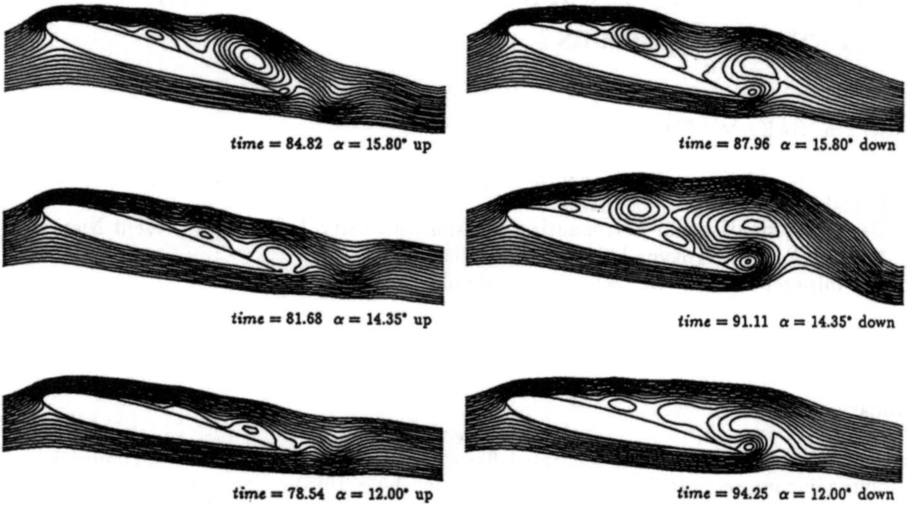

(a) Time series of instantaneous stream lines

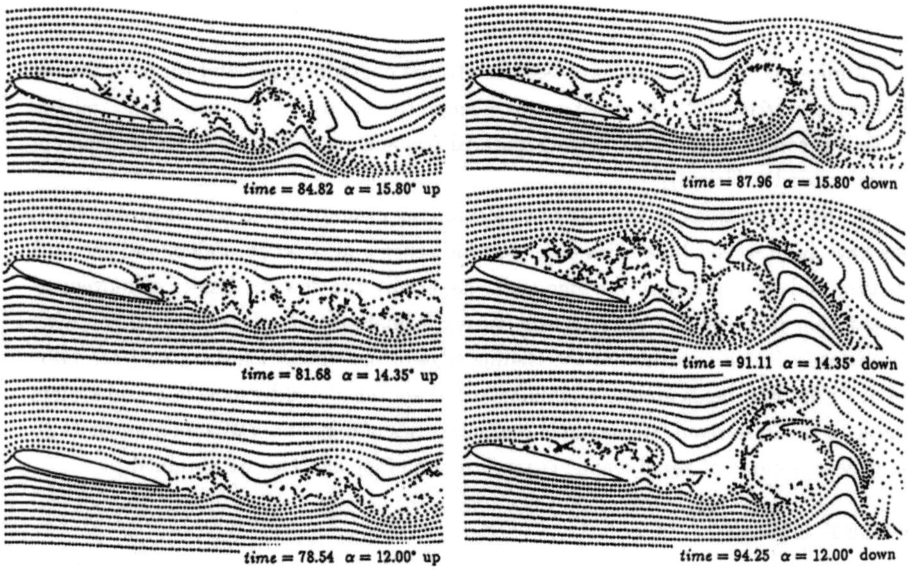

(b) Time series of numerical smoke wire diagrams

Fig.1 Flow patterns at $\alpha=12^0+4^0 \sin \omega$ t (k=0.2) (NACA0012, Re=70000).

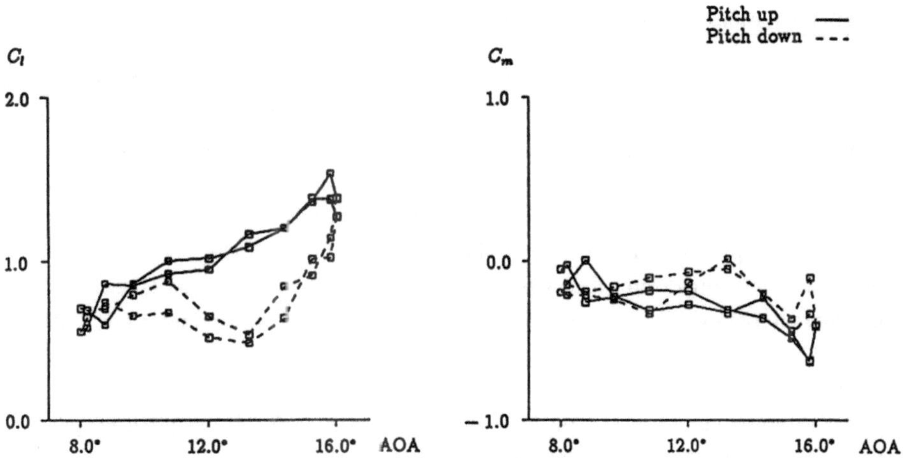

Fig.2 Hysteresis of C_L and C_M at $\alpha=12^0+4^0 \sin \omega$ t (k=0.2) (NACA0012, Re=70000).

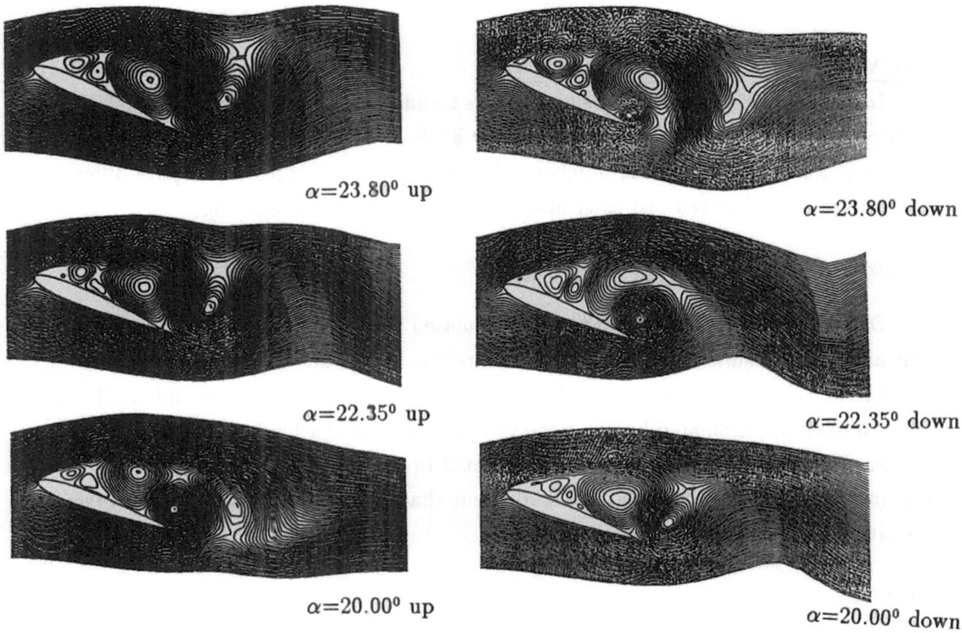

Fig.3 Flow patterns at $\alpha=20^0+4^0 \sin \omega$ t (k=0.2) (NACA0012, Re=70000).

Numerical Simulations on Separated Flows around an Oscillating Airfoil in Pitch

The representative instantaneous stream lines and numerical smoke-wire pictures of oscillating airfoil[8] are shown in Fig. 1. For the calculations of numerical smoke-wire picture numerous fluid particles are introduced at equi-spacing points along a perpendicular axis in front of an airfoil and their temporal traces convected by the velocity fields are calculated. Quite different characteristics of flow patterns between in pitching-up and pitching-down processes are obtained. The significant change of the flow patterns in pitching-up and pitching-down process show the existence of the Hysteresis of the aerodynamic characteristics. The results show the numerical smoke- wire pictures are quite useful in order to understand the structure of the flow fields and compare the calculated results with experiments.

The Hysteresis curves of C_L and C_M are showin in Fig. 2. C_L in pitch-up process is higher than that in pitch-down process. Those Hysteresis curves are remarkable characterestics of dynamic stall.

In Fig.3 flow patterns at higher mean attack angle are shown. As mean attack angle is large, the flow separates from the surface of the wing even in pitch-up process. Also the process of vortex shedding from trailing edge is quite different from that for smaller mean attack angle.

Flow Visualization around Oscillating Wing

In Fig.4 flow visualization by a smoke-wire technique is shown for pitch-up and pitch-down process of NACA0012 at smallest attack angle with $\alpha=16^0+4^0 \sin \omega t$. Freestream velocity is 2 m/s and Reynolds number is 3.7×10^4. The pictures show quite different flow patterns are observed just after and before the minimum attack angle.

Conclusions

Dynamic stall phenomena have been simulated numerically by viscous flow calculations and major results of visualization of numerical results are discussed.

In the calculations by viscous flow calculations separated flows around oscillating airfoil in pitch are simulated by using the third-order upwind scheme and a moving mesh system. The calculated separated region is small in pitching-up process and it becomes large in a pitching-down process. Quite different characteristics of flow patterns between in a pitching-up and pitching-down processes.

References

1) McCrosky, W. J. and Pucci, S.L.: AIAA paper 81-0051 (1981).
2) Aihara, Y., Koyama, H. and Murashige, A.: J. Aircraft, 2 (1985), pp.661-668.
3) Carr, L. W.: J. Aircraft, 25 (1988), pp.6-17.

4) Oshima, K and Oshima, Y.: Proc. 8th ICNMFD (1982), pp.94-106.
5) Kawamura, T. and Kuwahara, K.: AIAA paper 84-0340 (1984).
6) Steger, J. L. and Sorenson, R. L.: J. Computational Physics, 33 (1979), pp.405-410.
7) Nakamichi, J.: AIAA paper 87-1158 (1987).
8) ASO, S. and SAKAMOTO, A.: Tech. Rep. Kyushu Univ., 64 (1991), pp.249-255.

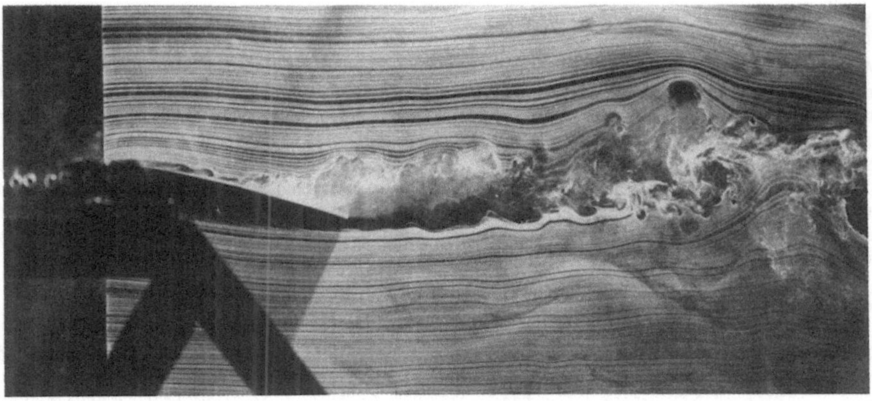

(a) Pitch-up process near $\alpha=12^0$

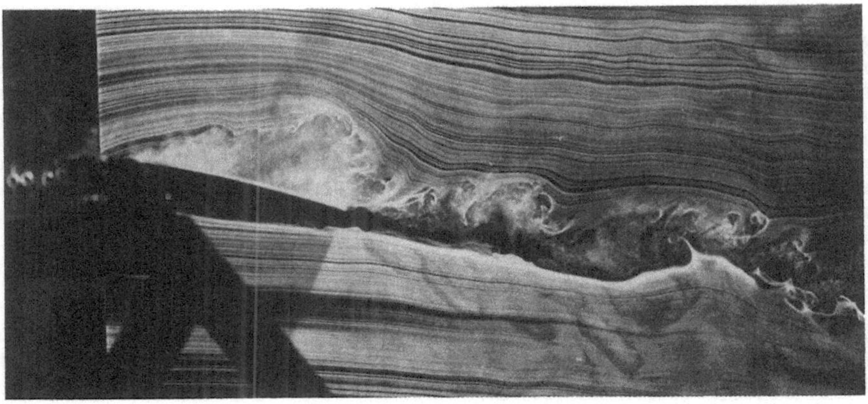

(b) Pitch-down process near $\alpha=12^0$

Fig.4 Flow patterns at $\alpha=16^0+4^0 \sin \omega t$ (k=0.2) (NACA0012, Re=37000).

Visualization Technique in Computational Fluid Dynamics for Rational Comparison with Experiment

Kozo Fujii and Yoshiaki Tamura

Institute of Space and Astronautical Science, Yoshinodai 3-1-1, Sagamihara, Kanagawa, 229, JAPAN

SUMMARY

A newly-developed computational method that simulates some of the flow visualization techniques, such as Schlieren photograph in compressible flow experiments is applied to several flow fields. By applying this method to the numerical data from the computational fluid dynamics simulations, same visual output as what is obtained in the experiment can be produced and the computational and experimental results can be fairly compared. Some of the flow features that are observed in the experiment but do not appear in the density contour plots of the computational results appear in the visual output in this method. It is shown that present method allows the correct comparison of the numerical data with the visual output of experiments and is useful for the validation of the computational approach to the fluid dynamics problems. In addition, accuracy and reliability of the present method are clarified by the theoretical evaluation of the assumptions used in the present approach.

INTRODUCTION

One of the important topics of the computational approach to fluid dynamics research is validation and verification of the simulation. One way for the validation and verification is comparison with the experiment. Visualized image obtained in the experiment can be frequently used for this purpose. Density contour plots from the computed result are most likely compared with the experimental visualized image such as Schlieren photographs and Shadowgraphs because these experimental visualization techniques use the deflection of light due to the nonuniform density distribution in the flow field. However, density contour pots are not the same as such visualized images and thus the comparison is not strict although good qualitative comparison can be made for many cases.

We have recently proposed a computational method that simulates the visualization techniques used in the experiment as one of the important visualization tools for the computational fluid dynamics research[1]. Using this method, strict comparison with the experimental visual image becomes possible. In the present paper, some of the recent applications of the proposed method are given. These examples indicate the importance of comparing the same output both from the computational result and experimental result. Also presented in the paper is the analysis of the errors introduced in the method. Two types of errors may be introduced in the method. One is the error due to the discretized integration of the light path. It is shown that this error can be negligible. Second type of error is introduced by the assumption that the light deflection is linear function of density gradient. Nonlinear effect is discussed and the visualization image including this effect is also presented.

PRINCIPLES OF OPTICAL VISUALIZATION METHODS

The basis of optical visualization method in the experimental study is that the deflection rate of light is different at each local point because the velocity of light changes according to the density of media through which the lights passes through. Basically, it can be shown that the deviation of light is proportional to the local density gradient in its direction. Mechanism of experimental visualization techniques can be found in the text book such as Ref. 1. Based on these knowledges, a method to simulate the visualization techniques in the experiments can be formulated. The detail of the formulation has already been described in Ref. 2. Here, only a brief description of the theoretical feature is presented for Schlieren photographs, Shadowgraph patterns and Interferograms.

1) Principle of Schlieren photographs and its computational simulation

In two-dimensional flows, intensity of the image depends on the density gradient. The normalized intensity can be written as

$$I = F_n + F_0 \frac{\partial \rho}{\partial y} \qquad (1)$$

Here, F_n corresponds to the background intensity that appears where no density gradient occurs in the test section. F_0 is the sensitivity factor for the density gradient. This factor corresponds to the effectiveness of the density gradient. The intensity of the light on the image is proportional to the density gradient as is shown in this equation. In the case of color Schlieren photographs, hue plays the role for the intensity in black and white images. Color filters are usually made of three strips of filters being red, green and blue. The rate of mixture now is proportional to the density gradient instead of light intensity and the resultant color is linearly proportional to the color function of hue. Hue is a periodic function and generally red corresponds to 0 and 1, green to 1/3 and blue to 2/3. Here F_n is set to be 1/3 and F_0 is set such that $0 \leq H \leq 2/3$ when the sequence of the color of the filter becomes red, green and blue. The derivatives are numerically evaluated using finite difference approximation.

In the case of three-dimensional flows, the local density gradient should be integrated along the light path all the way through to the image screen to obtain the same image as the Schlieren picture in the experiment. This integration is also numerically carried out by assuming linear variation of the density within the five tetrahedra that compose one computational cell for the flow computation.

2) Principle of shadowgraph patterns and its computational simulation

Shadowgraph technique is similar to the Schlieren Photograph. The major difference is that no knife edges are used and the screen is placed out of the focal point. Thus, the light does not come to the same point on the screen when the light is deflected. The brightness on the screen is now not uniform when the deflection of the light is not constant even without knife edges. Here, the intensity I is proportional to the deviation of the deflection of the light. In other words, the intensity (or brightness) is proportional to the Laplacian of the density.

$$I = I_0[1+F_s(\frac{\partial^2 \rho}{\partial x^2} + \frac{\partial^2 \rho}{\partial y^2})] \qquad (2)$$

F_s is a parameter determined by the distance between the screen and the focal point and is set to be negative. Note that the image could be distorted since the position of the image on the screen does not exactly correspond to that in the test section. Similar integration process as is in the simulation of Schlieren photograph is necessary for the three-dimensional shadowgraph simulation.

3) Principle of interferogram and its computational simulation

The principle of interferogram is that the shift of the stripes is proportional to the difference of density. In the finite fringe, the drift of stripes ΔN can be written as

$$\Delta N = N_s(x,y) + F_i(\rho/\rho_s - 1) \qquad (3)$$

where N_s is the stripes of the uniform flow, and ρ_s is the density at the reference area. F_i is the sensitivity parameter of the interferogram and denotes the correspondence of the density difference and the stripe width. The brightness of the final image corresponds to the fraction of ΔN. Again, in the case of three dimensional flows, numerical integration is necessary to create the final image.

APPLICATION EXAMPLES

First application example is the simulation of three-dimensional Schlieren photograph. In Fig.1(a), the Schlieren image created by the present method from the Navier-Stokes solution of the supersonic flow over an experimental wing-fuselage model is presented. The model is called AGARD-B model that is frequently used for the calibration of the wind tunnels. Freestream Mach number is 2.2 and the angle of attack is 10 degrees. The Schlieren pictures obtained in the experiment at the ISAS (Institute of Space and Astronautical Science) wind tunnel is shown in Fig. 1(b). Both original pictures are colored although it does not come out on the manuscript unfortunately. All the essential flow features such as shock wave from the nose, shock wave from the wing leading edge, expansion waves at the tail edge of the fuselage and rip shock wave at the sting region are well simulated. Important to note is that color (brightness in the manuscript) of the oblique shock wave from the apex of the fuselage is not the same in these figures especially in the region away from the fuselage. This is due to the lack of the grid resolution in the flow simulation. Since the Schlieren picture reflects the density gradient, the result indicates that the flow derivative is not well evaluated quantitatively at the shock wave in the flow simulation although the shock location is well simulated. This would never be noticed either in the black-and-white line contour plots or color image of density itself.

Second example shown in Fig. 2(a), (b) and (c) is shock-induced combustion in front of the blunt body. The flow field is axisymmetric and the wavy reaction front is observed in addition to the bow shock wave[3]. Density contour plots in the symmetry plane which is usually used for the comparison with the experiment is shown in Fig. 2(a). Figure 2(b) is the image obtained by the present method. Here the image is obtained by the light-path integration through the flow field. Figure 2(c) is the experimental shadowgraph pattern. In the experiment, so-called striation (straight lines shown between the reaction front and the body) is observed. On the other hand, in Fig. 2(a), no striations are observed. However, it appears in Fig. 2(b). This is a good example to demonstrate the importance of the comparison of the same output from the computation and the experiment. For relatively simple flow field such as this, it might be possible to imagine the integrated image from the computational result, but it is difficult and dangerous to do so for the complex three-dimensional flow field. Even the wrong interpretation might be deduced from the computational output.

Finally, simulation of the interferogram is presented in Fig. 3(a) and (b). These are one of the instantaneous images of the shock wave propagation through the curved duct. Small discrepancy exists in the vortical structure because the computational simulation is based on the Euler equations[4]. Same density increment level is used for both results and thus the quantitative comparison can be made.

DISCUSSIONS

As the interpolation and numerical integration are required to create the simulated image for three-dimensional flow fields, there may be accumulated errors. Here accuracy of the numerical integration is evaluated. Note that the error discussed here is different from the error of flow simulations. Therefore, a certain density distributions (which are not physical) are given analytically and the final images that are created by the analytical and numerical light-path integration are compared. Only the result is presented here in Figs. 4(a) and (b). The detail will be found in Ref. 5. Figure 4(a) shows the color Schlieren photograph of analytically integrated density gradient and Figure 4(b) shows the same photograph of numerically integrated density gradient. No difference is visible and the errors due to light-path integration is negligible at least in this particular case.

In the previous section, light path is considered to be straight under the assumption that the deviation of density and thus the deflection angle is small. The deflection of light path due to the secondary effect that light path can be bent can be analytically evaluated and the result indicates that the deflection of light path is in the order of β which is a non-dimensional parameter according to wavelength. Since β is in the order of 10^{-4} , this effect is negligible. However, this introduces the drift of focal point on the screen in Schlieren photograph when the light is deflected not linearly. As a result, the effect similar to the shadowgraph occurs on the Schlieren image on the screen. In the case of color Schlieren photograph, the brightness should be same on the image. If non-linear effect introduces the shadowgraph effect, the light intensity changes and the brightness becomes different locally. Figures 5(a) and (b) are the comparison of the ideal Schlieren image and non-linear Schlieren image for the supersonic intake. As can be seen, non-linear deflection of light induces the dark region in front of the oblique shock waves. We often see such effect on the experimental Schlieren images. The cause of such darkness near the shock wave could not be the out-of-range of the color filter but be the shadowgraph effect. Most important non-linear effect appears as a shadowgraph effect on the Schlieren image on the screen.

The results shown in the previous section indicate that the present method can be a good engineering tool for the validation and verification of the computational results. At the same time, missing features may come out on the images in the present method, which is useful for the better understanding of the flow fields. Line contour plots have been used for displaying the numerical results for long time before graphic workstation appeared. Now, people tend to show color image of the functions (such as density and pressure) that have been used for the line contour plots. It is important to note that color image of the function is different from the line contour plots. Line contour plots essentially show where gradient of the flow is large or small. If there is no gradient such as uniform flow, no contour lines are plotted. The level of the function (high pressure or low pressure, for instance) is not shown on the line contour plots. Color image of the function, on the other hand, shows the distribution of the absolute level of the plotted function.

Color image of the function is not necessarily better than the line contour plots. In the case of shock waves, line contour plots are better for the recognition of the strength of the shock wave rather than the color image of the function although the absolute level of the function can not be recognized from the plots without the labels of contour levels. It is important to consider what is good to draw on the screen for the purpose of fluid dynamics research. The method presented here to simulate the experimental visualization techniques is one example. With the operators such as Laplacian and certain integrals, feature of the flow field may appear on the screen. Finding the singular points in the flow field would show the stagnation points and else.

This will be more important when the amount of data obtained in the computational flow simulation is increased. Think about the three-dimensional unsteady flow simulation. One way for the plot is to take several planes and show the flow function on those planes timewisely on the screen. We have to decide in advance which planes are good to show for the understanding of the flow field. We may miss important flow features if such choice is not good. We need the visualization software that tells us where to plot, what to plot and so on. This is going to be a future research in the computational flow visualization.

CONCLUSIONS

A newly-developed computational method that simulates some of the flow visualization techniques, such as Schlieren photograph in compressible flow experiments was applied to several flow fields. By applying this method to the numerical data from the computational fluid dynamics simulation, same visual output as what is obtained in the experiment can be produced and the computational and experimental results can be fairly compared. Some of the flow features that are observed in the experiment but do not appear in the density contour plots of the computational results appear in the visual output in this method. It was shown that present method allows the correct comparison of the numerical data with the visual output of experiments and is useful for the validation of the computational approach. Non-linear effect of the light deflection is evaluated to clarify the errors in the present method and it is shown to have no predominant effect on the visualized image. The error analysis presented in this paper is also useful for understanding the visualized image obtained in the experiments.

REFERENCES

1. Tamura, Y. and Fujii, K.: Visualization for Computational Fluid Dynamics and the Comparison with Experiment, AIAA Paper 90-3031 (1990).
2. Liepmann, H. W. and Roshko, A : Element of Gas Dynamics, John Wiley, New York, 1956, pp. 153-179.
3. Matsuo, A. and Fujiwara, T.,: Numerical Investigation of Oscillatory Instability Mechanism in Shock-Induced Combustion around an Axisymmetric Blunt Body, AIAA Paper 91-1414 (1990) to appear as AIAA Journal article.
4. Fursenko, A. A. et al: Numerical Simulation of Propagation of Shock Waves through Channel Bends using TVD and ENO schemes, Proc. ISCFD Davis, Sept., (1991).
5. Tamura, Y. and Fujii, K.: Numerical Simulation of Experimental Visualization for Computational Fluid Dynamics Research, to appear in ISAS Report.

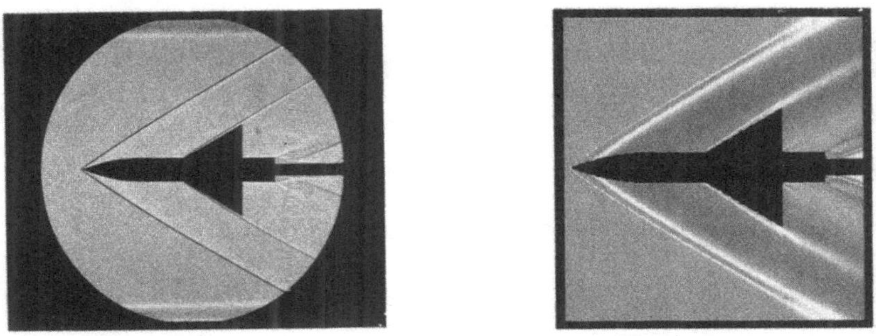

(a) Computational Schlieren photograph

(b) Experimental Schlieren Photograph

Figure 1 Supersonic flow over AGARD-B experimental wing-fuselage model.

(a) Density contour plots (b) Computational Shadowgraph (c) Experimental Shadowgraph

Figure 2 Shock-induced combustion in front of the axisymmetric blunt body.

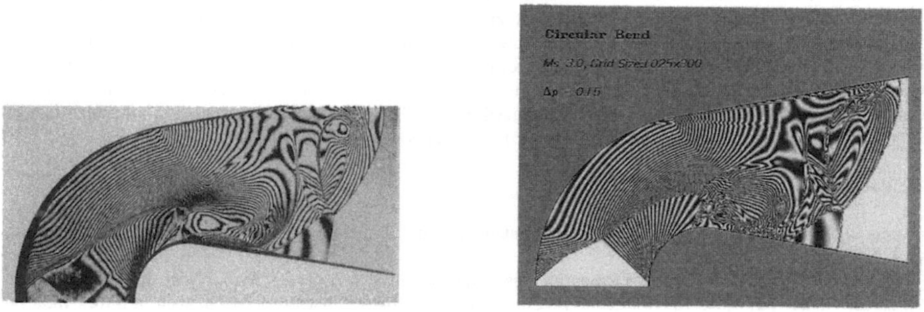

(a) Interferogram pattern in the experiment (b) Computational Interferogram pattern

Figure 3 Shock wave propagation through a curved duct.

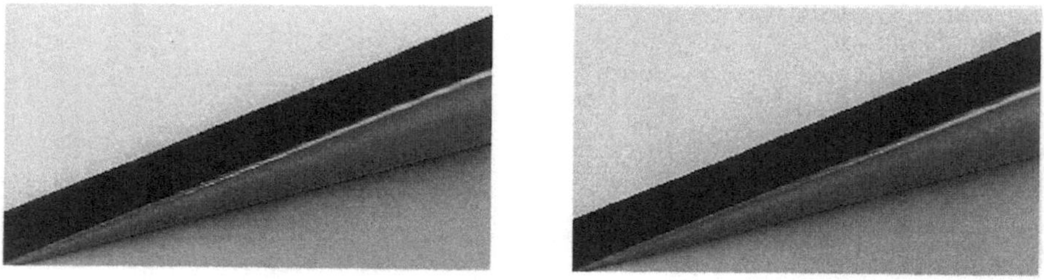

(a) Schlieren image by the analytical integration (b) Schlieren image by the numerical integration

Figure 4 Supersonic flow over a cone with zero degrees angles of attack.

(a) Simulated color Schlieren: linear evaluation (b) Simulated color Schlieren: nonlinear evaluation

Figure 5 Supersonic flow through two-dimensional intake duct.

Visualization of Molecular Motions by MD Method

Y. Matsumoto, J. Matsui and H. Ohashi

Department of Mechanical Engineering
The University of Tokyo
Hongo, Bunkyo-ku, Tokyo 113, Japan

Summary

Various flows in micro structure have significant influences on the performance of so-called micro machines, the manufacturing processes of semiconductors, and so on. In some cases, such flows are very sensitive to the boundary conditions on the solid wall, namely, the behavior of gas molecule near the surface. In this paper, molecular motions are visualized by the molecular dynamics methods, where the substance is modeled as an aggregate of particles that simulate the molecules. A solid thin film, which is consisted of monatomic molecules, is formed and then a monatomic gas molecule has collide on the surface. The molecular motions are well observed by this visualization and the numerical results reveal that the scattering behavior of the gas molecule is neither specular nor diffuse reflection.

Introduction

Flow in micro structure plays an important roles in the various situations [1], especially in the manufacturing processes of semiconductors [2]. In some cases, such flows are influenced very much by the boundary conditions on the solid wall, namely, the behavior of gas molecule near the surface. This gas-surface interaction is also significant for rarefied gas flows, such as flow around a space vehicle and flow between hard disk and disk head. Although considerable number of researches have been done [3,4,5], the micro-scale energy and momentum transfer in gas-surface interface has not been understood enough yet.

In this paper, molecular motions are simulated and visualized by the Molecular Dynamics method, where the substance is modeled as an aggregate of particles that simulate the molecules. A platinum thin film is formed and then a xenon gas molecule has collide on the surface. The molecular motions are visualized by the motion pictures and the numerical results reveal that the scattering behavior of the gas molecule is neither specular nor diffuse reflection.

Governing Equations and Numerical Method

The interaction between gas molecule and solid surface is simulated by Molecular Dynamics method. The equation of motion of molecule is written as,

$$m\frac{d^2\mathbf{x}_i}{dt} = -\sum_{i\neq j}\frac{\partial U(r_{i,j})}{\partial \mathbf{x}},$$

where m is the mass of molecule, $\mathbf{x}$ is the position of it, $r_{i,j}$ is the distance between two molecules, i and j are molecular index, and $U(r)$ is the interaction potential. The interaction potential between solid and gas is assumed to be the Morse potential,

$$U(r) = \varepsilon[\exp\{-2\sigma(r - r_0)\} - 2\exp\{-\sigma(r - r_0\}].$$

The potential between the solid molecules is assumed to be the Lennard-Jones inverse 12-6 power potential,

$$U(r) = 4\varepsilon\{(\sigma/r)^{12} - (\sigma/r)^6\}.$$

The parameters for the potentials are shown in table 1. The above equation of motion is integrated numerically, using multiple time-step method [6].

Table 1 Interaction potential parameters

	σ	r_0	ε
Xe - Pt	1.05 [$Å^{-1}$]	3.20 [Å]	0.0275 [eV]
Pt - Pt	2.523 [Å]	—	0.325 [eV]

The simulation is performed according to the following procedures:
(1) Thin platinum film is formed, which has (111) surface and is in thermal equilibrium. The thickness of the film is 6 molecular layers in z direction. The periodic condition is applied to the x and y directions of the model solid which are consisted of 864 molecules. The molecules are located to form a face-centered lattice. The kinetic energy which corresponds to a certain temperature is given to each molecule. The direction of the velocity is chosen randomly. The solid temperature is controlled to be the certain temperature.
(2) The gas molecule has collide onto the clean solid surface. The parameters, the initial translational kinetic energy and the incident angle are varied. The gas molecule is started far from the surface with certain kinetic energy and incident angle. The location from which the molecule is started is chosen randomly. The calculation for one case is continued for 10,000 steps. The locations of gas molecule and solid molecules are recorded on the memories from which motion pictures are produced.
(3) After the collision of gas molecule, the solid condition is set back before the collision and the next gas molecule is started.

Numerical Results

Trajectories of molecules of gas and solid are shown for several cases in the figures. The temperature of solid is set to 300[K], and other simulated conditions are shown in table2. Only a part of solid molecules are shown in the figures. The larger circle shows the molecules in the first layer, and the smaller shows those in the second layer. The locus of the gas molecule is shown with the gas molecule. The molecular motion is observed mainly from bird's-eye view.
Figure 1 shows the molecular motions for the case 1, where the gas temperature is 2967K and the incident angle is 40 degree. In the case where the initial gas temperature is high,

Table 2 Condition of calculation

case	Initial Gas Energy	Incident Angle [degree]
1	0.383 [eV] (2967 [K])	40.0
2	0.383 [eV] (2967 [K])	0.0
3	0.076 [eV] (592 [K])	40.0
4	0.014 [eV] (108 [K])	20.0

the gas molecule tends to reflect immediately from the surface. Two runs are shown from the bird's-eye view and the top view. In each run, the normal speed of the molecule changes rather gradually at the rebounds on the surface and also the tangential speeds in two directions along the surface are changed.

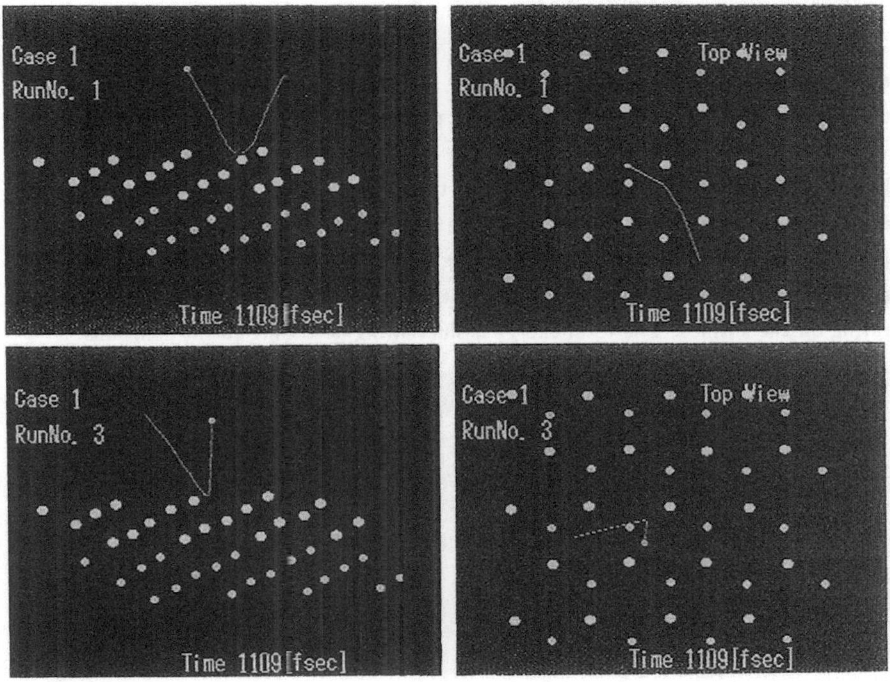

Fig.1. Traces of gas and solid molecules in case 1.

The molecular motions in the case 2 are shown in Fig.2 at 508 fs and 1109 fs. When the gas molecule collides just on the solid molecule, the solid molecule is observed to move downward sharply. Considerable amount of energy is transfered from the gas molecule to the solid one. The gas molecule reflects perpendicularly from the surface.

In the case 3 where the gas temperature is rather low, the gas molecules is accelerated to approach to the surface and then it rebounds. The molecule rebounds a few times and it

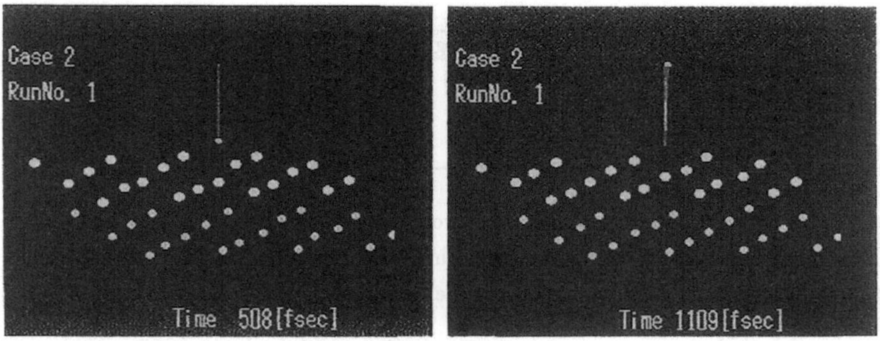

Fig.2. Traces of gas and solid molecules in case 2.

departs from the surface according to the solid molecular motions as shown in Fig.3. In Run 1, the gas molecule rebounds twice. The position of the second rebound is higher than the first one because the gas molecule has higher energy at the first collision.

In the case 4 where the gas temperature is low, the gas molecule is accelerated very much toward the solid surface and the gas molecule collide almost perpendicularly with the surface. The molecular motion is shown in Fig.4 from the bird's-eye view and the side view. Fairly amount of total energy is lost at the first collision and the gas molecule is migrating on the surface. The total energy is fluctuating according to the collisions.

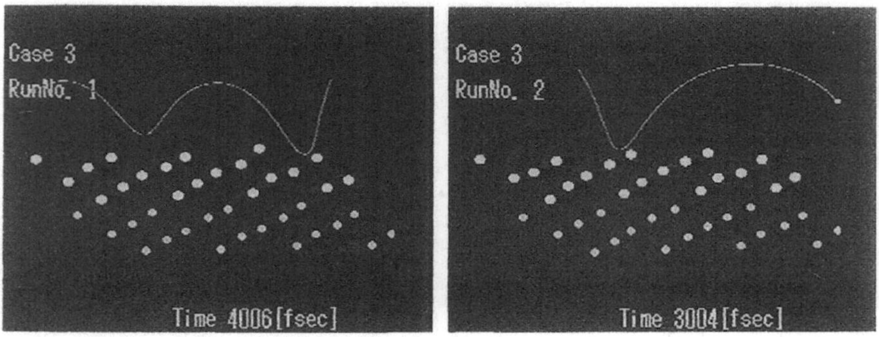

Fig.3. Traces of gas and solid molecules in case 3.

Concluding Remarks

The behavior of gas molecule, which collides on the solid surface is simulated by Molecular Dynamics method and visualized. The molecular motions are observed in detail and the characteristics of the gas-surface interactions are understood. The numerical results reveal that the scattering behavior of the gas molecule is neither specular nor diffuse reflection.

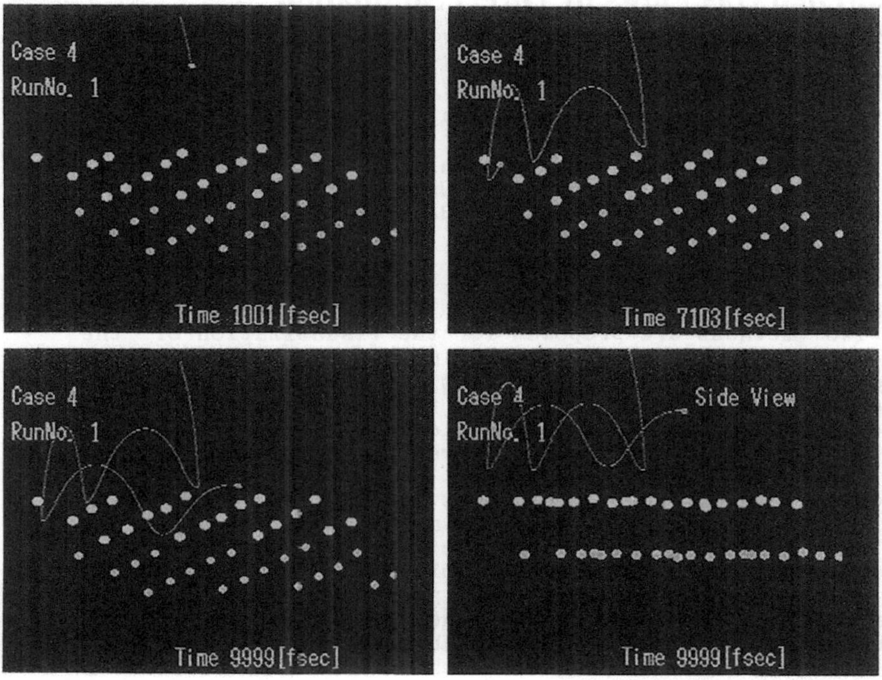

Fig.4. Traces of gas and solid molecules in case 4.

References

[1] Trefethen,L.M., *et al.*; "Some problems and Examples in Micro Fluid Dynamics" Forum on Micro Fluid Mechanics -1991-, ASME,FED-Vol.113,pp.29-42,1991

[2] Ikegawa,M and Kobayashi,J.; "Deposition Profile Simulation Using the Direct Simulation Monte Carlo Method" J. Electrochem. Soc.,136(10),pp.2982-2986,1989

[3] Oman,R.A., *et al.*; "Interactions of Gas Molecules with an Ideal Crystal Surface", AIAA Journal, Vol 2, pp.1722-1730,1964

[4] Cercignani,C. and Lampis,M.; "Kinetic models for gas-surface interactions ", Transport Theory and Statistical Physics,1(2),pp.101-114, 1971

[5] Goodmann, F.O. and Wachman, H.Y.; *"DYNAMICS OF GAS-SURFACE SCATTERING"*, Academic Press, 1976

[6] Streett,W.B., *et al.*; "Multiple time-step methods in molecular dynamics" , Mol. Phys., Vol 35, p639,1978

Numerical Visualization of Molecular Motion in Rarefied Gas Flows by Direct Simulation of Monte Carlo Method

Masahiro OTA

Department of Mechanical Engineering,
Tokyo Metropolitan University,
1-1,Minami-Oosawa, Hachi-Oji,
Tokyo,192-03,
JAPAN.

Summary

Two methods are proposed for the visualization of the gas molecule motion by Direct Simulation of Monte Carlo method. By the first one,the sizes and diameters of the gas molecules are neglected, and all the positions of the gas molecules are traced.Then,the loci of the molecular positions are visualized by drawing continuously the positions. By the second one, the diameters of the molecules are enlarged into about 10^7 times, comparing with the real diameters of the gas molecules . Then,the loci of the molecular positions are visualized by drawing sequentially expressions of the molecular positions at an each time step. Therefore the positions are expressed discretely. The motion of the gas moleclues in a computational cell are visualized by the first method. The traced curves are dependent with Knudsen number. With decreasing the Knudsen number,the curves fold a number of times. The rarefied gas flows through a circular tube installed a cicular disk, held perpendicularly and coaxially with the tube are visualized by the second method.Very few molecules intrude into the rear region of the disk.Rarefied gas flows through an thruster nozzle for an artificial satellite are simulated by the DSMC method.

Direct Simulation of Monte Carlo Method[1,2,3]

The Direct Simulation of Monte Carlo method (hereafter, abbreviated by DSMC method)applied to the molecular gas dynamics requires the solution of a transient problem wherein the initial condition is an appropriate disribution of gas molecules OF and the desired final state is a steady state. Since the steady state solution is the desired result, the initial condition should be chosen to permit the most rapid approach to the steady state. In addition, an efficient algorithm should be used to detect the onset of the steady state.

The steady state detection algorithm can be divided into three consecutive stage. In the first stage, the transient flow is simulated for a period of time equal to that which is required for incoming molecules to transverse the calculation-domain at a free-stream velocity with the stagnation condition at inlet boundaries. The transient simulation continues, in the second stage, until the monotonic drift in the number of gas molecules within the domain ceases.The number of the incoming gas molecules from the inlet boundaries minus the number of outgoing gas molecules to outlet boundaries is the number of gas molecules within the computational domain.

The accurate sampling of inter-molecular collisions is essential to DSMC simulations. Trasitionally, the time counter technique[1] has been used to simulate efficiently the collision processes. However, the technique introduces significant statistical errors as a result of an insufficient number samplings per time interval. Therefore, in this paper, the null collision technique[3] is used.The schematic flow chart of the DSMC method is shown in Fig.1.

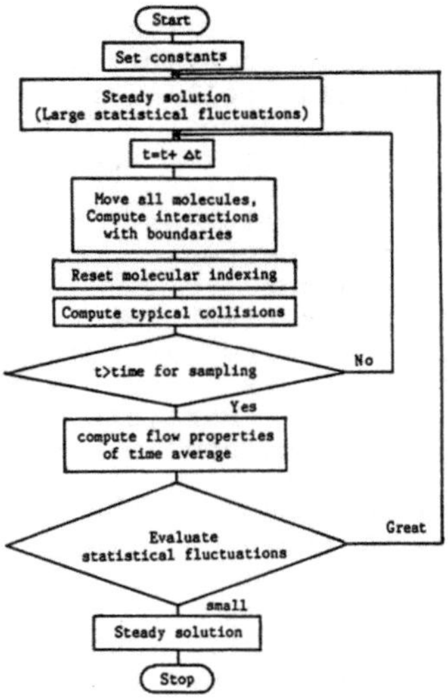

Fig.1 Schematic Flow Chart of Direct Simulation Monte Carlo Method

Results and Discussion

The motions of the gas molecules in one computational cell for the DSMC method are visualized in Fig.2 by the first method, the Traced Method.By the Traced Method the sizes and the diameters of the gas molecules are neglected, and all the positions of the gas molecules are traced even if the collisions of the molecules themselves are occured by the DSMC method.Then,the loci of the molecular positions are visualized by drawing continuously the positions on the CRT and so forth.The cell boundaries are made of solid walls.The gas molecules collided against the cell boundaries are reflected with a diffused reflection condition. The number of gas molucules is 100.The gas molecules do not have macroscopic velocities,only at first with Maxwellian microscopic velocity distribution.Therefore,the gas molecules collide against the solid walls and with the gas molecules themselves. The traced curves are dependent with Knudsen number. The Knudsen number is defined as the ratio of the mean free path length of the gas molecules to the length of the cell.With decreasing the

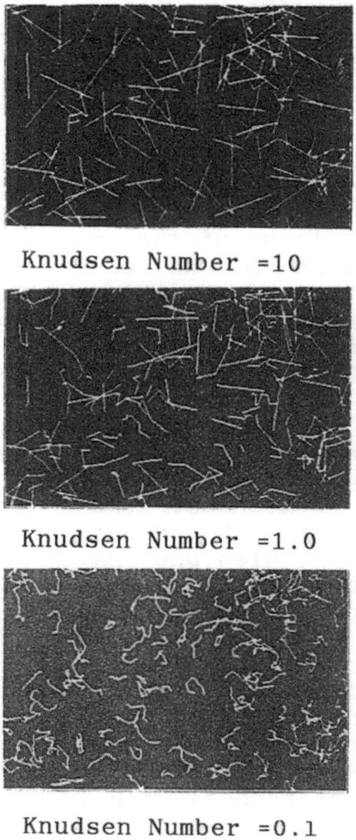

Fig.2 Numerical Visualization of Molecular Motion by Traced Method

Knudsen number, the curves fold a number of times. Even if the Knudsen number is 10, the molecular collisions occur but the number of the collisions is very small. It is also visualized that the molecular collisions increase with the decrease of the Knudsen number.

The rarefied gas flows through a circular tube are visualized by the second method, the Particle Method with DSMC method in Fig.3. The time steps by the DSMC method in Figures 3a,3b,3c,and 3d are 10 times,50 times, 100 times ,and 200 times of the mean free time of the gas molecules, Δt,respectively. The Knudsen number defined as the ratio of the mean free path length of the gas molecules with a stagnation condition at an inlet boundary to the diameter of the tube is 0.05.By the Particle Method,the diameters of the molecules are enlarged into larger than 10 times, comparing with the real diameters of the gas molecules molecular positions are shown by drawing sequential expressions of the molecular positions at an each time step. Therefore, the positions are expressed discretely.Some of the gas molecules flowing through the tube are impinged on a circular disk in the tube, held perpendicularly and coaxially with the tube. Very few molecules intrude into the rear region of the disk. Therefore, the number of the gas molecules per unit area behind the disk are more smaller than the number at other regions even if gases flows with macroscopic velocities.Very lower pressure regions exist behind the obstacles in the gas-flows.

Rarefied gas flows through a thruster nozzle for an artificial satellite are simulated by the DSMC method.

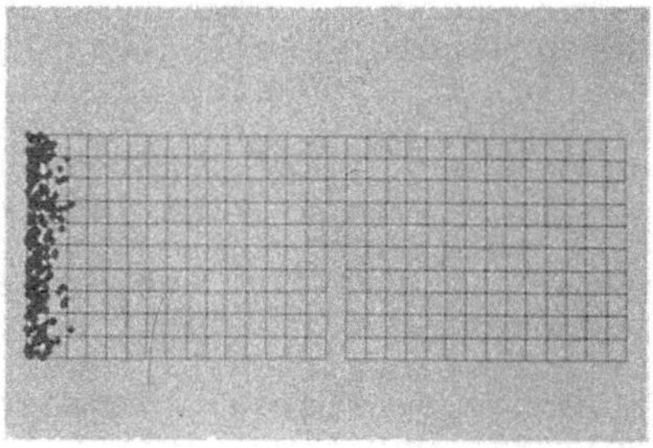

Fig.3a 10*time step

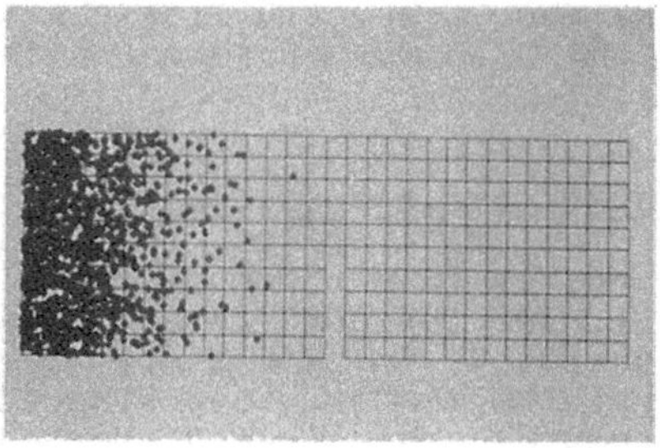

Fig.3b 50*time step

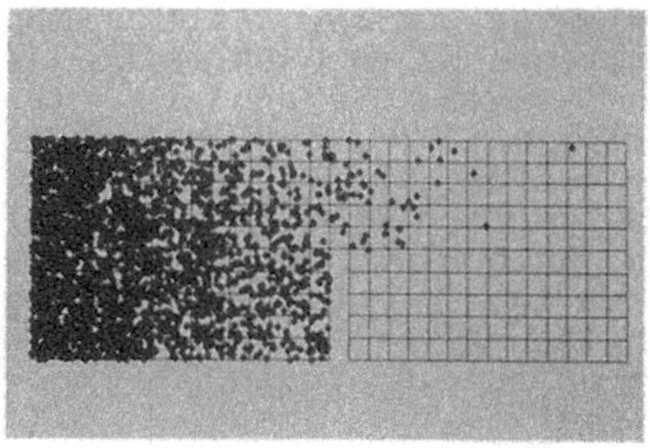

Fig.3c 100*time step

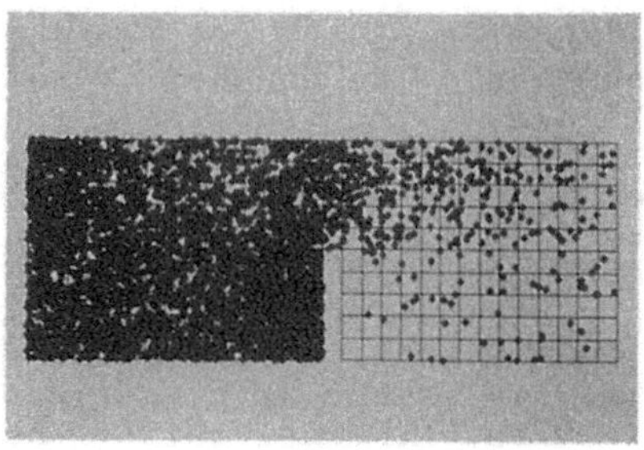

Fig.3d 200*time step

Fig.3 Numerical Visualization of Molecular Motion by Particle Method

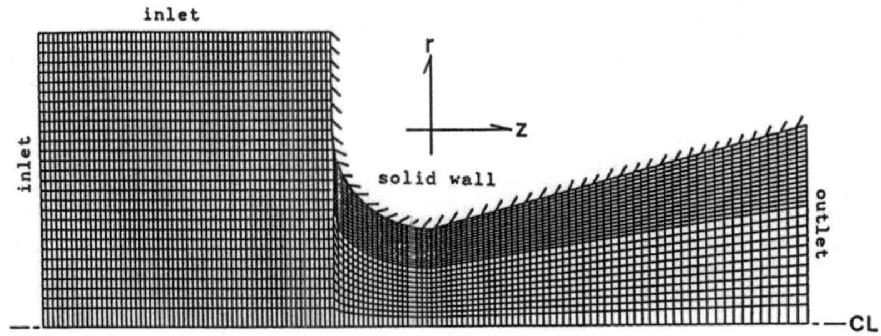

Fig.4 Cell Structure in Thruster Nozzle

The computational cell structure are shown in Fig.4. At close region to the inner surface of the nozzle, fine cell structures are used, since the macroscopic physical quantities are changed drastically. The hard sphere gas model is used for a mono-atomic gas molecule,argon. The molecules are incoming from inlet boundaries with a thermal velocity at an equilibrium condition and outgoing to outlet boundaries. Macroscopic axial velocity distibutions in a radial direction inside of the nozzle are shown in Fig.5a and 5b.The Knudsen numbers in Fig.5a and 5b are 0.025 and 0.25,respectively. With decreasing the Knudsen numbers, macroscopoic velocity distributions with non-constant velocities are built up close to the inner surface of the nozzle. The macroscopic velocities are not zero on the nozzle surface. Fig.6a and 6b show Mach number and number density distributions on the horizontal axis of the nozzle, respectively. The solid curves in the Fig.6 are the theoretical curves by one-dimensional invisid gas dynamics theory. With decreasing the Knudsen numbers the DSMC results approaches to the theoretical curves , but at the divergent region in the nozzle the deviation between the theory and DSMC results are great.

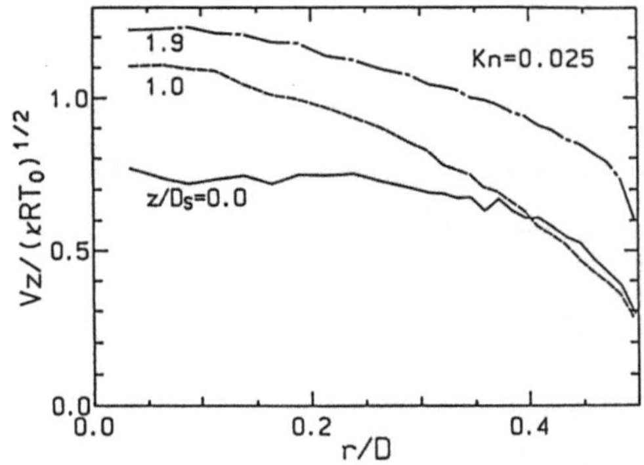

Fig.5a Knudsen Number=0.025

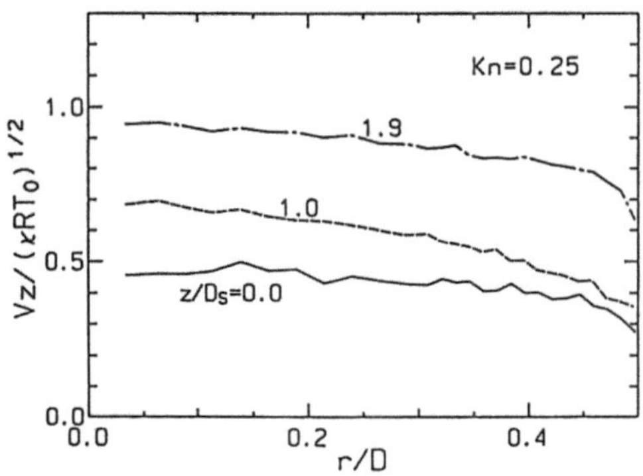

Fig.5b Knudsen Number=0.25

Fig.5 Effects of Knudsen Number on Axial Velocity Ditributions

Concluding Remarks

The motion of the gas molecules in rarefied gas flows is numerically visualized by sequential expressions of molecular positions at an each time step with the DSMC method. Two methods are proposed for the expression of the gas molecule motion. The motion of the gas moleclues in one computational cell are visualized by the first method. With decreasing the Knudsen number,the curves fold a number of times. The rarefied gas flows through a circular tube installed a cicular disk, held perpendicularly and coaxially with the tube are visualized by the second method.Very few molecules intrude into the rear region of the disk.Rarefied gas flows through a thruster nozzle for an artificial satellite are simulated by the DSMC method.

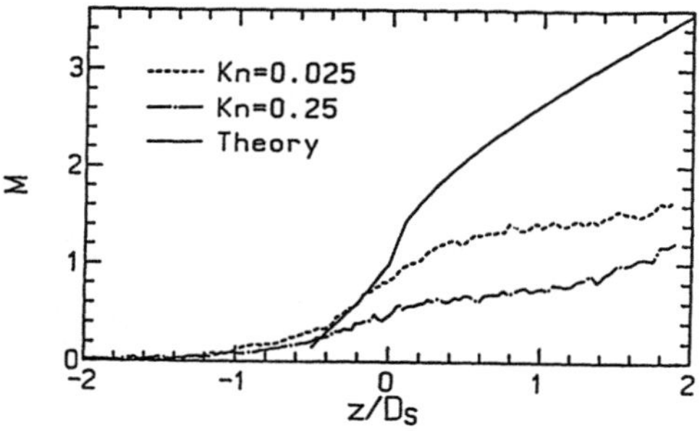

Mach Number Distributions

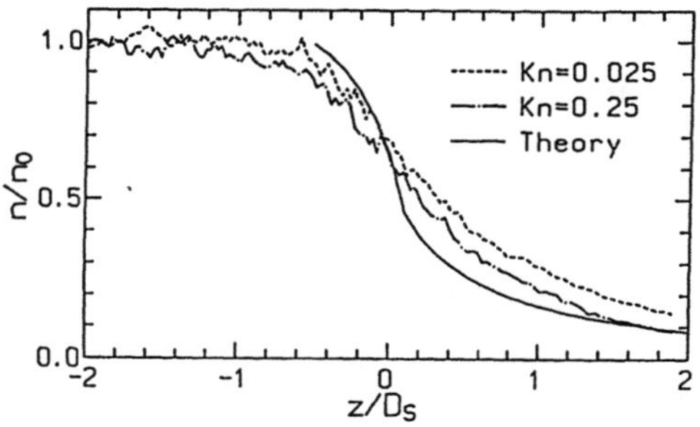

Number Density Distributions

Fig.6 Mach and Number Density Distributions on Horizontal Axis

References

1. Bird,G.A.: Molecular Gas Dynamics, Clarendon Press, Oxford,1976
2. Nanbu,K. : Proceedings of the 15th International Symposium on Rarefied Gas Dynamics, B.G.Teubner Stuttegart, Vol.1(1986) 369-383.
3. Koura,K. : Physics of Fluids,29(1986),3509-3511.

Three-Dimensional Computation of a Fluid Flow with the Method of the Lattice Boltzmann Equation

G. ALFONSI

Dipartimento di Ingegneria Idraulica,
Ambientale e del Rilevamento, Politecnico di Milano
Piazza Leonardo da Vinci 32, 20133 Milano, ITALY

Summary
A computational analysis is performed with the use of the method of the Lattice Boltzmann Equation (LBE). The case of a viscous fluid moving in a square duct at low values of the Reynolds number is considered and the calculations are compared with results of theoretical nature. The main characteristics of the new computational technique are also presented.

Introduction

It has recently been shown ([1],[2],[3],[4] among others) that it is theoretically possible to compute the flow field generated by a viscous incompressible fluid, by using a novel technique, generally known as Lattice Gas Automata (LGA). The method is based upon the idea of reproducing the behaviour of a viscous fluid by considerig the movements and the related collisions of a large number of discrete entities - the fluid particles - in the nodes of a lattice in which the flow field is ideally discretized. The particles are characterized by discrete properties, such as mass and velocity, and the microscopic collision rules must satisfy locally the mass and momentum conservation principles. An exclusion principle holds, so that, at each node, no two particles can have the same velocity.
Starting from a microdinamic point of view, it is possible to recover non linear partial differential equations with the same structure of the momentum conservation equations of fluid dynamics - the Navier-Stokes equations - in the incompressible limit, usually obtained by considering the hypotesis of fluid as a continuum and imposing the conservation principles. This result becomes possible by following an approach analogous to the one of the kinetic theory of gases, in particular for what the Chapman-Eskog functional expansion is concerned. Moreover, the method of the Lattice Boltzmann Equation (LBE) ([5],[6] among others), represents an even more recent development of the still novel Lattice Gas Automata theory, and it is mainly characterized by the fact that the Boltzmann transport equation is solved directly in its finite difference form, by noting that this equation gives an a priori average description of the lattice gas behaviour.

On the basis of the characteristics of the computational technique, the related computer codes are particularly suitable to be implemented on multiprocessor parallel computers and the calculations to be executed by following the parallel computing procedures.
In this paper, results of fluid flow calculations performed with the method of the Lattice Boltzmann Equation are presented. The computational analysis is performed in three dimensions and the case of a viscous incompressible fluid moving in a square duct is considered. The computer that has been used is a fine grain multiprocessor Meiko Computing Surface, with 20 processors Transputer T-800 with 2 or 4 Mbyte of memory each.

Lattice gas methods

Any lattice gas model is characterizeed by the rules according to which the model itself evolves. In general

$$n_i(t+1, \mathbf{r}+\mathbf{c}_i) = n_i(t, \mathbf{r}) + \Delta_i(\mathbf{n}) \tag{1}$$

where

$$\Delta_i(\mathbf{n}) = \sum_{s,s'} (s_i' - s_i)\, \xi_{s,s'} \prod_{j=1}^{b} n_j^{s_j} (1-n_j)^{(1-s_j)} \tag{2}$$

is the collision function. Since in each collision mass and momentum are conserved at each node, one has

$$\sum_{i=1}^{b} \Delta_i(\mathbf{n}) = 0 \tag{3}$$

$$\sum_{i=1}^{b} c_i\, \Delta_i\,(\mathbf{n}) = 0 \tag{4}$$

for each possible state **n**. As a consequence the conservation laws for the Boolean field can be written from equation (1) as

$$\sum_{i=1}^{b} n_i(t+1, \mathbf{r}+\mathbf{c}_i) = \sum_{i=1}^{b} n_i(t, \mathbf{r}) \tag{5}$$

$$\sum_{i=1}^{b} \mathbf{c}_i n_i(t+1, \mathbf{r}, \mathbf{c}_i) = \sum_{i=1}^{b} \mathbf{c}_i n_i(t, \mathbf{r}) \tag{6}$$

and the microdynamical equations can be used to study the process of evolution of a lattice gas model. In order to do that, a probabilistic point of view has to be adopted. The average macroscopic quantities can be calculated by means of the single particle distribution function; in particular, the total particle density n is given as

$$n = \sum_{i=1}^{b} f_i \tag{7}$$

while the average fluid velocity **u** is defined as

$$nu = \sum_{i=1}^{b} c_i f_i \quad . \tag{8}$$

The evolution in time and the variation in space of the distribution function are determined by the motions of the particles and the related collisions. This contributions can be kept into account by means of a proper master equation, which, in the absence of collisions, has the form

$$f_i\,(t+1, r+c_i) = f_i(t,r) \tag{9}$$

which is a finite difference equation and can be transformed in differential form in order to obtain a continuum macroscopic description. By performing a Taylor expansion, one has

$$f_i(\tau+\epsilon, x+\epsilon c_i) = f_i(\tau, x) + \epsilon\partial_t f_i + \epsilon c_i \cdot \nabla f_i + \tfrac{1}{2}\epsilon^2\,\partial_{tt}\,f_i + \ldots\ldots. \tag{10}$$

and retaining only the first order terms, the collisionless Boltzmann equation for the lattice gas is obtained

$$\partial_t f_i + c_i \cdot \nabla f_i = 0 \quad . \tag{11}$$

Collisions however do contribute to determine the behavoir of the distribution function. By introducing a collision term in (11), one has

$$\partial_t f_i + c_i \cdot \nabla f_i = \Omega_i \quad . \tag{12}$$

and, whatever the exact form of the collision term in (12) is, the mass an momentum principles at the microscopic level imply that

$$\sum_{i=1}^{b} \Omega_i = 0 \tag{13}$$

$$\sum_{i=1}^{b} c_i \Omega_i = 0 \quad . \tag{14}$$

By following the kinetic theory approach, after some manipulations from equation (12) the following expression for the conservation of mass is obtained

$$\partial_t n + \nabla \cdot (nu) = 0 \tag{15}$$

The same procedure applied for what the momentum conservation principle is concerned, gives

$$\partial_t \left(\sum_{i=1}^{b} c_i f_i \right) + \sum_{i=1}^{b} c_i (c_i \cdot \nabla f_i) = 0 \tag{16}$$

in which the distribution function still appears.
A more extended review of Lattice Gas Automata theory and models can be found in [7] and [8].

Results

Calculations of velocity profiles developed by a viscous fluid moving in a three-dimensional square duct have been performed. Two cases have been examined: the square ducts rispectively of 8*8*8 (Re = 17.14) and of 32*32*32 (Re = 68.55). In both cases the Mach number value has been M = 0.28, which assures the incompressible character of the fluid. The nucleous of the program is the one published in [9] and a minimum degree of parallellism has been implemented on the Meiko computing machine. The computing technique reflects the theory of the Lattice Boltzmann Equation. The results of the calculations are reported in Fig.1. The calculated velocity profiles for the steady state condition in the center of the square duct, are plotted and compared with results of analytical nature, corresponding to the theoretical solution for the square duct reported, among others, in [10].

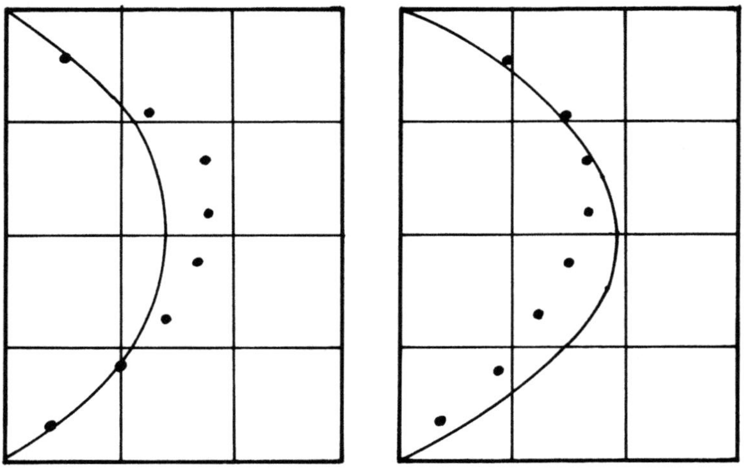

a) b)

Fig.1. Comparison between calculated velocity profiles (•••) and theoretical values (——) (steady state condition) in the case of the square duct of a) 8*8*8 and b) 32*32*32.

Conclusions

Numerical results obtained with the method of the Lattice Boltzmann Equation in the case of a viscous incompressible fluid moving in a three-dimensional square duct, are presented and compared with solutions of analytical nature. Two different cases are presented and the comparisons are shown when the steady state condition in the calculations is reached.

<u>References</u>

1. Clavin, P.; Lallenand, P.; Pomeau, Y.; Searby, G.: Simulation of free boundaries in flow systems by lattice gas models. Journal of Fluid Mechanics, 188 (1988) 437-464.
2. Frisch, U.; d'Humières, D.; Hasslacher, B.; Lallemand, P.; Pomeau, Y.; Rivet, J., P.: Lattice gas hydrodynamics in two and three dimensions. Complex Systems, 1 (1987) 649-707.
3. D'Humières, D.; Lallemand, P.: Numerical simulation of hydrodynamics with lattice gas automata in two dimensions. Complex Systems, 1 (1987) 599-632.
4. D'Humières, D.; Lallemand, P.; Frisch, U.: Lattice gas models for 3D hydrodynamics. Europhysics Letters, 2 (1986) 291-297.
5. Higuera, F.; Jimenez, J.: Boltzmann approach to lattice gas simulations. Europhysics Letters, 9 (1989) 663-668.
6. Mc Namara, G.; Zanetti, G.: Use of the Boltzmann equation to simulate lattice gas automata. Physical Revue Letters, 61 (1988) 2332- 2335.
7. Alfonsi, G.; Zampaglione, D.: Il metodo del gas reticolare nella meccanica dei fluidi incomprimibili. Atti XXII Convegno di Idraulica e Costruzicni Idrauliche, Cosenza (1990).
8. Alfonsi, G.; Mazzotti, M.; Zampaglione, D.: Modelli di gas reticolare e loro applicazioni in idrodinamica. Idrotecnica, 3 (1991) 223-230.
9. Foti, E.: Automi cellulari per la simulazione fluidodinamica nei mezzi porosi. Tesi di Laurea, Università di Catania (1989).
10. White, F.: Viscous fluid flow. McGraw Hill (1974).

Velocimetry by Image Processing

Particle Tracking Method

3D Particle Tracking Velocimetry-
Its Possibilities and Limitations

Themistocles Dracos and Nadeem A. Malik
Department of Hydromechanics and Water Resources Management,
The Swiss Federal Institute of Technology, ETH Zürich, CH-8093 Zürich, Switzerland.

1. Summary

Particle Tracking Velocimetry (PTV) is an evolving technique for 3D velocity field measurements in an observation volume. When the observation is made from a platform moving with the flow velocity it also yields Lagrangian trajectories of individual particles. It is one of the techniques which can cover the increasing need for measurements in a Lagrangian frame. The principles of the technique have been published in Papantoniou & Maas (1990) and Kasagi & Nishino (1991). Here some recent developments as well as the testing of the tracking scheme on the Kinematic Simulation Inertial Model (KSIM) test flow field will be presented.

2. Determination of 3D coordinates of particles

The PTV procedure developed at the ETH is similar to the one described by Kasagi & Nishino (1991). It consists of two steps. In a first step particle coordinates are determined by a photgrammetric method. Small neutrally bouyant particles seeded in a flowing liquid are illuminated and observed by two or more cameras. Details on particle size, number density of particles, illumination etc. are given in Kasagi & Nishino (1991), Maas *et al.* (1992) and Malik *et al.* (1992). To achieve a satisfactory accuracy a three camera system and good calibration of the sysytem are necessary. The coordinates of individual particles are in our case determined using the photogrammetric concept of epipolar lines (Maas (1992)). In the course of application of this method ambiguities arise which must be resolved if the yield of particle coordinates is to be high. One of the ambiguities is due to the overlapping of the images of individual particles. This problem has been solved by developing an anisotropic thresholding operator. An other ambiguity arises when correpondences of the coordinates of the particles in the three frames are established. It can be shown that for about 2000 particles in the observation volume, the number of ambiguities of this kind is reduced by an order of magnitude when three cameras instead of two are used and to nearly zero when four cameras are used. At the same time the accuracy of the coordinates is improved by a factor of about 1.4 when four instead of three cameras are used.

We call noise the spontaneous appearance or disappearance of particles due to unresolved ambiguities in the coordinate determination or to particles entering or due to leaving the observation volume through its boundaries.

3. The tracking procedure

An important parameter for tracking is the spacing-displacement ratio $p = \Delta_0/u'\Delta t$. Δ_0 is the average particle spacing $\Delta_0 = \rho_N^{-1/3}$, where $\rho_N = N/V$ is the particle number density and N is the number of particles in the observation volume V. The displacement is given by the rms turbulent velocity fluctuations u' and the imaging time step Δt. The tracking procedure used in our scheme is based on a four frame method, illustrated in figure 1. Let $[f]$ be the current frame and $[f-1]$ the previous frame. We assume that all possible links between particles in frame $[f-1]$ and $[f]$ have been established. $[f+1]$ is the next (search) frame to which links from particles in $[f]$ are now desired.

The essential idea of the PTV tracking scheme is to limit the search volume within which candidates for the continuation of particle trajectories are sought in the next time frame $[f+1]$. The tracking becomes easy when $p \gg 1$ and practically impossible when $p \ll 1$. Assuming that the turbulent velocity fluctuations are nearly Gaussian, the threshold between easy and difficult tracking is expected to be $p = 3$, since with 99.7% confidence particles will move less than $3u'\Delta t$ from their most current position.

Consider now a particle with 3D coordinate $\mathbf{x}_i^f$ in frame $[f]$ (figure 1). We define a correlation neighbourhood $\mathcal{N}_i^f$ of this particle in $[f]$. This neighbourhood is a sphere centred at $\mathbf{x}_i^f$ with radius $r_0 = F_r\lambda$ which is a fraction of the Taylor microscale λ. For $V^{1/3} \gg r_0$, the obvious choice is $F_r = 1$. For $V^{1/3} \approx O(r_0)$, $F_r = (n^{1/3}/\rho_N^{1/3})/\lambda$, where n is the number of particles sought in n_i^f. Next the expected location $\mathbf{e}_i^{f+1} \in [f+1]$ of the particle $\mathbf{x}_i^f \in [f]$ is determined. There are three possible cases:

(i) if the particle $\mathbf{x}_i^f$ is linked to a particle in frame $[f-1]$, say $\mathbf{x}_{i'}^{f-1}$, then its expected location in $[f+1]$ is the linear continuation $\mathbf{e}_i^{f+1} = \mathbf{x}_i^f + \mathbf{u}_i^f\Delta t$, and $\mathbf{u}_i^f = (\mathbf{x}_i^f - \mathbf{x}_{i'}^{f-1})/\Delta t$;

(ii) if the particle $\mathbf{x}_i^f$ has no previous link but n_j^f of the particles in $\mathcal{N}_i^f$ have; then $\mathbf{u}_i^f$ in (i) is replaced by $\overline{\mathbf{u}}_i^f = \left(1/n_j^f\right)\sum u_j^f$;

(iii) none of the particles in $\mathcal{N}_i^f$ are previously linked. $\mathbf{u}_i^f$ is replaced by $\overline{\mathbf{u}}_i^f = U_l - U_c$. U_c is the local averaged velocity of the flow and U_c is the carriage velocity of the imaging platform. This is always the case in the first step of the procedure.

The radius r_1 of the search neighbourhood $\mathcal{N}_i^f$ centred at $\mathbf{e}_i^{f+1}$ is related to the decorrelation of the particle velocity in the time Δt. It is obtained either (1) from the Lagrangian time structure function, viz (Malik *et al.* 1992) $r_1 = \left(A\pi C_L/3)^{1/2}\right)(\delta x_{max}/L)^{1/2}\,\delta x_{max}$. $\delta x_{max} = 3u'\Delta t$ is 3 times the standard deviation of the fluctuating movement of particles, $A \approx 1$, $C_L \approx 2.4$, and L is the Lagrangian integral length scale; or (2) from the mean nearest neighbour spacing of particles, which for a uniformly distributed set of particles of mean spacing Δ_0 gives $r_1 \approx \Delta_0/3$. r_1 is taken as the smaller of (1) and (2).

If only one particle is found in $\mathcal{N}_i^f$ an unambiguous connection is established. If more than one particle are found in $\mathcal{N}_i^{f+1}$ more than one connections can be made. To resolve the ambiguity we consider the movement of the particles in $\mathcal{N}_i^f$ in to the acceleration frame $[f+2]$. The expected location of these particles $\mathbf{e}_j^{f+2} \in [f+2]$ as well as the corresponding neighbourhoods r_2 are determined and the set $\mathcal{T}_i^f$ of all possible tracks $\mathbf{x}_i^f \rightarrow \mathbf{x}_j^{f+1} \rightarrow \mathbf{x}_k^{f+2}$ is formed. To this set of tracks are added all tracks that start from *any* particle $\mathbf{y}_i^f \in [f]$ and which intersect any of the links in the set $\mathcal{T}_i^f$. Then:

(i) if one of the links in these tracks coincides with the continuation of a track starting in frame $[f-1]$, then the track containing this link is taken as the true link;

(ii) if more than one track contains this link, then the track which shows the *minimum change in acceleration* is favoured as the true link.

Usually there remain some unlinked particles in $[f]$ and in $[f+1]$. If the search neighbourhood $\mathcal{N}_i^f$ of an unlinked particle in $[f]$ contains only one unlinked particle in $[f+1]$, the connection between the two particles is made.

The procedure decribed above is conservative; it is prefered not to make a connection rather than risk an erroneous connection.

4. Validation of the tracking procedure against simulated trajectories

A flow field was generated by the Kinematic Simualtion Inertial Model (KSIM) (Malik (1991), Fung *et al.* (1992)). The number of particles in each frame was fixed at $n = 1000$, and the spacing-displacement ratio was varied from $p = 2.31$ to $p = 7.5$. Noise was introduced by scrambling a per-centage of randomly chosen particle coordinates in each frame. Noise levels of $n = 0.0, 0.5, 1.0, 1.5, 2.0, 2.5, 3.0$ % were introduced in separate experiments. For complete details see Malik *et al.* (1992). Two per-centage yields of quantities were obtained: the absolute per-centage yield, i.e. the ratio of the total yield of a quantity to the maximum theoretically possible yield, times 100; and the relative per-centage yield (or quality), i.e. the ratio of the yield of the correct quantity to the total yield of the quantity, times 100.

First some results without noise are presented. Figure 2 shows the absolute and relative yield (quality) of links per frame as a function of p. Both yields are high. The absolute yield drops rapidly for $p < 3$ indicating that $p = 3$ is indeed a threshold between easy and difficult tracking. For $p > 3$ the absolute yield is higher than 98% and for $p > 5$ it is nearly 100%. It is remarkable that the quality is always very high. For the range of p investigated it is higher than 98% and becomes 100% for $p > 5$.

For Lagrangian statistics, it is necessary to have long trajectories. In the present simulation the longest possible trajectory contains 13 links for $p = 7.5$, and 10 links for $p = 2.31$. The difference is because a certain number of frames must be dropped initially before the tracking asymptotes to a steady yield of links between frames. The smaller the p value, the longer it takes to reach steady conditions (see Malik *et al.* 1992). The absolute yield of trajectories of these lengths drops rapidly for $p < 5$ (figure 3). For $p = 2.31$ this yield is reduced to nearly 40%. The quality of the obtained trajectories is, however, still higher than 95% in all the cases and nearly 100% for $p > 5$. This indicates that true links have not always been established but the links that are established are almost always good.

Tracking parameter $p = \Delta_0/u'\Delta t$	N, Length of track: 100	200	300	400	500	600	700	800	900	1000
2.31	00.0	00.0	00.0	00.0	00.0	00.0	00.0	00.0	00.0	00.0
2.50	00.3	00.0	00.0	00.0	00.0	00.0	00.0	00.0	00.0	00.0
3.00	08.0	00.6	00.1	00.0	00.0	00.0	00.0	00.0	00.0	00.0
3.75	22.2	04.9	01.1	00.2	00.1	00.0	00.0	00.0	00.0	00.0
5.00	52.2	27.4	14.4	07.6	04.0	02.1	01.1	00.6	00.3	00.2
7.00	57.5	33.3	19.3	11.1	06.4	03.7	02.2	01.2	00.7	00.4

(Noise level $n = 0.5\%$)

Table 1. Absolute per-centage yield of complete tracks.

Noise strongly reduces the tracking efficiency. Figure 4 shows that the absolute yield of links as a function of noise level. The yield drops nearly proportionally to $2n\%$ as expected.

Again, the quality of the obtained trajectories is still high and it shows that the quality of tracking is not strongly affected by the noise.

Noise affects strongly the yield of complete trajectories. Since the tracking error and the noise error are independant, the absolute yield of complete trajectories for N time steps can be estimated from $T(N;p,n') = (1-n')^{N+1}C_0^N(p)$ where $n' = 0.01n$ is the fractional level of noise and $C_0(p)$ is the fractional yield of links per frame without noise. These curves are plotted as dashed lines in figure 5 together with the simulation results. The agreement is good. The formula can thus be used to estimate the influence of noise on longer trajectories. The results are presented in Table 1 for a noise level of $n = 0.5\%$

5. Application on real data

Particle trajectories determined by KSIM (Malik 1991) as well as recent theoretical predictions (Hunt *et al.* 1991) show sharp changes in the direction of motion. A tracking scheme must be able to follow such sharp changes. To demonstrate the ability of the tracking scheme in this respect, trajectories of particles have been observed in a flow field produced by manual stirring of water in a small glass tank. The observation volume was $5 \times 20 \times 30$ cm^3 and a 3-camera imaging system was used. The average number of particles abserved in each frame was about 1000, the number of particle coordinates determined in the overlapping part of the images was about 850, and the number of vectors determined by the tracking scheme per frame was about 800. The particle tracks determined over 31 time steps are shown in figure 6. Sharp chnges in particle trajetories are clearly visible which the tracking scheme can easily follow.

6. Concluding remarks

The tracking procedure developed is reliable. The quality of the obtained links and trajectories is high and is not strongly influenced by noise. The number of vectors per frame is mainly limited by the ambiguities arising in the photogrammetric part of the procedure. It is expected that the use of four cameras will resolve most of these ambiguities. This may allow us to increase the number of vectors per frame from at present $\sim$ 900 to $\sim$ 1500 or more.

When only vector fields with good resolution are sought then the p ratio may be chosen to be small, say $2 < p < 3$. But for long trajectories, noise influences its absolute yield very strongly. In this case, when Lagrangian statistics are sought, a four camera system is necessary to reduce the noise level to a minimum, and p must be large, say $p > 10$. The tracking procedure is able to follow sudden changes in the direction of the movement of particles

This project has been funded by the Swiss National Science Foundation grant nos. 2-5316 and 2-26441.

References

FUNG, J.C.H., HUNT, J.C.R., MALIK, N.A. & PERKINS, R.J. (1992). Kinematic simulation of homogeneous turbulence by unsteady random Fourier modes. *J. Fluid Mech.* **236**, 281–317.

HUNT J.C.R., FUNG J.C.H., MALIK N.A., PERKINS R.J., VASSILICOS J.C., WRAY A.A., BUELL J.C. & BERTOGLIO J.P. (1991). Kinematics of small scale motions in homogeneous isotropic turbulence. *Advances in Turb.* **3**, 422–434.

KASAGI N. & NISHINO K. (1991). Probing turbulence with three-dimensional particle-tracking velocimetry. *Exptl. Thermal and Fluid Sci.* **4**, 601–612.
MALIK N.A. (1991). PhD. Thesis. University of Cambridge.
MALIK N.A., DRACOS T. & PAPANTONIOU D. (1992). Particle tracking velocimetry in three-dimensional flows. Part II: particle tracking. ***submitted to Experiments In Fluids.***
MAAS H.G. (1992). Ph.D Thesis, ETH (Swiss Federal Institute of Technology), Zürich.
MAAS H.G., GRUEN A. & PAPANTONIOU D. (1992). Particle tracking velocimetry in three-dimensional flows. Part I: the imaging technique. ***submitted to Experiments In Fluids.***
PAPANTONIOU, D. & MAAS, H.G. (1990). Recent Advances In 3D-Particle Tracking Velocimetry. *Proc. 5th Int. Symp. Applic. Laser Techniques in Fluid Mech.*, Lisbon.

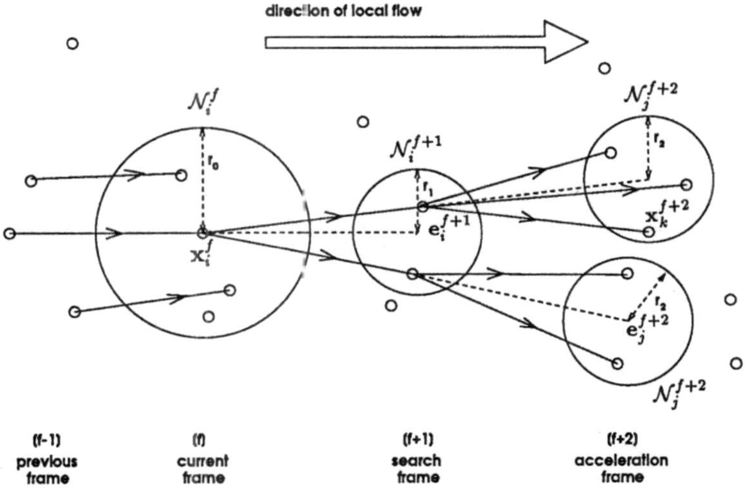

Figure 1. Schematic of Particle Tracking.

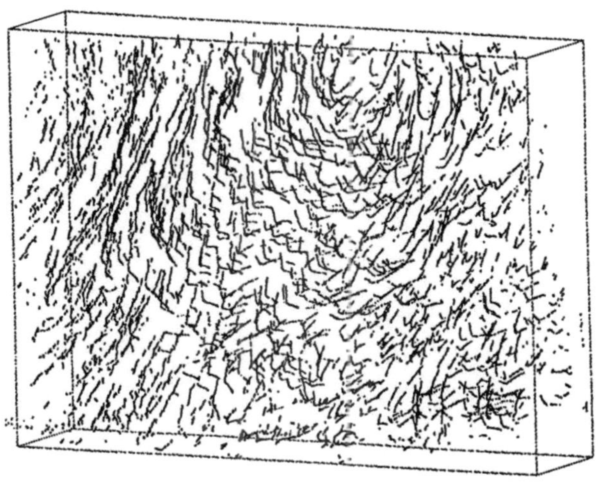

Figure 6. Particle trajectories observed in stirred water tank.

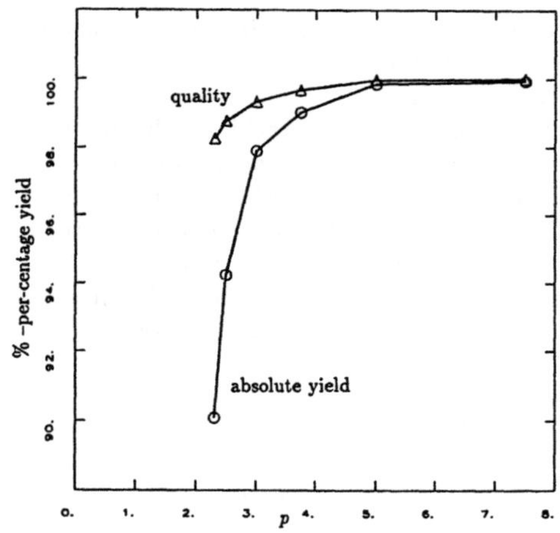

Figure 2. The absolute and relative (quality) percentage yield of links per frame as a function of p, the spacing-displacement parameter. From KSIM data, with no noise.

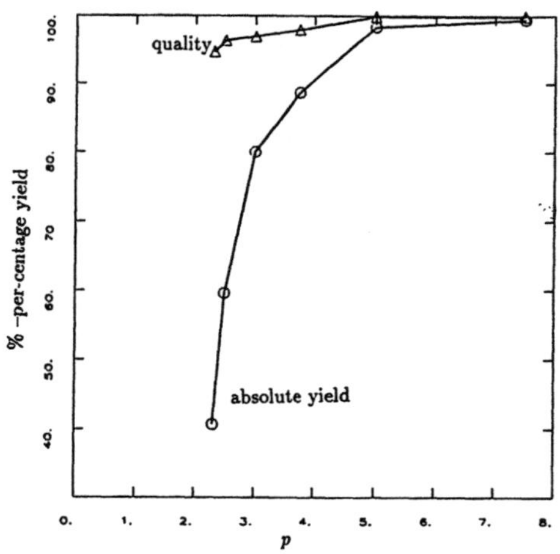

Figure 3. The absolute and relative (quality) percentage yield of complete trajectories as a function of p, the spacing-displacement parameter. From KSIM data, with no noise.

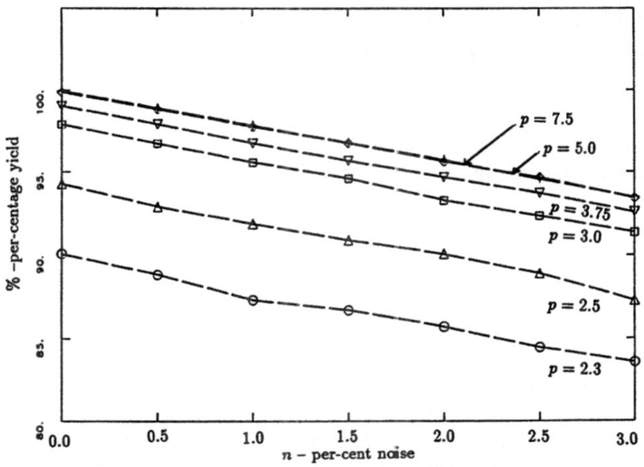

Figure 4. The absolute per-centage yield of links per frame as a function of noise n%. From KSIM data, for different values of p as shown.

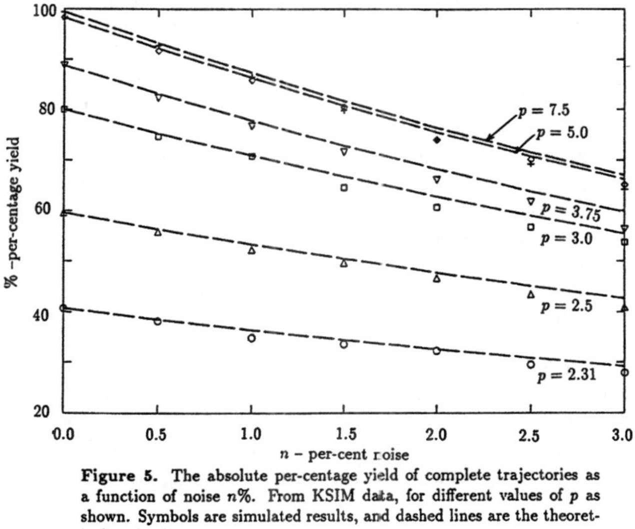

Figure 5. The absolute per-centage yield of complete trajectories as a function of noise n%. From KSIM data, for different values of p as shown. Symbols are simulated results, and dashed lines are the theoretical projections.

Improvement Toward High Measurement Resolution in Three-Dimensional Particle Tracking Velocimetry

Yutaka SATA and Nobuhide KASAGI
Department of Mechanical Engineering, The University of Tokyo, Japan

Summary

With new algorithm of 3-D particle tracking over multiple time steps, the three-dimensional particle tracking velocimetry (3-D PTV, hereafter) system originally developed by Nishino et al. [1-3] was improved in order to achieve higher measurement resolution. It newly includes three laser disc recorders and also an engineering workstation for faster data reduction. The technique tracks fine particles suspended in a flow over two to four consecutive time steps; a particle's new position, around which a search region for tracking is defined, is estimated from the particle's motion at the preceding time steps. With the size of the search region statistically optimized, the tracking algorithm is evaluated on the basis of modeled Lagrangian time spectra of velocity fluctuations. A computer simulation is performed to quantitatively evaluate effectiveness of the present particle tracking technique under a condition of nearly isotropic decaying turbulence.

Introduction

Three-dimensional particle tracking velocimetry has become a powerful flow measurement technique with which instantaneous spatial distributions of velocity vector can be measured [1-5]. The field information thus obtained should be particularly valuable for turbulence research; statistical and structural features of a turbulent flow can be studied in detail [3]. Up to the present, several 3-D PTV techniques were developed and applied to the measurement of various turbulent flows. However, there still remain demanding requirements for the future PTV technique; e.g., improvement in accuracy and spatial resolution of measurement, simplification of camera calibration, and application to gaseous flow measurement [3]. In particular, improvement in spatial resolution to acquire a larger number of velocity vectors is highly desirable for capturing and analyzing instantaneous flow structures. To do this, it is important to reduce ambiguity in identifying the 3-D positions and trajectories of many tracer particles. Possible strategies are, for example, to provide additional information for particle identification (say, size and color), to employ a high definition TV system and to develop effective particle tracking algorithm.

In the present study, the 3-D PTV system originally developed by Nishino et al. [1-3] was reconstructed and at the same time 3-D particle tracking procedures over multiple (up to four) time steps were introduced for higher measurement resolution. The size of a search region for particle tracking is analytically evaluated on the basis of Lagrangian time spectra of velocity fluctuations. In addition, the effectiveness of these improvements is confirmed by a computer simulation with modeled Lagrangian spectra of nearly isotropic decaying turbulence.

Measurement system

The present technique is based on tracking 3-D motions of the fine tracer particles suspended uniformly in a fluid and its basic measurement principle is the same as that described in Ref. [1-3].

As shown in Fig. 1, the new measurement system consists of a light source, three TV-cameras, three laser disk recorders (SONY LVR-5000, LVS-5000), a digital image processor (NEXUS 6810; 512 x 480 pix., 8-bit gray level), a micro computer, and an engineering

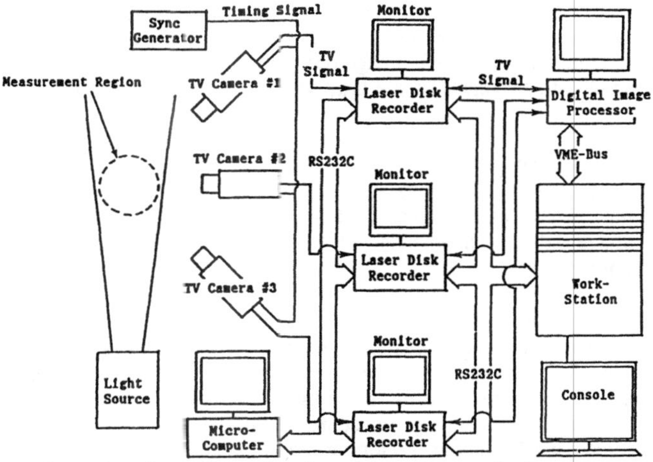

Figure 1. Three-dimensional particle tracking velocimetry system.

workstation (Titan 3020). When compared to the previous system, the present one is characterized by its use of the three laser disc recorders; this results in the fact that three instantaneous images taken simultaneously can be consecutively recorded at a time interval of either 1/30 s or 1/60 s. The replayed images are A/D converted by the image processor and transferred to the workstation, where image coordinates of particles and 3-D particle trajectories are calculated sequentially. The laser disk recorders and the image processor are interfaced with the workstation so that fully automated data reduction is achieved. Together with integration of the workstation, the computer programs for data reduction have been thoroughly revised, and as a result the data reduction speed in this system is five times faster than in the previous system; i.e., only about 30 min. are required to process a hundred sets of images.

Particle trajectory identification

The data reduction procedure is classified as camera calibration and particle tracking [1-3]. In the latter procedure, particles are three-dimensionally tracked by the following series of data reduction procedures, (i) calculation of particle coordinates on three image planes, (ii) calculation of 3-D particle positions, (iii) tracking of 3-D particle trajectories over multiple time steps, (iv) identification of independent trajectories. New algorithm is introduced in procedures (i) and (iii) to increase the number of instantaneous velocity vectors obtained, and thus improve resolution of measurement. Each of these procedures are briefly described below.

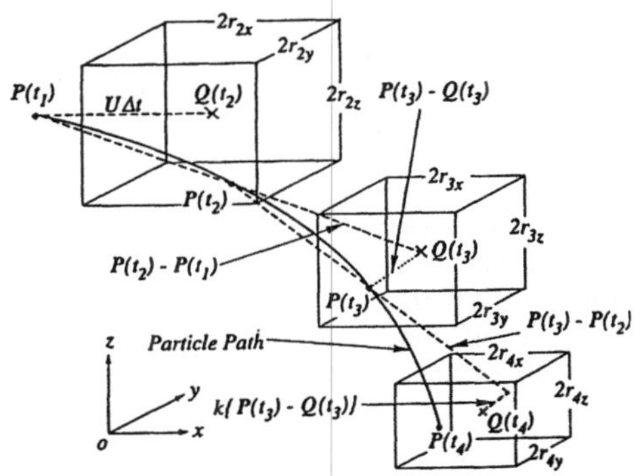

Figure 2. Multiple time step tracking

(i) Calculation of particle coordinates on image planes

A particle image should be distinguished from those lying close around it and this is vital when the particle number density is high. In this procedure, a pixel whose gray level is locally maximum is first searched for and then its neighbors whose gray levels exceed a certain fraction of the maximum are recognized as parts of the particle image. Note that the particle image occupies 4-9 pixels on average. The image coordinates of the particle are calculated as the gray level-weighted averages of all constituent pixels' coordinates.

(ii) Calculation of 3-D particle positions

This is done by extracting all the perspective rays of particle images and then searching for a set of three rays which intersect each other. Since measurement uncertainties are associated with the image data, the intersection of three rays are established when they pass through a common minute volume.

(iii) Tracking of 3-D particle trajectories

The 3-D particle tracking is accomplished in the following three steps; i.e., estimation of a future particle position, definition of a search region around it, and search for possible particle trajectories. The present algorithm performs these tasks over two to four consecutive time steps, say, t_1, t_2, t_3 and t_4. The tracking procedure is illustrated in Fig. 2 where $\boldsymbol{P}$ and $\boldsymbol{Q}$ denote true and estimated particle position vectors at each time step, respectively. The particle positions at later time steps can be estimated by the following equations:

$$Q_i(t_2) = P_i(t_1) + U_i(P(t_1))\Delta t, \tag{1}$$

$$Q_i(t_3) = 2P_i(t_2) - P_i(t_1), \tag{2}$$

$$Q_i(t_4) = (k+2)P_i(t_3) - (2k+1)P_i(t_2) + kP_i(t_1), \tag{3}$$

where $\boldsymbol{U}$ and Δt are an expected mean velocity and an interval between successive time steps, respectively, and subscript i represents the i-th direction. A constant k in Eq. (3) is a parameter whose value should be equal to 1 when the time interval is sufficiently shorter than the time scale of a flow measured, while approaches 0 as the time interval becomes relatively longer.

After the particle position at t_N is estimated, a search region is established around it as:

$$Q_i(t_N) - r_{Ni} < x_i < Q_i(t_N) + r_{Ni}, \tag{4}$$

where r_i is the half size of the search region in the i-th direction. Then, every particle position in the search region is stored as a candidate terminal position and the resultant trajectory is judged in the next procedure.

(iv) Identification of independent particle trajectories

Spurious particle trajectories are unavoidably included in the candidate trajectories recognized in procedure (iii). Hence, a true trajectory is identified only when all the participating particle positions are independent of other trajectories. The particle velocity vector is calculated by dividing the particle displacement by the time interval.

Analysis of the size of search region

The size of a search regions r_i should be carefully determined. It must be large enough to capture all particle movement and, at the same time, small enough compared with the mean distance between particles.

In the following analysis, a particle is assumed to trace fluid motion perfectly, so that its velocity $\boldsymbol{v}(t)$ is equal to the Lagrangian velocity of the fluid. The velocity is given in Fourier transform as:

$$v_i(t) = \int_{-\infty}^{\infty} V_i(f) exp(2\pi jft) df. \tag{5}$$

Integrating eq. (5), we obtain the particle position $\boldsymbol{P}$ at a time step t_N,

$$P_i(t_N) = \int_{-\infty}^{\infty} \frac{1}{2\pi jf} V_i(f) exp[2\pi jf\{t_1 + (N-1)\}\Delta t] + P_i(0). \tag{6}$$

By substituting eqs. (1), (2) and (3) into eq. (6), the standard deviation of the difference between the true and estimated particle positions is written as follows.

$$d_{Ni} = \left\{\overline{(P_i - Q_i)^2}\right\}^{0.5} = \left\{2\int_0^\infty G_N(f) X_{ii}(f) df\right\}^{0.5} \Delta t, \tag{7}$$

$$G_2 = \{2 - 2\cos(2\pi f\Delta t)\} / (2\pi f\Delta t)^2, \tag{8}$$

$$G_3 = \{2 - 2\cos(2\pi f\Delta t)\}^2 / (2\pi f\Delta t)^2, \tag{9}$$

$$G_4 = \{2 - 2\cos(2\pi f\Delta t)\}^2 \{k^2 + 1 - 2k\cos(2\pi f\Delta t)\} / (2\pi f\Delta t)^2, \tag{10}$$

where $X_{ii}(f)$ denotes the Lagrangian time spectrum of the velocity fluctuations measured. Thus, the size of a search region should be statistically determined as being proportional to the above standard deviation, e.g., four to six times, when the coverage is prescribed. It is noted that the constant k in eq. (3) should be optimized so that d_4 would be the smallest.

The distributions of G_2, G_3 and G_4 are shown for $k = 0.5$ and 1 in Fig. 3. These gain functions should preferably be small. Tracking over three and four time steps can certainly predict low frequency components of the particle motion. The accuracy for the components which exceed 1/6 of the imaging frequency $(\Delta t)^{-1}$, however, becomes worse compared to tracking over two time steps. Hence, in measurement of a turbulent flow whose characteristic time scale is sufficiently larger than the imaging frequency, the search region can be made quite small by employment of the tracking over three or four time steps.

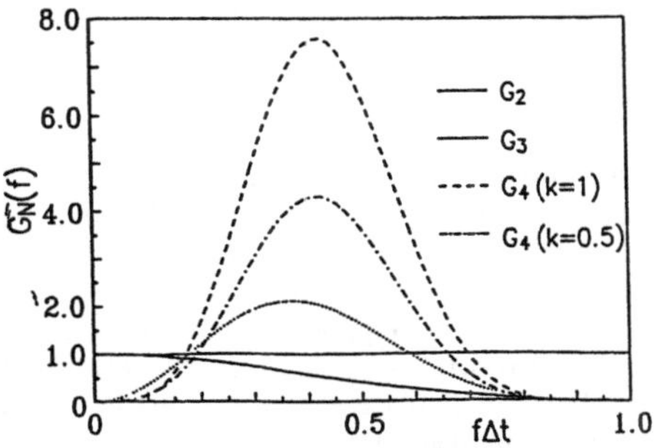

Figure 3. Gain functions.

Evaluation of multiple time step tracking by computer simulation

Effectiveness of the particle tracking procedures in the present study is evaluated by a computer simulation. The simulation is carried out as follows: (i) model a Lagrangian time spectrum of velocity fluctuations, (ii) determine the size of the search region by referring to eq. (7), (iii) numerically generate particles in space which move randomly according to the modeled spectrum, (iv) calculate image coordinates of the particles, and (v) track the particles through the above mentioned procedures.

The Lagrangian spectra shown in Fig. 4 are assumed under the condition of nearly isotropic decaying turbulence measured by Sata et al. [2]; cases 1 and 2 roughly correspond to the initial and terminal periods of the decay, respectively. The turbulent Reynolds number based on the turbulent kinematic energy and the integral scale is 358 for case 1 and 258 for case 2, while the time interval Δt is set to be 1/60 s. Table 1 summarizes the values of d_N in these two cases which is non-dimensionalized by the rms velocity fluctuation

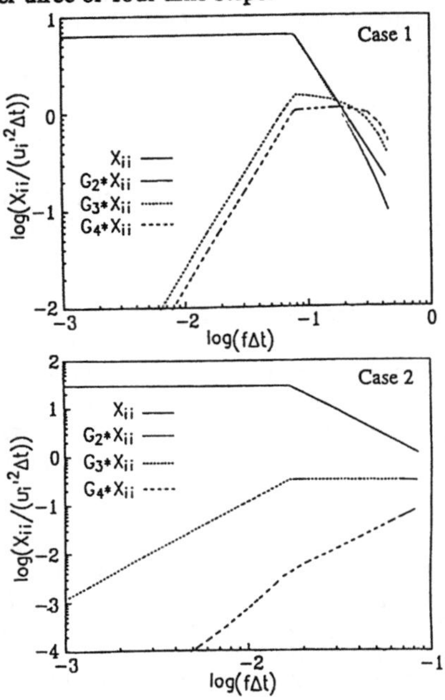

Figure 4. Lagrangian time spectra.

tracking time steps (N)	Case 1	Case 2
2	0.96	0.98
3	0.66	0.16
4	0.64 (k=0.2)	0.05 (k=0.95)

Table 1. Values of dimensionless standard deviation; $d_N/(u'\Delta t)$.

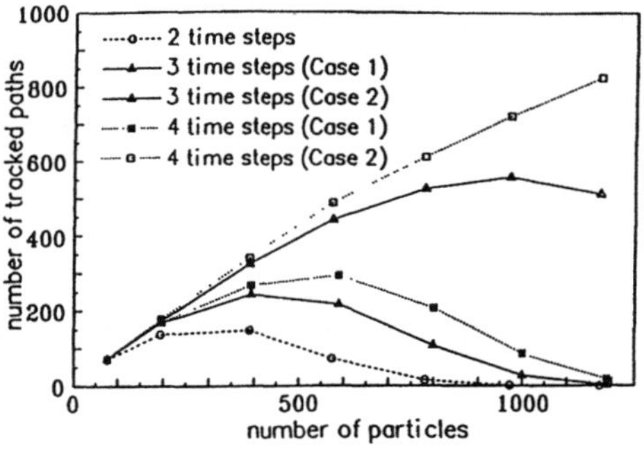

Figure 5. Number of tracked trajectories.

and the time interval. Note that k is optimized so that d_4 becomes the smallest. In case 1, the increase in tracking time steps does not make the search region size much smaller, because the characteristic time scale of the velocity fluctuation is not small compared to Δt. In case 2, however, it is markedly effective in reducing the search region size.

In the following, additional conditions are employed, i.e., (i) the search region sizes r_2, r_3 and r_4 are $4d_2$, $5d_3$, $5d_4$, respectively, (ii) the turbulent intensities are given so that the particle's mean displacement during the time interval is about 1/80 of the image plane size, (iii) in order to simulate overlaps of the particle images, two particles are recognized as one when the distance between them is less than 1/200 of the image plane size, (iv) the number of particles on the image plane is 5/4 of the number of particles which can be tracked. Figure 5 shows the number of velocity vectors tracked versus that of the particles given on the image. It is found that under the above conditions the tracking over two time steps can not track more than 200 particles, while the tracking over more than two time steps yields a larger number of velocity vectors. In measurement of turbulent flows whose time scale is sufficiently longer than the imaging interval (as in case 2 above), the tracking procedures over three and four time steps work effectively and result in more than 800 velocity vectors at most. It is noted, however, that these multiple time step tracking would make computational load heavier and necessitate larger computer memory.

Conclusions

The 3-D PTV system originally developed by Nishino et al. [1] was reconstructed with three laser disc recorders and an engineering workstation and, in addition, new particle tracking algorithm over multiple time steps was introduced for higher measurement resolution. The improved 3-D PTV system achieves five times faster data reduction than the previous one. The tracking algorithm is analytically evaluated on the basis of Lagrangian time spectra of velocity fluctuations. With the increase in tracking time steps, the accuracy in prediction of the particle's motion becomes better for the fluctuating velocity components less than 1/6 of the imaging frequency. Finally, a computer simulation is carried out to examine effectiveness of the tracking procedures proposed and it is found that more than 800 velocity vectors can be tracked with the present 3-D PTV system under practical experimental conditions.

References

[1] Nishino, K., Kasagi, N., and Hirata, M., *ASME J. Fluids Eng.*, **111**, 1989, pp. 384-391.
[2] Sata, Y., Nishino, K., and Kasagi, N., *Flow Visualization V*, 1990, Hemisphere, pp. 248-253.
[3] Kasagi, N., and Nishino, K., *Exp. Therm. Fluid Sci.*, **4**, 1992, pp. 601-613.
[4] Papantoniou, D., and Dracos, T., *Engineering Turbulence Modelling and Measurements*, 1990, Elsevier, pp. 315-324.
[5] Adrian, R. J., *Ann. Rev. Fluid Mech.*, **23**, 1991, pp. 261-304.
[6] Tennekes, H. and Lumley, J. L., *A First Course in Turbulence*, The MIT Press., 1972.

Application of a 3-D PTV to the Measurement of Turbulent Air Flows

H. Kato, K. Nishino and K. Torii

Department of Mechanical Engineering and Materials Science
Yokohama National University
Tokiwadai 156, Hodogaya-ku, Yokohama 240, Japan

Abstract

A three-dimensional particle tracking velocimeter (3-D PTV) is applied to the measurement of an air turbulent boundary layer on a flat plate with a free-stream velocity of 4 m/s. To capture particles' images that are relevant for PTV measurement, a unique illumination technique is developed by using three strobe lamps and a multi-channel signal retarder. Tracer particles used are plastic micro-balloons, whose average diameter and weight density are 50 μm and 36 kg/m^3, respectively. Examination of particles' traceability confirms that they can follow periodic velocity variation of the surrounding fluid up to 100 Hz, and that the traceability is good enough to resolve streamwise velocity fluctuations under the present flow conditions. Measured turbulence statistics, including the mean velocities, the turbulence intensities and the Reynolds stress, compare reasonably well with hot-wire measurement, previous LDV measurement and DNS data, hence verifying applicability of the present 3-D PTV to the measurement of turbulent air flows. It is, however, found that near wall measurement needs further refinement to deal with adhesion of the particles to the wall.

Introduction

Advent of high-performance digital image processors has promoted rapid development of various PTVs. Among them a 3-D PTV has attracted particular attention because of its capability of making whole field measurement of all three velocity components (e.g., see review paper by Kasagi and Nishino [1]). Although the 3-D PTVs developed so far have successfully been applied to several turbulent flows, most of those experiments were performed in water flows at relatively low velocities. Since measurement of air flows at higher velocities is of great importance from practical viewpoint, detailed examination on applicability of the 3-D PTV to turbulent air flows is strongly encouraged.

Experimental Apparatus

The measurement system was basically the same as that reported by Nishino et al. [2]. The present system consisted of three CCD cameras, an analogue laser disk recorder, a stroboscope equipped with three lamp houses, a multi-channel signal retarder, a digital image processor and others.

To visualize tracer particles in air flows, an illumination technique was devised as illustrated in Fig. 1(a). Strobe flashes were triggered by a vertical blanking signal of the sync-generator in such a way that the first flash was pulsed during an odd TV field and the second and third ones were done during the subsequent even TV field. The time interval between the first and second flashes, Δt_1, was short in order to facilitate particle tracking between two consecutive TV fields, while the time interval between the first and third flashes, Δt_2, was comparatively long so as to achieve satisfactory measurement accuracy. In the present measurement, Δt_1 and Δt_2 were respectively 100 μs and 500 μs and average particles' displacement over Δt_2 was about 40 pixels. A magnified picture of particles' images is shown in Fig. 1(b), in which three images of an identical particle are recognized: one in the odd TV field and the other two in the even TV field. Velocity vectors were calculated from the first and third positions, resulting in measurement uncertainties at 95 % coverage of 59, 125 and 51 mm/s in the streamwise, wall normal and spanwise directions, respectively.

Tracer particles chosen were plastic micro–balloons (Expancel, Japan Fillite Co., Ltd.), whose average diameter was 50 μm and weight density was 36 kg/m^3. Their frequency response to the surrounding fluid motions was analyzed with a procedure described by Hjemfelt and Mockros [3]. Fig. 2 gives an estimated amplitude ratio between a particle and fluid velocities as a function of frequency. Also included is power spectrum of the streamwise velocity fluctuations measured with the hot–wire anemometry in the present boundary layer flow. Most of fluctuation energy was contained in the frequency range lower than 100 Hz; that is the range in which particles' traceability was fairly good. Actually the frequency response would cause only 2 % decrease in the root–mean–square value of the velocity fluctuations.

The measurements were performed in a wind tunnel facility schematically shown in Fig. 3. The flow going through a settling chamber entered a square duct at a uniform free–stream velocity of 4 m/s with turbulence level of less than 0.6%. The duct 2 m in length had a square cross section of 30 x 30 cm^2. At the entrance of the duct, the flow was tripped with a 5 mm in diameter cylinder installed on the bottom wall, while the ceiling walls were adjusted to remove streamwise pressure gradient. The measuring station was located 1.1 m downstream of the cylinder, where the measuring volume of about 25 x 25 x 25 mm^3 was illuminated by the strobe lamps. Tracer particles were blown vertically upward into the flow through a circular hole of 12 mm in diameter. The hole was installed in the bottom wall, 35 mm upstream of the tripping cylinder, and the blowing velocities were 1–2 m/s.

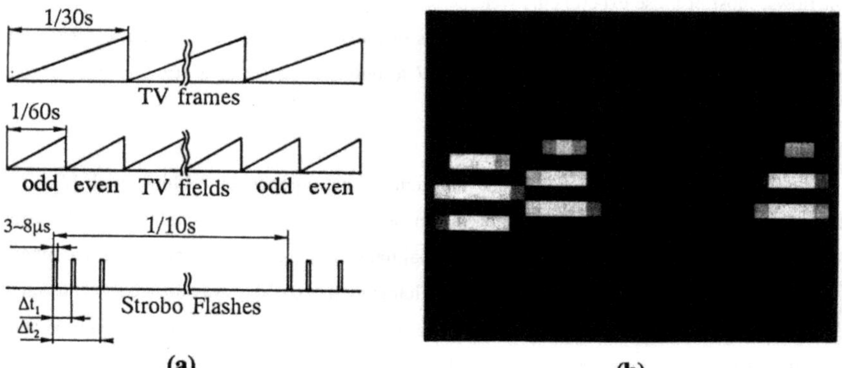

Fig. 1. (a) Timing chart of illumination of strobe flashes, (b) magnified picture of particles' images.

For comparison hot-wire measurement was carried out for the streamwise velocity component. Boundary layer parameters in the case of the 3-D PTV and hot-wire measurements were as follows:

	U_∞(m/s)	u_τ(m/s)	ν(mm^2/s)	δ^*(mm)	θ(mm)	Re_θ
PTV	4.03	0.176	14.89	10.08	7.11	1920
HW	3.98	0.181	14.73	8.17	5.79	1560

where U_∞ is the free-stream velocity, u_τ the friction velocity, ν the kinematic viscosity, δ^* the displacement thickness, θ the momentum thickness and Re_θ the Reynolds number ($=U_\infty\theta/\nu$).

Results and Discussion

Turbulence statistics were calculated from measured velocity vectors, which distribute randomly in the measuring volume. To take ensemble averages, the velocity data were classified into groups depending on their position from the wall. A distribution of the number of data is shown in Fig. 4(a). In the vicinity of the wall ($y^+<100$), image quality was considerably deteriorated by adhesion of the particles to the wall, resulting in decrease in the number of data. Nevertheless, image improvement based on background image subtraction was successful in retaining some velocity vectors. Note that four discontinuities in the distribution are due to changes in the grouping cell sizes, which are varied from 0.1 to 1.0 mm as going away from the wall.

Mean velocity distributions are presented in Fig. 4(b), where the streamwise, wall normal and spanwise directions are respectively denoted by x, y and z, and corresponding velocity components as u, v and w. The velocities are normalized with u_τ, which is determined so that the distribution of $\overline{u}$ in the logarithmic region fits the well-established profile indicated in Fig. 4(b). The distribution of $\overline{u}^+$ is in good agreement with those of the hot-wire measurement and DNS data of Spalart [4], excluding the near wall region ($y^+<6$) where influence of the particle adhesion becomes significant. The distribution of $\overline{v}^+$ is consistent with the boundary layer development while that of $\overline{w}^+$ is nearly zero as is required from symmetry of the flow field.

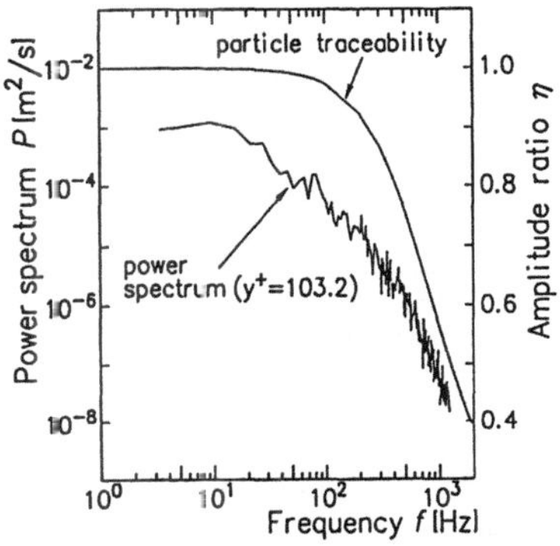

Fig. 2. Estimated amplitude ratio between particles and fluid motions, and power spectrum of the streamwise velocity fluctuations measured with the hot-wire anemometry.

Turbulence intensities are shown in Fig. 4(c). The measurement uncertainties mentioned above are subtracted as random noise from originally measured turbulence intensities. Their distributions are in general agreement with LDV measurements of Karlsson and Johansson [5]. However, there are notable differences in the peak value of u'^{+}_{rms} and values of w'^{+}_{rms} in the outer region (y^{+}>300). Although the reason for this is not yet clear, it is conjectured that the particle blowing at the entrance of the duct might affect the flow field. This influence is also noticed from the differences in δ^{*} and θ between the 3-D PTV and hot-wire measurements, both of which were performed under the same free-stream velocity. Finally the Reynolds stress is shown in Fig. 4(d); reasonable agreement with other data is recognized despite scatter of data.

Conclusions

A 3-D PTV was applied to the measurement of an air turbulent boundary layer on a flat plate. A unique illumination technique was devised by using three strobe lamps and a multi-channel signal retarder. It was verified that the plastic micro-balloons employed exhibited sufficient traceability under the present flow conditions. Measured turbulence statistics compare reasonably well with hot-wire measurement, previous LDV measurement and DNS data. From these results it was confirmed that the present 3-D PTV was applicable to turbulent air flows. Further improvement was, however, required for the near wall measurement to resolve the problem of particle adhesion to the wall.

Acknowledgements

The authors are grateful to Professor N. Kasagi, The University of Tokyo, for his cooperation. Assistance of Messrs. H. Shibazaki, T. Adachi and K. Murakoshi are gratefully acknowledged. This work was supported through the Grant-in-Aids for Co-operative Research (No. 03352018) by the Ministry of Education, Science and Culture.

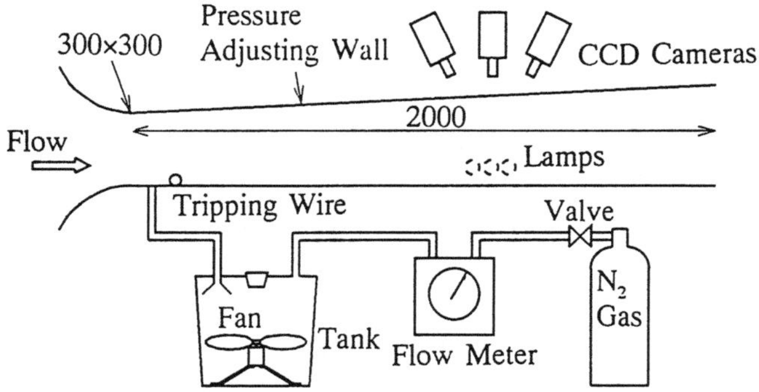

Fig. 3. Schematic of wind tunnel facility and seeding apparatus.

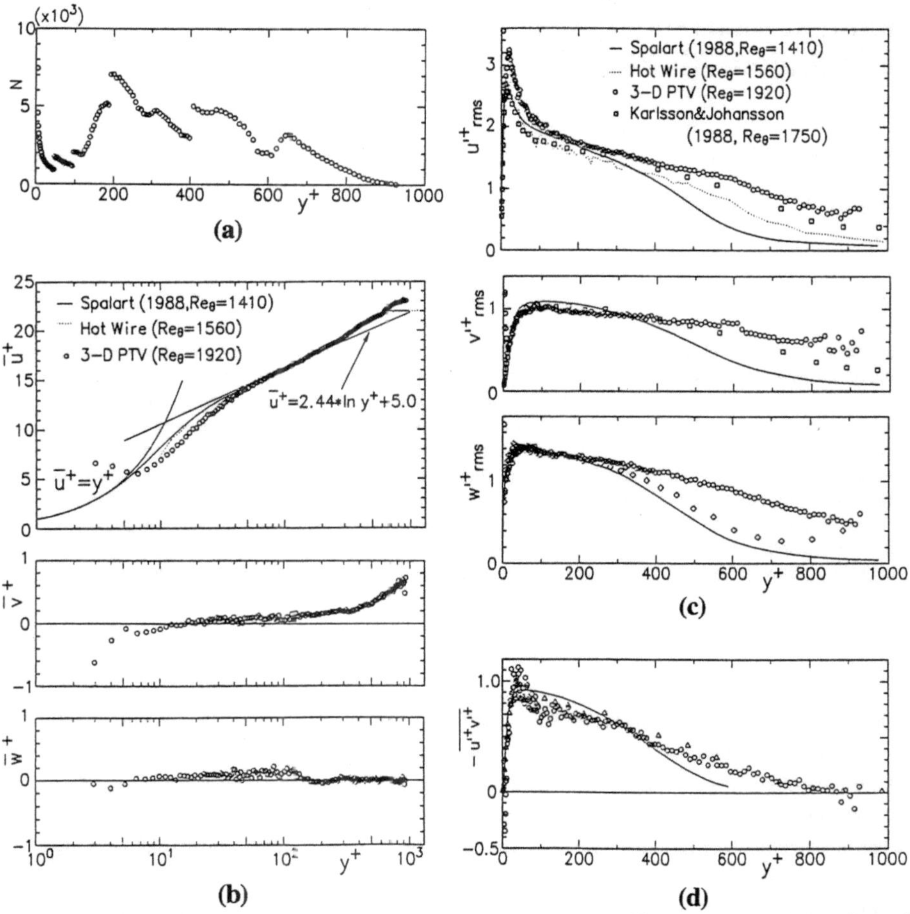

Fig. 4. (a) The number of velocity vectors, (b) mean velocities, (c) root–mean–square values of the velocity fluctuations, (d) the Reynolds stress. The superscript + denotes normalization with the friction velocity and kinematic viscosity.

References

1. Kasagi, N.; Nishino, K.: Probing Turbulence with Three–Dimensional Particle–Tracking Velocimetry. Experimental Thermal and Fluid Science, **4**–5 (1991) 601–612.

2. Nishino, K.; Kasagi, N.; Hirata, M.: Three–Dimensional Particle Tracking Velocimetry Based on Automated Digital Image Processing. Trans. ASME, J. Fluids Eng., **111**–4 (1989) 384–391.

3. Hjelmfelt, A. T.; Mockros, L. F.: Motion of Discrete Particles in a Turbulent Fluid. Appl. Sci. Res., **16** (1966) 149–161.

4. Spalart, P. R.; Direct Simulation of a Turbulent Boundary Layer up to R_θ=1410. J. Fluid Mech., **187** (1988) 61–98.

5. Karlsson, R. I.; Johansson, T. G.: LDV Measurements of Higher Order Moments of Velocity Fluctuations in a Turbulent Boundary Layer. Laser Anemometry in Fluid Mech. III, Adrian R. J. et al. (eds), 1988, 273–289.

Fractional Analysis of the Motion of Solid Particles in a Turbulent Air Flow

B.ŠIROK, I.NANČOVSKA, F.TRDIČ

TURBOINSTITUT Ljubljana, Slovenija

Summary

In the paper the fractional characteristics of the motion of a finite number of solid particles in a fully developed turbulent air flow are presented. The position of particles in space and time form patterns which can be observed by means of computer-aided visualization. By successive digitalization of shots the distribution of the particles in a restricted area a comparision can be made between the correlational dimension D_2 and the fractional dimension D_F. The correlational dimension was derived by estimating the time series by means of which the distribution of particles in a selected segment of space is described. The fractional dimension was obtained using the "Fractional Brownion motion" model of calculation. A qualitative relation between the above described dimensions has been established,which has shown that they are anisotropic. Therefore similarity between methods of forming topological patterns in space, as well as in time, may be supposed. The assumption may also be made that the patterns depend on the characteristics of turbulent flow in the observed process.

Introduction

Different patterns of motion of particles in space and time, are created by nature. The degree of order inside a pattern in a certain space and at a certain time may, in general, vary. The essential information about the dimension of the multiphase space status can be given by a fractional dimension. The fractional dimension can be derived from the topological structures of the turbulent flow [1] and from a time series based on measurements of the turbulent flow [2]. The aim of this paper is to present an experiment carried out in order to obtain an assessment of the dimension of image structures, as well as of the time series. It was possible to compare the two dimensions since both were obtained from the same experiment. During the experiment the motion of a limited number of particles in a turbulent air flow was observed. The observations were carried out using a computer-aided visualization in the observed plane. The assessment of the dimension was obtained using the correlational integral algorithm [2] and an algorithm for calculation of the fractional dimension [3].

Description of the experiment

The experiment was carried out in the vertical test volume shown in Fig. 1. Air flow enters the test volume thrugh an opening, located at the bottom of the test volume. The exit is located on the top of the tunnel. The inlet velocity of the air flow was measured and taken into account as an integral parameter of the process. Based on the diameter of the inlet pipe, a Reynold's number of $2{,}8 \, . \, 10^5$ to $3{,}5 \, . \, 10^5$ was maintained. In the observed space 5 particles of spherical shape, each having a diameter of dk = 12 mm, moved freely. The motion of the particles was observed by means of a camera, which was held perpendicular to the side wall of the test volume.

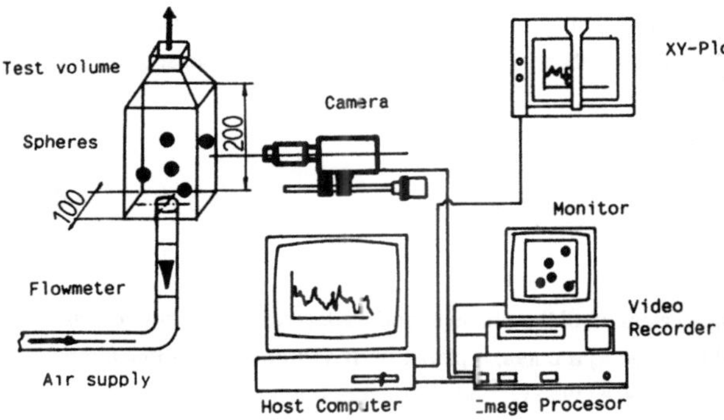

Fig. 1: The experiment

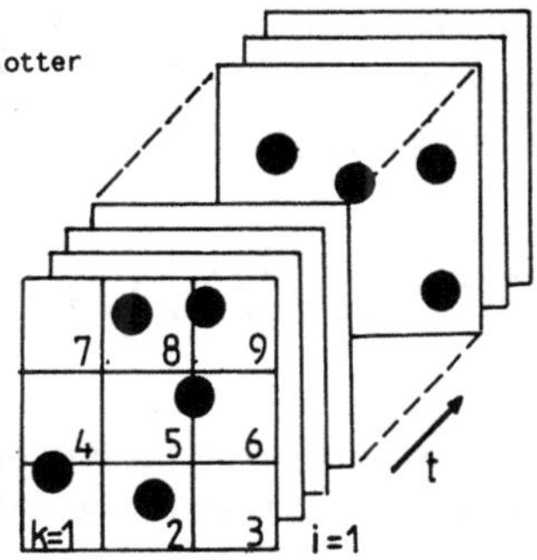

Fig. 2: Visualized image data

The exposure time of the CCD camera was 1/1000 sec. The camera was adjusted to a standard television signal velocity value, i.e. f=25 Hz. Sequential television signals were digitalized, with a resolution of 512 X 256 pixels. Digitalization was performed with a 8-bit A/D converter at an adjustable crest of grey images. The sequence of memorizing of the pictures by a computer is shown in Fig. 2.

The memorized digitalized pictures were used to calculate the fractional and correlational dimensions. In the plane of the experiment a field of 100 X 100 mm has been observed. This corresponded to digitalized picture boundaries defined by 150 X 150 pixels. The field was then divided up into 9 sections of square shape, each having 50 X 50 pixels.

For the assessment of the correlational dimension an integer type variable A(t,K) was introduced:

$$A(t,k) = \sum_{l=1}^{50}\sum_{m=1}^{50} E(l,m) \qquad E(l,m) = \begin{cases} 0 & \text{dark} \\ 1 & \text{light} \end{cases} \tag{1}$$

Using a variable A(t,k), a partion of dark surface of an observed window k at time t is given. The time interval of the observation is defined by the present camera velocity and the number of sequentially taken pictures in a time interval of t=40 sec.

The correlation dimension

Based on the general dynamic law [4], the status of an observed process at a time t+T can be defined by the status at a time T. If using the time series A(t,k) to define the components of vector $\underline{Y}$.

$$y_1 = A(t,k), y_2 = A(t+T,k), ..., y_{n+1} = A(t+nT,k), \quad T = 1/f$$

Then the supposition can be made that the characteristics of the time series A(t,k) are incorporated in a new vector $\underline{Y}(t,k)$. The dimensions of a newly established phase space can be assessed by the correlational integral [2]:

$$C(r) = \lim_{m \to \infty} \frac{1}{m^2} \sum_i \sum_{j \neq i} \Phi(r - |\underline{Y_i} - \underline{Y_j}|) \tag{2}$$

In C(r) the correlated pairs of points inside an element of the phase space having typical dimension r are counted up. If the points are evenly distributed along a curve, a plane, or in a three-dimensional metric space, then the number of correlated pairs increases with r, r^2, r^3. The exponent based on r therefore represents the dimension of the phase space. It is the lower value of the fractional dimension:

$$D_2 = \frac{\log(C(r))}{\log(r)} \tag{3}$$

The number of components n of the vector $\underline{Y}$ should be iteratively increased until the dimension D_2 becomes constant.

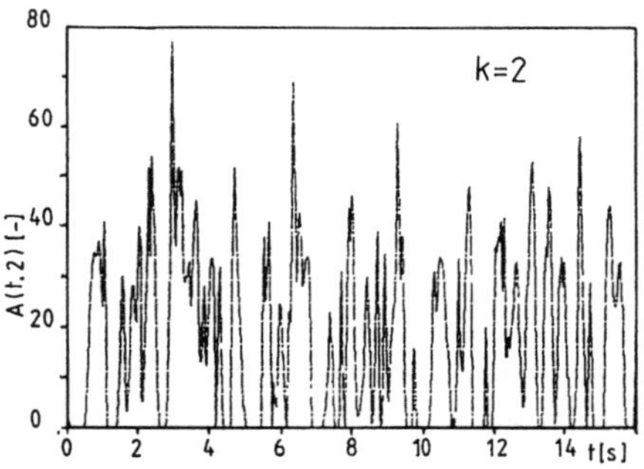

Fig. 3: The time signal A(t,k)

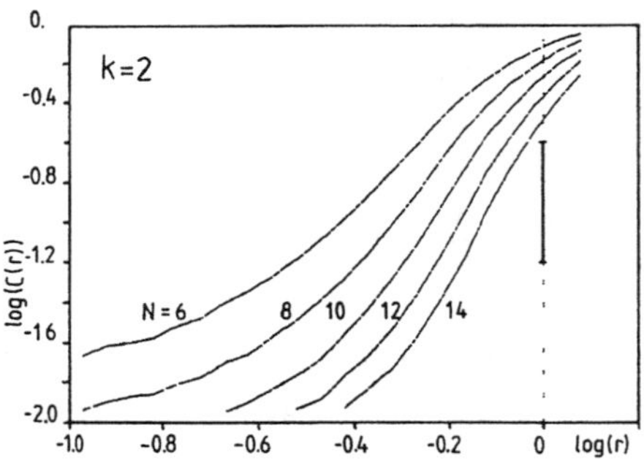

Fig. 4:The correlational integral C(r)

<u>The fractal dimension estimate by means of fBm model</u>

The matematical model of fractional Brownian motion - fBm [5] is a generalized concept of a random function which describes Brownian motion. It is a good basis for understanding of random chance phenomena involved in many phisical processes. fBm is a random single-valued function of one variable, wich is usually time t. The increments of fBm are stationary and statistically self-similar.

Pentland [6], Chen, Keller and Cronover [5] generalized the fBm function to a two-parametric space, by replacing the parameter t by a vector s(x,y). Using $B_H(s)$, the distribution of a grey colour on the image plane was described. Therefore it was possible to obtain increments of grey colour, they have Gaussian distribution with standard deviation.

$$\sigma[B_H(s+S)-B_H(s)] = C_H\|S\|^H, \tag{4}$$

$$s=(x,y), S=(\Delta x,\Delta y), \|S\| = \sqrt{(\Delta x)^2+(\Delta y)^2} \tag{5}$$

This property is called S^H law of fBm's. The parameter H is directly related to the fractal dimension of the intensity image by $D_F = 3\text{-}H$.
In this paper we use fBm for estimating the fractal dimensions of the gray images. Intensity images have been generated from successive black and white shots obtained during the experiment. The fundamental idea of gray image formation was to take into account time. This method we called "a method of forgetting", and is based on the addition of the successive grey images. The intensity of the grey colour of the i+1 image is described by:

$$GRAY_{i+1}(\omega) = \omega GRAY_i(\omega) + (1-\omega)B\&W_{i+1} \tag{6}$$

W is a coefficient of a forgetting and in our case it is constant. The expression B&W(i+1) is a black and white shot at the i+1 experiment step. It's supposed that the number of needed shots is limited and comparable with the number of components of the vector $\underline{Y}$.

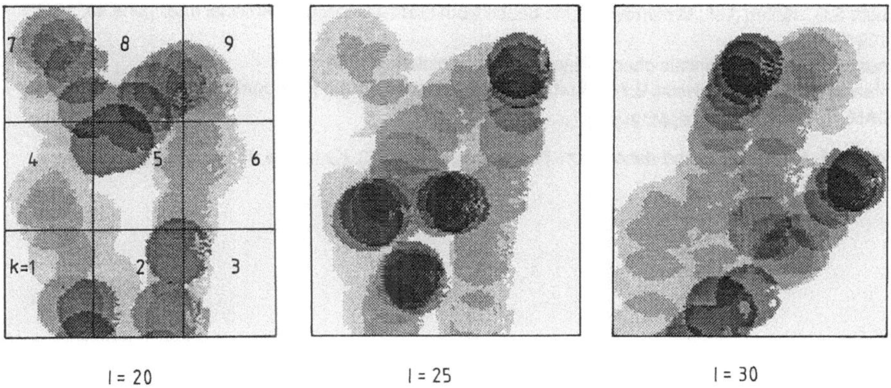

Fig. 5. The addition of the successive grey images

$D_2 = 2.61$ $N = 12$ $D_F = 2.53$ $k = 7$	$D_2 = 2.75$ $N = 12$ $D_F = 2.46$ $k = 8$	$D_2 = 2.62$ $N = 12$ $D_F = 2.55$ $k = 9$
$D_2 = 3.06$ $N = 14$ $D_F = 2.95$ $k = 4$	$D_2 = 3.16$ $N = 14$ $D_F = 2.50$ $k = 5$	$D_2 = 2.59$ $N = 14$ $D_F = 2.56$ $k = 6$
$D_2 = 2.28$ $N = 12$ $D_F = 1.86$ $k = 1$	$D_2 = 3.54$ $N = 14$ $D_F = 2.83$ $k = 2$	$D_2 = 2.65$ $N = 12$ $D_F = 2.55$ $k = 3$

Table 1: Comparision of the dimensions D_F and D_2

Results

The obtained correlational and fractal dimensions for the observed image segments are given in the Table 1. If the values of these dimensions are compared the anizotropy of the dimensions D_F and D_2, as well as their qualitative similarity , can be seen. The stability of a fractional dimension D_F is achived by means of the addition of successive grey images if a condition of $n=2XN+1$ is maintained. The above characteristic satisfies Taken's theorem [4].

References

1. Takeno, T.; Murayama, M.; Tanida, Y.: Fractal analysis of turbulent premixed flame surface. Experiments in fluids 10, 1990, (61-70)
2. Bonetti, M.; Meynart, R.; Boon, J.B.; Olivari, D.: Chaotic Dynamics in a Periodically Excited Air Jet. Physical Review Letters, Vol.55, num.5, 1985
3. Chen, S.S.; Keller, J.M.; Crownover, R.M.: Shape from Fractal Geometry, Artificial Intelligene 43, 1985, (199-218)
4. Shuster, H.G.: Deterministic chaos, Physik-Verlag, Weinheim, 1984
5. Mandelbrot, B.; Van Ness, J.W.: Fractional Brownian motions, fractional noises and aplications, SIAM Rev. 10(4), 1986, (422-437)
6. Pentland, A.: Fractal-based description, Proceedings IJCA I-83, Karlsruhe, FRG, 1983, (973-981)

Spectrum-Correlation Method and Filters Technique Application for the Velocity Field Estimation

TOMONAO KOBAYASHI

Department of Civil Engineering, Tokyo Rika University,
2641 Yamazaki, Noda city, Chiba, 278 Japan.

and

MIKIO HINO

Department of Civil Engineering, Tokyo Institute of Technology,
2-12-1 O-okayama, Meguro-ku, Tokyo, 152 Japan.

Abstract

A new image processing procedure by using Spectrum-Correlation has been developed to estimate an instantaneous velocity distribution. This has been applied to analyze velocity fields under breaking waves. As in an image-correlation method, this procedure finds out similar patterns from two successive pictures and estimates velocity vectors. The used pictures were taken with very short time lag between them. Therefore the procedure does not require to trace fine particles, and can be applied under various conditions. The processing time for evaluating the velocity distribution with a personal computer is relatively short. To improve the accuracy of the evaluated velocity distribution, a Laplacian filter and a smoothing filter are introduced. The Laplacian filter emphasizes image patterns near the water surface, and the smoothing filter reduces noises due to the flow rotating motions and turbulence.

1. Introduction

Understanding of velocity distribution in a flow field is very important to analyze phenomena in the flow. However this task is very difficult to realize with ordinary velocitymeters, especially in the case of unsteady flow. Recently, computer technology and advanced image processing technology with high speed have been progressing rapidly. Nowadays, with the help of both technologies, we can challenge to estimate an instantaneous velocity distribution from pictures by visualizing the flow field with an engineering work station or a personal computer.

Among the methods to estimate an instantaneous velocity distribution by using image processing technique, the tracer-particles method estimates velocity distributions with higher accuracy. However, this method cannot be applied to analyze an inner region under breaking waves, because this region involves innumerable air bubbles, and it is very difficult to trace particles from the image data of this flow field. On the other hand, the image-correlation method is useful in this case since it can be applied without tracing fine particles (*c.f.* Nishimura and Takewaka[1], Mano and Kamio[2]). Nevertheless, the analyzing time of this method is relatively long for a personal computer

In this study, an image processing technique by applying Spectrum-Correlation is employed to obtain an instantaneous velocity distribution of an inner region under breaking waves.

2. Method of Image Processing technique

2.1. Estimation of Velocity Distribution using Spectrum-Correlation

Spectrum-Correlation is applied to estimate an instantaneous velocity distribution in a flow field from two successive pictures of a visualized flow. The pictures were taken with very short time lag between them. This method is based on Okuno and Kinoshita[3]'s study.

In this method, we assume that the first picture of two successive ones has a brightness distribution expressed as a function $f(x, y)$. And if the flow moved $(x, y) = (a, b)$ in a very short period between two pictures, then the brightness distribution in the next picture is shown as $f(x-a, y-b)$. Results from the Fourier Transformation of the functions $f(x, y)$ and $f(x-a, y-b)$ are written as follows;

$$\begin{aligned} \mathcal{F}[f(x,y)] &= \int_{-\infty}^{\infty}\int_{-\infty}^{\infty} f(x,y)\, e^{-2\pi i k_x x} e^{-2\pi i k_y y}\, dx\, dy \\ &= F(k_x, k_y) \end{aligned} \qquad (1)$$

$$\begin{aligned} \mathcal{F}[f(x-a,y-b)] &= \int_{-\infty}^{\infty}\int_{-\infty}^{\infty} f(x-a,y-b)\, e^{-2\pi i k_x x} e^{-2\pi i k_y y}\, dx\, dy \\ &= F(k_x, k_y)\, e^{-2\pi i (a k_x + b k_y)}, \end{aligned} \qquad (2)$$

where $\mathcal{F}[\,]$ means Fourier Transformation; k_x and k_y are wave numbers of x and y directions respectively; $F(k_x, k_y)$ is a Power-Spectrum density function of $f(x, y)$. Afterwards, we calculate a ratio of the results of this transformation, and apply the Fourier Transformation again.

$$\begin{aligned} \mathcal{F}\left[\frac{\mathcal{F}[f(x,y)]}{\mathcal{F}[f(x-a,y-b)]}\right] &= \mathcal{F}[e^{2\pi i (a k_x + b k_y)}] \\ &= \delta(x-a, y-b), \end{aligned} \qquad (3)$$

where $\delta(x, y)$ is a delta function. One of the results obtained from Eq. 3 is shown in Fig. 1. From Eq. 3 and Fig. 1, it is clear that a location of a peak point of the delta function indicates a displacement of the velocity vector in a very short period between two pictures. Hence, we can estimate the velocity vector at each point from the location of the peak point in Eq. 3. In addition, the value of this peak indicates the accuracy of the estimated velocity vector. Generally, the evaluation of the instantaneous velocity distribution by applying the image-correlation method is accelerated if a First Fourier Transformation technique is used.

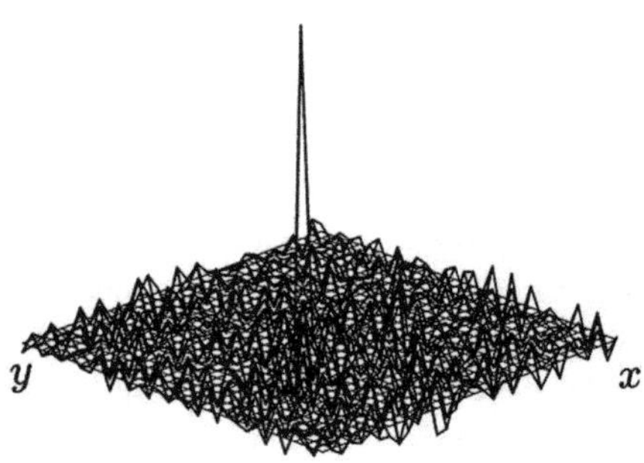

Figure 1: Distribution of a delta function, corresponding to Eq. 3.

2.2. Laplacian Filter

A picture of a flow field in a wave breaking zone contains a water surface which is the boundary between the fluid and its outer region; the fluid and its outer region; the fluid moves with certain velocity and the outer region remains at rest. Since the flow velocity cannot be exactly evaluated due to the different velocities of the outer region and fluid, we emphasize the movement of the surface boundary to estimate the velocity vector in its vicinity with the Laplacian filter. This filter is applied to processed data $f'(x,y)$ as follows;

$$f'(x,y) = \nabla^2 f(x,y). \tag{4}$$

The Laplacian filter is generally applied to emphasize image outlines.

2.3. Smoothing Filter

The Spectrum-Correlation method takes only a shifting motion of the flow into account, same as in the image-correlation method. Then the flow rotating motion and turbulence are regarded as noises. However, the flow field under breaking waves is strongly turbulent, and its motion involves many vortices. In this study, to diminish the influence of such motion and to get a fine result, a smoothing filter is introduced as a preprocessing for the evaluation of the velocity distribution. A processed data $\bar{f}(x,y)$ with this smoothing filter is computed as follows;

$$\bar{f}(x,y) = \frac{1}{A}\int_A f(x,y)\,dA, \tag{5}$$

where A is a small area around the point (x,y).

Sometimes for ordinary image processings, a smoothing filter causes aliasing, and instead of this one, a medium filter is usually introduced to reduce noises. However, since in this study the purpose to use this filter is different from general image processings, a smoothing filter is applied.

3. Application of the Method to Measurements of a Velocity Field under Breaking Waves

An instantaneous velocity distribution of a flow field under breaking waves is evaluated using the method and the filters explained above. Two cases of velocity fields are analyzed; one is a velocity field of a wave just breaking, and the other is the one around an obliquely descending eddy which is a large-scale structure under breaking waves and involves innumerable air bubbles.

First, the instantaneous velocity distribution of a wave field when a wave is just breaking, as shown in Fig. 2, is estimated This image data had been taken by Nishimura and Takewaka[1] with a high speed digital video tape recorder that can recorded 220×190 pixels frames with the speed of 1,000 frames per second. Figure 3 shows a velocity distribution estimated from two frames of image data which have 2 milliseconds time lag between them, by applying the method

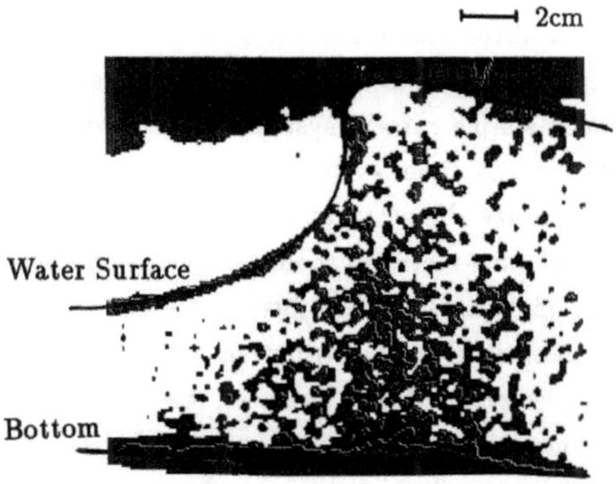

Figure 2: One of the successive image data of a just breaking wave, taken by Nishimura and Takewaka[1] with a high-speed digital VTR.

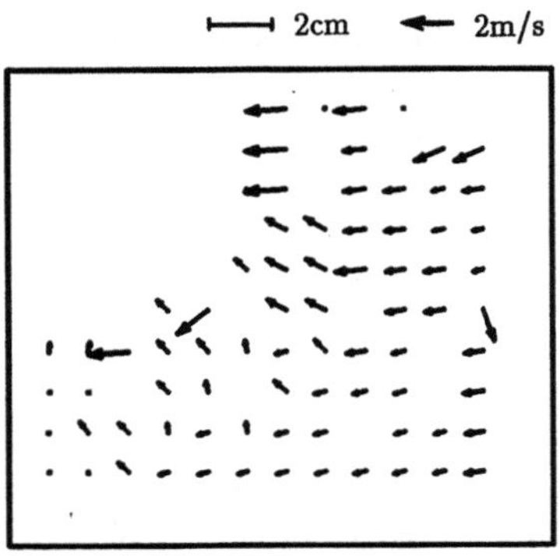

Figure 3: Evaluated instantaneous velocity distribution corresponding to Fig. 2 with a Spectrum-Correlation method.

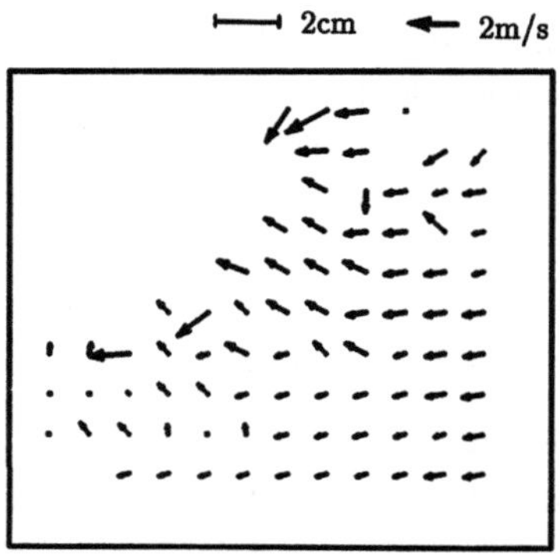

Figure 4: Combined results of an instantaneous velocity distribution from Fig. 3 and the estimated one by applying the Laplacian filter.

explained in Eqs. 1 to 3. As is observed in Fig. 3 the velocity distribution in the inner region is well estimated. However, the velocity vectors near the water surface are not. Therefore we emphasized the surface boundary image with the Laplacian filter, evaluating the velocity distribution again with the emphasized image data, and combining it with the distribution shown in Fig. 3, using the values of the peak in Fig. 1 as indices of accuracy of the evaluated velocity vectors. The result with this process is shown in Fig. 4, where it is observed that the velocity distribution is well estimated in the whole flow field, especially around the wave crest. This result indicates that the Spectrum-Correlation method is useful to estimate the instantaneous velocity distribution in a breaking wave field, and that the Laplacian filter improves its estimation near the surface boundary.

Next, a velocity distribution around an obliquely descending eddy is evaluated with the Spectrum-Correlation method. Since this eddy is caused by a wave breaking, it is unsteady, has a three dimensional large-scale complex structure and involves innumerable air bubbles. Figure 5 is one of the successive pictures of this eddy. The pictures were taken with a hand-made simple camera system which consists of a single-lens reflex camera, a strobe scope and mirrors. A time lag between two successive pictures is 6 milliseconds. These pictures involve such three dimensional rotations of flow motion and strong turbulence, that the Spectrum-Correlation method cannot be applied directly. Then, to obtain the velocity distribution around the eddy, a smoothing filter is applied as a preprocessing of the method. Figure 6 shows an evaluated velocity distribution from the flow field shown in Fig. 5. Global features of the eddy are observed in Fig. 6, however, it seems that the velocity distribution is not well estimated. This estimation with low accuracy is not indeed induced by the method but caused by the pictures. Three reasons which diminish the worth of the pictures are pointed out; a long time lag between pictures compared with the speed of the flow, the strong three dimensional pattern of the flow field, and the severe enhanced turbulence. Settling these problems, improvement of

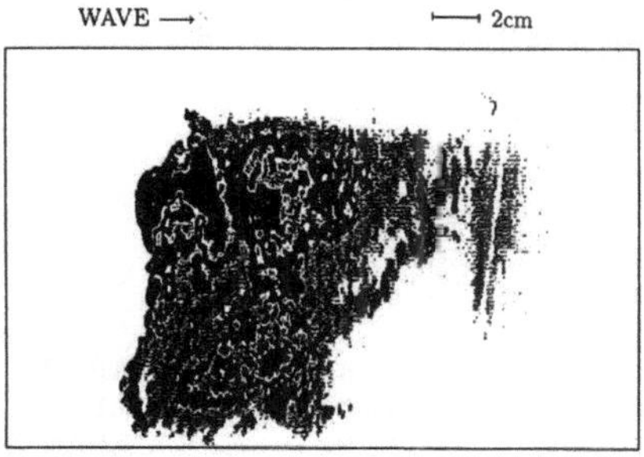

Figure 5: One of the successive pictures of a obliquely descending eddy under a breaking wave.

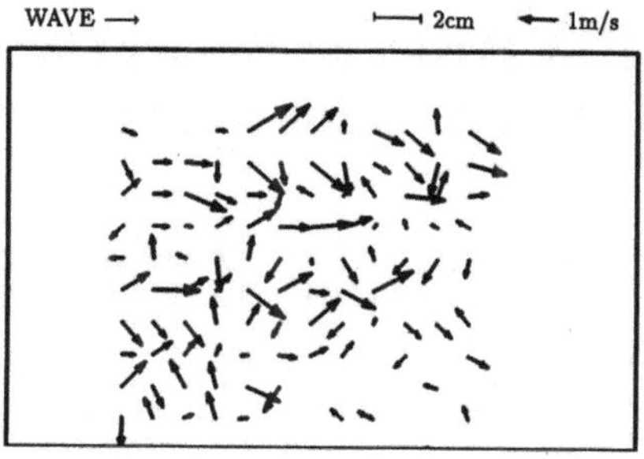

Figure 6: Evaluated instantaneous velocity distribution corresponding to Fig. 5, with a Spectrum-Correlation method and a smoothing filter.

the following points is required for the camera system; shortening of a time lag between pictures, and narrowing down a slit's width of the strobe scope. In other words, for some phenomena moving rapidly, a high speed camera and a strong slit of light are needed to take the pictures.

4. Conclusions

In this paper, a Spectrum-Correlation method to obtain an instantaneous velocity distribution was presented, and it was applied to analyze a flow field under breaking waves. This method has the merits that traced particles are unnecessary, and a processing time for estimation of a velocity distribution is relatively short. With the introduction of a Laplacian filter and a smoothing filter, as preprocessings of the estimation, the velocity distribution estimation is considerable improved for the flow near the boundaries and complex flow fields.

For more detailed analysis of the complex and air-entrained flow fields under breaking waves with this method, improvements in the camera system used in this study are required.

Acknowledgments: Thanks are given to Dr. Takewaka who supplied some of the image data used in this study, and to Nissan Science Foundation, for the support of this study.

References

1. Nishimura, Hitoshi and Satoshi Takewaka (1988): Estimation of a velocity field in a breaking wave region from VTR-image data using a correlation analysis (Japanese), Proc. of Coastal Eng., JSCE, No. 35, pp.45~48.
2. Mano, Akira and Shigeya Kamio (1990): Flow field analysis on air-entrained region of breaking waves by correlation method (Japanese), Proc. of JSCE, No.423/II -14, pp.171~180.
3. Okuno, Taketoshi and Taminori Kinoshita (1987): Development of a velocity measurement method using image processing technique (Japanese), Assoc. for the Study of Flow Measurement, Osaka, 4, p.177~184.
4. Hino, Mikio (1991): Some Improvements on the Velocity Estimation by FFT and Spectrum-Correlation Method, Tech. Rep., Dept. of Civil Eng., Tokyo Institute of Technology, No.44.

2D PTV Based on Binary Correlation Method and Its Application to Fluctuating Flows

Fujio YAMAMOTO*1, Tomomasa UEMURA*2, Mitsuo KOUKAWA*1, Manabu IGUCHI*3, Junichi OHTA*1 and Yoshihide SAKAI*4

*1 Fukui University, Dept. of Mech. Engg., Fukui, 910 JAPAN
*2 Osaka University, Fac. of Engg. Sci., Toyonaka,560 JAPAN
*3 Osaka University, Fac. of Engg., Suita Osaka, 565 JAPAN
*4 PFU Software Laboratory Co.,Unoke Ishikawa, 929-11 JAPAN

Summary
A two-dimensional particle tracking velocimetry (2D-PTV) based on binary correlation method which has been proposed by the authors before is applied to measure fluctuating flows around two circular cylinders. The experiments are carried out in the three types of tow cylinder arrangements: (a) tandem, (b) side by side, and (c) staggered. Analyzing the pictures of the flows visualized by the hydrogen bubble technique, instantaneous velocity distribution, root mean square (RMS) values of the fluctuating parts of the flows, streamlines and vorticity contours are obtained. From these experimental data the flow mechanism are discussed.
Key Words: Flow Visualization, Particle Tracking Velocimetry, Two Circular Cylinders, Fluctuating Flow

INTRODUCTION

Many types of digital image processing techniques for fluid flow measurement have been developed for these ten and several years. First, a method which computed flow velocities from the lengths of pathlines of tracer particle images was devised [1]. Then a method of particle tracking velocimetry (PTV) was proposed to identify automatically each particle in consecutive video pictures and compute the velocity of each particle. It has succeeded in solving the problem of vector directions [2]. A correlation method which computes the particle velocity vectors based on similarity of density distribution patterns of particle clouds between two pictures at a short time interval has been developed [3]. The authors have proposed so called a binary correlation method, which has a high speed algorithm for particle identification based on similarity of particle distribution patterns between two binarized pictures [4,5]. Recently the techniques of three-dimensional PTV have been also developed [6].

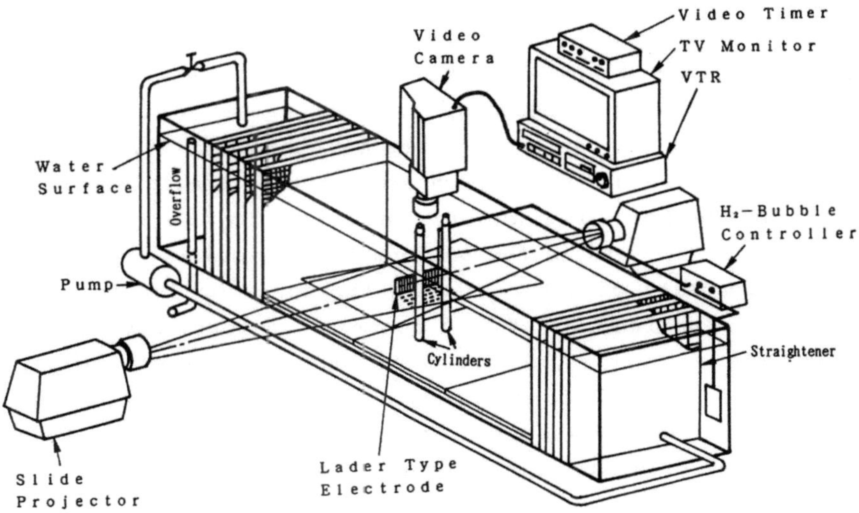

Fig. 1 Outline of Experimental Apparatus

The present study deals with fluctuating flows around two circular cylinders arrayed in the arrangement of (a) tandem, (b) side by side, and (c) staggered. After the flows are visualized by aluminum or hydrogen bubble tracer technique, 2D-PTV based on the binary correlation method is applied to measure the flow velocities of the whole field of the fluctuating flows. Using the results of instantaneous velocity vectors, vorticity contours, streamlines and root mean square values of the fluctuating velocities, the analysis and discussions of the flow mechanism are made.

EXPERIMENTAL APPARATUS AND PROCEDURE

Figure 1 shows the whole view of the experimental apparatus. It consists of uniform flow water channel, two circular cylinders, hydrogen bubble generator and its electrodes, lighting and video system, and digital image processing system including a personal computer. The image processor is of 8bit-256*256 pixel type and has the capacity of saving 16 picture frames. The two cylinder interaction fluctuating flow patterns are visualized and depicted using aluminum tracer technique in each case. A hydrogen bubble tracer technique is applied to the flow visualization and image processing. The velocity vectors, streamlines, vorticity contours, and RMS values of the flow velocity fluctuating parts are computed though fifty to one hundred consecutive pictures which are taken at the interval of 0.1s. Here the Reynolds number referred to the circular cylinder, Re = Ud/ν (U is the uniform flow velocity, ν is the kinematic viscosity of water), is in the range of 192 to 598. Before streamline function ϕ and contours of vorticity ζ are computed, the velocities at the locations of particle images are rearranged at grid points using an interpolation technique based on the inverse value of distance of four points in four quadrants whose original point is put at a grid point.

EXPERIMENTAL RESULTS AND DISCUSSION

The flow patterns around the two circular cylinders in tandem arrangement are classified into the three types based on the value of L/d. The classification agrees well with the Ohmi's results [7]. Figure 2 (a), (b) and (c) show the examples of instantaneous velocity vectors rearranged at grid points at a time, vorticity contours and streamlines for the tandem arrangement in the case of L/d = 5, respectively. In this case regularly alternating vortex shedding from the front cylinder can be seen. The fluctuation intensity of the flow takes the maximum at the front part of the rear cylinder. In addition, when L/d = 3, dead zone between the two cylinders are seen and the strong fluctuating flows are seen in the part of the rear cylinder. When L/d = 8, the strong fluctuating flows are observed between the two cylinders and expand along with the main flow. The vorticity is highest just behind the front cylinder and becomes lower flow approaches to the rear cylinder. Figure 3 shows the case of side by side for L/d = 3. From the nearly symmetric vortex shedding flow patterns about the downstream center line are seen. Figure 4 show the case of the staggered arrangement for L/d = $3\sqrt{2}\,d$ with the stages angle of 45°. The vortex shed from the front cylinder is inclined due to the interference of the rear cylinder.

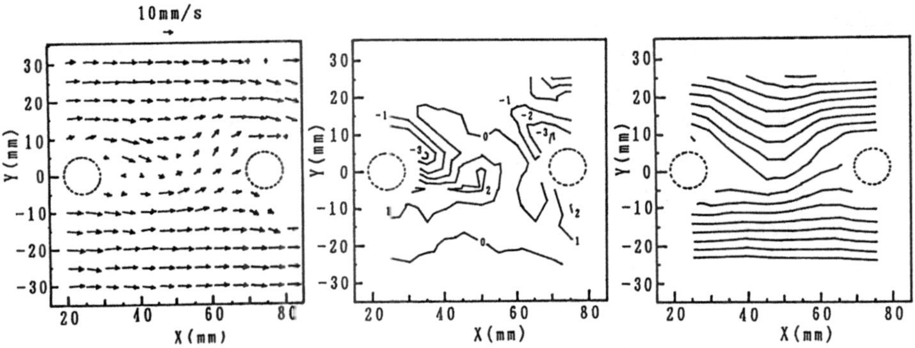

(a) Velocity Vectors (b) Vorticity Contours (c) Streamlines

Fig. 2 Measurement Results (Tandem, L/d = 5, Re = 208)

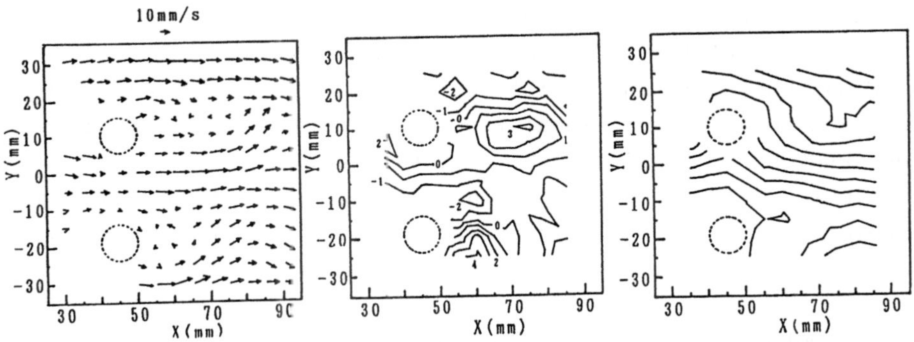

(a) Velocity Vectors (b) Vorticity Contours (c) Streamlines

Fig. 3 Measurement Results (Side by Side, L/d = 3, Re = 194)

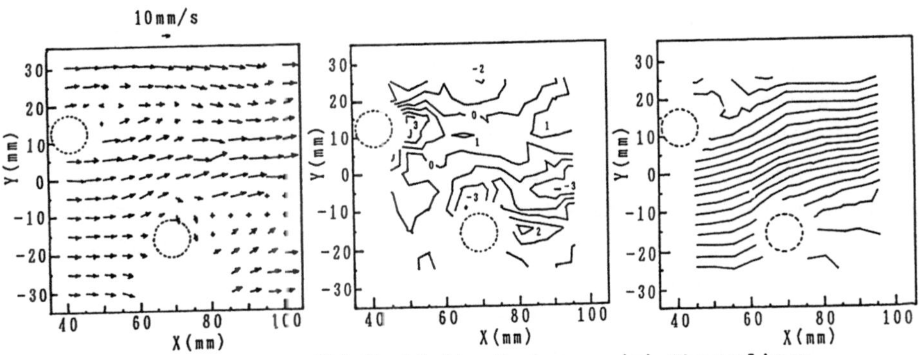

(a) Velocity Vectors (b) Vorticity Contours (c) Streamlines

Fig. 4 Measurement Results (Staggered, L/d = $3\sqrt{2}$, Re = 196)

CONCLUSION

The fluctuating interaction flows around two circular cylinders arrayed in the three cases of (a) tandem, (b) side by side, (c) staggered arrangement were measured using a two-dimensional particle tracking velocimetry (2D-PTV) based on binary correlation method. From the results of instantaneous velocity vectors, vorticity contours, streamlines and root mean square values of the fluctuating velocities the flow mechanisms are discussed.

REFERENCES

1. Dimotakis, P.E.; Debussy,F.D.; Koochesfahani,M.M.: Particle streak velocity field measurement in a two-dimensional mixing layers, Phys. Fluids, 24 (1981),995-999.

2. Kobayashi, T.; Saga, T.: A real-time velocity measurement algorithm for two-dimensional flow flow fields, Proc. 2nd Intl. Symp. FLUCOME'88, Sheffield U., (1988),174-178.

3. Kimura, I.; Takamori,T.: Image processing of flow around a circular cylinder by using correlation technique,Proc.4th Intl. Symp. on flow Visualization, Paris, Hemisphere Publ.Co.,(1986), 221-226.

4. Yamamoto,F.; Uemura,T.; Koukawa,M.; Itoh,M.; Teranishi,A.: Application of flow visualization and digital image processing techniques to unsteady viscous diffusing free doublet flow, Proc. 2nd Intl. Symp. FLUCOME'88, Sheffield U.,(1988), 184-188.

5. Uemura, T.; Yamamoto, F.; Koukawa, M.: High speed algorithm for particle tracking velocimetry using binary, J.Visualization Soc. of JAPAN.,Vol.10,No.38(1990),58-64. (in Japanese)

6. Nishino,k.;Kasagi,N.; Hirata,M.: Three-dimensional particle tracking velocimetry based on automated digital image processing, Trans.ASME, J.Fluids Eng., Vol.111,No.4(1989),384-391.

7. Ohmi,k.:A study on vortex formation around two-dimensional body using flow visualization techniques, Dr.Thesis at Osaka University, (1991). (in Japanese)

News 3D-PTV Algorithms Based on Correlation Between Binary Images

Tomomasa UEMURA*1, Fujio YAMAMOTO*2, Manabu IGUCHI*3

*1: Fac. of Engng. Sci., Osaka Univ., Toyonaka
*2: Fac. of Engng., Fukui Univ., Fukui
*3: Fac. of Engng., Osaka Univ., Suita

Summery

Two new algorithms of three-dimensional binary correlation method(BCM) are proposed for particle tracking velocimetry. The principal part of the method is an algorithm to evaluate a similarity between two local patterns of distributed particles in a three-dimensional space. Introducing a function, which represents a degree of closeness of two particles, the similarity with a reduced consumption of computer resources. Another method compares distributions on a two-dimensional plane by tranforming the coordinates of original patterns. The performances of the methods are examined by computer simulations.

Introduction

A method of particle tracking velocimetry(PTV) has been applied successfully to the measurements of three-dimensional flow 1] for these years. In contrast to the other methods, the most measurements has been made by so called the four-time-step-tracing method. In this paper, authors are going to propose a new three-dimensional PTV algorithms, which are improved from the two-dimensional binary correlation method(BCM) [2][3]. Their performances are expected as follows: a high analyzing speed, an acceptable memory consumption to a personal computer system, and sufficient reliability. Some results of preparatory experiments and simulations are shown for discussions.

Three-Dimensional Binary Correlation Method

The principle of BCM is based on pursuing the conformity of local distribution patterns of particles between sequential images. The concept is similar to identifying a specific star by using constellations as a clue. The proposed algorithms are designed making both requirement for memory consumptions and calculation time minimum. The former one achieved the purpose through simplifying the calculation of the coefficient by introducing a function named a proximity function. The second one solved the issue by transforming particle distributions in the three-dimensional space to a two-dimensional plane.

Definition of Cross Correlation Coefficient

A cross correlation coefficient C_{fg} between two three-dimensional functions $f(x,y,z)$ at time t and $g(x,y,z)$ at time $t+dt$ can be defined by the following equation;

$$C_{fg}=\frac{\int f(x,\ y,\ z)\cdot g(x,\ y,\ z)\,dV}{\sqrt{\int f^2(x,\ y,\ z)\,dV\cdot\int g^2(x,\ y,\ z)\,dV}} \tag{1}$$

When the above equation is applied to two binary digital images "i" and "j", a coefficient between the two patterns is calculated by superposing them so that the centroids of the center particles coincide. The numerator means the volume of overlapped parts, and the denominator means total volume of particles contained in each pattern. The combination having the maximum C_{ij} implies that the two particles at the center of the distributions are the same. Motions of particles can be traced by applying the above process to sequential pictures.

Introduction of Proximity Function

In order to simplify the calculation of the coefficient, actual particle images of various shapes and sizes are replaced by one imaginary particle with an appropriate shape and size. Then, the equation (1) is reformed as follows;

$$C_{ij}=\sum_{s} f\,(S_{st}\,/\,d)\,/\sqrt{n_i\cdot m_j} \tag{2}$$

where S_{st} is the distance between gravity centers of two adjacent member particles 's' and 't' in each pattern. 'd' is a characteristic diameter of the imaginary particle. n_i and m_j are numbers of particles contained in the local patterns, respectively.

The function *f(s/d)* is named a proximity function [4] which represents the overlapping volume ratio of two particles. The function used in the present example is shown in figure 1.

The merits brought by the function are as follows [5]:

1. A shape of the imaginary particle can be defined independent of the pixels composing the digital image.
2. The calculation of C_{ij} in the three dimensional space can be performed as efficiently as in the two dimensional space.
3. By giving the adequate form of the function, the resemblance between the patterns could be evaluated better and easier.

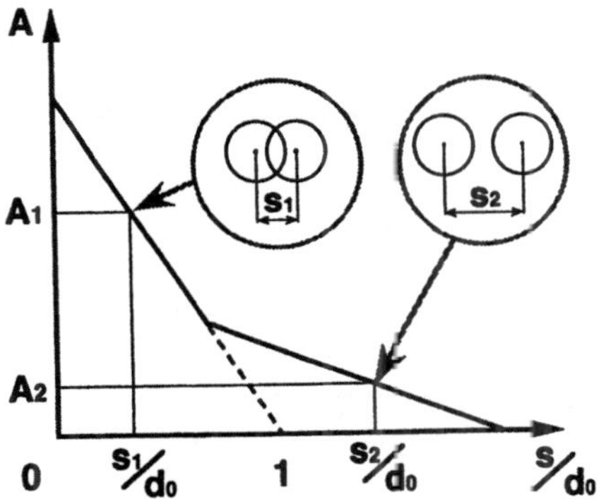

Fig.1 An example of proximity function used in the present analysis.

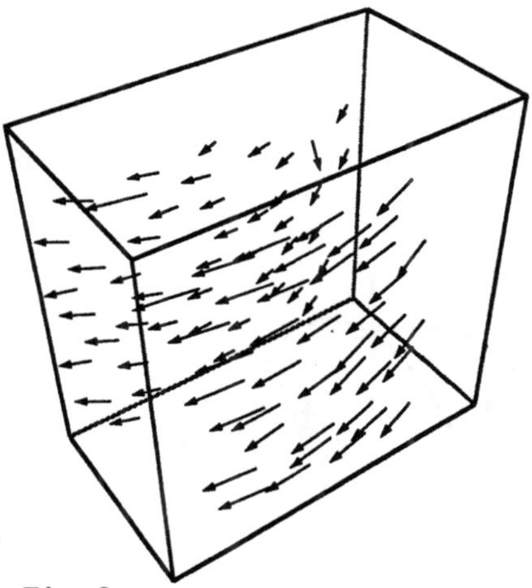

Fig.2
Three-dimensional vectors measured by the first method.

Results of Tracking Particles

In the experiment of particle tracking, before and after the motion distributed particles, three pictures are taken simultaneously from different directions. In the case of the parallel motion, 93 vectors are obtained from about 110 particles. When rotation is combined to the parallel motion, 84 correct vectors are obtained from 103 particle images(figure 2). In contrast to the fact that the double checking process is indispensable in the two-dimensional PTV by binary correlation method [2], particle patterns in a three-dimensional space happen to rarely become similar each other.

Method of Spherical Surface Projection

The second algorithm is based on the coordinates transformation of three-dimensional coordinates of particles. Taking a centroid of the center particle as an origin, the coordinates of the other particles are projected radial direction on a spherical surface (figure 3), then from the surface to a flat *Z*-plane (figure 4) following Sanson's map-projection formula.

$$x=R_O\cdot\phi\cdot\cos\theta,$$
$$y=R_O\cdot\phi. \qquad (3)$$

where, (x,y) are coordinates on the *Z*-plane, R_O is a radius of the sphere, θ and ϕ are the latitudinal and longitudinal angle, respectively.

On the *Z*-plane, a diameter of a projected particle is inversely proportional to the radial coordinate. The patterns on the *Z*-

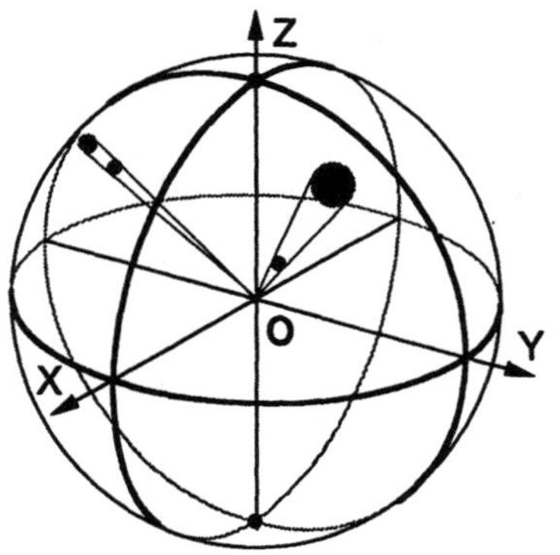

Fig.3
A polar coordinates system and projection of particles onto a spherical surface.

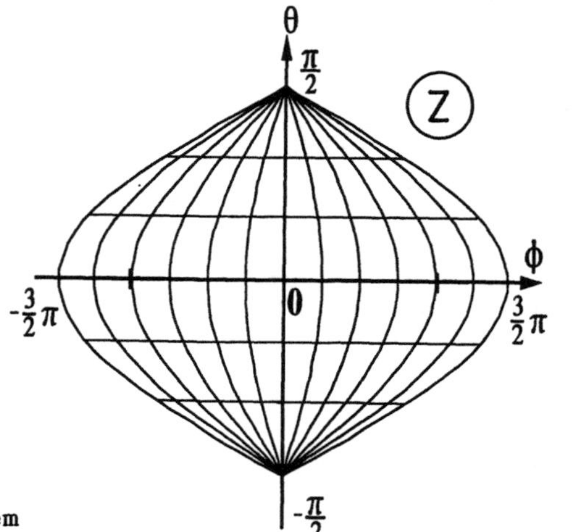

Fig.4 Coordinates of Sanson's nongeometric map-projection.

plane can be compared by making use of the proximity function. The rates of overlapping area (*A*) are estimated in the following manner for every member particle;

1. find a closest particle on the *Z*-plane from another distribution,
2. check the distance between the two particles,
3. find *A* using the proximity function.

In the present method, the first step in the above procedure can be made two-dimensionally, and the resemblance between two distributions can be recognized by visual inspection easily.

Conclusions

Two algorithms of binary correlation method for three-dimensional PTV are proposed. The performances are examined by both simulations and basic experiments.

The study is partly supported by the Scientific Research Fund, Grant No.02650484 of Ministry of Education, Science and Culture, and also Research Development Corporation of Japan No.3-33.

References

1. Nishino,K.; Kasagi,S.; Sata,Y.: Three-Dimensional Particle Tracking Velocimetry Based on Automated Digital Image Processing. Trans. ASME, J.Fluids Eng. 111-4 (1989) 384-391.

2. Uemura,T.; Yamamoto,F.: A High Speed Algorithm of Image .lm4 Analysis for Real Time Measurement of Two Dimensional Velocity Distribution. ASME FED-85 (1989) 129-134.

3. Uemura T.; Yamamoto F.; Koukawa M.: A High Speed Algorithm for Particle Tracking Velocimetry Using Binary Picture Correlation(in Japanese). J. Visualization Soc. Japan, 10-38 (1990) 196-202.

4. Uemura T.; Yamamoto F.; Nakajima H., Mizukoshi, W.: Study on the Application of Binary Correlation Method to 3-D Velocity Measurement(in Japanese). J. Visualization Soc. Japan, 10-Suppl.2 (1991) 77-80.

5. Yamamoto,F.; Uemura,T.; Iguchi,M.; Morita,Z.: A Binary Correlation Method Using a Proximity Function for 2-D and 3-D PTV. FED-128 (1991) 23-28.

Holographic Particle Image Velocimetry(HPIV)- An Extension of PIV to a 3-D Measuring Method

F. Dreesen, H. Hinrichs, K. Hinsch, A. Roshop

Fachbereich 8-Physik
Carl von Ossietzky Universität
D-2900 Oldenburg, Germany

Introduction

Particle Image Velocimetry is now a well-established, powerful tool for 2-D flow diagnostics. In its most common applications, a thin sheet of laser light illuminates a portion of the flow. A double exposure photograph is taken to record small light scattering particles seeding the flow. The 2-D flow field data obtained by this technique constitute great progress in comparison with single point measurements by conventional anemometry. Opposite to other whole-field techniques like smoke wire visualisation, for example, PIV is a quantitative measuring method. The opportunity to provide velocity data at one instant of time from a large field of view supports studies in the rapidly growing field of non stationary flows. This support would greatly gain, could PIV be extended in two respects:

1. Simultaneous recording over a certain region in depth
2. Registration of out-of-plane-components of the velocity vectors.

We present independent solutions to each of the above mentioned problems which, however, could be combined to provide the complete velocity information on a complex flow field. The solution to the first part makes use of a principle described by Abramson [1,2], the so called light-in-flight recording. The second problem is tackled by stereoscopic viewing. The deep-volume performance is shown in a wind tunnel study of vortex shedding behind a cylinder. The out-of-plane velocity measurement is demonstrated for model particles enclosed in a block of plastic. Both techniques retain the standard interrogation methods and demonstrate the advantages of fully optical processing.

Recording a voluminous flow field by holographic storing of a batch of light sheets

A holographic setup is used to register instantaneous data from a 3-D flow field. The application of holography seems straightforward: the illuminating light sheet is expanded in depth and a reference beam is added. Upon reconstruction an extended volume of the flow is available for inspection. Difficulties arise, however, if reliable PIV techniques shall be used for interrogation. When a PIV record is taken from the reconstructed image of the particle field by focusing on a certain plane in depth the many defocused images of particles from other regions contribute a considerable noise background. To overcome this problem a method has been developed that makes use of the limited coherence of laser light to select single sheets in depth in the evaluation step. The basic idea is illustrated in the example of Fig. 1., showing the setup used for an experiment on a real flow in a wind tunnel. A single light sheet will be redirected repeatedly through the flow at a spacing smaller than the flow scales to be resolved. The example illustrates this for just two passes. Special beam shaping optics are required to guarantee good quality light sheets at all passes through the flow. In this way a batch of light sheets can be illuminated by a doubly pulsed ruby laser at one instant of time. Light scattered by particles filling all the sheets is reaching a holographic plate. A number of reference beams arrive at the plate under different angles. Each reference beam has been designed in optical path such that it produces interference with light from a single sheet only. This takes into account the limited laser coherence and optical detours between different sheets and reference beams. By shifting through the set of reference beams during reconstruction, particles in one sheet can be evaluated without disturbing information from any other plane in depth. Taking a photo of each sheet results in an ordinary PIV record which can be interrogated with well-known techniques.

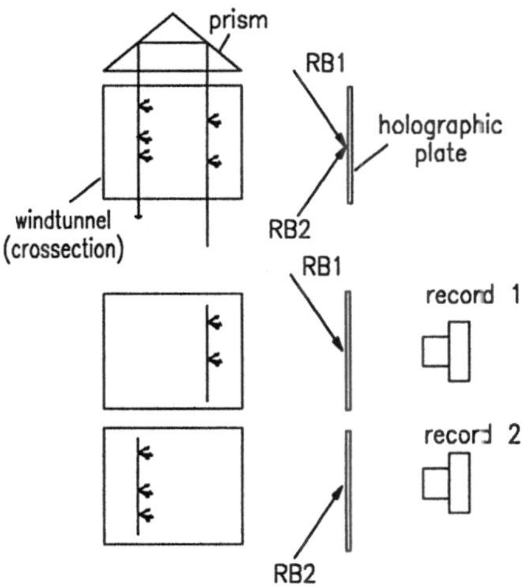

Fig. 1: Simultaneous holographic recording of two light sheets in a flow and their separate reconstruction

There are some restrictions to this method: First, the number of sheets is limited by the holographic recording process. The quality of the reconstructed images decreases with the number of holograms stored on a single plate. Secondly, the coherence function of the laser repeats with a period equal to twice the resonator length. The ratio of this length to the coherence length gives the number of sheets that can be reconstructed without interference. If a higher resolution in depth asks for more sheets, those sheets that are reconstructed simultaneously should be spaced in depth far enough to minimize cross talk.

3-D velocity registration by stereoscopic recording

In flows with a depth component of the velocity vector the light sheet must be shaped in such a way that the majority of particles will not leave the illuminated field due to their out-of-plane motion. The depth displacement of particles can then be recovered from a stereoscopic recording setup [3]. Our recording scheme is outlined in Fig. 2a . Double exposure particle images are observed with two cameras at different angles. The depth displacement is coded in differing pair separations.

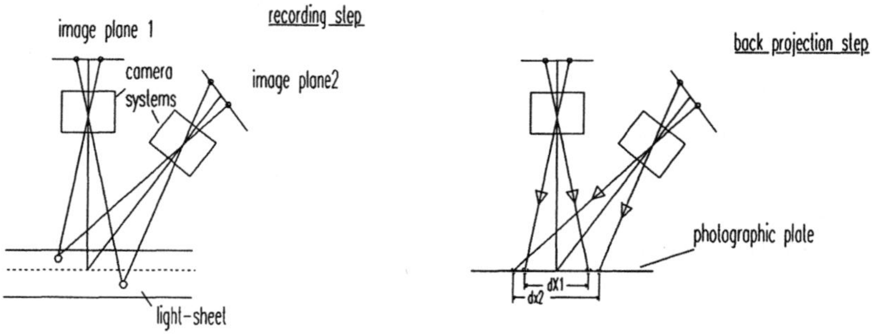

Fig. 2. Stereoscopic recording setup
a) particle recording b) back projection step

The larger the angle between the viewing directions, the better the resolution. When the optical axes of the cameras point at the centre of the field of view, however, only a limited region will be in focus. The small F-numbers needed for small particle images result also in a small depth of focus. A whole-field focused image requires to tilt the back plane of the camera in such a way that the planes of object, lens and image meet in a common line (Scheimpflug condition). Now there is a new problem: Magnification becomes position dependent and the image of the light sheet is thus distorted. This distortion problem is solved by a backprojection technique (Fig. 2b). Positive copies of the particle images recorded are repositioned in the image plane, illuminated from the back and imaged onto a photographic plate placed in the plane of the former light sheet. This backprojection is done with a white light source to avoid speckle noise.

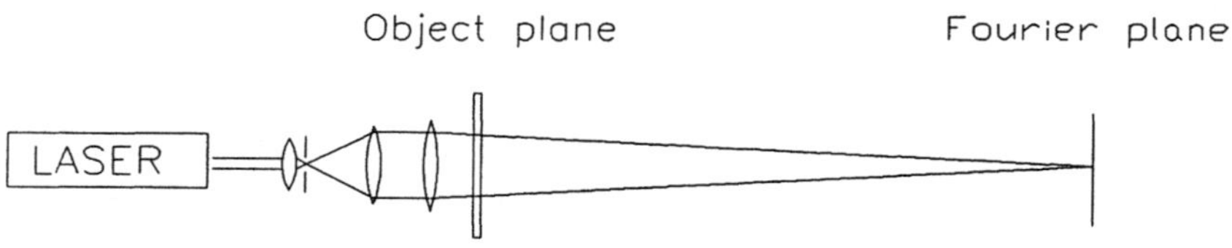

Fig. 3. Setup for optical Fourier transformation

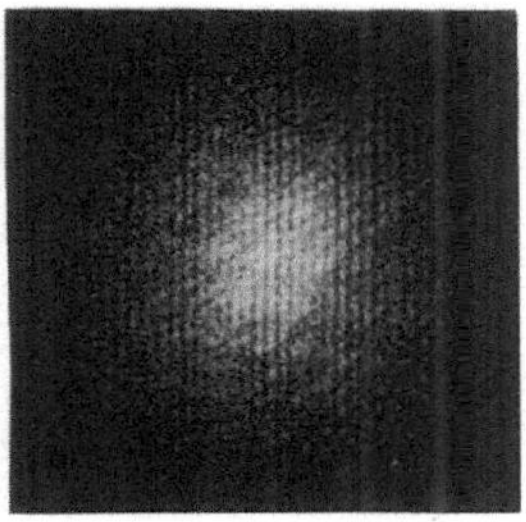

Fig. 4a. Youngs fringe system from a stereoscopic particle image input showing visibility modulation.

Fig. 4b. Autocorrelation by optical Fourier transformation Fof fringe system showing additional peaks caused by out-of-plane velocity.

Interrogation of a contact copy of this plate is done in the familiar setup for optical Fourier transformation (Fig. 3): A small area of the input record is illuminated by a focused laser beam producing the Fourier transform in the output plane. Due to the two different particle separations the resulting fringe system shows a beat modulation in fringe visibility (Fig. 4a). From the recording geometry and the beat period the out-of-plane velocity component in the observed area can be calculated. When the modulated fringe system is recorded on a transparency it also can be Fourier transformed optically to produce the autocorrelation function of the particle arrangement. In extension of the results obtained by ordinary 2-D PIV applications this output contains additional peaks (Fig. 4b) that are separated by a distance directly proportional to the depth component of the displacement.

Experimental results

A flow which requires a 3-D measuring method is irregular vortex shedding behind a cylinder. Wind tunnel measurements were carried out in a region, where the vortex shedding is influenced by statistical fluctuations. Two sheets are placed downstream behind a cylinder at a spacing in depth of 4cm. The experimental results are shown in Fig. 5: Recall that both vectorplots are obtained at the same instant of time. Thus, we see how the flow distribution varies in depth. Obviously , due to instabilities the flow is not merely two-dimensional. It is known that in some cases there arise obliquely shedded vortices. For a systematic study a statistical appoach has to be made with a large number of flow records. The holographic technique presented would provide a useful tool for such an investigation. In addition the depth velocity component could be measured by stereoscopic evaluation of the holographic recordings.

Fig. 5. Flow configuration behind a cylinder (Diameter 6 mm, Re=400) in two planes normal to the cylinder axis spaced 4 cm in depth. Size of the flow field 5.7cm X 2.4cm.

Conclusion

A concept has been presented that retains the well-established processing schemes of particle image velocimetry for the study of 3-D flow configurations. Results obtained from wind tunnel experiments of vortex shedding have been presented. By adding a stereoscopic recording scheme the whole 3-D structure of vortex shedding could be measured.

References

1. Abramson, N.: Light-in-flight recording by holography. Opt. Lett. 3 (1978) 121.

2. Hinrichs, H.; Hinsch, K.; Kuhfahl, G.; Meinlschmidt, P.: Holography with controlled coherence for 3-D particle image velocimetry (PIV). In: Tschudi, T. (Ed.). Holographics 1990. Mesago Publishing GmbH.

3. Arroyo M. P.; Greated, C. A.: Stereoscopic particle image velocimetry. Meas. Si. Technol. 2 (1991) 1181

Three-Dimensional Particle Image Velocimetry Using Velocity Vector Histogram

Koji OKAMOTO

Nuclear Engineering Research Laboratory,
The Faculty of Engineering, University of Tokyo,
Tokai-mura, Ibaraki, 319-11, JAPAN

Summary

In the three-dimensional Particle Image Velocimetry, three-dimensional positions of the tracer particles are needed. Because the obtained image is two dimensional, at least two images taken from different angles are used to determine the three-dimensional particle positions. Therefore, the identification of the same particles in both images is needed. Using the VVH method, three-dimensional particle identification could easily be carried out only from two different angle images. The effects of the particle concentration ratio and the image resolution are clarified. Using this method, the three-dimensional velocity vector can be obtained with high accuracy.

Introduction

In the three-dimensional Particle Image Velocimetry (PIV), three-dimensional positions of the tracer particles are needed. To determine the three-dimensional particle position, at least two images taken from different angles are used, because the image is two dimensional. Therefore, the identification of the same particles in both images is needed. The PIV methods are based on the identification technique of the same particle or the same particle groups between two different time images. The calculation of three-dimensional particle position is thought to be the identification of the same particle between two different angle images.

The author proposed the new PIV called VVH (Velocity Vector Histogram) method[1], in which the particle identification between different angle images and that between different time images could be operated simultaneously.

In the VVH method, when the particle concentration rate in the images increases, the miss identification rate of the particle three-dimensional position also increases. In this study, the effects of the particle concentration ratio on the accuracy of the measured velocities were discussed.

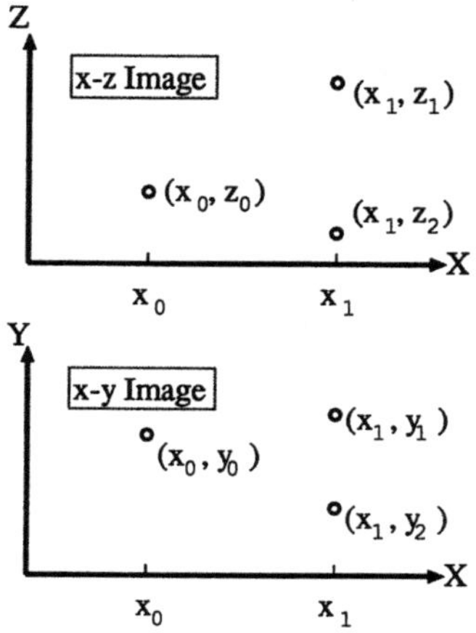

Figure 1. Three-dimensional particle position

Three-dimensional VVH method

Figure 1 shows an example of the images taken from the different angles. From the upper image the x and z coordinate position are determined, and from the lower image the x and y position are determined. If only one particle exists at the same x coordinate in both images, the three-dimensional position is determined uniquely [fig. 1 (x_0)]. While some particles exist at the same or very close x coordinate in both images, e.g., (x_1,y_1),(x_1,y_2) for x-y image and (x_1,z_1),(x_1,z_2) for x-z image, [fig. 1 (x_1)], the three-dimensional position can not be determined uniquely. In this case, to determine the position uniquely, another third image taken from different angle should be needed. Therefore, three cameras are used to record the three images taken from different angles, causing the measurement system to be a very complex one[2].

In the VVH method[1], the correct velocity vector was selected from the velocity vector histogram for all considerable velocity vectors. Using this method, highly accurate velocity could be obtained even from noisy images. In this method only two images taken from different angles are used to identify the particle three-dimensional positions. The uncertainty of the three-dimensional particle identification could be abandoned in the velocity vector histogram. For example, in the case of the fig. 1 (x_1), all positions are considered to be the particle positions, e.g., the considered particle positions are (x_1,y_1,z_1),(x_1,y_1,z_2), (x_1,y_2,z_1) and (x_1,y_2,z_2). Two positions out of these four positions are the correct positions. Though the remainder of them are miss identification, they can be considered as the noise particle, and can be omitted from the histogram like any other noise. Therefore, the selection of the correct positions from the all considerable positions is not needed. Using the VVH method, three-dimensional velocity vectors could be easily obtained from the two directional views.

Miss identification ratio

In the VVH method, the miss identification particle positions, i.e., noise particles, can be omitted in the velocity vector histogram. Because the noise particles are distributed at random in the three-dimensional histogram, the correct vectors can be obtained even in the relatively high noise condition. However, when the particle concentration rate in the flow is very large, the number of miss identification particles, i.e., noise particle rate, become also large, causing a lower reliability of the obtained vectors.

The noise particle ratio (N_e/N) can be calculated as follows,

$$\frac{N_e}{N} = 2\frac{S}{P}N \tag{1}$$

where N is the number of the tracer particles, N_e is the number of noise particles, P is the pixel number of the images and S is the pixel width in which the particles are considered to be at the same location.

For the simple measurement system, usually P is about 200 pixels and S is $1 \sim 3$ pixels. Therefore the equation (1) can be rewritten as, $N_e/N \approx 0.02N$. This equation shows that when the particle number is above 50, the number of the noise particles is greater than that of the correct particles.

Effects of miss identification

The noise particle effects on the VVH method were evaluated by a computer simulation. Figure 2 shows the simulated geometry that is the same geometry of Chen et al.[3]. The flow in the cubic region was solved by the three-dimensional finite element method code for laminar flow, which was called the reference velocity distributions here. Small particles were inserted into the reference flow with 5% artificial random noise, and the particle position was numerically simulated. The bottom and side view images were used to determine the three-dimensional particle position as shown in fig. 2. In the simulation, the area of the image was 200×200 pixels, and the resolution of the image was 1%, i.e., $P = 200$ and $S = 2$. The target region was divided into 125 small sub-region, i.e., $5 \times 5 \times 5$, in which the velocities were considered to be similar. If the particles were distributed at random in the target region, only $N/125$ particles existed in the small sub-region. Therefore, to obtain the enough particles many images should be needed.

The number of the simulated particles was changed from 50 to 200. In the case of $N = 100$, the noise particle number was calculated about 200 from eq. (1). The results of the obtained vectors are shown in fig. 3, which agreed well with the reference velocity vectors, except the outlet region. In this region, because the particles out of the outlet were not recorded on the images, the velocities could not be calculated correctly. The measurement technique should be improved.

Figure 4 shows the velocity distributions at the cross section. Because the y-z plane was parallel to the epipolar line of two images, the correct particles and noise particles were distributed on the same y-z plane, thus, it was thought to be difficult to obtain the correct vector. However, the direction of the obtained vectors was similar to that of the reference vectors, resulting in the correct vectors to be obtained.

Table 1 shows the simulated errors for several numbers of particles with the calculated noise particles' number from eq. (1). In every cases, the errors were about 8%, concluding the effects of the number of the noise particles on the obtained vectors to be relatively small. The accuracy of the obtained vectors could be improved by using the higher resolution images and by increasing the number of sampling images.

Conclusion

Using the VVH method, three-dimensional particle identification could easily be carried out only from two different angle images. The effects of the particle concentration ratio and the image resolution are clarified. Using this method, the three-dimensional velocity vector can be obtained with high accuracy.

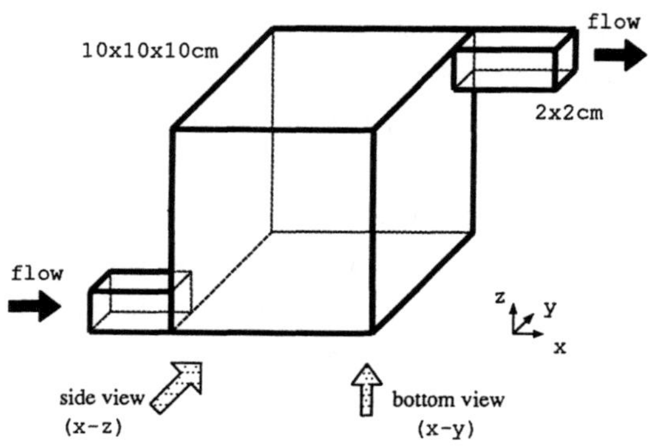

Figure 2. Test section

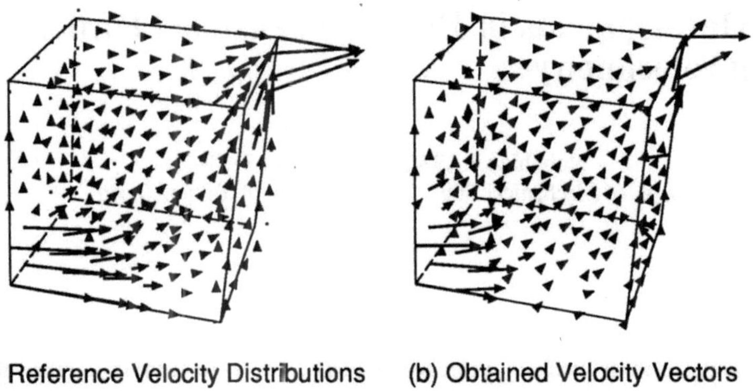

Figure 3. Results of VVH method

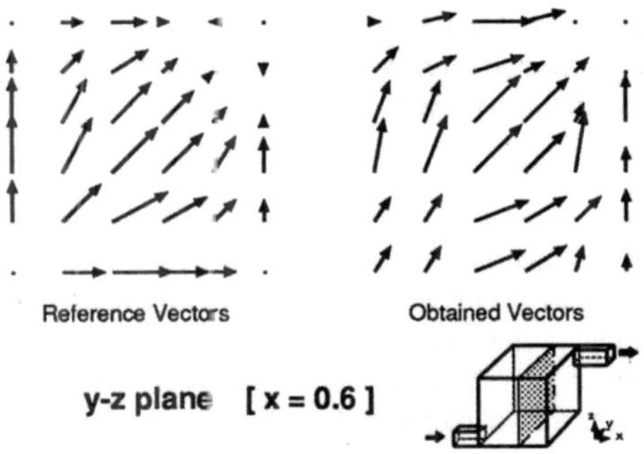

Figure 4. Cross sectional velocity distributions

Table 1. Effects of noise particles

Particles (N)	Noise (N_e)	error [%]
50	50	7.8
100	200	7.1
200	800	7.9

References

[1] Okamoto, K. et al., FLUCOME'91, (1991), pp.601

[2] Nishino K., et al., Trans. ASME, J. Fluid Eng., **111**, (1989), pp.384

[3] Chen L.J. et al., FLUCOME'91, (1991), pp.407

Recent Developments in Three-Dimensional Particle Tracking Velocimetry

Nobuhide KASAGI and Yutaka SATA
Dept. of Mechanical Eng., The University of Tokyo, Japan

Summary

Three-dimensional particle tracking velocimetry (3-D PTV, hereafter) is a relatively new measurement technique which can provide information on all three instantaneous velocity components in three-dimensional space. In this paper, the 3-D PTV technique originally developed by Nishino et al. [1] and later improved by the present authors is presented. Three TV-cameras and sophisticated camera calibration are adopted in order to achieve high accuracy and high spatial resolution of measurement. The present technique was applied to various fundamental turbulent flows and proved to be reliable and quite effective in both water and air flow measurements. Some examples of the turbulence measurements, e.g., in a fully developed channel flow, a separated and reattaching flow over a backward-facing step, and a flow in a square cross-section duct, are presented.

Introduction

As modern engineering turbulence measurement tends to be more complex, it has become obvious that the conventional flow measurement techniques such as hot-wire and laser-Doppler velocimetry have a serious disadvantage, i.e., difficulty in simultaneous measurement of all three velocity components at multiple locations. This fact prevents one from obtaining important turbulence statistics, e.g., Reynolds stress tensors, turbulence length scales, and two-points correlations tensors. In contrast to the above techniques, the three-dimensional particle tracking velocimetry can readily provide instantaneous spatial velocity vector distributions in a flow field [1-5]. The 3-D PTV herein described [1-3] is characterized by its employment of three TV-cameras and sophisticated camera calibration which results in high accuracy and high spatial resolution of the measurement. Furthermore, fully automated fast data reduction procedures are established to obtain a sufficient amount of data for calculation of turbulent statistics. In this paper, the latest 3-D PTV system at the University of Tokyo is briefly introduced and some examples of its application to turbulent flows are presented.

3-D Particle Tracking Velocimeter

The present technique is based on 3-D tracking of tracer particles suspended in a flow. They are observed from arbitrary viewing directions by three TV-cameras simultaneously and consecutive TV frames are recorded. From the recorded images, instantaneous 3-D particle positions are calculated, and then the particles are tracked over consecutive time steps.

The measurement system shown in Fig. 1 contains a light source, three TV-cameras, three laser disk recorders, a digital image processor (512x480 pix., 8 bit gray level), and an engineering workstation. Three TV frames simultaneously acquired by the cameras are respectively recorded on the laser disks. The time interval between consecutive images can be set to be either 1/30 s or 1/60 s by recording images onto TV frames or alternative odd and even fields, respectively. At the data reduction stage, replayed images are A/D converted by the image processor and transferred to the work station in which particle coordinates on the image

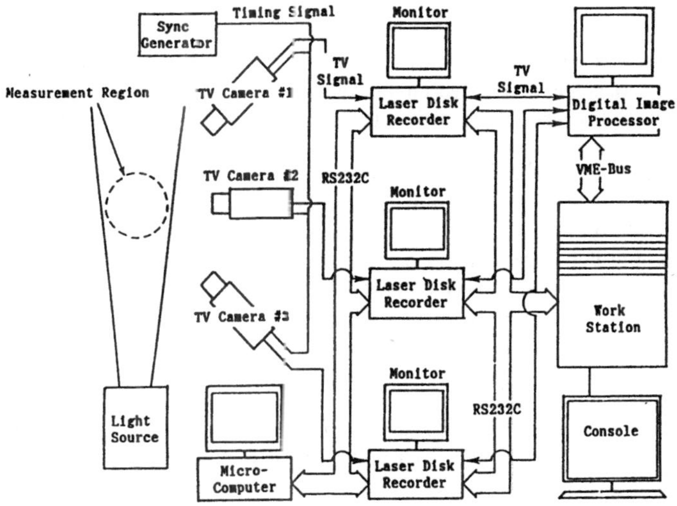

Figure 1. Three-dimensional particle tracking velocimeter system.

plane and 3-D particle trajectories are calculated.

Figure 2 summarizes data reduction procedures, which are classified into camera calibration and particle tracking [1,3]. In the camera calibration, eleven camera parameters, which define the orientation of each camera, are calculated first by taking images of calibration points, and then numerically solving the coplanarity equations which relate absolute and image coordinates of the calibration points in terms of the camera parameters. After acquisition of the calibration images, tracer particles are introduced into the flow field and their images are continuously recorded. In the particle tracking procedure, the particles' 3-D trajectories are calculated from the replayed images by the following sequences, (i) calculation of particle coordinates on three image planes, (ii) calculation of 3-D particle positions, (iii) tracking of 3-D particle trajectories over multiple (e.g., four) time steps, (iv) identification of independent trajectories. These procedures are described in detail by Sata and Kasagi [6]. Note that the use of three TV cameras with the above mentioned camera calibration is quite effective in calculating the particle positions both accurately and efficiently and that the particle tracking over four consecutive time steps is favorable for avoiding false particle trajectories [2,6]. As a result, the present 3-D PTV system has tracked a maximum of 630 particles at a time in the measurement of an axisymmetric turbulent free jet [7], and it is estimated by numerical simulations that more than 1000 particles can be tracked under a favorable condition. The above procedures are fully automated by the computer software developed, and only 15 s is needed to process a set of image

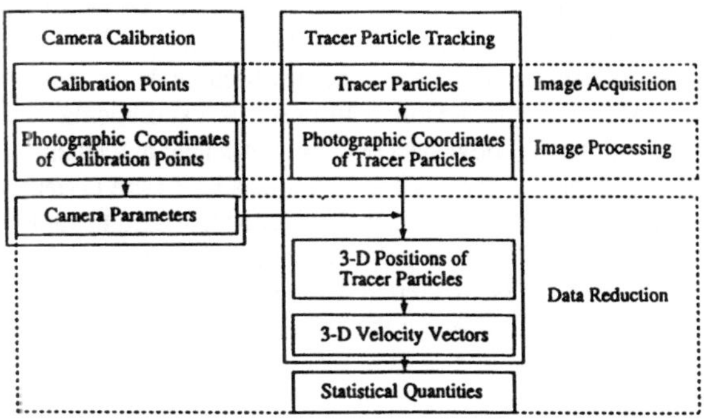

Figure 2. Data reduction procedures [3].

data for yielding a spatial vector distribution.

The tracer particles used in water flows are made of Nylon 12 of which density is 1.02 and their average diameter, which is optimized according to flow conditions, is typically about 200 μm. From numerical analyses based on the equation of particle motion, it is ensured that those particles' traceability should be satisfactorily good [1]. In air flow measurement, selection of tracers still remains to be a main technical problem. The tracers that the authors have adopted in a turbulence measurement described later are plastic microcapsules and tiny helium-filled bubbles. The former is about 100 μm in diameter and its density is about 30 times that of air, while the latter is 0.5-1.0 mm in diameter and almost neutrally buoyant. Some of these specifications are not suitable for air flow measurement when the flow is turbulent, but the 3-D PTV is yet capable of producing useful turbulence data as long as they are limited to statistics of low-order moment such as turbulent kinetic energy and Reynolds stresses.

Range of applicability of 3-D PTV

In turbulence measurement, it is important to examine whether the dynamic range of a measurement technique is wide enough to resolve turbulent fluctuations of various scales ranging from microscales to macroscales. Since the present technique is based on the measurement of particles' displacement, the following general requirements must be satisfied [5].

(a) The particle's displacement, which is caused by the microscale turbulence during the time interval, is sufficiently larger than the spatial resolution of the image processing. At the same time, the displacement caused by the macroscale turbulence are sufficiently smaller than the size of measuring volume.
(b) The tracking time interval is shorter than the smallest time scale of turbulence, while the measurement timespan is long enough compared to the largest time scale.
(c) The tracer particle diameter is smaller than the microscale of turbulence.

From these requirements, the uppermost Reynolds number of the 3-D PTV's applicability range is roughly estimated to be smaller than an order of 10^5 (see, Kasagi & Nishino [5]). To the authors' experience, sometimes all the requirements for the measurement of small scale turbulence may not be strictly satisfied, and in such a case proper evaluation and compensation of the turbulence statistics data should be indispensable.

Turbulence measurement with 3-D PTV

The 3-D PTV has been applied to various fundamental turbulent flows by Kasagi and his co-workers as listed in Table 1. In most of these measurements, mean velocities as well as turbulence statistics such as turbulent kinetic energy, Reynolds stresses and triple correlations have been obtained. Their accuracy has been evaluated and confirmed to be sufficiently good through extensive comparison with previous experimental and numerical data. For instance, the turbulent channel flow data [8] are in excellent agreement with a direct numerical simulation [15], as described below. Those PTV data bases are currently being used to evaluate various turbulence models and numerical simulations and also to study turbulent transport mechanisms. In some cases [7, 16], two-point correlation tensors of turbulence have been further calculated to investigate the quasi-coherent structures in turbulent free and wall shear flows. In the following, typical examples of the turbulent flow measurements are briefly introduced.

Table 1. Applications of the 3-D PTV at the University of Tokyo.

Unsteady laminar Couette flow between two concentric cylinders (water)	Nishino et al. (1989) [1]
Nearly isotropic decaying turbulence (water)	Nishino et al. (1989) [1] Sata et al. (1989) [2]
Fully developed turbulent channel flow (water)	Nishino & Kasagi (1989) [8]
Axi-symmetric turbulent free jet (water)	Kasagi et al. (1988)[9] Ninomiya (1992) [7]
Turbulent separated and reattaching flow over a backward-facing step (water)	Itoh & Kasagi (1989) [10] Kasagi et al. (1992) [11]
Turbulence near a free surface (water)	Sata & Kasagi (1990) [12]
Turbulent flow along a riblet surface (water)	Suzuki & Kasagi (1993) [13]
Turbulent flow in a square cross-section duct (air)	Takamura et al. (1992)[14]

Figure 3 shows the distribution of rms velocity fluctuations in a fully developed turbulent channel flow of water [8]. Here, u', v' and w' denote the streamwise, wall-normal and spanwise velocity components, respectively, and both the velocities and the wall distance are non-dimensionalized by the friction velocity and the kinematic viscosity. The Reynolds number based on the center-line velocity and the channel width is 7510. By processing about 48,000 TV frames, a total of 4.37 million velocity vectors were obtained; this sufficient large sample size allowed calculating well converged turbulence statistics data. Also included in Fig. 3 are the X-shape hot-film probe data of Kreplin and Eckelmann [17] and the direct simulation (DNS) data of Kim et al. [15]. The PTV and DNS data show excellent agreement between them, but poor agreement with the probe data, particularly in the v' and w' components. The probe measurement is found to be erroneous owing to a decrease in turbulence scales near the wall [18]. Note that the other PTV data such as mean velocities and higher order moments of the fluctuating components also compare well with the DNS data. Hence, the present technique is quite suitable to turbulent shear flow measurements even in the region close to a wall.

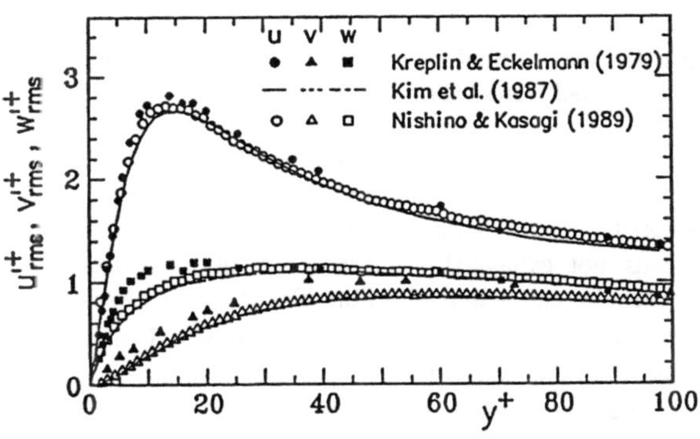

Figure 3. Distributions of rms velocity fluctuations in a turbulent channel flow [8].

Recently, the 3-D PTV measurement has been extended to more complex turbulent flows with separation and reattachment where the conventional techniques meet serious difficulty. For example, a detailed turbulent flow measurement was made in a plane channel with a one-sided 50% abrupt expansion, which acted as a backward-facing step, by Kasagi et al. [11]. The turbulent channel flow reached a fully developed state well upstream of the step; the Reynolds number based on the center-line velocity U_c and the step height H was 5541. The measurement region ranged from -2*H* upstream to 12*H* downstream of the step, and it was divided into about 300x120 sub-regions when the turbulence statistics were calculated. All the measured distributions of mean and fluctuating velocities well demonstrated characteristics of the separated flow. Figure 4 shows the wall-normal distributions of turbulent kinetic energy *k* at typical streamwise locations. The turbulence energy has been obtained as the sum of three turbulent normal stresses by processing of total of 192,000 TV fames. The mean reattachment point, where the forward flow fraction near the wall equaled to 0.5, was found to be located at 6.51*H* downstream of the step. Although not shown here, a peculiar phenomenon has been found that the spanwise turbulent normal stress is the largest among the three components in the reattaching region even without any direct production mechanism.

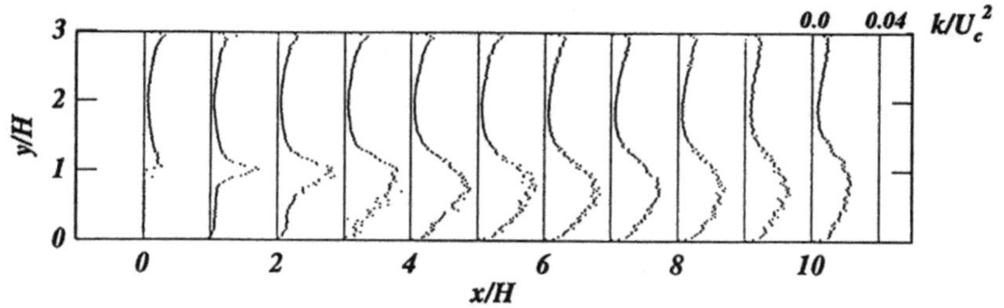

Figure 4. Wall-normal distributions of turbulent kinetic energy in a separated flow over a backward-facing step [11].

Gaseous flow measurement with the 3-D PTV is a challenging target [3] and has rarely been attempted. To the authors' knowledge, only Kent and Eaton [19] used helium-filled bubbles of 2 mm in diameter in a turbulent flow inside an engine cylinder, but accuracy of their data was not evaluated. Takamura et al. [14] recently used the aforementioned plastic microcapsules and helium-filled bubbles in a fully developed turbulent air flow in a straight duct of square cross-section 71x71 mm^2. The Reynolds number based on the center-line velocity and the hydraulic diameter is 2.3x10^4. Under the experimental condition, the Kolmogorov time scale is estimated to be 1.1 ms and hence the particle tracking was made over only two time steps with a time interval of 0.96 ms. It was anticipated that the plastic capsules would not sufficiently follow the fluctuating fluid motion at high frequencies although a major part of turbulent kinetic energy could be acquired and that the diameter of bubbles would not be small enough to measure fluctuations near the wall. The experimental results were compared with available experimental data obtained by hot-wire anemometry, and it turned out that, with both tracers, the measurement of mean velocities as well as second-order statistics could be successfully carried out. As a typical example, the

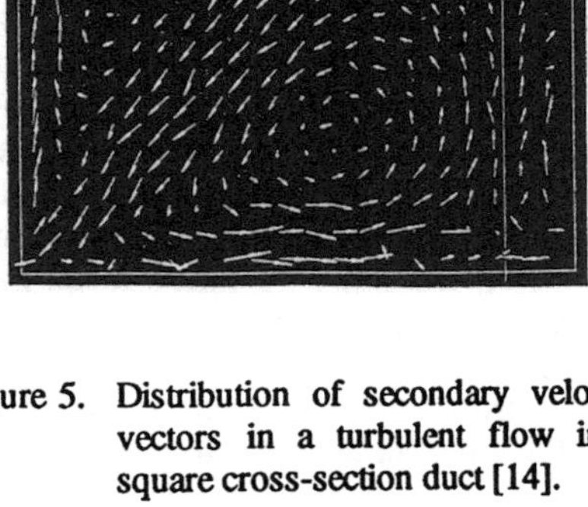

Figure 5. Distribution of secondary velocity vectors in a turbulent flow in a square cross-section duct [14].

velocity vector distribution on a quadrant of the cross-stream plane obtained with the He-filled bubbles is shown in Fig. 5, where the secondary flow of the second kind is clearly captured.

Conclusions

It is demonstrated that the 3-D PTV system developed should be a very powerful tool for measurement of various complex turbulent flows. The main advantage of the 3-D PTV over conventional measurement techniques lies in simultaneous measurement of all three velocity components in 3-D space, and thus it can provide crucial information on turbulent flow structures and mechanisms. There remain, of course, several urgent research needs for further development of 3-D PTV, e.g., (i) extension of the Reynolds number range in its application to turbulence measurement with improved resolution of digital image processing, (ii) increase in the yield of velocity vectors measured, and (iii) simplification of the camera calibration procedure for user-friendly PTV systems.

References

[1] Nishino, K., Kasagi, N., and Hirata, M., *ASME J. Fluids Eng.*, **111**, 1989, pp. 384-391.

[2] Sata, Y., Nishino, K., and Kasagi, N., *Flow Visualization V*, 1990, Hemisphere, pp. 248-253.

[3] Kasagi, N., and Nishino, K., *Exp. Therm. Fluid Sci.*, **4**, 1992, pp. 601-613.

[4] Papantoniou, D., and Dracos, T., *Engineering Turbulence Modelling and Measurements*, 1990, Elsevier, pp. 315-324.

[5] Adrian, R. J., *Ann. Rev. Fluid Mech.*, **23**, 1991, pp. 261-304.

[6] Sata, Y., and Kasagi, N., to be presented at *6th Int. Symp. Flow Vis.*, Yokohama, 1992.

[7] Ninomiya, N., *Dr. Eng. Thesis*, University of Tokyo, 1992.

[8] Nishino, K., and Kasagi, N., *Proc. 7th Symp. Turbulent Shear Flows*, Stanford, 1989, pp. 22.1.1-22.1.6.

[9] Kasagi, N., Ninomiya, N., and Hirata, M., *Exp. Heat Transfer, Fluid Mech. & Thermodyn.*, 1988, Elsevier, pp. 1502-1509.

[10] Itoh, N., and Kasagi, N., *Nagare-no Kashika (J. Flow Vis. Soc. Japan)*, **9**, 1989, pp. 245-248.

[11] Kasagi, N., Matsunaga, A., and Kawara, S., to be presented at *1st Int. Symp. Comp. Wind Eng.*, Tokyo, 1992.

[12] Sata, Y., and Kasagi, N., *Proc. Heat Transfer Symp. Japan*, **1**, 1990, pp. 55-57.

[13] Suzuki, Y., and Kasagi, N., to be presented at *Int. Conf. Near-Wall Turbulent Flows*, Tempe, 1993.

[14] Takamura, N., Sata, Y., Kasagi, N., and Sato, K., to appear in *Kasika Jyoho (J. Flow Vis. Japan)*, 1992.

[15] Kim, J., Moin, P., and Moser, R., *J. Fluid Mech.*, **177**, 1987, pp. 133-166.

[16] Nishino, K., and Kasagi, N., *Proc. 8th Symp. Turbulent Shear Flows*, Munich, 1991, pp. 28.3.1-28.3.6.

[17] Kreplin, H. P., and Eckelmann, H., *Phys. Fluids*, **22**, 1979, pp. 1233-1239.

[18] Suzuki, Y., and Kasagi, N., *Exp. Therm Fluid Sci.*, **5**, 1992, pp. 69-77.

[19] Kent, J. C., and Eaton, A. R., *Appl. Opt.*, **21**, 1982, pp. 904-912.

The Particle Tracking Method Using the Kalman Filter

Katsunori YAGOH, Kakuji OGAWARA, and Sei–ichi IIDA
Department of Mechanical Engineering II
Hokkaido University
N13 W8, Kita–ku, Sapporo, 060, Japan

An alternative particle tracking method for PIV(Particle Imaging Velocimetry) was developed by applying the Kalman filter. Numerical simulations to track particles in a 2–D cavity flow were performed to investigate the effects of the present method. It was proved that the method can considerably improve the accuracy and the dynamic range of the PIV measurement.

Key Words: Kalman Filter, Particle Tracking, Particle Imaging Velocimetry

1. INTRODUCTION

There are many advantages in PIV to study turbulence and chaos in fluid flows. Nishino and Kasagi used this technique for 3–D quantitative measurements of a turbulent channel flow and showed the effectiveness of this approach/1/. However, the dynamic range and accuracy of PIV need to be improved further in order to extend its applicability to various fluid flows.

While uncertainty in measured velocities is mainly caused by noises in observed positions of tracer particles, it is possible to estimate more accurate velocity values by using the Kalman filter at the particle tracking process. In our previous work, it was found that the predicted state of particles based on the Kalman filter algorithm is quite reliable, so that the step–by–step particles' identification becomes more successful/2/. In the present paper, the particle tracking method using the Kalman filter is applied to the measurement of a 2–D cavity flow which is obtained numerically for the purpose of both qualitative and quantitative investigation of this method.

2. APPLICATION OF THE KALMAN FILTER TO TRACKING PARTICLES

The position ξ, velocity $\dot{\xi}$ and acceleration $\ddot{\xi}$ of particles are considered as the state variables, and the state vector of particles at a certain time step k is described as $\xi_k = [\xi_k \ \dot{\xi}_k \ \ddot{\xi}_k]^T$ in X direction. Then the state equation which expresses the dynamics of particles in a short time interval Δt is approximately given as follows.

$$\xi_{k+1} = A\xi_k + B\alpha_k \ , \ A = \begin{bmatrix} 1 & \Delta t & \Delta t^2/2 \\ 0 & 1 & \Delta t \\ 0 & 0 & 1 \end{bmatrix} , \ B = \begin{bmatrix} 0 \\ 0 \\ 1 \end{bmatrix} , \tag{1}$$

where α is the system noise which means variation of the acceleration.

When only the particle positions with noise β are observed, the observation equation is written as

$$x_k = C\xi_k + \beta_k \ , \ C = [1 \ 0 \ 0] . \tag{2}$$

If the particle velocities with noise γ are also observed in addition to the particle positions, eq.(2) can be replaced by following equation.

$$\begin{bmatrix} x_k \\ \dot{x}_k \end{bmatrix} = C\xi_k + \begin{bmatrix} \beta_k \\ \gamma_k \end{bmatrix} , \ C = \begin{bmatrix} 1 & 0 & 0 \\ 0 & 1 & 0 \end{bmatrix} . \tag{3}$$

Supposing the noises are statistically independent and are steady zero-mean Gaussian, the Kalman filter algorithm is expressed as follows.

$$\begin{aligned} \hat{\xi}_k &= \tilde{\xi}_k + K\{x_k - C\tilde{\xi}_k\} \ , \\ x_k &= x_k \quad or \quad [x_k, \ \dot{x}_k]^T , \\ \tilde{\xi}_k &= A\hat{\xi}_{k-1} \ , \ \tilde{\xi}_0 = \bar{\xi}_0 \ , \end{aligned} \tag{4}$$

where K is the Kalman gain matrix which is separately obtained by solving the Ricatti equation/2/. The two kinds of the Kalman filter based on the eq.(2) and the eq.(3) are referred to as "type P" and "type PV", respectively, for convenience.

A particle is tracked by comparing its predicted state at the next step $\tilde{\xi}_k$ and the estimated states of particles $\hat{\xi}_k$ around the forecasted position, similarly to Chang et al./3/. The particle combination which yields a smaller difference in the two of state variables, $\tilde{\xi}_k$ and $\hat{\xi}_k$, is identified. In the present method, this tracking algorithm is executed for four successive time steps in every processing step to raise the accuracy of this treatment. Consequently, the estimated states of each particle are obtained for three steps ahead and are recorded as the measured data.

3. SIMULATED FLOW FOR ACCURACY TESTS

The 2-D flow in a square cavity, in which a number of particles are scattered, is used for producing sample data for the particle tracking simulations. All of the parameters involved in the flow calculations are appropriately normalized, e.g., the maximum velocity of the moving upper wall and the length of sides of the cavity are equal to unity. The ratio of the maximum to minimum velocity of particles in this flow was found to be order of 10^5. Therefore, a sufficiently wide dynamic range is needed in order to measure the velocity distribution of this flow in the entire flow field.

4. RESULTS AND DISCUSSION

Fig.1(a) shows the exact numerical solution of a particle path in the cavity flow during 500 time steps(Δt =0.02). When we observe this transition of a particle position, it is generally unavoidable that the noise gets into its observation. In order to illustrate this situation, we add the zero-mean Gaussian noise with the standard deviation 0.01 to each of the particle positions. Fig.1(c) shows the particle trajectory tracked by the present method; here the type P is used. It is clear that the estimated path line is quite close to the one in Fig.1(a). A temporal change of u-component of the particle velocity is plotted in Fig.2. The signal after filtering displays a fairly good agreement with the true velocity signal. Thus, it is verified that the state estimation based on the Kalman filter proposed in this paper runs successfully to obtain the accurate values of both positions and velocities of particles.

In the case that some precise information other than particle positions is observable, the present method can include the additional information in the state estimation. For

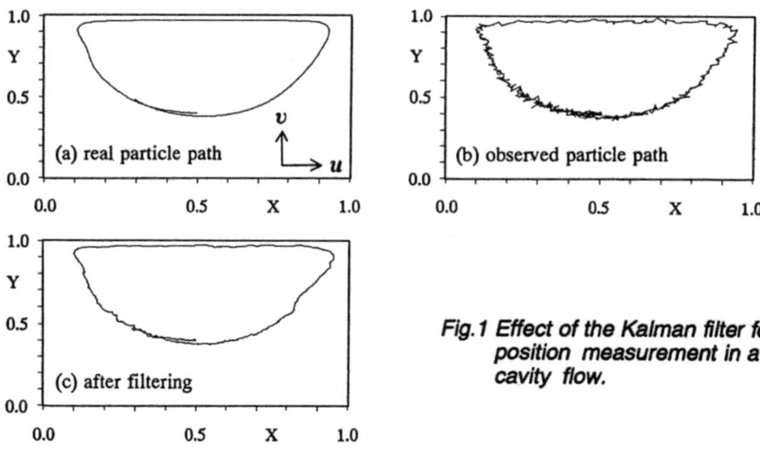

Fig.1 Effect of the Kalman filter for the position measurement in a 2-D cavity flow.

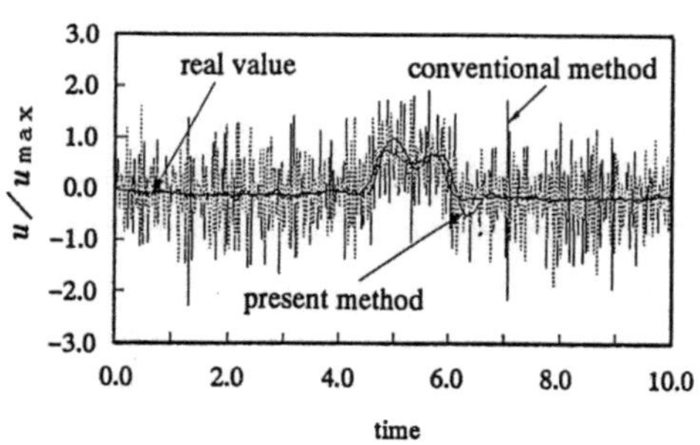

Fig.2 Comparison of time series of velocity outputs; the Kalman filter estimate, real value, and conventional evaluation.

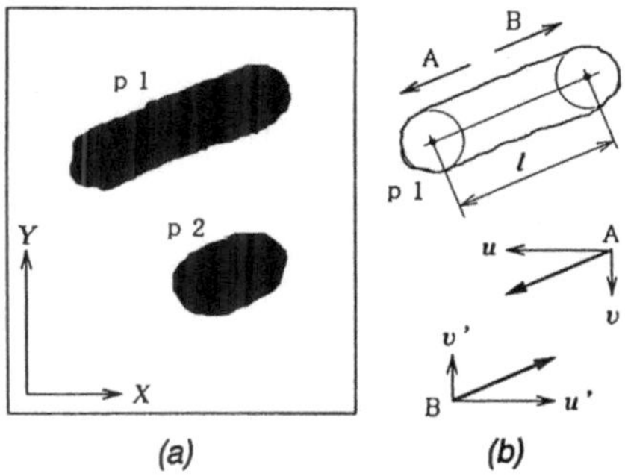

*Fig.3 (a) Example of particle path on a image,
(b) Displacement and moving direction
of a particle.*

example, let us consider the case that the absolute values of particle velocities can be evaluated from the paths on images and the velocity of a particle is either A or B, as shown in Fig.3. For this case, the measuring equation (3), i.e., type PV, is useful. Both the cases A and B are to be employed as the observed velocity components, and the one which has smaller difference between the predicted and estimated states for four successive time steps is selected to determine the sense of the vectors/4/.

In order to examine the effect of including this kind of information, the particle tracking simulations using the two types, P and PV, for 200 time steps(Δt =0.06) have been carried out in the cavity flow including 200 particles. Here, Δt is close to the critical value at which the average particle displacement during one time step exceeds the mean distance of their distributions $\sqrt{1/D}$; D denotes the particle density in the cavity. Hence, the difference in the displacement between high speed particles and low speed ones is significantly large. This implies that the wide dynamic range of tracking method is required for this case. Fig.4 compares the rates of the correctly tracked particles in the three ranges of particle velocity, as a function of the standard deviation of error in the observed velocity components normalized by the maximum velocity. Particles of lower speed(0 to 1/3), medium (1/3 to 2/3) and higher speed(2/3 to 1) belong to "range 1", "range 2" and "range 3" respectively. The type PV is superior to the type P in all the ranges of velocity, as far as the velocity components are observed sufficiently precise. Fig.5 shows the standard deviation of the error in velocity measurements versus that of the observation error. The velocity components are properly estimated owing to the reliability of the observed position and velocity. It can be concluded that even if the sense of vectors are unknown, the observation of the precise velocity information brings further improvement of the dynamic range and accuracy of the PIV measurement.

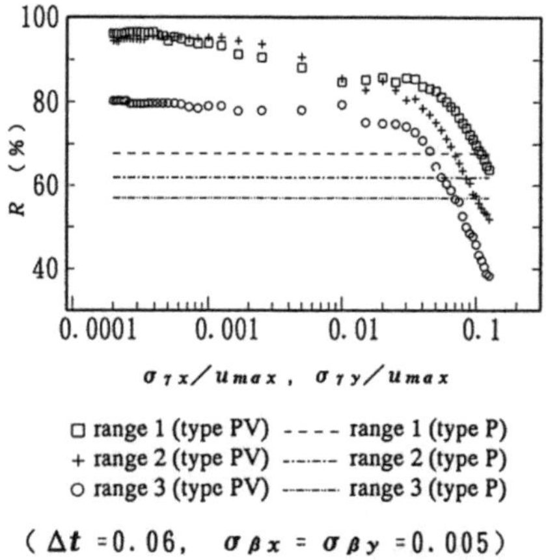

Fig.4 Rates of correctly tracked particles in each range of particle velocity.

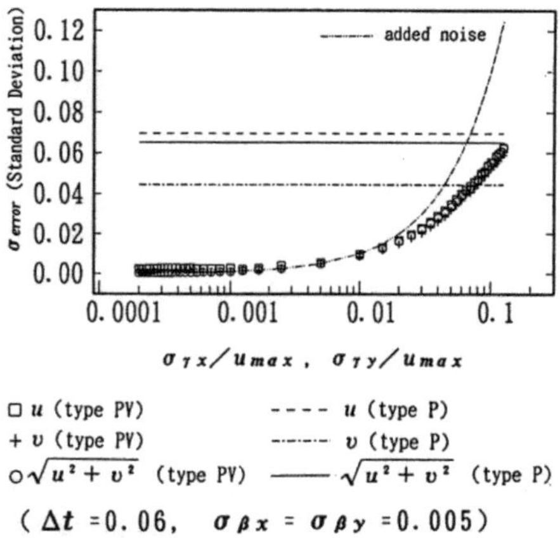

Fig.5 Standard deviation of errors in measured values of particles' velocity components.

5. SUMMARY

The particle tracking method using the Kalman filter was developed. The excellent performance of this method was demonstrated by simulating the PIV measurement of a 2-D cavity flow which was acquired numerically. The results showed that (1)the state estimation based on the Kalman filter algorithm ran successfully so that the present method made velocity measurements more accurate and practical, (2)both the dynamic range and the precision of velocity measurements were improved by observing absolute values of velocity components as well as the particle positions.

This method can be improved by including other information, like the particle acceleration, and also by taking account of the dynamics of fluid motion into the system equations.

REFERENCES

/1/ Nishino, K., and Kasagi, N., Trans. Jpn. Soc. Mech. Eng., (1990), 56-525, p.1338.
/2/ Yagoh, K., Ogawara, K., and Iida, S., JSME Int. J., (1992-Nov. or 1993-Feb.).
/3/ Chang, T. P., Wilcox, N. A., and Tatterson, G. B., Optical. Eng., 23-3, (1984), p.283.
/4/ Yagoh, K., Ogawara, K., and Iida, S., Trans. Jpn. Soc. Mech. Eng., B, (1992-Jun.).

Model-Based Velocity Estimation Using the Kalman Filter

Kohtaro OHBA, Tadashi ISHIHARA and Hikaru INOOKA

Department of Mechanical Engineering
Tohoku University
Aramaki aza Aoba, Aoba-ku, Sendai, 980, Japan

Summary

In this paper, we propose a model-based method for estimating the velocities of unsteady viscous 2-D flow from a series of images. The proposed method utilizes the Kalman filtering technique[1]. The observation equation is constructed by modeling the image as a distributed parameter system[2]. It is possible to construct a dynamic model for the flow by using the FEM(Finite Element Method). However, the resulting model is too complex to apply the Kalman filtering technique directly. We propose an adaptive Kalman filter algorithm which includes the identification of the dynamic model from the observed data. Experimental results are presented to show the effectiveness of the proposed algorithm.

Image Equation

Let $f(X,t)$ denote the image intensity of the location $X(x,y)$ at time t as a distributed parameter system. The image intensity of the location $X(x,y)$ at time t is equal to that of a location $X_0(x_0,y_0)$ at the initial time t_0. Let $\xi(X,t,t_0)$ denote a time-varying geometric transformation satisfying

$$\begin{aligned} f(X,t) &= f(X_0,t_0) \\ &= f(\xi(X,t,t_0),t_0). \end{aligned} \tag{1}$$

We assume that $\xi(X,t,t_0)$ is expressed as

$$\xi(X,t,t_0) = X + B(t,t_0), \tag{2}$$

where B is a 2×1 vector denoting translation.

Taking the partial derivatives of (1) with respect to the location X and time t, we get

$$\frac{\partial f(X,t)}{\partial t} = \frac{\partial f(X,t)}{\partial X}^T \dot{B}, \tag{3}$$

where $\dot{B}$ is the partial derivatives of B with respect to t. This is a model of the image as a distributed parameter system.

We can rewrite (3) as

$$f_{\mathcal{T}} = \begin{pmatrix} f_{\mathcal{X}} & f_{\mathcal{Y}} \end{pmatrix} \begin{pmatrix} u \\ v \end{pmatrix}, \tag{4}$$

where $f_{\mathcal{T}}, f_{\mathcal{X}}$ and $f_{\mathcal{Y}}$ denote the partial derivatives of the image intensity f with respect to t, x and y, respectively, and u, v denote the velocities of each axis.

We use the relation (4) as a fundamental relation to estimate the vector (u, v) from the image data. We set an $n \times n$ pixels window in the image plane and assume that the parameter vector (u, v) is constant within the window. Then, we have the following simultaneous equation.

$$\begin{pmatrix} f_{\mathcal{T}}^1 \\ f_{\mathcal{T}}^2 \\ \vdots \\ f_{\mathcal{T}}^{n^2} \end{pmatrix} = \begin{pmatrix} f_{\mathcal{X}}^1 & f_{\mathcal{Y}}^1 \\ f_{\mathcal{X}}^2 & f_{\mathcal{Y}}^2 \\ \vdots & \vdots \\ f_{\mathcal{X}}^{n^2} & f_{\mathcal{Y}}^{n^2} \end{pmatrix} \begin{pmatrix} u \\ v \end{pmatrix} \tag{5}$$

Model Equation

The unsteady viscous 2-D flow equation is described by the Navier-Stokes' equation of motion and the equation of continuity as follows.

$$-\rho(\frac{\partial u}{\partial t} + u\frac{\partial u}{\partial x} + v\frac{\partial u}{\partial y}) - \frac{\partial p}{\partial x} + \mu(\frac{\partial^2 u}{\partial x^2} + \frac{\partial^2 u}{\partial y^2}) + f_x = 0 \tag{6}$$

$$-\rho(\frac{\partial v}{\partial t} + u\frac{\partial v}{\partial x} + v\frac{\partial v}{\partial y}) - \frac{\partial p}{\partial y} + \mu(\frac{\partial^2 v}{\partial x^2} + \frac{\partial^2 v}{\partial y^2}) + f_y = 0 \tag{7}$$

$$\frac{\partial u}{\partial x} + \frac{\partial v}{\partial y} = 0 \tag{8}$$

These equations are the partial differential equations of distributed parameter system. Using the FEM (Finite Element Method), we can get the model equation as follows.

$$\begin{pmatrix} \mathbf{m} & 0 & 0 \\ 0 & \mathbf{m} & 0 \\ 0 & 0 & \mathbf{m} \end{pmatrix} \begin{pmatrix} \dot{\mathbf{u}} \\ \dot{\mathbf{v}} \\ \dot{\mathbf{p}} \end{pmatrix} + \begin{pmatrix} \mathbf{k}+\mathbf{k}_c & 0 & \mathbf{H}_x \\ 0 & \mathbf{k}+\mathbf{k}_c & \mathbf{H}_y \\ \mathbf{H}_x^T & \mathbf{H}_y^T & 0 \end{pmatrix} \begin{pmatrix} \mathbf{u} \\ \mathbf{v} \\ \mathbf{p} \end{pmatrix} = \begin{pmatrix} \mathbf{q}_x \\ \mathbf{q}_y \\ -\mathbf{q}_p \end{pmatrix} \tag{9}$$

Considering the node of element as the observation point, we define the extended state variable as,

$$\mathbf{X} = (\mathbf{u}|\mathbf{v}|\mathbf{p})^T \tag{10}$$

$$= (u_1 u_2 \cdots u_n | v_1 v_2 \cdots v_n | p_1 p_2 \cdots p_m)^T, \tag{11}$$

where subscripts denote the number of observation points. From (10), the differential equation (9) can be written as

$$\mathbf{G}\dot{\mathbf{X}} + (\mathbf{K} + \mathbf{K}_c)\mathbf{X} = \mathbf{Q}. \tag{12}$$

Using the Crank-Nicholson method, we get the descrete-time system equation as follows.

$$\mathbf{X}_{t+1} = \left(\mathbf{G} + \frac{\Delta t}{2}\mathbf{K}\right)^{-1} \left(\mathbf{G} - \frac{\Delta t}{2}\mathbf{K} - \Delta t \mathbf{K}_c \mathbf{X}_{t-1}\right) \mathbf{X}_t + \left(\mathbf{G} + \frac{\Delta t}{2}\mathbf{K}\right)^{-1} \Delta t \mathbf{Q} \tag{13}$$

Kalman Filter Algorithm

Adding system noise W_t to (13), we can get the system equation.

$$\mathbf{X}_{t+1} = F_t\mathbf{X}_t + W_t. \tag{14}$$

W_t is a zero-mean white noise sequence with a known covariance matrix Q_t. From (5), we get the observation equation.

$$Y_t = H_t\mathbf{X}_t + V_t, \tag{15}$$

where the observation noise V_t is a zero-mean white noise sequence with a known covariance matrix R_t.

Applying the Kalman filtering theory, we can obtain the following algorithm.

$$\begin{aligned}
\hat{\mathbf{X}}_{t+1/t} &= F_t\hat{\mathbf{X}}_{t/t}, \\
\hat{\mathbf{X}}_{t/t} &= \hat{\mathbf{X}}_{t/t-1}, + K_t\left(Y_t - H_t\hat{\mathbf{X}}_{t/t-1}\right), \\
K_t &= P_{t/t-1}H_t^T, \left(H_tP_{t/t-1}H_t^T + R_t\right)^{-1}, \\
P_{t+1/t} &= F_tP_{t/t}F_t^T + Q_t, \\
P_{t/t} &= P_{t/t-1} - K_tH_tP_{t/t-1}.
\end{aligned} \tag{16}$$

In the above algorithm, K_t is the Kalman gain matrix, and P is the covariance matrix of the estimation error. The subscript $t+1/t$ denotes the estimate at time $t+1$ based on the data available at time t. Using this algorithm, we can estimate the velocities and pressures at any point from the image data.

Model Identification

The matrix F_t in (14) derived from FEM is too complex to apply the Kalman filter algorithm and it takes a lot of time to estimate this matrix. To overcome this difficulty, we propose an adaptive technique which identifies the system matrix F_t.

The proposed method is based on simple observation that, if we can estimate the variable X accurately, then the relation

$$\hat{X}_{t+1/t} = F_t\hat{X}_{t/t-1} \tag{17}$$

holds. Assuming that (17) holds, we can estimate the system matrix F_t by the least square method.

Estimation Results

In this paper, we estimate the Kármán vortex flow velocity and identify the system matrix F_t at one point in the flow area as shown in Fig.1. The data is sampled with $1/30s$ and 100th sheets of serial image data. The image is digitized into 64th grades of intensities and 256×256 pixels data.

Comparing with this proposed method, we calculate the flow velocities with the correlation method[3] shown in Fig.2. In this method, we get the two image data with $1/15s$ sampling time.

With the proposed Kalman filter methods, we get the results and identified matrix parameter shown in Fig.3. This velocity is positioned at the top column and 5th row from left in Fig.2.

The results are shown to be influenced by noise of image data, and confirmed to be estimated the same value estimated with the correlation method. But the estimation time to caliculate is slightly shorter shown in Table.1.

References

[1] R.E.Kalman and R.S.Bucy, *"New Results in Linear Filtering and Prediction Theory", Trans. ASME, J. Basic Eng., 82D-1(1961), 95-108*

[2] R.J.Schalkoff, *"Dynamic Imagery Modeling and Motion Estimation Using Weak Formulation", IEEE Trans. Pattern Analysis and Machine Intelligence. Vol.PAMI9-4(1987), 578-583*

[3] I.Kimura, T.Takamori and T.Inoue, *"Image Processing Instrumentation of Flow Velocity Vector Distribution by Using Correlation Technique", Trans.SICE. 23-2(1987), 101-107.*

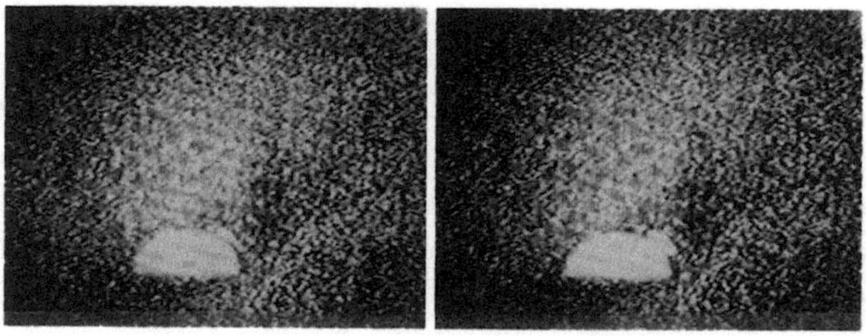

(a) t = 0 sec (b) t = 1/15 sec

Fig.1 Kármán vortex image.

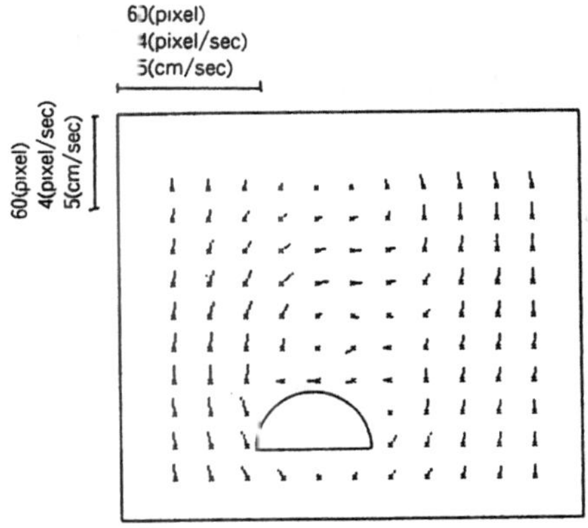

Fig.2 Optical Flow obtained by the correlation methods.

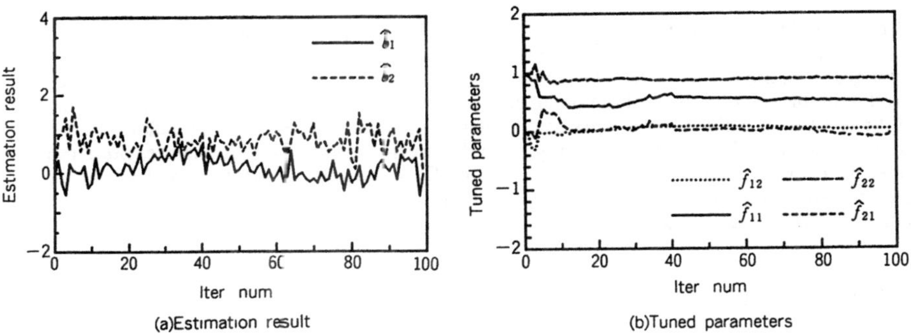

Fig.3 Velocity estimation obtained by Kalman filter.
(Window:m=7,n=7 ; Position:(130,50))

Table.1 Compensation of calculation time.

	Correlation Method (n=15, m=35)	Kalman Filter (n=7, m=7)
Addition	$5n^2m^2$ 1378125	$13n^4 + 150n^2 + 144 + 3n^2m^2$ 45910
Subtraction	$4n^2m^2$ 1102500	$\frac{n(n-1)^2}{2} + n^2 + 6^2 + 3n^2m^2$ 7414
Multiplication	$(3n^2+1)m^2$ 828100	$\frac{n(n-1)^2}{2} + 12n^4 + 150n^2 + 488$ 36776
Division	$3m^2$ 3675	$\frac{n(n+1)}{2} + 3m^2$ 175
	n: Window Size m: Research Width	n: Window Size m: Averaged Area Size

Development of Digital Vector Velocimetry Method and Its Application to Lid Driven Rotating Flow

You-Gon Kim* and Ching-Jen Chen**

* Senior Researcher, Fast Breeder Reactor Coolant Department,
Korea Atomic Energy Research Institute (KAERI), Daejon, Korea 305 -353

** Professor and Chair, Dept. of Mechanical Engineering,
The University of Iowa, Iowa City, Iowa 52242.

Summary

The lid-driven cylindrical cavity flows in the laminar region are quantitatively visualized by a new flow visualization method called the digital vector velocimetry (DVV) method. Particle streak images from the particle streak velocimetry (PSV) method are also taken and qualitatively compared with the velocity vector field obtained from the DVV method. Because of the convex wall of the test cell, the side view images of the rotating flows are distorted in the radial direction. These optical distortions are analyzed and corrected using calibration curves prepared to recover the original, accurate axisymmetric rotating flow. The DVV system provides measurement with subpixel accuracy. The dynamic range of the velocity is expanded using feedback analysis that selects consistent velocity vectors.

Digital Vector Velocimetry Method

In the digital vector velocimetry (DVV) method [2,3], a time sequence of images is captured by a high speed CCD (Charge-Coupled Device) video camera with each frame containing a single-exposure image. All possible velocity vectors between two temporally successive images in the interrogating window are digitally constructed on the digital vectorgram. Then the most probable velocity vector can be determined by estimating the position of the maximum intensity in the digital vectorgram. Unlike most of the quantitative methods that need optical postprocessing or complicated statistical method to analyze the images taken, the DVV method invokes the idea of probability with a direct count of the most probable state of the digital vector, or the most intense digital vector to determine the velocity vector. Therefore, it can be implemented without much computational effort and expensive optical system.

A digital vector is defined as a vector formed between two time-sequential particle images. The digital vector field sampled within an interrogating window on the digital vectorgram is given by

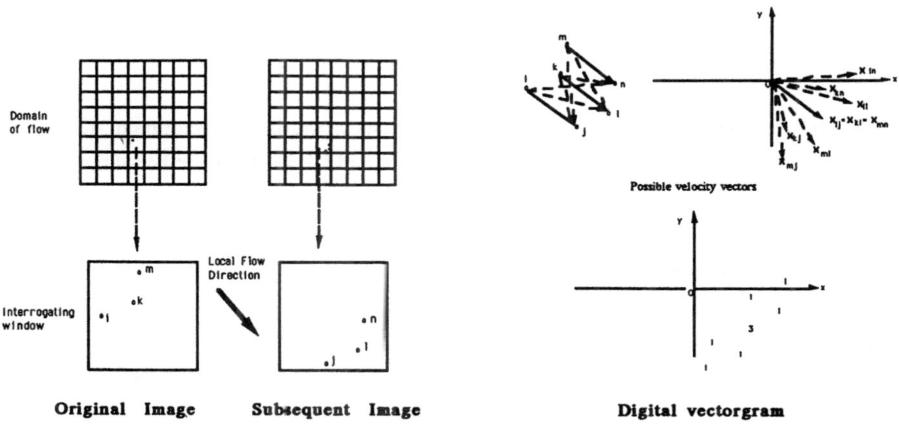

Figure 1. Construction of digital vectorgram

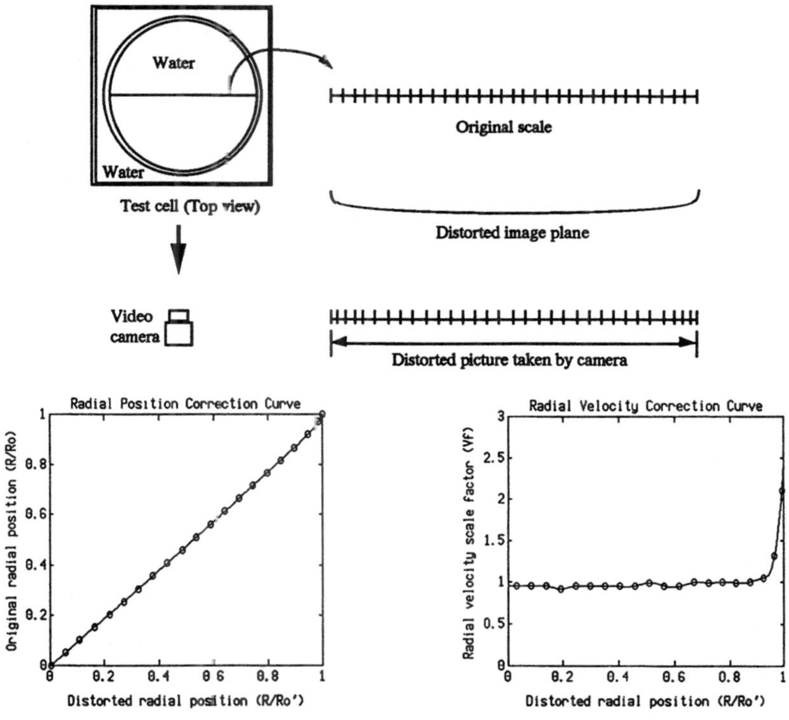

Figure 2. Calibration curves for lid-driven cylindrical cavity flow

$$I(\mathbf{x}) = \sum_{k=1}^{K} \sum_{l=1}^{L} Q_{kl} \cdot \delta[\mathbf{x}-\mathbf{x}_{kl}] \quad (1)$$

Here $I(\mathbf{x})$ is the distribution of the intensities of digital vectors on the digital vectorgram. K is the total number of the image particles in the original image, and L is the total number of the image particles in the subsequent image. $\mathbf{x}_{kl}$ (= $\mathbf{x}_l$ - $\mathbf{x}_k$) is the digital vector from k to l. Q_{kl} = $Q_k \times Q_l$, and $\delta[.]$ is Dirac delta function. Figure 1 illustrates the distribution of $I(\mathbf{x})$ from the two time-sequential images with particle image intensity set to be either 0 or 1.

In summary the quantitative flow visualization is achieved by the DVV method which interrogate a pair of single-exposure images element by element with 31×31 pixels of elements. From the first and second images within the interrogating element all possible digital vectors are projected to the digital vectorgram. The velocity vector possessing the maximum intensity is identified as the instantaneous velocity vector between the time interval of the images. By taking weighted average of the optical intensity of the digital vectors on the digital vectorgram, the subpixel accuracy of measurement is achieved. In order to achieve a broader dynamic range of the velocity field measurement, an analysis is performed using more than one pair of digital images in the present DVV method. As a post-treatment of velocity field of the side view obtained from the DVV method, optical distortion due to the presence of the convex geometry of the plexiglass wall is corrected. The correction includes two steps. The first step is to recover the axisymmetric flow by taking average of the two velocity fields with opposite directions of rotation of the lid. The second step is to achieve the radial distortion-free axisymmetric flow field using the correction curves, as shown in Figure 2, prepared for the rotating flows.

Results and Discussions

The digital vector velocimetry (DVV) method [2,3] has been developed to obtain the full-field velocity data of the lid-driven cylindrical cavity flow at angular speed of 10 rpm. In Figure 3 Particle streak images from the particle streak velocimetry (PSV) method [4] are compared with the velocity vector field obtained from the DVV method. The resultant flow fields obtained from the side view are observed to be strongly dependent on the rotating direction of the lid. Laminar numerical result obtained by the finite analytic (FA) method [1] is compared with the experimental data obtained by the digital vector velocimetry (DVV) method in Figure 4. Figure 5 shows numerical and experimental axial and radial velocity profiles. Here W_o is the edge velocity of the rotating lid, i.e., $R_o\omega$. The comparison reveals good agreement except in the steep velocity gradient region of the lower domain.

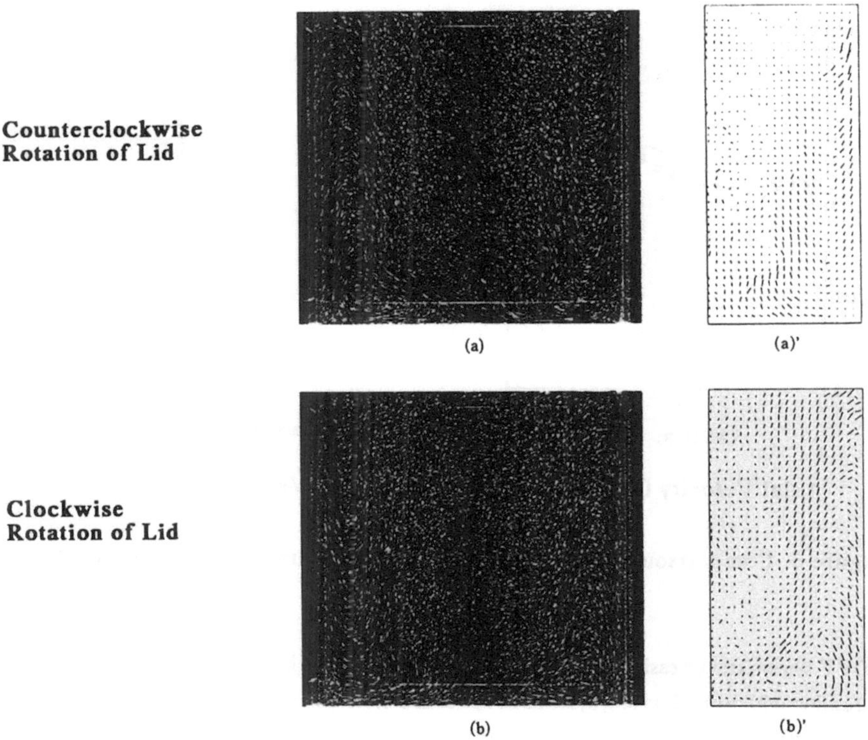

Figure 3. Strong dependence of DVV results on rotating direction

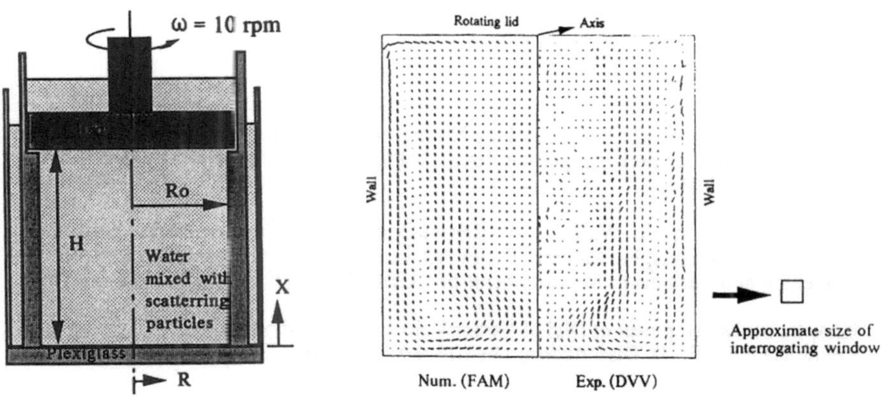

Figure 4. Schematic test cell and comparison of experimental (DVV) data and computational (FAM) result

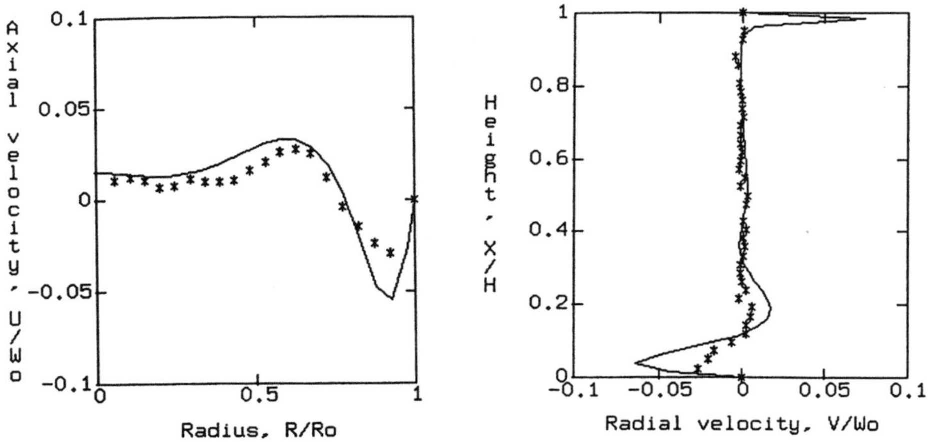

Axial Velocity (X=H/2) **Radial Velocity (R=Ro/2)**

Figure 5. Comparison of experimental (DVV) and numerical (FAM) results

The DVV system can be easily extended and improved to achieve higher accuracy in quantitative visualization with better imaging systems in optics and electronics. Higher power of the laser, higher acquisition speed system, and finer resolution of the CCD (Charge-Coupled Device) video camera are needed to provide better accuracy of measurement and broader dynamic range of the velocity in the high Reynolds number flow including turbulent flow.

References

1. Chen, C. J.; Chen, H. C.: Finite Analytic Numerical Method for Unsteady Two Dimensional Navier-Stokes Equation. Journal of Computational Physics, Vol. 53, No. 2, pp. 209-226, 1984.

2. Chen, C. J.; Kim, Y. G.; Walter, J. A.: Recent Developments in Quantitative Flow Visualization and Imaging Process. The Fourth International Symposium on Refined Flow Modelling and Turbulence Measurements of IAHR, Wuhan, China, pp. KA15-26, 1990.

3. Kim, Y. G.: Development of Digital Vector Velocimetry Method and Its Application to Rotating Flows. Ph.D. Dissertation, The University of Iowa, December 1991.

4. Walter, J. A.; Chen, C. J.: Flow Visualization of Particle Streaks in Offset Channel Flow by a Direct CCD Imaging Process. The Winter Annual Meeting of ASME, San Francisco, CA, FED-Vol. 85, pp. 115-120, 1989.

Application of a Fast Algorithm for Pattern Tracking on Airflow Measurements

Akikazu KAGA, Yoshio INOUE and Katsuhito YAMAGUCHI

Department of Environmental Engineering,
Osaka University, Osaka

Summary

A fast algorithm avoiding the calculation of correlation coefficients is applied to the pattern tracking for an airflow measurement. The pattern similarity is defined by the reputation number of the summations of gray level difference between corresponding pixels until the accumulation exceeds a critical value. An example of the sensitivity examination of parameters included are also described.

Introduction

For low velocity airflow measurement using tracer particles, the pattern tracking method is superior to the particle tracking method in the availability of fine particles. The former utilize gray level values on a whole image, and thus requires much more computation time than the latter. In the pattern tracking method, one has to find out the correspondence between a pair of image fragments, one is on the image at time $t = t_0$ and the other at $t = t_0 + \Delta t$ as shown in Fig.1. When the average gray levels of a pair of images do not differ significantly, the accumulation of gray level difference between corresponding pixels over the fragment is available as an index of similarity [1]; moreover, the calculations for dissimilar fragment's pairs can be interrupted on the way by adopting the reputation number of summations until the accumulation exceeds a critical value [2]. In this work the pattern similarity defined by reputation number is applied to airflow measurement. The pattern tracking using this definition has the same tracking performances as the method using correlation coefficient and saves computation time of one order or more when applied to airflow measurement [1].

Algorithm for Pattern Tracking

The algorithm used in this work is as follow (Fig.2. and Fig.3.):

(1) The absolute values of gray level difference between pixels in an image fragment at $t=t_0$ and one of its candidates at $t=t_0+\Delta t$ are accumulated n_S times in random order. These accumulations are executed for all candidates, and the minimum value of accumulation, $T_{S,min}$, is looked for.

(2) The minimum value of accumulation over the fragment, $T_{M,min}$, is estimated by Eq.(1).

$$T_{M,min}=T_{S,min}\times(n_M^2/n_S) \tag{1}$$

in which n_M is the dimension of fragment represented in pixels. The critical value $T_{M,cr}$ is determined by Eq.(2).

$$T_{M,cr}=\alpha\, T_{M,min} \tag{2}$$

in which α is a properly estimated parameter and has a value less than unity.

(3) The candidates already exceeding the critical value(e.g. ① in Fig.3) are eliminated here. For the remaining candidates, the summations are continued until the accumulation exceed the critical value. The candidate that gives the maximum reputation number is selected as the corresponding fragment.

Example of Measurement

A measured result of a vortex flow generated at the corner of a rectangular prism placed in a full-scale model of down flow type cleanroom is shown in Fig.4. A reasonable velocity distribution was obtained.

Optimum Value of Parameters

In this algorithm, the fragment dimension n_M, and the number of sampling pixels n_S and α-value in Eq.(2) for determining the critical value have to be selected properly. Although the optimum value of these parameters depends on flow characteristics, visualization method, sampling intervals Δt etc., it will be useful in the application to examine the optimum value of parameters for a simpler flow under the conditions similar to the objective measurement.

The effects of parameters in the measurement of uniform flow visualized with smoke particles in a wind tunnel were examined as an example. The visualized flow is shown in Fig.5. The relative frequency of gray level difference between corresponding pixels (gray level at $t=t_0+\Delta t-$ gray level at $t=t_0$), which was observed in the coordinate system carried by the uniform airflow, is shown in Fig.6. The gray level increased with the time proceeded since the flow was illuminated from a point downstream of the visualized zone. The distribution of gray level differences spread with time due to the pattern distortions.

The effects of parameters examined by tracking about 500 vectors in each case for Δt=1/30 s and 2/30 s are shown in Fig.7 to Fig.9. In these figures R_T represents the ratio of tracked points to the total points, R_C represents the ratio of correctly tracked vectors to the total vectors and N represents the average number of summations of gray level differences for each vector. R_T depends on n_M because a set point P in Fig.1 should not be tracked in this algorithm if the fragment surrounding P has a flat gray level distribution. N_s represents the number of summations to determine the critical value and N_r to n_r for each vector. Therefore N_s equals to {the number of candidates}× n_s . The number of candidates was 25×25 for Δt=1/30 s and 50×50 for Δt=2/30 s. The total computation time is proportional to N_s+N_r . The results represented in Fig.7 to Fig.9 are summarized as follows:

(1) The dimension of fragment n_M had to increase as the time step Δt increased to keep a high tracking performance because the gray level pattern was distorted significantly as the time proceeded. The increase of computation time was proportional to the first power of n_M rather than the second power.

(2) The critical value was properly determined with the sampling pixel number n_s lower than n_M.

(3) The critical value was properly determined with the α-value of 0.2 to 0.3.

References

1. Kaga,A;Y.Inoue;A.Yoshikawa: Velocity Distribution Measurement through digital image Processing, Proceedings of ROOMVENT'90 in Oslo(1990)22
2. Daniel,I.B.; F.S.Harvey: A Class of Algorithms for Fast Digital Image Registration, IEEE Trans. on Computers,1972,c-21,2,179-186

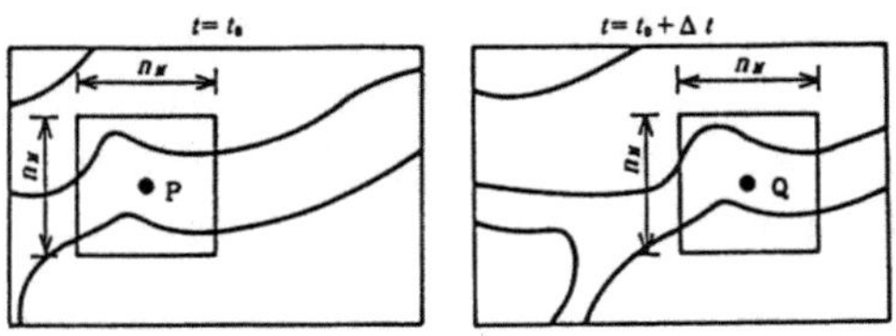

Fig.1. Pair of image fragments

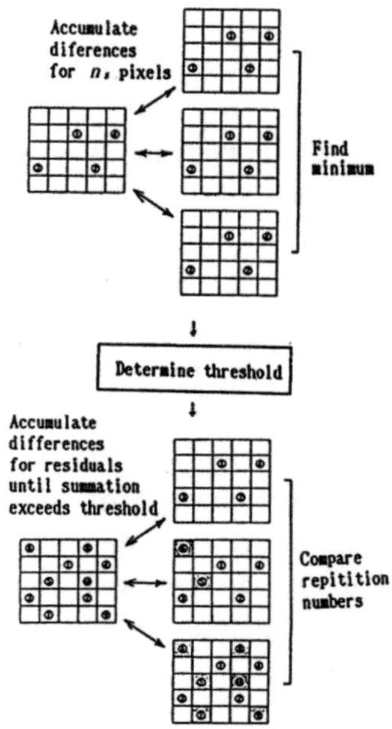

Fig.2. Algorithm for pattern tracking

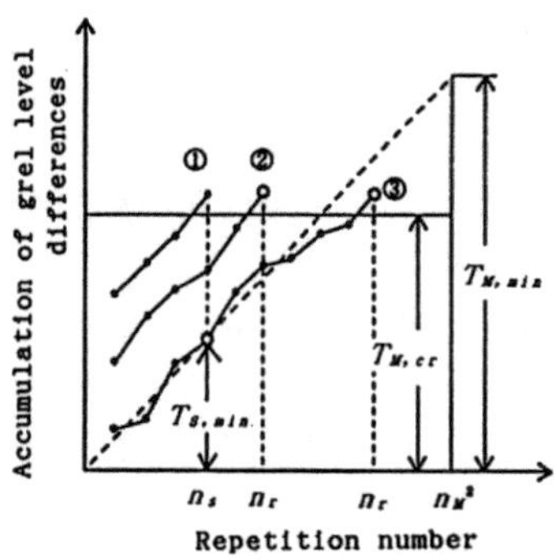

Fig.3. Growth of accumulation with reputation

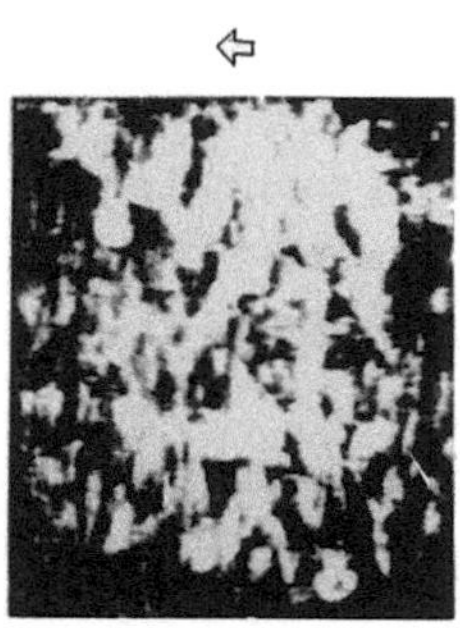

Fig.5. Uniform airflow

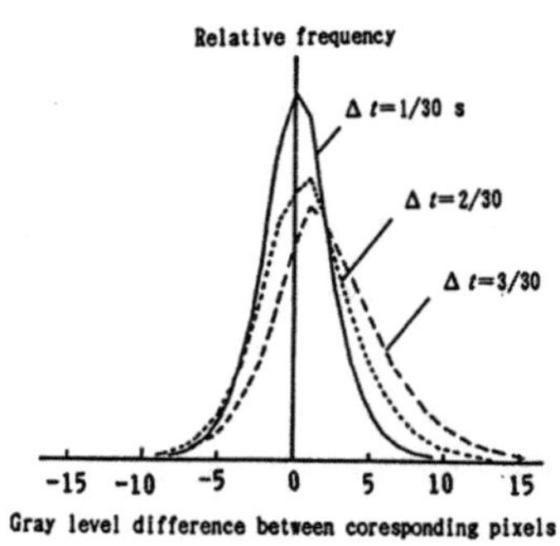

Fig.6. Distribution of gray level differences

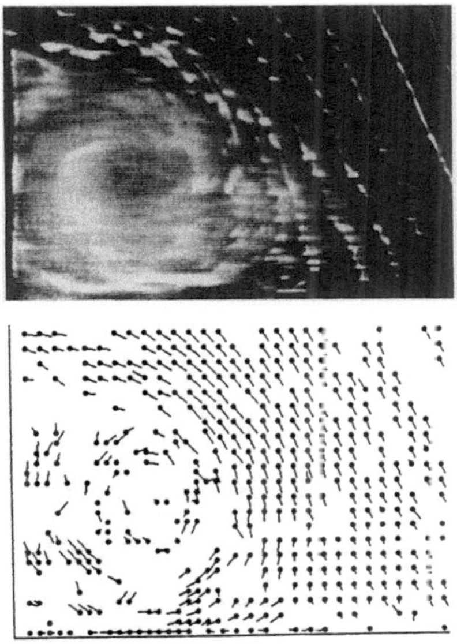

Fig.4. Velocity distribution of vortex flow

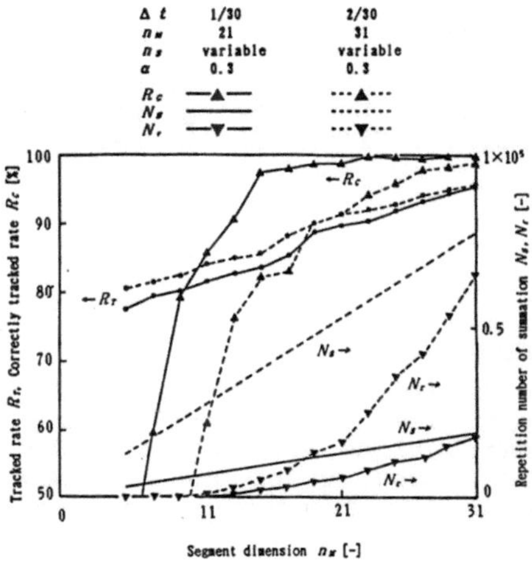

Fig.7. Effect of n_M on tracking performances

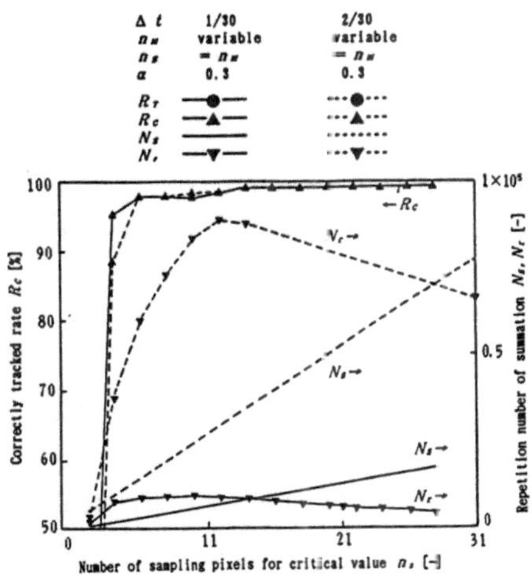

Fig.8. Effect of n_s on tracking performances

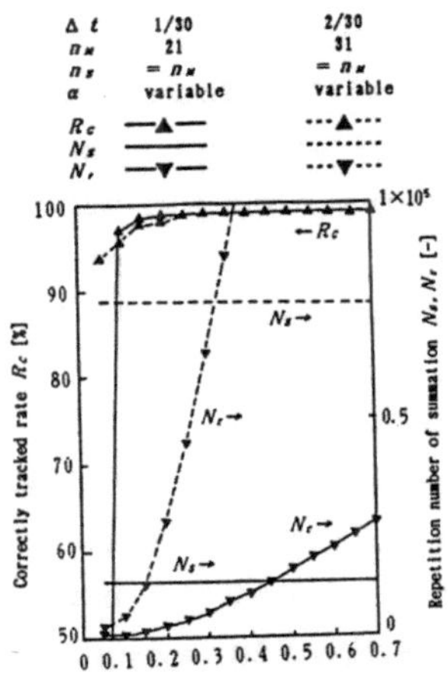

Fig.9. Effect of α on tracking performances

On the Accuracy of the Correlation Method

Ichiro FUJITA and Saburo KOMURA

Department of Civil Engineering
Faculty of Engineering, Gifu University
Yanagido, Gifu, 501-11, JAPAN

Summary

The correlation method, which is one of the image processing techniques, have been widely used for obtaining velocity distributions from visualized images of flow. The accuracy of this method, which has not been discussed in detail so far, is examined by simulating tracer-particle images of some simplified flows.

Simulation of Particle Images

Numerically created tracer-particle images are used for the simulation. The tracer particles, assumed to have elliptic shapes and the maximum intensities at their centers, are distributed on an imaginary rectangular area with specified average spacings, and then each of them is moved according to the given velocity distributions of the flows such as uniform flow, uniformly shearing flow, free shear-layer flow, and flow of a sinusoidal velocity profile. The digitized particle images before and after the particle movements are obtained by locally averaging the intensity distribution within a tiny area for every pixel. The correlation method is applied to a rectangular area composed of 150x150 pixels and about 400 velocity vectors are used for the analysis. The peak intensity of each particle is randomly chosen between 50 and 255.

The two parameters characterizing particle images are used in this research: the average area for one particle to exist, A_p, and the maximum number of particles possible to be packed in the above area, N_p. An example of the digitized images and the relative comparison of velocity profiles and particle distributions are schematically shown in Fig.1 and Fig.2, respectively.

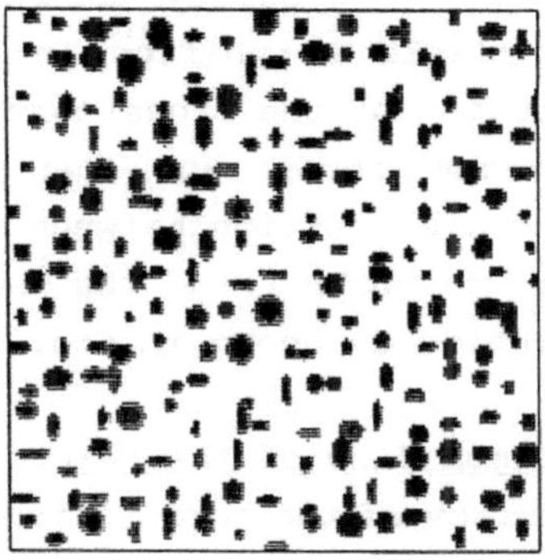

Fig.1. Example of digitized image (150x150 pixels ,Ap= 100,Np=3.3)

Method for Improving the Accuracy

In the correlation method, the cross-correlation coefficient of gray level intensities within a small area, referred to as a reference frame in this paper, in the first and the second images is systematically calculated to obtain the position having the maximum value of the coefficients, as is shown by Kimura and Takamori [1].

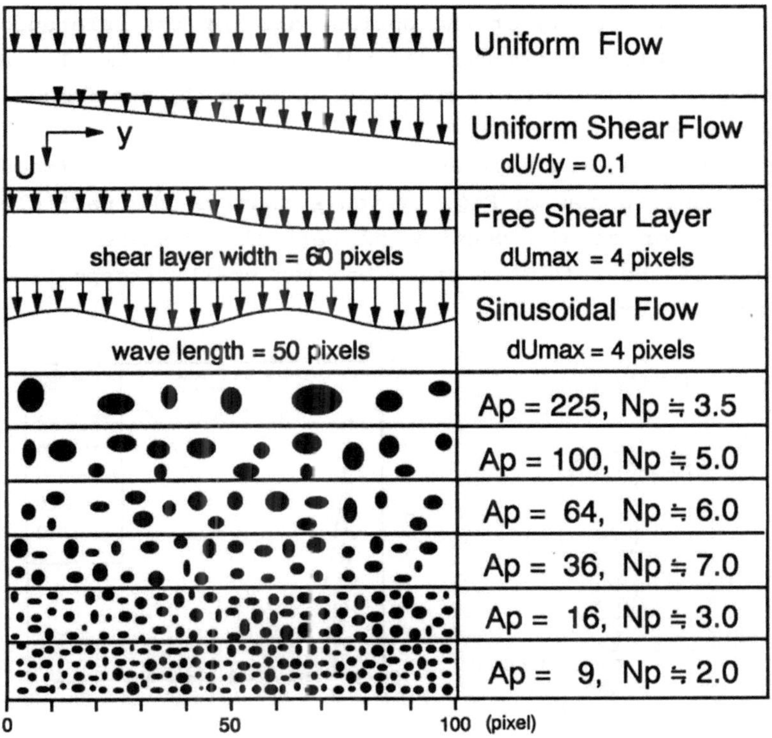

Fig.2. Comparison of types of flow and particle images

To improve the accuracy, at first, the reference frame composed of MxM pixels is shifted in pixel units to obtain the local distribution of the cross-correlation coefficients, Rab, around its maximum value, and then the true maximum value and its intermediate position between the pixels are estimated by best-fitting a surface of the second order to the coefficient distribution as shown in Fig.3.

The Optimum Size of the Reference Frame

The accuracy of the correlation method depends on the size of the reference frame; when the size is too small a lot of erroneous vectors are calculated, while the spatial resolution becomes lower when it is too large. Fig.4 indicates for the case of uniformly shearing flow that the production rate of erroneous vectors, Rerr, nearly vanishes or has a minimum value at a certain size of M, namely at the optimum size of the reference frame Mopt. Fig.4 also shows that the value of Rerr becomes large as the gradient of the particle movements between the two images, dU/dy, increases.

The relationship between Mopt and the product of the two image parameters, ApNp, shown in Fig.5 reveals that as the particle size decreases or as the density of particles increases the optimum size of the reference frame becomes small irrespective of the shear of flow.

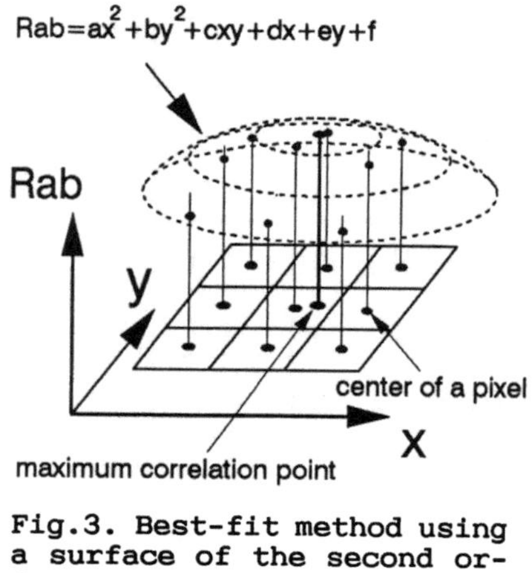

Fig.3. Best-fit method using a surface of the second order

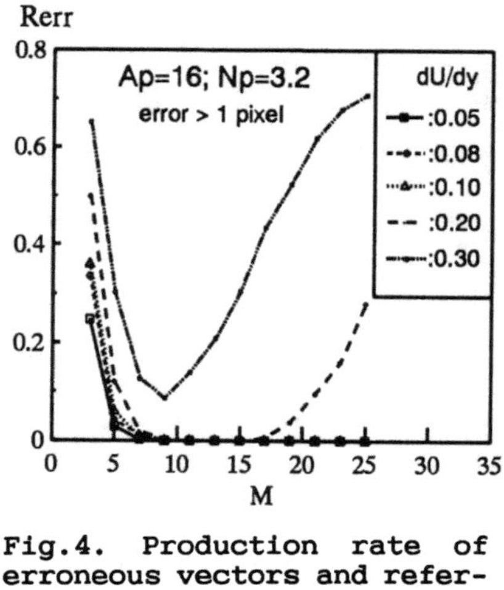

Fig.4. Production rate of erroneous vectors and reference frame size for uniform shear flows

Comparison of Accuracy

An example of the calculated distributions of the mean velocity and their standard deviation of errors, DD, for the optimum reference frame size is shown in Fig.6. Obviously, the accuracy becomes lower for the region of larger velocity gradients. The relationships between the standard deviation of errors for the all vectors obtained under the optimum conditions, Eopt, and the image parameters are shown in Fig.7. For uniform flows, the errors are always less than 0.1 pixel in any case. It supports the advantage of the previously mentioned the best-fit method.

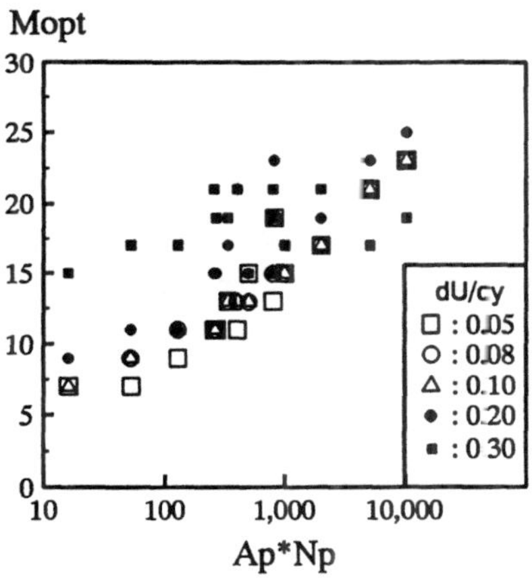

Fig.5. Distribution of Mopt for uniform shear flows

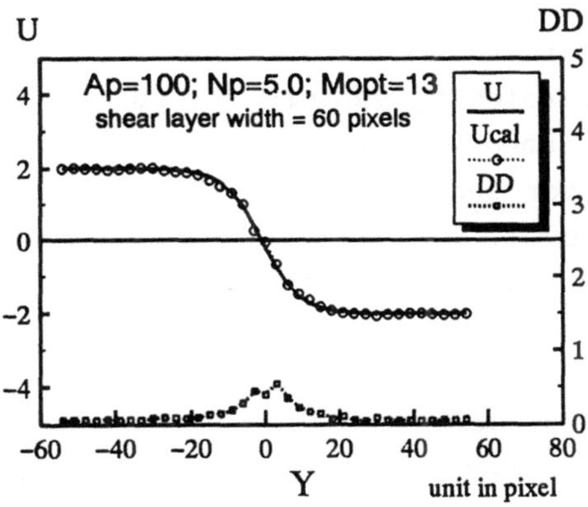

Fig.6. Calculated velocity and error distributions

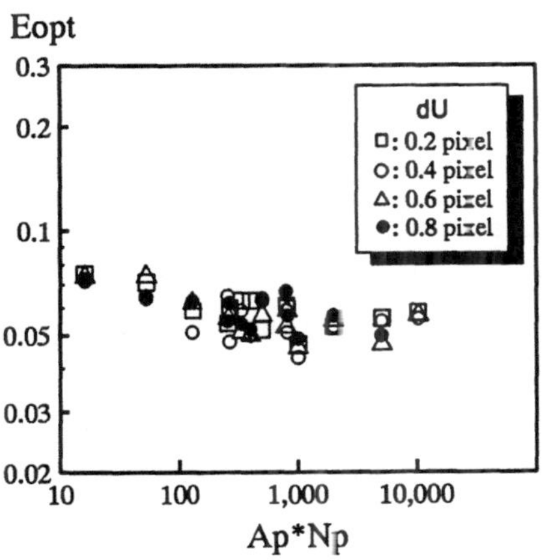

(a) Uniform flow

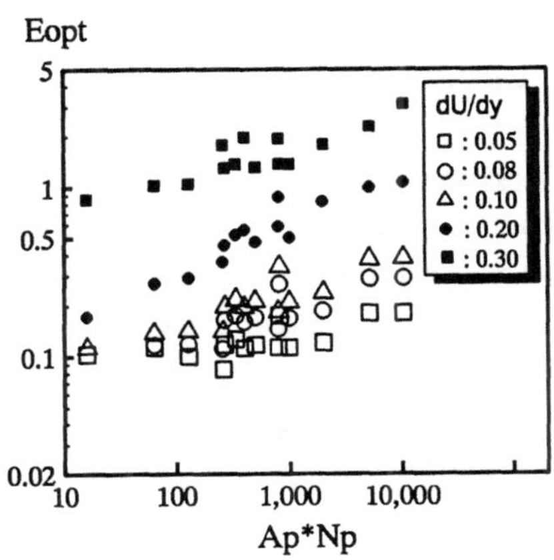

(b) Uniform shear flow

Fig.7. Distributions of average standard deviation of errors

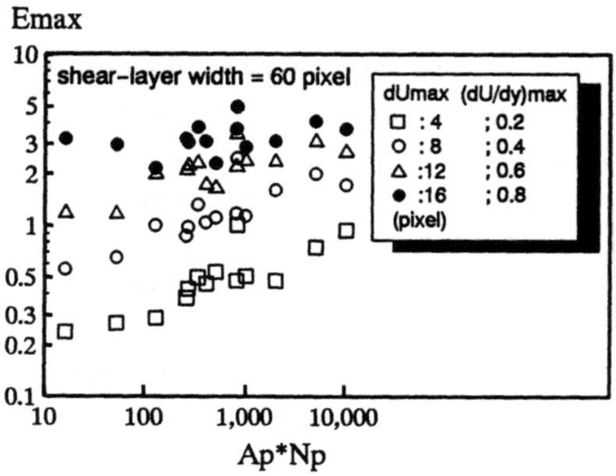

Fig.8. Distribution of the maximum standard deviation of errors for free-shear layers

In the case of uniform shear flows, the accuracy depends on the gradient of particle movements, i.e., the shear-gradient. To get the velocity vectors including the errors less than 0.5 pixel, the shear-gradient must at least be smaller than 0.1; for the larger shear-gradient the images of smaller ApNp must be used. However, for the shear-gradient greater than 0.3, we cannot get vectors having the errors less than one pixel for any particle image. The result for the free-shear layer shown in Fig.8 indicates the trend similar to the case for the uniform shear flows. Here, the maximum standard deviation of errors, Emax, is used instead of Eopt. Clearly, the accuracy depends on dUmax, the maximum relative movement of particles in the uniform flow regions. It should be noted that for the case of dUmax = 4 pixels, including the maximum shear-gradient of 0.2, the accuracy curve is almost the same as the case for the uniform shear flow with the same shear-gradient, 0.2.

Conclusions

By the simulation of particle images the accuracy of the correlation method was observed to depend on the image parameters and the shear-gradient between the two images. At least the time interval between the two images must be small enough, so that the gradient of the particle movements becomes smaller than 0.2 to get reliable velocity distributions.

References

1. Kimura, I. and Takamori, T.: Image Processing of Flow around a Circular Cylinder by Using Correlation Technique, Flow Visualization IV, Hemisphere pub. (1986) 221-226.

PIV for Lagrangian Scale Evaluation in a Convective Boundary Layer

A. Cenedese G.Querzoli
Mechanics and Aeronautics Department - University of Rome "La Sapienza"

1.Introduction

The Lagrangian description of the turbulence is the natural approach to the study of all the phenomena in which one want recognize the behaviour of the fluid particles during their motion within the field. A typical example of such phenomena is the pollutants dispersion in the Atmosphere, as the pollutant particles behave as tags of the fluid. The knowledge of the Lagrangian statistics gives a full description of the behaviour of a single particle since it was released from the source; this informations are very useful to calibrate the Lagrangian numerical models of dispersion, based on simulations of time histories of the pollutant particles released from a source (Thomson 1984). In the present experiment, the atmospheric Convective Boundary Layer (CBL), occurring in late morning and early afternoon of the sunny days above a uniform and plain terrain is simulated. Above this layer, a region in which the temperature grows with the height (the Capping Inversion), almost completely avoids the air-mass exchange between the CBL and the overlaying Free Atmosphere (Wyngaard 1985).

In the past, both theoretical and euristic relations have been suggested between Lagrangian and Eulerian characteristics of the turbulence since the classical experimental investigations give an Eulerian description of the field. Unfortunately they are affected by errors because not all the physical phenomena can be kept in account (Pasquill and Smith 1983, Lundgen and Pointin 1975). A recently developed technique, the Particle Image Velocimetry (PIV), gives the chance to obtain a Lagrangian description of the velocity field. The technique is based on the acquisition of series of images of the flow at constant time intervals. The working fluid, water in this case, is seeded by non-buoyant particles, so that their images (spots) are kept during the acquisition. A computer program individuates the positions of the particles at the successive times through the calculation of the centroids of the spots; the trajectories of the particles are then recognized and the velocities evaluated and recorded in Lagrangian coordinates: X(x,t) (where x is the position of the fluid particle at the initial time t_0, that is: $x = X(x,t_0)$). The function X(x,t) describes, as t varies, the trajectory of the fluid particle that at the initial time was at the position x. Furthermore, the velocity is given by:

$$V(x,t) = \frac{\partial X(x,t)}{\partial t} = u[X(x,t),t]$$

where u is the Eulerian velocity

2.Experimental set-up

The CBL is simulated in laboratory by a convection chamber, filled with water, in which the thermal stratification of the atmosphere is reproduced. Its horizontal dimensions are 41.0cm x 41.0cm; the side walls are of glass, while the lower surface is an aluminium plate for optimizing the thermal exchange. The chamber is heated from below by means of a heat exchanger placed under the aluminium plate, simulating the solar radiation. The upper surface of the water is insulated by means of a floating polystyrene slab to prevent the convection connected to the evaporation. The water is seeded by non-buoyant particles, 500 μm in diameter, and it is illuminated by a 1000W arc-lamp. The measuring volume is bounded by two parallel planes, orthogonal to the lower surface at 5.0cm distance. Its height is 10.0 cm, from 1.0 cm up to 11.0 cm above the lower plate. A 3-CCD video-camera takes the images, stored by a frame grabber in a buffer 512 x 512 pixels x 8 bits. The mean temperature of the CBL during the experiment is sensed by a thermocouple placed at 4.0 cm height.

3.Experimental procedure

The tank is filled, with ambient temperature water, up to a depth of 10.0 cm. Then a layer of 10.0 cm of hot water (20°C hotter than ambient temperature) is carefully stratified in order to obtain, after half an hour of rest, a layer between 7.0 and 13.0 cm of height in which the lapse rate is nearly constant. This layer simulates the Capping Inversion, usually present above the CBL (Young 1988). At this time the heating is applied to the lower surface and the convection starts mixing the layer at constant temperature. 240 s later, when the convection is completely developed, the acquisition starts.

4.Image Analysis

Starting 240 s after the heating is applied, 10 series of images are taken at fixed time intervals (110 s). Each series consist of 100 images taken at 1.1 Hz sampling rate. The images are thresholded in real time, through the use of the Look Up Tables (LUT), with the following law: the pixels whose intensity is lower than a fixed value are set to zero, while the remaining pixels are set to a value proportional to the sequential number of image. As a result, each particle of seeding gives a train of spots with increasing grey levels. The analysis of the images, that is carried out by a dedicated software, can be ideally parted in two steps:

i) The centroids of the spots are detected and stored in memory together with the grey levels of the pixels belonging to them (that is the time at which each spot is acquired);
ii) The list of centroids and grey levels is examined in order to recognize the trains of spots originated by the same particle of seeding, that is the trajectories.

The trajectories recognition criterion is based on the assumption that the spots of the same train are characterized by an increasing grey level, and match two requirements: first, the distance between two subsequent centroids have to be less than a fixed parameter ϑ; second, the difference between two successive displacements of a particle have to be, in absolute value, less than a parameter ε. These two conditions are equivalent to assume a maximum velocity and a maximum acceleration for the seeding particles. The parameters ϑ and ε have been chosen as large as possible to maximize the number of validated trajectories but small enough so that the number of errors in the recognition does not affect the velocity statistics. When the position of a particle, at fixed time intervals, is known, the non-dimensional velocity w/w_* is evaluated along the vertical axis, and the u/w_* along the horizontal axis, where w_* is the *convective velocity* (Deardorff 1970):

$$w_* = \sqrt[3]{g \cdot \alpha \cdot q_s \cdot z_i}$$

g: acceleration of gravity	α: thermal expansivity
z_i: inversion height	q_s: surface heat flux

The surface heat flux is evaluated from the time derivative of the mean temperature of the CBL by the relationship:

$$q_s = k \cdot z_i \frac{\partial \overline{\vartheta}}{\partial t}$$

obtained from the integration of the equation of heat balance, from the surface to the height at which the heat flux vanishes. Finally, k is a parameter stating the height $z_o = k\, z_i$ at which q_s=0.

5.Results

The probability distribution and the auto-correlation of both the vertical and horizontal velocity component have been evaluated from the analysis of 110 image series, that have been acquired during 11 different experiments, providing 82303 velocity samples.

5.1. Probability distribution

In figure 1, the contour levels of the vertical-velocity probability distributions are presented as function of the non-dimensional height. The plot shows the behaviour of the main turbulence statistics: the most

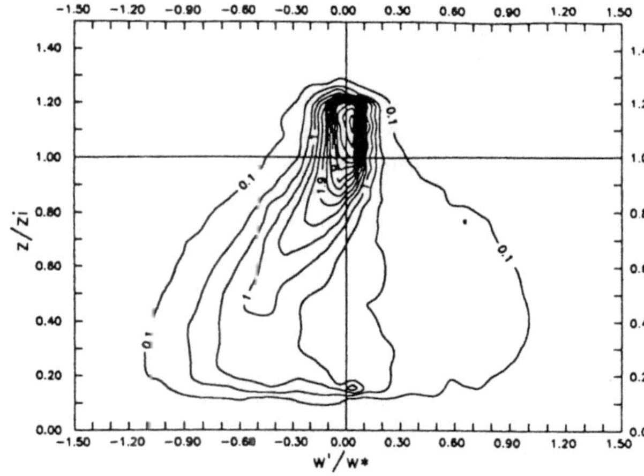

FIG.1 - PROBABILITY DISTRIBUTION OF THE VERTICAL VELOCITY

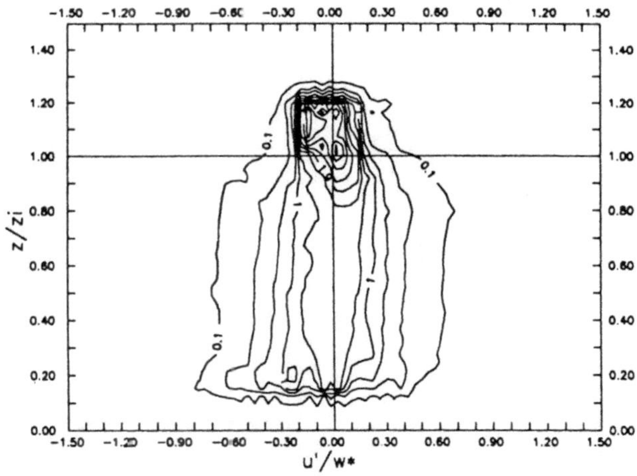

FIG.2 - PROBABILITY DISTRIBUTION OF THE HORIZONTAL VELOCITY

probable value is nearly zero at the lower surface and within the Capping Inversion, while it is negative inside the CBL with a minimum at z/z_i=0.3. This is due to the skewness that increases as far as the most probable value get negative because the mean value have however to be equal to zero(Stull 1988). Moreover, the horizontal spread of the contour lines is a qualitative measurement of the variance, that is a maximum at z/z_i=0.3 and it get very low in the Capping Inversion, where the velocity fluctuations are damped by the thermal stability; finally it vanishes at the lower surface because of the boundary conditions.

The probability distribution of the horizontal velocity (figure 2) is a maximum in a neighbourhood around the z-axis because of the horizontal symmetry of the phenomena occurring above a uniform terrain if no mean wind is present. Like for the vertical component, the variance of the horizontal component reduces in presence of the thermal stability of the Capping Inversion.

5.2. Lagrangian auto-correlation

The more extensive definition of Lagrangian auto-correlation is:

$$c = \overline{V(x,X_1,t_1)\cdot V(x,X_2,t_2)}$$

which is the average product of the velocity of the fluid particles that at the time t_1 are in X_1 and at the time t_2 are in the point X_2 (Monin and Yaglom 1971). Even if this formulation is theoretically more proper in the study of a non-homogeneous field and unsteady field, some assumption have to be made because, in point of fact, it is impossible to have a sufficient number of seeding particles matching those conditions. Actually, the phenomenon is assumed to be steady within the time of acquisition of each image series, so that the

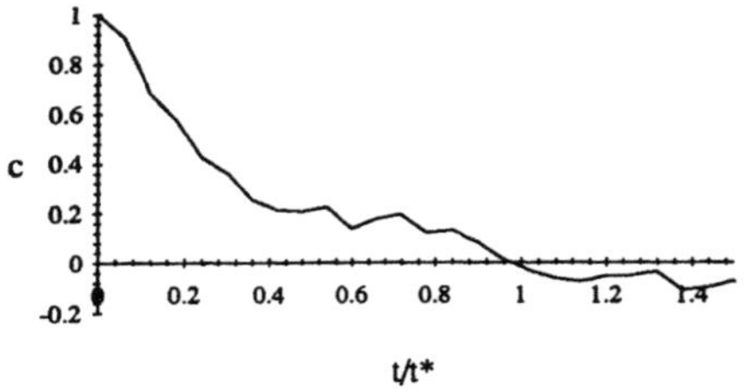

FIG.3 - U AUTO-CORRELATION AT Z/Z_1=0.2

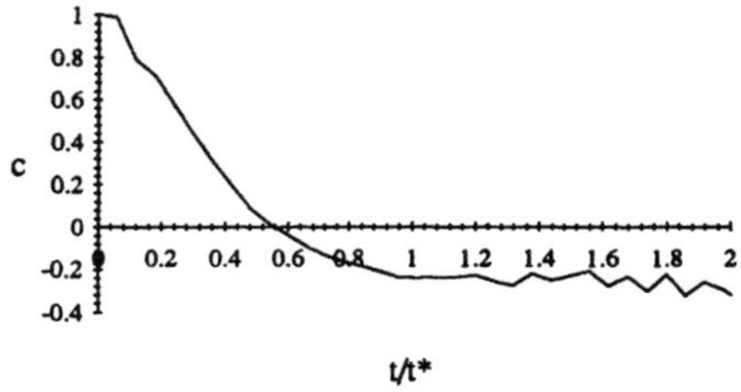

FIG. 4 - W AUTO-CORRELATION AT Z/Z_1=0.2

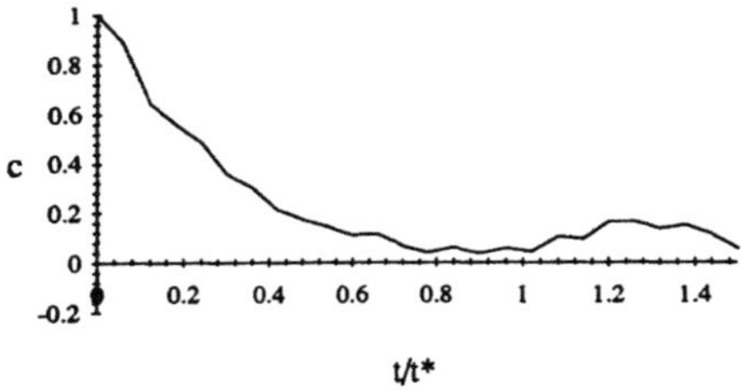

FIG. 5 - U AUTO-CORRELATION AT Z/Z_1=0.4

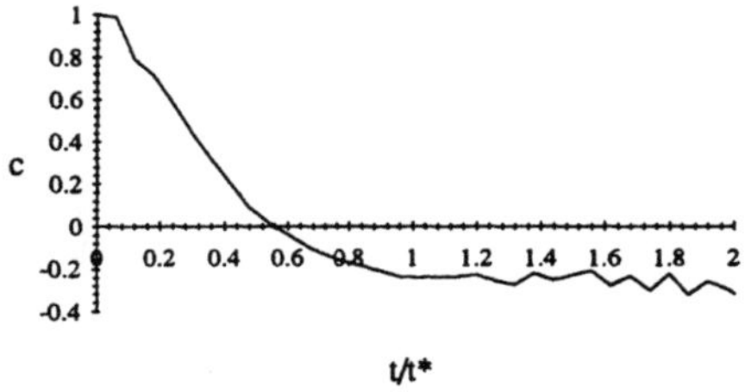

FIG. 6 - W AUTO-CORRELATION AT Z/Z_1=0.4

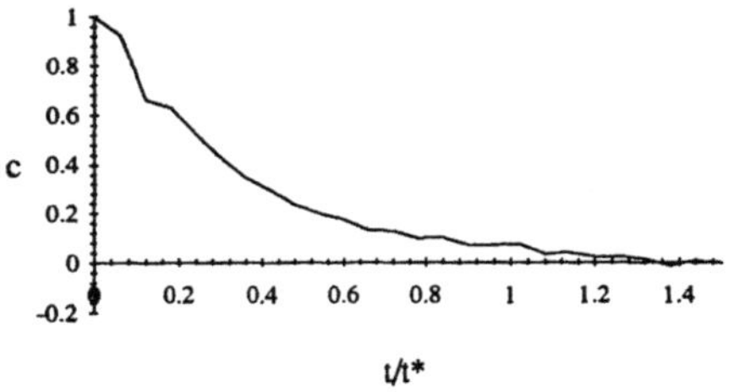

FIG. 7 - U AUTO-CORRELATION AT Z/Z_1=0.6

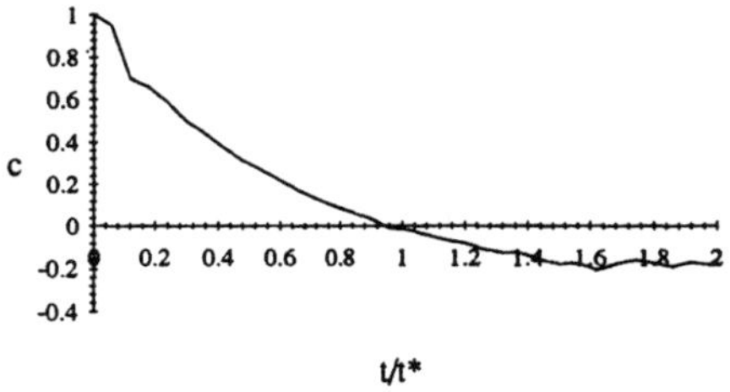

FIG. 8 - W AUTO-CORRELATION AT Z/Z_1=0.6

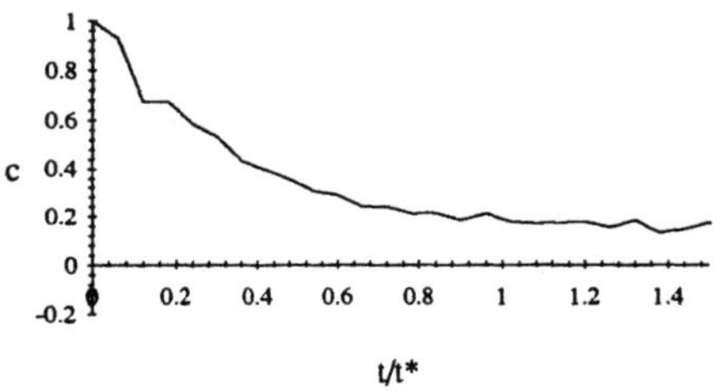

FIG. 9 - U AUTO-CORRELATION AT Z/Z_1=0.8

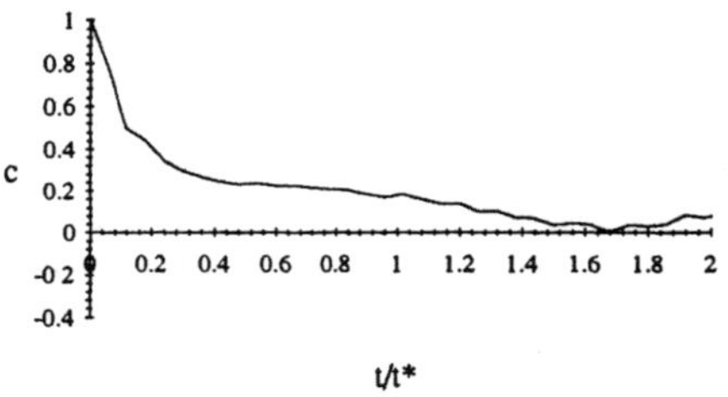

FIG. 10 - W AUTO-CORRELATION AT Z/Z_1=0.8

auto-correlation does not depends on the initial time t_1, the horizontal homogeneity is assumed and each auto-correlation is averaged on 0.1 intervals of the vertical variable. The Lagrangian auto-correlation of horizontal and vertical velocity, normalized by the variance, is shown in figures 3-10 as a function of the non-dimensional time lag t/t_*, at z/z_i=0.2, 0.4, 0.6, 0.8. The time scale is defined by $t_*=z_i/w_*$. An estimation of integral time scale is given by means of the 1/e-scale, which is the non-dimensional time lag at which the auto-correlation function decrease to the 1/e value (figures 11, 12). The 1/e-scale of the horizontal velocity increases with the height up to the top of the CBL, where the presence of the Capping Inversion cut the scale to low values. The behaviour of the vertical velocity scale is similar in the upper part of the CBL, while the 1/e scale exhibit a minimum at z/z_i=0.3, in accordance to the maximum of the vertical velocity variance. It should be noticed that the growth of the scale in the lower two points of the plot could be due to the influence of the lower border of the measurement volume, that can cause an undesirable conditional sampling. As a matter of fact, in that zone, the particles moving towards the surface (presumably with low vertical time-scale) go out of the measurement volume and do not give their contribution to the auto-correlation function, while the particles arising through the CBL (with a large vertical time-scale) are correctly tracked.

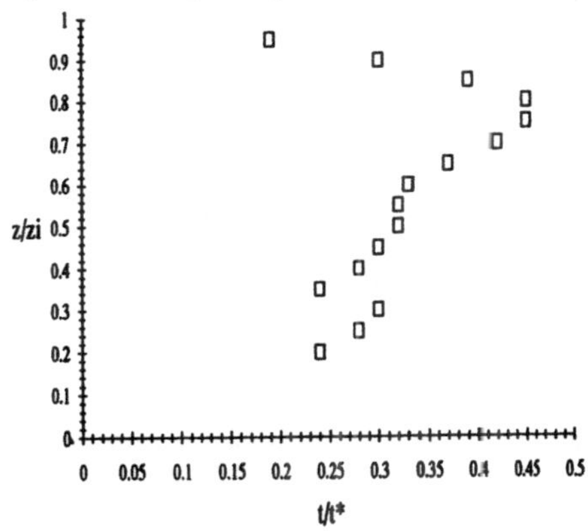

FIG.11 - 1/e SCALE OF THE HORIZONTAL VELOCITY COMPONENT

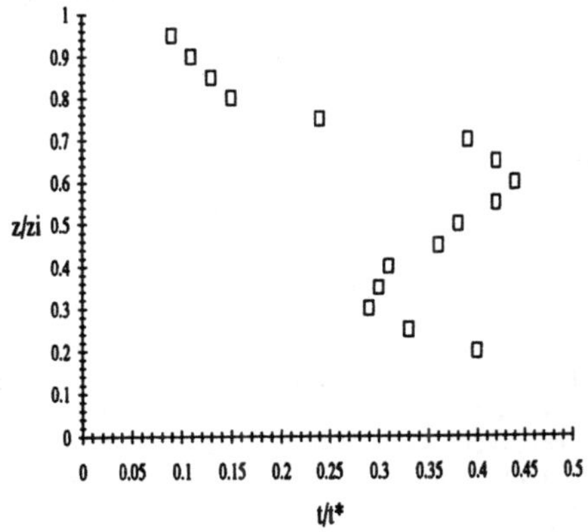

FIG.12 - 1/e SCALE OF THE VERTICAL VELOCITY COMPONENT

References

Deardorff, J.W.: Convective velocity and temperature scales for the unstable planetary boundary layer and for Rayleigh convection. J. Atmos. Sci. 27 (1970) 1211-1213

Lundgren, T.S.; Pointin, Y.B.: Turbulent self-diffusion. The Physics of Fluids 19 (1976) 355-358

Monin, A.S.; Yaglom, A.M.: Statistical Fluid Mechanics: Mechanics of turbulence. Vol 1. Cambridge Massachusetts, London: The MIT Press 1971

Thomson, D.J.: Random walk modelling of diffusion in inhomogeneous turbulence. Quart. J. R. Met. Soc. 100 (1984) 1107, 1120

Stull, R.:B.: An introduction to Boundary Layer Meteorology. Dordrecht. Kluwer Academic Publishers 1988

Pasquill, F.and Smith F.B.: Atmospheric diffusion. Chichester, Ellis Horwood Limited 1983

Wyngaard, J.C.: Structure of the planetary layer and implications for its modeling. J. Climate Appl. Meteorol. 41 (1985) 1959-1969

Young, G.S.: Convection in the atmospheric boundary layer. Earth-Sci. Rev. 25 (1988), 179-198

A High-Speed Data Acquisition Technique with Particle Image Velocimetry to Study Steam Bubble Collapse

Y.A. Hassan
W.D. Schmidl
T.K. Blanchat

Department of Nuclear Engineering
Texas A&M University
College Station, TX 77843-3133
Phone (409) 845-7090
FAX (409) 845-6443

Abstract

The use of a Particle Image Velocimetry (PIV) method (which uses digital cameras for data acquisition) for studying high speed fluid flows is usually limited by the camera frame acquisition rate. The velocity of the fluid under study must be limited to insure that particles suspended in the flow field remain in the camera's focal plane in successive images. However, the use of digital cameras for data acquisition is desirable to simplify and expedite the data analysis. A method is presented which will allow the measurement of fluid velocities with PIV techniques using two successive digital images and different framing rates, *simultaneously*. The first part of the method measures changes in the flow field which occur at the relatively slow framing rate of 53.8 ms. The second part measures changes which occur in the same flow field at the relatively fast framing rate of 100 to 320 μs. The effectiveness of this technique was tested by studying the collapse of steam bubbles, a relatively high-speed phenomena.

Introduction

Particle Image Velocimetry (PIV) is a two-dimensional, full-field, noninvasive flow visualization method. Its main advantage is that both qualitative and quantitative full field maps for fluid flow parameters (such as velocity and shear stress) can be produced over an extended region of interest [1].

The interest in the dynamics of steam or cavitation bubble collapse in liquids, especially water, mainly stems from their destructive action on solid surfaces [2]. Pictures are normally taken with high-speed photography and entered into a computer for evaluation with a graphical input device. Depending on the number of pictures to be studied, this step can be very time-consuming. A digital camera will not require this step, however, digital cameras are limited by the speed at which they can acquire images.

Description of the Experimental Arrangement

The setup for the data acquisition technique includes two Panasonic Charge Coupled Device (CCD) cameras (640 x 480 resolution), a beam splitter prism, a high-energy Nd-YAG pulsed laser, and two Epix frame grabber boards. The two cameras, set up perpendicular to the prism, look at the same focal plane through different faces of the prism. The setup to study steam bubble collapse is shown in Figure 1a. Two cameras are required to capture the microsecond separation between pulses in the double pulse mode of the laser. One camera takes a double exposure and the other takes a single exposure, as shown in Figure 1b. Since the second camera only captures the second pulse, it can be used to resolve the characteristic directional ambiguity of a double exposed image [3].

The light source is a double pulsed Spectra-Physics DCR-3G Nd-YAG laser. The time between each pulse in the set of double pulses can be varied from 100 μs to 320 μs. The time between sets of pulses is 53.8 ms or greater. Two EPIX frame grabber boards acquire the images. Each board can hold 13, sequential, images. The boards also send out a timed 5 volt TTL signal which is used to reset the cameras and trigger the laser.

To conduct the test, the plane of light was set above the tip of the nozzle. Steam for the experiment was generated in a glass flask and run though an insulated plastic hose into the tank, which was seeded with 6 μm diameter particle tracers (Expancel 551 DU, 1.02 specific gravity). The cameras were focused on the light sheet (camera view plane size 1.89 x 1.65 cm). Then the laser was fired and the data was acquired with the EPIX frame grabbers.

Data Analysis

Image processing techniques were used to convert the data into a binary format. Then a Cross-correlation Tracking Algorithm was used to determine particle correspondence between particles in two consecutive video frames. Every particle belongs to a characteristic group which has a specific local distribution pattern. This correspondence is obtained through the calculation of a correlation coefficient between a referenced pattern in the first binary image and a possible candidate pattern in the second binary image, where the latter is shifted so that the centroids of the possible particle pair coincide [4]. One particle in the first image will correspond to the particle in the second image which keeps the most similar pattern, providing the local pattern of the distributed particles changes little between sequential video frames.

Then an automated method was used to interpolate and clean the vector data. This method eliminates tedious, undesirable, manual, operator assistance in removing erroneous vectors [5]. The interpolation method that was used is based on the Hardy multiquadratic equations [6] as discussed by Narcowich and Ward [7]. Vorticity and flow streamlines can then be derived using the full-field equation for the interpolated vector field. For further details refer to Blanchat [5] and Hassan [4].

Results

This technique was tested by studying steam bubble collapse. The initial results demonstrated that it has the potential for providing useful quantitative and qualitative data. However, the following results which were obtained using the data analysis techniques previously described are not fully reliable because insufficient vectors were obtained. This problem will be corrected in future studies by increasing the seed concentration.

Long time frame (every 53.8 ms) results directly above the nozzle could not be obtained because it was very chaotic. The bubbles were collapsing every 10 ms. Therefore, results found in this area would not be accurate (since it depends on the local pattern between the two frames to be similar). However, the flow farther away from the nozzle was slower and could be tracked between the longer time frame.

Changes to the flow directly around the bubble may be studied by tracking between the shorter time frame (100 to 320 μs). The results obtained in this area did not show much of a change in the position of the tracers between the single and double exposure. However, there was a significant change in the bubble edge itself.

Figure 2a shows an enhanced image (grey levels multiplied by 3, zoomed by 5) double exposed image and Fig. 2b shows an enhanced single exposed (second pulse) image. A rough estimate of bubble collapse speed can be found by estimating the change in the steam bubble edge. Figure 2c is a tracing of the "double pulse" edge overlaid on the "single pulse" edge. The results estimate that the velocities close to the surface of the collapsing bubble were approximately 2 m/s. Figure 2d shows the resulting seed vectors in the water surrounding the steam bubbles (multiplied by 2 for emphasis) from the long time frame tracking results.

Conclusion

A method has been presented which can overcome the inherently slow frame acquisition rate of digital cameras, and allow the study of high speed phenomena (such as spray and steam bubble collapse) with Particle Image Velocimetry flow visualization techniques. The method can measure fast and slow fluid velocities simultaneously using successive digital images from two cameras. Standard PIV techniques can then be performed to quickly acquire flow information.

References

1. Hassan, Y. and Blanchat, T.: Flow Velocity Measurements using Digital Image Pulsed Laser Velocimetry. Optical Engineering. 30 (8) (1991) 1220-1227.
2. Blake, J. and Gibson, D.: Cavitation Bubbles Near Boundaries. Ann. Rev. Fluid Mech. 19 (1987) 97-123.
3. Adrian, R.J.: Image Shifting Techniques to Resolve Directional Ambiguity in Double Pulsed Laser Velocimetry. Appl. Optics. 25 (21) (1991) 3855-3858.
4. Hassan, Y.A.; Blanchat, T.K.; Seeley, C.H.: PIV Flow Visualization Using Particle Tracking Techniques. (1992) submitted to Measurement Science and Technology.
5. Blanchat, T.K.: Development of Pulsed Laser Velocimetry Techniques for Measurement of Two-Phase Interfacial Drag in a Horizontal Stratified Flow. Ph.D. Thesis. Texas A&M University, College Station, Texas 1991.
6. Hardy, R.L.: Multiquadric Equations of Topography and Other Irregular Surfaces. Journal of Geophysics Research. 76 (1971) 1905-1915.
7. Narcowich, F.J.; Ward, J.D.: Norms of Inverses and Condition Numbers for Matrices Associated with Scattered Data. Journal of Approximation Theory. 64 (1) (1990) 69-94.

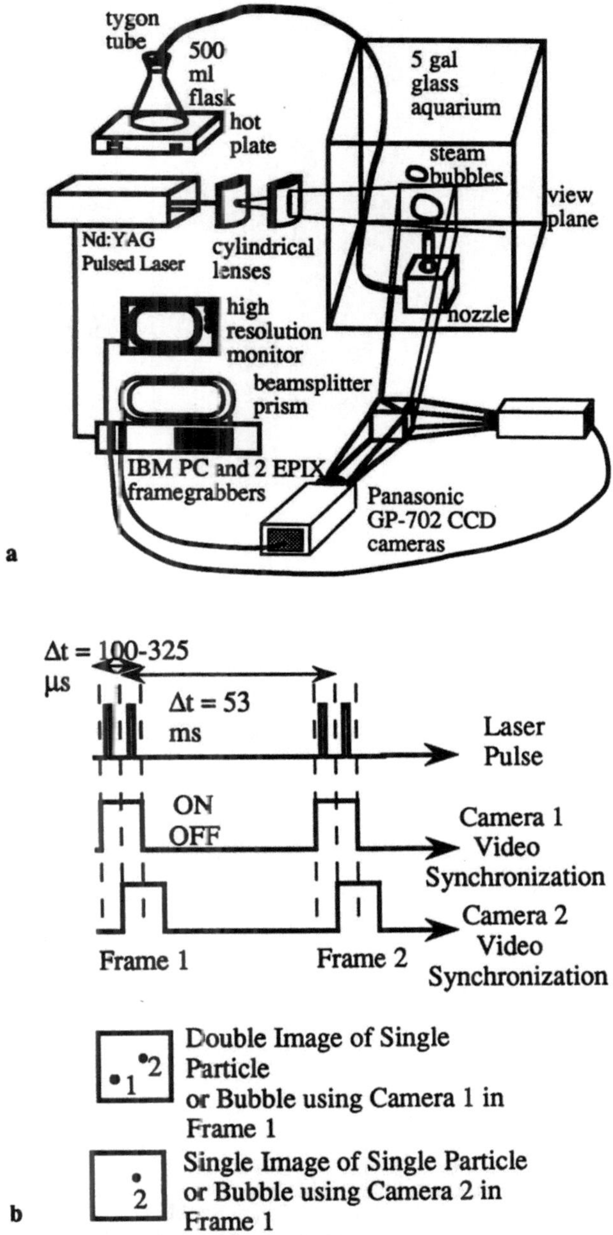

Fig. 1a-b. a Steam Bubble Collapse Experimental Setup.
b Sequential Image Acquisition of Laser Double Pulses.

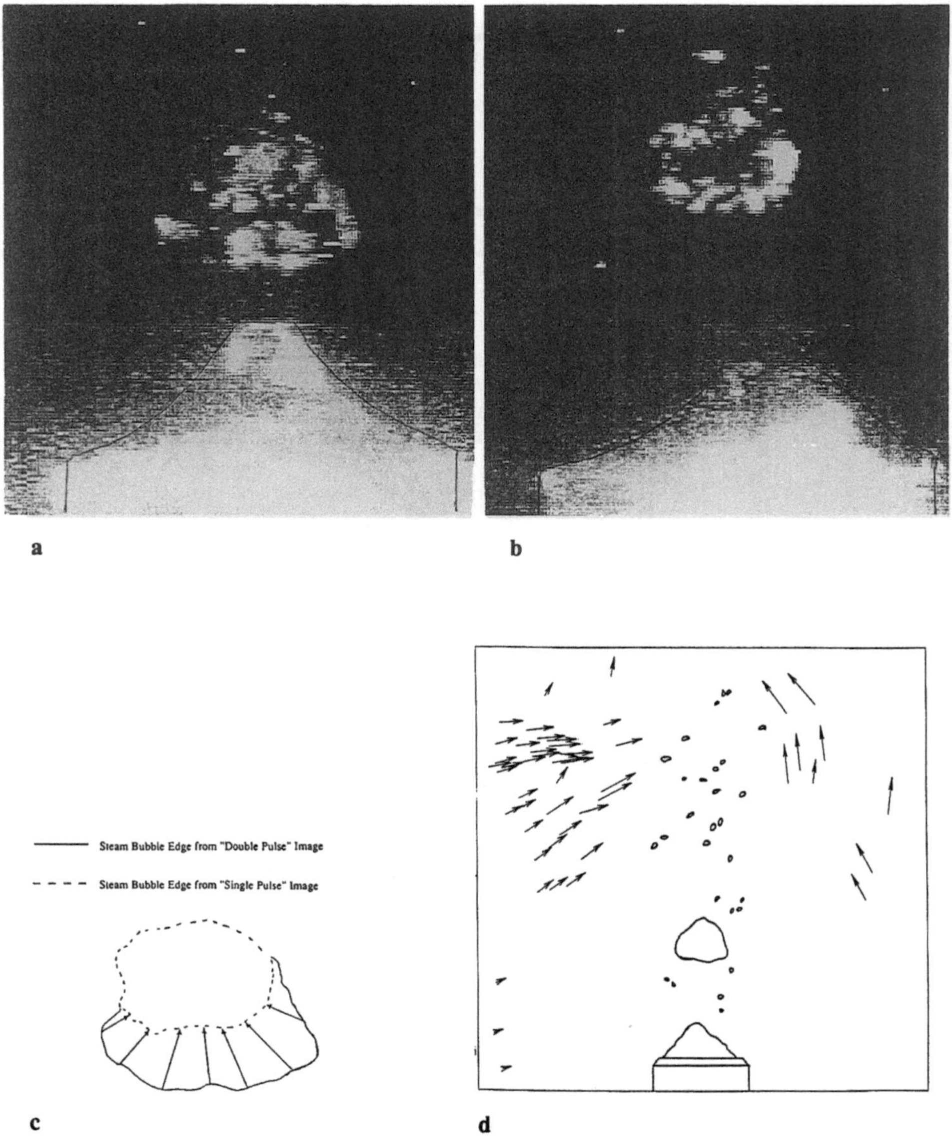

Fig. 2a-d. **a** Enhanced Steam Bubble Double Exposure. **b** Enhanced Steam Bubble Single Exposure. **c** Steam Bubble Edge Collapse. **d** Water Velocity Vectors Surrounding Steam Bubble.

Computer Aided Discrimination of Slow and Fast Tracer Paths

H.-H. Bartels*, W.J. Hiller, T.A. Kowalewski

Max-Planck-Institut für Strömungsforschung
Bunsenstr. 10, Göttingen, * Bayer A.G. Wuppertal-Elberfeld FRG

Summary
An optical method for the discrimination between slow and fast tracer particles conveyed with a flow is proposed. It is based on the dependence of the brightness of a trace from the velocity of the corresponding tracer. The images are captured by a CCD-camera and processed on-line. A practical application to the visualization of low velocity streamlines in a rotating duct is presented. Attempts to develop this technique into a quantitative tool for velocity measurements will be discussed.

Introductory Remarks

The visualization of streamline patterns in the vicinity of singular points or in complex flow geometries with large spatial velocity variations within the probe volume very often becomes difficult, as the field of view will be either crowded by a large number of traces due to fast moving particles, or the low speed tracers will become quite rare. This difficulty could be reduced considerably if there would be a fast and operative method to separate the low speed traces from the fast ones. One possibility to solve this problem is to classify the traces by their brightness, since the image of a track will become the darker the faster the particle moves.

Description of the Method

Let Φ_P be the luminant flux scattered continuously from a tracer that is collected by the imaging lens on a pixel of a CCD-camera. For a particle at rest and for displacements smaller than the dimensions of the particle image there will be locations on the trace that are illuminated during the entire integration time t_E of the camera. There, the energy collected will be $E_P = \Phi_P \cdot t_E$ (the size of a pixel be small compared to the size of the tracer image). For displacements larger than the dimensions of the tracer image the energy collected during time t_E by a pixel will be decreased by a factor of $L_P/(|v_P| \cdot t_E)$. This factor is the ratio between the length L_P of the particle image measured in the direction of its motion (e.g. a needle shaped object moving along its axis) and the displacement $|v_P|\cdot t_E$ of the image. v_P is the velocity of the particle image at the pixel concerned. This factor describes so to say the duty factor of the illumination for this pixel. For E_P that is directly proportional to the electric output of the CCD-camera we therefore get:

$$E_P = \Phi_P \cdot t_E \qquad \text{for} \quad L_P/(|v_P|\cdot t_E) \geq 1 \qquad [1]$$

$$E_P = \Phi_P \cdot L_P/|v_P| \qquad \text{for} \quad L_P/(|v_P|\cdot t_E) \leq 1 \qquad [2]$$

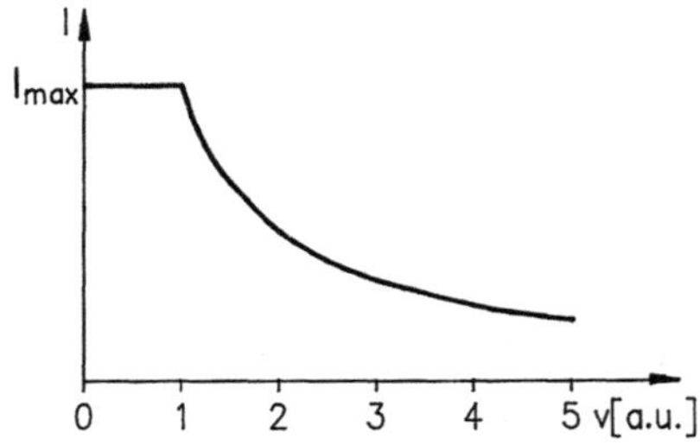

Fig. 1. Calculated maximum brightness I_{max} of a trace versus its velocity in arbitrary units. Tracers are observed against a black background.

This hyperbolic relation between the energy E_P and the modulus of v_P (Fig. 1) is optimally suited for the discrimination of velocities at the lower end of the operating range. The minimum detectable velocity v_{Pmin} is given by the condition of Eq.2. For spherical tracers of 100 µm and an exposure period of 40 ms we obtain 2.5 mm/s, if the centre region of a trace is analyzed. As for spherical tracer particles L_P decreases for the lateral regions the intensity across a trace will decrease, too. That means that the width of a trace becomes smaller, if the velocity of the particle image goes up. This effect can be used for a coarse visual estimation of the particle velocity. The maximum detectable velocity depends on the dynamic range of the camera output. By increasing the intensity of the scattered light the detectable velocity interval is shifted to higher values. Low velocity traces, of course, are overexposed. A detailed description of this method is given by Bartels-Lehnhoff et al. [1].

Capture of the Images and Classification of the Traces

The camera used for the present experiments is of the interline type (SONY XC-77CE). Its gain is kept fixed as light intensities are to be measured. The optical sensitive elements of 11 X 11 µm are arranged in a matrix of 581 lines and 756 columns. The frame is subdivided into two fields one containing the odd numbered lines and the other the even ones. On each of the two fields the

light is collected for 40 ms (CCIR) and the integration periods of the two fields overlap for 20 ms. During this time the image is captured in both fields, whereas for the first and last 20 ms only the odd and even fields are used, respectively. Depending on which of the two fields a common trace appears first, one can also detect the direction of motion. The images captured by the CCD-camera are acquired on-line by the 8-bit A/D-converter of a VFG-100 image processing board. The digitized images may be stored either directly in the memory or they can be classified online and afterwards stored.
Processing of the images is performed wiht look-up tables. The classification applied here consists in comparing the values of the intensity of the traces with an upper and lower threshold. The sections of the traces between these limits are preserved, and the constituent pixels are encoded with respect to their intensity, so that they become clearly visible. Detailed description of imaging processing may be found in the contributions of e.g. Bartels-Lehnhoff [1], Brodkey [2], and Jähne [2].
Generally a normal CCD-camera is expected to cover a dynamic range of 8 bits. From practical experience a resolution of 6 bits is more appropiate. For the above example of a 100 µm particle this corresponds to a velocity range from 0 to 160 mm/s divided into 64 classes. In the first class the particles with velocities of the image between 0 and 2.5 mm/s are contained and in the 64th class those with velocities higher than 160mm/s.

Tracer Particles and Seeding of the Flow

Normally tracers will not be uniform in size and shape. In these cases the brightness depends on the actual particle and the best, one can do is to divide the traces into two classes only. The cut should be chosen so that the low velocity class contains only the big particles. Disk-shaped particles, for instance, in a shear flow will generate dashed lines if properly viewed and illuminated.
The tracer density has to be matched with regard to the depth of the probe volume and the integration time selected. If it becomes too large the high velocity streaks will be buried in the noise. The extraction of the low velocity streaks is less sensitive to high particle concentrations.

Experiments and Results

The technique was applied to the visualization of the flow generated by a piston gliding inside a torus that is rotated in a vertical plane. The cross section is 1 X 1 cm and the mean radius of curvature of the channel amounts to 15 cm. A glycerol-water mixture was used as a flow medium and the tracers had a diameter of approx. 150 µm. The channel is illuminated within its plane of motion by a 1cm thick light sheet and observation is performed at right angles into the direction of the rotatory axis. Fig. 2a shows an unfiltered full frame of the particle tracks.

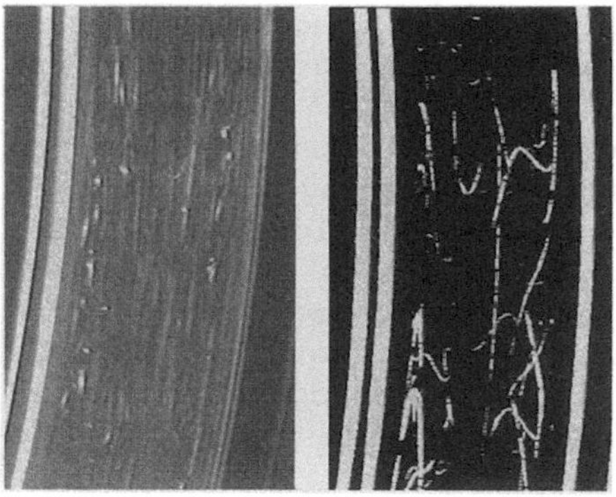

Fig. 2a (left). Unfiltered particle traces in a rotating duct. Duct walls move downwards.

Fig. 2b (right). A series of 24 successive frames of type 2a digitally filtered and superimposed. Depth of field: 10mm, field of view: 10x26 mm. Particle number density approx. 100/cm^3.

At first glance the image seems to contain only traces parallel to the side walls. For the region v_{ax} = 0 one expects due to secondary flow an inversion of the axial direction. To extract these slow but bright traces a series of 24 consecutive images (980 ms) was taken, properly filtered on-line, and superimposed (Fig. 2b). The shape of the streamlines crossing the surface v_{ax} = 0 becomes clearly visible, and also the ratio of the projection of the axial to the radial flow components into the plane of observation and the extension of v_{ax} into radial direction can be detected (outside this region no flow reversals will be observed). For velocity measurements, however, spherical tracers equal in size and of good optical scattering properties are

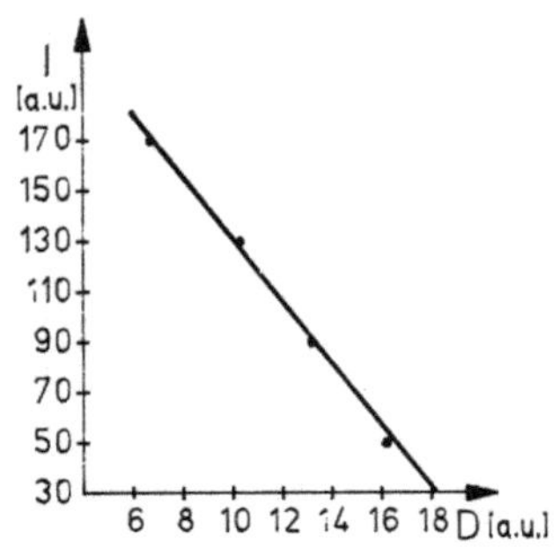

Fig. 3. Lateral distance D of isointensity line pairs versus intensity I of the trace. The distance D is measured in multiples of 11 µm (pixel size). I ranges from 0 to 255.

needed. To give an idea of the practicability of this technique a short trace was captured at a higher magnification, and the lateral distance D of corresponding isointensity lines was measured for a couple of values of the intensity (Fig. 3). As it can be deduced from Eq. 2, D really depends in good approximation linear on the intensity over a large range. For low intensities the trace disappears in the noise of the camera-system. For high values the deviation of our tracers from an ideal sphere becomes obvious, as they are strongly rotating. Fig. 4 gives a 3-D representation of a trace generated by such a coagulated particle.

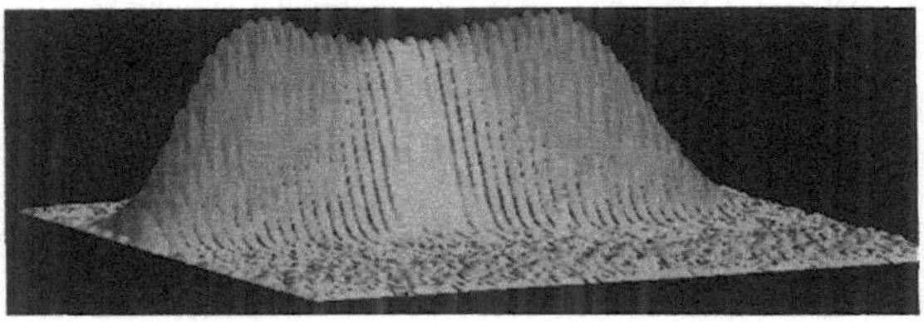

Fig. 4. Intensity distribution along a trace. The bumps at the beginning and at the end of the trace are accounted to irregularties of the particle shape. Exposure time 60 ms.

Conclusions

Despite the simplicity of this technique a lot of qualitative and quantitative results can be extracted from the traces of a single frame. Of course, there are still a lot of problems. Some are more of technical nature like the need for better CCD-cameras or suitable tracers. On the other hand automatic recognition and processing of the images is still at its infancy, further efforts are necessary, if a 3-D evaluation of the velocity field is intended.

References

1. Bartels-Lehnhoff, H.-H.; Hiller, W.J.; Kowalewski, T.A.: Computer aided Discrimination of slow and fast Tracer Paths. Accepted for publication in: Experiments in Fluids, 1992.

2. Brodkey, R.S.: Image processing and Analysis for Turbulence Research, Chem. Engr. Educ.,20 (1986) 202 - 207.

3. Jähne, B.: Digitale Bildverarbeitung. Berlin: Springer Verlag (1989)

4. Bartels-Lehnhoff. H.-H.: Erfassung dreidimensionaler Bahnlinien und Geschwindigkeitsfelder mittels automatischer Bildverarbeitung. Mitt. aus dem M.P.I. für Strömungsforschung, Göttingen (1991) Nr. 101.

Tracer Particles of Which Specific Gravity is Unity

T. ETOH and K. TAKEHARA

Department of Civil Engineering,
Kinki University

Summary

A method to produce microcapsules which satisfy the conditions that the specific gravity is unity and the specific heat is very close to that of water is developed. The microcapsule may be a near-ideal tracer of movement of water. It is filled with pure water and covered with a thin film of plastic material of which specific gravity is controlled to unity. The film material consists of polystyrene and ethylene-vinyl acetate copolymer(EVA), of which specific gravities are about 1.04 and 0.95, respectively.

Aim

A tracer particle is expected to accurately follow movement and temperature change of its surrounding water. Thus, the size should be smaller than the micro-scale of turbulence, and the specific gravity, specific heat, etc., should be as close as possible to those of water. The authors have developed a method to produce microcapsules which satisfy these conditions. The micro-capsules are filled with pure water and covered with a thin film of plastic material of which specific gravity is controlled to unity. The material consists of polystyrene and ethylene-vinyl acetate copolymer(EVA); their specific gravities are about 1.04 and 0.95, respectively. A micrograph of the micro-capsules is shown in Fig. 1.

Generally, a microcapsule has a diameter from several microns to several hundreds of microns, and the substance inside is protected by a thin film from the surrounding atmosphere or fluid. The film is made of a natural or synthetic macromolecule. The diameter and thickness of the film can be changed by different production methods.

Production Method

The method used to produce the microcapsules in this research is as follows (Kondoh and Koishi[1]). The film material is dissolved in a solvent which does not mix with water.

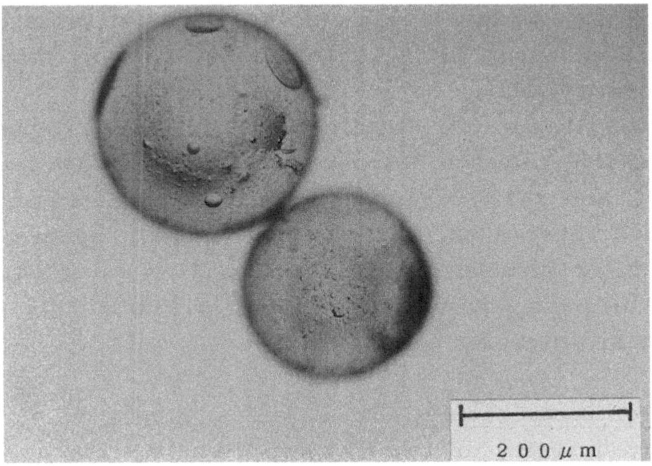

Fig.1. Micrograph of micro-capsules of which specific gravity is unity

Pure water is emulsified in the solvent. The emulsion is emulsified in water again. A particle in the triple-phase emulsion is basically made with oily solvent with the film material and has a very small sphere of pure water inside. The particles are suspended in water. The solvent is removed from the emulsion by heating and/or decompressing. Then the micro-capsules which are filled with pure water and covered with a thin film, are left suspended in the surrounding water.

In this research, mixture of polystyrene and EVA are chosen as the film material, and dichloromethane is chosen as solvent. The specific explanation of the production process is as follows:

1. Film material is made by mixing polystyrene and EVA with heat.
2. Film material is dissolved in dichloromethane.
3. Pure water which will be inside the microcapsule is emulsified in the dichloromethane. It is called a (W/O) type emulsion, where W means water, and O oil.
4. The (W/O) emulsion is emulsified again in water with a protective colloid which helps the (W/O) emulsion to be emulsified in the surrounding water. The emulsion is called a [(W/O)/W] type.

5. The [(W/O)/W] emulsion is heated to keep the temperature of the emulsion at about 40 °C, which is the boiling temperature of dichloromethane.

Then, the dichloromethane in the layer of the O-phase in a particle, which contains a sphere of pure water (W-phase), is evaporated, and film material dissolved in dichloromethane deposits and forms a thin film around a small sphere of pure water, which is suspended in the surrounding water. The small particles of pure water covered with plastic film are the resultant micro-capsules.

Condition for Productivity

The productivity of the microcapsules is shown in Table 1 as a function of mixing ratio of pure water to dichloromethane in which polystyrene and EVA are dissolved. Other conditions relating to the productivity are the same in all experiments. Experiments are repeated three times at each fixed condition. The number of microcapsules produced at each condition in the table is the average of the numbers for three experiments. It appears that the best mixing ratio lies in a range from 0.55 to 0.56.

Specific Gravity

Specific gravity of the microcapsules is very difficult to measure, because it is too close to that of water. However, after being left for more than one day in still water, the microcapsules are separated into three groups, i.e., one to float at the water surface, one to float still in water, and the rest to sink down to the bottom.

Size

Size distribution of the capsules is exemplified in Fig. 2. The average diameter and the standard deviation are 0.457 mm and 0.087 mm, respectively. It can be easily reduced by increasing rate or number of rotation of stirrer, when dichloromethane is emulsified in water.

Table 1. Productivity of microcapsules

Mixing ratio (pure water/dichloromethane)	number of microcapsules (number/cm^3)
0.12	24.8
0.14	25.8
0.16	31.1
0.18	31.4
0.20	34.9
0.22	51.2
0.24	46.5
0.26	27.6

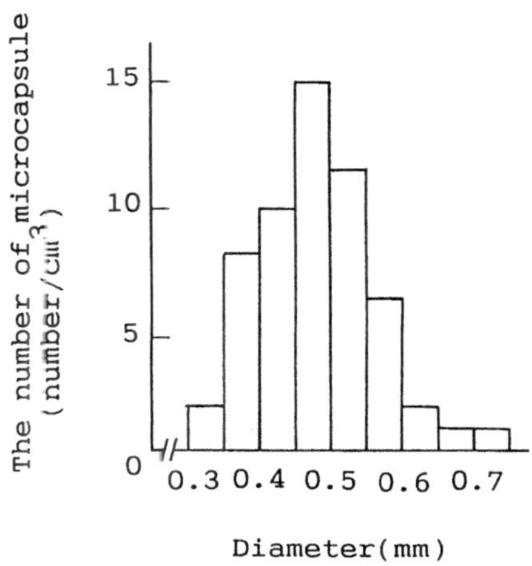

Fig.2. Frequency distribution of diameter of microcapsules

Concluding Remarks

Microcapsules whose specific gravity is equal to unity and heat-relating characteristics are very close to those of water is developed as tracer particles for PIV(particle-image velocimtry). They may be very usefully applied to measure velocity fields, slow turbulent motions of water with temperature gradient.

References

1. Kondoh, T., and Koishi, M., Microcapsule, Sunkyo Publishing Company(in Japanese), 1937.

Particle Imaging and Speckle Methods

Speckle Method and High Image Density PIV

Masaaki KAWAHASHI

Faculty of Engineering
Saitama University, Saitama

Motoo FUMIZAWA

Japan Atomic Energy Research Institute, Ibaragi

Summary

The method of full field velocity measurements by means of optical processing of double-exposed patterns of laser speckle or high density particle images obtained by a thin pulsed-light-sheet illumination in the fluid flow seeded by small tracer particles is known as speckle method or high-image-density PIV. This technique is superior to PIV by means of digital image processing in simplicity of processing procedure and applicable rage of velocity. Recent developments of devices for optical processing, such as liquid crystal display, BSO crystal or position sensitive device, are improving the system of velocity measurement by this technique.

Fundamental Technique

There are some different techniques to obtain velocity vector field of fluid flows by means of full field velocity measurments known as speckle method, particle image velocimetry or particle tracking velocimetry, as described by Adrian [1]. These are classified by some parameters determining the characteristics of images, such as the concentration of particles mixed in the fluid, the image density within a picture or the number of exposure times for recording images on each picture. The method of 2-D measurements by means of optical processing of double- or multi-exposure photographs recording high density images is called speckle method or high-image-density PIV. When a plane of velocity measurement in the fluid flow seeded densely with fine particles is illuminated successively by a thin sheet of pulsed laser light with a short time separation between pulses, laser speckle or particle image pattern is produced by the light scattered from moving particle field within the light sheet. A double- or a multi-exposures of the pattern is recorded on a photographic film by a still camera as "specklegram".

Optical Fourier's transformation of an interrogation spot in the specklegram illuminated by a laser beam gives Young's fringes in the image plane which indicates local information of particle displacement. The space and orientation of fringes are corresponding to the displacement of particle and its direction, respectively.

Fundamental technique of this method has been established, however, there still remains some problems to develop a system for practical usage. These are recording medium, image processing technique of Young's fringes, and so on.

Recent progress

Liquid crystal display can be used for recording medium in place of photographic film, but spatial resolution of it is not enough for the moment. Ferro-electric liquid crystal (FLC) and bismuth-silicone-oxide (BSO) crystal are useful devices to establish fast system of velocity measurement.

Two dimensional Fourier's transformation is an efficient method of Young's fringe analysis. The pattern of 2-D Fourier's transformation is obtained by digital image processing or by optical processing shown in fig.1. Optical processing systems composed by making use of FLC and BSO have been developed by Kobayashi[2] (shown in fig.2) and Takatsuji[3]. And a full analogue system combining with position sensitive device has investigated by Enomoto[4] as shown in fig.3.

To improve the accuracy of the measurement by this method, fine images are required. The choice of illumination technique is important for taking fine images. An efficient technique of illumination by dual-beam-sweep shown in fig.4 has been proposed by Kawahashi [5].

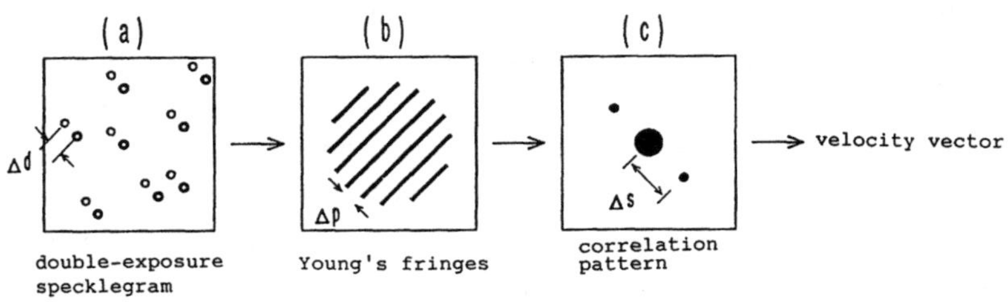

Fig.1 Optical processing of double-exposure specklegram and Young's fringes.

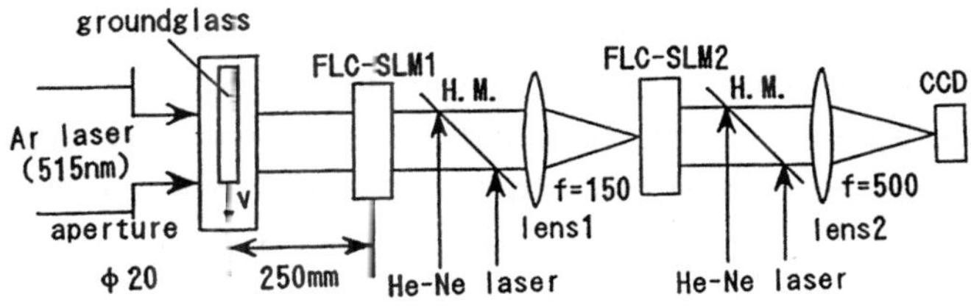

Fig.2 Optical processing system of specklegrams.

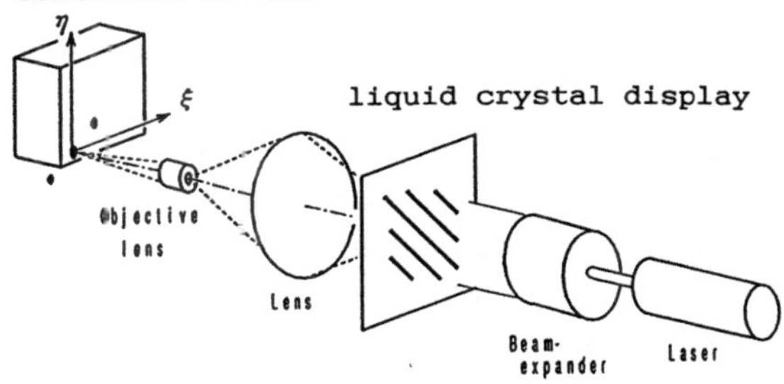

Fig.3 Optical system of 2-D Fourier's transformation combining with position sensitive device.

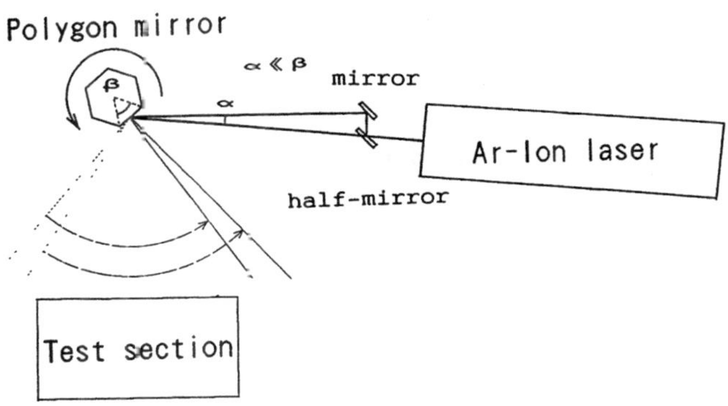

Fig.4 optical system of dual-beam-sweep illumination.

Application

The method of velocity measurements described in this paper was applied to the buoyancy-driven exchange flows through the inclined opening. This type flow might be caused by a pipe rupture accident in a high temperature gas cooled reactor.

The technique of single-beam-sweep illumination by making use of Ar-ion laser combining with a polygon mirror scanner is used for taking specklegrams. The optical arrangement and apparatus of this experiment are shown in fig.5. The diameter of the opening is 20mm. After helium gas had been filled in the test cylinder, the cap covered the top of the opening was abruptly removed. Then an ascending helium current issuing from the opening and an entering air current into the test cylinder were occurred. Flow patterns and the velocity at the opening of the exchange flows depended on inclinations of the test cylinder. When a vertical plane in the exchange flow field was periodically illuminated by sweeping beam scanned by a polygon mirror, multi-exposure specklegrams of the ascending helium current seeded with fine particles were recorded on the films. The scanner has 10 faces, and its revolution speed is fixed at 4000rpm. The particles mixed into the helium flow are finely powdered talc. The photographic film used is Kodak T-MAX P3200 and the shutter speed of camera is 1/15s, and the imaging magnification is 0.4. The velocity vector diagrams of the helium current were obtained from the specklegrams by means of optical method.

The specklegrams and the velocity vector diagrams obtained are shown in fig.6. The maximum velocity of helium current at the opening was obtained at the inclined angle 30°.

Conclusions

Speckle method (or high-image-density PIV) is useful to establish fast processing system for velocity measurement, and also it can be applied to wide range velocity measurement.

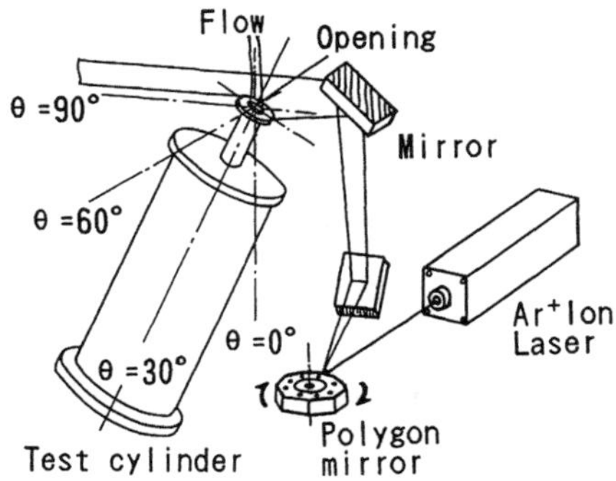

Fig.5 Experimental apparatus of air-helium exchange flow.

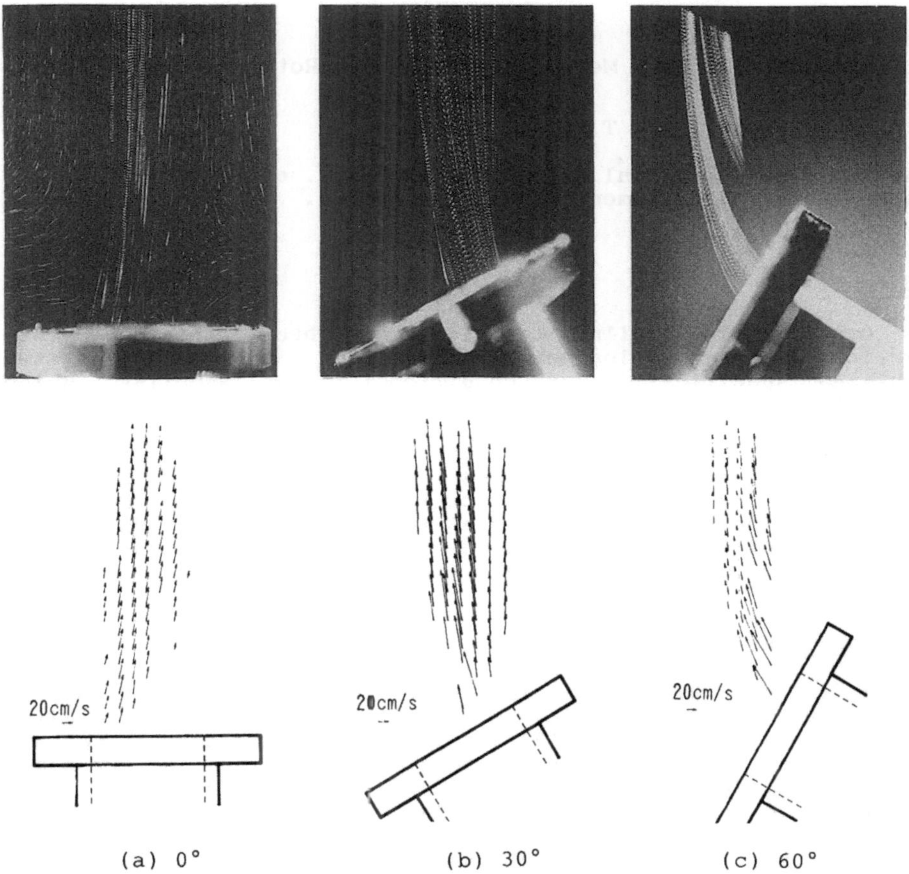

Fig.6 Specklegrams of helium currents issuing from small opening and velocity vector diagrams.

References

1. Adrian,R.J.:Particle-imaging-techniques for experimental fluid mechanics. Annu.Rev.Fluid Mech.23(1991)261-304.
2. Kobayashi,Y.;Mukozaki,M;Yoshida,N;Kurokawa,T:Displacement meter with FLC-SLM utilizing speckle photography. Abst.52nd meeting of J.S.A.P.,(1991)895.
3. Takatsuji,T;Nakagawa,K;Minemoto,T:Real time measurement of displacement distribution by speckle photography. Abst.50th Autumn meeting of J.S.A.P.,(1989)750.
4. Enomoto,H:A study on signal processing for particle image velocimetry. Master thesis of Keio Univ.,(1992).
5. Kawahashi,M;Hosoi,K:Dual-beam-sweep laser speckle velocimetry. Exp.Fluids 11,(1991)278-280.

Introduction of a Conditional Sampling to Particle-Imaging Velocimetry

-Conditional Velocity Measurement behind a Rotating Propeller-

D.H.Doh, T.Kobayashi, T.Saga, H.Kanamori *

Institute of Industrial Science of the Univ. of Tokyo
Satake Chemical Equipment Manufacturing Ltd.*

Abstract

A conditional sampling technique has been introduced to Particle-Imaging Velocimetry. The image processing system designed specifically for conditional image sampling which consists of a laser light sheet, a CCD camera, an image processor with a time difference detector allows to extract the image of the settled conditions from consecutive TV frame images. To confirm the availability of this technique, the system has been applied to the flows behind a rotating propeller, and the time-averaged velocity distributions at four discrete rotating angles of the propeller are measured. The difference of wake structures at each blade angle position was clearly extracted.

Introduction

Since Particle-Imaging Velocimetry(PIV) measures the simultaneous movements of many particles in flow fields through the digital image processing technique[1][2][3][4], this method would be very helpful for understanding the physical phenomena of unsteady flows. And its effectiveness will even more be shown when it is combined to a conditional sampling method which takes out a desired image on the consecutive images, because the data acquired by the conditional sampling method makes us easy to understand the spatial structure of flow fields under some

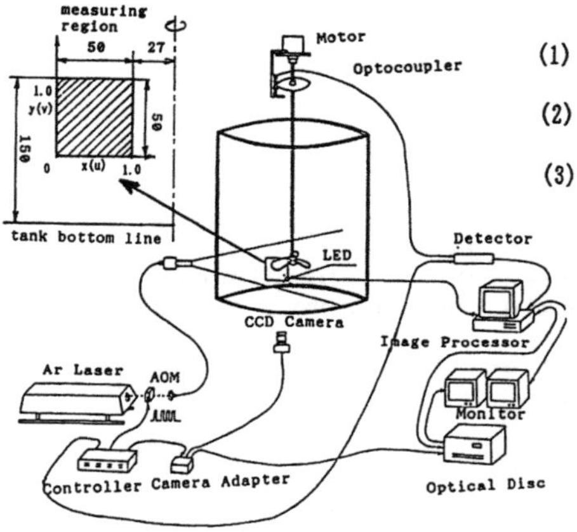

Fig. 1 Overall views of experimental apparatus and measuring region

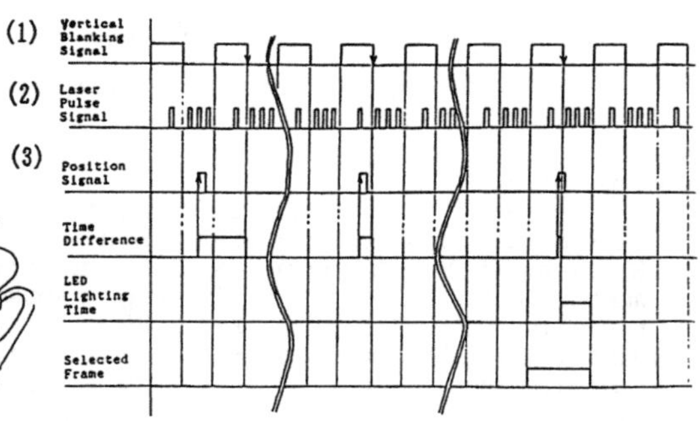

Fig. 2 System time relations

physical condition. On this view points, the conditional sampling method has been studied on PIV. After designing a system which is able to get a conditioned image automatically,it is applied to velocity measurements behind the flow field of the rotating propeller in a cylindrical water tank. The conditions considered here are the rotating speed and the rotating angle position of a blade. This paper aims at verifications of the effectiveness and the validity of the conditional sampling method on PIV.

Experimental Setup

A schematic arrangement of the experimental apparatus, the measured region and the definition of coordinate are shown in Fig.1. The diameter of the tank used is 490mm and its water level is 490mm. The propeller shaft is located on the center position of the tank and the vertical location of the propeller is 150mm from the tank bottom. An opto-coupler is set on the upper side of the rotating shaft for detecting the rotating speed and the blade angle positions of propeller. In the present experiment, the rotating speed of propeller is fixed with 60rpm. The output signal of the detector is sent to a time difference detector at which 12 bits data are generated, and the 12 bits data are sent to the host computer. The spherical particles used for visualization of the flow field are the nylon 12(mean diameter=300μm,specific gravity=1.02). Velocity measurement is made in a center vertical cross section of the vessel as shown in Fig.1. An argon laser light sheet passing through the measurement region(50mmx50mm,dimensionless length;1.0x1.0) with its width 20mm is illuminated intermittently four times per frame. Pictures of the flow field are recorded on the optical disc as analog signals and the instantaneous particles recorded on the image are transformed into digitized informations of particle sizes and positions through the image processor. The host computer works for both the image processing and the system controls. Up to the calculation of gravity centers, data are pre-processed on the image processor. The method of identification of the same particle among many particles is the four consecutive time tracking method established by T.Kobayashi et al.[1][5]

Selection of Conditioned Image

The time relations of this system are shown in Fig.2. The selection of desired image is worked out by finding the image that has density information of the LED. The LED attached on the out side of the tank is lighted when the blade comes to the presetted angle position of the blade. The time difference detector compares the position signal of blade and the vertical blanking signal of CCD camera, and send the difference as 12 bits digital data into the host computer. The host computer judges whether the rotating speed of propeller shaft is fluctuate and the received time difference Δt is in desired time difference (ΔT) or not. Figure 3 shows the algorithm of conditional sampling of image. The actual time relations between the vertical blanking signal and the blade angle position is shown in Fig.4. Here, the LLS and TV signals are synchronized with each other. The time difference Δt means the interval between the positive-going time of the position signal and the negative-going time of the first field vertical blanking signal. If the time difference, Δt, is less than ΔT(here, ΔT is set at 1386μs), the LED is lighted on the second field of frame and its density information is recorded on the image. The actual locations and movements of blade during measurements are shown in Fig.5. The analysis of the selected images is executed on four rotating angle positions. The positions a,b,c, and d are blade locations from the base angle. Each means the angle from the base angle 0°,30,°60°,and 90°,respectively. The velocity vectors are calculated from four consecutive pulsing times and the positions of four serial particles on the image. So the velocity vectors obtained are the results of tracking the four serial particles, which means that the blade moves between -2.2° and +5.1° from the base angle for four consecutive times.

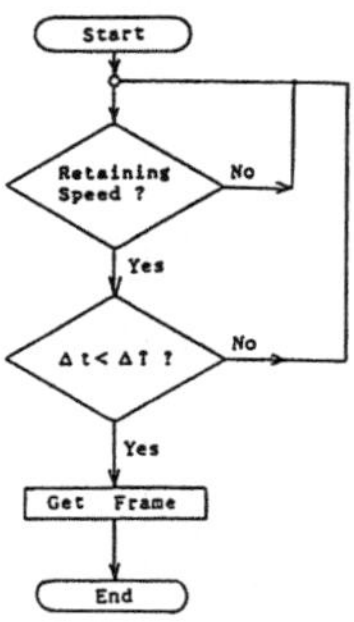

Fig.3 Algorithm of conditional sampling

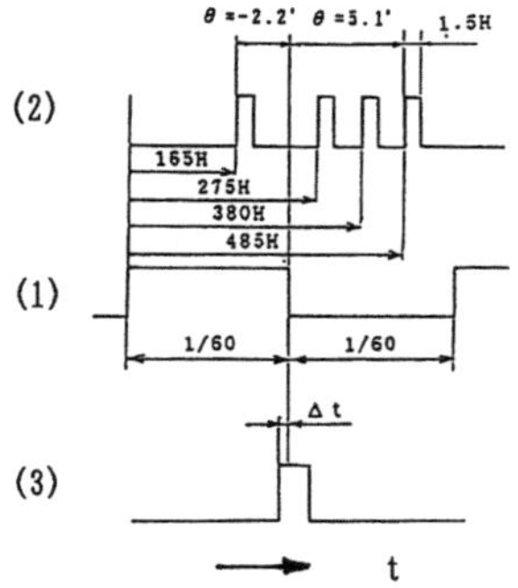

Fig.4 Time relations between the blade angle position and the vertical blanking signal of camera

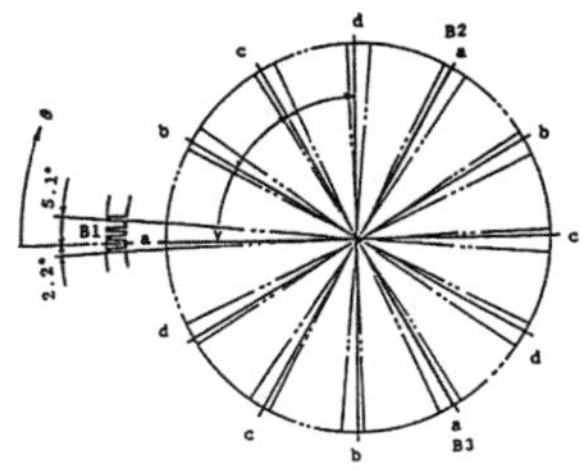

Fig.5 Sampling position

Image Analysis and Experimental Results

The selected images obtained through the conditional sampling are transferred from the optical disc into the image processor and are analyzed automatically. By repeating the above procedures and summing the resulted data on one angle position of blade, the time-averaged velocity vectors are obtained for the given blade angle position. The vectors are normalized with the tip peripheral velocity. The velocity vectors are also complemented on the grids 31x31, and the stream lines are calculated from the grid vectors. Final data are overlapped summations of the consecutive 80 separate selected frames and are shown in Fig.6. The velocity vectors of 80 frames without any condition are shown in Fig.6(a). And the velocity vectors on each blade angle position obtained through this method are shown in Figs.6(b),(c),(d), and (e). An arrow over Fig.6(a) means the tip peripheral velocity. Figure 7 shows the complemented velocity vectors on the grid for each case, and the stream lines are also shown in Fig.8. It can be seen from these figures that the structures of flow field behind the propeller have large quantitative difference with each other. The flow over the blade comes down on the surface of the blade and changes its direction suddenly and the flow goes down to the tank bottom with the increase of rotating angle of blade. To confirm this phenomenon,

(a) random

(b) $\theta=-2.2^\circ \sim 5.1^\circ$

(c) $\theta=27.8^\circ \sim 35.1^\circ$

(d) $\theta=57.8^\circ \sim 65.1^\circ$

(e) $\theta=87.8^\circ \sim 95.1^\circ$

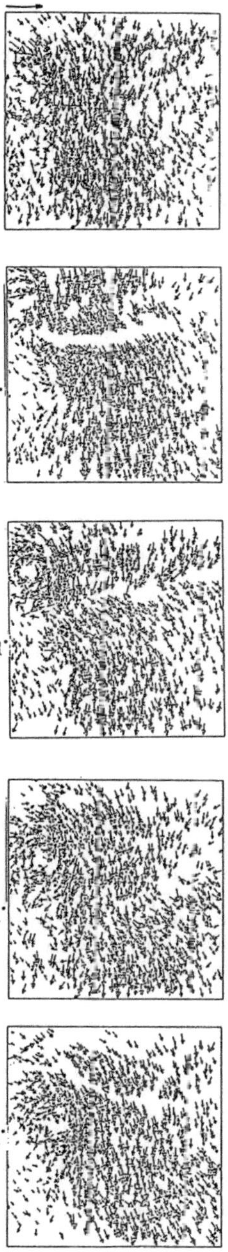

Fig. 6 Velocity vectors

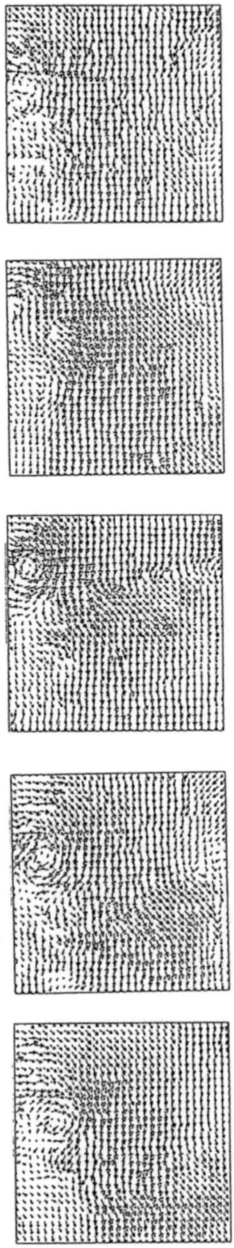

Fig. 7 Velocity vectors on grid

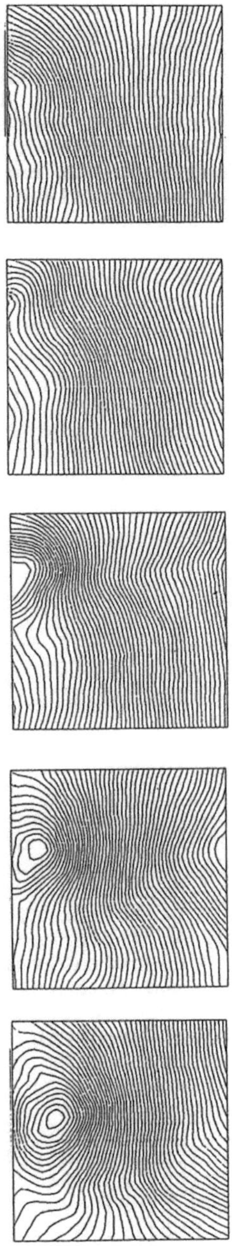

Fig. 8 Stream line

a comparison of velocity component, u, at x=0.4 is represented on Fig.9. Points, Pxx°, are the maximum inflection points of velocity component, u, on each blade angle position. After deciding the center of tip vortex by calculating the maximum vorticity, the distributions of the velocity components u and v around the vortex center are examined. Figure 10 shows that the velocity distributions around tip vortex on each blade angle position. It can be said that the velocity distribution around the vortex is able to be estimated quantitatively by the use of conditional sampling method.

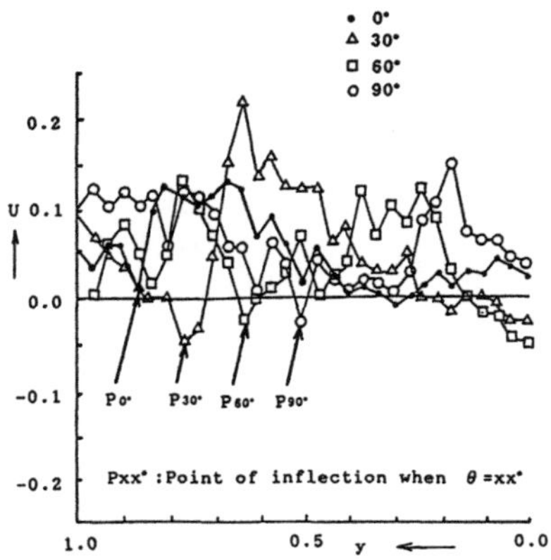

Fig. 9 Dispersions of u at x=0.4

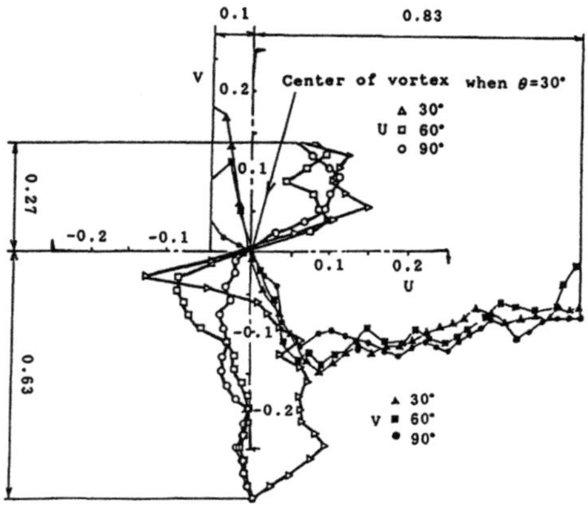

Fig. 10 Distributions of u and v around the vortex center

Conclusion

The conclusion of this study can be summed up as follows;

(1) A conditional sampling method has been introduced into PIV.

(2) The availability of this method has been identified by designing a conditional image processing system and applying it to the flow behind the propeller in a cylindrical tank.

(3) A difference of the flow structures has been illustrated quantitatively under the conditions for blade angle position and rotating speed of the propeller.

A three dimensional measuring method for this system is now being developed.

References

1. T.Kobayashi, T.Saga, T.Haeno. and N.Tsuda, ASME-FED128, 1991,pp9-14.
2. K.Nishino, N.Kasagi and M.Hirata, Trans.ASME-111,pp384-391.
3. Khaligi B. and Huebler M.S.,SAE Tech. Paper Series 880519, 1988,pp1-15.
4. Adamczyk A.A and Rimai L.,Exp. Fluids Vol.6,1989,pp380-389.
5. T.Kobayashi, T.Saga, S.Segawa, ISFV5,1990,pp197-202.

A High Accuracy and Fast Automatic Processing Method of Young's Fringes in PIV

Guangyun Ma , Gongxin Shen
Fluid Mechanic Institute (FMI),
Beijing University of Aeronautics and Astronautics (BUAA), Beijing CHINA

Abstract

A new method for the automatic processing of Young's fringes obtained from PIV photograph is presented. This processing method is not sensitive to the speckle noise in Young's fringe image, so that a high measuring accuracy could be obtained. The measuring error of fringe orientation and spacing are less than 0.5° and 1% respectively. Furthermore , only a less amount of computation is required.

Introduction

The particle image velocimetry(PIV) is an important nondestructive method to measuring flow velocity field in fluid mechanics [1,2] . The velocity distribution can be derived from the PIV photograph by optical Young's fringe method. A small region of PIV photograph is illuminated by a laser beam (Fig. 1) and the Young's fringe pattern could be observed (Fig. 2). The average displacement of particle in the interrogation spot is determined by the spacing and orientation of the fringes.The main problem in processing Young's fringes for the flow fleld is the high speckle noise level in Young's fringes. Various methods of automatic processing Young's fringe have been provided by many researcheres [3,4,5] . But most of them ,the reliablity and accuracy are low by some fast processing method, however, using some processing method with high measuring accuracy, the amount of computation is increased greatly.In this paper we propose a new Young's fringe processing method which gives the result of both high measuring accuracy and less amount of computation.

Automatic Processing of Young's Fringes

The procedure of the processing method in this paper is mainly as follows:

(1) Determining the fringe orientation

The fringe orientation is determined by searching the maximum of the orientation function $W(\alpha)$ which is constructed from the 2–D Young's fringe intensity distribution.

(a) Orientation function $W(\alpha)$

The coordinate systems (ξ_d,η_d) ,(ξ,η) and (ξ',η') are shown in Fig. 3. The ξ_d and η_d axis are the digital scanning coordinate axis for the Young's fringes. The η–axis parallels to the fringe orientation. The fringe angle is α_Y. The η'–axis has a arbitrary angle α (or called searching angle) with the ξ_d–axis. The 2–D Young's fringe intensity distribution in coordinate system (ξ',η') is expressed by $I_{DE}(\xi',\eta')$. The 1–D intensity distribution $I_\alpha(\xi')$ could be obtained by integrating the 2–D $I_{DE}(\xi',\eta')$ along the η'– axis, i.e.

$$I_\alpha(\xi')=\frac{1}{2r}\int_{-r}^{r} I_{DE}(\xi',\eta')d\eta' \qquad (-R_Y \leqslant \xi' \leqslant R_Y) \tag{1}$$

$$r=\sqrt{R_Y^2-\xi'^2}$$

Where R_Y is the radius of Young's fringes. Then the autocorrelation of the $I_\alpha(\xi')$ is given by

$$f_\alpha(u)=\frac{\int I_\alpha(\xi')\cdot I_\alpha(\xi'+u)d\xi'}{\int I_\alpha^2(\xi')d\xi'} \tag{2}$$

Fig. 4 gives a plot of typical $f_\alpha(u)$ distribution obtained from the Young's fringes shown in Fig. 2. As shown in Fig. 4, there are a series of maximum and minimum in $f_\alpha(u)$. The values of those maximum($u\neq0$) and minimum are expressed resprctively by $f_{max_i}(\alpha)$ *and* $f_{min_i}(\alpha)$ ($i=1,2,...,p$). The orientation function $W(\alpha)$ is given by

$$W(\alpha)=\sum_{i=1}^{p}(f_{max_i}(\alpha)-f_{min_i}(\alpha)) \qquad (0^\circ \leqslant \alpha \leqslant 180^\circ) \tag{3}$$

A typical $W(\alpha)$ distribution obtained from Fig. 4 is shown in Fig. 5.

(b) The characteristics of $W(\alpha)$

. The $W(\alpha)$ has the maximum value in the fringe orientation(i.e. $\alpha=\alpha_Y$).

. The $W(\alpha)$ is fairly sensitive to the fringe orientation when angle α approaches to the fringe angle α_Y.

. With the increase of fringe number, the $W(\alpha)$ distribution becomes sharper (Fig. 6), so that the corresponding measuring accuracy of the fringe orientation is increased.

.The $W(\alpha)$ is not sensitive to the speckle noise in the Young's fringes. Both the 1–D intensity compression (Eq. (1)) and the 1–D autocorrelation (Eq.(2)) could greatly suppress the influence of the speckle noise.

(c) Determining the fringe orientation according to the function $W(\alpha)$

The fringe orientation is determined by searching the maximum of $W(\alpha)$.

(2) Removing the influence of the diffraction halo

The 1–D fringe intensity distribution $I(\xi)$ is obtained by averaging the intensity along the lines which parallel to the fringe orentation. $I_0(\xi)$ is the 1–D average intensity distribution across the diffraction halo. The intensity of the $I_0(\xi)$ varies along the radial line and it results in the measuring error of the fringe spacing. So the influence of $I_0(\xi)$ must be removed before determining the spacing. The $I_0(\xi)$ is simulated by a function $\tilde{I}_0(\xi)$ which is given by

$$\tilde{I}_0(\xi)=e^{-(\frac{\xi}{R_Y})^2/\sigma^2}+C \tag{4}$$

Where σ、C are adjustable parmeters. So the 1–D intensity distribution $I(\xi)$ is modified by $\tilde{I}_0(\xi)$,i.e.

$$\tilde{I}(\xi)=I(\xi)/\tilde{I}_0(\xi) \tag{5}$$

Then, $\tilde{I}(\xi)$ is used to determine the fringe spacing. Fig. 7(a) shows a discrete distribution of the simulated diffraction halo intensity $\tilde{I}_0(\xi)$ with parameters $\sigma=0.5$, $C=-1.1$ and a typical discrete distribution of the 1–D fringe intensity $I(\xi)$ obtained from a

Young's fringes. Fig. 7(b) showns a discrete distribution of the $\tilde{I}(\xi)$ obtained from Fig. 7(a). As shown in Fig. 7(a)–7(b), a satisfactory simulation of the 1–D average halo intensity could be obtained by the function $\tilde{I}_0(\xi)$ expressed in Eq.(4).

(3) Determining fringe spacing

The fringe spacing is derived by calculating 1–D Fourier transformation of 1–D fringe intensity distribution $\tilde{I}(\xi)$. The Fourier transformation spectrum obtained from Fig. 7(b) is shown in Fig. 7(c). The location of the highest peak corresponds to the fringe spacing.

Discussion of the Accuracy and Computing Amount

The measuring error of fringe orientation is caused mainly by the speckle noise. The orientation function $W(\alpha)$ could strongly suppress the speckle noise and is very sensitive to the fringe orientation, so that a high measuring accuary of fringe orientation could be achived by searching the maximum of the $W(\alpha)$. The mearsuring error of fringe spacing is caused mainly by speckle noise and diffraction halo. In this paper the fringe spacing is determined by 1–D average fringe intensity distribution with the removel of diffraction halo, so that the influence of both the speckle noise and the diffraction halo is decreased. For a digital Young's fringes with 256×256 pixels, the measuring error of the fringe orientation and spacing here is less than 0.5° and 1% respectively.

This method has the same order of processing speed compareing with the fast processing method which determines the fringe orientation and spacing by calculating the correlation function of the fringe intensity along two lines in four dirrection [4], but this method could give the higher mearsuring accuracy than it. Compareing with the 2–D fringe intensity correlation method with high accuracy [5], the measuring accuracy in this paper is higher slightly, but the processing is speed up greatly.

Examples

Fig. 8 shows a examples of velocity distribution obtained from the PIV photographs by the Young's fringe method as mentioned above. The interrogation system by Young's fringe method is shown in Fig. 1. The PIV photographs corresponding to Fig. 8 are obtained using WLBIV techneque [2] for a starting flow around a 2–D backward step with $Re_a = 960(Re_a = \sqrt{8ha^3}/v$, h is the step height and a is the starting acceleration).

Conclusion

The Young's fringe processing method presented here could greatly suppress the speckle noise in the Young's fringes, so that a high measuring accuracy of fringe orientation and spacing could be obtained in fluid mechanic measurements. Also only a less amount of computation is required. It has been verified by the examples that the mentioned Young's fringe processing method is specially valid for PIV in the flow field.

References

[1] Reuss D. L.,Adrian R. J., et al.,″Instantaneous planar measurement of velocity and large–scale vorticity and strain rate in an engine using particle image velocimetry″,SAE Tech. Pater 890616,1989
[2] Gongxin Shen,Guangyun Ma,Fenggan Zhuang,″White light bubble image velocimetry″, Int. Con. on Exp. Fluids Mech.,1991,6,Chengdu,CHINA.
[3] G. W. Robinson,″Automatic fringe anslysis with a computer image processing system″, Appl. Opt.,Vol.22,No.14,PP2169–2175,1983
[4] Ralf Erbeok,″Fast image processing with a microcomputer applied to speckle photography″,Appl. Opt.,Vol.24,No.22,PP3838–3841,1985
[5] G. A. Reynolds,M. Short,M. C. Whiffen,″Automated reduction of instantaneous flow field images″,Opt. Eng.,Vol.24,No.3,PP475–479,1985

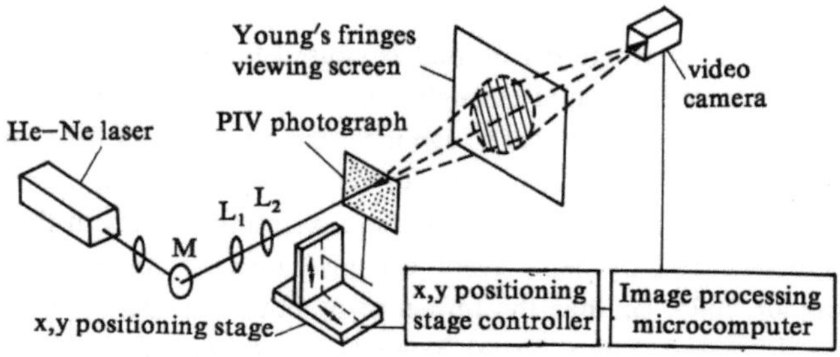

Fig.1 Schematic diagram of the interrogation system by Young′s fringe method

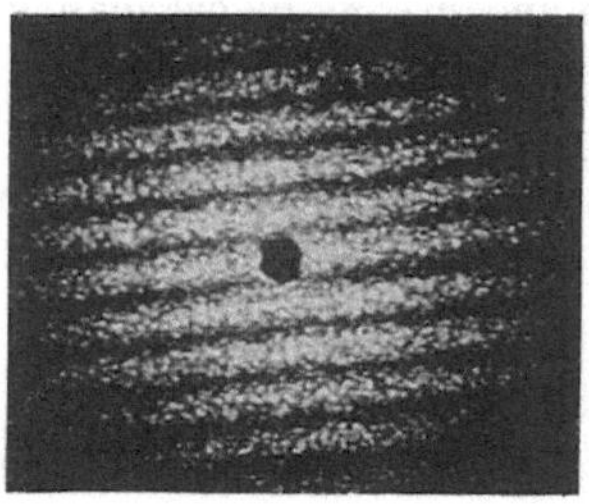

Fig.2 Young′s fringe pattern

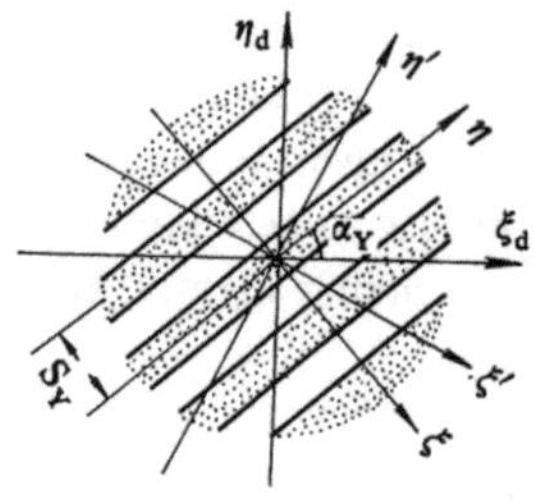

Fig.3 Schematic diagram of the coordinate system in Young′s fringes

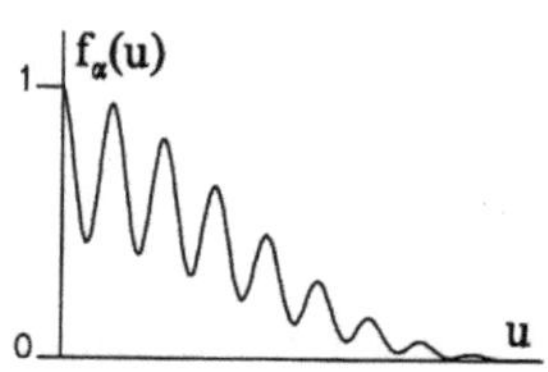

Fig.4 The $f_\alpha(u)$ distribution obtained from the Young′s fringes shown in Fig.2

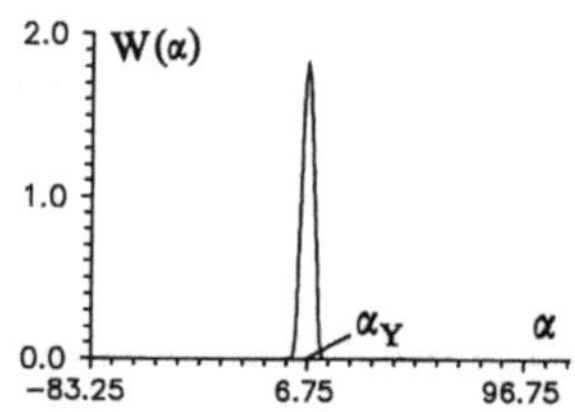

Fig.5 The $W(\alpha)$ distribution obtained from the $f_\alpha(u)$ shown in Fig.4

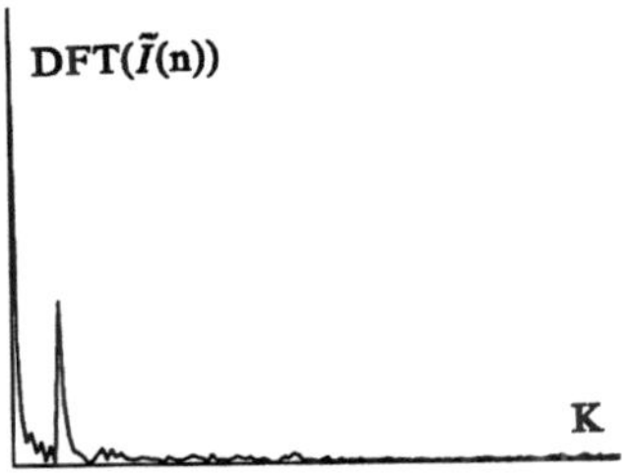

(c)

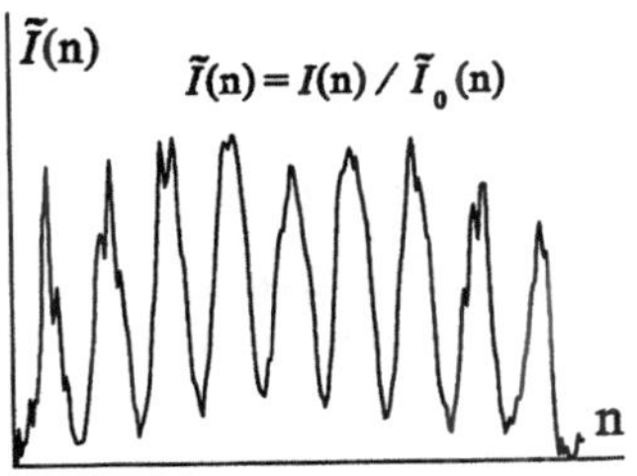

(b)

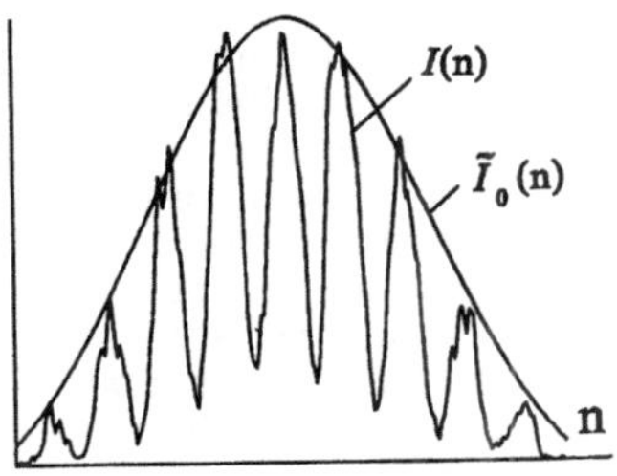

(a)

Fig.7 The 1–D discrete fringe intensity distribution before and after diffraction halo removal
(a) simumulated 1–D diffraction halo intensity $\tilde{I}_0(n)$ ($\sigma = 0.5$, $C = -1.1$) and 1–D fring intensity $I(n)$
(b) 1–D fring intensity $\tilde{I}(n)$ obtained from (a)
(c) Fourier transform spectrum of (b)

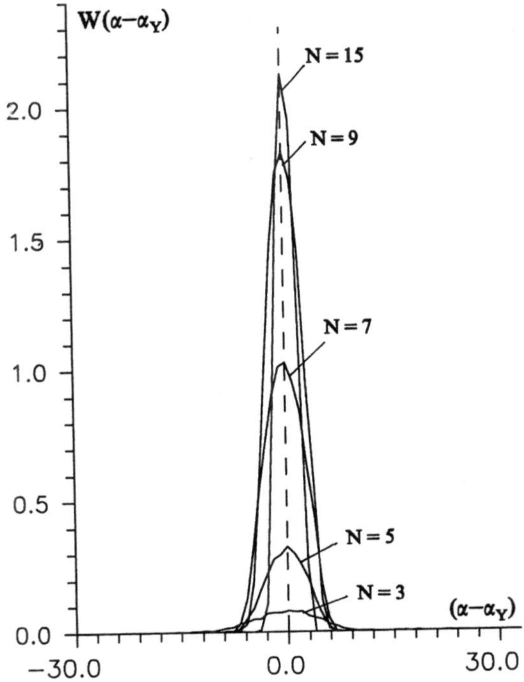

Fig.6 The W(α) distribution for different fringe number(N)

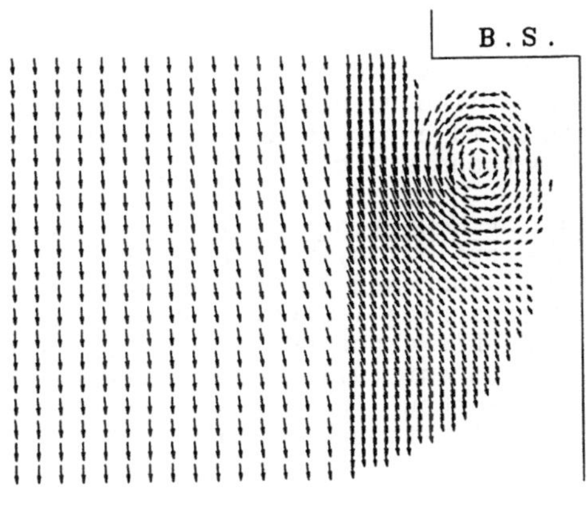

Fig.8 The velocity distribution derived from PIV photograph by Young's fringe method

Processing PIV Patterns by Holographic Correlation Technique

Noboru Nakatani and Takanori Oshio
Department of Industrial Education, Ashiya University
13-22, Rokurokuso-cho, Ashiya, Hyogo 659
Japan

Summary

This paper describes the methods for measuring iso-velocity distribution and iso-velocity direction from PIV pattern with a conventional holographic technique. Ring-shaped and rectangular patterns are used as master patterns for making holographic matched filters for iso-velocity distribution and iso-velocity-direction distribution, respectively. The capabilities of these methods are confirmed by model experiments.

Introduction

Particle Image Velocimetry (PIV) [1] has recently been developed into a powerful experimental technique for the measurement of instantaneous two dimensional velocity field in fluid flow. Local fluid velocities are extracted from the photographic recording image by measuring the spacing and orientation of parallel Young's-type fringes in the optical power spectrum of the image or by measuring the correlation between recorded images. However, these methods need sequential scanning for obtaining a complete two-dimensional velocity map. To eliminate this disadvantage, we developed some methods, which we proposed firstly in 1976 [2], for processing PIV patterns by the use of holographic correlation technique (holographic mached filtering technique). The conventional holographic correlation system is used. The methods for measuring the distributions of iso-velocity and iso-velocity-direction are described. The measured error is analyzed theoretically. The capabilities of these methods are confirmed by model experiments.

Method for Measuring Iso-velocity Distribution

The optical system of holographic correlation technique shown in Fig. 1 was used for processing a PIV pattern. We described method for measuring iso-velocity distributions from a PIV pattern.

A double-exposured photograph of a moving tracer particle is two transparent full circle as shown in Fig. 2. A master pattern of ring shape shown in Fig. 2 is set on P_1 in the optical system shown in Fig. 1 (a) to make a holographic filter. The double exposured recorded images of the moving tracer particles shown in Fig. 6 (a) (described in last section) is used as a test pattern. The test pattern is set on P_1 plane in Fig. 1 (b) and the holographic filter on P_2 plane to obtain cross-correlation between the test pattern and the master pattern. We assume the condition that the particle number is small and we can recognize particle moved distances during double exposures in distances among particle images. Under this condition, when the moved distances are below the diameter of the ring of the master pattern, the center between two points for the double exposured images has maximum intensity of cross-corre-

By using master patterns of various diameters, we can obtain iso-velocity distributions for each master pattern. The error of iso-velocity distribution measurement is considered by a model shown in Fig. 2. The intensity of cross-correlation normalized by maximum value is given by

$$I = \{[\pi r^2 - \theta r^2 + r^2 \phi - (a + \Delta a)(a + r)\sin\phi]/\pi r^2\}^2 \quad (1)$$

where θ: $\cos^{-1}(2ar - \Delta a^2 - 2a\Delta a)/2r(a + \Delta a)$, ϕ: $\cos^{-1}[(a + \Delta a)^2 + a(2r + a)]/2(a + r)(a + \Delta a)$, Δa: difference between radius a of a master pattern and half moved distance of a particle. The relation between I and Δa in various $\Delta a/r$ is shown in Fig. 3. The decay of the curve increases with decrease of a/r. Consequently, the moved distance is shorter and the measured error is smaller.

Measurement of Iso-Velocity Direction Distribution

A master pattern of rectangular shape shown in Fig. 4 is used to measure iso-velocity-direction distribution. When alignment of particle images for first exposure and for second exposure coincides to the direction of the master pattern, the maximum intensity of cross-correlation is at the particle images. The intensity of cross-correlation normalized by maximum value is given as

$$I = [\phi - (\sin 2\phi)/2]^2/\pi^2 \quad (2)$$

where ϕ: $\cos^{-1}[(a/r)\sin\Delta\theta - 1]$ in condition of $\sin\Delta\theta \leqq 2r/a$. The relation between I and $\Delta\theta$ in various 2r/a is shown in Fig. 5(b). The decay of the curve increases with decrease of 2r/a. Consequently, when the moved distance is larger, the measuring error of direction is smaller. From this theoretical analysis it is found that the measured error of the flow direction is below 2 ~3 degree in the region over the ratio of particle diameter to particle moved distance 0.1(Fig. 5).

Model Experiment

We made model experiments to cofirm the capabilities of our methods. To extract iso-velocity from the PIV pattern (Fig. 6 (a)), we used the holographic filter obtained from the master pattern (Fig. 6 (b)). One among the PIV double exposured images, which is matched to the diameter (moved distance) of the master pattern, is extracted as two large luminous spots shown in Fig 7. To extract a flow direction from the PIV pattern (Fig. 8 (a)), we used the holographic filter obtained from the master pattern (Fig. 8 (b)). One among the PIV double exposured images, which is matched to the direction of the master pattern, is extracted as two large luminous spots shown in Fig 9. The capabilities of these methods to processing patterns in high particle density will be presented in this symposium.

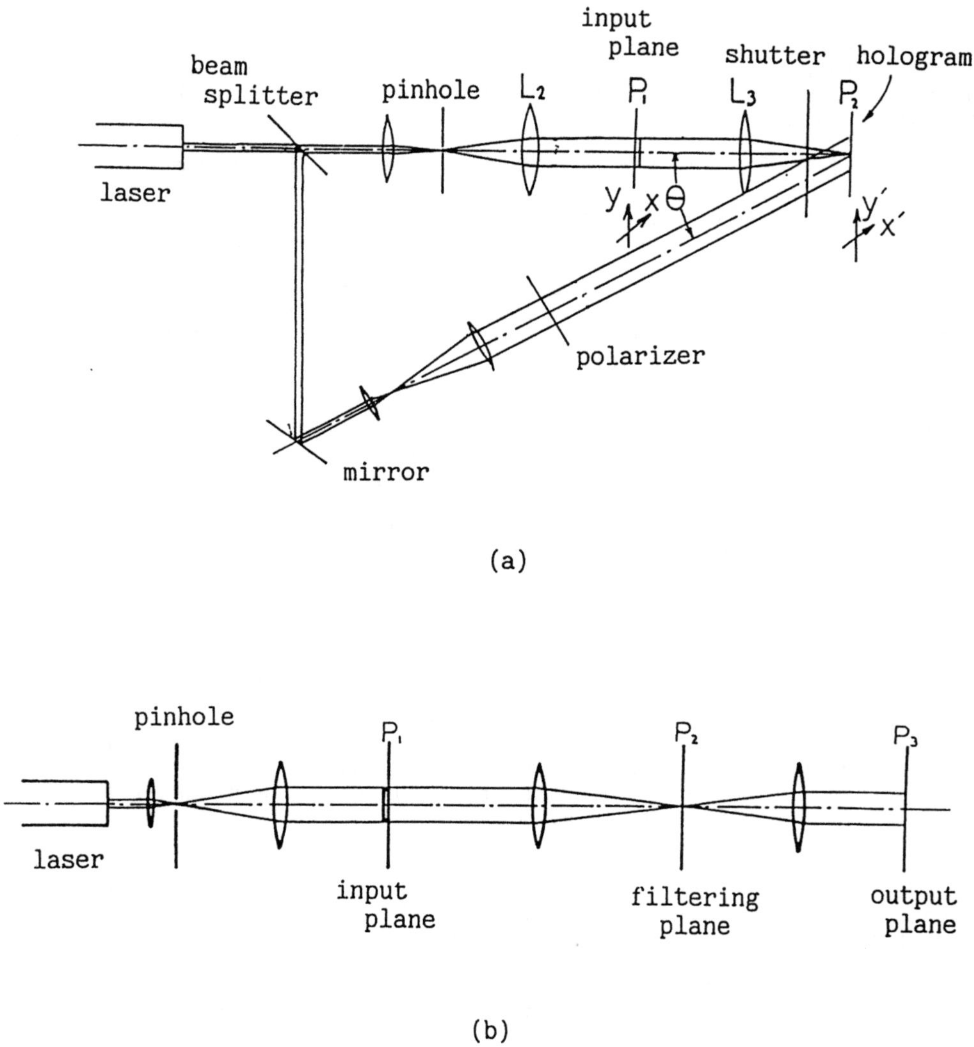

Fig. 1 Schematic diagram of optical system of holographic correlation technique. (a): making holographic filter from a master patterns, (b): correlation.

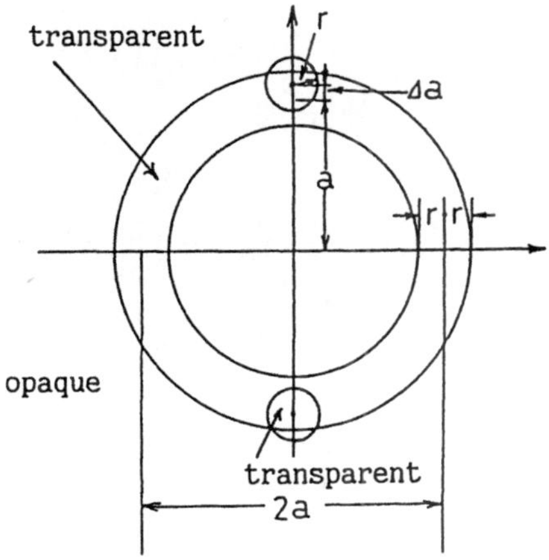

Fig. 2 Model of iso-velocity distribution measurement. Ring shaped pattern for a master pattern. Double-exposured particle images.

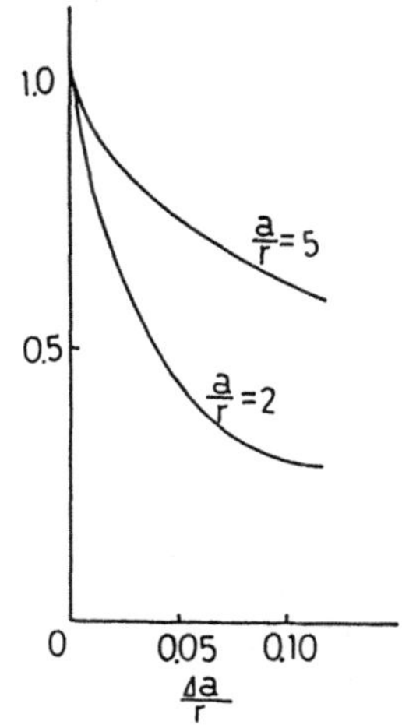

Fig. 3 The cross-correlation intensity I vs. Δa/r in various a/r for iso-velocity distribution measurement.

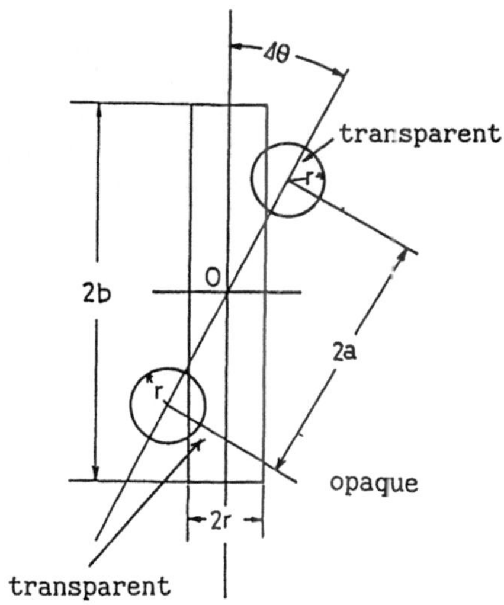

Fig. 4 Model of iso-velocity-direction distribution measurement. Rectangular pattern for a master pattern. Double-exposured particle images.

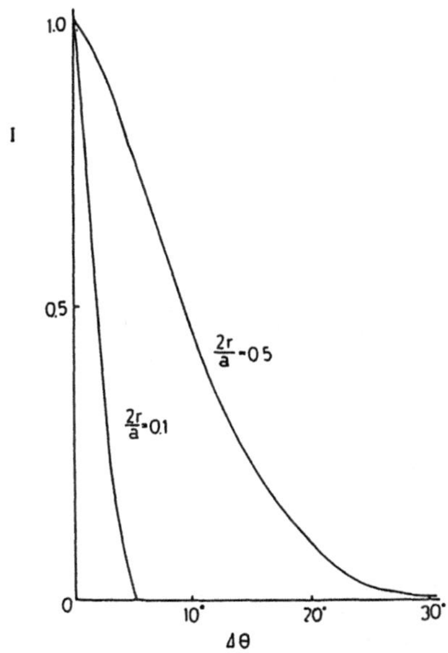

Fig. 5 The cross-correlation intensity I vs. Δa/r in various a/r for iso-velocity-direction distribution measurement.

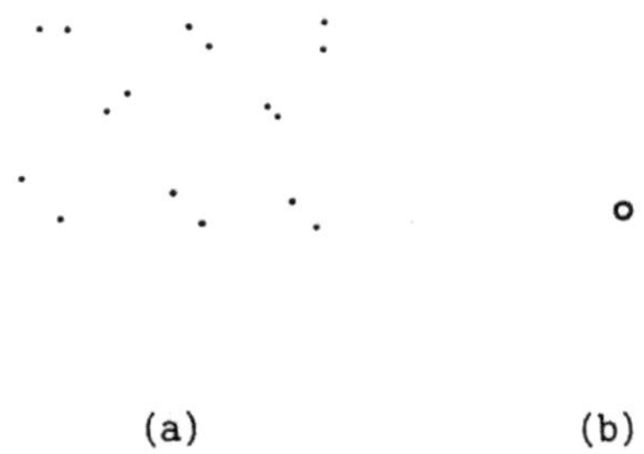

Fig. 6 The Model experiment for iso-velocity measurement. Dark parts are transparent. (a): Double exposured images, (b): a master pattern for making a holographic filter.

Fig. 7 The cross-correlation pattern showing iso-velocity.

Fig. 8 The model experiment for iso-velocity measurement. (a): Double exposured images, (b): a master pattern for making a holographic filter.

Fig. 9 The cross-correlation pattern showing iso-velocity-direction.

References

1) Adrian, R. J.;-Yao, C. S.: Developed of pulsed laser velocimetry for measurement of fluid flow. Proc. of 8th Biennial Symp. on Turbulence, 1983, University Missouri-Rolla (1983) 170-186.

2)Nakatani, N.; Matsumoto M.; Yamada, T.: Processing flow tracer patterns by holographic correlation technique. Proc. of Annual Meeting of Japan applied Physical Society, 28a-F-10 (1976) 169.

Subject Index

FSC
www.fsc.org
MIX
Papier aus verantwortungsvollen Quellen
Paper from responsible sources
FSC® C105338

If you have any concerns about our products,
you can contact us on
ProductSafety@springernature.com

In case Publisher is established outside the EU,
the EU authorized representative is:
Springer Nature Customer Service Center GmbH
Europaplatz 3, 69115 Heidelberg, Germany

Printed by Libri Plureos GmbH
in Hamburg, Germany